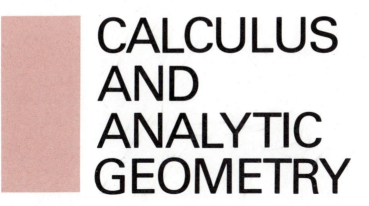

CALCULUS AND ANALYTIC GEOMETRY

J. DOUGLAS FAIRES
Youngstown State University

BARBARA TRADER FAIRES
Westminster College

PRINDLE, WEBER & SCHMIDT
Boston, Massachusetts

PWS PUBLISHERS

Prindle, Weber & Schmidt · ♣ · Willard Grant Press · **WG** · Duxbury Press · ♦
Statler Office Building · 20 Providence Street · Boston, Massachusetts 02116

To Mr. P.

PWS Publishers is a division of Wadsworth, Inc.

Printed and bound in the United States of America

83 84 85 86 87 — 10 9 8 7 6 5 4 3 2 1

Library of Congress Cataloging in Publication Data

Faires, J. Douglas.
    Calculus and analytic geometry.

    Includes index.
    1. Calculus. 2. Geometry, Analytic.
I. Faires, Barbara T. II. Title.
QA303.F294   1982   515'.15   82-18635

ISBN 0-87150-323-9

Cover photo © Erich Hartmann/MAGNUM

Photos of Escher prints courtesy of Vorpal Galleries, used by permission of Visual Artists and Galleries Assn., Inc.

Text designed and edited by Helen Walden. Composition by Weimer Typesetting. Printing and binding by Rand McNally. Covers printed by John Pow Company. Cover designed by Trisha Hanlon.

PREFACE

The study of calculus has been a core subject area in science and engineering for many years and is now included as an important topic in most other quantitative disciplines. Because of the broad range of students now taking calculus, diverse applications relating to the material have been included wherever possible.

Since calculus is primarily concerned with the behavior and applications of functions, the text begins with a discussion of functions and graphs that are used throughout the book. More material on graphing is presented than is seen in many other calculus books, since students understand new concepts more easily if they are given a familiar geometric representation. We also have found that students respond with more enthusiasm to the standard precalculus review topics when the relevance of those topics to calculus is shown, so this material has been incorporated into our discussion of functions. If further review of basic precalculus topics is needed, especially in trigonometry, students can refer to Appendix A.

We have made a concerted effort to present new concepts in small segments with reference to previously studied topics. This teaching philosophy can be seen in the following sequence. Chapter 1 contains exercises that ask the student to describe physical problems using functions. The same physical problems are reconsidered in Chapter 4 (Applications of the Derivative) and in Chapter 6 (Applications of the Integral). For example, Exercise 39 of Section 1.1 asks that the surface area of a cylindrical can of fixed volume be represented as a function of its radius. Exercise 15 of Section 4.2 uses this function to determine the dimensions of the can with minimal surface area.

The calculus applications can be difficult topics to master due to the fact that two relatively new concepts are used at one time; the mathematical description of the physical problems and the application of calculus techniques are required simultaneously. We have tried to separate these two concepts, enabling the students to comprehend and solve the applications.

Similarly, when vector-valued and multivariable functions are considered in Chapters 13, 15, and 16, we make frequent reference to the familiar single-variable concepts so that the students can compare the new concepts with the material previously studied.

The six standard trigonometric functions are presented early in the text. The availability of these nonalgebraic functions gives more meaning to the reciprocal, quotient and chain rules for differentiation. Introducing the trigonometric functions at this juncture also allows the consideration of more interesting applications.

The text has been written to appeal to a wide variety of curriculum requirements. Sections that might be omitted or discussed in a different order have been organized so that subsequent material can be presented easily. For example, conic sections (Chapter 11) can be presented at any time after Chapter 1. Although infinite limits and limits at infinity are discussed in Chapter 2 where their definitions can be easily compared to those of other limits, these topics can be postponed until the graphing techniques are considered in the final sections of Chapter 3. L'Hôpital's rule is presented in Chapter 4 as an application of the derivative. This topic could be postponed until Chapter 7. Separable differential equations are considered in Section 7.6 to provide applications emphasizing the natural exponential and logarithm functions. This topic could be discussed with the other forms of differential equations in Chapter 18.

The versatility of the text is also illustrated by the examples and the over 6,000 exercises, the majority of which are similar to the problems posed in the examples. However, there are throughout the text a large number of application problems from virtually every discipline commonly using calculus techniques. In addition, there are more exercises of a theoretical nature than we expect could be covered in a standard calculus course. Included among these are approximately fifty problems that have been posed on William Lowell Putnam examinations during the past forty years. These problems are generally very difficult but have been included primarily to show students the mathematical power of the topics they are studying and to introduce them in a minor way to the type of work that interests many professional mathematicians. Solutions to these problems can be found in the referenced issues of the *American Mathematical Monthly.*

There is a complete solutions manual available from the publisher for the instructors and a two-volume student supplement that contains solutions to more than one-third of the exercises.

ACKNOWLEDGMENTS

We would like to express our appreciation to the reviewers of the manuscript for this book:

Paul Bland, Eastern Kentucky University; C. Kenneth Bradshaw, San Jose State University; Thomas J. Brieske, Georgia State University; Robert Brooks, University of Utah; Arthur Dull, Diablo Valley College; Bruce Edwards, University of Florida; Frederick Gass, Miami University, Ohio; Beverly Gimmestad, Michigan Technologic University; Stuart Goldenberg, California Polytechnic State University; Paul Gormley, Villanova University; Edwin Haltar, University of Nebraska at Lincoln; Alan Heckenback, Iowa State University; Terry Herdman, Virginia Polytechnic Institute and State University; Louis Hoelzle, Bucks County Community College; Joseph Krebs, Boston College; John F. Lee, Oregon State University; Phil Locke, University of Maine, Orono; Richard Painter, Colorado State University; Jerome Paul, University of Cincinnati; V. Frederick Rickey, Bowling Green State University; Leon W. Rutland, Virginia Polytechnic Institute and State University; Phillip H. Schmidt, University of Akron; James R. Smith, Appalachian State University; Joel Stemple, CUNY, Queens College; David R. Stone, Georgia Southern College; and John Thorpe, SUNY, Stony Brook.

In particular Dave Stone, Fred Rickey, and Phil Schmidt did extensive reviews of the exercises, text, and historical notes, respectively. They deserve special thanks.

We would also like to thank the staff of PWS Publishers who worked diligently with us on this project; Theron Shreve, our editor; Mary LeQuesne, who provided excellent technical advice; and our production editor, Helen Walden, who turned our work into print. The typing of the text and solutions manuals was done by Kathy Bosak, who has our sincere gratitude.

Finally, we would like to express our appreciation to the students, faculty, and administrators at Westminster College and Youngstown State University for the encouragement we were given on this project.

Barbara and Doug Faires
October 17, 1982

CONTENTS

4
APPLICATION OF THE DERIVATIVE

5
THE INTEGRAL

6
APPLICATIONS OF THE DEFINITE INTEGRAL

7
THE CALCULUS OF INVERSE FUNCTIONS

8
TECHNIQUES OF INTEGRATION

9
SEQUENCES AND SERIES

10
POLAR COORDINATES AND PARAMETRIC EQUATIONS

11
CONIC SECTIONS

12
VECTORS

13
VECTOR-VALUED FUNCTIONS

14
MULTIVARIATE FUNCTIONS

15
THE DIFFERENTIAL CALCULUS OF MULTIVARIATE FUNCTIONS

1
FUNCTIONS

Calculus is concerned with the study and application of functions. For this reason, it is important that you fully understand this concept and can use it with ease. We have, therefore, devoted this first chapter to a study of functions—by definition, numerous examples, applications, and graphs.

1.1
DEFINITION OF A FUNCTION

The notion of a function is one of the most basic tools of mathematics. A function is a means of associating the elements of one set with those of another, in such a way that each element in the first set corresponds to precisely one element in the second (see Figure 1.1). For example, a function can be used to describe an association between a set of people and a set of numbers, by allowing each person (an element in the first set) to be associated with that person's age (an element in the second set).

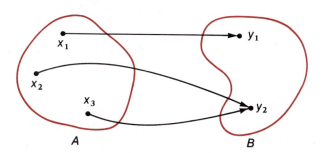

FIGURE 1.1

To represent a function completely three objects must be described: the set in which the function originates, the set in which the function terminates, and the rule that is used to describe the correspondence. The definition of a function given below contains each of these objects.

(1.1)
DEFINITION

A **function** from a set A into a set B is a rule of correspondence that assigns to each element in A precisely one element in B.

If f is used to denote a function from a set A into a set B, then $f(a)$ denotes the unique element in B that corresponds to the element a in A. The set A is called the **domain** of f. The **range** of f is the set of all elements in B that are associated with some element in A. To indicate that f is a function with domain A and range B, we often write $f: A \rightarrow B$ and say "f maps A into B."

EXAMPLE 1

The price of a symphony ticket is \$12. The amount of money received from the sale of n tickets is $R(n) = 12n$ dollars. If the symphony hall seats a maximum of 1000 people, what is the domain and range of R?

SOLUTION

The number of tickets sold is an integer between 0 and 1000 so the domain of R is $\{0, 1, 2, \ldots, 1000\}$.

The range of the function R is $\{0, 12, 24, \ldots 12000\}$, the set of all possible amounts that can be received. ☐

The study of calculus is primarily concerned with the analysis and application of functions whose domain and range are both contained in the set of real numbers $\mathbb{R}$. Functions of this type can often be described simply by giving their rule of correspondence. For example, $f(x) = x^3 + 1$ describes a function f. Unless specified otherwise, the domain of the function is assumed to be the largest subset of $\mathbb{R}$ for which the function is defined. The range is the set of real numbers that are associated with some number in the domain.

EXAMPLE 2

Find the domain and range of the function f whose rule of correspondence is $f(x) = \sqrt{x} + 1$.

SOLUTION

The square root of x is a real number if and only if $x \geq 0$. Consequently, the domain of f is the set of nonnegative real numbers.

The range of f is the set of real numbers greater than or equal to one, since the symbol $\sqrt{}$ indicates the principal or nonnegative square root. ☐

**HISTORICAL
NOTE**

This functional notation was first introduced by the Swiss mathematician **Leonhard Euler** (1707–1783). Euler also wrote the most complete calculus book of his day, including in the text many of his own original results. The volumes on differential calculus and differential equations were published in 1755, but his extensive volume on integral calculus did not appear in completed form until nineteen years later.

There is no special significance to the variable x; in fact, it might be better to describe the function f by writing $f(\underline{\hspace{1em}}) = \sqrt{\underline{\hspace{1em}}} + 1$. This form indicates more clearly that whatever value is used to fill the space on the left side of the equation is also used on the right side. For instance, 9 is in the domain of f, and $f(9) = \sqrt{9} + 1 = 4$.

If a real number x is in the domain of a function f, we say that f is defined at x or that $f(x)$ exists. The terminology "f is defined on A" means that A is a subset of the domain of f; that is, $f(x)$ exists for every number x in A.

In Chapter 3, we will consider in detail quotients of the form

$$\frac{f(x + h) - f(x)}{h},$$

where $h \neq 0$. For example, for the function f described by $f(x) = x^2 + x$, $f(x + h)$ is simply $(x + h)^2 + (x + h)$ and this quotient is:

$$\frac{f(x + h) - f(x)}{h} = \frac{(x + h)^2 + (x + h) - (x^2 + x)}{h}$$
$$= \frac{x^2 + 2xh + h^2 + x + h - x^2 - x}{h}$$
$$= \frac{h(2x + h + 1)}{h} = 2x + h + 1.$$

The domain and range of functions from $\mathbb{R}$ into $\mathbb{R}$ are often described using interval notation. The **open interval** (a, b) is defined by

$$(a, b) = \{x \mid a < x < b\},$$

read "the set of all real numbers x such that $a < x < b$." A geometric interpretation of an open interval is given by using the real number line as shown in Figure 1.2(a).

When the endpoints of the interval are included, the interval is called a **closed interval** and denoted

$$[a, b] = \{x \mid a \le x \le b\}.$$

(See Figure 1.2(b).) Half-open intervals contain one end point, but not the other. For example, $(a, b] = \{x \mid a < x \le b\}$ and $[a, b) = \{x \mid a \le x < b\}$ are half-open intervals.

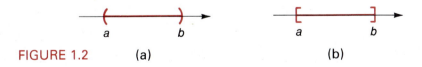

FIGURE 1.2   (a)   (b)

Unbounded intervals can be represented in a similar manner. (See Figure 1.3.)

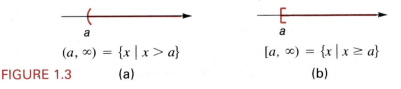

$(a, \infty) = \{x \mid x > a\}$ $[a, \infty) = \{x \mid x \ge a\}$

FIGURE 1.3   (a)   (b)

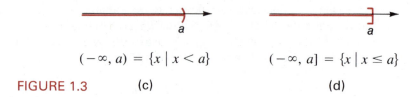

$$(-\infty, a) = \{x \mid x < a\} \qquad (-\infty, a] = \{x \mid x \leq a\}$$

FIGURE 1.3   (c)   (d)

The symbol ∞ is given the name "infinity" and is used to express that a set such as (a, ∞) has no greatest element. Similarly, $-\infty$ is used to express that a set such as $(-\infty, a)$ has no least element.

In general, the square bracket, [or], indicates that the element next to it is in the set while the round bracket, (or), indicates that the element next to it does not belong to the set. One obvious implication of this notation is that the symbols ∞ and $-\infty$ are never next to a square bracket, since they do not represent real numbers.

EXAMPLE 3 For the function f given by $f(x) = \dfrac{1}{x-1}$, find $f(0), f(-1)$, and the domain of f.

SOLUTION

$$f(0) = \frac{1}{0-1} = \frac{1}{-1} = -1 \qquad \text{and} \qquad f(-1) = \frac{1}{-1-1} = \frac{1}{-2} = -\frac{1}{2}.$$

The number 1 is not in the domain of f since $\dfrac{1}{1-1}$ is undefined. This is the only number for which f is undefined, so the domain of f is the set of all real numbers except 1, and is given in interval notation by $(-\infty, 1) \cup (1, \infty)$. ☐

EXAMPLE 4 Find $f(-2), f(0)$, and $f(5)$ for the function f defined by

$$f(x) = \begin{cases} x, & \text{if } x < 0 \\ 3, & \text{if } x \geq 0. \end{cases}$$

SOLUTION

Since $-2 < 0$ and $f(x) = x$ when $x < 0$, it follows that $f(-2) = -2$. When $x \geq 0, f(x) = 3$, so $f(0) = 3$ and $f(5) = 3$. ☐

EXAMPLE 5 Find the domain of the function described by

$$g(x) = \sqrt{x^2 - 4x + 3}.$$

SOLUTION

The domain of g is all real numbers x such that $x^2 - 4x + 3 \geq 0$. Since

$x^2 - 4x + 3 = (x - 3)(x - 1)$, it follows that x is in the domain of g if and only if the factors $(x - 3)$ and $(x - 1)$ are both nonnegative or both nonpositive.

In Figure 1.4, we see that both factors are nonnegative when $x \geq 3$ and that both are nonpositive when $x \leq 1$. So the domain of g is $(-\infty, 1] \cup [3, \infty)$. $\square$

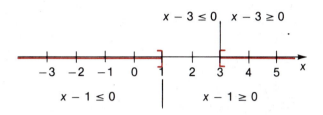

FIGURE 1.4

The **product set** $A \times B$ of the sets A and B consists of all ordered pairs (a, b) where a is in A and b is in B. Associated with any function from the set $\mathbb{R}$ into itself is a subset of the product set $\mathbb{R} \times \mathbb{R}$: the set consisting of all pairs $(x, f(x))$ where x is in the domain of f. This set is called the **graph** of f. To "sketch the graph of f" means to illustrate these points in the coordinate plane. For example, the graph of the function described in Example 4 is sketched in Figure 1.5.

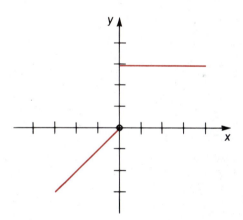

FIGURE 1.5

More generally, the graph of an equation in the variables x and y consists of all ordered pairs (x, y) that satisfy the equation. So the graph of a function is the graph of an equation describing the function. Usually the terminology is simplified by saying "graph the equation" instead of "sketch the graph of the equation" and the sketch is referred to as "the graph of the equation."

EXAMPLE 6   Determine which of the following equations describes a function in the form $y = f(x)$.

(a) $y = x^2$ (b) $y^2 = x$ (c) $x^2 + y^2 = 1$.

SOLUTION

The equation $y = x^2$ describes a function because each value of x corresponds to precisely one value of y: namely, the square of x.

The equation $y^2 = x$ does not describe a function since, in particular, the ordered pairs $(4, 2)$ and $(4, -2)$ both satisfy this equation, so $x = 4$ corresponds to both $y = 2$ and $y = -2$. Similarly, the equation $x^2 + y^2 = 1$ does not describe a function since, for example, the ordered pairs $(0, 1)$ and $(0, -1)$ both satisfy $x^2 + y^2 = 1$. $\square$

EXAMPLE 7   The graphs of the equations in Example 6 are shown in Figure 1.6. By looking at the graphs, determine which describe y as a function of x.

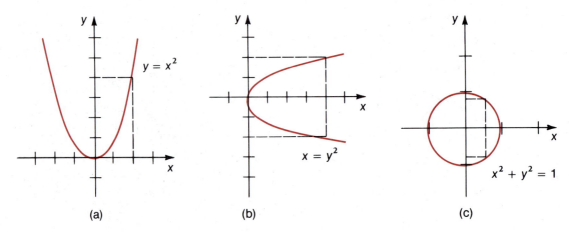

(a)   (b)   (c)

FIGURE 1.6

SOLUTION

In (a), each value of x is associated with exactly one value of y; (one such association is indicated by the dashed line). Thus the graph in (a) is the graph of a function. Note that any vertical line intersects this graph in at most one point.

In (b) and in (c) a value of x is shown that is associated with two different values of y and hence neither is the graph of a function. Observe that vertical lines exist which intersect these graphs in more than one point. $\square$

Example 7 illustrates a rule to distinguish the graphs of equations describing functions from those that do not:

An equation describes a function if and only if every vertical line intersects its graph in at most one point.
The domain of the function is described by those values on the horizontal axis through which a vertical line intersects the graph.
The range of the function is described by those values on the vertical axis through which a horizontal line intersects the graph.

Functional notation is used to express real-life situations in a form that can be handled mathematically.

EXAMPLE 8   A charter bus company charges $10 per person for a round trip to a ball game with a discount given for group fares. A group purchasing more than ten receives a reduction per ticket of twenty-five cents times the number of tickets purchased in excess of ten. If 11 people go to the game, each of the 11 tickets costs $9.75, if 16 people go, each ticket costs $8.50, and so on. Express the amount of money the company receives as a function of the number of tickets it sells to a group.

SOLUTION

Let x denote the number of tickets sold to a group and $R(x)$ the number of dollars received from the sale. When $0 \le x \le 10$, the amount received is $10x$ dollars. When $x > 10$, the price per ticket is reduced by $(.25)(x - 10)$ dollars. So, for $x > 10$ the price per ticket is $10 - .25(x - 10)$ dollars and the total amount received is $[10 - (.25)(x - 10)]x$ dollars. This equation simplifies to $(12.5 - .25x)x$ and

$$R(x) = \begin{cases} 10x & , \text{if } 0 \le x \le 10 \\ (12.5 - .25x)x, & \text{if } x > 10. \end{cases}$$

The domain of the function R is the set of nonnegative integers because x denotes the number of tickets sold.

In actual practice, the bus company would also place a restriction on the price reduction that would guarantee a minimal price per passenger. An extension of this example that includes this type of restriction is discussed in Exercise 33.   □

EXERCISE SET 1.1

1. If $f(x) = 4x^2 + 5$, find
    (a) $f(2)$ (b) $f(\sqrt{3})$ (c) $f(2 + \sqrt{3})$ (d) $f(2) + f(\sqrt{3})$

2. If $f(x) = 3x - 5x^2$, find
    (a) $f(7)$ (b) $f(\sqrt{2})$ (c) $f(7 + \sqrt{2})$ (d) $f(7) + f(\sqrt{2})$

In Exercises 3 through 6, a function is described. Find the domain of the function.

3. $f(x) = x^2$ **4.** $f(x) = \dfrac{1}{x - 3}$

5. $f(x) = \dfrac{x^2 - 1}{x^4 - x^2}$ **6.** $f(x) = \dfrac{1}{\sqrt{x - 3}}$

In Exercises 7 through 12, determine if the equation describes y as a function of x.

7. $3x + 2y = 5$ **8.** $y = x^2 + 3$

9. $x = y^2 + 3$ **10.** $x^2 + y^2 = 16$

11. $x^2 + 4x + y^2 = 12$ **12.** $x^2 - y^3 = 4$

In Exercises 13 through 16, the correspondence rule of a function f is given. Describe the function given by $f(-x)$, $-f(x)$, $f(1/x)$, $1/f(x)$, $f(\sqrt{x})$, and $\sqrt{f(x)}$.

13. $f(x) = x^2 + 2$ **14.** $f(x) = x^2 + 2x + 3$

15. $f(x) = \dfrac{1}{x}$ **16.** $f(x) = \sqrt{x}$

In Exercises 17 through 22, a function is described and a number in the range of the function is given. Find a number in the domain of the function corresponding to this value in the range.

17. $f(x) = x^2 - 1, 0$ **18.** $f(x) = x^2 - 1, 2$

19. $f(x) = x^2 - 1, \dfrac{1}{2}$ **20.** $h(x) = \sqrt{x}, \dfrac{3}{4}$

21. $g(x) = x^2 + 2x + 2, 1$ **22.** $f(x) = \dfrac{1}{x^2 + 2x + 2}, \dfrac{1}{2}$

Find $f(x + h)$ and $\dfrac{f(x + h) - f(x)}{h}$, if $h \neq 0$, for the functions described in Exercises 23 through 30.

23. $f(x) = 2x - 4$ **24.** $f(x) = 7x + 5$

25. $f(x) = \dfrac{3}{2}x - \dfrac{1}{4}$ **26.** $f(x) = -\dfrac{3}{2}x + 5$

27. $f(x) = x^2$ **28.** $f(x) = x^2 - 1$

29. $f(x) = 2x^2 - x$ **30.** $f(x) = 4x^2 + 3x + 1$

31. A function f is said to be **even** if for each number x in the domain of f, $-x$ is also in the domain of f and $f(-x) = f(x)$. A function f is said to be **odd** if for each number x in the domain of f, $-x$ is also in the domain of f and $f(-x) = -f(x)$. Determine which of the following functions are odd, which are even, and which are neither odd nor even.

(a) $f(x) = x^2$ (b) $f(x) = x^3$

(c) $f(x) = \sqrt{x}$ (d) $f(x) = x^2 + 1$

(e) $f(x) = x^2 + x^3$ (f) $f(x) = \dfrac{1}{x}$

32. Prove that if f is a function that is both odd and even, then $f(x) = 0$ for every x in its domain.

33. Example 8 discussed a problem involving a bus company offering group rates for a trip to a ball game. Reconsider this example with the added restriction that at no time will the price per person be lower than $6.25. Describe the function associated with this problem and sketch its graph.

34. A rectangular plot of ground containing 432 square feet is to be fenced off in a large lot. Express the perimeter of the plot as a function of the width. What is the domain of this function?

35. Two ships sail from the same port. The first ship leaves at 1:00 AM and travels eastward at 15 nautical miles per hour. The second ship leaves port at 2:00 AM and travels northward at 10 nautical miles per hour. Find the distance d between the ships as a function of the time after 2:00 AM.

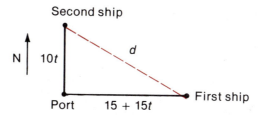

Second ship

N   10t

Port   15 + 15t   First ship d

36. Two tankers are traveling in the midst of the Atlantic Ocean. The first tanker is 100 miles due north of the second at 1:00 PM GMT (Greenwich Mean Time) and traveling due east at the rate of 20 miles per hour. The second tanker is traveling due north at 15 miles per hour. Express the distance d, shown in the accompanying figure, between the two tankers as a function of the number of hours after 1:00 PM GMT.

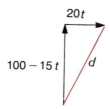

20t

100 − 15 t d

37. A rectangle is to be placed in a circle of radius 5 inches with its corners on the boundary of the circle. Express the area of the rectangle as a function of the width of one side and determine the domain of this function.

38. A rectangular beam is to be cut from a circular log of radius 5 inches in such a way that the corners of the beam all lie on the circumference of the circle. Define a function that describes the area of a cross section of the end of a beam of this type in terms of the width of the beam. What is the domain of this function?

39. A standard can contains a volume of 900 cm³. The can is in the shape of a right circular cylinder with a top and a bottom. Express the surface area of the can as a function of the radius of the bottom.

40. An open rectangular box is to be made from a piece of cardboard 8 inches wide and 11 inches long by cutting a square from each corner and bending up the sides. Express the volume of the box as a function of the length of a side of the square cut from each corner and determine the domain of this function.

41. A one-mile-long race track is to be built with two straight sides and with semicircles at the ends. Express the area enclosed by the track as a function of the diameter of the semicircles.

42. A house is being built with a straight driveway 800 feet long, as shown in the accompanying figure. A utility pole on a line perpendicular to the driveway and 200 feet from the end of the driveway is the closest point from which electricity can be furnished. The utility company will furnish power with underground cable at $2 per foot and with overhead lines at no charge. However, for overhead lines the company requires that a strip 30 feet wide

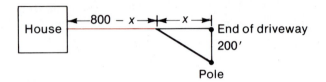

be cleared. The owner of the house estimates that to clear a strip this wide will cost \$3 for each foot of overhead wire used. How much will it cost to run the lines if they are run overhead to a point on the driveway x feet from the end of the driveway ($0 \le x \le 800$) and then run underground to the house?

1.2
LINEAR AND QUADRATIC FUNCTIONS

In this section we show some of the first steps of what we call *analyzing* a function or equation. The functions in these examples are used later to illustrate various concepts of calculus, so it is most important that you become completely familiar with their properties and graphs.

LINEAR FUNCTIONS

Functions whose graphs are straight lines are called **linear functions** and the equations describing them are known as **linear equations.**

Suppose l is a straight line that is not parallel to the y-axis and $P(x_1, y_1)$ and $Q(x_2, y_2)$ are two distinct points that lie on l, as illustrated in Figure 1.7.

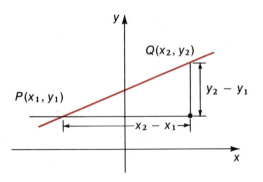

FIGURE 1.7

The **slope** m of the line l is defined to be

(1.2)
$$m = \frac{y_2 - y_1}{x_2 - x_1}.$$

A vertical line, one that is parallel to the y-axis, has the property that all points on the line have the same x-coordinate. Lines of this type have no slope (the calculation involved with the slope would require division by zero). Figure 1.8 shows various lines passing through a common point and lists their slopes.

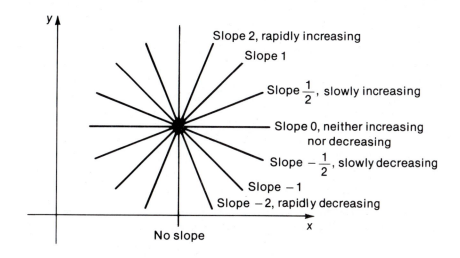

FIGURE 1.8

Notice that the slope determines the *direction* of a line, in the sense that:
1. A line with a positive slope is directed upward (looking from left to right). The values of y increase as the values of x increase, and the linear function is said to be increasing.
2. A line with a negative slope is directed downward (looking from left to right). The values of y decrease as the values of x increase, and the linear function is said to be decreasing.
3. A line with a large positive slope increases more rapidly than one with a smaller positive slope. A line having a negative slope that is large in magnitude decreases more rapidly than one with a negative slope of a smaller magnitude.

EXAMPLE 1   Find the slope of the line through the points $(2, 3)$ and $(-4, 0)$.

SOLUTION

This line has slope

$$m = \frac{3 - 0}{2 - (-4)} = \frac{3}{6} = \frac{1}{2}.$$

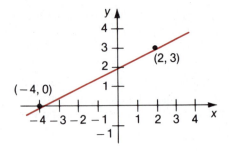

FIGURE 1.9

The graph of the line is shown in Figure 1.9. □

To determine an equation of the line l that passes through the point $P(x_1, y_1)$ and has slope m, suppose $Q(x, y)$ is any other point on l. Then

$$m = \frac{y - y_1}{x - x_1}$$

so the line has equation

(1.3) $y - y_1 = m(x - x_1).$

This equation is known as a **point-slope** form of the equation of a line.
     Equation (1.3) can be rewritten as

$$y = mx + y_1 - mx_1$$

or as

(1.4) $y = mx + b,$

where $b = y_1 - mx_1$. This equation is called the **slope-intercept** form of the equation of a line, since $(0, b)$ is the point where the line crosses the y-axis. The number b is called the **y-intercept** of the line.

EXAMPLE 2   Find a point-slope form and the slope-intercept form of the equation of the line passing through the points $(2, 3)$ and $(-4, 0)$.

SOLUTION
     From Example 1, we know that the slope of this line is 1/2. Using Equation (1.3) and the point $(2, 3)$, the point-slope form is

$$y - 3 = \frac{1}{2}(x - 2).$$

Using Equation (1.3) and the point $(-4, 0)$, the point-slope form is

$$y - 0 = \frac{1}{2}(x - (-4)).$$

When changed to the slope-intercept form, both equations reduce to

$$y = \frac{1}{2}x + 2.$$ □

EXAMPLE 3   Give the equations and sketch the graphs of the horizontal and vertical lines that pass through $(-1, 2)$.

SOLUTION

Horizontal lines have slope zero (all points on the line have the same y-coordinate), so the equation of the horizontal line passing through $(-1, 2)$ is

$$y = 2.$$

Since all points on a vertical line have the same x-coordinate, the equation of the vertical line that passes through $(-1, 2)$ is

$$x = -1.$$

The graphs of these lines are shown in Figure 1.10.   □

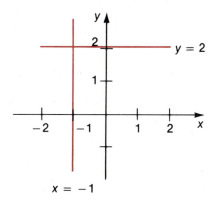

FIGURE 1.10

EXAMPLE 4   Sketch the graph of the linear function f described by $f(x) = 2x - 1$.

SOLUTION

Letting $y = f(x)$, we have

$$y = 2x - 1,$$

which is an equation of the line with slope 2 and y-intercept -1. Consequently, the point $(0, -1)$ lies on the line. To find another point, set $y = 0$. The x value of this point is called the **x-intercept** since this is where the line crosses the x-axis. When $y = 0, 0 = 2x - 1$, so

$$x = \frac{1}{2}.$$

The straight line through $(0, -1)$ and $(1/2, 0)$ is the graph of the function f and is shown in Figure 1.11.   □

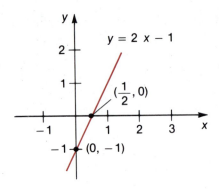

FIGURE 1.11

EXAMPLE 5   Determine the linear function that relates the temperature in degrees Celsius to the temperature in degrees Fahrenheit and use this function to determine the Fahrenheit temperature corresponding to 28° C.

SOLUTION

The freezing point of water at sea level corresponds to a temperature of 0°C and 32°F, while the boiling point of water at sea level corresponds to a temperature of 100°C and 212°F. If x denotes the temperature in degrees Celsius and y denotes the temperature in degrees Fahrenheit, then the linear function is described by the equation

$$y = mx + b$$

for some constants m and b.

Since $y = 32$ when $x = 0$, $b = 32$ and $y = mx + 32$. Substituting $y = 212$ and $x = 100$ implies that

$$212 = 100m + 32 \quad \text{and} \quad 180 = 100m;$$

so $m = 9/5$. Thus

$$y = \frac{9}{5}x + 32.$$

When $x = 28°C$,

$$y = \frac{9}{5}(28) + 32 = 82.4°F. \qquad \square$$

When two lines l_1 and l_2 intersect, two angles are determined. The angle generated by rotating l_1 counterclockwise about the point of intersection to coincide with l_2 is called the **angle from l_1 to l_2**. (See Figure 1.12.) The angle α from the x-axis to a line l is known as the **angle of inclination** of l. If l is parallel to the x-axis, then $\alpha = 0$.

The definition of the tangent function (see, in particular, equation (A.23) of Appendix A.3) implies that if a line has slope m and angle of inclination α, then $m = \tan \alpha$. Figure 1.12 illustrates the situation when $0 < \alpha < \pi/2$.

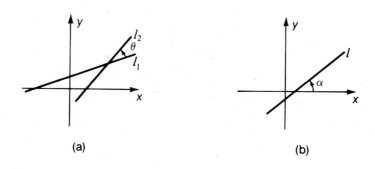

(a)   (b)

FIGURE 1.12

The slope of a line is used to determine when lines are parallel and when lines are perpendicular.

(1.5)
THEOREM
    (a) Two nonvertical lines are parallel if and only if they have the same slope.
    (b) Two nonvertical lines are perpendicular if and only if the product of their slopes is -1.

PROOF

Suppose the lines l_1 and l_2 have, respectively, slopes m_1 and m_2 and angles of inclination α_1 and α_2.

    (a) The lines are parallel if and only if $\alpha_1 = \alpha_2$. Since $0 \le \alpha_1 < \pi$ and $0 \le \alpha_2 < \pi$, $\alpha_1 = \alpha_2$ precisely when $\tan \alpha_1 = \tan \alpha_2$. Consequently, the lines are parallel if and only if $m_1 = \tan \alpha_1 = \tan \alpha_2 = m_2$.

    (b) If the lines are not parallel then $\alpha_1 < \alpha_2$ or $\alpha_2 < \alpha_1$. We will assume that $\alpha_1 < \alpha_2$ and let $\theta = \alpha_2 - \alpha_1$ denote the angle from l_1 to l_2 (see Figure 1.13).

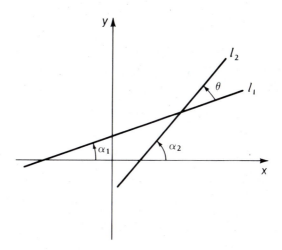

FIGURE 1.13

The lines l_1 and l_2 are perpendicular if and only if $\theta = \pi/2$. Using the trigonometric identity (A.19), $\theta = \pi/2$ precisely when

$$\tan \theta = \tan (\alpha_2 - \alpha_1) = \frac{\tan \alpha_2 - \tan \alpha_1}{1 + \tan \alpha_2 \tan \alpha_1} = \frac{m_2 - m_1}{1 + m_2 m_1}$$

is undefined. This implies that the lines are perpendicular if and only if $m_2 m_1 = -1$. ☐

EXAMPLE 6   Find an equation of the line that passes through $(1, 4)$ and is parallel to the line with equation $y = 2x - 1$.

SOLUTION

The slope of the line with equation $y = 2x - 1$ is 2, so the slope of the required line is 2. Consequently, for some constant b, the line has equation $y = 2x + b$. Since the line passes through $(1, 4)$, $4 = 2(1) + b$, and $b = 2$. The required line has equation $y = 2x + 2$. ☐

EXAMPLE 7   Find an equation of the line that passes through $(1, 4)$ and is perpendicular to the line with equation $y = 2x - 1$.

SOLUTION

If m is the slope of the line perpendicular to the line with equation $y = 2x - 1$, then m must satisfy the equation $2m = -1$. Hence $m = -1/2$. Therefore, the required line has an equation of the form $y = -(1/2)x + b$ for some constant b. Since the line passes through $(1, 4)$, $4 = -(1/2)1 + b$, so $b = 9/2$.

The desired equation is

$$y = -\frac{1}{2}x + \frac{9}{2}$$

or

$$x + 2y - 9 = 0. \qquad ☐$$

The equation of the line in Example 7 was expressed in the form

$$Ax + By + C = 0,$$

where A, B, and C are constants. Any linear equation can be written in this form. When $B = 0$ and $A \neq 0$, the equation describes a vertical line. When $B \neq 0$ and $A = 0$, the equation describes a horizontal line. When $B \neq 0$, the slope of the line is $-A/B$.

QUADRATIC FUNCTIONS

Quadratic functions are described by **quadratic equations** of the form

$$y = ax^2 + bx + c, \text{ where } a \neq 0.$$

The most basic of these equations occurs when $b = c = 0$ and $a = 1$. It is called the *squaring function* and is considered in the following example.

EXAMPLE 8   Sketch the graph of the function described by $f(x) = x^2$.

SOLUTION

The only point where the graph of the equation $y = x^2$ intersects either axis is $(0, 0)$. The domain of f is the set $\mathbb{R}$. Since $y = x^2 \geq 0$, the graph cannot lie below the x-axis; in fact, the range of f is $[0, \infty)$.

For values of x that are large in magnitude we see that $y = x^2$ is large and positive. Adding to this the knowledge that the points $(1, 1)$, $(-1, 1)$, $(2, 4)$, and $(-2, 4)$ all satisfy $y = x^2$ gives the graph shown in Figure 1.14. This graph is an example of a **parabola.**

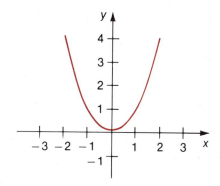

FIGURE 1.14

Because this graph has the same appearance on each side of the y-axis, we say it is *symmetric* with respect to the y-axis.   □

We have drawn the graph of f as a smooth and connected curve. To show that this is the case requires methods of calculus discussed in Chapter 3. At this stage, to sketch a graph, determine all the information you can, plot a few points to give a scale to the graph and then draw the simplest curve that agrees with the information. In most cases the graph will be reasonably correct.

The graph of the function described by $f(x) = x^2$ can be used to determine the graphs of other quadratic equations.

EXAMPLE 9   Analyze and sketch the graph of the function given by $g(x) = (x - 1)^2$.

SOLUTION

The correspondence described by g is similar to that of $f(x) = x^2$, but one unit is subtracted before the squaring operation is performed. Consequently, $g(1) = (1 - 1)^2 = 0^2 = f(0)$, $g(2) = 1^2 = f(1)$, $g(3) = f(2)$, and so on. For this reason, the graph of g is the same as the graph of f except that it is moved one unit to the right. (See Figure 1.15 on page 18).   □

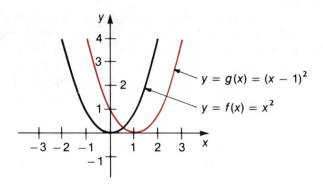

FIGURE 1.15

EXAMPLE 10   Sketch the graph of the function given by $h(x) = x^2 - 2x + 3$.

SOLUTION

To see the connection between h and the functions f and g in the two previous examples, we write $h(x)$ as a perfect square plus a real number, by "completing the square." That is, add and subtract the quantity

$$\left[\frac{1}{2}(\text{coefficient of } x)\right]^2.$$

Thus,

$$h(x) = x^2 - 2x + 3 = x^2 - 2x + \left[\frac{1}{2}(-2)\right]^2 - \left[\frac{1}{2}(-2)\right]^2 + 3$$
$$= x^2 - 2x + 1 + 2 = (x - 1)^2 + 2.$$

Written in this form, $h(x)$ is seen to be equal to $g(x) + 2$; and the graph of h is simply the graph of g moved up 2 units. Consequently, the graph of h has precisely the same "shape" as the graph of the squaring function f, as shown in Figure 1.16.   □

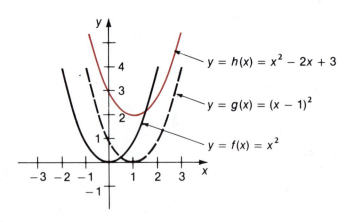

FIGURE 1.16

The graph of a quadratic function with equation of the form $y = ax^2$, for an arbitrary constant $a \neq 0$, has a shape similar to the graph of $y = x^2$. The particular value of a determines whether the graph will open upward, as in the case of the graphs shown in Figure 1.17 (a) where $a > 0$, or open downward, as in the case of the graphs shown in Figure 1.17 (b) where $a < 0$. As can be seen from the figures, the smaller the magnitude of a, the wider the opening of the graph.

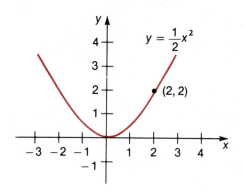
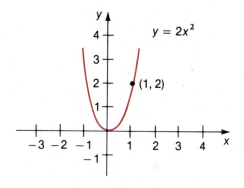

(a)

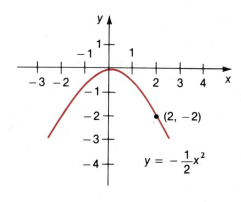
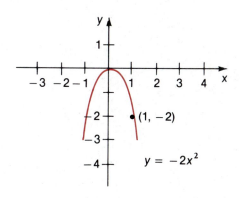

(b)

FIGURE 1.17

EXAMPLE 11   Sketch the graph of f if $f(x) = 2x^2 + 8x$.

SOLUTION

To complete the square on the right side of this equation, first factor the coefficient of x^2 from the expression. Then

$$f(x) = 2(x^2 + 4x) = 2(x^2 + 4x + 4) - 8$$
$$= 2(x + 2)^2 - 8.$$

Consequently, the graph has the same shape as the graph of $g(x) = 2x^2$ and can be obtained by shifting the graph of g two units to the left and eight units downward, as shown in Figure 1.18.   ☐

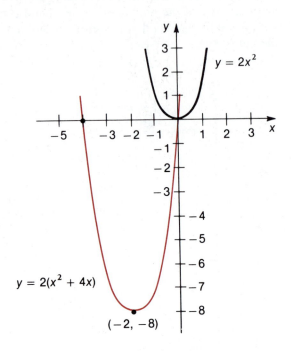

FIGURE 1.18

EXERCISE SET 1.2

1. Plot the pair of points given in each of the following and sketch the straight line determined by these points. Find the slope of the line.

   (a) $(0, 0)$, $(1, 2)$ (b) $(-1, 2)$, $(3, -2)$

   (c) $(1, -3)$, $(5, 1)$ (d) $(-3, -1)$, $(1, 2)$

2. Find an equation of the line determined by the pair of points in each part of Exercise 1.

3. Sketch the graph of the line associated with each of the following linear equations. Tell which pairs of lines are perpendicular.

   (a) $y = -1$ (b) $x = -2$ (c) $y = 2x + 3$

   (d) $y = 3x + 5$ (e) $y = \dfrac{1}{3}x - \dfrac{1}{3}$ (f) $y = -x - 3$

   (g) $y = -2x - 5$ (h) $y = -3x - 7$ (i) $y = -\dfrac{1}{3}x - \dfrac{5}{3}$

4. Tell which of the following pairs of linear equations describe lines that are parallel.

(a) $y = 2$ (b) $y = x + 1$ (c) $y = -x + 1$

(d) $y = x + 4$ (e) $y = 2x - 5$ (f) $y = -4$

(g) $x + y = 0$ (h) $4x - 2y - 4 = 0$ (i) $-x + y + 4 = 0$

In Exercises 5 through 12 the equation of a line is given together with a point not on the line. Find the slope-intercept form of the equation of the line that passes through the given point and is

    (a) parallel to the given line,   (b) perpendicular to the given line.

5. $y = 2x + 1, (0, 0)$ **6.** $y = 4x + 3, (0, 0)$

7. $y = 3x - 2, (1, 2)$ **8.** $y = \dfrac{1}{2}x + 2, (1, 1)$

9. $y = -2x + 3, (-1, 2)$ **10.** $y = -7x - 5, (-1, -3)$

11. $x + y + 1 = 0, (0, 0)$ **12.** $2x - 3y - 4 = 0, (1, -1)$

In Exercises 13 through 36 a quadratic equation is given. Sketch the graphs of these equations and find the range of the function described by the equation.

13. $y = x^2 + 1$ **14.** $y = x^2 - 2$

15. $y = (x - 3)^2$ **16.** $y = (x + 3)^2$

17. $y = x^2 - 4x + 4$ **18.** $y = x^2 - 4x$

19. $y = x^2 - 4x + 3$ **20.** $y = x^2 - 4x + 5$

21. $y = x^2 + 4x$ **22.** $y = x^2 + 4x + 5$

23. $y = -x^2$ **24.** $y = -x^2 + 1$

25. $y = -x^2 - 1$ **26.** $y = -(x - 1)^2$

27. $y = 2x^2$ **28.** $y = 2x^2 - 6$

29. $y = 2(x - 3)^2$ **30.** $y = 3x^2 + 6x$

31. $y = 2x^2 + 8x + 6$ **32.** $y = 2x^2 + 5x + 2$

33. $y = \dfrac{1}{2}x^2$ **34.** $y = \dfrac{1}{2}x^2 + 2$

35. $y = \dfrac{1}{2}x^2 - 2x + 2$ **36.** $y = \dfrac{1}{2}x^2 - 3x + 1$

37. Sketch the graph of f if   **38.** Sketch the graph of g if

$$f(x) = \begin{cases} x + 2, & \text{if } x < -1 \\ x^2, & \text{if } x \geq -1 \end{cases}$$

$$g(x) = \begin{cases} 0, & \text{if } x < 0 \\ x^2, & \text{if } 0 \leq x \leq 2 \\ -x + 6, & \text{if } x > 2 \end{cases}$$

39. The function defined by $s(t) = 640 - 16t^2$ describes the height, in feet, of a rock t seconds after being dropped from the top of a certain 60 story building. Sketch the graph of s and determine physically reasonable definitions for the domain and range of s. How long does it take the rock to reach the bottom of the building?

40. The function defined by $v(t) = -32t$ describes the velocity in feet per second of the rock t seconds after it was dropped from the building described in Exercise 39. Sketch the graph of v and determine physically reasonable definitions for the domain and range of v. What is the velocity of the rock when it reaches the bottom of the building?

41. A sweet potato farmer in North Carolina finds that he can produce 200 bushels of yams at an average cost of $3 per bushel and can produce 1000 bushels at an average cost of $2 per bushel. Assuming that the function describing the average cost is linear, find the average cost function and sketch its graph.

42. A hotel normally charges $40 for a room; however, special group rates are advertised: If the group requires more than 5 rooms, the price for each room is decreased by one dollar times the number of rooms exceeding five, with the restriction that the minimum price per room is $20.

(a) Express the charge per room as a function of the number of rooms used by a group and sketch the graph of this function.

(b) Express the revenue received from a group as a function of the number of rooms used by the group and sketch the graph of this function.

43. The average weight W (in grams) of a fish in a particular pond is assumed to depend on the total number n of fish in the pond in accordance with the model

$$W(n) = 500 - .5\,n.$$

(a) Sketch the graph of the function W.

(b) Express the total fish weight production (in grams) as a function of the number of fish in the pond. Sketch the graph of this function.

(c) What do you think happens when $n \geq 1000$?

44. When a solid rod is heated, its length increases by a certain amount depending on its coefficient of linear expansion. This coefficient, α, is assumed to be a constant, depending only on the material of the rod. The amount of increase in length is the product of the length, the change in temperature, and α. Suppose a steel rod has a length of 2 meters at 0°C and that the coefficient of linear expansion for this material is $\alpha = 11 \times 10^{-6}$. Find a function that describes the length of the rod in terms of its temperature above 0°C and determine its length when the temperature is 1000°C.

45. A function f has the property that $[f(x_1 + x_2)]^2 = [f(x_1)]^2 + [f(x_2)]^2$ for all real numbers x_1 and x_2. Tell as much as you can about f.

1.3
GRAPHS OF COMMON EQUATIONS

In this section we consider the graphs of some other basic equations often used in the study of calculus.

ABSOLUTE VALUE FUNCTION

The **absolute value** function, denoted by $f(x) = |x|$, associates with each real number x the distance of x from the origin. The function is defined by:

(1.6)
$$|x| = \begin{cases} x, & \text{if } x \geq 0 \\ -x, & \text{if } x < 0. \end{cases}$$

For example, $|3| = 3$, $|-2| = 2$, and $|0| = 0$.

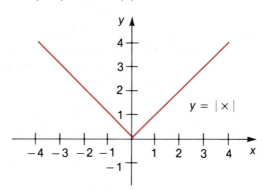

FIGURE 1.19

Graphing this function reduces to sketching the graph of the line $y = x$ when $x \geq 0$, and the graph of the line $y = -x$ when $x < 0$, as shown in Figure 1.19. Notice that this function is not "smooth" at the point $(0, 0)$; it turns abruptly at this point and forms a right angle.

EXAMPLE 1   Use the graph of $y = |x|$ to sketch the graph of $y = |x - 1| - 2$.

SOLUTION

Following the pattern discussed for $f(x) = x^2$, we see that the graph of $y = |x - 1|$ is the graph of $y = |x|$ shifted one unit to the right. To graph $y = |x - 1| - 2$, we move the graph of $y = |x - 1|$ two units downward as shown in Figure 1.20.   □

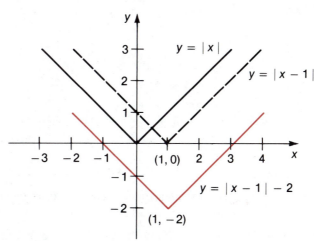

FIGURE 1.20

The absolute value can be used to express the distance between two points on the x-axis. If $x_2 > x_1$, then the distance $d(x_1, x_2)$ between x_1 and x_2 is given by $x_2 - x_1$. If $x_2 < x_1$, then $d(x_1, x_2) = x_1 - x_2 = -(x_2 - x_1)$. In both cases, the distance between x_1 and x_2 can be expressed by

$$d(x_1, x_2) = |x_2 - x_1|.$$

CIRCLES

A **circle** is a set of points that are equidistant from a fixed point. The equidistance is the **radius** of the circle and the fixed point is the **center** of the circle. The Pythagorean theorem implies that the distance between two points (x_1, y_1) and (x_2, y_2) in the plane is given by

(1.7) $$d((x_1, y_1), (x_2, y_2)) = \sqrt{(x_1 - x_2)^2 + (y_1 - y_2)^2}.$$

Hence, a circle with center (h, k) and radius r, as shown in Figure 1.21, has equation

$$\sqrt{(x-h)^2 + (y-k)^2} = r.$$

Squaring both sides gives

(1.8) $$(x-h)^2 + (y-k)^2 = r^2,$$

called the standard equation of a circle with center (h, k) and radius r. The special case when $r = 1$ is called a **unit circle.**

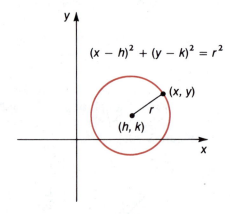

FIGURE 1.21

EXAMPLE 2   Sketch the graph of the circle with equation

$$x^2 + y^2 - 4x - 6y + 9 = 0.$$

SOLUTION

First rewrite the equation in standard form by completing the square in both variables x and y.

$$(x^2 - 4x) + (y^2 - 6y) + 9 = 0$$
$$(x^2 - 4x + 4) + (y^2 - 6y + 9) + 9 - 4 - 9 = 0.$$

Hence the original equation is equivalent to the equation

$$(x-2)^2 + (y-3)^2 = 4,$$

which describes a circle with center (2, 3) and radius 2. The graph of this circle is given in Figure 1.22.

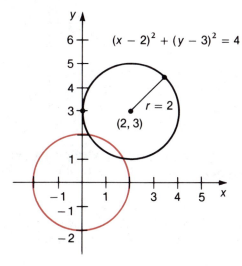

FIGURE 1.22

Notice that this is the graph of the circle with center (0, 0) and radius 2 shifted in a manner similar to that used for the quadratic and absolute value functions.   □

GREATEST INTEGER FUNCTION

The greatest integer function is denoted $f(x) = [\![x]\!]$ and defined by

$[\![x]\!]$ = the integer m that satisfies $m \le x < m + 1$.

In essence, the greatest integer function "rounds down" a real number to the next lowest integer. For example,

$$[\![1.2]\!] = 1, \qquad [\![-1.5]\!] = -2, \qquad [\![2]\!] = 2, \qquad [\![.3]\!] = 0.$$

This function assumes the constant integer value m on each interval $[m, m + 1)$, which implies that the graph of f consists of horizontal line segments beginning at each of the integers. For example, if x is in the interval $[-2, -1)$, then $-2 \le x < -1$ and $f(x) = -2$, if $0 \le x < 1$, $f(x) = 0$ and so on. The graph of the greatest integer function is given in Figure 1.23.

When	$[\![x]\!]$ is
$-2 \le x < -1$	-2
$-1 \le x < 0$	-1
$0 \le x < 1$	0
$1 \le x < 2$	1
$2 \le x < 3$	2

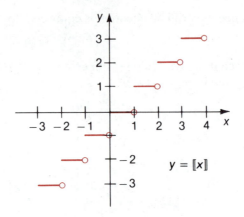

FIGURE 1.23

Since the graph of the greatest integer function "jumps" at each of the integers, this function is important for describing certain real-world situations.

EXAMPLE 3   A cab driver charges two dollars up to the first mile and one dollar for each succeeding mile. Describe graphically the relationship between the distance traveled and the cost. Find a function that relates the cost to the distance traveled.

SOLUTION

The cost for some distance ranges are listed as follows:

Miles, x	Cost
$0 < x < 1$	$2
$1 \leq x < 2$	$3
$2 \leq x < 3$	$4
$3 \leq x < 4$	$5

The charge "jumps" from $2 to $3 when the meter registers 1 mile and remains at $3 until the meter registers 2 miles. Graphically, we have the situation shown in Figure 1.24. Using the greatest integer function, we can express the cost c as a function of the number of miles x.

$$c(x) = \$2 + \$1 \, [\![x]\!] \qquad \text{for } x > 0.$$ □

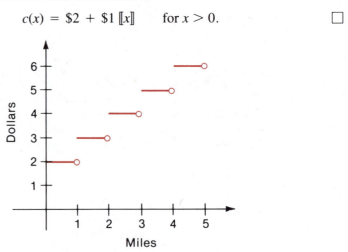

FIGURE 1.24

EXERCISE SET 1.3

Find the distance between the pairs of points given in Exercises 1 and 2.

1. $(0, 0), (1, -2)$ **2.** $(-1, 4), (0, -5)$

Use the graph of $y = |x|$ to sketch the graphs of the functions described in Exercises 3 through 8.

3. $f(x) = |x - 3|$ **4.** $f(x) = |x + 1|$

5. $f(x) = |x - 3| + 3$ **6.** $f(x) = |x + 1| - 1$

7. $f(x) = |x - 2| + 2$ **8.** $f(x) = |x + 2| - 2$

Use the graph of $y = [\![x]\!]$ to sketch the graphs of the functions described in Exercises 9 through 14.

9. $h(x) = [\![x - 2]\!]$ **10.** $h(x) = [\![x + 2]\!]$

11. $h(x) = [\![x]\!] - 2$ **12.** $h(x) = [\![x]\!] + 2$

13. $h(x) = [\![x + 2]\!] - 2$ **14.** $h(x) = -[\![x]\!]$

The equation of a circle is given in each of Exercises 15 through 22. Find the center and radius of each circle and sketch its graph.

15. $x^2 + y^2 = 9$ **16.** $x^2 + y^2 = 16$

17. $x^2 + (y - 1)^2 = 1$ **18.** $(x - 1)^2 + y^2 = 1$

19. $(x - 2)^2 + (y - 1)^2 = 9$ **20.** $(x + 2)^2 + (y - 1)^2 = 9$

21. $x^2 - 4x + y^2 - 2y = 4$ **22.** $x^2 + y^2 + 4x + 2y = 4$

In each of Exercises 23 through 26, the center and radius of a circle are given. Write the equation of the circle in each case and sketch its graph.

23. $(0, 0), r = 3$ **24.** $(1, 1), r = 1$

25. $(-1, 2), r = \dfrac{1}{2}$ **26.** $(1, -3), r = 2$

27. The sketch of the graph of the function described by $f(x) = x^3$ is shown below.

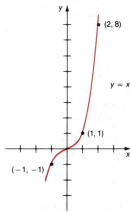

Use this graph to sketch the graph of g if g is defined by

(a) $g(x) = x^3 + 1$ (b) $g(x) = (x + 1)^3$

(c) $g(x) = x^3 - 1$ (d) $g(x) = (x - 1)^3$

(e) $g(x) = 2x^3$ (f) $g(x) = -2x^3$

28. Some functions have the property that for each element in the range there corresponds exactly one element in the domain. Functions with this property are called **one-to-one**. Determine which of the following functions are one-to-one.

(a) $f(x) = x^2$ (b) $f(x) = 2x + 3$

(c) $f(x) = |x|$ (d) $f(x) = [\![x]\!]$

(e) $g(x) = x^3$ (f) $h(x) = x^3 + 1$

29. Find the points (x, y) in the plane that are 5 units from $(0, 0)$ and 3 units from the x-axis.

30. What geometric figure is described by the equation

$$d((x, y), (1, 1)) = d((x, y), (3, 3))?$$

31. Express the distance from the point $(0, 1)$ to a point on the parabola $y = x^2$ as a function of the x-coordinate of the point on the curve.

32. Express the distance from the point $(0, 1)$ to a point on the circle $x^2 + y^2 = 1$ as a function of the y-coordinate of the point on the curve.

33. A cab driver charges five dollars for the first two miles and one dollar for each succeeding mile. Describe a function that relates cost to the distance traveled and sketch the graph of this function.

34. A cab driver charges $1.15 for the first fifth of a mile and 25 cents for each succeeding fifth of a mile. Describe a function that relates cost to the distance traveled. Suppose the driver tells you that it costs about $10.00 to go from your hotel to the airport and that you know the distance is 6 miles. How much tip is included in this approximation?

35. The Ohio Turnpike is 241 miles in length and has service plazas located 75 and 160 miles from Eastgate, the entrance to the Turnpike at the Pennsylvania line. Express the maximum distance of a car from the closest service plaza as a function of the car's distance from Eastgate and sketch the graph of this function.

36. Whooping cranes spend the winter at feeding grounds in the Aransas Bay in Texas and the summer at feeding grounds 2500 miles away in Wood Buffalo National Park, Northwest Territories, Canada. When migrating north in the spring, they stop to rest at feeding grounds in Valentine, Nebraska, a distance 1200 miles from Aransas. Find a function that describes the minimal distance of the cranes from one of these feeding grounds as a function of their distance from Aransas.

Putnam exercise:

37. If f and g are real-valued functions of one real variable, show that there exist numbers x and y such that $0 \le x \le 1, 0 \le y \le 1$, and $|xy - f(x) - g(y)| \ge 1/4$. (This exercise was problem 4, part I of the twentieth William Lowell Putnam examination given on November 21, 1959. The examination and its solution can be found in the January 1961 issue of the *American Mathematical Monthly*, pages 27–33.)

1.4
COMBINING FUNCTIONS

A profit function for a business firm is defined by a difference,

$$P(x) = R(x) - C(x),$$

where R is the revenue function, C is the cost function, and x is the number of items sold. In this case, functions have been combined using an arithmetic operation. We discuss in this section the combining of functions by both arithmetic and non-arithmetic methods.

(1.9)
DEFINITION

If f and g are functions, then the functions $f + g$, $f - g$, $f \cdot g$ and $\dfrac{f}{g}$ are defined by:

$$(f + g)(x) = f(x) + g(x)$$

$$(f - g)(x) = f(x) - g(x)$$

$$(f \cdot g)(x) = f(x) \cdot g(x)$$

$$\left(\frac{f}{g}\right)(x) = \frac{f(x)}{g(x)}.$$

For any of these operations to be defined at a number x, both $f(x)$ and $g(x)$ must be defined. The domains of $f + g$, $f - g$, and $f \cdot g$ consist of those real numbers that are common to both the domain of f and the domain of g. The domain of f/g consists of those real numbers x that are in both the domain of f and the domain of g, and that also satisfy $g(x) \neq 0$.

EXAMPLE 1

Find $f + g$, $f - g$, $f \cdot g$, f/g, and their domains if $f(x) = 1/(x^2 - 1)$ and $g(x) = x/\sqrt{x+4}$.

SOLUTION

$$(f + g)(x) = f(x) + g(x) = \frac{1}{x^2 - 1} + \frac{x}{\sqrt{x+4}},$$

$$(f - g)(x) = f(x) - g(x) = \frac{1}{x^2 - 1} - \frac{x}{\sqrt{x+4}},$$

$$(f \cdot g)(x) = f(x) \cdot g(x) = \frac{1}{x^2 - 1} \cdot \frac{x}{\sqrt{x+4}} = \frac{x}{(x^2 - 1)\sqrt{x+4}}$$

$$\left(\frac{f}{g}\right)(x) = \frac{f(x)}{g(x)} = \frac{\dfrac{1}{x^2 - 1}}{\dfrac{x}{\sqrt{x+4}}} = \frac{\sqrt{x+4}}{x(x^2 - 1)}.$$

Since the domain of f is all real numbers except ± 1 and the domain of g is the interval $(-4, \infty)$, the domain of $f + g$, $f - g$, and $f \cdot g$ is the set of real numbers that satisfy $x > -4$ and $x \neq \pm 1$. The domain of f/g is all real numbers in this set for which $g(x) \neq 0$. Since $g(x) = 0$ implies $x = 0$, the domain of f/g is all real numbers that satisfy $x > -4$, $x \neq \pm 1$, and $x \neq 0$.   ☐

Notice that although the simplified form of $(f/g)(x)$ in Example 1 seems to imply that -4 is in the domain of f/g, this is not true because -4 is not in the domain of g.

The arithmetic operations can be used to express complicated functions as a combination of more easily handled ones.

Consider the function described by $f(x) = |x| + x/2$. One method of sketching the graph of this function is to observe that $f(x) = g(x) + h(x)$ where $g(x) = |x|$ and $h(x) = x/2$. The graphs of g and h are shown in Figure 1.25 as dashed lines. We can graph f by observing that for each x, the value of $f(x)$ can be found by first finding the value of $g(x)$ and then adding the value of $h(x)$, as shown in Figure 1.25.

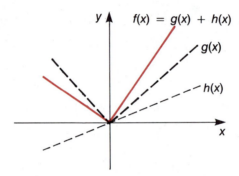

FIGURE 1.25

Although the product and quotient of functions are not generally useful as graphing techniques, special cases of these operations can be applied. In particular, if f is defined by $f(x) = c$, then $(f \cdot g)(x) = f(x) \cdot g(x) = cg(x)$. Graphing the equation $y = cg(x)$, if the graph of the function g is known, employs the same technique used in Section 1.2 to sketch the graph of $y = ax^2$ given the graph of $y = x^2$. The graphs of a typical function g with various values of c are shown in Figure 1.26.

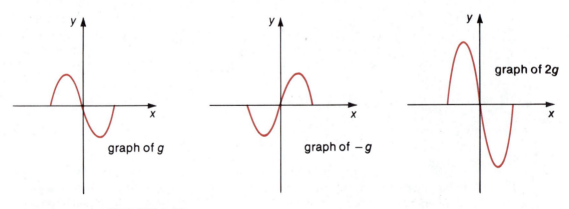

graph of g

graph of $-g$

graph of 2g

FIGURE 1.26

The reciprocal of a function g, defined by $(1/g)(x) = 1/g(x)$, is a special case of the quotient of two functions. A graphing technique using the reciprocal of a function is illustrated in the next example. The technique is elaborated on in Exercise 32.

EXAMPLE 2   Use the graph of $g(x) = x$ to determine the graph of $h(x) = (1/g)(x) = 1/x$.

SOLUTION

The graph of g is the straight line through $(0, 0)$ with slope 1. Consequently, the function h is undefined at $x = 0$ and its values increase in magnitude as x approaches zero. In addition, $h(x) = 1/x$ approaches zero as x increases in magnitude. Noting that $h(x)$ and $g(x)$ always have the same sign and that both graphs pass through the points $(1, 1)$ and $(-1, -1)$ leads to the graph shown in Figure 1.27. ☐

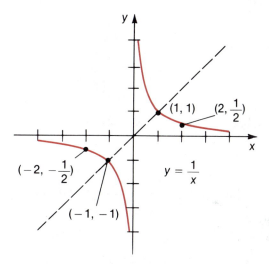

FIGURE 1.27

Another method of combining functions is called the **composition** of functions. It can be used as a type of building process that makes complicated functions from more elementary ones. The reverse process of taking a complicated function and decomposing it into a sequence of familiar, less complicated functions is extremely useful in the study of calculus.

(1.10)
DEFINITION

If g is a function mapping a set A into a set B and f is a function mapping the set B into a set C, then the **composition** of f and g, written $f \circ g$, is a function mapping the set A into the set C defined for each a in A by

$$(f \circ g)(a) = f(g(a)).$$

The composition of a pair of functions f and g is illustrated in Figure 1.28.

The domain of $f \circ g$ is defined to be those values x in the domain of g for which $g(x)$ is in the domain of f.

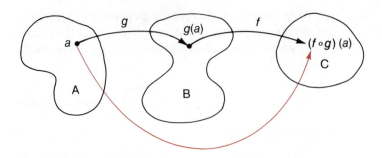

FIGURE 1.28

EXAMPLE 3   Find $(f \circ g)(2)$ and $(g \circ f)(2)$ if $f(x) = 2x + 3$ and $g(x) = x^2$.

SOLUTION

$$(f \circ g)(2) = f(g(2)) = f(4) = 11$$
$$(g \circ f)(2) = g(f(2)) = g(7) = 49.$$

Note that

$$(f \circ g)(2) \neq (g \circ f)(2). \qquad \square$$

In general, $(f \circ g)(x) \neq (g \circ f)(x)$.

EXAMPLE 4   Find $f \circ g$, $g \circ f$, and their domains if $f(x) = \sqrt{x-1}$ and $g(x) = 1/x^2$.

SOLUTION

$$(f \circ g)(x) = f(g(x)) = f\left(\frac{1}{x^2}\right) = \sqrt{\frac{1}{x^2} - 1}$$

$$(g \circ f)(x) = g(f(x)) = g(\sqrt{x-1}) = \frac{1}{(\sqrt{x-1})^2} = \frac{1}{x-1}.$$

The domain of g is the set of nonzero real numbers and the domain of f is the interval $[1, \infty)$. Consequently, the domain of $f \circ g$ is the set of nonzero real numbers x such that $g(x) = 1/x^2 \geq 1$, that is, $x \neq 0$ and $x^2 \leq 1$. In interval notation, the domain of $f \circ g$ is $[-1, 0) \cup (0, 1]$.

The domain of $g \circ f$ is the set of real numbers x such that $x \geq 1$ and $f(x) = \sqrt{x-1} \neq 0$, the open interval $(1, \infty)$. $\qquad \square$

EXAMPLE 5   Write the function described by $h(x) = |x^2 - 4x|$ as the composition of two functions.

SOLUTION

One way to decompose h is to observe the following chain:

$$x \longrightarrow (x^2 - 4x) \longrightarrow |x^2 - 4x|.$$

If we let $f(x) = |x|$ and $g(x) = x^2 - 4x$, then $(f \circ g)(x) = f(g(x)) = f(x^2 - 4x) = |x^2 - 4x|$ and $h = f \circ g$.

This representation of h is not unique. For example, the composition $f \circ g$ will also give h where the functions f and g are given by $f(x) = |x - 4|$ and $g(x) = (x - 2)^2$. ☐

Composition of functions will cause no difficulty if you remember that functional notation such as $f(x) = x^2 - 1$ means that the element x in the domain of f is mapped into the element obtained by squaring x and then subtracting 1. In particular, keep in mind that there is nothing special about the variable x, only the relation of the variable to its image. Using the notation $f(\_\_) = (\_\_)^2 - 1$ can be helpful when considering the composition since it emphasizes the fact that $f(g(x)) = (g(x))^2 - 1$.

Composition of functions is particularly useful in applications involving changes with respect to time.

EXAMPLE 6   A boat is leaking oil into a lake. Suppose the shape of the oil spill is approximately circular and at any time t minutes after the leak begins, the radius of the circle is $r(t) = \sqrt{t} + 1$. Find the area of the spill at any time t after the leak begins.

SOLUTION

The area A of a circle with radius r is given by $A(r) = \pi r^2$. The area of the spill depends on the radius, and the radius depends on time. Hence, A is a function of t given by

$$(A \circ r)(t) = A(r(t)) = A(\sqrt{t} + 1) = \pi(\sqrt{t} + 1)^2.$$ ☐

EXERCISE SET 1.4

In Exercises 1 through 8 a pair of functions f and g is described. Determine the equation describing each of the following functions and the domain in each case.

(a) $f + g$ (b) $f - g$ (c) $f \cdot g$

(d) $\left(\dfrac{f}{g}\right)$ (e) $f \circ g$ (f) $g \circ f$

1. $f(x) = x$, $g(x) = x + 1$ **2.** $f(x) = x^2$, $g(x) = x + 1$

3. $f(x) = \dfrac{1}{x}$, $g(x) = \sqrt{x - 1}$ **4.** $f(x) = x^2 + 1$, $g(x) = \dfrac{1}{\sqrt{x}}$

5. $f(x) = \dfrac{x}{x-1}$, $g(x) = \sqrt{x}$ **6.** $f(x) = \dfrac{x}{x-1}$, $g(x) = \dfrac{x}{x+1}$

7. $f(x) = x^2 + 1$, $g(x) = x - 2$ **8.** $f(x) = |x|$, $g(x) = 2x - 1$

Use the graphs of the absolute value and greatest integer functions to sketch the graphs of the functions described in Exercises 9 through 22.

9. $f(x) = -|x|$ **10.** $f(x) = 2|x|$

11. $f(x) = |x| + |x + 1|$ **12.** $f(x) = 2|x| + |x + 1|$

13. $f(x) = |x| - |x + 1|$

14. $f(x) = |x + 1| - |x|$

15. $f(x) = \dfrac{1}{|x|}$

16. $f(x) = \dfrac{1}{|x + 1|}$

17. $f(x) = -[\![x]\!]$

18. $f(x) = 2[\![x]\!]$

19. $f(x) = [\![x]\!] + [\![x + 1]\!]$

20. $f(x) = [\![x]\!] - [\![x + 1]\!]$

21. $f(x) = |x| - [\![x]\!]$

22. $f(x) = [\![x]\!] - |x|$

23. Express $F(x) = \sqrt{x^2 + 2x}$ as a composition of two functions, one of which is defined by $g(w) = \sqrt{w}$.

24. Express $G(x) = (x^3 - 5x + 1)^5$ as a composition of two functions, one of which is defined by $f(x) = x^5$.

25. Express $F(x) = (\sqrt{x^2 + 1} + 2)^3$ as a composition of three functions.

26. Functions f and g are defined by $f(x) = x^2 - 9$ and $g(x) = x + 3$. Sketch the graph of f/g. Sketch the graph of h if $h(x) = x - 3$. How do the two graphs differ?

27. Show that for any function f, $g(x) = (f(x) + f(-x))/2$ describes an even function. (An even function is defined in Exercise 31 of Exercise Set 1.1).

28. Show that for any function f, $h(x) = (f(x) - f(-x))/2$ describes an odd function. (An odd function is defined in Exercise 31 of Exercise Set 1.1).

29. Use Exercises 27 and 28 to show that any function f can be written as the sum of an odd and an even function, provided that when x is in the domain of f, $-x$ is also in the domain of f.

30. Use the results in Exercises 27, 28, and 29 to write each of the following functions as the sum of an odd and an even function.

(a) $f(x) = x^2 + x$

(b) $f(x) = |x + 1|$

(c) $f(x) = |x| + 1$

(d) $f(x) = \dfrac{1}{x} + 1$

31. A polynomial of degree n, where n is a positive integer, is a function P_n with domain $\mathbb{R}$ that is of the form

$$P_n(x) = a_n x^n + a_{n-1} x^{n-1} + a_{n-2} x^{n-2} + \cdots + a_2 x^2 + a_1 x + a_0$$

where $a_0, a_1, a_2, \ldots, a_n$ are all constants and $a_n \neq 0$. If P_n and Q_n are both polynomials of degree n, what can you say about the following functions?

(a) $P_n + Q_n$

(b) $P_n - Q_n$

(c) $P_n \cdot Q_n$

(d) $3P_n - 2Q_n$

(e) f described by $f(x) = xP_n(x) - Q_n(x)$

32. Answer the following concerning a function g and its reciprocal $1/g$.

(a) $\dfrac{1}{g(x)}$ is undefined when $g(x) = $ _____ .

(b) $g(x)$ and $\dfrac{1}{g(x)}$ have the same value when $g(x) = $ _____ or when $g(x) = $ _____ .

(c) Do $g(x)$ and $\dfrac{1}{g(x)}$ always have the same sign?

(d) When the magnitude of $g(x)$ is small, what is the magnitude of $\dfrac{1}{g(x)}$?

(e) When the magnitude of $g(x)$ is large, what is the magnitude of $\dfrac{1}{g(x)}$?

33. Use the results of Exercise 32 and the graph of g to sketch the graph of $h(x) = 1/g(x)$, when

(a) $g(x) = 2x - 1$ (b) $g(x) = |x|$

(c) $g(x) = x^2 - 1$ (d) $g(x) = x^2 - 4x + 3$

34. The graphs of the sine and cosine functions are shown in the accompanying figure.

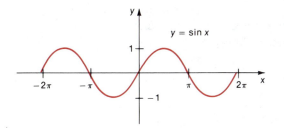 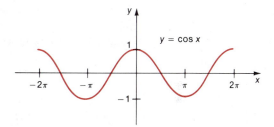

Use these graphs and the results of Exercise 32 to sketch the graphs of the cosecant and secant functions, defined by: csc $x = 1/\sin x$ and sec $x = 1/\cos x$.

35. A metal sphere is heated so that at any time t seconds after the heat is applied, the radius $r(t)$ is given by $r(t) = 3 + .01t$ centimeters. Express the volume of the sphere as a function of t.

36. Sand is being poured onto a conical pile whose radius and height are always equal, although both increase with time. If the height of the pile at any time t seconds after the pour begins is given by $h(t) = 10 + .25t$ feet, express the volume of the pile as a function of t.

37. In Exercise 39 of Exercise Set 1.2, the height of a rock above the ground t seconds after it was dropped from the top of a building was stated to be $s(t) = 640 - 16t^2$. The domain of the function s is the closed interval whose left endpoint is zero and whose right endpoint is the time it takes the rock to reach the ground. Define a function $\bar{s}$ that describes the height of the rock above the ground, assuming that the rock is dropped at $t = 2$ instead of at $t = 0$. What is the domain and range of $\bar{s}$?

38. Newton's law of gravitational attraction states that the attraction between an object of mass m_1 and an object of mass m_2 is directly proportional to the product of the masses m_1 and m_2 of the objects and inversely proportional to the square of the distance r between the centers of mass of the objects. Write a functional relationship describing this force in terms of the distance r, assuming that the masses remain constant. What restrictions must be put on the domain of this function if it is to describe the physical situation? Sketch the graph of the function.

39. A balloon is being inflated so that its radius at the end of t seconds is $r(t) = 3\sqrt{t} + 5$ cm, $0 \le t \le 4$. Express the volume and surface area as a function of time. What are the units of these quantities?

Putnam exercises:

40. Determine all polynomials P such that $P(x^2 + 1) = (P(x))^2 + 1$ and $P(0) = 0$. (This exercise was Problem A–2 of the thirty-second William Lowell Putnam examination given on December 4, 1971. The examination and its solution can be found in the February 1973 issue of the *American Mathematical Monthly*, pages 172–179.)

41. Let F be a real valued function defined for all real x except for $x = 0$ and $x = 1$ and satisfying the functional equation $F(x) + F\{(x-1)/x\} = 1 + x$. Find all functions F satisfying these conditions. (This exercise was Problem B–2 of the thirty-second William Lowell Putnam examination given on December 4, 1971. The examination and its solution can be found in the February 1973 issue of the *American Mathematical Monthly*, pages 172–179.)

1.5
ADDITIONAL TECHNIQUES FOR GRAPHING

In this section we consider additional topics associated with graphing equations and demonstrate some ways to approach the subject of graphing in a systematic manner.

SYMMETRY

A graph is symmetric with respect to a line if the portion of the graph on one side of the line is the mirror image of the portion on the other side. The importance of symmetry in sketching graphs is that it cuts the work in half. For example, if a graph is symmetric with respect to the y-axis, we need determine only the portion of the graph for $x \ge 0$; the portion for $x < 0$ is then obtained by symmetry.

Although there are many types of symmetry, we will restrict our discussion to symmetry with respect to the y-axis, the x-axis, and the origin.

(1.11)
DEFINITION

The graph of an equation is **symmetric with respect to the y-axis** provided that whenever (a, b) is on the graph, $(-a, b)$ is also on the graph. (See Figure 1.29(a).)

(1.12)
DEFINITION

The graph of an equation is **symmetric with respect to the x-axis** provided that whenever (a, b) is on the graph, $(a, -b)$ is also on the graph. (See Figure 1.29(b).)

(1.13)
DEFINITION

The graph of an equation is **symmetric with respect to the origin** provided that whenever (a, b) is on the graph, $(-a, -b)$ is also on the graph. (See Figure 1.29(c).)

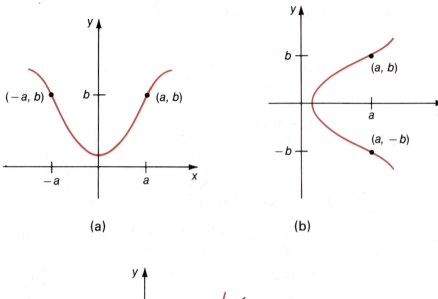

(a)   (b)

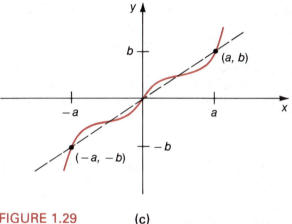

FIGURE 1.29   (c)

Two observations should be made concerning these symmetry definitions:

(1.14)

The graph of a function is **symmetric with respect to the *x*-axis** if and only if it is identically zero. Thus, we need not check for symmetry with respect to the *x*-axis when graphing a function.

(1.15)

The graph of an equation that is **symmetric with respect to either axis** is symmetric with respect to the origin if and only if it is symmetric with respect to both axes. That is, we need not check a graph for symmetry with respect to the origin unless we have determined that the graph is not symmetric with respect to either axis.

In the following examples we develop a systematic method of graphing that includes determining:
1. where the graph intersects the coordinate axes,
2. symmetry properties of the graph, and
3. excluded regions for the graph.

EXAMPLE 1　　　Sketch the graph of the equation

$$4x^2 + 9y^2 = 36.$$

SOLUTION

1. Intercepts: The intercepts occur when $4x^2 + 9 \cdot 0 = 36$, $x = \pm 3$, and when $4 \cdot 0 + 9y^2 = 36$, $y = \pm 2$. Consequently, the x-intercepts are at $(3, 0)$ and $(-3, 0)$, and the y-intercepts are at $(0, 2)$ and $(0, -2)$.

2. Symmetry: Both x and y appear only as squares in the equation $4x^2 + 9y^2 = 36$; hence, if (a, b) satisfies this equation, so do $(-a, b)$ and $(a, -b)$. This implies that the graph is symmetric with respect to both the x-axis and the y-axis. Knowing this, we need determine the graph only in the first quadrant. Symmetry will provide the graph in the remaining quadrants.

3. Excluded regions: To determine the regions of the plane in which the graph cannot lie, we solve the equation for both x and y.

   (i) $x^2 = \dfrac{1}{4}(36 - 9y^2)$ implies $x = \pm \dfrac{1}{2}\sqrt{36 - 9y^2}$

   (ii) $y^2 = \dfrac{1}{9}(36 - 4x^2)$ implies $y = \pm \dfrac{1}{3}\sqrt{36 - 4x^2}$

   From (i), we see that $36 - 9y^2 \geq 0$ so that $y^2 \leq 4$. Thus, $-2 \leq y \leq 2$.

   From (ii), we see that $36 - 4x^2 \geq 0$; that is, $x^2 \leq 9$ and $-3 \leq x \leq 3$.

The graph of this equation must lie in the region described by $-2 \leq y \leq 2$ and $-3 \leq x \leq 3$.

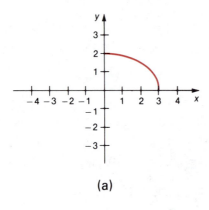

(a)

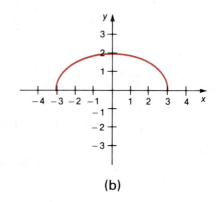

(b)

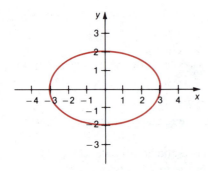

FIGURE 1.30 (c)

To determine the basic shape of the graph in the first quadrant, we plot the point when $x = 2$. This point, approximately $(2, 1.49)$, and the intercept points $(0, 2)$, $(3, 0)$ enable us to see that the graph in the first quadrant appears as shown in Figure 1.30(a). Symmetry with respect to the y-axis yields the graph shown in Figure 1.30(b). Using symmetry with respect to the x-axis, we complete the graph, as illustrated in Figure 1.30(c). ☐

EXAMPLE 2   Sketch the graph of the equation

$$x^2y^2 - y^2 + x^2 = 0.$$

SOLUTION

1. Intercepts: Since $y = 0$ implies $x^2 = 0$ and $x = 0$ implies $-y^2 = 0$, we see that this graph intersects the axes only at the origin $(0, 0)$.
2. Symmetry: The equation $x^2y^2 - y^2 + x^2 = 0$ involves only squares of both x and y, so the graph is symmetric with respect to both axes. Consequently, we need determine the graph only in the first quadrant.
3. Excluded regions: Solving the equation $x^2y^2 - y^2 + x^2 = 0$ for x and for y yields:

   (i) $x^2 = \dfrac{y^2}{y^2 + 1}$ or $x = \pm\sqrt{\dfrac{y^2}{y^2 + 1}}$

   (ii) $y^2 = \dfrac{-x^2}{x^2 - 1} = \dfrac{x^2}{1 - x^2}$ or $y = \pm\sqrt{\dfrac{x^2}{1 - x^2}}.$

We can see from the equations in either (i) or (ii) that $x^2 < 1$ and hence that $-1 < x < 1$. For example, in (ii) $y = \pm\sqrt{x^2/(1 - x^2)}$, which implies that $1 - x^2$ must be positive, and that $x^2 < 1$. From the equation $x = \sqrt{y^2/(y^2 + 1)}$, we see that as y becomes increasingly large, x approaches 1.

Plotting the points $(0, 0)$ and $(1/\sqrt{2}, 1)$ and drawing the curve in the first quadrant that agrees with our analysis gives the graph in this region as shown in Figure 1.31(a).

Applying the fact that the graph is symmetric with respect to both the y-axis and the x-axis, we obtain the final result shown in Figure 1.31(b). ☐

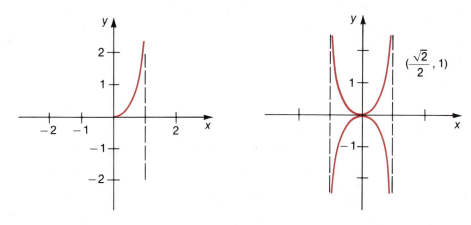

FIGURE 1.31 (a)   (b)

EXAMPLE 3   Sketch the graph of the equation

$$x^2 y + 3y - 6 = 0.$$

SOLUTION

This equation describes a function, since $y = 6/(x^2 + 3)$ and each value of x determines a unique value of y.

1. Intercepts: If $x = 0$, $y = 2$; however, if $y = 0$, we have $-6 = 0$, a false statement. Consequently, the graph intersects the y-axis at $(0, 2)$, but does not intersect the x-axis.

2. Symmetry: If (a, b) is on the graph of this function, $a^2 b + 3b - 6 = 0$ and since $(-a)^2 b + 3b - 6 = a^2 b + 3b - 6 = 0$, $(-a, b)$ is also on the graph. The graph is therefore symmetric with respect to the y-axis. By (1.14), the graph is not symmetric with respect to the x-axis.

3. Excluded regions: Solving for both x and y gives:

   (i) $y = \dfrac{6}{x^2 + 3}$

   and

   (ii) $x^2 = \dfrac{3(2 - y)}{y}$ or $x = \pm\sqrt{\dfrac{3(2 - y)}{y}}.$

From equation (i), we see that y must always be positive. Moreover, the smallest possible value for the denominator, and hence the largest possible value for y, occurs when $x = 0$ and $y = 6/(0 + 3) = 2$. As x increases in magnitude, the value of $y = 6/(x^2 + 3)$ becomes close to zero.

By plotting a few points, say $(1, 3/2)$ and $(2, 6/7)$, and using the symmetry with respect to the y-axis, we see that the graph of $y = 6/(x^2 + 3)$ is as shown in Figure 1.32.   □

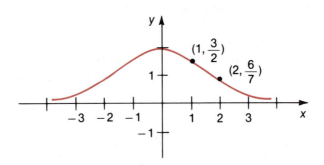

FIGURE 1.32

The exact shape of a graph can be more easily determined once the concepts of calculus are introduced in Chapter 3. However, if you understand the procedures outlined in this section, you can use them to give a rough sketch of the graph and then use the calculus concepts for fine tuning.

We close this section with a brief visual review of some of the more important aspects of graphing in this chapter.

SOME COMMON GRAPHS

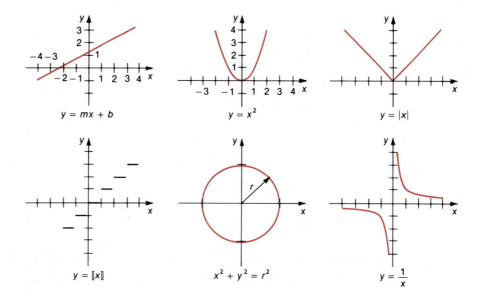

$y = mx + b$ $y = x^2$ $y = |x|$

$y = [\![x]\!]$ $x^2 + y^2 = r^2$ $y = \dfrac{1}{x}$

FIGURE 1.33

TRANSLATION

——— the graph of $y = f(x)$
- - - - - the graph of $y = f(x - a)$
——— the graph of $y = f(x) + b$
- - - - - the graph of $y = f(x - a) + b$

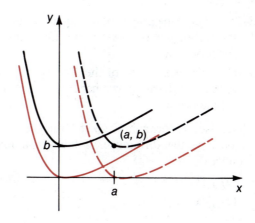

FIGURE 1.34

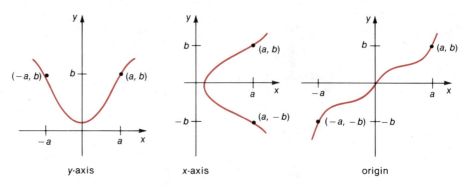

FIGURE 1.35

EXERCISE SET 1.5

Find the x- and y-intercepts of the equations in Exercises 1 through 6.

1. $x^2y^2 - x + 2y = 0$ **2.** $xy - y^2 + 1 = 0$

3. $xy - 2y^2 = 2$ **4.** $x^3 + 2y^2 = 3$

5. $x^2y^2 = 1$ **6.** $xy = 0$

In Exercises 7 through 22, determine whether the equations are symmetric with respect to (a) the y-axis, (b) the x-axis, (c) the origin.

7. $x^2y^2 - x^2 - y^2 - 0$ **8.** $xy - y^2 - 1 = 0$

9. $x^2y^2 + x^2 - 1 = 0$ **10.** $yx^2 + 4x + y = 0$

11. $x^2y^2 = 1$ **12.** $yx^2 + 4x + y = 1$

13. $xy^2 + 3x - 6 = 0$ **14.** $x^2 - xy + 1 = 0$

15. $x^2 - y^2 = x^2y^2$ **16.** $x^2y^2 + x^2 + y^2 = 0$

17. $(x-y)^2 = 0$ **18.** $x^2y^2 + 4x^2 - y^2 = 0$

19. $(x-y)^2 = 1$ **20.** $xy - 2y^2 - 2 = 0$

21. $yx^2 + 4y - x = 0$ **22.** $x^2 + 2xy + y^2 = 0$

23–38. Sketch the graphs of the equations in Exercises 7 through 22 by applying the techniques of this section. Determine if the equations describe y as a function of x.

39. Consider the graphs of the six trigonometric functions given in Appendix A.3, and determine whether each is symmetric with respect to (a) the y-axis, (b) the origin.

In Exercises 40 through 42, sketch the graphs of the functions in the order given and observe the difference in the graph that each successive complication introduces.

40. (a) $f_1(x) = x - 1$ (b) $f_2(x) = (x - 1)^2$

    (c) $f_3(x) = x^2 - 2x$ (d) $f_4(x) = |x^2 - 2x|$

    (e) $f_5(x) = \dfrac{1}{|x^2 - 2x|}$ (f) $f_6(x) = \dfrac{-1}{|x^2 - 2x|}$

41. (a) $f_1(x) = x - 2$
    (b) $f_2(x) = (x - 2)^2$
    (c) $f_3(x) = x^2 - 4x + 2$
    (d) $f_4(x) = |x^2 - 4x + 2|$

    (e) $f_5(x) = \dfrac{1}{|x^2 - 4x + 2|}$
    (f) $f_6(x) = \dfrac{2}{|x^2 - 4x + 2|}$

42. (a) $f_1(x) = x + 1$
    (b) $f_2(x) = (x + 1)^3$
    (c) $f_3(x) = x^3 + 3x^2 + 3x$
    (d) $f_4(x) = |x^3 + 3x^2 + 3x|$

    (e) $f_5(x) = \dfrac{1}{|x^3 + 3x^2 + 3x|}$
    (f) $f_6(x) = \dfrac{-2}{|x^3 + 3x^2 + 3x|}$

43. Consider the functions described in Exercise 40. Find a function g_1 with the property that $g_1 \circ f_1 = f_2$, a function g_2 with the property that $g_2 \circ f_2 = f_3$, etc. Show that if the functions g_1, g_2, g_3, g_4 and g_5 are chosen in this manner then we can write $f_5 = g_5 \circ g_4 \circ g_3 \circ g_2 \circ g_1 \circ f_1$.

44. Repeat the instructions given in Exercise 43 for the functions described in Exercise 41.

45. Repeat the instructions given in Exercise 43 for the functions described in Exercise 42.

46. Show that a function is even if and only if its graph is symmetric with respect to the y-axis.

47. Show that a function is odd if and only if its graph is symmetric with respect to the origin.

REVIEW EXERCISES

For the functions in Exercises 1 through 4, sketch the graph of f and find

    (a) $f(\sqrt{2})$
    (b) $f(\sqrt{2} + 1)$
    (c) $f(\sqrt{2}) + f(1)$

    (d) $f(\sqrt{x})$
    (e) $\sqrt{f(x)}$
    (f) $\dfrac{f(x + h) - f(x)}{h}$

1. $f(x) = 2x - 3$
    2. $f(x) = x^2 + 1$
3. $f(x) = x^2 - 7x + 10$
    4. $f(x) = x^2 + 2x + 4$

Find an equation of each of the lines described in Exercises 5 through 14. Sketch the graph of each line.

5. The line passes through the point $(0, 2)$ and has slope -1.

6. The line passes through the point $(1, -2)$ and is parallel to the line with equation $y - x - 1 = 0$.

7. The line passes through the point $(1, -2)$ and is perpendicular to the line with equation $y - x - 1 = 0$.

8. The line passes through the points $(-2, 0)$ and $(1, 3)$.

9. A horizontal line through the point $(1, -3)$.

10. A vertical line through the point $(1, -3)$.

11. A line with x-intercept 2 and slope -3.

12. A line with y-intercept -3 and slope -2.

13. A line with x-intercept 1 and y-intercept 2.

14. A line perpendicular to the line joining $(-1, 2)$ and $(2, 0)$ and passing through $(3/2, 1)$.

Sketch the graphs of the functions described in each of the Exercises 15 through 22.

15. $f(x) = x^2 - 2x - 1$

16. $f(x) = x^2 - 2x + 4$

17. $f(x) = x^2 - 2x - 4$

18. $f(x) = x^2 - 2x$

19. $f(x) = -x^2 + 6x$

20. $f(x) = 2x^2 + 4x + 3$

21. $f(x) = |x^2 - 2x|$

22. $f(x) = \dfrac{1}{x^2 - 2x}$

The graph of $f(x) = \dfrac{x^2}{x^2 + 1}$ is shown here.

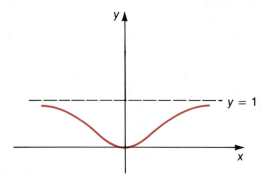

Use this graph to sketch the graph of each of the functions described in Exercises 23 through 28.

23. $f(x) = \dfrac{x^2}{x^2 + 1} + 1$

24. $h(x) = \dfrac{x^2}{x^2 + 1} - 1$

25. $k(x) = \dfrac{-x^2}{x^2 + 1}$

26. $l(x) = \dfrac{3x^2}{x^2 + 1}$

27. $u(x) = \dfrac{x^2 + 1}{x^2}$

28. $v(x) = -\dfrac{x^2 + 1}{x^2}$

For the functions described in Exercises 29 through 32, find $f \circ g$ and $g \circ f$ and the domain of each.

29. $f(x) = \sqrt{x^2 - 4}$, $g(x) = \dfrac{1}{x}$

30. $f(x) = |x|$, $g(x) = x^2 - 2x$

31. $f(x) = \dfrac{1}{x^2 - 1}$, $g(x) = x + 1$

32. $f(x) = \dfrac{x}{x + 1}$, $g(x) = \sqrt{x}$

33. Express the function described by $f(x) = (x^2 - 7x + 1)^3$ as a composition of two functions, one of which is given by $h(w) = w^3$.

34. Express the function described by $f(x) = \sqrt{x^3 + 1}$ as a composition of two functions, one of which is given by $h(w) = \sqrt{w}$.

Analyze each of the equations in Exercises 35 through 44 and sketch its graph.

35. $y = \dfrac{x}{x^2 + 1}$ **36.** $y = \dfrac{x - 1}{x + 1}$

37. $x^2 + y^2 = 9$ **38.** $x^2 + y^2 + 2x - 2y - 2 = 0$

39. $xy + y - 1 = 0$ **40.** $x^2 + 8x - 4y + 12 = 0$

41. $y = x(x - 2)(x + 1)$ **42.** $x^3 - y^2 = 0$

43. $xy^2 + x - y = 0$ **44.** $x^2 - 2xy + y^2 = 0$

45. A number of factors are considered when an anesthetic is administered. One factor is the age of the patient. A common formula gives a dosage $D(x) = D_0 x/(x + 12)$ to a patient of age x, where D_0 is the maximum allowable dosage. Sketch the graph of D.

2

LIMITS AND CONTINUITY OF FUNCTIONS

Calculus is distinguished from algebra, geometry, and other precalculus topics by the concept of the limit. One reason calculus was developed was a need to understand ideas such as velocity at a particular time, the tangent to a curve at a point, and the area under a curve. At the center of each of these problems lies the limit concept.

2.1
THE LIMIT OF A FUNCTION: THE INTUITIVE NOTION

To obtain some intuitive feeling for the concept of the limit, consider the following problems.

Suppose an empty 55-gallon barrel is supported by a wooden frame and water is added to the barrel at the rate of .75 gallons per minute. Precisely 56 minutes after the water begins to flow into the barrel the frame breaks and the water is spilled. The question we would like to answer is: If we rebuild the frame in exactly the same way, how much water can be put into the barrel without breaking the frame?

Since the water ran into the barrel for 56 minutes at the rate of .75 gallons per minute, the amount of water in the barrel when it spilled was

$$(56 \text{ min}) \, (.75 \text{ gal/min}) = 42 \text{ gal}.$$

Consequently, if the system is rebuilt exactly as before, 42 gallons is the *limit* of the amount of water we can put into the barrel. The best answer to the question is that we can put in any amount *less* than this limiting value.

Now suppose a bar of gold bullion, initially at 200° F, is placed in a freezing unit kept at a constant temperature of 0° F. At the end of one hour the temperature of the bar is found to be 100° F, at the end of two hours the temperature is 50° F, and so on. At the end of each hour interval the temperature is reduced to one half its previous value, as shown in Figure 2.1.

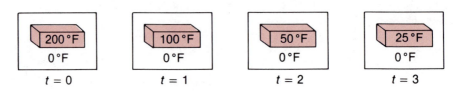

FIGURE 2.1

As time passes the temperature of the metal decreases, but does it ever reach 0° F?

Regardless of your answer to this question, it is clear that the temperature of the bar approaches 0° F as *t* increases. In other words, the *limiting* temperature of the bar is zero.

This example is somewhat different from the problem of the water barrel. By the limiting temperature of the gold bar we mean the temperature as the time becomes increasingly large, that is, as time approaches infinity. The water barrel problem involves a limit at a finite value (56 minutes). We will consider only limits at finite values in the first four sections of this chapter, deferring the discussion of limits involving infinity to Sections 2.5 and 2.6.

The limit of a function *f* at a number *a* describes the behavior of *f*(*x*) when *x* "approaches" or "is close to" *a*, as opposed to the particular value of *f* at *a*. The intuitive notion of the limit is expressed by saying that a function *f* has the limit *L* at the number *a* provided that *f*(*x*) becomes and remains close to *L* as *x* becomes close, but not equal, to *a*. This concept is expressed by writing

$$\lim_{x \to a} f(x) = L.$$

If such an *L* can be found, we say the limit of *f* at *a* exists or that *f* has a limit at *a*.

EXAMPLE 1   Find $\lim_{x \to 2} f(x)$ for the function defined by $f(x) = x^2$.

SOLUTION

The questions are: Is there a number *L* with the property that x^2 becomes and remains close to *L* as *x* approaches 2? And, if such a number *L* exists, what is it? To answer these questions, consider the values of *f* at some numbers close to 2.

x	$f(x)$		x	$f(x)$
1.9	3.61		2.1	4.41
1.99	3.9601		2.01	4.0401
1.999	3.996001		2.001	4.004001
1.9999	3.99960001		2.0001	4.00040001
1.99999	3.9999600001		2.00001	4.0000400001

It seems reasonable to conclude from this table that the limit of this function at $x = 2$ is 4. ☐

EXAMPLE 2   Find $\lim\limits_{x \to 2} g(x)$ for the function g defined by

$$g(x) = \frac{x^3 - 2x^2}{x - 2}.$$

SOLUTION

The graph of the function is given in Figure 2.2. Notice that the limit of the function f considered in Example 1 agrees with the value of f at $x = 2$. For the function g in this example, 2 is not in the domain, so this is certainly not the case.

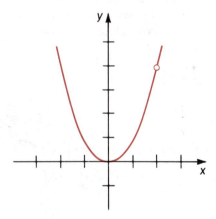

FIGURE 2.2

However,

$$g(x) = \frac{x^3 - 2x^2}{x - 2}$$

$$= \frac{x^2(x - 2)}{x - 2}$$

$$= x^2, \text{ when } x \neq 2.$$

So, the value of g at any number $x \neq 2$ agrees with the value of f at x. Since we are specifically excluding the value of the function at the point when finding the limit, the limits of f and g at the number 2 are the same and $\lim\limits_{x \to 2} g(x) = 4$. ☐

EXAMPLE 3   Determine $\lim\limits_{x \to 2} h(x)$ for the function h defined by

$$h(x) = \begin{cases} x^2 \, , \text{ if } x \neq 2 \\ 0 \; , \text{ if } x = 2. \end{cases}$$

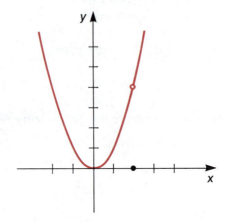

FIGURE 2.3

SOLUTION

   Since h and f agree at each number $x \neq 2$, the table listing the values of $f(x)$ in Example 1 also lists the values of $h(x)$ for $x \neq 2$. Thus, as x becomes "close" to 2, $h(x)$ becomes "close" to 4 and $\lim\limits_{x \to 2} h(x) = 4$ even though $h(2) = 0$.   □

EXAMPLE 4   Find $\lim\limits_{x \to 2} [\![x]\!]$.

SOLUTION

   In Chapter 1 we saw that the graph of the greatest integer function consists of "steps" at the integers, as shown in Figure 2.4.

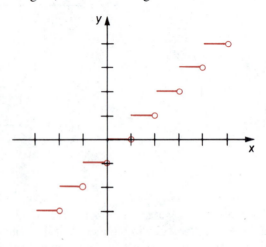

FIGURE 2.4

Examining the graph, we see that this function has a limit at certain points but not at others. For instance, all values of x sufficiently close to 3/2 have $[\![x]\!] = 1$, so we expect that $\lim\limits_{x\to 3/2} [\![x]\!] = 1$. However, for x close to 2, $[\![x]\!]$ can be either 1 or 2 depending on whether $x < 2$ or $x > 2$. Since no unique number exists that is close to *all* values of $[\![x]\!]$ when x is close, but not equal, to 2, $\lim\limits_{x\to 2} [\![x]\!]$ does not exist. In fact, this same reasoning indicates that the limit of the greatest integer function does not exist at any integer.   ☐

   Examples 2, 3, and 4 illustrate why we use the words "x becomes close, but not equal, to a" in the intuitive definition of the limit. Remember, when finding $\lim\limits_{x\to a} f(x)$, we are not interested in what happens *at a,* only *near a.*
   The next examples illustrate how a calculator can be used to gain an intuitive idea about limits.

EXAMPLE 5   Find $\lim\limits_{x\to 0} \sin x$.

SOLUTION

   The table below lists some values of $\sin x$ for x close to 0. If you are verifying the values of $\sin x$ by using a calculator, be sure that you use radians rather than degrees for the values of x.

x	$\sin x$	x	$\sin x$
.1	.099833417	$-.1$	$-.099833417$
.01	.009999833	$-.01$	$-.009999833$
.001	.001000000	$-.001$	$-.001000000$
.0001	.000100000	$-.0001$	$-.000100000$

The table suggests that $\lim\limits_{x\to 0} \sin x = 0$.   ☐

EXAMPLE 6   Find $\lim\limits_{x\to 0} \dfrac{\sin x}{x}$.

SOLUTION

   This problem is different from our previous examples in that both the numerator and denominator of the quotient approach zero yet have no common factor. The entries given in the tables in Example 5 suggest that
$$\lim_{x\to 0} \frac{\sin x}{x} = 1.$$ ☐

   The final example gives an important geometrical application of the limiting process.

EXAMPLE 7   Consider the function described by $f(x) = x^2$.

If $x \neq 1$, a unique line can be drawn joining the distinct points $(1, 1)$ and (x, x^2). This is called a **secant line** of the graph of f. (One secant line is drawn in Figure 2.5).

(a) Find the slope of the secant line joining the points $(1, 1)$ and (x, x^2) when $x \neq 1$.

(b) Find the limit of the slopes of the secant lines as x approaches 1.

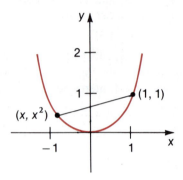

FIGURE 2.5

SOLUTION

(a) The slope of the secant line joining $(1, 1)$ and (x, x^2) is

$$\text{slope} = \frac{x^2 - 1}{x - 1} = \frac{(x - 1)(x + 1)}{x - 1}$$

$$= x + 1, \text{ when } x \neq 1.$$

(b) Since the slope of the secant line is $x + 1$ when $x \neq 1$, the limit of the slopes of the secant lines as x approaches 1 is

$$\lim_{x \to 1} \frac{x^2 - 1}{x - 1} = \lim_{x \to 1} (x + 1) = 2.$$

This limit is the slope of the line that will be defined in Chapter 3 to be tangent to the graph of f at $(1, 1)$. Figure 2.6 illustrates secant lines joining (x, x^2) and $(1, 1)$ for various values of x, together with the tangent line at $(1, 1)$. □

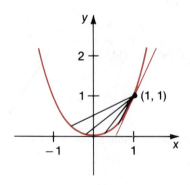

FIGURE 2.6

EXERCISE SET 2.1

In Exercises 1 and 2, the graph of a function f is given. Use the graph to find the indicated limits, if they exist.

1.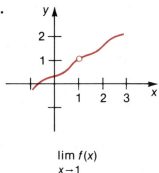

$$\lim_{x \to 1} f(x)$$

2.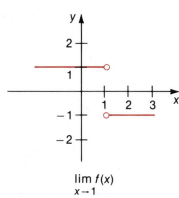

$$\lim_{x \to 1} f(x)$$

In Exercises 3 through 32, a function f is described and a number a is given. Determine the limit of f at a, if it exists.

3. $f(x) = 5x - 7, a = 2$

4. $f(x) = 3 - 2x^2 + 5x, a = 4$

5. $f(x) = \dfrac{1}{x^2}, a = \dfrac{1}{2}$

6. $f(x) = \dfrac{x^2 + 1}{x^3 + 2}, a = -1$

7. $f(x) = \dfrac{x^2 - 9}{x - 3}, a = 3$

8. $f(x) = \dfrac{x^2 - 9}{x + 3}, a = -3$

9. $f(x) = \dfrac{x^2 - 16}{x + 4}, a = -4$

10. $f(x) = \dfrac{x^2 - 16}{x - 4}, a = 4$

11. $f(x) = x + \dfrac{x^2 - 4}{x - 2}, a = 2$

12. $f(x) = x - \dfrac{x^2 - 4}{x - 2}, a = 2$

13. $f(x) = \dfrac{x^3 - 1}{x - 1}, a = 1$

14. $f(x) = \dfrac{x^3 - 27}{x - 3}, a = 3$

15. $f(x) = \dfrac{x^3 - 3x^2 - 4x + 12}{x^2 - 5x + 6}, a = 3$

16. $f(x) = \dfrac{x^3 - 3x^2 - 4x + 12}{x^2 - 5x + 6}, a = 2$

17. $f(x) = \dfrac{|x|}{x}, a = 1$

18. $f(x) = \dfrac{|x|}{x}, a = -\sqrt{3}$

19. $f(x) = \dfrac{|x|}{x}, a = 0$

20. $f(x) = [\![x]\!], a = 4$

21. $f(x) = \dfrac{[\![x]\!]}{x}, a = 1$

22. $f(x) = \dfrac{x}{[\![x]\!]}, a = 1$

23. $f(x) = \dfrac{1}{x - 1}, a = 1$

24. $f(x) = \dfrac{x - 4}{x - 2}, a = 2$

25. $f(x) = \sin x, a = \dfrac{\pi}{2}$

26. $f(x) = \cos x, a = 3\pi$

27. $f(x) = \sec x$, $a = 0$

28. $f(x) = \tan x$, $a = \dfrac{\pi}{4}$

29. $f(x) = \dfrac{\tan x}{\sec x}$, $a = 0$

30. $f(x) = \dfrac{\tan x}{\sec x}$, $a = \dfrac{\pi}{2}$

31. $f(x) = \tan x \cos x$, $a = \dfrac{\pi}{2}$

32. $f(x) = \cot x \sin x$, $a = 0$

In Exercises 33 through 36, a function f is described and a number a is given. Use a calculator to find the values of the function for x near a and determine the limit of f at a, if it exists.

33. $f(x) = \dfrac{\cos x - 1}{x}$, $a = 0$

34. $f(x) = \dfrac{\tan x}{x}$, $a = 0$

35. $f(x) = \dfrac{\cos x}{x - \frac{\pi}{2}}$, $a = \dfrac{\pi}{2}$

36. $f(x) = \dfrac{\sin 2x}{x}$, $a = 0$

37. Consider the function f described by $f(x) = x^2$.
  (a) Find an equation for the slope of the secant line joining the points $(2, 4)$ and (x, x^2) when $x \neq 2$.
  (b) Find the limit of the slopes of the secant lines as x approaches 2.

38. Consider the function described by $f(x) = x^2$.
  (a) Find an equation for the slope of the secant line joining the points (a, a^2) and (x, x^2) when $x \neq a$.
  (b) Find the limit of the slopes of the secant lines as x approaches a.

39. Consider the function f described by $f(x) = x^3$. If $x \neq 1$, a unique line can be drawn joining the points $(1, 1)$ and (x, x^3).
  (a) Find an equation for the slope of the secant line joining the points $(1, 1)$ and (x, x^3) when $x \neq 1$.
  (b) Find the limit of the slopes of the secant lines as x approaches 1.

40. Air is pumped into a spherical balloon at the rate of 30 cubic inches per second. Precisely 75 minutes after the pumping begins, the balloon bursts. What is the limiting amount of air that was pumped into the balloon?

41. Suppose $f(x) = 1000/(x + 1)$ gives the reading of a thermometer in degrees Fahrenheit when the thermometer is x inches from a flame and that the thermometer will immediately burst if it touches the flame. What is the limiting temperature reading as the thermometer approaches the flame?

2.2
THE LIMIT OF A FUNCTION: THE DEFINITION

We have seen the properties the limit concept must possess and have discussed situations in which the limit does and does not exist. The task now is to present the definition of the limit in a manner that can be used analytically, without referring to such vague terms as "close to" and "approaching." In this section we discuss a definition that precisely describes $\lim\limits_{x \to a} f(x) = L$ when both a and L are real numbers.

The difficulty in defining the limit lies in the fact that we are dealing with a relative concept: $f(x)$ being close to the limit L generally depends upon what is meant by x being close to the value a. The definition presented below resolves this difficulty by using tolerances about both the number a and the value of the limit L.

(2.1)
DEFINITION

Suppose f is a function that is defined on an open interval containing a, except possibly at a itself. The **limit of the function f at a is L** provided that for every number $\varepsilon > 0$, a number $\delta > 0$ can be found with the property that

$$|f(x) - L| < \varepsilon$$

whenever

$$0 < |x - a| < \delta.$$

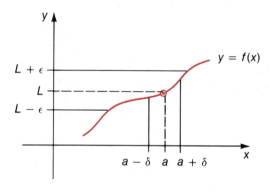

FIGURE 2.7

To see why we need a definition of this type, observe how this definition evolves from our intuitive notion of the limit. Consider the phrase "$f(x)$ is close to L"; "close" can be expressed by choosing an appropriate tolerance and stating that the value of $f(x)$ differs from L by less than this tolerance. The tolerance about L is usually denoted by ε (epsilon), and represents an arbitrary, but generally small, positive real number. Figure 2.8(a) shows an ε tolerance interval

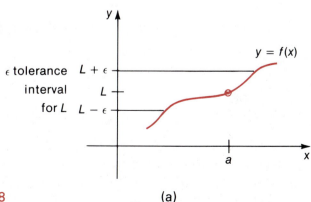

FIGURE 2.8   (a)

about L. The phrase "$f(x)$ is close to L" can then be replaced by the phrase "the difference between $f(x)$ and L is less than ε" and concisely written

(2.2)
$$|f(x) - L| < \varepsilon.$$

The next step is to express the statement "x is close, but not equal, to a" by an analytical expression and provide a link with the statement "$f(x)$ is close to L." We again use a tolerance, denoted by δ (delta), representing a positive real number and illustrated in Figure 2.8(b). The phrase "x is close, but not equal, to a" can be replaced by the phrase "the difference between x and a is less than δ and is not zero," or by writing

(2.3)
$$0 < |x - a| < \delta.$$

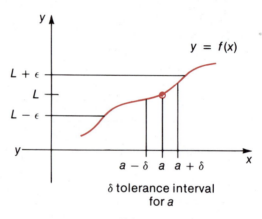

FIGURE 2.8   **(b)**

The link between statements (2.2) and (2.3) is provided by requiring that for *every* tolerance $\varepsilon > 0$, a tolerance $\delta > 0$ can be found with the property that "$f(x)$ differs from L by less than ε" for every number x provided "x differs from a by less than δ, and $x \neq a$." Using statements (2.2) and (2.3) to express this requirement leads to Definition 2.1.

Notice in the definition that for a given tolerance $\varepsilon > 0$, the number δ is not unique. In fact, once a specific value of δ is found, any number $\bar{\delta} > 0$ that is smaller than δ will also suffice because $0 < |x - a| < \bar{\delta}$ implies that $0 < |x - a| < \delta$.

EXAMPLE 1   Figure 2.9 shows the graph of $f(x) = 2x - 1$. Prove that

$$\lim_{x \to 2} f(x) = 3.$$

SOLUTION

Given the tolerance $\varepsilon > 0$ we must show that there is a number $\delta > 0$ so that

$$|(2x - 1) - 3| < \varepsilon$$

for all x satisfying $0 < |x - 2| < \delta$.

We can see from Figure 2.9 that such a δ can be found. In fact, given a horizontal strip of width ε on each side of the line $y = 3$, we can find a vertical

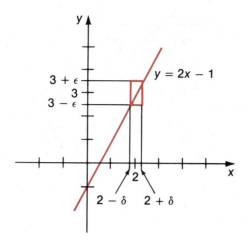

strip of width δ on each side of the line $x = 2$, with the property that the graph of f lies entirely within the rectangle formed by the intersection of the two strips.

The graph in Figure 2.10 shows that for $\varepsilon = 1$, we can choose $\delta = 1/2$; for $\varepsilon = 1/2$, we can choose $\delta = 1/4$; for $\varepsilon = 1/4$, $\delta = 1/8$; and so on. The graph convinces us that for any tolerance $\varepsilon > 0$, no matter how small, a number $\delta > 0$ can be found. However, we cannot rely on pictures to *prove* a result; we must do this analytically.

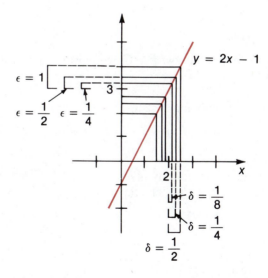

Suppose x satisfies the inequality $0 < |x - 2| < \delta$; then

$$|f(x) - 3| = |(2x - 1) - 3| = 2\,|x - 2| < 2\delta.$$

Thus, for any given number $\varepsilon > 0$, choose δ to satisfy $2\delta = \varepsilon$, that is, $\delta = \varepsilon/2$ (or any smaller positive number). If

$$0 < |x - 2| < \delta,$$

then $$|f(x) - 3| < 2\delta = 2(\varepsilon/2) = \varepsilon.$$

This completes the proof. The value of $\delta = \varepsilon/2$ could be anticipated because the graph of f is a line with slope 2.   □

The graph of the function defined by $f(x) = -3x + 3$ is given in Figure 2.11. Show that $\lim\limits_{x \to 2} f(x) = -3$.

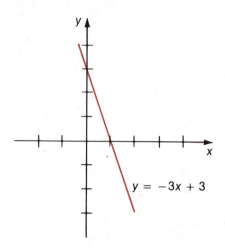

$y = -3x + 3$

FIGURE 2.11

SOLUTION

Given a number $\varepsilon > 0$, we must show that there is a number $\delta > 0$ so that

$$|(-3x + 3) - (-3)| < \varepsilon$$

for all x satisfying $0 < |x - 2| < \delta$. If $0 < |x - 2| < \delta$, then

$$|f(x) - (-3)| = |(-3x + 3) - (-3)|$$
$$= |-3(x - 2)| = 3|x - 2| < 3\delta.$$

Consequently, for any given $\varepsilon > 0$, choose δ so that $0 < \delta \le \varepsilon/3$. If x satisfies $0 < |x - 2| < \delta$, then $|f(x) - (-3)| < \varepsilon$.

This shows that $\lim\limits_{x \to 2} f(x) = -3.$ □

Notice that to prove $\lim\limits_{x \to a} f(x) = L$ requires three steps: simplification of $|f(x) - L|$ to involve $|x - a|$, choosing δ, and verifying that the choice of δ suffices.

In Examples 1 and 2 the functions were linear, and the application of the limit definition was straightforward. However, for more complex functions, the procedure is usually difficult to apply. The next example illustrates the problem.

EXAMPLE 3   Consider the function given by $f(x) = x^2$. Show that $\lim\limits_{x \to 2} f(x) = 4$.

SOLUTION

For a given number $\varepsilon > 0$, we must find a number $\delta > 0$ with the property that

$$|x^2 - 4| < \varepsilon,$$

whenever

$$0 < |x - 2| < \delta.$$

Proceeding as in the previous examples, if

$$0 < |x - 2| < \delta,$$

then

$$|x^2 - 4| = |(x - 2)(x + 2)| = |x - 2|\,|x + 2| < \delta\,|x + 2|.$$

However, we cannot solve for a *constant* δ by using the equation $\delta|x + 2| = \varepsilon$.

The problem can be solved by recalling the freedom we have in choosing δ. The only stipulation on δ is that if $0 < |x - 2| < \delta$, then $|x^2 - 4| < \varepsilon$. In our present situation, we choose δ small enough so that when $0 < |x - 2| < \delta$, x is not near -2. To ensure this and to simplify the problem, we first require that $\delta \leq 1$. If x satisfies

$$0 < |x - 2| < \delta \leq 1,$$

then

$$-1 \leq x - 2 \leq 1 \quad \text{and} \quad x \neq 2$$

so

$$1 \leq x \leq 3 \quad \text{and} \quad x \neq 2.$$

This keeps x within a bounded interval and ensures that $x \neq -2$. But we must still satisfy $|f(x) - 4| < \varepsilon$. If $0 < |x - 2| < \delta$, then

$$\begin{aligned}
|f(x) - 4| = |x^2 - 4| &= |x - 2|\,|x + 2| \\
&< \delta\,(|x + 2|) \leq \delta\,(|x| + |2|) \\
&\leq \delta\,(3 + 2) = 5\delta.
\end{aligned}$$

This implies that we should choose $0 < \delta \leq \varepsilon/5$. To satisfy both $0 < \delta \leq 1$ and $0 < \delta \leq \varepsilon/5$, we choose δ to be the minimum of $\varepsilon/5$ and 1. This will ensure that $|f(x) - 4| < \varepsilon$ for all values of x satisfying $0 < |x - 2| < \delta$. $\square$

The proof of the limit result in Example 3 is complicated even though the function is relatively simple. It should be clear that before we can work efficiently with limits of functions, we must develop some machinery for determining and demonstrating limits without actually applying the limit definition. This topic is considered in Section 2.3.

Having given the definition of the limit of a function at a point and considered some examples using this definition, it is important to realize that the definition ensures that there is *at most* one limit for any function at a number. This was one of the important intuitive conditions for the limit.

(2.4)
THEOREM

If $\lim_{x \to a} f(x) = L$ and $\lim_{x \to a} f(x) = M$, then $L = M$.

PROOF

To prove this theorem we use proof by contradiction. We assume that $L \neq M$ and show that this assumption leads to a contradiction. If $L \neq M$, then $\dfrac{|L - M|}{2} > 0$. Since $\lim\limits_{x \to a} f(x) = L$, for $\varepsilon = \dfrac{|L - M|}{2}$, there is a number

$\delta_1 > 0$ such that $|f(x) - L| < \varepsilon = \dfrac{|L - M|}{2}$ whenever $0 < |x - a| < \delta_1$.

Since $\lim\limits_{x \to a} f(x) = M$, for $\varepsilon = \dfrac{|L - M|}{2}$, there is a number $\delta_2 > 0$ such that

$|f(x) - M| < \varepsilon = \dfrac{|L - M|}{2}$ whenever $0 < |x - a| < \delta_2$.

If δ is the smaller of δ_1 and δ_2, then whenever $0 < |x - a| < \delta$,

$$|L - M| = |L - f(x) + f(x) - M| \le |L - f(x)| + |f(x) - M|$$

$$< \frac{|L - M|}{2} + \frac{|L - M|}{2} = |L - M|.$$

That is, whenever $0 < |x - a| < \delta, |L - M| < |L - M|$. This is a contradiction and our assumption is false. The only possibility is that $L = M$. $\square$

EXERCISE SET 2.2

In Exercises 1 through 12, a function f is described and values of a and L are given with $\lim\limits_{x \to a} f(x) = L$. For the given value of ε, find a number $\delta > 0$ with the property that $|f(x) - L| < \varepsilon$ whenever $0 < |x - a| < \delta$.

1. $f(x) = 2x + 3, a = 2, L = 7, \varepsilon = 1$

2. $f(x) = 2x + 3, a = 2, L = 7, \varepsilon = .5$

3. $f(x) = x - 2, a = 2, L = 0, \varepsilon = .005$

4. $f(x) = x + 1, a = 1, L = 2, \varepsilon = 10^{-4}$

5. $f(x) = 2x + 1, a = 1, L = 3, \varepsilon = .1$

6. $f(x) = 3x + 1, a = 1, L = 4, \varepsilon = .1$

7. $f(x) = 3 - 2x, a = 2, L = -1, \varepsilon = .1$

8. $f(x) = 3 - 2x, a = 2, L = -1, \varepsilon = 10^{-10}$

9. $f(x) = \dfrac{x^2 - 4}{x + 2}, a = 2, L = 0, \varepsilon = .01$

10. $f(x) = \dfrac{x^2 - 4}{x - 2}, a = 2, L = 4, \varepsilon = .01$

11. $f(x) = x^2, a = 3, L = 9, \varepsilon = .1$

12. $f(x) = x^2 + 1, a = -1, L = 2, \varepsilon = .1$

In Exercises 13 through 24, use the definition of the limit to prove each assertion.

13. $\lim\limits_{x \to 3} 2x + 5 = 11$

14. $\lim\limits_{x \to 2} 2x + 3 = 7$

15. $\lim\limits_{x \to -1} 3x + 4 = 1$

16. $\lim\limits_{x \to -2} -3x + 4 = 10$

17. $\lim\limits_{x \to -2} -2x + 5 = 9$

18. $\lim\limits_{x \to 2} |x - 2| = 0$

19. $\lim_{x \to 2} |x + 3| = 5$

20. $\lim_{x \to 3/2} [\![x]\!] = 1$

21. $\lim_{x \to 0} x^2 = 0$

22. $\lim_{x \to -1} x^2 + 1 = 2$

23. $\lim_{x \to 2} 1/x = 1/2$

24. $\lim_{x \to 4} \sqrt{x} = 2$

25. Suppose that $\lim_{x \to a} f(x) = L$. Does this imply that $\lim_{x \to a} |f(x)| = |L|$? If so, prove it; if not, find a counterexample to the statement. (A *counterexample* is an example for which the statement is false.)

26. Suppose that $\lim_{x \to a} |f(x)| = |L|$. Does this imply that $\lim_{x \to a} f(x) = L$? If so, prove it; if not, find a counterexample to the statement.

27. Show that $\lim_{x \to a} f(x) = L$ if and only if $\lim_{x \to a} (f(x) - L) = 0$.

28. Suppose that numbers $\delta > 0$ and M exist with $f(x) \leq M$ for all values of x in $(a - \delta, a + \delta)$. Show that $\lim_{x \to a} f(x) \leq M$, if this limit exists.

29. Consider the function described by

$$f(x) = \begin{cases} 1, & \text{if } x = \pm\dfrac{1}{n}, \ n = 1, 2, 3, 4, \ldots \\ 0, & \text{if } x \neq \pm\dfrac{1}{n}. \end{cases}$$

Does $\lim_{x \to 0} f(x)$ exist?

30. Consider the function described by

$$f(x) = \begin{cases} 1, & \text{if } x \text{ is rational} \\ -1, & \text{if } x \text{ is irrational}. \end{cases}$$

(a) Describe the graph of f.

(b) For which numbers a does $\lim_{x \to a} f(x)$ exist?

31. Consider the function described by

$$f(x) = \begin{cases} x, & \text{if } x \text{ is rational} \\ 1 - x, & \text{if } x \text{ is irrational}. \end{cases}$$

(a) Describe the graph of f.

(b) For which numbers a does $\lim_{x \to a} f(x)$ exist?

32. Show that if $\lim_{x \to a} f(x) = 0$ and $\varepsilon > 0$ is given, an interval containing a exists with $|f(x)| < \varepsilon$ for all $x \neq a$ in the interval.

33. Show that if $\lim_{x \to a} f(x) > 0$, then an interval containing a exists with $f(x) > 0$ for all $x \neq a$ in the interval.

34. Show that if $\lim_{x \to a} f(x) < 0$, then an interval containing a exists with $f(x) < 0$ for all $x \neq a$ in the interval.

2.3
LIMITS AND CONTINUITY

In this section we first discuss what is meant by a continuous function and then develop some results that can be used to simplify the limit process in many cases.

First let us distinguish between two types of functions that have limits at a number, by reviewing some of the functions (and their graphs) discussed in Section 2.1. (See Figure 2.12.)

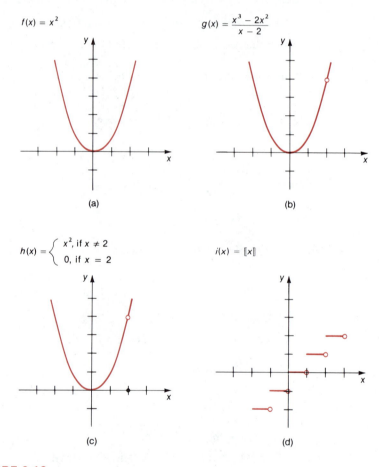

FIGURE 2.12

The limit at 2 exists in examples (a), (b), and (c). In (a) the limit at 2 agrees with the value of the function at 2, $\lim_{x \to 2} f(x) = 4 = f(2)$. This is not the case, however, in either (b), where $\lim_{x \to 2} g(x) = 4$, but $g(2)$ is undefined; or in part (c), where $\lim_{x \to 2} h(x) = 4 \neq 0 = h(2)$. To distinguish between these situations we have the following definition.

(2.5)
DEFINITION

A function f is **continuous** at the number a provided:
(i) $f(a)$ is defined,
(ii) $\lim\limits_{x \to a} f(x)$ exists,
(iii) $\lim\limits_{x \to a} f(x) = f(a)$.

If a function f is not continuous at a, then f is said to be **discontinuous** at a. So, if any one of the three criteria (i), (ii), (iii), is not met, then f is discontinuous at a.

A function f is called a **continuous function** provided f is continuous at every real number.

Reconsidering the previous examples, we see that:
(a) For $f(x) = x^2$, $\lim\limits_{x \to 2} f(x) = \lim\limits_{x \to 2} x^2 = 4 = f(2)$. Since all three of the conditions for continuity at $x = 2$ are satisfied, f is continuous at 2. In fact, f is continuous at each real number. So f is a continuous function.
(b) For $g(x) = (x^3 - 2x^2)/(x - 2)$, $g(2)$ is not defined. This violates part (i) of the definition of continuity, and g is discontinuous at 2. The manner in which g is defined, however, implies that g is continuous at every real number except 2.

(c) For $h(x) = \begin{cases} x^2, & \text{if } x \neq 2 \\ 0, & \text{if } x = 2 \end{cases}$, $\lim\limits_{x \to 2} h(x) = \lim\limits_{x \to 2} x^2 = 4$, but $h(2) = 0 \neq 4$.

Condition (iii) of the definition of continuity is not satisfied at 2, so h is not continuous at 2. The function h is, however, continuous at every real number except 2.
(d) For $i(x) = [\![x]\!]$, $\lim\limits_{x \to 2} i(x) = \lim\limits_{x \to 2} [\![x]\!]$ does not exist. Thus, i is discontinuous at 2 because part (ii) of the definition is violated. In fact, i is discontinuous at every integer, but continuous at every other real number.

Looking again at the graphs of these functions, we see that only the graph of f does not "jump" or is not "interrupted" at $x = 2$. Our mathematical definition of continuous is, in this sense, equivalent to the dictionary definition of continuous behavior as that behavior which is "without cessation or interruption."

Many physical properties have continuous behavior. For example, temperature, the height of a person, and the flight of a baseball are continuous in nature. On the other hand, the cost of postage, the number of hamburgers consumed per day, and the population of a certain town are discontinuous in nature since each of these measurements is given in integers only.

At the end of Section 2.2 we said that the $\varepsilon - \delta$ process involved with limits is too difficult and tedious to employ in most situations. The proofs of the following results depend on the $\varepsilon - \delta$ definition of the limit given in Section 2.2, but once the results have been proved, most limits can be found without referring to the definition.

The first result concerns limits and continuity of functions whose graphs are straight lines.

(2.6)
THEOREM

For any constants m and b, and any real number a,

$$\lim_{x \to a} (mx + b) = ma + b.$$

Consider the graphs of the straight lines with positive and negative slopes shown in Figures 2.13(a) and (b). Notice that for a given value of ε, the number δ can be chosen as ε divided by the magnitude of the slope of the line, provided that the slope is not zero. When the slope is zero any value of δ will suffice because the line is horizontal, as shown in Figure 2.13(c).

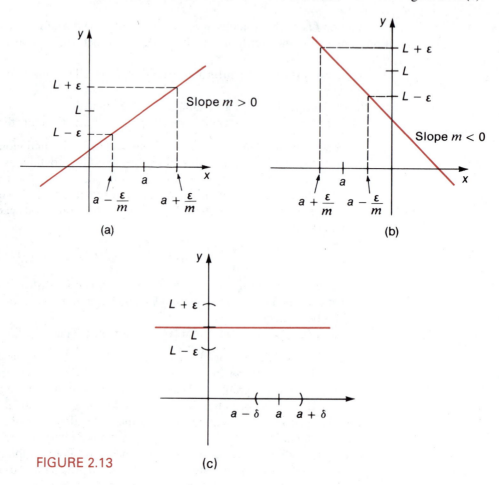

FIGURE 2.13   (c)

PROOF

Suppose $\varepsilon > 0$ is given. To find a number δ with the property that

$$|(mx + b) - (ma + b)| < \varepsilon$$

whenever $0 < |x - a| < \delta$, it is necessary to consider two cases:
(i) If $m \neq 0$, then $|(mx + b) - (ma + b)| = |m| \cdot |x - a| < |m| \, \delta$ and the desired condition can be obtained by choosing δ to satisfy $0 < \delta \leq \varepsilon/|m|$. If

$$0 < |x - a| < \delta \leq \frac{\varepsilon}{|m|},$$

then

$$|(mx + b) - (ma + b)| = |mx - ma| = |m| \cdot |x - a|$$
$$< |m| \cdot \delta \leq |m| \cdot \frac{\varepsilon}{|m|} = \varepsilon.$$

(ii) If $m = 0$, then $|(mx + b) - (ma + b)| = 0 < \varepsilon$, regardless of the value of x. In this case, any number $\delta > 0$ will suffice. □

As an immediate corollary we obtain a corresponding result about the continuity of linear functions.

(2.7)
COROLLARY

If f is a function described by $f(x) = mx + b$, then f is a continuous function.

A special case of Corollary 2.7 is that if $f(x) = b$, then f is a continuous function; that is, constant functions are continuous functions.

The next theorem is an extremely powerful result for determining limits of complicated functions. This result provides us with a constructive process to split a complicated problem into more manageable parts. Although the proof is not difficult, it is rather technical and is given in Appendix B.

(2.8)
THEOREM

If f and g are functions with $\lim\limits_{x \to a} f(x) = L$ and $\lim\limits_{x \to a} g(x) = M$, then

(a) $\lim\limits_{x \to a} [f(x) + g(x)] = L + M$,

(b) $\lim\limits_{x \to a} [f(x) - g(x)] = L - M$,

(c) $\lim\limits_{x \to a} [f(x) \cdot g(x)] = L \cdot M$,

(d) $\lim\limits_{x \to a} cf(x) = cL$, for any constant c,

(e) $\lim\limits_{x \to a} \dfrac{1}{f(x)} = \dfrac{1}{L}$, provided $L \neq 0$,

(f) $\lim\limits_{x \to a} \dfrac{g(x)}{f(x)} = \dfrac{M}{L}$, provided $L \neq 0$,

(g) $\lim\limits_{x \to a} \sqrt[n]{f(x)} = \sqrt[n]{L}$ if n is an odd positive integer or if n is an even

positive integer and $L > 0$.

EXAMPLE 1

Show that $\lim\limits_{x \to 2} x^2 = 4$.

SOLUTION

By Theorem 2.6, $\lim\limits_{x \to 2} x = 2$. An application of part (c) of Theorem 2.8 with $f(x) = x$ and $g(x) = x$ yields:

$$\lim_{x \to 2} x^2 = \lim_{x \to 2} x \cdot x = 2 \cdot 2 = 4.$$ □

The ease with which we found $\lim\limits_{x \to 2} x^2$ in Example 1 using the results of Theorem 2.8 contrasts sharply with the method used in Example 3 of Section 2.2.

EXAMPLE 2   Show that $\lim\limits_{x \to 8} x^{2/3} = 4$.

SOLUTION

By Theorem 2.6, $\lim\limits_{x \to 8} x = 8$ and by part (c) of Theorem 2.8,

$$\lim_{x \to 8} x^2 = \lim_{x \to 8} x \cdot x = 8 \cdot 8 = 64.$$

Applying part (g) of Theorem 2.8 implies that

$$\lim_{x \to 8} x^{2/3} = \lim_{x \to 8} \sqrt[3]{x^2} = \sqrt[3]{64} = 4. \qquad \square$$

EXAMPLE 3   Determine $\lim\limits_{x \to 3} f(x)$, when $f(x) = \dfrac{x^3 - x^2 - 10x + 12}{x^2 - 9}$.

SOLUTION

The denominator of $f(x)$ is zero at $x = 3$, so Theorem 2.8 cannot be directly applied to determine this limit. However, since the numerator of $f(x)$ is also zero at $x = 3$, $(x - 3)$ must be a factor of both the numerator and denominator of $f(x)$. Consequently, for $x \neq 3$,

$$\begin{aligned}
f(x) &= \frac{x^3 - x^2 - 10x + 12}{x^2 - 9} \\
&= \frac{(x^2 + 2x - 4)\,(x - 3)}{(x + 3)\,(x - 3)} \\
&= \frac{x^2 + 2x - 4}{x + 3}
\end{aligned}$$

and using Theorem 2.8, part (f), we have

$$\lim_{x \to 3} f(x) = \lim_{x \to 3} \frac{x^2 + 2x - 4}{x + 3} = \frac{\lim\limits_{x \to 3}(x^2 + 2x - 4)}{\lim\limits_{x \to 3}(x + 3)} = \frac{11}{6}. \qquad \square$$

(2.9)
COROLLARY

If f and g are functions that are continuous at a, then

(a) $f + g$ is continuous at a,

(b) $f - g$ is continuous at a,

(c) $f \cdot g$ is continuous at a,

(d) cf is continuous at a, for any constant c,

(e) $\dfrac{1}{f}$ is continuous at a, provided $f(a) \neq 0$,

(f) $\dfrac{g}{f}$ is continuous at a, provided $f(a) \neq 0$,

(g) $f^{1/n}$, defined by $f^{1/n}(x) = \sqrt[n]{f(x)}$, is continuous at a if n is an odd positive integer or if n is an even positive integer and $f(a) > 0$.

One of the many applications of Corollary 2.9 is considered in the following example.

EXAMPLE 4   Show that the function described by $f(x) = 7x^2 - 7x + 1$ is a continuous function.

SOLUTION

The function can be expressed as

$$f(x) = 7(g_1(x))^2 + g_2(x)$$

where $g_1(x) = x$ and $g_2(x) = -7x + 1$. It follows from Corollary 2.7 that both g_1 and g_2 are continuous functions. Part (c) of Corollary 2.9 implies that $g_1^2 = g_1 \cdot g_1$ is continuous at every real number and part (d) implies that $7g_1^2$ also has this property. Now, part (a) of Corollary 2.9 can be applied to see that $f = 7g_1^2 + g_2$ is continuous at each real number. So f is a continuous function. ☐

The reasoning used in the last example can be extended to any polynomial.

(2.10)
THEOREM   If $f(x) = a_n x^n + a_{n-1}x^{n-1} + \ldots + a_1 x + a_0$ where the a_is are real numbers, then f is a continuous function.

To see how this result can be applied to a larger class of functions, consider the following example.

EXAMPLE 5   Determine the numbers at which the function described by

$$f(x) = \frac{x^2 + 3x + 2}{x^2 - 2x - 8}$$

is continuous.

SOLUTION

The function f is the quotient of two continuous functions g and h, where $g(x) = x^2 + 3x + 2$ and $h(x) = x^2 - 2x - 8$. Part (f) of Corollary 2.9 implies that $f = g/h$ is continuous at every real number except when

$$x^2 - 2x - 8 = (x - 4)(x + 2) = 0.$$

The conclusion is that f is continuous except at $x = 4$ and $x = -2$. ☐

The procedure used to demonstrate the continuity of the function in Example 5 can be applied to any function that is a quotient of polynomials. A function of this type is called a **rational function.**

(2.11)
THEOREM

A rational function f described by

$$f(x) = \frac{a_n x^n + a_{n-1} x^{n-1} + \ldots + a_1 x + a_0}{b_m x^m + b_{m-1} x^{m-1} + \ldots + b_1 x + b_0},$$

is continuous at all x for which $b_m x^m + b_{m-1} x^{m-1} + \ldots + b_1 x + b_0 \neq 0$.

For composite functions we might *conjecture* (a conjecture is an educated guess) that if $\lim_{x \to a} g(x) = b$ and $\lim_{x \to b} f(x) = L$, then $\lim_{x \to a} f(g(x)) = L$. Unfortunately our intuition in this case is incorrect. (See Exercise 56.) However, if f is continuous at b, the result is valid.

(2.12)
THEOREM

If $\lim_{x \to a} g(x) = b$ and f is continuous at b, then $\lim_{x \to a} f(g(x)) = f(b)$.

(2.13)
COROLLARY

If g is continuous at a and f is continuous at $b = g(a)$, then $f \circ g$ is continuous at a.

EXAMPLE 6

Show that the function h described by

$$h(x) = \sqrt{x^2 + 2}$$

is a continuous function.

SOLUTION

The function h is the composition, $h = f \circ g$, of the functions g and f defined by

$$g(x) = x^2 + 2 \qquad \text{and} \qquad f(x) = \sqrt{x}.$$

The function g is the sum of two continuous functions and is therefore continuous. By Corollary 2.9, f is continuous for all $x > 0$. The previous corollary tells us that $h = f \circ g$ is continuous for all x such that $g(x) = x^2 + 2 > 0$; that is, h is continuous at every real number.

Another way to show that h is continuous is to apply part (g) of Corollary 2.9 directly to the function h. □

EXERCISE SET 2.3

1. Which of the following illustrate continuous behavior?
   (a) Number of dogs in a household.
   (b) Weight of a cat.
   (c) Flight of a golfball.
   (d) Sales tax rate.
2. Which of the following illustrate continuous behavior?
   (a) Interest rate.
   (b) Grade point average.
   (c) Growth of a tomato plant.
   (d) The number of grains of sand in the bottom half of an hourglass.

In each of Exercises 3 through 6 the graph of a function is given. From the graph, determine the points at which the function is discontinuous and give the criterion of Definition 2.5 that is not met.

3.

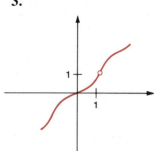

4.

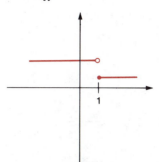

5.

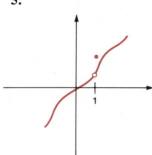

6.

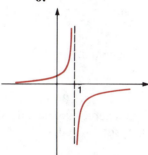

Sketch the graph of each function described in Exercises 7 through 14, and determine any numbers at which the function is discontinuous. For each discontinuity, give the criterion of Definition 2.5 that is not met.

7. $f(x) = \dfrac{x^2 - 4}{x - 2}$

8. $f(x) = \begin{cases} \dfrac{x^2 - 4}{x - 2}, & \text{if } x \neq 2 \\ 1 & \text{, if } x = 2 \end{cases}$

9. $f(x) = \dfrac{|x|}{x}$

10. $f(x) = \dfrac{|x + 4|}{x + 4}$

11. $f(x) = \begin{cases} \dfrac{x^3 - 3x^2 - 4x + 12}{x^2 - 5x + 6}, & \text{if } x \neq 2,3 \\ 0 & \text{,if } x = 2 \\ 1 & \text{,if } x = 3 \end{cases}$

12. $f(x) = [\![x - 1]\!]$

13. $f(x) = \tan x$ (Note: Graphs of the trigonometric functions are given in Appendix A.3.)

14. $f(x) = \sec x$

Functions f and g are described in Exercises 15 through 18. Find $f \circ g$ and $g \circ f$ and determine where these functions are continuous.

15. $f(x) = \dfrac{1}{x}, g(x) = x - 2$

16. $f(x) = x^3, g(x) = \sqrt{x}$

17. $f(x) = \dfrac{x - 1}{x + 1}, \ g(x) = \sqrt[4]{x}$ **18.** $f(x) = \dfrac{x - 1}{x + 1}, \ g(x) = \sqrt[3]{x}$

Employ the limit theorems in this section to determine the limit at the specified number in Exercises 19 through 30. Determine whether the function is continuous at the number.

19. $\lim\limits_{x \to -3} (x^4 - 5x^3 + 7)$ **20.** $\lim\limits_{x \to 4} (x^3 - 3x^2 + 5)$

21. $\lim\limits_{x \to 4} (x^2 + 3)(x - 2)$ **22.** $\lim\limits_{x \to 2} (\sqrt[3]{x^2} + 4x + 1)$

23. $\lim\limits_{x \to 2} \dfrac{1}{\sqrt[3]{x^2 - 3}}$ **24.** $\lim\limits_{x \to -1} (3x + 2x^2 + 1)(4x^2 - 3)$

25. $\lim\limits_{x \to 0} \dfrac{x^2 - 4x + 5}{2x - 7}$ **26.** $\lim\limits_{x \to -2} \sqrt[3]{4x^3 - 2}$

27. $\lim\limits_{x \to -2} \dfrac{1}{\sqrt{x^2 + 12} - \sqrt[3]{x^2 + 4}}$ **28.** $\lim\limits_{x \to -2} \dfrac{3x^2 + 2x + 1}{5x - 4}$

29. $\lim\limits_{x \to 4} \sqrt{x + 5} \ \sqrt[3]{\sqrt{x} - 1}$ **30.** $\lim\limits_{x \to 0} \dfrac{1}{\sqrt{x^2 + 10} - \sqrt{5 - 3x^2}}$

In Exercises 31 through 46, determine the limits or tell why they do not exist. (These exercises generally require an algebraic simplification before the limit can be determined.)

31. $\lim\limits_{x \to -1} \dfrac{x^2 - 3x - 4}{5x + 5}$ **32.** $\lim\limits_{x \to 2} \dfrac{x^3 - 8}{x - 2}$

33. $\lim\limits_{x \to -1} \dfrac{x^2 - 3x - 4}{3x + 5}$ **34.** $\lim\limits_{x \to 2} \dfrac{x^2 - 5x + 6}{x^3 - 2x^2 + 4x - 8}$

35. $\lim\limits_{x \to -5/2} \dfrac{3x^2 - 2x + 1}{2x + 5}$ **36.** $\lim\limits_{x \to -1} \dfrac{3x + 5}{x^2 - 3x - 4}$

37. $\lim\limits_{x \to 4} \dfrac{2x + 5}{3x^2 - 2x + 1}$ **38.** $\lim\limits_{x \to -5/2} \dfrac{2x + 5}{3x^2 - 2x + 1}$

39. $\lim\limits_{x \to 0} \dfrac{(2 + x)^2 - 4}{x}$ **40.** $\lim\limits_{x \to 0} \dfrac{(2 + x)^3 - 8}{x}$

41. $\lim\limits_{x \to 2} \dfrac{(1/x) - (1/2)}{x - 2}$ **42.** $\lim\limits_{x \to 0} \dfrac{1/(2 + x) - (1/2)}{x}$

43. $\lim\limits_{x \to 1} f(x)$, if $f(x) = \begin{cases} x^2, & \text{when } x < 1 \\ -x, & \text{when } x \geq 1 \end{cases}$

44. $\lim\limits_{x \to -1} f(x)$, if $f(x) = \begin{cases} x^2, & \text{when } x < -1 \\ 0, & \text{when } x \geq -1 \end{cases}$

45. $\lim\limits_{x \to 4} \dfrac{(x - 4)^2}{\sqrt{x + 12} - 4}$ **46.** $\lim\limits_{x \to 3} \dfrac{\sqrt{6 + x} - 3}{x - 3}$

(*Hint:* Rationalize the denominator.)

47. If $f(x) = \begin{cases} 2x, & \text{if } x < 2 \\ 0, & \text{if } x = 2 \\ x^2, & \text{if } x > 2 \end{cases}$ find

(a) $f(2)$ (b) $\lim\limits_{x \to 2} f(x)$ (c) $\lim\limits_{x \to 1} f(x)$

48. Find the constant a that will make

$$f(x) = \begin{cases} \dfrac{x^3 - 8}{x - 2}, & \text{if } x \neq 2 \\ a, & \text{if } x = 2 \end{cases}$$

continuous at $x = 2$.

49. Find constants a and b that will make

$$g(x) = \begin{cases} x^2, & \text{if } x \leq 1 \\ ax + b, & \text{if } 1 < x < 2 \\ x^3, & \text{if } x \geq 2 \end{cases}$$

continuous for all x.

50. When a function f is discontinuous at a number a but $\lim\limits_{x \to a} f(x)$ exists, the discontinuity is called a *removable discontinuity*. A discontinuity that is not removable is called an *essential discontinuity*. Classify the discontinuities in Exercises 3 through 14 as removable or essential.

51. Suppose the functions f, g, and h are all continuous at a. Prove that

(a) $f + g + h$ is continuous at a.

(b) $f \cdot g \cdot h$ is continuous at a.

(c) $f \cdot (g + h)$ is continuous at a.

What must be true if $f/(g \cdot h)$ is to be continuous at a?

52. Use Corollary 2.9 to show that any polynomial of degree 2 is continuous at every real number.

53. Use Corollary 2.9 to show that any rational function of the form $f(x) = \dfrac{ax^2 + bx + c}{dx^2 + ex + f}$ is continuous at any number x for which $dx^2 + ex + f \neq 0$.

54. What must be true about a rational function f of the form $f(x) = \dfrac{ax^2 + bx + c}{dx^2 + ex + f}$ if it is known that f has a limit at x_0 and $dx^2_0 + ex_0 + f = 0$?

55. Use Theorem 2.12 to show that $\lim\limits_{x \to a} |f(x)| = |L|$, whenever $\lim\limits_{x \to a} f(x) = L$.

56. Let functions f and g be defined by $g(x) = 2$ and $f(x) = \begin{cases} x^2, & \text{if } x \neq 2 \\ 0, & \text{if } x = 2 \end{cases}$. Show that $\lim\limits_{x \to 1} g(x) = 2$ and $\lim\limits_{x \to 1} f(g(x)) \neq f(2)$. Why does this not contradict Theorem 2.12?

In Exercises 57 through 60, find functions f and g that satisfy the specified conditions.

57. The function $f + g$ is continuous at 1, but neither f nor g is continuous at 1.

58. The function $f \cdot g$ is continuous at 1, but neither f nor g is continuous at 1.

59. Both of the functions $f + g$ and $f \cdot g$ are continuous at 1, but neither f nor g is continuous at 1.

60. The function f/g is continuous at 1, but neither f nor g is continuous at 1.

61. An $\varepsilon - \delta$ definition of continuity can be given as follows:

A function f is continuous at the number a provided that $f(a)$ exists and that whenever a number $\varepsilon > 0$ is given, a number $\delta > 0$ can be found with the property that $|f(x) - f(a)| < \varepsilon$ whenever $|x - a| < \delta$.

Use this definition to prove that the function f defined by $f(x) = 2x - 3$ is continuous at each real number.

62. Prove that a function f is continuous at a number a if and only if $\lim_{h \to 0} f(a + h) = f(a)$.

2.4
ONE-SIDED LIMITS AND CONTINUITY

For the limit of the function f at the point a to exist, f must be defined at every point in some open interval containing a, except possibly at a itself. At times it is useful to analyze the limiting behavior of a function on only one side of a given number. To describe this situation we need the notion of the *one-sided limit*. The water barrel problem given in Section 2.1 involves a one-sided limit. A function describing the amount of water in a barrel t minutes after the water began to flow is given by:

$$f(t) = \begin{cases} .75\,t, & \text{if } 0 \le t < 56 \\ 0, & \text{if } t \ge 56. \end{cases}$$

The limit of this function at $t = 56$ does not exist. However, there is a *physical* limit to the amount of water that can run into the barrel. As t approaches 56 and $t < 56$, $f(t)$ approaches 42. This implies that the limit of $f(t)$ from the lower or left side of 56 is 42.

The function defined by $f(x) = \sqrt{x} + 1$ has no negative real numbers in its domain, so $\lim_{x \to 0} f(x)$ is not defined. However, for $x > 0$, $f(x) = \sqrt{x} + 1$ approaches 1 as x approaches 0, so there is a limiting value of f at zero when only values of x on the positive side of zero are considered. (See Figure 2.14.)

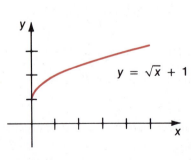

FIGURE 2.14

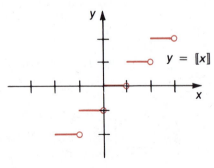

FIGURE 2.15

Another one-sided limit situation arises in the greatest integer function graphed in Figure 2.15. In Section 2.1, we saw that the limit of this function does not exist at 2. However, a limit could be defined if we were permitted to restrict the values of x to be entirely on one side of 2. In order to discuss this type of situation and to distinguish between some of the various cases when the limit does not exist, we make a definition relaxing part of the limit requirements.

(2.14)
DEFINITION

The **limit from the right of the function f at a is L,** written

$$\lim_{x \to a^+} f(x) = L,$$

provided that for every number $\varepsilon > 0$, a number $\delta > 0$ can be found with the property that

$$|f(x) - L| < \varepsilon$$

whenever $0 < x - a < \delta$.

Similarly, the **limit from the left of the function f at a is L,** written

$$\lim_{x \to a^-} f(x) = L,$$

provided that for every number $\varepsilon > 0$ a number $\delta > 0$ can be found with the property that

$$|f(x) - L| < \varepsilon$$

whenever $-\delta < x - a < 0$.

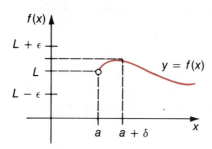

FIGURE 2.16

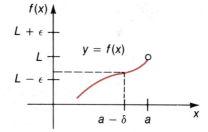

FIGURE 2.17

The next theorem follows immediately from these definitions and the definition of $\lim_{x \to a} f(x)$.

(2.15)
THEOREM

$\lim_{x \to a} f(x)$ exists and is L if and only if the following conditions hold:

(i) $\lim_{x \to a^+} f(x)$ exists,

(ii) $\lim_{x \to a^-} f(x)$ exists,

(iii) $\lim_{x \to a^+} f(x) = \lim_{x \to a^-} f(x) = L$.

For the function defined by $f(x) = \sqrt{x} + 1$, find $\lim\limits_{x \to 0^+} f(x)$, $\lim\limits_{x \to 0^-} f(x)$ and $\lim\limits_{x \to 0} f(x)$.

SOLUTION

As x approaches 0 and $x > 0$, $\sqrt{x}$ approaches 0, so $\lim\limits_{x \to 0^+} (\sqrt{x} + 1) = 1$. As x approaches 0 and $x < 0$, $\sqrt{x}$ is not defined, so $\lim\limits_{x \to 0^-} (\sqrt{x} + 1)$ does not exist.

Therefore, by Theorem 2.15, $\lim\limits_{x \to 0} f(x)$ does not exist since condition (ii) fails. □

EXAMPLE 2   For the function h defined by $h(x) = [\![x]\!]$, find $\lim\limits_{x \to 2^+} h(x)$, $\lim\limits_{x \to 2^-} h(x)$ and $\lim\limits_{x \to 2} h(x)$.

SOLUTION

When x is close to 2 and $x > 2$, $[\![x]\!] = 2$ so $\lim\limits_{x \to 2^+} h(x) = 2$. When x is close to 2 and $x < 2$, $[\![x]\!] = 1$ so $\lim\limits_{x \to 2^-} h(x) = 1$. So $\lim\limits_{x \to 2} h(x)$ does not exist because $\lim\limits_{x \to 2^+} h(x) \neq \lim\limits_{x \to 2^-} h(x)$. □

Since the definition of continuity is based directly upon that of the limit, corresponding definitions can be made for continuity from the right and from the left.

(2.16)
DEFINITION

A function f is continuous from the right at the number a provided
(i) $f(a)$ exists,
(ii) $\lim\limits_{x \to a^+} f(x)$ exists,
(iii) $\lim\limits_{x \to a^+} f(x) = f(a)$.

Likewise, a **function f is continuous from the left at the number a** provided
(i) $f(a)$ exists,
(ii) $\lim\limits_{x \to a^-} f(x)$ exists,
(iii) $\lim\limits_{x \to a^-} f(x) = f(a)$.

As might be expected, this definition leads to a theorem similar to Theorem 2.15.

(2.17)
THEOREM

A function f is continuous at a if and only if both of the following statements hold:
(i) f is continuous from the right at a,
(ii) f is continuous from the left at a.

The concepts of continuity from the right and left can be used to define continuity of a function on an interval.

(2.18)
DEFINITION

(i) A function f **is said to be continuous on the open interval** (a, b) if f is continuous at every number in (a, b).

(ii) A function f **is said to be continuous on the closed interval** $[a, b]$ if f is continuous on (a, b), is continuous from the right at a, and is continuous from the left at b.

For example, the function described by $f(x) = \sqrt{x} + 1$ is continuous at every positive real number and $\lim_{x \to 0^+} f(x) = 1 = f(0)$, so f is continuous on the interval $[0, \infty)$. See Figure 2.18.

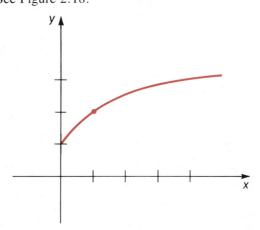

FIGURE 2.18

The results in Theorems 2.6 and 2.8 are equally valid when the limits are replaced by one-sided limits. Similarly the results in Corollaries 2.7 and 2.9 hold for continuity from the right or from the left. We will make use of these results even though they have not been stated explicitly.

EXAMPLE 3   Determine $\lim_{x \to 4^-} (5 - \sqrt{4 - x})$.

SOLUTION

$$\lim_{x \to 4^-} (5 - \sqrt{4 - x}) = \lim_{x \to 4^-} 5 - \lim_{x \to 4^-} \sqrt{4 - x} = 5 - \lim_{x \to 4^-} \sqrt{4 - x}.$$

Since $x < 4$, $4 - x > 0$ and $\sqrt{4 - x}$ is defined. Consequently, $\lim_{x \to 4^-} \sqrt{4 - x} = 0$ and $\lim_{x \to 4^-} (5 - \sqrt{4 - x}) = 5$. Note that $\lim_{x \to 4^+} (5 - \sqrt{4 - x})$ does not exist, since $\sqrt{4 - x}$ is undefined when $x > 4$. As a consequence of Theorem 2.15, $\lim_{x \to 4} (5 - \sqrt{4 - x})$ does not exist.

The function described by $f(x) = 5 - \sqrt{4 - x}$ is continuous at every number less than 4 and $\lim_{x \to 4^-} f(x) = 5 = f(4)$, so f is continuous on the interval $(-\infty, 4]$.   □

We conclude this section with an important result concerning functions that are continuous on closed intervals. The proof of this theorem is quite difficult and is not included.

(2.19)
THEOREM

Intermediate Value Theorem   Suppose f is a function that is continuous on the interval $[a, b]$, and M is any number between $f(a)$ and $f(b)$. There exists at least one number c in (a, b) with $f(c) = M$.

Geometrically, the intermediate value theorem implies that if f is continuous on $[a, b]$ and M is a number between $f(a)$ and $f(b)$, then the graph of f intersects the line $y = M$ at some number between a and b. Figure 2.19 shows that the intersection can occur at more than one number; the theorem guarantees that there is at least one.

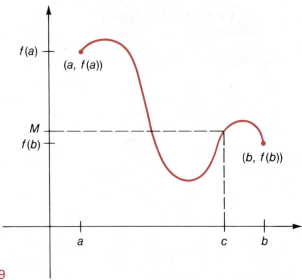

FIGURE 2.19

EXAMPLE 4   Use the intermediate value theorem to show that if $f(x) = x^3 + x - 1$, then there is a number c in the interval $(0, 1)$ with $f(c) = 0$.

SOLUTION

Since this function is a polynomial, it is a continuous function; in particular, f is continuous on $[0, 1]$. We find that $f(0) = -1$ and $f(1) = 1$, so $f(0) < 0 < f(1)$. The intermediate value theorem implies that there is a number c in $(0, 1)$ with $f(c) = 0$. This means that the graph of f crosses the x-axis at least once on the interval $[0, 1]$. ☐

EXERCISE SET 2.4

1. The graph of a function f is shown in the figure below.

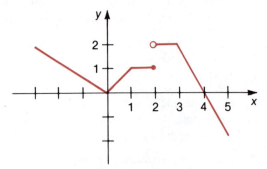

(a) Find $\lim\limits_{x \to 2^+} f(x)$.   (b) Find $\lim\limits_{x \to 2^-} f(x)$.   (c) Find $\lim\limits_{x \to 2} f(x)$.

2. Use the graph of the function shown in Exercise 1 to determine the following.

    (a) $\lim\limits_{x \to 1^+} f(x)$ (b) $\lim\limits_{x \to 1^-} f(x)$ (c) $\lim\limits_{x \to 1} f(x)$

Evaluate the limits in Exercises 3 through 22, provided that the limit exists.

3. $\lim\limits_{x \to 0^-} (2 - \sqrt{x})$ 4. $\lim\limits_{x \to 0^+} (2 - \sqrt{x})$

5. $\lim\limits_{x \to 2^+} \sqrt{x^2 - 4}$ 6. $\lim\limits_{x \to 3^+} \sqrt{x^2 - 4}$

7. $\lim\limits_{x \to -3^-} \sqrt{x^2 - 9}$ 8. $\lim\limits_{x \to -3^+} \sqrt{x^2 - 9}$

9. $\lim\limits_{x \to 3^+} \sqrt{x^2 - 4}$ 10. $\lim\limits_{x \to 3^-} \sqrt{x^2 - 4}$

11. $\lim\limits_{x \to 0^+} \dfrac{|x|}{x}$ 12. $\lim\limits_{x \to 0^-} \dfrac{|x|}{x}$

13. $\lim\limits_{x \to 0} \dfrac{|x|}{x}$ 14. $\lim\limits_{x \to 0} \dfrac{\sqrt{x^2}}{|x|}$

15. $\lim\limits_{x \to 0^+} \left(\dfrac{1}{x} - \dfrac{1}{|x|} \right)$ 16. $\lim\limits_{x \to 0^-} \left(\dfrac{1}{x} + \dfrac{1}{|x|} \right)$

17. $\lim\limits_{x \to 3^-} (\llbracket x \rrbracket + \llbracket -x \rrbracket)$ 18. $\lim\limits_{x \to 3^+} (\llbracket x \rrbracket + \llbracket -x \rrbracket)$

19. $\lim\limits_{x \to 2^+} \sqrt{x - \llbracket x \rrbracket}$ 20. $\lim\limits_{x \to 2^-} \dfrac{1}{\sqrt{x - \llbracket x \rrbracket}}$

21. $\lim\limits_{x \to 0^+} \dfrac{\sqrt{x^2 + 4} - 2}{x}$ 22. $\lim\limits_{x \to 4^-} \dfrac{(1/\sqrt{x}) - (1/2)}{x - 4}$

In Exercises 23 through 26 the graph of a function is given. Use the graph to find the intervals on which the function is continuous.

23.

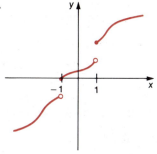

24.

25.

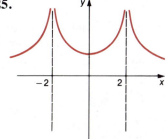

26.

Determine the intervals on which the functions described in Exercises 27 through 36 are continuous.

27. $f(x) = x^3 + 2$

28. $f(x) = 2x^2 - 1$

29. $f(x) = \dfrac{1}{x}$

30. $f(x) = \sqrt{x}$

31. $f(x) = \sqrt{x - 3}$

32. $g(x) = \dfrac{1}{\sqrt{x - 3}}$

33. $f(x) = |x|$

34. $g(x) = \dfrac{1}{|x|}$

35. $h(x) = -[\![x]\!]$

36. $g(x) = \dfrac{1}{[\![x]\!]}$

37. Suppose $f(x) = \begin{cases} x^2 + 1, & \text{if } x < -1 \\ 2x - 3, & \text{if } -1 \le x \le 2 \\ x + 2, & \text{if } x > 2. \end{cases}$

(a) Sketch the graph of f.

(b) Find $\lim\limits_{x \to 2^+} f(x)$, $\lim\limits_{x \to 2^-} f(x)$, and determine whether f is continuous from the right or left at 2. Is f continuous at 2?

(c) Find $\lim\limits_{x \to -1^+} f(x)$, $\lim\limits_{x \to -1^-} f(x)$ and determine whether f is continuous from the right or left at -1. Is f continuous at -1?

(d) Determine the intervals on which f is continuous.

38. Suppose $f(x) = \begin{cases} -\dfrac{1}{3}x - 1, & \text{if } x \le 0 \\ x^2 - 1, & \text{if } 0 < x < 1 \text{ or } 1 < x < 2 \\ \dfrac{1}{2}x + 2, & \text{if } x \ge 2. \end{cases}$

(a) Sketch the graph of f.

(b) Find $\lim\limits_{x \to 0^+} f(x)$, $\lim\limits_{x \to 0^-} f(x)$, and determine if f is continuous from the right or left at 0. Is f continuous at 0?

(c) Find $\lim\limits_{x \to 1^+} f(x)$, $\lim\limits_{x \to 1^-} f(x)$, and determine if f is continuous from the right or left at 1. Is f continuous at 1?

(d) Find $\lim\limits_{x \to 2^+} f(x)$, $\lim\limits_{x \to 2^-} f(x)$, and determine if f is continuous from the right or left at 2. Is f continuous at 2?

(e) Determine the intervals on which f is continuous.

In Exercises 39 through 42 a function f, an interval $[a, b]$, and a number M between $f(a)$ and $f(b)$ are given. Verify that the conclusion of the intermediate value theorem holds by finding a number c in $[a, b]$ with $f(c) = M$.

39. $f(x) = 7x - 4$, $[1, 3]$, $M = 15$

40. $f(x) = x^2 - 1$, $[-1, 2]$, $M = 2$

41. $f(x) = x^3 + 1$, $[-2, 3]$, $M = 9$

42. $f(x) = x^2 + 5x + 3$, $[-3, 1]$, $M = 1$

In Exercises 43 through 46, a function f and an interval are given. Use the intermediate value theorem to show that there is a number c in the interval for which $f(c) = 0$.

43. $f(x) = x^3 - 3x - 1$, $[1, 2]$

44. $f(x) = x^4 + x^3 - 3$, $[-1, 2]$

45. $f(x) = \cos x - x$, $\left[0, \dfrac{\pi}{2}\right]$

46. $f(x) = \sin x + \cos x$, $\left[\dfrac{\pi}{2}, \pi\right]$

47. Find a function f with the property that $\lim\limits_{x\to 1^+} f(x)$, $\lim\limits_{x\to 1^-} f(x)$, and $\lim\limits_{x\to 1} |f(x)|$ all exist, but $\lim\limits_{x\to 1} f(x)$ does not exist.

48. If a function f has an essential discontinuity (see Exercise 50 of Section 2.3) at the number a and both $\lim\limits_{x\to a^+} f(x)$ and $\lim\limits_{x\to a^-} f(x)$ exist, then f is said to have a *jump discontinuity* at a. Which, if any, of the discontinuities in Exercises 7 through 14 of Section 2.3 are jump discontinuities?

49. Find a function f that is continuous on the interval $(0, 1)$ and has the property that for any positive number M, numbers x_1 and x_2 exist in $(0, 1)$ with $f(x_1) < -M$ and $f(x_2) > M$.

50. Suppose g is a continuous function on $[a, b]$ and that $a \le g(x) \le b$ for all x in $[a, b]$. Show that there is a number p in $[a, b]$ such that $g(p) = p$. (p is called a *fixed point* of g.) [*Hint:* Define $h(x) = g(x) - x$ and apply the intermediate value theorem to h.]

51. Prove Theorem 2.15 (from page 73).

52. Prove Theorem 2.17 (from page 74).

2.5
LIMITS AT INFINITY: HORIZONTAL ASYMPTOTES

In Section 2.1 we discussed a problem involving a heated gold bar placed in a freezing unit. The question concerned the limiting temperature of the bar as t (the time after the bar is placed in the unit) increases. In this section, we show how the limit can describe situations in which quantities increase or decrease without bound.

As mentioned in Section 1.1, it is general practice in mathematics to use the symbols ∞, called infinity, and $-\infty$ to describe unbounded behavior. The notation

$$\lim_{x\to\infty} f(x) = L$$

is used to express that the values of f approach L as x increases without bound. Although the symbol ∞ does not represent a real number, $\lim\limits_{x\to\infty} f(x) = L$ is often read "$f(x)$ approaches L as x approaches ∞" or "the limit of f at ∞ is L."

In a similar manner, $\lim\limits_{x\to -\infty} f(x) = L$ is used to express that the values of f

approach L as x decreases without bound; and is read "the limit of f at $-\infty$ is L."

In the gold bar problem, the temperature of the bar (initially 200° F) was 100° F at the end of 1 hour, 50° F at the end of 2 hours, and so on. The function T defined by

$$T(t) = \frac{200}{2^t}, \qquad t \geq 0$$

describes the bar's temperature, in degrees Fahrenheit, t hours after being placed in the freezing unit. The limiting temperature of the bar is given by

$$\lim_{t \to \infty} T(t).$$

As t increases, 2^t increases without bound, so $\dfrac{1}{2^t}$ approaches zero. Consequently,

$$\lim_{t \to \infty} T(t) = \lim_{t \to \infty} \frac{200}{2^t} = 0.$$

Note, however, that no matter how large t becomes, $T(t) = \dfrac{200}{2^t} > 0$; so $T(t)$

approaches zero as $t \to \infty$, but is not equal to zero for any value of t.

To define precisely the limit at infinity, we need to express "increase without bound" in an analytical manner. To say that "x increases without bound" means that the values of x exceed any specified bound.

(2.20)
DEFINITION

Suppose f is a function defined on an interval (a, ∞). The **limit of f at ∞ is** the number L, written

$$\lim_{x \to +\infty} f(x) = L,$$

provided that for every number $\varepsilon > 0$, a number M can be found with the property that $|f(x) - L| < \varepsilon$ whenever $x > M$.

A function with limit L at ∞ is illustrated in Figure 2.20. Observe that when $x > M$, the graph of f stays between the lines $y = L + \varepsilon$ and $y = L - \varepsilon$.

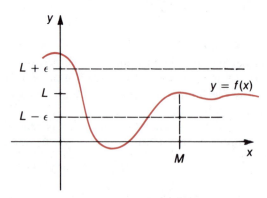

FIGURE 2.20

If f is defined on an interval $(-\infty, a)$, $\lim\limits_{x \to -\infty} f(x) = L$ provided that for every number $\varepsilon > 0$, a number M can be found with the property that $|f(x) - L| < \varepsilon$ whenever $x < M$.

In Section 1.4 (Example 2), we considered the graph of $h(x) = 1/x$ and saw that $h(x) = 1/x$ approaches zero as x increases in magnitude. For example, if we let $x > 100$, then $1/x < 1/100 = .01$; if $x > 10,000$, then $1/x < 1/10,000 = .0001$. In general, if $\varepsilon > 0$ and $x > 1/\varepsilon$, then $|1/x - 0| = 1/x < \varepsilon$. By Definition 2.20, $\lim_{x \to \infty} 1/x = 0$. Similarly, $\lim_{x \to -\infty} 1/x = 0$.

The following result is useful in determining the limits at infinity of rational functions. The proof is considered in Exercise 46.

(2.21)
THEOREM

If r is a positive rational number and c is any real number, then

(i) $\lim_{x \to \infty} \dfrac{c}{x^r} = 0$,

(ii) $\lim_{x \to -\infty} \dfrac{c}{x^r} = 0$, provided x^r is defined for $x < 0$.

Limit theorems analogous to those given in Section 2.3 concerning limits of sums, products, quotients, and roots are true for the cases $x \to \infty$ and $x \to -\infty$. In particular, all the results of Theorem 2.8 remain valid if we assume that a is replaced by the symbol ∞ or by $-\infty$.

EXAMPLE 1   Determine $\lim\limits_{x \to \infty} f(x) = \dfrac{3x^2 + 7x + 2}{x^2 - 2x - 8}$.

SOLUTION
Since we are interested in large values of x, we may assume $x \neq 0$ and divide the numerator and denominator of $(3x^2 + 7x + 2)/(x^2 - 2x - 8)$ by x^2 (the largest power of x in the expression). Using the limit theorems, we have:

$$\lim_{x \to \infty} f(x) = \lim_{x \to \infty} \frac{3x^2 + 7x + 2}{x^2 - 2x - 8}$$

$$= \lim_{x \to \infty} \frac{3 + \dfrac{7}{x} + \dfrac{2}{x^2}}{1 - \dfrac{2}{x} - \dfrac{8}{x^2}}$$

$$= \frac{\lim\limits_{x \to \infty} 3 + \lim\limits_{x \to \infty} \dfrac{7}{x} + \lim\limits_{x \to \infty} \dfrac{2}{x^2}}{\lim\limits_{x \to \infty} 1 - \lim\limits_{x \to \infty} \dfrac{2}{x} - \lim\limits_{x \to \infty} \dfrac{8}{x^2}}$$

$$= \frac{3 + 0 + 0}{1 - 0 - 0} = 3. \qquad \square$$

EXAMPLE 2   Determine $\lim\limits_{x \to -\infty} \dfrac{\sqrt{x^2 + 3}}{\sqrt[3]{x^4 - 4x^3 + 5}}$.

SOLUTION

If we divide the numerator and denominator by $x^{4/3}$, the largest power of x in the expression, then

$$\lim_{x \to -\infty} \frac{\sqrt{x^2 + 3}}{\sqrt[3]{x^4 - 4x^3 + 5}} = \lim_{x \to -\infty} \frac{\sqrt{\dfrac{x^2 + 3}{x^{8/3}}}}{\sqrt[3]{\dfrac{x^4 - 4x^3 + 5}{x^4}}}$$

$$= \lim_{x \to -\infty} \frac{\sqrt{\dfrac{1}{x^{2/3}} + \dfrac{3}{x^{8/3}}}}{\sqrt[3]{1 - \dfrac{4}{x} + \dfrac{5}{x^4}}}$$

$$= \frac{\sqrt{\displaystyle\lim_{x \to -\infty} \dfrac{1}{x^{2/3}} + \lim_{x \to -\infty} \dfrac{3}{x^{8/3}}}}{\sqrt[3]{1 - \displaystyle\lim_{x \to -\infty} \dfrac{4}{x} + \lim_{x \to -\infty} \dfrac{5}{x^4}}}$$

$$= \frac{\sqrt{0 + 0}}{\sqrt[3]{1 - 0 + 0}} = 0. \qquad \square$$

EXAMPLE 3   The monthly sales s (in dollars) of a new product depend on the time t (in months) after the product has been introduced according to the rule:

$$s(t) = \frac{2000\,(1 + t)}{2 + t}.$$

What is $\lim_{t \to \infty} s(t)$? What information concerning the sales of the new product does this limiting value provide?

SOLUTION

Dividing both the numerator and denominator by t and proceeding as in Example 1, we have

$$\lim_{t \to \infty} s(t) = \lim_{t \to \infty} \frac{2000\,(1 + t)}{2 + t}$$

$$= \lim_{t \to \infty} \frac{\dfrac{2000}{t} + 2000}{\dfrac{2}{t} + 1}$$

$$= 2000.$$

The graph of the function s for $t > 0$ is shown in Figure 2.21.

The limiting value of $2000 in sales per month is approximately the monthly sales expected, once the product has been accepted by the public. $\qquad \square$

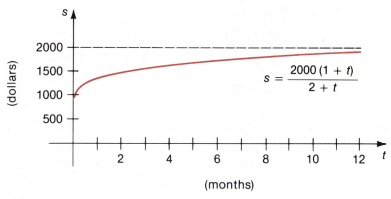

FIGURE 2.21

In the previous example the horizontal line, $s = 2000$, gives a graphical illustration of the amount that the sales differ from the limiting value at any particular time. This line is called a *horizontal asymptote* of the graph.

(2.22)
DEFINITION   A line with equation $y = b$ is called a **horizontal asymptote** of the graph of the function f if either $\lim\limits_{x \to \infty} f(x) = b$ or $\lim\limits_{x \to -\infty} f(x) = b$.

EXAMPLE 4   Find any horizontal asymptotes of the graph of the function defined by

$$f(x) = \frac{x^2}{x^2 + 1}.$$

SOLUTION
Since the equation describing f involves only squares of x, $f(x) = f(-x)$ (the graph of f is symmetric with respect to the y-axis) and $\lim\limits_{x \to \infty} f(x) = \lim\limits_{x \to -\infty} f(x)$. To find $\lim\limits_{x \to \infty} f(x)$, we divide the numerator and denominator of the quotient by x^2:

$$\lim_{x \to \infty} f(x) = \lim_{x \to \infty} \frac{x^2}{x^2 + 1} = \lim_{x \to \infty} \frac{1}{1 + \dfrac{1}{x^2}} = 1.$$

Thus, the only horizontal asymptote of the graph of f is the line $y = 1$. The graph of f is sketched in Figure 2.22. In Chapter 3 we will consider how the exact shape of this graph is determined. ☐

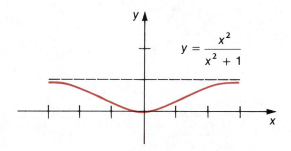

FIGURE 2.22

EXERCISE SET 2.5

Find the limits in Exercises 1 through 38 if they exist.

1. $\displaystyle\lim_{x \to -\infty} \frac{2x - 3}{3x + 5}$

2. $\displaystyle\lim_{x \to -\infty} \frac{5x - 7}{x + 300}$

3. $\displaystyle\lim_{x \to \infty} \frac{1}{x - 2}$

4. $\displaystyle\lim_{x \to -\infty} \frac{1}{x - 2}$

5. $\displaystyle\lim_{x \to -\infty} \frac{x^2}{x^3 + 1}$

6. $\displaystyle\lim_{x \to \infty} \frac{x + 2}{x^2}$

7. $\displaystyle\lim_{x \to \infty} \frac{x^2 + 3}{x^2 + 1}$

8. $\displaystyle\lim_{x \to -\infty} \frac{x^3 - 2}{x^3 + 4}$

9. $\displaystyle\lim_{x \to \infty} \frac{3x^2 + 2x + 1}{4x^2 - 3}$

10. $\displaystyle\lim_{x \to \infty} \frac{142x^3 + 1}{.1x^4}$

11. $\displaystyle\lim_{x \to -\infty} \frac{50 - x^3}{-x^3}$

12. $\displaystyle\lim_{x \to \infty} \frac{50 - x^3}{-x^3}$

13. $\displaystyle\lim_{x \to \infty} \frac{\sin x}{x} \left(\textit{Hint: } \left|\frac{\sin x}{x}\right| \leq \frac{1}{|x|}\right)$

14. $\displaystyle\lim_{x \to \infty} \frac{1 + \sin x}{x}$

15. $\displaystyle\lim_{x \to \infty} \frac{\cos x}{x + 2}$

16. $\displaystyle\lim_{x \to \infty} \frac{x^2 \cos x}{x^3 + 2}$

17. $\displaystyle\lim_{x \to -\infty} \frac{x^2 + \sin x \cos x}{x^2 + 2x - 1}$

18. $\displaystyle\lim_{x \to \infty} \frac{1 + \sqrt[4]{x}}{1 - \sqrt[4]{x}}$

19. $\displaystyle\lim_{x \to -\infty} \frac{1 + \sqrt[3]{x}}{1 - \sqrt[3]{x}}$

20. $\displaystyle\lim_{x \to \infty} \frac{4 + \sqrt[3]{8x}}{4 - \sqrt[3]{x}}$

21. $\displaystyle\lim_{x \to \infty} \frac{\sqrt{x^2 - 4}}{5x + 2}$

22. $\displaystyle\lim_{x \to -\infty} \frac{\sqrt{x^2 - 4}}{2x + 5}$

23. $\displaystyle\lim_{x \to \infty} \frac{\sqrt{4x^2 + 3x - 9}}{x - 2}$

24. $\displaystyle\lim_{x \to \infty} \frac{3x^2 - 4}{\sqrt{x^5 - 4x^2}}$

25. $\displaystyle\lim_{x \to -\infty} \frac{3x^2 - 4}{\sqrt{x^5 - 4x^2}}$

26. $\displaystyle\lim_{x \to -\infty} \frac{|1 - x|}{\sqrt{x^2 - 3x + 4}}$

27. $\displaystyle\lim_{x \to \infty} \frac{(2x + 5)(x^2 - 2x + 4)}{(9x^2 + 3)(2x - 7)}$

28. $\displaystyle\lim_{x \to -\infty} \frac{(x^3 + 3x^2 + 2)(5x - 6)}{(x^2 + 2x + 1)(4 - x^2)}$

29. $\displaystyle\lim_{x \to -\infty} \frac{(4 - x^2)(\sqrt{x^2 + 2})}{3x^3 - 9x + \pi}$

30. $\displaystyle\lim_{x \to \infty} \frac{(3x^2 - 5)\sqrt{x^2 + x + 1}}{(4x - 3)\sqrt{9x^4 + 3x^2 + 1}}$

31. $\displaystyle\lim_{x \to \infty} (\sqrt{x^2 + 1} - x)$

$\left(\textit{Hint: } \text{Multiply by } \dfrac{\sqrt{x^2 + 1} + x}{\sqrt{x^2 + 1} + x}\right)$

32. $\displaystyle\lim_{x \to \infty} (\sqrt{4x^2 + 1} - 2x)$

33. $\displaystyle\lim_{x \to \infty} (\sqrt{x^2 + x} - x)$

34. $\displaystyle\lim_{x \to \infty} (\sqrt{x^2 + 2x + 2} - x)$

35. $\displaystyle\lim_{x \to \infty} \frac{\sqrt{x^2 + \sqrt{x^2 + 1}}}{x}$

36. $\displaystyle\lim_{x \to -\infty} \frac{\sqrt{x^2 + \sqrt{x^2 + 1}}}{\sqrt{x^2 + 1}}$

37. $\lim\limits_{x\to\infty} \dfrac{x}{[\![x]\!]}$

38. $\lim\limits_{x\to-\infty} \dfrac{[\![x]\!]}{x}$

In Exercises 39 through 42, find the horizontal asymptotes of the graph of the function defined and sketch the graph.

39. $f(x) = \dfrac{1}{x^2 + 1}$

40. $f(x) = \dfrac{x}{x^2 + 1}$

41. $f(x) = \dfrac{x}{\sqrt{x^2 + 1}}$

42. $f(x) = \dfrac{\sin x}{x^2 + 1}$

43. We can see that $\lim\limits_{x\to\infty} \dfrac{1}{x^3} = 0$. How large must x be to ensure that $\dfrac{1}{x^3}$ is less than

(a) .001?

(b) 10^{-6}?

44. We can see that $\lim\limits_{x\to\infty} \dfrac{1}{x + 1} = 0$. How large must x be to ensure that $\dfrac{1}{x + 1}$ is less than

(a) .01?

(b) .001?

45. Use Definition 2.20 to prove that $\lim\limits_{x\to\infty} 1/x^2 = 0$.

46. If r is a positive rational number and c is any real number, prove that $\lim\limits_{x\to\infty} c/x^r = 0$.

47. Suppose that f is an even function. Show that $\lim\limits_{x\to\infty} f(x) = \lim\limits_{x\to-\infty} f(x)$, provided these limits exist.

48. Suppose that f is an odd function. Show that $\lim\limits_{x\to\infty} f(x) = -\lim\limits_{x\to-\infty} f(x)$, provided these limits exist.

49. Suppose that f is a rational function of the form

$$f(x) = \frac{a_n x^n + a_{n-1} x^{n-1} + \ldots + a_1 x + a_0}{b_n x^n + b_{n-1} x^{n-1} + \ldots + b_1 x + b_0}$$

where $a_n \neq 0$ and $b_n \neq 0$. What can be said about $\lim\limits_{x\to\infty} f(x)$ and $\lim\limits_{x\to-\infty} f(x)$?

50. Suppose that P is a polynomial. What possibilities exist for $\lim\limits_{x\to\infty} P(x)$ and $\lim\limits_{x\to-\infty} P(x)$, and when do these possibilities occur?

51. The Dull calculator company estimates that the cost of producing x calculators is

$$C(x) = 2000 + 10x + \frac{100}{x} \text{ dollars.}$$

If the average cost per calculator is $C(x)/x$, find the limit of the average cost per calculator as x approaches ∞. What does this limit indicate?

52. The intensity of illumination at a point is proportional to the product of the strength of the light source and the inverse of the square of the distance from the source. If a light source has strength s, what happens to the illumination as the distance from the source increases?

53. The relationship between the gross photosynthetic rate P of a leaf and the light intensity I is given by

$$P(I) = \left(a + \frac{b}{I} \right)^{-1}$$

where a and b are positive constants that depend on the species and levels of the other extremal factors. Find the maximum attainable photosynthetic rate. Sketch the graph of P when $a = b = 1$.

54. A test of one-hour duration has a reliability r $(0 < r < 1)$. If the test is performed for t hours, $t > 0$, the reliability of the new test is given by

$$R(t) = \frac{tr}{1 + (t - 1)r}.$$

Find $\lim\limits_{t \to \infty} R(t)$ and interpret the result. Find $\lim\limits_{t \to 0} R(t)$ and interpret the result.

2.6
INFINITE LIMITS: VERTICAL ASYMPTOTES

The limit concept is also used to express the fact that the values of a function f increase without bound as x gets close to a number a. For this we write

$$\lim_{x \to a} f(x) = \infty,$$

which is read "$f(x)$ becomes infinite as x approaches a."

The notation $\lim\limits_{x \to a} f(x) = -\infty$ is used to express that the values of f decrease without bound as x gets close to a and is read "$f(x)$ becomes negatively infinite as x approaches a."

One-sided infinite limits are defined similarly and are often the form in which infinite limits are employed.

An application of an infinite limit is associated with the intensity of the illumination of a light at its source. If a particular light source has strength s, then the illumination as a function of the distance d from the source is given by

$$I(d) = \frac{s}{d^2}.$$

As the distance from the source of the light decreases (d approaches 0), the intensity increases (when $d = .01$, $I = 10,000s$; when $d = .0001$, $I = 100,000,000s$, and so forth) and

$$\lim_{d \to 0^+} I(d) = \infty$$

($d > 0$, since d describes a distance).

EXAMPLE 1   Find $\lim\limits_{x \to 1^+} \dfrac{1}{x - 1}$, $\lim\limits_{x \to 1^-} \dfrac{1}{x - 1}$, and sketch the graph of $f(x) = \dfrac{1}{(x - 1)}$.

SOLUTION

As x approaches 1 and $x > 1$, $x - 1$ approaches 0 and $x - 1 > 0$. Thus, as x approaches 1 from the right, $\dfrac{1}{(x - 1)}$ increases without bound;

$$\lim_{x \to 1^+} \frac{1}{x - 1} = \infty.$$

As x approaches 1 and $x < 1$, $x - 1$ approaches 0 and $x - 1 < 0$, so $\dfrac{1}{(x - 1)}$ is negative and increasing in magnitude.

That is,
$$\lim_{x \to 1^-} \frac{1}{x - 1} = -\infty.$$

The function f is not defined at 1, the y-intercept is -1, and the graph does not intersect the x-axis. Also

$$\lim_{x \to \infty} \frac{1}{x - 1} = 0 \qquad \text{and} \qquad \lim_{x \to -\infty} \frac{1}{x - 1} = 0.$$

Plotting the points $(2, 1)$ and $(0, -1)$ and using the preceding analysis gives the graph shown in Figure 2.23. Notice that the graph of this function is simply the graph of $g(x) = 1/x$ moved one unit to the right. $\square$

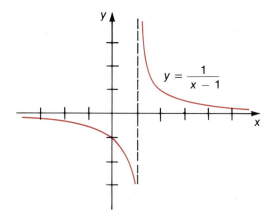

FIGURE 2.23

To define $\lim\limits_{x \to a} f(x) = \infty$, we need a precise way to express that as x approaches a, the values of f become and remain larger than any prescribed bound. Such a description is given in Definition 2.23.

(2.23)
DEFINITION

Let f be a function that is defined on an open interval containing a, except possibly at a. The **values of f are said to become positively infinite as x approaches a,** written

$$\lim_{x \to a} f(x) = \infty,$$

provided that, for any number M, a number $\delta > 0$ can be found with the property that $f(x) > M$ whenever $0 < |x - a| < \delta$.

This definition is illustrated in Figure 2.24(a).

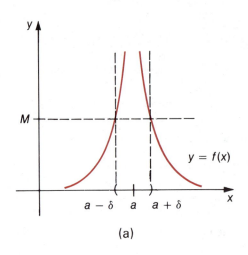

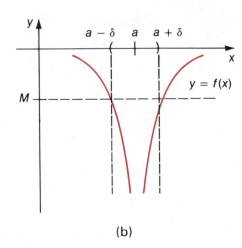

(a)   (b)

FIGURE 2.24

Similarly, the **values of f are said to become negatively infinite as x approaches a,** written

$$\lim_{x \to a} f(x) = -\infty,$$

provided that for any number M, a number $\delta > 0$ can be found with the property that $f(x) < M$ whenever $0 < |x - a| < \delta$. (See Figure 2.24(b).)

One-sided infinite limits can be defined similarly. For example, we write $\lim_{x \to a^+} f(x) = \infty$ provided that given any number M, a number $\delta > 0$ can be found with the property that $f(x) > M$ whenever $0 < x - a < \delta$. Some of the other definitions are considered in the exercises.

A number of theorems can be established concerning the arithmetic of infinite limits. A typical example of this type of theorem is the following:

(2.24)
THEOREM

If a is any real number and f and g are functions with $\lim_{x \to a} f(x) = \infty$ and $\lim_{x \to a} g(x) = L$ where L is a nonzero real number, then

(i) $\lim_{x \to a^-} (f(x) \pm g(x)) = \infty,$

(ii) $\lim_{x \to a} (f(x) \cdot g(x)) = \begin{cases} \infty, & \text{if } L > 0 \\ -\infty, & \text{if } L < 0, \end{cases}$

(iii) $\lim_{x \to a} \dfrac{f(x)}{g(x)} = \begin{cases} \infty, & \text{if } L > 0 \\ -\infty, & \text{if } L < 0, \end{cases}$

(iv) $\lim_{x \to a} \dfrac{g(x)}{f(x)} = 0.$

This theorem also holds if in each case a is replaced by a^+, a^-, ∞, or $-\infty$; a similar result holds when $\lim_{x \to a} f(x) = -\infty$.

EXAMPLE 2   Find $\lim\limits_{x \to 0} h(x)$ for the function h defined by $h(x) = \dfrac{x - 1}{x^2}$.

SOLUTION

   Observe that $h(x) = f(x) \cdot g(x)$ where $f(x) = 1/x^2$ and $g(x) = x - 1$. Since $\lim\limits_{x \to 0} f(x) = \infty$ and $\lim\limits_{x \to 0} g(x) = -1$, we can use part (ii) of Theorem 2.24 to show that $\lim\limits_{x \to 0} h(x) = -\infty$.

   Another way to see that this limit is $-\infty$ is: The numerator $x - 1$ is approximately -1 when x is close to 0, but the denominator x^2 is a positive number close to 0. Consequently, the quotient $\dfrac{x - 1}{x^2}$ becomes negatively infinite as x approaches 0.   ☐

   The graph of $h(x) = (x-1)/x^2$ approaches the line $x = 0$ as x approaches 0, as shown in Figure 2.25. The line $x = 0$ is called a *vertical asymptote* of the graph of h.

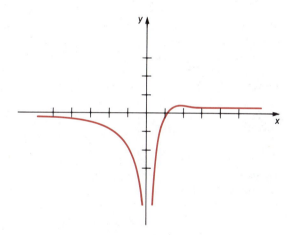

FIGURE 2.25

(2.25)   A line with equation $x = a$ is called a **vertical asymptote** of the graph of the
DEFINITION   function f, if $\lim\limits_{x \to a^+} f(x) = \infty$, $\lim\limits_{x \to a^+} f(x) = -\infty$, $\lim\limits_{x \to a^-} f(x) = \infty$, or $\lim\limits_{x \to a^-} f(x) = -\infty$.

EXAMPLE 3·   Find any horizontal and vertical asymptotes and sketch the graph of the function defined by

$$f(x) = \frac{x^2 + 3x + 2}{x^2 - 2x - 8}.$$

SOLUTION

To find the horizontal asymptotes, consider

$$\lim_{x \to \infty} f(x) = \lim_{x \to \infty} \frac{x^2 + 3x + 2}{x^2 - 2x - 8} = \lim_{x \to \infty} \frac{1 + \dfrac{3}{x} + \dfrac{2}{x^2}}{1 - \dfrac{2}{x} - \dfrac{8}{x^2}} = 1.$$

Also $\lim_{x \to -\infty} f(x) = 1$; so the only horizontal asymptote is the line with equation $y = 1$, shown as a horizontal dashed line in Figure 2.26.

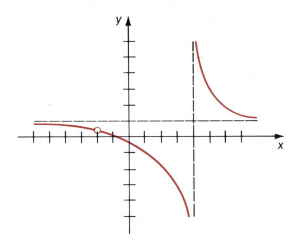

FIGURE 2.26

Since

$$f(x) = \frac{x^2 + 3x + 2}{x^2 - 2x - 8} = \frac{(x + 2)(x + 1)}{(x + 2)(x - 4)},$$

f is undefined at $x = -2$ and $x = 4$. The lines $x = -2$ and $x = 4$ are thus likely candidates for vertical asymptotes. To determine if they are indeed vertical asymptotes, we consider the one-sided limits at $x = -2$ and $x = 4$.

$$\lim_{x \to -2^+} f(x) = \lim_{x \to -2^+} \frac{(x + 2)(x + 1)}{(x + 2)(x - 4)}$$

$$= \lim_{x \to -2^+} \frac{x + 1}{x - 4} = \frac{-1}{-6} = \frac{1}{6}.$$

Similarly, $\lim_{x \to -2^-} f(x) = 1/6$, so $\lim_{x \to -2} f(x) = 1/6$ and $x = -2$ is not a vertical asymptote of the graph of f.

However,

$$\lim_{x \to 4^+} f(x) = \lim_{x \to 4^+} \frac{(x + 2)(x + 1)}{(x + 2)(x - 4)} = \lim_{x \to 4^+} \frac{x + 1}{x - 4}.$$

Since $\lim_{x \to 4^+} (x + 1) = 5$ and $\lim_{x \to 4^+} \dfrac{1}{x - 4} = \infty$, $\lim_{x \to 4^+} \dfrac{x + 1}{x - 4} = \infty$, so $x = 4$ is a vertical asymptote of the graph of f (indicated by the vertical dashed line in Figure 2.26).

Moreover, $$\lim_{x \to 4^-} f(x) = \lim_{x \to 4^-} \frac{x + 1}{x - 4} = -\infty.$$

When $x \neq -2$, $f(x) = (x + 1)/(x - 4)$, which describes a continuous function on the intervals $(-\infty, 4)$ and $(4, \infty)$ and has intercepts at $(0, -1/4)$ and $(-1, 0)$. The graph is shown in Figure 2.26. To indicate that -2 is not in the domain of f, a circle has been drawn on the graph at $(-2, 1/6)$.

We have again made assumptions about the graph that we cannot presently justify: that the graph has no oscillatory behavior and moves smoothly on the intervals $(-\infty, -2)$, $(-2, 4)$ and $(4, \infty)$. The validity of these assumptions can be established after we study Section 3.11. ☐

EXERCISE SET 2.6

Find the limits in Exercises 1 through 20 if they exist.

1. $\lim\limits_{x \to 2^+} \dfrac{1}{x - 2}$ **2.** $\lim\limits_{x \to 2^-} \dfrac{1}{x - 2}$

3. $\lim\limits_{x \to 2} \dfrac{1}{x - 2}$ **4.** $\lim\limits_{x \to 2} \dfrac{1}{(x - 2)^2}$

5. $\lim\limits_{x \to 2^+} \dfrac{x + 2}{x - 2}$ **6.** $\lim\limits_{x \to 2^-} \dfrac{x + 2}{x - 2}$

7. $\lim\limits_{x \to 1^+} \dfrac{x + 1}{x^2 - 1}$ **8.** $\lim\limits_{x \to 1} \dfrac{x + 1}{x^2 - 1}$

9. $\lim\limits_{x \to 1} \dfrac{x^2 - 1}{x - 1}$ **10.** $\lim\limits_{x \to -1} \dfrac{x^2 - 1}{x + 1}$

11. $\lim\limits_{x \to -3} \dfrac{x^2 + 2x - 8}{x^2 + x - 6}$ **12.** $\lim\limits_{x \to 2} \dfrac{x^2 + 2x - 8}{x^2 + x - 6}$

13. $\lim\limits_{x \to 5^+} \dfrac{3x^2 + 1}{(x - 2)(x - 5)}$ **14.** $\lim\limits_{x \to 5^-} \dfrac{3x^2 + 1}{(x - 2)(x - 5)}$

15. $\lim\limits_{x \to 2^+} \dfrac{3x^2 + 1}{(x - 2)(x - 5)}$ **16.** $\lim\limits_{x \to 2^-} \dfrac{3x^2 + 1}{(x - 2)(x - 5)}$

17. $\lim\limits_{x \to 0^-} \dfrac{|x|}{x^2}$ **18.** $\lim\limits_{x \to 2} \dfrac{x}{|x - 2|}$

19. $\lim\limits_{x \to 1^+} \sqrt{\dfrac{x}{x - 1}}$ **20.** $\lim\limits_{x \to 1^+} \dfrac{x}{x^2 - 1}$

Determine the equations of any vertical and horizontal asymptotes to the graphs of the functions and equations described in Exercises 21 through 40. Sketch each of these graphs.

21. $f(x) = \tan x$ **22.** $f(x) = \cot x$

23. $f(x) = \sec x$

24. $f(x) = \csc x$

25. $f(x) = \dfrac{x - 1}{x + 1}$

26. $f(x) = \dfrac{2x - 3}{1 - x}$

27. $f(x) = \dfrac{1}{x^2 - 4x + 6}$

28. $f(x) = \dfrac{1}{x^2 - 4x + 5}$

29. $f(x) = \dfrac{x}{\sqrt{x^2 - 4}}$

30. $f(x) = \dfrac{x}{\sqrt{4 - x^2}}$

31. $f(x) = \dfrac{1}{\sqrt{1 - x^2}}$

32. $f(x) = \dfrac{x^2 - 5x + 6}{x^2 - 9}$

33. $f(x) = \dfrac{x^2 - 1}{x^2 - 2x + 1}$

34. $f(x) = \dfrac{1 + x^2}{x^2}$

35. $f(x) = \dfrac{4x^2}{9 - x^2}$

36. $f(x) = \dfrac{x^2 - 4}{9 - x^2}$

37. $x^2y^2 - x^2 - y^2 = 0$

38. $xy^2 + 3x - 6 = 0$

39. $yx^2 + 4y - x = 0$

40. $xy - y^2 - 1 = 0$

Exercises 41 through 46 describe conditions that the graph of a function is to satisfy. In each case, sketch such a graph.

41. A vertical asymptote at $x = 0$ and a horizontal asymptote at $y = -1$.

42. Vertical asymptotes at $x = 1$ and $x = -1$ and a horizontal asymptote at $y = 1$.

43. Vertical asymptotes at $x = 1$ and $x = -1$ and no horizontal asymptotes.

44. Horizontal asymptotes at $y = 1$ and $y = -1$ and a vertical asymptote at $x = 1$.

45. Horizontal asymptotes at $y = 1$ and $y = -1$ and no vertical asymptotes.

46. Horizontal asymptotes at $y = 1$ and $y = -1$ and vertical asymptotes at $x = 1$ and $x = -1$.

47. We can see that $\lim\limits_{x \to 0^+} 1/x = \infty$. How small must x be to ensure that $1/x$ is greater than

(a) 100?   (b) 10,000?

48. We can see that $\lim\limits_{x \to 0} 1/x^4 = \infty$. How small in absolute value must x be to ensure that $1/x^4$ is greater than

(a) 100?   (b) 10,000?

In Exercises 49 through 56, construct functions f, g, and h with $\lim\limits_{x \to \infty} f(x) = 0$, $\lim\limits_{x \to \infty} g(x) = \infty$, and $\lim\limits_{x \to \infty} h(x) = -\infty$ that satisfy the given condition.

49. $\lim\limits_{x \to \infty} f(x)g(x) = 1$

50. $\lim\limits_{x \to \infty} f(x)g(x) = \infty$

51. $\lim\limits_{x \to \infty} f(x)g(x) = 0$

52. $\lim\limits_{x \to \infty} (g(x) + h(x)) = 1$

53. $\lim\limits_{x \to \infty} (g(x) + h(x)) = 0$

54. $\lim\limits_{x \to \infty} (g(x) + h(x)) = -1$

55. $\lim\limits_{x \to \infty} \dfrac{g(x)}{h(x)} = -\infty$

56. $\lim\limits_{x \to \infty} \dfrac{g(x)}{h(x)} = 0$

57. Give a definition for $\lim\limits_{x \to a^+} f(x) = -\infty$ and use this definition to show that $\lim\limits_{x \to 0^+} -1/x = -\infty$.

58. Give a definition for $\lim\limits_{x \to \infty} f(x) = \infty$ and use this definition to show that $\lim\limits_{x \to \infty} x^2 = \infty$.

59. In Example 2 we proved that $\lim\limits_{x \to 0} \dfrac{x - 1}{x^2} = -\infty$ by showing that $\lim\limits_{x \to 0} (x - 1) = -1$ while $\lim\limits_{x \to 0} 1/x^2 = \infty$. Suppose instead that we had rewritten $(x - 1)/x^2 = 1/x - 1/x^2$ and considered $\lim\limits_{x \to 0} 1/x$ and $\lim\limits_{x \to 0} 1/x^2$. Why can we not use these results to show that $\lim\limits_{x \to 0} (x - 1)/x^2 = -\infty$?

60. Living tissue can be excited by an electric current only if the current reaches or exceeds a certain threshold. A function describing the dependence of the threshold on the duration t of current flow is given by Weiss' Law:

$$f(t) = \frac{a}{t} + b,$$

where a and b are positive constants. Describe the behavior of the threshold when t approaches zero and when t approaches ∞.

61. Van der Waals equation of state for one mole of a gas is

$$\left(P + \frac{a}{V^2}\right)(V - b) = RT,$$

where P is the pressure, V is the volume, and T is the temperature of the gas, and R, a, and b are positive constants. What does this law predict will happen to the temperature of the gas as V approaches zero and as V approaches ∞, assuming that the pressure remains constant?

62. The rate at which an enzyme-catalysed reaction proceeds depends on the concentration of the enzyme and substrate. This rate can be described by the equation

$$v(x) = \frac{Vx}{x + K}$$

where $v(x)$ is the velocity of the reaction, x is the substrate concentration, V is the maximum value of the velocity, and K is a constant, called the *Michaelis constant*, that depends on the reaction. This equation is known variously as the *Michaelis-Menton* and the *Briggs-Haldane* equation. Sketch the graph of v and label any asymptotes. Find the value of x corresponding to $v(x) = V/2$.

63. The special theory of relativity states that the mass of an object relative to a system depends on its velocity relative to that system. If we let m_0 denote the mass of an object when it is at rest and $m(v)$ denote the mass of the object when it has velocity v, then

$$m(v) = m_0 \left(1 - \frac{v^2}{c^2}\right)^{-1/2},$$

where c is the speed of light, a constant. Sketch the graph of the function m. Describe physically what must occur as an object approaches the speed of light by considering $\lim\limits_{v \to c^-} m(v)$.

REVIEW EXERCISES

Evaluate the limits in Exercises 1 through 20.

1. $\lim\limits_{x \to 5} \dfrac{x^2 - 25}{x^2 - 6x + 5}$

2. $\lim\limits_{x \to 5^-} \dfrac{1}{x - 5}$

3. $\lim\limits_{x \to 6^+} \sqrt{x - 6}$

4. $\lim\limits_{x \to \infty} \dfrac{1}{x^2 + 2}$

5. $\lim\limits_{x \to -\infty} \dfrac{1}{x^2 - 1}$

6. $\lim\limits_{x \to 1^-} \dfrac{|x - 1|}{x - 1}$

7. $\lim\limits_{x \to 1^+} \dfrac{|x - 1|}{x - 1}$

8. $\lim\limits_{x \to 1} \dfrac{|x - 1|}{x - 1}$

9. $\lim\limits_{x \to 2} \dfrac{\sqrt{2 + x} - 2}{x - 2}$

10. $\lim\limits_{x \to 1^+} \dfrac{x}{x - 1}$

11. $\lim\limits_{x \to 3} \dfrac{\sqrt{10} - \sqrt{x^2 + 1}}{x - 3}$

12. $\lim\limits_{x \to \infty} \dfrac{x^3 - 17x}{4x^3 + 1}$

13. $\lim\limits_{x \to \infty} \dfrac{2 - x}{x^2}$

14. $\lim\limits_{x \to -\infty} \dfrac{x^3 - 7}{x^2 + 2x}$

15. $\lim\limits_{x \to 0} \dfrac{1}{|x|}$

16. $\lim\limits_{x \to 1} \dfrac{\sqrt{x} - 1}{x - 1}$

17. $\lim\limits_{x \to 2^-} \dfrac{1}{x^2 - 4}$

18. $\lim\limits_{x \to 2^+} \dfrac{1}{x^2 - 4}$

19. $\lim\limits_{h \to 0} \dfrac{\dfrac{1}{\sqrt{x + h}} - \dfrac{1}{\sqrt{x}}}{h}$

20. $\lim\limits_{h \to 0} \dfrac{(x + h)^2 - x^2}{h}$

In Exercises 21 through 26, determine any numbers at which the function is discontinuous and give the reason for the discontinuity.

21. $f(x) = \dfrac{1}{x^2 - 4}$

22. $f(x) = \dfrac{|x - 1|}{x - 1}$

23. $f(x) = \begin{cases} x^2, & \text{if } x \le 0 \\ 2x, & \text{if } x > 0 \end{cases}$

24. $f(x) = \begin{cases} x + 1, & \text{if } x < 0 \\ x^3, & \text{if } x \ge 0 \end{cases}$

25. $f(x) = \dfrac{x^2 - 5x + 6}{x^2 - 7x + 12}$

26. $f(x) = \dfrac{x^2 - 1}{(x^2 + 4)(x - 1)}$

Determine the intervals on which the functions described in Exercises 27 through 30 are continuous.

27. $f(x) = \dfrac{1}{4x - 3}$

28. $g(x) = \sqrt{2x - 1}$

29. $h(x) = \sqrt{x^2 - 1}$

30. $f(x) = \dfrac{1}{1 - x^2}$

In Exercises 31 through 38 find the horizontal and vertical asymptotes of the graph of the function and sketch the graph.

31. $f(x) = \dfrac{1}{(x - 1)^2}$

32. $f(x) = \dfrac{1}{x^2 + 2x}$

33. $f(x) = \dfrac{x^2 - 2x + 1}{x^2 - 2x}$

34. $f(x) = \dfrac{x^3 + x^2}{x^2 - 4}$

35. $f(x) = \dfrac{3x^2}{x^2 + 3}$

36. $f(x) = \sqrt{\dfrac{x^3 + x^2}{x^2 + 1}}$

37. $f(x) = \dfrac{x - 2}{x^2 - 2x}$

38. $f(x) = \dfrac{x}{|x - 4|}$

39. Use the intermediate value theorem to show that a number x, $-1 < x < 2$, exists for which $2x^3 - x - 2 = 0$.

40. Use the intermediate value theorem to show that $f(x) = x^2 - 2 \cos \pi x$ is zero for some number x in $(0, 1)$.

41. Use the definition of the limit to show that $\lim\limits_{x \to 1} (3x + 5) = 8$.

42. Use the definition of the limit to show that $\lim\limits_{x \to 3} 1/(x - 1) = 1/2$.

43. Suppose f and g are functions and that $\lim\limits_{x \to a} g(x) = \infty$ and $\lim\limits_{x \to a} f(x) = 2$. Find:

(a) $\lim\limits_{x \to a} (f(x) + g(x))$

(b) $\lim\limits_{x \to a} (f(x) - g(x))$

(c) $\lim\limits_{x \to a} f(x)g(x)$

(d) $\lim\limits_{x \to a} \dfrac{f(x)}{g(x)}$

44. If $f(x) = \begin{cases} x + 2, & \text{if } x < -2 \\ 2, & \text{if } -2 \le x \le 2, \\ 2 - x, & \text{if } x > 2 \end{cases}$ find:

(a) $\lim\limits_{x \to -2^-} f(x)$

(b) $\lim\limits_{x \to -2^+} f(x)$

(c) $\lim\limits_{x \to 0} f(x)$

45. Find a constant a that will make the function f continuous at $x = 1$, if

$$f(x) = \begin{cases} \dfrac{x^3 - 3x^2 + 2}{x^2 - 1}, & \text{for } x \ne 1 \\ a, & \text{for } x = 1. \end{cases}$$

46. If $\lim\limits_{x \to a} (f(x) + g(x))$ exists, does this imply that both $\lim\limits_{x \to a} f(x)$ and $\lim\limits_{x \to a} g(x)$ exist?

47. If $\lim\limits_{x \to a} f(x)g(x)$ exists, does this imply that both $\lim\limits_{x \to a} f(x)$ and $\lim\limits_{x \to a} g(x)$ exist?

48. Give an example of two functions f and g and a number a with the property that $\lim\limits_{x \to a} f(x)/g(x)$ exists but neither $\lim\limits_{x \to a} f(x)$ nor $\lim\limits_{x \to a} g(x)$ exists.

49. Sketch the graph of a function that has a vertical asymptote $x = 1$ and no horizontal asymptote.

50. Sketch the graph of a function that has vertical asymptotes $x = -1$ and $x = 3$ and a horizontal asymptote $y = 2$.

51. We can see that $\lim\limits_{x \to \infty} 1/x^2 = 0$. How large must x be to ensure that $1/x^2$ is less than

(a) .01?   (b) .001?

52. We can see that $\lim\limits_{x \to 0} 1/x^2 = \infty$. How small in absolute value must x be in order that $1/x^2 > 10{,}000$?

3

THE DERIVATIVE

The derivative is a fundamental concept of calculus, with applications occurring in virtually every area of scientific study. Physicists use the derivative to study the velocity and acceleration of particles. Biologists use the derivative to study the growth rate of organisms. Engineers use the concept to study a multitude of subjects, including heat flow, circuit theory, and the effects of chemical reactions. Economists analyze marginal cost and revenue using the derivative, and psychologists use it to study the response to stimuli.

Indeed, any area that depends on methods of approximation or statistics frequently uses the derivative. This list of applications only touches on a few of its uses. Although the applications come from diverse areas, they are linked by the common need to measure the change in a certain quantity relative to the change in another quantity.

3.1
THE SLOPE OF A CURVE AT A POINT

The slope of a line is constant and describes the direction of the line. If a line with equation $y = mx + b$ has a positive slope, the values of y increase as the values of x increase. If the line has a negative slope, the values of y decrease as the values of x increase. (See page 11.)

Describing the increasing and decreasing behavior of an arbitrary curve is more difficult than for a straight line, since this behavior varies with the point on the curve, as illustrated in Figure 3.1.

To determine the slope or direction of the graph of an arbitrary function at a point, we consider the line that best approximates the graph of the function at

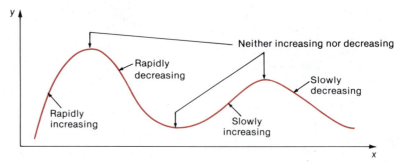

FIGURE 3.1

the point. This line is called the **tangent line** to the graph at the point. The slope of the tangent line describes the slope or direction of the graph. Some tangent lines to the graph of a function are shown in Figure 3.2.

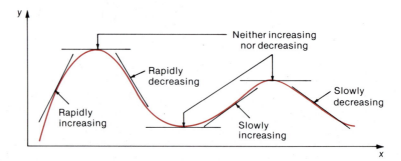

FIGURE 3.2

To determine the tangent line to the graph of a function f at a point $(a, f(a))$, we must find a line that:

1. passes through the point $(a, f(a))$, and
2. has a slope that indicates the direction of the graph of f at $(a, f(a))$.

The first condition is easy to satisfy; to see how the second condition can be satisfied we return to an example discussed in Section 2.1.

Consider the function described by $f(x) = x^2$ at the point $(1, 1)$. Lines that approximate the graph of f at $(1, 1)$ can be obtained by choosing x "close to" 1 and considering the *secant line* that passes through (x, x^2) and $(1, 1)$.

As x approaches 1, the slopes of the secant lines approach the slope of the tangent line, as shown in Figure 3.3. Consequently, the slope of the tangent line at $(1, 1)$ is defined to be the limit of the slopes of the secant lines joining (x, x^2) and $(1, 1)$ as x approaches 1. Since the slope of the line joining (x, x^2) and $(1, 1)$ is

$$\frac{x^2 - 1}{x - 1},$$

the slope of the tangent line at $(1, 1)$ is

$$\lim_{x \to 1} \frac{x^2 - 1}{x - 1} = \lim_{x \to 1} \frac{(x - 1)(x + 1)}{x - 1} = \lim_{x \to 1} x + 1 = 2.$$

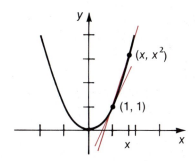

FIGURE 3.3

(3.1)
DEFINITION

The **slope of the tangent line to the graph of a function f at the point $(a, f(a))$** is defined to be

$$\lim_{x \to a} \frac{f(x) - f(a)}{x - a},$$

provided this limit exists. This is also called the slope of the curve $y = f(x)$ at $(a, f(a))$.

Figure 3.4 gives an illustration of this definition.

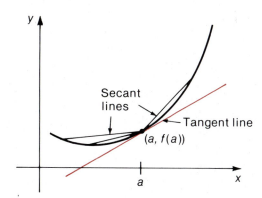

FIGURE 3.4

EXAMPLE 1   Find the slope and an equation of the tangent line to the graph of the function described by $f(x) = x^2 + x$ at the point $(-2, 2)$.

SOLUTION

The slope is

$$\lim_{x \to -2} \frac{f(x) - f(-2)}{x - (-2)} = \lim_{x \to -2} \frac{(x^2 + x) - ((-2)^2 + (-2))}{x + 2} = \lim_{x \to -2} \frac{x^2 + x - 2}{x + 2}.$$

Both the numerator and denominator have the limit zero at -2, so the limit is evaluated by factoring the numerator. This implies that the slope of the tangent line is

$$\lim_{x \to -2} \frac{(x - 1)(x + 2)}{x + 2} = \lim_{x \to -2} x - 1 = -3.$$

Since $(-2, 2)$ is a point on this line, the line has equation $y - 2 = -3(x + 2)$ or $y = -3x - 4$. $\square$

EXAMPLE 2   Find the slope of the line tangent to the graph of the function described by $f(x) = x^2$ at an arbitrary point (a, a^2).

SOLUTION

This problem is no more difficult than if the value of a was specified; the slope of the curve at (a, a^2) is

$$\lim_{x \to a} \frac{f(x) - f(a)}{x - a} = \lim_{x \to a} \frac{x^2 - a^2}{x - a} = \lim_{x \to a} \frac{(x - a)(x + a)}{x - a}$$

$$= \lim_{x \to a} (x + a)$$

$$= 2a.$$

When $a < 0$, the slope of the curve is negative. When $a > 0$, the slope is positive. This relationship can be seen in Figure 3.5(a).   ☐

EXAMPLE 3   Find the slope of the tangent line to the graph of the function described by $f(x) = 1/x$ at $(a, f(a))$, $a \neq 0$.

SOLUTION

The slope is given by

$$\lim_{x \to a} \frac{f(x) - f(a)}{x - a} = \lim_{x \to a} \frac{1/x - 1/a}{x - a} = \lim_{x \to a} \frac{(a - x)/xa}{x - a}$$

$$= \lim_{x \to a} \frac{a - x}{xa(x - a)}$$

$$= \lim_{x \to a} \frac{-1}{xa} = -\frac{1}{a^2}.$$

Consequently, the slope of the graph of $f(x) = 1/x$ is always negative. (See Figure 3.5(b).)   ☐

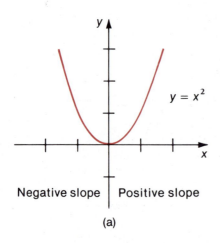

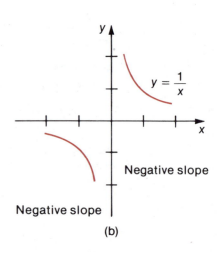

(a)   (b)

FIGURE 3.5

The **normal line** to the graph of a function at a point is the line perpendicular to the tangent line at the point. As shown in Theorem 1.5, the slopes m_1 and m_2 of nonvertical lines that are perpendicular satisfy the equation $m_1 = -1/m_2$.

EXAMPLE 4   Find an equation of the line normal to the graph of the function described by $f(x) = x^2 + x$ at $(-2, 2)$.

SOLUTION

The tangent line at $(-2, 2)$ was shown in Example 1 to have slope -3. The slope of the normal line to the graph of f at $(-2, 2)$ is consequently $-1/-3 = 1/3$. The point $(-2, 2)$ lies on the normal line, so the line has equation

$$\frac{y - 2}{x + 2} = \frac{1}{3} \qquad \text{or} \qquad y = \frac{1}{3}x + \frac{8}{3}.$$

(See Figure 3.6.)   □

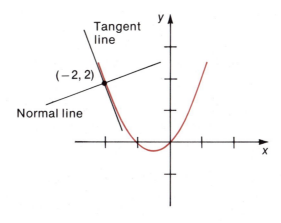

FIGURE 3.6

EXERCISE SET 3.1

In Exercises 1 through 12, a function f is described and a point on its graph is given.

   (a) Sketch the graph of the function.

   (b) Find the slope of the tangent line to the graph at the given point.

   (c) Find an equation of the tangent line to the graph at the point.

   (d) Find an equation of the normal line to the graph at the point.

1. $f(x) = 2x^2 + 1$, $(2, 9)$ **2.** $f(x) = 2x^2 - 1$, $(2, 7)$

3. $f(x) = x - 3x^2$, $(1, -2)$ **4.** $f(x) = 3 - 4x^2$, $(2, -13)$

5. $f(x) = x^2 + 2x + 3$, $(-1, 2)$ **6.** $f(x) = 2x^2 + 4x + 1$, $(2, 17)$

7. $f(x) = x^3$, $(1, 1)$ **8.** $f(x) = x^3$, $(2, 8)$

9. $f(x) = 1/x$, $(2, 1/2)$ **10.** $f(x) = 1/x$, $(-2, -1/2)$

11. $f(x) = x^2 + 2x + 2$, $(a, f(a))$ **12.** $f(x) = 2x^2 - 3x + 4$, $(a, f(a))$

13. Find a point on the graph of $f(x) = x^2 + 2x + 2$ where the tangent line has slope 1.

14. Find a point on the graph of $f(x) = 2x^2 - 3x + 4$ where the tangent line has slope -1.

15. Find a point on the graph of $f(x) = 2x^2 + 3x$ where the tangent line is parallel to the line with equation $x + y = 4$.

16. Find a point on the graph of $f(x) = 3x^2 + 2x - 1$ where the tangent line is parallel to the line with equation $y - 4x = 3$.

17. Find a point on the graph of $f(x) = 2x^2 + 3x$ where the normal line is parallel to the line with equation $x - y = 4$.

18. Find a point on the graph of $f(x) = 3x^2 + 2x - 1$ where the normal line is parallel to the line with equation $y - x = 2$.

19. Find a number a with the property that the line tangent to the graph described by $f(x) = x^2$ at $(a, f(a))$ passes through $(-1, 0)$.

20. Show that no number a exists with the property that the line tangent to the graph described by $f(x) = x^2$ at the point $(a, f(a))$ passes through the point $(0, 1)$.

21. Consider the function described by

$$f(x) = \begin{cases} x, & \text{if } x \text{ is a rational number} \\ -x, & \text{if } x \text{ is an irrational number.} \end{cases}$$

Show that the graph of f does not have a tangent line at the point $(0, 0)$.

22. Consider the function g described by

$$g(x) = \begin{cases} x^2, & \text{if } x \text{ is a rational number} \\ -x^2, & \text{if } x \text{ is an irrational number.} \end{cases}$$

Show that the graph of g has a tangent line at the point $(0, 0)$.

3.2
THE DERIVATIVE OF A FUNCTION

The slope of a tangent line to the graph of a function is the limiting value of a quotient. This quotient describes the change in the values of the function relative to the change in the independent variable. The need to determine the rate of change of a certain quantity relative to the change in another quantity occurs in many other situations. The physicist studying the velocity of a particle is concerned with the change in distance relative to the change in time. The biologist studying the growth rate of a colony of bacteria is interested in the change in the number of colony members relative to the change in time. The economist studying a marginal revenue problem is studying the change in the amount of money received from the sale of items relative to the change in the demand for the items, and so on. Mathematically, all these problems have the same format:

1. find a function that associates one quantity with another, and
2. determine the rate of change of the function with respect to the change in the independent variable.

To discuss this rate of change we introduce one of the most fundamental concepts of calculus: the **derivative**.

(3.2)
DEFINITION

The derivative of a function f at a number a is denoted $f'(a)$ and defined by

$$f'(a) = \lim_{x \to a} \frac{f(x) - f(a)}{x - a},$$

provided this limit exists. When the limit exists, f is said to be **differentiable** at a.

By comparing Definitions 3.1 and 3.2, we see that the slope of the tangent line to the graph of f at $(a, f(a))$ is simply $f'(a)$.

EXAMPLE 1   Find $f'(a)$ if $f(x) = 3x - 2$.

SOLUTION

Before applying Definition 3.2, consider the graph of f shown in Figure 3.7 to see what you think the derivative should be.

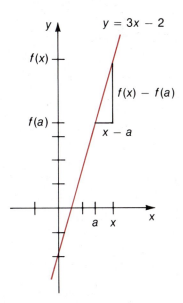

FIGURE 3.7

Because of the relationship between the derivative and the slope of the tangent line to the curve, you should expect that the derivative of f is the slope of the line. The application of Definition 3.2 shows that this is indeed the case.

$$f'(a) = \lim_{x \to a} \frac{f(x) - f(a)}{x - a}$$

$$= \lim_{x \to a} \frac{(3x - 2) - (3a - 2)}{x - a}$$

$$= \lim_{x \to a} \frac{3(x - a)}{x - a} = 3.$$ □

The quotient

$$\frac{f(x) - f(a)}{x - a}$$

is called the **difference quotient** and gives the **average rate of change** of the function values with respect to the change in the independent variable *from x to a*. The derivative, which is the limit of the difference quotient as *x* approaches *a*, is the **instantaneous rate of change** of the function values with respect to the change in the independent variable *at a*. It is this property of the derivative that makes the concept so useful for applications.

In some cases, an equivalent limit definition of the derivative is more convenient to use. If we let *h* denote the change in the independent variable, then $h = x - a$, and we can rewrite $f'(a)$ (see Figure 3.8) as

$$f'(a) = \lim_{h \to 0} \frac{f(a + h) - f(a)}{h}.$$

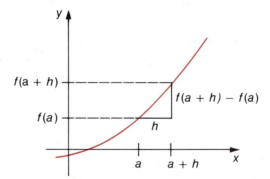

FIGURE 3.8

The notation for the derivative of a function can also assume different forms, depending on the application and on historical practice. Common alternative notations for $f'(a)$ are $df(a)/dx$, and $D_x f(a)$, and when the independent variable represents time, $\dot{f}(a)$. When *f* is defined by $y = f(x)$, it is also common to use

---

HISTORICAL NOTE   The notation $f'(a)$ first appeared in *Mécanique Analytique*, a treatise applying calculus techniques to mechanics, published in 1788 by the German mathematician **Joseph-Louis Lagrange** (1736–1813). This book has the interesting feature of containing no diagrams or drawings, just text and formulas. The dy/dx notation for the derivative has its basis with one of the mathematicians credited with founding calculus, Gottfried Liebniz. Isaac Newton, the other founder of calculus, used a notation on which $\dot{f}(a)$ is based.

dy/dx and y' to denote the derivative. We will use these alternate expressions and notations involving the derivative interchangeably to ensure that they are familiar when encountered in other sources.

The process of finding the derivative of a function is called **differentiation** and the phrases, "differentiate the function" or "find $D_x f(x)$" are often used in place of the longer expression "find the derivative of the function f."

EXAMPLE 2   Differentiate the function defined by $f(x) = x^3$.

SOLUTION

For any real number a,

$$f'(a) = \lim_{h \to 0} \frac{f(a + h) - f(a)}{h}$$

$$= \lim_{h \to 0} \frac{(a + h)^3 - a^3}{h}$$

$$= \lim_{h \to 0} \frac{a^3 + 3a^2h + 3ah^2 + h^3 - a^3}{h}$$

$$= \lim_{h \to 0} \frac{h(3a^2 + 3ah + h^2)}{h}$$

$$= \lim_{h \to 0} 3a^2 + 3ah + h^2 = 3a^2.$$

Observe that for all nonzero numbers this derivative is positive; that is, the slope of the graph of $f(x) = x^3$ is always positive. (See Figure 3.9.) ☐

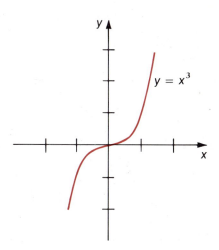

FIGURE 3.9

EXAMPLE 3   Find the derivative of the function defined by $f(x) = x^3$ at $x = -1$. What is the slope of the tangent line to the graph of f at $(-1, -1)$?

SOLUTION

From Example 2,

$$f'(a) = 3a^2.$$

So at -1, $$f'(-1) = 3(-1)^2 = 3.$$

The slope of the tangent line to the graph of f at $(-1, -1)$ is the derivative of f at $x = -1; f'(-1) = 3.$ ☐

EXAMPLE 4   Find the instantaneous rate of change of $f(x)$ with respect to x at $x = 2$ if $f(x) = 1/x$.

SOLUTION

The instantaneous rate of change of $f(x)$ with respect to x at $x = 2$ is given by $f'(2)$.

$$f'(2) = \lim_{x \to 2} \frac{f(x) - f(2)}{x - 2}$$

$$= \lim_{x \to 2} \frac{(1/x) - (1/2)}{x - 2} = \lim_{x \to 2} \frac{2 - x}{2x(x - 2)}$$

$$= \lim_{x \to 2} \frac{-1}{2x}, \text{ since } x \neq 2$$

$$= -\frac{1}{4}.$$ ☐

You should be able to recognize and use each form of the definition of the derivative. The next example gives an illustration of using both forms to solve the same problem.

EXAMPLE 5   Find $D_x f(a)$ if $f(x) = \sqrt{x}$ and $a > 0$.

SOLUTION

By Definition 3.2:

$$D_x f(a) = \lim_{x \to a} \frac{f(x) - f(a)}{x - a} = \lim_{x \to a} \frac{\sqrt{x} - \sqrt{a}}{x - a}.$$

To evaluate this limit we factor the term $x - a$ into:

$$x - a = (\sqrt{x} - \sqrt{a})(\sqrt{x} + \sqrt{a}).$$

Then, $D_x f(a) = \lim_{x \to a} \dfrac{\sqrt{x} - \sqrt{a}}{(\sqrt{x} - \sqrt{a})(\sqrt{x} + \sqrt{a})}$

$$= \lim_{x \to a} \frac{1}{\sqrt{x} + \sqrt{a}} = \frac{1}{2\sqrt{a}}.$$

Notice that $f(0)$ is defined, but $f'(0)$ is not defined.

Using the alternate definition, $f'(a) = \lim\limits_{h \to 0} \dfrac{f(a + h) - f(a)}{h}$,

$$f'(a) = \lim_{h \to 0} \frac{f(a + h) - f(a)}{h}$$
$$= \lim_{h \to 0} \frac{\sqrt{a + h} - \sqrt{a}}{h}.$$

To simplify this expression, we multiply both numerator and denominator by $(\sqrt{a + h} + \sqrt{a})$, which, in effect, "rationalizes" the numerator.

$$f'(a) = \lim_{h \to 0} \left(\frac{\sqrt{a + h} - \sqrt{a}}{h} \right) \left(\frac{\sqrt{a + h} + \sqrt{a}}{\sqrt{a + h} + \sqrt{a}} \right)$$
$$= \lim_{h \to 0} \frac{a + h - a}{h(\sqrt{a + h} + \sqrt{a})}$$
$$= \lim_{h \to 0} \frac{h}{h(\sqrt{a + h} + \sqrt{a})}$$
$$= \lim_{h \to 0} \frac{1}{\sqrt{a + h} + \sqrt{a}}$$
$$= \frac{1}{2\sqrt{a}},$$

the same result, of course, as that obtained by the original definition.   □

The derivative f' of a function f is also a function. The rule of correspondence for f' is given by

$$f'(x) = \lim_{h \to 0} \frac{f(x + h) - f(x)}{h}.$$

The domain of f' is the set of real numbers in the domain of f for which this limit exists.

The function considered in Example 5 illustrates that the domain of a function and the domain of its derivative need not be equal. For $f(x) = \sqrt{x}$, $f'(x) = 1/(2\sqrt{x})$; 0 is in the domain of f, but not in the domain of f'.

An important connection existing between the concepts of differentiability and continuity is established in the following theorem.

(3.3)
THEOREM

If $f'(a)$ exists, then f is continuous at a.

PROOF

If $f'(a)$ exists, then a is in the domain of f'. Since the domain of f' is contained in the domain of f, $f(a)$ exists. To complete the proof that f is continuous at a, we must show that $\lim\limits_{x \to a} f(x) = f(a)$. This is equivalent to showing that

$$\lim_{x \to a} (f(x) - f(a)) = 0.$$

Since $\lim_{x \to a} (x - a) = 0$, we have

$$\lim_{x \to a} (f(x) - f(a)) = \lim_{x \to a} \frac{f(x) - f(a)}{x - a} \cdot (x - a)$$

$$= \lim_{x \to a} \frac{f(x) - f(a)}{x - a} \cdot \lim_{x \to a} (x - a)$$

$$= f'(a) \cdot 0$$

$$= 0.$$

This implies that $\lim_{x \to a} f(x) = f(a)$ and continuity at a is established.   □

The converse of this theorem does not hold; that is, there are functions continuous at places where they are not differentiable. For example, the absolute value function is continuous at zero (see Figure 3.10), but is not differentiable there. To see this, consider the definition of $f'(0)$.

$$f'(0) = \lim_{x \to 0} \frac{|x| - |0|}{x - 0} = \lim_{x \to 0} \frac{|x|}{x},$$

provided this limit exists. However,

$$\lim_{x \to 0^+} \frac{|x|}{x} = 1 \quad \text{and} \quad \lim_{x \to 0^-} \frac{|x|}{x} = -1,$$

so the limit does not exist. In fact,

$$D_x |x| = \begin{cases} 1, \text{ if } x > 0 \\ -1, \text{ if } x < 0 \\ \text{undefined, if } x = 0. \end{cases}$$

Notice that there is no unique tangent line to the graph of $f(x) = |x|$ when $x = 0$.

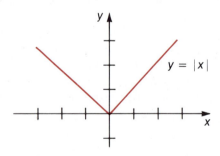

FIGURE 3.10

Figure 3.10 illustrates one of the common geometric features the graph of a function may assume when the derivative fails to exist. The graph changes direction abruptly to form a "corner" at $(0, 0)$.

In Figure 3.11 a graph is shown for a function f that has no derivative at x_1, x_2, x_3, or x_4. The derivative cannot exist at x_1 since f is discontinuous at x_1. A

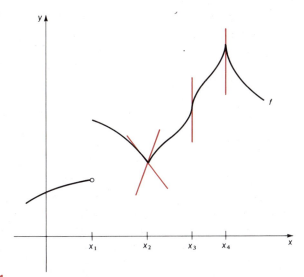

FIGURE 3.11

"corner" occurs at $(x_2, f(x_2))$, which implies that the one-sided limits of the difference quotient differ at x_2 (as they did for the absolute value function at zero).

The derivative fails to exist at both x_3 and x_4 because the limit of the difference quotient becomes infinite. The tangent lines at $(x_3, f(x_3))$ and at $(x_4, f(x_4))$ are vertical and consequently have no slope.

We close this section by mentioning another notation that is historically associated with the derivative. It is common to let Δx (read delta x) denote a change in the variable x and, if $y = f(x)$, to let Δy denote the corresponding change in the function values. Consequently, if $\Delta x = x - a$, then $\Delta y = f(x) - f(a)$, and we can write $f'(a)$ as

(3.4)
$$f'(a) = \lim_{\Delta x \to 0} \frac{\Delta y}{\Delta x}$$

or

(3.5)
$$f'(a) = \lim_{\Delta x \to 0} \frac{f(a + \Delta x) - f(a)}{\Delta x}.$$

Equation (3.4) is the basis for the $\dfrac{dy}{dx}$ notation for the derivative.

EXERCISE SET 3.2

Use the formula $D_x f(a) = \lim\limits_{x \to a} \dfrac{f(x) - f(a)}{x - a}$ to find $D_x f(a)$ in Exercises 1 through 16.

1. $f(x) = 3x, a = 1$ **2.** $f(x) = 5x + 2, a = 1$

3. $f(x) = x^2 + x, a = 3$ **4.** $f(x) = x^2 - x, a = 3$

5. $f(x) = 2x^2 + 4, a = 1$ **6.** $f(x) = -2x^2 + 3x, a = -1$

7. $f(x) = x^2 + 2x + 1, a = -1$ **8.** $f(x) = 4x^2 - 2x + 3, a = .003$

9. $f(x) = x^3, a = -4$ **10.** $f(x) = -x^3, a = 4$

11. $f(x) = \dfrac{1}{\sqrt{x}}, a = 4$ **12.** $f(x) = \dfrac{1}{\sqrt{x}}, a = 9$

13. $f(x) = \dfrac{1}{\sqrt{x+2}}, a = 2$ **14.** $f(x) = \dfrac{1}{\sqrt{x+1}}, a = 8$

15. $f(x) = x^2 + \dfrac{1}{x}, a = 2$ **16.** $f(x) = x + \sqrt{x}, a = 4$

Use the formula $D_x f(a) = \lim\limits_{h \to 0} \dfrac{f(a+h) - f(a)}{h}$ to find $f'(a)$ in Exercises 17 through 22. Compare this method with the corresponding exercises in 1 through 16 and decide which method you prefer in each case.

17. $f(x) = x^2 + x, a = 3$ **18.** $f(x) = x^2 - x, a = 3$

19. $f(x) = x^2 + 2x + 1, a = -1$ **20.** $f(x) = x^3, a = -4$

21. $f(x) = \dfrac{1}{\sqrt{x+2}}, a = 2$ **22.** $f(x) = \dfrac{1}{\sqrt{x+1}}, a = 8$

In Exercises 23 through 34, a function f is described. Find f' and its domain, and sketch the graphs of f and f'.

23. $f(x) = 4x - 2$ **24.** $f(x) = 3x + 2$

25. $f(x) = x^2 + 2x + 1$ **26.** $f(x) = x^2 + 2x - 1$

27. $f(x) = x^3$ **28.** $f(x) = x^3 + 1$

29. $f(x) = (x + 1)^3$ **30.** $f(x) = x^4$

31. $f(x) = \dfrac{1}{x}$ **32.** $f(x) = \dfrac{1}{x^2}$

33. $f(x) = |x|$ **34.** $f(x) = [\![x]\!]$

35. Find the instantaneous rate of change of the function described by $f(x) = x^2 + x$ at $x = 1$.

36. Find the instantaneous rate of change of the function described by $f(x) = x^3 + x$ at $x = 1$.

37. The area of a circle with radius r is given by $A(r) = \pi r^2$. Find the instantaneous rate of change of the area with respect to the radius.

38. The volume of a cube with side s is given by $V(s) = s^3$. Find the instantaneous rate of change of the volume with respect to s.

The right- and left-hand derivatives of a function can be defined by using the definitions for right- and left-hand limits. The **right-hand derivative,** f'_+, of the function f at a is defined by

$$f'_+(a) = \lim_{x \to a^+} \frac{f(x) - f(a)}{x - a},$$

provided this limit exists and the **left-hand derivative,** f'_-, of the function f at a is defined by

$$f'_-(a) = \lim_{x \to a^-} \frac{f(x) - f(a)}{x - a},$$

provided this limit exists. (See the figure below.) Use these definitions to find $f'_+(a)$ and $f'_-(a)$ in Exercises 39 through 42.

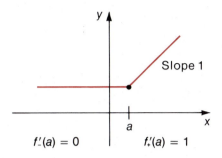

$f'_-(a) = 0$ $f'_+(a) = 1$

39. $f(x) = x^2 + x$, $a = 3$ **40.** $f(x) = |x|$, $a = 1$

41. $f(x) = |x|$, $a = 0$ **42.** $f(x) = [\![x]\!]$, $a = 1$

43. Prove that $f'(a)$ exists if and only if $f'_+(a)$ and $f'_-(a)$ exist and $f'_+(a) = f'_-(a)$.

44. The volume of a yeast starter for making sourdough bread can be approximated by

$$V = \begin{cases} 1, \text{ if } t \leq 2 \\ 1 + \sqrt{t - 2}, \text{ if } 2 < t < 6 \\ 3, \text{ if } 6 \leq t \end{cases}$$

where t is measured in days and V in cups. What is the instantaneous rate of change of the starter with respect to time when $t = 3$? Is V a differentiable function of t when $t = 2$ and when $t = 6$?

45. A simple model for population growth utilizes the statement that the rate of change of the population with respect to time is proportional to the population. Express this statement as an equation involving a derivative.

46. A theory of group behavior presented by G. C. Homans is based on I, the intensity of interaction among group members; F, the level of friendliness among members; A, the amount of activity by members; and E, the amount of activity imposed on group members by external environmental forces. Each of these quantities represents an average over the group members and may change with time. H. A. Simon, recipient of the 1979 Nobel prize in Economics, expressed this theory mathematically by:

$$I(t) = a_1 F(t) + a_2 A(t)$$
$$F'(t) = b[I(t) - \beta F(t)]$$
$$A'(t) = c_1[F(t) - \gamma A(t)] + c_2[E(t) - A(t)]$$

where a_1, a_2, b, β, c_1, γ, c_2 are positive constants. What do these relations indicate?

47. Find the y-intercept and x-intercept of the tangent line to the graph of f at a point $(a, f(a))$.

48. Find the y-intercept and x-intercept of the normal line to the graph of f at a point $(a, f(a))$.

49. Show that a function f is differentiable at a number x_0 if and only if a constant m and a function ξ exist with the property that

$$f(x) = f(x_0) + m(x - x_0) + \xi(x)(x - x_0)$$

and

$$\lim_{x \to x_0} \xi(x) = 0.$$

50. Suppose f is a non-constant function with the property that for all real numbers x_1 and x_2, $f(x_1 + x_2) = f(x_1) f(x_2)$. Show that $f(0) = 1$. Use this to show that if $f'(0)$ exists, then $f'(a)$ exists for all a in $\mathbb{R}$ and $f'(a) = f(a) \cdot f'(0)$.

51. Suppose f is a function with the property that for all positive real numbers x_1 and x_2, $f(x_1 x_2) = f(x_1) + f(x_2)$. Show that $f(1) = 0$ and use this to show that if $f'(1)$ exists, then $f'(a) = f'(1)/a$ for any positive real number a.

3.3
FORMULAS FOR DIFFERENTIATION

Before we can effectively study the concept of the derivative we need a way to quickly determine derivatives of a relatively large class of functions that can be used for examples and applications. In this section we will present results that eliminate the need for using the limit process when finding the derivatives of many common functions.

Consider a linear function, one whose graph is a nonvertical line such as that shown in Figure 3.12. For any real numbers a and x, $(f(x) - f(a))/(x - a)$ is the slope of the line. Consequently, the derivative of a linear function at any point is the slope of the line.

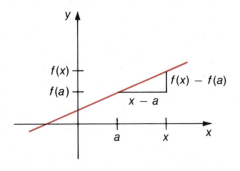

FIGURE 3.12

(3.6)
THEOREM   If f is a function described by $f(x) = mx + b$, then $f'(x) = m$ for any real number x.

PROOF

$$f'(x) = \lim_{h \to 0} \frac{f(x + h) - f(x)}{h}$$

$$= \lim_{h \to 0} \frac{(m(x + h) + b) - (mx + b)}{h}$$

$$= \lim_{h \to 0} \frac{mh}{h}$$

$$= m. \qquad \qquad \square$$

Since this theorem is true for each pair of constants m and b, a special case of this result is:

(3.7)
COROLLARY

(a) If $f(x) = mx$ for some constant m, then $f'(x) = m$.
(b) If $f(x) = b$ for some constant b, then $f'(x) = 0$.

The next theorem generalizes two results that we have seen in our examples. In Example 2 of Section 3.1 we saw that for $f(x) = x^2$, $D_x x^2 = f'(x) = 2x$. In Example 2 of Section 3.2 we found that $D_x x^3 = 3x^2$. A reasonable conjecture from these results is that for any positive integer n, $D_x x^n = nx^{n-1}$. This result can be established using the **binomial theorem**, which states that for any constants a and b and any positive integer n:

$$(a + b)^n = a^n + na^{n-1} b + \frac{n(n - 1)}{2} a^{n-2}b^2 + \cdots + nab^{n-1} + b^n.$$

(3.8)
THEOREM

The Power Rule If $f(x) = x^n$, where n is a positive integer, then $f'(x) = nx^{n-1}$.

PROOF

$$f'(x) = \lim_{h \to 0} \frac{f(x + h) - f(x)}{h}$$

$$= \lim_{h \to 0} \frac{(x + h)^n - x^n}{h}$$

By letting $a = x$ and $b = h$ in the binomial theorem,

$$f'(x) = \lim_{h \to 0} \frac{(x^n + nx^{n-1} h + \frac{n(n - 1)}{2} x^{n-2} h^2 + \cdots + nx\, h^{n-1} + h^n) - x^n}{h}$$

$$= \lim_{h \to 0} \frac{h(nx^{n-1} + \frac{n(n - 1)}{2} x^{n-2} h + \cdots + nx\, h^{n-2} + h^{n-1})}{h}$$

$$= \lim_{h \to 0} \left(nx^{n-1} + \frac{n(n - 1)}{2} x^{n-2} h + \cdots + nx\, h^{n-2} + h^{n-1} \right).$$

All terms except the first include a factor of h and approach 0 as h approaches 0. Thus, $f'(x) = nx^{n-1}$. $\square$

EXAMPLE 1   If $f(x) = x^9$, find $f'(-1)$.

SOLUTION

From Theorem 3.8,

$$f'(x) = 9x^{9-1} = 9x^8,$$

so,

$$f'(-1) = 9(-1)^8 = 9. \qquad \square$$

In Chapter 2 we developed some results about the limit of the sum, difference, product, and quotient of two functions. Since the derivative involves the limit, these results can be used to simplify the derivative process.

(3.9)
THEOREM

If f and g are differentiable at x, then

$$D_x(f+g)(x) = D_x f(x) + D_x g(x),$$

and

$$D_x(f-g)(x) = D_x f(x) - D_x g(x).$$

PROOF

$$
\begin{aligned}
D_x(f+g)(x) &= \lim_{h \to 0} \frac{(f+g)(x+h) - (f+g)(x)}{h} \\
&= \lim_{h \to 0} \frac{f(x+h) + g(x+h) - f(x) - g(x)}{h} \\
&= \lim_{h \to 0} \left(\frac{f(x+h) - f(x)}{h} + \frac{g(x+h) - g(x)}{h} \right) \\
&= \lim_{h \to 0} \frac{f(x+h) - f(x)}{h} + \lim_{h \to 0} \frac{g(x+h) - g(x)}{h} \\
&= D_x f(x) + D_x g(x).
\end{aligned}
$$

The proof that $D_x(f-g)(x) = D_x f(x) - D_x g(x)$ can be handled in the same manner (see Exercise 42). $\qquad \square$

(3.10)′
THEOREM

If f is differentiable at x, then for any constant c,

$$D_x(cf)(x) = cD_x f(x).$$

PROOF

$$
\begin{aligned}
D_x(cf)(x) &= \lim_{h \to 0} \frac{cf(x+h) - cf(x)}{h} \\
&= \lim_{h \to 0} \frac{c[f(x+h) - f(x)]}{h} \\
&= c \lim_{h \to 0} \frac{f(x+h) - f(x)}{h} \\
&= cD_x f(x). \qquad \square
\end{aligned}
$$

EXAMPLE 2   Find $D_x f(x)$ for $f(x) = 3x^2 + 2x - 1$.

SOLUTION

The theorems just presented can be used to show that

$$\begin{aligned}
D_x f(x) &= D_x(3x^2 + 2x - 1) \\
&= D_x(3x^2) + D_x(2x - 1) \\
&= 3D_x(x^2) + 2 \\
&= 3(2x) + 2 \\
&= 6x + 2.
\end{aligned}$$

$\square$

By repeated application of Theorems 3.8 and 3.9, it can be seen that any polynomial

$$P(x) = a_n x^n + a_{n-1} x^{n-1} + \ldots + a_1 x + a_0$$

will have as its derivative

$$P'(x) = na_n x^{n-1} + (n-1) a_{n-1} x^{n-2} + \ldots + 2a_2 x + a_1.$$

We have established that when two functions are differentiable the derivative of the sum or difference can be obtained by simply taking the sum or difference of their derivatives. The natural question to ask is whether the same type of result is true for products and quotients. To see that it is not, let $f(x) = 1$ and $g(x) = x^2$. Then $D_x(fg)(x) = D_x(x^2) = 2x$, while $[D_x f(x)][D_x g(x)] = 0 \cdot 2x = 0$.
A rule for multiplication does exist and is easily applied. It is called the **product rule.**

(3.11)
THEOREM

The Product Rule   If f and g are differentiable at x, then
$$D_x(fg)(x) = f(x) D_x g(x) + g(x) D_x f(x).$$

PROOF

$$(fg)'(x) = \lim_{h \to 0} \frac{f(x+h) g(x+h) - f(x) g(x)}{h}.$$

To change this into a form involving the quotients associated with $f'(x)$ and $g'(x)$,

$$\frac{f(x+h) - f(x)}{h} \quad \text{and} \quad \frac{g(x+h) - g(x)}{h},$$

we add and subtract the quantity $f(x+h) g(x)$ in the numerator.

$$\begin{aligned}
(fg)'(x) &= \lim_{h \to 0} \frac{f(x+h) g(x+h) - f(x+h) g(x) + f(x+h) g(x) - f(x)g(x)}{h}. \\
&= \lim_{h \to 0} \left[f(x+h) \frac{g(x+h) - g(x)}{h} + \frac{f(x+h) - f(x)}{h} g(x) \right] \\
&= \lim_{h \to 0} f(x+h) \lim_{h \to 0} \left[\frac{g(x+h) - g(x)}{h} \right] + \lim_{h \to 0} \left[\frac{f(x+h) - f(x)}{h} \right] \lim_{h \to 0} g(x) \\
&= \lim_{h \to 0} f(x+h) g'(x) + f'(x) \lim_{h \to 0} g(x).
\end{aligned}$$

Since f has a derivative at x, Theorem 3.3 implies that f is continuous at x; so $\lim_{h \to 0} f(x + h) = f(x)$. Moreover, $g(x)$ is independent of h, so $\lim_{h \to 0} g(x) = g(x)$.

Therefore, $(fg)'(x) = f(x) g'(x) + f'(x) g(x)$.   ☐

EXAMPLE 3   Calculate $\dfrac{df(x)}{dx}$ if $f(x) = (2x^2 + x - 1)(2x + 4)$.

SOLUTION

$$\frac{df(x)}{dx} = \frac{d}{dx} [(2x^2 + x - 1)(2x + 4)]$$

$$= (2x^2 + x - 1) \frac{d}{dx} (2x + 4) + (2x + 4) \frac{d}{dx} (2x^2 + x - 1)$$

(by Theorem 3.11)

$$= (2x^2 + x - 1)(2) + (2x + 4) \left[\frac{d}{dx}(2x^2) + \frac{d}{dx}(x - 1) \right]$$

(by Theorems 3.6 and 3.9)

$$= 4x^2 + 2x - 2 + (2x + 4) \left[2\frac{d}{dx}(x^2) + 1 \right] \quad \text{(by Theorem 3.10)}$$

$$= 4x^2 + 2x^2 - 2 + (2x + 4) [2(2x) + 1] \quad \text{(by Theorem 3.8)}$$

$$= 4x^2 + 2x - 2 + 8x^2 + 18x + 4$$

$$= 12x^2 + 20x + 2.$$

The result in this example could also be obtained by performing the multiplication $(2x^2 + x - 1)(2x + 4) = 4x^3 + 10x^2 + 2x - 4$ and then taking the derivative of each term in succession.   ☐

The formula for the derivative of the product of three or more differentiable functions can be derived by applying the product rule as many times as necessary. This is illustrated in the next example.

EXAMPLE 4   Find $D_x [(x^2 + 1)(x^3 + 2x - 1)(4x - 2)]$.

SOLUTION

Using the product rule,

$$D_x[(x^2 + 1)(x^3 + 2x - 1)(4x - 2)] = (x^2 + 1)(x^3 + 2x - 1) D_x(4x - 2)$$
$$+ (4x - 2) D_x[(x^2 + 1)(x^3 + 2x - 1)]$$

$$= (x^2 + 1)(x^3 + 2x - 1) D_x(4x - 2)$$
$$+ (4x - 2)(x^2 + 1) D_x(x^3 + 2x - 1)$$
$$+ (4x - 2)(x^3 + 2x - 1) D_x(x^2 + 1)$$

$$= (x^2 + 1)(x^3 + 2x - 1)4$$
$$+ (4x - 2)(x^2 + 1)(3x^2 + 2)$$
$$+ (4x - 2)(x^3 + 2x - 1)2x.$$ ☐

In general,

(3.12) $D_x(f \cdot g \cdot h)(x) = f(x)g(x)D_xh(x) + f(x)h(x)D_xg(x) + g(x)h(x)D_xf(x),$

provided all derivatives exist. This rule can be expanded in a similar manner to include the products of more than three functions.

Before stating the theorem concerning the derivative of the quotient of two functions, we will prove a result about the derivative of the reciprocal of a function.

**(3.13)
THEOREM**

The Reciprocal Rule   If g is differentiable at x and $g(x) \neq 0$, then

$$D_x\left(\frac{1}{g}\right)(x) = -\frac{D_x g(x)}{[g(x)]^2}$$

PROOF

$$D_x\left(\frac{1}{g}\right)(x) = \lim_{h \to 0} \frac{\left(\frac{1}{g}\right)(x + h) - \left(\frac{1}{g}\right)(x)}{h}$$

$$= \lim_{h \to 0} \frac{\dfrac{1}{g(x + h)} - \dfrac{1}{g(x)}}{h}$$

$$= \lim_{h \to 0} \frac{g(x) - g(x + h)}{g(x + h)g(x)\, h}$$

$$= \lim_{h \to 0} \left(\frac{g(x + h) - g(x)}{h}\right)\left(\frac{-1}{g(x + h)g(x)}\right)$$

$$= \lim_{h \to 0} \frac{g(x + h) - g(x)}{h} \lim_{h \to 0} \frac{-1}{g(x + h)g(x)}$$

$$= g'(x) \frac{-1}{\lim\limits_{h \to 0} g(x + h) \cdot g(x)}.$$

Since $g'(x)$ exists, g is continuous at x and $\lim\limits_{h \to 0} g(x + h) = g(x)$. With this,

$$D_x\left(\frac{1}{g}\right)(x) = g'(x) \frac{-1}{g(x)g(x)} = \frac{-g'(x)}{[g(x)]^2}. \qquad \square$$

The product and reciprocal rules can now be combined to give the quotient rule.

**(3.14)
THEOREM**

The Quotient Rule   If f and g are differentiable at x and $g(x) \neq 0$, then

$$D_x\left(\frac{f}{g}\right)(x) = \frac{g(x)D_xf(x) - f(x)\,D_xg(x)}{[g(x)]^2}.$$

PROOF

Using Theorems 3.11 and 3.13

$$\left(\frac{f}{g}\right)'(x) = \left(f \cdot \frac{1}{g}\right)'(x)$$

$$= f(x)\left(\frac{1}{g}\right)'(x) + f'(x)\left(\frac{1}{g}\right)(x)$$

$$= f(x)\left(\frac{-g'(x)}{[g(x)]^2}\right) + f'(x)\left(\frac{1}{g(x)}\right)$$

$$= \frac{-f(x)g'(x) + f'(x)g(x)}{[g(x)]^2}$$

$$= \frac{f'(x)g(x) - f(x)g'(x)}{[g(x)]^2}.$$ □

EXAMPLE 5   Find $D_x f(x)$ if

$$f(x) = \frac{x^4 + 2x - 1}{3x^2 + 5}.$$

SOLUTION

$$D_x f(x) = D_x\left(\frac{x^4 + 2x - 1}{3x^2 + 5}\right)$$

$$= \frac{(3x^2 + 5)D_x(x^4 + 2x - 1) - (x^4 + 2x - 1)D_x(3x^2 + 5)}{(3x^2 + 5)^2}$$

$$= \frac{(3x^2 + 5)(4x^3 + 2) - (x^4 + 2x - 1)(6x)}{(3x^2 + 5)^2}$$

$$= \frac{6x^5 + 20x^3 - 6x^2 + 6x + 10}{(3x^2 + 5)^2}.$$ □

EXAMPLE 6   Use the quotient rule and then the product rule to find $f'(t)$ if
$$f(t) = \frac{(t + 1)(t + 2)}{t - 1}.$$

SOLUTION

$$f'(t) = \frac{(t - 1)D_t[(t + 1)(t + 2)] - (t + 1)(t + 2)D_t(t - 1)}{(t - 1)^2}$$

$$= \frac{(t - 1)[(t + 1)D_t(t + 2) + (t + 2)D_t(t + 1)] - (t + 1)(t + 2)(1)}{(t - 1)^2}$$

$$= \frac{(t - 1)[(t + 1) + (t + 2)] - (t + 1)(t + 2)}{(t - 1)^2}$$

$$= \frac{(t - 1)(2t + 3) - (t + 1)(t + 2)}{(t - 1)^2}$$

$$= \frac{2t^2 + t - 3 - t^2 - 3t - 2}{(t - 1)^2}$$

$$= \frac{t^2 - 2t - 5}{t^2 - 2t + 1}.$$ □

EXAMPLE 7   Find the derivative of

$$h(x) = \frac{1}{x^2}.$$

SOLUTION

While we could apply the result of Theorem 3.14 to the function h, with $f(x) = 1$ and $g(x) = x^2$, it is easier to apply Theorem 3.13 with $h = 1/g$.

$$h'(x) = D_x\left(\frac{1}{g}\right)(x) = \frac{-g'(x)}{[g(x)]^2}$$

$$= \frac{-2x}{(x^2)^2} = \frac{-2}{x^3}. \qquad \square$$

Example 7 implies that $D_x(x^{-2}) = -2x^{-3}$, so the power rule holds for $n = -2$. In fact the rule holds for all negative as well as positive integers.

(3.15)
COROLLARY   If $f(x) = x^n$, where n is any nonzero integer, then $f'(x) = nx^{n-1}$.

PROOF

The result has already been shown in the case when n is a positive integer. If n is a negative integer, then $-n$ is positive and since $x^n = 1/x^{-n}$, we can apply Theorem 3.8 and Theorem 3.13 to deduce that

$$f'(x) = D_x\left(\frac{1}{x^{-n}}\right) = \frac{-D_x(x^{-n})}{(x^{-n})^2}$$

$$= \frac{-(-nx^{-n-1})}{x^{-2n}} = nx^{-n-1}x^{2n}$$

$$= nx^{n-1}. \qquad \square$$

EXAMPLE 8   Find

$$D_x\left(\frac{2}{x^3} + \frac{1}{x^5}\right).$$

SOLUTION

$$D_x\left(\frac{2}{x^3} + \frac{1}{x^5}\right) = 2D_x(x^{-3}) + D_x(x^{-5})$$

$$= 2(-3x^{-4}) + (-5x^{-6})$$

$$= -6x^{-4} - 5x^{-6}. \qquad \square$$

EXAMPLE 9   Find $D_x(x+1)^{-2}$.

SOLUTION

Rewrite $(x+1)^{-2}$ as:

$$(x+1)^{-2} = \frac{1}{(x+1)^2} = \frac{1}{x^2 + 2x + 1}.$$

Then,
$$D_x(x + 1)^{-2} = D_x\left(\frac{1}{x^2 + 2x + 1}\right)$$
$$= \frac{-D_x(x^2 + 2x + 1)}{(x^2 + 2x + 1)^2}$$
$$= \frac{-(2x + 2)}{(x^2 + 2x + 1)^2} = \frac{-2(x + 1)}{(x + 1)^4}$$
$$= \frac{-2}{(x + 1)^3}.$$
□

EXERCISE SET 3.3

Use the results of this section to find the derivatives of the functions described in Exercises 1 through 28.

1. $f(x) = x^2 - 2x + 7$

2. $f(x) = 2x^2 - 3x + 4$

3. $f(x) = \frac{1}{2}x^2 + 4x + 9$

4. $f(x) = \frac{1}{4}x^4 + \frac{1}{3}x^3 + \frac{1}{2}x^2 + x + 1$

5. $f(x) = 17x^{31} + 14x^{29} + x^4$

6. $f(x) = 21x^{16} + 14x^{12} + 11x^9$

7. $f(t) = \pi t^2 + (3 + 4\sqrt{2})t + 9$

8. $f(s) = (\pi^2 + 1)s^2 + \sqrt{3}(s + 1)$

9. $h(x) = (x^2 + 4)(x^2 - 4)$

10. $r(u) = (u^2 + u + 2)(u^4 + u^3)$

11. $h(t) = (t^4 - 3t^2 + 5)(3t^4 - t^{-2})$

12. $g(z) = (z^3 - 3z^{-2} + z)(5 + z^{-6})$

13. $f(x) = \frac{1}{3}x^2 + \frac{3}{x^2}$

14. $f(x) = \frac{1}{4}x^4 + \frac{4}{x^4}$

15. $f(z) = \frac{z}{z + 1}$

16. $r(z) = \frac{z + 1}{z}$

17. $H(x) = \frac{3x + 4}{5x^2 + 1}$

18. $g(w) = \frac{w^2 + 3w - 2}{w^2 + 3w + 2}$

19. $H(t) = \frac{t^2 - t - 6}{t^2 + 3t - 4}$

20. $f(x) = \frac{x^3 - x^2 + 2x - 2}{3x + 4}$

21. $F(s) = \left(\frac{s^2 + 1}{s^2 + 2}\right)(s^2 + 3)$

22. $g(t) = \left(\frac{t^2 + 1}{t^2 + 2}\right)\left(\frac{t^2 + 3}{t^2 + 4}\right)$

23. $g(x) = \frac{x(x^2 - 1)}{(x^2 + 1)}$

24. $h(t) = \frac{t^2 + 1}{(t^2 + 2)(t^2 + 3)}$

25. $f(x) = (x + 1)(x + 2)(x + 3)$

26. $f(x) = (x^2 - 1)(x^2 - 2)(x^2 - 3)$

27. $f(x) = x(x^2 + 2x + 1)(x - 5)$

28. $g(x) = x(x^2 + 1)(x^3 + 2)$

29. Find a point on the graph of $f(x) = x^2 + 3x - 4$ at which the slope of the tangent line is zero.

30. Find the point on the graph of $y = x^2 - x$ at which the tangent line is horizontal.

31. Find an equation of the line tangent to the graph of $y = (x^2 - 1)/(x^2 + 1)$ at $(2, 3/5)$.

32. Find an equation of the line normal to the graph of $y = (x^2 - 1)/(x^2 + 1)$ at $(2, 3/5)$.

33. Water is pumped into a reservoir so that the volume V (in gallons) t minutes after the pumping begins is given by $V = 50t^2$. (a) Find the average rate of change of V with respect to t from $t = 1$ to $t = 2$ minutes. (b) Find the instantaneous rate of change of V with respect to t when $t = 1$ minute.

34. The size in square centimeters of a bacteria population residing on a nutrient agar at time t measured in hours is given by $P(t) = 1000 + 100t - 10t^2$. Find the instantaneous rate of growth when $t = 2$ hours, when $t = 5$ hours, and when $t = 7$ hours.

35. The size of an insect population at time t (measured in days) is given by $P(t) = 10{,}000 - 9000/(t + 1)$. (a) What is the initial population $P(0)$? (b) Find the instantaneous rate of growth at any time t, assuming that P is a continuous function of time.

36. The heat capacity of oxygen depends on temperature according to the formula

$$C_p = 8.27 + 2.6 \times 10^{-4}T - 1.87 \times 10^{-5}T^2 \frac{\text{cal}}{\text{deg-mole}},$$

where T is measured in °C, and $25 < T < 650$. Find the instantaneous rate of change of the heat capacity with respect to temperature when $T = 500$°C.

37. For temperatures close to absolute zero $(-273$°C$)$ the atomic heat of aluminum can be accurately described by Debye's equation

$$C = 464.6 \left(\frac{T}{375}\right)^3 \frac{\text{cal}}{\text{deg-mole}},$$

where T is the temperature in degrees Kelvin; °K $=$ °C $+ 273$. Find the instantaneous rate of change in the heat with respect to T when the temperature is -250°C.

38. A liquid flowing through a cylindrical tube has varying velocity depending on its distance r from the center of the tube:

$$v(r) = k(R^2 - r^2),$$

where R is the radius of the tube and k is a constant that depends on the length of the tube and the velocity of the fluid at the ends. Find the rate of change of $v(r)$ with respect to the distance from the center of the tube (a) at the center of the tube, (b) halfway to the wall of the tube, and (c) at the wall of the tube.

39. In Example 5 of Section 3.2 we showed that $D_x\sqrt{x} = 1/(2\sqrt{x})$, which is equivalent to stating that $D_x x^{1/2} = (1/2)x^{-1/2}$. Use this fact and the product or reciprocal rule to find
(a) $D_x x^{3/2}$ (b) $D_x x^{5/2}$ (c) $D_x x^{-1/2}$

40. By considering the results of Exercise 39, find a formula for the derivative of the function described by $f(x) = x^{n/2}$, where n is a positive odd integer. Derive this formula by writing $n = 2k + 1$ for some positive integer k and expressing $x^{n/2}$ as $x^k \cdot x^{1/2}$.

41. Use the formula you deduced in Exercise 40 to find a formula for $D_x x^{-n/2}$ that is valid for any positive integer n.

42. Prove that the second assertion in Theorem 3.9 is true.

43. Suppose $f'(a)$ exists. Show that for any positive integer n

$$\lim_{x \to a} \frac{[f(x)]^n - [f(a)]^n}{x - a} = n[f(a)]^{n-1} f'(a).$$

Use this to find $D_x (2x + 1)^4$.

44. What is faulty about the following reasoning?

$$\overbrace{x^2 = x + x + \ldots + x}^{x \text{ times}}$$

so

$$D_x x^2 = D_x(x + \ldots + x).$$

But then

$$\overbrace{2x = 1 + 1 + \ldots + 1,}^{x \text{ times}}$$

which implies

$$2x = x, \text{ and } 1 = 2.$$

3.4
DIFFERENTIATION OF TRIGONOMETRIC FUNCTIONS

In this section we determine the derivatives of the trigonometric functions. A review of the definitions and basic properties of these functions is provided in Appendix A.3. In the study of calculus, the trigonometric functions are usually described using radian measure. We will assume this representation here and in other sections, unless specified otherwise.

The calculation of the derivative of the sine function involves finding

$$D_x \sin x = \lim_{h \to 0} \frac{\sin (x + h) - \sin x}{h}.$$

Expanding $\sin (x + h)$ and using the limit theorems gives

$$D_x \sin x = \lim_{h \to 0} \frac{\sin x \cos h + \sin h \cos x - \sin x}{h}$$

$$= \lim_{h \to 0} \left(\frac{\sin x (\cos h - 1)}{h} + \cos x \frac{\sin h}{h} \right)$$

$$= \sin x \lim_{h \to 0} \frac{\cos h - 1}{h} + \cos x \lim_{h \to 0} \frac{\sin h}{h}.$$

Determining the derivative of this basic trigonometric function hinges on finding values for

$$\lim_{h \to 0} \frac{\sin h}{h} \quad \text{and} \quad \lim_{h \to 0} \frac{\cos h - 1}{h}.$$

To find these limits we need the squeeze or **sandwiching theorem,** so named because it gives results about a function that is squeezed or sandwiched between two other functions. (See Figure 3.13.)

(3.16)
THEOREM

Sandwiching Theorem   Suppose f, g, and h are functions with the property that

$$f(x) \leq g(x) \leq h(x)$$

for each $x \neq a$ in an open interval containing a. If $\lim_{x \to a} f(x) = L$ and $\lim_{x \to a} h(x) = L$, then $\lim_{x \to a} g(x) = L$.

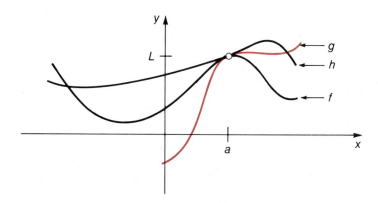

FIGURE 3.13

The proof of the sandwiching theorem follows from the definition of the limit. Similar results hold when the limits are all the same one-sided limits. These results provide an important step in the proof of the following theorem.

(3.17)
THEOREM

$$\lim_{t \to 0} \frac{\sin t}{t} = 1.$$

PROOF

First we will show that $\lim_{t \to 0^+} \frac{\sin t}{t} = 1$. The demonstration of this result is geometric and requires comparing the areas of triangles OBP and OBA to the area of the sector OBP, shown in Figure 3.14. This sector is the portion of the unit circle bounded by the rays of the angle with radian measure t. Since we are interested in a limit at 0 from the right, we can restrict t to $0 < t < \pi/2$.

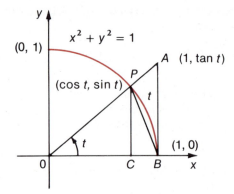

FIGURE 3.14

$$\text{Area of triangle } OBP = \frac{1}{2} \cdot \text{base} \cdot \text{altitude} = \frac{1}{2} d(O, B) \, d(C, P)$$

$$= \frac{1}{2} \cdot 1 \cdot \sin t = \frac{1}{2} \sin t$$

and

$$\text{Area of triangle } OBA = \frac{1}{2} d(O, B) \, d(B, A)$$

$$= \frac{1}{2} \cdot 1 \cdot \tan t$$

$$= \frac{1}{2} \tan t = \frac{1}{2} \frac{\sin t}{\cos t}.$$

To determine the area of sector OBP we use the fact that the area of the unit circle is π and that this sector constitutes the portion $t/2\pi$ of the circle.

$$\text{Area of sector } OBP = \frac{t}{2\pi} \cdot \pi = \frac{1}{2} t.$$

From Figure 3.14, we see that

Area of triangle $OBP <$ Area of sector $OBP <$ Area of triangle OBA,

so

(3.18)
$$\frac{1}{2} \sin t < \frac{1}{2} t < \frac{1}{2} \frac{\sin t}{\cos t}.$$

Multiplying (3.18) by $\dfrac{2}{\sin t}$ produces the inequality

$$1 < \frac{t}{\sin t} < \frac{1}{\cos t}$$

and taking reciprocals gives

$$1 > \frac{\sin t}{t} > \cos t.$$

However, $\cos t = \sqrt{1 - (\sin t)^2}$ and by (3.18) $\sin t < t$. Therefore,

(3.19) $$1 > \frac{\sin t}{t} > \cos t > \sqrt{1 - t^2}.$$

The sandwiching theorem for limits from the right implies that

(3.20) $$1 = \lim_{t \to 0^+} 1 \geq \lim_{t \to 0^+} \frac{\sin t}{t} \geq \lim_{t \to 0^+} \cos t \geq \lim_{t \to 0^+} \sqrt{1 - t^2} = 1$$

so $$\lim_{t \to 0^+} \frac{\sin t}{t} = 1.$$

Since $\sin(-t) = -\sin t$,

$$\lim_{t \to 0^-} \frac{\sin t}{t} = \lim_{t \to 0^+} \frac{\sin(-t)}{-t} = \lim_{t \to 0^+} \frac{\sin t}{t} = 1.$$

By Theorem 2.15, $$\lim_{t \to 0} \frac{\sin t}{t} = 1. \qquad \square$$

An additional result from line (3.20) is that

(3.21) $$\lim_{t \to 0^+} \cos t = 1.$$

Since $\cos(-t) = \cos t$ for all values of t this is sufficient to show that $\lim_{t \to 0} \cos t = 1$, which implies that the cosine function is continuous at zero.

(3.22)
COROLLARY
$$\lim_{t \to 0} \frac{\cos t - 1}{t} = 0.$$

PROOF

$$\lim_{t \to 0} \frac{\cos t - 1}{t} = \lim_{t \to 0} \frac{\cos t - 1}{t} \cdot \frac{\cos t + 1}{\cos t + 1}$$

$$= \lim_{t \to 0} \frac{(\cos t)^2 - 1}{t(\cos t + 1)} = \lim_{t \to 0} \frac{-(\sin t)^2}{t(\cos t + 1)}$$

$$= \lim_{t \to 0} \frac{-\sin t}{t} \cdot \frac{\sin t}{\cos t + 1}$$

$$= -1 \cdot \frac{0}{2} = 0. \qquad \square$$

The basic result concerning the derivatives of trigonometric functions can now be derived.

(3.23)
THEOREM

For any real number x,

$$D_x \sin x = \cos x \qquad \text{and} \qquad D_x \cos x = -\sin x.$$

PROOF

Using Theorem 3.17 and Corollary 3.22:

$$D_x \sin x = \lim_{h \to 0} \frac{\sin (x + h) - \sin x}{h}$$

$$= \lim_{h \to 0} \frac{\sin x \cos h + \sin h \cos x - \sin x}{h}$$

$$= \sin x \lim_{h \to 0} \frac{\cos h - 1}{h} + \cos x \lim_{h \to 0} \frac{\sin h}{h}$$

$$= \sin x \cdot 0 + \cos x \cdot 1 = \cos x.$$

And,

$$D_x \cos x = \lim_{h \to 0} \frac{\cos (x + h) - \cos x}{h}$$

$$= \lim_{h \to 0} \frac{\cos x \cos h - \sin x \sin h - \cos x}{h}$$

$$= \lim_{h \to 0} \left[\cos x \left(\frac{\cos h - 1}{h} \right) - \sin x \left(\frac{\sin h}{h} \right) \right]$$

$$= \cos x \lim_{h \to 0} \frac{\cos h - 1}{h} - \sin x \lim_{h \to 0} \frac{\sin h}{h}$$

$$= \cos x \cdot 0 - \sin x \cdot 1 = - \sin x. \qquad \square$$

EXAMPLE 1   Find $D_x(x \sin x - \cos x)$.

SOLUTION

$$D_x(x \sin x - \cos x) = x D_x \sin x + \sin x D_x x - D_x \cos x$$
$$= x \cos x + \sin x - (- \sin x)$$
$$= x \cos x + 2 \sin x. \qquad \square$$

EXAMPLE 2   Find $D_x \left(\dfrac{x^2 + 2}{\sin x} \right)$.

SOLUTION

$$D_x \left(\frac{x^2 + 2}{\sin x} \right) = \frac{\sin x D_x(x^2 + 2) - (x^2 + 2) D_x \sin x}{(\sin x)^2}$$

$$= \frac{2x \sin x - (x^2 + 2) \cos x}{(\sin x)^2}. \qquad \square$$

The derivatives of the remaining trigonometric functions are obtained by using Theorem 3.23 and the reciprocal and quotient rules.

(3.24)
COROLLARY

For each real number x for which the functions are defined:

(a) $D_x \tan x = (\sec x)^2$

(b) $D_x \sec x = \sec x \tan x$

(c) $D_x \cot x = - (\csc x)^2$

(d) $D_x \csc x = - \csc x \cot x.$

PROOF

(a)
$$D_x \tan x = D_x \left(\frac{\sin x}{\cos x} \right)$$

$$= \frac{\cos x \, D_x \sin x - \sin x \, D_x \cos x}{(\cos x)^2}$$

$$= \frac{(\cos x)^2 + (\sin x)^2}{(\cos x)^2}.$$

But $(\cos x)^2 + (\sin x)^2 = 1$ and $\sec x = 1/\cos x$, so

$$D_x \tan x = (\sec x)^2.$$

(b)
$$D_x \sec x = D_x \left(\frac{1}{\cos x} \right) = \frac{-D_x \cos x}{(\cos x)^2}$$

$$= \frac{-(-\sin x)}{(\cos x)^2}$$

$$= \frac{1}{\cos x} \cdot \frac{\sin x}{\cos x}$$

$$= \sec x \tan x.$$

Parts (c) and (d) can be verified in a similar manner (see Exercises 33 and 34). ☐

EXAMPLE 3   Find the slope of the line tangent to the graph of $f(x) = x^2 \tan x + \csc x$ at $(\pi/4, f(\pi/4))$.

SOLUTION
$$D_x(x^2 \tan x + \csc x) = D_x(x^2 \tan x) + D_x(\csc x)$$

$$= \tan x \, D_x x^2 + x^2 D_x \tan x + (-\csc x \cot x)$$

$$= 2x \tan x + x^2(\sec x)^2 - \csc x \cot x.$$

So when $x = \dfrac{\pi}{4}$, the slope of the tangent line is

$$2\left(\frac{\pi}{4}\right) \tan \frac{\pi}{4} + \left(\frac{\pi}{4}\right)^2 \left(\sec \frac{\pi}{4}\right)^2 - \csc \frac{\pi}{4} \cot \frac{\pi}{4} = \frac{\pi}{2}(1) + \frac{\pi^2}{16}(\sqrt{2})^2 - \sqrt{2}(1)$$

$$= \frac{\pi}{2} + \frac{\pi^2}{8} - \sqrt{2}. \qquad ☐$$

EXERCISE SET 3.4

Use the results of this section to find the derivatives of the functions described in Exercises 1 through 26.

1. $f(x) = \sin x + \tan x$

2. $f(x) = \cos x - \cot x$

3. $f(x) = x^3 - \cos x$

4. $f(x) = 2x^2 + \tan x$

5. $g(t) = t^3 \cos t$

6. $h(z) = 2z^4 \sin z$

7. $h(x) = \sin x \cos x$

8. $f(x) = \sec x \csc x$

9. $r(t) = \tan t \cot t$

10. $h(x) = \sec x \cos x$

11. $f(x) = (\sin x)^2$

12. $f(x) = (\tan x)^2$

13. $f(x) = \dfrac{2x}{\sin x}$

14. $g(r) = \dfrac{2r^2 + 1}{\cos r}$

15. $f(x) = x \tan x + x^2 \cot x$

16. $f(x) = (x^2 + 1)\sin x + 2x \sec x$

17. $f(\theta) = \dfrac{\theta^2}{1 + \cos \theta}$

18. $g(\theta) = \dfrac{2\theta^2}{1 + \tan \theta}$

19. $h(\theta) = \dfrac{\sec \theta + 1}{\tan \theta}$

20. $h(\theta) = \dfrac{\theta + \sin \theta}{\cos \theta}$

21. $f(t) = \dfrac{\sin t \tan t}{\sin t + \cos t}$

22. $y(t) = \dfrac{\sin t + 1}{\cos t \cot t}$

23. $g(x) = (\csc x + \sec x)\sin x \cos x$

24. $h(x) = (x^2 - 1)\sin x \cos x$

25. $g(x) = x(\tan x)^2$

26. $g(x) = (2x^2 - 1)(\sin x)^2$

Use Theorem 3.17 and Corollary 3.22 to find the limits in Exercises 27 through 32.

27. $\lim\limits_{t \to 0} \dfrac{\sin t}{2t}$

28. $\lim\limits_{t \to 0} \dfrac{\sin 2t}{2t}$

29. $\lim\limits_{t \to 0} \dfrac{(\sin t)^2}{t}$

30. $\lim\limits_{t \to 0} \dfrac{(\sin t)^2}{t^2}$

31. $\lim\limits_{t \to 0} \dfrac{\tan t}{t}$

32. $\lim\limits_{t \to 0} \dfrac{(\cos t)^2 - 1}{t}$

33. Use Theorem 3.23 and the quotient rule to show that $D_x \cot x = -(\csc x)^2$.

34. Use Theorem 3.23 and the quotient rule to show that $D_x \csc x = -\csc x \cot x$.

35. Find an equation of the line tangent to the graph of $y = \sin x$ when $x = \pi/3$.

36. Find an equation of the line normal to the graph of $y = \sin x$ when $x = \pi/3$.

37. Find the point on the graph of $y = \sin x + \cos x$, $0 \le x \le \pi$, at which the tangent line is horizontal.

38. Find the numbers x in the interval $[0, 2\pi]$ at which the slope of the tangent line to the graph of $y = \sin x$ is zero.

39. Show that at no value of x is the line tangent to $y = \sin x$ perpendicular to the line tangent to $y = \cos x$.

40. The accompanying figure shows a triangle in which x, y, and θ are changing, but the third side remains fixed. Express x in terms of θ and determine the instantaneous rate of change of x with respect to θ when $\theta = \dfrac{\pi}{4}$.

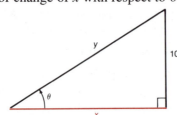

41. Using the figure for Exercise 40, express y in terms of θ and determine the instantaneous rate of change of y with respect to θ when $\theta = \dfrac{\pi}{4}$.

42. Suppose f, g, and h are functions with the property that $f(x) \le g(x) \le h(x)$ for each $x \ne a$ in the open interval $(a, a+\delta)$, for some constant $\delta > 0$. If

$$\lim_{x \to a^+} f(x) = L \qquad \text{and} \qquad \lim_{x \to a^+} h(x) = L,$$

prove that $\lim\limits_{x \to a^+} g(x) = L$.

43. Suppose f, g, and h are functions with the property that $f(x) \le g(x) \le h(x)$ for each $x \ne a$ in the open interval $(a - \delta, a)$, for some constant $\delta > 0$. If

$$\lim_{x \to a^-} f(x) = L \qquad \text{and} \qquad \lim_{x \to a^-} h(x) = L,$$

prove that $\lim\limits_{x \to a^-} g(x) = L$.

44. Consider the function described by $f(x) = \sin \dfrac{1}{x}$.

(a) If n is an even positive integer, what is

$$f\!\left(\frac{2}{(2n - 1)\,\pi}\right)?$$

(b) If n is an odd positive integer, what is

$$f\!\left(\frac{2}{(2n - 1)\,\pi}\right)?$$

(c) Find $\lim\limits_{x \to 0^+} f(x)$ if this limit exists.

3.5
THE DERIVATIVE OF A COMPOSITE FUNCTION: THE CHAIN RULE

In Section 3.3 we found that $D_x(x + 1)^{-2} = -2(x + 1)^{-3}$. It is natural to ask whether there is a connection between this result and the power rule: $D_x x^n = nx^{n-1}$, for nonzero integers n. To examine this possibility further, consider

$$D_x(2x^2 + 3)^2 = D_x[(2x^2 + 3)(2x^2 + 3)]$$
$$= (2x^2 + 3)(4x) + (2x^2 + 3)(4x)$$
$$= 8x(2x^2 + 3).$$

This is perhaps not quite the result that was expected. However, rewriting this as $D_x(2x^2 + 3)^2 = 2(2x^2 + 3)^1(4x)$, we see that it follows somewhat the pattern of $D_x x^n = nx^{n-1}$. The difference is that it is multiplied by the factor $4x$, the derivative of $2x^2 + 3$.

The pattern of the preceding illustrations leads us to expect that $D_x(\sin x)^2 = 2(\sin x)^1 \cdot \cos x$. Using the product rule, we can verify that this is true:

$$D_x(\sin x)^2 = D_x[(\sin x) \cdot (\sin x)]$$
$$= (\sin x)(\cos x) + (\cos x)(\sin x)$$
$$= 2 \sin x \cos x.$$

The general rule that applies to these examples is called the **chain rule.**

(3.25)
THEOREM

Chain Rule If g is differentiable at a and f is differentiable at $g(a)$, then $f \circ g$ is differentiable at a and $(f \circ g)'(a) = f'(g(a)) \cdot g'(a)$.

To see why we expect the chain rule to apply, consider

$$(f \circ g)'(a) = \lim_{x \to a} \frac{(f \circ g)(x) - (f \circ g)(a)}{x - a}$$

$$= \lim_{x \to a} \frac{f(g(x)) - f(g(a))}{x - a}$$

$$= \lim_{x \to a} \frac{f(g(x)) - f(g(a))}{x - a} \frac{g(x) - g(a)}{g(x) - g(a)}, \text{ provided that } g(x) - g(a) \ne 0$$

$$= \lim_{x \to a} \frac{f(g(x)) - f(g(a))}{g(x) - g(a)} \cdot \frac{g(x) - g(a)}{x - a}$$

$$= \lim_{x \to a} \frac{f(g(x)) - f(g(a))}{g(x) - g(a)} \cdot \lim_{x \to a} \frac{g(x) - g(a)}{x - a}.$$

Since g is differentiable at a, g is continuous at a and $\lim_{x \to a} g(x) = g(a)$. Thus,

$$(f \circ g)'(a) = \lim_{x \to a} \frac{f(g(x)) - f(g(a))}{g(x) - g(a)} \cdot \lim_{x \to a} \frac{g(x) - g(a)}{x - a}$$

$$= \lim_{g(x) \to g(a)} \frac{f(g(x)) - f(g(a))}{g(x) - g(a)} \cdot \lim_{x \to a} \frac{g(x) - g(a)}{x - a}$$

$$= f'(g(a)) \cdot g'(a).$$

The difficulty with constructing a proof based on this motivation is that we cannot proceed beyond the first step unless we know that $g(x) - g(a) \ne 0$ when $x \ne a$. In general this is not true; for example, if g is a constant function $g(x) - g(a) = 0$ for all values of x. Consequently, the proof of this result must proceed on a different course. It has been deferred to Appendix B.

EXAMPLE 1

Use the chain rule to find $D_x(x + 1)^{-2}$.

SOLUTION

Let $g(x) = x + 1$ and $f(x) = x^{-2}$. Then

$$D_x(x + 1)^{-2} = (f \circ g)'(x) = f'(g(x)) \cdot g'(x)$$
$$= -2(x + 1)^{-3} \cdot 1$$
$$= -2(x + 1)^{-3}. \qquad \square$$

An easy way to remember the chain rule is to use Leibniz notation for the derivative. Suppose $u = g(x)$ and $y = (f \circ g)(x) = f(g(x)) = f(u)$. Then,

$$\frac{dy}{dx} = (f \circ g)'(x), \qquad \frac{dy}{du} = f'(g(x)), \qquad \text{and} \qquad \frac{du}{dx} = g'(x).$$

Consequently, the chain rule can be expressed in the form

(3.26)
$$\frac{dy}{dx} = \frac{dy}{du} \cdot \frac{du}{dx},$$

a form that makes the rule seem quite natural.

EXAMPLE 2   Find $D_x(2x^2 + 3)^2$ by using the form of the chain rule $\dfrac{dy}{dx} = \dfrac{dy}{du} \cdot \dfrac{du}{dx}$.

SOLUTION

Let $u = 2x^2 + 3$ and $y = (2x^2 + 3)^2 = u^2$.

Then $\dfrac{dy}{du} = 2u$ and $\dfrac{du}{dx} = 4x$ so

$$
\begin{aligned}
D_x(2x^2 + 3)^2 &= \frac{dy}{dx} = \frac{dy}{du} \cdot \frac{du}{dx} \\
&= (2u)(4x) \\
&= 2(2x^2 + 3)(4x) \\
&= 8x(2x^2 + 3).
\end{aligned}
$$
$\square$

EXAMPLE 3   Differentiate $h(x) = \dfrac{1}{(x^2 + x + 1)^2}$.

SOLUTION

Letting $y = h(x)$ and $u = x^2 + x + 1$, we have $y = u^{-2}$, and the chain rule implies that

$$
\begin{aligned}
h'(x) &= \frac{dy}{dx} = \frac{dy}{du} \cdot \frac{du}{dx} \\
&= -2u^{-3}(2x + 1) = -2(x^2 + x + 1)^{-3}(2x + 1).
\end{aligned}
$$
$\square$

Note that in all of these examples, we have used the special case of the chain rule that if $y = f(x)$, then

(3.27)
$$
D_x y^n = n y^{n-1} \cdot D_x y
$$

for n any nonzero integer. The following examples illustrate the chain rule when other types of compositions are involved.

EXAMPLE 4   Find $D_x \sin (3x^2 + 1)$.

SOLUTION

Let $u = 3x^2 + 1$ and $y = \sin (3x^2 + 1) = \sin u$. Then,

$$
\begin{aligned}
D_x \sin (3x^2 + 1) &= \frac{dy}{dx} = \frac{dy}{du} \cdot \frac{du}{dx} \\
&= (\cos u)(6x) \\
&= 6x \cos (3x^2 + 1).
\end{aligned}
$$
$\square$

EXAMPLE 5   Find $D_x(\cos(x^2 + 1))^3$.

SOLUTION

This function involves two compositions that can be expressed as

$$
x \to x^2 + 1 \to \cos (x^2 + 1) \to (\cos (x^2 + 1))^3,
$$

so the chain rule must be applied twice. First

$$D_x (\cos (x^2 + 1))^3 = 3(\cos (x^2 + 1))^2 D_x \cos (x^2 + 1).$$

Then since

$$D_x \cos (x^2 + 1) = -\sin (x^2 + 1)D_x (x^2 + 1) = (-\sin (x^2 + 1))(2x),$$

$$D_x (\cos (x^2 + 1))^3 = 3(\cos (x^2 + 1))^2(-\sin (x^2 + 1))(2x)$$

$$= -6x \sin (x^2 + 1)(\cos (x^2 + 1))^2. \qquad \square$$

EXAMPLE 6   Determine $f'(x)$ if $f(x) = \left(\dfrac{\sin (2x + 1)}{\cos (3x^2 - 4)} \right)^4 .$

SOLUTION

$f(x)$ can be considered as being constructed as follows:

$$x \rightarrow 2x + 1 \rightarrow \sin (2x + 1) \searrow \dfrac{\sin (2x + 1)}{\cos (3x^2 - 4)} \rightarrow \left(\dfrac{\sin (2x + 1)}{\cos (3x^2 - 4)} \right)^4$$

$$x \rightarrow 3x^2 - 4 \rightarrow \cos (3x^2 - 4) \nearrow$$

The last operation used to construct $f(x)$ is a composition, so the first operation used to determine $f'(x)$ is the chain rule.

$$f'(x) = 4 \left(\dfrac{\sin (2x + 1)}{\cos (3x^2 - 4)} \right)^3 D_x \left(\dfrac{\sin (2x + 1)}{\cos (3x^2 - 4)} \right).$$

Applying first the quotient rule and then the chain rule

$$D_x \left(\dfrac{\sin (2x + 1)}{\cos (3x^2 - 4)} \right) = \dfrac{\cos (3x^2 - 4)D_x \sin (2x + 1) - \sin(2x + 1)D_x \cos (3x^2 - 4)}{(\cos (3x^2 - x))^2}$$

$$= \dfrac{\cos (3x^2 - 4)\cos (2x + 1)D_x (2x + 1) - \sin (2x + 1)(-\sin (3x^2 - 4))D_x (3x^2 - 4)}{(\cos (3x^2 - 4))^2}$$

$$= \dfrac{\cos (3x^2 - 4)\cos (2x + 1)(2) + \sin (2x + 1)\sin (3x^2 - 4)(6x)}{(\cos (3x^2 - 4))^2} .$$

So

$$f'(x) = 4 \left(\dfrac{\sin (2x + 1)}{\cos (3x^2 - 4)} \right)^3 \left[\dfrac{2 \cos(3x^2 - 4)\cos(2x + 1) + 6x \sin(2x + 1)\sin (3x^2 - 4)}{(\cos (3x^2 - 4))^2} \right] .$$

$\square$

The chain rule is often used in applications involving the rate of change of quantities with respect to time.

EXAMPLE 7   The area of a circle with radius r is given by $A = \pi r^2$. Suppose the radius of the circle depends on time t in accordance with the formula $r = t^3 + 1$.
(a) Find dA/dt, the instantaneous rate of change of the area with respect to time.
(b) Find dA/dt when $t = 2$.

SOLUTION

(a) The chain rule can be applied to find

$$\frac{dA}{dt} = \frac{dA}{dr} \cdot \frac{dr}{dt} = (2\pi r)(3t^2).$$

Expressing dA/dt in terms of t, we have

$$\frac{dA}{dt} = 2\pi(t^3 + 1)(3t^2) = 6\pi t^2(t^3 + 1).$$

(b) $\quad\quad\quad\dfrac{dA}{dt}$ (when $t = 2$) $= 6\pi 2^2(2^3 + 1) = 216\pi.$ $\square$

EXERCISE SET 3.5

Find $D_x f(x)$ in Exercises 1 through 32.

1. $f(x) = (x + 1)^4$

2. $f(x) = (x^2 + 1)^4$

3. $f(x) = (x^2 - 3x + 4)^7$

4. $f(x) = (x^3 - x^2 + 1)^3$

5. $f(x) = (6x + 4)^{-2}$

6. $f(x) = (6x)^{-2} + 4^{-2}$

7. $f(x) = \dfrac{1}{(6x^2 - 2x + 1)^2}$

8. $f(x) = \dfrac{5}{(3x^2 + 1)^4}$

9. $f(x) = \dfrac{(3x^2 + 1)^4}{5}$

10. $f(x) = (x^2 + 3x + 5)^{-2}$

11. $f(x) = (3x^2 + 13)^{-7}$

12. $f(x) = (3x^2 + 13x)^{-7}$

13. $f(x) = \left(\dfrac{x - 1}{x + 1}\right)^3$

14. $f(x) = \left(\dfrac{8x + 3}{1 + 9x}\right)^4$

15. $f(x) = \dfrac{(x - 1)^2}{(x + 1)^3}$

16. $f(x) = (4x - 5)(3x^2 + 2x + 1)^2$

17. $f(x) = \sin \pi x$

18. $f(x) = 3 \tan 2x$

19. $f(x) = (\cos x - 1)^3$

20. $f(x) = \cos(x^3 - 1)$

21. $f(x) = (\cos x)^3 - 1$

22. $f(x) = \cos x^3 - 1$

23. $f(x) = \tan\left(\dfrac{x + 1}{x - 1}\right)$

24. $f(x) = \left(\dfrac{x}{\sin x}\right)^3$

25. $f(x) = x^2 \cos(x^2 + 1)$

26. $f(x) = 2x^3 \sec(x^2 - 1)$

27. $f(x) = (x + x^{-1})^{-1}$

28. $f(x) = (x + x^{-1})^4$

29. $f(x) = (\tan 2x)^3 \sin(1 - x^2)$

30. $f(x) = [\sin (2x + 1)]^3 \cos(x^2 + 1)$

31. $f(x) = [(\sin x)^3 + 1]^2$

32. $f(x) = [(\sec x)^2 + 2]^3$

33. Find the derivative of each of the following functions.
    (a) $f_1(x) = (4x - 5)^3$
    (b) $f_2(x) = (4x - 5)^3 + 2x + 1$
    (c) $f_3(x) = [(4x - 5)^3 + 2x + 1]^2$
    (d) $f_4(x) = \{[(4x - 5)^3 + 2x + 1]^2 - 7x\}^4$

34. Find the derivative of each of the following functions.
    (a) $f_1(x) = (x^2 - 1)^2$
    (b) $f_2(x) = (x^2 - 1)^2 + 4x^3 - 3$
    (c) $f_3(x) = [(x^2 - 1)^2 + 4x^3 - 3]^3$
    (d) $f_4(x) = \{[(x^2 - 1)^2 + 4x^3 - 3]^3 - 3x^4\}^2$

35. Find the derivative of each of the following functions.
    (a) $f_1(x) = x^2 - x$
    (b) $f_2(x) = \sin(x^2 - x)$
    (c) $f_3(x) = [x^3 + \sin(x^2 - x)]^2$

36. Find the derivative of each of the following functions.
    (a) $f_1(x) = x^2 - x$
    (b) $f_2(x) = (x^2 - x)^2$
    (c) $f_3(x) = \sin[x^3 + (x^2 - x)^2]$

37. Find an equation of the line tangent to the graph of

$$f(x) = \left(\frac{(x - 1)}{(x + 2)}\right)^3 \text{ at } (2, 1/64).$$

38. Find an equation of the line normal to the graph of

$$f(x) = \left(\frac{(x - 1)}{(x + 2)}\right)^3 \text{ at } (2, 1/64).$$

39. Find an equation of a line tangent to the graph of $f(x) = \left(\dfrac{x + 1}{x - 1}\right)^2$ and parallel to the line with equation $y = 4x - 1$.

40. Find an equation of a line normal to the graph of $f(x) = \left(\dfrac{x + 1}{x - 1}\right)^2$ and perpendicular to the line with equation $x = 1$.

41. The volume V of a sphere with radius r is $V(r) = 4\pi r^3/3$. Suppose that the radius of the sphere depends on time t in accordance with the formula $r(t) = t^2 + 3$. Find $D_t V$, the derivative of the volume of the sphere with respect to time.

42. The volume V of a cone with radius r and altitude h is $V = \pi r^2 h/3$. Suppose that the radius and altitude of the cone are always equal and depend on time t according to the formula $r = h = 3t$. Find dV/dt, the derivative of the volume of the cone with respect to time.

43. Suppose the radius of a circle is increasing at the rate of 5 cm/sec, that is, $dr/dt = 5$ cm/sec. Find dA/dt, the instantaneous rate of change of its area

with respect to time. How fast is the area increasing when the radius is 30 cm?

44. Suppose a metal sphere is heated so that its radius increases at the rate of 2 millimeters per second. Find dV/dt, the instantaneous rate of change of its volume with respect to time. How fast is the volume increasing when the radius is 6 cm?

45. Sand is being poured onto a conical pile whose radius and height are always equal, although both increase with time. The height of the pile is increasing at the rate of 3 in/sec. Find dV/dt, the rate at which the volume is increasing.

46. A spherical ice ball is melting so that its radius decreases at the rate of .1 cm/hr. Find dV/dt and the rate at which the volume is decreasing when the diameter is 10 cm.

47. The monthly sales s (in dollars) of a new product depend on the time t (in months) after the product has been introduced according to the rule:

$$s(t) = \frac{2000(1 + t)}{2 + t}.$$

If the monthly profit is given by $P(s) = .25s$, find the rate of change of the profit with respect to the time t, for any time t after the new product is introduced.

48. The volume of a solid varies with temperature. For pure silver the volume of a 1-gram bar is

$$V = .095 \ [1 + 5.83 \times 10^{-5} \ (T - 25.0)] \ \text{cm}^3$$

at T degrees Celsius. Suppose that the temperature of a bar of silver with a volume of 100 cm^3 at 25°C depends on time t according to the formula

$$T = 7.5 + 3t + .04t^2 + \frac{17.5}{t + 1} \frac{°C}{\text{hour}}.$$

Find the rate of change of the volume at $t = 3$ hours.

The derivative of the absolute value function is

$$D_x|x| = \begin{cases} 1, \text{ if } x > 0 \\ \text{undefined, if } x = 0 \\ -1, \text{ if } x < 0. \end{cases}$$

Use the chain rule to find $f'(x)$ for the function f described in each of Exercises 49 through 52.

49. $f(x) = |2x|$ **50.** $f(x) = |x + 3|$

51. $f(x) = |x^2 - 1|$ **52.** $f(x) = |1 - x^2|$

53. Suppose f and g are differentiable functions with the property that $(f \circ g)(x) = x$. Use the chain rule to show that $f'(g(x)) = 1/g'(x)$.

54. Use the result of Exercise 53 to find $g'(x)$ when $g(x) = \sqrt[3]{x}$.

55. To connect a circular segment of a roadbed for a highway or railroad to a straight segment, engineers often use a cubic polynomial whose coefficients are chosen to match the tangents of the circle and line segments. Find the equation of the polynomial in the accompanying figure.

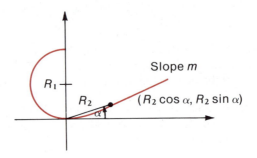

3.6
IMPLICIT DIFFERENTIATION

Consider the problem of finding the slope of the tangent line to the circle described by $x^2 + y^2 = 4$ at $(1, -\sqrt{3})$. See Figure 3.15.

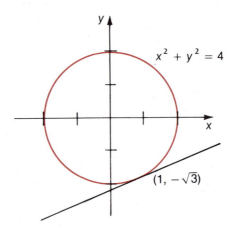

FIGURE 3.15

The graph of $x^2 + y^2 = 4$ is not the graph of a function. However, $x^2 + y^2 = 4$, $y \leq 0$, describes a function: $f(x) = -\sqrt{4 - x^2}$, and $(1, -\sqrt{3})$ lies on the graph of f. In Example 5 of Section 3.2, we found that $D_x\sqrt{x} = 1/(2\sqrt{x})$. Consequently, the chain rule can be used to show that

$$y' = f'(x) = \frac{-1}{2\sqrt{4 - x^2}} D_x (4 - x^2) = \frac{x}{\sqrt{4 - x^2}}.$$

When $x = 1$, $y' = 1/\sqrt{3} = \sqrt{3}/3$, the slope of the tangent line to the circle at $(1, -\sqrt{3})$.

The chain rule gives an alternate approach to finding the slope of the curve $x^2 + y^2 = 4$ at a point, an approach that does not require that we first express y explicitly as a function of x. If y is a function of x, the chain rule implies that

$$D_x \, y^n = n y^{n-1} D_x \, y.$$

Using this result and differentiating both sides of $x^2 + y^2 = 4$ with respect to x:

$$D_x(x^2 + y^2) = D_x(4),$$

$$D_x(x^2) + D_x(y^2) = 0,$$

$$2x + 2y \cdot D_x \, y = 0.$$

Solving for $D_x \, y$,

$$D_x \, y = \frac{-2x}{2y} = -\frac{x}{y},$$

and evaluating at $(1, -\sqrt{3})$

$$D_x \, y \, (\text{at } (1, -\sqrt{3})) = \frac{-1}{-\sqrt{3}} = \frac{1}{\sqrt{3}} = \frac{\sqrt{3}}{3}.$$

This procedure is called **implicit differentiation** since it involves differentiating a function that is given by an implied, rather than an explicit, relation. We assume that the given equation defines y as a function of x, if correctly restricted, and that $D_x \, y$ exists.

In the preceding example, we originally solved for y in terms of x. This is not always possible or desirable and in these situations implicit differentiation finds the most application.

EXAMPLE 1   Use implicit differentiation to find $\dfrac{dy}{dx}$ if $xy^2 + 6y + x = 0$.

SOLUTION

To use implicit differentiation, we assume that $xy^2 + 6y + x = 0$ implicitly defines y as a differentiable function of x. Then,

$$D_x(xy^2 + 6y + x) = D_x(0)$$

$$xD_x(y^2) + y^2 D_x \, x + 6D_x \, y + D_x \, x = 0$$

$$x(2y D_x \, y) + y^2 + 6D_x \, y + 1 = 0$$

$$2xy \, D_x \, y + y^2 + 6D_x \, y + 1 = 0$$

$$y^2 + (2xy + 6)D_x \, y + 1 = 0$$

so

$$\frac{dy}{dx} = D_x y = \frac{-1 - y^2}{2xy + 6}. \qquad \square$$

EXAMPLE 2   The graph of the equation $3(x^2 + y^2)^2 = 25(x^2 - y^2)$ is a **lemniscate** and is shown in Figure 3.16. Find an equation of the line tangent to this curve at $(2, 1)$.

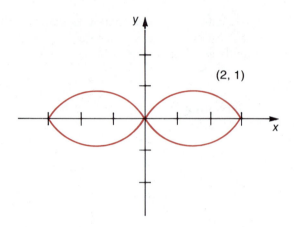

FIGURE 3.16

SOLUTION

Assuming that the given equation defines y as a differentiable function of x, we take the derivative of both sides with respect to x:

$$D_x(3(x^2 + y^2)^2) = D_x(25(x^2 - y^2))$$
$$3 \cdot 2(x^2 + y^2) D_x(x^2 + y^2) = 25(D_x(x^2 - y^2))$$
$$6(x^2 + y^2)(2x + 2y D_x y) = 25(2x - 2y D_x y).$$

Dividing both sides by 2 and solving for $D_x y$, we have:

$$6x(x^2 + y^2) + 6y(x^2 + y^2)D_x y = 25x - 25y D_x y$$
$$(6y(x^2 + y^2) + 25y)D_x y = 25x - 6x(x^2 + y^2)$$

so
$$D_x y = \frac{25x - 6x(x^2 + y^2)}{25y + 6y(x^2 + y^2)}.$$

The slope of the tangent line at $(2, 1)$ is

$$D_x y \text{ (at } (2, 1)) = \frac{(25)(2) - 6(2)(4 + 1)}{25 + 6(4 + 1)} = \frac{50 - 60}{55} = -\frac{2}{11}.$$

An equation of the tangent line to the lemniscate at $(2, 1)$ is

$$y - 1 = -\frac{2}{11}(x - 2) \qquad \text{or} \qquad 2x + 11y = 15. \qquad \square$$

EXAMPLE 3   Find $D_x y$ if $\cos xy + \sin x = 1$.

SOLUTION

We assume that the equation $\cos xy + \sin x = 1$ defines y as a differentiable function of x. Then

$$D_x (\cos xy + \sin x) = D_x(1)$$
$$- \sin xy\, (D_x\, xy) + \cos x = 0$$
$$- \sin xy\, (y + xD_x\, y) + \cos x = 0$$
$$- x \sin xy\, D_x\, y = y \sin xy - \cos x,$$

so

$$D_x\, y = \frac{y \sin xy - \cos x}{-x \sin xy}. \qquad \square$$

Implicit differentiation can also be used to extend the power rule.

(3.28)
THEOREM

If $f(x) = x^n$ where n is any nonzero rational number, then $f'(x) = nx^{n-1}$, whenever the derivative exists.

PROOF

The result has been shown to be true whenever n is a nonzero integer.

If n is a nonzero rational number, then n can be written as $n = p/q$ where p and q are nonzero integers. Letting

$$y = x^n = x^{p/q}$$

and taking y to the qth power, we have

$$y^q = x^p.$$

Since both y and x now have integral exponents, we can use implicit differentiation to obtain

$$D_x\, y^q = D_x\, x^p,$$
$$qy^{q-1}\, D_x\, y = px^{p-1},$$

and

$$D_x\, y = \frac{px^{p-1}}{qy^{q-1}}.$$

Simplifying this expression,

$$D_x\, y = \frac{p}{q}\frac{x^{p-1}}{(x^{p/q})^{q-1}} = \frac{p}{q}\frac{x^{p-1}}{x^{p-p/q}} = \frac{p}{q}x^{p-1}\, x^{-p+p/q} = \frac{p}{q}x^{p/q-1}$$

so $D_x\, x^n = nx^{n-1}$ for all nonzero rational numbers n. $\qquad \square$

EXAMPLE 4

Find $f'(x)$ for $f(x) = \sqrt{x} + \dfrac{1}{\sqrt{x}}$.

SOLUTION

$f(x) = \sqrt{x} + \dfrac{1}{\sqrt{x}} = x^{1/2} + x^{-1/2}$. Using Theorem 3.28,

$$f'(x) = \frac{1}{2}x^{-1/2} - \frac{1}{2}x^{-3/2}$$

$$= \frac{1}{2\sqrt{x}} - \frac{1}{2(\sqrt{x})^3}. \qquad \square$$

Theorem 3.28 is so frequently used in conjunction with the chain rule that we list the following result for ease of reference.

(3.29)
COROLLARY

If $g(x)$ is a differentiable function and n is any nonzero rational number, then

$$D_x[g(x)]^n = n[g(x)]^{n-1} \cdot D_x g(x).$$

EXAMPLE 5   Find $D_x \sqrt[3]{(x^2 + 1)^2}$.

SOLUTION

$$D_x \sqrt[3]{(x^2 + 1)^2} = D_x(x^2 + 1)^{2/3}$$

$$= \frac{2}{3}(x^2 + 1)^{-1/3} D_x(x^2 + 1)$$

$$= \frac{2}{3}(x^2 + 1)^{-1/3}(2x)$$

$$= \frac{4x}{3\sqrt[3]{x^2 + 1}}. \qquad \square$$

EXERCISE SET 3.6

Use implicit differentiation to find $D_x y$ in Exercises 1 through 12.

1. $x^3 + y^3 = 0$ **2.** $x^4 - y^4 = 0$

3. $xy = 1$ **4.** $x^2 = y^3$

5. $x^2 - 2xy + y^2 = 0$ **6.** $x^2 + 2xy + y^2 = 0$

7. $2x^2 - x^2y^2 + y^{-3} = 4$ **8.** $4x^3 - 5x^2y^2 + 4y^3 = 3y$

9. $\sin x + \cos y = 1$ **10.** $\sin xy = \cos x$

11. $y^2 - \sin x = 0$ **12.** $\tan x + xy + y^2 = 0$

Find $D_x y$ in Exercises 13 through 38.

13. $y = x^{1/3}$ **14.** $y = (x + 1)^{1/3}$

15. $y = 3x^{4/5}$ **16.** $y = (3x)^{4/5}$

17. $y = (x + 1)^{2/3} + x$ **18.** $y = ((x + 1)^{2/3} + x)^{1/3}$

19. $y = \sqrt[4]{x^2 + 2x + 2}$ **20.** $y = \sqrt[3]{\dfrac{3x - 1}{2x + 5}}$

21. $y = \sqrt[3]{(3x^2 + 4x + 1)^2}$ **22.** $y = \sqrt[3]{x} + \dfrac{1}{\sqrt[3]{x}}$

23. $y = x^{1/2} + x^{3/4} + x^{-6/5}$ **24.** $y = [x^2 + (x^2 + 4)^{1/2}]^{1/2}$

25. $y = (x^2 + 2)^{1/3}(x^2 - 2)^{1/4}$ **26.** $y = (x^2 - 2)^{1/3}(x^2 - 2)^{1/4}$

27. $y = 6x^2 + \dfrac{3}{x} - \dfrac{2}{3\sqrt[3]{2x^2}}$

28. $y = x\sqrt{4x^2 + 1} + \dfrac{5x^{5/3}}{4x - 2}$

29. $y = \sqrt{x^2 + \sqrt{x^2 + \sqrt{x + 1}}}$

30. $y = \dfrac{(x - 1)^{1/2}\,(x + 1)^{1/2}}{(x - 2)^{1/2}\,(x + 2)^{1/2}}$

31. $x^{1/3} + y^{2/3} = 5$

32. $\sqrt{xy} = x + y$

33. $y^2 = \dfrac{x}{xy + 1}$

34. $x^2y + xy^2 = 6(x^2 + y^2)$

35. $\sin\dfrac{x}{y} = x$

36. $\sin\dfrac{x}{y} = y$

37. $y = \tan xy$

38. $y = \sec xy$

39. Find an equation of the line tangent to the graph of $f(x) = 3x^{1/3}$ at $(1, 3)$.

40. Find an equation of the line normal to the graph of $f(x) = 3x^{1/3}$ at $(1, 3)$.

41. Find the slope of the tangent line to the graph of $x^2y - 3x + y = 4$ at $(1, 7/2)$.

42. Find the slope of the line normal to the graph of $x^2y^2 - 2xy + x = 1$ at $(1, 2)$.

43. Find an equation of the line tangent to the graph of $x^3 + y^3 = 2$ at $(1, 1)$.

44. Find an equation of the line normal to the graph of $x^3 + y^3 = 2$ at $(1, 1)$.

45. The graph of the equation $x^2 + y^2 = r^2$ for a positive constant r is a circle of radius r with center at $(0, 0)$. Find the slope of the tangent line to the circle at an arbitrary point (x_0, y_0) on the circle.

46. Use the result of Exercise 45 to prove that the tangent line to a circle at any point on the circle is perpendicular to the line joining the point and the center of the circle.

47. Find an equation of the line tangent to the circle $x^2 + y^2 = 4$ at $(\sqrt{3}, 1)$.

48. Sketch the graph of the equation $x^4 + y^4 = 1$. Then find all points on the graph that have the property that the tangent line to the graph is perpendicular to the line joining the point and $(0, 0)$.

49. The graph of $x^{2/3} + y^{2/3} = 1$ is called an *astroid*. Find an equation of the line tangent to the astroid at $(-1/8, 3\sqrt{3}/8)$.

50. The graph of $x^3 + y^3 = 9xy$ is called a *folium of Descartes*. Find an equation of the line tangent to the folium of Descartes at $(4, 2)$.

51. Suppose that for each positive rational number a function f_r is defined by

$$f_r(x) = \begin{cases} 0, & \text{if } x \le 0 \\ x^r, & \text{if } x > 0. \end{cases}$$

For which values of r is f_r differentiable at every real number?

52. Show that $|x| = \sqrt{x^2}$ for any real number x and use this representation to find $f'(x)$ if

(a) $f(x) = |x|$

(b) $f(x) = |x^2 - 1|$

(c) $f(x) = \sqrt{|x + 3|}$

(d) $f(x) = |\sin x|$

53. The production cost x and profit $p(x)$ of a firm are related by the equation $3x + .1x\,p(x) + (p(x))^2 = 130,\ 0 \leq x \leq 40$ where x and $p(x)$ are measured in hundreds of dollars per week. Find the instantaneous rate of change of the profit with respect to the production cost when the production cost is $3000 per week and the profit is $500 per week.

54. Van der Waals equation of state for a gas is

$$\left(P + \frac{a}{V^2}\right)(V - b) = RT,$$

written for one mole of the gas, where R, a, and b are constants depending on the gas and T, V, and P represent, respectively, temperature, volume, and pressure. Suppose that the volume and temperature vary while the pressure is held constant. Use implicit differentiation to find the rate of change of the volume with respect to the temperature.

3.7
HIGHER DERIVATIVES

Since a differentiable function f determines a new function f', it is reasonable to suspect that there will be instances when the function f' is also differentiable and its derivative, the function $(f')'$, can be defined. Similarly, the function $(f')'$ may be differentiable and we could define $((f')')'$ and so on, continuing this process as long as the functions are differentiable.

To express this concept more concisely we use the notations f'', $D_x^2 f$, and d^2f/dx^2 to denote $(f')'$ and call this function the **second derivative** of f. Likewise, f''', $D_x^3 f$ and d^3f/dx^3, are used to represent $((f')')'$, the third derivative of f, and so on. The notation $f^{(n)}$ is also used to denote the **nth derivative** of f, especially when n is greater than 3. For example, the 10th derivative of f at x would be written $f^{(10)}(x)$, $D_x^{10} f(x)$, or $\dfrac{d^{10}f(x)}{dx^{10}}$.

When a function f is defined by an equation $y = f(x)$ we also use notation such as $y^{(n)}$, $D_x^n y$ and $\dfrac{d^n y}{dx^n}$ to denote the nth derivative.

EXAMPLE 1   If $f(x) = x^3 + 3x^2 - 2x + \sin x$, find the first four derivatives of f.

SOLUTION

$$f'(x) = 3x^2 + 6x - 2 + \cos x$$
$$f''(x) = 6x + 6 - \sin x$$
$$f'''(x) = 6 - \cos x$$
$$f^{(4)}(x) = \sin x.$$

$\square$

EXAMPLE 2   If $f(x) = 1/x$, find the first three derivatives of f and establish a general form for the nth derivative of f.

SOLUTION

Since

$$f(x) = \frac{1}{x} = x^{-1},$$
$$f'(x) = -x^{-2}$$
$$f''(x) = 2x^{-3}$$
$$f'''(x) = -2 \cdot 3x^{-4} = -6x^{-4}.$$

The general form for the nth derivative of f is given by

$$f^{(n)}(x) = (-1)^n (1 \cdot 2 \cdot 3 \cdots n) x^{-n-1},$$

provided $x \neq 0$. The product $1 \cdot 2 \cdot 3 \cdots n$ is denoted $n!$ and read n factorial. Using this notation,

$$f^{(n)}(x) = (-1)^n n! x^{-n-1}. \qquad \square$$

The next example demonstrates finding higher derivatives for functions expressed implicitly.

EXAMPLE 3   Find $D_x y$, $D^2_x y$, and $D^3_x y$ if $x^2 + y^2 = 1$.

SOLUTION

Using implicit differentiation gives:

$$2x + 2y\, D_x y = 0$$

so

$$D_x y = \frac{-x}{y}.$$

Thus, $D^2_x y = D_x\left(\frac{-x}{y}\right) = -\frac{y\, D_x x - x\, D_x y}{y^2}.$

Substituting $D_x y = \frac{-x}{y}$, we have

$$D^2_x y = -\frac{y - x\left(\dfrac{-x}{y}\right)}{y^2} = -\frac{y^2 + x^2}{y^3} = -\frac{1}{y^3},$$

since $x^2 + y^2 = 1$.
   Similarly,

$$D^3_x y = D_x\left(-\frac{1}{y^3}\right) = \frac{3}{y^4} D_x y = \frac{3}{y^4}\left(-\frac{x}{y}\right) = \frac{-3x}{y^5}. \qquad \square$$

EXERCISE SET 3.7

Find $D_x y$ and $D^2_x y$ in Exercises 1 through 16.

1. $y = 3x^2 + 4x$

2. $y = 3x^3 - 2x^2 + x - 1$

3. $y = x^4 + 1 + x^{-4}$

4. $y = \dfrac{x^3}{3} - x^{-1}$

5. $y = 3\sqrt{x} + 2x^2$

6. $y = 7x^5 - 2\sqrt[3]{x}$

7. $y = \sin 2x$

8. $y = \sin x \cos x$

9. $y = \dfrac{1}{\sqrt{2x + 5}}$

10. $y = \sqrt{x} + \dfrac{1}{\sqrt{x}}$

11. $y = \dfrac{x}{x + 1}$

12. $y = \dfrac{x + 1}{x}$

13. $y = \dfrac{\sqrt{x} - 1}{\sqrt{x} + 1}$

14. $y = \dfrac{1}{\sqrt{x^2 + 3}}$

15. $y = \sin(x^2 + 1)$

16. $y = (\sin x)^3$

In Exercises 17 through 22, a function f is described. Find f' and f''.

17. $f(x) = \sqrt{x} + \sqrt{x^3}$

18. $f(x) = \dfrac{1}{\sqrt{x}} + \dfrac{1}{\sqrt{x^3}}$

19. $f(x) = \tan 2x$

20. $f(x) = \cos \pi x$

21. $f(x) = \sqrt{x + x^3}$

22. $f(x) = \dfrac{1}{\sqrt{x + x^3}}$

23–26. Find $f^{(3)}$ and $f^{(4)}$ for the functions described in Exercises 17 through 20.

Find a general formula expressing $f^{(n)}(x)$ for any positive integer n for the functions described in Exercises 27 through 34.

27. $f(x) = \dfrac{1}{x + 1}$

28. $f(x) = \dfrac{1}{x - 1}$

29. $f(x) = \dfrac{1}{2x + 1}$

30. $f(x) = \dfrac{1}{(x - 1)^2}$

31. $f(x) = x^{3/2}$

32. $f(x) = x^{1/3}$

33. $f(x) = \sin x$

34. $f(x) = \cos x$

Use implicit differentiation to find $D^2_x y$ in Exercises 35 through 42.

35. $x^2 + y^2 = 1$

36. $x^{1/2} + y^{1/2} = 1$

37. $x^2 + 2xy + y^2 = 16$

38. $x^2 + xy + y^2 = 4$

39. $x^2 + xy = y$

40. $x^2 y = 1 + xy$

41. $xy + 1 = \sqrt{y}$

42. $\dfrac{x}{y} + 1 = \dfrac{y}{x}$

43. Find h'' in terms of the derivatives of f and g if $h = f \circ g$.

44. Find a function f with the property that the domain of f is not equal to the domain of f' and the domain of f' is not equal to the domain of f''.

45. Suppose $P(x) = a_2 x^2 + a_1 x + a_0$, where a_0, a_1, a_2 are constants. Find $P(0)$, $P'(0)$ and $P''(0)$.

46. Find a polynomial P of degree two or less such that $P(0) = 1$, $P'(0) = 4$, and $P''(0) = 3$. Is this polynomial unique?

47. Suppose P is a polynomial of degree n and that

$$P(x) = a_n x^n + a_{n-1} x^{n-1} + \ldots + a_1 x + a_0.$$

Find a relationship between the coefficients $a_0, a_1, \ldots, a_n$ of P and the evaluation of P and its derivatives at $x = 0$.

3.8
MAXIMA AND MINIMA OF FUNCTIONS

We initiated the study of derivatives by considering the problem of determining when the graph of a function is increasing or decreasing. To continue the study of this problem, we need to present some definitions and theorems.

The first result we will discuss is the *extreme value theorem* for continuous functions. Although the statement of this theorem is quite elementary, its proof involves theory generally deferred to a course in advanced calculus.

(3.30)
THEOREM

Extreme Value Theorem   If f is continuous on a closed interval $[a,b]$, then there exist numbers c_1 and c_2 in $[a,b]$ with the property that for all x in $[a,b]$

$$f(c_1) \leq f(x) \leq f(c_2).$$

The value $f(c_2)$ in Theorem 3.30 is the absolute maximum of the function f on $[a,b]$ and $f(c_1)$ is the absolute minimum of f on $[a,b]$. In general, a function f is said to have an **absolute maximum** at c on an interval I if $f(x) \leq f(c)$ for all x in I. Similarly, f is said to have an **absolute minimum** at c on I if $f(x) \geq f(c)$ for all x in I. Maximum and minimum values are also called **extrema**. The function f whose graph is shown in Figure 3.17 has an absolute minimum at c_1 and an absolute maximum at c_2.

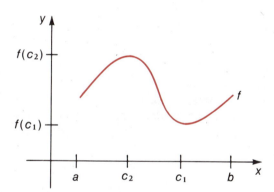

FIGURE 3.17

Although the extreme value theorem ensures that absolute maximum and minimum values exist for functions continuous on closed intervals, it does not provide a constructive method for determining these values. Methods of calculus can be used to determine these values for many functions. The following definition is needed to describe these methods.

(3.31)
DEFINITION

A function f is said to have a relative, or local, maximum at c if there exists an open interval I, containing c, with the property that $f(x) \leq f(c)$ for all x in I. Similarly, f is said to have a relative, or local, minimum at c if an open interval I, containing c, exists with the property that $f(x) \geq f(c)$ for all x in I.

The function f whose graph is shown in Figure 3.18 has relative minima at c_1 and c_3 and relative maxima at c_2 and c_4. The graph suggests that the tangent lines at c_1, c_2 and c_3 have slope zero and that there is no unique tangent line to the graph at c_4; that is, $f'(c_1) = f'(c_2) = f'(c_3) = 0$ and $f'(c_4)$ does not exist.

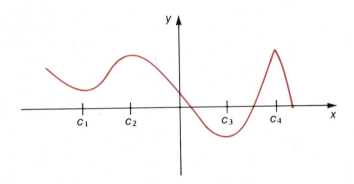

FIGURE 3.18

The following theorem gives an important connection between the derivative of a function and the location of the numbers at which the function has a relative maximum or relative minimum value. This result provides the basis for many applications involving the derivative.

(3.32)
THEOREM
Suppose f is a function that has a relative minimum or relative maximum at c. If $f'(c)$ exists, then $f'(c) = 0$.

PROOF

We will show the result for the case when f has a relative maximum at c; the case when f has a relative minimum at c can be handled similarly. Since a relative maximum occurs at c, there is an open interval I containing c with the property that

$$f(x) \le f(c) \text{ for all } x \text{ in } I.$$

Assume that $f'(c)$ does exist; if this is false, we have nothing to prove. Since $f'(c)$ exists, both one-sided limits of the difference quotient exist and are equal to $f'(c)$; that is,

$$f'(c) = \lim_{x \to c^+} \frac{f(x) - f(c)}{x - c} \quad \text{and} \quad f'(c) = \lim_{x \to c^-} \frac{f(x) - f(c)}{x - c}.$$

If the values of x are restricted so that x is in I, then $f(x) - f(c)$ is always nonpositive. Thus, when $x > c$,

$$\frac{f(x) - f(c)}{x - c} \le 0$$

and when $x < c$,

$$0 \le \frac{f(x) - f(c)}{x - c}.$$

This implies that

$$0 \le \lim_{x \to c^-} \frac{f(x) - f(c)}{x - c} = f'(c) = \lim_{x \to c^+} \frac{f(x) - f(c)}{x - c} \le 0$$

and that $f'(c) = 0$. ☐

It follows from Theorem 3.32 that to determine relative maximum and minimum values of a function, we need consider only numbers in the domain of f at which the derivative is undefined or is zero. These numbers are called **critical numbers** or, more commonly, **critical points.** (The name is derived from the fact that these points are critical in sketching the graph of a function.) Consequently, the only numbers at which a function can have relative maximum or minimum values are the critical points.

EXAMPLE 1   Find the relative maximum and minimum values of the function given by $f(x) = 2x^3 - 9x^2 + 12x - 2$ and sketch the graph of f.

SOLUTION

The derivative given by

$$f'(x) = 6x^2 - 18x + 12$$

is defined for all values of x, so relative maximum or minimum values of f can occur only when $f'(x) = 0$.
Since

$$f'(x) = 6x^2 - 18x + 12 = 6(x^2 - 3x + 2)$$
$$= 6(x - 1)(x - 2),$$

$f'(x) = 0$ at $x = 1$ and $x = 2$.

Evaluating the function at the critical points, we have $f(1) = 3$ and $f(2) = 2$. Using the knowledge that $\lim_{x \to +\infty} f(x) = \infty$ and $\lim_{x \to -\infty} f(x) = -\infty$, we see that the graph of f must be similar to that shown in Figure 3.19. The function f therefore has a relative maximum at $x = 1$ and a relative minimum at $x = 2$. ☐

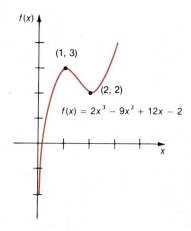

FIGURE 3.19

EXAMPLE 2   Determine the relative maximum and minimum values of the function f, where $f(x) = x^2 - 3x^{2/3}$. Sketch the graph of f.

SOLUTION

This function has a derivative given by

$$f'(x) = 2x - 2x^{-1/3}$$

which is undefined at $x = 0$. To determine any remaining critical points, we set $f'(x) = 0$ and solve for x.

$$0 = 2x - 2x^{-1/3}$$
$$= 2x^{-1/3}(x^{4/3} - 1),$$

so

$$x^{4/3} = 1.$$

Solutions to this equation are $x = \pm 1$.

Evaluating f at the critical points, we see that $f(0) = 0$, $f(-1) = -2$, $f(1) = -2$.

The graph is symmetric with respect to the y-axis and intersects the x-axis only when

$$x^2 - 3x^{2/3} = x^{2/3}(x^{4/3} - 3) = 0,$$

that is, when $x = 0$ and $x = \pm 3^{3/4} \approx \pm 2.3$.

Since $\lim_{x \to \infty} f(x) = \infty$, the graph is as shown in Figure 3.20. The shape of the graph at $(0, 0)$ reflects the fact that $f'(x)$ does not exist at $x = 0$. The function f has relative (and absolute) minima at $x = \pm 1$ and a relative maximum at $x = 0$. □

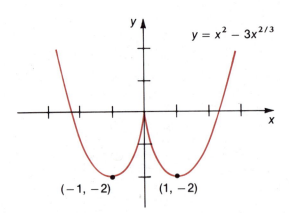

$$y = x^2 - 3x^{2/3}$$

$(-1, -2)$ $(1, -2)$

FIGURE 3.20

It is *not* true that a relative maximum or minimum occurs at every critical point, as we see in the next example.

EXAMPLE 3   Find the critical points of the function described by $f(x) = x^3$ and verify that there are no relative maximum or relative minimum values for f.

SOLUTION

The derivative of this function is $f'(x) = 3x^2$, for all real numbers x. The only critical point of f occurs at $x = 0$. However, if $x < 0$, then $f(x) < 0$ and if $x > 0$, then $f(x) > 0$, so f has neither a relative maximum nor a relative minimum at $x = 0$. This can be seen by considering the graph in Figure 3.21.   □

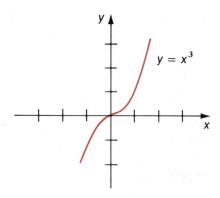

FIGURE 3.21

The first result of this section, the extreme value theorem, asserts that a function continuous on a closed interval assumes absolute maximum and minimum values on the interval. Relative extreme values may or may not be the absolute extreme values of the function. The absolute extreme values can occur at the endpoints of the interval.

EXAMPLE 4   Determine the absolute maximum and absolute minimum values of $f(x) = 2x^3 - 9x^2 + 12x - 2$, on the interval $[0, 4]$.

SOLUTION

In the solution of Example 1 it was determined that there are critical points at $x = 1$ and $x = 2$. A relative maximum, $f(1) = 3$, occurs at $x = 1$ and a relative minimum, $f(2) = 2$, occurs at $x = 2$. However, at the endpoints of the interval $[0, 4]$, $f(0) = -2$ and $f(4) = 30$. Consequently, the absolute maximum value is 30 and the absolute minimum value is -2, as can be seen in Figure 3.19. Both absolute extrema occur at the *endpoints* of the interval, not at the critical points.   □

EXAMPLE 5   Determine the absolute maximum and absolute minimum values of $f(x) = x^2 - 3x^{2/3}$, on $[0, 8]$.

SOLUTION

In Example 2, we found that f has a relative minimum at $x = \pm 1$ and $f(1) = f(-1) = -2$. Also f has a relative maximum at 0 and $f(0) = 0$. Considering the values of f at the endpoints we see that $f(0) = 0$ and $f(8) = 52$. So f has an absolute minimum of -2 at $x = 1$ and an absolute maximum of 52 at the endpoint $x = 8$.   □

Examples 4 and 5 illustrate the method for determining the absolute maximum and minimum values of a function f that is continuous on a closed interval:

Find the values of f at the critical points and at the endpoints of the interval. The absolute maximum is the largest of these values, and the absolute minimum the smallest of these values.

Absolute maximum and minimum values may not exist for functions defined on intervals that are not closed. For example, the function considered in Example 5, $f(x) = x^2 - 3x^{2/3}$, has no absolute maximum on the interval $(0, 8)$ since $\lim_{x \to 8^-} f(x) = 52$, but $f(x) < 52$ for all x in $(0,8)$. It does have an absolute minimum at $x = 1$, however.

Maximizing and minimizing quantities finds applications in many areas. Most often, the interest is in absolute maximum or absolute minimum values. The following example illustrates one such application in the area of economics. Applications involving extrema are considered in more detail in Section 4.2.

EXAMPLE 6   The profit realized from selling x tons of sand per week, for $0 \le x \le 50$, is given by $P(x) = 150 + 22x - x^2/2$ dollars. What is the number of tons of sand that should be sold in order to maximize the profit?

SOLUTION

The mathematical interpretation of this question is: at what number in the closed interval $[0, 50]$ does the function P have an absolute maximum?

First find the critical points of P.

$$P'(x) = 22 - x,$$

so P has a critical point at $x = 22$. Since $P(22) = 392$, $P(0) = 150$, and $P(50) = 0$, P has an absolute maximum at $x = 22$. The graph of P is given in Figure 3.22. Notice that if more than 50 tons of sand are sold, the profit has a negative value and a loss results. This explains why the domain of P is restricted to $[0, 50]$.

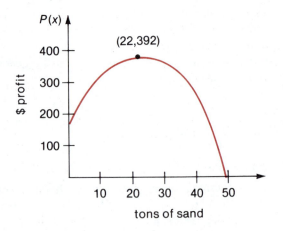

FIGURE 3.22

EXERCISE SET 3.8

Find the critical points of the functions described in Exercises 1 through 24.

1. $f(x) = x^2 - 4x + 4$

2. $f(x) = 9 - 3x^2$

3. $f(x) = x^3 - 6x^2 + 9x + 1$

4. $f(x) = x^3 - 7x^2 + 14x - 8$

5. $f(x) = x^3 + 3x^2 - 24x + 7$

6. $f(t) = t^3 + t^2 + t + 1$

7. $h(s) = s^3 + s^2 - s - 1$

8. $f(z) = z^3 + z^2 + z - 1$

9. $f(x) = \dfrac{1}{x}$

10. $f(x) = \dfrac{1}{x + 1}$

11. $f(x) = \dfrac{x}{x - 1}$

12. $g(x) = \dfrac{x + 1}{x - 1}$

13. $h(t) = \dfrac{t}{t^2 + 4}$

14. $f(t) = \dfrac{t}{t^2 - 4}$

15. $g(s) = \dfrac{s - 2}{s^2 - 4}$

16. $r(t) = \dfrac{t + 4}{t^2 - 16}$

17. $f(x) = \sqrt{x + 2}$

18. $h(x) = \sqrt{x^2 - 4}$

19. $f(x) = \sin x + \cos x$

20. $g(x) = x - \sin x$

21. $h(x) = (\sin x)^2 + \cos x$

22. $f(x) = \tan x - \sec x$

23. $f(t) = \cos 2t + \cos t$

24. $f(x) = \sin x - \dfrac{x}{2} + 1$

Find the critical points of the functions described in Exercises 25 through 46 and use this information to sketch the graph of each function. Where do relative and absolute extrema occur in each case?

25. $f(x) = x^2 - 6x + 4$

26. $f(x) = x^2 - 6x + 3$

27. $f(x) = 4 + 6x - x^2$

28. $h(x) = x^3 - 3x^2 + 3$

29. $v(t) = t^4 - 6t^2 + 8t$

30. $g(s) = s^3 - 12s + 4$

31. $f(x) = \dfrac{x}{x - 1}$

32. $f(t) = t + \dfrac{1}{t}$

33. $f(x) = x^2 + \dfrac{1}{x^2}$

34. $g(x) = \dfrac{x + 1}{x - 1}$

35. $g(x) = \dfrac{x + 1}{x^2 - 1}$

36. $g(x) = \dfrac{x^2 + 1}{x^2 - 1}$

37. $f(x) = \sin\left(x + \dfrac{\pi}{2}\right)$

38. $f(x) = \cos\left(x - \dfrac{\pi}{2}\right)$

39. $f(x) = x - \cos x$

40. $h(x) = 1 + \sin x$

41. $g(x) = \sin 4x$

42. $g(x) = \sin x + \cos x$

43. $f(x) = \sqrt{4 - x^2}$

44. $f(x) = \dfrac{1}{\sqrt{4 - x^2}}$

45. $f(x) = |x^3 - x^2|$

46. $f(x) = [\![x]\!]$

In Exercises 47 through 62 the domain of the function is restricted to the interval given. Sketch the graph of the function on this interval and determine the absolute maximum and absolute minimum values of the function, if these values exist.

47. $f(x) = x^2 - 4x + 3; [0, 4]$ **48.** $f(x) = x^3 - 3x; [-2, 2]$

49. $f(x) = \dfrac{1}{x}; [-4, -2]$ **50.** $f(x) = \dfrac{1}{x}; [-1, 1], x \neq 0$

51. $f(x) = \dfrac{2}{(x-3)}; [4, 5]$ **52.** $f(x) = \dfrac{x}{(x-2)^2}; [-3, 1]$

53. $f(x) = 2x^3 - 3x^2 - 12x + 15; [0, 3]$

54. $f(x) = 2x^3 - 3x^2 - 12x + 15; [-2, 4]$

55. $f(x) = \sin x; [-\pi, \pi]$ **56.** $f(x) = \sin 2x; [-\pi, \pi]$

57. $f(x) = \cos x; [-1, 1]$ **58.** $f(x) = \sin x, \left[-\dfrac{\pi}{4}, \dfrac{\pi}{4}\right)$

59. $f(x) = |x^3 - 4x|; [-2, 2]$ **60.** $f(x) = |x^4 - x|; [-2, 2)$

61. $f(x) = \begin{cases} 2x + 7, & \text{if } -2 \le x < -1 \\ 4 - x, & \text{if } -1 \le x \le 5 \end{cases}$

62. $f(x) = \begin{cases} x^2 + 2x + 1, & \text{if } -3 \le x \le 0 \\ 2x - x^2 - 3, & \text{if } 0 < x < 2 \\ 3 - x, & \text{if } 2 \le x \le 3 \end{cases}$

63. Use the result of Exercise 32 to find the minimum value possible for the sum of a positive number and its reciprocal.

64. Prove Theorem 3.32 for the case when f has a local minimum at c.

65. Use the fact that a polynomial has no more distinct roots than its degree to prove that a polynomial of degree n has at most $n - 1$ critical points.

66. Suppose f is defined by

$$f(x) = \frac{ax + b}{cx + d}$$

for some constants a, b, c, and d and that f has a critical point. What can be said about the constants a, b, c, and d?

67. The extreme value theorem states that any function that is continuous on a closed interval $[a, b]$ assumes both its maximum and minimum values on that interval. This result need not be true unless the interval is closed.

(a) Find a function that is continuous on the interval $(0, 1]$ and does not assume a maximum on this interval.

(b) Find a function that is continuous on $(0, 1]$ and does not assume a minimum on this interval.

(c) Find a function that is continuous on $(0, 1)$ and does not assume either a maximum or a minimum on this interval.

(d) Find a function that is continuous on $(0, 1]$ and assumes neither a maximum nor a minimum on this interval. (One possibility: modify the function $f(x) = \sin x$ and use the fact that the sine function is continuous.)

68. What is the maximum slope of a line tangent to $f(x) = x^3 - 3x^2 + 4x - 5$ for x in $[-1, 2]$?

69. The cost of producing $1000x$ square yards of carpet per week is given by $C(x) = x^2 - 10x + 30$, where $C(x)$ is given in hundreds of dollars. Because of factory size, at most 15,000 square yards can be produced in one week. Sketch the graph of this function. Find the absolute maximum and absolute minimum weekly costs.

70. The size (in square centimeters) of a bacteria population residing on a nutrient agar at time t (in hours) is given by $P(t) = 1000 + 100t - 10t^2$. Give the domain of P and find the absolute maximum and absolute minimum size of the bacteria population.

71. The solubility of sucrose in water is temperature-dependent according to the formula

$$S = 64.542 + 7.982 \times 10^{-2}T + 1.9658 \times 10^{-3}T^2 - 9.691 \times 10^{-6}T^3$$

where T is measured in degrees Celsius and S is given in grams of sucrose per 100 grams of solution. Find the lowest temperature that produces a maximum solubility.

72. Another equation that describes the solubility of sucrose in water is

$$S = \begin{cases} 64.53 + 9.37 \times 10^{-2}T + 1.2 \times 10^{-3}T^2, & \text{if } T \le 38.04 \\ 61.15 + 2.25 \times 10^{-1}T + 8.4 \times 10^{-5}T^2, & \text{if } 38.04 < T. \end{cases}$$

Find the lowest temperature that produces a maximum solubility.

73. The power P of a steam turbine depends upon the peripheral speed of the wheel surrounding the turbine blades. If S_1 is the speed of the steam entering the turbine and S_2 is the peripheral speed of the wheel, then

$$P = kS_2 (S_1 - S_2)$$

for some constant k. Suppose that the turbine blades can be tilted to vary S_2 while S_1 remains constant. What value of S_2 produces the maximum power?

74. The velocity of a particular class of chemical reactions obeys the law $v = a(b + x)(c - x)$ where a, b, and c are positive constants and x is the amount of substrate that has decomposed. Find the value of x at which the velocity is a maximum.

75. The solubility of cuprous chloride in solutions containing excess chloride is

$$S = \frac{1.8 \times 10^{-7}}{[\text{Cl}^-]} + 1.0 \times 10^{-5} + 7.7 \times 10^{-2} [\text{Cl}^-],$$

where $[\text{Cl}^-]$ is the concentration of free chloride moles per liter. Find the concentration that minimizes S.

Putnam exercises:

76. Let $f(x) = a_1\sin x + a_2\sin 2x + \cdots + a_n\sin nx$, where $a_1, a_2 \cdots , a_n$ are real numbers and where n is a positive integer. Given that $|f(x)| \le |\sin x|$ for all real x, prove that

$$|a_1 + 2a_2 + \cdots + na_n| \le 1.$$

(This exercise was problem A–1 of the twenty-eighth William Lowell Putnam examination given on December 2, 1967. The examination and its solution can be found in the September 1968 issue of the *American Mathematical Monthly,* pages 734–739.)

77. Consider a polynomial $f(x)$ with real coefficients having the property $f(g(x)) = g(f(x))$ for every polynomial $g(x)$ with real coefficients. Determine and prove the nature of $f(x)$.

(This exercise was problem 5, part I of the twenty-first William Lowell Putnam examination given on December 3, 1960. The examination and its solution can be found in the September 1961 issue of the *American Mathematical Monthly,* pages 632–637.)

3.9
THE MEAN VALUE THEOREM

The mean value theorem has applications at various stages in the study of calculus. The proof of this important theorem is facilitated by first considering the following lemma. (A lemma is a result that is important in its own right, but its primary importance is to prove a major theorem.)

(3.33)
LEMMA

Rolle's Theorem   Suppose f is continuous on the interval $[a, b]$ and differentiable on (a, b). If $f(a) = f(b)$, then there exists at least one number c, $a < c < b$, with $f'(c) = 0$.

PROOF

There are two possibilities for the function f.

(i) f is a constant function on $[a, b]$. (See Figure 3.23(a).) In this case, $f'(x) = 0$ for all x in (a, b) and c can be chosen to be any number between a and b.

(ii) f is not constant on $[a, b]$. (See Figure 3.23(b).) In this case there must exist an x in (a, b) with either $f(x) > f(a)$ or $f(x) < f(a)$. For the sake of argument, suppose $f(x) > f(a)$. The extreme value theorem guarantees that there is a number c in $[a, b]$ at which f assumes its maximum value. Since $f(x) > f(a)$, $f(c) > f(a) = f(b)$, so c cannot be either a or b. But an absolute maximum occurs at c, and an absolute maximum can occur only at an endpoint of the interval or at a critical point of f. Therefore,

HISTORICAL NOTE   **Michel Rolle** (1652–1719) was a respected member of the Académie of Science in France who criticized the "new" calculus proposed by Isaac Newton and Gottfried Liebniz. The theorem that bears his name appeared in a little known treatise on geometry and algebra entitled *Méthode pour résoudre les égalités* published in 1691. It is ironic that one of the most basic results in the theory of calculus was proved by a person who was vigorously opposed to the calculus methods of his contemporaries. In later life, Rolle acknowledged that the calculus techniques were of value and basically sound.

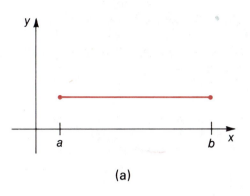

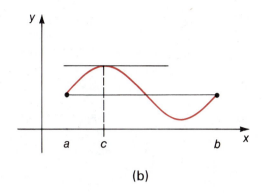

(a)   (b)

FIGURE 3.23

c must be a critical point of f. Since f' exists at each number in (a, b), Theorem 3.32 implies that $f'(c) = 0$.   □

Rolle's theorem says that if f is continuous on $[a, b]$, differentiable on (a, b), and $f(a) = f(b)$, then at some point $(c, f(c))$, $a < c < b$, the tangent line to the graph of f is horizontal. That is, the tangent line is parallel to the line joining the points $(a, f(a))$ and $(b, f(b))$. The mean value theorem makes a similar statement when $f(a)$ and $f(b)$ are not necessarily equal.

(3.34)
THEOREM

The Mean Value Theorem   If a function f is continuous on the interval $[a, b]$ and differentiable on (a, b), then there exists at least one number c, with $a < c < b$, and

$$f'(c) = \frac{f(b) - f(a)}{b - a}.$$

Before presenting the proof of the mean value theorem, consider the graph given in Figure 3.24 of a function f that is continuous on $[a, b]$ and differentiable on (a, b).

Since $(f(b) - f(a))/(b - a)$ is the slope of the line joining the points $(a, f(a))$ and $(b, f(b))$, and $f'(c)$ is the slope of the tangent line to the graph of f

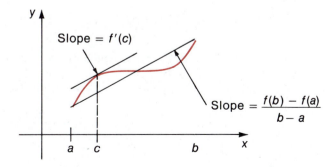

Slope $= f'(c)$

Slope $= \dfrac{f(b) - f(a)}{b - a}$

FIGURE 3.24

at $(c, f(c))$, the mean value theorem asserts that a number c exists between a and b where these two slopes are equal. That is, the tangent line to the graph of f at $(c, f(c))$ is parallel to the line joining the points $(a, f(a))$ and $(b, f(b))$. (In our sketch, there are actually two choices for c.)

PROOF

We define a function g that satisfies the hypothesis of Rolle's theorem on $[a, b]$. Let $g(x)$ be the difference between $f(x)$ and the y-coordinate of the point on the line joining $(a, f(a))$ and $(b, f(b))$ whose first coordinate is x. See Figure 3.25.

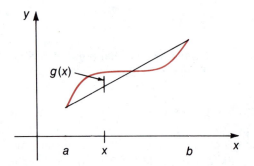

FIGURE 3.25

Since the line joining $(a, f(a))$ and $(b, f(b))$ has slope $(f(b) - f(a))/(b - a)$, its equation can be expressed as

$$y = \frac{f(b) - f(a)}{b - a} (x - a) + f(a).$$

Thus, g is defined by

$$g(x) = f(x) - \left[\frac{f(b) - f(a)}{b - a} (x - a) + f(a) \right],$$

for x in $[a, b]$.
It is easily verified that $g(a) = 0 = g(b)$.

Since f is continuous on $[a, b]$, g is also continuous on $[a, b]$. Moreover,

$$g'(x) = f'(x) - \frac{f(b) - f(a)}{b - a},$$

so g' exists whenever f' exists, which implies that g is differentiable on (a, b). Therefore, g satisfies the hypothesis of Rolle's theorem so a number c in (a, b) exists with $g'(c) = 0$. That is, for some c in (a, b)

$$0 = g'(c) = f'(c) - \frac{f(b) - f(a)}{b - a}.$$

Consequently,

$$f'(c) = \frac{f(b) - f(a)}{b - a}. \qquad \square$$

EXAMPLE 1   Given $f(x) = x^{2/3}$, find a value of c in $(0, 8)$ that is ensured by the conclusion of the mean value theorem.

SOLUTION

The graph of f is shown in Figure 3.26. We will first show that the hypotheses of the mean value theorem are satisfied on the interval $[0, 8]$.

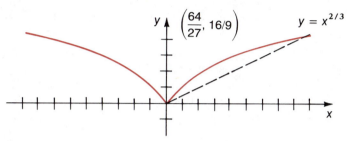

FIGURE 3.26

The function given by $f(x) = x^{2/3}$ is continuous on $[0, 8]$. Its derivative, $f'(x) = (2/3)x^{-1/3}$, exists except when $x = 0$, so f is differentiable on $(0, 8)$. We must find c, $0 < c < 8$, which satisfies

$$f'(c) = \frac{f(8) - f(0)}{8 - 0}.$$

That is,

$$\frac{2}{3} c^{-1/3} = \frac{4 - 0}{8} = \frac{1}{2}.$$

Although the mean value theorem does not give a constructive method of finding c, the value of c in this example can be obtained by solving the equation

$$\frac{2}{3} c^{-1/3} = \frac{1}{2},$$

so,

$$c = \left(\frac{3}{4}\right)^{-3} = \left(\frac{4}{3}\right)^{3} = \frac{64}{27}. \qquad \square$$

Note that the conclusion of the mean value theorem does not hold for $f(x) = x^{2/3}$ when we consider the interval $[-1, 1]$ (see Figure 3.27). At no number in this interval is the derivative equal to zero.

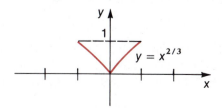

FIGURE 3.27

This is not contradictory, however, since there is a number, $x = 0$, in the interval $(-1, 1)$ at which the derivative, $f'(x) = (2/3)x^{-1/3}$, does not exist. Consequently, the hypotheses of the mean value theorem are not satisfied for f on $[-1, 1]$.

There are many important applications of the mean value theorem in the development of the calculus. We conclude this section with two of these.

(3.35)
THEOREM

If f is differentiable on an interval $[a, b]$ and $f'(x) = 0$ at each number x in $[a, b]$, then f is a constant function on $[a, b]$.

PROOF

Let x denote an arbitrary number in $(a, b]$. Since f is differentiable on $[a, b]$ and $a < x \le b$, f is differentiable and, hence, continuous on $[a, x]$. The hypotheses of the mean value theorem are satisfied on $[a, x]$, so there is a number c, $a < c < x$, satisfying

$$f'(c) = \frac{f(a) - f(x)}{a - x}.$$

However, $f'(x) = 0$ at each x in $[a, b]$ and $a < c < x$, so $f'(c) = 0$. Thus,

$$0 = \frac{f(a) - f(x)}{a - x},$$

which implies that

$$f(x) = f(a).$$

That is, f is constant on $[a, b]$. □

(3.36)
COROLLARY

Suppose f and g are differentiable on the interval $[a, b]$. Then, $f'(x) = g'(x)$ for each x in $[a, b]$ if and only if $f(x) = g(x) + C$ for some constant C.

PROOF

If $f(x) = g(x) + C$ for some constant C, then $D_x f(x) = D_x g(x) + D_x C = D_x g(x) + 0$, so $f'(x) = g'(x)$.

To show the converse, consider the function $h = f - g$. h is differentiable on $[a, b]$ and $h'(x) = f'(x) - g'(x) = 0$ for each x in $[a, b]$. It follows from Theorem 3.35 that h is a constant function on $[a, b]$; that is, there is a constant C such that $h(x) = C$ for each x in $[a, b]$. Consequently, $f(x) - g(x) = C$ and $f(x) = g(x) + C$. □

EXAMPLE 2

Find a function f such that $f(0) = 4$ and $f'(x) = 3$ for all x.

SOLUTION

If $g(x) = 3x$, then $g'(x) = 3 = f'(x)$. From Corollary 3.36, it follows that $f(x) = 3x + C$ for some constant C. Since $f(0) = 4$, $4 = f(0) = 3 \cdot 0 + C$ and $C = 4$. Consequently, the only function with the required properties is given by

$$f(x) = 3x + 4.$$ □

EXERCISE SET 3.9

In Exercises 1 through 10, determine whether the hypotheses of Rolle's theorem are satisfied for the given function and interval. If the hypotheses are satisfied, find a number c in the interval with $f'(c) = 0$.

1. $f(x) = x^2 - 3x + 2, [1, 2]$ **2.** $f(x) = x^2 + 9, [-3, 3]$

3. $f(x) = x^3 - 2x^2 - x + 2, [-1, 2]$ **4.** $f(x) = x^3 - 3x, [0, 3]$

5. $f(x) = \dfrac{x^2 - 1}{x}, [-1, 1]$ **6.** $f(x) = \dfrac{x^2 - 1}{x^2 + 1}, [-1, 1]$

7. $f(x) = \sin x, [0, \pi]$ **8.** $f(x) = \tan x, [0, \pi]$

9. $f(x) = \sin x - \cos x, \left[\dfrac{\pi}{4}, \dfrac{5\pi}{4}\right]$ **10.** $f(x) = 1 - \cos x, [0, 2\pi]$

In Exercises 11 through 22, determine whether the hypotheses of the mean value theorem are satisfied for the given function and interval. If the hypotheses are satisfied, find a number c in the interval that is ensured by the conclusion of the mean value theorem.

11. $f(x) = x^2 + 4x - 5, [-1, 2]$ **12.** $f(x) = 2x^2 + 3x - 5, [1, 4]$

13. $f(x) = x + \dfrac{1}{x}, [1, 2]$ **14.** $f(x) = x + \dfrac{1}{x}, [-2, -1]$

15. $f(x) = x^{3/2}, [0, 8]$ **16.** $f(x) = x^{3/2}, [-8, 8]$

17. $f(x) = \sqrt{1 - x^2}, [-1, 1]$ **18.** $f(x) = \sqrt{1 - x^2}, [-1, 0]$

19. $f(x) = \dfrac{x^2 + 3x}{x - 1}, [-1, 0]$ **20.** $f(x) = \dfrac{x^2 + 3x}{x - 1}, [0, 2]$

21. $f(x) = \sin x, \left[0, \dfrac{\pi}{2}\right]$ **22.** $f(x) = \cos x, [0, \pi]$

In Exercises 23 through 30, find a function f that satisfies the given conditions.

23. $f(0) = 3$ and $f'(x) = 0$ for all x.

24. $f(0) = -1$ and $f'(x) = 2$ for all x.

25. $f(0) = 5$ and $f'(x) = 2x$ for all x.

26. $f(1) = 3$ and $f'(x) = -2$ for all x.

27. $f(0) = 2$ and $f'(x) = \cos x$ for all x.

28. $f(\pi/2) = 3$ and $f'(x) = 2 \cos x$ for all x.

29. $f(0) = 2, f'(0) = 3$, and $f''(x) = 2$ for all x.

30. $f(0) = 1, f'(0) = 2$, and $f''(x) = 6x$ for all x.

31. Explain why the hypotheses of the mean value theorem are not satisfied for the given function and interval in each of the following.

  (a) $f(x) = |x|, [-1, 1]$ (b) $f(x) = \tan x, [0, \pi]$

  (c) $f(x) = [\![x]\!], [0, 2]$ (d) $f(x) = \sec x, [0, \pi]$

32. Use Rolle's theorem to show that the graph of f intersects the x-axis exactly once in the given interval.

(a) $f(x) = x^3 - 3x - 1$, $[1, 2]$

(b) $f(x) = \cos x - x$, $\left[0, \dfrac{\pi}{4}\right]$

(c) $f(x) = 2x^3 - 3x^2 - 12x$, $[3, 4]$

(d) $f(x) = x^7 - 4x^3 + 1$, $[0, 1]$

33. Use Rolle's theorem to show that the graph of $f(x) = x^3 + 2x + k$ crosses the x-axis exactly once, regardless of the value of the constant k.

34. Suppose that f is a function whose second derivative is defined at each number in the interval $[a, b]$ and for which three distinct numbers x_1, x_2, and x_3 exist in $[a, b]$ with $f(x_1) = f(x_2) = f(x_3) = 0$. Use Rolle's theorem to show that $f''(c) = 0$ for some value of c in (a, b).

35. Give an example of a function that is continuous and differentiable on the interval $(0, 1]$, but for which the conclusion of the mean value theorem does not hold for the interval $[0, 1]$.

36. Give an example of a function that is continuous on $[0, 1]$ and differentiable except at $x = 1/2$, but for which the conclusion of the mean value theorem does not hold.

37. Suppose f is a quadratic function described by $f(x) = a_2x^2 + a_1x + a_0$, for arbitrary constants a_0, a_1, and a_2. Show that the hypotheses of the mean value theorem hold for any interval $[a, b]$, that the number c ensured by the theorem is unique, and is equal to $(a + b)/2$.

38. Use the mean value theorem to show that if $f'(x) \neq 0$ for all values of x, then $a \neq b$ implies that $f(a) \neq f(b)$.

39. Use the mean value theorem to show that $\left|\sin a - \sin b\right| \leq \left| a - b \right|$ and deduce from this result that $\left|\sin a + \sin b\right| \leq \left|a + b\right|$.

40. In Exercise 50 of Section 2.4, we found that if g is continuous and $a \leq g(x) \leq b$ whenever $a \leq x \leq b$, then a fixed-point p of g exists in $[a, b]$, that is, $a \leq p \leq b$ and $g(p) = p$. Suppose, in addition, that $\left|g'(x)\right| < 1$ on $[a, b]$. Show that there is exactly one fixed point in $[a, b]$.

41. Suppose $f'(x) > 0$ on (a, b) and $f'(x) < 0$ on (b, c). Show that if x_1 is in (a, b) there is at most one number x_2 in (b, c) with $f(x_1) = f(x_2)$.

3.10
INCREASING AND DECREASING FUNCTIONS: THE FIRST DERIVATIVE TEST

With the help of the mean value theorem we are ready to consider in detail the increasing and decreasing property of functions.

(3.37)
DEFINITION

A function f **is increasing on I** if f has the property that $f(x_1) < f(x_2)$ for each pair of numbers x_1 and x_2 in I with $x_1 < x_2$.

A function f **is decreasing on** I if f has the property that $f(x_1) > f(x_2)$ for each pair of numbers x_1 and x_2 in I with $x_1 < x_2$.

By examining the graph shown in Figure 3.28, it can be seen that on the intervals where the slopes of the tangent lines are positive the function is increasing, and when the slopes of the tangent lines are negative the function is decreasing.

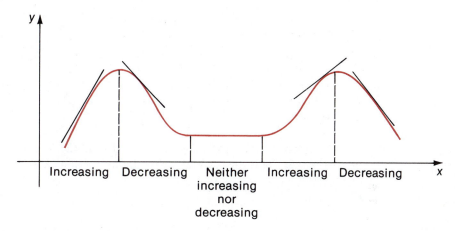

FIGURE 3.28

(3.38)
THEOREM

If I is any interval and
(i) $f'(x) > 0$ for all x in I, then f is increasing on I;
(ii) $f'(x) < 0$ for all x in I, then f is decreasing on I.

PROOF

(i) Suppose x_1 and x_2 are in I with $x_1 < x_2$. Since $f'(x) > 0$ for all x in I, f' exists on I and consequently on $[x_1, x_2]$. Therefore, f is continuous and satisfies the hypotheses of the mean value theorem on $[x_1, x_2]$. Hence, a number c, $x_1 < c < x_2$, exists with

$$f'(c) = \frac{f(x_2) - f(x_1)}{x_2 - x_1}.$$

Since $x_2 - x_1 > 0$ and $f'(c) > 0$, this implies that

(3.39) $f(x_2) - f(x_1) > 0$

so

(3.40) $f(x_1) < f(x_2).$

Thus f is increasing on I.
(ii) The proof of case (ii) is the same as (i) except $f'(c) < 0$, so the inequalities in (3.39) and (3.40) are reversed.   ☐

EXAMPLE 1

If $f(x) = 2x^3 - 9x^2 + 12x - 2$, find the intervals on which f is increasing and the intervals on which f is decreasing.

SOLUTION

In Example 1 of Section 3.8, we found critical points of this function at $x = 1$ and $x = 2$ and sketched its graph. Since f' is continuous, the critical points are the only places at which $f'(x) = 6(x - 1)(x - 2)$ can change sign. If we consider the intervals $(-\infty, 1)$, $(1, 2)$, $(2, \infty)$ and use Theorem 3.38, we can determine when f is increasing and when f is decreasing. Our work is summarized in the following table:

When x is in the interval	$x - 1$	$x - 2$	$f'(x) = $ $6(x-1)(x-2)$	f is
$(-\infty, 1)$	negative	negative	positive	increasing
$(1, 2)$	positive	negative	negative	decreasing
$(2, \infty)$	positive	positive	positive	increasing

Thus f is increasing on the intervals $(-\infty, 1)$ and $(2, \infty)$ and decreasing on the interval $(1, 2)$. This analysis agrees with our previous knowledge of the graph of f. (See Figure 3.19 on p. 147.)   □

EXAMPLE 2   For $f(x) = x + 1/x$, find the intervals on which f is increasing and the intervals on which f is decreasing. Sketch the graph of f.

SOLUTION

First observe that f is discontinuous at $x = 0$. The derivative of f at x, when $x \neq 0$, is

$$f'(x) = 1 - \frac{1}{x^2} = \frac{x^2 - 1}{x^2},$$

which is zero if $x = 1$ or $x = -1$. Partition the real line into intervals determined by the critical points, $x = -1$ and $x = 1$, and the point of discontinuity $x = 0$. These are the only values at which f' can change sign. The behavior of f in each interval can be determined by considering the value of f' at any arbitrary number in the interval.

When x is in the interval	$x^2 - 1$	x^2	$f'(x) = \dfrac{x^2 - 1}{x^2}$	f is
$(-\infty, -1)$	positive	positive	positive	increasing
$(-1, 0)$	negative	positive	negative	decreasing
$(0, 1)$	negative	positive	negative	decreasing
$(1, \infty)$	positive	positive	positive	increasing

Note, in addition, that the graph of f is symmetric with respect to the origin and that both $\lim\limits_{x \to \infty} x + 1/x = \infty$ and $\lim\limits_{x \to 0^+} x + 1/x = \infty$. The graph of f must consequently appear as shown in Figure 3.29.   □

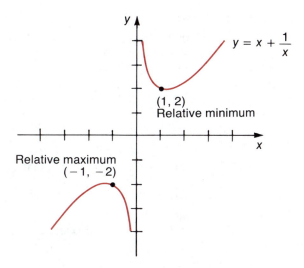

FIGURE 3.29

In Section 3.8 we saw that if a function has a relative extrema, then it must occur at a critical point; however, not every critical point gives a relative extrema. A relative minimum occurs when the function is decreasing to the left of a critical point and increasing to the right of the critical point. A relative maximum occurs when the function is increasing to the left of the critical point and decreasing to the right. The sign of the derivative on intervals containing the critical point can be used to determine when this occurs, as illustrated in Figure 3.30.

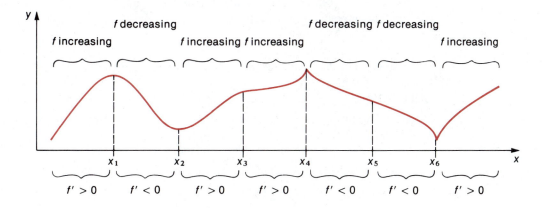

FIGURE 3.30

(3.41)
THEOREM

The First Derivative Test   Suppose c is a critical point of f and (a, b) is an open interval containing c on which f is continuous.

(i)   If $f'(x) > 0$ on (a, c) and $f'(x) < 0$ on (c, b), then f has a relative maximum at $x = c$.

(ii) If $f'(x) < 0$ on (a, c) and $f'(x) > 0$ on (c, b), then f has a relative minimum at $x = c$.

(iii) If $f'(x) > 0$ on both (a, c) and (c, b), or if $f'(x) < 0$ on both (a, c) and (c, b), then f has neither a relative maximum nor a relative minimum at $x = c$.

EXAMPLE 3   Using the first derivative test, find relative maxima and minima of f if $f(x) = x^2 - 3x^{2/3}$.

SOLUTION

We found in Section 3.8 (Example 2) that

$$f'(x) = 2x - 2x^{-1/3} = 2x^{-1/3}(x^{4/3} - 1).$$

So the critical points are $x = 0$ (since $f'(0)$ is undefined), $x = 1$, and $x = -1$ (since $f'(1) = f'(-1) = 0$). These are the only values of x at which f' can change sign.

When x is in the interval	$x^{-1/3}$	$x^{4/3} - 1$	$f'(x) = 2x^{-1/3}(x^{4/3} - 1)$
$(-\infty, -1)$	negative	positive	negative
$(-1, 0)$	negative	negative	positive
$(0, 1)$	positive	negative	negative
$(1, +\infty)$	positive	positive	positive

Since $f'(x) < 0$ on $(-\infty, -1)$ and $f'(x) > 0$ on $(-1, 0)$, the first derivative test implies that f has a relative minimum at $x = -1$. A relative minimum also occurs at $x = 1$, while a relative maximum occurs at $x = 0$. This agrees with the conclusion reached in Section 3.8. (See Figure 3.20 on page 148.) □

EXAMPLE 4   Use the first derivative test to sketch the graph of $f(x) = x(4 - x^2)^{1/2}$.

SOLUTION

The domain of f is the interval $[-2, 2]$, since $4 - x^2$ must be nonnegative. Using the product rule,

$$f'(x) = (4 - x^2)^{1/2} + x\left(\frac{1}{2}\right)(4 - x^2)^{-1/2}(-2x)$$

$$= (4 - x^2)^{1/2} - x^2(4 - x^2)^{-1/2}$$

$$= 2(2 - x^2)(4 - x^2)^{-1/2}.$$

Critical points occur at $x = 2$, $x = -2$, where $f'(x)$ is undefined, and at $x = \sqrt{2}$, $x = -\sqrt{2}$, where $f'(x) = 0$.

When x is in the interval	$(2 - x^2)$	$(4 - x^2)^{-1/2}$	$f'(x) = 2(2-x^2)(4-x^2)^{-1/2}$
$(-2, -\sqrt{2})$	negative	positive	negative
$(-\sqrt{2}, \sqrt{2})$	positive	positive	positive
$(\sqrt{2}, 2)$	negative	positive	negative

The first derivative test implies that a relative minimum occurs when $x = -\sqrt{2}$ and a relative maximum when $x = \sqrt{2}$. The graph of f is shown in Figure 3.31.

The graph is symmetric with respect to the origin. It has a vertical tangent at $x = 2$ and $x = -2$ because $\lim\limits_{x \to 2^{-}} f'(x) = \lim\limits_{x \to 2^{-}} 2(2 - x^2)(4 - x^2)^{-1/2} = -\infty.$ □

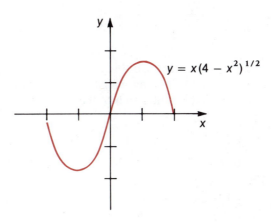

$y = x(4 - x^2)^{1/2}$

FIGURE 3.31

EXERCISE SET 3.10

In Exercises 1 through 6, the graph of a function f is given. Use the graph to determine the intervals on which the function is increasing and those on which it is decreasing. Where do relative and absolute extrema occur? What is the derivative at these points? Find the absolute extrema when they exist.

1.

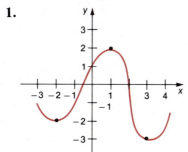

2.

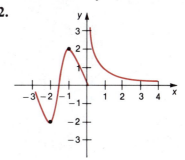

3.

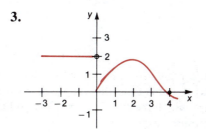

4.

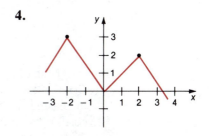

5.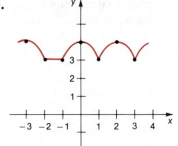

6.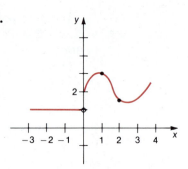

In Exercises 7 through 48, determine the intervals on which the function is increasing and those on which the function is decreasing. Find the relative extrema and sketch the graph of the function.

7. $f(x) = 4x - 3$

8. $f(x) = -2x + 3$

9. $f(x) = x^2 - 4x + 4$

10. $f(x) = 4 + 4x - x^2$

11. $f(x) = 2x^2 - 9x + 3$

12. $f(x) = x^3 - 6x^2 + 9x - 4$

13. $f(x) = x^3 - 3x$

14. $f(x) = x^3 - x^2$

15. $f(x) = x^4 - 2x^2$

16. $f(x) = x^4 + 2x^2$

17. $f(x) = 2x + \dfrac{1}{2x}$

18. $f(x) = 2x - \dfrac{1}{2x}$

19. $f(x) = x^3 (1 - x)$

20. $f(x) = x^2 (1 - x)^2$

21. $f(x) = \sin x$

22. $f(x) = \cos x$

23. $f(x) = \tan x$

24. $f(x) = \sec x$

25. $f(x) = x^2 + \dfrac{1}{x^2}$

26. $f(x) = x + \dfrac{1}{\sqrt{x}}$

27. $f(x) = \dfrac{x + 1}{x - 1}$

28. $f(x) = \dfrac{2x - 3}{x - 2}$

29. $f(x) = \dfrac{x^2 + 1}{x^2 - 1}$

30. $f(x) = \dfrac{x^2 - 1}{x^2 + 1}$

31. $f(x) = x\sqrt{9 - x^2}$

32. $f(x) = x\sqrt{x^2 - 9}$

33. $f(x) = (x - 2)^2 (x + 1)^2$

34. $f(x) = (3x - 5)^2 (4 - x)^3$

35. $f(x) = (x - 1)^2 (x - 2)^3 (x - 3)^4$

36. $f(x) = (x^2 - 1)^2 (x^2 + 1)^2$

37. $f(x) = |x^2 - 2|$

38. $f(x) = x |x^2 - 2|$

39. $f(x) = x^{3/2} - 3x^{1/2}$

40. $f(x) = x^{3/2} + 3x^{1/2}$

41. $f(x) = x^{7/6} - x^{5/3}$

42. $f(x) = x^{7/3} + x^{4/3} - 3x^{1/3}$

43. $f(x) = \sqrt[4]{x^4 - 2x^2}$

44. $f(x) = \sqrt[3]{x^3 - 3x}$

45. $f(x) = \sin x - \dfrac{x}{2}$

46. $f(x) = 3 \sin 2x$

47. $f(x) = x - \sin x$

48. $f(x) = \sin x - \cos x$

49. The cost of producing x gallons, $500 \le x \le 4000$, of maple syrup is $C(x) = 15x - .002x^2$ dollars. Find the values of x for which the cost is increasing and those for which the cost is decreasing. What is the maximum cost? What is the minimum cost?

50. The amount of sales of a new product t months after the product is introduced is given by $s(t) = 2000(1 + t)/(10 + t)$. Show that the amount of sales is increasing. Is there a maximum amount of sales?

In Exercises 51 through 54, find values of a, b, and c that ensure that the function f is increasing on the interval $(-\infty, -1)$, decreasing on the interval $(-1, 1)$ and increasing on the interval $(1, \infty)$, or show that no such constants exist.

51. $f(x) = ax^2 + bx + c$ **52.** $f(x) = ax^3 + bx + c$

53. $f(x) = ax^3 + bx^2 + cx$ **54.** $f(x) = ax^3 + bx^2 + c$

55. Suppose f and g are functions defined on $[a, b]$ and both $f'(x) > 0$ and $g'(x) > 0$ for every x in $[a, b]$. Must the function $h = f \cdot g$ be increasing on $[a, b]$?

56. Suppose f and g are increasing functions on their respective domains and $f \circ g$ exists for all x in $[a, b]$. Show that $f \circ g$ is increasing on $[a, b]$.

57. Prove part (ii) of Theorem 3.38. If $f'(x) < 0$ for all x in an interval I, then f is decreasing on I.

3.11
THE SECOND DERIVATIVE OF A FUNCTION

Another test for determining the relative maxima and minima of a function uses the second derivative of the function. This result is more restrictive in its application than the first derivative test, but is easy to apply if the second derivative of the function can be found. Figure 3.32 gives an illustration of this result.

(3.42)
THEOREM

The Second Derivative Test   Suppose f is differentiable on an open interval containing c and $f'(c) = 0$.
(i)   If $f''(c) < 0$, then f has a relative maximum at c.
(ii)   If $f''(c) > 0$, then f has a relative minimum at c.

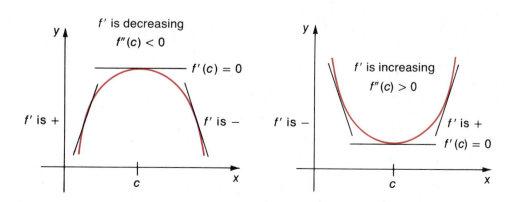

FIGURE 3.32

PROOF

(i) If $f''(c) = \lim\limits_{x \to c} \dfrac{f'(x) - f'(c)}{x - c} < 0$, then an interval (a, b) about c exists with

(3.43)
$$\frac{f'(x) - f'(c)}{x - c} < 0$$

whenever x is in (a, b) and $x \neq c$. For x satisfying $a < x < c$, $x - c < 0$ and (3.43) implies that $f'(x) - f'(c) > 0$. Therefore $f'(x) > f'(c) = 0$ and f is increasing on (a, c). For x satisfying $c < x < b$, $x - c > 0$ and (3.43) implies that $f'(x) - f'(c) < 0$. Therefore $f'(x) < f'(c) = 0$ and f is decreasing on (c, b). By the first derivative test, a relative maximum must occur at c.

Part (ii) is proved in a similar manner. The proof is considered in Exercise 64. □

EXAMPLE 1   If $f(x) = x^4 - 4x^3 - 2x^2 + 12x + 1$, use the second derivative test to find the relative maxima and minima of f.

SOLUTION

Since

$$f'(x) = 4x^3 - 12x^2 - 4x + 12,$$

the only critical points are when $f'(x) = 0$. Noting that $f'(1) = 0$, we can factor to find:

$$f'(x) = 4(x^3 - 3x^2 - x + 3) = 4(x - 3)(x - 1)(x + 1).$$

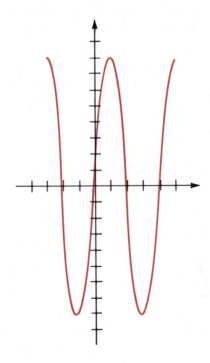

FIGURE 3.33

The critical points are $x = 3$, $x = 1$, and $x = -1$. The second derivative is

$$f''(x) = 12x^2 - 24x - 4.$$

So $f''(3) = 32 > 0$ and f has a relative minimum at 3; $f''(1) = -16 < 0$ and f has a relative maximum at 1; $f''(-1) = 32 > 0$ and f has a relative minimum at -1. The graph of f is shown in Figure 3.33.   □

    Nothing can be implied from Theorem 3.42 when both $f'(c) = 0$ and $f''(c) = 0$. This is evident from considering the sequence of functions shown in Figure 3.34. Each of these functions has $x = 0$ as its only critical point and the second derivative at $x = 0$ is also zero. However, while f_1 and f_2 have neither a relative maximum nor a relative minimum at $x = 0$, f_3 has a relative minimum at $x = 0$ and f_4 has a relative maximum there.

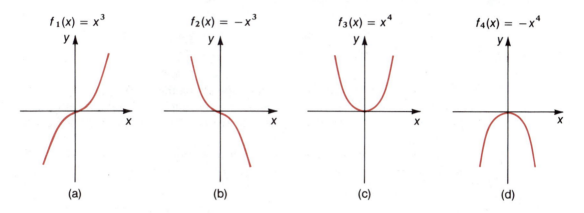

FIGURE 3.34

    The second derivative of a function can give more information about the shape of the graph of a function. It can be used to determine when a curve joining two points $(a, f(a))$ and $(b, f(b))$ has the shape shown in Figure 3.35(a) or that shown in Figure 3.35(b).

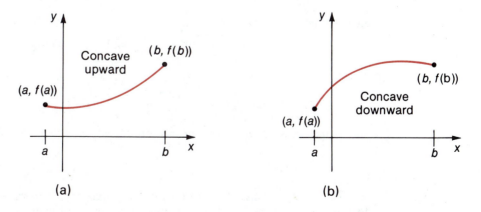

FIGURE 3.35

The graph of the function in Figure 3.35(a) is called **concave upward.** This graph has the property that any tangent line lies below the curve. If the graph of a function is concave upward, then the derivative of the function is increasing, and the tangent lines increase in slope as x increases.

The graph of the function in Figure 3.35(b) is called **concave downward.** This graph has the property that any tangent line lies above the curve. If the graph of a function is concave downward, then the derivative of the function is decreasing, and the tangent lines decrease in slope as x increases.

Rather than define concavity in terms of the position of the tangent line relative to the curve, it is convenient to express the definition in terms of the behavior of the first derivative.

(3.44)
DEFINITION

The graph of a function f is called concave upward at $(c, f(c))$ if there exists an open interval I containing c, with the property that f' is increasing on I.

The graph of f is called concave downward at $(c, f(c))$ if there exists an open interval I containing c, with the property that f' is decreasing on I.

The graph of f is said to be concave upward on the interval I if f is concave upward at $(x, f(x))$ for every x in I. Similarly, the graph of f is said to be concave downward on the interval I if f is concave downward at $(x, f(x))$ for every x in I.

The following theorem concerning the concavity of a function follows immediately from the first derivative test (Theorem 3.41) when applied to the function f'.

(3.45)
THEOREM

If I is an open interval and
(i) $f''(x) > 0$ for all x in I, then the graph of f is concave upward on I;
(ii) $f''(x) < 0$ for all x in I, then the graph of f is concave downward on I.

EXAMPLE 2   For $f(x) = x^3 - 3x^2 + x + 2$, find intervals on which the graph is concave upward and intervals on which the graph is concave downward.

SOLUTION

$$f'(x) = 3x^2 - 6x + 1 \quad \text{and} \quad f''(x) = 6x - 6.$$

Since

$$f''(x) > 0 \text{ when } x > 1 \quad \text{and} \quad f''(x) < 0 \text{ when } x < 1$$

the graph of f is concave upward on $(1, \infty)$ and concave downward on $(-\infty, 1)$. $\quad\square$

The points at which the concavity of the graph changes from concave upward to concave downward (and conversely) are naturally of interest.

(3.46)
DEFINITION

A point $(c, f(c))$ is called a point of inflection for the graph of f if an interval (a, b) containing c exists with either of the following properties:
(i) f is concave upward on (a, c) and concave downward on (c, b);
or
(ii) f is concave downward on (a, c) and concave upward on (c, b).

This definition is illustrated in Figure 3.36.

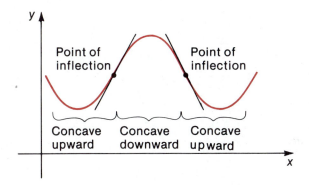

FIGURE 3.36

In light of Theorem 3.45 and the fact that a function changes sign only when the function is zero or discontinuous, we have the result:

(3.47)
COROLLARY

If $(c, f(c))$ is a point of inflection for the graph of a function f and f'' is continuous at c, then $f''(c) = 0$.

EXAMPLE 3

Find any points of inflection for the graph of the function described by $f(x) = 2x^3 - 9x^2 + 12x - 2$.

SOLUTION

Since $f'(x) = 6x^2 - 18x + 12$, the second derivative of f is given by

$$f''(x) = 12x - 18 = 6(2x - 3).$$

By Corollary 3.47, the only possible point of inflection is $(3/2, f(3/2))$. Since $f''(x) > 0$ if $x > 3/2$ and $f''(x) < 0$ if $x < 3/2$, the graph is concave upward on $(3/2, \infty)$ and concave downward on $(-\infty, 3/2)$. Consequently, $(3/2, f(3/2)) = (3/2, 5/2)$ is a point of inflection for the graph of f.

The graph of this function was considered as Example 1 of Section 3.10. If we add the information about the concavity of the graph of f to the knowledge we obtained in that example, we have the graph shown in Figure 3.37.   □

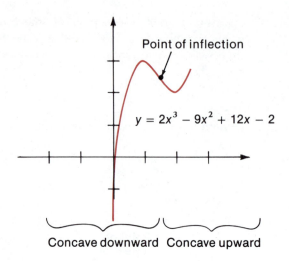

Point of inflection

$y = 2x^3 - 9x^2 + 12x - 2$

Concave downward Concave upward

FIGURE 3.37

EXAMPLE 4   Sketch the graph of the function described by

$$f(x) = x^4 + 2x^3 - 1.$$

SOLUTION

$$f'(x) = 4x^3 + 6x^2 = 2x^2(2x + 3).$$

Critical points occur at $x = 0$ and $x = -3/2$.

$$f''(x) = 12x^2 + 12x = 12x(x + 1).$$

Since $f''(x) = 0$ when $x = 0$ and when $x = -1$, possible points of inflection are $(0, f(0))$ and $(-1, f(-1))$. Partitioning the real line using the critical points and possible points of inflection gives the information in the following table.

When x is in the interval	$f'(x)$	$f''(x)$	The graph of f is
$\left(-\infty, -\dfrac{3}{2}\right)$	negative	positive	decreasing, concave upward
$\left(-\dfrac{3}{2}, -1\right)$	positive	positive	increasing, concave upward
$(-1, 0)$	positive	negative	increasing, concave downward
$(0, \infty)$	positive	positive	increasing, concave upward

The table shows that there is a relative minimum when $x = -3/2$ and points of inflection occur when $x = -1$ and $x = 0$. Evaluating f at these numbers we find that the points $(-3/2, -43/16)$, $(-1, -2)$, and $(0, -1)$ lie on the graph of f. Since $f(-3) > 0$ and $f(-2) < 0$, the graph has an x-intercept in the interval $(-3, -2)$. It also has an x-intercept in $(0, 1)$. Using this information with the observation that $\lim\limits_{x \to -\infty} f(x) = \infty$ and $\lim\limits_{x \to \infty} f(x) = \infty$ gives the graph shown in Figure 3.38.

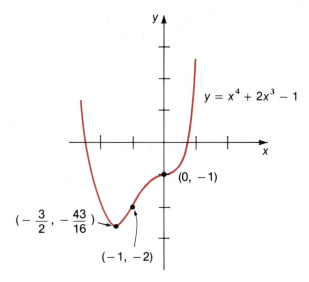

$$y = x^4 + 2x^3 - 1$$

$(0, -1)$

$\left(-\dfrac{3}{2}, -\dfrac{43}{16}\right)$

$(-1, -2)$

FIGURE 3.38

EXAMPLE 5   Sketch the graph of the function described by

$$f(x) = \sin x + \cos x.$$

SOLUTION

Since $f(x + 2n\pi) = f(x)$ for each integer n, we need to analyze the graph only on the interval $[0, 2\pi]$.

$$f'(x) = \cos x - \sin x,$$

so critical points occur when $\cos x = \sin x$; that is, when $x = \pi/4$ or $x = 5\pi/4$. Since

$$f''(x) = -\sin x - \cos x,$$

possible points of inflection occur when $\sin x = -\cos x$; that is, when $x = 3\pi/4$ or $x = 7\pi/4$. We use the critical points and possible points of inflection to partition the interval $[0, 2\pi]$ and summarize the results in the table given below.

When x is in the interval	$f'(x)$	$f''(x)$	The graph of f is
$\left(0, \dfrac{\pi}{4}\right)$	positive	negative	increasing, concave downward
$\left(\dfrac{\pi}{4}, \dfrac{3}{4}\pi\right)$	negative	negative	decreasing, concave downward
$\left(\dfrac{3}{4}\pi, \dfrac{5}{4}\pi\right)$	negative	positive	decreasing, concave upward
$\left(\dfrac{5}{4}\pi, \dfrac{7}{4}\pi\right)$	positive	positive	increasing, concave upward
$\left(\dfrac{7}{4}\pi, 2\pi\right)$	positive	negative	increasing, concave downward

The table indicates that f has a relative maximum at $x = \pi/4$ and a relative minimum at $x = 5\pi/4$. Points of inflection occur when $x = 3\pi/4$ and when $x = 7\pi/4$. The graph of f is shown in Figure 3.39.

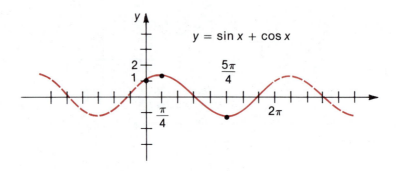

FIGURE 3.39   □

EXAMPLE 6   Sketch the graph of

$$f(x) = \frac{x^2}{x^2 + 1} + 1.$$

SOLUTION

Since $1 \le \dfrac{x^2}{x^2 + 1} + 1 < 2$, the values of f are in the interval $[1, 2)$ and the only axis intercept for the graph is at $(0, 1)$. Moreover, $\displaystyle\lim_{x \to \infty} \dfrac{x^2}{x^2 + 1} + 1 = 2$, so $y = 2$ is a horizontal asymptote to the graph of f. The graph of f is symmetric with respect to the y-axis since $f(x) = f(-x)$. Since

$$f'(x) = \frac{(x^2 + 1)2x - x^2(2x)}{(x^2 + 1)^2} = \frac{2x}{(x^2 + 1)^2},$$

$f'(x) = 0$ when $x = 0$, the only critical point.

To determine the concavity of the graph, we need f''.

$$f''(x) = \frac{2(x^2 + 1)^2 - 2x[2(x^2 + 1)(2x)]}{(x^2 + 1)^4}$$

$$= \frac{2(x^2 + 1)(x^2 + 1 - 4x^2)}{(x^2 + 1)^4}$$

$$= \frac{2(1 - 3x^2)}{(x^2 + 1)^3}.$$

Since $(x^2 + 1)^3 > 0$ for all values of x, the sign of $1 - 3x^2$ determines the sign of $f''(x)$. When $-\sqrt{3}/3 < x < \sqrt{3}/3$, $1 - 3x^2 > 0$, so $f''(x) > 0$ and the graph is concave upward. When $x < -\sqrt{3}/3$ or $x > \sqrt{3}/3$, $1 - 3x^2 < 0$, so $f''(x) < 0$ and the graph is concave downward. Consequently, both

$$\left(-\frac{\sqrt{3}}{3}, f\left(-\frac{\sqrt{3}}{3}\right)\right) = \left(-\frac{\sqrt{3}}{3}, \frac{5}{4}\right)$$

and

$$\left(\frac{\sqrt{3}}{3}, f\left(\frac{\sqrt{3}}{3}\right)\right) = \left(\frac{\sqrt{3}}{3}, \frac{5}{4}\right)$$

are points of inflection.

   We can use the critical points and possible points of inflection to partition the domain of f. Our findings are summarized in the table given below.

If x is in the interval	$f'(x)$	$f''(x)$	The graph of f is
$\left(-\infty, -\dfrac{\sqrt{3}}{3}\right)$	negative	negative	decreasing, concave downward
$\left(-\dfrac{\sqrt{3}}{3}, 0\right)$	negative	positive	decreasing, concave upward
$\left(0, \dfrac{\sqrt{3}}{3}\right)$	positive	positive	increasing, concave upward
$\left(\dfrac{\sqrt{3}}{3}, +\infty\right)$	positive	negative	increasing, concave downward

Our analysis implies that the graph of f must appear as shown in Figure 3.40.

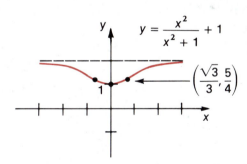

$$y = \frac{x^2}{x^2 + 1} + 1$$

$$\left(\frac{\sqrt{3}}{3}, \frac{5}{4}\right)$$

FIGURE 3.40

EXERCISE SET 3.11

In Exercises 1 through 12, use the second derivative test, when applicable, to determine relative extrema.

1. $f(x) = x^2 - 4x + 2$

2. $f(x) = -x^2 + x$

3. $f(x) = x^3 + 3x$

4. $f(x) = 2 - 4x + x^3$

5. $f(x) = x^3 + 6x^2 + 2$ **6.** $f(x) = x^4 + 6x^2$

7. $f(x) = \sin x$ **8.** $f(x) = \cos x$

9. $f(x) = x + \dfrac{1}{x}$ **10.** $f(x) = 2x^2 + \dfrac{1}{2x^2}$

11. $f(x) = x^{3/2} - 3x^{1/2}$ **12.** $f(x) = 2x^{5/3} - 5x^{2/3}$

In Exercises 13 through 28, determine intervals on which the graph of the given function is concave upward, where it is concave downward, and any points of inflection of the function.

13. $f(x) = x^4 - 2x^3 - 12x^2 + 3x + 2$

14. $f(x) = x^4 + 4x^3 + 6x^2 - 2x - 1$

15. $f(x) = \dfrac{x}{1 - x}$ **16.** $f(x) = \dfrac{x}{1 - x^2}$

17. $f(x) = \tan x$ **18.** $f(x) = \cot x$

19. $f(x) = \dfrac{x + 1}{x - 1}$ **20.** $f(x) = \dfrac{x^2 + 1}{x^2 - 1}$

21. $f(x) = x\sqrt{x^2 - 4}$ **22.** $f(x) = \dfrac{x}{\sqrt{x^2 - 4}}$

23. $f(x) = (x^2 - 9)^2$ **24.** $f(x) = (x^2 - 16)^2$

25. $f(x) = (x - 2)^2(x + 1)^2$ **26.** $f(x) = (x - 2)^2(x + 3)^3$

27. $f(x) = \sin 2x + 8 \sin x$ **28.** $f(x) = \cos 2x + 4 \cos x$

For the functions described in Exercises 29 through 48, find the critical points, intervals on which the function is decreasing and increasing, relative extrema, intervals on which the function is concave upward or downward, and points of inflection.

29. $f(x) = x^3 - 6x^2 + 9x - 4$ **30.** $f(x) = x^3 - 3x$

31. $f(x) = x^4 - 2x^2$ **32.** $f(x) = x^4 + 2x^2$

33. $f(x) = x(x - 2)^2$ **34.** $f(x) = x^2(x - 2)$

35. $f(x) = x^2 + \dfrac{1}{x^2}$ **36.** $f(x) = x + \dfrac{1}{\sqrt{x}}$

37. $f(x) = x\sqrt{x - 1}$ **38.** $f(x) = x\sqrt{x + 1}$

39. $f(x) = (x - 2)^2(x + 1)^2$ **40.** $f(x) = x^2(x - 2)^2$

41. $f(x) = x^{2/3} + x^{5/3}$ **42.** $f(x) = x^{7/6} - x^{5/3}$

43. $f(x) = \dfrac{x^2 - 1}{x^3}$ **44.** $f(x) = \dfrac{x - 1}{x^2}$

45. $f(x) = x + \cos x$ **46.** $f(x) = x - \sin x$

47. $f(x) = |x^2 - 2|$ **48.** $f(x) = x|x^2 - 2|$

Sketch the graphs of the equations in Exercises 49 through 56 as completely as possible. When appropriate, describe symmetry, intercepts, asymptotes, relative extrema, points of inflection, concavity, and so on. The graphs of these equations were considered previously in Exercise Set 1.5.

49. $x^2y^2 - x^2 - y^2 = 1$

50. $xy^2 + 3x - 6 = 0$

51. $x^2 - y^2 = x^2y^2$

52. $yx^2 + 4y - x = 0$

53. $xy - y^2 - 1 = 0$

54. $x^2y + 4x + y = 0$

55. $x^2 - xy + 1 = 0$

56. $xy - 2y^2 - 2 = 0$

57. Show that the graph of a quadratic polynomial cannot have a point of inflection.

58. Show that the graph of a cubic polynomial must have exactly one point of inflection.

59. Show that the graph of a quartic (fourth degree) polynomial can have two points of inflection or no points of inflection, but it cannot have just one point of inflection.

60. Suppose the function described by $f(x) = a_3x^3 + a_2x^2 + a_1x + a_0$ for arbitrary constants a_0, a_1, a_2, and a_3, has critical points at x_1 and x_2. Show that the graph of f has a point of inflection at $(x_0, f(x_0))$ where $x_0 = (x_1 + x_2)/2$.

61. Find a cubic polynomial that passes through the point $(0, 0)$, has a relative maximum when $x = 1$, and whose graph has a point of inflection when $x = 0$.

62. Suppose f is a cubic polynomial that has both a horizontal tangent and point of inflection at $(1, 1)$. Find the most general form of the equation of f.

63. A function f has the property that $f''(x) > 0$ if $x < 0$ and $f''(x) > 0$ if $x > 0$. In addition, $f(-1) = 0$, $f(0) = 2$, and $f(1) = 1$. What must be true about f' at $x = 0$?

64. Prove part (ii) of Theorem 3.42.

Putnam exercise:

65. On the domain $0 \le \theta \le 2\pi$: Prove that $(\sin \theta)^2 \cdot \sin (2\theta)$ takes its maximum at $\pi/3$ and $4\pi/3$ (and hence its minimum at $2\pi/3$ and $5\pi/3$).
(This exercise was problem B–6, part (a) of the thirty-fourth William Lowell Putnam examination given on December 1, 1973. The examination and its solution can be found in the December 1974 issue of the *American Mathematical Monthly*, pages 1089–1095.)

REVIEW EXERCISES

Find the derivative of the functions described in Exercises 1 through 20.

1. $f(x) = 3x^4 - 2x^3 + x - 5$

2. $f(x) = \sqrt{3x} + \dfrac{1}{\sqrt{3x}}$

3. $g(t) = \dfrac{16t^2 - 32}{t}$

4. $f(w) = (w^3 - w + 1)^4$

5. $f(x) = (x^2 - 7)^3 (x^4 + 1)$

6. $h(x) = \dfrac{\sqrt{x^2 - 4}}{x + 2}$

7. $g(x) = [(x^2 + 1)^3 - 7x]^5$

8. $f(x) = [(2x + 1)^{1/3} + x^3]^4$

9. $h(u) = u^3 - \dfrac{1}{u^2} + \dfrac{1}{u} - 2$

10. $f(x) = \sin(2x + 3)$

11. $f(x) = (x^3 - 1)\sqrt{3x^2 + 4}$

12. $h(t) = \sqrt{\cos t}$

13. $g(x) = \left(\dfrac{x^3 - 8}{x^2 + 4}\right)^5$

14. $h(x) = \left(\dfrac{x^4 - 1}{x^2 + 2x + 1}\right)^{-3}$

15. $f(x) = \left(\dfrac{x}{x^3 + 1}\right)^{-4}$

16. $g(w) = (w^3 + 1)^2(w + 2)^4 \, 2w^5$

17. $f(x) = (\sin x + \cos x)^2$

18. $h(t) = [\sin(t^3 - 7t)]^4$

19. $f(t) = \tan 2t + \sec 3t$

20. $f(x) = \csc x^2 \cot x$

21–24. Find the second and third derivatives of the functions described in Exercises 1 through 4.

25. Use the definition of the derivative to show that:
$$D_x(x^2 - 7x + 1) = 2x - 7.$$

26. Give an example of a function that is continuous at $x = 1$ but not differentiable there.

27. Find an equation of the tangent line to the graph of $f(x) = x(4 - x^2)^{1/2}$ at $(0, 0)$.

28. Find an equation of the tangent line to the graph of $f(x) = \sqrt[3]{4 - x}$ that is parallel to the line with equation $x + 12y - 13 = 0$.

29. Find an equation of the tangent line to the graph of $f(x) = x^3 - 3x + 2$ that passes through $(0, 0)$.

In Exercises 30 through 33, find $\dfrac{dy}{dx}$.

30. $x^2y + xy^2 + y^2 = 2$

31. $(x^2 + y^2)^{1/2} = xy$

32. $y \sin x + x \sin y = 0$

33. $\sqrt{xy} + \dfrac{1}{\sqrt{xy}} = 3$

In Exercises 34 and 35, find (a) $\dfrac{dy}{dx}$ and (b) $\dfrac{d^2y}{dx^2}$.

34. $y^3 + x^3 = 3xy$

35. $x^2 + 2xy + y^2 = 1$

36. Find the first five derivatives of f if

(a) $f(x) = x^4 - 7x^2$

(b) $f(x) = (x + 1)^{1/2}$

For the functions described in Exercises 37 through 46, find the critical points, intervals on which the function is decreasing and increasing, relative extrema, intervals on which the function is concave upward or downward, and points of inflection. Find any horizontal or vertical asymptotes to the graph and sketch the graph of the function.

37. $f(x) = 2x^3 - 3x^2 - 12x + 13$

38. $f(x) = x^3 - x$

39. $f(x) = x^{3/2} - 3x^{1/2}$

40. $f(x) = \dfrac{x^2}{8} - \dfrac{1}{x}$

41. $f(x) = 2\sqrt{x} - x$

42. $f(x) = 3x^4 - 4x^3$

43. $f(x) = \dfrac{1}{x^2 - 2x}$

44. $f(x) = \dfrac{2x^2 - 1}{x^2 - 1}$

45. $f(x) = \dfrac{x + 1}{\sqrt{x} - 1}$

46. $f(x) = \dfrac{x - 1}{\sqrt{x} + 1}$

47. Find the absolute maximum and minimum values of $f(x) = 2 - |1 - x|$ on the interval $[0, 2]$.

48. Find the absolute extrema of $f(x) = (x - 2)^{1/3} (2x - 2)^{2/3}$ on the interval $[0, 2]$.

49. Show that if $k > 0$, then the function described by $f(x) = x^3 + 3kx - 5$ has no relative extrema.

50. Sketch the graph of a function that has a relative maximum at $(0, 2)$, a relative minimum at $(2, 0)$, and a point of inflection at $(1, 1)$.

51. Sketch the graph of a function that is increasing on the intervals $(-\infty, -1)$ and $(3, \infty)$, and satisfies $\lim\limits_{x \to 6^+} f(x) = -\infty, \ \lim\limits_{x \to 6^-} f(x) = \infty$.

52. Let f be such that $f(0) = 1$ and $f'(x) = 2$ for all x. What is $f(x)$?

53. Let f be such that $f(0) = 1$ and $f'(x) = 3x^2$ for all x. What is $f(x)$?

4

APPLICATIONS OF THE DERIVATIVE

In the introduction to Chapter 3, we stated that the derivative is used in virtually every area of scientific study. In this chapter we consider these applications in more detail.

The first part of the chapter concerns physical applications of the derivative, primarily in scientific areas. Sections 4.4, 4.5, and 4.6 concern topics that are important in your subsequent study of mathematics, as well as having applications in their own right. Section 4.7 is devoted to an introduction of calculus techniques applied to business problems.

4.1
RECTILINEAR MOTION

One of the most important applications of the derivative concerns the motion of a particle that moves in a straight line. This motion is called **rectilinear motion.** An interest in methods for studying the nature of continuous motion was one of the primary reasons for developing the calculus.

Suppose a rock is dropped from a height of 1600 feet and takes 10 seconds to fall to the ground. We will show later in this section that the height of the rock at any time t, $0 \leq t \leq 10$, is given by the equation

$$s(t) = 1600 - 16t^2,$$

where t is measured in seconds and $s(t)$ in feet. The question we wish to consider is, "How fast is the rock moving at the end of 5 seconds?"

The average rate at which an object travels is found by dividing the distance traveled by the time required to travel that distance. **Velocity** describes this rate and also the direction in which the object is moving. The **speed** of an object is defined to be the magnitude or absolute value of its velocity. Since our problem involves an object that is falling to earth, the distance from the ground is decreasing with time, so the velocity at any time will be negative. For example, the average velocity of the rock from $t = 5$ seconds to $t = 6$ seconds is given by:

$$\begin{aligned}
\text{Average velocity} &= \frac{s(6) - s(5) \text{ ft}}{(6 - 5) \text{ sec}} \\
\text{(from } t = 5 \text{ to } t = 6) \\
&= \frac{(1600 - 16(6)^2) - (1600 - 16(5)^2) \text{ ft}}{1 \text{ sec}} \\
&= 1024 - 1200 \frac{\text{ft}}{\text{sec}} = -176 \frac{\text{ft}}{\text{sec}}.
\end{aligned}$$

This calculation gives the average velocity of the rock for the one-second interval from $t = 5$ seconds to $t = 6$ seconds. But we want to find the *instantaneous* velocity *at* the time $t = 5$ seconds, a somewhat different notion. The calculation shown above approximates this velocity. A better approximation is expected if the time interval is decreased. For example,

$$\begin{aligned}
\text{Average velocity} &= \frac{s(5.5) - s(5) \text{ ft}}{5.5 - 5 \text{ sec}} \\
\text{(from } t = 5 \text{ to } t = 5.5) \\
&= \frac{[1600 - 16(5.5)^2] - [1600 - 16(5)^2] \text{ ft}}{.5 \text{ sec}} \\
&= -168 \frac{\text{ft}}{\text{sec}},
\end{aligned}$$

a result that is likely to be closer to the instantaneous velocity.

To obtain increasingly better approximations to the instantaneous velocity at $t = 5$ seconds, we can consider successively smaller intervals of time and compute the average velocity on those intervals.

For the interval $t = 5$ to $t =$	The average velocity is
6.0	$\dfrac{s(6) - s(5)}{6 - 5} \dfrac{\text{ft}}{\text{sec}} = -176 \dfrac{\text{ft}}{\text{sec}}$
5.5	$\dfrac{s(5.5) - s(5)}{5.5 - 5} \dfrac{\text{ft}}{\text{sec}} = -168 \dfrac{\text{ft}}{\text{sec}}$
5.25	$\dfrac{s(5.25) - s(5)}{5.25 - 5} \dfrac{\text{ft}}{\text{sec}} = -164 \dfrac{\text{ft}}{\text{sec}}$
5.125	$\dfrac{s(5.125) - s(5)}{5.125 - 5} \dfrac{\text{ft}}{\text{sec}} = -162 \dfrac{\text{ft}}{\text{sec}}$
5.0625	$\dfrac{s(5.0625) - s(5)}{5.0625 - 5} \dfrac{\text{ft}}{\text{sec}} = -161 \dfrac{\text{ft}}{\text{sec}}$
5.03125	$\dfrac{s(5.03125) - s(5)}{5.03125 - 5} \dfrac{\text{ft}}{\text{sec}} = -160.5 \dfrac{\text{ft}}{\text{sec}}$

This table suggests that the concept of the limit can be used to define instantaneous velocity at $t = 5$ seconds:

$$\lim_{t \to 5} \frac{s(t) - s(5)}{t - 5} \frac{\text{ft}}{\text{sec}}.$$

This, however, is precisely the derivative of the function s at $t = 5$. Since $s(t) = 1600 - 16t^2$, the instantaneous velocity, denoted by $v(t)$, is:

$$v(t) = s'(t) = -32t \frac{\text{ft}}{\text{sec}}.$$

When $t = 5$, $v(5) = -160$ ft/sec, so the rock is traveling 160 ft/sec at the end of 5 seconds and the negative sign indicates correctly that the rock is falling.

The rate at which the instantaneous velocity of an object changes with respect to time is also of interest in physical problems. The derivative of the velocity gives this rate of change. We call $v'(t)$ the instantaneous acceleration of the object and denote $v'(t)$ by $a(t)$. For the problem we have been considering,

$$a(t) = v'(t) = -32 \frac{\text{ft/sec}}{\text{sec}} = -32 \frac{\text{ft}}{(\text{sec})^2}.$$

(4.1)
DEFINITION

Suppose $s(t)$ describes the rectilinear motion of an object. The instantaneous **velocity** of the object at time t is denoted $v(t)$ and defined by $v(t) = s'(t)$, provided that $s'(t)$ exists.

The instantaneous **acceleration** at time t is denoted $a(t)$ and defined by $a(t) = v'(t) = s''(t)$, provided $s''(t)$ exists.

When the terms velocity and acceleration are used without reference to either average or instantaneous, we will always assume that instantaneous is intended.

In rectilinear motion problems it is necessary to choose a positive and negative direction for the motion. We generally assume that the positive direction along a vertical line is upward. The direction to the right is usually taken as positive on a horizontal line.

EXAMPLE 1

Suppose a particle moves along a straight line in a manner described by $s(t) = t^3 - 12t^2 + 36t$, for $t \geq 0$, where t is measured in minutes (min) and $s(t)$ in centimeters (cm). Find the velocity and acceleration of the particle and describe its motion.

SOLUTION

The velocity at any time t is

$$v(t) = s'(t) = 3t^2 - 24t + 36 \frac{\text{cm}}{\text{min}}$$

and the acceleration is

$$a(t) = v'(t) = s''(t) = 6t - 24 \frac{\text{cm}}{\text{min}^2}.$$

We can use $v(t)$ to determine the direction of the particle. Since $v(t) = 3(t^2 - 8t + 12) = 3(t - 6)(t - 2)$, $v(2)$ and $v(6)$ are both zero and the particle is instantaneously stopped at $t = 2$ and at $t = 6$. Between $t = 0$ and $t = 2$ min, the particle travels in a positive direction; between $t = 2$ min and $t = 6$ min, the particle travels in a negative direction; and when $t > 6$, the particle again travels in a positive direction. Calculating $s(t)$ for $t = 0$, $t = 2$, $t = 6$, $t = 8$, we can sketch the pattern of motion of the particle shown in Figure 4.1. Although the actual path of the particle lies along the line s, we show the motion above this line so that overlapping paths can be distinguished.

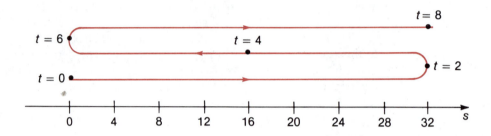

FIGURE 4.1

Since $a(t) = 6t - 24$ is negative when $0 \le t < 4$, the velocity of the particle is decreasing during that time. The velocity is increasing when $t > 4$ minutes. At $t = 4$, the velocity is not changing.   □

EXAMPLE 2   A ball is thrown upward from the earth with an initial velocity of 88 ft/sec. With the positive direction of motion assumed to be upward, the equation of distance from the earth is $s(t) = 88t - 16t^2$ feet. This equation is valid from time $t = 0$ until the ball returns to earth.
Find
(a) the velocity and acceleration of the ball at any time t.
(b) how many seconds it takes the ball to reach its highest point;
(c) how high the ball will go;
(d) how many seconds it takes the ball to reach the ground; and
(e) the velocity of the ball when it hits the ground.

SOLUTION

(a) $v(t) = s'(t) = 88 - 32t \dfrac{\text{ft}}{\text{sec}}$ and $a(t) = s''(t) = -32 \dfrac{\text{ft}}{\text{sec}^2}$.

(b) At the ball's highest altitude, $s(t)$ is a maximum and, consequently, a critical point of s must occur at that time. Therefore, the highest altitude occurs at the time when

$$0 = s'(t) = v(t) = 88 - 32t \qquad \text{or when} \qquad t = \frac{88}{32} = \frac{11}{4} \text{ sec.}$$

This implies that the ball is instantaneously stopped when it reaches its maximum height.

(c) The highest altitude is found by evaluating $s(t)$ at $t = 11/4$ sec,

$$s\left(\frac{11}{4}\right) = 88\left(\frac{11}{4}\right) - 16\left(\frac{11}{4}\right)^2 = 121 \text{ feet.}$$

(d) The ball is on the ground precisely when $s(t) = 0$.
Solving $0 = s(t) = 88t - 16t^2 = 8t(11 - 2t)$
for t, we have $t = 0$, the time when the ball was thrown, and $t = 11/2$, the time when it returns to earth.

(e) The velocity when the ball hits the ground is

$$v\left(\frac{11}{2}\right) = 88 - 32\left(\frac{11}{2}\right) = -88\,\frac{\text{ft}}{\text{sec}},$$

the negative of the velocity at which the ball was thrown.
A sketch of the motion of the ball is shown in Figure 4.2. □

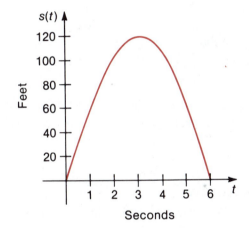

FIGURE 4.2

The acceleration of the ball in Example 2 is the same constant obtained in the initial example of dropping a rock. This constant, -32 ft/sec^2 (-9.8 meters/sec^2 if metric units are used), is the value generally assumed to be the acceleration of an object near the surface of the earth due to the gravitational force of the earth. The assumption $a(t) = -32$ ft/sec^2 can be used to determine equations for the velocity of an object, $v(t)$, and its distance, $s(t)$, above the earth.

Consider how the equation for distance given in Example 2 is obtained. Suppose at time $t = 0$ the ball is on the ground, that is, $s(0) = 0$, the initial velocity is 88 ft/sec, and $a(t) = -32$ ft/sec^2.

In Corollary 3.36 we saw that two functions with the same derivative can differ only by a constant. Since

$$D_t v(t) = a(t) = -32 = D_t(-32t),$$

a constant C exists with

$$v(t) = -32t + C.$$

But the initial velocity, $v(0)$, is 88 ft/sec, so

$$88 = v(0) = -32 \cdot 0 + C = C.$$

Thus,
$$v(t) = -32t + 88\frac{\text{ft}}{\text{sec}}.$$

Similarly,

$$D_t s(t) = v(t) = -32t + 88 = D_t(-16t^2 + 88t),$$

so a constant K exists with

$$s(t) = -16t^2 + 88t + K.$$

Since $s(0) = 0$,

$$0 = s(0) = -16 \cdot 0^2 + 88 \cdot 0 + K = K$$

and

$$s(t) = -16t^2 + 88t.$$

The process we used to recover $s(t)$ from its derivative is called **antidifferentiation.** This very important topic will be considered in detail in Chapter 5.

EXAMPLE 3   A child stands at the top of a cliff and throws a rock straight down into a pond 200 feet below. The rock hits the pond in 2 seconds. How fast was the rock thrown?

SOLUTION

We assume again that the earth's gravitational force is constant, $a(t) = -32\,\dfrac{\text{ft}}{\text{sec}^2}$. Since

$$v'(t) = a(t) = -32,$$
$$v(t) = -32t + C$$

for some constant C. In fact, the constant C is the initial velocity $v(0)$. In this case, however, $v(0)$ is our unknown, we will solve for its value later. Since

$$s'(t) = v(t) = -32t + v(0) = D_t(-16t^2 + v(0)t),$$
$$s(t) = -16t^2 + v(0)t + K$$

for some constant K. The rock is 200 ft above the pond at $t = 0$, so $s(0) = 200 = K$. Consequently,

$$s(t) = -16t^2 + v(0)t + 200.$$

The rock hits the pond in 2 seconds, so $s(2) = 0$.
Thus,

$$0 = -16(2)^2 + v(0) \cdot 2 + 200.$$

Solving this equation for $v(0)$ yields $v(0) = -68$ ft/sec.   □

The procedure used in Example 3 can be used to derive the velocity and distance functions for any free-falling object. If the object has initial velocity v_0 feet per second and height s_0 feet, its velocity at time t is

$$v(t) = v_0 - 32t \frac{\text{ft}}{\text{sec}}$$

and its height at time t is

$$s(t) = s_0 + v_0 t - 16t^2 \text{ ft.}$$

EXERCISE SET 4.1

In Exercises 1 through 8, $s(t)$ describes the distance of an object from a specified point at the end of time t. Find the instantaneous velocity and acceleration of the object at the time given.

1. $s(t) = 32t^2 + 3, t = 4$

2. $s(t) = 16t^2 + 300, t = 4$

3. $s(t) = t + \dfrac{1}{t}, t = 2$

4. $s(t) = \sqrt{t^2 + 1}, t = 6$

5. $s(t) = \sqrt[3]{t^2 + 2t + 1}, t = 0$

6. $s(t) = \dfrac{t + 1}{\sqrt{t^2 + 1}}, t = 1$

7. $s(t) = \dfrac{\sin t}{t + 1}, t = \pi$

8. $s(t) = \sin t + t \cos t, t = \pi/4$

In Exercises 9 through 16, $s(t)$ describes the distance of an object from a specified point at the end of time $t \geq 0$. Sketch a figure of the motion of the object similar to the figure given in Example 1. Determine if the object is ever momentarily stopped.

9. $s(t) = t^2 - 2t + 2$

10. $s(t) = t^2 - t$

11. $s(t) = t^3 - 3t^2 + 4$

12. $s(t) = t^3 + 2$

13. $s(t) = t^4 - 4t^3 + 4t^2 + 1$

14. $s(t) = 2t^3 - 9t^2 + 12t + 12$

15. $s(t) = \dfrac{t}{t + 1}$

16. $s(t) = \sin t, 0 \leq t \leq 2\pi$

17. A ball rolling down an inclined plane from a height of 100 feet is a distance $s(t) = 100 - 4t^2$ feet above the ground at any time t (in seconds) before it reaches the ground. Find the velocity and acceleration of the ball when it reaches the ground, and when it is 50 feet above the ground.

18. A ball thrown from the roof of a building is $s(t) = 245 + 24.5t - 4.9t^2$ meters above the ground t seconds after it is thrown. Determine the height of the building, the time it takes the ball to reach the ground, and the velocity when it hits the ground.

19. A rock is dropped from a height of 1600 feet and takes 10 seconds to fall to the ground. Find an equation for the distance of the rock above the ground at any time t. (Compare this equation to the equation on p. 181.)

20. What is the minimal initial velocity necessary to throw a ball 100 feet high? How long does it take the ball to reach 100 feet when thrown at this minimal initial velocity?

21. The child described in Example 3 throws another rock in a vertical direction so that it lands in the pool. This rock takes 4 seconds to reach the water. At what velocity was the rock thrown?

22. An unintelligent archer fires an arrow directly upward with a velocity of 20 meters/second. How much time does the archer have to find cover if he does not wish to be skewered?

23. A cat jumps vertically upward to reach the top of a ledge that is 5 feet above the original position of the cat. What is the minimal initial velocity the cat must have to reach the ledge?

24. A pole vaulter can clear a bar at a height of 16 feet. What must his initial velocity be to clear this height? What is the minimal time he must be in the air?

25. The acceleration due to gravity of an object on the moon is approximately one-sixth of the acceleration due to gravity on earth. If the pole vaulter in Exercise 24 can clear a height of 16 feet on earth, how high can he clear with the same vault if he is on the moon? How long does it take him to return to the moon's surface if he makes this vault?

26. Suppose that the pole vaulter described in the previous exercises is on the surface of a planet with approximately the same mass as the sun and whose gravitational force is twenty-four times that of the earth. Assuming again that the vaulter can clear a height of 16 feet on the earth, how high can he vault on this planet? How long would it take him to complete this vault?

27. A railroad company needs to install a signal that can be used to instruct a railway engineer whether to stop at an upcoming station. What is the least distance from the station the signal can be placed if it is to be used to stop a train traveling at 60 miles per hour, which has brakes that will slow the train down at 2 ft/sec^2 (that is, the acceleration is -2 ft/sec^2)?

28. Suppose the train in Exercise 27 travels at a maximum of 30 miles per hour. By what amount does this decrease the distance required for the placement of the signal?

29. A track star figures that he accelerates for the first 16 yards at a constant rate and then runs the remainder of a 100-yard dash with a constant velocity. He knows he can run the 100 yards in 10 seconds. What is his acceleration for the first 16 yards and how long does it take to run the first 16 yards?

30. The track star described in Exercise 29 has set a goal to run a mile at the same pace that he runs his 100 yard dash; that is, he will run the first 100 yards of the mile in 10 seconds and will cover the remaining 1660 yards with the same constant velocity that he ran the last 84 yards of the dash. If he succeeds, what will be his time for the mile? How likely is he to succeed?

31. An open construction elevator is ascending on the outside of a building at the rate of 8 ft/sec. A ball is dropped from the floor of the elevator when the elevator's floor is 80 feet above the ground. How long does it take the ball to reach the ground?

32. An open construction elevator is ascending on the outside of a building at the rate of 8 ft/sec. When the bottom of the elevator is 80 feet above the ground a ball is dropped from 160 feet above the floor of the elevator to the floor of the elevator. How long does it take the ball to reach the floor of the elevator?

33. A building consists of 140 stories, each story being the same height. An object is dropped from the roof of the building and is observed to take two seconds to pass from the 105th story to the 70th story. What is the height of the building?

34. A child shoots a ball bearing in a slingshot directly up in the air. A man on the third floor of a building sees the ball bearing ascend past his 4 foot window in 1/4 second. How long from the time he first sees the projectile is it safe for him to put his head out the window to yell at the child?

Putnam exercises:

35. A particle moving on a straight line starts from rest and attains a velocity v_0 after traversing a distance s_0. If the motion is such that the acceleration was never increasing, find the maximum time for the traverse.
(This exercise was problem B–2 of the thirty-third William Lowell Putnam examination given on December 2, 1972. The examination and its solution can be found in the November 1973 issue of the *American Mathematical Monthly,* pages 1019–1028.)

36. A circle stands in a plane perpendicular to the ground and a point A lies in this plane exterior to the circle and higher than its bottom. A particle starting from rest at A slides without friction down an inclined straight line until it reaches the circle. Which straight line allows descent in the shortest time? (The starting point A and the circle are fixed; the stopping point B is allowed to vary over the circle.)

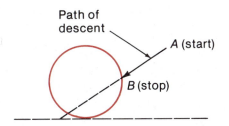

(This exercise was problem A–2 of the thirty-fifth William Lowell Putnam examination given on December 7, 1974. The examination and its solution can be found in the November 1975 issue of the *American Mathematical Monthly,* pages 907–912.)

4.2
APPLICATIONS INVOLVING RELATIVE MAXIMA AND MINIMA

One of the most important applications of the derivative concerns the solution of problems that call for minimizing or maximizing a quantity. For example, businesses are concerned with maximizing profit and minimizing loss; a motorist might want a route that minimizes the distance traveled or perhaps the travel time; a homeowner wants to minimize the amount of heating fuel used.

By introducing variables to represent certain quantities, each of these problems becomes a particular case of a general type of mathematical problem in which it is required to find an absolute maximum or an absolute minimum. The methods involved in the solutions of problems of this type can best be outlined after considering some examples.

EXAMPLE 1   Find two positive real numbers whose product is 16 and whose sum is a minimum.

SOLUTION

We first translate this verbal problem into mathematical language by attaching labels (assigning variables) to the quantities we wish to find. In this example we let x denote one of the numbers to be found and y the other. Then $xy = 16$. The quantity we wish to minimize is the sum $S = x + y$. Since $y = 16/x$, this sum can be expressed as a function of x,

$$S(x) = x + \frac{16}{x}.$$

The essential elements of this problem are:

Know	Find
(a) $y = \dfrac{16}{x}$, and	(1) the value of x that minimizes $S(x)$, and
(b) $S = x + y$	(2) the value of y corresponding to the value of x found in (1).
From (a) and (b):	
(c) $S(x) = x + \dfrac{16}{x}$.	

The problem has been reduced to finding where the absolute minimum of the function S occurs. An absolute minimum can occur only at a critical point or at an endpoint of the domain of S. Since

$$S'(x) = 1 - \frac{16}{x^2},$$

the only positive critical point of S is $x = 4$.

The second derivative of S,

$$S''(x) = \frac{32}{x^3},$$

shows that S has a relative minimum at $x = 4$, since $S''(4) = 1/2 > 0$.

Recalling that x must be a positive real number, we see that the domain of S is $(0, \infty)$ and there are no endpoint extrema. Hence, $x = 4$ is also an absolute minimum for S. The pair of numbers whose product is 16 and sum is a minimum is $x = 4$ and $y = 16/4 = 4$. □

EXAMPLE 2   Farmer MacDonald has 300 feet of chicken wire to use for constructing a rectangular pen for the chickens. The edge of a 300 foot long barn is to be used as one side of the pen, so the wire is needed only to fence in the remaining three sides. How can this be accomplished in a way that will maximize the amount of space the birds have to roam?

SOLUTION

We construct a picture of the situation confronting Farmer MacDonald, labeling the unknown quantities. See Figure 4.3.

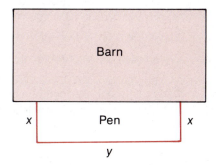

FIGURE 4.3

In this way we can write precisely what is known and the mathematical problem that needs to be solved. It also gives us time to reflect on the problem and determine any logical flaws it might have.

Know	Find
(a) $2x + y = 300$, and	(1) the value of x that maximizes $A(x)$, and
(b) area of pen is $A = xy$. From (a) and (b) (c) $A(x) = x(300 - 2x)$.	(2) the value of y corresponding to the value of x found in (1).

Notice that for a pen to be formed the values of x must lie between 0 and 150.

Since
$$A(x) = x(300 - 2x),$$
$$A'(x) = 300 - 2x + x(-2) = 300 - 4x$$

and $A'(x) = 0$ implies $x = 75$. The second derivative $A''(x) = -4$ is always negative so A has a relative maximum at $x = 75$. The absolute maximum value of A can occur only at this relative maximum or at the endpoints of its domain.

At the endpoints, 0 and 150, $A = 0$, so A has an absolute maximum at $x = 75$.

Consequently, Farmer MacDonald can maximize the space inside the pen by making the sides adjacent to the barn 75 feet long and the remaining side 150 feet long. □

After these two examples you have probably noticed some common steps in the solutions.

1. If possible, draw a picture to illustrate the problem and label the pertinent parts.
2. Write down any relationships between the variables and constants in the problem.
3. Express the quantity to be maximized or minimized as a function of one of the variables.
4. Find the critical points of the function.
5. Use the first derivative test or second derivative test to determine whether the critical points are relative maxima or relative minima.
6. If the domain of the function is a closed interval, determine whether extrema occur at the endpoints.

EXAMPLE 3   A rectangle is to be inscribed in a semicircle with radius 2 inches. Find the dimensions of the rectangle that encloses the maximum area.

SOLUTION

By drawing a picture of the physical situation and labeling the various parts, we can summarize the problem's essential elements. See Figure 4.4.

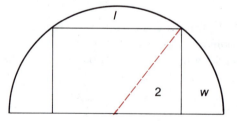

FIGURE 4.4

Know	Find
(a) area of the rectangle is: $A = lw$, and	(1) the value of w that maximizes $A(w)$, and
(b) $\left(\frac{1}{2}l\right)^2 + w^2 = 4$ so $l = 2\sqrt{4 - w^2}$.	(2) the value of l corresponding to the value of w found in (1).

From (a) and (b):

(c) $A(w) = 2w\sqrt{4 - w^2}$.

To maximize A, we first find any critical points of A. Since

$$A'(w) = 2\sqrt{4 - w^2} + 2w\left(\frac{1}{2}\right)(4 - w^2)^{-1/2}(-2w)$$

$$= \frac{8 - 4w^2}{\sqrt{4 - w^2}},$$

$A'(w)$ is undefined when $w = \pm 2$ and $A'(w) = 0$ when $w = \pm\sqrt{2}$. However, the domain of A is [0, 2], so the only critical points are $w = 2$ and $w = \sqrt{2}$.

The first derivative test ensures that A has a relative maximum at $w = \sqrt{2}$: when $0 < w < \sqrt{2}$, $A'(w) > 0$ and when $\sqrt{2} < w < 2$, $A'(w) < 0$. To check endpoint extrema, we need only note that when $w = 0$ and when $w = 2$, $A = 0$. Consequently A has an absolute maximum when $w = \sqrt{2}$.

The value of l corresponding to this value of w is $l = 2\sqrt{4 - w^2} = 2\sqrt{2}$. $\square$

EXAMPLE 4   Find the ratio of the radius to the height of a right circular cylinder that has minimal surface area, assuming that the volume of the cylinder is fixed and that the cylinder has a top and a bottom.

SOLUTION

With the notation and drawing given in Figure 4.5, the problem can be summarized as follows:

Know	Find
(a) $V = \pi r^2 h$, (V is a constant)	(1) the value of r that minimizes $S(r)$,
(b) $S = 2\pi rh + 2\pi r^2$, and	(2) the value of h corresponding to the
From (a) and (b):	value of r found in (1), and
(c) $S(r) = 2\pi r\left(\dfrac{V}{\pi r^2}\right) + 2\pi r^2$	(3) the ratio of the values of r and h found in (1) and (2).
$\quad = \dfrac{2V}{r} + 2\pi r^2.$	

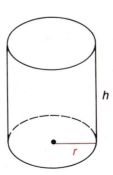

FIGURE 4.5

Calculating

$$S'(r) = \frac{-2V}{r^2} + 4\pi r,$$

we see that $S'(r) = 0$ implies

$$\frac{-2V}{r^2} + 4\pi r = 0,$$

$$4\pi r^3 = 2V$$

so

$$r = \sqrt[3]{\frac{V}{2\pi}}.$$

Applying the second derivative test ensures that S has a relative minimum at

$$r = \sqrt[3]{\frac{V}{2\pi}}:$$

$$S''(r) = \frac{4V}{r^3} + 4\pi$$

$$S''\left(\sqrt[3]{\frac{V}{2\pi}}\right) = \frac{4V}{V/(2\pi)} + 4\pi = 8\pi + 4\pi = 12\pi > 0.$$

Since the domain of S is $(0, \infty)$, there are no endpoint extrema and S has an absolute minimum at $r = \sqrt[3]{V/(2\pi)}$. The value of h corresponding to this value of r is

$$h = \frac{V}{\pi r^2} = \frac{V}{\pi\left(\sqrt[3]{V/(2\pi)}\right)^2} = \frac{V}{\pi\left(V/(2\pi)\right)^{2/3}} = \sqrt[3]{\frac{4V}{\pi}},$$

and the ratio of the radius to the height is

$$\frac{r}{h} = \frac{\sqrt[3]{V/(2\pi)}}{\sqrt[3]{4V/\pi}} = \sqrt[3]{\frac{V}{2\pi} \cdot \frac{\pi}{4V}} = \sqrt[3]{\frac{1}{8}} = \frac{1}{2}. \qquad \square$$

EXAMPLE 5   A beer distributor has determined that 200 gallons of Old Horse Light beer can be sold to Leo's Tavern if the price is $2 per gallon. For each cent per gallon that the price is lowered, 10 more quarts of beer are sold. At what price should beer be sold in order to maximize the money received by the distributor?

SOLUTION

Let x denote the number of cents per gallon by which the price is lowered. Then, the price is $(2.00 - .01x)$ dollars/gallon and the amount of Old Horse sold is

$$200 \text{ gal} + 10 \, x \text{ qt} = (200 + 2.5x) \text{ gallons}.$$

Assuming that the beer is not given away, $0 \le x < 200$. Summarizing this information, we have:

Know	Find
(a) price is $(2.00 - .01x)$ dollars/gallon,	(1) the value of x that maximizes $R(x)$, and
(b) amount sold is $(200 + 2.5x)$ gallons, and	(2) the price corresponding to the value of x found in (1).
(c) revenue earned is $R(x) = (2.00 - .01x)(200 + 2.5x)$ dollars.	

To maximize R, we need the critical points of R.

Since
$$\begin{aligned}
R'(x) &= -.01(200 + 2.5x) + 2.5(2.00 - .01x) \\
&= -2 - .025x + 5 - .025x \\
&= 3 - .05x,
\end{aligned}$$

the only critical point occurs when

$$0 = 3 - .05x$$

or at
$$x = 60.$$

Since $R''(x) = -.05$, the second derivative test implies that R has a relative maximum at $x = 60$.

The price corresponding to $x = 60$ is

$$2.00 - (.01)(60) = \$1.40,$$

and the revenue produced at that price is

$$R(60) = (2.00 - .60)[200 + 2.5(60)] = \$490.$$

Checking for endpoint extrema we find that the revenue at $x = 0$ is \$400. Therefore, the maximum revenue occurs when the price is \$1.40 per gallon. □

EXAMPLE 6   The problem considered in Example 5 would be more realistic if instead of maximizing the revenue received, we could determine the price that would maximize the profit. Suppose the distributor must pay \$.90 per gallon to the Old Horse brewery. At what price should the beer be sold to maximize the profit, assuming that the other conditions in Example 5 are the same?

SOLUTION

Let x denote the number of cents per gallon by which the price is lowered. The profit per gallon is the price per gallon minus the cost per gallon to the distributor; that is, the profit earned per gallon is $[(2.00 - .01x) - .90]$ dollars.

Know	Find
(a) price is $(2.00 - .01x)$ dollars/gallon,	(1) the value of x that maximizes $P(x)$, and
(b) profit earned is $[(2.00 - .01x) - .90]$ dollars/gallon,	(2) the price corresponding to the value of x found in (1).
(c) amount sold is $(200 + 2.5x)$ gallons, and	
(d) total profit earned is $P(x) = (1.10 - .01x)(200 + 2.5x)$ dollars.	

Since
$$P'(x) = -.01(200 + 2.5x) + 2.5(1.10 - .01x)$$
$$= -2 - .025x + 2.75 - .025x$$
$$= .75 - .05x,$$

a critical point of P occurs when

$$0 = .75 - .05x$$

or $$x = 15.$$

The price corresponding to $x = 15$ is

$$(2.00 - .01(15)) \text{ dollars/gallon} = \$1.85 \text{ per gallon}$$

and the corresponding total profit earned is

$$P(15) = (1.10 - .15)(200 + 37.50) = \$225.62.$$

The domain of P is $[0, 200)$. (Note that $x > 110$ produces negative profit.) The profit at $x = 0$ is $220, which is less than the profit at the critical point $x = 15$. Consequently, the selling price to maximize the profit is $1.85 per gallon.   □

EXAMPLE 7   The mathematically inclined Farmer MacDonald needs to build a trough to slop hogs. The trough is to be made from three oak 1×10's (1 inch by 10 inches), each eight feet long. At what angle should the sides of the trough be sloped to enclose the maximum quantity of swill? How much will this trough hold?

SOLUTION

A cross-section of a typical trough is shown in Figure 4.6. The capacity of the trough is maximized when the cross-sectional area, A, is maximized.

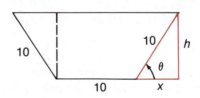

FIGURE 4.6

The problem is expressed concisely as:

Know	Find
(a) $A = 10h + 2\left(\frac{1}{2}xh\right),$	(1) the value of θ that maximizes $A(\theta)$,
(b) $h = 10 \sin \theta,$	(2) the maximum value of A, and
(c) $x = 10 \cos \theta,$ and	(3) the maximum volume.
(d) $0 < \theta \le \dfrac{\pi}{2}.$	

From (a), (b), (c) and (d)

(e) $A(\theta) = 100 \sin \theta + 100 \sin \theta \cos \theta.$

$$
\begin{aligned}
A'(\theta) &= 100 \cos \theta + 100((\cos \theta)^2 - (\sin \theta)^2) \\
&= 100 \cos \theta + 100(\cos \theta)^2 - 100(1 - (\cos \theta)^2) \\
&= 200(\cos \theta)^2 + 100 \cos \theta - 100,
\end{aligned}
$$

so critical points of A occur when

$$
\begin{aligned}
0 &= 200(\cos \theta)^2 + 100 \cos \theta - 100 \\
&= [2(\cos \theta)^2 + \cos \theta - 1](100) \\
&= [(2 \cos \theta - 1)(\cos \theta + 1)](100).
\end{aligned}
$$

Thus,

$$
\cos \theta = \frac{1}{2}, \qquad \text{so } \theta = \frac{\pi}{3} \text{ or } \frac{2\pi}{3},
$$

or

$$
\cos \theta = -1, \qquad \text{so } \theta = \pi.
$$

Only $\theta = \dfrac{\pi}{3}$ is in the domain of A and

$$
\begin{aligned}
A\left(\frac{\pi}{3}\right) &= 100 \sin \frac{\pi}{3} + 100 \sin \frac{\pi}{3} \cos \frac{\pi}{3} \\
&= 100\left(\frac{\sqrt{3}}{2}\right) + 100\left(\frac{\sqrt{3}}{2}\right)\left(\frac{1}{2}\right) \\
&= 75\sqrt{3}.
\end{aligned}
$$

Considering the value of A at the endpoint $\theta = \dfrac{\pi}{2}$, we see that

$$
A\left(\frac{\pi}{2}\right) = 100 < A\left(\frac{\pi}{3}\right).
$$

Hence,

$$
A\left(\frac{\pi}{3}\right) = 75\sqrt{3} \text{ in}^2 = \frac{75\sqrt{3}}{144} \text{ ft}^2
$$

is the maximum cross-sectional area that can be produced. The maximum volume is

$$V = \frac{75\sqrt{3}}{144} \cdot 8 \approx 7.22 \text{ ft}^3.$$

Since one cubic foot is approximately 7.48 gallons, the maximum volume is approximately 54 gallons. (This is a quantity sufficient to feed 71 adult hogs, if filled to capacity once a day).   □

EXERCISE SET 4.2

1. Find the positive number with the property that the sum of the number and its reciprocal is a minimum.

2. Find the positive number with the property that the sum of the square of the number and the square of its reciprocal is a minimum.

3. A rectangular plot of ground containing 432 square feet is to be fenced off in a large lot. Find the dimensions of the plot that requires the least amount of fence.

4. A rectangular plot of ground containing 432 square feet is to be fenced off in a large lot and a fence is to be constructed down the middle of the lot to separate it into equal parts. Find the dimensions of the plot that requires the minimal amount of fencing.

5. Suppose the fence being used to enclose the plot of ground described in Exercise 4 costs $10 per foot and the fence used to divide the plot into parts costs $5 per foot. Find the dimensions of the plot that requires the least expense for fencing.

6. A rectangular dog run is to contain 864 square feet. If the dog's owner must pay for the fencing, what should be the dimensions of the run to minimize cost? Suppose a neighbor has agreed to let the owner use an already constructed fence for one side of the run. What should the dimensions of the run be in this situation if the owner's cost is to be a minimum?

7. A rectangular box with no top is to contain 2250 in^3. Find the dimensions to minimize the amount of material used to construct the box if the length of the base is three times the width.

8. Suppose the box described in Exercise 7 is to be constructed with a top. What dimensions would minimize the amount of material required?

9. The United States Postal Service has decreed that no rectangular-shaped parcel can be mailed if the total of its length and girth exceeds 84 inches. Excepted are post offices serving an area with less than 600 delivery units, where packages can be mailed whose length and girth do not exceed 100 inches. Find the maximum volume that can be mailed from each type of post office in a rectangular parcel if the cross-section of the parcel is a square.

10. The speed of a point on the rim of a flywheel t seconds after the flywheel starts to turn is given by the formula $v = 36t^2 - t^3$ feet per second. Find its greatest speed. How long was it running before it reached this speed?

11. The turning effect of a ship's rudder is known to be $T = k \cos \theta (\sin \theta)^2$, where k is a positive constant and θ is the angle that the direction of the rudder makes with the keel line of the ship ($0 < \theta < \pi/2$). For what value of θ is the rudder most effective?

12. A one-mile race track is to be built with two straight sides and semicircles at the ends. What is the maximum amount of area needed to construct the track? What is the minimum amount of area that can be used to construct the track?

13. The following problem was described in Example 8 of Section 1.1. A charter bus company charges $10 per person for a round trip to a ball game with a discount given for group fares. A group purchasing more than ten tickets at one time receives a reduction per ticket of twenty-five cents times the number of tickets purchased in excess of ten. Determine the maximum revenue that can be received by the bus company.

14. The following problem was described in Exercise 42 of Section 1.2. A hotel normally charges $40 for a room; however, special group rates are advertised: If the group requires more than 5 rooms, the price for each room is decreased by one dollar times the number of rooms exceeding five, with the restriction that the minimum price per room is $20. Find the maximum revenue that the hotel can receive from a group.

15. A standard can contains a volume of 900 cm^3. The can is in the shape of a right circular cylinder with a top and bottom. Find the dimensions of the can that will minimize the amount of material needed for construction.

16. In constructing a can in the shape of a right circular cylinder, little waste is produced when the side of the can is cut, but the top and bottom are each stamped from a square sheet and the remainder is wasted. Find the relative dimensions of the can that uses the least amount of material under this construction method.

17. An open rectangular box is to be made from a piece of cardboard 8 inches wide and 8 inches long by cutting a square from each corner and bending up the sides. Find the dimensions of the box with the largest volume.

18. A rectangular plot is to be laid out that will contain a vineyard of one acre in area (43,560 square feet). The vineyard must have a boundary of 8 feet on all sides for equipment to pass and an 8-foot pathway down the middle. What is the minimal acreage required for this plot?

19. Suppose that (x_0, y_0) is a point that does not lie on the circle $x^2 + y^2 = 1$. Show that the shortest distance from (x_0, y_0) to the circle is along a line that passes through the center of the circle.

20. Show that the shortest distance from the point (x_0, y_0) to the line with equation $Ax + By + C = 0$ is

$$\frac{Ax_0 + By_0 + C}{A^2 + B^2}.$$

21. A rectangle is to be placed inside a circle of radius r with its corners on the boundary of the circle. Of all such rectangles, find the dimensions of the one that encloses the maximum area.

22. A rectangle is to be placed inside a circle of radius r with its corners on the boundary of the circle. What dimensions should be given to the rectangle to maximize the sum of its perimeter and the length of its two diagonals? What dimensions would minimize this sum?

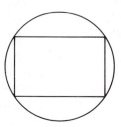

23. An isosceles triangle is to be placed inside a circle of radius r with its vertices on the boundary of the circle. How should this be accomplished if the area of the triangle is to be maximized?

24. "The isosceles triangle with two fixed equal sides and maximum area is not an equilateral triangle." Show that this statement is true by finding the length of the base of an isosceles triangle that maximizes the area over all such triangles.

25. The area of the print on a book page is 42.5 in^2. The margins are 1 inch on the sides and bottom and 1/2 inch at the top. What should the dimensions of a page of this book have been if the only object was to use the minimal amount of paper?

26. A field is to be fenced off in the form of a rectangle containing 10,000 square feet. In addition to the fencing required for the perimeter of the field, an isosceles triangle is to be fenced off in one corner by running a fence from the midpoint on the shortest side to the adjacent side enclosing the corner of the rectangle. Find the dimensions of the field that minimizes the amount of fencing required.

27. A wire 1-foot long is to be cut in two pieces: one piece is used to construct a square, the other to construct an equilateral triangle. Where should the cut be made to minimize the sum of the areas of the figures?

28. A wire 1-foot long is to be cut in two pieces: one piece is used to construct a square, the other to construct a circle. Where should the cut be made to minimize the sum of the areas of the figures?

29. Find the volume of the largest right circular cylinder that can be placed inside a sphere of radius 1.

30. Find the volume of the largest right circular cone that can be placed inside a sphere of radius 1.

31. The strength of a rectangular beam is directly proportional to the product of the width of the beam and the square of its depth. Find the dimensions of the strongest beam that can be cut from a log with radius r inches.

32. The stiffness of a rectangular beam is directly proportional to the product of the width of the beam and the cube of its depth. Find the dimensions of the stiffest beam that can be cut from a log with radius r inches.

33. Refer to Problem 42 in Exercise Set 1.1, on page 9. This problem involved running a power line from a utility pole to a house. At what point on the driveway should the switch from overhead to underground be made in order to minimize the cost?

34. A crew of painters is assigned to paint a second floor wall on the outside of a building along a busy sidewalk. They must leave a corridor, for unsuperstitious pedestrians, between the wall and their ladders. The corridor is 6 feet wide and 8 feet high. What is the minimal length of ladder they can use to reach the wall and how far from the base of the wall should it be placed?

35. A mugger armed with a knowledge of calculus is planning an attack on his next victim. The attack must be made on a sidewalk between two lights that are 200 feet apart, one of which is twice as bright as the other. Before dropping out of high school the mugger took a physics course and recalls that the intensity of illumination from a light varies inversely as the square of the distance from the light. Where will he attack if he always attacks at the darkest point between the lights?

36. Refer to Problem 36 in Exercise Set 1.1, on page 9. This problem involved two tankers in the Atlantic ocean. At what time are the tankers closest together and what is the minimal distance separating them?

37. A warehouse is to be built beside a long straight highway running north and south. This warehouse will house equipment produced in two factories and sent there by air for storage. The northern factory lies 80 miles east of the highway; the other lies on the highway. The point on the highway that is closest to the northern factory is 100 miles north of the second factory. Where should the warehouse be located if it is to minimize the sum of the air distances from the warehouse to the two factories?

38. Consider the problem described in Exercise 37 with the following modification: The southern factory lies 80 miles west of the highway instead of beside the highway. Where should the warehouse now be located along the highway if the sum of the air distances from the warehouse to the two factories is to be minimized?

39. The Boondockia Outfitting Corporation has decided to produce and market a small backpacking tent. Their first task is to determine the dimensions of the tent requiring the minimal amount of material to construct that satisfies the following conditions: (a) the tent must have two sides, two ends and a bottom; (b) the volume of the tent must be at least 100 cubic feet; (c) the tent must be eight feet long; (d) the cross-section of the tent parallel to each end must be in the shape of a triangle. Find the dimensions of the tent using the minimal amount of material that satisfies these requirements.

40. Boondockia Inc. has decided that the tent described in Exercise 39 using the minimal amount of material is not marketable because it does not have enough floor space. They have decided instead to design the tent so that it is six feet wide at the bottom and has a cross section in the form of a rectangle topped by a triangle. The minimal height of the rectangle is to be one foot and the total height to the center of the tent at least 30 inches. Assuming all other specifications are the same as those in Exercise 39, how should the tent be designed to minimize the amount of material required?

41. The law of reflection proposed by Euclid in *Catoptrics*, states that when light strikes a flat reflecting surface, the angle of incidence θ_i and angle of reflection θ_r are equal. Prove this law by using Fermat's principle that light will travel from point P to point Q along the path that requires the least time. See the accompanying figure.

4.3
RELATED RATES

The derivative has been applied to problems that describe the slope of a graph of a function at a point and that determine the instantaneous velocity of an object traveling in a straight line. The common factor in these two applications is that the derivative gives a measure of the instantaneous change in one quantity with respect to the change in another. The slope of the graph of a function at a point measures the change in the function values with respect to the change in its independent variable. The velocity of an object measures the change in distance the object has traveled with respect to time.

In a large class of physical problems, functions are involved in which both the independent variable and the value of the function are dependent upon time. Suppose y is a function of x and t is a time variable on which both x and y depend; problems involving relationships between dx/dt and dy/dt are called **related rate problems**.

EXAMPLE 1   If the radius of a circle is increasing at the rate of 7 cm/sec, how fast is its area increasing when the radius is 20 cm?

SOLUTION

If r denotes the radius of the circle, then the area is $A(r) = \pi r^2$. The essential information concerning this problem is concisely presented below.

Know	Find
(a) $A(r) = \pi r^2$, and	(1) $\dfrac{dA}{dt}$, and
(b) $\dfrac{dr}{dt} = 7 \dfrac{\text{cm}}{\text{sec}}$.	(2) $\dfrac{dA}{dt}$ when $r = 20$ cm·

Since we know $\dfrac{dr}{dt}$, we will use the chain rule to find $\dfrac{dA}{dt}$:

$$\frac{dA}{dt} = \frac{dA}{dr} \cdot \frac{dr}{dt}$$

$$= (2\pi r)\,(7)\,\frac{cm^2}{sec}$$

$$= 14\pi r\,\frac{cm^2}{sec}.$$

Thus, $\dfrac{dA}{dt}$ (when $r = 20$ cm) $= 14\pi(20) = 280\,\dfrac{cm^2}{sec}.$ □

EXAMPLE 2   A painter is painting a house using a ladder 15 feet long. A dog runs by the ladder dragging a leash that catches the bottom of the ladder and drags it directly away from the house at 22 ft/sec. Assuming that the ladder continues to be pulled away from the wall at this speed, how fast is the top of the ladder moving down the wall when the top is 5 feet from the ground?

SOLUTION
Figure 4.7 describes this situation.

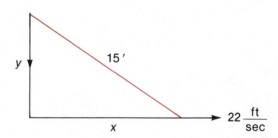

FIGURE 4.7

The information given in the problem can be expressed as follows:

Know	Find
(a) $x^2 + y^2 = (15)^2$,	(1) $\dfrac{dy}{dt}$, and
(b) $\dfrac{dx}{dt} = 22\,\dfrac{ft}{sec}$, and	(2) $\dfrac{dy}{dt}$ when $y = 5$ ft.
(c) when $y = 5$ ft,	
$\quad x = \sqrt{225 - 25} = \sqrt{200}$ ft.	

From (a) we have $y = \sqrt{225 - x^2}$ and using the chain rule,

$$\frac{dy}{dt} = \frac{dy}{dx}\frac{dx}{dt}$$

$$= \frac{1}{2}(225 - x^2)^{-1/2}(-2x)\frac{dx}{dt}$$

$$= \frac{-x}{\sqrt{225 - x^2}}\frac{dx}{dt},$$

so

$$\frac{dy}{dt} = \frac{-x}{y}\, 22\,\frac{ft}{sec}.$$

The velocity of the top of the ladder when it is 5 ft above the ground is therefore

$$\frac{dy}{dt}\text{ (when } y = 5) = \frac{-\sqrt{200}}{5}\,(22)\,\frac{ft}{sec} \approx -62.23\,\frac{ft}{sec}.$$

The negative value for the velocity indicates correctly that the distance is decreasing with time.   □

Implicit differentiation could also have been used to determine $\dfrac{dy}{dt}$ in Example 2. From (a)

$$\frac{d}{dt}(x^2 + y^2) = \frac{d}{dt}(225)$$

so

$$2x\frac{dx}{dt} + 2y\frac{dy}{dt} = 0$$

and

$$\frac{dy}{dt} = \frac{-x}{y}\frac{dx}{dt}.$$

Implicit differentiation is a useful tool to employ in many problems involving related rates. It is required in the next example.

EXAMPLE 3   For the situation described in Example 2, find how fast the angle between the ladder and the ground is changing when the ladder is 5 feet from the ground.

SOLUTION

Let θ denote the angle between the ladder and ground. The problem is to find $d\theta/dt$ given dx/dt, so we need an equation that relates θ and x.

Know	Find
(a) $\cos \theta = \dfrac{x}{15}$ and	(1) $\dfrac{d\theta}{dt}$, and
(b) $\dfrac{dx}{dt} = 22 \dfrac{\text{ft}}{\text{sec}}$.	(2) $\dfrac{d\theta}{dt}$ when $y = 5$ ft.

Taking the derivative of both sides of the equation in (a) with respect to t, we have

$$\frac{d \cos \theta}{dt} = \frac{d\left(\dfrac{x}{15}\right)}{dt},$$

$$-\sin \theta \frac{d\theta}{dt} = \frac{1}{15} \cdot \frac{dx}{dt}.$$

Consequently,

$$\frac{d\theta}{dt} = -\frac{1}{15 \sin \theta} \frac{dx}{dt}$$

$$= \frac{-22}{15 \sin \theta}.$$

When $y = 5$ feet, $\sin \theta = \dfrac{1}{3}$, so

$$\frac{d\theta}{dt} = \frac{-22}{15(1/3)} = -\frac{22}{5}.$$ □

Since θ is given in radians, the units for $d\theta/dt$ are radians/sec.

EXAMPLE 4   Two cars approach an intersection at right angles. One car is traveling 50 mph and the other is traveling 40 mph. How fast are the cars approaching each other when the first car is 30 feet from the intersection and the second is 40 feet from the intersection?

SOLUTION

Notice that the speed of the cars is given in terms of miles per hour, while the distances are given in feet. In order to have consistent units, we will transform all the dimensions into either feet or miles. Using

$$1 \frac{\text{mi}}{\text{hr}} = \frac{5280 \text{ ft}}{3600 \text{ sec}} = \frac{22}{15} \frac{\text{ft}}{\text{sec}},$$

we will translate the terms involving miles per hour into feet per second.

In Figure 4.8, x denotes the distance from the intersection for the first car, y the distance from the intersection for the second car, and z the distance between the cars.

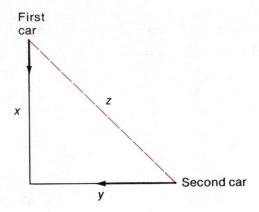

FIGURE 4.8

Consequently, we have the following outline:

Know	Find
(a) $z^2 = x^2 + y^2$, and	(1) $\dfrac{dz}{dt}$, and
(b) $\dfrac{dx}{dt} = -50\dfrac{\text{mi}}{\text{hr}} = -50\left(\dfrac{22}{15}\right)\dfrac{\text{ft}}{\text{sec}}$	(2) $\dfrac{dz}{dt}$ when $x = 30$ ft
$\qquad = \dfrac{-220}{3}\dfrac{\text{ft}}{\text{sec}}$, and	$\qquad$ and $y = 40$ ft.
(c) $\dfrac{dy}{dt} = -40\dfrac{\text{mi}}{\text{hr}} = -40\left(\dfrac{22\text{ ft}}{15\text{ sec}}\right) = \dfrac{-176}{3}\dfrac{\text{ft}}{\text{sec}}$.	

The derivatives of x and y are negative since these distances are decreasing with time.

We will apply implicit differentiation to the equation $z^2 = x^2 + y^2$ to find dz/dt. It is possible to find dz/dt without using implicit differentiation, but more work is required.

Taking the derivative of both sides of the equation $z^2 = x^2 + y^2$ with respect to t and using the chain rule:

$$2z\frac{dz}{dt} = 2x\frac{dx}{dt} + 2y\frac{dy}{dt}$$

and

$$\frac{dz}{dt} = \frac{x}{z}\frac{dx}{dt} + \frac{y}{z}\frac{dy}{dt}$$

$$= -\frac{220}{3}\frac{x}{z} - \frac{176}{3}\frac{y}{z}\frac{\text{ft}}{\text{sec}}.$$

When $x = 30$ ft and $y = 40$ ft,

$$z = \sqrt{x^2 + y^2} = \sqrt{900 + 1600} = 50 \text{ ft}$$

and $$\frac{dz}{dt} = -\frac{220}{3}\left(\frac{30}{50}\right) - \frac{176}{3}\left(\frac{40}{50}\right) = -90.93\,\frac{\text{ft}}{\text{sec}}.$$

Translating back into miles per hour, if desired, we find that the cars are approaching each other at 90.93 (15/22) ≈ 62 mph.   □

EXAMPLE 5   A revolving beacon located 3 miles from a straight shoreline makes 2 revolutions per minute. Find the speed of the spot of light along the shore when it is two miles from the point on the shore nearest the light.

SOLUTION

The problem is illustrated in Figure 4.9.

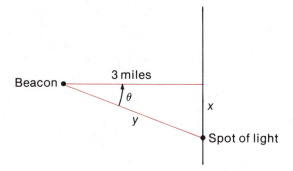

FIGURE 4.9

Since the beacon makes 2 revolutions per minute and one revolution is 2π radians, $d\theta/dt = 4\pi$ radians/minute. The rate at which the light is moving along the shore is given by $\dfrac{dx}{dt}$, the instantaneous rate of change of the distance x with respect to time t.

Know	Find
(a) $\dfrac{d\theta}{dt} = 4\pi$, and	(1) $\dfrac{dx}{dt}$, and
(b) $\tan \theta = \dfrac{x}{3}$.	(2) $\dfrac{dx}{dt}$, when $x = 2$.

Taking the derivative of both sides of the equation in (b), we have

$$(\sec \theta)^2 \frac{d\theta}{dt} = \frac{1}{3} \frac{dx}{dt}.$$

Solving for $\dfrac{dx}{dt}$ yields:

$$\frac{dx}{dt} = 3(\sec \theta)^2 \frac{d\theta}{dt} = 3(\sec \theta)^2 \, 4\pi.$$

To find dx/dt when $x = 2$, we must first find $(\sec \theta)^2$ when $x = 2$. Since $\tan \theta = x/3 = 2/3$ and $(\tan \theta)^2 + 1 = (\sec \theta)^2$, $(\sec \theta)^2 = (2/3)^2 + 1 = 13/9$.

Consequently,

$$\frac{dx}{dt} = 3\left(\frac{13}{9}\right)4\pi = \frac{52}{3}\pi \, \frac{\text{mi}}{\text{min}}. \qquad \square$$

EXAMPLE 6   A straw is used to drink water from a straight-sided cup that has a diameter of 3 inches at the bottom, 4 inches at the top, and a height of 8 inches. The liquid is being consumed at the rate of 4 cubic inches per second. How fast is the level of the water dropping when there is a 3-inch depth of water left in the cup?

SOLUTION

The shape of the cup is called the frustum of a right circular cone. The volume of such a solid is $V = \pi h (r_1^2 + r_1 r_2 + r_2^2)/3$, where r_1 and r_2 are the radii of the two ends of the frustum and h is the height. (This formula is derived in Example 6 of Section 6.2.) In our situation r_2 is always 3/2, the radius of the bottom of the cup, but the radius $r_1 = r$ of the top of the frustum depends on the depth of the water, h, and hence on time. The relationship between r and h can be obtained by considering Figure 4.10 and extending this into a triangle as shown in Figure 4.11.

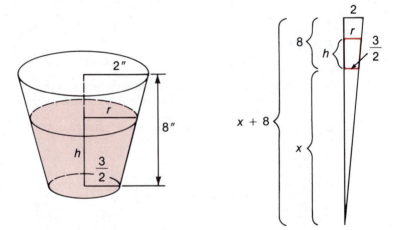

FIGURE 4.10   FIGURE 4.11

First, use similar triangles to find x:
Since

$$\frac{2}{8 + x} = \frac{3/2}{x}, \qquad 2x = 12 + \frac{3x}{2} \qquad \text{and} \qquad x = 24.$$

Using similar triangles again, we see that

$$\frac{r}{2} = \frac{24 + h}{24 + 8} = \frac{24 + h}{32},$$

so

$$r = \frac{3}{2} + \frac{h}{16}.$$

The information appearing in the problem is:

Know	Find

(a) $V = \dfrac{1}{3}\pi h \left[r^2 + r\left(\dfrac{3}{2}\right) + \left(\dfrac{3}{2}\right)^2 \right]$,

(1) $\dfrac{dh}{dt}$, and

(b) $r = \dfrac{3}{2} + \dfrac{h}{16}$, and

(2) $\dfrac{dh}{dt}$, when $h = 3$.

(c) $\dfrac{dV}{dt} = -4\,\dfrac{in^3}{sec}$.

Combining (a) and (b) we have

$$V = \frac{1}{3}\pi h \left[\left(\frac{3}{2} + \frac{h}{16}\right)^2 + \left(\frac{3}{2} + \frac{h}{16}\right)\left(\frac{3}{2}\right) + \left(\frac{3}{2}\right)^2 \right]$$

$$= \frac{1}{3}\pi h \left(\frac{9}{4} + \frac{3}{16}h + \frac{h^2}{256} + \frac{9}{4} + \frac{3}{32}h + \frac{9}{4} \right)$$

$$= \frac{1}{3}\pi \left(\frac{h^3}{256} + \frac{9}{32}h^2 + \frac{27}{4}h \right).$$

Thus,

$$\frac{dV}{dt} = \frac{1}{3}\pi \left(\frac{3h^2}{256} + \frac{9}{16}h + \frac{27}{4} \right)\frac{dh}{dt}$$

and

$$\frac{dh}{dt} = \frac{\dfrac{dV}{dt}}{\dfrac{1}{3}\pi \left(\dfrac{3h^2}{256} + \dfrac{9}{16}h + \dfrac{27}{4} \right)}$$

$$= \frac{-4}{\dfrac{1}{3}\pi \left(\dfrac{3h^2}{256} + \dfrac{9}{16}h + \dfrac{27}{4} \right)}.$$

When $h = 3$,

$$\frac{dh}{dt} = \frac{-4}{\dfrac{1}{3}\pi \left(\dfrac{27}{256} + \dfrac{27}{16} + \dfrac{27}{4} \right)}$$

$$= \frac{-1024}{729\pi} \approx -.45\,\frac{in}{sec}.$$

The conclusion is that the depth of the water is decreasing at the rate of .45 in/sec when the depth of the water in 3 inches.   □

EXERCISE SET 4.3

1. A metal sphere is heated so that its radius increases at the rate of 1 mm/sec. How fast is its volume changing when its radius is 30 mm?

2. The radius of a circle is increasing at the constant rate of 3 cm/sec. At what rate is the area of the circle increasing when the radius is 20 cm?

3. The sides of a square are increasing at the constant rate of 2 in/min. At what rate is the area of the square increasing when the sides are 4 in?

4. The length of a rectangle is three times its width and the length is increasing at the rate of 9 in/sec. How fast is the area of the rectangle changing?

5. The edges of a cube are increasing at the constant rate of 2 in/min. At what rate is the volume of the cube increasing when the edges are 4 in?

6. Reconsider the cube described in Exercise 5. At what rate is the total surface area of the cube increasing when the edges are 4 in?

7. The sides of an equilateral triangle are increasing at the constant rate of 1 cm/sec. At what rate is the area of the triangle increasing when the sides are 4 cm?

8. The two equal sides of an isosceles triangle are increasing at the constant rate of 1 cm/sec while the third side is held at 4 cm. At what rate is the area of the triangle increasing when the sides are all 4 cm?

9. A stone is dropped into a pool of still water from a height of 150 feet. Circular ripples radiate at the rate of 3 in/sec from the spot the stone hits the water. What is the area of the disturbed water four seconds after the stone hits? How fast is the area changing at this time?

10. A certain yeast culture grows in a circular colony. As it grows the surface area it covers is directly proportional to its population and contains 10^5 members when the area is 1 cm^2. How fast is the population increasing when the radius of the circle is 12 cm if the radius of the circle is increasing at the rate of 3 cm/hr?

11. Gas is being pumped into a spherical balloon at the rate of 1 ft^3/min. How fast is the diameter of the balloon increasing when the balloon contains 36 ft^3 of gas?

12. Sand is being poured onto a conical pile whose radius and height are always equal, although both increase with time. The height of the pile is increasing at the rate of 3 in/sec. How fast is the volume of the pile increasing when the height of the pile is 10 feet?

13. Find the rate of change of the area of a circle with respect to its radius. Compare this with the circumference of the circle.

14. Find the rate of change of the volume of a sphere with respect to its radius. Compare this with the surface area of a sphere.

15. A single-engine airplane passes over a beacon and heads east at 100 miles per hour. Two hours later a jet passes over the beacon at the same altitude traveling north at 400 miles per hour. Assuming that the planes stay on these courses, how fast are they separating one hour after the jet passes over the beacon?

16. Two ships meet at a point in the ocean with one of the ships traveling south at 15 miles per hour and the other traveling west at 20 miles per hour. At what rate are the ships separating two hours after they meet?

17. A kite is flying at an altitude of 80 meters and is being carried horizontally at this altitude by the wind at the rate of 2 meters/second. At what rate is the string being released to maintain this flight when 100 meters of string have been released?

18. A kite is being carried horizontally at 1.5 meters/second and rising at 2.0 meters/second. How fast is the string being released to maintain this flight when 100 meters of string have been released and the kite is at an altitude of 80 meters?

19. An object moves along the graph described by the equation $y = \sqrt{x}$ with its x-coordinate increasing at the rate of 4 cm/sec. How fast is the y-coordinate of the object changing when the object is at the point (16, 4)?

20. An object moves along the graph described by the equation $y = \sin x$ with its x-coordinate increasing at the rate of 2 cm/sec. How is the y-coordinate of the object changing when the object is at the point $(2\pi/3, \sqrt{3}/2)$?

21. An object moves counterclockwise along a circle of radius r from the point $(r, 0)$ to the point $(-r, 0)$. How does the y-coordinate of the object change if the x-coordinate decreases at the constant rate of 1 cm/sec?

22. An object moves counterclockwise along a circle of radius r from the point $(r, 0)$ to the point $(-r, 0)$. Find the rate of change of the angle θ shown in the accompanying figure if the x-coordinate decreases at the constant rate of 1 cm/sec.

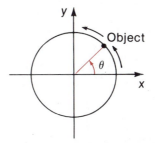

23. A rectangular swimming pool 50 feet long and 30 feet wide is being filled with water to a depth of 8 feet at the rate of 3 ft³/min. How long does it take to fill the pool? At what rate is the depth of water in the pool increasing when the pool is half full of water?

24. A rectangular swimming pool 50 feet long and 30 feet wide has a depth of 8 feet for the first 20 feet of its length, a depth of 3 feet on the last 20 feet of its length and tapers linearly for the 10 feet in the middle of its length. The pool is being filled with water at the rate of 3 ft³/min. How long does it take to fill the pool? At what rate is the depth of water in the pool increasing when the pool is half full of water?

25. A woman on a dock is using a rope to pull in a canoe at the rate of 2 ft/sec. The woman's hands are 3 feet above the level point where the rope is attached to the canoe. How fast is the canoe approaching the dock when the length of rope from her hands to the canoe is 10 feet?

26. A fisherman sitting on the end of a pier with his pole 3 meters above the water snags what he assumes to be a large fish, perhaps even a whale. He reels in his line at the steady rate of 1 meter/second. He does not realize that the object is actually an old log lying just below the surface, until the log is 5 meters from the pier. How fast is the log approaching the pier at this time?

27. An 8-foot 2 × 4 is leaning against a 10-foot wall. The lower end of the board is pulled away from the wall at the constant rate of 1 ft/sec. How fast is the top of the board moving toward the ground when it is five feet from the ground? When it is four feet from the ground?

28. An 8-foot 2 × 4 is leaning against a 5-foot wall with the remainder of the board hanging over the wall. The lower end of the board is pulled away from the wall at the constant rate of 1 ft/sec. How fast is the top of the board moving toward the ground when this end is five feet from the ground, that is, when the upper end just reaches the wall? How fast is the top of the board approaching the ground when it is four feet from the ground?

29. A camera televising the return of the opening kickoff of a football game is located 5 yards from the east edge of the field and in line with the goal line. The player with the football runs down the east edge (just in bounds) for a touchdown. When he is 10 yards from the goal line, the camera is turning at a rate of .5 radians per second. How fast is the player running?

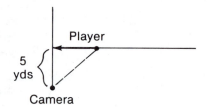

30. A revolving beacon located 1 mile from a straight shoreline turns at 1 revolution per minute. Find the speed of the spot of light along the shore when it is two miles away from the point on the shore nearest the light.

31. A metal cylinder contracts as it cools, the height of the cylinder decreasing at 4.5×10^{-4} cm/sec and the radius decreasing at 3.75×10^{-5} cm/sec. At what rate is the volume of the cylinder decreasing when its height is 200 cm and its radius is 10 cm?

32. A woman 5 feet 6 inches tall walks at the rate of 6 ft/sec toward a street light that is 16 feet above the ground. At what rate is the tip of her shadow moving? At what rate is the length of her shadow changing when she is 10 feet from the base of the light?

33. A horse trough 10 feet long has a cross section in the shape of an inverted equilateral triangle with an altitude of two feet. When filled with water it is found that the trough leaks water through a crack in the bottom at the rate of 1 ft³/hour. At what rate is the height of the water in the trough decreasing when the depth of water is 1 foot? At what rate is the height of water in the trough decreasing when the trough is half full of water?

34. A 3-foot-high picture is placed on a wall with its base 3 feet above an observer's eye level. If the observer approaches the wall at the rate of 1 ft/sec, how fast is the angle subtended at her eye by the picture changing when she is 10 feet from the wall?

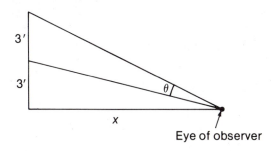

35. An object that weighs w_0 lb on the surface of the earth weighs approximately

$$w(r) = w_0 \left(\frac{4000}{4000 + r} \right)^2 \text{ lb}$$

when lifted a distance of r miles from the earth's surface. Find the rate at which the weight of an object weighing 2000 lb on the earth's surface is changing when it is 100 miles above the earth's surface and being lifted at the rate of 10 mi/sec.

36. The owner of a dog kennel reads in *Dog's Life* that the surface area of a dog is approximately related to its weight by the equation

$$s = .1 \, w^{2/3},$$

where the weight w of the dog is measured in kilograms and the surface area s of the dog is measured in square meters. Since the amount of flea powder the owner must purchase is directly proportional to the surface area of his dogs, he can use this equation to determine the rate at which he must increase his purchase of powder as his dogs mature. He knows that the average pup in his kennel gains weight at approximately .8 kg/wk. At what rate does he have to increase his purchase of powder if he has 23 dogs, the average dog weighs 20 kg, and a can of powder covers 3 square meters?

37. Oil is leaking from an ocean tanker at the rate of 5000 liters per second. The leakage results in a circular oil slick with a depth of 5 cm. (*Note:* 1 liter $=$ 1000 cm³).
   (a) How fast is the radius of the oil slick increasing when the radius is 300 meters?
   (b) How fast is the radius of the oil slick increasing 4 hours after the leakage begins?

38. In actual practice an oil slick like the one in Exercise 37 does not have a constant depth; the depth of the slick decreases as the oil moves from the point of spillage and depends primarily on the turbulence of the water and the viscosity of the oil. Suppose that the depth of the oil varies linearly from a maximum of 5 cm at the point of leakage to a minimum of .5 cm at the outside edge of the slick, and that oil is leaking from the tanker at the rate of 5000 liters per second. How is the radius of the slick increasing four hours after the leakage begins?

4.4 DIFFERENTIALS

The derivative of a function can be used to estimate the effect that a small change in the independent variable makes in the value of the function. Consider a function defined by $y = f(x)$. If x is changed by the amount Δx, from x to $x + \Delta x$, the corresponding change in y is given by

$$\Delta y = f(x + \Delta x) - f(x).$$

(See Figure 4.12.)

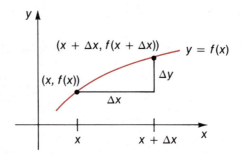

FIGURE 4.12

Let us return briefly to the definition of the derivative; since $y = f(x)$,

$$f'(x) = \lim_{\Delta x \to 0} \frac{f(x + \Delta x) - f(x)}{\Delta x} = \lim_{\Delta x \to 0} \frac{\Delta y}{\Delta x}.$$

This implies that $\Delta y/\Delta x$ is close to $f'(x)$ whenever Δx is sufficiently close to zero. Rewritten, we have

$$\Delta y \approx f'(x)\, \Delta x,$$

whenever Δx is close to zero. The relationship between Δy and $f'(x)\, \Delta x$ is illustrated in Figure 4.13(a) (where $\Delta x > 0$) and in Figure 4.13(b) (where $\Delta x < 0$).

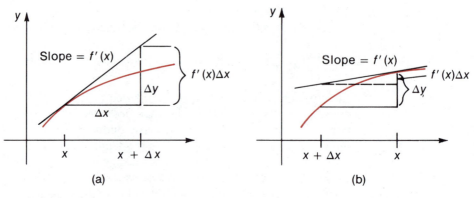

(a)   (b)

FIGURE 4.13

(4.2)
DEFINITION

If f is differentiable on an interval containing x and $x + \Delta x$, where Δx represents a change in the variable x, and if $y = f(x)$, then
(i) the **differential of x** is denoted dx and defined by: $dx = \Delta x$;
(ii) the **differential of y** is denoted dy and defined by:
$dy = f'(x) \Delta x = f'(x)dx$.

Part (ii) of this definition implies that $f'(x)$ is the quotient of the differentials dy and dx. This is the reason for using the Leibniz notation dy/dx for the derivative.

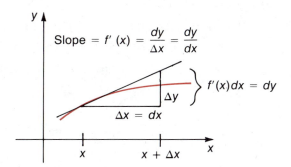

FIGURE 4.14

Figure 4.14 illustrates the relationship between the differential dy and the value of Δy corresponding to the change Δx. Δy represents the change in the function values, while the differential dy represents the change along the tangent line. When dy is used to approximate Δy, the tangent line is being used to approximate the curve.

EXAMPLE 1

Suppose the function f is described by $y = f(x) = 3x^2 - 4x + 5$. Find dy and Δy when $x = 3$ and $\Delta x = .01$.

SOLUTION

Since
$$\frac{dy}{dx} = f'(x) = 6x - 4,$$
$$dy = (6x - 4)dx.$$

When $x = 3$ and $\Delta x = .01$, $dx = .01$ and $dy = (18 - 4)(.01) = .14$.
Finding Δy requires the following calculation:

$$\Delta y = f(x + \Delta x) - f(x)$$
$$= 3(x + \Delta x)^2 - 4(x + \Delta x) + 5 - (3x^2 - 4x + 5)$$
$$= 3x^2 + 6x\Delta x + 3(\Delta x)^2 - 4x - 4\Delta x + 5 - 3x^2 + 4x - 5$$
$$= (6x - 4)\Delta x + 3(\Delta x)^2$$

When $x = 3$ and $\Delta x = .01$,

$$\Delta y = [6(3) - 4] (.01) + 3(.01)^2$$
$$= .14 + .0003 = .1403.$$ $\square$

As shown in Example 1, the differential dy can be used as an approximation to Δy. Since $\Delta y = f(x + \Delta x) - f(x)$,

$$f(x + \Delta x) = f(x) + \Delta y \approx f(x) + dy$$

whenever Δx is small.

EXAMPLE 2   Use differentials to approximate the value of $\sqrt{80}$.

SOLUTION

The function described by $y = f(x) = \sqrt{x}$ is easily evaluated at $x = 81$, so the plan is to approximate $\sqrt{80}$ by using the following:

$$\begin{aligned} \sqrt{80} = f(80) &= f(81 + \Delta x) \approx f(81) + dy \\ &= \sqrt{81} + dy \\ &= 9 + dy. \end{aligned}$$

Consequently, we need to find dy when $x = 81$ and $\Delta x = 80 - 81 = -1$. With $f(x) = \sqrt{x}$, we see that

$$f'(x) = \frac{1}{2} x^{-1/2} = \frac{1}{2\sqrt{x}},$$

so $dy = f'(x)\, \Delta x = \dfrac{1}{2\sqrt{x}}\Delta x$; and

$$dy \ (\text{when } x = 81 \text{ and } \Delta x = -1) = \frac{1}{2\sqrt{81}}(-1) = -\frac{1}{18}.$$

The approximation to $\sqrt{80}$ is therefore

$$\sqrt{80} \approx 9 + dy = 9 - \frac{1}{18} \approx 8.9444,$$

which compares favorably to the correct four-decimal value 8.9443.   □

EXAMPLE 3   The diameter of a sphere is determined to be 16 cm, with a maximal measurement error of $\pm .001$ cm. Use differentials to determine an estimate for the maximum error in computing the volume of this sphere.

SOLUTION

If r denotes the radius and D the diameter of the sphere, then the volume is

$$V = \frac{4}{3}\pi r^3 = \frac{1}{6}\pi D^3.$$

Hence,

$$\frac{dV}{dD} = \frac{1}{2}\pi D^2 \quad \text{and} \quad dV = \frac{1}{2}\pi D^2\, dD.$$

With $D = 16$ cm and $|dD| \leq .001$ cm, we find that the maximum error for the volume is

$$|dV| = \left| \frac{1}{2} \pi D^2 \, dD \right| = \frac{1}{2} \pi D^2 \left| \, dD \, \right| \leq \frac{1}{2} \pi (16)^2 (.001) \approx .40 \text{ cm}^3.$$

The actual volume, V, is bounded approximately by

$$\left[\frac{1}{6} \pi (16)^3 - .40 \right] \text{cm}^3 \leq V \leq \left[\frac{1}{6} \pi (16)^3 + .40 \right] \text{cm}^3,$$

so

$$2144.26 \text{ cm}^3 \leq V \leq 2145.06 \text{ cm}^3. \qquad \square$$

EXERCISE SET 4.4

2-12 (even)   23,24,27 (8,24).

Find dy and Δy in terms of x and Δx in Exercises 1 through 12.

1. $y = 3x - 5$ **2.** $y = 3 - 2x$

3. $y = x^2 - 2x + 3$ **4.** $y = 1 - x + 2x^2$

5. $y = x^3 - x$ **6.** $y = x^3 - 3x^2 + 3x - 1$

7. $y = \sqrt{x + 1}$ **8.** $y = \dfrac{1}{x - 1}$

9. $y = x + \dfrac{1}{x}$ **10.** $y = \dfrac{1}{\sqrt{x}}$

11. $y = \tan x$ **12.** $y = x + \sin x$

Find dy and Δy in Exercises 13 through 16 if $y = f(x)$ and
(a) $x = 2, \Delta x = .1$ (b) $x = -1, \Delta x = -.2$.

13. $f(x) = x^2 - 4x + 1$ **14.** $f(x) = 3 - 2x^2$

15. $f(x) = x^3 - 2x$ **16.** $f(x) = x^3 - 2x^2$

Find dy and Δy in Exercises 17 through 22.

17. $f(x) = \sin x, x = \dfrac{\pi}{2}, \Delta x = \dfrac{\pi}{4}$ **18.** $f(x) = \cos x, x = \dfrac{\pi}{2}, \Delta x = \dfrac{\pi}{4}$

19. $f(x) = \dfrac{1}{x}, x = 1, \Delta x = -.02$ **20.** $f(x) = \dfrac{x}{x + 1}, x = 0, \Delta x = .1$

21. $f(x) = \sqrt{x}, x = 9, \Delta x = .03$

22. $f(x) = \dfrac{1}{\sqrt[3]{x + 1}}, x = 7, \Delta x = -.1$

Use differentials to approximate the quantities in Exercises 23 through 30. Check your answers by using a calculator.

23. $\sqrt{9.03}$ **24.** $\sqrt{50}$

25. $\sqrt{6560}$ **26.** $\sqrt[3]{65}$

27. $\sqrt[3]{2200}$

28. $\sqrt[3]{.0011}$

29. $\sqrt{\dfrac{1}{9.03}}$

30. $\dfrac{1}{\sqrt[6]{1000001}}$

31. Use differentials to approximate the maximum possible error that can be produced when calculating the area of a circle, if it is known that the radius of the circle is $2 \pm .001$ feet.

32. Use differentials to approximate the maximum possible error that can be produced when calculating the circumference of a circle, if it is known that the radius of the circle is $2 \pm .001$ feet.

33. Use differentials to approximate the maximum possible error that can be produced when calculating (a) the volume, and (b) the surface area of a sphere, if the radius of the sphere is $2 \pm .001$ feet.

34. Use differentials to approximate the maximum possible error that can be produced when calculating (a) the volume, and (b) the surface area of a cube, if the length of an edge is $1 \pm .0005$ feet.

35. A hole 1/4 inch in diameter has been drilled through a wooden 4 × 4. The hole is to be enlarged to 5/16 inch in diameter. Use differentials to approximate how much additional wood has to be removed. (The dimensions of a wooden 4 × 4 are 3.5 inches by 3.5 inches.)

36. An open cylindrical pipe for pumping fuel oil has an outside diameter of 3 inches and a length of 8 feet. The outside of the pipe is to be painted with a coating 1/16 inch thick. Use differentials to determine the number of gallons of paint needed to paint 100 pipes of this type. (One U.S. gallon is equivalent to 231 cubic inches.)

37. In Exercise 31 of Exercise Set 4.2 the dimensions were found for the strongest rectangular beam that can be cut from a circular log of radius r inches. Use this result and differentials to approximate the percentage of increase in the strength of the strongest rectangular beam that can be cut from a circular log if the radius of the log increases from r inches to $r + \Delta r$ inches.

38. In Exercise 32 of Exercise Set 4.2 the dimensions were found for the stiffest rectangular beam that can be cut from a circular log of radius r inches. Use this result and differentials to approximate the percentage increase in the stiffness of the stiffest rectangular beam that can be cut from a circular log if the radius of the log is increased from r inches to $r + \Delta r$ inches.

39. A metal spherical ball with a volume of 36 cm^3 is given a chrome plating .2 cm thick. Use differentials to approximate the volume of the chromium required to plate the ball.

40. A string is placed around the earth at the equator. (Assume that the earth's surface at the equator is a circle with radius 4000 miles.) We cut the string and raise it above the surface of the earth exactly one foot at each point (so that small dogs can pass beneath, perhaps). By some magical means the string will stay in this position. In raising the string by one foot the ends of the string no longer meet. How much string is needed to complete the circle one foot above the earth?

(a) Follow the procedure of Exercise 32 to find a solution to this problem.

(b) Solve the problem directly by determining the circumference of each of the circles and subtracting to find their difference.

41. Temperature on weather forecasts is generally given to the nearest degree. Suppose you are told that the present temperature is 20°C, which means that it is somewhere between 19.5°C and 20.5°C. Use differentials to approximate the maximum error in degrees Fahrenheit that can be produced if the 20°C figure is converted to 68°F.

42. Two thermometers are placed in a room. Both read the temperature in Celsius with an error of no more than $\pm$ .51°C. The thermometers are read simultaneously to the nearest degree Celsius and then converted to a Fahrenheit scale. Use differentials to approximate the maximum number of degrees Fahrenheit by which the thermometers can disagree.

43. A stick is placed in the ground 100 feet from the base of a tree. A line joining the top of the tree and the top of the stick touches the ground at a point 10 feet from the base of the stick. How high is the tree (in terms of the stick height)? Suppose that an error of no more than one inch is made in determining the spot where the line joining the top of the tree and the top of the stick hits the ground; the other measurements are assumed to be exact. Use differentials to approximate the maximum error that can be produced in calculating the height of the tree.

44. Reconsider the situation in Exercise 43 but assume instead that an error of no more than one inch is made in measuring the distance from the base of the tree to the base of the stick; the other measurements are assumed to be exact. Use differentials to approximate the maximum error that can be produced in calculating the height of the tree.

4.5
L'HÔPITAL'S RULE

Early in the study of the limit we established that

$$\lim_{x \to a} \frac{f(x)}{g(x)} = \frac{\lim_{x \to a} f(x)}{\lim_{x \to a} g(x)},$$

provided $\lim_{x \to a} f(x)$ and $\lim_{x \to a} g(x)$ both exist and $\lim_{x \to a} g(x) \neq 0$. However, this result was not sufficient to answer many of the questions concerning the limit of a quotient of two functions. Indeed, the derivative itself involves the limit of a quotient in which both the numerator and denominator have the limit zero:

$$D_x f(a) = \lim_{x \to a} \frac{f(x) - f(a)}{x - a}.$$

In this section we will see how the derivative can be applied to resolve limits of quotients in which both the numerator and denominator have limit zero or both have infinite limits.

To simplify the discussion, we say that the quotient $f(x)/g(x)$ has the **indeterminate form 0/0** at a if both $\lim_{x \to a} f(x) = 0$ and $\lim_{x \to a} g(x) = 0$ (a can be a real number or $\pm \infty$). For example, $(x^2 - 4)/(x - 2)$ has indeterminate form 0/0 at 2, while $\sin x/x$ has indeterminate form 0/0 at 0.

Similarly, we say that $f(x)/g(x)$ has the **indeterminate form** ∞/∞ at a if the limits at a of both f and g are infinite (∞ or $-\infty$).

Previously we have resorted to various techniques to resolve indeterminate forms. In Section 2.1 we examined $\lim\limits_{x \to 2} (x^2 - 4)/(x - 2)$. This problem was resolved by factoring the numerator to give

$$\lim_{x \to 2} \frac{x^2 - 4}{x - 2} = \lim_{x \to 2} (x + 2) = 4.$$

The first time we encountered an indeterminate form that an algebraic manipulation would not resolve was in evaluating $\lim\limits_{x \to 0} \sin x/x = 1$. A geometric technique was used to determine this limit.

To introduce a general method for evaluating indeterminate forms, we consider a special case of the form $0/0$. Suppose that $f(x)/g(x)$ assumes this form (so $\lim\limits_{x \to a} f(x) = 0$ and $\lim\limits_{x \to a} g(x) = 0$) and that both f and g are differentiable at a. Differentiability at a implies continuity at a, so

$$\lim_{x \to a} f(x) = f(a), \ \lim_{x \to a} g(x) = g(a) \quad \text{and hence} \quad f(a) = g(a) = 0.$$

So

$$\lim_{x \to a} \frac{f(x)}{g(x)} = \lim_{x \to a} \frac{\dfrac{f(x) - f(a)}{x - a}}{\dfrac{g(x) - g(a)}{x - a}}, \qquad \text{provided } g(x) \neq g(a) \text{ near } a$$

$$= \frac{\lim\limits_{x \to a} \dfrac{f(x) - f(a)}{x - a}}{\lim\limits_{x \to a} \dfrac{g(x) - g(a)}{x - a}} = \frac{f'(a)}{g'(a)} \cdot$$

In the problem $\lim\limits_{x \to 2} (x^2 - 4)/(x - 2)$, $D_x(x^2 - 4) = 2x$ and $D_x(x - 2) = 1$, so

$$\lim_{x \to 2} \frac{x^2 - 4}{x - 2} = \frac{2 \cdot 2}{1} = 4.$$

Also $D_x \sin x = \cos x$ and $D_x x = 1$, so

$$\lim_{x \to 0} \frac{\sin x}{x} = \frac{\cos 0}{1} = 1.$$

While this is the basic procedure we will apply to evaluate indeterminate forms, the requirement that both f and g be differentiable at a is an overly restrictive condition. For example, the condition could never be satisfied when considering the value of an indeterminate form at infinity.

To establish a more general result we first need to consider the following extension of the mean value theorem.

(4.3)
THEOREM

Cauchy Mean Value Theorem   Suppose f and g are continuous on $[a, b]$ and differentiable on (a, b). If g' is nonzero on (a, b), then a number c exists, $a < c < b$, with

$$\frac{f(b) - f(a)}{g(b) - g(a)} = \frac{f'(c)}{g'(c)}.$$

PROOF

The proof of this theorem is similar to the proof of the mean value theorem— a new function h is defined on $[a, b]$ that satisfies the hypotheses of Rolle's theorem.

Let $h(x) = [f(b) - f(a)][g(x) - g(a)] - [g(b) - g(a)][f(x) - f(a)]$.

Then h is continuous on $[a, b]$ and differentiable on (a, b). Additionally,

$h(a) = [f(b) - f(a)][g(a) - g(a)] - [g(b) - g(a)][f(a) - f(a)] = 0$

and

$h(b) = [f(b) - f(a)][g(b) - g(a)] - [g(b) - g(a)][f(b) - f(a)] = 0$,

so Rolle's theorem implies that a number c, $a < c < b$, exists with

$0 = h'(c) = [f(b) - f(a)]g'(c) - [g(b) - g(a)]f'(c)$.

Since g' is nonzero on (a, b), both $g'(c)$ and $g(b) - g(a)$ are nonzero. Consequently,

$$\frac{f(b) - f(a)}{g(b) - g(a)} = \frac{f'(c)}{g'(c)}. \qquad \square$$

We can now prove the basic theorem involving indeterminate forms. This theorem and the various modifications we will discuss later are known collectively as **L'Hôpital's rule.**

(4.4)
THEOREM

Suppose f and g are differentiable on an open interval containing a number a, except possibly at a, and that g' is nonzero on the interval. If $f(x)/g(x)$ has the indeterminate form 0/0 at a and $\displaystyle\lim_{x \to a} \frac{f'(x)}{g'(x)} = L$, then $\displaystyle\lim_{x \to a} \frac{f(x)}{g(x)} = L$.

HISTORICAL
NOTE

Guillaume Francois Antoine Marquis de L'Hôpital (1661–1704) introduced this method in the first calculus textbook *Analyse des Infiniments Petits* published in Paris in 1696. The rule was evidently a discovery of Jean Bernoulli, who sent all his mathematical discoveries to L'Hôpital in exchange for a regular salary. Controversy developed regarding whether this arrangement was clandestine, with L'Hôpital attempting to receive credit for Bernoulli's work.

In his preface, L'Hôpital takes special note of the contribution of all the members of the prolific Bernoulli family, particularly Jean. It was only after L'Hôpital's death that Bernoulli accused him of plagiarism, an accusation that at the time was generally dismissed, but now seems to be well founded. In any case, the book was a major contribution to the spread of the knowledge of calculus throughout Europe, particularly in France.

PROOF

A minor technical difficulty arises because neither f nor g is assumed to be continuous at a. To alleviate this, functions F and G are defined by

$$F(x) = \begin{cases} f(x), & \text{if } x \neq a \\ 0, & \text{if } x = a \end{cases} \quad \text{and} \quad G(x) = \begin{cases} g(x), & \text{if } x \neq a \\ 0, & \text{if } x = a. \end{cases}$$

Since $\lim_{x \to a} F(x) = \lim_{x \to a} f(x) = 0 = F(a)$ and $\lim_{x \to a} G(x) = \lim_{x \to a} g(x) = 0 = G(a)$, both F and G are continuous on the open interval containing a and differentiable on this interval, except possibly at a. For both cases $x < a$ and $x > a$, the Cauchy mean value theorem can be applied to F and G on the interval with endpoints x and a. The conclusion is that there is a number c between x and a with

$$\frac{F(x) - F(a)}{G(x) - G(a)} = \frac{F'(c)}{G'(c)}$$

and consequently, that

$$\frac{f(x)}{g(x)} = \frac{f'(c)}{g'(c)}.$$

Since c is between x and a,

$$\lim_{x \to a} \frac{f(x)}{g(x)} = \lim_{x \to a} \frac{f'(c)}{g'(c)} = \lim_{c \to a} \frac{f'(c)}{g'(c)} = L. \qquad \square$$

In the examples that follow, we have not included the verification that f and g are differentiable and that $g'(x) \neq 0$. This is usually straightforward and is left to the reader.

EXAMPLE 1   Evaluate $\lim_{x \to -1} \dfrac{x^6 - x^4}{x^5 + 2x^2 - 1}$.

SOLUTION

$\lim_{x \to -1} (x^6 - x^4) = 0 = \lim_{x \to -1} (x^5 + 2x^2 - 1)$, so L'Hôpital's rule can be applied.

$$\lim_{x \to -1} \frac{x^6 - x^4}{x^5 + 2x^2 - 1} = \lim_{x \to -1} \frac{6x^5 - 4x^3}{5x^4 + 4x} = \frac{-6 + 4}{5 - 4} = -2. \qquad \square$$

EXAMPLE 2   Evaluate $\lim_{x \to 9} \dfrac{\sqrt{x} - 3}{x - 9}$.

SOLUTION

$\lim_{x \to 9} \sqrt{x} - 3 = 0 = \lim_{x \to 9} x - 9$, so L'Hôpital's rule can be applied.

$$\lim_{x \to 9} \frac{\sqrt{x} - 3}{x - 9} = \lim_{x \to 9} \frac{D_x(\sqrt{x} - 3)}{D_x(x - 9)}$$

$$= \lim_{x \to 9} \frac{\frac{1}{2\sqrt{x}}}{1} = \frac{1}{6}.$$

Try to find two other methods of determining this limit.   □

EXAMPLE 3   Evaluate $\lim\limits_{x \to \pi} \dfrac{1 + \cos x}{\sin x}$.

SOLUTION

$\lim\limits_{x \to \pi} (1 + \cos x) = 0 = \lim\limits_{x \to \pi} \sin x$, so

$$\lim_{x \to \pi} \frac{1 + \cos x}{\sin x} = \lim_{x \to \pi} \frac{D_x (1 + \cos x)}{D_x (\sin x)}$$

$$= \lim_{x \to \pi} \frac{- \sin x}{\cos x} = 0.   □$$

The case of the indeterminate form ∞/∞ at a can be deduced from Theorem 4.3, but the details are more complicated. The proof can be found in most advanced calculus books.

(4.5)
THEOREM   Suppose f and g are differentiable on an open interval containing a number a, except possibly at a, and that g' is nonzero on the interval. If $f(x)/g(x)$ has the indeterminate form ∞/∞ at a and $\lim\limits_{x \to a} \dfrac{f'(x)}{g'(x)} = L$, then $\lim\limits_{x \to a} \dfrac{f(x)}{g(x)} = L$.

Theorems 4.4 and 4.5 also hold for one-sided limits and for the case when a is ∞ or $-\infty$. We will assume these results.

EXAMPLE 4   Evaluate $\lim\limits_{x \to \frac{\pi}{2}^-} \dfrac{2x + \sec x}{3 + \tan x}$.

SOLUTION

This limit has the indeterminate form ∞/∞, since $\lim\limits_{x \to \frac{\pi}{2}^-} (2x + \sec x) = \infty$ and $\lim\limits_{x \to \frac{\pi}{2}^-} (3 + \tan x) = \infty$.

Consequently,

$$\lim_{x \to \frac{\pi}{2}^-} \frac{2x + \sec x}{3 + \tan x} = \lim_{x \to \frac{\pi}{2}^-} \frac{2 + \sec x \tan x}{(\sec x)^2}$$

$$= \lim_{x \to \frac{\pi}{2}^-} [2(\cos x)^2 + \sin x] = 1.   □$$

EXAMPLE 5   Evaluate $\lim\limits_{x \to 0} \dfrac{x \cos x - \sin x}{x - \sin x}$.

SOLUTION

$$\lim_{x \to 0} (x \cos x - \sin x) = 0 = \lim_{x \to 0} (x - \sin x)$$

so

$$\lim_{x \to 0} \frac{x \cos x - \sin x}{x - \sin x} = \lim_{x \to 0} \frac{\cos x - x \sin x - \cos x}{1 - \cos x}$$

$$= \lim_{x \to 0} \frac{-x \sin x}{1 - \cos x}.$$

But, $\lim\limits_{x \to 0} - x \sin x = 0 = \lim\limits_{x \to 0} (1 - \cos x)$, so we must apply L'Hôpital's rule again.

$$\lim_{x \to 0} \frac{x \cos x - \sin x}{x - \sin x} = \lim_{x \to 0} \frac{-x \sin x}{1 - \cos x}$$

$$= \lim_{x \to 0} \frac{-\sin x - x \cos x}{\sin x}.$$

Once again we have the form 0/0; this time L'Hôpital's rule resolves the problem.

$$\lim_{x \to 0} \frac{x \cos x - \sin x}{x - \sin x} = \lim_{x \to 0} \frac{-\sin x - x \cos x}{\sin x}$$

$$= \lim_{x \to 0} \frac{-\cos x - \cos x + x \sin x}{\cos x}$$

$$= \frac{-2}{1} = -2. \qquad \square$$

EXAMPLE 6   Evaluate $\lim\limits_{x \to 0} \dfrac{x}{1 + \sin x}$.

SOLUTION

This is *not* a problem requiring L'Hôpital's rule, since

$$\lim_{x \to 0} \frac{x}{1 + \sin x} = \frac{\lim\limits_{x \to 0} x}{\lim\limits_{x \to 0} (1 + \sin x)} = \frac{0}{1} = 0.$$

In fact, if L'Hôpital's rule is mistakenly applied to this limit problem, an incorrect answer results:

$$\lim_{x \to 0} \frac{D_x x}{D_x(1 + \sin x)} = \lim_{x \to 0} \frac{1}{\cos x} = 1. \qquad \square$$

Beware of the pitfalls that arise from applying L'Hôpital's rule to functions that are not in one of the indeterminate forms $0/0$ or $\pm \infty/\infty$; it generally leads to incorrect results. L'Hôpital's rule can be applied only when needed, that is, when the problem is one of the indeterminate forms.

Other types of limits lead to indeterminate forms. If $\lim\limits_{x \to a} f(x) = 0$ and $\lim\limits_{x \to a} g(x) = \infty$, then $f(x) \cdot g(x)$ has the indeterminate form $0 \cdot \infty$ at a. Quite often this limit can be resolved by reexpressing $f(x)\, g(x)$ as

$$f(x)g(x) = \frac{f(x)}{\left(\dfrac{1}{g(x)}\right)}$$

or

$$f(x)g(x) = \frac{g(x)}{\left(\dfrac{1}{f(x)}\right)}$$

and applying L'Hôpital's rule to the functions f and $1/g$ or to g and $1/f$.

EXAMPLE 7   Evaluate $\lim\limits_{x \to \infty} x \sin \dfrac{1}{x}$.

SOLUTION

This limit has indeterminate form $0 \cdot \infty$, since $\lim\limits_{x \to \infty} x = \infty$ and $\lim\limits_{x \to \infty} \sin \dfrac{1}{x} = 0$.

However, $\lim\limits_{x \to \infty} x \sin \dfrac{1}{x} = \lim\limits_{x \to \infty} \dfrac{\sin \dfrac{1}{x}}{\dfrac{1}{x}}$, which is of the form $0/0$. Thus,

$$\lim\limits_{x \to \infty} x \sin \dfrac{1}{x} = \lim\limits_{x \to \infty} \frac{\cos \dfrac{1}{x}\left(-\dfrac{1}{x^2}\right)}{-\dfrac{1}{x^2}}$$

$$= \lim\limits_{x \to \infty} \cos \dfrac{1}{x} = 1. \qquad \square$$

An indeterminate form of the type $\infty - \infty$ occurs when evaluating $\lim\limits_{x \to a}(f(x) - g(x))$, where $\lim\limits_{x \to a} f(x) = \infty = \lim\limits_{x \to a} g(x)$. Problems of this type can often be manipulated into a form appropriate for applying L'Hôpital's rule.

EXAMPLE 8   Evaluate $\lim\limits_{x \to 0^+} (\csc x - \cot x)$.

SOLUTION

$\lim\limits_{x \to 0^+} \csc x = \infty = \lim\limits_{x \to 0^+} \cot x$, so this limit has the indeterminate form $\infty - \infty$.

However,

$$\lim_{x \to 0^+} (\csc x - \cot x) = \lim_{x \to 0^+} \left(\frac{1}{\sin x} - \frac{\cos x}{\sin x} \right)$$

$$= \lim_{x \to 0^+} \frac{1 - \cos x}{\sin x}$$

$$= \lim_{x \to 0^+} \frac{D_x (1 - \cos x)}{D_x \sin x}$$

$$= \lim_{x \to 0^+} \frac{\sin x}{\cos x} = 0. \qquad \square$$

Other indeterminate forms, 1^∞, 0^0, and ∞^0 will be considered in Chapter 7.
    We end this section with another caution against attempting to apply
L'Hôpital's rule when it is not required. It almost always leads to an erroneous
conclusion, especially when prearranged by devious calculus instructors.

EXERCISE SET 4.5

Determine the limits in Exercises 1 through 40.

1. $\lim\limits_{x \to 1} \dfrac{x^3 - 2x^2 + 2x - 1}{x^2 - 1}$

2. $\lim\limits_{x \to -2} \dfrac{x^3 + 2x^2 - x - 2}{3x^2 + 10x + 8}$

3. $\lim\limits_{x \to 3} \dfrac{x^4 - 6x^2 - 5x - 12}{3x^2 - 7x - 6}$

4. $\lim\limits_{x \to 1} \dfrac{x^4 - x^3 + 3x^2 + x - 4}{x^3 + x^2 - 7x + 5}$

5. $\lim\limits_{x \to -2} \dfrac{2 + 9x - 2x^2 - 3x^3}{2x^4 + 7x^3 + 10x^2 + 7x - 2}$

6. $\lim\limits_{x \to -4} \dfrac{x^3 + 8x^2 + 11x - 20}{2x^3 + 5x^2 - 8x + 16}$

7. $\lim\limits_{x \to 0} \dfrac{1 - \cos x}{2x}$

8. $\lim\limits_{x \to 0} \dfrac{2x}{\tan x}$

9. $\lim\limits_{x \to 0} \dfrac{\sin 2x}{3x}$

10. $\lim\limits_{x \to 0} \dfrac{\cos x + x - 1}{3x}$

11. $\lim\limits_{x \to -2} \dfrac{\sqrt{x + 6} - 2}{x^2 - 4}$

12. $\lim\limits_{x \to 2} \dfrac{\sqrt{x + 7} - 3}{x^2 - 4}$

13. $\lim\limits_{x \to -2} \dfrac{\sqrt{6 + x} - \sqrt{2 - x}}{x + 2}$

14. $\lim\limits_{x \to 1} \dfrac{2 - \sqrt{x + 3}}{1 - x}$

15. $\lim\limits_{x \to 2} \dfrac{\sqrt{4x + 1} - \sqrt[3]{13x + 1}}{x + 2}$

16. $\lim\limits_{x \to 4} \dfrac{2 - \sqrt[3]{x + 4}}{4 + \sqrt{12 + x}}$

17. $\lim\limits_{x\to\infty} \dfrac{x^2 + 5}{3x^2 + 1}$

18. $\lim\limits_{x\to\infty} \dfrac{3x^2 - 7}{2x^2 + x + 3}$

19. $\lim\limits_{x\to\infty} \dfrac{3x^2 + 4x + 5}{2x^2 + 3x - 2}$

20. $\lim\limits_{x\to-\infty} \dfrac{2x^3 + 7x - 5}{7x^3 + 2x - 5}$

21. $\lim\limits_{x\to-\infty} \dfrac{5 - 2x^3 - 3x^4}{3 + 2x + 7x^4}$

22. $\lim\limits_{x\to\infty} \dfrac{3x + 4x^2 - 5x^4}{2 + x + 4x^2 - 3x^4}$

23. $\lim\limits_{x\to\frac{\pi}{2}} \dfrac{1 - \sin x}{\cos x}$

24. $\lim\limits_{x\to0} \dfrac{1 - \sin x}{\cos x}$

25. $\lim\limits_{x\to0} \dfrac{\csc x}{\cot x}$

26. $\lim\limits_{x\to\frac{\pi}{2}^-} \dfrac{\sec x}{\tan x}$

27. $\lim\limits_{x\to0} \dfrac{\sin x - x}{x^3}$

28. $\lim\limits_{x\to0} \dfrac{\sin x^2}{(\sin x)^2}$

29. $\lim\limits_{x\to0} \dfrac{4x^2 \cos 3x}{(\sin 3x)^2}$

30. $\lim\limits_{x\to0} \dfrac{1 - \cos x^2}{x^3}$

31. $\lim\limits_{x\to0} \left(\dfrac{1}{x} - \dfrac{1}{\sin x} \right)$

32. $\lim\limits_{x\to0} \left(\dfrac{1}{1 - \cos x} - \dfrac{1}{x^2} \right)$

33. $\lim\limits_{x\to\infty} \left(\dfrac{1}{x^2 + 2} - \dfrac{1}{x} \right)$

34. $\lim\limits_{x\to\frac{\pi}{2}^+} \left(\dfrac{1}{\sec x} - \dfrac{1}{x - \pi/2} \right)$

35. $\lim\limits_{x\to-\infty} \left(\dfrac{x}{x^2 + 1} - \dfrac{x}{x^2 - 1} \right)$

36. $\lim\limits_{x\to\frac{\pi}{2}} (\tan x - \sec x)$

37. $\lim\limits_{x\to\infty} x^2 \sin \dfrac{1}{x}$

38. $\lim\limits_{x\to0^-} x^2\cot x$

39. $\lim\limits_{x\to\frac{\pi}{2}^-} \left(x - \dfrac{\pi}{2} \right)\sec x$

40. $\lim\limits_{x\to0^+} 4x^2\cot 3x$

In Exercises 41 through 44, find all values of c that satisfy the conclusions of the Cauchy mean value theorem, if the hypotheses of the theorem hold for the given function and interval.

41. $f(x) = x$, $g(x) = x^2$, $[1, 2]$ **42.** $f(x) = x - 2$, $g(x) = x^3$, $[-1, 0]$

43. $f(x) = x^2$, $g(x) = x^3$, $[-1, 1]$

44. $f(x) = x^2 - 3x$, $g(x) = x^2 - 1$, $[0, 2]$

45. What must a be if

$$g(x) = \begin{cases} \dfrac{x^3}{\sin x - x}, & \text{if } x \neq 0 \\ a, & \text{if } x = 0 \end{cases}$$

is continuous at zero?

46. Determine

$$\lim_{x \to \infty} \frac{\sin x - x}{\sin x + x}.$$

Does L'Hôpital's rule resolve this problem?

47. Explain why L'Hôpital's rule could not have been used to evaluate $\lim_{x \to 0} \dfrac{\sin x}{x}$ in Section 3.4, even if the rule had been known at that time.

48. Find a pair of differentiable functions f and g and a real number a with the property that $\lim_{x \to a} f(x)/g(x)$ exists and is finite, but $\lim_{x \to a} f'(x)/g'(x)$ does *not* exist.

49. Find a pair of differentiable functions f and g with the property that $\lim_{x \to \infty} f(x)/g(x)$ exists and is finite, but $\lim_{x \to \infty} f'(x)/g'(x)$ does *not* exist.

50. Show that all the hypotheses of the Cauchy mean value theorem are necessary. (a) Find a pair of functions that are not continuous on an interval $[a, b]$ and show that the conclusion does not hold. (b) Find a pair of functions that are not differentiable on (a, b), but are continuous on $[a, b]$, and show that the conclusion does not hold. (c) Find a pair of functions that are continuous on $[a, b]$, differentiable on (a, b), but for which the conclusion does not hold.

51. Suppose $f'(a)$ and $\lim_{x \to a} f'(x)$ exist. Show that f' must be continuous at a.

52. Suppose f'' exists at a. Show that

$$f''(a) = \lim_{h \to 0} \frac{f(a + h) + f(a - h) - 2f(a)}{h^2}.$$

53. In the proof of the Cauchy mean value theorem it is stated that since g' is nonzero on (a, b), $g(b) - g(a) \neq 0$. Show that this statement is true.

4.6
NEWTON'S METHOD

The derivative can also be applied to find approximate solutions to an equation expressed in the form $f(x) = 0$, where f is a differentiable function.

If f is a linear or quadratic function, the solution to this problem is elementary. In fact, solutions to this equation for polynomials of degree three and four can be found exactly by algebraic formulas, although these formulas are seldom used. In most instances, however, it is either necessary or much easier to find an approximate, rather than an exact, solution to the equation $f(x) = 0$. The most popular and generally efficient procedure for handling such problems is the **Newton-Raphson**, or simply **Newton's method**.

Suppose we have determined that the graph of a function f crosses the x-axis precisely once in the interval $[a, b]$, and that its derivative exists and does not change sign in $[a, b]$. The graph of such a function might take a form similar to one of those sketched in Figure 4.15.

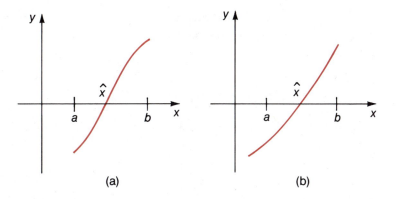

FIGURE 4.15

To approximate the point $\hat{x}$, where the graph of f crosses the x-axis, we assume an initial approximation x_0 to $\hat{x}$. We then construct the line tangent to the graph of f at $(x_0, f(x_0))$, and call the x intercept of this line x_1, as shown in Figure 4.16.

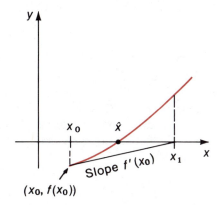

FIGURE 4.16

The tangent line passes through $(x_0, f(x_0))$ and has slope $f'(x_0)$, so its equation is

$$\frac{y - f(x_0)}{x - x_0} = f'(x_0) \quad \text{or} \quad y - f(x_0) = f'(x_0)(x - x_0).$$

Since the line crosses the x-axis at $(x_1, 0)$,

$$0 - f(x_0) = f'(x_0)(x_1 - x_0).$$

Solving this equation for x_1 gives

$$x_1 = x_0 - \frac{f(x_0)}{f'(x_0)}.$$

Newton's method is based on the assumption that each new value constructed in this manner is a better approximation to $\hat{x}$ than the previous. The procedure can be repeated using x_1 in place of x_0 to obtain another approximation x_2, as shown in Figure 4.17. In general,

(4.6)
$$x_{n+1} = x_n - \frac{f(x_n)}{f'(x_n)}$$

for each integer $n \geq 0$, provided $f'(x_n) \neq 0$.

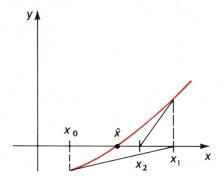

FIGURE 4.17

EXAMPLE 1   Find an approximate solution to the equation
$$x^3 - 2x - 5 = 0.$$

SOLUTION

If $f(x) = x^3 - 2x - 5$, then f is a continuous function. Since $f(2) = -1$ and $f(3) = 16$, it follows from the intermediate value theorem (2.19) that there is a number $\hat{x}$ in [2, 3] with $f(\hat{x}) = 0$. Since $f'(x) = 3x^2 - 2 > 0$ for x in [2, 3], f is increasing on the interval [2, 3] and its graph intersects the x-axis at most once. So there is precisely one number $\hat{x}$ in [2, 3] with $f(\hat{x}) = 0$. Choosing $x_0 = 2$, we have

$$x_1 = x_0 - \frac{f(x_0)}{f'(x_0)}$$

$$= x_0 - \frac{x_0^3 - 2x_0 - 5}{3x_0^2 - 2} = 2 - \frac{2^3 - 2(2) - 5}{3(2)^2 - 2} = 2 - \frac{-1}{10} = 2.1.$$

Reapplying Newton's method with x_1 replacing x_0 gives the new approximation:

$$x_2 = x_1 - \frac{f(x_1)}{f'(x_1)}$$

$$= 2.1 - \frac{(2.1)^3 - 2(2.1) - 5}{3(2.1)^2 - 2} \approx 2.0946.$$

Applying the method once more leads to the approximation

$$x_3 = x_2 - \frac{f(x_2)}{f'(x_2)}$$

$$\approx 2.0946 - \frac{(2.0946)^3 - 2(2.0946) - 5}{3(2.0946)^2 - 2}$$

$$\approx 2.09455148,$$

a result that can be shown to be correct to the places listed.   □

EXAMPLE 2   Find an approximate value for $\sqrt{3}$.

SOLUTION

To use Newton's method to find a solution to this problem we need to determine a function that is zero at $\hat{x} = \sqrt{3}$. One such function is described by

$$f(x) = x^2 - 3.$$

To approximate the positive solution to the equation $f(x) = 0$, we choose $x_0 = 1.75$ as our initial approximation. Since $f'(x) = 2x$,

$$x_1 = x_0 - \frac{f(x_0)}{f'(x_0)}$$

$$= 1.75 - \frac{(1.75)^2 - 3}{2(1.75)} \approx 1.732.$$

Reapplying the method,

$$x_2 = x_1 - \frac{f(x_1)}{f'(x_1)}$$

$$\approx 1.732 - \frac{(1.732)^2 - 3}{2(1.732)}$$

$$\approx 1.732050808,$$

which is the correct value of $\sqrt{3}$ to the places listed.   □

As these examples indicate, Newton's method can be very effective. Often, if the initial approximation is sufficiently accurate, each application of Newton's method will essentially double the number of correct decimal places in the approximation. (The precise accuracy result will be considered in Section 9.10.) Another advantage of Newton's method is that, since it is a recursive procedure, the approximations $x_1, x_2, x_3, \ldots$ are easily generated with a simple computer program. However, Newton's method is not successful in all situations and in others it will work, but the approximations will not improve as rapidly as our examples have shown.

HISTORICAL NOTE   In 1685, **John Wallis** (1616–1703) used Example 1 in his book, *Algebra,* to illustrate the method Isaac Newton devised for approximating solutions of equations. A special case of Newton's method had been known from Babylonian times for approximating square roots and was also known to apply to problems involving polynomials. Newton showed in his *Method of Fluxions* that the technique works equally well for more general equations.

The first systematic account of the method was provided by Joseph Raphson (1648–1715) in 1690.

EXERCISE SET 4.6

For these exercises, assume that if two consecutive approximations agree to a certain number of decimal places, the later approximation is accurate to at least that number of places.

In Exercises 1 through 10, approximate to four decimal places the value of x lying in the given interval for which $f(x) = 0$.

1. $f(x) = x^3 - 2x^2 - 5, [2, 3]$ **2.** $f(x) = x^3 - x - 1, [1, 2]$

3. $f(x) = x^3 - 9x^2 + 12, [-2, 0]$ **4.** $f(x) = x^3 - 9x^2 + 12, [0, 2]$

5. $f(x) = x^3 - 9x^2 + 12, [8, 9]$ **6.** $f(x) = x^3 + 3x^2 - 1, [-4, -1]$

7. $f(x) = \cos x - x, \left[0, \dfrac{\pi}{2}\right]$ **8.** $f(x) = x - .2\sin x - .8, \left[0, \dfrac{\pi}{2}\right]$

9. $f(x) = x^4 - 2x^3 - 5x^2 + 12x - 5, [0, 1]$

10. $f(x) = 2x^4 - 8x^3 + 8x^2 - 1, [1, 2]$

11. Use Newton's method to approximate $\sqrt{5}$ to four decimal places.

12. Use Newton's method to approximate $\sqrt[3]{25}$ to five decimal places.

13. Use Newton's method to approximate $\sqrt[4]{1000}$ to five decimal places.

14. Newton's method applied to $f(x) = x + \sqrt{x} - 1 = 0$ can be expressed as

$$x_{n+1} = x_n - \frac{f(x_n)}{f'(x_n)} = x_n - \frac{2x_n\sqrt{x_n} + 2x_n - 2\sqrt{x_n}}{2\sqrt{x_n} + 1}.$$

Apply Newton's method to this function on the interval $[0, 1]$ with (a) $x_0 = 1$, (b) $x_0 = 0$. Explain any apparent difficulties.

15. Use Newton's method to find an approximation, accurate to three decimal places, to the point on the graph of $y = x^2$ that is closest to $(1, 0)$. (*Hint:* Let $d(x)$ denote the distance from (x, x^2) on the graph of $y = x^2$ to $(1, 0)$. Then use the methods of Section 4.2 to minimize $[d(x)]^2$.)

16. Use Newton's method to find an approximation, accurate to three decimal places, to the point on the graph of $y = 1/x$ that is closest to the point $(1, 0)$.

17. Exercise 16 of Exercise Set 4.2 describes the process involved in constructing a can with a fixed volume in the shape of a right circular cylinder. The actual construction of such a can requires that the circular top and bottom be cut with a radius greater than the radius of the can so that the excess material can be used to form a seal with the side of the can. Assume that circles of diameter $2r + .5$ cm are required to form the top and bottom of a can with radius r. Approximate, to three decimal places, the minimal amount of material to construct such a can containing 1500 cm^3.

18. The **secant method** approximates the solution of an equation of the form $f(x) = 0$ without using the derivative of f. The method proceeds by assuming two initial approximations, x_0 and x_1, to the solution of the equation and computing a new approximation x_2, by determining where the line joining $(x_0, f(x_0))$ and $(x_1, f(x_1))$ crosses the x-axis. The points $(x_1, f(x_1))$ and

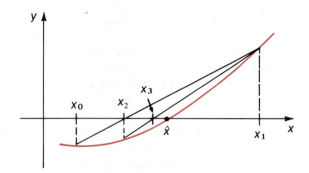

$(x_2, f(x_2))$ can then be used to construct another line that crosses the x-axis at the perhaps better approximation x_3, and so on.

(a) Show that the value for x_2 is

$$x_2 = x_1 - \frac{f(x_1)\,(x_1 - x_0)}{f(x_1) - f(x_0)}.$$

(b) How could the secant method be derived from Newton's method?

19. Use the secant method described in Exercise 18 to find an approximation, accurate to two decimal places, to the solution in the interval $[1, 2]$ of $x^3 - 3x - 1 = 0$.

20. Use the intermediate value theorem to show that $\sqrt{17}$ lies in the interval $[4, 5]$. Use the secant method described in Exercise 18 to find an approximation to $\sqrt{17}$ that is accurate to two decimal places.

4.7
APPLICATIONS TO BUSINESS AND ECONOMICS

Determining the rate at which certain quantities change is a subject of primary interest to financial planners. Company executives need to know how a price change relates to the change in the demand, how a demand change relates to the change in revenue and profit, and so on. In economics, the adjective "marginal" is used to express the instantaneous rate of change of a quantity. The reason for using the term marginal is more easily seen if you consider it in light of its synonym, "extra." The marginal total cost, for example, is the rate at which the total cost changes relative to the change in the number of units produced. As such, it provides an estimate of the cost of producing one extra, or marginal, unit of the item.

Before considering the application of the derivative to economic problems, we need the definitions of some common economic functions.

Cost, or **total cost, function** (denoted C or TC) describes the cost $C(x)$ of producing x units of an item.

Average cost function (denoted by c or AC) describes the cost $c(x)$ of producing a single unit of an item if x units are produced: $c(x) = C(x)/x$.

Demand function (denoted p) describes the price $p(x)$ per unit of an item when x units are sold.

Revenue, or **total revenue, function** (denoted R or TR) describes the revenue $R(x)$ produced when x units of an item are sold: $R(x) = xp(x)$.

Profit, or **total profit, function** (denoted P or TP) describes the profit $P(x)$ realized from the sale of x units of an item: $P(x) = R(x) - C(x)$.

To see how the marginal concepts are developed, consider the case of marginal cost. If $C(x)$ is the cost of producing x units of an item, then the cost of producing one additional item is $\Delta C = C(x + 1) - C(x)$. In this case $\Delta x = dx = 1$. For large values of x it is reasonable to approximate ΔC by the differential, dC; thus,

$$C'(x) = \frac{dC}{dx} \approx \frac{C(x + 1) - C(x)}{1} = C(x + 1) - C(x).$$

The marginal cost at x is consequently defined to be $C'(x)$. In fact, associated with each differentiable economic function is a marginal function.

Marginal cost function (denoted C', or by economists MC) describes the rate of change of the cost relative to the change in the number of units produced.

Marginal average cost function (denoted c' or MAC) describes the rate of change of the cost of a single unit of an item relative to the change in the number of units produced.

Marginal demand function (denoted p' or Mp) describes the rate of change of the demand for an item relative to the change in the number of units produced.

Marginal revenue function (denoted R' or MR) describes the rate of change of the revenue produced from the sale of an item relative to the number of units sold.

Marginal profit function (denoted P' or MP) describes the rate of change of the profit produced from the sale of an item relative to the number of units sold.

EXAMPLE 1

Maple Leaf Moccasins Ltd. can produce between 100 and 1000 pairs of moccasins per day. The cost of producing x pairs of moccasins, $100 \leq x \leq 1000$, is given by

$$C(x) = 295 + 3.28x + .003x^2 \text{ dollars.}$$

The demand function for the company is:

$$p(x) = 7.47 + \frac{321}{x} \text{ dollars per pair.}$$

(a) Describe the marginal cost, marginal average cost, marginal demand, marginal revenue, and marginal profit functions.
(b) Determine the number of pairs that should be produced if the average cost per unit is to be a minimum, and the number of pairs that should be produced in order to maximize the profit.

SOLUTION

(a) The average cost is given by

$$c(x) = \frac{C(x)}{x} = \frac{295}{x} + 3.28 + .003x,$$

the revenue by

$$R(x) = xp(x) = 7.47x + 321$$

and the profit by

$$P(x) = R(x) - C(x) = 4.19x + 26 - .003x^2.$$

The marginal functions are consequently:

Marginal cost: $C'(x) = 3.28 + .006x$

Marginal average cost: $c'(x) = \frac{-295}{x^2} + .003$

Marginal demand: $p'(x) = \frac{-321}{x^2}$

Marginal revenue: $R'(x) = 7.47$

Marginal profit: $P'(x) = 4.19 - .006x.$

To see the significance of these marginal functions, suppose we let $x = 200$. Then the 201st pair of moccasins will cost approximately $C'(200) = \$4.48$ to produce, lower the average production cost per pair by $c'(200) = \$-.004375$, decrease the price per pair by approximately $p'(200) = \$-.008025$, produce approximately $R'(200) = \$7.47$ in revenue, and generate approximately $P'(200) = \$2.99$ profit.

(b) The minimal average cost can occur only at the endpoints of the domain of c (when $x = 100$ or $x = 1000$) or when $c'(x) = 0$. If $c'(x) = 0$, then

$$\frac{295}{x^2} = .003$$

and

$$x^2 = \frac{295}{.003};$$

so

$$x = \sqrt{\frac{295}{.003}} \approx 314.$$

Since $c(314) = 5.16$, $c(100) = 6.53$, and $c(1000) = 6.58$, to obtain the minimal average cost, 314 pairs should be produced.

The maximum profit can occur only when $x = 100$, $x = 1000$ or when $P'(x) = 0$. If $P'(x) = 0$, $.006x = 4.19$, so $x = 4.19/.006 \approx 698$ and $P(698) = 1489$. Since $P(100) = 415$ and $P(1000) = 1216$, the production of 698 pairs produces the maximum profit.   □

Business executives frequently use charts and graphs to determine trends and as a basis for business decisions. Shown in Figure 4.18 are the graphs of the average cost and marginal cost functions from Example 1. Note that in Example 1 the minimal average cost of production occurs at the intersection of the average and marginal cost graphs. This fact is true in general.

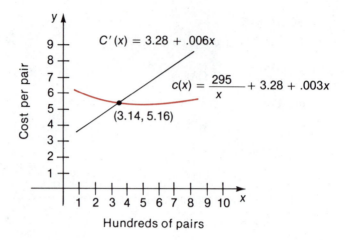

FIGURE 4.18

Since

$$c(x) = \frac{C(x)}{x},$$

$$c'(x) = \frac{xC'(x) - C(x)}{x^2},$$

which is zero precisely when $xC'(x) - C(x) = 0$ so

$$C'(x) = \frac{C(x)}{x} = c(x).$$

Consequently, a relative minimum for $c(x)$ can occur only when the marginal cost and average cost functions coincide.

In a practical business situation, it may be unrealistic to determine explicit cost functions. However, as production increases it is relatively easy to determine values of $c(x)$ and estimates for $C'(x)$ based on the approximation

$$C'(x) \approx C(x + 1) - C(x).$$

In general, $C'(x)$ is less than $c(x)$ and is increasing while $c(x)$ is decreasing. When $c(x)$ and $C'(x)$ agree, production should increase no further if production levels are based on minimal average cost of production.

In a similar manner, the number of units that should be produced and sold to maximize profit can be estimated, since

$$P'(x) = R'(x) - C'(x).$$

If $P'(x) = 0$, then $R'(x) = C'(x)$, that is, the profit function P can have a local maximum only if the marginal revenue is equal to the marginal cost. Both of these quantities can be estimated from sales and production data by

$$R'(x) \approx R(x + 1) - R(x)$$

and $$C'(x) \approx C(x + 1) - C(x).$$

The graphs of the marginal cost and marginal revenue functions from Example 1 intersect when $x = 698$ pairs, which is the production that yields a maximum profit. See Figure 4.19.

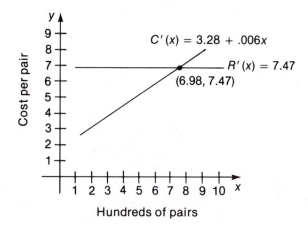

FIGURE 4.19

The next example illustrates how a relatively minor change in a firm's cost function with no accompanying change in the demand function can influence the profitability of the company.

EXAMPLE 2   Suppose the government levies a tax of 6% on the cost of producing the moccasins in Example 1 and that the demand function for the firm remains

$$p(x) = 7.47 + \frac{321}{x} \text{ dollars per pair.}$$

Determine the number of pairs that should be produced if the average cost per unit is to be a minimum and the number of pairs that should be produced to maximize the profit.

SOLUTION

The cost function after the government tax is

$$C(x) = 1.06(295 + 3.28x + .003x^2)$$

so $$c(x) = 1.06\left(\frac{295}{x} + 3.28 + .003x\right) \text{ dollars per pair.}$$

Consequently,

$$c'(x) = 1.06\left(\frac{-295}{x^2} + .003\right)$$

and the only critical point again occurs at $x = 314$. Since $c(314) = 5.47$, $c(100) = 6.92$, and $c(1000) = 6.97$, the minimal average cost, while increasing by 6%, still occurs at $x = 314$.

The profit function is now

$$\begin{aligned}
P(x) &= xp(x) - C(x) \\
&= 7.47x + 321 - 1.06(295 + 3.28x + .003x^2) \\
&= 3.9932x + 8.30 - .00318x^2.
\end{aligned}$$

Thus,

$$P'(x) = 3.9932 - .00636x$$

and a critical point occurs when

$$x = \frac{3.9932}{.00636} \approx 628.$$

Since $P(628) \approx 1262$, $P(100) \approx 376$, and $P(1000) \approx 822$, the maximum profit occurs when 628 pairs are produced.

By looking at the maximum profit of $1489 found in Example 1, we see that the 6% increase in the cost of production decreased the maximum profit by $227, or over 15%. The increase in cost also caused a decrease in production of over 10%, if the maximum profit is to be realized.   □

A generally accepted business principle is that an increase in the price of an object is accompanied by a decrease in its demand while a price decrease leads to a demand increase. If the unit price is increased, more revenue per unit will result; but if the price increase leads to a significant decrease in demand the total revenue can decrease. To make sound pricing decisions, a quantitative means is needed to measure the public's responsiveness to a price change. A measure of this responsiveness is given by the ratio of the percentage of change in demand to the percentage of change in price. The ratio when a units are sold is:

(4.7)
$$-\frac{\left(\dfrac{x - a}{a}\right)}{\left(\dfrac{p(x) - p(a)}{p(a)}\right)} = -\frac{p(a)}{a} \cdot \frac{1}{\dfrac{p(x) - p(a)}{x - a}},$$

where $p(x)$ denotes the selling price when x units are sold. (The minus sign is used so that this ratio will in general be positive; if $p(x) - p(a) > 0$, then $x - a$ will usually be negative.)

When x is close to a, $(p(x) - p(a))/(x - a)$ can be approximated by $p'(a)$, so the ratio in (4.7) can be approximated by $-p(a)/ap'(a)$.
The function E defined by

(4.8)
$$E(x) = \frac{-p(x)}{xp'(x)}$$

measures the change in demand relative to a change in the selling price when x units are sold, and is called the **price elasticity of demand**.

Elasticity can be used to measure the responsiveness of total revenue to price changes. Since $R(x) = xp(x)$,

$$R'(x) = p(x) + xp'(x)$$

$$= p(x)\left(1 + \frac{xp'(x)}{p(x)}\right)$$

$$= p(x)\left(1 - \frac{1}{E(x)}\right).$$

(i) When $E(x) > 1$, for x within an interval, $1/E(x) < 1$, $R'(x) > 0$ and $R(x)$ is increasing. The demand function is said to be **elastic** on the interval, indicating that an increase in the price produces an increase in total revenue. The larger the value of $E(x)$, the more dramatic the change.

(ii) When $E(x) < 1$, for x within an interval, $1/E(x) > 1$, $R'(x) < 0$ and $R(x)$ is decreasing. The demand function is said to be **inelastic** on the interval, indicating that an increase in price is accompanied by a decrease in total revenue.

(iii) When $E(x) = 1$, the demand function is said to have **unit elasticity**; changes in price do not influence total revenue.

It is important to realize that elasticity measures the relative change in total revenue with respect to the change in price. Elasticity does not involve the cost function, so it does not measure the change in profit with respect to the change in price. Only when profit can be assumed to be a fixed proportion of the revenue can elasticity be used to determine pricing strategies for maximizing profit.

EXAMPLE 3

Suppose that a demand function for food processors is $p(x) = 200 - .02x$ dollars per unit. Will a rise in price increase or decrease the revenue received if 1000 food processors are currently sold? At what level of sales will it not make any difference whether the price is increased or decreased?

SOLUTION

$$E(x) = \frac{-p(x)}{xp'(x)} = \frac{-(200 - .02x)}{x(-.02)} = \frac{200 - .02x}{.02x}$$

$$E(1000) = \frac{200 - 20}{20} = 9 > 1.$$

Since $E(1000) > 1$, there will be an increase in revenue if prices rise.

To determine the level of sales at which a price change will not influence the total revenue, we determine the number of units x for which $E(x) = 1$.

$$E(x) = \frac{200 - .02x}{.02x} = 1 \text{ if } .04x = 200, \qquad \text{that is, if } x = 5000.$$

If less than 5000 units are currently sold, an increase in price produces an increase in total revenue. A decrease in revenue is produced if more than 5000 units are currently sold and prices rise. There is unit elasticity at $x = 5000$. $\square$

EXERCISE SET 4.7

For each of the cost functions given in Exercises 1 through 4 find (a) the average cost function, (b) the marginal cost function, (c) the marginal average cost function, (d) the average cost per unit if 1000 units of an item are produced, and (e) the cost of the one-thousand-and-first unit produced.

1. $C(x) = 1500 + 2x - .0003x^2$

2. $C(x) = 875 + 2.6x + .01x^2$

3. $C(x) = 15500 + 77x - .00001x^3$

4. $C(x) = 38 + 4.75x - .01\sqrt{x}$

In Exercises 5 through 8, $R(x)$ denotes the revenue realized from the sale of x units of an item. Find the marginal revenue function.

5. $R(x) = 3.49 - .0002x$

6. $R(x) = 43.52 - .01x + .00002x^2$

7. $x^2R(x) - 500x + 300R(x) - 1900 = 0$

8. $x^2R(x) + 10R(x) + .01\,R(x)^2 - 25 = 0$

Trasho Manufacturing Inc. figures that the total cost of producing x plastic wastebaskets per week is

$$C(x) = 1700 + .76x + .0001x^2 \text{ dollars}$$

and that the total revenue from their sale is

$$R(x) = 2.85x - .00008x^2 \text{ dollars}.$$

9. Determine the marginal cost, marginal revenue, and marginal profit functions for the Trasho corporation.

10. Determine the number of wastebaskets that Trasho should produce if the average cost per basket is to be minimized.

11. (a) How many wastebaskets per week should be produced if Trasho wishes to maximize profit? (b) How much profit is made if this number is produced?

12. Sketch the graphs of the marginal cost and marginal revenue functions and show that they intersect when the maximum profit is produced.

The employees of the Trasho company are given a raise that adds $.09x$ to the weekly cost function. This raise is passed on to the distributor, which changes the revenue function to

$$R(x) = 2.94x - .00008x^2.$$

13. How does this change modify the maximum profit the company can realize and the number of wastebaskets that should be produced to maximize the profit?

14. Suppose that the total wages paid to all Trasho employees is proportional to the maximum profit that can be realized. Have the employees collectively benefited from the wage increase?

Suppose that instead of adding $.09x$ to the weekly Trasho cost function, the management determines that the amount of the raise should be incorporated into the fixed cost. The new cost function in this case is

$$C(x) = 2400 + .76x + .0001x^2.$$

The revenue function remains

$$R(x) = 2.94x - .00008x^2.$$

15. What is the maximum profit that can now be realized and how many wastebaskets should be produced? Compare this result with those in Exercises 11 and 13.

16. Suppose that the total wages paid to all Trasho employees is proportional to the maximum profit that can be realized. Should the employees prefer this cost-function approach to the one described before Exercise 13?

17. The cost of producing x genuine, right-off-the-tree, birchbark canoes is

$$C(x) = 300x + 500 \text{ dollars.}$$

Each canoe is sold for $1000. Find the revenue function, profit function, and marginal profit function.

18. The Dull calculator company estimates that the cost of producing x calculators is

$$C(x) = 2000 + 10x - \frac{10000}{x} \text{ dollars.}$$

Find (a) the marginal cost function, (b) the average cost function, (c) the marginal cost when $x = 1000$, (d) the approximate cost of producing the one-thousand-and-first calculator, and (e) the number of calculators that should be produced to minimize the average cost. Sketch the graphs of the marginal cost and average cost functions.

19. Let $p(x) = \dfrac{600 - x^2}{25}$ be the selling price when x units of an item are sold.

    (a) Find the price elasticity of demand.
    (b) Find the value of x that gives unit elasticity.
    (c) Will a rise in price increase or decrease revenue if 10 units are sold? If 20 units are sold?

20. Consider a demand function

$$p(x) = \frac{36}{x - 36} + 12,$$

where $p(x)$ is the selling price when x units of an item are sold. Find the price elasticity of demand. Will a rise in price increase or decrease revenue if 40 units are in demand?

21. The demand function for a certain item is

$$p(x) = 80 - .01x \text{ dollars per unit.}$$

Find the number of units at which a price change will not influence total revenue.

REVIEW EXERCISES

1. Consider the function described by $f(x) = x^3/3 - x^2 - 3x + 4$.
    (a) Find the slope of the tangent line to the graph of f at any point $(x, f(x))$.
    (b) If $f(x)$ describes the distance of an object from a fixed point at the end of time x, find the instantaneous velocity of the object at time x.
    (c) If $f(x)$ gives the profit realized from the sale of x units of an item, find the marginal profit function.
    (d) If $y = f(x)$, find dy.

2. The distance of an object from a specified point at the end of time t is given by $s(t) = 4t - t^2$. (a) Find the average velocity from $t = 0$ to $t = 2$. (b) Find the instantaneous velocity and acceleration of the object when $t = 2$.

3. A ball is thrown upward from a point 3 feet above the ground with an initial velocity of 88 ft/sec. How long will the ball remain in the air?

4. Find the dimensions of the largest rectangle that can be inscribed in a right triangle with sides 5, 12, and 13 inches, if one vertex of the rectangle is on the longest side of the triangle.

5. A can in the shape of a right circular cylinder with a bottom but no top contains a volume of 900 cm^3. Find the dimensions of the can that will minimize the amount of material needed for construction.

6. An angler has a fish at the end of a line. The line is reeled in at 2 feet per second from a bridge 30 feet above the water. At what rate is the fish moving through the water when the length of the line is 50 feet?

7. Find the shortest distance from the point $(0, 5)$ to the parabola $4y = x^2$.

8. An open water tank in the shape of an inverted right circular cone is to be designed to hold 576 ft^3. Find the dimensions of the cone that has the least lateral surface area.

9. Find the dimensions of the right circular cylinder of greatest volume that can be inscribed in a right circular cone with a radius of 6 inches and a height of 12 inches.

10. A car traveling 60 mph northward on a straight road crosses a railroad track that is perpendicular to the road. A train going 80 mph directly eastward crosses the road 15 minutes later. At what rate are the car and train separating 30 minutes after the train crosses the road?

11. Water flows at the rate of 2 ft^3/min into a tank in the shape of an inverted right circular cone of altitude 6 feet and radius 2 feet. At what rate is the surface of the water rising when the tank is half full?

12. A baseball diamond is a 90 foot square. A player hits a ball along the third-base line at 100 feet per second and runs to first base at 25 feet per second.
    (a) At what rate is the distance between the ball and first base changing when the ball is halfway to third base?
    (b) At what rate is the distance between the ball and the player changing when the ball is halfway to third base?

13. An object moves along the graph described by the equation $y = x^2$. Find the point on the graph at which the y-coordinate is changing at the same rate as the x-coordinate.

14. When a circular metal plate is heated, its radius increases at the rate of 1 millimeter per second. At what rate is the area of the plate increasing when the radius is 4 centimeters?

15. A spherical water tank has many coats of old paint and a radius, with the paint, of exactly 3 feet. The paint is ground off and the tank repainted with a net decrease of .2 inches in the radius. Use differentials to approximate the volume of paint removed.

16. Use Newton's method to find an approximation, accurate to three decimal places, to the value of x for which $f(x) = 0$ in the interval given.
    (a) $f(x) = x^3 - 2x - 1$, $[1, 2]$
    (b) $f(x) = \sin x - \left(\dfrac{x}{2}\right)^2$, $(0, \infty)$
    (c) $f(x) = x^4 - x^3 - 1$, $[0, 2]$

17. Use Newton's method to find an approximation, accurate to three decimal places, to a solution of $x = \tan x$, $x > 0$.

18. A potter estimates that the cost of making x mugs is

$$C(x) = 500 + 3x + \frac{10}{x} \text{ dollars.}$$

Find:
(a) the marginal cost function,
(b) the average cost function,
(c) the cost of producing the one-hundred-and-first mug.

Use L'Hôpitals rule, if applicable, to determine the limits in Exercises 19 through 28.

19. $\displaystyle\lim_{x \to 1} \frac{x^3 - 3x^2 + 3x - 1}{x^2 - 1}$

20. $\displaystyle\lim_{x \to -1} \frac{x^3 - 3x^2 + 3x - 1}{x^2 - 1}$

21. $\displaystyle\lim_{x \to 3} \frac{\sqrt{x + 1} - 2}{x^2 - 9}$

22. $\displaystyle\lim_{x \to 2} \frac{3 - \sqrt{x + 7}}{x^2 - 4x + 4}$

23. $\displaystyle\lim_{x \to 0^+} \frac{\sin x}{\sqrt{x}}$

24. $\displaystyle\lim_{x \to 0} \frac{\sin x - x}{x^3}$

25. $\displaystyle\lim_{x \to \frac{\pi}{2}^-} \frac{\cot x}{x - \dfrac{\pi}{2}}$

26. $\displaystyle\lim_{x \to 0} \frac{x \cos x}{\sin x}$

27. $\displaystyle\lim_{x \to 0^+} \left(\frac{1}{x} - \frac{1}{\tan x} \right)$

28. $\displaystyle\lim_{x \to 0^-} \left(\frac{1}{\sin x} - \frac{1}{\tan x} \right)$

29. Two trees 21 feet apart are to be reinforced by a nylon rope connected to the trunks and tied at ground level to a stake in the ground between the trees. If the rope is to be tied 15 feet above the ground on one tree and 20 feet above the ground on the other tree, find the location of the stake that will minimize the amount of rope to be used.

5

THE INTEGRAL

Calculus is divided into two main areas. One is differential calculus, in which the derivative is the principal concept. The other is integral calculus, which involves the integral.

In this chapter the integral is introduced and studied. We begin with the problem of finding the area under a given curve and show how the solution of this problem leads to the notion of the integral.

5.1
AREA

Consider the problem of determining the area enclosed in the region bounded by the graph of a nonnegative continuous function f, the x-axis, and the lines $x = a, x = b$. (See Figure 5.1.)

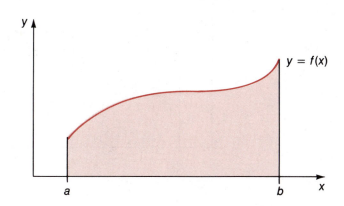

FIGURE 5.1

There are a limited number of geometric figures for which the area can easily be determined. We can calculate the area of a rectangle, trapezoid, triangle, or circle, and other geometric figures such as a pentagon or octagon. In general there is no algebraic method for calculating the area of a region such as that shown in Figure 5.1. We can, however, approximate this area.

A crude approximation A_1 to the area A of the region is the area of the rectangle whose base is the interval $[a, b]$ and whose height is the value of the function f at b, as shown in Figure 5.2.

$$A_1 = f(b) (b - a).$$

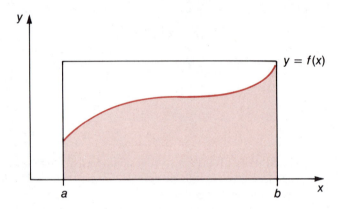

FIGURE 5.2

To find a better approximation to A, divide the interval $[a, b]$ into two equal parts as in Figure 5.3. Since the width of $[a, b]$ is $b - a$, the two subintervals of equal width are $[a, a + (b - a)/2]$ and $[a + (b - a)/2, b]$. The height of the rectangle on the left is $f\left(a + \dfrac{b - a}{2}\right) = f\left(\dfrac{a + b}{2}\right)$ and the height of the rectangle on the right is $f(b)$.

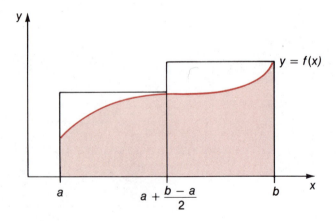

FIGURE 5.3

Letting A_2 denote the sum of the area of the two rectangles in Figure 5.3, we have

$$A_2 = f\left(a + \frac{b-a}{2}\right) \cdot \left(\frac{b-a}{2}\right) + f(b) \cdot \left(\frac{b-a}{2}\right).$$

Likewise, dividing the interval $[a, b]$ into three equal parts as in Figure 5.4 produces the approximation

$$A_3 = f\left(a + \frac{b-a}{3}\right) \cdot \left(\frac{b-a}{3}\right) + f\left(a + 2\frac{(b-a)}{3}\right) \cdot \left(\frac{b-a}{3}\right)$$

$$+ f(b) \cdot \left(\frac{b-a}{3}\right).$$

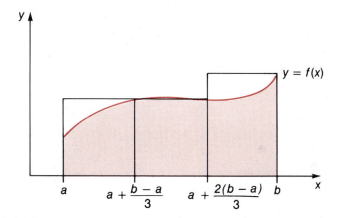

FIGURE 5.4

At the nth stage of approximation, that is, when $[a, b]$ is divided into n subintervals, the approximating area

$$A_n = f\left(a + \frac{b-a}{n}\right)\left(\frac{b-a}{n}\right) + f\left(a + \frac{b-a}{n}2\right)\left(\frac{b-a}{n}\right) + \ldots$$

$$+ f\left(a + \frac{b-a}{n}(n-1)\right)\left(\frac{b-a}{n}\right) + f(b)\left(\frac{b-a}{n}\right)$$

is the sum of the area of n rectangles. (See Figure 5.5.)

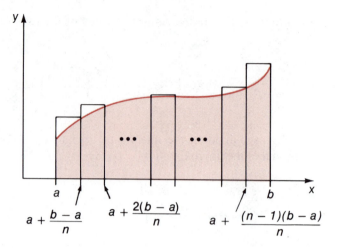

FIGURE 5.5

If n is large, the notation involved with this summation is cumbersome. To condense this notation, we use the summation symbol, Σ (the Greek capital letter sigma), to represent consecutive sums.

(5.1)
DEFINITION   The notation $\displaystyle\sum_{i=p}^{n} a_i$ denotes the sum of the quantities a_i starting with the index p and ending with the index n; that is,

$$\sum_{i=p}^{n} a_i = a_p + a_{p+1} + a_{p+2} + \ldots + a_{n-1} + a_n.$$

EXAMPLE 1   Find the values of $\displaystyle\sum_{i=1}^{10} i$, $\displaystyle\sum_{i=2}^{5} \frac{2i-1}{3}$, and $\displaystyle\sum_{i=1}^{4} \sqrt{3i+2}$.

SOLUTION

$$\sum_{i=1}^{10} i = 1 + 2 + 3 + 4 + 5 + 6 + 7 + 8 + 9 + 10 = 55.$$

$$\sum_{i=2}^{5} \frac{2i-1}{3} = \frac{2\cdot 2 - 1}{3} + \frac{2\cdot 3 - 1}{3} + \frac{2\cdot 4 - 1}{3} + \frac{2\cdot 5 - 1}{3}$$

$$= \frac{3}{3} + \frac{5}{3} + \frac{7}{3} + \frac{9}{3} = \frac{24}{3} = 8,$$

and

$$\sum_{i=1}^{4} \sqrt{3i+2} = \sqrt{3\cdot 1 + 2} + \sqrt{3\cdot 2 + 2} + \sqrt{3\cdot 3 + 2} + \sqrt{3\cdot 4 + 2}$$

$$= \sqrt{5} + \sqrt{8} + \sqrt{11} + \sqrt{14} \approx 12.123. \qquad \square$$

The following results concerning summation notation will be used to solve area problems. In general, the technique of mathematical induction discussed in Appendix A.4 is needed to prove these results.

(5.2) $$\sum_{i=1}^{n} 1 = n$$

(5.3) $$\sum_{i=p}^{n} ca_i = c \sum_{i=p}^{n} a_i, \text{ for any constant } c$$

(5.4) $$\sum_{i=p}^{n} (a_i + b_i) = \sum_{i=p}^{n} a_i + \sum_{i=p}^{n} b_i$$

(5.5) $$\sum_{i=1}^{n} i = \frac{n(n + 1)}{2}$$

(5.6) $$\sum_{i=1}^{n} i^2 = \frac{n(n + 1)(2n + 1)}{6}$$

(5.7) $$\sum_{i=1}^{n} i^3 = \frac{n^2(n + 1)^2}{4}$$

EXAMPLE 2   Find the value of the sum $\sum_{i=1}^{10} (3i + 5)$.

SOLUTION

Utilizing the properties listed above, we have:

$$\sum_{i=1}^{10} (3i + 5) = \sum_{i=1}^{10} 3i + \sum_{i=1}^{10} 5$$

$$= 3\sum_{i=1}^{10} i + 5\sum_{i=1}^{10} 1$$

$$= 3\left(\frac{10 \cdot 11}{2}\right) + 5 \cdot 10 = 215. \qquad \square$$

Returning to the problem of calculating approximations to the area A, we see now that approximations A_1, A_2 and A_3 can be written as

$$A_1 = f(b)\,(b - a) = \sum_{i=1}^{1} f(a + (b - a)\,i)\,(b - a),$$

$$A_2 = f\left(\frac{b - a}{2}\right)\left(\frac{b - a}{2}\right) + f(b)\left(\frac{b - a}{2}\right) = \sum_{i=1}^{2} f\left(a + \frac{b - a}{2}\,i\right)\left(\frac{b - a}{2}\right),$$

$$A_3 = f\left(a + \frac{b - a}{3}\right)\left(\frac{b - a}{3}\right) + f\left(a + \frac{b - a}{3}\,2\right)\left(\frac{b - a}{3}\right) + f(b)\left(\frac{b - a}{3}\right)$$

$$= \sum_{i=1}^{3} f\left(a + \frac{b - a}{3}\,i\right)\left(\frac{b - a}{3}\right),$$

and in general,

$$A_n = f\left(a + \frac{b-a}{n}\right)\left(\frac{b-a}{n}\right) + f\left(a + \frac{b-a}{n}2\right)\left(\frac{b-a}{n}\right) + \ldots$$

$$+ f\left(a + \frac{b-a}{n}(n-1)\right)\left(\frac{b-a}{n}\right) + f(b)\left(\frac{b-a}{n}\right)$$

$$= \sum_{i=1}^{n} f\left(a + \frac{b-a}{n}i\right)\left(\frac{b-a}{n}\right).$$

As the number of rectangles is increased, we would expect the approximation to the actual area to improve. Thus to find A, the actual value of the area, we need to determine the value that A_n approaches as n becomes increasingly large. That is, we need to find

(5.8) $$A = \lim_{n \to \infty} A_n = \lim_{n \to \infty} \sum_{i=1}^{n} f\left(a + \frac{b-a}{n}i\right)\left(\frac{b-a}{n}\right),$$

assuming that this limit exists.

EXAMPLE 3   Calculate the area of the region bounded by the graph of $f(x) = x^2$, the x-axis, and the lines $x = 1$ and $x = 3$.

SOLUTION

Some approximations to this area are shown in Figure 5.6.

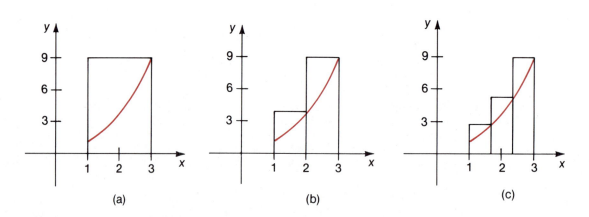

(a)   (b)   (c)

FIGURE 5.6

When $n = 1$ (as in Figure 5.6(a)):

$$A_1 = f(3)(3 - 1) = 9(2) = 18.$$

When $n = 2$ (as in Figure 5.6(b)):

$$A_2 = f(2) \cdot \frac{(3-1)}{2} + f(3) \cdot \frac{(3-1)}{2}$$
$$= 4(1) + 9(1) = 13.$$

When $n = 3$ (as in Figure 5.6(c)):

$$A_3 = f\left(\frac{5}{3}\right)\left(\frac{2}{3}\right) + f\left(\frac{7}{3}\right)\left(\frac{2}{3}\right) + f(3)\left(\frac{2}{3}\right)$$
$$= \left(\frac{25}{9}\right)\left(\frac{2}{3}\right) + \left(\frac{49}{9}\right)\left(\frac{2}{3}\right) + (9)\left(\frac{2}{3}\right) \approx 11.5.$$

In the general case shown in Figure 5.7,

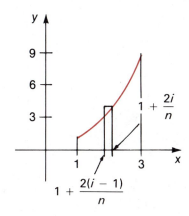

FIGURE 5.7

$$A_n = \sum_{i=1}^{n} f\left(1 + \frac{2}{n}i\right)\left(\frac{2}{n}\right) = \sum_{i=1}^{n} \left(1 + \frac{2i}{n}\right)^2\left(\frac{2}{n}\right)$$

$$= \sum_{i=1}^{n} \left(1 + \frac{4i}{n} + \frac{4i^2}{n^2}\right)\left(\frac{2}{n}\right) = \sum_{i=1}^{n} \left(\frac{2}{n} + \frac{8i}{n^2} + \frac{8i^2}{n^3}\right)$$

$$= \sum_{i=1}^{n} \frac{2}{n} + \sum_{i=1}^{n} \frac{8}{n^2}i + \sum_{i=1}^{n} \frac{8}{n^3}i^2$$

$$= \frac{2}{n}\sum_{i=1}^{n} 1 + \frac{8}{n^2}\sum_{i=1}^{n} i + \frac{8}{n^3}\sum_{i=1}^{n} i^2.$$

By properties (5.2), (5.5), and (5.6),

$$A_n = \frac{2}{n}(n) + \frac{8}{n^2} \cdot \frac{n(n+1)}{2} + \frac{8}{n^3} \cdot \frac{n(n+1)(2n+1)}{6}$$

$$= 2 + \frac{4}{n^2}(n^2 + n) + \frac{4}{3n^3}(2n^3 + 3n^2 + n)$$

$$= \frac{26}{3} + \frac{8}{n} + \frac{4}{3n^2}.$$

The area of the region bounded by the x-axis, the lines $x = 1$ and $x = 3$, and the graph of f is therefore

$$A = \lim_{n \to \infty} A_n = \lim_{n \to \infty} \left(\frac{26}{3} + \frac{8}{n} + \frac{4}{3n^2} \right) = \frac{26}{3}. \qquad \square$$

EXAMPLE 4   Suppose $f(x) = (x^3 + 2)/2$. Find the area of the region bounded by the graph of f, the x-axis, and the lines $x = -1$ and $x = 2$.

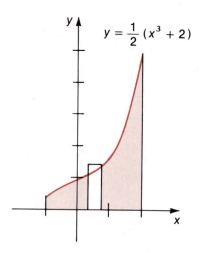

$$y = \frac{1}{2}(x^3 + 2)$$

FIGURE 5.8

SOLUTION

The region is shown in Figure 5.8. If we divide the interval $[-1, 2]$ into n subintervals of equal width, the common width of the subintervals is

$$\frac{2 - (-1)}{n} = \frac{3}{n}.$$

Since the left boundary of the region is the line $x = -1$, the first subinterval is $[-1, -1 + 3/n]$, the second subinterval $[-1 + 3/n, -1 + (3/n)2]$ and so on. The right endpoint of the ith subinterval is $-1 + (3/n)i$, so

$$A_n = \sum_{i=1}^{n} f\left(-1 + \frac{3}{n}i \right)\left(\frac{3}{n} \right)$$

$$= \sum_{i=1}^{n} \frac{1}{2}\left[\left(-1 + \frac{3}{n}i \right)^3 + 2 \right]\left(\frac{3}{n} \right).$$

Expanding and simplifying this expression for A_n leads to:

$$A_n = \frac{3}{2n} \sum_{i=1}^{n} \left[(-1)^3 + 3(-1)^2 \left(\frac{3i}{n}\right) + 3(-1)\left(\frac{3i}{n}\right)^2 + \left(\frac{3i}{n}\right)^3 + 2 \right]$$

$$= \frac{3}{2n} \sum_{i=1}^{n} \left[-1 + \frac{9i}{n} - \frac{27i^2}{n^2} + \frac{27i^3}{n^3} + 2 \right]$$

$$= \frac{3}{2n} \left[\sum_{i=1}^{n} 1 + \frac{9}{n} \sum_{i=1}^{n} i - \frac{27}{n^2} \sum_{i=1}^{n} i^2 + \frac{27}{n^3} \sum_{i=1}^{n} i^3 \right]$$

$$= \frac{3}{2n} \left[n + \frac{9}{n} \frac{n(n+1)}{2} - \frac{27}{n^2} \frac{n(n+1)(2n+1)}{6} + \frac{27}{n^3} \frac{n^2(n+1)^2}{4} \right]$$

$$= \frac{3}{2} + \frac{27}{4} \left(\frac{n+1}{n}\right) - \frac{27}{4} \frac{(n+1)(2n+1)}{n^2} + \frac{81}{8} \frac{(n+1)^2}{n^2}.$$

Thus, if A denotes the area of our region,

$$A = \lim_{n \to \infty} A_n$$

$$= \lim_{n \to \infty} \left[\frac{3}{2} + \frac{27}{4} \left(\frac{n+1}{n}\right) - \frac{27}{4} \frac{(n+1)(2n+1)}{n^2} + \frac{81}{8} \frac{(n+1)^2}{n^2} \right]$$

$$= \lim_{n \to \infty} \left[\frac{3}{2} + \frac{27}{4} \left(1 + \frac{1}{n}\right) - \frac{27}{4} \left(2 + \frac{3}{n} + \frac{1}{n^2}\right) + \frac{81}{8} \left(1 + \frac{2}{n} + \frac{1}{n^2}\right) \right]$$

$$= \frac{3}{2} + \frac{27}{4} - \frac{27}{4}(2) + \frac{81}{8}$$

$$= \frac{39}{8}. \qquad \square$$

EXAMPLE 5   Let $f(x) = \sin x$. Find the area of the region bounded by the graph of f and the x-axis on the interval $[0, \pi]$.

SOLUTION

Divide the interval $[0, \pi]$ into n subintervals as shown in Figure 5.9. The width of each subinterval is $(\pi - 0)/n = \pi/n$. The right endpoint of the ith subinterval is therefore $0 + (\pi/n)i = \pi i/n$.

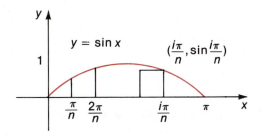

FIGURE 5.9

The area enclosed in the ith rectangle is

$$\left(\frac{\pi}{n}\right)f\left(\frac{\pi i}{n}\right) = \frac{\pi}{n}\sin\frac{\pi i}{n},$$

so the total area in the region is

$$\lim_{n\to\infty}\sum_{i=1}^{n}\frac{\pi}{n}\sin\frac{\pi i}{n} = \lim_{n\to\infty}\frac{\pi}{n}\sum_{i=1}^{n}\sin\frac{\pi i}{n}.$$

At this time we have no way to evaluate the limit. The table below lists values of $\dfrac{\pi}{n}\displaystyle\sum_{i=1}^{n}\sin\dfrac{\pi i}{n}$ obtained using a programmable calculator. This table suggests that the limit is 2, which is in fact so, as we shall see after we study the fundamental theorem of calculus in Section 5.4.

n	A_n
1	0
5	1.933766
10	1.983524
20	1.995886
50	1.999342
100	1.999836
1000	1.999998

EXERCISE SET 5.1

Find the values of the sums in Exercises 1 through 12.

1. $\displaystyle\sum_{i=1}^{4}(i + 2)$

2. $\left(\displaystyle\sum_{i=1}^{4}i\right) + 2$

3. $\displaystyle\sum_{i=1}^{7}(i^2 - 5)$

4. $\displaystyle\sum_{i=2}^{5}(3 - 2i^2)$

5. $\displaystyle\sum_{j=3}^{7}(j^2 + 2j + 2)$

6. $\left(\displaystyle\sum_{j=3}^{7}j^2 + j\right) + 2$

7. $\displaystyle\sum_{i=1}^{6}(-1)^i\, i$

8. $\displaystyle\sum_{j=3}^{7}(-1)^j\, j^2$

9. $\displaystyle\sum_{i=1}^{7}(i + 1)^2$

10. $\displaystyle\sum_{i=1}^{7}(i^2 + 2i)$

11. $\displaystyle\sum_{j=2}^{4}(ij + kj^2)$

12. $\displaystyle\sum_{i=2}^{4}(ij + kj^2)$

Use Formulas 5.2 through 5.7 to evaluate the sums in Exercises 13 through 18.

13. $\displaystyle\sum_{i=1}^{11}(i^2 + 2i)$

14. $\displaystyle\sum_{i=1}^{15}(3i^3 - i^2)$

15. $\displaystyle\sum_{i=1}^{n} (i^2 + i)$

16. $\displaystyle\sum_{i=1}^{n} i(i - 1)$

17. $\displaystyle\sum_{i=10}^{20} (i^3 - i^2)$

18. $\displaystyle\sum_{i=5}^{15} (i^3 + i^2 + 1)$

Express the sums given in Exercises 19 through 24 in closed form by using the summation notation.

19. $2 + 5 + 8 + \ldots + 23$

20. $1 + 3 + 5 + 7 + \ldots + 15$

21. $4 + 9 + 16 + \ldots + 64$

22. $2 - 4 + 6 - 8 + \ldots + 1002$

23. $1 - \dfrac{1}{2} + \dfrac{1}{3} - \dfrac{1}{4} + \ldots + \dfrac{1}{11}$

24. $\dfrac{1}{2} + \dfrac{2}{3} + \dfrac{3}{4} + \ldots + \dfrac{9}{10}$

In Exercises 25 through 36, find the area of the region bounded by the x-axis, the graph of f, and the given lines.

25. $f(x) = 2x + 3, x = -1, x = 1$

26. $f(x) = x + 1, x = 0, x = 1$

27. $f(x) = x^2, x = 0, x = 1$

28. $f(x) = x^2, x = 2, x = 4$

29. $f(x) = x^2 + x, x = 2, x = 4$

30. $f(x) = 2x^2, x = -1, x = 1$

31. $f(x) = (x + 1)^2, x = -1, x = 1$

32. $f(x) = (x + 1)^2, x = -3, x = -1$

33. $f(x) = 4 - x^2, x = -2, x = 0$

34. $f(x) = x^2 + 3x + 2, x = 0, x = 2$

35. $f(x) = x^3 + 1, x = 0, x = 2$

36. $f(x) = 3x^3 - 1, x = 1, x = 2$

37. Find the area of the region bounded by the lines $x = 1$ and $x = 3$, the x-axis, and the graph of f, where f is described below. Compare the answers in (c) and (d) to those in (a) and (b).
(a) $f(x) = 3$
(b) $f(x) = x$
(c) $f(x) = 3 + x$
(d) $f(x) = 3x$

38. The following sums have the "telescoping" property that each term cancels a part of the next term. Compute the sum by writing out the sums and cancelling where possible.

(a) $\displaystyle\sum_{k=5}^{15} \left(\dfrac{1}{k} - \dfrac{1}{k + 1} \right)$

(b) $\displaystyle\sum_{i=1}^{4} (2^i - 2^{i+1})$

(c) $\displaystyle\sum_{i=10}^{100} (10^{i+1} - 10^i)$

(d) $\displaystyle\sum_{j=1}^{10} (\sqrt{2j + 1} - \sqrt{2j - 1})$

39. Derive the formula for $\displaystyle\sum_{i=1}^{n} i$ by using the fact that the sum can be written as both

(a) $\displaystyle\sum_{i=1}^{n} i = 1 + 2 + 3 + \ldots + (n - 2) + (n - 1) + n$ and

(b) $\displaystyle\sum_{i=1}^{n} i = n + (n - 1) + (n - 2) + \ldots + 3 + 2 + 1$, and then add equations (a) and (b).

40. Derive a formula for $\displaystyle\sum_{i=1}^{n} r^i$ by using the fact that

$$\sum_{i=1}^{n} r^i = r + r^2 + r^3 + \ldots + r^n \text{ and}$$

$$r\sum_{i=1}^{n} r^i = r^2 + r^3 + \ldots + r^{n+1}.$$

41. The brothers of Gamma Gamma Gamma fraternity are planning an empty beer can stacking contest. They are going to require that the cans be stacked in a triangular fashion: the top row will have one can, the second row two cans, and so on.

(a) How many cans will they have to empty if they want to provide enough to reach 15 rows?

(b) Find a formula that will tell them how many cans they need to reach n rows, for any integer n.

42. A grocer stacks oranges in pyramid form with 150 on the bottom level in 15 rows of 10 each. How many oranges are in this pyramid if they are stacked until only one row is on top?

43. Suppose that $a_1, a_2, \ldots, a_n$ are constants and that f is a function defined by $f(x) = \displaystyle\sum_{i=1}^{n} (x - a_i)^2$. Show that f has an absolute minimum when $x = \dfrac{1}{n} \displaystyle\sum_{i=1}^{n} a_i$, the average of the numbers $a_1, a_2, \ldots, a_n$.

5.2
THE DEFINITE INTEGRAL

The repetitive summation process used in Section 5.1 occurs in many other applications. The volume of a solid can be approximated by summing the volumes of discs that resemble slices of the solid. (See Figure 5.10.)

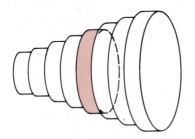 

FIGURE 5.10

The length of a curve can be approximated by summing the lengths of line segments joining points on the curve. The mass of a rod with varying density can be approximated by summing the mass of small segments of the rod on which the density is assumed to be constant.

Each of these applications and many others involve the same mathematical concept: find the limiting value of approximating sums. This concept involves the **definite integral**.

Let f be defined on the closed interval $[a, b]$. To define the definite integral of f we must introduce some new terminology.

We say that the finite set of numbers $\mathcal{P} = \{x_0, x_1, \ldots, x_n\}$ is a **partition** of the interval $[a, b]$ provided that $a = x_0 < x_1 < x_2 < \ldots < x_{n-1} < x_n = b$.

If Δx_1 denotes the quantity $x_1 - x_0$, Δx_2 denotes $x_2 - x_1$, and in general $\Delta x_i = x_i - x_{i-1}$, for each $i = 1, 2, \ldots, n$ we say that the **norm** of the partition $\mathcal{P}$, denoted by $\|\mathcal{P}\|$, is the largest of the values $\Delta x_1, \Delta x_2, \ldots, \Delta x_n$, that is, $\|\mathcal{P}\| = \text{maximum } \{\Delta x_1, \Delta x_2, \ldots, \Delta x_n\}$. (See Figure 5.11.)

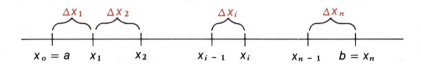

FIGURE 5.11

For example, $\mathcal{P} = \{0, 1/4, 3/4, 1\}$ is a partition of the interval $[0, 1]$; $\Delta x_1 = 1/4$, $\Delta x_2 = 1/2$ and $\Delta x_3 = 1/4$. So the norm of $\mathcal{P}$, $\|\mathcal{P}\|$, is $1/2$.

(5.9)
DEFINITION

Let f be defined on $[a, b]$, and $\mathcal{P} = \{x_0, x_1, \ldots, x_n\}$ be a partition of $[a, b]$. Suppose $z_1, z_2, \ldots, z_n$ are any numbers in $[a, b]$ with the property that $x_{i-1} \le z_i \le x_i$, for each $i = 1, 2, \ldots, n$. The **Riemann sum** of f with respect to the partition $\mathcal{P}$ and $z_1, z_2, \ldots, z_n$ is defined by

$$S(f, \mathcal{P}, \{z_i\}) = \sum_{i=1}^{n} f(z_i)\Delta x_i.$$

EXAMPLE 1

Suppose $f(x) = x^3$ and $\mathcal{P}$ is the partition $\{-1, 0, .4, 1, 1.25, 2\}$ of $[-1, 2]$. Find the norm of $\mathcal{P}$ and the Riemann sum associated with f and $\mathcal{P}$, if $z_1 = -.6$, $z_2 = .2$, $z_3 = .5$, $z_4 = 1$, and $z_5 = 1.7$.

SOLUTION

For the partition $\mathcal{P}$, $\Delta x_1 = 1$, $\Delta x_2 = .4$, $\Delta x_3 = .6$, $\Delta x_4 = .25$, and $\Delta x_5 = .75$; so

$$\|\mathcal{P}\| = \text{maximum } \{1, .4, .6, .25, .75\} = 1.$$

The Riemann sum is

$$\sum_{i=1}^{5} f(z_i)\Delta x_i = (-.6)^3 (1) + (.2)^3 (.4) + (.5)^3 (.6) + (1)^3 (.25) + (1.7)^3 (.75)$$

$$= -.216 + .0032 + .075 + .25 + 3.68475$$

$$= 3.79695.$$

The graph of $f(x) = x^3$ and a geometric illustration of the Riemann sum is shown in Figure 5.12. □

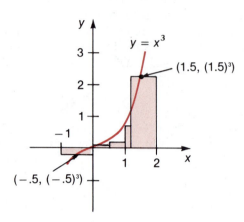

FIGURE 5.12

Suppose f is continuous and nonnegative on $[a, b]$, $\mathscr{P}$ is the partition with $x_i = a + \dfrac{b - a}{n} i$, and $z_i = x_i$ for each $i = 1, 2, \ldots, n$. The Riemann sum of f is identical to the sum A_n that approximates the area of the region bounded by the graph of f, the x-axis, and the lines $x = a$, and $x = b$. (See Equation (5.8) page 250.)

Notice, however, that the general definition of a Riemann sum of a function f differs from the discussion of area in a number of ways.

For a Riemann sum:
(1) the function need not be continuous on $[a, b]$;
(2) the function can assume negative values;
(3) the definition of a partition allows the interval $[a, b]$ to be divided into unequal parts; and
(4) the value z_i at which the function is to be evaluated can be arbitrarily chosen within the ith interval.

The concept of the Riemann sum is used to give the definition of the definite integral of a function.

HISTORICAL NOTE   **Georg Friedrich Bernhard Riemann** (1826–1866) made many of the important discoveries classifying functions that have integrals. He constructed a function having an infinite number of points of discontinuity in an interval on which its integral exists, and determined precise conditions for bounded functions to be integrable. He also did fundamental work in geometry and complex function theory.

(5.10)
DEFINITION

If f is defined on $[a, b]$ and L is a number with the property that

$$L = \lim_{\|\mathcal{P}\| \to 0} \sum_{i=1}^{n} f(z_i) \, \Delta x_i,$$

then L is called the **definite integral** of f on $[a, b]$ and is denoted

$$\int_{a}^{b} f(x) \, dx.$$

When $\int_{a}^{b} f(x)dx$ exists, we say f is **integrable** on $[a, b]$.

The limit in the definition of the definite integral is somewhat more complicated than the limit discussed in Chapter 2. The difference is that we want this limit to be independent of both the partition $\mathcal{P}$ and the manner in which the numbers $z_1, z_2, \ldots, z_n$ are chosen from that partition. To be precise we define

$$L = \lim_{\|\mathcal{P}\| \to 0} \sum_{i=1}^{n} f(z_i)\Delta x_i$$

provided that, for any number $\varepsilon > 0$, a number $\delta > 0$ can be found with the property that for any partition $\mathcal{P} = \{x_1, x_2, \ldots, x_n\}$, with $\|\mathcal{P}\| < \delta$, and any set of numbers $z_1, z_2, \ldots, z_n$, chosen with $x_{i-1} \leq z_i \leq x_i$ for each $i = 1, 2, \ldots, n$,

$$\left| L - \sum_{i=1}^{n} f(z_i)\Delta x_i \right| < \varepsilon.$$

EXAMPLE 2

Evaluate the definite integral $\int_{0}^{2} -3 \, dx$.

SOLUTION

In this case, all associated Riemann sums have the same value:

$$\sum_{i=1}^{n} f(z_i)\Delta x_i = \sum_{i=1}^{n} -3\Delta x_i = -3 \sum_{i=1}^{n} \Delta x_i$$

$$= -3[(x_1 - x_0) + (x_2 - x_1) + (x_3 - x_2) + \ldots + (x_n - x_{n-1})]$$

$$= -3(x_n - x_0) = -3(2) = -6.$$

Consequently,

$$\int_{0}^{2} -3 \, dx = \lim_{\|\mathcal{P}\| \to 0} \sum_{i=1}^{n} f(z_i)\Delta x_i = -6. \qquad \square$$

The definite integral of a function on an interval is a number that depends solely on the function f, called the **integrand**, and the interval $[a, b]$. The numbers a and b are called the *limits of integration: a* is the lower limit, *b* the upper. The symbol $\int$ used in representing the definite integral is derived from an elongated S and indicates the connection between the definite integral and the notion of the Riemann sum.

The definition of the definite integral is complicated and may seem to impose rather restrictive conditions for a function to be integrable on an interval. The next theorem, however, shows that the class of such functions is quite large. The proof of this theorem requires the introduction of concepts we will not otherwise need and consequently is not presented here. It can be found in any standard advanced calculus textbook.

(5.11)
THEOREM   If f is continuous on $[a, b]$, then f is integrable on $[a, b]$.

The scope of this theorem is quite broad, since most of the functions we have considered are continuous on every closed interval in their domain. The theorem is not conclusive, however; many functions are integrable on intervals containing points of discontinuity. The greatest integer function, for example, is discontinuous at every integer and yet is integrable on every closed interval $[a, b]$. (See Exercise 33.)

The discontinuities of an integrable function must be of a particular type, however, as the following result indicates. Like Theorem 5.11, the proof of this result is quite complicated and not presented.

(5.12)
THEOREM   If f is integrable on $[a, b]$, then f is bounded on $[a, b]$, that is, a constant M exists with $|f(x)| \leq M$ for all x in $[a, b]$.

The definition of the definite integral was specifically chosen to ensure that when a function f is integrable on an interval $[a, b]$ the limit

$$\lim_{\|\mathcal{P}\| \to 0} \sum_{i=1}^{n} f(z_i)\Delta x_i$$

is independent of the way in which the Riemann sums are chosen. For calculation purposes it is usually most convenient to let the partition consist of numbers that divide the interval into subintervals of equal width and, in each subinterval, to select the number z_i to be the right endpoint of the subinterval. In this case, the norm of a partition $\mathcal{P}$ is

$$\|\mathcal{P}\| = \frac{b-a}{n},$$

and the specification that $\|\mathcal{P}\| \to 0$ can be replaced by the specification $n \to \infty$. Consequently, when f is integrable on $[a, b]$,

$$\int_a^b f(x)\, dx = \lim_{n \to \infty} \sum_{i=1}^{n} f\left(a + \frac{b-a}{n} i\right)\left(\frac{b-a}{n}\right).$$

This is the same limit that we considered in Section 5.1 in the discussion of area. Consequently,

when f is continuous and nonnegative on $[a, b]$, the area of the region bounded by the graph of f, the x-axis and the lines $x = a$ and $x = b$ is $\int_a^b f(x)\,dx.$

EXAMPLE 3   Find $\int_1^3 x^2 \, dx$.

SOLUTION

The function described by $f(x) = x^2$ is continuous and nonnegative on $[1, 3]$ so $\int_1^3 x^2 \, dx$ is the area of the region bounded by the graph of f, the x-axis, and the lines $x = 1$ and $x = 3$. In Example 3 of Section 5.1 we found that this area is $26/3$. Thus,

$$\int_1^3 x^2 \, dx = \frac{26}{3}.$$

☐

EXAMPLE 4   Find $\int_0^2 x^3 \, dx$.

SOLUTION

Since the function described by $f(x) = x^3$ is continuous on $[0, 2]$, the integral exists. Choose $\Delta x_i = (2 - 0)/n = 2/n$ for each i and $z_i = 0 + (2/n) i = 2i/n$. Applying the summation results discussed in Section 5.1:

$$\int_0^2 x^3 \, dx = \lim_{\|\mathcal{P}\| \to 0} \sum_{i=1}^n f(z_i)\Delta x_i$$

$$= \lim_{n \to \infty} \sum_{i=1}^n \left(\frac{2i}{n}\right)^3 \left(\frac{2}{n}\right)$$

$$= \lim_{n \to \infty} \frac{16}{n^4} \sum_{i=1}^n i^3$$

$$= \lim_{n \to \infty} \frac{16}{n^4} \frac{n^2(n + 1)^2}{4}$$

$$= 4.$$

Since f is nonnegative on $[0, 2]$, the value of this integral also gives the area of the region bounded by the graph of $y = x^3$, the x-axis, and the lines $x = 0$ and $x = 2$. (See Figure 5.13.)   ☐

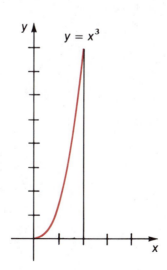

FIGURE 5.13

EXAMPLE 5   Find $\displaystyle\int_{-1}^{3/2} 2x \, dx$.

SOLUTION

   The function described by $f(x) = 2x$ is continuous on $[-1, \ 3/2]$ so the integral exists. Suppose the interval $[-1, \ 3/2]$ is partitioned into n subintervals of equal width:

$$\Delta x_i = \frac{(3/2) - (-1)}{n} = \frac{5}{2n}.$$

The right endpoint of the ith subinterval is $z_i = -1 + \dfrac{5}{2n} i$. Thus,

$$\int_{-1}^{3/2} 2x \, dx = \lim_{n \to \infty} \sum_{i=1}^{n} 2\left[-1 + \frac{5}{2n} i \right] \left(\frac{5}{2n} \right)$$

$$= \lim_{n \to \infty} \frac{5}{n} \left[-\sum_{i=1}^{n} 1 + \frac{5}{2n} \sum_{i=1}^{n} i \right]$$

$$= \lim_{n \to \infty} \frac{5}{n} \left[-n + \frac{5}{2n} \cdot \frac{n(n+1)}{2} \right]$$

$$= \lim_{n \to \infty} \left[-5 + \frac{25}{4} \frac{n+1}{n} \right]$$

$$= \frac{5}{4}.$$

   The function f is negative on a portion of the interval $[-1, \ 3/2]$ as shown in Figure 5.14. We cannot interpret this result as the area of a region, as we did in Examples 3 and 4. $\square$

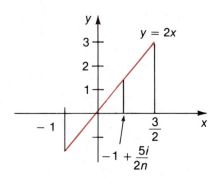

FIGURE 5.14

EXERCISE SET 5.2

Find the value of the Riemann sum associated with the function, partition, and points described in Exercises 1 through 10.

1. $f(x) = 2x + 3$, $\mathcal{P} = \{0, 1, 2, 3\}$, $z_1 = 0$, $z_2 = 1$, $z_3 = 2$

2. $f(x) = 2x + 3$, $\mathcal{P} = \{0, 1, 2, 3\}$, $z_1 = 1$, $z_2 = \dfrac{3}{2}$, $z_3 = 2$

3. $f(x) = x^2$, $\mathcal{P} = \left\{0, \dfrac{1}{4}, \dfrac{1}{2}, \dfrac{3}{4}, 1\right\}$, $z_1 = 0$, $z_2 = \dfrac{1}{4}$, $z_3 = \dfrac{1}{2}$, $z_4 = \dfrac{3}{4}$

4. $f(x) = x^2$, $\mathcal{P} = \left\{0, \dfrac{1}{4}, \dfrac{1}{2}, \dfrac{3}{4}, 1\right\}$, $z_1 = \dfrac{1}{4}$, $z_2 = \dfrac{1}{2}$, $z_3 = \dfrac{3}{4}$, $z_4 = 1$

5. $f(x) = x^2$, $\mathcal{P} = \left\{0, \dfrac{1}{4}, \dfrac{1}{2}, \dfrac{3}{4}, 1\right\}$, $z_1 = 0$, $z_2 = \dfrac{1}{2}$, $z_3 = \dfrac{1}{2}$, $z_4 = 1$

6. $f(x) = x^2$, $\mathcal{P} = \left\{0, \dfrac{1}{4}, \dfrac{1}{2}, \dfrac{3}{4}, 1\right\}$, $z_1 = \dfrac{1}{8}$, $z_2 = \dfrac{3}{8}$, $z_3 = \dfrac{5}{8}$, $z_4 = \dfrac{7}{8}$

7. $f(x) = x^2 - x$, $\mathcal{P} = \{-1, -.5, .5, 1\}$, $z_1 = -1$, $z_2 = 0$, $z_3 = .75$

8. $f(x) = \dfrac{1}{x}$, $\mathcal{P} = \{-2, -1, 0, 1, 2\}$, $z_1 = -1.5$, $z_2 = -.5$, $z_3 = .5$, $z_4 = 1.5$

9. $f(x) = \sin x$, $\mathcal{P} = \{0, \dfrac{\pi}{4}, \dfrac{2\pi}{3}, \pi\}$, $z_1 = \dfrac{\pi}{4}$, $z_2 = \dfrac{\pi}{3}$, $z_3 = \dfrac{3\pi}{4}$

10. $f(x) = \tan x$, $\mathcal{P} = \{0, \dfrac{\pi}{4}, \dfrac{2\pi}{3}, \pi\}$, $z_1 = \dfrac{\pi}{6}$, $z_2 = \dfrac{\pi}{4}$, $z_3 = \dfrac{5\pi}{6}$

Find the value of the definite integrals given in Exercises 11 through 18.

11. $\displaystyle\int_0^1 (2x + 1)dx$

12. $\displaystyle\int_{-1}^1 (2x + 3)dx$

13. $\displaystyle\int_0^1 x^2\,dx$

14. $\displaystyle\int_0^3 (x^2 + x)dx$

15. $\displaystyle\int_0^1 x^3\,dx$

16. $\displaystyle\int_{-1}^1 (x^2 + x)dx$

17. $\displaystyle\int_{-1}^1 x^3\,dx$

18. $\displaystyle\int_{-2}^3 -x^2\,dx$

In Exercises 19 and 20, use the results of Exercises 15 and 14 to find the area enclosed in the region bounded by the x-axis, the graph of f, and the given lines.

19. $f(x) = x^3$, $x = 0$, $x = 1$ **20.** $f(x) = x^2 + x$, $x = 0$, $x = 3$

Use the application of the definite integral as an area to evaluate the definite integrals in Exercises 21 through 24.

21. $\displaystyle\int_0^2 |x|\,dx$

22. $\displaystyle\int_{-2}^2 |x|\,dx$

23. $\displaystyle\int_{-2}^{-1} 4\,dx$

24. $\displaystyle\int_{-1}^1 \sqrt{1 - x^2}\,dx$

In Exercises 25 through 30 a sum is given. In each case, determine a function for which the sum is a Riemann sum.

25. $\sum_{i=1}^{n} \pi(x_i)^2 \, \Delta x_i$

26. $\sum_{i=1}^{n} \sqrt{1 + x_i} \, \Delta x_i$

27. $\sum_{i=1}^{n} (\cos x_i)^2 \, \Delta x_i$

28. $\sum_{i=1}^{n} \dfrac{1}{x_i} \, \Delta x_i$

29. $\sum_{i=1}^{n} \sqrt{1 + (f'(x_i))^2} \, \Delta x_i$

30. $\sum_{i=1}^{n} 2\pi \, x_i f(x_i) \Delta x_i$

31. Find the area bounded by the graph of $f(x) = x^3$, the x-axis, and the lines $x = 1$ and $x = 3$ using a procedure similar to that outlined in Section 5.1. Take A_n to be the sum of the areas of the regions described below.
   (a) Rectangles whose heights are the values of f at the left endpoint of each subinterval.
   (b) Rectangles whose heights are the values of f at the right endpoint of each subinterval.
   (c) Rectangles whose heights are the values of f at the midpoint of each subinterval.

32. Theorem 5.11 states that a function continuous on a closed interval has a definite integral on that interval. Show that the converse of this theorem is not true by showing that the function described by $f(x) = \text{sgn } x$ has a definite integral on the interval $[-1, 1]$ but is not continuous at $x = 0$. Recall that

$$\text{sgn } x = \begin{cases} 1 & \text{if } x \geq 0 \\ -1 & \text{if } x < 0. \end{cases}$$

33. Show that the discontinuous function described by $f(x) = [\![x]\!]$ has a definite integral on $[1/2, 3/2]$.

34. The graph of a function f is shown below.

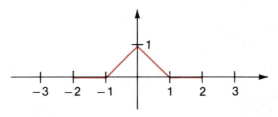

Suppose that for x in $[-2, 2]$, a function g is defined by $g(x) = \displaystyle\int_{-2}^{x} f(t)dt$. Sketch the graph of g.

5.3
PROPERTIES OF THE DEFINITE INTEGRAL

Arithmetic formulas similar to those for the derivative can be obtained to help us more easily calculate the definite integral of a function. The proofs of these results are relatively straightforward applications of the definition of the definite integral, although in some cases the details become rather complicated. We present the proof of the first result, and refer the reader to Appendix B for verification of the others.

(5.13)
THEOREM

If f is a constant function described by $f(x) = k$ on the interval $[a, b]$, then
$$\int_a^b f(x)dx = k(b - a), \text{ that is}$$

$$\int_a^b k\, dx = k(b - a).$$

The case when $k \geq 0$ is illustrated in Figure 5.15.

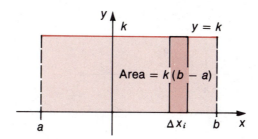

FIGURE 5.15

PROOF

Let $\mathcal{P} = \{x_0, x_1, \ldots, x_n\}$ be a partition of $[a, b]$ and $z_1, z_2, \ldots, z_n$ be any collection of numbers satisfying $x_{i-1} \leq z_i \leq x_i$ for each $i = 1, 2, \ldots, n$. Since $f(x) = k$ for each x in $[a, b]$, the Riemann sum $S(f, \mathcal{P}, \{z_i\})$ is

$$\sum_{i=1}^n f(z_i)\Delta x_i = \sum_{i=1}^n k\Delta x_i.$$

Thus, $\displaystyle\sum_{i=1}^n f(z_i)\Delta x_i = k \sum_{i=1}^n \Delta x_i = k \sum_{i=1}^n (x_i - x_{i-1})$

$$= k[(x_1 - x_0) + (x_2 - x_1) + (x_3 - x_2) + \ldots$$
$$+ (x_{n-2} - x_{n-3}) + (x_{n-1} - x_{n-2}) + (x_n - x_{n-1})]$$
$$= k(-x_0 + x_n) = k(b - a).$$

All Riemann sums of f have the same value $k(b - a)$, which implies that

$$\int_a^b k\, dx = \lim_{\|\mathcal{P}\| \to 0} \sum_{i=1}^n f(z_i)\Delta x_i = k(b - a). \qquad \square$$

The next theorem is analogous to the result about the derivative of a function multiplied by a constant.

(5.14)
THEOREM

If f is integrable on $[a, b]$ and k is a constant, then kf is also integrable on $[a, b]$ and

$$\int_a^b kf(x)dx = k \int_a^b f(x)dx.$$

In the special case when f is continuous and nonnegative on $[a, b]$, and k is positive, this theorem implies that the area of the region bounded by the graph of $y = kf(x)$, the x-axis, and the lines $x = a$ and $x = b$ is k times the area bounded by the graph of $y = f(x)$, the x-axis, and the lines $x = a$ and $x = b$. The case when $k = 2$ is illustrated in Figure 5.16.

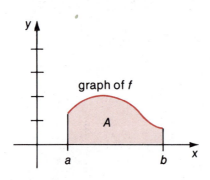

FIGURE 5.16

EXAMPLE 1

Find $\int_1^3 - 2x^2 \, dx$.

SOLUTION

According to Theorem 5.14,

$$\int_1^3 - 2x^2 \, dx = -2 \int_1^3 x^2 \, dx.$$

It was found in Example 3 of Section 5.2 that $\int_1^3 x^2 \, dx = \dfrac{26}{3}$.

Consequently,

$$\int_1^3 - 2x^2 \, dx = -2 \left(\frac{26}{3} \right) = -\frac{52}{3}. \qquad \square$$

(5.15)
THEOREM

If f and g are integrable on $[a, b]$, then
(i) $f + g$ is integrable on $[a, b]$ and

$$\int_a^b (f + g)(x) \, dx = \int_a^b f(x)dx + \int_a^b g(x)dx,$$

(ii) $f - g$ is integrable on $[a, b]$ and

$$\int_a^b (f - g)(x) \, dx = \int_a^b f(x)dx - \int_a^b g(x)dx.$$

In the special case when f and g are continuous and nonnegative, the result in (i) has the geometrical interpretation shown in Figure 5.17.

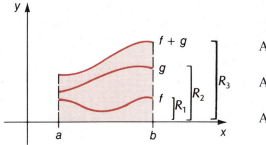

Area of $R_1 = \displaystyle\int_a^b f(x)dx$

Area of $R_2 = \displaystyle\int_a^b g(x)dx$

Area of $R_3 = \displaystyle\int_a^b [f(x) + g(x)]dx$

FIGURE 5.17

EXAMPLE 2   Find $\displaystyle\int_1^3 (2x^2 - 5)dx$.

SOLUTION
By Theorem 5.15 and Theorem 5.14,

$$\int_1^3 (2x^2 - 5)dx = \int_1^3 2x^2\,dx - \int_1^3 5\,dx = 2\int_1^3 x^2\,dx - \int_1^3 5\,dx.$$

In Example 2 of Section 5.2 we found that $\int_1^3 x^2\,dx = 26/3$. Theorem 5.13 tells us that $\int_1^3 5\,dx = 5(3 - 1) = 10$. Combining these facts, we have

$$\int_1^3 (2x^2 - 5)dx = 2\int_1^3 x^2\,dx - \int_1^3 5\,dx$$

$$= 2\left(\frac{26}{3}\right) - 10 = \frac{22}{3}. \qquad \square$$

(5.16)
THEOREM   If f is integrable on the intervals $[a, c]$ and $[c, b]$, then f is integrable on $[a, b]$ and

$$\int_a^b f(x)dx = \int_a^c f(x)dx + \int_c^b f(x)dx.$$

A geometrical interpretation of this theorem when f and g are continuous and nonnegative is illustrated in Figure 5.18.

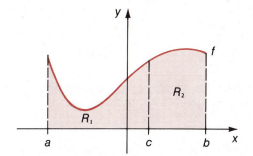

Area of $R_1 = \displaystyle\int_a^c f(x)dx$

Area of $R_2 = \displaystyle\int_c^b f(x)dx$

Area of R_1 and $R_2 = \displaystyle\int_a^b f(x)dx$

FIGURE 5.18

The definition of the definite integral $\int_a^b f(x)dx$ requires that $a < b$. We would like to remove this restriction and define this integral when $b \leq a$. When a and b are the same, the area interpretation of the integral requires that the integral be zero, so we make the definition that whenever $f(a)$ exists

$$\int_a^a f(x)dx = 0.$$

If we want Theorem 5.16 to hold regardless of the values of a and b, this definition requires that

$$0 = \int_a^a f(x)dx = \int_a^b f(x)dx + \int_b^a f(x)dx.$$

For this reason, for $b < a$ we define

(5.17) $$\int_a^b f(x)dx = -\int_b^a f(x)dx.$$

Theorem 5.16 can then be reexpressed as:
If f is integrable on an interval that contains the numbers a, b, and c, then

$$\int_a^b f(x)dx = \int_a^c f(x)dx + \int_c^b f(x)dx$$

regardless of the ordering of the numbers a, b, and c on the interval.

(5.18)
THEOREM

If f is integrable on $[a, b]$ and $f(x) \geq 0$ for each x in $[a, b]$, then

$$\int_a^b f(x)dx \geq 0.$$

Theorem 5.18 and part (ii) of Theorem 5.15 can be used to show the next result.

(5.19)
THEOREM

If f and g are integrable on $[a, b]$ and $f(x) \geq g(x)$ for each x in $[a, b]$, then

$$\int_a^b f(x)dx \geq \int_a^b g(x)dx.$$

The special case when f and g are continuous and nonnegative has the area interpretation shown in Figure 5.19. In this figure

$$\int_a^b g(x)\, dx = \text{area of } R_1 \leq \text{area of } R_2 = \int_a^b f(x)dx.$$

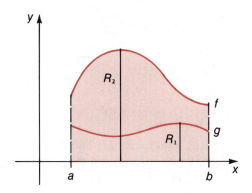

FIGURE 5.19

EXERCISE SET 5.3

In Exercises 1 through 6, find the definite integrals.

1. $\int_1^3 2\, dx$ **2.** $\int_{-2}^5 \sqrt{3}\, dx$ **3.** $\int_0^2 -3\, dx$

4. $\int_{-1}^3 dx$ **5.** $\int_1^5 c\, dx$ **6.** $\int_{-1}^3 (c+1)dx$

Use the information that $\int_0^1 x^2\, dx = 1/3$, $\int_0^2 x^2\, dx = 8/3$, $\int_0^2 x\, dx = 2$,

$\int_0^1 x^3\, dx = 1/4$, $\int_0^2 x^3\, dx = 4$, and the theorems in this section to find the definite integrals in Exercises 7 through 14.

7. $\int_0^2 3x^2\, dx$ **8.** $\int_0^2 (3x^2 + x)dx$

9. $\int_0^2 -2x^3\, dx$ **10.** $\int_0^2 (x^3 - 4)dx$

11. $\int_0^1 (3x^3 - x^2 + 4)dx$ **12.** $\int_0^2 (2x^3 + x^2 - x - 1)dx$

13. $\int_1^2 x^2\, dx$ **14.** $\int_1^2 (x^3 + 1)dx$

15. Use the fact that $\int_0^\pi \sin x\, dx = 2$ to find each of the following:

(a) $\int_0^\pi 2 \sin x\, dx$ (b) $\int_0^\pi -\sin x\, dx$

(c) $\int_0^\pi (\sin x - 3)dx$ (d) $\int_\pi^0 \sin x\, dx$

16. Use the fact that $\int_0^{\pi/2} \cos x \, dx = 1$ to find each of the following:

(a) $\int_0^{\pi/2} 3 \cos x \, dx$

(b) $\int_0^{\pi/2} - \cos x \, dx$

(c) $\int_0^{\pi/2} (2 - \cos x) dx$

(d) $\int_{\pi/2}^0 \cos x \, dx$

In Exercises 17 through 22, find the area of the region bounded by the x-axis, the graph of the function, and the given lines. Make a sketch to show the region in each case and use the fact that

$$\int_0^{\pi/2} \cos x \, dx = 1, \qquad \int_0^{\pi} \sin x \, dx = 2.$$

17. $f(x) = 2 \cos x, \, x = 0, \, x = \dfrac{\pi}{2}$ **18.** $f(x) = \cos x + 1, \, x = 0, \, x = \dfrac{\pi}{2}$

19. $f(x) = \sin x + 1, \, x = 0, \, x = \pi$ **20.** $f(x) = \cos x, \, x = -\dfrac{\pi}{2}, \, x = \dfrac{\pi}{2}$

21. $f(x) = - \cos x, \, x = 0, \, x = \dfrac{\pi}{2}$ **22.** $f(x) = \sin x, \, x = 0, \, x = \dfrac{\pi}{2}$

23. Find the definite integrals given below. (*Hint:* Use the fact that if f is continuous and nonnegative on $[a, b]$, then $\int_a^b f(x) \, dx$ is the area of the region bounded by the graph of f, the x-axis, and the lines $x = a$ and $x = b$.)

(a) $\int_{-4}^3 |x| \, dx$

(b) $\int_{-4}^3 - 2|x| \, dx$

(c) $\int_{-4}^3 3|x| \, dx$

(d) $\int_{-4}^3 (|x| + 2) dx$

24. Without evaluating the integrals, show that $\int_0^1 x \, dx \geq \int_0^1 x^2 dx$ and that $\int_1^2 x \, dx \leq \int_1^2 x^2 dx.$

25. Without evaluating the integral, show that $0 \leq \int_0^1 (3x^5 + 1) dx \leq 4.$

26. If $f(x) \geq m$ for all x in $[a, b]$ and f is integrable on $[a, b]$, use Theorem 5.19 to prove that $\int_a^b f(x) dx \geq m(b - a)$. Illustrate this result graphically.

27. If $f(x) \leq M$ for all x in $[a, b]$ and f is integrable on $[a, b]$, use Theorem 5.19 to prove that $\int_a^b f(x) dx \leq M(b - a)$. Illustrate this result graphically.

28. Use Theorem 5.19 to show that $\left| \int_a^b f(x) dx \right| \leq \int_a^b |f(x)| dx$ for every continuous function f and any interval $[a, b]$.

29. Construct a simple example showing that

$$\int_a^b f(x)g(x)dx \neq \left(\int_a^b f(x)dx\right)\left(\int_a^b g(x)dx\right).$$

30. Suppose that f is a continuous function and a and b are constants with $0 < a \leq f(x) \leq b$ when $0 \leq x \leq 1$. Show that

$$2a \leq \int_0^1 f(x)dx + ab\int_0^1 \frac{dx}{f(x)} \leq 2b.$$

Putnam exercises:

31. Find all continuous positive functions $f(x)$, for $0 \leq x \leq 1$, such that

$$\int_0^1 f(x)dx = 1$$

$$\int_0^1 f(x)x\, dx = \alpha$$

$$\int_0^1 f(x)x^2 dx = \alpha^2$$

where α is a given real number. (This exercise was problem 2, part I of the twenty-fifth William Lowell Putnam examination given on December 5, 1964. The examination and its solution can be found in the September 1965 issue of the *American Mathematical Monthly*, pages 734–739.)

32. Prove that if a function f is continuous on the closed interval, $[0, \pi]$ and if

$$\int_0^\pi f(\theta)\cos\theta\, d\theta = \int_0^\pi f(\theta)\sin\theta\, d\theta = 0$$

then there exist points α and β such that

$$0 < \alpha < \beta < \pi \quad \text{and} \quad f(\alpha) = f(\beta) = 0.$$

(This exercise was problem 5, part I of the twenty-fourth William Lowell Putnam examination given on December 7, 1963. The examination and its solution can be found in the June–July 1964 issue of the *American Mathematical Monthly*, page 636.)

5.4
THE FUNDAMENTAL THEOREM OF CALCULUS

This section has a rather imposing title, but one that is completely justified. It will be shown that the seemingly distinct concepts of the derivative and the definite integral are in fact intimately related. This connection can be used to calculate the definite integral of many functions without referring to Riemann sums. The fundamental theorem of calculus climaxes a series of beautiful results concerning the definite integral. In order to prove the theorem we need to develop some additional mathematical tools. Be assured, however, that the final result makes this development worthwhile.

(5.20)
THEOREM

Mean Value Theorem for Integrals   If f is continuous on $[a, b]$, then at least one number z in (a, b) exists with

$$f(z) = \frac{1}{b - a} \int_a^b f(x)dx.$$

PROOF

Since f is continuous on $[a, b]$, the extreme value theorem (3.30) implies that numbers c_1 and c_2 exist in $[a, b]$ at which f assumes its maximum and minimum values; that is, $f(c_1) \leq f(x) \leq f(c_2)$ for each value of x in $[a, b]$. Theorem 5.19 implies that

$$\int_a^b f(c_1)dx \leq \int_a^b f(x)dx \leq \int_a^b f(c_2)dx.$$

However, $f(c_1)$ and $f(c_2)$ are constants, so

$$\int_a^b f(c_1)dx = f(c_1)(b - a) \quad \text{and} \quad \int_a^b f(c_2)dx = f(c_2)(b - a).$$

Hence,

$$f(c_1)(b - a) \leq \int_a^b f(x)dx \leq f(c_2)(b - a)$$

and

$$f(c_1) \leq \frac{1}{b - a} \int_a^b f(x)dx \leq f(c_2).$$

The number $(b - a)^{-1} \int_a^b f(x)dx$ lies between two values of the function f. Since f is continuous, the intermediate value theorem (2.19) ensures that there is a number z between c_1 and c_2 with

$$f(z) = \frac{1}{b - a} \int_a^b f(x)dx.$$

Since both c_1 and c_2 belong to $[a, b]$, z must be in (a, b). $\square$

In the special case when the function f is continuous and nonnegative, the definite integral $\int_a^b f(x)dx$ gives the area of the region bounded by the graph of f, the x-axis, and the lines $x = a$ and $x = b$. The mean value theorem for integrals implies that there is a number z in (a, b) such that this area is equal to the area of the rectangle with base of width $b - a$ and height $f(z)$. That is,

$$\int_a^b f(x)dx = f(z)(b - a).$$

(See Figure 5.20.) Because of this geometrical interpretation, the number $(b - a)^{-1} \int_a^b f(x)dx$ is called the **average value** of the function f over the interval $[a, b]$.

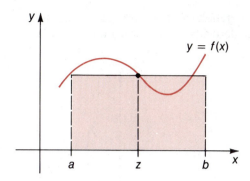

FIGURE 5.20

EXAMPLE 1   For the function given by $f(x) = 2x^2 - 5$ and the interval $[1, 3]$, find a number z guaranteed by the mean value theorem for integrals.

SOLUTION

The function f is continuous on $[1, 3]$ so the hypothesis of the mean value theorem for integrals is satisfied. We need to find a number z in $(1, 3)$ with

$$f(z) = 2z^2 - 5 = \frac{1}{3 - 1} \int_1^3 (2x^2 - 5)dx.$$

In Example 2 of Section 5.3, we found that $\int_1^3 (2x^2 - 5)dx = 22/3$. Consequently, to find z, we can solve the equation

$$2z^2 - 5 = \left(\frac{1}{2}\right)\left(\frac{22}{3}\right)$$

to obtain $z = \pm\dfrac{\sqrt{39}}{3}$.

Since only one of these values, $\sqrt{39}/3$, is in the interval $(1, 3)$, this must be the value guaranteed by the theorem. $\square$

This example illustrates the mean value theorem for integrals, but our reason for presenting this theorem is not associated with solving such problems. Rather, we use the theorem as a connecting link in the proof of our next result.

This final preliminary result to the fundamental theorem of calculus needs a few introductory remarks. A function that is integrable on an interval $[a, b]$ will also be integrable on the subinterval $[a, x]$ for each $a \leq x \leq b$. Therefore, when f is integrable on $[a, b]$, the expression

(5.21) $G(x) = \displaystyle\int_a^x f(t)dt$

defines a function whose domain contains the interval $[a, b]$. The variable t has been used in this integral to eliminate any confusion with the upper limit of integration. (See Figure 5.21.) The most important result in calculus concerns the derivative of this function.

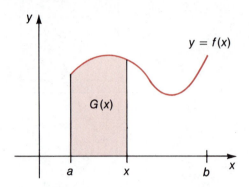

FIGURE 5.21

(5.22)
LEMMA

Fundamental Lemma of Calculus. If f is continuous on $[a, b]$ and G is defined by

$$G(x) = \int_a^x f(t)dt,$$

for each x in $[a, b]$, then G is differentiable on $[a, b]$ and $G'(x) = f(x)$ for each x in (a,b).

Additionally, the derivative of G from the right at a, $G'_+(a)$, is $f(a)$ and the derivative of G from the left at b, $G'_-(b)$, is $f(b)$.

PROOF

Let x be an arbitrary number in (a, b). By the definition of the derivative of G at x,

$$G'(x) = \lim_{h \to 0} \frac{G(x + h) - G(x)}{h}$$

$$= \lim_{h \to 0} \frac{1}{h}\left[\int_a^{x+h} f(t)dt - \int_a^x f(t)dt\right]$$

$$= \lim_{h \to 0} \frac{1}{h}\left[\int_a^x f(t)dt + \int_x^{x+h} f(t)dt - \int_a^x f(t)dt\right]$$

$$= \lim_{h \to 0} \frac{1}{h}\int_x^{x+h} f(t)dt.$$

If h is restricted so that $x + h$ is also in the interval $[a, b]$, then the mean value theorem for integrals implies that a number z between x and $x + h$ exists with

$$f(z) = \frac{1}{h}\int_x^{x+h} f(t)dt.$$

See Figure 5.22.

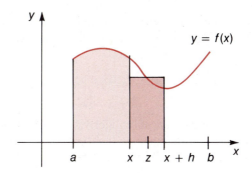

FIGURE 5.22

Thus,

$$G'(x) = \lim_{h \to 0} \frac{1}{h} \int_x^{x+h} f(t)dt = \lim_{h \to 0} f(z).$$

Since z is between x and $x + h$ and f is continuous, $\lim_{h \to 0} f(z) = f(x)$. This implies that

$$G'(x) = f(x).$$

Since x is an arbitrary number in (a,b), we have the conclusion that $G'(x) = f(x)$ for each x in (a, b).

The conclusion listed for $x = a$ and $x = b$ is obtained by the same method, restricting $x + h$ to lie within $[a, b]$. These results are considered in Exercise 50. □

EXAMPLE 2   Find $D_x \int_1^x (t^5 - 3t^3 + 4t + 1)dt$.

SOLUTION

Since the function defined by $f(t) = t^5 - 3t^3 + 4t + 1$ is a continuous function, the fundamental lemma of calculus tells us that

$$D_x \int_1^x (t^5 - 3t^3 + 4t + 1)dt = x^5 - 3x^3 + 4x + 1. □$$

Functions F with the property that $F' = f$ are called **antiderivatives** of f. According to the fundamental lemma, if f is a continuous function and

$$G(x) = \int_a^x f(t)dt,$$

then $G' = f$; that is, G is an antiderivative of f.

We first used antiderivatives in Section 4.1 where the motion $s(t)$ of an object was derived from its acceleration at time t, $a(t) = s''(t)$.

Functions do not have unique antiderivatives. For example, $D_x(x^2) = D_x(x^2 + 2) = 2x$, so two antiderivatives of $f(x) = 2x$ are $F(x) = x^2$ and

$G(x) = x^2 + 2$. An application of the mean value theorem, Corollary 3.36, showed that if $F'(x) = G'(x)$, then $F(x) = G(x) + C$. Stated using antiderivative terminology, this becomes:

(5.23)   Functions F and G are antiderivatives of the same function if and only if a constant C exists with $F = G + C$.

Consequently, all antiderivatives of a function f can be obtained by determining one function F with $F'(x) = f(x)$ and considering the class of functions of the form $F + C$ for all constants C.

EXAMPLE 3   Find all antiderivatives of the function described by $f(x) = 4x^3 - 3x^2 + 3$.

SOLUTION

Since $D_x x^4 = 4x^3, D_x x^3 = 3x^2$ and $D_x 3x = 3$,

$$D_x(x^4 - x^3 + 3x) = 4x^3 - 3x^2 + 3.$$

A function F is an antiderivative of f if and only if

$$F(x) = x^4 - x^3 + 3x + C$$

for some constant C.   □

(5.24) THEOREM   **The Fundamental Theorem of Calculus**   If f is continuous on $[a, b]$ and F is any antiderivative of f, then

$$\int_a^b f(t)dt = F(b) - F(a).$$

PROOF

The fundamental lemma implies that the function defined by

$$G(x) = \int_a^x f(t)dt$$

is an antiderivative of f. It follows from (5.23) that if F is any antiderivative of f, then the functions F and G differ by a constant. Therefore, a constant C exists with

$$F(x) = G(x) + C = \int_a^x f(t)dt + C$$

for all x in $[a, b]$.
Evaluating F when $x = a$ implies that

$$F(a) = \int_a^a f(t)dt + C = 0 + C = C.$$

Hence, $C = F(a)$ and

$$F(x) = \int_a^x f(t)dt + F(a).$$

Evaluating F when $x = b$ implies that

$$F(b) = \int_a^b f(t)dt + F(a)$$

and that

$$\int_a^b f(t)dt = F(b) - F(a). \qquad \square$$

The fundamental theorem gives the important connection between the definite integral and the concept of the derivative. We can now find the value of many definite integrals without referring to Riemann sums.

EXAMPLE 4   Find $\int_{-2}^1 2x \, dx$.

SOLUTION

To apply the fundamental theorem, we observe that $D_x x^2 = 2x$, so $F(x) = x^2$ defines an antiderivative that can be used in the evaluation of the definite integral.

$$\int_{-2}^1 2x \, dx = F(1) - F(-2)$$
$$= 1^2 - (-2)^2 = -3. \qquad \square$$

EXAMPLE 5   Use the fundamental theorem of calculus to find

$$\int_1^3 (2x^2 - 5)dx.$$

SOLUTION

To apply the fundamental theorem, we must first have an antiderivative of $f(x) = 2x^2 - 5$. Since $D_x(x^3) = 3x^2$, $D_x\left(\frac{2}{3}x^3\right) = 2x^2$. Also $D_x(5x) = 5$, so if $F(x) = 2x^3/3 - 5x$, then

$$F'(x) = D_x\left(\frac{2}{3}x^3 - 5x\right) = 2x^2 - 5 = f(x).$$

Consequently,

$$\int_1^3 (2x^2 - 5)dx = F(3) - F(1)$$
$$= \left[\frac{2}{3}(3)^3 - 5(3)\right] - \left[\frac{2}{3}(1)^3 - 5(1)\right]$$
$$= \frac{54}{3} - 15 - \frac{2}{3} + 5 = \frac{22}{3},$$

a result consistent with our previous work. (See Example 2 of Section 5.3.) $\square$

Note that if $F(x) + C$ is used in place of $F(x)$ in the fundamental theorem, the same result is obtained,

$$\int_a^b f(x)dx = (F(b)+C) - (F(a)+C) = F(b) - F(a).$$

This agrees with the statement in the theorem that *any* antiderivative of f can be used in the evaluation of $\int_a^b f(x)dx$.

If F is any antiderivative of a continuous function f, a notation that is commonly used to summarize the fundamental theorem is

$$(5.25) \qquad \int_a^b f(x)dx = F(x) \Big]_a^b = F(b) - F(a).$$

For example,

$$\int_{-2}^1 2x \, dx = x^2 \Big]_{-2}^1 = 1 - 4 = -3.$$

To conclude this section we return to an area problem that was considered at the end of Section 5.1.

EXAMPLE 6   Let $f(x) = \sin x$. Find A, the area of the region bounded by the graph of f, the x-axis, and the lines $x = 0$, $x = \pi$.

SOLUTION

Since $y = \sin x \geq 0$ for x in the interval $[0, \pi]$, the area is given by the definite integral

$$A = \int_0^\pi \sin x \, dx.$$

An antiderivative of $f(x) = \sin x$ is given by $F(x) = -\cos x$, so

$$A = \int_0^\pi \sin x \, dx = -\cos x \Big]_0^\pi$$
$$= -\cos \pi - (-\cos 0)$$
$$= 1 + 1 = 2. \qquad \square$$

EXERCISE SET 5.4

In Exercises 1 through 4, find a number z guaranteed by the mean value theorem for integrals. Use the results and instructions from Exercises 7 through 14 in Exercise Set 5.3 to evaluate the integrals.

1. $\displaystyle\int_0^2 3x^2 \, dx$

2. $\displaystyle\int_0^2 x \, dx$

3. $\displaystyle\int_0^2 -2x^3 \, dx$

4. $\displaystyle\int_0^2 (3x^2 + x)dx$

Use the fundamental theorem of calculus to evaluate the definite integrals in Exercises 5 through 20.

5. $\int_2^5 2x\, dx$

6. $\int_{-2}^1 2x\, dx$

7. $\int_{-1}^1 3x^2\, dx$

8. $\int_1^3 4x^3\, dx$

9. $\int_1^3 (3x^2 - 2x)dx$

10. $\int_4^5 (3x^2 + 2)dx$

11. $\int_0^{\pi/2} \cos x\, dx$

12. $\int_{-\pi/2}^{\pi/2} \cos x\, dx$

13. $\int_0^{\pi/2} \sin x\, dx$

14. $\int_{-\pi/2}^0 \sin x\, dx$

15. $\int_{-\pi/4}^{\pi/4} (\sec x)^2\, dx$

16. $\int_0^{\pi/4} \sec x \tan x\, dx$

17. $\int_0^3 x^2\, dx$

18. $\int_{-2}^3 (x^2 + x)dx$

19. $\int_0^2 (6x^2 - 5)dx$

20. $\int_{-1}^2 (2x^2 - 5)dx$

Use the fundamental lemma of calculus to find $D_x F(x)$ for the function F described in each of Exercises 21 through 26.

21. $F(x) = \int_0^x t\sqrt{t^2 + 9}\, dt$

22. $F(x) = \int_0^x \sqrt{t^2 + 9}\, dt$

23. $F(x) = \int_0^3 t\sqrt{t^2 + 9}\, dt$

24. $F(x) = \int_1^4 \sqrt{t^2 + 9}\, dt$

25. $F(x) = x\int_1^x \sqrt{t^2 + 9}\, dt$

26. $F(x) = x + \int_0^x \sqrt{t^2 + 9}\, dt$

In Exercises 27 through 36, use the fundamental theorem of calculus to find the area of the region bounded by the x-axis, the graph of f, and the given lines. Make a sketch to show the region in each case.

27. $f(x) = 3, x = 0, x = 4$

28. $f(x) = -3, x = 0, x = 4$

29. $f(x) = 3x^2 + 1, x = -1, x = 1$

30. $f(x) = 3x^2 + 1, x = 0, x = 1$

31. $f(x) = 3x^2 + 2x, x = 0, x = 3$

32. $f(x) = 4x^3, x = 0, x = 2$

33. $f(x) = \cos x, x = -\dfrac{\pi}{2}, x = \dfrac{\pi}{2}$

34. $f(x) = \cos x, x = 0, x = \dfrac{\pi}{2}$

35. $f(x) = \sin x, x = 0, x = \dfrac{\pi}{2}$

36. $f(x) = \cos x, x = \dfrac{3\pi}{2}, x = 2\pi$

In Exercises 37 through 40, find the average value of the function on the given interval.

37. $f(x) = 2x + 1, [0, 3]$

38. $f(x) = 2x + 3, [-1, 1]$

39. $f(x) = 3x^2, [-1, 1]$

40. $f(x) = x^2, [1, 4]$

41. Sketch the graph of the function described by $f(x) = x$. Find the average value of this function over the interval $[0, 10]$. Does this agree with your intuitive idea of the average value of a function over an interval?

42. Use the mean value theorem for integrals to show that if f is continuous and $\int_a^b f(x)dx = 0$, then $f(c) = 0$ for some number c in (a, b).

43. Suppose f is a continuous function and $g(x) = x \int_0^x f(t)dt$. Show that $g'(x) = \dfrac{g(x)}{x} + xf(x)$, provided $x \neq 0$.

44. Use L'Hôpital's rule and the fundamental lemma of calculus to find

$$\lim_{x \to 0} \frac{\int_0^x t\sqrt{3 + t^2}\, dt}{x^2}.$$

45. Use L'Hôpital's rule and the fundamental lemma of calculus to find

$$\lim_{x \to \infty} \frac{\int_1^x (t^2 + t)dt}{x^3 + 1}.$$

46. Suppose f' is continuous on $[0, 2]$. What is

$$\lim_{x \to 1} \frac{\int_1^x f'(t)dt}{x - 1}?$$

47. Suppose f' is continuous on $[-a, a]$. What is $\int_{-a}^a f'(x)dx$ if
(a) f is an odd function?
(b) f is an even function?

48. Show that if f is continuous and positive on $[a, b]$, then the function G defined in Lemma 5.22 is an increasing function on $[a, b]$. What conditions on f would imply that G is decreasing on $[a, b]$?

49. Suppose f is a function that is differentiable on $[a, b]$ with $f'(x) > 0$ for all x in $[a, b]$. Show that the graph of the function G defined in Lemma 5.22 is concave upward on $[a, b]$.

50. Prove the two additional statements in the fundamental lemma of calculus:

$$G'_+ (a) = f(a) \qquad \text{and} \qquad G'_- (b) = f(b).$$

51. Show that if f' is continuous on $[a, b]$, then

$$\int_a^b f(x)f'(x)dx = \frac{(f(b))^2 - (f(a))^2}{2}.$$

52. Suppose that f' and g' are continuous on $[-1, 1]$. What is

$$\lim_{x \to 0} \frac{\int_0^x f(t)g'(t)dt + \int_0^x f'(t)g(t)dt}{x}?$$

53. The specific heat of oxygen depends on temperature according to the formula

$$C_p = 8.27 + 2.6 \times 10^{-4}T - 1.87 \times 10^{-5}T^2 \frac{\text{cal}}{\text{deg-mole}},$$

where T is measured in °C. Find the average specific heat for T in the interval [25°C, 650°C].

5.5
THE INDEFINITE INTEGRAL

Before the fundamental theorem of calculus can be used to any great extent, we need an efficient procedure for constructing antiderivatives of functions. This section contains results and terminology that will make the discussion and evaluation of antiderivatives easier.

(5.26)
DEFINITION
The class of all antiderivatives of the function f is denoted $\int f$ and called the **indefinite integral of f.**

If F is an antiderivative of f we can write $\int f = F + C$, where C is an arbitrary constant called the **constant of integration**. Usually this equation is written with reference to a variable, such as

$$\int f(x)dx = F(x) + C \qquad \text{or} \qquad \int f(u)du = F(u) + C.$$

EXAMPLE 1   Find $\int 3(x + 1)^2 dx$ and $\int_0^2 3(x + 1)^2 dx$.

SOLUTION
Since $D_x (x + 1)^3 = 3(x + 1)^2$,

$$\int 3(x + 1)^2 dx = (x + 1)^3 + C.$$

The fundamental theorem of calculus states that *any* antiderivative of f can be used to evaluate $\int_a^b f(x)dx$,

so

$$\int_0^2 3(x + 1)^2 \, dx = (x + 1)^3 \Big]_0^2$$

$$= (2 + 1)^3 - (0 + 1)^3$$

$$= 27 - 1 = 26. \qquad \square$$

Many of the formulas involving derivatives can be reexpressed using the concept of the indefinite integral and the fact that:

$$D_x F(x) = f(x) \qquad \text{if and only if} \qquad \int f(x)dx = F(x) + C.$$

For example, for any rational number $n \neq -1$,

$$\int x^n \, dx = \frac{x^{n+1}}{n+1} + C$$

because

$$D_x\left(\frac{x^{n+1}}{n+1} + C\right) = x^n.$$

These results are summarized in the following table:

TABLE 5.1

Derivative Results	Corresponding Integral Results
1. $D_x\,(f + g)(x) = D_x f(x) + D_x g(x)$	$\int (f + g)(x)dx = \int f(x)dx + \int g(x)dx$
2. $D_x(kf)(x) = k\,D_x f(x)$ for any constant k	$\int k\,f(x)\,dx = k \int f(x)\,dx$
3. $D_x\,x^{n+1} = (n + 1)x^n$, for any rational number $n \neq -1$	$\int x^n \, dx = \dfrac{x^{n+1}}{n+1} + C$, for any rational number $n \neq -1$
4. $D_x \sin x = \cos x$	$\int \cos x \, dx = \sin x + C$
5. $D_x \cos x = -\sin x$	$\int \sin x \, dx = -\cos x + C$
6. $D_x \tan x = (\sec x)^2$	$\int (\sec x)^2 \, dx = \tan x + C$
7. $D_x \cot x = -(\csc x)^2$	$\int (\csc x)^2 \, dx = -\cot x + C$
8. $D_x \sec x = \sec x \tan x$	$\int \sec x \tan x \, dx = \sec x + C$
9. $D_x \csc x = -\csc x \cot x$	$\int \csc x \cot x \, dx = -\csc x + C$

Notice that we can apply the fundamental theorem of calculus only to a function whose antiderivative is known. For example, we can evaluate $\int (\sec x)^2 \, dx$ using property (6) and $\int \sec x \tan x \, dx$ using property (8), but until we determine a function whose derivative is $\sec x$ we cannot evaluate $\int \sec x \, dx$.

It is important to realize that there is a significant difference between finding the derivative and antiderivative of a function. Finding the derivative of a function is generally straightforward, even though at times it becomes quite complicated. For example, if $f(x) = (3x^2 + 4)^6 (2x^2 + 1)^{3/2}$, $f'(x)$ can be found by applying the elementary arithmetic and composition rules. Determining the antiderivative of a function such as f is a much more difficult problem. Now we are searching for a function F such that $F'(x) = (3x^2 + 4)^6 (2x^2 + 1)^{3/2}$, and we have no elementary rules to apply. Keep in mind that the functions used for examples in the next few chapters have been specifically chosen because their integrals can be readily obtained. More difficult integration problems will be considered later, particularly in Chapter 8.

EXAMPLE 2   Find $\int (x^3 - \sqrt[3]{x^2})dx$.

SOLUTION

$$\int (x^3 - \sqrt[3]{x^2})dx = \int (x^3 - x^{2/3})dx$$

$$= \int x^3\, dx - \int x^{2/3}\, dx$$

$$= \frac{x^{3+1}}{3+1} - \frac{x^{(2/3)+1}}{(2/3)+1} + C$$

$$= \frac{1}{4}x^4 - \frac{3}{5}x^{5/3} + C.$$

To check the solution, note that

$$D_x\left[\frac{1}{4}x^4 - \frac{3}{5}x^{5/3} + C\right] = x^3 - x^{2/3}. \qquad \square$$

EXAMPLE 3   Find $\int_1^4 \left(t\sqrt{t} + \frac{3}{t^2}\right)dt$.

SOLUTION

We first find the indefinite integral:

$$\int \left(t\sqrt{t} + \frac{3}{t^2}\right)dt = \int (t^{3/2} + 3t^{-2})dt$$

$$= \int t^{3/2}\, dt + \int 3t^{-2}\, dt$$

$$= \frac{t^{5/2}}{5/2} + 3\frac{t^{-1}}{-1} + C$$

$$= \frac{2}{5}t^{5/2} - 3t^{-1} + C.$$

We can verify this solution since

$$D_t\left[\frac{2}{5}t^{5/2} - 3t^{-1} + C\right] = t^{3/2} + 3t^{-2}.$$

Consequently,

$$\int_1^4 \left(t\sqrt{t} + \frac{3}{t^2}\right)dt = \frac{2}{5}t^{5/2} - \frac{3}{t}\Bigg]_1^4$$

$$= \left[\frac{2}{5}(4)^{5/2} - \frac{3}{4}\right] - \left(\frac{2}{5} - 3\right)$$

$$= \frac{64}{5} - \frac{3}{4} - \frac{2}{5} + 3$$

$$= \frac{293}{20}. \qquad \square$$

EXAMPLE 4   Find $\int (\tan x)^2 \, dx$.

SOLUTION

The indefinite integral of $(\tan x)^2$ is not listed in Table 5.1. However, $(\tan x)^2 = (\sec x)^2 - 1$, so

$$\int (\tan x)^2 \, dx = \int [(\sec x)^2 - 1] \, dx$$

$$= \int (\sec x)^2 \, dx - \int 1 \, dx$$

$$= \tan x - x + C. \qquad \square$$

Notice in Example 3 that the constant of integration is omitted in the evaluation of the definite integral, because it does not contribute to the final answer. However, *the constant cannot be omitted when finding indefinite integrals*. This is emphasized in the final examples of this section.

EXAMPLE 5   Find an equation of the curve that passes through $(0, 1)$ and whose tangent line at any point (x, y) has slope $x^2 - 2$.

SOLUTION

The slope of the tangent line to a curve $y = f(x)$ at any point (x, y) is given by $f'(x)$. Hence, we need to find a function f that satisfies $f'(x) = x^2 - 2$ and $f(0) = 1$.

Using antidifferentiation, we see that

$$f(x) = \int (x^2 - 2) \, dx = \frac{x^3}{3} - 2x + C.$$

Since $f(0) = 1$, $C = 1$ and the desired equation is

$$y = f(x) = \frac{x^3}{3} - 2x + 1. \qquad \square$$

The concluding example is of the type considered in Section 4.1. Here we will use the indefinite integral to solve the problem.

EXAMPLE 6   A ball is thrown upward from the top of a building 96 feet high with an initial velocity of 80 ft/sec. Find
(a) $s(t)$, the distance of the ball from the ground at time t.
(b) the time it takes the ball to hit the ground.
(c) the height the ball goes before descending.

SOLUTION

The ball is under the constant force of acceleration due to the earth's gravity, $a(t) = -32$ ft/sec^2. Its velocity at the end of t seconds is given by

$$v(t) = \int a(t) \, dt = \int -32 \, dt = -32t + C.$$

Since $v(0) = 80$, $C = 80$ and consequently

$$v(t) = -32t + 80.$$

The distance above the ground at any time t after the ball is thrown is

$$s(t) = \int v(t)dt = \int (-32t + 80)dt$$
$$= -16t^2 + 80t + K,$$

for some constant K. At $t = 0$, $s(t) = 96$, so $K = 96$. Hence the answer to (a) is

$$s(t) = -16t^2 + 80t + 96.$$

When the ball hits the ground $s(t) = 0$. The only positive solution to

$$0 = -16t^2 + 80t + 96 = -16(t^2 - 5t - 6) = -16(t - 6)(t + 1)$$

is $t = 6$, so the answer to part (b) is 6 seconds.

At its maximum height the velocity is zero. This occurs when

$$0 = -32t + 80, \text{ so } t = \frac{80}{32} = \frac{5}{2} \text{ seconds.}$$

Consequently the answer to part (c) is

$$s\left(\frac{5}{2}\right) = -16\left(\frac{5}{2}\right)^2 + 80\left(\frac{5}{2}\right) + 96 = 196 \text{ feet.} \qquad \square$$

When finding indefinite integrals, you should check your result by taking the derivative of the indefinite integral. If the derivative is the integrand, the result is correct; if it is not the integrand, some error has been made.

It is important that you check your answers in this manner, and not rely on the answers in the back of the book. In the long run this practice will save time and effort, since two correct answers can often assume forms that appear at first to be different. For example, in Example 1 we found that

$$\int 3(x + 1)^2 \, dx = (x + 1)^3 + C, \qquad \text{since } D_x(x + 1)^3 = 3(x + 1)^2.$$

However, this indefinite integral can also be evaluated by first expanding $(x + 1)^2$ to give

$$\int 3(x + 1)^2 \, dx = \int 3(x^2 + 2x + 1)dx$$
$$= \int 3x^2 \, dx + \int 6x \, dx + \int 3 \, dx$$
$$= x^3 + 3x^2 + 3x + C.$$

This answer is also easily verified to be correct, since

$$D_x(x^3 + 3x^2 + 3x) = 3x^2 + 6x + 3 = 3(x^2 + 2x + 1) = 3(x + 1)^2.$$

The two seemingly different antiderivatives $(x + 1)^3$ and $x^3 + 3x^2 + 3x$ are both valid because they differ by a constant:

$$(x + 1)^3 = (x^3 + 3x^2 + 3x) + 1.$$

EXERCISE SET 5.5

Find the indefinite integral in Exercises 1 through 26. Check the answers to these problems by differentiating your result.

1. $\displaystyle\int 4x^3\,dx$

2. $\displaystyle\int 3x^2\,dx$

3. $\displaystyle\int x^3\,dx$

4. $\displaystyle\int x^5\,dx$

5. $\displaystyle\int (2x^3 + 3x^2)dx$

6. $\displaystyle\int (5x^4 - 4x^5)dx$

7. $\displaystyle\int (3x^2 + 4x - 2)dx$

8. $\displaystyle\int (4x^3 + 7x^2 - 6x + 1)dx$

9. $\displaystyle\int (x^{2/3} + 3x^{1/3} + 4x^2)dx$

10. $\displaystyle\int (x^{3/4} - 2x^{1/2} + x^3)dx$

11. $\displaystyle\int (\cos x + 1)dx$

12. $\displaystyle\int (\sin x - x)dx$

13. $\displaystyle\int (\cot u)^2\,du$

14. $\displaystyle\int \frac{1}{(\cos x)^2}\,dx$

15. $\displaystyle\int 3t(t^3 + 1)\,dt$

16. $\displaystyle\int x\sqrt{x}\,dx$

17. $\displaystyle\int 3(3x + 1)^2\,dx$

18. $\displaystyle\int 3t(t^3 + 4)^2\,dt$

19. $\displaystyle\int (\sqrt[5]{t} - \sqrt[5]{t^2})\,dt$

20. $\displaystyle\int \left(x^3\sqrt{x} + \frac{1}{x^{-3}}\right)dx$

21. $\displaystyle\int \frac{y^2 - 3y}{y}\,dy$

22. $\displaystyle\int \frac{3x^3 - 2x^2 + x}{x}\,dx$

23. $\displaystyle\int \frac{x^3 - 2x^2 + 4}{x^5}\,dx$

24. $\displaystyle\int \frac{z^{1/2} + z^3 - z^4}{z}\,dz$

25. $\displaystyle\int (x^3 - x^5)^2\,dx$

26. $\displaystyle\int x^3(x + 1)(3x + 2)dx$

Find the definite integral in Exercises 27 through 36.

27. $\displaystyle\int_2^4 (4x^2 + 3x + 1)dx$

28. $\displaystyle\int_0^1 (4x^5 + 5x^3)dx$

29. $\displaystyle\int_{-1}^1 2(3x + 4)^3\,dx$

30. $\displaystyle\int_{-1}^1 (x^3 - x^2)^2\,dx$

31. $\displaystyle\int_0^\pi \sin x\,dx$

32. $\displaystyle\int_0^\pi \cos x\,dx$

33. $\displaystyle\int_1^2 \frac{\sqrt{x}+1}{\sqrt{x}}\, dx$

34. $\displaystyle\int_1^4 \frac{t^{1/4}-t^2}{t}\, dt$

35. $\displaystyle\int_0^{\pi/4} (\sec x)^2\, dx$

36. $\displaystyle\int_0^{\pi/4} (\tan t)^2\, dt$

In Exercises 37 through 44, the derivative of a function and the value of the function at one point is given. Determine the function.

37. $f'(x) = 3, f(0) = 2$

38. $f'(x) = 2x, f(0) = 1$

39. $f'(x) = x, f(1) = 0$

40. $f'(x) = \cos x, f(0) = 0$

41. $f'(x) = x^2 - x, f(0) = 0$

42. $f'(x) = x^4, f(1) = 1$

43. $f'(x) = \sin x - x, f(0) = -1$

44. $f'(x) = x + \dfrac{1}{x^2}, f(1) = 2$

45. Find an equation of the curve that passes through $(0, 0)$ and whose tangent line at any point (x, y) has slope $x + 2$.

46. Find an equation of the curve that passes through $(1, 0)$ and whose tangent line at any point (x, y) has slope $1 - 3x^2$.

47. Find an equation of the curve passing through the points $(0, 2)$ and $(1, 3)$ with the property that $y'' = 6x - 2$.

48. Find an equation of the curve passing through $(0, 1)$ and $(1, 0)$ with the property that $y'' = 12x^2 - 4$.

49. By observing production and sales data, the Trasho Company has found that its marginal profit from selling wastebaskets is given by

$$P'(x) = 2.8 - .006x^2$$

when x units are sold. It is known that Trasho loses \$1000 when nothing is sold. Find the profit function of this company.

50. The marginal cost of a certain company is given by $C'(x) = 5.25 + .02x$ when x units are sold. Find the cost function of this company if the cost is \$100 when no units are sold.

51. An object is moving in a straight line and has acceleration described by $a(t) = 3t^2$ cm/sec^2. Find its velocity at the end of 2 seconds if it has an initial velocity of 4 cm/sec.

52. Suppose the velocity of a ball t seconds after it is thrown into the air is given by $v(t) = 80 - 32t$ and that when the ball is thrown, its distance above the ground is 64 feet. Find its distance $s(t)$ above the ground at any time t after it is thrown.

53. A rock is thrown vertically into the air with an initial velocity of 60 mph. How high will the rock go? What is the velocity of the rock at the end of 3 seconds?

54. An open construction elevator is rising at the rate of 3 feet per second. A hammer falls from the elevator when it is 100 feet from the ground. How long does it take the hammer to reach the ground?

55. The CN tower in Toronto, Ontario is the tallest free-standing structure in the world. This tower is 1815 feet high and the main concrete structure supporting the tower is 1464 feet high. Suppose an object is dropped from the top of the tower. How much later should another object be dropped from the top of the concrete supporting structure if the objects are to reach the ground simultaneously?

5.6
INTEGRATION BY SUBSTITUTION

The chain rule for derivatives plays an important role in finding indefinite integrals. Suppose F is an antiderivative of f and g is a differentiable function whose range is contained in the domain of F. The chain rule states that

$$D_x(F(g(x))) = F'(g(x))g'(x)$$
$$= f(g(x))g'(x).$$

The corresponding result for indefinite integrals is that if f is continuous, then

(5.27) $$\int f(g(x))g'(x)dx = F(g(x)) + C.$$

The reason for using the differential-like notation in the integral can now be appreciated. Letting $u = g(x)$, the differential du, with respect to x, is $du = g'(x)dx$. With this notation, equation (5.27) becomes

(5.28) $$\int f(g(x))g'(x)dx = \int f(u)du = F(u) + C = F(g(x)) + C.$$

The process of choosing the appropriate value for u in a given problem is called **integration by substitution**, an important technique that will be used early and often.

EXAMPLE 1   Find $\int (3x + 4)^4 \, 3dx$.

SOLUTION
        To apply the integration by substitution technique, let $u = 3x + 4$ (which is the first part of the composition). Then $du = 3\,dx$. Substituting into the original problem and integrating, we have

$$\int (3x + 4)^4 \, 3dx = \int u^4 \, du$$

$$= \frac{u^5}{5} + C$$

$$= \frac{(3x + 4)^5}{5} + C.$$

The final answer is expressed in terms of x, the variable involved in the original integral.   □

EXAMPLE 2   Find $\int 2x \cos (x^2 + 2)dx$.

SOLUTION

If

$$u = x^2 + 2, \text{ then } du = 2x\,dx.$$

So

$$\int 2x \cos (x^2 + 2)dx = \int \cos u\,du$$
$$= \sin u + C$$
$$= \sin (x^2 + 2) + C.$$

Notice that again the substitution u is made for the first part of a composition:

$$x \rightarrow \overbrace{x^2 + 2}^{u} \rightarrow \cos (x^2 + 2).$$

The solution can be verified by differentiating:

$$D_x\,[\sin (x^2 + 2) + C] = \cos (x^2 + 2)(2x) = 2x \cos (x^2 + 2). \qquad \square$$

EXAMPLE 3   Find $\int (4x^2 - 3)^9 \, x\,dx$.

SOLUTION

Let $u = 4x^2 - 3$, then $du = 8x\,dx$. Since the exact form of du does not appear in the integral $\int (4x^2 - 3)^9\,x\,dx$, the substitution is made as follows: $du = 8x\,dx$ implies $du/8 = x\,dx$, so

$$\int (4x^2 - 3)^9\,x\,dx = \int u^9 \frac{1}{8}\,du$$

$$= \frac{1}{8} \int u^9\,du$$

$$= \frac{1}{8}\left(\frac{u^{10}}{10}\right) + C$$

$$= \frac{(4x^2 - 3)^{10}}{80} + C. \qquad \square$$

EXAMPLE 4   Find $\int_{.5}^{1} (4x^2 - 3)^9 \, x\,dx$.

SOLUTION

In Example 3, we found that

$$\int (4x^2 - 3)^9\,x\,dx = \frac{(4x^2 - 3)^{10}}{80} + C.$$

Consequently,

$$\int_{.5}^{1} (4x^2 - 3)^9 \, x \, dx = \left. \frac{(4x^2 - 3)^{10}}{80} \right]_{.5}^{1}$$

$$= \frac{(4 - 3)^{10}}{80} - \frac{(1 - 3)^{10}}{80}$$

$$= \frac{1}{80} - \frac{1024}{80} = \frac{-1023}{80}.$$ □

There is another method for finding the value of a definite integral such as the one given in Example 4. This involves changing limits of integration to correspond to the new variable after the substitution is made. The values of $u = 4x^2 - 3$ that correspond to $x = .5$ and $x = 1$ are, respectively, $u = 4(.5)^2 - 3 = -2$ and $u = 4(1)^2 - 3 = 1$. Thus,

$$\int_{x=.5}^{x=1} (4x^2 - 3)^9 \, x \, dx = \frac{1}{8} \int_{u=-2}^{u=1} u^9 \, du$$

$$= \left. \frac{u^{10}}{80} \right]_{-2}^{1}$$

$$= \frac{1}{80} - \frac{(-2)^{10}}{80} = \frac{1}{80} - \frac{1024}{80} = \frac{-1023}{80},$$

which, of course, agrees with the result in Example 4.

The following problems are somewhat more complicated than those in the previous examples. Notice, however, that the method of solution is the same—a substitution is made for the first part of a composition.

EXAMPLE 5   Find $\displaystyle\int \frac{x + 1}{\sqrt{x^2 + 2x}} \, dx$.

SOLUTION

The term $x^2 + 2x$ is the first part of the composition described by $\dfrac{1}{\sqrt{x^2 + 2x}}$. If $u = x^2 + 2x$, then $du = (2x + 2)dx = 2(x + 1)dx$, and

$$\int \frac{x + 1}{\sqrt{x^2 + 2x}} \, dx = \int \frac{1}{2} \frac{du}{\sqrt{u}}$$

$$= \frac{1}{2} \int u^{-1/2} \, du$$

$$= \frac{1}{2} \frac{u^{1/2}}{1/2} + C$$

$$= \sqrt{x^2 + 2x} + C.$$ □

EXAMPLE 6   Find $\int (\tan x)^3 (\sec x)^2 \, dx$.

SOLUTION

Two compositions are present in this problem, one involving the term $(\tan x)^3$, the other involving $(\sec x)^2$. Consider the resulting expression if we make the substitution $u = \tan x$. In this case, $du = (\sec x)^2 \, dx$ and

$$\int (\tan x)^3 (\sec x)^2 \, dx = \int u^3 \, du$$

$$= \frac{u^4}{4} + C$$

$$= \frac{(\tan x)^4}{4} + C.$$

The problem could also be solved by making the substitution $u = \sec x$, but the solution is more difficult and the trigonometric identity $(\tan x)^2 + 1 = (\sec x)^2$ must be employed. It is instructive to carry out this substitution, however, and for this reason it is included as Exercise 47.   □

EXAMPLE 7   Find $\int (x + 1)^2 \cos(x^3 + 3x^2 + 3x) dx$.

SOLUTION

Two compositions are present in this problem, one involved in $(x + 1)^2$, the other in $\cos(x^3 + 3x^2 + 3x)$. If we make the substitution $u = x + 1$, $du = dx$, it does not reduce the complication in the integrand. On the other hand, if we let $u = x^3 + 3x^2 + 3x$, then $du = (3x^2 + 6x + 3)dx$, and

$$\frac{1}{3} du = (x^2 + 2x + 1)dx = (x + 1)^2 \, dx.$$

So,

$$\int (x + 1)^2 \cos(x^3 + 3x^2 + 3x)dx = \frac{1}{3} \int \cos u \, du$$

$$= \frac{1}{3} \sin u + C$$

$$= \frac{1}{3} \sin(x^3 + 3x^2 + 3x) + C.$$ □

EXAMPLE 8   Find $\int x \sqrt{x + 1} \, dx$.

SOLUTION

If we follow our rule that a substitution be made for the first part of a composition, then

$$u = x + 1; \, du = dx \qquad \text{and} \qquad x = u - 1.$$

This substitution simplifies the radical without significantly complicating the rest of the integrand.

$$\int x \sqrt{x + 1}\, dx = \int (u - 1) \sqrt{u}\, du$$

$$= \int (u^{3/2} - u^{1/2})du$$

$$= \frac{2}{5} u^{5/2} - \frac{2}{3} u^{3/2} + C$$

$$= \frac{2}{5} (x + 1)^{5/2} - \frac{2}{3} (x + 1)^{3/2} + C.$$

We verify this solution by differentiating:

$$D_x \left[\frac{2}{5} (x + 1)^{5/2} - \frac{2}{3} (x + 1)^{3/2} + C \right] = (x + 1)^{3/2} - (x + 1)^{1/2}$$

$$= (x + 1)^{1/2} [(x + 1) - 1]$$

$$= x \sqrt{x + 1}. \qquad \square$$

EXERCISE SET 5.6

Evaluate the integrals in Exercises 1 through 46.

1. $\displaystyle\int 2(2x - 3)^2\, dx$

2. $\displaystyle\int 5(5x + 4)^5\, dx$

3. $\displaystyle\int 3t^2(t^3 + 4)dt$

4. $\displaystyle\int 3t^2(t^3 + 4)^5\, dt$

5. $\displaystyle\int 3t^2(t^3 + 4)^{1/2}\, dt$

6. $\displaystyle\int 3t^2(t^3 + 4)^{-1/2}\, dt$

7. $\displaystyle\int \sqrt{4x - 5}\, dx$

8. $\displaystyle\int (x + 1) \sqrt{x^2 + 2x}\, dx$

9. $\displaystyle\int \sin \pi x\, dx$

10. $\displaystyle\int \cos (2x + 1)dx$

11. $\displaystyle\int (\sin x)^{10} \cos x\, dx$

12. $\displaystyle\int \frac{\sin x}{(\cos x)^5}\, dx$

13. $\displaystyle\int \frac{(\sqrt{x} - 1)^2}{\sqrt{x}}\, dx$

14. $\displaystyle\int \frac{(\sqrt{x} - 1)^{1/2}}{\sqrt{x}}\, dx$

15. $\displaystyle\int (x^3 + x^2)^4(3x^2 + 2x)\, dx$

16. $\displaystyle\int \sqrt{x^3 + x^2}\, (3x^2 + 2x)dx$

17. $\int_{-1}^{1} \frac{x+1}{(x^2+2x+2)^3} \, dx$

18. $\int_{0}^{1} 5x^3(x^4+1)^4 \, dx$

19. $\int \sin 3x \, (\cos 3x)^3 \, dx$

20. $\int_{0}^{\pi} \cos x \sqrt{\sin x} \, dx$

21. $\int \frac{x}{(\cos x^2)^2} \, dx$

22. $\int (x+1) \sin (x^2+2x+3) dx$

23. $\int_{0}^{\pi/4} \tan x(\sec x)^2 \, dx$

24. $\int \cot 3x \csc 3x \, dx$

25. $\int \left(1+\frac{1}{t}\right)^3 \frac{1}{t^2} \, dt$

26. $\int \frac{x^2}{(x^3+1)^4} \, dx$

27. $\int_{0}^{2} (t+2) \sqrt{t^2+4t+1} \, dt$

28. $\int_{-1}^{1} x^2 \sqrt{x^3+1} \, dx$

29. $\int \frac{2}{\sqrt{3t-7}} \, dt$

30. $\int_{1}^{4} \frac{1}{\sqrt{x}\,(\sqrt{x}+1)^2} \, dx$

31. $\int (z+1)\sqrt{z-1} \, dz$

32. $\int_{1}^{2} (z+1) \sqrt{z-1} \, dz$

33. $\int_{0}^{1} \frac{x}{\sqrt{x+1}} \, dx$

34. $\int x\sqrt{2x+1} \, dx$

35. $\int_{1}^{2} \frac{\sqrt{\sqrt{x}+1}}{\sqrt{x}} \, dx$

36. $\int \sqrt{x\sqrt{x}+1} \, dx$

37. $\int (x^2+1)\sqrt{x-2} \, dx$

38. $\int x^3\sqrt{x^2-1} \, dx$

39. $\int \frac{x^2+2x}{x^2+2x+1} \, dx$

40. $\int \frac{x^3-3x^2+3x}{x^3-3x^2+3x+1} \, dx$

(*Hint*: First divide the denominator into the numerator.)

41. $\int \frac{1}{x^2+6x+9} \, dx$

42. $\int_{0}^{2} \frac{1}{x^2+2x+1} \, dx$

43. $\int_{1}^{2} \left(1+\frac{1}{t^2}\right)^3 \frac{1}{t^3} \, dt$

44. $\int \left(2+\frac{1}{\sqrt{z}}\right)^4 \frac{1}{z\sqrt{z}} \, dz$

45. $\int \frac{\sin x}{(2+3\cos x)^2} \, dx$

46. $\int \frac{(\sec x)^2}{(1+\tan x)^3} \, dx$

47. Find $\int (\tan x)^3 \, (\sec x)^2 \, dx$ by making the substitution $u = \sec x$. What is the difference between this solution and the one given in Example 6?

48. Find the area of the region bounded by the graph of $f(x) = \sin x \cos x$, the x-axis, and the lines $x = 0$, $x = \pi/2$.

In Exercises 49 through 52, the derivative of a function and its value at one point are given. Find the function.

49. $f'(x) = \sin x \cos x, f(0) = 0$

50. $f'(x) = x \cos (x^2 + \pi), f(0) = 1$

51. $f'(x) = (x + 2) \sqrt{x^2 + 4x}, f(0) = 0$

52. $f'(x) = \dfrac{x}{\sqrt{x^2 + 1}}, f(0) = 1$

5.7
THE NATURAL LOGARITHM FUNCTION

The first integrals we studied involved functions of the form $f(x) = x^n$ for some rational number n. The derivative rule for these functions, $D_x x^n = n x^{n-1}$, led immediately to the integration formula

$$\int x^n \, dx = \frac{x^{n+1}}{n + 1} + C,$$

provided $n \neq -1$. We cannot extend this formula to include $n = -1$, since this would involve division by zero.

The function described by $f(x) = x^{-1}$ has an antiderivative, but it is a different type of function from those we have discussed thus far. This function, called the natural logarithm function, is most easily defined in terms of an integral of the function f. The reason for calling this function a logarithm will become clear after we have studied some of its properties.

(5.29)
DEFINITION

The **natural logarithm** function ln is defined by

$$\ln x = \int_1^x \frac{1}{t} \, dt$$

for all real numbers $x > 0$.

We have used the variable t in the integrand of the definition to avoid confusion between the variable of integration and the variable that describes ln.

When $x > 1$, $\ln x$ is the positive number that describes the area of the region shown in Figure 5.23. When $0 < x < 1$, $\ln x$ is the negative number that describes the negative of the area of the region shown in Figure 5.24.

HISTORICAL NOTE   **Nicolaus Mercator** (1620–1687), a Danish mathematician who lived most of his life in London, first used the description of the natural logarithm in 1668 in his manuscript *Logarithmotechnia*. Mercator's given name was Nicolaus Kaufmann—he was not related to the famous Flemish cartographer Gerhardus Mercator.

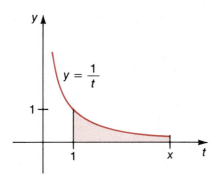

FIGURE 5.23

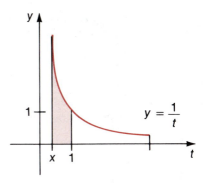

FIGURE 5.24

Two properties of ln that follow immediately from the definition are:

(5.30) $\ln 1 = 0$

and

(5.31) $D_x \ln x = D_x \left(\int_1^x \frac{1}{t}\, dt \right) = \frac{1}{x}$,   for all $x > 0$.

Property (5.30) follows because

$$\ln 1 = \int_1^1 \frac{1}{t}\, dt,$$

and any definite integral with the same upper and lower limits has been defined to be zero. Property (5.31) is a consequence of the fundamental lemma of calculus.

It follows immediately from the chain rule that if u is a differentiable function and $u > 0$, then

(5.32) $D_x \ln u = \frac{1}{u} D_x u.$

Other properties of the natural logarithm function are given in the next theorem. These properties are common to all logarithm functions and should be familiar to you if you have studied logarithms with base 10.

(5.33)
THEOREM

If a and b are positive real numbers and r is a rational number, then

(a) $\ln ab = \ln a + \ln b$ (b) $\ln \frac{a}{b} = \ln a - \ln b$

(c) $\ln a^r = r \ln a$

PROOF

(a) $\ln ab = \int_1^{ab} \frac{1}{t}\, dt = \int_1^a \frac{1}{t}\, dt + \int_a^{ab} \frac{1}{t}\, dt$

$= \ln a + \int_a^{ab} \frac{1}{t}\, dt.$

It remains to show that $\int_a^{ab} 1/t \, dt = \ln b$. To do this we introduce a substitution $u = t/a$. Then $du = dt/a$, so $dt = a \, du$ and

$$\int_{t=a}^{t=ab} \frac{1}{t} \, dt = \int_{u=1}^{u=b} \frac{a \, du}{u \, a} = \int_1^b \frac{du}{u} = \ln b.$$

Thus, $\ln ab = \ln a + \ln b$.

(b) This property follows from (a) by noting that since $a = \dfrac{a}{b} \cdot b$,

$$\ln a = \ln \frac{a}{b} \cdot b = \ln \frac{a}{b} + \ln b.$$

Therefore,

$$\ln \frac{a}{b} = \ln a - \ln b.$$

(c) This property can be proved in a manner similar to the proof of (a), but we will use a different method. It follows from (5.32) and the chain rule that for all $x > 0$ and rational numbers r,

$$D_x(\ln x^r) = \frac{1}{x^r} r x^{r-1} = \frac{r}{x} = D_x(r \ln x).$$

Consequently, $\ln x^r$ and $r \ln x$ are antiderivatives of the same function and

$$\ln x^r = r \ln x + C, \text{ for some constant } C.$$

In particular, when $x = 1$,

$$0 = \ln 1 = \ln 1^r = r \ln 1 + C = r \cdot 0 + C, \qquad \text{so } C = 0,$$

and $\ln x^r = r \ln x$.

Consequently, for any positive real number a and rational number r

$$\ln a^r = r \ln a. \qquad \qquad \square$$

EXAMPLE 1   Find $\displaystyle\int_2^4 \frac{1}{t} \, dt$.

SOLUTION

$$\int_2^4 \frac{1}{t} \, dt = \int_1^4 \frac{1}{t} \, dt - \int_1^2 \frac{1}{t} \, dt$$

$$= \ln 4 - \ln 2 = \ln \frac{4}{2}$$

$$= \ln 2 \approx .69315.$$

The value for $\ln 2$ was found using a calculator. (The fact that $\ln$ is a statement function on most scientific calculators should give a clue to its importance.) $\square$

EXAMPLE 2   Find $D_x \ln (\sqrt{x^2 + 3x})$.

SOLUTION

We can find this derivative by using (5.32):

$$D_x \ln (\sqrt{x^2 + 3x}) = \frac{1}{\sqrt{x^2 + 3x}} D_x \sqrt{x^2 + 3x},$$

but it is easier to first use property (c) of Theorem 5.33 to simplify the expression.

$$D_x \ln (\sqrt{x^2 + 3x}) = D_x \ln (x^2 + 3x)^{1/2} = D_x \frac{1}{2} \ln (x^2 + 3x)$$

$$= \frac{1}{2} \frac{1}{x^2 + 2x} D_x (x^2 + 3x)$$

$$= \frac{2x + 3}{2(x^2 + 2x)}.$$ □

The natural logarithm function can be used to differentiate a function that involves combinations of exponents, quotients, and products. This procedure, known as **logarithmic differentiation,** is illustrated in the next example.

EXAMPLE 3   Find $D_x y$ if $y = \dfrac{x \sqrt[3]{x + 4}}{\sqrt{x^2 - 1}}$.

SOLUTION

Taking the natural logarithm of each side and simplifying using the properties of ln, we have

$$\ln y = \ln x + \ln \sqrt[3]{x + 4} - \ln \sqrt{x^2 - 1}$$

$$= \ln x + \frac{1}{3} \ln (x + 4) - \frac{1}{2} \ln (x^2 - 1).$$

Now, differentiate both sides with respect to x to obtain:

$$\frac{1}{y} D_x y = \frac{1}{x} + \frac{1}{3(x + 4)} - \frac{2x}{2(x^2 - 1)}$$

and solve for $D_x y$:

$$D_x y = y \left(\frac{1}{x} + \frac{1}{3(x + 4)} - \frac{2x}{2(x^2 - 1)} \right)$$

$$= \frac{x \sqrt[3]{x + 4}}{\sqrt{x^2 - 1}} \left(\frac{1}{x} + \frac{1}{3(x + 4)} - \frac{x}{x^2 - 1} \right).$$ □

If $y \le 0$ in Example 3, $\ln y$ is not defined. However, for $y < 0$, we could use $\ln |y|$ and obtain the same result, as the following discussion shows.

If $x < 0$, then $-x > 0$, so $\ln (-x)$ exists. By the chain rule,

$$D_x \ln (-x) = \frac{1}{-x} (-1) = \frac{1}{x}.$$

Consequently, for any $x \neq 0$, $D_x \ln |x| = 1/x$. The corresponding integration formula is

(5.34)
$$\int \frac{1}{x} \, dx = \ln |x| + C.$$

EXAMPLE 4   Determine $\int \frac{x}{x^2 + 3} \, dx$.

SOLUTION

Let $u = x^2 + 3$. Then $du = 2x \, dx$ and $\frac{1}{2} du = x \, dx$.

$$\int \frac{x}{x^2 + 3} \, dx = \frac{1}{2} \int \frac{1}{u} \, du$$
$$= \frac{1}{2} \ln |u| + C$$
$$= \frac{1}{2} \ln (x^2 + 3) + C.$$

Notice that the absolute value can be omitted in the final result since $x^2 + 3 > 0$ for all values of x. □

Property (5.34) can be used to find integration formulas for the trigonometric functions that could not be determined in Section 5.5.

To integrate the tangent function we write

$$\int \tan x \, dx = \int \frac{\sin x}{\cos x} \, dx$$

and use the substitution $u = \cos x$. Then $du = -\sin x \, dx$ and

$$\int \tan x \, dx = -\int \frac{du}{u}$$
$$= -\ln |u| + C$$
$$= -\ln |\cos x| + C.$$

Since $-\ln |\cos x| = \ln |\cos x|^{-1} = \ln |\sec x|$, we can rewrite the formula as

(5.35)
$$\int \tan x \, dx = \ln |\sec x| + C.$$

In a similar manner, we can show that

(5.36)
$$\int \cot x \, dx = \ln |\sin x| + C.$$

To integrate either the secant or cosecant functions requires a "trick":

$$\int \sec x \, dx = \int \sec x \left(\frac{\sec x + \tan x}{\sec x + \tan x} \right) dx$$
$$= \int \frac{(\sec x)^2 + \sec x \tan x}{\sec x + \tan x} \, dx.$$

Let $u = \sec x + \tan x$, then $du = [\sec x \tan x + (\sec x)^2]dx$ and

$$\int \sec x \, dx = \int \frac{1}{u} \, du$$
$$= \ln |u| + C,$$

so

(5.37) $$\int \sec x \, dx = \ln |\sec x + \tan x| + C.$$

In a similar manner, it can be shown that

(5.38) $$\int \csc x \, dx = \ln |\csc x - \cot x| + C.$$

EXAMPLE 5   Find $\int \tan(x + 2)dx$.

SOLUTION
Let $u = x + 2$, then $du = dx$. Using (5.35),

$$\int \tan(x + 2)dx = \int \tan u \, du$$
$$= \ln |\sec u| + C$$
$$= \ln |\sec(x + 2)| + C. \qquad \square$$

EXAMPLE 6   Determine $\int x \csc(3x^2 + 4)dx$.

SOLUTION
Let $u = 3x^2 + 4$, then $du = 6x \, dx$, $x \, dx = (1/6) \, du$ and

$$\int x \csc(3x^2 + 4)dx = \int \csc u \, \frac{du}{6}$$
$$= \frac{1}{6} \ln |\csc u - \cot u| + C$$
$$= \frac{1}{6} \ln |\csc(3x^2 + 4) - \cot(3x^2 + 4)| + C. \qquad \square$$

We complete this section by giving additional facts and observations that will enable us to sketch the graph of the natural logarithm function.

(5.39) $1/2 < \ln 2 < 1 < \ln 4$.

(5.40) A number, denoted e, exists with the property that $2 < e < 4$ and $\ln e = 1$. (In fact, e is an irrational number with an approximate value of 2.71828.)

(5.41) ln is an increasing function whose graph is concave downward.

(5.42) $\lim\limits_{x \to \infty} \ln x = \infty$ and $\lim\limits_{x \to 0^+} \ln x = -\infty$.

HISTORICAL NOTE   The symbol e was first used for this purpose in 1727 by **Leonhard Euler** (1707–1783), who also introduced the $f(x)$ notation for function representation and the Σ notation for sums.

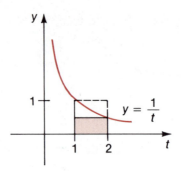

FIGURE 5.25

To show (5.39), consider the graph of $f(t) = 1/t$ shown in Figure 5.25. The area of the region bounded by the graph of f, the t-axis, and the lines $t = 1$, $t = 2$ is $\ln 2$. This area is greater than the area inside the inscribed rectangle, which is $(2 - 1)(1/2) = 1/2$, and less than the area inside the circumscribed rectangle, $(2 - 1)(1) = 1$. Thus, $1/2 < \ln 2 < 1$. Since $\ln 4 = \ln 2^2 = 2 \ln 2$, $\ln 4 > 2 (1/2) = 1$. This establishes (5.39).

To show (5.40), note that $\ln$ is differentiable on $[2, 4]$, so $\ln$ is continuous on $[2, 4]$. By the intermediate value theorem, $\ln$ takes on every value between $\ln 2 < 1$ and $\ln 4 > 1$. Therefore, a number in $[2, 4]$ exists whose natural logarithm is 1.

To verify (5.41), recall that $D_x \ln x = 1/x > 0$, so $\ln$ is increasing. Since the second derivative, $D_x^2 \ln x = -1/x^2$, is negative, the graph of $\ln$ is concave downward.

To verify that $\lim_{x \to \infty} \ln x = \infty$, let M be any positive real number and n be an integer with $n > M$. Then $\ln e^n = n \ln e = n \cdot 1 = n > M$. Since $\ln$ is increasing, if $x > e^n$, $\ln x > \ln e^n = n > M$. This implies that $\lim_{x \to \infty} \ln x = \infty$.

To see that $\lim_{x \to 0^+} \ln x = -\infty$, let $z = 1/x$. Then

$$\lim_{x \to 0^+} \ln x = \lim_{z \to \infty} \ln \frac{1}{z} = \lim_{z \to \infty} \ln z^{-1} = - \lim_{z \to \infty} \ln z = -\infty.$$

The graph shown in Figure 5.26 is the natural conclusion to the observations in (5.39) through (5.42).

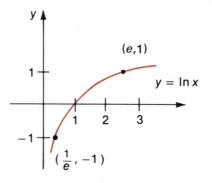

FIGURE 5.26

EXERCISE SET 5.7

Find the derivative of the functions described in Exercises 1 through 24.

1. $f(x) = \ln(x^2 + 3)$
2. $f(x) = \ln(x^2 + 2x - 1)$

3. $f(x) = \ln x^{1/4}$
4. $f(x) = (\ln x)^{1/4}$

5. $f(x) = \ln(\ln x)$
6. $f(x) = \ln\left(\dfrac{1}{\ln x}\right)$

7. $f(x) = \ln(x + \sqrt{x^2 - 1})$
8. $f(x) = \ln(x - \sqrt{x^2 - 1})$

9. $f(x) = \ln\left(\dfrac{1+x}{1-x}\right)$
10. $f(x) = \ln\left(\dfrac{x^3 + 4x + 5}{x^2 + 4x + 1}\right)$

11. $f(x) = \left(\ln\dfrac{1+x}{1-x}\right)^{1/2}$
12. $f(x) = \ln\sqrt[4]{\dfrac{x^3 + 4x + 5}{x^2 + 4x + 1}}$

13. $f(x) = \ln(\sin x)$
14. $f(x) = \ln(\csc x)$

15. $f(x) = x\ln(x^2 + 1)$
16. $f(x) = \dfrac{\ln(x^2 + 1)}{x}$

17. $f(x) = \dfrac{x^2 + 1}{\ln x}$
18. $f(x) = x\ln x - x$

19. $f(x) = \ln|2x + 1|$
20. $f(x) = \ln|9x^2 - 4|$

21. $f(x) = (\ln\sqrt{x^2 + 2})^3$
22. $f(x) = (\ln\sqrt{x} + 1)^{1/2}$

23. $f(x) = \cos(\ln x)$
24. $f(x) = \tan(\ln x)$

Find $D_x y$ in Exercises 25 through 30.

25. $\ln xy = 1$
26. $\ln(x + y) = x$

27. $\ln(x + y) = y$
28. $\ln\dfrac{x}{y} + \ln\dfrac{y}{x} = y$

29. $\ln\left(\dfrac{x}{y}\right) + x^2 + y^2 = 3$
30. $\ln(x + y) = \ln(x - y) + 1$

Evaluate the integrals in Exercises 31 through 44.

31. $\displaystyle\int \dfrac{dx}{x - 3}$
32. $\displaystyle\int \dfrac{dx}{2x + 5}$

33. $\displaystyle\int_{-3}^{-1} \dfrac{dx}{x - 1}$
34. $\displaystyle\int_{-5}^{-1} \dfrac{dx}{2x + 1}$

35. $\displaystyle\int \dfrac{x + 1}{x^2 + 2x - 3}\,dx$
36. $\displaystyle\int \dfrac{2x}{(x^2 + 4)^2}\,dx$

37. $\displaystyle\int_{2}^{3} \dfrac{\ln x}{x}\,dx$
38. $\displaystyle\int \dfrac{1}{x}(\ln x)^3\,dx$

39. $\displaystyle\int \dfrac{1}{x}(\ln x^2)^3\,dx$
40. $\displaystyle\int_{1}^{4} \dfrac{\cos(\ln x)}{x}\,dx$

41. $\displaystyle\int_0^3 \frac{dx}{\sqrt{x + 1}}$ **42.** $\displaystyle\int \frac{dx}{\sqrt{x}\,(\sqrt{x} + 1)}$

43. $\displaystyle\int \frac{\cos x}{1 + \sin x}\,dx$ **44.** $\displaystyle\int \frac{(\sec x)^2}{2 + \tan x}\,dx$

Use logarithmic differentiation to find $D_x y$ in Exercises 45 through 50.

45. $y = (x^2 + 8)^3\,(3x + 7)^5$ **46.** $y = \dfrac{(x^2 + 7x)^5\,\sqrt{x^3 + 1}}{(x^3 - 2x + 1)^{1/5}}$

47. $y = \sqrt{x}\,\sqrt{(x + 1)\sqrt{x}}$ **48.** $y = \dfrac{(x^3 - 5)^{2/3}\,(x^2 + 2)^{1/3}}{\sqrt{x}}$

49. $y = \sqrt{x^2 + 1}\,(x^3 - 1)^5\,(2x^2 + 1)^7$ **50.** $y = \sqrt{x^2\,\sqrt{(3x - 7)\sqrt{x^3 + 1}}}$

51. Sketch the graphs of the functions described below.
   (a) $f(x) = \ln(x + 1)$ (b) $f(x) = \ln(x - 1)$
   (c) $f(x) = 1 + \ln x$ (d) $f(x) = \ln(-x)$
   (e) $f(x) = -\ln x$ (f) $f(x) = \ln|x|$

52. Use L'Hôpital's rule to determine the following limits:
   (a) $\displaystyle\lim_{x\to\infty} \frac{\ln x}{x}$ (b) $\displaystyle\lim_{x\to\infty} \frac{\ln x^5}{x}$
   (c) $\displaystyle\lim_{x\to\infty} \frac{\ln x}{x^5}$ (d) $\displaystyle\lim_{x\to 1} \frac{\ln x}{x - 1}$
   (e) $\displaystyle\lim_{x\to 0} \frac{\ln(1 + x)}{x}$ (f) $\displaystyle\lim_{x\to 0} 2x \ln x^2$

53. Find the area of the region bounded by the graph of $f(x) = 2/(x - 1)$, the x-axis, and the lines $x = 2$ and $x = 3$.

54. Find an equation of the line tangent to the graph of $y = 3 \ln x + x$ at the point $(1, 1)$.

55. Use the fact that $\dfrac{1}{1 - x^2} = \dfrac{1}{2}\left[\dfrac{1}{1 - x} + \dfrac{1}{1 + x}\right]$ to find $\displaystyle\int \frac{1}{1 - x^2}\,dx$.

56. If r is a nonzero rational number, use the substitution $u = t^{1/r}$ to show that $\ln a^r = r \ln a$ in a manner similar to that used in proving property (a) of Theorem 5.33.

57. Use Newton's method to find an approximation, accurate to within 10^{-3}, to a number that satisfies

$$\frac{x}{10} = \ln x.$$

58. Suppose $f(x) > 0$ and $f'(x) = f(x)g(x)$ for x in $[0, 1]$. Show that
$$\int_0^1 g(x)\,dx = \ln\frac{f(1)}{f(0)}.$$

59. The vapor pressure P of water measured in millimeters of mercury depends upon the temperature of the water according to the equation

$$\ln P = 21.020 - \frac{5.319 \times 10^3}{T}, \quad 100°C < T.$$

Find an expression for $\dfrac{dP}{dT}$.

60. The molar absorption coefficient in spectroscopy, α, can be defined by

$$\alpha = \frac{-.4343}{cx} \ln \left(\frac{I(x)}{I_0} \right),$$

where c is the molar concentration of the substance in solution, x is the thickness of the solution layer that absorbs the light, I_0 is the initial intensity of the light beam and $I(x)$ is the intensity of the light after passing through the location. Find an expression for the rate at which $I(x)$ changes with respect to x.

61. When a gas is expanded or compressed from a volume V_1 to a volume V_2, the work W done is $W = \displaystyle\int_{V_1}^{V_2} P(V)\, dV$, where P is the pressure exerted by the gas as a function of the volume V. Suppose that a gas satisfies the equation of state $P(V)V = RT + \alpha P(V)$, where α is a function of the temperature T. Show that if the gas expands isothermally (without changing T) from V_1 to V_2, then

$$W = RT \ln \left(\frac{P(V_1)}{P(V_2)} \right).$$

62. A company must make a decision whether to purchase a new machine for shredding scrap metal. It is estimated that the machine will have a useful life of 5 years and be worth \$500 at the end of that time. The machine costs \$4500 new and the savings from the machine during its useful life are given by $s(t) = 3500/(t + 2)$, where t is time in years. Will the machine pay for itself?

63. Show that for all $x > 0$, $x > \sqrt{1 + x} \ln (1 + x)$.

5.8
IMPROPER INTEGRALS

The functions described by

$$f(x) = \frac{1}{x} \quad \text{and} \quad g(x) = \frac{1}{x^2}$$

have similar properties when restricted to the interval $[1, \infty)$. Both functions assume the value one when $x = 1$, both are positive for $x > 1$, and the values of each approach zero as x approaches infinity. This can be seen in Figures 5.27 and 5.28.

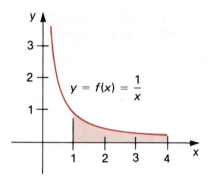

FIGURE 5.27

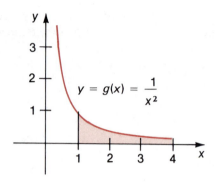

FIGURE 5.28

If we calculate the definite integrals of f and g on an interval $[1, M]$, where $M > 1$, then

$$\int_1^M f(x)dx = \int_1^M \frac{1}{x}\, dx = \ln M$$

and

$$\int_1^M g(x)dx = \int_1^M \frac{1}{x^2}\, dx = -\frac{1}{x}\bigg|_1^M = 1 - \frac{1}{M}.$$

We can now see a significant difference between f and g on $[1, \infty)$:

$$\lim_{M \to \infty} \int_1^M \frac{1}{x}\, dx = \lim_{M \to \infty} \ln M = \infty$$

while

$$\lim_{M \to \infty} \int_1^M \frac{1}{x^2}\, dx = \lim_{M \to \infty} \left(1 - \frac{1}{M}\right) = 1.$$

This implies that the area of the region bounded by the x-axis, the graph of f, and the lines $x = 1$ and $x = M$ becomes unbounded as M approaches infinity. However, we could logically define the area of the region bounded by the x-axis, the graph of g, and the line $x = 1$ to be one. For certain functions, then, it is reasonable to extend the concept of the definite integral to infinite intervals.

If f is continuous on $[a, \infty)$, then

(5.43)
$$\int_a^\infty f(x)dx = \lim_{M \to \infty} \int_a^M f(x)dx$$

whenever this limit is finite. Similarly, if f is continuous on $(-\infty, a]$, then

(5.44)
$$\int_{-\infty}^a f(x)dx = \lim_{M \to -\infty} \int_M^a f(x)dx$$

if this limit is finite. We define

(5.45)
$$\int_{-\infty}^\infty f(x)dx = \int_{-\infty}^a f(x)dx + \int_a^\infty f(x)dx$$

if both $\int_a^\infty f(x)dx$ and $\int_{-\infty}^a f(x)dx$ are finite for some real number a.

Integrals of this type are called **improper integrals**. They are said to **converge** if the limit is finite and said to **diverge** otherwise.

EXAMPLE 1   Evaluate $\int_2^\infty \frac{1}{\sqrt{x}}\, dx$, if this improper integral converges.

SOLUTION

$$\int_2^\infty \frac{1}{\sqrt{x}}\, dx = \lim_{M \to \infty} \int_2^M x^{-1/2}\, dx$$

$$= \lim_{M \to \infty} 2x^{1/2} \Big]_2^M$$

$$= \lim_{M \to \infty} (2\sqrt{M} - 2\sqrt{2}) = \infty.$$

The improper integral diverges.   ☐

EXAMPLE 2   Evaluate $\int_0^\infty \frac{x}{(1 + x^2)^2}\, dx$, if this integral converges.

SOLUTION

$$\int_0^\infty \frac{x}{(1 + x^2)^2}\, dx = \lim_{M \to \infty} \int_0^M \frac{x}{(1 + x^2)^2}\, dx.$$

To evaluate this integral let $u = 1 + x^2$, so $du = 2x\, dx$. Then, $(1/2)\, du = x\, dx$ and

$$\int \frac{x}{(1 + x^2)^2}\, dx = \int \frac{1}{u^2} \frac{du}{2}$$

$$= \frac{1}{2} \int u^{-2}\, du$$

$$= -\frac{1}{2u} + C$$

$$= -\frac{1}{2(1 + x^2)} + C.$$

Thus,

$$\int_0^\infty \frac{x}{(1 + x^2)^2}\, dx = \lim_{M \to \infty} -\frac{1}{2(1 + x^2)} \Big]_0^M$$

$$= \lim_{M \to \infty} \left(-\frac{1}{2(1 + M^2)} + \frac{1}{2} \right)$$

$$= \frac{1}{2}.$$ ☐

EXAMPLE 3   Evaluate $\int_{-\infty}^{\infty} \dfrac{x}{(1 + x^2)^2}\, dx$.

SOLUTION

We can rewrite this integral as

$$\int_{-\infty}^{\infty} \frac{x}{(1 + x^2)^2}\, dx = \int_{-\infty}^{0} \frac{x}{(1 + x^2)^2}\, dx + \int_{0}^{\infty} \frac{x}{(1 + x^2)^2}\, dx.$$

The second integral in this sum was found in Example 2 to have the value 1/2. It was also found that

$$\int \frac{x}{(1 + x^2)^2}\, dx = \frac{-1}{2(1 + x^2)} + C.$$

Hence, the first integral can be evaluated by:

$$\int_{-\infty}^{0} \frac{x}{(1 + x^2)^2}\, dx = \lim_{M \to -\infty} \int_{M}^{0} \frac{x}{(1 + x^2)^2}\, dx$$

$$= \lim_{M \to -\infty} \frac{-1}{2(1 + x^2)} \Bigg]_{M}^{0}$$

$$= \lim_{M \to -\infty} \left(-\frac{1}{2} + \frac{1}{2(1 + M^2)} \right)$$

$$= -\frac{1}{2}.$$

Consequently $\int_{-\infty}^{\infty} \dfrac{x}{(1 + x^2)^2}\, dx = -\dfrac{1}{2} + \dfrac{1}{2} = 0.$ □

Consider the improper integral $\int_{0}^{\infty} \dfrac{1}{(x - 1)^2}\, dx$. If we apply the fundamental theorem of calculus, we conclude that

$$\int_{0}^{M} \frac{dx}{(x - 1)^2} = -\frac{1}{M - 1} - 1,$$

and consequently that

$$\int_{0}^{\infty} \frac{dx}{(x - 1)^2} = \lim_{M \to \infty} \left[-\frac{1}{M - 1} - 1 \right] = -1.$$

This is obviously a fallacy since the integrand $1/(x - 1)^2$ is never negative. The error is that we applied the fundamental theorem of calculus on an interval over which it does not apply. The integrand is undefined at $x = 1$ and hence not continuous on any interval that includes this number. The fundamental theorem cannot be applied on an interval containing $x = 1$.

This problem brings us to the second type of improper integral: the extension of the concept of the definite integral to an interval on which a function is unbounded.

If f is continuous on $(a, b]$ and has an infinite discontinuity at a, that is, $\lim\limits_{x \to a^+} f(x) = \pm \infty$, then

(5.46)
$$\int_a^b f(x)dx = \lim_{M \to a^+} \int_M^b f(x)dx$$

provided this limit is finite.

Similarly, if f is continuous on $[a, b)$ and has an infinite discontinuity at b, then

(5.47)
$$\int_a^b f(x)dx = \lim_{M \to b^-} \int_a^M f(x)dx$$

if this limit is finite.

If f is continuous on $[a, c)$ and on $(c, b]$, but has an infinite discontinuity at c, then

(5.48)
$$\int_a^b f(x)dx = \int_a^c f(x)dx + \int_c^b f(x)dx,$$

provided both $\int_a^c f(x)dx$ and $\int_c^b f(x)dx$ are finite.

An integral involving an infinite discontinuity is also called an **improper integral**; it is said to **converge** if the limit is finite and is said to **diverge** otherwise.

EXAMPLE 4   Evaluate $\int_0^1 \dfrac{dx}{(x-1)^2}$, if this improper integral converges.

SOLUTION

An infinite discontinuity occurs at $x = 1$ so we must examine

$$\lim_{M \to 1^-} \int_0^M \frac{dx}{(x-1)^2} = \lim_{M \to 1^-} \left[-\frac{1}{x-1} \right]_0^M = \lim_{M \to 1^-} \left[-\frac{1}{M-1} - 1 \right] = \infty.$$

The improper integral diverges and is the cause of the difficulty in the problem we discussed before presenting (5.46). □

EXAMPLE 5   Evaluate $\int_0^4 \dfrac{dx}{\sqrt{x}}$, if this integral converges.

SOLUTION

The function $f(x) = \dfrac{1}{\sqrt{x}}$ has an infinite discontinuity at $x = 0$.

$$\lim_{M \to 0^+} \int_M^4 \frac{dx}{\sqrt{x}} = \lim_{M \to 0^+} 2\sqrt{x} \Big]_M^4$$
$$= \lim_{M \to 0^+} [2\sqrt{4} - 2\sqrt{M}] = 4.$$

Hence, the improper integral converges and

$$\int_0^4 \frac{dx}{\sqrt{x}} = 4. \qquad □$$

EXAMPLE 6   Evaluate $\int_{-1}^{1} x^{-2/3} \, dx$, if this integral converges.

SOLUTION

Since $f(x) = x^{-2/3}$ has an infinite discontinuity at 0, the integral must be considered as

$$\int_{-1}^{1} x^{-2/3} \, dx = \int_{-1}^{0} x^{-2/3} \, dx + \int_{0}^{1} x^{-2/3} \, dx.$$

Evaluating each improper integral, we have

$$\int_{-1}^{0} x^{-2/3} \, dx = \lim_{M \to 0^-} \int_{-1}^{M} x^{-2/3} \, dx$$

$$= \lim_{M \to 0^-} 3x^{1/3} \Big]_{-1}^{M} = \lim_{M \to 0^-} [3M^{1/3} - 3(-1)] = 3$$

and

$$\int_{0}^{1} x^{-2/3} \, dx = \lim_{M \to 0^+} \int_{M}^{1} x^{-2/3} \, dx = \lim_{M \to 0^+} 3x^{1/3} \Big]_{M}^{1} = \lim_{M \to 0^+} (3 - 3M^{1/3}) = 3.$$

So,

$$\int_{-1}^{1} x^{-2/3} \, dx = 3 + 3 = 6. \qquad \square$$

EXERCISE SET 5.8

Determine whether the integrals in Exercises 1 through 34 converge or diverge and evaluate those that converge.

1. $\int_{1}^{\infty} \dfrac{1}{x^{3/4}} \, dx$

2. $\int_{-\infty}^{-1} \dfrac{dx}{x^3}$

3. $\int_{1}^{\infty} \dfrac{x^2}{(x^3 + 1)^2} \, dx$

4. $\int_{100}^{\infty} \dfrac{1}{x + 5} \, dx$

5. $\int_{3}^{\infty} \dfrac{1}{x(\ln x)^2} \, dx$

6. $\int_{0}^{\infty} \dfrac{x}{1 + x^2} \, dx$

7. $\int_{1}^{\infty} \dfrac{x}{(x^2 + 1)^3} \, dx$

8. $\int_{1}^{\infty} \dfrac{1}{x\sqrt{x}} \, dx$

9. $\int_{1}^{\infty} \dfrac{1}{\sqrt{x}\,(\sqrt{x} + 1)} \, dx$

10. $\int_{-\infty}^{0} \dfrac{x}{\sqrt{x^2 + 1}} \, dx$

11. $\int_{1}^{\infty} \left(\dfrac{1}{x^3} + \dfrac{1}{x^2} \right) dx$

12. $\int_{1}^{\infty} \left(\dfrac{1}{x^3} + \dfrac{1}{x^4} \right) dx$

13. $\int_0^\infty \cos x \, dx$

14. $\int_5^\infty x \sin x^2 \, dx$

15. $\int_{-1}^0 \frac{1}{x^2} \, dx$

16. $\int_0^1 \frac{dx}{(x-1)^{2/3}}$

17. $\int_1^2 \frac{1}{(x-1)^2} \, dx$

18. $\int_0^1 \frac{1}{\sqrt{1-x}} \, dx$

19. $\int_0^{\pi/2} (\sec x)^2 \, dx$

20. $\int_0^{\pi/2} (\tan x)^2 \, dx$

21. $\int_0^{\pi/2} \sec x \, dx$

22. $\int_0^{\pi/2} \tan x \, dx$

23. $\int_0^2 \frac{1}{(x-1)^2} \, dx$

24. $\int_{-1}^1 \frac{1}{x^3} \, dx$

25. $\int_0^1 \frac{1}{\sqrt[3]{3x-1}} \, dx$

26. $\int_{-1}^0 \frac{dx}{\sqrt[3]{2x+1}}$

27. $\int_{-\infty}^\infty \frac{x}{(1+x^2)^3} \, dx$

28. $\int_{-\infty}^\infty \frac{x^2}{1+x^3} \, dx$

29. $\int_0^2 \left(\frac{1}{x} + \frac{1}{x^2} \right) dx$

30. $\int_1^\infty \left(\frac{1}{x} + \frac{1}{x^2} \right) dx$

31. $\int_0^\infty \frac{1}{\sqrt[3]{x-1}} \, dx$

32. $\int_{-1}^\infty \frac{1}{\sqrt[3]{x}} \, dx$

33. $\int_{-\infty}^0 \frac{1}{(x+2)^2} \, dx$

34. $\int_{-\infty}^0 \frac{x}{(x^2-4)^3} \, dx$

In Exercises 35 through 38, the boundaries of a region are given. Determine whether the region has finite area and if so, calculate the area.

35. $y = \frac{1}{x^2}$, x-axis, $x = 0$, $x = 1$.

36. $y = \tan x$, x-axis, $x = 0$, $x = \frac{\pi}{2}$.

37. $y = \frac{1}{x^3}$, x-axis, $x = 1$, $x \geq 1$.

38. $y = \frac{1}{\sqrt{x+1}}$, x-axis, $x = 3$, $x \geq 3$.

39. Show that the integral $\int_1^\infty \frac{1}{x^p} \, dx$ converges if $p > 1$ and diverges if $p \leq 1$.

40. Find all values of p for which the integral $\int_{-\infty}^{-1} \frac{1}{x^p} \, dx$ converges.

41. (a) Show that $\lim\limits_{M \to \infty} \int_{-M}^{M} x\,dx$ exists.

(b) Show that $\int_{-\infty}^{\infty} x\,dx$ diverges.

Parts (a) and (b) show the reason for defining $\int_{-\infty}^{\infty} f(x)\,dx$ as in Equation (5.45) instead of as $\lim\limits_{M \to \infty} \int_{-M}^{M} f(x)\,dx$.

42. Show that if $\int_{1}^{\infty} f(x)dx$ converges and c is a constant, then $\int_{1}^{\infty} cf(x)dx$ converges.

43. Suppose f and g are continuous on $[1, \infty)$ and $0 \le g(x) \le f(x)$ for $x \ge 1$. Show that if $\int_{1}^{\infty} f(x)dx$ converges, then $\int_{1}^{\infty} g(x)dx$ converges.

5.9
THE DISCOVERY OF CALCULUS

Gottfried Liebniz (1646–1716) and Isaac Newton (1642–1727) are generally credited with discovering the theory of calculus independently during the last third of the 17th century. But which part of calculus did they discover: the differential calculus, the integral calculus, or the important link between these two concepts, the fundamental theorem of calculus? The answer to this question is one that appears all too frequently on multiple choice examinations—none of the above. To place this statement in perspective, we must consider some important mathematical results that led to the unified theory of calculus.

The underlying concept of integral calculus was employed by Greek mathematicians at least as early as the time of Eudoxus of Cnidus (408–355 B.C.) and Archimedes (~287–212 B.C.). Archimedes determined the area of a circle by computing the area of inscribed and circumscribed polygons of increasing numbers of sides, as shown in Figure 5.29.

FIGURE 5.29

In this and similar applications, he was using the concept behind Riemann sums. Archimedes also devised ways to determine the tangent lines to certain curves, including the curve that bears his name: the spiral of Archimedes, shown in Figure 5.30.

FIGURE 5.30

In some sense, then, Archimedes could be considered the founder of calculus. He did not, however, have a notion of a unified theory that could be applied to more than a few specific cases, nor did he recognize a connection between the differential and integral concepts of calculus.

Little progress was made toward the discovery of calculus until the beginning of the 17th century, when the techniques of Eudoxus and Archimedes were employed by Bonaventura Cavalieri (1598–1647) and in the work of Galileo Galilee (1564–1642). Cavalieri was able to compute the areas and volumes of a number of common figures; his techniques were extended in Italy by Evangelista Toricelli (1608–1647), in France by Pierre Fermat (1601–1665), Christiaan Huygens (1629–1695), and Giles Personne de Roberval (1602–1675), and in England by John Wallis (1616–1703) and Isaac Barrow (1630–1677).

The problem of constructing tangents to curves was also revived during this period, first by Fermat and his French colleague René Descartes (1596–1650) and later by Roberval and Barrow. With such active research, it was only a matter of time until the unified theory was uncovered.

The first published statement of the fundamental theorem of calculus appears in *Lectiones Geometricae,* a treatise published by Barrow in 1667. The theorem, however, is believed to have been intuitively recognized by Galileo fifty years earlier in connection with his study of motion.

This brings us to the time of Newton, a young student of Barrow at Cambridge in the 1660's and to Liebniz, the Leipzig-born student of Huygens in Paris. Newton developed most of his calculus, called the "method of fluxions," during the period of 1664–1669 and compiled his results in the manuscript *De Analysi* in 1669. Although this manuscript was circulated and reviewed by a number of his English contemporaries, it did not appear in print until 1711, over forty years later. In fact, Newton used his calculus to develop many important discoveries regarding gravitation and the motion of objects, but his treatise on the subject, *Philosophiae Naturalis Principia Mathematica,* contains only classical geometrical demonstrations.

It is difficult to determine precisely when Liebniz first became interested in the calculus, but it was probably shortly before he traveled to England in 1672 as a political envoy. While visiting the London home of John Collins (1625–1683), he became acquainted with *De Analysi* and later communicated with Newton regarding the discoveries that Newton had made. The two exchanged a series of letters during 1676 and 1677 by which time Liebniz had developed his own theory of calculus. The letters generally describe the extent of their work, but often omit crucial details necessary for the methods of discovery. Liebniz understandably expected that Newton would soon publish a treatise on his method of fluxions and, when it became obvious that this work was not forthcoming, began in 1684 to publish his own version in a work entitled *Nova methodus pro maximus et minimus, itemque tangentibus, qua nec irrationales quantitates moratur* (A new method for finding maxima and minima, and also for tangents, which is not obstructed by irrational quantities).

Because of Liebniz's prior publication, his calculus became the version known to the mathematical public of the time, particularly on the scientifically active continent of Europe. We use his differential notation, dy/dx, and his elongated S symbol $\int$ to represent integration. Newton's notation was generally more cumbersome, although his symbol $\dot{y}$ to denote the derivative of y is still commonly used to indicate differentiation with respect to time.

The most generally accepted reason for Newton's failure to capitalize on his discovery of calculus lies in the lack of a rigorous method of proving the results. We have seen that the limit concept is basic to the study of both the differential and integral calculus. While this concept is intuitively clear, its definition is quite sophisticated. It was not until almost two centuries later, in 1872, that the German mathematician Eduard Heine (1821–1881), influenced by the work of Karl Weierstrass (1815–1897), published in his *Elements* the definition for the limit of a function that is used today. With this definition, the results of calculus can be proved as rigorously as the geometric demonstrations required for a result to be accepted by mathematicians in Newton's time.

REVIEW EXERCISES

Find the values of the sums in Exercises 1 through 6.

1. $\displaystyle\sum_{i=1}^{4} 2^i$

2. $\displaystyle\sum_{i=1}^{5} \frac{1}{2}\left(1 + \frac{i}{2}\right)^2$

3. $\displaystyle\sum_{i=3}^{6} (2i + 1)$

4. $\displaystyle\sum_{j=1}^{3} (j^2 + 2j - 1)$

5. $\displaystyle\sum_{i=1}^{n} (3i - 5)$

6. $\displaystyle\sum_{i=1}^{n} (3i^2 + 4i - 5)$

7. Use the definition of the definite integral to show that $\displaystyle\int_{1}^{3} (x^2 - 2x)\, dx = \frac{2}{3}$.

8. Use the definition of the definite integral to find the value of

$$\int_{2}^{5} (3x^2 - 1)\, dx.$$

9. Express each of the following limits as a definite integral over the interval $[0, 2]$ and find its value.

(a) $\displaystyle\lim_{n\to\infty} \sum_{i=1}^{n} 3(x_i)^2 \frac{2}{n}$

(b) $\displaystyle\lim_{n\to\infty} \sum_{i=1}^{n} (2x_i + 1)\frac{2}{n}$

10. Determine a function for which each of the following is a Riemann sum.

(a) $\displaystyle\sum_{i=1}^{n} (x_i^3 + x_i - 1)\Delta x_i$

(b) $\displaystyle\sum_{i=1}^{n} \sqrt{1 + x_i^2}\,\Delta x_i$

Evaluate the integrals in Exercises 11 through 48.

11. $\int x^3 \, dx$

12. $\int (2x - 15) \, dx$

13. $\int_{-3}^{5} (2x^3 + 5) \, dx$

14. $\int \left(\sqrt{x} - 2x + \frac{1}{x^2} \right) dx$

15. $\int 4(4x + 7)^3 \, dx$

16. $\int (4x + 7)^3 \, dx$

17. $\int \left(x\sqrt{x} + \frac{1}{x} \right) dx$

18. $\int (x^3 - 2)^2 \, dx$

19. $\int \sqrt[3]{8x^8} \, dx$

20. $\int (x^2 + 2x)(x^3 + 3x^2 + 1)^{1/2} dx$

21. $\int \frac{1}{x^4} + \frac{1}{\sqrt[4]{x}} \, dx$

22. $\int_{-2}^{2} 3w\sqrt{4 - w^2} \, dw$

23. $\int_{0}^{\pi} (\cos t - t) \, dt$

24. $\int x \cos (x^2 - 1) \, dx$

25. $\int \frac{x}{\sqrt{4 - x^2}} \, dx$

26. $\int_{2}^{3} \frac{(x + 1)}{x^2 + 2x + 3} \, dx$

27. $\int w^3\sqrt{w^2 + 1} \, dw$

28. $\int x^2\sqrt{x - 1} \, dx$

29. $\int \frac{1}{\sqrt{x}(1 - \sqrt{x})^2} \, dx$

30. $\int_{2}^{5} \frac{1}{x} \, dx$

31. $\int \frac{1}{x - 2} \, dx$

32. $\int \frac{x}{x^2 + 2} \, dx$

33. $\int \frac{x - 3}{x^2 - 6x + 5} \, dx$

34. $\int \frac{1}{x^2 + 2x + 1} \, dx$

35. $\int_{0}^{2} 2u^2\sqrt{u^3 + 1} \, du$

36. $\int x^5\sqrt{x^2 + 4} \, dx$

37. $\int_{1}^{2} \frac{s}{(1 + 2s)^3} \, ds$

38. $\int \frac{t^3 - 7t^2 + 6t + 1}{t^2} \, dt$

39. $\int \frac{x^2}{x^2 - 2x + 2} \, dx$

40. $\int \frac{x^2 - 2x}{x^2 - 5x + 6} \, dx$

41. $\int \cos 2x \, dx$

42. $\int_{0}^{1} \sin \pi x \, dx$

43. $\int \tan 4x \, dx$

44. $\int \frac{\sin x}{1 - \cos x} \, dx$

45. $\int (\sin x)^2 \cos x \, dx$

46. $\int \frac{\sin \sqrt{x}}{\sqrt{x}} \, dx$

47. $\int \frac{\cos (\ln x)}{x} \, dx$

48. $\int [(\sin x)^2 + \sin x] \cos x \, dx$

Determine whether the integrals in Exercises 49 through 62 converge or diverge and evaluate those that converge.

49. $\int_{1}^{\infty} \dfrac{3}{x^2}\, dx$

50. $\int_{0}^{2} \dfrac{1}{\sqrt{x}}\, dx$

51. $\int_{0}^{\infty} \dfrac{1}{(x+1)^2}\, dx$

52. $\int_{0}^{\infty} \dfrac{1}{10x}\, dx$

53. $\int_{2}^{\infty} \dfrac{1}{x(\ln x)^2}\, dx$

54. $\int_{-\infty}^{\infty} \dfrac{1}{x^2 - 2x + 1}\, dx$

55. $\int_{3}^{\infty} \dfrac{1}{\sqrt{3x}}\, dx$

56. $\int_{0}^{9} \dfrac{1}{(x-1)^{2/3}}\, dx$

57. $\int_{-3}^{3} \dfrac{x\, dx}{\sqrt{9 - x^2}}$

58. $\int_{0}^{1} \dfrac{1}{\sqrt{1 - x}}\, dx$

59. $\int_{-1}^{8} x^{-2/3}\, dx$

60. $\int_{0}^{\infty} \dfrac{x}{\sqrt{x^2 + 1}}\, dx$

61. $\int_{-\infty}^{\infty} \cos x\, dx$

62. $\int_{-\infty}^{\infty} \dfrac{1}{1 - x}\, dx$

Find the derivative of the functions described in Exercises 63 through 70.

63. $f(x) = (\ln x)^3$

64. $f(x) = x \ln x - x$

65. $f(x) = \ln (x^3 - 7x + 1)$

66. $g(t) = \dfrac{\ln (\sqrt{t} + 1)}{\sqrt{t}}$

67. $h(x) = [\ln(x^{1/3} + 1)]^{1/3}$

68. $h(x) = (\ln x^4)(x^3 - 2x)$

69. $f(x) = x \displaystyle\int_{1}^{x} \ln t\, dt$

70. $f(x) = \displaystyle\int_{1}^{x} \dfrac{1}{t^2 + 1}\, dt$

71. Find an equation of the curve that passes through the point $(4, 0)$ and whose tangent line at any point (x, y) has slope $2x - 1$.

72. Find an equation of the curve that has a horizontal tangent line at the point $(0, 1)$ and for which $y'' = 3x^2 - 1$.

73. Find the area of the region bounded by the curve $y = (x - 2)^{2/3}$, the x-axis, and the lines $x = 1$ and $x = 10$. Make a sketch of the region.

74. Find the area of the region bounded by the curve $y = 9 - x^2$ and the x-axis. Make a sketch of the region.

75. Find the area of the region bounded by the graph of $f(x) = x^3 - 3x^2 - x + 3$, the lines $x = -1$, $x = 1$, and the x-axis. Make a sketch of the region.

76. Verify that $\int \ln x\, dx = x \ln x - x + C$ and use this to find each of the following:

(a) $\displaystyle\int_{1}^{3} \ln x\, dx$

(b) $\displaystyle\int (3 \ln x - 2x)\, dx$

(c) $\displaystyle\int_{2}^{4} (\ln x - 2x)\, dx$

(d) $\displaystyle\int 2x \ln x^2\, dx$

(e) The area of the region bounded by the graph of $f(x) = \ln x$, the x-axis, and the line $x = 3$.

(f) An equation of the curve that passes through the point $(1, 1)$ and whose tangent line at any point (x, y) has slope $\ln x$.

6

APPLICATIONS OF THE DEFINITE INTEGRAL

To introduce the definite integral, we considered the problem of determining the area of a region bounded by the x-axis and the graph of a nonnegative function defined on a closed interval. In this chapter we consider some other applications of the definite integral. The common thread in these applications is that in each problem finite sums are used as approximations. Each of these sums is a Riemann sum and leads to a definite integral as the number of terms increases without bound.

6.1
AREAS OF REGIONS IN THE PLANE

In Section 5.1, we found that if f is a continuous nonnegative function defined on an interval $[a, b]$, then the area A of the region bounded by the graph of f, the x-axis, and the lines $x = a$, $x = b$, is $A = \lim_{n \to \infty} A_n$, where $A_n = \sum_{i=1}^{n} f(x_i) \dfrac{b - a}{n}$ is the sum of the areas of n rectangles. Then in Section 5.2, we saw that for each n, A_n is actually a Riemann sum for f on $[a, b]$ with $\Delta x_i = (b - a)/n$ for each i. Consequently,

$$(6.1) \qquad A = \lim_{n \to \infty} A_n = \int_a^b f(x)dx.$$

315

We call this the area under the curve from a to b, as in Figure 6.1.

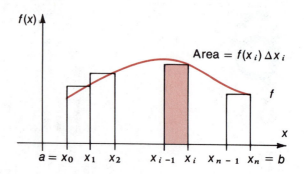

FIGURE 6.1

EXAMPLE 1   Find the area of the region bounded by the x-axis, the line $x = 2$, and the graph of $f(x) = x^3 - 2x^2 + 3$.

SOLUTION

First we sketch enough of the graph of f to show the region whose area we wish to find. This region and an approximating rectangle are shown in Figure 6.2.

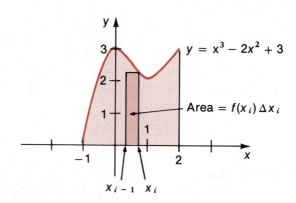

FIGURE 6.2

Since f is nonnegative on the interval $[-1, 2]$, the area is

$$\int_{-1}^{2} f(x)dx = \int_{-1}^{2} (x^3 - 2x^2 + 3)dx = \left(\frac{x^4}{4} - \frac{2x^3}{3} + 3x\right)\Bigg]_{-1}^{2}$$

$$= \left[\frac{16}{4} - \frac{16}{3} + 6\right] - \left[\frac{1}{4} + \frac{2}{3} - 3\right] = \frac{27}{4}.$$

☐

When f is a function that is continuous and nonpositive on an interval $[a, b]$, the area of the region bounded by $x=a$, $x=b$, the x-axis, and the graph of f can be found by considering the continuous nonnegative function $-f$ on $[a, b]$. See Figure 6.3.

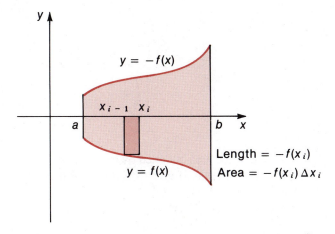

FIGURE 6.3

Since the area of the region below the x-axis bounded by the graph of f is the same as the area of the region above the x-axis bounded by the graph of $-f$, this area is

$$\int_a^b -f(x)dx = -\int_a^b f(x)dx.$$

EXAMPLE 2   Find the area of the region bounded by the x-axis, the lines $x = 1$, $x = 3$, and the graph of $f(x) = x^2 - 3x$.

SOLUTION

The graph of f intersects the x-axis at $x = 0$ and $x = 3$ as shown in Figure 6.4.

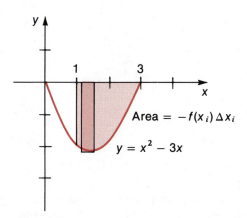

FIGURE 6.4

Since f is nonpositive on the interval $[1, 3]$, the desired area is found by calculating the integral of $-f(x)$:

$$\int_1^3 -(x^2 - 3x)dx = -\int_1^3 (x^2 - 3x)dx$$

$$= -\left(\frac{x^3}{3} - \frac{3x^2}{2}\right)\Bigg]_1^3$$

$$= -\left[\left(9 - \frac{27}{2}\right) - \left(\frac{1}{3} - \frac{3}{2}\right)\right] = -\left(-\frac{10}{3}\right) = \frac{10}{3}. \quad \square$$

To find the area bounded by the graph of a continuous function f that changes sign on an interval, divide the interval at the points where the graph crosses the x-axis.

EXAMPLE 3   Find the area of the region bounded by the x-axis, the lines $x = 0$, $x = 3$ and the graph of $f(x) = x^2 - 3x + 2$.

SOLUTION

Since $x^2 - 3x + 2 = (x - 2)(x - 1)$, $f(x)$ is positive when $x < 1$ and when $x > 2$ and $f(x)$ is negative when $1 < x < 2$. The graph of f is shown in Figure 6.5.

To find the area, divide the region into three parts, R_1, R_2, and R_3, as shown in Figure 6.5. The area A is the sum of the areas of these three regions.

$$A = \int_0^1 (x^2 - 3x + 2)dx - \int_1^2 (x^2 - 3x + 2)dx + \int_2^3 (x^2 - 3x + 2)dx$$

$$= \left(\frac{x^3}{3} - \frac{3x^2}{2} + 2x\right)\Bigg]_0^1 - \left(\frac{x^3}{3} - \frac{3x^2}{2} + 2x\right)\Bigg]_1^2 + \left(\frac{x^3}{3} - \frac{3x^2}{2} + 2x\right)\Bigg]_2^3$$

$$= \left[\left(\frac{1}{3} - \frac{3}{2} + 2\right) - 0\right] - \left[\left(\frac{8}{3} - 6 + 4\right) - \left(\frac{1}{3} - \frac{3}{2} + 2\right)\right]$$

$$+ \left[\left(9 - \frac{27}{2} + 6\right) - \left(\frac{8}{3} - 6 + 4\right)\right]$$

$$= \frac{11}{6}. \quad \square$$

Now we can extend the area concept to the situation in which the region is bounded by the graphs of two distinct functions.

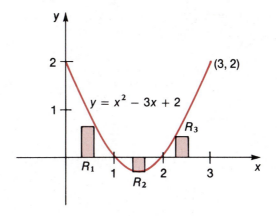

FIGURE 6.5

Suppose f and g are continuous functions on $[a, b]$ and $f(x) \geq g(x)$ for all x in $[a, b]$. The region bounded by the graphs of f and g and the lines $x = a$, $x = b$ is shown in Figure 6.6.

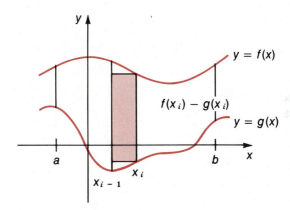

FIGURE 6.6

The area of this region is approximated by

$$\sum_{i=1}^{n} [f(x_i) - g(x_i)]\Delta x_i,$$

which is a Riemann sum for the function $f - g$. Consequently, the total area between the graphs of f and g is given by

(6.2) $A = \lim\limits_{n \to \infty} \sum\limits_{i=1}^{n} [f(x_i) - g(x_i)]\Delta x_i = \int_{a}^{b} [f(x) - g(x)]dx.$

EXAMPLE 4   Find the area of the region bounded by $x = 1$, $x = 4$, the graph of $f(x) = \sqrt{x}$, and the graph of $g(x) = -x$.

SOLUTION

First we draw the graphs of f and g on the same set of coordinate axes and observe that $f(x) > g(x)$ for all x in $[1, 4]$ as shown in Figure 6.7. The area of the shaded region is:

$$A = \int_1^4 [f(x) - g(x)]dx$$

$$= \int_1^4 (\sqrt{x} - (-x))dx = \int_1^4 (\sqrt{x} + x)dx$$

$$= \left(\frac{2}{3}x^{3/2} + \frac{x^2}{2} \right) \Big]_1^4$$

$$= \left[\left(\frac{16}{3} + 8 \right) - \left(\frac{2}{3} + \frac{1}{2} \right) \right] = \frac{73}{6}. \qquad \square$$

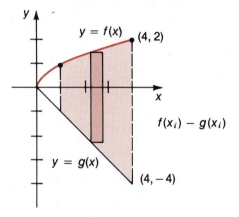

FIGURE 6.7

EXAMPLE 5   Find the area of the region bounded by the graphs of $f(x) = 2x^3 - 9x^2 + 9x$ and $g(x) = (x^2 - 3x)/2$.

SOLUTION

The interval boundaries for the region are the points of intersection of the two graphs. To find these points, we solve the equations simultaneously.

$$2x^3 - 9x^2 + 9x = \frac{x^2 - 3x}{2}$$

so

$$4x^3 - 19x^2 + 21x = 0,$$

$$x(4x^2 - 19x + 21) = 0,$$

$$x(x - 3)(4x - 7) = 0,$$

and the graphs intersect when $x = 0$, $x = 3$, and $x = 7/4$. The points of intersection are $(0, 0)$, $(3, 0)$ and $(7/4, -35/32)$. The region bounded by the graphs of f and g is shown in Figure 6.8.

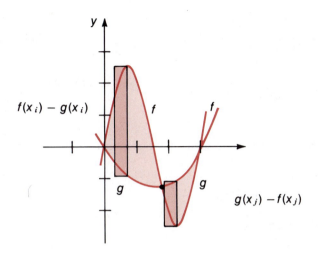

FIGURE 6.8

On $[0, 7/4]$, $f(x) \geq g(x)$ so the height of a representative rectangle is $[f(x) - g(x)]$. On $[7/4, 3]$, $g(x) \geq f(x)$ so the height of a representative rectangle is $[g(x) - f(x)]$. Consequently,

$$A = \int_0^{7/4} [f(x) - g(x)]dx + \int_{7/4}^3 [g(x) - f(x)]dx$$

$$= \int_0^{7/4} \left(2x^3 - 9x^2 + 9x - \frac{x^2}{2} + \frac{3x}{2} \right) dx$$

$$+ \int_{7/4}^3 \left(\frac{x^2}{2} - \frac{3x}{2} - 2x^3 + 9x^2 - 9x \right) dx$$

$$= \left(\frac{2x^4}{4} - \frac{9x^3}{3} + \frac{9x^2}{2} - \frac{x^3}{6} + \frac{3x^2}{4} \right) \Bigg]_0^{7/4}$$

$$+ \left(\frac{x^3}{6} - \frac{3x^2}{4} - \frac{2x^4}{4} + \frac{9x^3}{3} - \frac{9x^2}{2} \right) \Bigg]_{7/4}^3 \approx 5.34. \qquad \square$$

Notice that the area considered in the previous example can be expressed as

$$\int_0^3 |f(x) - g(x)| \, dx.$$

The comprehensive statement that gives the area between the graphs of any two continuous functions is:

If f and g are continuous on $[a, b]$, then the area of the region bounded by $x = a$, $x = b$, and the graphs of f and g is

(6.3)
$$\int_a^b |f(x) - g(x)|\, dx.$$

The final example involves a region that is bounded on different intervals by distinct curves.

EXAMPLE 6   Find the area of the region bounded by the graphs of $y^2 = x + 1$ and $x + y = 1$.

SOLUTION
The region is shown in Figure 6.9.

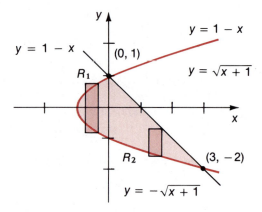

FIGURE 6.9

The equation $y^2 = x + 1$ does not describe y as a function of x since $y = \pm\sqrt{x + 1}$. However, the region R_1 to the left of the y-axis is bounded by the graphs of $y = \sqrt{x + 1}$ and $y = -\sqrt{x + 1}$, so the area of R_1 is

$$\int_{-1}^0 (\sqrt{x + 1} - (-\sqrt{x + 1}))dx = 2\int_{-1}^0 (x + 1)^{1/2}dx$$

$$= 2\left(\frac{2}{3}(x + 1)^{3/2}\right)\Bigg]_{-1}^0 = \frac{4}{3}.$$

The region R_2 to the right of the y-axis is bounded by the graphs of $y = 1 - x$ and $y = -\sqrt{x + 1}$, so the area of R_2 is

$$\int_0^3 [(1 - x) - (-\sqrt{x + 1})]dx = \int_0^3 (1 - x + (x + 1)^{1/2})\, dx$$

$$= \left(x - \frac{x^2}{2} + \frac{2}{3}(x + 1)^{3/2} \right) \Big]_0^3$$

$$= \left[3 - \frac{9}{2} + \frac{2}{3}(8) - \frac{2}{3} \right] = \frac{19}{6}.$$

The sum of the areas of R_1 and R_2 gives the total area, $A = 4/3 + 19/6 = 9/2$.

This area can also be found by treating x as a function of the variable y. The equations $x = 1 - y$ and $x = y^2 - 1$ both describe x as a function of y, and the region R is bounded by the graphs of these two equations, as shown in Figure 6.10.

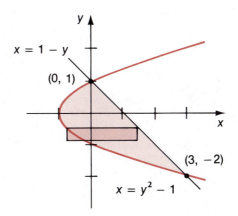

FIGURE 6.10

Therefore, the area of R is

$$A = \int_{-2}^1 [(1 - y) - (y^2 - 1)]dy = \left(y - \frac{y^2}{2} - \frac{y^3}{3} + y \right) \Big]_{-2}^1$$

$$= \left[\left(2 - \frac{1}{2} - \frac{1}{3} \right) - \left(-4 - 2 + \frac{8}{3} \right) \right]$$

$$= \frac{9}{2}.$$

EXERCISE SET 6.1

Find the area bounded by the regions described in Exercises 1 through 38. Sketch the graph of each region.

1. $y = x^2 - 2x$, $x = 2$, $x = 4$, $y = 0$

2. $y = 6x - x^2 - 5$, $x = 2$, $x = 4$, $y = 0$

3. $y = x^2 - 2x$, $y = 0$

4. $y = 6x - x^2 - 5$, $x = 0$, $x = 1$, $y = 0$

5. $y = \dfrac{1}{x}$, $x = 1$, $x = 5$, $y = 0$

6. $y = \dfrac{1}{x}$, $x = .1$, $x = 1$, $y = 0$

7. $y = \dfrac{1}{x}$, $x = -1$, $x = -5$, $y = 0$

8. $y = \dfrac{1}{x}$, $x = -.1$, $x = -1$, $y = 0$

9. $y = x^2$, $y = \sqrt{x}$

10. $y = x^2$, $y = \sqrt{x}$, $x = 2$, $x = 3$

11. $y = x^3 - x$, $y = 0$

12. $y = x - x^5$, $y = 0$

13. $y = \tan x$, $x = -\dfrac{\pi}{4}$, $x = \dfrac{\pi}{4}$, $y = 0$

14. $y = |\sin x|$, $x = 0$, $x = 2\pi$, $y = 0$

15. $y = x^3$, $y = x^2$

16. $y = x^3$, $y = \sqrt{x}$

17. $y = 1 - x^2$, $y = -3$

18. $y = x^2 - 3$, $y = 4$

19. $y = x^2 - 1$, $y = 1 - x^2$

20. $y = x^2 - 2$, $y = x$

21. $y = x^{3/2}$, $y = x$

22. $y = -x^3$, $y = x$, $y = 1$

23. $y = \cos x$, $y = \sin x$, $x = \dfrac{-\pi}{2}$, $x = \dfrac{\pi}{2}$

24. $y = (\sec x)^2$, $y = \cos x$, $x = -\dfrac{\pi}{4}$, $x = \dfrac{\pi}{4}$

25. $y = x$, $y = 1 - x$, $y = -\dfrac{x}{2}$

26. $y = x + 1$, $y = \dfrac{2}{3}x + \dfrac{1}{3}$, $y = \dfrac{2}{5}x + \dfrac{8}{5}$

27. $y = 2x - x^2$, $y = 2 - x$

28. $y = x^2$, $y = 4x - x^2$

29. $y = \sin 2x$, $y = \sin x$, $x = 0$, $x = \pi$

30. $y = \sin 2x$, $y = \sin x$, $x = 0$, $x = 2\pi$

31. $y = x^3 - x$, $y = 2x^3 - 2x$

32. $y = x^3 - 3x^2 - x + 3$, $y = x^2 - 1$

33. $y = |x|$, $y = x + 2$, $y = 2 - x$

34. $y = |x|$, $y = 2 - x^2$

35. $x = y^2$, $x + y = 2$

36. $x = y^2$, $y = x - 6$

37. $y^2 = x + 2$, $y = -x$

38. $y^2 = x + 1$, $y^2 = 1 - x$

39. Use integration to find the area of the region inside the triangle with vertices $(0, 0)$, $(1, 1)$ and $(1, 0)$.

40. Use integration to find the area of the region inside the triangle with vertices (1, 1), (3, 2) and (2, 4).

41. A line through (0, 0) intersects the curve $y = x^2$ at a point (a, a^2). The area of the region bounded above by the line and below by the curve is 27. Find a.

42. Consider the region bounded by $y = x^2 - 1$ and $y = 1 - x^2$.
(a) Show that the line $y = x$ divides the region into two equal parts.
(b) Show that any line through the origin divides the region into two equal parts.

43. Suppose $A(a)$ denotes the area of the region bounded by $y = x^2 - a^2$ and $y = a^2 - x^2$. Find the rate of change of this area with respect to the change in a.

44. Suppose that $A(a)$ denotes the area of the region bounded by $y = x^2 - a$ and $y = a - x^2$ for $a > 0$. Find the rate of change of this area with respect to the change in a.

45. Farmer MacDonald has a pasture bounded on the south and east by a pair of perpendicular dirt roads that meet at the southeast corner of the pasture. The pasture runs 390 feet along the south-bounding road and 1430 feet along the east-bounding road. The remaining boundary can be described by a parabola that has its vertex at the north point of the east-bounding road and is symmetric with respect to the line made by this road. Each cow requires 1/2 acre of this pasture and an acre is 43,560 ft^2. How many cows can MacDonald pasture in this field?

46. Find the area of the region described in Example 4 by the following procedure:
(i)   Find the area of the region R_1 bounded by the graph of f, the x-axis, and the lines $x = 1$ and $x = 4$.
(ii) Find the area of the region R_2 bounded by the graph of g, the x-axis, and the lines $x = 1$ and $x = 4$.
(iii) Add the area of R_1 and the area of R_2. Compare your answer to that of Example 4.

6.2
VOLUMES OF SOLIDS OF ROTATION: DISKS

A **solid of rotation** is generated when a region in the plane is rotated about a line that does not intersect the interior of the region. One of the many applications of the definite integral is to find the volume of such solids.

Suppose R is a region in the plane with boundaries $x = a$, $x = b$, the x-axis, and the graph of a continuous function f. (See Figure 6.11.)

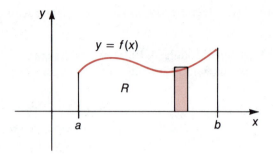

FIGURE 6.11

If R is rotated about the x-axis, the volume of the solid of rotation can be approximated in a manner similar to that of approximating the area of a region.

We begin by dividing $[a, b]$ into n subintervals with endpoints

$$a = x_0 < x_1 < \cdots < x_n = b$$

and choosing numbers z_i with

$$x_{i-1} \leq z_i \leq x_i$$

for each $i = 1, 2, \ldots, n$.

Rotating the rectangles with height $f(z_i)$ about the x-axis produces a collection of disks, which together approximate the solid generated by rotating R. A representative disk is shown in Figure 6.12.

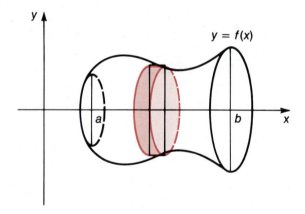

FIGURE 6.12

An approximation to the volume of this solid can be obtained by summing the volumes of these disks.

Rotating a rectangle with base $\Delta x_i = x_i - x_{i-1}$ and height $f(z_i)$ results in a disk with radius $f(z_i)$ and height Δx_i. Since the volume of a cylinder with radius r and height h is $\pi r^2 h$, the volume generated is $\Delta V_i = \pi [f(z_i)]^2 \Delta x_i$. Summing these volumes yields

$$\sum_{i=1}^{n} \pi [f(z_i)]^2 \Delta x_i,$$

which is a Riemann sum for the function $\pi(f)^2$. As the norms of the partitions approach zero, the number of approximating rectangles increases, leading naturally to a definite integral.

(6.4)
DEFINITION

If f is continuous on $[a, b]$, then the volume of the solid generated by rotating about the x-axis the region bounded by the graph of f, the lines $x = a$, $x = b$, and the x-axis is

$$V = \pi \int_a^b [f(x)]^2 \, dx.$$

EXAMPLE 1

Find the volume generated by rotating about the x-axis the region bounded by the graph of $f(x) = x^2/2$, the lines $x = 1, x = 3$, and the x-axis.

SOLUTION

Setting up a volume problem is easier if you first give a rough sketch of the solid that is generated. The solid described in this example and an approximating disk are shown in Figure 6.13. The volume of a typical approximating disk is

$$\Delta V_i = \pi[f(z_i)]^2 \, \Delta x_i = \pi \left(\frac{1}{2} z_i^2 \right)^2 \Delta x_i.$$

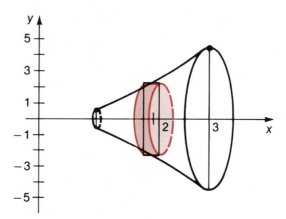

FIGURE 6.13

Consequently, the volume of the solid is

$$V = \pi \int_1^3 \left[\frac{1}{2} x^2 \right]^2 dx = \frac{1}{4} \pi \int_1^3 x^4 \, dx = \frac{1}{4} \pi \left(\frac{x^5}{5} \right) \Big]_1^3$$

$$= \frac{1}{4} \pi \left[\left(\frac{243}{5} \right) - \left(\frac{1}{5} \right) \right] = \frac{242}{20} \pi = \frac{121}{10} \pi. \quad \square$$

EXAMPLE 2   Find the volume generated by rotating about the y-axis the region bounded by the graph of $y = x^2$, the y-axis, and the lines $y = 1$, $y = 9/2$.

SOLUTION

Since we are rotating about the y-axis in this example, we first express the variable x as a function of y, $x = f(y) = \sqrt{y}$. The resulting solid is shown in Figure 6.14. A typical approximating disk has radius $f(w_i)$, where $y_{i-1} \le w_i \le y_i$, and height Δy_i. Hence the volume of the disk is $\pi \, [f(w_i)]^2 \, \Delta y_i$.

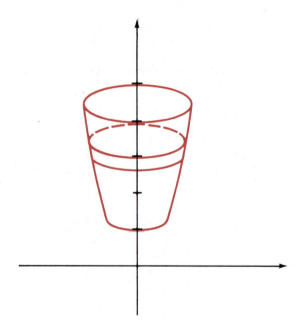

FIGURE 6.14

The volume of the solid is

$$V = \pi \int_1^{9/2} (y^{1/2})^2 \, dy = \pi \int_1^{9/2} y \, dy = \pi \left(\frac{y^2}{2}\right)\Bigg]_1^{9/2}$$

$$= \pi \left(\frac{81}{8} - \frac{1}{2}\right) = \frac{77}{8}\,\pi. \qquad \square$$

EXAMPLE 3   Find the volume of the solid generated by rotating about the x-axis the region bounded by the graph of $y = \tan x$, the x-axis, and the lines $x = -\pi/4$, $x = \pi/4$.

SOLUTION

First we make a sketch indicating the region to be rotated, as shown in Figure 6.15.

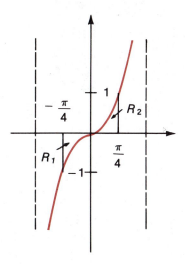

FIGURE 6.15

The volume is given by

$$V = \int_{-\pi/4}^{\pi/4} \pi \, [\tan x]^2 \, dx$$

$$= \pi \int_{-\pi/4}^{\pi/4} [(\sec x)^2 - 1] \, dx = \pi \, (\tan x - x) \Big]_{-\pi/4}^{\pi/4}$$

$$= \pi \left[\left(1 - \frac{\pi}{4} \right) - \left(-1 + \frac{\pi}{4} \right) \right] = 2\pi - \frac{\pi^2}{2}.$$

Since the graph of the tangent function is symmetric with respect to the origin, the volume generated by rotating R_1 about the x-axis is the same as the volume obtained by rotating R_2 about the x-axis. (See Figure 6.15). Thus, another way to find the volume described in this example is to double the volume of the solid generated by rotating R_1 about the x-axis.   □

The volume generated by rotating the area of a region bounded by two vertical lines and the graphs of two functions can be obtained in a similar manner. Suppose f and g are continuous with $f(x) \geq g(x) \geq 0$ for each x in $[a, b]$ and R is the region bounded by the lines $x = a$, $x = b$, and the graphs of f and g. If R is rotated about the x-axis, the solid generated is of the type shown in Figure 6.16.

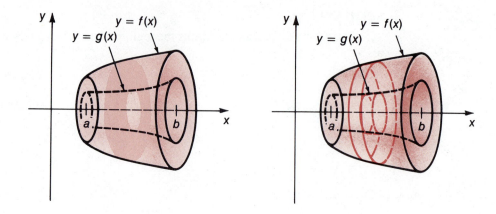

FIGURE 6.16   FIGURE 6.17

To find the volume of this solid, first determine the volume of the solid generated by rotating about the x-axis the region bounded by the graph of f, the lines $x = a$, $x = b$, and the x-axis. Then subtract the volume of the solid generated by rotating about the x-axis the region bounded by the graph of g, $x = a$, $x = b$, and the x-axis. Thus,

(6.5)
$$V = \pi \int_a^b [f(x)]^2 \, dx - \pi \int_a^b [g(x)]^2 \, dx$$
$$= \pi \int_a^b [(f(x))^2 - (g(x))^2] \, dx$$

Formula (6.5) can also be derived by considering a typical rectangular segment bounded by the graphs of f and g. (See Figure 6.17). Rotating the rectangle about the x-axis produces a washer whose volume

$$\Delta V_i = \pi [f(z_i)]^2 \, \Delta x_i - \pi [g(z_i)]^2 \, \Delta x_i = \pi \{ [f(z_i)]^2 - [g(z_i)]^2 \} \Delta x_i$$

is the difference between the two approximating disks. Summing these volumes and taking the limit as the norm of the partition approaches zero gives (6.5).

EXAMPLE 4   Find the volume generated by rotating about the x-axis the region in the first quadrant bounded by the graphs of $y = x^3 + 1$ and $y = x + 1$.

SOLUTION

The region to be rotated and an approximating washer are shown in Figures 6.18 (a) and (b).

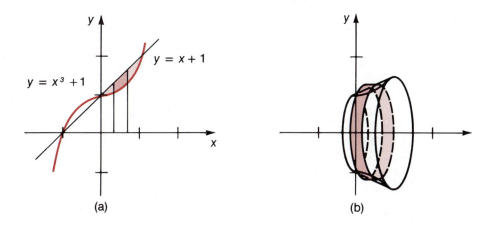

FIGURE 6.18

Since the points of intersection in the first quadrant occur when $x = 0$ and $x = 1$, the volume generated is

$$V = \pi \int_0^1 [(x + 1)^2 - (x^3 + 1)^2]\, dx$$

$$= \pi \frac{(x + 1)^3}{3} \Bigg]_0^1 - \pi \int_0^1 (x^6 + 2x^3 + 1)\, dx$$

$$= \pi \left[\frac{8}{3} - \frac{1}{3} \right] - \pi \left[\frac{x^7}{7} + \frac{2x^4}{4} + x \right]_0^1$$

$$= \frac{7}{3}\pi - \pi \left[\frac{1}{7} + \frac{1}{2} + 1 \right] = \frac{29}{42}\pi. \qquad \Box$$

EXAMPLE 5   Find the volume generated by rotating the region described in Example 4 about the y-axis.

SOLUTION

Since we are rotating about the y-axis, the first step is to write x in each instance as a function of y; $y = x^3 + 1$ implies that $x = (y - 1)^{1/3}$ and $y = x + 1$ implies that $x = y - 1$. The region and an approximating washer are shown in Figure 6.19.

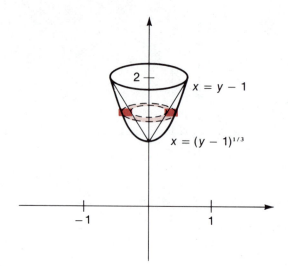

FIGURE 6.19

The resulting volume is

$$V = \int_{1}^{2} \{\pi[(y-1)^{1/3}]^2 - \pi[y-1]^2\}\, dy = \pi \int_{1}^{2} [(y-1)^{2/3} - (y-1)^2]\, dy$$

$$= \pi \left(\frac{3}{5}(y-1)^{5/3} - \frac{(y-1)^3}{3} \right) \Bigg]_{1}^{2}$$

$$= \pi \left(\frac{3}{5} - \frac{1}{3} \right) = \frac{4}{15}\pi. \qquad \square$$

EXAMPLE 6   The drinking cup shown in Figure 6.20(a) has radius r_1 at the top, radius r_2 at the bottom, and height h. Show that the volume contained in the cup is given by

$$V = \frac{1}{3}\pi h\,(r_1^2 + r_1 r_2 + r_2^2).$$

(a)

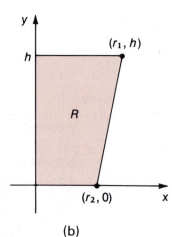

(b)

FIGURE 6.20

SOLUTION

   The shape of the cup is called the *frustum* of a right circular cone. This solid can be considered as a solid of rotation in which the region R shown in Figure 6.20(b) is rotated about the y-axis. To find the volume of this solid of rotation we need the equation of the line joining the points $(r_2, 0)$ and (r_1, h). This is

$$\frac{y}{x - r_2} = \frac{h}{r_1 - r_2} \qquad \text{or} \qquad x = \frac{r_1 - r_2}{h} y + r_2.$$

The volume is

$$V = \pi \int_0^h \left[\left(\frac{r_1 - r_2}{h} \right) y + r_2 \right]^2 dy$$

$$= \pi \left(\frac{h}{r_1 - r_2} \right) \frac{1}{3} \left[\left(\frac{r_1 - r_2}{h} \right) y + r_2 \right]^3 \Bigg]_0^h$$

$$= \pi \left(\frac{h}{r_1 - r_2} \right) \frac{1}{3} [(r_1 - r_2 + r_2)^3 - (r_2)^3]$$

$$= \frac{1}{3} \pi \frac{h}{r_1 - r_2} (r_1{}^3 - r_2{}^3) = \frac{1}{3} \pi h (r_1{}^2 + r_1 r_2 + r_2{}^2).$$

   The formula for the volume of a cone with radius r and height h is the special case of this result when $r_1 = 0$ and $r_2 = r$:

$$V = \frac{1}{3} \pi r^2 h.$$

$\square$

   The final example illustrates that the methods of this section can be used to find the volume generated when a plane region is rotated about a line other than the x- or y-axis.

EXAMPLE 7   Find the volume generated if the region bounded by the graph of $y = x^2 + 2$, the x-axis, and the lines $x = 0$, $x = 1$ is rotated about the line $y = 3$.

SOLUTION

   The region to be rotated is shown in Figure 6.21(a), and the solid of rotation in Figure 6.21(b). The volume of an approximating washer is

$$\Delta V_i = \pi(3)^2 \, \Delta x_i - \pi[3 - (x_i^2 + 2)]^2 \, \Delta x_i = \pi[9 - (1 - x_i^2)^2]\Delta x_i$$

so the total volume generated is

$$V = \pi \int_0^1 [9 - (1 - x^2)^2] \, dx = \pi \int_0^1 (8 + 2x^2 - x^4) \, dx$$

$$= \pi \left(8x + \frac{2x^3}{3} - \frac{x^5}{5} \right) \Bigg]_0^1 = \frac{127}{15} \pi. \qquad \square$$

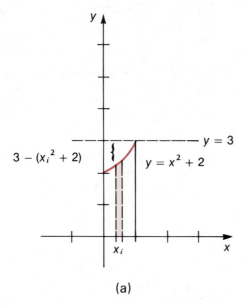

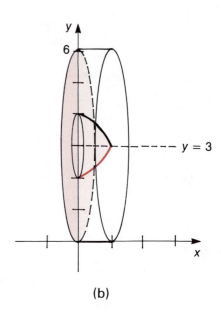

(a)

(b)

FIGURE 6.21

EXERCISE SET 6.2

Find the volume generated if the region bounded by the curves described in Exercises 1 through 26 is rotated about the x-axis. Sketch the region to be rotated.

1. $y = x^2$, $y = 0$, $x = 3$ **2.** $y = x^3$, $y = 0$, $x = 2$

3. $y = x^3$, $y = 0$, $x = 1$, $x = 2$ **4.** $y = x^2$, $y = 0$, $x = 1$, $x = 3$

5. $y = x^2 + 2x + 1$, $x = 1$, $y = 0$ **6.** $y = 4x - x^2$, $y = 0$

7. $y = \dfrac{1}{x}$, $y = 0$, $x = 1$, $x = 2$ **8.** $y = \dfrac{1}{\sqrt{x}}$, $x = 1$, $x = 5$, $y = 0$

9. $y = x^2$, $y = x^3$ **10.** $y = x^2$, $y = \sqrt{x}$

11. $y = x^3$, $y = 1$, $x = 2$ **12.** $y = x^2$, $y = 8 - x^2$

13. $y = 2x^2$, $y = 3 - x^2$ **14.** $y = x^2$, $y = 1$, $x = 3$

15. $y = x^2 - x$, $y = 0$ **16.** $y = x^3 - x$, $y = 0$

17. $y = \sec x$, $x = \dfrac{-\pi}{4}$, $x = \dfrac{\pi}{4}$, $y = 0$ **18.** $y = \cot x$, $x = \dfrac{\pi}{4}$, $x = \dfrac{3\pi}{4}$, $y = 0$

19. $y = \sin x$, $x = 0$, $x = \pi$, $y = 0$
$\left(Hint: (\sin x)^2 = \dfrac{1 - \cos 2x}{2} \right)$

20. $y = \cos x$, $x = 0$, $x = \pi$, $y = 0$
$\left(Hint: (\cos x)^2 = \dfrac{1 + \cos 2x}{2} \right)$

21. $y = \sin x$, $x = -\pi$, $x = 0$,
$y = 0$

22. $y = \sin x$, $x = 0$, $x = 2\pi$,
$y = 0$

23. $y = \dfrac{1}{x}$, $y = x$, $x = 2$, $y = 0$

24. $y = \dfrac{1}{\sqrt{x}}$, $y = x$, $y = 0$, $x = 5$

25. $y = x + 1$, $y = 3x - 5$, $y = 1$

26. $y = x + 1$, $y = 3x - 5$,
$y = \dfrac{1}{2}x + 1$

Find the volume generated if the region bounded by the curves described in Exercises 27 through 36 is rotated about the y-axis.

27. $y = x^3$, $x = 0$, $y = 1$

28. $y = \sqrt{x}$, $x = 0$, $y = 2$

29. $y = x^2$, $y = x$

30. $y = x^2$, $y = x^3$

31. $y = \dfrac{1}{x}$, $y = 1$, $y = 2$

32. $y = 2x - 1$, $y = -2x + 7$,
$y = 1$

33. $y = x^3$, $y = x$

34. $y = x^2 - x$, $y = 0$

35. $y = 2x - 1$, $y = -2x + 7$,
$y = x$

36. $y = x^2 - 2x + 1$, $y = x + 1$.

37. Rotate about the x-axis the region bounded by the x-axis and the upper portion of the circle with equation $x^2 + y^2 = r^2$ to generate a sphere of radius r. Show that the volume of this sphere is $4\pi r^3/3$.

38. Rotate about the x-axis the region bounded by the x-axis and the lines $x = 0$, $x = h$, and $y = r$ to generate a right circular cylinder of radius r and height h. Show that the volume of this cylinder is $\pi r^2 h$.

39. Rotate about the x-axis the region bounded by the x-axis, the line $x = h$ and the line segment joining the point $(0, 0)$ with the point (h, r) to generate a cone of radius r and height h. Show that the volume of this cone is $\pi r^2 h/3$.

40. A water tower is in the shape of a sphere with radius 50 feet. The height of the water in the tank is 75 feet. How many gallons of water are in the tank? (One U.S. gallon is equivalent to 231 cubic inches.)

41. Find the volume of the solid generated if the region bounded by $y = x^2$ and the lines $x = 0$ and $y = 1$ is rotated about the line

(a) $x = 1$ (b) $y = 1$

(c) $x = -1$ (d) $y = 2$

42. Find the volume of the solid generated if the region bounded by $y = \sqrt{x}$ and the lines $x = 0$ and $y = 1$ is rotated about the line

(a) $x = 1$ (b) $y = 1$

(c) $x = -1$ (d) $y = 2$

43. Explain why it is unnecessary to add the conditional phrase "provided this integral exists" to Definition 6.4.

6.3
VOLUMES OF SOLIDS OF ROTATION: SHELLS

In this section we present another method for finding the volume of a solid of rotation. Although the method is derived for rotations about the y-axis, the examples illustrate that it can also be used when the rotation is about other lines. We will show that the volume of the solid generated by rotating a region R about the y-axis can be found using cylindrical shells of the type shown in Figure 6.22.

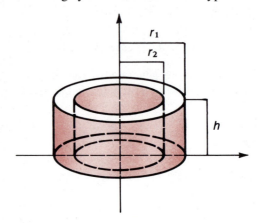

FIGURE 6.22

The volume of a cylindrical shell of this type with outer radius r_1, inner radius r_2, and height h is

$$\pi r_1^2 h - \pi r_2^2 h = \pi(r_1^2 - r_2^2) h$$

(6.6)
$$= \pi(r_1 + r_2)(r_1 - r_2) h$$

$$= 2\pi\left(\frac{r_1 + r_2}{2}\right)(r_1 - r_2) h.$$

Let f be continuous and nonnegative on $[a, b]$ and suppose R is the region bounded by the graph of f, the x-axis, and the lines $x = a$ and $x = b$. To find the volume of the solid generated by rotating R about the y-axis, we begin by partitioning $[a, b]$ with $a = x_0 < x_1 < \ldots < x_n = b$. For each $i = 1, 2, \ldots n$, let z_i be the average of the endpoints of the interval $[x_{i-1}, x_i]$:

$$z_i = \frac{x_{i-1} + x_i}{2}.$$

See Figure 6.23(a). If the approximating rectangle with base $\Delta x_i = x_i - x_{i-1}$ and height $f(z_i)$ is rotated about the y-axis, the resulting solid is the cylindrical shell shown in Figure 6.23(b).

This shell has outer radius x_i, inner radius x_{i-1}, and height $f(z_i)$. It follows from (6.6) that the volume ΔV_i of this shell is

(6.7)
$$\Delta V_i = 2\pi\left(\frac{x_i + x_{i-1}}{2}\right)(x_i - x_{i-1}) f(z_i),$$

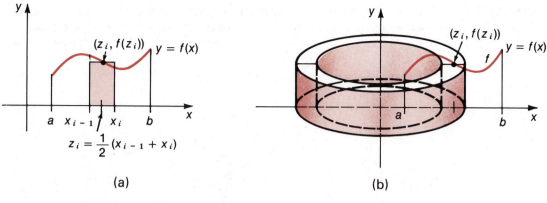

FIGURE 6.23

so

(6.8) $$\Delta V_i = 2\pi z_i \, f(z_i)\Delta x_i.$$

One way to remember this volume is to imagine that the cylindrical shell has been cut along the side and flattened as shown in Figure 6.24. The resulting solid is approximately a flat plate with length $f(z_i)$, width $2\pi z_i$ (the circumference of the cylinder), and thickness Δx_i. Consequently, its volume is

$$2\pi z_i \, f(z_i)\Delta x_i.$$

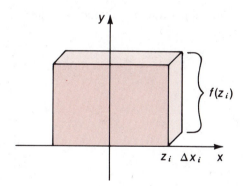

FIGURE 6.24

The total approximating volume for the solid is

$$\sum_{i=1}^{n} 2\pi z_i \, f(z_i)\Delta x_i,$$

which is a Riemann sum for the function described by $2\pi x f(x)$ on $[a, b]$. Consequently,

if f is continuous on $[a, b]$, then the volume of the solid generated by rotating about the y-axis the region bounded by the graph of f, the lines $x = a$, $x = b$, and the x-axis is

(6.9) $$V = \int_a^b 2\pi x f(x)dx.$$

EXAMPLE 1   Find the volume of the solid obtained by rotating about the y-axis the region bounded by the graph of $f(x) = x^2$, the lines $x = 1$, $x = 3$, and the x-axis.

SOLUTION

The region to be rotated is shown in Figure 6.25(a).

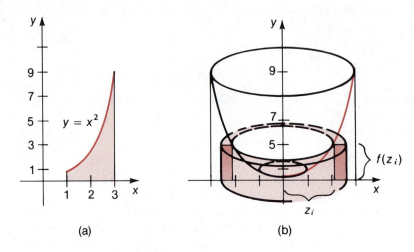

FIGURE 6.25

The approximating cylindrical shell shown in Figure 6.25(b) has volume $2\pi z_i\, f(z_i)\Delta x_i$, so

$$V = 2\pi \int_1^3 x f(x)dx = 2\pi \int_1^3 x\,(x^2)dx$$

$$= 2\pi \int_1^3 x^3 dx = 2\pi \left(\frac{x^4}{4}\right)\Bigg]_1^3$$

$$= 2\pi \left[\frac{81}{4} - \frac{1}{4}\right] = 40\pi. \qquad \square$$

To find the volume in Example 1 by using the method of disks would require breaking the region into two subregions: R_1 (the rectangle bounded by the x-axis, $y = 1$, $x = 1$, and $x = 3$); and R_2 (the area bounded by the graph of $f(x) = x^2$, and the lines $y = 1$ and $x = 3$).

The volume of the solid generated by rotating a region bounded by the graph of two functions and two vertical lines can also be found using the method of shells. Suppose f and g are continuous on $[a, b]$ and $f(x) \geq g(x)$ for all x in $[a, b]$.

The approximating cylindrical shell has volume

$$2\pi z_i\,[f(z_i) - g(z_i)]\Delta x_i.$$

Consequently, the total volume is

(6.10) $$V = 2\pi \int_a^b x[f(x) - g(x)]dx.$$

EXAMPLE 2   Find the volume of the solid generated by rotating about the y-axis the region bounded by the graphs of

$$f(x) = 2x - x^2 \text{ and } g(x) = x^2 - x.$$

SOLUTION

First we sketch the graphs of f and g to show the region that is to be rotated (shown in Figure 6.26(a)).

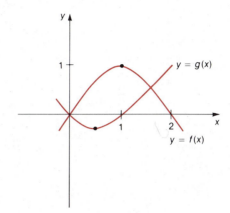

FIGURE 6.26(a)

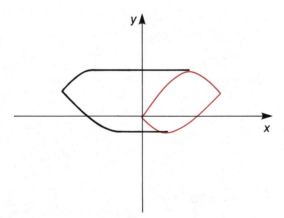

FIGURE 6.26(b)

The graphs of f and g intersect at $x = 0$ and at $x = 3/2$ and for x in $[0, 3/2]$, $f(x) \geq g(x)$. The volume of the solid generated (shown in Figure 6.26(b)) is

$$V = 2\pi \int_0^{3/2} x[(2x - x^2) - (x^2 - x)]dx$$

$$= 2\pi \int_0^{3/2} x(3x - 2x^2)dx = 2\pi \int_0^{3/2} (3x^2 - 2x^3)dx$$

$$= 2\pi \left(x^3 - \frac{x^4}{2} \right) \Bigg]_0^{3/2}$$

$$= 2\pi \left[\frac{27}{8} - \frac{81}{32} \right] = \frac{27\pi}{16}.$$ □

EXAMPLE 3   Find the volume of the solid obtained by rotating about the line $x = 1$ the region bounded by the graph of $f(x) = x^2 - 4$, the line $x = 4$, and the x-axis.

SOLUTION

First we sketch the region to be rotated (Figure 6.27(a)) and an approximating cylindrical shell (Figure 6.27(b)).

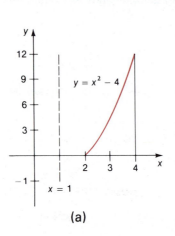

(a)

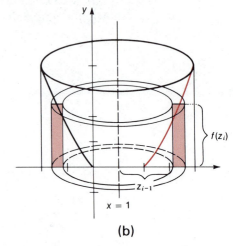

(b)

FIGURE 6.27

Using the notation given in Figure 6.27(b), we see that the volume of an approximating cylindrical shell is

$$\Delta V_i = 2\pi(z_i - 1) f(z_i)\Delta x_i.$$

Consequently, the total volume is

$$V = 2\pi\int_2^4 (x - 1)(x^2 - 4)dx$$

$$= 2\pi\int_2^4 (x^3 - x^2 - 4x + 4)dx$$

$$= 2\pi\left(\frac{x^4}{4} - \frac{x^3}{3} - 2x^2 + 4x\right)\Bigg|_2^4$$

$$= 2\pi\left[\left(64 - \frac{64}{3} - 32 + 16\right) - \left(4 - \frac{8}{3} - 8 + 8\right)\right] = \frac{152\pi}{3}. \qquad \square$$

EXERCISE SET 6.3

Use the method of shells to find the volume generated if the region bounded by the curves described in Exercises 1 through 14 is rotated about the y-axis. Sketch the region to be rotated.

1. $y = x^2$, $x = 1$, $y = 0$ **2.** $y = x$, $x = 5$, $y = 0$

3. $y = x^3$, $x = 1$, $x = 2$, $y = 0$ **4.** $y = \sqrt{x^2 + 1}$, $x = 1$, $x = 3$, $y = 0$

5. $y = x^2$, $y = \sqrt[3]{x}$ **6.** $y = x^3$, $y = x$

7. $y = \cos x^2, x = 0, y = 0,$

$x = \sqrt{\dfrac{\pi}{2}}$

8. $y = x - x^2, y = 0$

9. $y = x - x^3, y = 0$

10. $y = x^2, x = 1, x = 3,$

$y = \dfrac{1}{2}x + \dfrac{1}{2}$

11. $y = \dfrac{1}{2}x + 1, x + y = 4, y = 1$

12. $y = \dfrac{1}{2}x + 1, y + x = 4,$

$x + 4y = 4$

13. $y = x + \dfrac{2}{x}, x = 1,$

$y = 0, y = \dfrac{11}{6}(x - 1)$

14. $y = x + \dfrac{2}{x}, x = 1,$

$y = \dfrac{1}{6}(x + 23)$

Use the method of shells to find the volume generated if the region bounded by the curves described in Exercises 15 through 26 is rotated about the x-axis. (This requires that x be expressed in terms of y.)

15. $x = \sqrt{y}, x = 0, y = 4$

16. $y = x^{2/3}, x = 0, y = 4$

17. $y = \dfrac{1}{x}, y = 1, y = 2, x = 0$

18. $x = \sqrt{4 - y}, x = 0, y = 0$

19. $y = x^2, y = x$

20. $y = \sqrt[3]{x}, y = x^2$

21. $y = x^3, y = x^2$

22. $y = \dfrac{2}{x}, x + y = 3$

23. $x = y - y^2, x = y^2 - y$

24. $x = 2y - y^2, x + y = 0$

25. $y = 2x - x^2, x = 0, x + y = 2$

26. $y = x + 1, 2x + y = 4, x = 0$

27. Find the volume of the solid generated by rotating the region bounded by $x = y^2 - 2y$ and $x = y - y^2$ about the x-axis. Compare this region and volume with those given in Example 2.

28. (a) Find the volume of the solid generated by rotating the region bounded by the graphs of $f(x) = x^2$ and $g(x) = \sqrt{x}$ about the y-axis.

(b) Find the volume of the solid generated by rotating the region described in (a) about the x-axis.

(c) Compare the results of parts (a) and (b).

29. Find the volume of a right circular cylinder of height h and radius r by using the method of shells to determine the volume generated by rotating about the y-axis the region bounded by the line $x = r$, the line $y = h$, and the x- and y-axes.

30. Find the volume of a sphere of radius r by using the method of shells to determine the volume generated by rotating about the y-axis a portion of the region bounded by the circle $x^2 + y^2 = r^2$.

31. Find the volume of a cone of height h and radius r by using the method of shells to determine the volume generated by rotating about the y-axis the region bounded by the x- and y-axes and the line segment joining the points $(r, 0)$ and $(0, h)$.

32. A cylindrical hole is bored through the center of a sphere of radius 4. Find the volume removed from the sphere if the height of the cylinder is 4.

33. Use the method of shells to find the volume of the solid generated if the region bounded by $y = x^2$ and the lines $x = 0$ and $y = 1$ is rotated about the line

(a) $x = 1$ (b) $y = 1$ (c) $x = -1$ (d) $y = 2$

34. Use the method of shells to find the volume of the solid generated if the region bounded by $y = \sqrt{x}$ and the lines $x = 0$ and $y = 1$ is rotated about the line

(a) $x = 1$ (b) $y = 1$ (c) $x = -1$ (d) $y = 2$

35. When a liquid flows through a cylindrical tube, friction at the walls of the tube tends to slow its motion. This results in the liquid flowing faster near the center of the tube than near the walls. In fact, the liquid can be considered as flowing in circular layers, or laminae, which have the constant velocity in each layer given by *Poiseuille's law*:

$$v(r) = K(R^2 - r^2),$$

where R is the radius of the tube, r is the distance from the center of the tube, and K is a constant. Partition the radius of the tube and use this partition to construct a Riemann sum that will lead to the formula $F = K\pi R^4/2$, called *Poiseuille's equation*, that describes the volume of flow through the tube per unit time.

6.4
VOLUMES OF SOLIDS WITH KNOWN CROSS SECTIONS

The geometric figures used to approximate the volume of solids in the disk method have easily determined cross-sectional areas. The volume of any solid whose cross-sectional area is known can be found in a similar manner.

Suppose S is a solid lying between the planes perpendicular to the x-axis that pass through $x = a$ and $x = b$. (See Figure 6.28(a).) In addition, suppose A is a continuous function on $[a, b]$ that describes the area of the cross section of S perpendicular to the x-axis at any number in $[a, b]$.

Let $a = x_0 < x_1 < \ldots < x_n = b$ partition the interval $[a, b]$ and z_i be a number with $x_{i-1} \leq z_i \leq x_i$ for each $i = 1, 2, \ldots, n$. An approximation to the volume of that portion of the solid intersecting the interval $[x_{i-1}, x_i]$ is the product of the area of the cross section at z_i and the width of the interval $[x_{i-1}, x_i]$.

(6.11) $$\Delta V_i = A(z_i)\Delta x_i.$$

(See Figure 6.28(b)).

Summing n such elements of volume gives a Riemann sum approximating the volume of the solid S:

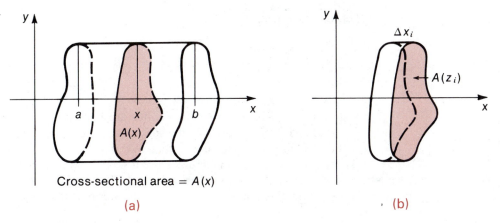

Cross-sectional area $= A(x)$

(a)

(b)

FIGURE 6.28

(6.12)
$$\sum_{i=1}^{n} A(z_i)\Delta x_i.$$

Allowing the norm of the partition to approach zero and hence n to increase leads to the definite integral

(6.13)
$$V = \int_{a}^{b} A(x)dx.$$

EXAMPLE 1   The base of a solid is bounded in the xy-plane by the graphs of $x = y^2$ and $x = 4$ and each cross section perpendicular to the x-axis is an isosceles triangle with altitude one. Find the volume of the solid.

SOLUTION
The base of the solid and a typical cross section are shown in Figure 6.29. Since $A(x)$ is the area of the triangle with height 1 and base $2\sqrt{x}$,

$$A(x) = \frac{1}{2}(2\sqrt{x}) \cdot 1 = \sqrt{x}.$$

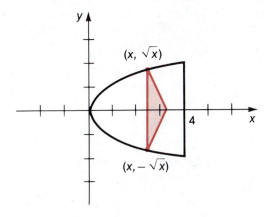

FIGURE 6.29

By (6.13), the volume of the solid is

$$V = \int_0^4 \sqrt{x}\, dx = \frac{2}{3} x^{3/2} \Big]_0^4 = \frac{2}{3}(8 - 0) = \frac{16}{3}. \qquad \square$$

Equation (6.13) implies that if two solids are bounded between a pair of parallel planes and if the sections of the solids cut by any plane parallel to and between the bounding planes are equal, then the volumes of the solids are equal. This is known as **Cavalieri's principle,** since it was first expressed by the Italian mathematician Bonaventura Cavalieri (1598–1647) in 1635. Exercise 17 gives an example of how this principle can be used to determine the volume of certain solids.

EXAMPLE 2   Find the volume of the solid shown in Figure 6.30, where each cross section perpendicular to the x-axis is a square.

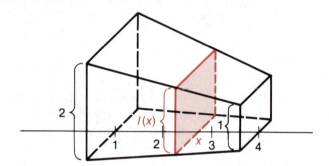

FIGURE 6.30

SOLUTION

The sides of the square cross sections of this figure taper linearly from a length of 2 when $x = 1$ to a length of 1 when $x = 4$. Let l denote the linear function describing the length of the side. Since $l(1) = 2$ and $l(4) = 1$,

$$\frac{l(x) - 2}{x - 1} = \frac{1 - 2}{4 - 1}$$

and

$$l(x) = -\frac{1}{3}x + \frac{7}{3}.$$

Each cross section perpendicular to the x-axis is a square, so the area of this cross section for each x in [1, 4] is

$$A(x) = [l(x)]^2 = \frac{1}{9}(x^2 - 14x + 49).$$

The volume of the solid is

$$V = \int_1^4 A(x)dx$$

$$= \frac{1}{9} \int_1^4 (x^2 - 14x + 49)dx$$

$$= \frac{1}{9} \left(\frac{x^3}{3} - 7x^2 + 49x \right) \Bigg]_1^4 = 7.$$ $\square$

EXAMPLE 3   Find the volume of a wedge removed from a red oak (*Quercus borealis*) tree of radius 2 feet. The wedge is removed by making one cut halfway through the trunk parallel to the ground and another starting 2 feet above this cut and meeting it in the center. (See Figure 6.31(a).)

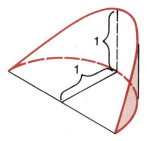

FIGURE 6.31(a)

SOLUTION

First we place a rectangular coordinate system on the plane made by the first cut, with the origin of the system at the center of the circle made by the intersection of the tree trunk with the plane. The boundary of the circle can then be expressed as $x^2 + y^2 = 4$, and the shaded portion of Figure 6.31(b) represents the base of the wedge that is removed.

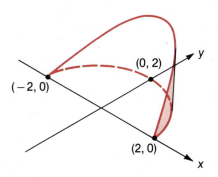

FIGURE 6.31(b)

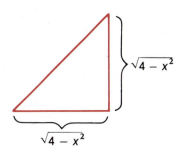

FIGURE 6.31(c)

The wedge has the shape of an isosceles triangle when viewed from the side, as shown in Figure 6.31(c).

Each cross section perpendicular to the x-axis is a triangle whose area is

$$A(x) = \frac{1}{2} \sqrt{4 - x^2} \sqrt{4 - x^2} \text{ ft}^2,\ 0 \le x \le 2,$$

since $\sqrt{4 - x^2}$ is the height of the triangular cross section and also the length of its base. The volume of this wedge is consequently

$$V = 2 \int_0^2 \frac{1}{2} (4 - x^2)dx = \left. \left(4x - \frac{x^3}{3} \right) \right]_0^2$$

$$= 8 - \frac{8}{3} = \frac{16}{3} \text{ ft}^3. \qquad \square$$

EXERCISE SET 6.4

1. The base of a solid is bounded by the circle in the xy-plane with equation $x^2 + y^2 = r^2$ and each cross section perpendicular to the x-axis is a square. Find the volume of the solid.

2. The base of a solid is bounded by the circle in the xy-plane with equation $x^2 + y^2 = r^2$ and each cross section perpendicular to the x-axis is an equilateral triangle. Find the volume of the solid.

$\backslash$**3.** A solid has its base in the xy-plane bounded by $y = x^2$, and $y = 1$. Find the volume of the solid if each cross section perpendicular to the x-axis is in the shape of a square.

4. A solid has its base in the xy-plane bounded by $y = x^2$ and $y = 1$. Find the volume of the solid if each cross section perpendicular to the y-axis is in the shape of a square.

5. The base of a solid is bounded by the x-axis and the graph of $y = \sqrt{r^2 - x^2}$. Each cross section of the solid perpendicular to the x-axis is an isosceles triangle with altitude r. Set up the integral describing this volume.

$\backslash$**6.** The base of a solid is bounded by the y-axis and the graph of $y = \sqrt{r^2 - x^2}$. Each cross section of the solid perpendicular to the y-axis is an isosceles triangle with altitude r. Set up the integral describing this volume. How does this volume compare to the volume determined in Exercise 5?

7. A right circular cylinder of radius r and height h can be described as a solid whose base is contained in the circle $x^2 + y^2 = r^2$ and each of whose cross sections perpendicular to the xy-plane is a rectangle with height h. Set up the integral describing this volume.

8. A hemisphere of radius r can be described as a solid whose base is contained in the circle $x^2 + y^2 = r^2$ and each of whose cross sections perpendicular to the xy-plane is a semicircle with radius y. Set up the integral describing this volume.

9. The recommended procedure to follow when felling a tree is to first make a notch on the fall side of the tree, that is, the side on which the tree will fall. This notch should be made by first making a horizontal cut a third of the way through the tree. Then a cut is made about a third as far above the horizontal cut as the tree is thick, meeting the horizontal cut at its end. Set up the integral that describes the volume of wood removed from the tree by this notch, if the tree is 18 inches in diameter. Do not try to evaluate this integral.

10. A pyramid has a base in the form of a square of length r and has an altitude h. Find the volume of the pyramid.

11. The center of a universal joint on an automobile driveshaft is in the form of two right circular cylinders of the same radii, $r = .5$ inch, intersecting at right angles. What is the volume contained in this intersection?

12. Suppose that the circular cylinders that make up the universal joint described in Exercise 11 are both 2.5 inches in length. How much metal is required to make this universal joint?

13. A bull's horn can be described by drawing a curve $y = x^2/8$ inches for x in $[0, 8]$ and then drawing circles perpendicular to the x-axis whose centers are along this curve and where the radius of the circle with center $(x, x^2/8)$ is $r = (8 - x^2/8)/8 = 1 - x^2/64$ inches. Find the volume in the bull's horn.

14. A swimming pool is 30 feet wide and 50 feet long with straight sides. The depth of the water at one end is 8 feet and the pool continues at this depth for 20 feet. The depth of the pool then decreases linearly for the next 10 feet to a depth of 3 feet and continues at this depth for the final 20 feet of the pool. Find the volume of water in the pool.

15. A hole of radius 3 is bored through the center of a sphere with radius 4. How much volume has been removed?

16. A hole of radius r is bored through the center of a sphere of radius R, removing half the volume of the sphere. What is r?

17. Solve Exercise 16 by using Cavalieri's principle as follows: Suppose that a plane perpendicular to the axis of the cylindrical hole cuts both the original solid and the sphere with radius $[R^2 - r^2]^{1/2}$ a distance h from the center of each sphere. Show that the cross-sectional area cut from each solid is equal and use the volume of the smaller sphere to deduce the result.

6.5
ARC LENGTH OF A CURVE AND SURFACES OF REVOLUTION

A problem of equal interest with those of calculating area and volume is to determine the length of an arbitrary curve. In this section we consider a special case of this problem: when the curve is the graph of a differentiable function.

Suppose C denotes the graph of a differentiable function f from $(a, f(a))$ to $(b, f(b))$ (see Figure 6.32). The length of C is called the **arc length** of the graph of f from $(a, f(a))$ to $(b, f(b))$.

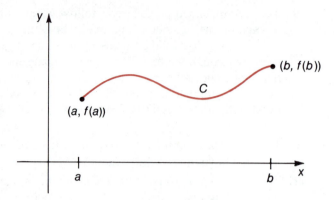

FIGURE 6.32

If $a = x_0 < x_1 < \ldots < x_n = b$ partitions $[a, b]$, an approximation to the length of C can be found by summing the straight-line distances from $(x_{i-1}, f(x_{i-1}))$ to $(x_i, f(x_i))$ for each $i = 1, 2, \ldots, n$. (See Figure 6.33.)

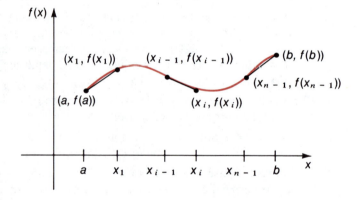

FIGURE 6.33

The length of the line segment from $(x_{i-1}, f(x_{i-1}))$ to $(x_i, f(x_i))$ is

$$\Delta L_i = \sqrt{(x_i - x_{i-1})^2 + (f(x_i) - f(x_{i-1}))^2}.$$

To demonstrate that finding arc length is an application of the definite integral, we rewrite this equation as

(6.14)
$$\Delta L_i = \sqrt{1 + \left[\frac{f(x_i) - f(x_{i-1})}{x_i - x_{i-1}}\right]^2} (x_i - x_{i-1}).$$

Since f is differentiable on the interval $[a, b]$, the mean value theorem guarantees the existence of a number z_i with $x_{i-1} < z_i < x_i$ and

$$f'(z_i) = \frac{f(x_i) - f(x_{i-1})}{x_i - x_{i-1}}.$$

Using this result and the fact that $\Delta x_i = x_i - x_{i-1}$, Equation (6.14) can be rewritten

$$\Delta L_i = \sqrt{1 + [f'(z_i)]^2}\, \Delta x_i.$$

The sum of the lengths of these straight line segments approximates the length of C:

$$L \approx \sum_{i=1}^{n} \Delta L_i = \sum_{i=1}^{n} \sqrt{1 + [f'(z_i)]^2}\, \Delta x_i.$$

This is a Riemann sum for the function described by $\sqrt{1 + [f'(x)]^2}$ and leads to the definition of the length of the graph of f.

(6.15)
DEFINITION

If f is differentiable on $[a, b]$, the **arc length** of the graph of f from $(a, f(a))$ to $(b, f(b))$ is

$$L_{[a,\, b]}(f) = \int_a^b \sqrt{1 + [f'(x)]^2}\, dx,$$

provided this integral exists.

When this integral exists, f is said to be **rectifiable** on the interval $[a, b]$.

A condition sufficient to guarantee that a function f is rectifiable on an interval is that f' is continuous on the interval. Such a function is said to be **smooth** on that interval. When f is smooth on $[a, b]$, $\sqrt{1 + [f'(x)]^2}$ describes a continuous function, so it follows from Theorem 5.11 that the integral in Definition (6.15) exists.

EXAMPLE 1   Find the arc length of $f(x) = x^{3/2}$ from $(1, 1)$ to $(9, 27)$.

SOLUTION

Since $f'(x) = 3x^{1/2}/2$ is continuous, f is smooth on $[1, 9]$ and hence rectifiable, so

$$L_{[1,\, 9]}(f) = \int_1^9 \sqrt{1 + \left[\frac{3}{2}x^{1/2}\right]^2}\, dx$$

$$= \int_1^9 \sqrt{1 + \frac{9}{4}x}\, dx.$$

To determine $\int \sqrt{1 + 9x/4}\, dx$, let $u = 1 + 9x/4$. Then $du = \dfrac{9}{4}\, dx$, so

$$\int \sqrt{1 + \frac{9}{4}x}\, dx = \int \sqrt{u}\, \frac{4}{9}\, du$$

$$= \frac{4}{9} \cdot \frac{2}{3} u^{3/2} + C = \frac{8}{27}\left(1 + \frac{9}{4}x\right)^{3/2} + C.$$

Thus,
$$L_{[1,\,9]}(f) = \frac{8}{27}\left(1 + \frac{9}{4}x\right)^{3/2}\Bigg]_1^9$$

$$= \frac{8}{27}\left[\left(1 + \frac{81}{4}\right)^{3/2} - \left(1 + \frac{9}{4}\right)^{3/2}\right]$$

$$\approx 27.3.$$

This result appears reasonable from the graph shown in Figure 6.34.   ☐

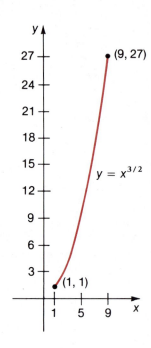

FIGURE 6.34

EXAMPLE 2   Show that the function described by $f(x) = x^2$ is rectifiable on the interval $[0, 1]$.

SOLUTION

Since $f'(x) = 2x$ describes a continuous function on the interval $[0, 1]$, f is rectifiable on $[0, 1]$. The obvious question to ask in return is: What is the arc length of the graph of $f(x) = x^2$ from $(0, 0)$ to $(1, 1)$?

The arc length is given by

$$L_{[0,\,1]}(f) = \int_0^1 \sqrt{1 + [2x]^2}\, dx$$

$$= \int_0^1 \sqrt{1 + 4x^2}\, dx,$$

an integral we cannot yet evaluate.   ☐

The previous example uncovers a significant difficulty associated with determining the arc length of a curve: even for relatively elementary functions, an integrand of the form $\sqrt{1 + [f'(x)]^2}$ can be difficult or impossible to evaluate. In Chapter 8 we will study ways of evaluating some integrals of this type and show methods for approximating the integrals of those we cannot evaluate exactly.

AREA OF A SURFACE OF ROTATION

When a curve is rotated about a line, a surface is generated. For example, if a right triangle is rotated about one of the legs forming the right angle, the surface of a cone is obtained, as shown in Figure 6.35.

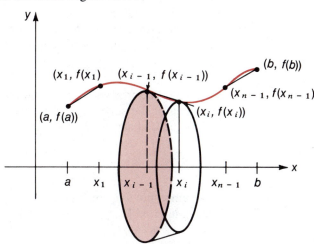

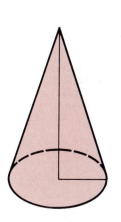

FIGURE 6.35   FIGURE 6.36

Deriving the formula for the area of a surface generated by rotating a curve about the x-axis is similar to deriving the formula for the arc length of a curve, with a few added complications.

Suppose f is nonnegative and differentiable on $[a, b]$ and the graph of f between $(a, f(a))$ and $(b, f(b))$ is rotated about the x-axis. To find the area of the surface obtained by this rotation, let $a = x_0 < x_1 < \ldots < x_n = b$ partition the interval $[a, b]$ and consider the straight line segments joining the points $(x_{i-1}, f(x_{i-1}))$ and $(x_i, f(x_i))$ for each $i = 1, 2, \ldots, n$. Rotating the straight line segments joining $(x_{i-1}, f(x_{i-1}))$ and $(x_i, f(x_i))$ about the x-axis gives an approximation to the surface generated by rotating about the x-axis the portion of the graph of f that lies between these two points. See Figure 6.36.

The surface generated by rotating each of these line segments is the frustum of a cone and has surface area equal to $\pi(f(x_{i-1}) + f(x_i))$ (slant height). (This formula is given on the inside cover of the book.) Consequently,

(6.16) $\Delta SA_i = \pi[f(x_{i-1}) + f(x_i)]\sqrt{(x_i - x_{i-1})^2 + [f(x_i) - f(x_{i-1})]^2}.$

Following the procedure used to develop the arc length formula, we rewrite Equation (6.16) as

$$\Delta SA_i = \pi[f(x_{i-1}) + f(x_i)]\sqrt{1 + \left[\frac{f(x_i) - f(x_{i-1})}{x_i - x_{i-1}}\right]^2}\,\Delta x_i$$

and employ the mean value theorem to deduce that there is a number z_i in the interval (x_{i-1}, x_i) with

(6.17) $\Delta SA_i = \pi[f(x_{i-1}) + f(x_i)]\sqrt{1 + [f'(z_i)]^2}\,\Delta x_i.$

Note that (6.17) can be rewritten as

(6.18) $\Delta SA_i = 2\pi\left[\dfrac{f(x_{i-1}) + f(x_i)}{2}\right]\sqrt{1 + [f'(z_i)]^2}\,\Delta x_i,$

where $(f(x_{i-1}) + f(x_i))/2$ is simply the average of the numbers $f(x_{i-1})$ and $f(x_i)$. Since f is a continuous function, the intermediate value theorem implies that there is a number w_i in the interval $[x_{i-1}, x_i]$ with $f(w_i) = (f(x_{i-1}) + f(x_i))/2$. Equation (6.18) can now be rewritten

$$\Delta SA_i = 2\pi f(w_i)\sqrt{1 + [f'(z_i)]^2}\,\Delta x_i.$$

Summing the surface area on each of the frustums gives an approximation to the total area of the surface generated by rotating the graph of f between $(a, f(a))$ and $(b, f(b))$ about the x-axis.

$$SA \approx \sum_{i=1}^{n} \Delta SA_i = \sum_{i=1}^{n} 2\pi f(w_i)\sqrt{1 + [f'(z_i)]^2}\,\Delta x_i.$$

This sum is not a Riemann sum, however, since it involves two values w_i and z_i in each interval $[x_{i-1}, x_i]$. Fortunately, even if $w_i \neq z_i$,

$$\lim_{\|\mathscr{P}\|\to 0} \sum_{i=1}^{n} 2\pi f(w_i)\sqrt{1 + [f'(z_i)]^2}\,\Delta x_i = \int_a^b 2\pi f(x)\sqrt{1 + [f'(x)]^2}\,dx.$$

This is a result of a theorem stated by George A. Bliss (1876–1961) in 1914 in a paper entitled ''A Substitute for Duhamel's Theorem.'' (The more general Duhamel's theorem is considered in Chapter 17.)

(6.19)
THEOREM

Bliss's Theorem   Suppose f and g are integrable on $[a, b]$. If $\mathscr{P} = \{x_0, x_1,$ $\ldots, x_n\}$ is a partition of $[a, b]$ and w_i and z_i are both in $[x_{i-1}, x_i]$ for each i, then $\displaystyle\lim_{\|\mathscr{P}\|\to 0} \sum_{i=1}^{n} f(w_i)\, g(z_i)\Delta x_i = \int_a^b f(x)g(x)dx.$

(6.20)
DEFINITION

If f is nonnegative and differentiable on $[a, b]$, the area of the surface generated by rotating the graph of f between $(a, f(a))$ and $(b, f(b))$ about the x-axis is

$$SA_{[a,b]}(f) = 2\pi\int_a^b f(x)\sqrt{1 + [f'(x)]^2}\,dx,$$

provided this integral exists.

EXAMPLE 3   Find the area of the surface generated by rotating about the x-axis the graph of $f(x) = x^3$, for x in $[0, 2]$.

SOLUTION

Since f is differentiable on $[0, 2]$ and $f'(x) = 3x^2$, Definition 6.20 gives the surface area as

$$SA_{[0,\,2]}(f) = 2\pi \int_0^2 x^3 \sqrt{1 + [3x^2]^2}\, dx = 2\pi \int_0^2 x^3 \sqrt{1 + 9x^4}\, dx.$$

If $u = 1 + 9x^4$, then $du = 36x^3 dx$ and

$$SA_{[0,\,2]}(f) = 2\pi \int_{x=0}^{x=2} x^3 \sqrt{1 + 9x^4}\, dx$$

$$= 2\pi \int_{u=1}^{u=145} \frac{1}{36} u^{1/2}\, du = \frac{\pi}{18} \left(\frac{2}{3} u^{3/2} \right) \Bigg]_1^{145}$$

$$= \frac{\pi}{27} [(145)^{3/2} - 1] \approx 203. \qquad \square$$

EXAMPLE 4   Use Definition 6.20 to show that the surface area of an open right circular cylinder of height h and radius r is $2\pi rh$.

SOLUTION

A cylinder of this type can be obtained by rotating a rectangle about the x-axis, as can be seen in Figure 6.37.

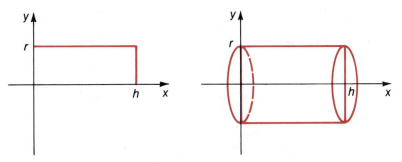

FIGURE 6.37

With $f(x) = r$ for x in $[0, h]$, Definition 6.20 implies that the surface area is

$$SA = 2\pi \int_0^h r \sqrt{1 + 0}\, dx = 2\pi rx \Bigg]_0^h = 2\pi rh. \qquad \square$$

The formula for surface area has the same complications as the arc length formula. The integral is difficult to evaluate except in special cases, like the example above. Even for $f(x) = x^2$, the determination of the surface area obtained by rotating the graph of f about the x-axis involves an integral,

$$2\pi \int x^2 \sqrt{1 + 4x^2}\, dx,$$

which we cannot evaluate at this time. For more complicated functions the difficulties generally increase.

EXERCISE SET 6.5

Find the length of the curves described in Exercises 1 through 18 for x lying in the given interval.

1. $y = 3x + 4$, $[2, 4]$ **2.** $y = -2x + 1$, $[0, 2]$

3. $y = x^{3/2}$, $[3, 15]$ **4.** $y = (x + 1)^{3/2}$, $[0, 1]$

5. $y = 3x^{3/2}$, $[0, 1]$ **6.** $y = (x - 4)^{3/2}$, $[4, 7]$

7. $y = \left(x^2 - \dfrac{2}{3}\right)^{3/2}$, $[1, 2]$ **8.** $y = \dfrac{x^3}{6} + \dfrac{1}{2x}$, $[1, 2]$

9. $y = \dfrac{x^4 + 12}{12x}$, $[1, 3]$ **10.** $y = \dfrac{x^6 + 32}{32x^2}$, $[2, 3]$

11. $y = x^{3/2} - \dfrac{1}{3}\sqrt{x}$, $[1, 4]$ **12.** $y = \dfrac{1}{3}\sqrt{x}\,(1 - 3x)$, $[1, 4]$

13. $y = x^{2/3}$, $[1, 8]$. (*Hint:* Write x in terms of y.)

14. $x = 3y^{3/2} - 1$, $[2, 23]$

15. $x^{2/3} + y^{2/3} = 1$, $\left[\dfrac{1}{8}, 1\right]$ **16.** $x^{2/3} + y^{2/3} = 1$, $\left[-1, -\dfrac{1}{8}\right]$

17. $(y + 1)^3 = (x - 1)^2$, $[1, 2]$ **18.** $(y + 1)^2 = (x - 1)^3$, $[1, 2]$

19. $y = x^3$, $[1, 3]$ **20.** $y = x^{1/2}$, $[2, 6]$

21. $y = x + 1$, $[0, 2]$ **22.** $y = 2x + 3$, $[1, 3]$

23. $y = \sqrt{x + 1}$, $[1, 5]$ **24.** $y = \sqrt{3x - 1}$, $[3, 12]$

25. $y = \dfrac{x^3}{6} + \dfrac{1}{2x}$, $[1, 2]$ **26.** $y = \sqrt{4 - x^2}$, $[0, 1]$

Find the area of the surface generated by rotating about the x-axis the curve described in Exercises 19 through 26 for x lying in the given interval.

27. Find a formula for the surface area of a cone by rotating about the x-axis the line segment joining $(0, 0)$ and (h, r).

28. Find a formula for the surface area of a sphere of radius r by rotating the upper portion of the circle $x^2 + y^2 = r^2$ about the x-axis.

29. Find the coordinates of the point on the graph of $f(x) = x^{3/2}$ that is midway on this graph from the points $(0, 0)$ and $(1, 1)$.

30. Find the perimeter of the area bounded by the curves $y = x^{3/2}$, $y = x + 4$, and $x = 0$.

31. A local Boy Scout troop wants to make a canvas tepee. The tepee will be conical with a base of diameter 12 feet and a vertex 10 feet above the ground. They want to leave the tepee open at the top from one foot below the vertex for supporting poles to extrude. How much canvas must they purchase?

32. The arc length of a curve from $(0, 1)$ to $(1, 0)$ is

$$\int_0^1 \sqrt{\frac{x^2}{1 - x^2} + 1} \, dx = \int_0^1 \sqrt{\frac{1}{1 - x^2}} \, dx.$$

Find a possible equation for the curve. Is the curve uniquely defined by these conditions?

33. Show that

$$\int_{-1}^1 \sqrt{1 - x^2} \, dx = \frac{1}{2} \int_{-1}^1 \frac{1}{\sqrt{1 - x^2}} \, dx$$

without evaluating either integral. (*Hint:* Consider Exercise 32.)

6.6
FLUID PRESSURE

A somewhat surprising (but easily verified) physical fact is that the pressure exerted by a liquid on the bottom of a container depends only on the height of the liquid in the container, not on the volume. A simple experiment used to demonstrate this fact consists of placing water in a glass apparatus of the type shown in Figure 6.38. Since the water level in each of the containers above the reservoir is the same, the pressure of the water on the bottom of the containers must be equal, even though the containers vary in size and shape.

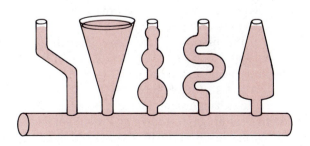

FIGURE 6.38

The pressure p exerted by a liquid on the bottom of a container, or on any horizontal surface in the liquid, is the product of the density w (weight per unit volume) of the liquid and the height h of the liquid above the surface; that is,

$$p = wh.$$

Since pressure is force per unit area, the total force F exerted by a liquid on a horizontally submerged plate of area A is

(6.21) $$F = pA = whA,$$

where w is the density of the liquid and h is the depth of the plate below the surface.

EXAMPLE 1   A 55-gallon cylindrical barrel half full of water is sitting on an end. The barrel is 3 feet high with a radius of .88 feet. Find the total force exerted by the water on the bottom of the barrel.

SOLUTION

The force at the bottom of the barrel depends on the pressure exerted by the water and the area of an end of the barrel. The area is

$$A = \pi r^2 \approx 2.43 \text{ ft}^2.$$

Since the density of water is approximately 62.4 pounds per cubic foot and the depth of the water in the half full barrel is 1.5 feet, the total force exerted by the water on the bottom of the barrel is

$$F \approx \left(62.4 \frac{\text{lb}}{\text{ft}^3}\right)(1.5 \text{ ft})(2.43 \text{ ft}^2) \approx 227 \text{ lb}. \qquad \square$$

When a plate is submerged other than horizontally the situation is more complicated because the depth of the liquid does not remain constant. We can handle this situation in a manner similar to that used in the previous applications of the definite integral:

Divide the plate into narrow strips, calculate the force on each strip, sum to obtain an approximation, and then use the definite integral as the limit of these approximations.

Suppose that a plate is submerged vertically in a liquid with density w from a depth $x = a$ to a depth $x = b$. See Figure 6.39(a). To find the total force on the plate, first we divide the interval $[a, b]$ into subintervals with endpoints $a = x_0 < x_1 < \ldots < x_n = b$. We then determine the force exerted on a typical rectangle with area ΔA_i which is at a depth of z_i, where $x_{i-1} \le z_i \le x_i$. (See Figure 6.39(b).)

The calculation of the force on a typical rectangle is simplified by applying a physical principle known as **Pascal's law.** (See the Historical Note.) This law states that each point in a stationary liquid is subject to forces of the same magnitude in every direction, or the point would move in the direction of least force. Hence the force of the liquid at a point on a vertical (or any other) plate at a depth h is the same as the force if the point was on a horizontal plate at this depth.

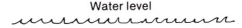

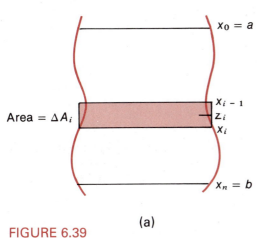

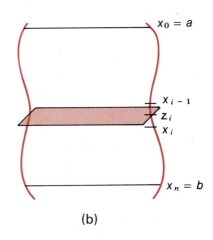

FIGURE 6.39

(a)   (b)

Applying Pascal's law to our situation, if $\Delta x_i = x_i - x_{i-1}$ is sufficiently small, the force exerted by the liquid on the vertical rectangle with area ΔA_i is approximately the same as if the rectangle were horizontal. This force is given by

$$\Delta F_i = wz_i \Delta A_i.$$

Let f denote the function with domain $[a, b]$ that describes the width of the plate as a function of its distance below the surface of the liquid. Thus, for each $i = 1, 2, \ldots, n,$

$$\Delta A_i = f(z_i)(x_i - x_{i-1}) = f(z_i)\Delta x_i$$

and

$$\Delta F_i = \text{(density) (depth) (area)}$$
$$= wz_i f(z_i)\Delta x_i.$$

The total force exerted on the plate can therefore be approximated by the Riemann sum

$$F \approx \sum_{i=1}^{n} \Delta F_i = \sum_{i=1}^{n} wz_i f(z_i)\Delta x_i$$

and the reasonable definition of the total force on the plate is the integral of the function described by $wxf(x)$.

HISTORICAL NOTE

Blaise Pascal (1623–1662) began his mathematical research career at age 16 by publishing an important result concerning the geometry of conic sections. Together with Pierre Fermat (1601–1665), he is considered to be a founder of the theory of probability. He initiated his study of probability at the request of Chevalier de Méré, who needed a method for computing the payoff for an interrupted dice game.

Pascal had a religious experience on November 23, 1654 that turned his interest from science and mathematics to philosophy and religion. He returned to mathematics only briefly in 1658–1659, when he did fundamental work on the study of cycloids and in the theory of hydrostatics.

(6.22)
DEFINITION

Suppose a plate is submerged vertically in a liquid whose density is w. If f is a function with domain $[a, b]$ that describes the width of the plate as a function of its depth below the surface of the liquid, then the total force F exerted on the plate by the liquid is

$$F = \int_a^b wxf(x)dx,$$

provided this integral exists.

EXAMPLE 2

A four-by-eight sheet of plywood is submerged vertically in a five-foot-high tank filled with paint remover. One of the eight-foot sides of the sheet lies on the bottom of the tank. The density of the paint remover is 56.7 lb/ft^3. Find the total force exerted on the plywood.

SOLUTION

The plywood extends from a depth of 1 foot below the surface of the liquid to a depth of 5 feet, and the width of the sheet at any depth is 8 feet. So

$$F = \int_1^5 (56.7)(8)x \, dx = 453.6 \frac{x^2}{2}\Bigg]_1^5 = 5443.2 \text{ lb.} \qquad \square$$

EXAMPLE 3

Suppose the 55-gallon barrel considered in Example 1 is half full of water and lying on its side. What is the total force exerted by the water on an end of the barrel?

SOLUTION

The radius of the barrel is .88 ft. To use Definition 6.22 we need to know the width of the barrel at each value of x in $[0, .88]$. By using the right triangle shown in Figure 6.40(b), we see that the width of the barrel z_i feet

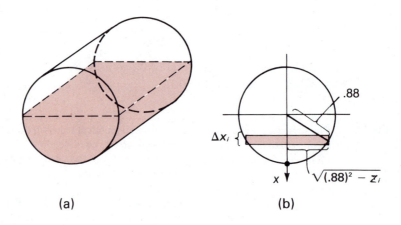

(a)   (b)

FIGURE 6.40

below the surface of the water is $2\sqrt{(.88)^2 - z_i^2}$ feet. Consequently $\Delta A_i = 2\sqrt{(.88)^2 - z_i^2}\,\Delta x_i$. Since the density of water is 62.4 pounds per cubic foot, $\Delta F_i = (62.4)z_i\,(2\sqrt{(.88)^2 - z_i^2}\,\Delta x_i)$ and the total force exerted on an end of the barrel is

$$F = \int_0^{.88} 2(62.4)\,x\sqrt{(.88)^2 - x^2}\,dx$$

$$= \frac{-124.8}{3}\,[(.88)^2 - x^2]^{3/2}\bigg]_0^{.88}$$

$$\approx 28.3 \text{ pounds.} \qquad \square$$

It is often more convenient to reorient the coordinate system when solving fluid pressure problems. The technique is illustrated in the following example.

EXAMPLE 4   A vertical gate on a dam has the shape of a parabola with a height of 8 feet and a width of 8 feet at its top. Find the force of the water on the dam if the level of the water is at the top of the gate.

SOLUTION

The region describing the gate is sketched in Figure 6.41 with a rectangular coordinate system placed so that the depth of the water is measured along the y-axis and the origin is at the bottom of the gate. The gate is parabolic, so the boundary of the region must have the form $y = ax^2$. Since $y = 8$ when $x = 4$, $a = y/x^2 = 8/16 = 1/2$. The gate is consequently described by $y = x^2/2$, $y = 8$.

At a point y on the y-axis, the depth of the water is $(8 - y)$ feet and the width of the dam is $2\sqrt{2y}$ feet. The force is

$$\int_0^8 w(8 - y)2\sqrt{2y}\,dy = 2\sqrt{2}\,w \int_0^8 (8y^{1/2} - y^{3/2})\,dy$$

$$= 2\sqrt{2}w\left(8y^{3/2}\cdot\frac{2}{3} - y^{5/2}\cdot\frac{2}{5}\right)\bigg]_0^8$$

$$= 4\sqrt{2}w\left(\frac{8}{3}(8)^{3/2} - \frac{1}{5}(8)^{5/2}\right)$$

$$= 4\sqrt{2}\cdot64\sqrt{8}w\left(\frac{1}{3} - \frac{1}{5}\right) = 1024\,\frac{2}{15} = \frac{2048}{15}\,w.$$

For water $w \approx 62.4$ lb/ft^3, so

$$F = \frac{2048}{15}(62.4) \approx 8520 \text{ lb.} \qquad \square$$

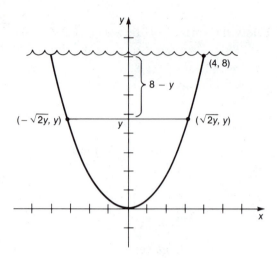

FIGURE 6.41

EXERCISE SET 6.6

1. A plate containing 6 square feet of area is lying flat on the bottom of a tank of water 8 feet deep. What is the pressure on the plate when the plate is in the shape of (a) a rectangle 2 feet wide and 3 feet long? (b) a circle? (c) an equilateral triangle? What is the total force exerted on each of the plates by the water?

2. A metal plate in the shape of a rectangle 2 feet high and 3 feet wide is placed vertically in a tank of water 8 feet deep, with the base of the rectangle at the bottom of the tank. What is the total force exerted by the water on the plate?

3. What would be the force on the plate described in Exercise 2 if it is rotated 90°, so that the 2-foot side is on the bottom of the tank?

4. Suppose the plate described in Exercise 2 is placed in the tank with its 3-foot width lying on the surface of the water. What is the force on the plate in this situation?

5. A vertical gate on a dam is in the shape of a rectangle 10 feet wide and 5 feet deep with its width parallel to the surface of the water. Find the force of the water on the gate when the level of water in the dam is at the top of the gate.

6. Reconsider the dam gate discussed in Exercise 5. How much force is exerted by the water on this gate when the top of the gate is 5 feet below the surface of the water?

7. Reconsider the dam gate discussed in Exercise 5. How much force is exerted by the water on this gate when the water is 1 foot below the top of the gate?

8. A vertical gate on a dam is in the shape of the region described in feet by $y = x^2$, $y = 9$. The top of the dam is described by $y = 9$. Find the force of the water on the gate when the level of water is 3 feet below the top of the dam.

9. A metal plate in the shape of the region $y = 4 - x^2$, $y = 0$, is submerged vertically in a tank of water 6 feet deep so that the flat edge of the plate is on the bottom of the tank. The dimensions of the plate are given in feet. Find the force on the plate.

10. Suppose that a plate in the shape of an equilateral triangle with area $4\sqrt{3}/3$ square feet is placed vertically at the bottom of an 8 ft tank of water. What is the force exerted by the water on this plate if one side of the triangle is on the bottom of the tank?

11. What is the force on the plate described in Exercise 10 if a vertex of the triangle is touching the bottom and the side opposite this vertex is parallel to the bottom?

12. A circular plate with an area of 6 square feet is placed vertically on the bottom of a tank of water 8 feet deep. Set up the integral that describes the force on this plate.

13. In Exercise 14 of Section 6.4, we discussed some facts concerning a swimming pool with dimensions 30 feet by 50 feet that was 8 feet deep for 20 feet on one end, sloped linearly upward for 10 feet, and was 3 feet deep at the other end. How much force does the water exert on each of the sides of the pool when the pool is filled with water?

14. How much force is exerted on each side of the pool described in Exercise 13 when the pool is half-full of water?

15. How much force is acting on the bottom of the swimming pool described in Exercise 13 when the pool is filled with water?

16. How much force is acting on the bottom of the pool described in Exercise 13 when the pool is half-full of water?

17. A pig trough 8 feet long has a cross section in the shape of an isosceles trapezoid with a lower base of one foot, an upper base of two feet, and an altitude of one foot. The trough is filled with swill of density 64 lb/ft^3. (a) What is the total force on an end of the trough? (b) What is the total force on the bottom of the trough?

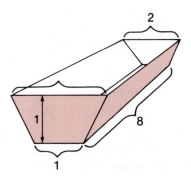

18. A glass in the shape of a right circular cylinder with height 6 inches and radius 2 inches is filled with water. What is the total force acting on the side of the glass?

19. Suppose that the filled glass described in Exercise 18 is placed in a basin of water and that the water in the basin just reaches the top of the glass. What is the total force acting on the sides of the glass?

6.7
WORK

The concept of work is used to describe the transfer of energy. When a constant force F acts through a distance x the work W done by the force is defined as

$$W = F \cdot x.$$

Suppose we compute the work required to lift an object weighing 10 pounds from the ground to a point 3 feet above the ground. The weight of the object describes the force exercised on the object by the earth's gravity; an equivalent force must be applied if the object is to be lifted. Consequently, the work W done in lifting the object is the product of the force required, $F = 10$ pounds, and the distance moved, 3 feet, or

$$W = 10 \cdot 3 = 30 \text{ ft lb}.$$

Energy also needs to be transferred in order to stretch or compress a spring. From your own experience, you no doubt realize that the farther a spring is stretched from its natural or unstressed length, the more difficult it is to stretch further. This indicates that the force required to stretch a spring is not constant. In fact, from physical experiments, it can be shown that the force required to stretch a spring x units beyond its natural length l is

$$F(x) = kx.$$

This formula is known as **Hooke's law** and is illustrated in Figure 6.42. (See the Historical Note.) The value k is called the **spring constant** and depends only on the particular spring (and the units of measurement in which x and $F(x)$ are expressed.)

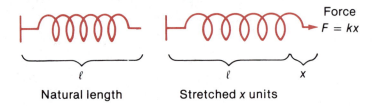

Natural length   Stretched x units

FIGURE 6.42

There are many other situations in which a force is variable and depends on the position of the object on which it is applied. Pulling a cart on an unlevel road and pumping water from a cylindrical barrel are two examples.

HISTORICAL NOTE   **Robert Hooke** (1635–1703) was an English contemporary of Isaac Newton. His work was primarily in mathematical physics where he did fundamental research in the applications of balance springs and pendulums. He hypothesized Newton's law of gravitation, but was unable to establish the relationship mathematically.

We can determine the work W done by a variable force by partitioning the interval over which the force is acting, assuming the force is constant over each subinterval, and determining approximating sums for W. The procedure is similar to others in this chapter.

Suppose F is a function describing the force acting on an object as the object moves along a straight line from point a to point b. To find the total work done by F, we partition the interval $[a, b]$ into subintervals with endpoints $a = x_0 < x_1 < \ldots < x_n = b$ and assume that the force F is constant on each interval $[x_{i-1}, x_i]$. If $x_{i-1} \le z_i \le x_i$ for each i, the work done by F on $[x_{i-1}, x_i]$ is approximately

$$\Delta W_i = F(z_i)\Delta x_i.$$

The total work done by F in moving an object from a to b is approximated by the Riemann sum

$$\sum_{i=1}^{n} F(z_i)\Delta x_i,$$

which leads to a definite integral.

(6.23)
DEFINITION

If F is a function describing the force acting on an object as the object moves linearly from point a to point b, then the work done by the force is

$$W = \int_a^b F(x)dx,$$

provided this integral exists.

EXAMPLE 1   Use Definition 6.23 to determine the work required to lift a 10-pound weight 3 feet above the ground.

SOLUTION

The force in this example is a constant 10 lb so the work required is

$$W = \int_0^3 10\, dx = 30 \text{ ft lb,}$$

which agrees with the result in the first illustration of this section. □

EXAMPLE 2   Use Definition 6.23 to determine the work required to stretch a spring with natural length 1 foot and spring constant 12 a distance 3 feet from its natural length.

SOLUTION

Since the force of the spring is given by $F(x) = 12x$, the work required is

$$W = \int_0^3 12x\, dx = 6x^2 \Big]_0^3 = 54 \text{ ft lb.}$$ □

EXAMPLE 3   A 5 pound hammer is lying on the bottom of a well 16 feet deep. How much work is required to raise the hammer to the surface by a rope weighing 1 oz/ft?

SOLUTION

When the hammer is x feet from the surface, the combined weight of the hammer and the rope is

$$F(x) = 5 \text{ lb} + (x \text{ ft})(1 \text{ oz/ft}) = (5 + x/16)\text{lb}.$$

So the work required is

$$W = \int_0^{16} \left(5 + \frac{x}{16}\right) dx = 5x + \frac{x^2}{32}\Bigg]_0^{16} = 80 + 8 = 88 \text{ ft lb.} \qquad \square$$

The previous examples show that it is easy to determine the work done by a force when a functional representation of the force is known. Unfortunately, important problems for determining work involve forces whose functional representations are not immediately recognized. This is true, for example, when the force is due to liquid pressure. In the remaining examples of this section we consider problems of this type, and use the approximation process to develop the definite integrals associated with work.

EXAMPLE 4   Suppose a well 10 feet deep with a uniform diameter of 6 feet is filled with water. How much work is required to pump all the water to the top of the well?

SOLUTION

The situation is sketched in Figure 6.43.

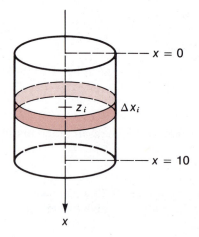

FIGURE 6.43

Our plan is to approximate the work done in pumping a thin layer of water to the top and then sum these approximations; this will lead to a definite integral.

First divide the interval $[0, 10]$ into subintervals with endpoints $0 = x_0 < x_1$

$< \ldots < x_n = 10$ and for each i let z_i be a number with $x_{i-1} \leq z_i \leq x_i$. The force that must be applied to move the portion of water between levels x_{i-1} and x_i to the top is

$$F_i = (\text{density})(\text{volume})$$
$$= 62.4 \text{ lbs/ft}^3 \ (\text{cross-sectional area})(\text{thickness})$$
$$= (62.4)\pi(3)^2 \Delta x_i \text{ lbs.}$$

The work required to pump the ith layer of water to the top is approximately

$$\Delta W_i = (F_i)(\text{distance})$$
$$= [(62.4)(9\pi)\Delta x_i] \, z_i,$$

and the work required to pump all the water to the top is approximated by

$$W \approx \sum_{i=1}^{n} \Delta W_i = \sum_{i=1}^{n} (62.4)(9\pi z_i)\Delta x_i.$$

Consequently, the total work is

$$W = \int_0^{10} (62.4)9\pi x \, dx = 561.6\pi \frac{x^2}{2}\bigg]_0^{10} = 28,080\pi \text{ ft lb.} \qquad \square$$

In some cases it is more convenient to represent a work situation so that the distance over which the force is applied is measured along the y-axis. This is illustrated in our next example.

EXAMPLE 5   A reservoir in the shape of a hemisphere with radius 10 feet is full of water. What is the work required to pump this water to the top of the reservoir?

SOLUTION

The hemisphere is shown in Figure 6.44. A center cross section of the hemisphere is sketched in Figure 6.45 with a rectangular coordinate system placed so that the origin is at the center of the sphere and the direction in which the water is to be pumped is along the y-axis.

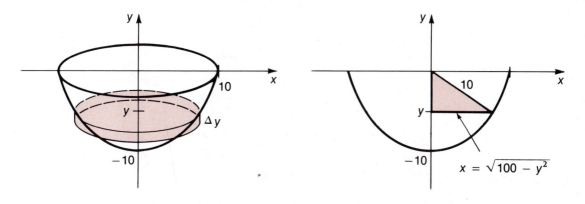

FIGURE 6.44   FIGURE 6.45

The distance through which a thin layer of water at a point on the y-axis, $y \le 0$, must be pumped is $(0 - y)$. Since the force that must be applied is density times volume, we need the volume of a thin layer of water. This volume is approximated by the cross-sectional area times the thickness. We see from Figure 6.45 that this volume is $\pi x^2 \Delta y$ so the work is

$$\int_{-10}^{0} 62.4(0 - y)\pi x^2 \, dy = -62.4\pi \int_{-10}^{0} y(100 - y^2) \, dy$$

$$= -62.4\pi \left[\frac{100y^2}{2} - \frac{y^4}{4} \right]_{-10}^{0} = 62.4\pi \left[\frac{(100)^2}{2} - \frac{(100)^2}{4} \right]$$

$$= 62.4\pi(25)(100) = 156,000\pi \text{ ft lb.} \qquad \square$$

EXAMPLE 6   How much work is required to pump the water in the reservoir described in Example 5 to a point 5 feet above the reservoir?

SOLUTION

The only difference between this example and the preceding one is the distance that the water must be pumped. With the same notation as in Example 5, we see from Figure 6.46 that this distance is $(5 - y)$ and

$$W = \int_{-10}^{0} 62.4\pi(5 - y)(100 - y^2) \, dy$$

$$= 62.4\pi \int_{-10}^{0} (5000 - 100y - 5y^2 + y^3) \, dy$$

$$= 62.4\pi \left[5000y - 50y^2 - \frac{5y^3}{3} + \frac{y^4}{4} \right]_{-10}^{0}$$

$$= -62.4\pi \left[5000(-10) - 50(100) - \frac{5(-1000)}{3} + \frac{10^4}{4} \right]$$

$$= 62.4\pi \left(\frac{17,500}{3} \right) = 364,000\pi \text{ ft lb.} \qquad \square$$

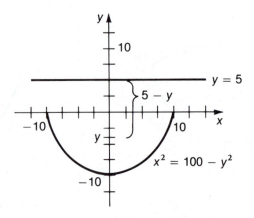

FIGURE 6.46

EXERCISE SET 6.7

1. Determine the work required to lift a 50-pound bag of dog food 3 feet above the ground.

2. How much work is done in lifting a 110-pound barbell from the ground to a point 14 inches above your head?

3. An automobile weighing 3256 lb is driven on a 458-mile trip from Cheyenne, Wyoming to Salt Lake City, Utah. How much work is done by the automobile on this trip?

4. A diesel locomotive on a freight train exerts the constant force of 8 tons on a train while moving the train at 40 mph. How much work is done by the locomotive when the train travels one mile?

5. A force of 6 newtons stretches a spring 40 centimeters. How much work is done if the spring is stretched 80 centimeters? (*Hint:* First use the given information to find the spring constant.)

6. A force of 3 pounds stretches a spring 4 inches. How much work is done when the spring is stretched one foot?

7. A force of 5 newtons stretches a spring 20 centimeters. How far will a force of 10 newtons stretch the spring? How much work is done when the force of 5 newtons is applied? How much work is done when the force of 10 newtons is applied?

8. A spring is 6 inches long and is stretched 1 inch by a 5-pound weight.
   (a) What is the work done in stretching this spring from 6 to 8 inches?
   (b) What is the work done in stretching this spring from 8 to 10 inches?

9. A spring is 10 inches long and a force of 5 pounds is required to compress it to a length of 6 inches. How much work is required to stretch it from its natural length of 10 inches to a length of 20 inches?

10. A spring has a natural length of 20 centimeters, and a force of 4 newtons is required to compress it to a length of 10 centimeters. How much work is required to stretch the spring from 20 centimeters to 40 centimeters?

11. A 3-gallon bucket full of water is at the bottom of a 100-ft well. How much work is required to lift this bucket to the top of the well, assuming that the bucket weighs 3 pounds when empty and that the weight of the rope is insignificant? (1 gal $\approx$ .134 cu ft)

12. A 50-ft rope weighing 2 oz/ft is hanging from the top of a 100-ft bridge. How much work is required to pull this rope to the top of the bridge?

13. A 3-gallon bucket full of water is at the bottom of a 100-ft well. How much work is required to lift this bucket to the top of the well, assuming that the bucket weighs 3 pounds when empty and that the rope weighs 4 oz/ft?

14. A 50-ft rope weighing 2 oz/ft is hanging from the top of a 25-ft bridge with 25 feet of the rope coiled at the bottom. How much work is required to pull this rope to the top of the bridge?

15. A crane is used to move heavy equipment within a building. The cable of the crane weighs 1.5 lb/ft and the hook at the end of the cable weighs 35 lb. How much work is done by this crane in lifting a 500-lb object 6 feet off the ground?

16. The anchor on the *U.S.S. Constitution (Old Ironsides)* was forged in the 1790's and weighs about 5300 pounds. The ship is presently moored at Boston Naval Yard in Charlestown, Mass., under the care of the U.S. Navy. The depth in the harbor is approximately 40 feet (depending on the tide). (a) How much work would be required to raise the anchor from the bottom of the harbor to a point 16 feet above the water, assuming that the weight of the chain is negligible? (b) If the chain actually weighs 50 lb/ft, how much work does this add to the task?

17. A swimming pool with dimensions 30 feet by 50 feet and 8 feet deep is filled with water.
(a) How much work is required to pump all the water from the pool?
(b) How much work is required to pump all the water to a point 3 feet above its surface?

18. Suppose the water in the swimming pool described in Exercise 17 is 7 feet deep. How much work is required to pump all the water from the pool?

19. A cylindrical tank of radius 2 feet and height 6 feet is full of oil weighing 55 lb/ft^3. (a) How much work is necessary to pump the oil to the top of the tank? (b) Suppose the tank is placed on its side and the oil is pumped to the top of the tank. Set up the integral that gives the work required in this case.

20. Suppose a one-horsepower motor is used to pump the oil from the tank described in Exercise 19. (One horsepower equals 550 ft lb per second.) How long will it take to pump the oil into a tank truck 8 feet above the original level of the oil?

21. In Exercise 14 of Section 6.4, we described a swimming pool with dimensions 30 feet by 50 feet that was 8 feet deep for twenty feet, then sloped linearly upward for 10 feet, and was 3 feet deep for the remaining twenty feet. How much work is required to pump all the water from the pool to a point 3 feet above its original surface?

22. A 10,000-gallon tank full of water is in the shape of an inverted cone with equal height and diameter. The water is pumped into a cylindrical tank with 10,000-gallon capacity and equal diameter and height. The tops of the two tanks are in the same plane. How much work is done to pump the water from the inverted cone into the cylinder? If the procedure is reversed, with the cylindrical tank full and the inverted cone empty, how much work is necessary to pump the water? (1 gal $=$ .134 cu ft)

23. A service station has a 2000-gallon cylindrical tank whose top is buried 2 feet below the ground, with the axis of the cylinder parallel to the ground. Once a week this tank is serviced by the supplier. In one week, 800 gallons of gasoline are sold, so the service station has the choice of having the tank filled every other week or having it topped off each week. Which method of purchase will require the least work for the service station pumps?

24. Show that when a spring is stretched from a distance x_1 beyond its natural length to a distance x_2 beyond its natural length, the work required is the product of the distance $(x_2 - x_1)$ and the average of the forces $F(x_1)$ and $F(x_2)$; that is,

$$W = \frac{F(x_1) + F(x_2)}{2}(x_2 - x_1).$$

6.8
MOMENTS AND THE CENTER OF MASS

Suppose n objects lie on a coordinate line at the points $x_1, x_2, \ldots, x_n$ and that these objects have respective masses $m_1, m_2, \ldots, m_n$. There is precisely one point along the line where the weight of the system is balanced, in the sense that if the system is supported only at this point it will not rotate. This balance point $\bar{x}$ is called the **center of mass** of the system. (See Figure 6.47.)

FIGURE 6.47

To determine the center of mass of a system of this type we define the moment of the system about p, for each real number p:

(6.24)
$$M_p = \sum_{i=1}^{n} m_i(x_i - p).$$

Physically, M_p describes the tendency of the system to rotate when a pivot is placed at p. If $M_p > 0$, the rotation will be in the clockwise direction; if $M_p < 0$, the rotation will be in the counterclockwise direction.

Since the center of mass $\bar{x}$ is the point where no rotation occurs, the center of mass is distinguished by the condition $M_{\bar{x}} = 0$. Thus

$$0 = M_{\bar{x}} = \sum_{i=1}^{n} m_i(x_i - \bar{x}) = \sum_{i=1}^{n} m_i x_i - \bar{x}\sum_{i=1}^{n} m_i$$

and

(6.25)
$$\bar{x} = \frac{\sum_{i=1}^{n} m_i x_i}{\sum_{i=1}^{n} m_i}.$$

Equation (6.25) implies that the center of mass of the system is the moment of the system about the origin, M_0, divided by the total mass of the system,

$$M = \sum_{i=1}^{n} m_i.$$

EXAMPLE 1   Suppose that two children A and B of the same weight w are on opposite ends of a seesaw 12 feet in length. What is the moment of this system about the location of the pivot, p, if p is
(a) 2 feet from A?   (b) 6 feet from A?   (c) 9 feet from A?

SOLUTION

The solutions to (a), (b), and (c) are illustrated in Figure 6.48.   ☐

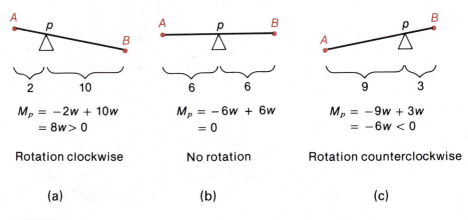

$M_p = -2w + 10w$	$M_p = -6w + 6w$	$M_p = -9w + 3w$
$= 8w > 0$	$= 0$	$= -6w < 0$
Rotation clockwise	No rotation	Rotation counterclockwise
(a)	(b)	(c)

FIGURE 6.48

The concepts of moments and center of mass can easily be extended to the situation involving a collection of objects in the plane. Suppose that n objects lie in a coordinate plane at the points with coordinates (x_1, y_1), (x_2, y_2), (x_3, y_3), . . . , (x_n, y_n) and that these objects have the respective masses $m_1, m_2, \ldots , m_n$ as shown in Figure 6.49.

Moments for the system are defined in both coordinate directions. The moment about a line $x = p$ is defined by

$$M_{x,p} = \sum_{i=1}^{n} m_i(x_i - p).$$

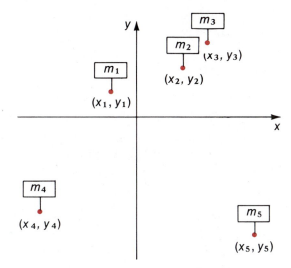

FIGURE 6.49

$M_{x,p}$ describes the tendency of the system to pivot about the line $x = p$. The number $\bar{x}$ is obtained by specifying that $M_{x,\bar{x}} = 0$, so

$$0 = \sum_{i=1}^{n} m_i(x_i - \bar{x})$$

and

(6.26)
$$\bar{x} = \frac{\displaystyle\sum_{i=1}^{n} m_i x_i}{\displaystyle\sum_{i=1}^{n} m_i}.$$

Since the line $x = 0$ is the y-axis, the moment, $M_{x,0}$ is generally called the **moment about the y-axis** and alternatively denoted M_y. With this notation,

(6.27)
$$\bar{x} = \frac{M_y}{M}.$$

Similarly, we can define the moment about a line $y = q$ by

$$M_{y,q} = \sum_{i=1}^{n} m_i(y_i - q),$$

which describes the tendency of the system to pivot about the line $y = q$. The number $\bar{y}$ is defined to satisfy $M_{y,\bar{y}} = 0$, so

$$0 = \sum_{i=1}^{n} m_i(y_i - \bar{y}).$$

Letting M_x denote the moment about the line $y = 0$, called the **moment about the x-axis,** we have

(6.28)
$$\bar{y} = \frac{\sum\limits_{i=1}^{n} m_i y_i}{\sum\limits_{i=1}^{n} m_i} = \frac{M_x}{M}.$$

The point with coordinates $(\bar{x}, \bar{y})$ is called the **center of mass of the system** and is the unique point at which a pivot can be placed so that the system is balanced.

EXAMPLE 2   Particles with masses 2, 3, 5, and 6 units are located at the points $(1, -1)$, $(2, 1)$, $(-1, 1)$, and $(-2, -2)$, respectively. (See Figure 6.50.) Find the center of mass of this system.

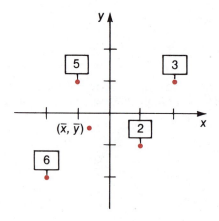

FIGURE 6.50

SOLUTION

The total mass of the system is $M = 2 + 3 + 5 + 6 = 16$, and the moments about the x- and y-axes are

$$M_x = 2(-1) + 3(1) + 5(1) + 6(-2) = -6$$
and
$$M_y = 2(1) + 3(2) + 5(-1) + 6(-2) = -9.$$

The center of mass occurs at $(\bar{x}, \bar{y})$ where

$$\bar{x} = \frac{M_y}{M} = -\frac{9}{16} \qquad \text{and} \qquad \bar{y} = \frac{M_x}{M} = -\frac{6}{16} = -\frac{3}{8}. \qquad \square$$

When we move from a finite system in the plane to compute the moments of a region in the plane, we need the definite integral. We cannot yet compute the moments and center of mass of a planar region of varying density. This involves a function of two variables and will not be discussed until Chapter 16.

We can, however, find the moments and center of mass for a region of constant density.

A thin sheet of material, called a **lamina**, can be described by a region in the plane. A lamina with constant or uniform density is called a **homogeneous lamina**, and its center of mass is known as the **centroid** of the lamina.

Suppose a lamina lies in the plane and is bounded by the x-axis, the lines $x = a$, $x = b$, and the graph of a continuous nonnegative function f.

Partition the interval $[a, b]$ by $a = x_0 < x_1 < \ldots < x_n = b$ and choose z_i in $[x_{i-1}, x_i]$ to be $z_i = (x_{i-1} + x_i)/2$. The center of mass of the rectangle with base the interval $[x_{i-1}, x_i]$ and height $f(z_i)$ is the geometric center of the rectangle:

$$\left(\frac{x_{i-1} + x_i}{2}, \frac{1}{2} f(z_i)\right) = \left(z_i, \frac{1}{2} f(z_i)\right).$$

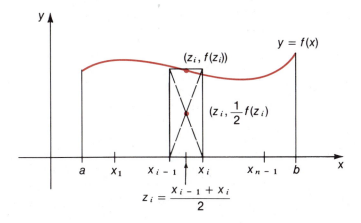

FIGURE 6.51

We make the simplifying, but correct, assumption that all mass of a rectangle can be concentrated at its center of mass without changing its moments. Since the area of the ith rectangle is $f(z_i)\Delta x_i$, the mass of this rectangle is $\rho f(z_i)\Delta x_i$, where ρ denotes the constant area density of the lamina. The moment about the x-axis of the ith rectangle is

$$(\rho f(z_i)\Delta x_i)\left(\frac{1}{2} f(z_i)\right) = \frac{\rho}{2} [f(z_i)]^2 \Delta x_i.$$

So the moment about the x-axis of the system consisting of n rectangles is

$$\sum_{i=1}^{n} \frac{\rho}{2} [f(z_i)]^2 \Delta x_i,$$

which is a Riemann sum for the function described by $\rho[f(x)]^2/2$ on the interval $[a, b]$.

Similarly, the moment of the system of n rectangles about the y-axis is

$$\sum_{i=1}^{n} [\rho f(z_i)\Delta x_i] z_i = \sum_{i=1}^{n} \rho z_i f(z_i)\Delta x_i,$$

a Riemann sum for the function described by $\rho x f(x)$ on the interval $[a, b]$.

Since the total mass of a homogeneous lamina is the product of its density ρ and its area $\int_a^b f(x)dx$, we have the following definition.

(6.29)
DEFINITION

Suppose R is a homogeneous lamina with density ρ, bounded by the x-axis, the lines $x = a$, $x = b$ and the graph of a continuous nonnegative function f. The **moment of R about the x-axis** is

$$M_x = \frac{\rho}{2} \int_a^b [f(x)]^2 \, dx;$$

the **moment of R about the y-axis** is

$$M_y = \rho \int_a^b xf(x)dx;$$

and the **center of mass** (or **centroid**) **of R** occurs at $(\bar{x}, \bar{y})$,

where

$$\bar{x} = \frac{M_y}{M} = \frac{\displaystyle\int_a^b xf(x)dx}{\displaystyle\int_a^b f(x)dx}$$

and

$$\bar{y} = \frac{M_x}{M} = \frac{\displaystyle\int_a^b [f(x)]^2 \, dx}{2\displaystyle\int_a^b f(x)dx}.$$

EXAMPLE 3   Find the centroid of the homogeneous lamina bounded by the x-axis, the line $x = 2$ and the graph of $f(x) = x^2$.

SOLUTION

Since the centroid of a homogeneous lamina is independent of the density of the region we can assume that $\rho = 1$.

By Definition 6.29, the moment about the x-axis is

$$M_x = \frac{1}{2} \int_0^2 (x^2)^2 \, dx = \frac{x^5}{10} \Big]_0^2 = \frac{16}{5},$$

the moment about the y-axis is

$$M_y = \int_0^2 x \, (x^2) \, dx = \frac{x^4}{4} \Big]_0^2 = 4,$$

and the mass of the lamina is the area of the region

$$M = \int_0^2 x^2 \, dx = \frac{x^3}{3} \Big]_0^2 = \frac{8}{3}.$$

The centroid of the lamina is the point $(\bar{x}, \bar{y})$ where

$$\bar{x} = \frac{M_y}{M} = \frac{4}{\dfrac{8}{3}} = \frac{3}{2}$$

and

$$\bar{y} = \frac{M_x}{M} = \frac{\dfrac{16}{5}}{\dfrac{8}{3}} = \frac{6}{5}.$$

See Figure 6.52.   ☐

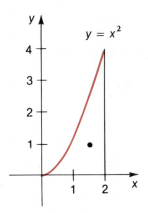

$$y = x^2$$

FIGURE 6.52

Suppose a homogeneous lamina with density ρ is described by a region R bounded by the lines $x = a$, $x = b$, and the graphs of the continuous functions f and g, where $g(x) \le f(x)$ on $[a, b]$. The moments about the axes and center of mass of the lamina can be found in a manner similar to that which led to Definition 6.29.

Partition $[a, b]$ by $a = x_0 < x_1 < \ldots < x_n = b$ and choose z_i in $[x_{i-1}, x_i]$ to be $z_i = (x_{i-1} + x_i)/2$. The midpoint of the rectangle with boundaries $x = x_{i-1}$, $x = x_i$, $y = f(z_i)$ and $y = g(z_i)$ (see Figure 6.53) occurs at

$$\left(z_i, g(z_i) + \frac{f(z_i) - g(z_i)}{2} \right) = \left(z_i, \frac{f(z_i) + g(z_i)}{2} \right),$$

and the area of this rectangle is

$$[f(z_i) - g(z_i)]\Delta x_i.$$

Consequently, the mass of the ith rectangle is

$$\rho[f(z_i) - g(z_i)]\Delta x_i,$$

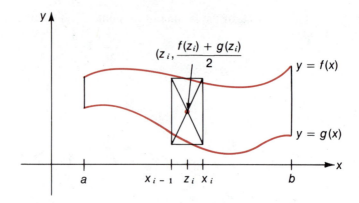

FIGURE 6.53

the moment about the x-axis is

$$\{\rho[f(z_i) - g(z_i)]\Delta x_i\}\left[\frac{f(z_i) + g(z_i)}{2}\right] = \frac{\rho}{2}\{[f(z_i)]^2 - [g(z_i)]^2\}\Delta x_i,$$

and the moment about the y-axis is

$$\{\rho[f(z_i) - g(z_i)]\Delta x_i\}z_i = \rho z_i[f(z_i) - g(z_i)]\Delta x_i.$$

Summing these values produces Riemann sums that lead to integrals for the mass of the lamina

$$M = \int_a^b \rho[f(x) - g(x)]dx,$$

the moment about the x-axis

$$M_x = \int_a^b \frac{\rho}{2}\{[f(x)]^2 - [g(x)]^2\}dx,$$

and the moment about the y-axis

$$M_y = \int_a^b \rho x[f(x) - g(x)]dx.$$

The center of mass occurs at $(\bar{x}, \bar{y})$ where

$$\bar{x} = \frac{M_y}{M} = \frac{\displaystyle\int_a^b x[f(x) - g(x)]dx}{\displaystyle\int_a^b [f(x) - g(x)]dx}$$

and

$$\bar{y} = \frac{M_x}{M} = \frac{\displaystyle\int_a^b \{[f(x)]^2 - [g(x)]^2\}dx}{2\displaystyle\int_a^b [f(x) - g(x)]dx}.$$

EXAMPLE 4   Find the centroid of the homogeneous lamina bounded by the y-axis and the graphs of $f(x) = 1 - x$ and $g(x) = -\sqrt{x} + 1$. See Figure 6.54.

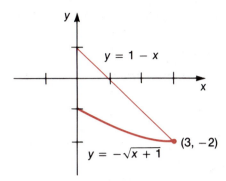

FIGURE 6.54

SOLUTION

Assuming that $\rho = 1$, the mass of the lamina is

$$M = \int_0^3 [(1 - x) + \sqrt{x + 1}] \, dx$$

$$= \left(\frac{-(1 - x)^2}{2} + \frac{2}{3} (x + 1)^{3/2} \right) \Bigg]_0^3 = \frac{19}{6},$$

the moment about the x-axis is

$$M_x = \frac{1}{2} \int_0^3 [(1 - x)^2 - (x + 1)] \, dx$$

$$= \frac{1}{2} \left(\frac{-(1 - x)^3}{3} - \frac{(x + 1)^2}{2} \right) \Bigg]_0^3 = -\frac{9}{4},$$

and the moment about the y-axis is

$$M_y = \int_0^3 x(1 - x + \sqrt{x + 1}) \, dx$$

$$= \left(\frac{x^2}{2} - \frac{x^3}{3} \right) \Bigg]_0^3 + \int_0^3 x\sqrt{x + 1} \, dx.$$

To evaluate $\int x\sqrt{x + 1} \, dx$, let $u = x + 1$. Then

$$\int x\sqrt{x + 1} \, dx = \int (u - 1)\sqrt{u} \, du$$

$$= \int (u^{3/2} - u^{1/2}) du = \frac{2}{5} u^{5/2} - \frac{2}{3} u^{3/2} + C$$

$$= \frac{2}{5} (x + 1)^{5/2} - \frac{2}{3} (x + 1)^{3/2} + C.$$

So

$$M_y = \left(\frac{x^2}{2} - \frac{x^3}{3} + \frac{2}{5} (x + 1)^{5/2} - \frac{2}{3} (x + 1)^{3/2} \right) \Bigg]_0^3 = \frac{97}{30}.$$

The centroid of the lamina is at $(\bar{x}, \bar{y})$, where

$$\bar{x} = \frac{M_y}{M} = \frac{\left(\dfrac{97}{30}\right)}{\left(\dfrac{19}{6}\right)} = \frac{97}{95} \quad \text{and} \quad \bar{y} = \frac{M_x}{M} = \frac{\left(\dfrac{-9}{4}\right)}{\left(\dfrac{19}{6}\right)} = -\frac{27}{38}. \qquad \square$$

EXERCISE SET 6.8

Objects are placed along a line with the coordinates and relative masses given in Exercises 1 through 4. Find the moment of the system with respect to the origin and the center of mass of the system.

1. $x_1 = 1, m_1 = 1; x_2 = 3, m_2 = 2; x_3 = 5, m_3 = 3$.

2. $x_1 = -2, m_1 = 1; x_2 = 1, m_2 = 1; x_3 = 4, m_3 = 1; x_4 = 6, m_4 = 1$.

3. $x_1 = -1, m_1 = 3; x_2 = 0, m_2 = 1; x_3 = 2, m_3 = 4; x_4 = 4, m_4 = 3$.

4. $x_1 = -5, m_1 = 3; x_2 = -1, m_2 = 4; x_3 = 0, m_3 = 5; x_4 = 3, m_4 = 3;$
$x_5 = 4, m_5 = 2$.

Objects are placed in the plane with the coordinates and relative masses given in Exercises 5 through 8. Find the moments of the system with respect to the axes and determine the center of mass of the system.

5. $P_1(1, 0), m_1 = 1; P_2(2, 3), m_2 = 3; P_3(-1, 1), m_3 = 2$.

6. $P_1(-2, 1), m_1 = 1; P_2(0, 0), m_2 = 1; P_3(1, 3), m_3 = 1; P_4(5, 7),$
$m_4 = 1$.

7. $P_1(3, 2), m_1 = 3; P_2(2, -1), m_2 = 2; P_3(5, -2), m_3 = 4; P_4(-2, 0),$
$m_4 = 1$.

8. $P_1(-4, -3), m_1 = 4; P_2(-2, 2), m_2 = 6; P_3(-5, 0), m_3 = 2; P_4(4, 4),$
$m_4 = 3; P_5(6, 2), m_5 = 1$.

9. Find the center of mass of the system shown in the accompanying figure, if the top rectangle has a uniform density of 2 lb/ft^2 and the bottom rectangle has a uniform density of 1 lb/ft^2.

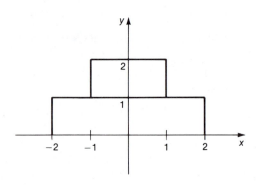

10. A square with sides 1-foot long and uniform density of 1 lb/ft^2 sits on the center of the top of a rectangle with height 3 feet, width 2 feet and a uniform density of 2 lb/ft^2. Find the center of mass of this system.

11. A 2-pound mass is located at a point with coordinates $(-1, 1)$ and a 5-pound mass is located at $(2, 4)$. Where should a 1-pound mass be located if the system is to be balanced at the origin?

12. Two children are playing on a seesaw 16 feet long. One child weighs 60 pounds, the other 67 pounds. Suppose the lighter child sits at one end of the seesaw. How far from the opposite end must the other child sit if the seesaw is to be balanced in the middle, assuming that the weight of the seesaw itself is insignificant?

13. Suppose that instead of moving the children on the seesaw to balance the weight in Exercise 12, we have the facility to add a weight to the seesaw at a point 3 feet from one end. How much weight would be required to balance the seesaw if the children sit at opposite ends and the weight of the seesaw itself is insignificant?

14. Suppose that the seesaw described in Exercise 12 weighs 2 lb/ft. Recalculate the position at which the child must sit.

Find the centroid of the homogeneous lamina bounded by the graphs of the equations given in Exercises 15 through 26.

15. $y = 4 - x^2, y = 0$ **16.** $y = x^2 + 1, x = -1, x = 1, y = 0$

17. $y = x^2, y = 1$ **18.** $y = x^2, y = x$

19. $y = x^3, y = x, x \geq 0$ **20.** $y = 1 - x, x = 0, y = 0$

21. $y = x^2, y = \sqrt{x}$ **22.** $y = x^3, y = x^2$

23. $y = \dfrac{1}{x^2}, x = 1, x = 4, y = 0$ **24.** $y = \dfrac{1}{x}, x = 1, x = 4, y = 0$

25. $y = x, y = 4 - x, y = 0$ **26.** $y = \dfrac{1}{2}x, y = 1 - x, y = 0$

27. Suppose that a rod has the dimension of length only and is of length l. In this case, the mass at each point is determined by a function that describes its linear density, that is, its mass per unit length. Let ρ be the continuous function with domain $[0, l]$ that describes the density at a point on the rod as a function of its distance from a fixed end of the rod. The total mass of the rod is

$$M = \int_0^l \rho(x)\, dx$$

and the moment of the rod about p for p in $[0, l]$ is defined by

$$M_p = \int_0^l (x - p)\rho \text{ see } (x)dx.$$

A straight rod has the density given by the function in each of the following. Find the moment of the rod with respect to the origin and its center of mass.

(a) $\rho(x) = x, [0, 2]$ (b) $\rho(x) = x^2, [0, 2]$

(c) $\rho(x) = x, [1, 3]$ (d) $\rho(x) = x - x^2, [0, 1]$

28. A solid pole 16 feet long has uniform density and tapers linearly from a diameter of 6 inches to a diameter of 4 inches. How far from the larger end of the pole is the center of mass?

29. Suppose that instead of being solid, the pole described in Exercise 28 is a hollow pipe made of half-inch-thick stock. Would this change the position of the center of mass?

30. The *first theorem of Pappus* (see Historical Note) states that the volume of the solid generated by rotating a plane region about a line that does not intersect the region is the product of the area of the region and the distance traveled by the centroid of the region as it is rotated about the line. Prove this theorem for the case when the line is the y-axis and the region is bounded by the x-axis, the graph of a continuous function f, and the lines $x = a$ and $x = b$, where $0 < a < b$.

31. Prove the first theorem of Pappus for the case when the line is the x-axis, and the region is bounded by the x-axis, the graph of a continuous function f, and the lines $x = a$ and $x = b$ where $a < b$.

32. Use the first theorem of Pappus to find the volume of the solid generated by rotating the region bounded by the graphs of $y = x^2$ and $y = x$ about

　(a) the x-axis　　　　　　　　　　(b) the y-axis

33. Use the first theorem of Pappus to find the volume of the solid generated by rotating a circle of radius r centered at $(R, 0)$ about the y-axis, where $r < R$. This solid is called a *torus*, and is the shape of a donut with a hole.

34. A rod of uniform density has the shape of the graph of a continuously differentiable function f on the interval $[a, b]$. Show that the center of mass of this rod occurs at $(\bar{x}, \bar{y})$ where

$$\bar{x} = \frac{\displaystyle\int_a^b x\sqrt{1 + [f'(x)]^2}\, dx}{\displaystyle\int_a^b \sqrt{1 + [f'(x)]^2}\, dx} \quad \text{and} \quad \bar{y} = \frac{\displaystyle\int_a^b f(x)\sqrt{1 + [f'(x)]^2}\, dx}{\displaystyle\int_a^b \sqrt{1 + [f'(x)]^2}\, dx}.$$

35. The *second theorem of Pappus* states that the surface area generated by rotating a plane curve about a line that does not intersect the curve is the product of the length of the curve and the distance traveled by the center of mass of the curve as it is rotated about the line. Use the result of Exercise 34 to prove that this theorem is correct when the line is the x-axis and the curve is described by the graph of a continuously differentiable function f on an interval $[a, b]$.

36. Use the second theorem of Pappus to find a formula for the surface area of a cone in terms of its radius r and its altitude h.

37. The surface area of a sphere of radius r is $4\pi r^2$. Use this fact and the second theorem of Pappus to find the center of mass of a rod of uniform mass that is bent in the shape of a semicircle of radius 1.

38. Show that the centroid of a circular homogeneous lamina is located at the center of the circle.

HISTORICAL NOTE　　**Pappus of Alexandria** (300 A.D.) compiled a set of 8 volumes that contain the bulk of the Greek mathematics done in his time. The first and second theorems of Pappus are included in this collection and appear to be his own contribution. He also did extensive work on conic sections and wrote valuable commentaries on Euclid's *Data* and *Elements* and on the *Almagest* by Ptolemy.

39. Show that a homogeneous lamina in the shape of a rectangle has its centroid at the same point as the center of mass of a system of four uniform masses placed at the vertices of the rectangle.

40. Show that a homogeneous lamina in the shape of a triangle has its centroid at the same point as the center of mass of a system of three uniform masses placed at the vertices of the triangle.

41. A sandwich is constructed of two identical pieces of bread and a piece of mozzarella cheese. The cheese extends beyond the boundary of the bread by varying amounts. Show that the sandwich can be divided by one cut in such a manner that both the cheese and the bread are split into equal parts. Is it still possible to divide evenly the various ingredients if we add a slice of ham to the sandwich?

42. In statistics, moments are used to describe the shape of the distribution of a random variable. If the distribution of a random variable is described by a continuous function f, then the moment about the origin of the distribution is called the *mean* of the distribution, and is given by

$$\mu = \int_{-\infty}^{\infty} xf(x)dx.$$

Find the mean of the distribution given by

$$f(x) = \begin{cases} \dfrac{1}{2}x, & \text{for } 0 < x < 2 \\ 0, & \text{elsewhere.} \end{cases}$$

43. The *second moment about the mean* of the distribution of a random variable described by a continuous function f is called the *variance*. This second moment is defined by

$$\sigma^2 = \int_{-\infty}^{\infty} (x - \mu)^2 f(x)dx.$$

This moment measures the spread or dispersion of the distribution. The square root of the variance, σ, is called the *standard deviation* of the distribution. Find the variance for the distribution in Exercise 42.

REVIEW EXERCISES

Find the area bounded by the regions described in Exercises 1 through 8.

1. $y = \sqrt{x}, y = x^3$

2. $y = \sqrt{x - 1}, y = \sqrt{2 - x}, y = 0$

3. $y^2 = 4 - x, y = x + 2$

4. $y = 4x - x^2, y = x$

5. $y = x^2 - 1, y = 7 - x^2$

6. $y = \sin \pi x, y = \cos \pi x, x = 0, x = 1$

7. $y = 2x^3 - 9x^2 + 12x - 2, y = -2, x = 3$

8. $y = 2x^3 - 3x^2 - 12x + 13, x = -1, x = 2, y = 0$

Find the volume generated if the region bounded by the curves described in Exercises 9 through 16 is rotated about the indicated line.

9. $y = \dfrac{6}{x}$, $x = 2$, $x = 4$, $y = 0$; about the x-axis

10. $y = \dfrac{6}{x}$, $x = 2$, $x = 4$, $y = 0$; about the y-axis.

11. $y^2 = 4x$, $x = 4$; about the x-axis   **12.** $y^2 = 4x$, $x = 4$; about the y-axis

13. $y^2 = 4x$, $x = 4$; about $x = 4$ **14.** $y^2 = 4x$, $x = 4$; about $x = 6$

15. $y = x^2 + 2x + 2$, $x = 0$, $x = 2$, $y = 0$; about the y-axis.

16. $y = 4 - 4x^2$, $y = 1 - x^2$

    (a) about the x-axis   (b) about $y = -1$ (c) about $y = 4$

17. The base of a solid is bounded by the circle in the xy-plane with equation $x^2 + y^2 = 25$ (with the radius given in inches) and each cross section perpendicular to the x-axis is an isosceles triangle with altitude 6 inches. Set up the integral that describes the volume of this solid.

18. The base of a solid is bounded by the circle in the xy-plane with equation $x^2 + y^2 = 25$ (with the radius given in inches) and each cross section perpendicular to the y-axis is an equilateral triangle. Set up the integral that describes the volume of this solid.

19. Find the length of the curve described by $y = \ln(\cos x)$ from $x = -\pi/4$ to $x = \pi/4$.

20. Find the length of the curve described by $y = (x^3 + 3/x)/6$ from $x = 1$ to $x = 3$.

21. Find the area of the surface generated by rotating about the x-axis the curve described by $y = \sqrt{2x - 1}$ for $1/2 \le x \le 1$.

22. Find the area of the surface generated by rotating about the x-axis the curve described by $y = (x^{3/2}/3) - x^{1/2}$ for $1 \le x \le 4$.

23. Find the centroid of the homogeneous lamina bounded by the parabola $y^2 = 4x$ and lines $x = 0$, $y = 4$.

24. A ship's anchor weighs 2000 lb and the anchor chain weighs 30 lb/ft. How much work is required to pull this anchor 50 feet vertically up to the deck?

25. How much work is done in stretching a spring 4 inches if it takes 250 pounds of force to stretch it that amount?

26. A force of 5 pounds stretches a spring one foot. (a) How much work is done in stretching the spring this distance? (b) How much work is done if the spring is stretched from one foot to two feet beyond its natural length?

27. A conical tank with the vertex at the bottom has radius 5 feet and altitude 20 feet. If the water in the tank is 12 feet deep, find the work required to pump the water (a) over the top, (b) to a point 10 feet above the top.

28. A floodgate has a parabolic shape described by $y = x^2$. If the gate is 4 feet deep and 4 feet wide at the top, find the force of the water on the gate when the level of water is at the top of the gate.

29. A right circular cylinder with base area A and height h is filled with fluid of density ρ. Show that the work done in pumping all the fluid to the top of the cylinder is given by $W = \rho h^2 A/2$.

7

THE CALCULUS OF INVERSE FUNCTIONS

We have studied many of the differential and integral calculus properties of algebraic and trigonometric functions. It is time to broaden our knowledge of elementary functions by considering a new and very important class: the exponential functions. The exponential functions are ideally suited for calculus study and applications. In particular, the natural exponential function defined in Section 7.3 is so widely used that you might find it difficult to imagine that you have survived, mathematically speaking, so long without it.

7.1
INVERSE FUNCTIONS

A function is described by a rule of correspondence between two sets, the domain A and the range B, that assigns to each element of A precisely one element in B. The definition permits elements of B to be the image of more than one element of A. For example, if $f(x) = x^2$, every nonzero element y in the range of f is the image of two elements, $\sqrt{y}$ and $-\sqrt{y}$, as shown in Figure 7.1.

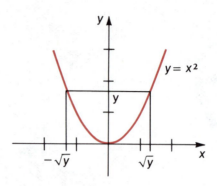

FIGURE 7.1

In this section we will study one-to-one functions.

(7.1)
DEFINITION

A function is **one-to-one** if each element in the range corresponds to precisely one element in the domain.

We can immediately determine whether a function is one-to-one by looking at its graph. *A function is one-to-one precisely when every horizontal line intersects its graph at most once.* From Figure 7.1, we can see that $f(x) = x^2$ is not one-to-one. Neither is $f(x) = \sin x$, whose graph is shown in Figure 7.2 (b). The functions graphed in Figures 7.2 (a) and (c) are one-to-one.

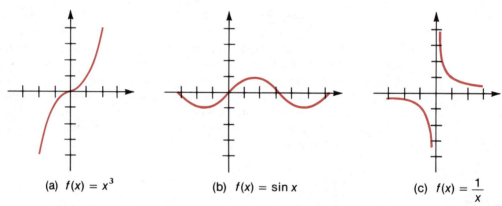

(a) $f(x) = x^3$ (b) $f(x) = \sin x$ (c) $f(x) = \dfrac{1}{x}$

FIGURE 7.2

(7.2)
DEFINITION

Suppose f is a one-to-one function with domain A and range B. The **inverse function of f, f^{-1}**, is the function with domain B and range A defined by

$$f^{-1}(y) = x \qquad \text{if and only if} \qquad y = f(x).$$

It follows from the definition of f^{-1} that

1. for every x in the domain of f, $f^{-1}(f(x)) = x$;
2. for every y in the domain of f^{-1}, $f(f^{-1}(y)) = y$.

EXAMPLE 1   Suppose $f(x) = 2x - 3$. Find f^{-1}.

SOLUTION

Let $y = f(x) = 2x - 3$;

then
$$x = \frac{y + 3}{2},$$

so
$$f^{-1}(y) = x = \frac{y + 3}{2}.$$

Since the variable describing a function is irrelevant, the function can also be described by

$$f^{-1}(x) = \frac{x + 3}{2},$$

which conforms with the usual notation.
The graphs of f and f^{-1} are shown in Figure 7.3.   □

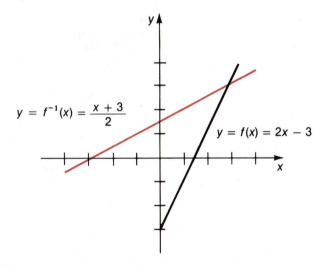

FIGURE 7.3

EXAMPLE 2   Show that the function described by $g(x) = x^2$ with domain $[0, \infty)$ is one-to-one, and find g^{-1}.

SOLUTION

By restricting the domain of the squaring function to $[0, \infty)$, we see that:

$$y = x^2 \text{ and } x \geq 0 \qquad \text{if and only if} \qquad x = \sqrt{y}$$

and y is the image of precisely one element in $[0, \infty)$. Thus, g is one-to-one and g^{-1} is described by

$$g^{-1}(y) = \sqrt{y}$$

or equivalently by,

$$g^{-1}(x) = \sqrt{x}.$$

Figure 7.4 shows the relationship between the graphs of g and g^{-1}.   ☐

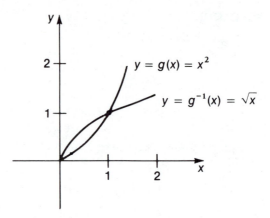

FIGURE 7.4

Figures 7.3 and 7.4 illustrate an important relationship between the graphs of a function and its inverse. If the functions are expressed in terms of the same variable x, then the graph of f^{-1} is the reflection of the graph of f about the line $y = x$, as illustrated in Figure 7.5.

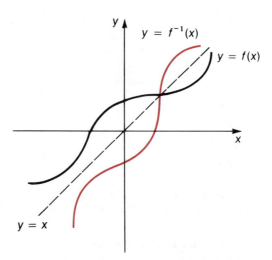

FIGURE 7.5

The following theorem is the first in a series of results about the inverse of increasing and decreasing functions.

(7.3)
THEOREM   If f is increasing on $[a, b]$, then f^{-1} exists and is increasing on $[f(a), f(b)]$.

PROOF

Suppose x_1 and x_2 are in $[a, b]$ and $x_1 \neq x_2$. If $x_1 < x_2$, then, since f is increasing, $f(x_1) < f(x_2)$. Similarly, if $x_2 < x_1$, then $f(x_2) < f(x_1)$. Consequently,

$x_1 \neq x_2$ implies $f(x_1) \neq f(x_2)$. This shows that f is one-to-one and hence that f^{-1} exists.

To show that f^{-1} is increasing, suppose $y_1 < y_2$ where y_1 and y_2 are in the domain of f^{-1}. Since the domain of f^{-1} is the range of f, there are numbers x_1 and x_2 in $[a, b]$ with $f(x_1) = y_1$ and $f(x_2) = y_2$. Suppose $x_1 = f^{-1}(y_1) \geq f^{-1}(y_2) = x_2$. Since f is increasing, $y_1 = f(x_1) \geq f(x_2) = y_2$, which is a contradiction. Consequently, it must be true that $f^{-1}(y_1) < f^{-1}(y_2)$. This implies that f^{-1} is increasing on its domain.

Finally, y is in the domain of f^{-1} if and only if for some x, $a \leq x \leq b$, $y = f(x)$. Since f is increasing, this is true if and only if

$$f(a) \leq f(x) = y \leq f(b).$$

The domain of f^{-1} is $[f(a), f(b)]$. ☐

An equivalent result is true for decreasing functions. The proof of this result follows from Theorem 7.3 and the fact that f is a decreasing function on $[a, b]$ if and only if $-f$ is increasing on $[a, b]$.

(7.4)
COROLLARY   If f is decreasing on $[a, b]$, then f^{-1} exists and is decreasing on $[f(b), f(a)]$.

By adding the condition of continuity to f in the hypothesis of Theorem 7.3 and Corollary 7.4, we gain the assurance of continuity of f^{-1}. The proof of this result, although not difficult, requires considerable manipulation between the function and its inverse, and is presented in Appendix B.

(7.5)
THEOREM   If f is increasing and continuous on $[a, b]$, then f^{-1} is increasing and continuous on $[f(a), f(b)]$.

(7.6)
COROLLARY   If f is decreasing and continuous on $[a, b]$, then f^{-1} is decreasing and continuous on $[f(b), f(a)]$.

Figure 7.6(a) illustrates the result in Theorem 7.5, and Figure 7.6(b) illustrates Corollary 7.6.

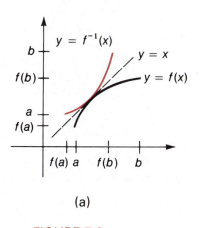

(a)

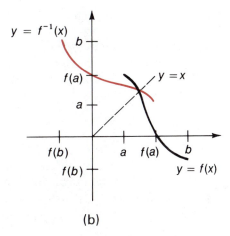

(b)

FIGURE 7.6

EXAMPLE 3   Verify Theorem 7.5 for $f(x) = x^2 - 2$ on the interval $[1, 8]$.

SOLUTION

Since

$$D_x f(x) = D_x(x^2 - 2) = 2x > 0$$

for x in $[1, 8]$, f is increasing and continuous on $[1, 8]$. To find f^{-1}, let $y = x^2 - 2$. Then $x = \sqrt{y + 2} = f^{-1}(y)$, so f^{-1} is described by $f^{-1}(x) = \sqrt{x + 2}$ for x in $[f(1), f(8)] = [-1, 62]$. Since

$$D_x f^{-1}(x) = D_x(\sqrt{x + 2}) = \frac{1}{2}(x + 2)^{-1/2} > 0,$$

f^{-1} is also an increasing and continuous function on its domain $[-1, 62]$.   □

The final result concerning increasing functions and their inverses adds the hypothesis of differentiability of the function to produce the conclusion of differentiability for the inverse function. This theorem and its counterpart for decreasing functions are used repeatedly throughout this chapter and together are called the *inverse function theorem*.

**(7.7)
THEOREM**

The Inverse Function Theorem   If f is increasing and differentiable on $[a, b]$ and $D_x f(x) \neq 0$, then f^{-1} is increasing and differentiable on $[f(a), f(b)]$. If $x = f^{-1}(y)$ on $[f(a), f(b)]$, then

$$D_y f^{-1}(y) = \frac{1}{D_x f(x)}.$$

PROOF

Fix y_0 in $[f(a), f(b)]$. To find $D_y f^{-1}(y_0)$ we must determine

$$\lim_{y \to y_0} \frac{f^{-1}(y) - f^{-1}(y_0)}{y - y_0}.$$

With $y = f(x)$ and $y_0 = f(x_0)$, this becomes

$$D_y f^{-1}(y) = \lim_{y \to y_0} \frac{f^{-1}(y) - f^{-1}(y_0)}{y - y_0}$$

$$= \lim_{y \to y_0} \frac{x - x_0}{f(x) - f(x_0)}.$$

But f is differentiable on $[a, b]$, so f is also continuous on $[a, b]$. By Theorem 7.5, this implies that f^{-1} is continuous on $[f(a), f(b)]$. Therefore,

$$\lim_{y \to y_0} x = \lim_{y \to y_0} f^{-1}(y) = f^{-1}(y_0) = x_0,$$

so $\qquad D_y f^{-1}(y_0) = \displaystyle\lim_{y \to y_0} \frac{x - x_0}{f(x) - f(x_0)}$

$$= \lim_{x \to x_0} \frac{x - x_0}{f(x) - f(x_0)}$$

$$= \frac{1}{\displaystyle\lim_{x \to x_0} \frac{f(x) - f(x_0)}{x - x_0}}$$

$$= \frac{1}{D_x f(x_0)}, \text{ provided } D_x f(x_0) \neq 0. \qquad \square$$

(7.8)
COROLLARY

If f is decreasing and differentiable on $[a, b]$ and $D_x f(x) \neq 0$, then f^{-1} is decreasing and differentiable on $[f(b), f(a)]$. If $x = f^{-1}(y)$, then

$$D_y f^{-1}(y) = \frac{1}{D_x f(x)}.$$

The equation in Theorem 7.7 and Corollary 7.8 is often expressed using differential notation as

$$\frac{dx}{dy} = \frac{1}{\dfrac{dy}{dx}}.$$

EXAMPLE 4   If $f(x) = x^3 + 1$, find $(f^{-1})'(9)$.

SOLUTION

Let $y = x^3 + 1$. It follows from the inverse function theorem that

$$(f^{-1})'(y) = D_y f^{-1}(y) = \frac{1}{D_x f(x)} = \frac{1}{3x^2} = \frac{1}{3(y - 1)^{2/3}},$$

so

$$(f^{-1})'(9) = \frac{1}{3(9 - 1)^{2/3}} = \frac{1}{12}.$$

To determine $(f^{-1})'(9)$, we did not need an explicit expression for $(f^{-1})'(y)$. Since $y = x^3 + 1$ and $y = 9$ imply that $x = 2$,

$$(f^{-1})'(y) = \frac{1}{3x^2}, \text{ so } (f^{-1})'(9) = \frac{1}{3 \cdot 2^2} = \frac{1}{12}. \qquad \square$$

EXERCISE SET 7.1

In Exercises 1 through 6, find an inverse function of f, if one exists, and state the domain of the inverse function. Sketch the graph of f and the inverse function on the same set of axes.

1. $f(x) = 3x + 4$ **2.** $f(x) = 1 - 2x$

3. $f(x) = x^3$ **4.** $f(x) = 2x^3 + 1$

5. $f(x) = x^2 + 4$ **6.** $f(x) = x^2 - 1$

In Exercises 7 through 12, a function is described that is not one-to-one, so no inverse function exists. Determine a new function that is one-to-one by modifying the domain of the given function. Sketch the graph of the new function and its inverse on the same set of axes.

7. $f(x) = x^2 - 4$ **8.** $f(x) = 2x^2 + 1$

9. $f(x) = |x + 1|$ **10.** $f(x) = \cos x$

11. $f(x) = \sin x$ **12.** $f(x) = \tan x$

In Exercises 13 through 20, a differentiable one-to-one function is described on a particular interval. Find $(f^{-1})'$ directly and also by using the inverse function theorem.

13. $f(x) = 3x - 1, \, (-\infty, \infty)$ **14.** $f(x) = -2x + 5, \, (-\infty, \infty)$

15. $f(x) = x^3, \, [1, 8]$ **16.** $f(x) = x^2 + 3, \, [1, 9]$

17. $f(x) = \dfrac{1}{x}, \, (0, \infty)$ **18.** $f(x) = \dfrac{1}{x}, \, (-\infty, 0)$

19. $f(x) = \dfrac{x + 1}{x}, \, [1, 2]$ **20.** $f(x) = \dfrac{x - 4}{x + 1}, \, [0, 1]$

In Exercises 21 through 26, use the inverse function theorem to find the slope of the line tangent to the graph of f^{-1} at the point $(a, f^{-1}(a))$.

21. $f(x) = x^3 + x + 1, \, (3, 1)$ **22.** $f(x) = 2x^5 + 3x^2, \, (5, 1)$

23. $f(x) = \sqrt{x + 4} + x^3, \, (2, 0)$ **24.** $f(x) = \dfrac{x^3 + 8}{x^5 - 1}, \, (0, -2)$

25. $f(x) = \cos x, \, \left(\dfrac{1}{2}, \dfrac{\pi}{3}\right)$

26. $f(x) = \sin x, \, \left(-\dfrac{\sqrt{3}}{2}, -\dfrac{\pi}{3}\right)$

Show that the functions f described in Exercises 27 through 32 have the property that $f = f^{-1}$.

27. $f(x) = x$ **28.** $f(x) = -x$

29. $f(x) = \dfrac{1}{x}$ **30.** $f(x) = -\dfrac{1}{x}$

31. $f(x) = \sqrt{1 - x^2}, \, 0 \le x \le 1$ **32.** $f(x) = -\sqrt{4 - x^2}, \, -2 \le x \le 0$

33. By examining the graphs of the functions described in Exercises 27 through 32, make a conjecture concerning the symmetry of a function f with the property that $f = f^{-1}$.

34. Suppose f is the linear function described by $f(x) = ax + b$, $a \neq 0$. Show that f^{-1} exists and is also a linear function.

35. Suppose f is a linear function described by $f(x) = ax + b$ and that $f = f^{-1}$. What can be said about a and b?

Each of the functions in Exercises 36 through 38 has an inverse. Sketch the graph of the function and its inverse and define $f^{-1}(x)$.

36. $f(x) = \begin{cases} x, & \text{if } x < 0 \\ x^2, & \text{if } 0 \leq x \leq 1 \\ \sqrt{x}, & \text{if } 1 < x \end{cases}$ **37.** $f(x) = \begin{cases} \dfrac{1}{x^2}, & \text{if } x < 0 \\ 0, & \text{if } x = 0 \\ -\dfrac{1}{x^2}, & \text{if } x > 0 \end{cases}$

38. $f(x) = \begin{cases} -x + 2, & \text{if } x < 1 \\ \dfrac{1}{x}, & \text{if } x \geq 1 \end{cases}$

39. Describe a function T that converts temperature readings given in Fahrenheit into readings in Celsius. Determine T^{-1}. What application does T^{-1} have? What do T' and $(T^{-1})'$ describe?

40. Suppose the lines tangent to the graph of f at (a, b) and to the graph of f^{-1} at (b, a) intersect. Show that the intersection point must lie on the line $y = x$.

41. Show that the following formula holds, provided that the necessary differentiability conditions on f and f^{-1} are satisfied and that $D_x f(x) \neq 0$:

$$D_y^2 f^{-1}(y) = \frac{-D_x^2 f(x)}{[D_x f(x)]^3}.$$

42. Suppose f is a function defined by $f(x) = x^k$ for some rational number k. What values are possible for k if f has an inverse and $f(x) = f^{-1}(x)$ for all x in the domain of f?

43. Show that an even function cannot be one-to-one unless its domain contains only the number zero.

44. Give an example of a one-to-one odd function with domain $(-\infty, \infty)$.

45. Give an example of a non one-to-one odd function with domain $(-\infty, \infty)$.

46. Find a function f with the property that whenever g is defined by $g(x) = f(x) + c$ for any constant c, then $g = g^{-1}$.

47. Find a function f with the property that whenever g is defined by $g(x) = cf(x)$ for any nonzero constant c, then $g = g^{-1}$.

7.2
INVERSE TRIGONOMETRIC FUNCTIONS

The graph of the sine function shown in Figure 7.2(b) on p. 384 indicates that this function has no inverse. However, if we restrict the domain of the sine function to the interval $[-\pi/2, \pi/2]$, the restricted function is increasing and has as its range the interval $[-1, 1]$, the range of the sine function. For this restricted function we can define an inverse called the **inverse sine** or **arcsine function**, denoted arcsin. (The inverse sine is denoted in some sources by $\sin^{-1}$.)

**(7.9)
DEFINITION**

The **arcsine function** is defined by

$$\arcsin x = y \qquad \text{for } -1 \leq x \leq 1$$

if and only if

$$x = \sin y \quad \text{and} \quad -\frac{\pi}{2} \leq y \leq \frac{\pi}{2}.$$

The graph of the arcsine function is simply the reflection about the line $y = x$ of the graph of the restricted sine function. These graphs are shown in Figure 7.7.

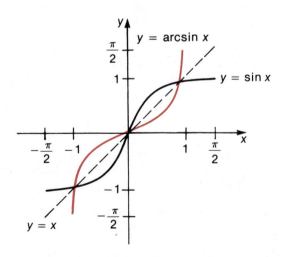

FIGURE 7.7

EXAMPLE 1   Find arcsin 1, arcsin (-1) and arcsin 1/2.

SOLUTION

$\sin \dfrac{\pi}{2} = 1$ and $\dfrac{\pi}{2}$ is in $\left[-\dfrac{\pi}{2}, \dfrac{\pi}{2} \right]$,   so arcsin $1 = \dfrac{\pi}{2}$.

$\sin -\dfrac{\pi}{2} = -1$ and $-\dfrac{\pi}{2}$ is in $\left[-\dfrac{\pi}{2}, \dfrac{\pi}{2} \right]$,   so arcsin $(-1) = -\dfrac{\pi}{2}$.

$\sin \dfrac{\pi}{6} = \dfrac{1}{2}$ and $\dfrac{\pi}{6}$ is in $\left[-\dfrac{\pi}{2}, \dfrac{\pi}{2} \right]$,   so arcsin $\dfrac{1}{2} = \dfrac{\pi}{6}$. $\square$

The inverse function theorem provides the means for determining the derivative of the arcsine function. Since

$$y = \arcsin x \quad \text{if and only if} \quad x = \sin y \quad \text{and} \quad -\frac{\pi}{2} \le y \le \frac{\pi}{2},$$

$$D_x \arcsin x = \frac{dy}{dx} = \frac{1}{\dfrac{dx}{dy}} = \frac{1}{D_y \sin y} = \frac{1}{\cos y}, \quad -\frac{\pi}{2} \le y \le \frac{\pi}{2}.$$

However, $(\cos y)^2 + (\sin y)^2 = 1$

so $\cos y = \pm \sqrt{1 - (\sin y)^2} = \pm \sqrt{1 - x^2}.$

But y is in $[-\pi/2, \pi/2]$ and in this interval $\cos y \ge 0$. Thus,

(7.10) $D_x \arcsin x = \dfrac{1}{\sqrt{1 - x^2}},$

provided $|x| < 1$.

The indefinite integral formula associated with this derivative result is

(7.11) $\displaystyle\int \frac{dx}{\sqrt{1 - x^2}} = \arcsin x + C.$

EXAMPLE 2   Evaluate $\displaystyle\int \frac{dx}{\sqrt{4 - x^2}}$.

SOLUTION

$$\int \frac{dx}{\sqrt{4 - x^2}} = \int \frac{dx}{\sqrt{4 \left(1 - \dfrac{x^2}{4} \right)}} = \frac{1}{2} \int \frac{dx}{\sqrt{1 - \dfrac{x^2}{4}}}.$$

If $u = x/2$, then $du = dx/2$ and

$$\int \frac{dx}{\sqrt{4 - x^2}} = \frac{1}{2} \int \frac{2 \, du}{\sqrt{1 - u^2}} = \int \frac{du}{\sqrt{1 - u^2}}$$

$$= \arcsin u + C$$

$$= \arcsin \left(\frac{x}{2} \right) + C. \qquad \square$$

By making the same type of variable substitution as illustrated in Example 2, it follows that for any $a > 0$,

(7.12)
$$\int \frac{dx}{\sqrt{a^2 - x^2}} = \arcsin\left(\frac{x}{a}\right) + C.$$

General integration formulas will not always be stated. They are, however, the type of formulas given in integral tables such as those appearing on the inside covers of this book.

By using similar domain restrictions on the other trigonometric functions we can construct additional inverse trigonometric functions. The domain of the cosine function is restricted to $[0, \pi]$, because the function is one-to-one on this interval and has range $[-1, 1]$.

(7.13)
DEFINITION

The **inverse cosine** or **arccosine function,** arccos, is defined by

arccos $x = y$ for $-1 \le x \le 1$ if and only if $x = \cos y$ and $0 \le y \le \pi$.

The graphs of the arccosine and restricted cosine functions are shown in Figure 7.8.

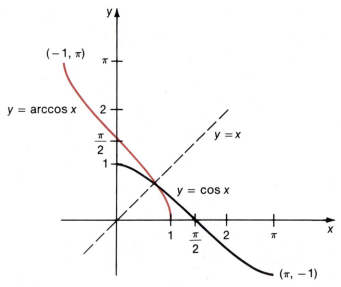

FIGURE 7.8

EXAMPLE 3   Find $\cos [\arcsin (1/2)]$ and $\sin [\arccos (-1/2)]$.

SOLUTION

Let $y = \arcsin (1/2)$. Then, since $\sin y = 1/2$ and $-\pi/2 \le y \le \pi/2$, $y = \pi/6$. Thus,

$$\cos\left[\arcsin\left(\frac{1}{2}\right) \right] = \cos\left(\frac{\pi}{6}\right) = \frac{\sqrt{3}}{2}.$$

If $y = \arccos(-1/2)$, then, since $\cos y = -1/2$ and $0 \le y \le \pi$, $y = 2\pi/3$. Thus,

$$\sin\left[\arccos\left(-\frac{1}{2}\right)\right] = \sin\frac{2\pi}{3} = \frac{\sqrt{3}}{2}. \qquad \square$$

EXAMPLE 4 Find $\cos[\arccos(4/5) + \arcsin(3/5)]$.

SOLUTION

The trigonometric identity $\cos(a + b) = \cos a \cos b - \sin a \sin b$ implies that

$$\cos[\arccos(4/5) + \arcsin(3/5)] = \cos[\arccos(4/5)] \cdot \cos[\arcsin(3/5)]$$
$$- \sin[\arccos(4/5)] \cdot \sin[\arcsin(3/5)].$$

Letting $y = \arcsin(3/5)$ implies that $\sin y = 3/5$. Thus,

$$\cos\left[\arcsin\left(\frac{3}{5}\right)\right] = \cos y = \sqrt{1 - (\sin y)^2} = \sqrt{1 - \left(\frac{3}{5}\right)^2} = \frac{4}{5}.$$

Similarly,

$$\sin\left[\arccos\left(\frac{4}{5}\right)\right] = \frac{3}{5}.$$

Thus, $\cos\left[\arccos\left(\frac{4}{5}\right) + \arcsin\left(\frac{3}{5}\right)\right] = \left(\frac{4}{5}\right)\left(\frac{4}{5}\right) - \left(\frac{3}{5}\right)\left(\frac{3}{5}\right) = \frac{7}{25}.$ $\square$

The inverse function theorem can also be used to show that

$$D_x \arccos x = -\frac{1}{\sqrt{1 - x^2}}.$$

The integral formula associated with this result is

$$\int \frac{1}{\sqrt{1 - x^2}}\, dx = -\arccos x + C.$$

Since we have already established that $\int \dfrac{1}{\sqrt{1 - x^2}}\, dx = \arcsin x + C$,

$$D_x \arcsin x = D_x(-\arccos x).$$

Consequently,

$$\arcsin x = -\arccos x + C$$

for some constant C. In fact, it is not difficult to show that $C = \pi/2$.

The main application of the inverse trigonometric functions is in the evaluation of integrals. Since the arcsine and the arccosine are used to evaluate the same type of integral, we need not consider the arccosine function further. The other inverse trigonometric functions that will be needed for integral evaluation are the arctangent and arcsecant functions.

(7.14)
DEFINITION

The **inverse tangent** or **arctangent function,** arctan, is defined by

$$\arctan x = y \qquad \text{for } -\infty < x < \infty$$

if and only if

$$x = \tan y \quad \text{and} \quad -\pi/2 < y < \pi/2.$$

The graphs of the arctangent and restricted tangent functions are shown in Figure 7.9.

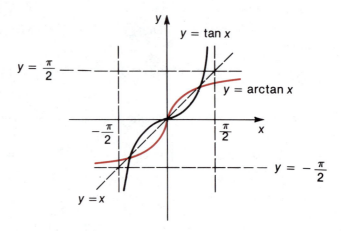

FIGURE 7.9

To find the derivative of the arc tangent function, we again use the inverse function theorem. Let $y = \arctan x$. Then $x = \tan y$ and $-\pi/2 < y < \pi/2$,

$$D_x \arctan x = \frac{dy}{dx} = \frac{1}{\dfrac{dx}{dy}} = \frac{1}{(\sec y)^2} \qquad \text{for } -\frac{\pi}{2} < y < \frac{\pi}{2}.$$

However, $(\sec y)^2 = 1 + (\tan y)^2 = 1 + x^2$; so

(7.15)
$$D_x \arctan x = \frac{1}{1 + x^2} .$$

The associated integral formula is

(7.16)
$$\int \frac{1}{1 + x^2} \, dx = \arctan x + C.$$

The next example illustrates how a variable substitution leads to the general integration formula

(7.17)
$$\int \frac{1}{a^2 + x^2} \, dx = \frac{1}{a} \arctan \left(\frac{x}{a} \right) + C, \, a > 0.$$

EXAMPLE 5

Find the area of the region bounded by the graph of $f(x) = \dfrac{1}{x^2 + 4}$, the x-axis, the y-axis, and the line $x = 2$.

SOLUTION

The region described is shown in Figure 7.10.

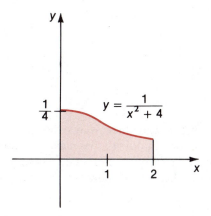

FIGURE 7.10

The area is given by

$$\int_0^2 \frac{dx}{x^2 + 4} = \frac{1}{4} \int_0^2 \frac{dx}{\left(\dfrac{x}{2}\right)^2 + 1} \, .$$

Let $u = \dfrac{x}{2}$; then $du = \dfrac{1}{2} dx$.

So
$$\int \frac{1}{x^2 + 4} \, dx = \frac{1}{2} \int \frac{1}{u^2 + 1} \, du$$
$$= \frac{1}{2} \arctan u + C = \frac{1}{2} \arctan \frac{x}{2} + C,$$

which agrees with formula (7.17) when $a = 2$. The area is

$$\int_0^2 \frac{1}{x^2 + 4} \, dx = \frac{1}{2} \arctan \frac{x}{2} \Big|_0^2 = \frac{1}{2}[\arctan 1 - \arctan 0]$$
$$= \frac{1}{2}\left[\frac{\pi}{4} - 0\right] = \frac{\pi}{8} \, . \qquad \square$$

EXAMPLE 6   Determine $\displaystyle \int \frac{dx}{x^2 + 2x + 2} \, .$

SOLUTION
$$\int \frac{dx}{x^2 + 2x + 2} = \int \frac{dx}{(x^2 + 2x + 1) + 1}$$
$$= \int \frac{dx}{(x + 1)^2 + 1} \, .$$

If $u = x + 1$, then $du = dx$ and

$$\int \frac{dx}{x^2 + 2x + 2} = \int \frac{dx}{(x + 1)^2 + 1}$$

$$= \int \frac{du}{u^2 + 1}$$

$$= \arctan u + C$$

$$= \arctan (x + 1) + C. \qquad \square$$

EXAMPLE 7   A night watchman wants the greatest possible view of the second story windows of a building as he makes his rounds. The windows are 4 feet high and begin 11 feet from the ground. Assuming the watchman's eyes are precisely 6 feet from the ground, how far from the building should the watchman walk?

SOLUTION

The situation is described in Figure 7.11. The problem is to find the value of x that will maximize the angle θ.

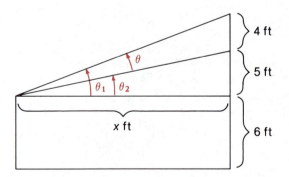

FIGURE 7.11

From the triangles in Figure 7.11, we see that

$$\tan \theta_1 = \frac{9}{x} \quad \text{and} \quad \tan \theta_2 = \frac{5}{x}.$$

Thus,

$$\tan \theta = \tan (\theta_1 - \theta_2) = \frac{\tan \theta_1 - \tan \theta_2}{1 + \tan \theta_1 \tan \theta_2}$$

$$= \frac{\dfrac{9}{x} - \dfrac{5}{x}}{1 + \dfrac{9}{x} \cdot \dfrac{5}{x}} = \frac{4x}{x^2 + 45}.$$

Consequently, $\quad \theta(x) = \arctan\left(\dfrac{4x}{x^2 + 45}\right).$

Since

$$\theta'(x) = \dfrac{1}{1 + \left(\dfrac{4x}{x^2 + 45}\right)^2} \; D_x\left(\dfrac{4x}{x^2 + 45}\right)$$

$$= \dfrac{(x^2 + 45)^2}{(x^2 + 45)^2 + 16x^2} \; \dfrac{4(x^2 + 45) - 4x(2x)}{(x^2 + 45)^2}$$

$$= \dfrac{4(45 - x^2)}{(x^2 + 45)^2 + 16x^2},$$

the only critical points occur when $x = \pm\sqrt{45}$.

When $-\sqrt{45} < x < \sqrt{45}$, $\theta'(x) > 0$, when $x > \sqrt{45}$, $\theta'(x) < 0$. A relative and absolute maximum occurs at $x = \sqrt{45}$ so the optimal distance from the wall for the watchman to walk is

$$x = \sqrt{45} \approx 6.71 \text{ feet.} \qquad\qquad \square$$

(7.18)
DEFINITION

The **inverse secant** or **arcsecant function**, arcsec, is defined by

$$\text{arcsec } x = y \quad \text{for } |x| > 1 \quad \text{if and only if} \quad \sec y = x \quad \text{and} \quad \begin{cases} 0 \le y \le \pi. \\ y \ne \dfrac{\pi}{2}. \end{cases}$$

The graphs of the arcsecant and restricted secant functions are shown in Figure 7.12.

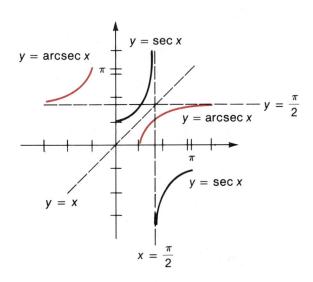

FIGURE 7.12

The graph of the restricted secant function shows clearly the reason for excluding the value of $\pi/2$ from the domain of the arcsecant function.

The derivative and general integral formulas for the arcsecant function are:

(7.19) $$D_x \operatorname{arcsec} x = \frac{1}{|x|\sqrt{x^2 - 1}}$$

and

(7.20) $$\int \frac{dx}{x\sqrt{x^2 - a^2}} = \frac{1}{a} \operatorname{arcsec} \left| \frac{x}{a} \right| + C, \qquad \text{provided that } a > 0.$$

EXAMPLE 8   Find $\displaystyle\int_{3\sqrt{2}}^{6} \frac{dx}{x\sqrt{x^2 - 9}}$.

SOLUTION

$$\int \frac{dx}{x\sqrt{x^2 - 9}} = \frac{1}{3} \operatorname{arcsec} \left| \frac{x}{3} \right| + C,$$

so

$$\int_{3\sqrt{2}}^{6} \frac{dx}{x\sqrt{x^2 - 9}} = \frac{1}{3} (\operatorname{arcsec} 2 - \operatorname{arcsec} \sqrt{2})$$

$$= \frac{1}{3} \left(\frac{\pi}{3} - \frac{\pi}{4} \right) = \frac{\pi}{36} . \qquad \square$$

Since the inverse cotangent and inverse cosecant functions do not add to our knowledge of integration techniques, we have deferred their definition and discussion to the exercises.

EXERCISE SET 7.2

Find the exact value of the expressions given in Exercises 1 through 14.

1. $\arccos \left(\dfrac{\sqrt{2}}{2} \right)$

2. $\arcsin (1)$

3. $\arcsin \left(\dfrac{\sqrt{3}}{2} \right)$

4. $\arccos \left(-\dfrac{\sqrt{2}}{2} \right)$

5. $\arctan (1)$

6. $\operatorname{arcsec} \left(\dfrac{2\sqrt{3}}{3} \right)$

7. arcsec (-1)

8. arctan $(\sqrt{3})$

9. $\sin\left[\arccos\left(\dfrac{5}{13}\right)\right]$

10. $\cos\left[\arcsin\left(\dfrac{1}{3}\right)\right]$

11. $\tan[\arccos(1)]$

12. $\sec\left[\arcsin\left(-\dfrac{\sqrt{2}}{2}\right)\right]$

13. $\sin\left[\arccos\left(\dfrac{1}{2}\right) + \arctan\left(\dfrac{3}{4}\right)\right]$

14. $\tan\left[\arctan(2) + \arccos\left(\dfrac{1}{4}\right)\right]$

15. If $y = \arcsin(1)$, find

    (a) $\sin y$ (b) $\cos y$ (c) $\sec y$ (d) $\cot y$

16. If $y = \arccos\left(\dfrac{1}{2}\right)$, find

    (a) $\sin y$ (b) $\sec y$ (c) $\tan y$ (d) $\csc y$

17. If $y = \arctan(-1)$, find

    (a) $\sin y$ (b) $\cos y$ (c) $\cot y$ (d) $\csc y$

18. If $y = \text{arcsec}(\sqrt{2})$, find

    (a) $\cos y$ (b) $\csc y$ (c) $\cot y$ (d) $\tan y$

Find the derivative of the functions described in Exercises 19 through 26.

19. $f(x) = \arcsin(x + 1)$

20. $f(x) = \arccos(x + 1)$

21. $f(x) = \arctan(x^2 - x + 3)$

22. $f(x) = \dfrac{\arcsin(3x + 5)}{x + 2}$

23. $f(x) = \arcsin(x \arccos x)$

24. $f(x) = \dfrac{\arcsin x}{\arccos x}$

25. $f(x) = \sin x \arcsin x$

26. $f(x) = \sin(\arcsin x)$

Evaluate the integrals in Exercises 27 through 44.

27. $\displaystyle\int \dfrac{dx}{x^2 + 36}$

28. $\displaystyle\int \dfrac{dx}{\sqrt{1 - 9x^2}}$

29. $\displaystyle\int_0^1 \dfrac{dx}{\sqrt{4 - x^2}}$

30. $\displaystyle\int \dfrac{dx}{x\sqrt{x^2 - 16}}$

31. $\displaystyle\int \dfrac{dx}{4x^2 + 9}$

32. $\displaystyle\int \dfrac{dx}{(x + 3)^2 + 1}$

33. $\displaystyle\int \dfrac{dx}{x\sqrt{16x^2 - 1}}$

34. $\displaystyle\int_3^4 \dfrac{dx}{\sqrt{x^4 - x^2}}$

35. $\displaystyle\int \dfrac{dx}{x\sqrt{x^4 - 1}}$

36. $\displaystyle\int \dfrac{2x\,dx}{\sqrt{1 - x^4}}$

37. $\displaystyle\int_{1/4}^{1/2} \dfrac{dx}{\sqrt{x - x^2}}$

38. $\displaystyle\int_2^\infty \dfrac{dx}{x\sqrt{x^2 - 1}}$

39. $\displaystyle\int_0^1 \frac{x}{x^2 + 1} \, dx$

40. $\displaystyle\int \frac{2x + 3}{x^2 + 4} \, dx$

41. $\displaystyle\int \frac{5 - x}{\sqrt{25 - x^2}} \, dx$

42. $\displaystyle\int \frac{2x + 1}{x^2 + x + 3} \, dx$

43. $\displaystyle\int \frac{dx}{x^2 + 4x + 8}$

44. $\displaystyle\int \frac{x \, dx}{x^2 + 4x + 8}$

Show that the inverse trigonometric functions satisfy the identities listed in Exercises 45 through 48.

45. $\arcsin(-x) = -\arcsin x$

46. $\arccos(-x) = \pi - \arccos x$

47. $\cos(\arcsin x) = \sqrt{1 - x^2}$

48. $\arcsin x = \arctan \dfrac{x}{\sqrt{1 - x^2}}$, if $-1 < x < 1$

Sketch the graphs of the functions described in Exercises 49 through 56.

49. $y = 2 \arcsin x$

50. $y = \dfrac{1}{\pi} \operatorname{arccot} x$

51. $y = \arcsin(2x)$

52. $y = -\arctan x$

53. $y = \dfrac{\pi}{2} + \arcsin x$

54. $y = \dfrac{\pi}{2} + \arctan x$

55. $y = \arctan \dfrac{1}{x}$

56. $y = \arctan x + \arctan \dfrac{1}{x}$

Use L'Hôpital's rule to find the limits in Exercises 57 through 60.

57. $\displaystyle\lim_{x \to 0} \frac{\arctan x}{x}$

58. $\displaystyle\lim_{x \to 0} \frac{\arcsin x}{\dfrac{\pi}{2} - \arccos x}$

59. $\displaystyle\lim_{x \to 0^+} \csc x \arcsin x$

60. $\displaystyle\lim_{x \to \infty} \left(\frac{\pi}{2} - \arctan x\right) \cot\left(\frac{1}{x}\right)$

61. (a) Find $D_x \left[\arctan \dfrac{1}{x} \right]$.

(b) Use the result of (a) and $D_x[\arctan x]$ to show that if $x \neq 0$, then

$$\arctan x + \arctan \frac{1}{x} = C$$

for some constant C.

(c) Use the graph in Exercise 56 to determine the value of C.

62. Find the area of the region bounded by the y-axis, the x-axis, and the graph of $f(x) = \dfrac{1}{x^2 + 1}$.

63. In high school geometry one is told that the circumference of a circle of radius r is $2\pi r$. Verify the truth of this statement by finding the length of the curve that is the graph of $x^2 + y^2 = r^2$.

64. A Coast Guard cutter is sitting 100 meters from a straight shoreline. A searchlight on the ship revolves at 8 revolutions per minute. Find the rate at which the light beam is moving down the shore when the beam is 200 meters from the point on shore that is closest to the cutter.

65. A museum plans to hang a rare masterpiece for public viewing. The painting is 5 feet high and the viewers pass the picture six feet from the wall on which the painting will be mounted. How high should the painting be mounted in order that the average 5-foot 6-inch eyelevel viewer will be afforded the best view? (The best view occurs when the angle formed by the viewer's eyes and the top and bottom of the picture is a maximum.)

66. Suppose that the painting discussed in Exercise 65 is mounted with its base 8 ft above the floor. How far should the average viewer stand from the painting to ensure the best view of the painting if there is no restriction on the viewer's distance from the painting?

67. Farmer MacDonald has three pieces of straight fencing each 10 feet long, to use for fencing pigs along the side of a 120-foot barn. One side of the fence is to be parallel to the side of the barn. What angle should the fence make with the side of the barn to maximize the area enclosed?

68. Show that $\sin (\cos x) < \cos (\sin x)$ but that $\cos (\arcsin x) < \arcsin (\cos x)$, $0 \le x \le 1$.

69. Show that $\displaystyle\int_0^{\pi/2} \sin x \, dx + \int_0^1 \arcsin x \, dx = \frac{\pi}{2}$, without evaluating the integrals.

Putnam exercise:

70. Prove

$$\frac{22}{7} - \pi = \int_0^1 \frac{x^4(1 - x)^4}{1 + x^2} \, dx.$$

(This exercise was problem A–1 of the twenty-ninth William Lowell Putnam examination given on December 7, 1968. The examination and its solution can be found in the October 1969 issue of the *American Mathematical Monthly*, pages 911–915.)

7.3
EXPONENTIAL FUNCTIONS

The natural logarithm function, ln, is increasing and differentiable on its domain $(0, \infty)$. The inverse function theorem implies that ln has an inverse function that is differentiable and increasing on $(-\infty, \infty)$, the range of the natural logarithm function. This inverse function, called the *natural exponential function,* is a likely candidate for the function most important for calculus applications.

Before beginning the study of the exponential function, we summarize the properties of the natural logarithm function, considered in Section 5.7.
1. Domain of ln is $(0, \infty)$, range of ln is $(-\infty, \infty)$
2. $\ln 1 = 0$
3. $D_x \ln |x| = \dfrac{1}{x}$, for $x \neq 0$

If a and b are positive real numbers and r is a rational number, then
4. $\ln ab = \ln a + \ln b$
5. $\ln \dfrac{a}{b} = \ln a - \ln b$
6. $\ln a^r = r \ln a$.

(7.21)
DEFINITION

The **natural exponential function**, exp, is the inverse function for the natural logarithm function, ln. As such:
(i) $\exp x = y$ if and only if $\ln y = x$;
(ii) the domain of exp is $(-\infty, \infty)$;
(iii) the range of exp is $(0, \infty)$.

Immediate implications of this definition are that

(7.22) $\exp(\ln x) = x,$ whenever $x > 0$

and

(7.23) $\ln(\exp x) = x,$ for all real numbers x.

Because of the relationship between the graph of a function and its inverse, the graph of exp is readily obtained and shown in Figure 7.13.

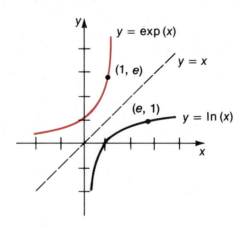

FIGURE 7.13

EXAMPLE 1   Find exp 0 and exp 1.

SOLUTION

Using (7.23), ln (exp 0) = 0. Thus exp 0 is the unique number whose natural logarithm is zero:

$$\exp 0 = 1.$$

Also, ln (exp 1) = 1, and in Section 5.7 we found that the real number whose natural logarithm is 1 is denoted $e \approx 2.71828$. So

$$\exp (1) = e.$$

The inverse relationship with the natural logarithm function can be used to show that the function exp satisfies the properties usually associated with exponentials.

(7.24)
THEOREM

If a and b are any real numbers, then

(i) $\exp (a + b) = \exp a \cdot \exp b$

(ii) $\exp (a - b) = \dfrac{\exp a}{\exp b}$

(iii) $\exp (ar) = [\exp a]^r$, for any rational number r.

PROOF

Let $M = \exp a$ and $N = \exp b$; then $\ln M = a$ and $\ln N = b$.

(i)   Since $a + b = \ln M + \ln N = \ln MN$, Definition 7.21 implies that

$$\exp(a + b) = MN = \exp a \cdot \exp b.$$

(ii) Similarly, $a - b = \ln M - \ln N = \ln \left(\dfrac{M}{N}\right)$

so $\exp(a - b) = \dfrac{M}{N} = \dfrac{\exp a}{\exp b}.$

(iii) Also,

$$ar = r \ln M = \ln M^r,$$

so $\exp(ar) = M^r = [\exp a]^r.$ □

In algebra courses we learn that for any positive real number a and any rational number $x = p/q$, we can define a^x by

$$a^x = a^{\frac{p}{q}} = \sqrt[q]{a^p},$$

where p and q are integers and $q \neq 0$. This definition, however, is inappropriate for defining a^x when x is an irrational number. (For example, what is meant by 2^π or $3^{\sqrt{2}}$?) Until the definition of a^x is extended to include irrational as well as rational numbers, calculus cannot be used to study exponentials. Without this extension, an exponential function is not defined on an interval, so the concepts of continuity and differentiability cannot be applied. The natural exponential function is used to define this extension.

(7.25)
DEFINITION

For any positive real number a and any real number x, we define $a^x = \exp(x \ln a)$. For each fixed positive real number $a \neq 1$, $f(x) = a^x$ describes a function with domain $(-\infty, \infty)$ and range $(0, \infty)$. This function is called the **exponential function with base** a. (When $a = 1$, $a^x \equiv 1$, which describes a constant function.)

The general shape of the graph of an exponential function with base a is shown in Figure 7.14.

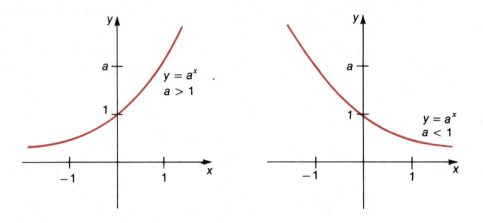

FIGURE 7.14

Notice that when $x = \dfrac{p}{q}$ is a rational number, this definition reduces to the original definition.

$$a^{\frac{p}{q}} = \exp\left(\frac{p}{q} \ln a\right) = \exp\left(\frac{1}{q} \ln a^p\right)$$

$$= \exp\left(\ln \sqrt[q]{a^p}\right) = \sqrt[q]{a^p}.$$

We can now show that the logarithm property $\ln a^r = r \ln a$ holds for irrational as well as rational numbers r. If x is any real number and a any positive real number, then

$$\ln a^x = \ln \left(\exp\left(x \ln a\right)\right) = x \ln a.$$

The following theorem shows that the exponential functions satisfy the usual properties associated with exponentials.

(7.26)
THEOREM

For any real numbers x, y, and a with $a > 0$,

(i) $a^{x+y} = a^x a^y$

(ii) $a^{x-y} = \dfrac{a^x}{a^y}$

(iii) $(a^x)^y = a^{xy}$

PROOF

These properties follow almost immediately from the relations in Theorem 7.24. We will prove only part (i) and leave the proofs of parts (ii) and (iii) as exercises.

$$
\begin{aligned}
\text{(i) } a^{x+y} &= \exp\left((x+y)\ln a\right) \\
&= \exp\left(x\ln a + y\ln a\right) \\
&= \exp\left(x\ln a\right)\exp\left(y\ln a\right) \quad \text{(by Theorem 7.24(i))} \\
&= a^{x}a^{y}.
\end{aligned}
$$
$\square$

For the particular case when the base in the exponential function is e (the real number whose natural logarithm is 1),

$$
e^{x} = \exp(x\ln e) = \exp x.
$$

The notation e^{x} is the most common way of describing the natural exponential function and the one we will generally use in the remainder of this text. However, most computer programming languages use the exp notation for expressing the natural exponential function, so you should keep this original notation in mind.

Using the e^{x} notation for the natural exponential function we can rewrite equations (7.22) and (7.23) as

(7.27) $$e^{\ln x} = x, \qquad \text{whenever } x > 0$$

and

(7.28) $$\ln(e^{x}) = x, \qquad \text{for all real numbers } x.$$

The results in Theorem 7.24 can be rewritten as

$$
e^{a+b} = e^{a}e^{b}
$$

$$
e^{a-b} = \frac{e^{a}}{e^{b}}
$$

and

$$
e^{ar} = (e^{a})^{r}.
$$

The exponential function with base a can be written

$$
a^{x} = e^{x\ln a}.
$$

The result given by the next theorem is one of the primary reasons for the importance of the natural exponential function in calculus applications.

(7.29)
THEOREM

For any real number x, $D_{x}e^{x} = e^{x}$.

PROOF

Let $y = e^{x}$, then $x = \ln y$. It follows from the inverse function theorem that

$$
\frac{dy}{dx} = \frac{1}{\dfrac{dx}{dy}} = \frac{1}{\dfrac{1}{y}} = y = e^{x}.
$$

Consequently, $D_{x}e^{x} = e^{x}$.
$\square$

EXAMPLE 2   Find $D_x e^{x^2}$.

SOLUTION
$$D_x e^{x^2} = e^{x^2} D_x x^2 = 2x e^{x^2}. \qquad \square$$

EXAMPLE 3   Find $D_x e^{x \sin x}$.

SOLUTION
$$\begin{aligned} D_x e^{x \sin x} &= e^{x \sin x} \cdot D_x (x \sin x) \\ &= e^{x \sin x}(\sin x + x\cos x). \qquad \square \end{aligned}$$

EXAMPLE 4   Find $D_x \sin(e^{x^2})$.

SOLUTION
$$\begin{aligned} D_x \sin(e^{x^2}) &= \cos(e^{x^2}) \cdot D_x e^{x^2} \\ &= \cos(e^{x^2}) \cdot e^{x^2} \cdot 2x = 2x e^{x^2} \cos(e^{x^2}). \qquad \square \end{aligned}$$

The integral formula associated with the derivative result for the natural exponential function is

(7.30)
$$\int e^x \, dx = e^x + C.$$

EXAMPLE 5   Evaluate $\int e^{4x-1} dx$.

SOLUTION
Let $u = 4x - 1$, then $du = 4dx$ and
$$\begin{aligned} \int e^{4x-1} dx &= \int e^u \frac{1}{4} \, du \\ &= \frac{1}{4} e^u + C \\ &= \frac{1}{4} e^{4x-1} + C. \qquad \square \end{aligned}$$

EXAMPLE 6   Evaluate $\int_0^\pi e^{3 \sin x} \cos x \, dx$.

SOLUTION

Let $u = 3 \sin x$, then $du = 3 \cos x \, dx$ so

$$\int e^{3 \sin x} \cos x \, dx = \int e^u \frac{1}{3} \, du$$

$$= \frac{1}{3} e^u + C$$

$$= \frac{1}{3} e^{3 \sin x} + C.$$

Thus,

$$\int_0^\pi e^{3 \sin x} \cos x \, dx = \frac{1}{3} [e^{3 \sin \pi} - e^{3 \sin 0}]$$

$$= \frac{1}{3} [e^0 - e^0] = 0. \qquad \square$$

Theorem 7.29 can also be used to deduce information about the derivative and integral of the general exponential functions.

(7.31)
COROLLARY

For any $a > 0$ and any real number x,

(i) $D_x a^x = a^x \ln a$

(ii) $\displaystyle\int a^x dx = \frac{a^x}{\ln a} + C$, when $a \neq 1$.

PROOF

(i)

$$D_x a^x = D_x e^{x \ln a}$$

$$= e^{x \ln a} D_x (x \ln a) \quad \text{(by Theorem 7.29)}$$

$$= e^{x \ln a} \ln a$$

$$= a^x \ln a.$$

(ii) Since $\ln a$ is a constant, part (i) implies that $\dfrac{a^x}{\ln a}$ is an antiderivative of a^x so,

$$\int a^x dx = \frac{a^x}{\ln a} + C. \qquad \square$$

We have seen that if n is any rational number, then

$$D_x(x^n) = nx^{n-1}.$$

Now that we can define x^n for all real numbers n, provided $x > 0$, we can extend this derivative result.

(7.32)
THEOREM

For any real number n,

$$D_x x^n = nx^{n-1}, \text{ for all } x > 0.$$

PROOF

Since $x^n = \exp(n \ln x) = e^{n \ln x}$,

$$D_x x^n = D_x(e^{n \ln x})$$
$$= e^{n \ln x} D_x(n \ln x)$$
$$= e^{n \ln x} \cdot \frac{n}{x}$$
$$= x^n \cdot \frac{n}{x} = nx^{n-1}. \qquad \square$$

EXAMPLE 7   Find $D_x x^{\sqrt{2}}$.

SOLUTION

$$D_x x^{\sqrt{2}} = \sqrt{2} x^{\sqrt{2} - 1}. \qquad \square$$

EXAMPLE 8   Find $D_x(\sin x)^\pi$.

SOLUTION

Using the chain rule, we obtain

$$D_x(\sin x)^\pi = \pi (\sin x)^{(\pi - 1)} D_x \sin x$$
$$= \pi (\sin x)^{(\pi - 1)} \cos x. \qquad \square$$

Definition (7.25) also enables us to consider functions of the form $[f(x)]^{g(x)}$, provided $f(x) > 0$, since

$$[f(x)]^{g(x)} = e^{g(x) \ln f(x)}.$$

EXAMPLE 9   Find $D_x x^x$, when $x > 0$.

SOLUTION

Since $x^x = e^{x \ln x}$,

$$D_x x^x = e^{x \ln x} \cdot D_x(x \ln x)$$
$$= x^x \left(\ln x + x \cdot \frac{1}{x} \right) = x^x(\ln x + 1). \qquad \square$$

EXERCISE SET 7.3

Find $D_x y$ in Exercises 1 through 26.

1. $y = e^{3x}$ **2.** $y = e^{x^2 + x}$

3. $y = e^{\sqrt{x}}$ **4.** $y = e^{-3x^2}$

5. $y = xe^{-x}$

6. $y = (e^x + e^{-x})^2$

7. $y = e^x \ln x$

8. $y = e^{2 \ln x}$

9. $y = e^x \sin x + e^x \cos x$

10. $y = (e^{x^2+1})^{1/2}$

11. $y = 2^{x^2}$

12. $y = 3^x \, 2^{x^2}$

13. $y = e^{\sin x}$

14. $y = (\sin x)^e$

15. $y = (3 + \pi)^x$

16. $y = x^{3 + \pi}$

17. $y = \ln(e^{x^2})$

18. $y = \ln(2e^x)$

19. $y = x^{\pi + \sqrt{2}}$

20. $y = (\sin x)^x$

21. $y = x^{\sin x}$

22. $y = (\sin x)^{\sin x}$

23. $ye^x + xe^y = 1$

24. $x^3 e^y + \sin y \, e^x = 2$

25. $e^{xy} = \ln x$

26. $e^{xy} = e^x e^{-y} + e^y e^{-x}$

Evaluate the integrals in Exercises 27 through 38.

27. $\displaystyle\int e^{3x-2} dx$

28. $\displaystyle\int xe^{x^2} dx$

29. $\displaystyle\int_0^1 \frac{e^x + e^{-x}}{2} dx$

30. $\displaystyle\int \frac{e^x + 1}{e^{2x}} dx$

31. $\displaystyle\int \frac{e^x}{e^x + 1} dx$

32. $\displaystyle\int \frac{e^x - e^{-x}}{e^x + e^{-x}} dx$

33. $\displaystyle\int \frac{e^x + e^{-x}}{e^x - e^{-x}} dx$

34. $\displaystyle\int_0^1 xe^{2-x^2} dx$

35. $\displaystyle\int e^{2x} e^{5x} dx$

36. $\displaystyle\int \frac{e^x}{e^{2x} + 1} dx$

37. $\displaystyle\int \frac{e^x}{\sqrt{1 - e^{2x}}} dx$

38. $\displaystyle\int \frac{dx}{\sqrt{e^{2x} - 1}}$

Sketch the graphs of the functions described in Exercises 39 through 42.

39. $f(x) = e^x + e^{-x}$

40. $f(x) = e^x - e^{-x}$

41. $f(x) = 2^x + 3^x$

42. $f(x) = 2^x - 3^x$

Sketch the graph of each pair of functions f and g described in Exercises 43 through 48, and determine when $f = g^{-1}$.

43. $f(x) = \ln x^2, \; g(x) = \exp\left(\dfrac{x}{2}\right)$

44. $f(x) = \ln \dfrac{x}{2}, \; g(x) = \exp(2x)$

45. $f(x) = \ln |x|, \; g(x) = \exp|x|$

46. $f(x) = -\ln x, \; g(x) = \exp(-x)$

47. $f(x) = 1 + \ln x, \; g(x) = \exp(x - 1)$

48. $f(x) = 2 \ln x, \; g(x) = \dfrac{1}{2} \exp x$

Use L'Hôpital's rule to evaluate the limits in Exercises 49 through 54.

49. $\lim\limits_{x \to 0} \dfrac{e^x - 1}{x}$

50. $\lim\limits_{x \to \infty} \dfrac{e^x}{x^2}$

51. $\lim\limits_{x \to \infty} \dfrac{x^2}{e^x}$

52. $\lim\limits_{x \to \infty} \dfrac{\ln x}{x}$

53. $\lim\limits_{x \to 0} \dfrac{2^x - x - 1}{x^2}$

54. $\lim\limits_{x \to \infty} \dfrac{x}{3^x}$

55. Find an equation of the line tangent to the graph of $f(x) = \exp x$ that passes through the point $(3, 0)$.

56. Find an equation of the line tangent to the graph of $f(x) = \exp 2x$ that passes through the point $(2, 0)$.

57. Find the area of the region in the plane that is bounded by the x-axis, the y-axis, the line $x = 1$, and the graph of $y = e^x$.

58. Find the area of the region bounded by the graphs of $x = 0$, $y = e^x$ and $y = e^{2x} - 2$.

59. Find the area of the region bounded by the x-axis, the y-axis, the line $x = a$ and the graph of $f(x) = \dfrac{e^x}{e^{2x} + 1}$.

60. Show that the area of the region bounded by the x-axis, the lines $x = -a$ and $x = a$, and the graph of $f(x) = \dfrac{e^{|x|}}{e^{2|x|} + 1}$ is twice the area of the region found in Exercise 59.

61. Find the volume of the solid generated when the region bounded by the graph of $f(x) = e^{-x^2}$, the lines $x = 0$ and $x = 1$, and the x-axis is rotated about the y-axis. Attempt, for a short while, to find the area of this region.

62. Use the derivatives of the function described by $f(x) = e^x - \ln x$ to demonstrate that the graphs of $y = \ln x$ and $y = e^x$ cannot intersect.

63. Suppose $f(x) = x^x$ and $g(x) = f(2x)$. Find $g'(x)$.

64. Suppose $f'(x) = x^x$ and $g(x) = f(2x)$. Find $g'(x)$.

65. Does the graph of $f(x) = e^x(1 - x) + x^2$ cross the x-axis between $x = 0$ and $x = 1$?

66. Show that $f(x) = \left(1 + \dfrac{1}{x}\right)^x$ is an increasing function when x is positive. [*Hint*: Consider $\ln f(x)$.]

67. The Laplace transform of a continuous function f, described in the variable t, is a new function, described in the variable s, defined by

$$L(f) = F(s) = \int_0^\infty e^{-st}f(t)\, dt,$$

provided that this improper integral exists. Show that if the Laplace transforms of f and g exist and if a and b are constants, then

$$L\{af + bg\} = aL(f) + bL(g).$$

68. Use the definition of the Laplace transform given in Exercise 67 to find $L(e^{at})$, where a is a constant.

69. The study of how the time rate of a chemical reaction changes with respect to temperature was first done by Svante August Arrhenius (1859–1927) in 1889. It is generally assumed that the time rate k has the form

$$k = z e^{-Q/RT},$$

where T is the temperature, Q is the activation energy of the chemical reaction, R is the gas constant of the chemical substance, and z is a constant. Show that k satisfies the equation

$$\frac{d \ln k}{dT} = \frac{Q}{RT^2}.$$

70. Only a small proportion of seedlings planted on a piece of bare ground will reach maturity. In an experiment with foxglove plants, the number y surviving at time t (measured in months from the emergence of the seedlings) was found to be

$$y = 100e^{-0.2310t}.$$

(a) What was the original number of foxglove seedlings?

(b) What was the half-life of this group of seedlings? (That is, the time when half the original number of seedlings have died.)

(c) If it takes 15 months for the seedlings to flower, how many seedlings are likely to survive to flower?

71. The number n of *Drosophila Melanogaster* (fruit flies) in a colony after t days of breeding is given by the logistic function:

$$n = \frac{230}{1 + 6.9310e^{-0.1702t}}.$$

(a) What is the maximum number of flies in the colony?

(b) How many flies were originally in the colony?

(c) When would the number of flies be twice the original number?

72. The concentration $C(t)$ of a drug in the bloodstream of a patient is described by

$$C(t) = .03t\, e^{-.01t} \text{ milligrams per cubic centimeter,}$$

where t is measured in minutes after an injection. Use Newton's method to determine, to the nearest second, when the concentration is .001 mg/cm^3.

73. In 1916, Dubois and Dubois suggested that the surface area S, in square numbers, of a human body is related to the height H, in centimeters, and weight W, in kilograms, by the equation

$$S = .007184\, W^{.425} H^{.725}.$$

Suppose that a person 5 feet 6 inches tall is on a diet, losing 3 pounds per week. At what rate is her skin area changing when she weighs 118 pounds?

74. A drug is infused into the circulation system of a cat at time $t = 0$. The concentration $C(t)$ of the drug for $t > 0$ is given by

$$C(t) = c_1 e^{-k_1 t} + c_2 e^{-k_2 t},$$

where k_1, k_2, c_1, and c_2 are positive constants. Suppose that the volume V of the circulating fluid is constant. Determine the rate at which the amount of the drug is changing with respect to time. What is the limit of this rate as t approaches ∞?

Putnam exercise:

75. The graph of the equation $x^y = y^x$ in the first quadrant (i.e. the region where $x > 0$ and $y > 0$) consists of a straight line and a curve. Find the coordinates of the intersection point of the line and the curve.
(This exercise was problem 1, part I of the twenty-second William Lowell Putnam examination given on December 2, 1961. The examination and its solution can be found in the October 1962 issue of the *American Mathematical Monthly*, pages 762–767.)

7.4
GENERAL LOGARITHMIC FUNCTIONS

In the previous section we used the fact that the natural logarithm function, ln, has an inverse in order to define the natural exponential function, exp. From this definition, we defined the exponential function a^x for any positive base $a \neq 1$. We also found that $D_x a^x = a^x \ln a$, so

when $0 < a < 1$, $\ln a < 0$ and $D_x a^x < 0$;
when $a > 1$, $\ln a > 0$ and $D_x a^x > 0$.

The function f described by $f(x) = a^x$ is decreasing if $0 < a < 1$ and increasing when $a > 1$. In either case, $f(x) = a^x$ describes a differentiable function that, by the inverse function theorem, has a differentiable inverse. We define this inverse function to be the logarithm function with base a. (See Figure 7.15.)

(7.33)
DEFINITION

If a is a positive real number, $a \neq 1$, the logarithm function with base a, $\log_a$, is defined to be the inverse of the exponential function with base a:
(i) $\log_a x = y$ if and only if $x = a^y$;
(ii) the domain of $\log_a$ is $(0, \infty)$;
(iii) the range of $\log_a$ is $(-\infty, \infty)$.

The special case when $a = 10$ is called the *common logarithm* and is often considered in high school algebra and trigonometry courses.

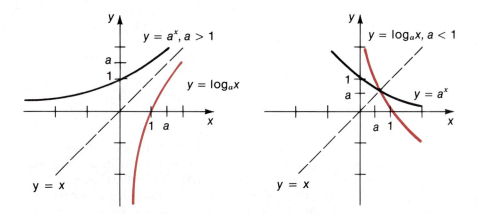

FIGURE 7.15

Since $\ln x = y$ if and only if $x = e^y$, Definition 7.33 implies that for all $x > 0$,

$$(7.34) \qquad \log_e x = \ln x.$$

An interesting relation exists between the logarithm function with base a and the natural logarithm, $\ln$. Let $M = \log_a x$; then $x = a^M = e^{M \ln a}$. Thus, $\ln x = M \ln a$, and

$$(7.35) \qquad \log_a x = \frac{\ln x}{\ln a}.$$

Properties of the natural logarithm function can now be used to derive similar properties for the general logarithm function. For any positive real number a, $a \neq 1$, and any $x > 0$,

$$(7.36) \qquad D_x \log_a x = \frac{1}{x \ln a}.$$

If x and y are any positive real numbers, then

(i) $\log_a(x\,y) = \log_a x + \log_a y,$

(ii) $\log_a \left(\dfrac{x}{y}\right) = \log_a x - \log_a y,$

(iii) $\log_a x^y = y \log_a x.$

EXAMPLE 1 Find $D_x \log_{10}(x^2 + 4x + 4)$.

SOLUTION

$$D_x \log_{10}(x^2 + 4x + 4) = D_x \left(\frac{\ln(x^2 + 4x + 4)}{\ln 10} \right)$$

$$= \frac{1}{(x^2 + 4x + 4)\ln 10} \frac{1}{\ln 10} D_x(x^2 + 4x + 4)$$

$$= \frac{1}{(x + 2)^2 \ln 10}(2x + 4)$$

$$= \frac{2}{(x + 2)\ln 10}. \qquad \square$$

We have seen a rather interesting transition through the logarithm and exponential functions. We began in Section 5.7 by defining a particular logarithm function, the natural logarithm, and used indirect methods to transfer information to a particular exponential function, the natural exponential. Having defined the natural logarithm and natural exponential functions, we used them to define general exponential functions. The general exponential functions were then used to define the general logarithm functions. The only explicit definitions in this entire procedure are those given by $\ln x = \int_1^x 1/t\, dt$ and $a^x = \exp(x \ln a)$. Figure 7.16 lists the steps that were performed and a few of the more important results obtained. The symbol $\leftrightarrow$ in this diagram is used in place of the phrase "if and only if."

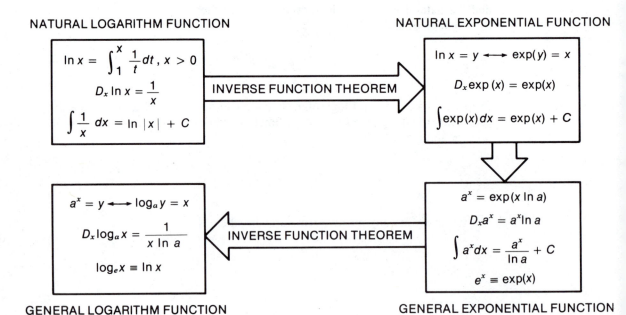

NATURAL LOGARITHM FUNCTION

$$\ln x = \int_1^x \frac{1}{t} dt, \ x > 0$$

$$D_x \ln x = \frac{1}{x}$$

$$\int \frac{1}{x} dx = \ln |x| + C$$

INVERSE FUNCTION THEOREM

NATURAL EXPONENTIAL FUNCTION

$$\ln x = y \longleftrightarrow \exp(y) = x$$

$$D_x \exp(x) = \exp(x)$$

$$\int \exp(x) dx = \exp(x) + C$$

GENERAL LOGARITHM FUNCTION

$$a^x = y \longleftrightarrow \log_a y = x$$

$$D_x \log_a x = \frac{1}{x \ln a}$$

$$\log_e x \equiv \ln x$$

INVERSE FUNCTION THEOREM

GENERAL EXPONENTIAL FUNCTION

$$a^x = \exp(x \ln a)$$

$$D_x a^x = a^x \ln a$$

$$\int a^x dx = \frac{a^x}{\ln a} + C$$

$$e^x \equiv \exp(x)$$

FIGURE 7.16

We close this section with an interesting representation of e that also introduces some new indeterminant forms. We first show that

$$e = \lim_{x \to \infty} (1 + 1/x)^x.$$

As x increases without bound, $1 + 1/x$ approaches 1, while the exponent x becomes infinite. This produces an indeterminate form of the type 1^∞. However

$$\left(1 + \frac{1}{x} \right)^x = e^{x \ln \left(1 + \frac{1}{x} \right)}$$

and the natural exponential function is continuous,

so
$$\lim_{x \to \infty} \left(1 + \frac{1}{x} \right)^x = \lim_{x \to \infty} e^{x \ln \left(1 + \frac{1}{x} \right)}$$
$$= e^{\lim_{x \to \infty} x \ln \left(1 + \frac{1}{x} \right)}.$$

Since $x \ln \left(1 + \frac{1}{x} \right)$ has the indeterminate form $\infty \cdot 0$, we rewrite this as

$$\frac{\ln \left(1 + \frac{1}{x} \right)}{\frac{1}{x}}$$

and apply L'Hôpital's rule.

$$\lim_{x \to \infty} x \ln \left(1 + \frac{1}{x} \right) = \lim_{x \to \infty} \frac{\ln \left(1 + \frac{1}{x} \right)}{\frac{1}{x}} = \lim_{x \to \infty} \frac{D_x \ln \left(1 + \frac{1}{x} \right)}{D_x \frac{1}{x}}$$

$$= \lim_{x \to \infty} \frac{\left(1 \Big/ \left(1 + \frac{1}{x} \right) \right) \left(-\frac{1}{x^2} \right)}{-\frac{1}{x^2}}$$

$$= \lim_{x \to \infty} \frac{x}{x + 1} = 1.$$

Hence,

(7.37) $$\lim_{x \to \infty} \left(1 + \frac{1}{x} \right)^x = e^{\lim_{x \to \infty} x \ln \left(1 + \frac{1}{x} \right)} = e^1 = e.$$

Approximations to e can consequently be found by evaluating $(1 + 1/x)^x$ for large values of x. The following table lists some approximations. Notice that x must be quite large before the approximation $(1 + 1/x)^x$ produces accurate results. (The correct value of e to the places listed is 2.7182818.)

x	$\left(1 + \dfrac{1}{x}\right)^x$
100	2.7048138
1,000	2.7169239
10,000	2.7181459
100,000	2.7182682
1,000,000	2.7182805

In the preceding discussion, we showed how to handle an indeterminate form of the type 1^∞. Other indeterminate forms that arise from expressions of the type $[f(x)]^{g(x)}$ are 0^0 and ∞^0. These can be handled similarly, as illustrated in the next examples.

EXAMPLE 2   Find $\lim\limits_{x \to \infty} x^{1/x}$.

SOLUTION

This illustrates the indeterminate form ∞^0. Since $x^{1/x} = e^{(1/x)\ln x}$ and the natural exponential function is continuous,

$$\lim_{x \to \infty} x^{1/x} = e^{\lim\limits_{x \to \infty}\left(\frac{1}{x}\ln x\right)}.$$

Seeing that $\dfrac{1}{x}\ln x$ has an indeterminate form ∞/∞, we apply L'Hôpital's rule:

$$\lim_{x \to \infty} \frac{\ln x}{x} = \lim_{x \to \infty} \frac{1/x}{1} = 0.$$

Consequently

$$\lim_{x \to \infty} x^{1/x} = e^0 = 1. \qquad \square$$

EXAMPLE 3   Find $\lim\limits_{x \to 0^+} (e^x - 1)^x$.

SOLUTION

We use the procedure illustrated in Example 2 for this indeterminate form 0^0. First

$$\lim_{x \to 0^+} (e^x - 1)^x = e^{\lim\limits_{x \to 0^+} [x \ln (e^x - 1)]}.$$

Since $x \ln (e^x - 1)$ has an indeterminate form $0 \cdot \infty$, we rewrite it as

$$\frac{\ln (e^x - 1)}{1/x}$$

and apply L'Hôpital's rule:

$$\lim_{x \to 0^+} x \ln (e^x - 1) = \lim_{x \to 0^+} \frac{\ln (e^x - 1)}{1/x}$$

$$= \lim_{x \to 0^+} \frac{\left(\dfrac{1}{e^x - 1}\right) e^x}{-1/x^2}$$

$$= \lim_{x \to 0^+} \frac{-e^x x^2}{e^x - 1}.$$

Since this has an indeterminate form of the type 0/0, we can apply L'Hôpital's rule again.

$$\lim_{x \to 0^+} x \ln (e^x - 1) = \lim_{x \to 0^+} \frac{-e^x x^2 - 2xe^x}{e^x} = 0,$$

so

$$\lim_{x \to 0^+} (e^x - 1)^x = e^0 = 1. \qquad \square$$

EXERCISE SET 7.4

Find $D_x y$ in Exercises 1 through 12.

1. $y = \log_{10}(x^2 + 1)$

2. $y = \log_3(x^2 + 2x + 3)$

3. $y = \log_2\left(\dfrac{x^2 + 1}{x^2 + 2}\right)$

4. $y = \log_{10}(\log_{10}x + 3)$

5. $y = \log_{10}(\ln x)$

6. $y = \ln(\log_{10}x)$

7. $y = \log_\pi (\sqrt{x^2 + 3x - 1})$

8. $y = \log_2(x^{-1} + 2x^{-3})$

9. $y = \log_4 \dfrac{\sqrt{x^2 + 2x}}{\sqrt[3]{x^4 - 5}}$

10. $y = \log_{10}(\sqrt{x^2 + 3x} \, \sqrt[3]{3x - 5})$

11. $y \log_{10}x = x \log_{10}y$

12. $\log_{10}(xy) = \log_9(xy)$

Find the limits in Exercises 13 through 26.

13. $\lim\limits_{x \to 0^+} x^x$

14. $\lim\limits_{x \to 0^+} (1 + x^2)^{1/x^2}$

15. $\lim\limits_{x \to 0^+} x^{\sin x}$

16. $\lim\limits_{x \to 0^+} (\sin x)^x$

17. $\lim\limits_{x \to 0^+} (x + e^x)^{1/x}$

18. $\lim\limits_{x \to \infty} (x + e^x)^{1/x}$

19. $\lim\limits_{x \to 0^+} (1 + x)^{\ln x}$

20. $\lim\limits_{x \to \infty} \left(\sin \dfrac{1}{x} \right)^x$

21. $\lim\limits_{x \to \frac{\pi^-}{2}} (1 + \tan x)^{\cos x}$

22. $\lim\limits_{x \to \frac{\pi^-}{2}} (1 + \cos x)^{\tan x}$

23. $\lim\limits_{x \to \infty} (e^{-x})^{e^x}$

24. $\lim\limits_{x \to \infty} (e^x)^{e^{-x}}$

25. $\lim\limits_{x \to 0^+} (x^x)^x$

26. $\lim\limits_{x \to 0^+} x^{(x^x)}$

Sketch the graphs of the functions described in Exercises 27 through 32.

27. $f(x) = \log_2 x$

28. $f(x) = \log_{1/2} x$

29. $f(x) = \log_3 x - \log_2 x$

30. $f(x) = x \log_{10} x$

31. $f(x) = \log_2 (\log_2 x)$

32. $f(x) = \dfrac{\log_2 x}{x}$

Use the fact that $\lim\limits_{x \to \infty} \left(1 + \dfrac{1}{x} \right)^x = e$ to determine the limits in Exercises 33 through 36.

33. $\lim\limits_{x \to 0^+} (1 + x)^{1/x}$

34. $\lim\limits_{x \to \infty} \left(1 + \dfrac{1}{x} \right)^{2x}$

35. $\lim\limits_{x \to \infty} \left(1 + \dfrac{1}{2x} \right)^x$

36. $\lim\limits_{x \to \infty} \left(1 + \dfrac{2}{x} \right)^x$

37. Show that if a and b are positive real numbers with $a \neq 1$ and $b \neq 1$, then
$$\log_a b = \frac{1}{\log_b a}.$$

38. Find $\lim\limits_{x \to e} \log_x e$.

39. Consider the function $f(x) = \log_x a$, where a is a constant and $x > 0$.
(a) Use Exercise 37 to find $f'(x)$.
(b) Sketch the graph of f for $a > 1$.
(c) Sketch the graph of f for $0 < a < 1$.

40. If the interest on an investment P is compounded m times a year and the interest rate is i, the total amount of investment at the end of t years is

$$P(t) = P\left(1 + \frac{i}{m}\right)^{mt}.$$

Find $P(t)$ if the compounding is continuous (that is, if $m \to \infty$).

41. Suppose that $a > 1$ is a constant. Sketch the graphs of $y = \log_a x$ and $y = \log_{1/a} x$ on the same set of axes.

42. Show that the graphs of $y = \log_a x$ and $y = \log_b x$, for $a \neq b$, intersect only at $(1, 0)$.

43. Show that a solution to the equation

$$x + \log_a x = a$$

is $x = \dfrac{a^a}{c}$, where $c^c = a^{a^a}$.

44. Evaluate $\displaystyle\lim_{t \to 0^+} \left[\int_0^1 [bx + a(1 - x)]^t \, dx \right]^{1/t}$, where a and b are constants.

7.5
EXPONENTIAL GROWTH AND DECAY

The natural exponential function is extremely important for use in applications. These applications include population growth, financial problems involving a continuous rate of interest, radioactive decay, the change in the dilution or contamination of a liquid, and many other subjects.

In this section we examine a few of these applications and see that the mathematical problem in each case is the same: to find a function f with positive values satisfying

(7.38) $f'(t) = kf(t)$, where k is a constant.

To find $f(t)$ we rewrite (7.38) as

$$\frac{f'(t)}{f(t)} = k,$$

and then integrate with respect to t:

$$\int \frac{f'(t)}{f(t)} \, dt = kt + C$$

for some constant C. Since $f(t) > 0$ and $D_t \ln f(t) = \dfrac{f'(t)}{f(t)}$,

$$\ln f(t) = kt + C.$$

Thus

$$f(t) = e^{kt+C} = e^C e^{kt}.$$

Moreover,

$$f(0) = e^C e^{k \cdot 0} = e^C,$$

so the function f that satisfies (7.38) is given by

(7.39) $f(t) = f(0)e^{kt}.$

The first type of application we will consider involves the decay of a radio-active substance. The rate at which such a substance decays changes with time and is proportional to the amount of substance present. If $A(t)$ denotes the amount of material present at time t, the situation can be described by

$$\frac{dA(t)}{dt} = kA(t),$$

where k is the constant of proportionality.

The half-life of a radioactive substance is the length of time required for the substance to decay to half of its original amount. This concept is used in the following example to determine the constant k.

EXAMPLE 1   It is assumed that radioactive carbon 14, denoted ^{14}C, has a half-life of 5570 years. If 100 milligrams of ^{14}C are present at time $t = 0$, find an expression for the amount present at any time t. Determine the amount of ^{14}C present when $t = 2000$ years.

SOLUTION

Let $A(t)$ denote the amount of ^{14}C after t years. Then $A(0) = 100$ and $A(5570) = 50$. The problem can be summarized as:

Know	Find
(a) $\dfrac{dA(t)}{dt} = kA(t),$	(1) $A(t)$, and
(b) $A(0) = 100$, and	(2) $A(2000)$.
(c) $A(5570) = 50$.	

Since

$$\frac{dA(t)}{dt} = kA(t),$$

it follows from (7.39) that

$$A(t) = A(0)e^{kt} = 100e^{kt}.$$

To find the constant k, we use the half-life of ^{14}C.

$$50 = A(5570) = 100e^{5570k}$$

so

$$\frac{1}{2} = e^{5570k}.$$

Taking the natural logarithm of both sides,

$$\ln \frac{1}{2} = 5570k$$

and

$$k = \frac{-\ln 2}{5570}.$$

Thus

$$A(t) = 100e^{-t \ln 2/5570}$$
$$A(2000) = 100e^{-2000 \ln 2/5570}$$
$$\approx 77.97 \text{ milligrams} \qquad \square$$

Lending institutions usually pay interest on savings accounts at a certain rate compounded daily. Assuming instead that the interest is compounded continuously leads to an accurate approximation of the actual growth of a savings account. The following example involves a problem of this type.

EXAMPLE 2   Suppose $1000 is deposited in a savings account that pays $7\frac{1}{4}\%$ per annum compounded daily. How much will be in the account at the end of four years? When will the account have a balance of $1500?

HISTORICAL NOTE   Archaeologists have relied heavily on the decay property of ^{14}C to establish the age of relics and artifacts since the Nobel laureate, **William Libby** (1908–1980) introduced the technique in the early 1950's. The assumption is made that the percentage of ^{14}C and ^{12}C, the common nondecaying carbon isotope, in all *living* things has remained constant throughout history. The date when an organism ceased to live can be estimated by determining the current ^{14}C and ^{12}C proportions and comparing these to the original proportions of approximately 1 part ^{14}C to 10^5 parts ^{12}C. Since the original proportion of ^{14}C is small compared to that of ^{12}C, this procedure has limitations, but current laboratory equipment can accurately detect 1 part ^{14}C to 10^8 parts ^{12}C.

In recent years, the assumption that the ^{14}C to ^{12}C proportions in living matter has remained constant throughout history has been challenged. In particular, some scientists maintain that the industrial revolution in the mid-19th century modified the ^{14}C and ^{12}C proportions, and consequently, that these proportions are invalid for dating artifacts from antiquity.

SOLUTION

Let $A(t)$ denote the amount of money in the account at the end of t years and assume that the interest is compounded continuously, instead of daily. The change in the amount at any time t is proportional to the amount accumulated at that time, the constant of proportionality being the interest rate. Stated concisely, the problem is:

Know	Find
(a) $\dfrac{dA(t)}{dt} = .0725A(t)$, and	(1) $A(t)$, and
(b) $A(0) = \$1000.$	(2) $A(4).$
	(3) t when $A(t) = \$1500.$

It follows from (7.39) that

$$A(t) = A(0)e^{.0725t} = 1000e^{.0725t},$$

so

$$A(4) = 1000e^{(.0725)(4)} = 1000e^{.29} = \$1336.43.$$

The actual amount that would have accumulated if compounded daily is $1336.32.

When $A(t) = 1500$,

$$1500 = 1000e^{.0725t}$$

or

$$1.5 = e^{.0725t}.$$

So

$$\ln 1.5 = .0725t$$

and

$$t = \frac{\ln 1.5}{.0725} \approx 5.6.$$

The account will reach $1500 in approximately 5.6 years. ☐

In the two previous examples the same type of equation

$$\frac{dA(t)}{dt} = kA(t)$$

described each situation. The primary difference between these two problems is in the sign of k. In the decay problem, $k < 0$. In the deposit problem, a growth situation, $k > 0$.

The growth (or attrition) rate of an unrestricted population is often assumed to be proportional to the number of individuals in the population. Although the actual population is given in integer values, it is much easier to solve the mathematical problem associated with population change if we allow nonintegral values for the population. In this case we have the same type of problem as discussed previously.

EXAMPLE 3

Suppose a population of 10,000 people lives in a certain Arizona community. Census figures show that the influx rate (rate at which people enter) in this community is 55 persons per 1000 and the attrition rate (rate at which people leave) is 14 persons per 1000. The sewage treatment facilities are to be expanded to ensure that they will handle the needs of the community for the next fifteen years. How many people should the facility support? How many should the facility support if it is designed to handle the needs for the next 100 years?

SOLUTION

The net rate of increase each year in the population is $55 - 14 = 41$ persons per 1000. If $P(t)$ represents the population in t years, the problem can be summarized by:

Know	Find
(a) $\dfrac{dP(t)}{dt} = \dfrac{41}{1000} P(t)$, and	(1) $P(t)$,
(b) $P(0) = 10{,}000$.	(2) $P(15)$, and
	(3) $P(100)$.

Since
$$P(t) = P(0)e^{.041t},$$
$$P(15) = 10{,}000e^{.041(15)} = 18{,}496$$

and
$$P(100) = 10{,}000e^{.041(100)} = 603{,}403. \qquad \square$$

A city planner might accept the population prediction in the previous example for fifteen years, but would hardly be expected to accept the hundred-year figure. While both figures were calculated by the same mathematical formula, the formula is probably only accurate for relatively small values of t. When t becomes large, the assumption of unrestricted growth is no longer valid and the equation $dP(t)/dt = kP(t)$ does not accurately model the population growth problem.

Newton's law of cooling states that the rate of change of the temperature of an object is proportional to the difference between the temperature of the object and the temperature of its surrounding medium. If $T(t)$ denotes the temperature of the object at time t and L denotes the temperature of the surrounding medium, this law of cooling can be expressed as

$$\frac{dT(t)}{dt} = k(T(t) - L).$$

Letting $S(t)$ denote the difference between the temperature of the object and that of the surrounding medium at time t, $S(t) = T(t) - L$, implies that

$$\frac{dS(t)}{dt} = kS(t),$$

which has the solution

$$S(t) = S(0)\, e^{kt}.$$

Restating the solution in terms of $T(t)$ implies that

$$T(t) = (T(0) - L)e^{kt} + L$$

describes the temperature of the object at any time t.

EXAMPLE 4   A metal outdoor thermometer reading $-3°C$ is brought into a $20°C$ room. One minute later the thermometer reads $5°C$. How long will it take for the temperature to reach $19.5°C$?

SOLUTION

   Let $T(t)$ represent the temperature of the sending unit of the thermometer, and hence the reading on the thermometer, t minutes after the thermometer is brought into the room. It follows from the preceding discussion that

$$T(t) = (T(0) - L)\,e^{kt} + L$$
$$= -23\,e^{kt} + 20.$$

Since
$$5 = T(1) = -23\,e^{k} + 20,$$

$$k = \ln\!\left(\frac{15}{23}\right) \approx -.427.$$

Thus,
$$T(t) = -23\,e^{-.427t} + 20.$$

   The temperature reaches $19.5°C$ when t satisfies

$$19.5 = -23\,e^{-.427t} + 20,$$

that is, when

$$t = -\frac{1}{.427}\ln\!\left(\frac{.5}{23}\right) \approx 9.0 \text{ minutes.} \qquad \square$$

EXERCISE SET 7.5

1. The radioactive isotope thorium $_{90}\text{Th}^{234}$ has a half-life of approximately 590 hours. If there were 50 milligrams present at time $t = 0$, find
   (a) an expression for the amount of thorium present at any time t, and
   (b) how much thorium will remain at the end of 100 hours.

2. Suppose that 100 milligrams of the radioactive isotope thorium 234 described in Exercise 1 are present initially. How much thorium will remain at the end of one week?

3. Radioactive radiothorium $_{90}\text{Th}^{228}$ has a half-life of approximately 1.90 years. If 100 milligrams of radiothorium are present today how much was present one year ago?

4. The rate of growth of the number of bacteria present in a certain culture is proportional to the number present. If there are 500 present at time $t = 0$ and 1000 present at the end of 5 hours, find an expression for the number present at any time t.

5. The number of bacteria present in a certain culture doubles every four hours.
(a) Assuming that there were 1000 present at time $t = 0$, find an expression for the number present at any time t.
(b) How many bacteria will be present after 7 hours?

6. Under ideal conditions, a cell of the bacteria *Escherichia coli* will divide in approximately 22 minutes. Suppose that the initial amount $A(0)$ of bacteria is large enough to assume that the bacteria multiply continuously. Find an expression for $A(t)$, where t is measured in minutes. Determine the rate at which the bacteria is increasing at the end of one hour.

7. One bank advertises that it pays an interest rate of $5\frac{1}{4}\%$ compounded continuously while a second bank advertises that it pays $5\frac{1}{2}\%$ compounded semiannually. Which rate is most advantageous to the investor?

8. An amount of money doubles at the end of 9 years when invested at a certain rate of interest that is compounded continuously. What is that rate of interest?

9. A census of the United States population is taken every ten years. Given is a table listing, in thousands of people, the population from 1930 to 1980:

Year	1930	1940	1950	1960	1970	1980
Population (in thousands)	123,203	131,669	150,697	179,323	203,212	226,505

Assume that population changes by an amount proportional to the amount of population present. Predict the population in the year 2000 based on the population in the years
(a) 1930 and 1940
(b) 1970 and 1980
Compare your results and make conclusions.

10. Consider the population table given in Exercise 9. Predict the population in 1980 based on the population in the years
(a) 1930 and 1940
(b) 1960 and 1970
Do these results lead you to different conclusions about your prediction in Exercise 9?

11. The rate at which sugar dissolves in a liquid kept at a constant temperature is proportional to the amount that remains to be dissolved. Suppose that a home brewer adds 10 lbs of sugar to a vat of water and that 5 lbs dissolves in the first five minutes. How long does it take for the next four pounds to dissolve?

12. In a chemical reaction, one chemical X is converted into another chemical at a rate proportional to the amount of X present at any time. Suppose there were initially 50 g of X and 1 hour later there were 12g.
(a) What percent of X will have been converted at the end of 30 minutes?
(b) When will 90% of X have been converted into the other chemical?

13. An oven thermometer is taken from a drawer that is at a temperature of 65°F and placed in an oven that has been set at 375°F. At the end of two minutes the thermometer reads 300°F and at the end of three minutes it reads 345°F. Assuming the thermometer is accurate, is the oven temperature actually 375°? If not, is the actual oven temperature higher or lower than 375°F?

14. A cup of tea is made by pouring boiling water (212°F) into a cup. Three minutes later the tea is brewed, but at 180°F is too hot to drink. Assuming that the air temperature is a constant 68°, how long will it take for the tea to become a drinkable temperature of 160° or less?

15. A thermometer reading 0°C is placed in a room where the temperature is 20°C. In three minutes the thermometer reads 12°C. How long will it take for this thermometer to read 19°C?

16. An indoor/outdoor thermometer has two sensing units, one placed outside, the other inside. The outdoor thermometer reads 5°F and the indoor thermometer reads 65°F when the thermometers and their sensing units are placed in a wine cellar that is held at 45°F. One minute later the outdoor thermometer reads 25°F and the indoor thermometer reads 55°F. Which of the two thermometers will first be within five degrees of the correct temperature? Can you conclude anything significant about the thermometers from this problem?

17. A thermometer reading 0°C is brought into a 20°C room. In three minutes the thermometer reads 12°C. (See Exercise 15). The thermometer is then placed in a freezer for three minutes where the temperature is 0°C and immediately brought back into the room where the temperature is 20°C. What is the reading on the thermometer three minutes after it has been brought back into the room? Does this agree with your intuition?

18. Suppose the cycle described in Exercise 17 is repeated: for three minutes the thermometer is placed in the freezer and then for three minutes it is placed in the room. What is the temperature at the end of this time?

19. A towel hung on a line loses its moisture at a rate directly proportional to its moisture content. If the towel loses one half of its moisture content in two hours, how long will it take to be 95% dry?

20. Bankers often approximate the amount of time it takes to double the amount of an investment made at a fixed interest rate by dividing the percent of annual interest into 70. For example, $10,000 invested at 8.75% per year will become $20,000 in approximately $70/8.75 = 8$ years. Show that this formula has mathematical validity.

21. In the text it was stated that laboratory equipment can accurately detect as little as 1 part of ^{14}C in 10^8 parts of ^{12}C. Assuming that a living organism contains 1 part ^{14}C to 10^5 parts ^{12}C, determine the approximate range of the carbon-14 dating technique.

7.6
SEPARABLE DIFFERENTIAL EQUATIONS

In Section 7.5 we discussed some applications of the exponential function involving the equation $f'(x) = kf(x)$ (or, in the notation we will use in this section, $y' = ky$). This equation is an example of a **first-order differential equation:** an equation that involves a function and its first derivative.

In this section we consider the problem of finding the solution to a broader class of first-order differential equations. These equations are called **separable,** and have the form

(7.40)
$$y' = \frac{p(x)}{q(y)},$$

where p and q are continuous functions and $q(y) \neq 0$. The equation $y' = ky$ considered in Section 7.5 is an example of a separable equation where $p(x) = k$ and $q(y) = 1/y$.

The separable equation (7.40) can be rewritten

$$q(y) \frac{dy}{dx} = p(x)$$

or
$$0 = q(y) \frac{dy}{dx} - p(x).$$

Integrating both sides of the equation with respect to x implies that for an arbitrary constant C:

$$C = \int q(y) \frac{dy}{dx} dx - \int p(x) dx$$

and

(7.41)
$$C = \int q(y) dy - \int p(x) dx.$$

This produces the complete or *general* solution to the separable differential equation.

EXAMPLE 1   Find the general solution to the separable equation

$$y' = 3x^2 y.$$

SOLUTION

$$\frac{1}{y} \frac{dy}{dx} = 3x^2$$

so
$$0 = \frac{1}{y} \frac{dy}{dx} - 3x^2.$$

Thus
$$C = \int \frac{1}{y} dy - \int 3x^2 dx = \ln|y| - x^3$$

so
$$\ln|y| = x^3 + C$$

and
$$|y| = e^{x^3 + C} = e^C e^{x^3}.$$

Since e^C can assume any positive value, the general solution has the form $y = \hat{C} e^{x^3}$ for an arbitrary constant $\hat{C}$. $\square$

EXAMPLE 2   Find the solution of the differential equation $y' = \dfrac{1}{x^2 y^2}$ that satisfies the condition $y(1) = 3$.

SOLUTION

Since
$$\frac{dy}{dx} = \frac{1}{x^2 y^2},$$

$$0 = y^2 \frac{dy}{dx} - \frac{1}{x^2}.$$

Thus,
$$C = \int y^2 dy - \int \frac{1}{x^2} dx = \frac{y^3}{3} + \frac{1}{x}.$$

Since $y = 3$ when $x = 1$,

$$C = \frac{27}{3} + \frac{1}{1} = 10.$$

Consequently,
$$10 = \frac{y^3}{3} + \frac{1}{x},$$

$$y^3 = 30 - \frac{3}{x},$$

and
$$y = \left(30 - \frac{3}{x}\right)^{1/3}. \qquad \square$$

In Section 7.5 we considered the growth of an unrestricted population, a population $P(t)$ described by the equation

$$\frac{dP(t)}{dt} = kP(t),$$

where k is the growth rate. In Example 3 of Section 7.5 we found that this model of population growth is likely to give an unreasonable prediction unless the time period is quite small.

A modification of this differential equation produces better approximations. Suppose certain known community considerations make it reasonable to assume that the maximum population that can be supported is a constant L. The differential equation

(7.42)
$$\frac{dP(t)}{dt} = k(L - P(t))\, P(t)$$

expresses the more realistic situation that:
1. the rate of growth depends directly on the amount of population, and
2. as $P(t)$ approaches the limiting population L, the rate at which the population changes approaches zero.

Differential equation (7.42) is called the **logistic equation** for population growth. To solve the logistic equation, let $y = P(t)$ and consider the equivalent equation

$$0 = \frac{1}{(L - y)y} \frac{dy}{dt} - k.$$

Integrating with respect to t gives

$$C = \int \frac{1}{(L - y)y} dy - kt.$$

However, it is easily verified that

$$\frac{1}{(L-y)y} = \frac{1}{L}\left[\frac{1}{y} + \frac{1}{L-y}\right],$$

so

$$C = \frac{1}{L}\int\left[\frac{1}{y} + \frac{1}{L-y}\right]dy - kt$$

$$= \frac{1}{L}\left[\ln y - \ln(L-y)\right] - kt = \frac{1}{L}\ln\frac{y}{L-y} - kt$$

$$= \frac{1}{L}\ln\frac{P(t)}{L-P(t)} - kt.$$

When $t = 0$,

$$C = \frac{1}{L}\ln\frac{P(0)}{L-P(0)},$$

so

$$\frac{1}{L}\ln\frac{P(0)}{L-P(0)} = \frac{1}{L}\ln\frac{P(t)}{L-P(t)} - kt.$$

Thus

$$kLt = \ln\frac{P(t)}{L-P(t)} - \ln\frac{P(0)}{L-P(0)} = \ln\frac{P(t)}{L-P(t)}\frac{L-P(0)}{P(0)}$$

and

$$\frac{P(t)(L-P(0))}{P(0)(L-P(t))} = e^{kLt}.$$

Solving this equation for $P(t)$ gives

$$P(t) = \frac{P(0)Le^{kLt}}{(L-P(0)) + P(0)e^{kLt}}$$

$$= \frac{P(0)L}{P(0) + (L-P(0))e^{-kLt}}.$$

The graph of the solution to the logistic equation is shown in Figure 7.17.

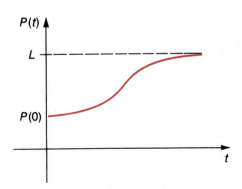

FIGURE 7.17

Initially the graph is concave upward and $P(t)$ appears to grow in an exponential manner; then it becomes concave downward and approaches the limiting population. Some features of this graph are discussed in Exercise 18.

EXAMPLE 3   Consider the population problem in Example 3 of Section 7.5, with the restriction that the limiting population in the community is 100,000 persons. In that example the initial population was 10,000 and the increase rate per year was .041. Use the logistic equation to predict the population in fifteen years and one hundred years.

SOLUTION

The constant k in the logistic equation is chosen so that the term $k(L - P(t))$ is the initial growth rate, that is, the rate when $t = 0$.

Thus $.041 = k(L - P(0)) = k(100,000 - 10,000)$

so $$k = \frac{.041}{90,000}.$$

The logistic equation predicts that the population in t years will be

$$P(t) = \frac{P(0)L}{P(0) + (L - P(0))e^{-.41t/9}}$$

$$= \frac{(10,000)(100,000)}{(10,000) + (90,000)e^{-.41t/9}}.$$

Evaluating when $t = 15$ and $t = 100$ gives

$$P(15) = 18,036 \qquad \text{and} \qquad P(100) = 91,359.$$

The predicted population in fifteen years is quite close to our previous prediction of 18,496 persons. A significant difference occurs in the 100-year figure; our previous prediction was 603,403. □

EXAMPLE 4   A tank originally contains 500 gal of brine in which 150 lb of salt has been dissolved. Fresh water runs into the tank at the rate of 5 gal/min and the well-stirred mixture is drained from the tank at the same rate. Find the amount of salt remaining in the tank at the end of one hour and the amount of time required to reduce the salt in the tank to 10 lb.

SOLUTION

Let $A(t)$ denote the number of pounds of salt in the tank t minutes after the fresh water is introduced to the tank. Then $A'(t)$, the rate of change in the amount of salt with respect to time, is the difference between the rate at which salt enters the tank, 0 lb/min, and the rate at which salt leaves the tank,

$$A'(t) = \text{Rate in} - \text{Rate out} = 0\,\frac{\text{lb}}{\text{gal}} \cdot 5\,\frac{\text{gal}}{\text{min}} - \frac{A(t)}{500}\,\frac{\text{lb}}{\text{gal}} \cdot 5\,\frac{\text{gal}}{\text{min}} = \frac{A(t)}{100}\,\frac{\text{lb}}{\text{min}}.$$

This problem can be expressed concisely as

Know	Find
(a) $\dfrac{dA(t)}{dt} = -\dfrac{A(t)}{100}$, and	(1) $A(t)$,
(b) $A(0) = 150$.	(2) $A(60)$, and
	(3) t, with $A(t) = 10$.

Since
$$A(t) = A(0)e^{-t/100} = 150e^{-t/100},$$
$$A(60) = 150e^{-60/100} \approx 82.3 \text{ lb.}$$

To find t with $A(t) = 10$ requires solving for t in the equation
$$10 = 150e^{-t/100}.$$

Thus
$$-\frac{t}{100} = \ln\frac{10}{150} = -\ln 15$$

and
$$t = 100 \ln 15 \approx 270.8 \text{ minutes.} \qquad \square$$

EXAMPLE 5   Suppose an object is propelled from the surface of the earth with an initial velocity v_0. What is the minimal value of v_0 that will ensure that the object does not return to earth?

SOLUTION

This problem requires two physical assumptions not previously discussed. First, the earth's gravitational force on an object a distance of x miles from the surface of the earth is

$$F = -\frac{mgR^2}{(x + R)^2},$$

where m is the mass of the object, $g = 32$ ft/sec^2 is the gravitational force on the earth's surface, and $R \approx 4000$ miles is the earth's radius. This assumption is a result of Newton's law of gravitation, a result that will be considered in Section 13.6.

The other assumption is Newton's second law of motion, which states that the product of the mass and acceleration of an object is equal to the sum of the forces acting on the object. In this case

$$m\frac{dv}{dt} = -\frac{mgR^2}{(x + R)^2}.$$

Since
$$\frac{dv}{dt} = \frac{dv}{dx}\frac{dx}{dt} = \frac{dv}{dx} \cdot v,$$

this equation can be rewritten as

$$v\frac{dv}{dx} = -\frac{gR^2}{(x + R)^2},$$

a separable first-order differential equation in x and v. Consequently

$$C = \int v \, dv + \int \frac{gR^2}{(x + R)^2} \, dx = \frac{v^2}{2} - \frac{gR^2}{(x + R)}$$

and since $v = v_0$ when $x = 0$,

$$C = \frac{v_0^2}{2} - gR.$$

Thus

$$\frac{v_0^2}{2} - gR = \frac{v^2}{2} - \frac{gR^2}{(x + R)}$$

and

$$\frac{(v(x))^2}{2} = \frac{v_0^2}{2} - gR + \frac{gR^2}{(x + r)}.$$

To reverse direction and return to earth, $v(x)$ would have to be zero, so the object will not return to earth if $v(x) > 0$ for all values of x. To ensure that $v(x) > 0$, we choose v_0 so that the right side of the preceding equation is always positive. Since $\lim_{x \to \infty} gR^2/(x + R) = 0$, this will be assured when

$$\frac{v_0^2}{2} - gR > 0,$$

that is, when

$$v_0 > \sqrt{2gR} = \left[2\left(32 \, \frac{\text{ft}}{\text{sec}^2} \right)\left(\frac{1 \text{ mi}}{5280\text{ft}} \right)(4000 \text{ mi}) \right]^{1/2} \approx 6.97 \, \frac{\text{mi}}{\text{sec}},$$

a rather large initial velocity.   □

EXERCISE SET 7.6

Find the general solution to the differential equations in Exercises 1 through 12.

1. $y' = 7y$ **2.** $y' = y^2$

3. $y' = y^2 + 1$ **4.** $y' = x(y^2 + 1)$

5. $y' = \dfrac{\sin x + x^2}{\cos y - 1}$ **6.** $y' = e^{x + 3y}$

7. $y' + y^2\cos x = 0$ **8.** $xy' - \sqrt{1 - y^2} = 0$

9. $e^x dy + e^y dx = 0$ **10.** $\tan x \, dx + (\sec y)^2 \, dy = 0$

11. $y \, dx + e^x dy = 0$ **12.** $\cos y \csc x \, dy - e^{\sin y}\sin x \, dx = 0$

In Exercises 13 through 16, find the solution to the differential equation that satisfies the given condition.

13. $y' = x(y + 1)$, $y(0) = 1$ **14.** $y' = \dfrac{\sin x}{y}$, $y(0) = \pi$

15. $y' = e^y \cos x, \; y(\pi) = 0$

16. $y(1 - x^2)dy + x(y^2 + 1)dx = 0, \; y(0) = 0$

17. An automobile cooling system contains 17 quarts of liquid. Initially, the system contains a mixture of 2 gallons of antifreeze and 9 quarts of water. Water runs into the system at the rate of 1 gallon per minute and the homogeneous mixture runs out a petcock at the same rate. How much antifreeze remains in the system after 5 minutes?

18. A tank with a 500-gallon capacity contains 300 gallons of salt-free water. A brine containing .5 pound of salt per gallon of water runs into the tank at 2 gallons per minute and the well-stirred mixture runs out at 2 gallons per minute. What is the concentration of salt in the tank at the end of 10 minutes?

19. It is estimated that the limiting population the United States can support is five hundred million people. Assume that the population of the United States satisfies the logistic equation with $L = 5 \times 10^8$. Predict the population in the year 2000 based on the population in the years
(a) 1930 and 1940   (b) 1970 and 1980
Compare these predictions with the corresponding predictions in Exercise 9 of Section 7.5.

Year	1930	1940	1950	1960	1970	1980
Population (in thousands)	123,203	131,669	150,697	179,323	203,212	226,505

20. The dispersion of information among a population can be modeled by the differential equation

$$\frac{dP}{dt} = k(1 - P),$$

where P denotes the proportion of the population aware of the information at time t, and k is a positive constant. Suppose that 10% of the population learns of a tax rebate on the 7:00 AM news and that 50% is aware of the rebate by noon. What percentage will know about the rebate before the 6:00 PM news?

21. A body is found floating face down in Lake Gotchaheny (a lake having a relatively constant temperature of 62°F). When taken from the water at 11:50 AM, the temperature of the body was 66°F. The temperature of the body when first found at 11:00 AM was 67°F. When did the victim meet his demise? (Assume that the victim was a normal 98.6°F before going to the watery grave.)

22. The differential equation specified by Newton's law of cooling,

$$\frac{dT(t)}{dt} = k(T(t) - L),$$

is a separable differential equation. Find the general solution to this equation directly, without using the variable substitution made in Section 7.5.

23. Early work in quantitative psychology by Ernst Heinrich Weber (1795–1878) and Gustav Fechner (1801–1887) suggested that the reaction R of an organism to a stimulus S is inversely proportional to the intensity of the stimulus. The differential equation

$$\frac{dR}{dS} = \frac{k}{S}$$

expressing this relationship is known as the *Weber-Fechner law*. The threshold level of a stimulus S_0 is defined to be the minimal level at which a reaction is obtained. Find the solution to the differential equation describing the Weber-Fechner law.

24. *Fick's first law* states that the diffusion of a substance across a cellular membrane is directly proportional to the difference in the fixed concentration C outside the cell and the varying concentration $c(t)$ inside the cell. In addition, the diffusion is directly proportional to the surface area of the membrane and inversely proportional to the volume of the cell. Derive the differential equation described by Fick's law and find its general solution.

25. The rate at which an enzyme reaction converts a substrate into a product is described by the *Michaelis-Menten equation*

$$\frac{dS(t)}{dt} = -\frac{A\,S(t)}{B\,+\,S(t)},$$

where $S(t)$ is the amount of substrate at time t, and A and B are constants depending on the maximum velocity of the reaction and the experimental conditions. Find the general solution to the Michaelis-Menten equation.

26. Suppose a population satisfies the logistic equation

$$\frac{dP(t)}{dt} = k(L - P(t))P(t).$$

Show that the graph of P must be as shown in Figure 7.17. Find the values of t and $P(t)$ that correspond to a point of inflection.

27. Another mathematical model for describing the restricted growth of a population uses the *Gompertz function*, given by

$$f(t) = a\,e^{-be^{-kt}}, \quad t \geq 0,$$

for positive constants a, b, and k. Sketch the graph of f noting, in particular, any points of inflection. What are the initial and limiting populations in this model?

28. Show that the Gompertz function described in Exercise 27 satisfies the differential equation

$$f'(t) = kf(t) \ln \frac{a}{f(t)}.$$

Interpret this differential equation as a rate of change and decide what the constant a represents.

29. Use the Gompertz function to predict the population of the United States in the year 2000 based on the assumptions and data described in Exercise 19.

30. A culture of Heriff Dansk Ølgær yeast is introduced into a mixture of malt, hops, sugar, and water. The yeast converts the sugar to alcohol, producing a liquid commonly known as beer. This good beer yeast will produce a 9% alcohol beverage; at this point, the alcohol retards the growth of the yeast and there is no further activity. The logistic differential equation is an excellent model for yeast and alcohol production in beer. Suppose that there are initially 10 grams of yeast introduced, that three days later this amount has increased to 320 grams and that the limiting amount of yeast produced is 500 grams. In order to ensure a certain amount of foam, or head, the beer is to be bottled when 98.5% of the fermentation is complete. When should the beer be bottled?

7.7
HYPERBOLIC AND INVERSE HYPERBOLIC FUNCTIONS

Certain arithmetic combinations of the natural exponential function occur sufficiently often that special functions are defined to more easily discuss them. These are known as the **hyperbolic functions**.

The most common hyperbolic functions are the **hyperbolic sine**, denoted sinh, and the **hyperbolic cosine**, denoted cosh, defined by:

(7.43)
$$\sinh x = \frac{e^x - e^{-x}}{2}$$

and

(7.44)
$$\cosh x = \frac{e^x + e^{-x}}{2}.$$

The graphs of the hyperbolic sine and cosine functions can be readily obtained by referring to the graphs of $y = e^x/2$ and $y = e^{-x}/2$, shown in Figure 7.18.

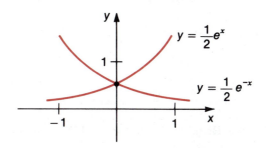

FIGURE 7.18

The graph of $y = \cosh x$ is the sum of these graphs (shown in Figure 7.19(a)), while the graph of $y = \sinh x$ is the difference (see Figure 7.19(b)).

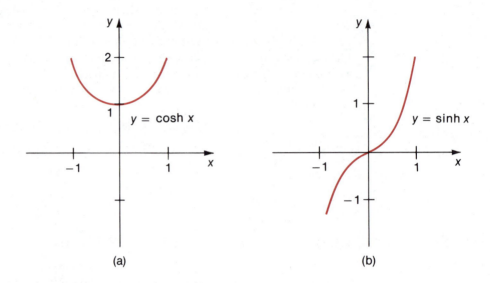

FIGURE 7.19

The names suggest that the hyperbolic functions have properties in common with the trigonometric functions. Some of these properties are listed below:

(i) The trigonometric sine and cosine functions can be derived by mapping the real line onto the unit circle having equation $x^2 + y^2 = 1$. (See Figure 7.20(a).) The hyperbolic sine and cosine functions can be derived by mapping the real line onto the portion of the hyperbola $x^2 - y^2 = 1$ corresponding to $x \geq 1$. (See Figure 7.20(b).)

(ii) The area of the region within the unit circle $x^2 + y^2 = 1$ bounded by the rays through the origin intersecting the circle at $P(0)$ and at $P(t)$ is $A = \pi(t/2\pi) = t/2$. If you solve Exercise 54, you will find that the region bounded by the hyperbola $x^2 - y^2 = 1$ and the rays through the origin intersecting the hyperbola at $P(0)$ and at $P(t)$ also has area $t/2$.

(iii) The relation of $\cosh t$ and $\sinh t$ to the unit hyperbola produces an identity similar to the trigonometric identity $(\cos t)^2 + (\sin t)^2 = 1$:

(7.45) $(\cosh t)^2 - (\sinh t)^2 = 1.$

HISTORICAL NOTE   The Italian mathematician **Vincenzo Riccati** (1707–1775) introduced the notions of hyperbolic functions in 1757, but the notation we use to describe these functions is due to **Johann Heinrich Lambert** (1728–1777), who used this notation in a paper published in 1771.

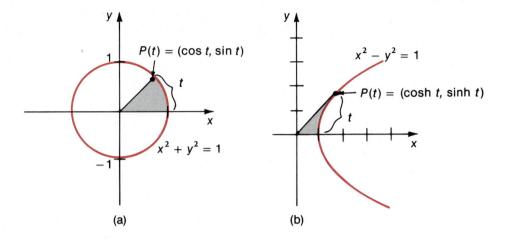

FIGURE 7.20

Identity (7.45) can also be established by using the definitions of these functions:

$$(\cosh t)^2 - (\sinh t)^2 = \left[\frac{e^t + e^{-t}}{2}\right]^2 - \left[\frac{e^t - e^{-t}}{2}\right]^2$$

$$= \frac{e^{2t} + 2e^t e^{-t} + e^{-2t}}{4} - \frac{e^{2t} - 2e^t e^{-t} + e^{-2t}}{4}$$

$$= \frac{4e^t e^{-t}}{4} = 1.$$

Another relationship of the hyperbolic sine and cosine that is reminiscent of the trigonometric functions is the property of their derivatives and integrals:

$$D_x \sinh x = D_x \frac{e^x - e^{-x}}{2} = \frac{e^x + e^{-x}}{2} = \cosh x,$$

$$D_x \cosh x = D_x \frac{e^x + e^{-x}}{2} = \frac{e^x - e^{-x}}{2} = \sinh x,$$

$$\int \cosh x \, dx = \sinh x + C,$$

$$\int \sinh x \, dx = \cosh x + C.$$

The other hyperbolic functions are called the hyperbolic tangent, cotangent, secant, and cosecant functions. These functions are defined in a manner similar to their trigonometric counterparts:

$$\tanh x = \frac{\sinh x}{\cosh x} = \frac{e^x - e^{-x}}{e^x + e^{-x}};$$

$$\coth x = \frac{\cosh x}{\sinh x} = \frac{e^x + e^{-x}}{e^x - e^{-x}}, \text{ provided } x \neq 0;$$

$$\text{sech } x = \frac{1}{\cosh x} = \frac{2}{e^x + e^{-x}};$$

$$\text{csch } x = \frac{1}{\sinh x} = \frac{2}{e^x - e^{-x}}, \text{ provided } x \neq 0.$$

The graphs of these functions are shown in Figure 7.21.

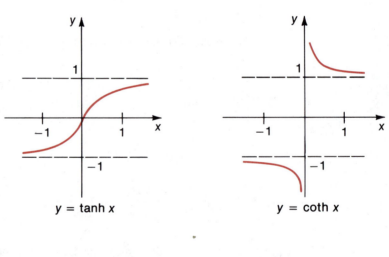

$y = \tanh x$ $y = \coth x$

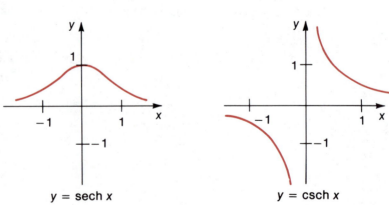

$y = \text{sech } x$ $y = \text{csch } x$

FIGURE 7.21

The derivatives and corresponding indefinite integral formulas for these functions follow. It is interesting to note the similarities, as well as the differences, between these results and the corresponding results for the trigonometric functions.

$$D_x \tanh x = (\text{sech } x)^2, \text{ so } \int (\text{sech } x)^2 \, dx = \tanh x + C$$

$$D_x \coth x = -(\text{csch } x)^2, \text{ so } \int (\text{csch } x)^2 \, dx = -\coth x + C$$

$$D_x \text{ sech } x = -\text{sech } x \tanh x, \text{ so } \int \text{sech } x \tanh x \, dx = -\text{sech } x + C$$

$$D_x \text{ csch } x = -\text{csch } x \coth x, \text{ so } \int \text{csch } x \coth x \, dx = -\text{csch } x + C$$

There are identities relating the other hyperbolic functions analogous to the trigonometric identities. Two of these are:

$$(\text{sech } x)^2 + (\tanh x)^2 = 1$$
$$(\coth x)^2 - (\text{csch } x)^2 = 1.$$

EXAMPLE 1   Evaluate $\int (\cosh x)^3 \sinh x \, dx$.

SOLUTION

Letting $u = \cosh x$, we see that $du = \sinh x \, dx$ and

$$\int (\cosh x)^3 \sinh x \, dx = \int u^3 \, du$$

$$= \frac{u^4}{4} + C$$

$$= \frac{(\cosh x)^4}{4} + C. \qquad \square$$

EXAMPLE 2   Find $D_x (\tanh x^3)(\text{sech } x^3)$.

SOLUTION

Applying the product and chain rules, we have:

$$D_x (\tanh x^3)(\text{sech } x^3) = (\text{sech } x^3)^2 (3x^2)(\text{sech } x^3) - (\tanh x^3)(\text{sech } x^3)(\tanh x^3)\, 3x^2$$

$$= 3x^2 \text{ sech } x^3 [(\text{sech } x^3)^2 - (\tanh x^3)^2]. \qquad \square$$

EXAMPLE 3   Evaluate $\int \tanh x \, dx$.

SOLUTION

$$\int \tanh x \, dx = \int \frac{\sinh x}{\cosh x} \, dx.$$

With $u = \cosh x$, $du = \sinh x \, dx$, so

$$\int \tanh x \, dx = \int \frac{du}{u} = \ln|u| + C$$

$$= \ln \cosh x + C. \qquad \square$$

The hyperbolic cosine function has an important application in the representation of a curve commonly occurring in physical problems. Suppose a flexible cable of uniform density w is supported at its ends and hangs under its own weight as shown in Figure 7.22.

FIGURE 7.22

If an xy-coordinate system is placed so that the x-axis lies parallel to the ground level and the y-axis is midway between the supports, then the position of the cable is described by the graph of the equation

$$y = \frac{T_0}{w} \cosh \frac{wx}{T_0},$$

where T_0 is the horizontal tension on the cable at its lowest point, a distance T_0/w above the x-axis. See Figure 7.23. A curve of this type is called a **catenary**. The name is derived from the Latin word *catena*, meaning chain. The shape assumed by a power line between two poles, a stationary jump rope, or an unweighted necklace is a catenary. Exercises 57 and 58 consider some practical applications of catenary design.

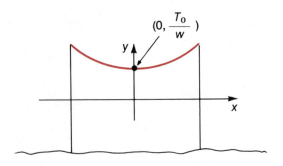

FIGURE 7.23

The graph of the hyperbolic sine function shown in Figure 7.19(b) on p. 438 indicates that this function is one-to-one and increasing. This can be easily verified by observing that

$$D_x \sinh x = \cosh x > 0$$

for all values of x.

It follows from the inverse function theorem that sinh has a differentiable inverse function that is also increasing. We call this function the **inverse hyperbolic sinh function**, denoted arcsinh:

(7.46) $y = \operatorname{arcsinh} x$ if and only if $x = \sinh y.$

The inverse function theorem also implies that

$$D_x \operatorname{arcsinh} x = \frac{1}{D_y \sinh y} = \frac{1}{\cosh y}.$$

Since $(\cosh y)^2 - (\sinh y)^2 = 1,$

$$(\cosh y)^2 = 1 + (\sinh y)^2,$$

and $\cosh y = \pm \sqrt{1 + (\sinh y)^2}.$

But the hyperbolic cosine function is always positive, so

$$\cosh y = \sqrt{1 + (\sinh y)^2} = \sqrt{1 + x^2}$$

and

(7.47) $$D_x \operatorname{arcsinh} x = \frac{1}{\sqrt{1 + x^2}}.$$

The corresponding indefinite integral formula is:

$$\int \frac{dx}{\sqrt{1 + x^2}} = \operatorname{arcsinh} x + C.$$

By making a variable substitution, this formula implies that

$$\int \frac{dx}{\sqrt{a^2 + x^2}} = \operatorname{arcsinh} \frac{x}{a} + C,   \text{for any constant } a > 0.$$

The graphs of the hyperbolic functions, tanh, coth, and csch, indicate that these functions are also one-to-one and consequently have inverse functions. If the domains of the hyperbolic cosine and secant functions are restricted to $[0, \infty)$, they also have inverse functions.

The graphs of the inverse hyperbolic functions are shown in Figure 7.24.

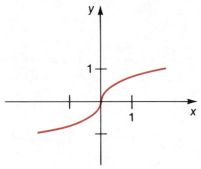

(i) $y = \text{arcsinh } x$

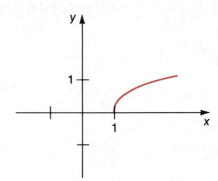

(ii) $y = \text{arccosh } x$
$x > 1$

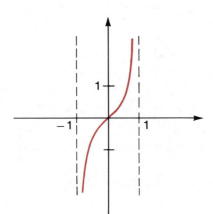

(iii) $y = \text{arctanh } x$
$|x| < 1$

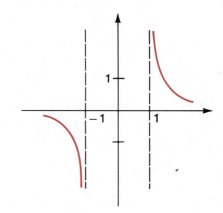

(iv) $y = \text{arccoth } x$
$|x| > 1$

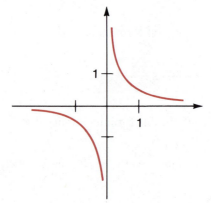

(v) $y = \text{arccsch } x$
$|x| > 0$

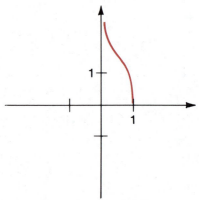

(vi) $y = \text{arcsech } x$
$0 < x \leq 1$
$y \geq 0$

FIGURE 7.24

A procedure similar to that used in finding the derivative of the arcsinh can be used to find the derivatives of the other inverse hyperbolic functions. The derivative formulas for these functions are given below.

$$D_x \text{ arccosh } x = \frac{1}{\sqrt{x^2 - 1}}, \quad \text{for } x > 1.$$

$$D_x \text{ arctanh } x = \frac{1}{1 - x^2}, \quad \text{for } |x| < 1.$$

$$D_x \text{ arccoth } x = \frac{1}{1 - x^2}, \quad \text{for } |x| > 1.$$

$$D_x \text{ arccsch } x = \frac{-1}{|x| \sqrt{1 + x^2}} x \neq 0.$$

$$D_x \text{ arcsech } x = \frac{-1}{x \sqrt{1 - x^2}}, \quad \text{for } 0 < x < 1.$$

EXAMPLE 4   Find $D_x \text{ arcsinh } (x^2 - 1).$

SOLUTION

Using Equation (7.47) and the chain rule, we have

$$D_x \text{ arcsinh } (x^2 - 1) = \frac{1}{\sqrt{1 + (x^2 - 1)^2}} \cdot D_x (x^2 - 1)$$

$$= \frac{2x}{\sqrt{x^4 - 2x^2 + 2}}. \qquad \square$$

The integral formulas associated with the inverse hyperbolic functions are:

$$\int \frac{1}{\sqrt{x^2 - a^2}} \, dx = \text{arccosh } \frac{x}{a} + C, \quad \text{for } x > a > 0,$$

$$\int \frac{1}{a^2 - x^2} \, dx = \begin{cases} \frac{1}{a} \text{ arctanh } \frac{x}{a} + C, & \text{for } a > |x| > 0, \\ \frac{1}{a} \text{ arccoth } \frac{x}{a} + C, & \text{for } |x| > a > 0. \end{cases}$$

$$\int \frac{1}{x \sqrt{a^2 + x^2}} \, dx = \frac{-1}{a} \text{ arccsch } \frac{|x|}{a} + C, \quad \text{for } |x| > 0,$$

$$\int \frac{1}{x \sqrt{a^2 - x^2}} \, dx = -\frac{1}{a} \text{ arcsech } \frac{|x|}{a} + C, \quad \text{for } 0 < |x| < a.$$

EXAMPLE 5   Evaluate $\int \frac{e^x}{\sqrt{1 + e^{2x}}} \, dx.$

SOLUTION

Let $u = e^x$, then $du = e^x\,dx$ and

$$\int \frac{e^x}{\sqrt{1 + e^{2x}}}\,dx = \int \frac{du}{\sqrt{1 + u^2}}$$

$$= \operatorname{arcsinh} u + C$$

$$= \operatorname{arcsinh} e^x + C. \qquad \square$$

EXAMPLE 6   Evaluate $\displaystyle\int_0^1 \frac{x}{4 - x^4}\,dx$.

SOLUTION

First we evaluate the indefinite integral. Let $u = x^2$, then $du = 2x\,dx$ and

$$\int \frac{x}{4 - x^4}\,dx = \frac{1}{2}\int \frac{du}{4 - u^2} = \frac{1}{4}\operatorname{arctanh}\frac{u}{2} + C$$

$$= \frac{1}{4}\operatorname{arctanh}\frac{x^2}{2} + C.$$

So

$$\int_0^1 \frac{x}{4 - x^4}\,dx = \frac{1}{4}\left(\operatorname{arctanh}\frac{1}{2} - \operatorname{arctanh} 0\right)$$

$$= \frac{1}{4}\operatorname{arctanh}\frac{1}{2} \approx .137. \qquad \square$$

EXAMPLE 7   Evaluate $\displaystyle\int \frac{dx}{(x - 1)\sqrt{2x - x^2}}$.

SOLUTION

The expression $2x - x^2$ can be written $1 - (x^2 - 2x + 1) = 1 - (x - 1)^2$. Let $u = x - 1$, then $du = dx$ and

$$\int \frac{dx}{(x - 1)\sqrt{2x - x^2}} = \int \frac{du}{u\sqrt{1 - u^2}}$$

$$= \operatorname{arcsech}|u| + C$$

$$= \operatorname{arcsech}|x - 1| + C. \qquad \square$$

EXERCISE SET 7.7

Differentiate the functions described in Exercises 1 through 18.

1. $f(x) = \sinh x^2$
2. $f(x) = \cosh(3x^2 + 2)$
3. $f(x) = \sinh\sqrt{x^2 + 2x}$
4. $f(x) = \tanh e^x$
5. $f(x) = \cosh(\ln x)$
6. $f(x) = \sinh x^3 - \cosh x^3$
7. $f(x) = \operatorname{sech}(e^{x^3 + 3x})$
8. $f(x) = \ln(\coth\sqrt{x})$
9. $f(x) = \operatorname{arcsinh} x^3$
10. $f(x) = \operatorname{arctanh}(3x^2 + 1)$
11. $f(x) = \operatorname{arccosh}\sqrt{x}$
12. $f(x) = \sqrt{\operatorname{arccosh} x}$

13. $f(x) = \sinh x^2 \operatorname{arcsinh} x^2$

14. $f(x) = \sinh (\operatorname{arcsinh} x^2)$

15. $f(x) = \operatorname{arctanh} (\tan x)$

16. $f(x) = \tan (\operatorname{arctanh} x)$

17. $f(x) = \dfrac{\operatorname{arcsinh} x}{\operatorname{arccosh} x}$

18. $f(x) = (\operatorname{arccosh} x)^2 - (\operatorname{arcsinh} x)^2$

Evaluate the integrals in Exercises 19 through 36.

19. $\displaystyle\int \dfrac{\cosh \sqrt{x}}{\sqrt{x}}\, dx$

20. $\displaystyle\int_0^1 e^x \sinh(e^x)\, dx$

21. $\displaystyle\int (\operatorname{sech} 2x)^2\, dx$

22. $\displaystyle\int (\tanh 2x)^2 dx$

23. $\displaystyle\int \sinh x \cosh x\, dx$

24. $\displaystyle\int x \tanh x^2\, dx$

25. $\displaystyle\int \dfrac{\cosh x}{(\sinh x)^2}\, dx$

26. $\displaystyle\int \sqrt{\tanh x}\,(\operatorname{sech} x)^2\, dx$

27. $\displaystyle\int_0^4 \dfrac{dx}{\sqrt{x^2 + 9}}$

28. $\displaystyle\int_0^1 \dfrac{dx}{\sqrt{9x^2 + 4}}$

29. $\displaystyle\int \dfrac{x\, dx}{\sqrt{x^4 + 9}}$

30. $\displaystyle\int \dfrac{\cos x\, dx}{\sqrt{1 + (\sin x)^2}}$

31. $\displaystyle\int \dfrac{dx}{\sqrt{9x^2 - 4}}$

32. $\displaystyle\int \dfrac{dx}{x\sqrt{4 - 9x^2}}$

33. $\displaystyle\int \dfrac{e^x dx}{1 - e^{2x}}$

34. $\displaystyle\int \dfrac{dx}{x\sqrt{x^2 + 9}}$

35. $\displaystyle\int \dfrac{dx}{4x - x^2}$

36. $\displaystyle\int \dfrac{dx}{\sqrt{x^2 + 4x + 5}}$

Verify the identities given in Exercises 37 through 48.

37. $\cosh x + \sinh x = e^x$

38. $\cosh x - \sinh x = e^{-x}$

39. $(\operatorname{sech} x)^2 + (\tanh x)^2 = 1$

40. $(\coth x)^2 - (\operatorname{csch} x)^2 = 1$

41. $\sinh(a \pm b) = \sinh a \cosh b \pm \cosh a \sinh b$

42. $\cosh (a \pm b) = \cosh a \cosh b \pm \sinh a \sinh b$

43. $\tanh(a \pm b) = \dfrac{\tanh a \pm \tanh b}{1 \pm \tanh a \tanh b}$

44. $\cosh \dfrac{x}{2} = \sqrt{\dfrac{\cosh x + 1}{2}}$

45. $\sinh \dfrac{x}{2} = \begin{cases} \sqrt{\dfrac{\cosh x - 1}{2}}, & \text{if } x \geq 0 \\[2ex] -\sqrt{\dfrac{\cosh x - 1}{2}}, & \text{if } x < 0. \end{cases}$

46. $\operatorname{arcsinh} x = \ln (x + \sqrt{x^2 + 1})$.

47. $\operatorname{arccosh} x = \ln (x + \sqrt{x^2 - 1}), \quad x \geq 1.$

48. $\operatorname{arctanh} x = \dfrac{1}{2} \ln \dfrac{1 + x}{1 - x}$, $|x| < 1$.

49. Show that $\operatorname{arctanh} \dfrac{3}{5} = \operatorname{arcsinh} \dfrac{3}{4} = \operatorname{arccosh} \dfrac{5}{4}$.

50. Find the volume generated by rotating $y = \cosh x$ from $(0, 1)$ to $(a, \cosh a)$ about
(a) the x-axis   (b) the y-axis

51. Find the volume generated by rotating $y = \sinh x$ from $(0, 0)$ to $(a, \sinh a)$ about
(a) the x-axis   (b) the y-axis

52. Show that the length of the curve $y = \cosh x$ from $(0, 1)$ to $(a, \cosh a)$ is $\sinh a$.

53. Show that the surface area generated by rotating $y = \cosh x$ from $(0, 1)$ to $(a, \cosh a)$ about the x-axis is $\dfrac{\pi}{2}(\sinh 2a + 2a)$.

54. Show that the area of the region bounded by the hyperbola $x^2 - y^2 = 1$ and the rays through the origin intersecting the hyperbola at $P(0) = (\cosh 0, \sinh 0)$ and $P(t) = (\cosh t, \sinh t)$ is $t/2$.

55. Show that $(\sinh x + \cosh x)^n = \sinh nx + \cosh nx$ for any positive integer n.

56. Show that a catenary is concave upward at each point.

57. A cable hangs between two 40-foot power poles whose bases are on the same level. The sag in the middle of the cable is 5 feet and the towers are 100 feet apart.
(a) Find the length of the cable in terms of the quotient T_0/w.
(b) Use Newton's method to solve for T_0/w and use this to find the length of the cable accurate to within .1 ft.

58. The Gateway Arch in St. Louis, Missouri was built in the shape of an inverted catenary with both height and width at ground level of 630 feet. The arch sits on the west bank of the Mississippi river and contains elevators that take visitors through the hollow arches to an observation post at the top of the arch.
(a) Use Newton's method to find an approximate equation for the arch. (*Hint*: Place an xy-coordinate system with the origin T_0/w units above the top of the arch and the positive y direction directed downward.)
(b) Find the length of this arch.

REVIEW EXERCISES

Find $\dfrac{dy}{dx}$ in Exercises 1 through 26.

1. $y = \arcsin\left(\dfrac{x}{3}\right)$ **2.** $y = \arctan(2x - 1)$

3. $y = e^{x^3}$

4. $y = e^x \arccos x$

5. $y = 2^{x^3}$

6. $y = \left(e^{\frac{1}{x}}\right)^2$

7. $y = e^{3\ln x}$

8. $y = 3^{x^2} 2^x$

9. $y = xe^{\tan x}$

10. $y = \log_{10} 4x$

11. $y = \log_{10} \dfrac{x+1}{x^2+1}$

12. $y = \arccos(e^x + 2^x)$

13. $y = \ln e^{x^3}$

14. $y = \dfrac{1}{10^{2x}}$

15. $y = x - \ln(1 + e^x)$

16. $y = x^{\sqrt{2}}$

17. $y = x^e$

18. $y = \arctan\left(e^{\frac{1}{x}}\right)$

19. $y = \dfrac{1 + e^{3x}}{\cosh x}$

20. $y = \cosh(x^2 - 1)$

21. $y = e^x \sinh(2x - 1)$

22. $y = (\tanh e^x)^3$

23. $y = x^{\cos x}$

24. $y = (\cos x)^x$

25. $\ln(x + y) = \arctan\left(\dfrac{x}{y}\right)$

26. $\cosh(x + y) = \sinh(xy)$

Evaluate the integrals in Exercises 27 through 40.

27. $\displaystyle\int \dfrac{dx}{x^2 + 9}$

28. $\displaystyle\int_0^7 \dfrac{dx}{x^2 + 49}$

29. $\displaystyle\int_1^2 x^2 e^{x^3}\, dx$

30. $\displaystyle\int \dfrac{xe^{x^2}}{e^{x^2} + 1}\, dx$

31. $\displaystyle\int \dfrac{dx}{\sqrt{9 - x^2}}$

32. $\displaystyle\int \dfrac{dx}{x\sqrt{x^2 - 4}}$

33. $\displaystyle\int \dfrac{e^x\, dx}{\sqrt{1 - e^{2x}}}$

34. $\displaystyle\int \dfrac{dx}{x\sqrt{4x^2 - 1}}$

35. $\displaystyle\int \dfrac{dx}{x^2 - 2x + 2}$

36. $\displaystyle\int_{-2}^{-1} \dfrac{dx}{x^2 + 4x + 5}$

37. $\displaystyle\int_{-1}^1 2^x\, dx$

38. $\displaystyle\int \dfrac{e^x}{e^{2x} + 1}\, dx$

39. $\displaystyle\int \dfrac{\sinh(\ln x)}{x}\, dx$

40. $\displaystyle\int e^x \cosh e^x\, dx$

41. Find the exact value of the following expressions.

(a) $\arcsin\left(-\dfrac{1}{2}\right)$ (b) $\arctan(-1)$ (c) $\arccos \dfrac{\sqrt{3}}{2}$

42. Find the exact value of the following expressions.

(a) $\sin\left[\arccos \dfrac{\sqrt{2}}{2}\right]$ (b) $\operatorname{arcsec}\sqrt{2}$ (c) $\arctan \dfrac{\sqrt{3}}{3}$

Find the limits in Exercises 43 through 50.

43. $\lim\limits_{x\to 0^+} x^{\ln x}$

44. $\lim\limits_{x\to 0^+} (2x)^x$

45. $\lim\limits_{x\to 0^+} x^{\tan x}$

46. $\lim\limits_{x\to -\infty} x \tan \dfrac{1}{x}$

47. $\lim\limits_{x\to\infty} (1 + x)^{\frac{1}{x}}$

48. $\lim\limits_{x\to\infty} \left(\dfrac{x+2}{x}\right)^x$

49. $\lim\limits_{x\to\infty} \left(\tan \dfrac{1}{x}\right)^x$

50. $\lim\limits_{x\to\infty} (e^x)^{e^{-x}}$

Sketch the graphs of the functions described in Exercises 51 through 56.

51. $f(x) = \arcsin(x - 1)$

52. $f(x) = x + \arcsin x$

53. $f(x) = 1 + 2^x$

54. $f(x) = e^{x-1}$

55. $f(x) = \sinh 2x$

56. $f(x) = \sinh(x - 1)$

Find the general solution to the differential equations in Exercises 57 through 62.

57. $y' = x^2 + 2x$

58. $y' = \dfrac{y}{2}$

59. $y' = e^{x+y}$

60. $y' = y \tan x$

61. $y' = \dfrac{1 + y^2}{1 + x^2}$

62. $y' = \dfrac{y}{x} \ln x$

In Exercises 63 through 66, find the solution to the differential equation that satisfies the given condition.

63. $y' = 2xy$, $y(0) = 5$

64. $y' = y(x + 1)$, $y(0) = 1$

65. $2x(y + 1)dx - y\,dy = 0$, $y(1) = 0$

66. $y' = x^2 y$, $y(0) = 3$

67. A bacteria culture is known to grow at a rate proportional to the amount present. After one hour, 1000 bacteria are present and after four hours, 3000 are present. Find (a) an expression for the number of bacteria present in the culture at any time t and (b) the number of bacteria originally in the culture.

68. A thermometer reading 70°F is taken outside, where the temperature is 0°F. After three minutes outside, the thermometer reading is 15°F. Express the reading on the thermometer as a function of time.

69. The radioactive isotope uranium $_{92}U^{235}$ has a half-life of 8.8×10^8 years. If 1 gram was initially present, find the amount decayed by the end of 1000 years.

70. A tank originally contains 100 gallons of pure water. Starting at time $t = 0$, brine containing 3 lb of dissolved salt per gallon runs into the tank at the rate of 4 gal/min. The mixture is kept uniform by stirring and the well-stirred mixture runs out of the tank at the same rate.
   (a) Find the amount of salt remaining in the tank at the end of 30 minutes.
   (b) Find the amount of time required to raise the amount of salt in the tank to 10 lb.

71. How long does it take for an amount of money to double if it is deposited at 6% compounded continuously?

8

TECHNIQUES OF INTEGRATION

Most of the definite integral problems discussed thus far can be solved by a relatively simple substitution technique. In Section 6.5, however, we found that, to determine the arc length of $y = x^2$, we needed to evaluate $\int \sqrt{1 + 4x^2}\, dx$, an integral for which no elementary substitution technique will work. This is not an isolated situation; we have avoided many natural integration problems that cannot be solved using integration by substitution.

In this chapter we examine some of the more frequently used techniques for evaluating integrals, and discuss the most common methods for approximating definite integrals.

8.1 INTEGRATION BY PARTS

When the integration formulas in Chapter 5 were developed, the formula corresponding to the product rule for differentiation was not given. This rule does, however, provide a very useful technique of integration.

Since the product rule for differentiation is

$$D_x\,[\,f(x)\,g(x)\,] = f(x)D_x\,g(x) + g(x)D_x f(x),$$

the corresponding indefinite integral equation is

$$f(x)g(x) = \int f(x)D_x\,g(x)dx + \int g(x)D_x f(x)dx.$$

For integration purposes this integral formula is expressed as

(8.1)
$$\int f(x) D_x\, g(x)\, dx = f(x)g(x) - \int g(x) D_x f(x)\, dx$$

and called the **integration by parts** formula. The name indicates that the application involves splitting the integrand into two parts: $f(x)$ and $D_x\, g(x)$.

EXAMPLE 1   Determine $\int x \cos x\, dx$.

SOLUTION

We split the integrand into two parts.

Let $f(x) = x$ and $D_x\, g(x) = \cos x$.

Then $D_x f(x) = 1$ and $g(x) = \sin x + C$.

The integration by parts formula implies that

$$\int x \cos x\, dx = x(\sin x + C) - \int (\sin x + C)\, dx$$
$$= x \sin x + Cx + \cos x - Cx + K$$
$$= x \sin x + \cos x + K. \qquad \square$$

Before considering more applications of the integration by parts technique, we introduce some notation to simplify the procedure. Let $u = f(x)$ and $v = g(x)$. Then $du = D_x f(x)\, dx$ and $dv = D_x\, g(x)\, dx$; so the integration by parts procedure is given by

(8.2)
$$\int u\, dv = uv - \int v\, du.$$

EXAMPLE 2   Determine $\int x \ln x\, dx$.

SOLUTION

Let $u = \ln x$ and $dv = x\, dx$.

Then $du = \dfrac{1}{x}\, dx$ and $v = \dfrac{x^2}{2} + C$.

By formula (8.2),

$$\int x \ln x\, dx = (\ln x)\left(\frac{x^2}{2} + C\right) - \int \left(\frac{x^2}{2} + C\right)\frac{1}{x}\, dx$$

$$= \frac{x^2}{2} \ln x + C \ln x - \int \frac{x}{2}\, dx - \int \frac{C}{x}\, dx$$

$$= \frac{x^2}{2} \ln x + C \ln x - \frac{x^2}{4} - C \ln x + K$$

$$= \frac{x^2}{2} \ln x - \frac{x^2}{4} + K. \qquad \square$$

Notice that the constant of integration C that comes from determining v from dv makes no contribution to the final result. This will always be the case when using integration by parts. If v is any solution to $dv = D_x g(x)\, dx$, then all solutions are of the form $v + C$ for some constant. But,

$$u(v + C) - \int (v + C)\, du = uv + uC - \int v\, du - C \int du$$

$$= uv + uC - \int v\, du - Cu$$

$$= uv - \int v\, du.$$

Consequently the constant C can be chosen to be zero, as we will do in the remainder of this section.

Making the right choice for u and dv depends on experience and a few general rules, the most basic of which is that *you must be able to integrate the choice for dv*. Other guidelines will be discussed as we proceed through the examples.

EXAMPLE 3   Determine $\int \ln x\, dx$.

SOLUTION

Let $$u = \ln x \qquad \text{and} \qquad dv = dx,$$

then $$du = \frac{1}{x}\, dx \qquad \text{and} \qquad v = x.$$

So, $$\int \ln x\, dx = x \ln x - \int x\, \frac{1}{x}\, dx$$

$$= x \ln x - x + C. \qquad \square$$

EXAMPLE 4   Evaluate $\displaystyle\int_0^1 x\, e^x\, dx$.

SOLUTION

We first determine the indefinite integral $\displaystyle\int x\, e^x\, dx$.

Let $$u = x \qquad \text{and} \qquad dv = e^x\, dx,$$

then $$du = dx \qquad \text{and} \qquad v = e^x.$$

So $$\int x\, e^x\, dx = x\, e^x - \int e^x\, dx = x\, e^x - e^x + C.$$

Hence, $$\int_0^1 x\, e^x\, dx = x\, e^x - e^x \Big]_0^1 = (e - e) - (0 - 1) = 1. \qquad \square$$

To illustrate the importance of making the right choice for u and dv, notice that if in Example 4 we had chosen $u = e^x$ and $dv = x\, dx$, then

$$du = e^x\, dx \qquad \text{and} \qquad v = \frac{x^2}{2}$$

and

$$\int x\, e^x\, dx = \frac{x^2}{2}\, e^x - \int \frac{x^2}{2}\, e^x\, dx.$$

Although this is a correct result, it is more complicated than the original problem.

EXAMPLE 5   Evaluate $\displaystyle\int_0^\pi x^2 \sin x\, dx$.

SOLUTION

Let $\qquad\qquad\qquad u = x^2 \qquad$ and $\qquad dv = \sin x\, dx,$

then $\qquad\qquad\qquad du = 2x\, dx \qquad$ and $\qquad v = -\cos x.$

So, $\qquad\qquad\displaystyle\int x^2 \sin x\, dx = -x^2 \cos x + 2 \int x \cos x\, dx.$

To evaluate $\displaystyle\int x \cos x\, dx$, we use the integration by parts technique again.

Let $\qquad\qquad\qquad u = x \qquad$ and $\qquad dv = \cos x\, dx;$

then $\qquad\qquad\qquad du = dx \qquad$ and $\qquad v = \sin x.$

So $\qquad\qquad\displaystyle\int x \cos x\, dx = x \sin x - \int \sin x\, dx$

$$= x \sin x + \cos x + C.$$

Thus,

$$\int x^2 \sin x\, dx = -x^2 \cos x + 2x \sin x + 2 \cos x + C.$$

The definite integral is

$$\int_0^\pi x^2 \sin x\, dx = \left[-x^2 \cos x + 2x \sin x + 2 \cos x \right]_0^\pi$$

$$= [(-\pi^2(-1) - 2) - (2)] = \pi^2 - 4. \qquad \square$$

Other problems that can be solved using the integration by parts technique are those that tend to "cycle" with repeated application of the procedure.

EXAMPLE 6   Evaluate $\displaystyle\int e^x \cos x\, dx$.

SOLUTION

Let $u = e^x \qquad$ and $\qquad dv = \cos x\, dx,$

then $\qquad\qquad\qquad du = e^x\, dx \qquad$ and $\qquad v = \sin x.$

So
$$\int e^x \cos x \, dx = e^x \sin x - \int e^x \sin x \, dx.$$

Apply the technique again to determine $\int e^x \sin x \, dx$.

Let $\qquad u = e^x \qquad$ and $\qquad dv = \sin x \, dx,$

then $\qquad du = e^x \, dx \qquad$ and $\qquad v = -\cos x.$

So
$$\int e^x \sin x \, dx = -e^x \cos x + \int e^x \cos x \, dx.$$

The original integral is:
$$\int e^x \cos x \, dx = e^x \sin x - \int e^x \sin x \, dx = e^x \sin x + e^x \cos x - \int e^x \cos x \, dx$$

or, since the term $\int e^x \cos x \, dx$ appears on each side of the equation,

$$2 \int e^x \cos x \, dx = e^x (\sin x + \cos x) + C.$$

Hence, $\qquad \int e^x \cos x \, dx = \dfrac{e^x}{2}(\sin x + \cos x) + C.$

Note the addition of the constant of integration to the right side of the equation. This was needed when the indefinite integral was deleted from that side. $\qquad\qquad\qquad\qquad\qquad\qquad\qquad\qquad\qquad\qquad\square$

The next example illustrates some pitfalls to be avoided when using integration by parts. In our previous examples the choices of u and dv have been relatively straightforward. This is not the case in the next example.

EXAMPLE 7 Evaluate $\int x^3 e^{x^2} \, dx.$

SOLUTION

Following the lead of Example 4, we first let

$$u = x^3 \qquad \text{and} \qquad dv = e^{x^2} \, dx,$$

then $\qquad du = 3x^2 \, dx \qquad$ and $\qquad v = \int e^{x^2} \, dx.$

But $\int e^{x^2} \, dx$ cannot be evaluated: *Pitfall #1.* Try again: let

$$u = e^{x^2} \qquad \text{and} \qquad dv = x^3 \, dx$$

then $\qquad du = 2xe^{x^2} \qquad$ and $\qquad v = \dfrac{x^4}{4}.$

So
$$\int x^3 e^{x^2} \, dx = \frac{x^4}{4} e^{x^2} - \frac{1}{2} \int x^5 e^{x^2} \, dx.$$

Now a more complicated integral results: *Pitfall #2*. Although there are a number of alternative choices for u and dv, all lead to the same type of difficulties, except:

let $$u = x^2 \quad \text{and} \quad dv = x\,e^{x^2}\,dx$$

then $$du = 2x\,dx \quad \text{and} \quad v = \int x\,e^{x^2} = \frac{1}{2}\,e^{x^2}.$$

So $$\int x^3\,e^{x^2}\,dx = \frac{x^2}{2}\,e^{x^2} - \int x\,e^{x^2}\,dx$$

$$= \frac{x^2}{2}\,e^{x^2} - \frac{1}{2}\,e^{x^2} + C. \qquad \square$$

The previous example shows how successful integration by parts substitutions are generally made. The substitution $dv = x\,e^{x^2}\,dx$ is the most complicated portion of the product $x^3\,e^{x^2}\,dx$ that can be readily integrated. Reviewing the other examples in this section you will see that this rule prevails in those problems as well. The priority of substitution is for dv rather than u, since it is more difficult to determine v from dv than to find du given u.

We close this section with an example of integration by parts applied to the evaluation of an improper integral.

EXAMPLE 8   Evaluate the improper integral $\displaystyle\int_0^e \frac{\ln x}{x^2}\,dx$, if it converges.

SOLUTION

First we determine the indefinite integral $\displaystyle\int \frac{\ln x}{x^2}\,dx$.

Let $$u = \ln x \quad \text{and} \quad dv = \frac{1}{x^2}\,dx,$$

then $$du = \frac{1}{x}\,dx \quad \text{and} \quad v = -\frac{1}{x}.$$

So $$\int \frac{\ln x}{x^2}\,dx = -\frac{\ln x}{x} + \int \frac{1}{x^2}\,dx$$

$$= -\frac{\ln x}{x} - \frac{1}{x} + C.$$

Since the integrand has an infinite discontinuity at zero,

$$\int_0^e \frac{\ln x}{x^2}\,dx = \lim_{M \to 0^+} \int_M^e \frac{\ln x}{x^2}\,dx = \lim_{M \to 0^+} \left. -\frac{\ln x}{x} - \frac{1}{x} \right]_M^e$$

$$= \lim_{M \to 0^+} \left[\left(-\frac{1}{e} - \frac{1}{e} \right) - \left(-\frac{\ln M}{M} - \frac{1}{M} \right) \right]$$

$$= -\frac{2}{e} + \lim_{M \to 0^+} \frac{\ln M + 1}{M} = -\infty.$$

The integral diverges. $\qquad \square$

EXERCISE SET 8.1

Evaluate the integrals in Exercises 1 through 46.

1. $\displaystyle\int x e^{-x}\, dx$

2. $\displaystyle\int_0^{\pi/2} x \sin x\, dx$

3. $\displaystyle\int_0^{\pi} x \sin 2x\, dx$

4. $\displaystyle\int_{-1}^0 3x\, e^{2x}\, dx$

5. $\displaystyle\int_1^e x \ln x\, dx$

6. $\displaystyle\int_1^e \sqrt{x}\, \ln x\, dx$

7. $\displaystyle\int \arcsin x\, dx$

8. $\displaystyle\int \arctan x\, dx$

9. $\displaystyle\int \arccos x\, dx$

10. $\displaystyle\int \operatorname{arc\,cot} x\, dx$

11. $\displaystyle\int x^2\, e^x\, dx$

12. $\displaystyle\int (x^2 + 1)e^{2x}\, dx$

13. $\displaystyle\int_{\pi/2}^{\pi} x^2 \cos x\, dx$

14. $\displaystyle\int_0^1 4x^2\, e^{-3x}\, dx$

15. $\displaystyle\int \ln x^2\, dx$

16. $\displaystyle\int (\ln x)^2\, dx$

17. $\displaystyle\int x^3 \ln x^2\, dx$

18. $\displaystyle\int x^3 (\ln x)^2 dx$

19. $\displaystyle\int_0^4 x\sqrt{2x + 1}\, dx$

20. $\displaystyle\int x(x - 1)^5\, dx$

21. $\displaystyle\int x(\sec x)^2\, dx$

22. $\displaystyle\int \sin x \ln \cos x\, dx$

23. $\displaystyle\int x^3(1 - x^2)^{-1/2} dx$

24. $\displaystyle\int x^3 \sin x^2\, dx$

25. $\displaystyle\int e^x \sin x\, dx$

26. $\displaystyle\int e^{2x} \cos x\, dx$

27. $\displaystyle\int e^{3x} \cos 2x\, dx$

28. $\displaystyle\int e^{2x} \cos 3x\, dx$

29. $\displaystyle\int (\sec x)^3\, dx$

30. $\displaystyle\int_0^{\pi/2} \sin x \cos 2x\, dx$

31. $\displaystyle\int_0^{\pi/6} \sin 2x \sin 3x\, dx$

32. $\displaystyle\int x \tan x \sec x\, dx$

33. $\displaystyle\int \cos \sqrt{x}\, dx$

34. $\displaystyle\int \sin \sqrt{2x}\, dx$

35. $\displaystyle\int e^{\sqrt{x}}\, dx$

36. $\displaystyle\int_0^4 e^{\sqrt{2x+1}}\, dx$

37. $\displaystyle\int_0^1 x \arctan x\,dx$

38. $\displaystyle\int_{\sqrt{2}}^2 x \operatorname{arcsec} x\,dx$

39. $\displaystyle\int x\, 3^x\,dx$

40. $\displaystyle\int x\, 2^{-x}\,dx$

41. $\displaystyle\int x^2\, 2^x\,dx$

42. $\displaystyle\int x^3\, 3^x\,dx$

43. $\displaystyle\int \sin \ln x\,dx$

44. $\displaystyle\int \cos \ln x\,dx$

45. $\displaystyle\int \ln(x + \sqrt{x^2 + 1})\,dx$

46. $\displaystyle\int (1 + x)e^x \ln x\,dx$

47. Use the method of shells to find the volume generated by rotating about the y-axis the region bounded by the x-axis, the line $x = \pi/2$, and the graph of $y = \sin x$.

48. Use the method of disks to find the volume requested in Exercise 47.

49. Determine the volume of the solid generated by rotating about the x-axis the region bounded by the x-axis, the line $x = e$, and the graph of $y = \ln x$.

50. Find the volume of the solid generated when the region described in Exercise 49 is rotated about the y-axis.

51. Find the area of the region bounded by the line $y = (x - 1)/(e - 1)$ and the graph of $y = \ln x$.

52. Find the volume generated when the region described in Exercise 51 is rotated about the x-axis.

53. A distribution in statistics known as the *gamma distribution* has probability density given by

$$f(x) = \begin{cases} kx^{\alpha-1}\, e^{-x/\beta}, & \text{for } x > 0 \\ 0, & \text{for } x \le 0 \end{cases}$$

where $\alpha > 0$, $\beta > 0$ and k must be such that $\displaystyle\int_{-\infty}^{\infty} f(x)\,dx = 1$. For $\alpha = \beta = 2$, find k.

54. An *exponential distribution* has density function described by

$$f(x) = \begin{cases} e^{-x}, & \text{for } x > 0 \\ 0, & \text{for } x \le 0. \end{cases}$$

Find the mean μ and standard deviation σ of this distribution. (*Hint:* See Exercises 42 and 43 in Section 6.8 for the definitions of μ and σ.)

55. The *gamma function* is defined by

$$\Gamma(x) = \int_0^{\infty} t^{x-1}\, e^{-t}\,dt \qquad \text{for } x > 0.$$

(a) Show that $\Gamma(x) = (x - 1)\Gamma(x - 1)$ for each $x > 1$.

(b) Show that $\Gamma(1) = 1$.

(c) Deduce from (a) and (b) that $\Gamma(n + 1) = n(n - 1) \ldots 2 \cdot 1 \equiv n!$ for each positive integer n.

56. The Laplace transform of a function f was defined in Exercise 67 of Section 7.3 by $L(f) = F(s) = \int_0^\infty e^{-st} f(t) dt$. Suppose f' is continuous on $[0, \infty)$ and $|f(t)| \le Ce^{kt}$ for all $t > 0$, where C and k are positive constants. Use integration by parts to show that $L(f') = sL(f) - f(0)$, whenever $s \ge k$.

57. Assume that a is constant. Determine

(a) $L(\sin a t)$ (b) $L(\cos a t)$ (c) $L(e^{at})$.

8.2
INTEGRALS OF PRODUCTS OF TRIGONOMETRIC FUNCTIONS

Integrals involving products and powers of trigonometric functions occur frequently in the evaluation of other integrals. The first type we consider involves products of sine and cosine functions.

The technique required to evaluate the integral

$$\int (\sin x)^n (\cos x)^m \, dx$$

for positive integers m and n depends on whether the integers are both even or at least one of the integers is odd.

(1) If n is odd, make the substitution $u = \cos x$. Use the identity $(\sin x)^2 = 1 - (\cos x)^2$ to change the expression $(\sin x)^{n-1}$ into one involving only powers of $\cos x$.

(2) If m is odd, make the substitution $u = \sin x$. Use the identity $(\cos x)^2 = 1 - (\sin x)^2$ to change the expression $(\cos x)^{m-1}$ into one involving only powers of $\sin x$.

(3) If both m and n are even, use the identities $(\sin x)^2 = \dfrac{1 - \cos 2x}{2}$ and $(\cos x)^2 = \dfrac{1 + \cos 2x}{2}$ to reduce the powers in the integrand.

(4) If both m and n are odd, choose (1) if $n < m$ and (2) if $m < n$. If $m = n$ the choice is irrelevant.

EXAMPLE 1   Evaluate $\int (\sin x)^3 (\cos x)^2 \, dx$.

SOLUTION

Since the power of $\sin x$ is odd, let $u = \cos x$. Then $du = -\sin x \, dx$ and

$$\int (\sin x)^3 (\cos x)^2 \, dx = \int (\sin x)^2 \sin x (\cos x)^2 \, dx$$

$$= \int [1 - (\cos x)^2] (\cos x)^2 \sin x \, dx$$

$$= \int [(\cos x)^2 - (\cos x)^4] \sin x \, dx$$

$$= \int (u^2 - u^4)(-du) = -\frac{u^3}{3} + \frac{u^5}{5} + C$$

$$= -\frac{(\cos x)^3}{3} + \frac{(\cos x)^5}{5} + C. \qquad \square$$

EXAMPLE 2   Evaluate $\int (\sin x)^2 (\cos x)^5 \, dx$.

SOLUTION

Since the power of $\cos x$ is odd, let $u = \sin x$. Then $du = \cos x \, dx$ and

$$\int (\sin x)^2 (\cos x)^5 \, dx = \int (\sin x)^2 (\cos x)^4 \cos x \, dx$$

$$= \int (\sin x)^2 [1 - (\sin x)^2]^2 \cos x \, dx$$

$$= \int u^2 (1 - u^2)^2 \, du$$

$$= \int u^2 (1 - 2u^2 + u^4) \, du$$

$$= \int (u^2 - 2u^4 + u^6) \, du = \frac{u^3}{3} - \frac{2u^5}{5} + \frac{u^7}{7} + C$$

$$= \frac{(\sin x)^3}{3} - \frac{2(\sin x)^5}{5} + \frac{(\sin x)^7}{7} + C. \qquad \square$$

EXAMPLE 3   Evaluate $\int (\sin x)^2 (\cos x)^2 \, dx$.

SOLUTION

Since both powers are even, we use the identities

$$(\sin x)^2 = \frac{1}{2} (1 - \cos 2x) \qquad \text{and} \qquad (\cos x)^2 = \frac{1}{2} (1 + \cos 2x).$$

$$\int (\sin x)^2 (\cos x)^2 \, dx = \frac{1}{4} \int (1 - \cos 2x)(1 + \cos 2x) \, dx$$

$$= \frac{1}{4} \int [1 - (\cos 2x)^2] \, dx.$$

However, $$(\cos 2x)^2 = \frac{1}{2} (1 + \cos 4x),$$

so $$\int (\sin x)^2 (\cos x)^2 \, dx = \frac{1}{4} \int \left[1 - \frac{1}{2} (1 + \cos 4x) \right] dx$$

$$= \frac{1}{4} \int \left(\frac{1}{2} - \frac{1}{2} \cos 4x \right) dx$$

$$= \frac{1}{4} \left(\frac{x}{2} - \frac{\sin 4x}{8} \right) + C$$

$$= \frac{1}{8} x - \frac{1}{32} \sin 4x + C. \qquad \square$$

EXAMPLE 4   Evaluate $\int \frac{(\cos x)^3}{\sqrt{\sin x}} \, dx$.

SOLUTION

Even though this integral contains a nonintegral power of $\sin x$, the technique is the same as in the previous examples. Let $u = \sin x$. Then $du = \cos x \, dx$ and

$$\int \frac{(\cos x)^3}{\sqrt{\sin x}} \, dx = \int \frac{(\cos x)^2}{\sqrt{u}} \, du$$

$$= \int \frac{1 - (\sin x)^2}{\sqrt{u}} \, du$$

$$= \int \frac{1 - u^2}{\sqrt{u}} \, du$$

$$= \int (u^{-1/2} - u^{3/2}) \, du = 2u^{1/2} - \frac{2}{5} u^{5/2} + C$$

$$= 2(\sin x)^{1/2} - \frac{2}{5} (\sin x)^{5/2} + C. \qquad \square$$

The technique involved in the integration of

$$\int (\tan x)^n (\sec x)^m \, dx$$

is similar to the powers of sines and cosines. In this instance three different approaches are needed depending on whether m and n are odd or even.

   (1) If m is even, make the substitution $u = \tan x$. Use the identity $(\sec x)^2 = 1 + (\tan x)^2$ to change the expression $(\sec x)^{m-2}$ into one involving only powers of $\tan x$.
   (2) If m and n are both odd, make the substitution $u = \sec x$. Use the identity $(\tan x)^2 = 1 - (\sec x)^2$ to change the expression $(\tan x)^{n-1}$ into one involving only powers of $\sec x$.
   (3) If n is even and m is odd, use integration by parts with $dv = \tan x \sec x \, dx$ to reduce the powers of the integrand.

EXAMPLE 5   Evaluate $\int \tan x \, (\sec x)^4 \, dx$.

SOLUTION

Since the power of $\sec x$ is even, let $u = \tan x$. Then $du = (\sec x)^2 \, dx$ and

$$\int \tan x \, (\sec x)^4 \, dx = \int u \, (\sec x)^2 \, du.$$

Since $(\sec x)^2 = 1 + (\tan x)^2 = 1 + u^2$,

$$\int \tan x \, (\sec x)^4 \, dx = \int u \, (1 + u^2) du$$

$$= \frac{u^2}{2} + \frac{u^4}{4} + C$$

$$= \frac{(\tan x)^2}{2} + \frac{(\tan x)^4}{4} + C. \qquad \square$$

EXAMPLE 6   Evaluate $\int (\tan x)^5 \sec x \, dx$.

SOLUTION

Since both powers are odd, let $u = \sec x$. Then $du = \sec x \tan x \, dx$ and

$$\int (\tan x)^5 \sec x \, dx = \int (\tan x)^4 \tan x \sec x \, dx$$

$$= \int [(\sec x)^2 - 1]^2 \tan x \sec x \, dx$$

$$= \int [u^2 - 1]^2 \, du$$

$$= \int (u^4 - 2u^2 + 1) \, du = \frac{u^5}{5} - \frac{2u^3}{3} + u + C$$

$$= \frac{(\sec x)^5}{5} - \frac{2(\sec x)^3}{3} + \sec x + C. \qquad \square$$

EXAMPLE 7   Evaluate $\int (\tan x)^2 \sec x \, dx$.

SOLUTION

Since the power of $\tan x$ is even while the power of $\sec x$ is odd, we must use integration by parts to reduce the powers in the integrand.

Let $\qquad\qquad u = \tan x \qquad$ and $\qquad dv = \tan x \sec x \, dx$.

Then $\qquad\qquad du = (\sec x)^2 \, dx \qquad$ and $\qquad v = \sec x,$

so $\qquad\int (\tan x)^2 \sec x \, dx = \sec x \tan x - \int (\sec x)^3 \, dx.$

But

$$(\sec x)^3 = (\sec x)^2 \sec x = [1 + (\tan x)^2] \sec x = \sec x + (\tan x)^2 \sec x.$$

Thus,

$$\int (\tan x)^2 \sec x \, dx = \sec x \tan x - \int \sec x \, dx - \int (\tan x)^2 \sec x \, dx$$

and

$$2 \int (\tan x)^2 \sec x \, dx = \sec x \tan x - \int \sec x \, dx.$$

Hence,

$$\int (\tan x)^2 \sec x \, dx = \frac{1}{2} \left(\sec x \tan x - \int \sec x \, dx \right)$$

$$= \frac{1}{2} (\sec x \tan x - \ln |\sec x + \tan x|) + C. \qquad \square$$

The last example of this section illustrates how to evaluate integrals of the type

$$\int (\cot x)^n (\csc x)^m \, dx.$$

Note the similarity between this procedure and that for evaluating

$$\int (\tan x)^n (\sec x)^m \, dx.$$

EXAMPLE 8   Evaluate $\int (\csc x)^4 (\cot x)^2 \, dx$.

SOLUTION

Since $D_x \cot x = -(\csc x)^2$ and $(\csc x)^2 = (\cot x)^2 + 1$, we make the substitution $u = \cot x$. Then $du = -(\csc x)^2 \, dx$ and

$$\int (\csc x)^4 (\cot x)^2 \, dx = \int (\csc x)^2 (\cot x)^2 (\csc x)^2 \, dx$$

$$= -\int (\csc x)^2 \, u^2 \, du$$

$$= -\int [(\cot x)^2 + 1] \, u^2 \, du$$

$$= -\int (u^2 + 1) u^2 \, du$$

$$= -\int (u^4 + u^2) \, du = -\left(\frac{u^5}{5} + \frac{u^3}{3}\right) + C$$

$$= -\frac{(\cot x)^5}{5} - \frac{(\cot x)^3}{3} + C. \qquad \square$$

EXERCISE SET 8.2

Evaluate the integrals in Exercises 1 through 40.

1. $\int (\sin x)^2 \cos x \, dx$

2. $\int (\cos x)^3 \, dx$

3. $\int_0^{\pi/2} (\sin 2x)^2 (\cos 2x)^3 \, dx$

4. $\int (\sin (x + 1))^5 \, dx$

5. $\int \sin 3x \, (\cos 3x)^4 \, dx$

6. $\int_{\pi/4}^{\pi/2} (\sin x)^2 (\cos x)^2 \, dx$

7. $\int_0^{\pi/6} (\sin 3x)^2 (\cos 3x)^4 \, dx$

8. $\int_{\pi/4}^{\pi/2} (\sin 2x)^4 (\cos 2x)^3 \, dx$

9. $\int_0^{\pi} (\cos 5x)^4 \, dx$

10. $\int (\sin 2x)^4 \, dx$

11. $\int \sqrt{\sin x} \, (\cos x)^3 \, dx$

12. $\int \sqrt{\sin x} \, \cos x \, dx$

13. $\int_{-\pi/2}^{\pi/2} (\sin x + \cos x)^3 \, dx$

14. $\int_{-\pi/2}^{\pi/2} (\sin x + \cos x)^2 \, dx$

15. $\displaystyle\int (\sec x)^4 (\tan x)^3 \, dx$ **16.** $\displaystyle\int (\tan x)^3 (\sec x)^2 \, dx$

17. $\displaystyle\int_{\pi/4}^{\pi/2} (\cot x)^2 \, dx$ **18.** $\displaystyle\int_0^{\pi/2} \left(\tan \frac{x}{2}\right)^2 dx$

19. $\displaystyle\int \cot x \, (\csc x)^4 \, dx$ **20.** $\displaystyle\int (\csc 5x)^2 (\cot 5x)^2 \, dx$

21. $\displaystyle\int (\cot x)^5 \csc x \, dx$ **22.** $\displaystyle\int \cot x \, (\csc x)^3 \, dx$

23. $\displaystyle\int (\cot 2x)^3 \, dx$ **24.** $\displaystyle\int x(\cot x^2)^2 \csc x^2 \, dx$

25. $\displaystyle\int (\tan x + \cos x)^2 \, dx$ **26.** $\displaystyle\int (\tan 2x + \cot 2x)^2 \, dx$

27. $\displaystyle\int \ln(\sin x)(\cos x)^3 \, dx$ **28.** $\displaystyle\int \ln(\tan x)(\sec x)^4 \, dx$

29. $\displaystyle\int e^{\sin x} (\cos x)^3 \, dx$ **30.** $\displaystyle\int e^{\tan x} (\sec x)^4 \, dx$

31. $\displaystyle\int (\cos x)^2 (\tan x) \, dx$ **32.** $\displaystyle\int (\cos x)^2 (\tan x)^5 \, dx$

33. $\displaystyle\int \frac{\tan x}{\sec x} \, dx$ **34.** $\displaystyle\int \frac{\tan x}{(\sec x)^3} \, dx$

35. $\displaystyle\int \frac{\sec x}{\tan x} \, dx$ **36.** $\displaystyle\int \frac{(\sec x)^2}{(\tan x)^3} \, dx$

37. $\displaystyle\int (\tan x)^3 \, dx$ **38.** $\displaystyle\int (\sin x)^2 (\cot x)^3 \, dx$

39. $\displaystyle\int \frac{\cot x}{(\csc x)^4} \, dx$ **40.** $\displaystyle\int (\csc x)^3 \, dx$

41. A particle is traveling with velocity $v(t) = \left(\cos \dfrac{\pi t}{3}\right)^2$ ft/sec. Find the distance traveled by the particle from $t = 0$ to $t = 1$.

42. Find the area bounded by $x = 0$, $x = \pi/4$ and the graphs of $y = (\sin x)^2$ and $y = (\cos x)^2$.

43. Use the method of disks to find the volume generated by rotating about the x-axis the region bounded by the x-axis, the line $x = \pi/4$, and the graph of $f(x) = \tan x$.

44. Find the volume generated when the region described in Exercise 43 is rotated about the x-axis.

Use the identities

$$\sin a \cos b = \frac{1}{2}[\sin (a + b) + \sin (a - b)]$$

$$\cos a \cos b = \frac{1}{2}[\cos (a + b) + \cos (a - b)]$$

$$\sin a \sin b = \frac{1}{2}[\cos (a - b) - \cos (a + b)]$$

to evaluate the integrals in Exercises 45 through 50.

45. $\displaystyle\int \sin x \cos 2x \, dx$ **46.** $\displaystyle\int \sin 2x \cos x \, dx$

47. $\displaystyle\int \sin 2x \cos 3x \, dx$ **48.** $\displaystyle\int \sin 2x \sin 4x \, dx$

49. $\displaystyle\int \cos 3x \cos 5x \, dx$ **50.** $\displaystyle\int \sin (-7x) \cos x \, dx$

51. Find a general formula for

    (a) $\displaystyle\int \sin ax \cos bx \, dx,$ (b) $\displaystyle\int \cos ax \cos bx \, dx,$

    (c) $\displaystyle\int \sin ax \sin bx \, dx,$

    when $a^2 \neq b^2$.

8.3
INTEGRATION USING TRIGONOMETRIC SUBSTITUTIONS

Many applications involve integrals containing an expression of the form $\sqrt{a^2 - x^2}$, $\sqrt{a^2 + x^2}$, or $\sqrt{x^2 - a^2}$, for a constant a. Integrals of this type can often be simplified by making an appropriate trigonometric substitution to eliminate the radical. The substitutions are based on the identity

$$a^2 = a^2 [(\sin \theta)^2 + (\cos \theta)^2]$$

or its equivalent form involving tangents and secants,

$$a^2 [1 + (\tan \theta)^2] = a^2 (\sec \theta)^2.$$

When an integral involves an expression of the form

$$\sqrt{a^2 - x^2}, \quad a > 0,$$

we make the substitution

$$x = a \sin \theta, \quad \text{where } -\frac{\pi}{2} \leq \theta \leq \frac{\pi}{2}.$$

This changes the radical to

$$\sqrt{a^2 - x^2} = \sqrt{a^2 - a^2 (\sin \theta)^2} = \sqrt{a^2 (\cos \theta)^2}.$$

Since $-\dfrac{\pi}{2} \leq \theta \leq \dfrac{\pi}{2}$, $\cos \theta \geq 0$, so $\sqrt{a^2 (\cos \theta)^2} = a \cos \theta$ and

$$\sqrt{a^2 - x^2} = a \cos \theta.$$

When an integral involves an expression of the form

$$\sqrt{a^2 + x^2}, \quad a > 0,$$

we make the substitution

$$x = a \tan \theta, \quad \text{where } -\frac{\pi}{2} < \theta < \frac{\pi}{2},$$

which simplifies this radical to

$$\sqrt{a^2 + x^2} = \sqrt{a^2 + a^2 (\tan \theta)^2} = \sqrt{a^2 (\sec \theta)^2} = a \sec \theta.$$

When the integral involves an expression of the form

$$\sqrt{x^2 - a^2}, \quad a > 0,$$

we make the substitution

$$x = a \sec \theta, \quad \text{where } 0 \leq \theta \leq \pi, \quad \theta \neq \frac{\pi}{2},$$

which simplifies this radical to

$$\sqrt{x^2 - a^2} = \sqrt{a^2 (\sec \theta)^2 - a^2} = \sqrt{a^2 [(\sec \theta)^2 - 1]}$$

$$= \sqrt{a^2 (\tan \theta)^2}$$

$$= \begin{cases} a \tan \theta, & \text{for } 0 \leq \theta < \dfrac{\pi}{2} \\ -a \tan \theta, & \text{for } \dfrac{\pi}{2} < \theta \leq \pi. \end{cases}$$

The difference in sign occurs because $\tan \theta \geq 0$ for $0 \leq \theta < \dfrac{\pi}{2}$ and $\tan \theta \leq 0$ for $\dfrac{\pi}{2} < \theta \leq \pi$.

The restrictions on θ in the trigonometric substitutions are made to ensure that x/a is in the domain of the appropriate inverse trigonometric function. Keeping these restrictions in mind, the substitutions are expressed concisely in the table below.

Term in the Original Integrand	Trigonometric Substitution	Simplified Term in the Integrand
$\sqrt{a^2 - x^2}, a > 0$	$x = a \sin \theta$	$a \cos \theta$
$\sqrt{a^2 + x^2}, a > 0$	$x = a \tan \theta$	$a \sec \theta$
$\sqrt{x^2 - a^2}, a > 0$	$x = a \sec \theta$	$\pm a \tan \theta$

EXAMPLE 1   Evaluate $\displaystyle\int \sqrt{4 - x^2} \, dx$.

SOLUTION

Let $x = 2 \sin \theta$; then $dx = 2 \cos \theta \, d\theta$ and

$$\int \sqrt{4 - x^2} \, dx = \int \sqrt{4 - 4(\sin \theta)^2} \, 2 \cos \theta \, d\theta$$

$$= \int (2 \cos \theta) \, 2 \cos \theta \, d\theta$$

$$= 4 \int (\cos \theta)^2 \, d\theta.$$

The trigonometric identity

$$(\cos \theta)^2 = \frac{1}{2}(1 + \cos 2\theta)$$

reduces the integral to

$$\int \sqrt{4 - x^2} \, dx = 4 \int \frac{1}{2}(1 + \cos 2\theta) d\theta$$

$$= 2\theta + \sin 2\theta + C.$$

The solution will be complete when we rewrite this result in terms of the variable x.

Since

$$x = 2 \sin \theta,$$

$$\theta = \arcsin\left(\frac{x}{2}\right).$$

Also

$$\sin 2\theta = 2 \sin \theta \cos \theta$$

$$= 2 \sin \theta \sqrt{1 - (\sin \theta)^2}$$

$$= 2 \left(\frac{x}{2}\right) \sqrt{1 - \left(\frac{x}{2}\right)^2}$$

$$= \frac{x}{2} \sqrt{4 - x^2}.$$

Thus,

$$\int \sqrt{4 - x^2} \, dx = 2 \arcsin\left(\frac{x}{2}\right) + \frac{x}{2} \sqrt{4 - x^2} + C. \qquad \square$$

EXAMPLE 2   Evaluate $\displaystyle\int \frac{1}{\sqrt{9 + x^2}} \, dx$.

SOLUTION

Let $x = 3 \tan \theta$; then $dx = 3(\sec \theta)^2 \, d\theta$ and

$$\int \frac{1}{\sqrt{9 + x^2}} \, dx = \int \frac{3(\sec \theta)^2}{3 \sec \theta} \, d\theta$$

$$= \int \sec \theta \, d\theta = \ln |\sec \theta + \tan \theta| + C.$$

Since $x = 3 \tan \theta$, $\tan \theta = \dfrac{x}{3}$ and

$$\sec \theta = \sqrt{1 + (\tan \theta)^2} = \sqrt{1 + \left(\dfrac{x}{3}\right)^2} = \dfrac{1}{3}\sqrt{9 + x^2}.$$

Thus

$$\int \dfrac{1}{\sqrt{9 + x^2}}\, dx = \ln \left| \dfrac{1}{3}\sqrt{9 + x^2} + \dfrac{x}{3} \right| + C$$

$$= \ln \left| \sqrt{9 + x^2} + x \right| + \ln \dfrac{1}{3} + C.$$

Since C is an arbitrary constant and $\ln (1/3)$ is also a constant, it is standard practice to replace the sum $\ln (1/3) + C$ simply by the arbitrary constant symbol C. In addition, $\sqrt{9 + x^2} + x$ is always positive so

$$\int \dfrac{1}{\sqrt{9 + x^2}}\, dx = \ln \left(\sqrt{9 + x^2} + x \right) + C. \qquad \square$$

EXAMPLE 3   Evaluate $\displaystyle\int_2^4 \dfrac{\sqrt{x^2 - 4}}{x}\, dx.$

SOLUTION

First we will evaluate the indefinite integral

$$\int \dfrac{\sqrt{x^2 - 4}}{x}\, dx.$$

Let $x = 2 \sec \theta$. Then $dx = 2 \sec \theta \tan \theta\, d\theta$ and

$$\int \dfrac{\sqrt{x^2 - 4}}{x}\, dx = \int \dfrac{\sqrt{4(\sec \theta)^2 - 4}}{2 \sec \theta}\, 2 \sec \theta \tan \theta\, d\theta$$

$$= 2 \int \sqrt{(\tan \theta)^2} \, \tan \theta\, d\theta$$

$$= \pm\, 2 \int (\tan \theta)^2 d\theta$$

$$= \pm\, 2 \int [(\sec \theta)^2 - 1] d\theta$$

$$= \pm\, 2(\tan \theta - \theta) + C.$$

The sign choice is needed in the indefinite integral because we do not know the values of θ. However, from the definite integral, we see that since $\sec \theta = \dfrac{x}{2}$ and $2 \leq x \leq 4$,

$$1 \leq \sec \theta \leq 2.$$

This implies $0 \le \theta < \dfrac{\pi}{2}$ and the positive sign applies. To express the indefinite integral in terms of x, we use

$$\tan \theta = \sqrt{(\sec \theta)^2 - 1} = \sqrt{\left(\frac{x}{2}\right)^2 - 1} = \frac{\sqrt{x^2 - 4}}{2} \quad \text{and} \quad \theta = \text{arcsec} \frac{x}{2}.$$

So

$$\int \frac{\sqrt{x^2 - 4}}{x}\, dx = 2(\tan \theta - \theta) + C$$

$$= \sqrt{x^2 - 4} - 2\, \text{arcsec}\, \frac{x}{2} + C$$

and

$$\int_2^4 \frac{\sqrt{x^2 - 4}}{x}\, dx = \sqrt{x^2 - 4} - 2\, \text{arcsec}\, \frac{x}{2}\Bigg]_2^4$$

$$= (\sqrt{12} - 2\, \text{arcsec}\, 2) - (0 - 2\, \text{arcsec}\, 1)$$

$$= \sqrt{12} - 2\, \text{arcsec}\, 2 = 2\sqrt{3} - 2\pi/3. \qquad \square$$

These examples demonstrate the basic procedure used to evaluate integrals requiring trigonometric substitution. In the next example we give an alternate method for simplifying the last step in the procedure. This method relies on the geometric representation of the trigonometric functions.

EXAMPLE 4   Evaluate $\displaystyle\int \frac{\sqrt{4 - x^2}}{x}\, dx$.

SOLUTION

Let $x = 2 \sin \theta$, then $dx = 2 \cos \theta\, d\theta$ and

$$\int \frac{\sqrt{4 - x^2}}{x}\, dx = \int \frac{\sqrt{4 - 4(\sin \theta)^2}}{2 \sin \theta}\, 2 \cos \theta\, d\theta$$

$$= \int \frac{\cos \theta}{\sin \theta}\, 2 \cos \theta\, d\theta$$

$$= 2 \int \frac{(\cos \theta)^2}{\sin \theta}\, d\theta = 2 \int \frac{1 - (\sin \theta)^2}{\sin \theta}\, d\theta$$

$$= 2 \int \csc \theta\, d\theta - 2 \int \sin \theta\, d\theta$$

$$= 2 \ln |\csc \theta - \cot \theta| + 2 \cos \theta + C.$$

Instead of using trigonometric identities to find $\csc \theta$, $\cot \theta$, and $\cos \theta$, we use the right triangle in Figure 8.1 to determine these expressions.

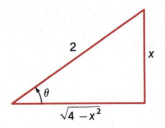

FIGURE 8.1

The condition $\sin \theta = x/2$ determines two sides of this triangle; the third is determined by the Pythagorean theorem. The other trigonometric functions can then be written in terms of the sides of this triangle:

$$\csc \theta = \frac{2}{x}, \quad \cot \theta = \frac{\sqrt{4 - x^2}}{x}, \quad \text{and} \quad \cos \theta = \frac{\sqrt{4 - x^2}}{2}.$$

Thus,

$$\int \frac{\sqrt{4 - x^2}}{x}\, dx = 2 \ln|\csc \theta - \cot \theta| + \cos \theta + C$$

$$= 2 \ln\left| \frac{2}{x} - \frac{\sqrt{4 - x^2}}{x} \right| + \sqrt{4 - x^2} + C. \qquad \square$$

EXAMPLE 5   Evaluate $\displaystyle\int \frac{dx}{(x^2 + 2x + 5)}.$

SOLUTION

First we complete the square in $x^2 + 2x + 5$.

$$x^2 + 2x + 5 = (x^2 + 2x + 1) + 4 = (x + 1)^2 + 4.$$

So $$\int \frac{dx}{x^2 + 2x + 5} = \int \frac{dx}{(x + 1)^2 + 4}.$$

Since the squared term involving x is $(x + 1)^2$ and $a = 2$, the substitution $x + 1 = 2 \tan \theta$ is used. Then $dx = 2(\sec \theta)^2 d\theta$ and

$$\int \frac{dx}{x^2 + 2x + 5} = \int \frac{dx}{(x + 1)^2 + 4}$$

$$= \int \frac{2(\sec \theta)^2\, d\theta}{4(\tan \theta)^2 + 4} = \int \frac{2(\sec \theta)^2\, d\theta}{4(\sec \theta)^2}$$

$$= \frac{1}{2}\int d\theta = \frac{1}{2}\theta + C$$

$$= \frac{1}{2} \arctan \frac{x + 1}{2} + C. \qquad \square$$

The preceding example shows that a knowledge of trigonometric substitutions eliminates the need for knowing the integration formulas involving inverse trigonometric functions. For example, the formula

$$\int \frac{1}{a^2 + x^2}\, dx = \frac{1}{a}\arctan\frac{x}{a} + C$$

can be quickly derived by substituting $x = a \tan \theta$.

EXERCISE SET 8.3

Evaluate the integrals in Exercises 1 through 28.

1. $\displaystyle\int \sqrt{1 - x^2}\, dx$

2. $\displaystyle\int \sqrt{100 - x^2}\, dx$

3. $\displaystyle\int_0^4 \sqrt{16 - x^2}\, dx$

4. $\displaystyle\int_{-3}^3 \sqrt{9 - x^2}\, dx$

5. $\displaystyle\int \frac{3}{x^2 + 9}\, dx$

6. $\displaystyle\int \frac{3}{9x^2 + 1}\, dx$

7. $\displaystyle\int \frac{1}{\sqrt{16 + x^2}}\, dx$

8. $\displaystyle\int_0^5 \frac{1}{\sqrt{25 + x^2}}\, dx$

9. $\displaystyle\int_0^1 \frac{x^2}{x^2 + 1}\, dx$

10. $\displaystyle\int \frac{x^2}{\sqrt{1 + x^2}}\, dx$

11. $\displaystyle\int_1^2 x\sqrt{x^2 - 1}\, dx$

12. $\displaystyle\int_2^3 2x\sqrt{x^2 - 4}\, dx$

13. $\displaystyle\int_{-2}^{-1} \frac{\sqrt{x^2 - 1}}{x}\, dx$

14. $\displaystyle\int_{-2}^{-1} \frac{x}{\sqrt{x^2 - 1}}\, dx$

15. $\displaystyle\int_{-1}^1 \sqrt{16 - 9x^2}\, dx$

16. $\displaystyle\int_{-9/2}^{9/2} \sqrt{81 - 4x^2}\, dx$

17. $\displaystyle\int \sqrt{4 - (x + 1)^2}\, dx$

18. $\displaystyle\int \frac{1}{1 + (2x - 3)^2}\, dx$

19. $\displaystyle\int \frac{1}{x^2 + 2x + 10}\, dx$

20. $\displaystyle\int \sqrt{5 - x^2 - 2x}\, dx$

21. $\displaystyle\int \sqrt{2x - x^2}\, dx$

22. $\displaystyle\int \sqrt{x^2 - 2x}\, dx$

23. $\displaystyle\int \frac{1}{\sqrt{x}(x+1)}\,dx$ **24.** $\displaystyle\int \frac{\sqrt{1-x}}{\sqrt{x}}\,dx$

(*Hint:* Let $\sqrt{x} = \tan\theta$.)

25. $\displaystyle\int e^x\sqrt{1-e^{2x}}\,dx$ **26.** $\displaystyle\int_0^{\ln\sqrt{3}} \frac{e^x}{e^{2x}+1}\,dx$

27. $\displaystyle\int \frac{e^{x/2}}{e^x+1}\,dx$ **28.** $\displaystyle\int \frac{1}{\sqrt{e^{2x}+1}}\,dx$

29. Find the area of the region bounded by $y = 1/(x^2+1)$, the x-axis, and the lines $x = -2$, $x = 2$.

30. Show that the area enclosed in the circle $x^2 + y^2 = r^2$ is πr^2.

31. Find the area of the region enclosed by the ellipse $x^2/16 + y^2/9 = 1$.

32. Show that the area of the region enclosed by the ellipse $x^2/a^2 + y^2/b^2 = 1$ is πab.

Find the arc length of the graph of each function given in Exercises 33 through 36.

33. $f(x) = \ln x$ from $x = 1$ to $x = e$.

34. $f(x) = x^2$ from $x = 0$ to $x = 2$.

35. $f(x) = \sqrt{4-x^2}$ from $x = -2$ to $x = 2$.

36. $f(x) = x^2 - 2x$ from $x = 0$ to $x = 2$.

37. Show that the circumference of the circle $x^2 + y^2 = r^2$ is $2\pi r$.

38. Find the volume of the solid obtained by rotating about the y-axis the region bounded by $y = x/\sqrt{1+x^2}$, $x = 3$, and the x-axis.

39. The velocity of an object moving along a straight line is given by $v(t) = \sqrt{25-t^2}$ cm/sec. Find the distance traveled by the object during the time interval $[0, 5]$.

40. A *torus* is a solid generated by rotating a circle about a line that does not intersect the circle. A baker has been making, and selling for 20¢ each, plain bagels in the shape of a torus. The bagel can be described by rotating a circle of diameter 1 inch about a line 1 inch from the center of the circle. This baker decides to make a giant bagel described by a torus generated by rotating a circle of radius 1 inch about a line 1.25 inches from the center of the circle. What should be the selling price of the giant bagels?

41. Show that if $u = \tan x/2$, $-\pi < x < \pi$, then

$$\sin x = \frac{2u}{1+u^2}, \quad \cos x = \frac{1-u^2}{1+u^2}, \quad \text{and} \quad dx = \frac{2du}{1+u^2}.$$

Use the substitution and relations in Exercise 41 to determine the integrals in Exercises 42 through 47.

42. $\displaystyle\int_0^{\pi/2} \frac{dx}{1 + \sin x}$

43. $\displaystyle\int_0^{\pi/2} \frac{dx}{1 + \cos x}$

44. $\displaystyle\int \frac{dx}{\cos x - \sin x + 1}$

45. $\displaystyle\int \frac{dx}{\cos x + \sin x + 1}$

46. $\displaystyle\int \frac{dx}{\sin x + \cos x}$

47. $\displaystyle\int \frac{dx}{\sin x + \tan x}$

8.4
INTEGRATION USING PARTIAL FRACTIONS

The method of partial fractions is used to integrate rational functions: functions that are quotients of polynomials. The method is based on the fact that any rational function can be written as a sum of a polynomial and fractions of the form

$$\text{(i)} \ \frac{1}{(x + a)^n}, \qquad \text{(ii)} \ \frac{1}{(x^2 + ax + b)^n}, \quad \text{and} \quad \text{(iii)} \ \frac{x}{(x^2 + ax + b)^n}.$$

Fractions having one of these forms are called **partial fractions**.

The method of partial fractions can be applied only to rational functions when the degree of the numerator is less than the degree of the denominator. Such functions are called **proper rational functions**. If the rational function is not of this type, we can use polynomial division to express it as a sum of a polynomial and a proper rational function.

EXAMPLE 1   Write the rational function described by

$$R(x) = \frac{2x^4 - x^3 - 5x^2 - 3x + 10}{x^3 - x^2 - 4x + 4}$$

as the sum of a polynomial and a proper fraction.

SOLUTION

We divide the numerator by the denominator

$$
\begin{array}{r}
2x + 1 \\
x^3 - x^2 - 4x + 4 \overline{\smash{)}\ 2x^4 - x^3 - 5x^2 - 3x + 10} \\
\underline{2x^4 - 2x^3 - 8x^2 + 8x } \\
x^3 + 3x^2 - 11x + 10 \\
\underline{x^3 - x^2 - 4x + 4} \\
4x^2 - 7x + 6
\end{array}
$$

Thus $R(x) = 2x + 1 + \dfrac{4x^2 - 7x + 6}{x^3 - x^2 - 4x + 4}.$ $\square$

The first step in the partial fraction procedure is to factor completely the denominator of the proper rational function in terms of linear and quadratic factors. Theoretically, the factorization can always be accomplished, but practically the procedure can be quite difficult, if indeed possible. The factorization relies on the fact that:

If P is a polynomial, then $x - a$ is a factor of $P(x)$ if and only if $P(a) = 0$.

EXAMPLE 2   Factor $P(x) = x^3 - x^2 - 4x + 4$ completely.

SOLUTION

Since $P(1) = 0$, $x = 1$ is a factor of $P(x)$. Dividing $P(x)$ by $x - 1$ gives

$$P(x) = (x - 1)(x^2 - 4) = (x - 1)(x - 2)(x + 2). \qquad \square$$

The procedure for decomposing a proper fraction into the sum of partial fractions can be explained by considering cases determined by the factors of the denominator.

Case 1. $R(x)$ is a proper fraction of the form $Q(x)/P(x)$ and $P(x) = (x - r_1)$ $(x - r_2) \ldots (x - r_n)$, where $r_1, r_2, \ldots, r_n$ are distinct real numbers.

Then $R(x)$ can be written

$$R(x) = \frac{Q(x)}{P(x)} = \frac{A_1}{x - r_1} + \frac{A_2}{x - r_2} + \cdots + \frac{A_n}{x - r_n}$$

for a unique set of constants $A_1, A_2, \ldots, A_n$.

The terms on the right side of this equation describe the most general collection of fractions with linear denominators that can contribute to a common denominator of the form $P(x)$. The procedure used to determine the specific values of the constants is demonstrated in the following example.

EXAMPLE 3   Evaluate $\displaystyle\int \frac{2x^4 - x^3 - 5x^2 - 3x + 10}{x^3 - x^2 - 4x + 4} \, dx$.

SOLUTION

We must first express

$$R(x) = \frac{2x^4 - x^3 - 5x^2 - 3x + 10}{x^3 - x^2 - 4x + 4}$$

as the sum of a polynomial and a proper fraction. We showed in Example 1 that

$$R(x) = 2x + 1 + \frac{4x^2 - 7x + 6}{x^3 - x^2 - 4x + 4}.$$

So $\displaystyle\int R(x)dx = \int (2x + 1)dx + \int \frac{4x^2 - 7x + 6}{x^3 - x^2 - 4x + 4} \, dx$

$$= x^2 + x + \int \frac{4x^2 - 7x + 6}{x^3 - x^2 - 4x + 4} \, dx.$$

To evaluate the last integral, express

$$\frac{Q(x)}{P(x)} = \frac{4x^2 - 7x + 6}{x^3 - x^2 - 4x + 4}$$

in partial fraction form. From Example 2,

$$\frac{Q(x)}{P(x)} = \frac{4x^2 - 7x + 6}{(x - 1)(x - 2)(x + 2)} = \frac{A_1}{x - 1} + \frac{A_2}{x - 2} + \frac{A_3}{x + 2}.$$

To determine the constants A_1, A_2, and A_3, we first multiply both sides of the equation by the original denominator $P(x) = (x - 1)(x - 2)(x + 2)$:

$$4x^2 - 7x + 6 = A_1(x - 2)(x + 2) + A_2(x - 1)(x + 2) + A_3(x - 1)(x - 2)$$

and then find the values of the constants that satisfy this polynomial equation. In particular, for this equation to be satisfied at the zeros of P:

$$x = 1: \quad 3 = A_1(-1)(3) + A_2(0)(3) + A_3(0)(-1), \qquad \text{so } A_1 = -1;$$

$$x = 2: \quad 8 = -1(0)(2) + A_2(1)(4) + A_3(1)(0), \qquad \text{so } A_2 = 2;$$

$$x = -2: \quad 36 = -1(-4)(0) + 2(-3)(0) + A_3(-3)(-4), \qquad \text{so } A_3 = 3.$$

This implies that the partial fractions form of $R(x)$ is

$$\frac{4x^2 - 7x + 6}{x^3 - x^2 - 4x + 4} = \frac{-1}{x - 1} + \frac{2}{x - 2} + \frac{3}{x + 2}$$

and

$$\int \frac{4x^2 - 7x + 6}{x^3 - x^2 - 4x + 4}\, dx = -\int \frac{dx}{x - 1} + 2\int \frac{dx}{x - 2} + 3\int \frac{dx}{x + 2}$$

$$= -\ln|x - 1| + 2\ln|x - 2| + 3\ln|x + 2| + C$$

$$= -\ln|x - 1| + \ln(x - 2)^2 + \ln|x + 2|^3 + C$$

$$= \ln\left|\frac{(x - 2)^2\,(x + 2)^3}{x - 1}\right| + C.$$

Thus

$$\int \frac{2x^4 - x^3 - 5x^2 - 3x + 10}{x^3 - x^2 - 4x + 4}\, dx = x^2 + x + \ln\left|\frac{(x - 2)^2(x + 2)^3}{x - 1}\right| + C. \quad \square$$

Case 2. $R(x)$ is a proper fraction of the form $Q(x)/P(x)$ and $P(x) = (x - r_1)$ $(x - r_2) \ldots (x - r_n)$, where $r_1, r_2, \ldots r_n$ are real numbers, not all of which are distinct.

We demonstrate the procedure for this case in the next example.

EXAMPLE 4   Evaluate

$$\int \frac{x^2 + 3x + 3}{x(x + 2)^2}\, dx.$$

SOLUTION

The form of the partial fractions is

(8.3)
$$\frac{x^2 + 3x + 3}{x(x + 2)^2} = \frac{A_1}{x} + \frac{A_2}{x + 2} + \frac{A_3}{(x + 2)^2},$$

because these terms involve the most general factors that can contribute to a common denominator of the form $x(x + 2)^2$. In this example, as in Example 3, the number of constants to be determined agrees with the degree of the polynomial in the denominator of the original rational function. This will always be the case, regardless of the form of the denominator.

Multiplying both sides of equation (8.3) by $x(x + 2)^2$ gives

(8.4)
$$x^2 + 3x + 3 = A_1(x + 2)^2 + A_2(x + 2)x + A_3x.$$

$$x = 0: \quad 3 = A_1(4) + A_2(0) + A_3(0), \qquad \text{so } A_1 = \frac{3}{4};$$

$$x = -2: \quad 1 = \frac{3}{4}(0) + A_2(0) + A_3(-2), \qquad \text{so } A_3 = -\frac{1}{2}.$$

There is no value of x that can be used to reduce the coefficients of both A_1 and A_3 to zero. To find A_2, evaluate equation (8.4) at any number other than 0 and -2, for example,

$$x = 1: \quad 7 = \frac{3}{4}(3)^2 + A_2(3)(1) - \frac{1}{2}(1), \qquad \text{so } A_2 = \frac{1}{4}.$$

Hence,

$$\frac{x^2 + 3x + 3}{x(x + 2)^2} = \frac{3}{4x} + \frac{1}{4(x + 2)} - \frac{1}{2(x + 2)^2},$$

and

$$\int \frac{x^2 + 3x + 3}{x(x + 2)^2}\,dx = \frac{3}{4}\int \frac{dx}{x} + \frac{1}{4}\int \frac{dx}{x + 2} - \frac{1}{2}\int \frac{dx}{(x + 2)^2}$$

$$= \frac{3}{4}\ln|x| + \frac{1}{4}\ln|x + 2| + \frac{1}{2}\left(\frac{1}{x + 2}\right) + C. \qquad \square$$

Case 3. $R(x)$ is a proper fraction of the form $Q(x)/P(x)$ and $P(x)$ has irreducible factors of the form $x^2 + ax + b$.

In this case the partial fraction decomposition contains fractions of the form

$$\frac{Bx}{x^2 + ax + b} \qquad \text{and} \qquad \frac{C}{x^2 + ax + b},$$

since these are the most general proper fractions that can contribute to a common denominator containing $x^2 + ax + b$.

EXAMPLE 5   Evaluate

$$\int \frac{4x^2 - x + 5}{(x - 5)(x^2 + 4x + 5)}\,dx.$$

SOLUTION
Since $x^2 + 4x + 5$ is irreducible, the partial fraction form of the integrand is

$$\frac{4x^2 - x + 5}{(x - 5)(x^2 + 4x + 5)} = \frac{A}{x - 5} + \frac{Bx}{x^2 + 4x + 5} + \frac{C}{x^2 + 4x + 5}.$$

As in the previous examples, the number of constants to be determined is the same as the degree of the original denominator. To find the constants, multiply by $(x - 5)(x^2 + 4x + 5)$ to obtain:

$$4x^2 - x + 5 = A(x^2 + 4x + 5) + Bx(x - 5) + C(x - 5).$$

If

$$x = 5: \quad 100 = 50A, \quad \text{so } A = 2;$$

$$x = 0: \quad 5 = 10 - 5C, \quad \text{so } C = 1.$$

To find B let x be any number other than 0 and 5, for example, $x = 1$.

$$x = 1: 8 = 20 - 4B - 4, \quad \text{so } B = 2.$$

The partial fraction form of the integrand is therefore

$$\frac{4x^2 - x + 5}{(x - 5)(x^2 + 4x + 5)} = \frac{2}{x - 5} + \frac{2x}{x^2 + 4x + 5} + \frac{1}{x^2 + 4x + 5}$$

and

$$\int \frac{4x^2 - x + 5}{(x - 5)(x^2 + 4x + 5)} \, dx = \int \left(\frac{2}{x - 5} + \frac{2x + 1}{x^2 + 4x + 5} \right) dx$$

$$= 2 \ln|x - 5| + \int \frac{2x + 1}{x^2 + 4x + 5} \, dx.$$

To evaluate $\int \dfrac{2x + 1}{x^2 + 4x + 5} \, dx$, let $u = x^2 + 4x + 5$. Then $du = (2x + 4)dx$, and

$$\int \frac{2x + 1}{x^2 + 4x + 5} \, dx = \int \frac{2x + 4 - 3}{x^2 + 4x + 5} \, dx$$

$$= \int \frac{2x + 4}{x^2 + 4x + 5} \, dx - \int \frac{3}{x^2 + 4x + 5} \, dx$$

$$= \int \frac{1}{u} \, du - \int \frac{3}{x^2 + 4x + 5} \, dx$$

$$= \ln|u| - \int \frac{3}{x^2 + 4x + 5} \, dx$$

$$= \ln|x^2 + 4x + 5| - \int \frac{3}{x^2 + 4x + 5} \, dx.$$

Finally, to evaluate $\int \dfrac{3}{x^2 + 4x + 5} \, dx$, we complete the square of the denominator

$$x^2 + 4x + 5 = x^2 + 4x + 4 + 1 = (x + 2)^2 + 1$$

and make the substitution $x + 2 = \tan \theta$, $dx = (\sec \theta)^2 \, d\theta$. Then

$$\int \frac{3}{x^2 + 4x + 5} \, dx = \int \frac{3 \, (\sec \theta)^2}{(\tan \theta)^2 + 1} \, d\theta$$

$$= 3 \int \frac{(\sec \theta)^2}{(\sec \theta)^2} \, d\theta$$

$$= 3\theta + C = 3 \arctan (x + 2) + C.$$

Thus,

$$\int \frac{4x^2 - x + 5}{(x - 5)(x^2 + 4x + 5)} \, dx = 2 \ln|x - 5| + \ln|x^2 + 4x + 5| - 3 \arctan (x + 2) + C.$$

$\square$

Case 4. $R(x)$ is a proper fraction of the form $Q(x)/P(x)$ and $P(x)$ has factors of the form $(ax^2 + bx + c)^n$, $n \geq 2$.

The partial fraction decomposition of $R(x)$ contains the terms

$$\frac{B_j x}{(x^2 + ax + b)^j} \quad \text{and} \quad \frac{C_j}{(x^2 + ax + b)^j}, \quad \text{for each } j = 1, 2, \ldots, n.$$

This naturally complicates the situation; however, the procedure for determining the constants B_j and C_j is the same as before.

EXAMPLE 6   Evaluate

$$\int \frac{dx}{x(1 + x^2)^2} .$$

SOLUTION

To find the partial fraction decomposition of

$$R(x) = \frac{1}{x(1 + x^2)^2} ,$$

we write

(8.5) $$\frac{1}{x(1 + x^2)^2} = \frac{A}{x} + \frac{B_1 x}{1 + x^2} + \frac{C_1}{1 + x^2} + \frac{B_2 x}{(1 + x^2)^2} + \frac{C_2}{(1 + x^2)^2}$$

and solve for the constants. Note, again, that the number of constants, five, agrees with the degree of the denominator. To find the constants, we write (8.5) as:

$$1 = A(1 + x^2)^2 + B_1 x^2 (1 + x^2) + C_1 x(1 + x^2) + B_2 x^2 + C_2 x.$$

When $x = 0$, $1 = A$, so

$$1 = (1 + x^2)^2 + B_1 x^2 (1 + x^2) + C_1 x(1 + x^2) + B_2 x^2 + C_2 x.$$

To find the remaining constants, we could evaluate this equation at four arbitrary nonzero values of x and solve the resulting system of equations. We choose, however, to demonstrate another method for finding the constants.

Expand the right side of the last equation and combine similar terms:

$$1 = 1 + 2x^2 + x^4 + B_1 x^2 + B_1 x^4 + C_1 x + C_1 x^3 + B_2 x^2 + C_2 x$$

or

$$1 = (1 + B_1)x^4 + C_1 x^3 + (2 + B_1 + B_2)x^2 + (C_1 + C_2)x + 1.$$

In order for both sides of this equation to be equal for all values of x, the coefficients of each of the terms involving the various powers of x must be the same on each side of the equation. Consequently,

$$x^4: \quad 1 + B_1 = 0 \text{ implies } B_1 = -1$$
$$x^3: \quad C_1 = 0$$
$$x^2: \quad 2 + B_1 + B_2 = 0 \text{ implies } B_2 = -1$$
$$x: \quad C_1 + C_2 = 0 \text{ implies } C_2 = 0.$$

The partial fraction decomposition is:

$$\frac{1}{x(1+x^2)^2} = \frac{1}{x} - \frac{x}{1+x^2} - \frac{x}{(1+x^2)^2}$$

and

$$\int \frac{dx}{x(1+x^2)^2} = \int \frac{dx}{x} - \int \frac{x}{1+x^2}\,dx - \int \frac{x}{(1+x^2)^2}\,dx$$

$$= \ln|x| - \frac{1}{2}\ln(1+x^2) + \frac{1}{2(1+x^2)} + C. \qquad \square$$

EXERCISE SET 8.4

Evaluate the integrals in Exercises 1 through 30.

1. $\displaystyle \int \frac{1}{x^2 - 1}\,dx$

2. $\displaystyle \int \frac{1}{x^2 - 5x + 6}\,dx$

3. $\displaystyle \int_3^4 \frac{1}{x^2 - 3x + 2}\,dx$

4. $\displaystyle \int_0^1 \frac{x}{x^2 - 2x - 3}\,dx$

5. $\displaystyle \int \frac{x}{x^2 - 1}\,dx$

6. $\displaystyle \int \frac{x^2}{x^2 - 1}\,dx$

7. $\displaystyle \int_{-1}^1 \frac{x^2 + 1}{4 - x^2}\,dx$

8. $\displaystyle \int \frac{x^3 + 2}{x^2 + 7x + 12}\,dx$

9. $\displaystyle \int \frac{x^3 + x^2 + x}{x^2 + 2x + 1}\,dx$

10. $\displaystyle \int \frac{x^2}{(x - 1)^2}\,dx$

11. $\displaystyle \int \frac{x^2 + 1}{x^3 + x^2 - 4x - 4}\,dx$

12. $\displaystyle \int \frac{1}{x^3 + 2x^2 - x - 2}\,dx$

13. $\displaystyle \int \frac{2x^2 - 9x + 12}{x^3 - 4x^2 + 4x}\,dx$

14. $\displaystyle \int \frac{5x^2 - 19x + 18}{x^3 - 5x^2 + 3x + 9}\,dx$

15. $\displaystyle \int \frac{3}{x^4 - 5x^2 + 4}\,dx$

16. $\displaystyle \int \frac{1}{x^4 - 3x^3 - 7x^2 + 27x - 18}\,dx$

17. $\displaystyle \int \frac{1}{x^2 + 2x + 2}\,dx$

18. $\displaystyle \int_0^2 \frac{x}{x^2 + 6x + 10}\,dx$

19. $\displaystyle\int \frac{1}{x^3 + x^2 + x + 1}\,dx$

20. $\displaystyle\int \frac{1}{x^3 - x^2 + x - 1}\,dx$

21. $\displaystyle\int_3^4 \frac{3x^2 - 2x - 1}{x^3 - x^2 - x - 2}\,dx$

22. $\displaystyle\int \frac{x^2 + 2x + 2}{x^3 + 3x^2 + 3x + 2}\,dx$

23. $\displaystyle\int \frac{x^2}{(x^2 + 1)^2}\,dx$

24. $\displaystyle\int \frac{2x^5 - 5x}{(x^2 + 2)^2}\,dx$

25. $\displaystyle\int \frac{x + 2}{(x^2 + x + 1)^2}\,dx$

26. $\displaystyle\int \frac{x^3}{(x^2 - 3x + 9)^2}\,dx$

27. $\displaystyle\int \frac{2x - 3}{(2x^2 + 5)(x^2 + 3)}\,dx$

28. $\displaystyle\int \frac{8x^3 + 3x^2 + 4x + 12}{(x^2 + 4)(2x^2 + 1)}\,dx$

29. $\displaystyle\int \frac{\cos x}{\sin x(\sin x - 1)}\,dx$

30. $\displaystyle\int \frac{(\sec x)^2}{\tan x(\tan x + 1)}\,dx$

31. Sketch the graph of the function described by

$$f(x) = \frac{1}{x^2 - x - 6}$$

for x in $[-1, 2]$. Find the area of the region bounded by the graph of f, the lines $x = -1$, $x = 2$, and the x-axis.

32. Let $f(x) = 1/(x^2 + x)$. Find, if possible, the area of the region bounded by the graph of f and the x-axis on the interval $[1, \infty)$.

33. Find the volume of the solid obtained if the region described in Exercise 31 is rotated about the x-axis.

34. Find, if possible, the volume of the solid obtained if the region described in Exercise 32 is rotated about the x-axis.

35. Find the volume of the solid obtained if the region bounded by $x = 2$, $x = 0$, $y = 0$, and the graph of $f(x) = \dfrac{1}{x^2 - x - 6}$ is rotated about the y-axis.

36. Find, if possible, the volume of the solid obtained if the region described in Exercise 32 is rotated about the y-axis.

37. A model for the dispersion of information was considered in Exercise 20 of Section 7.6. Another model uses the differential equation

$$\frac{dP}{dt} = kP(1 - P),$$

where P denotes the proportion of the population aware of the information at time t, and k is a positive constant. Suppose that 10% of the population learns of a tax rebate on the 7:00 AM news and that 50% is aware of the rebate by noon. What percentage will know about the rebate before the 6:00 PM news?

38. The production of the enzyme trypsin from its inactive precursor trypsinogen is an autocatalytic reaction; the production of trypsin from trypsinogen does not begin unless trypsin is present, and the production rate increases with the increased production of trypsin. The production rate can be described by a separable differential equation of the form

$$\frac{dA(t)}{dt} = k(A_1 + A(t)(A_2 - A(t)))$$

where $A(t)$ denotes the amount of trypsin present at time t, A_1 and A_2 are the amounts of trypsinogen and trypsin, respectively, initially present, and k is a positive constant.

(a) Find the general solution to this differential equation.

(b) When is the reaction rate a maximum?

39. Partial fraction methods appear frequently in various engineering areas, particularly electrical, because of their connection with Laplace transforms. The Laplace transform changes a differential equation into an algebraic equation whose solution can be found by ordinary algebraic manipulation. Reversing the transform procedure produces the solution to the differential equation. If L denotes the Laplace transform operation, then $L(\sin at) = s/(s^2 + a)$, $L(\cos at) = a/(s^2 + a)$ and $L(e^{at}) = 1/(s - a)$, where a is a constant. (This was discussed in Exercise 57 of Section 8.1.) Use these results together with the arithmetic result in Exercise 67 of Section 7.3 to find the function whose Laplace transform is

$$\frac{s^3 - 5s - 26}{(s^2 + s - 2)(s^2 + 4)}.$$

Use the substitution and relations in Exercise 41 of Section 8.3 to evaluate the integrals in Exercises 40 and 41.

40. $\displaystyle\int \frac{dx}{3 \cos x - 4 \sin x}$
   41. $\displaystyle\int_0^{\pi/2} \frac{dx}{(3 + 2 \cos x)^2}$

42. Use the substitution $z = \sqrt[6]{x}$ to evaluate $\displaystyle\int \frac{\sqrt{x}}{1 + \sqrt[3]{x}} \, dx$.

43. Use the substitution $z = \sqrt[12]{x}$ to evaluate $\displaystyle\int \frac{dx}{\sqrt[3]{x} + \sqrt[4]{x}}$.

44. Make a substitution based on the idea in Exercise 42 to evaluate
$$\int \frac{e^{x/2}}{1 + e^{x/3}} \, dx.$$

45. Make a substitution based on the idea in Exercise 43 to evaluate
$$\int \frac{dx}{e^{x/3} + e^{x/4}}.$$

8.5
NUMERICAL INTEGRATION

In this chapter we have examined a number of methods for evaluating indefinite integrals. These methods are a representative selection of standard integration techniques and each method applies to a particular class of functions important in the study of calculus and differential equations. Many other integration techniques are known, some of which were considered in the exercises in the previous sections. When an integral cannot be evaluated by one of the techniques we have discussed, the usual procedure is to consult a list of *integral tables*. A list of those integrals most commonly used is contained in the end pages of this text. More comprehensive lists are found in handbooks of mathematical tables, such as the *CRC Handbook of Mathematical Tables*, which contains over 700 definite and indefinite integrals.

In this section we examine some methods used to find approximate values of definite integrals. Since these methods do not require knowing an anti-derivative of the integrand, they can be applied to a wide variety of problems.

You might expect that approximation methods would be reserved as a last resort, used only for those problems to which none of the integration techniques can be applied. However, these methods are often applied to definite integrals of quite elementary functions, especially when those functions arise within a larger problem requiring a computer solution. To halt a computer program to evaluate an integral is generally both inconvenient and inefficient. The error in the approximation of the integrals seldom significantly affects the accuracy of the final result.

Another advantage of approximation methods is that values of the integrand are required only at certain specified values. In practical applications, this information might be obtained experimentally, in which case an explicit representation of the integrand is unknown.

The methods we discuss are based on determining polynomials that agree with the integrand at specified values. These polynomials are defined in the following theorem. (See Exercise 19 for the proof of this result.)

(8.6)
THEOREM

Suppose f is a function defined at the $n + 1$ numbers $x_0, x_1, \ldots, x_n$. The only polynomial of degree at most n whose graph passes through the points $(x_0, f(x_0)), (x_1, f(x_1)), \ldots (x_n, f(x_n))$ is given by

$$L_n(x) = \sum_{i=0}^{n} \frac{(x - x_0)(x - x_1) \ldots (x - x_{i-1})(x - x_{i+1})(x - x_n)}{(x_i - x_0)(x_i - x_1)(x_i - x_{i-1})(x_i - x_{i+1})(x_i - x_n)} f(x_i).$$

The function L_n is called the **Lagrange polynomial** that agrees with the function f at the numbers $x_0, x_1, \ldots, x_n$.

EXAMPLE 1

Find the Lagrange polynomial of degree at most two that agrees with $f(x) = e^x$ when $x_0 = -1$, $x_1 = 0$, and $x_2 = 1$. Use this polynomial to approximate $\int_{-1}^{1} e^x \, dx$.

SOLUTION

The polynomial L_2 is described by

$$L_2(x) = \frac{(x - x_1)(x - x_2)}{(x_0 - x_1)(x_0 - x_2)} f(x_0) + \frac{(x - x_0)(x - x_2)}{(x_1 - x_0)(x_1 - x_2)} f(x_1)$$

$$+ \frac{(x - x_0)(x - x_1)}{(x_2 - x_0)(x_2 - x_1)} f(x_2)$$

$$= \frac{(x - 0)(x - 1)}{(-1)(-2)} e^{-1} + \frac{(x + 1)(x - 1)}{(1)(-1)} e^0 + \frac{(x + 1)(x - 0)}{(2)(1)} e^1$$

$$\approx .54308 x^2 + 1.17520x + 1.$$

The graphs of f and L_2 are shown in Figure 8.2.

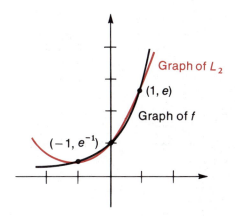

Graph of L_2

$(1, e)$

Graph of f

$(-1, e^{-1})$

FIGURE 8.2

From these sketches it can be seen that the polynomial is a reasonable approximation to f on the interval $[-1, 1]$ but deteriorates rapidly outside this interval.

To approximate the integral of f on $[-1, 1]$ we can integrate the approximating polynomial L_2:

$$\int_{-1}^{1} L_2(x)\, dx = \left[\frac{.54308}{3} x^3 + \frac{1.17520}{2} x^2 + x \right]_{-1}^{1} = 2.36205.$$

Since

$$\int_{-1}^{1} e^x\, dx = e^1 - e^{-1} \approx 2.35040,$$

the error in this approximation is

$$\left| \int_{-1}^{1} e^x\, dx - \int_{-1}^{1} L_2(x)\, dx \right| = .01165. \qquad \square$$

The previous example illustrates the essential idea behind the numerical integration techniques we will consider: we approximate an arbitrary function by a polynomial and then integrate the polynomial.

The first method we will discuss is called the **trapezoidal rule**. Suppose f is a function defined on an interval $[a, b]$ and that $\{x_0, x_1, \ldots, x_n\}$ is an equally spaced partition of $[a, b]$ with $h = x_i - x_{i-1}$ for each $i = 1, 2, \ldots, n$. We will approximate the integral of f on each subinterval $[x_{i-1}, x_i]$ by integrating over this interval the linear function that passes through $(x_{i-1}, f(x_{i-1}))$ and $(x_i, f(x_i))$. (See Figure 8.3.)

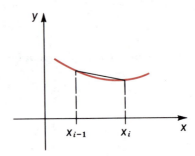

FIGURE 8.3

The Lagrange polynomial of degree at most 1 passing through $(x_{i-1}, f(x_{i-1}))$ and $(x_i, f(x_i))$ is

$$L_1(x) = \frac{x - x_i}{x_{i-1} - x_i} f(x_{i-1}) + \frac{x - x_{i-1}}{x_i - x_{i-1}} f(x_i)$$

$$= \frac{1}{h} [-(x - x_i)f(x_{i-1}) + (x - x_{i-1})f(x_i)].$$

Consequently,

$$\int_{x_{i-1}}^{x_i} f(x)dx \approx \int_{x_{i-1}}^{x_i} L_1(x)dx$$

$$= \frac{1}{h} \int_{x_{i-1}}^{x_i} [-(x - x_i)f(x_{i-1}) + (x - x_{i-1})f(x_i)]dx$$

$$= \frac{1}{h} \left[-\frac{(x - x_i)^2}{2} f(x_{i-1}) + \frac{(x - x_{i-1})^2}{2} f(x_i) \right]_{x_{i-1}}^{x_i}$$

$$= \frac{1}{h} \left[\frac{(x_{i-1} - x_i)^2}{2} f(x_{i-1}) + \frac{(x_i - x_{i-1})^2}{2} f(x_i) \right]$$

$$= \frac{h}{2} [f(x_{i-1}) + f(x_i)].$$

An approximation to the integral of f on $[a, b]$ is found by summing the approximations on each subinterval

$$\int_a^b f(x)dx = \int_{x_0}^{x_1} f(x)dx + \int_{x_1}^{x_2} f(x)dx + \cdots + \int_{x_{n-1}}^{x_n} f(x)dx$$

$$\approx \frac{h}{2} [f(x_0) + f(x_1)] + \frac{h}{2} [f(x_1) + f(x_2)] + \cdots + \frac{h}{2} [f(x_{n-1}) + f(x_n)].$$

Combining the approximations gives

(8.7) Trapezoidal Rule

$$\int_a^b f(x)dx \approx \frac{h}{2}[f(x_0) + 2f(x_1) + 2f(x_2) + \cdots + 2f(x_{n-1}) + f(x_n)].$$

We can see why this is called the trapezoidal rule. When f is positive on $[a, b]$, $\int_a^b f(x)dx$ is approximated by summing the areas of the trapezoids shown in Figure 8.4.

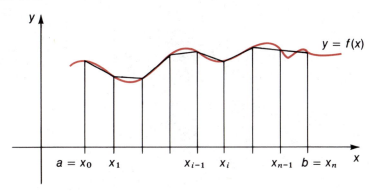

FIGURE 8.4

EXAMPLE 2   Approximate $\int_{-1}^{1} e^x\, dx$ by using the partition with $x_i = -1 + \frac{i}{4}$, for each $i = 0, 1, \ldots, 8$ and

  (a) the trapezoidal rule;
  (b) the Riemann sum with rectangles above the curve; and
  (c) the Riemann sum with rectangles below the curve.

SOLUTION

The graph of the function $f(x) = e^x$ is sketched in Figure 8.5.
(a) Since $h = 1/4$, the trapezoidal rule gives:

$$\int_{-1}^{1} e^x dx \approx \frac{\frac{1}{4}}{2}[e^{-1} + 2e^{-.75} + 2e^{-.5} + 2e^{-.25} + 2e^0$$
$$+ 2e^{.25} + 2e^{.5} + 2e^{.75} + e^1]$$
$$= 2.36263.$$

(b) Since the exponential function is increasing, the Riemann sum with rectangles above the curve gives

$$\int_{-1}^{1} e^x dx \approx \frac{1}{4}\left[e^{-.75} + e^{-.5} + e^{-.25} + e^0 + e^{.25} + e^{.5} + e^{.75} + e^1\right]$$
$$= 2.65643.$$

(c) The Riemann sum with rectangles below the curve gives

$$\int_{-1}^{1} e^x dx \approx \frac{1}{4}\left[e^{-1} + e^{-.75} + e^{-.5} + e^{-.25} + e^0 + e^{.25} + e^{.5} + e^{.75}\right]$$
$$= 2.06883.$$

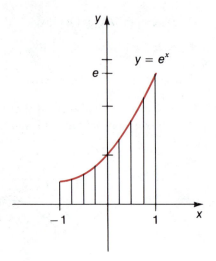

FIGURE 8.5

Since, to the accuracy listed, $\displaystyle\int_{-1}^{1} e^x \, dx = 2.35040$, the errors in the approximations are (a) .01223, (b) .30603, and (c) $-.28157$.   □

In Example 2 the trapezoidal rule is clearly superior to the rectangular approximations. Looking back to Example 1, however, we see that the quadratic polynomial in that example gives an even better approximation than the trapezoidal rule. Although this particular quadratic is unusually accurate, the use of quadratic, rather than linear, functions for approximation leads to generally superior results.

The method using quadratic polynomials and equally spaced points is called **Simpson's rule.** Suppose f is continuous on $[a, b]$ and $\{x_0, x_1, \ldots, x_n\}$ is a partition of $[a, b]$ of evenly spaced points for an even integer n. (See Figure 8.6.) Let L_2 be the Lagrange polynomial agreeing with f at x_{i-1}, x_i and x_{i+1}. Then

$$L_2(x) = \frac{(x-x_i)(x-x_{i+1})}{(x_{i-1} - x_i)(x_{i-1} - x_{i+1})} f(x_{i-1}) + \frac{(x - x_{i-1})(x - x_{i+1})}{(x_i - x_{i-1})(x_i - x_{i+1})} f(x_i)$$

$$+ \frac{(x - x_{i-1})(x - x_i)}{(x_{i+1} - x_{i-1})(x_{i+1} - x_i)} f(x_{i+1}).$$

Letting $h = x_{i+1} - x_i = x_i - x_{i-1}$ and integrating over $[x_{i-1}, x_{i+1}]$ gives (see Exercise 20):

$$\int_{x_{i-1}}^{x_{i+1}} f(x)dx \approx \int_{x_{i-1}}^{x_{i+1}} L_2(x)dx = \frac{h}{3}[f(x_{i-1}) + 4f(x_i) + f(x_{i+1})].$$

An approximation to the integral of f is found by summing the approximations on each subinterval:

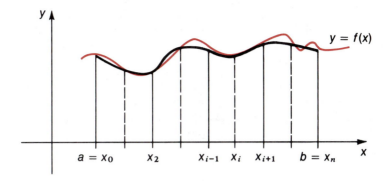

FIGURE 8.6

$$\int_a^b f(x)dx = \int_{x_0}^{x_2} f(x)dx + \int_{x_2}^{x_4} f(x)dx + \cdots + \int_{x_{n-2}}^{x_n} f(x)dx$$

$$\approx \frac{h}{3}[f(x_0) + 4f(x_1) + f(x_2)] + \frac{h}{3}[f(x_2) + 4f(x_3) + f(x_4)]$$

$$+ \cdots + \frac{h}{3}[f(x_{n-2}) + 4f(x_{n-1}) + f(x_n)].$$

Combining the approximations gives

(8.8) Simpson's Rule

$$\int_a^b f(x)dx \approx \frac{h}{3}[f(x_0) + 4f(x_1) + 2f(x_2) + 4f(x_3) + \cdots + 2f(x_{n-2})$$

$$+ 4f(x_{n-1}) + f(x_n)].$$

EXAMPLE 3   Use Simpson's rule with $n = 8$ to approximate $\int_{-1}^{1} e^x \, dx$. Compare this result to the trapezoidal rule approximation found in Example 2.

SOLUTION
Simpson's rule with $n = 8$ gives

$$\int_{-1}^{1} e^x dx \approx \frac{\left(\frac{1}{4}\right)}{3}[e^{-1} + 4e^{-.75} + 2e^{-.5} + 4e^{-.25}$$

$$+ 2e^0 + 4e^{.25} + 2e^{.5} + 4e^{.75} + e^1]$$

$$= 2.35045.$$

HISTORICAL NOTE   This formula (8.8) appeared in **Thomas Simpson's** (1710–1761) book *Mathematical Dissertations on Physical and Analytical Subjects* (1743), but can be found as early as 1668 in *Exercitationes Geometricæ* by James Gregory. Simpson was a self-taught mathematician, a genius who was originally trained as a weaver. He published textbooks on geometry, algebra, and trigonometry that became the best selling mathematical books of his time.

The error in this approximation is .00005 compared to an error of .01223 in the trapezoidal rule. Simpson's rule is significantly superior for this problem.   □

In most instances, Simpson's rule gives a better approximation than the trapezoidal rule, as the results of the following theorems show. These results give information about the maximum error occurring when the trapezoidal and Simpson's rules are used.

(8.9)
THEOREM

Trapezoidal Rule If f is defined on $[a, b]$ and $|f''(x)| \leq M$ for all x in $[a, b]$, then

$$\left| \int_a^b f(x)dx - \frac{h}{2}[f(x_0) + 2f(x_1) + 2f(x_2) + \ldots + 2f(x_{n-1}) + f(x_n)] \right| \leq \frac{(b-a)M}{12}h^2,$$

where $\{x_0, x_1, \ldots, x_n\}$ is a partition of $[a, b]$ and $h = x_i - x_{i-1}$ for each $i = 1, 2, \ldots, n$.

(8.10)
THEOREM

Simpson's Rule If f is defined on $[a, b]$ and $|f^{(4)}(x)| \leq M$ for all x in $[a, b]$, then

$$\left| \int_a^b f(x)dx - \frac{h}{3}[f(x_0) + 4f(x_1) + 2f(x_2) + \cdots + 2f(x_{n-2}) + 4f(x_{n-1}) + f(x_n)] \right| \leq \frac{(b-a)M}{180}h^4,$$

provided n is even, where $\{x_0, x_1, \ldots, x_n\}$ is a partition of $[a, b]$ and $h = x_i - x_{i-1}$ for each $i = 1, 2, \ldots, n$.

The expressions on the right-hand side of the inequalities in Theorems 8.18 and 8.19 are called *error bounds*.

EXAMPLE 4

Compare the error bounds for the trapezoidal rule and Simpson's rule when applied to approximate $\int_{-1}^1 e^x\, dx$ with $n = 8$.

SOLUTION

Since $f(x) = e^x, f''(x) = f^{(4)}(x) = e^x$, and both derivatives are bounded on $[-1, 1]$ by $M = e^1 = e$. The bound for the trapezoidal rule with $n = 8$ is, therefore,

$$\frac{2}{12}\, e\left(\frac{1}{4}\right)^2 \approx .02832,$$

while the bound for Simpson's rule with $n = 8$ is

$$\frac{2}{180} \, e \left(\frac{1}{4} \right)^4 \approx .00012.$$

Notice that the actual errors in approximation, .01223 for the trapezoidal rule and .00005 for Simpson's rule, are well within the error bounds.   □

EXAMPLE 5   Find an approximate value for the arc length of the curve described by $f(x) = \sin x$ from $(0, 0)$ to $(\pi/4, \sqrt{2}/2)$.

SOLUTION

The arc length of the curve is

$$L = \int_0^{\pi/4} \sqrt{1 + [D_x f(x)]^2} \, dx$$

$$= \int_0^{\pi/4} \sqrt{1 + (\cos x)^2} \, dx.$$

This integral belongs to the class of *elliptic integrals,* a class that cannot be evaluated by any integration technique. Using the trapezoidal and Simpson's rules, we can generate approximations to this integral, as shown in the following table.

n	2	4	6	8	10
Trapezoidal Rule	1.0527995	1.0567808	1.0575120	1.0577674	1.0578856
Simpson's Rule	1.0582938	1.0581079	1.0580979	1.0580963	1.0580958

Even without determining the error bounds we would conclude from the agreement of the Simpson's rule approximations that the approximation 1.0581 is accurate to the number of decimal places listed.   □

The study of statistics is concerned with the way in which randomly chosen values are distributed. To describe a continuous numerical distribution, statisticians use a **probability density function:** a non-negative function with the property that the area bounded by the curve and the x-axis is one. Integrating the probability density function over an interval $[a, b]$ gives the probability that a randomly chosen value will fall between a and b.

An important family of probability density functions have the form

(8.11) $$f(x) = \frac{1}{\sigma \sqrt{2\pi}} \, e^{-\frac{1}{2} \left(\frac{x - \mu}{\sigma} \right)^2},$$

where μ and σ are constants and $\sigma > 0$. The probability distribution described by this type of function is called a **normal distribution;** its values assume the form of a bell-shaped curve of the type shown in Figure 8.7.

Normal distributions are associated with problems involving the distribution of errors in experimental data, tolerances of standard machined parts, scores on

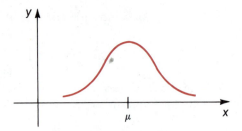

FIGURE 8.7

standardized examinations, and characteristic features of large populations, features such as the height, weight, or voting preferences of individuals. The constant μ in the probability density function describes the **mean,** or average, value of the distribution. The constant σ, called the **standard deviation,** describes the tendency of the values to cluster about the mean. A large value of σ indicates that the distribution is widely spread, while a small value of σ implies that the values are tightly clustered. (See Figure 8.8.)

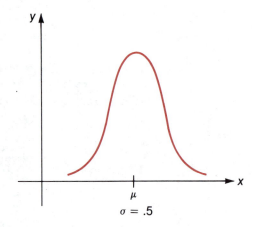

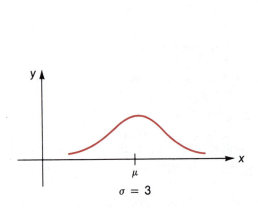

$\sigma = 3$

$\sigma = .5$

FIGURE 8.8

Integrals of normal probability density functions are of interest because of their applications to probability. Unfortunately, these functions have no elementary antiderivative, so the fundamental theorem of calculus cannot be applied. To find the probability that a randomly chosen value from a normal distribution lies in an interval $[a, b]$, we must determine numerically the integral

$$\int_a^b \frac{e^{-\frac{1}{2}\left(\frac{x-\mu}{\sigma}\right)^2}}{\sigma\sqrt{2\pi}}\, dx.$$

Actually this probability can be determined by consulting extensive tables associated with the probability density function of the *standard normal* distribution, the normal distribution with $\mu = 0$ and $\sigma = 1$. These tabulated values, however, were determined by numerical approximations.

EXAMPLE 6   The scores on a standard college mathematical aptitude examination are normally distributed with mean $\mu = 500$ and standard deviation $\sigma = 100$. Find the probability that a randomly chosen examination score will lie between 300 and 700.

SOLUTION

The probability that a score will lie in the interval [300, 700] is

$$P = \int_{300}^{700} \frac{e^{-\frac{1}{2}\left(\frac{x-500}{100}\right)^2}}{100\sqrt{2\pi}}\,dx.$$

Using the variable substitution $z = \dfrac{(x-500)}{100}$, $dz = \dfrac{dx}{100}$ transforms the problem into

$$P = \int_{x=300}^{x=700} \frac{e^{-\frac{1}{2}\left(\frac{x-500}{100}\right)^2}}{100\sqrt{2\pi}}\,dx = \int_{z=-2}^{z=2} \frac{e^{\frac{-z^2}{2}}}{\sqrt{2\pi}}\,dz.$$

Since e^{-z^2} is symmetric with respect to the vertical axis,

$$P = \int_{-2}^{2} \frac{e^{-\frac{z^2}{2}}}{\sqrt{2\pi}}\,dz = 2\int_{0}^{2} \frac{e^{-\frac{z^2}{2}}}{\sqrt{2\pi}}\,dz.$$

Using Simpson's rule with various choices of n produces the results in the following table.

n	2	4	6	8
P	.94721	.95440	.95448	.95449

From this table we conclude that the probability that a randomly chosen score lies between 300 and 700 is approximately .954, or over 95% of the time. □

EXERCISE SET 8.5

1. Use the trapezoidal and Simpson's rules to approximate $\int_{1}^{3} \ln x \, dx$, using 4 subintervals. Compare the approximations to the actual value.

2. Approximate $\int_{0}^{.1} x^{1/3}\,dx$ using the trapezoidal and Simpson's rules with 2 subintervals. Compare the approximations to the actual value.

Approximate the integrals in Exercises 3 through 8 using (a) the trapezoidal rule and (b) Simpson's rule, with the indicated number n of subintervals.

3. $\displaystyle\int_{0}^{1} e^{x^2}\,dx,\ n = 4$

4. $\displaystyle\int_{-1}^{0} \sqrt{1 + x^3}\,dx,\ n = 4$

5. $\displaystyle\int_{-1}^{1} \sin \pi x^2\,dx,\ n = 4$

6. $\displaystyle\int_{0}^{\pi/2} \sqrt{1 + \cos x}\,dx,\ n = 2$

7. $\displaystyle\int_{1}^{2} \frac{\sin x}{x}\,dx,\ n = 8$

8. $\displaystyle\int_{1}^{5} \frac{\ln x}{x}\,dx,\ n = 8$

9. Use the trapezoidal rule with $n = 4$ to find an approximation to the length of the cosine curve from $x = 0$ to $x = \pi/2$.

10. Use the trapezoidal rule with 8 subintervals to approximate the area of the region bounded by the normal curve

$$y = \frac{1}{\sqrt{2\pi}} e^{-x^2/2}$$

and the x-axis on the interval $[-1, 1]$.

11. Repeat Exercise 9 using Simpson's rule.

12. Repeat Exercise 10 using Simpson's rule.

13. Find an approximation to the area of the region bounded by the normal curve

$$y = \frac{1}{\sigma\sqrt{2\pi}} e^{-(x/\sigma)^2/2}$$

and the x-axis on the interval $[-\sigma, \sigma]$ by using the trapezoidal rule with $n = 8$.

14. Repeat Exercise 13 using instead Simpson's rule with $n = 8$.

15. Use Theorem 8.9 to find an integer n such that the error in the approximation of $\int_1^3 \ln x \, dx$ by the trapezoidal rule with $h = 2/n$ is less than 10^{-3}.

16. Use Theorem 8.9 to find an upper bound for the error in using the trapezoidal rule with $n = 4$, $h = 1/2$, to approximate $\int_1^3 \ln x \, dx$.

17. Use Theorem 8.10 to find an integer n such that the error in the approximation of $\int_1^3 \ln x \, dx$ by Simpson's rule with $h = 2/n$ is less than 10^{-3}.

18. Show that the polynomial $L_n(x)$ described in Theorem 8.6 has the property that $L_n(x_i) = f(x_i)$ for each $x = 0, 1, \ldots, n$.

19. The fundamental theorem of algebra implies that a polynomial of degree n has at most n distinct zeros. Use this fact to show that if P is a polynomial of degree at most n whose graph passes through $(x_0, f(x_0))$, $(x_1, f(x_1)) \ldots$ $(x_n, f(x_n))$, then $P \equiv L_n$.

20. Show that $\int_{x_{i-1}}^{x_{i+1}} L_2(x)dx = \frac{h}{3}[f(x_{i-1}) + 4f(x_i) + f(x_{i+1})]$.

[*Hint:* Use the substitution $x = x_{i-1} + th$.]

21. A *superegg* is described by Martin Gardner in his "Mathematical Games" column of the *Scientific American,* Sept. 1965 as the solid of rotation obtained by rotating the *super ellipse*

$$\frac{x^{2.5}}{a^{2.5}} + \frac{y^{2.5}}{b^{2.5}} = 1$$

about the x-axis. Use Simpson's rule with $n = 4$ to approximate the volume of the superegg when $a = 3$ and $b = 2$.

22. In an electrical circuit containing an impressed voltage ε, a capacitor with capacitance C farads, and a resistance of R ohms, the following relationship holds

$$\varepsilon = Ri + \frac{1}{C}\int_0^t i \, dt,$$

where i is the current, in amperes, in the circuit at time t, in seconds. Suppose that in a particular circuit $R = .1$ ohms, $C = 1$ microfarad (10^{-6} farads), and at time t, $i = \sqrt{1 + (\sin \pi t/4)^2}$ amperes. Use Simpson's rule with $n = 4$ to find the voltage after 2 seconds.

23. Suppose that the grade-point averages of all college students in the United States are normally distributed with mean equal to 2.4 and standard deviation equal to .8. Find, to within 10^{-3}, the probability that a randomly chosen student has a grade-point average (a) between 1.6 and 3.2; (b) between 2.4 and 3.2; (c) between 2.4 and 3.0; (d) greater than 3.6.

REVIEW EXERCISES

Evaluate the integrals in Exercises 1 through 50.

1. $\int xe^{-x}\,dx$

2. $\int (x-1)\sin x\,dx$

3. $\int (\sin 2x)^3\,(\cos 2x)^2\,dx$

4. $\int \dfrac{1}{x\sqrt{x^2+4}}\,dx$

5. $\int x^2 \ln x\,dx$

6. $\int x(x^2+1)^5\,dx$

7. $\int \sin \ln x\,dx$

8. $\int \dfrac{\sin \ln x \cos \ln x\,dx}{x}$

9. $\int \dfrac{x-1}{x^2+x}\,dx$

10. $\int \dfrac{x^3}{\sqrt{1-x^2}}\,dx$

11. $\int x2^x\,dx$

12. $\int \sqrt{x^2+4}\,dx$

13. $\int x^3(16-x^2)^{3/2}\,dx$

14. $\int \dfrac{1}{x^2-8x+25}\,dx$

15. $\int \dfrac{x}{x^2+6x+10}\,dx$

16. $\int \dfrac{1}{x^2+4x+5}\,dx$

17. $\int (\sin x)^e\,(\cos x)^3\,dx$

18. $\int (\sec x)^3\,(\tan x)^3\,dx$

19. $\int \dfrac{x^2}{(1-x^2)^{3/2}}\,dx$

20. $\int \dfrac{x^3+6x^2+3x+6}{x^3+2x^2}\,dx$

21. $\int \dfrac{x^5}{\sqrt{1-2x^3}}\,dx$

22. $\int \sin 3x \cos x\,dx$

23. $\int \dfrac{x\,dx}{\sqrt{x-x^2}}$

24. $\int (\tan x)^5\,dx$

25. $\int \dfrac{\sin 2x}{1+(\sin x)^2}\,dx$

26. $\int (\sin x)^3 \ln \cos x\,dx$

27. $\int (\tan x)^5\,(\sec x)^4\,dx$

28. $\int e^x(\cot e^x)^2\,dx$

29. $\int \dfrac{x+2}{x^3+2x^2-3x}\,dx$

30. $\int \dfrac{x^3+5x^2+2x-4}{x^4-1}\,dx$

31. $\int \dfrac{\sqrt{x^2-1}}{x}\,dx$

32. $\int \dfrac{x}{\sqrt{x^2-1}}\,dx$

33. $\displaystyle\int \cos x \ln \sin x \, dx$

34. $\displaystyle\int \frac{e^x}{1 + e^{2x}} \, dx$

35. $\displaystyle\int 2x\sqrt{x + 1} \, dx$

36. $\displaystyle\int \frac{x^3 - 4x}{(x^2 + 1)^2} \, dx$

37. $\displaystyle\int \frac{e^x}{1 + e^x} \, dx$

38. $\displaystyle\int \ln\sqrt{2x - 1} \, dx$

39. $\displaystyle\int \cos 2x \sin 3x \, dx$

40. $\displaystyle\int \frac{1}{(x + 1)(x^2 + x + 1)} \, dx$

41. $\displaystyle\int \frac{dx}{x - \sqrt{x}}$

42. $\displaystyle\int \frac{\cosh \ln x}{x} \, dx$

43. $\displaystyle\int \frac{e^{\tan x}}{1 - (\sin x)^2} \, dx$

44. $\displaystyle\int \operatorname{arcsinh} x \, dx$

45. $\displaystyle\int \frac{1}{\sqrt{7 - 6x - x^2}} \, dx$

46. $\displaystyle\int (\cos x)^5\sqrt{\sin x} \, dx$

47. $\displaystyle\int \frac{1}{1 - \sin x} \, dx$

48. $\displaystyle\int \frac{\sqrt{x^2 + 1}}{x^3} \, dx$

49. $\displaystyle\int \frac{1}{2 - \sqrt{3x}} \, dx$

50. $\displaystyle\int \frac{2}{\cos x + \cot x} \, dx$

51. Find the volume of the solid generated by rotating about the x-axis the region bounded by the x-axis, the lines $x = -2$ and $x = 2$, and the graph of $y = 4/(x^2 + 4)$.

52. Find the area of the region bounded by the x-axis, the line $x = e$, and the graph of $y = \ln x$.

53. Find the volume of the solid generated by rotating about the y-axis the region bounded by the x-axis, the lines $x = 1$ and $x = 3$, and the graph of $y = e^x$.

54. Sketch the graph of the function described by $f(x) = 1/(x^2 + 1)$. Find the volume of the solid obtained if the region bounded by the graph of f, the x-axis, and the lines $x = -2$, $x = 2$ is rotated about the x-axis.

55. Find the centroid of the homogeneous lamina bounded by the x-axis, the line $x = e$ and the graph of the function described $f(x) = \ln x$.

Approximate the integrals in Exercises 56 through 59 using (a) the trapezoidal rule and (b) Simpson's rule, with the indicated number n of subintervals.

56. $\displaystyle\int_0^1 \sqrt{1 - x^3} \, dx, \ n = 4$

57. $\displaystyle\int_0^{\pi/3} \sqrt{\tan x} \, dx, \ n = 2$

58. $\displaystyle\int_1^3 \frac{(\sin x)^2}{x} \, dx, \ n = 8$

59. $\displaystyle\int_{-\pi/2}^{\pi/2} \sqrt{1 + \sin x} \, dx, \ n = 4$

60. Use the trapezoidal rule with $n = 8$ to find an approximation to the arc length of the graph of the function described by $f(x) = x^3$ from $x = 0$ to $x = 2$.

61. Use Simpson's rule with $n = 8$ to find an approximation to the length of the sine curve from $x = -\pi/2$ to $x = \pi/2$.

9

SEQUENCES AND SERIES

In this chapter we discuss a new tool to use in our study of the calculus properties of functions. The topics considered here form an important background for future mathematical study in areas such as differential equations, numerical analysis, and complex analysis. In the latter sections of the chapter we will find that there is a connecting link between seemingly distinct classes of functions such as the trigonometric functions, the exponential functions, and the algebraic polynomials.

9.1 INFINITE SEQUENCES

Our study of calculus has examined functions that have both their domain and range in the set of real numbers. Generally, these domains consist of intervals, since only for functions of this type can we effectively discuss the concepts of continuity, differentiability and integrability. In this section we study the behavior of a class of functions whose domain is a set of nonnegative integers.

(9.1)
DEFINITION
A function whose domain consists of an infinite set of nonnegative integers is called an **infinite sequence**, or simply a **sequence**.

The image of a particular integer n is called the nth term of the sequence, often denoted using subscript notation, such as a_n. The sequence itself is then denoted $\{a_n\}$.

We saw an example of a sequence $\{A_n\}$ in Section 5.1 where for each positive integer n, A_n was the sum of areas of n rectangles:

$$A_n = \sum_{i=1}^{n} f\left(a + i\left(\frac{b-a}{n}\right)\right)\left(\frac{b-a}{n}\right).$$

The 1st, 2nd, 3rd, 4th, and nth terms of the sequence $\{A_n\}$ are illustrated in Figure 9.1.

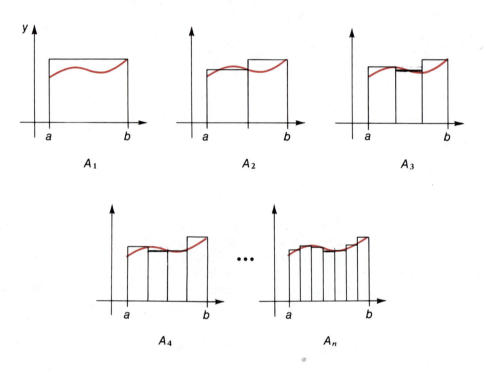

FIGURE 9.1

EXAMPLE 1   Give the first five terms of the sequence described by

$$f(n) = 2n - 1, \qquad n \geq 0.$$

SOLUTION

Using the notation $a_n = f(n)$, we have $a_0 = -1$, $a_1 = 1$, $a_2 = 3$, $a_3 = 5$, $a_4 = 7$. Since $f(x) = 2x - 1$ describes a line with slope 2, the terms of the sequence $\{a_n\} = \{2n - 1\}$ lie on this line at the points $(n, 2n - 1)$ for $n = 0$, 1, 2, 3, ... (See Figure 9.2.)

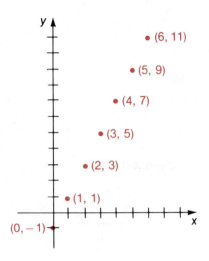

FIGURE 9.2

A sequence of this type, where $a_n = bn + c$ for some pair of constants b and c, is called an *arithmetic sequence*.   □

EXAMPLE 2   Describe the sequence $\left\{\dfrac{1}{2^n}\right\}$, $n \geq 1$.

SOLUTION

The successive terms of this sequence are $\dfrac{1}{2}, \dfrac{1}{4}, \dfrac{1}{8}, \dfrac{1}{16}, \cdots$. Figure 9.3 gives two geometric representations of the first terms of this sequence. In Figure 9.3(a), the graph of the sequence is given; and in Figure 9.3(b), the points corresponding to the first terms of the sequence are plotted on a horizontal axis.   □

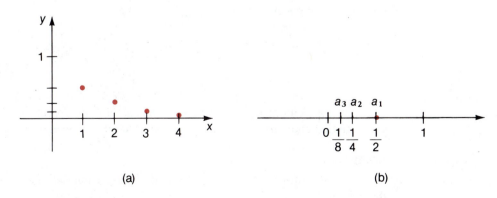

(a)   (b)

FIGURE 9.3

EXAMPLE 3   Describe the sequence $\left\{ \dfrac{2^n - 1}{2^n} \right\}$, $n \geq 1$.

SOLUTION

The successive terms of this sequence are

$$\frac{1}{2}, \frac{3}{4}, \frac{7}{8}, \frac{15}{16}, \frac{31}{32}, \cdots .$$

These are illustrated in Figure 9.4.

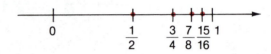

FIGURE 9.4

It is interesting to note the connection between this sequence and the sequence in Example 2. The nth term of this sequence, $(2^n - 1)/2^n$, is the sum of the first n terms of the sequence in Example 2:

$$\frac{1}{2}, \quad \frac{3}{4} = \frac{1}{2} + \frac{1}{4}, \quad \frac{7}{8} = \frac{1}{2} + \frac{1}{4} + \frac{1}{8}, \cdots .$$

Sequences of this type will be studied extensively beginning with Section 9.2.   □

EXAMPLE 4   Find the first six terms of the sequence

$$\left\{ (-1)^{n+1} \frac{2n}{n^2 + 1} \right\}, \qquad n \geq 1.$$

SOLUTION

The terms of this sequence alternate in sign because of the factor $(-1)^{n+1}$. The first six terms are

$$a_1 = 1, \quad a_2 = \frac{-4}{5}, \quad a_3 = \frac{3}{5}, \quad a_4 = \frac{-8}{17}, \quad a_5 = \frac{5}{13}, \quad a_6 = \frac{-12}{37}. \quad □$$

EXAMPLE 5   The nth term in the sequence $\{100(1 + .06)^n\}$ is the amount in a savings account at the end of n years if \$100 is deposited at 6 percent interest per year compounded annually. How much is in the account at the end of 5 years?

SOLUTION

Since the fifth term gives the amount in the account at the end of 5 years, this amount is

$$100(1 + .06)^5 = \$133.82. \qquad □$$

The terms of the sequence $\left\{\dfrac{1}{2^n}\right\}$ in Example 2 approach 0 as n increases, though no term has the value 0. In Example 3, the terms of the sequence $\left\{\dfrac{2^n - 1}{2^n}\right\}$ approach 1 as n increases, and in Example 4, the terms $(-1)^{n+1}\dfrac{2n}{n^2 + 1}$ approach 0. Situations of this type are described by using the limit concept. Since the statement "as n increases without bound" can be replaced by "n approaches ∞," a new definition is not required. We simply modify our original definition of the limit of a function at infinity.

Recall that a function f has the limit L at ∞, written $\lim\limits_{x \to \infty} f(x) = L$, provided that for every number $\varepsilon > 0$, a number M can be found with the property that $|f(x) - L| < \varepsilon$ whenever $x > M$. Definition 9.2 follows from this statement by replacing x by n and $f(x)$ by a_n.

(9.2)
DEFINITION

The limit of a sequence $\{a_n\}$ is a number L written

$$\lim_{n \to \infty} a_n = L,$$

provided that for every number $\varepsilon > 0$, a number M can be found with the property that

$$|a_n - L| < \varepsilon \text{ whenever } n > M.$$

This definition is illustrated in Figures 9.5(a) and (b).

A sequence that has a finite limit is called a **convergent sequence**; if $\lim\limits_{n \to \infty} a_n = \pm \infty$ or if $\lim\limits_{n \to \infty} a_n$ does not exist, then the sequence $\{a_n\}$ is called a **divergent sequence**.

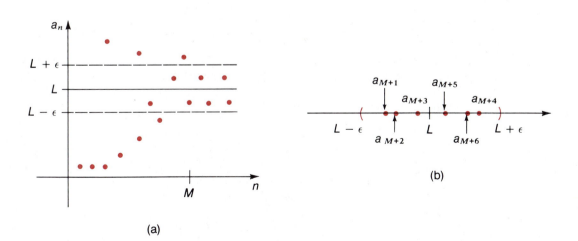

(a)

(b)

FIGURE 9.5

EXAMPLE 6   Prove that $\lim\limits_{n \to \infty} \dfrac{1}{n} = 0$.

SOLUTION

Given $\varepsilon > 0$, we must show that a number M can be found with the property that

$$\left| \frac{1}{n} - 0 \right| = \frac{1}{n} < \varepsilon \text{ whenever } n > M.$$

Suppose $n > M > 0$; then $\left| \dfrac{1}{n} - 0 \right| = \dfrac{1}{n} < \dfrac{1}{M}$. For any given $\varepsilon > 0$, choose M to satisfy $\dfrac{1}{M} \leq \varepsilon$, that is, $M \geq \dfrac{1}{\varepsilon}$. If $n > M \geq \dfrac{1}{\varepsilon}$, then

$$\left| \frac{1}{n} - 0 \right| = \frac{1}{n} < \frac{1}{M} \leq \varepsilon.$$

This completes the proof that $\lim\limits_{n \to \infty} \dfrac{1}{n} = 0$.   □

EXAMPLE 7   Determine whether the following sequences are convergent or divergent and give the limit of the sequence if it converges.

(a) $\left\{ \left(\dfrac{1}{5} \right)^n \right\}$ (b) $\left\{ \left(\dfrac{3}{2} \right)^n \right\}$ (c) $\left\{ (-1)^n \right\}$

SOLUTION

The sequence in (a) can be written as $\left\{ \dfrac{1}{5^n} \right\}$. As $n \to \infty$, $5^n \to \infty$, so $\dfrac{1}{5n} \to 0$. Consequently, $\lim\limits_{n \to \infty} \left(\dfrac{1}{5} \right)^n = 0$.

Since $\dfrac{3}{2} > 1$, $\lim\limits_{n \to \infty} \left(\dfrac{3}{2} \right)^n = \infty$, so the sequence in (b) is divergent. The terms of the sequence in (c) are $-1, 1, -1, 1, -1, 1, \ldots$, so the sequence does not converge. Consequently, $\lim\limits_{n \to \infty} (-1)^n$ does not exist and $\{(-1)^n\}$ is also a divergent sequence.   □

The sequences in Example 7 are of the type r^n, where r is a constant. In (a), $|r| < 1$ and the sequence converges. In (b), $|r| > 1$ and the sequence diverges. In fact, it is true in general that

$$\lim\limits_{n \to \infty} r^n = 0 \quad \text{if} \quad |r| > 1$$

and

$$\{r^n\} \text{ diverges} \quad \text{if} \quad |r| > 1.$$

When $r = -1$, the divergent sequence in Example 7(c) results. When $r = 1$, the series is the convergent constant sequence all of whose terms are 1.

In general, a constant sequence is a sequence $\{a_n\}$ for which $a_n = c$, for some constant c. For such a sequence, $\lim\limits_{n \to \infty} a_n = c$.

The limit theorems discussed in Chapter 2 are equally valid for sequences and eliminate the necessity for using the definition in most limit problems.

(9.3)
THEOREM

If $\{a_n\}$ and $\{b_n\}$ are sequences with $\lim\limits_{n \to \infty} a_n = L$ and $\lim\limits_{n \to \infty} b_n = M$, then

(a) $\lim\limits_{n \to \infty} (a_n \pm b_n) = L \pm M$;

(b) $\lim\limits_{n \to \infty} (a_n b_n) = LM$;

(c) $\lim\limits_{n \to \infty} \left(\dfrac{a_n}{b_n} \right) = \dfrac{L}{M}$, provided $M \neq 0$;

(d) $\lim\limits_{n \to \infty} ca_n = cL$, for any constant c;

(e) $\lim\limits_{n \to \infty} \sqrt[m]{a_n} = \sqrt[m]{L}$, if m is an odd positive integer, or if m is an even positive integer and $L > 0$.

EXAMPLE 8

Determine $\lim\limits_{n \to \infty} \dfrac{n^3 + n + 1}{4n^3 + 2n + 3}$.

SOLUTION

We can find this limit by using Theorem 9.3 and the result proved in Example 6, $\lim\limits_{n \to \infty} 1/n = 0$. We first divide the numerator and denominator by the highest power of n, in this case n^3.

$$\lim_{n \to \infty} \frac{n^3 + n^2 + 1}{4n^3 + 2n + 3} = \lim_{n \to \infty} \frac{1 + \dfrac{1}{n} + \dfrac{1}{n^3}}{4 + \dfrac{2}{n^2} + \dfrac{3}{n^3}}$$

$$= \frac{\lim\limits_{n \to \infty} 1 + \lim\limits_{n \to \infty} \dfrac{1}{n} + \lim\limits_{n \to \infty} \dfrac{1}{n^3}}{\lim\limits_{n \to \infty} 4 + \lim\limits_{n \to \infty} \dfrac{2}{n^2} + \lim\limits_{n \to \infty} \dfrac{3}{n^3}}$$

$$= \frac{1}{4}. \qquad \qquad \square$$

The procedure used in Example 8 should seem familiar; it was used to find limits of functions in Chapter 2.

The next theorem is analogous to the sandwiching theorem studied in Section 3.4.

(9.4)
THEOREM

Suppose $\{a_n\}$, $\{b_n\}$, and $\{c_n\}$ are sequences with $\lim\limits_{n \to \infty} a_n = L$ and $\lim\limits_{n \to \infty} b_n = L$. If for some $M > 0$, $a_n \le c_n \le b_n$ when $n > M$, then $\lim\limits_{n \to \infty} c_n = L$.

EXAMPLE 9   Find $\lim\limits_{n\to\infty} \dfrac{\sin n}{n}$.

SOLUTION

For all n, $-1 \le \sin n \le 1$, so $-\dfrac{1}{n} \le \dfrac{\sin n}{n} \le \dfrac{1}{n}$. However, $\lim\limits_{n\to\infty} -\dfrac{1}{n} = 0$ and $\lim\limits_{n\to\infty} \dfrac{1}{n} = 0$ so Theorem 9.4 implies that $\lim\limits_{n\to\infty} \dfrac{\sin n}{n} = 0$. $\square$

An interesting special case of Theorem 9.4 occurs when $a_n = c_n$ for $n > M$; in this case, the theorem says that sequences that differ by only a finite number of terms have the same limit.

(9.5)
COROLLARY

If $\{a_n\}$ and $\{c_n\}$ are sequences with $\lim\limits_{n\to\infty} a_n = L$ and for some $M > 0$, $a_n = c_n$ when $n > M$, then $\lim\limits_{n\to\infty} c_n = L$.

The close relationship between the limit of a function at infinity and the limit of a sequence produces the next result.

(9.6)
THEOREM

Suppose f is a function with the property that $\lim\limits_{x\to\infty} f(x) = L$. If for each positive integer n, $f(n) = a_n$, then $\lim\limits_{n\to\infty} a_n = L$.

PROOF

Let $\varepsilon > 0$ be given. Since $\lim\limits_{x\to\infty} f(x) = L$, there is a number M such that $|f(x) - L| < \varepsilon$ whenever $x > M$. Thus, if $n > M$,

$$|f(n) - L| = |a_n - L| < \varepsilon.$$

Consequently, $\lim\limits_{n\to\infty} a_n = L$. $\square$

EXAMPLE 10   Find $\lim\limits_{n\to\infty} \dfrac{\ln n}{n}$.

SOLUTION

Since $\lim\limits_{x\to\infty} \ln x = \infty$ and $\lim\limits_{x\to\infty} x = \infty$, L'Hôpital's rule can be applied to find $\lim\limits_{x\to\infty} \dfrac{\ln x}{x}$.

$$\lim\limits_{x\to\infty} \dfrac{\ln x}{x} = \lim\limits_{x\to\infty} \dfrac{D_x \ln x}{D_x x} = \lim\limits_{x\to\infty} \dfrac{\dfrac{1}{x}}{1} = 0.$$

Consequently, Theorem 9.6 implies that

$$\lim_{n \to \infty} \frac{\ln n}{n} = 0.$$

□

BOUNDED SEQUENCES

A sequence $\{a_n\}$ is said to be **bounded above** if there is a number M with $a_n \leq M$ for each n; M is called an **upper bound** for $\{a_n\}$. Similarly, $\{a_n\}$ is **bounded below** if there is a number m with $a_n \geq m$ for each n. In this case, m is called a **lower bound** for $\{a_n\}$. A sequence that is bounded both above and below is called a **bounded sequence**.

EXAMPLE 11   Give an upper bound and a lower bound for the sequence

$$\left\{ \frac{n}{n + 1} \right\}, \qquad n \geq 1.$$

SOLUTION

An upper bound for the sequence with terms 1/2, 2/3, 3/4, 4/5, . . . is 1 (or any number greater than 1) and a lower bound for this sequence is any number less than or equal to 1/2.

□

(9.7)
THEOREM   If $\{a_n\}$ is a convergent sequence, then $\{a_n\}$ is bounded.

PROOF

Suppose L is the limit of this sequence. Then for any $\varepsilon > 0$, and in particular for $\varepsilon = 1$, an integer M exists with the property that

$$|a_n - L| < \varepsilon = 1 \qquad \text{whenever } n > M.$$

Consequently, for $n > M$, all terms of the sequence satisfy the inequality

$$-1 < a_n - L < 1, \qquad \text{so } L - 1 < a_n < L + 1$$

and these terms are bounded. There are only a finite number of terms of the sequence $a_1, a_2, \ldots, a_M$ not covered by this inequality. If we let

$$M_1 = \text{maximum } \{L + 1, a_1, a_2, \ldots, a_M\}$$

and

$$M_2 = \text{minimum } \{L - 1, a_1, a_2, \ldots, a_M\},$$

then for any n, $M_2 \leq a_n \leq M_1$, so the sequence is bounded.

□

The converse of Theorem 9.7 is not true. For example, the sequence $\{(-1)^n\}$ is bounded above by 1 and below by -1 and yet not convergent. However, if a bounded sequence is always increasing or always decreasing, then the sequence is convergent.

Recalling the definitions of increasing and decreasing for functions, we can see that a sequence $\{a_n\}$ is:

(i) **increasing** if $a_n < a_{n+1}$ for each n
(ii) **decreasing** if $a_n > a_{n+1}$ for each n.

A sequence is called **monotonic** if it is either an increasing sequence or a decreasing sequence. The study of the limiting behavior of such sequences requires the **completeness property** of the real number system.

(9.8)
THE COMPLETENESS PROPERTY

If a sequence $\{a_n\}$ of real numbers has an upper bound, then it has a least upper bound $\hat{M}$; that is, if a number M exists with $a_n \leq M$ for all n, then a number $\hat{M}$ exists with $a_n \leq \hat{M}$ for all n and $\hat{M} \leq M$ for all upper bounds M of $\{a_n\}$.

The equivalent statement holds for sequences that have lower bounds: every sequence of real numbers that has a lower bound has a greatest lower bound.

EXAMPLE 12

Give the least upper bound and the greatest lower bound of each of the following sequences if the sequence is bounded.

(a) $\left\{\dfrac{1}{n}\right\}$, $n \geq 1$ (b) $\left\{\dfrac{n}{n+1}\right\}$, $n \geq 1$ (c) $\{(-1)^n n\}$, $n \geq 1$.

SOLUTION

(a) This sequence has terms $1, \dfrac{1}{2}, \dfrac{1}{3}, \dfrac{1}{4}, \dfrac{1}{5}, \cdots$. It has greatest lower bound 0 and least upper bound 1.

(b) This sequence has terms $\dfrac{1}{2}, \dfrac{2}{3}, \dfrac{3}{4}, \dfrac{4}{5}, \dfrac{5}{6}, \dfrac{6}{7}, \cdots$. It has greatest lower bound $\dfrac{1}{2}$ and least upper bound 1.

(c) The sequence has terms $-1, 2, -3, 4, -5, \ldots$. It is bounded neither below nor above. The sequence has no greatest lower bound or least upper bound. $\square$

The sequence $\{n/(n+1)\}$ is an increasing sequence that converges to 1, the least upper bound of the sequence. The sequence $\{1/n\}$ is a decreasing sequence that converges to its greatest lower bound 0. That bounded monotonic sequences always behave in this way is a result of the following theorem.

(9.9)
THEOREM

If $\{a_n\}$ is a bounded monotonic sequence, then $\{a_n\}$ is convergent. If $\{a_n\}$ is an increasing sequence, its limit is its least upper bound. If $\{a_n\}$ is a decreasing sequence, its limit is its greatest lower bound.

EXERCISE SET 9.1

The nth term of a sequence is given in Exercises 1 through 36. (a) List the first 3 terms of the sequence. (Assume $n \geq 0$ unless otherwise specified.) (b) Determine if the sequence is convergent. If the sequence is convergent, give its limit. (It might help to review the infinite limit problems in Section 2.6 and the L'Hôpital's rule exercises in Sections 4.5 and 7.4.) (c) Determine if the sequence is bounded.

1. $a_n = \dfrac{2}{n+1}$

2. $a_n = \dfrac{n}{n^2+3}$

3. $a_n = (-1)^n \dfrac{1}{n}, \; n \geq 1$

4. $a_n = (-1)^n \, n$

5. $a_n = \dfrac{n}{n+4}$

6. $a_n = (-1)^n \dfrac{n}{n+4}$

7. $a_n = \dfrac{n^2+2n+3}{n^2-3n+4}$

8. $a_n = \dfrac{n^2}{n^2-4}, \; n \geq 3$

9. $a_n = \dfrac{n^3+2n^2-1}{n^2+n}, \; n \geq 1$

10. $a_n = \dfrac{n^3+2n^2-1}{n^4+n}, \; n \geq 1$

11. $a_n = \dfrac{\pi}{n}, \; n \geq 1$

12. $a_n = \ln n, \; n \geq 1$

13. $a_n = \dfrac{e^n}{n}, \; n \geq 1$

14. $a_n = \dfrac{\ln n}{n^2} - 1, \; n \geq 1$

15. $a_n = (-1)^n \dfrac{n^2+3}{\sqrt[4]{n^9+3n^3+4}}$

16. $a_n = \dfrac{n^3+3}{\sqrt[4]{n^9+3n^3+4}}$

17. $a_n = \sin \dfrac{\pi}{n}, \; n \geq 1$

18. $a_n = \cos \dfrac{\pi}{n}, \; n \geq 1$

19. $a_n = \dfrac{e^n}{n^2+3n-1}$

20. $a_n = \dfrac{n^3-1}{n}, \; n \geq 1$

21. $a_n = \dfrac{1}{n} - \dfrac{1}{n+1}, \; n \geq 1$

22. $a_n = \dfrac{1}{n+1} - \dfrac{1}{n+2}$

23. $a_n = \sqrt{n^2+1} - n$

24. $a_n = \sqrt{n+1} - \sqrt{n}$

25. $a_n = \dfrac{1}{\sqrt{n^2+1} - n}$

26. $a_n = \dfrac{1}{\sqrt{n^2+n} - n}, \; n \geq 1$

27. $a_n = \dfrac{1}{n} \sin \dfrac{\pi}{n}, \; n \geq 1$

28. $a_n = n \sin \dfrac{\pi}{n}, \; n \geq 1$

29. $a_n = \dfrac{1}{n} \cos n\pi, \; n \geq 1$

30. $a_n = n \cos n\pi$

31. $a_n = \dfrac{n}{n^2+1} \cos n\pi$

32. $a_n = \dfrac{n}{n^2+1} \sin \dfrac{n\pi}{2}$

33. $a_n = \dfrac{e^n}{1 + e^n}$

34. $a_n = \dfrac{2^n}{1 + 2^n}$

35. $a_n = \left(1 + \dfrac{1}{n}\right)^n,\ n \geq 1$

36. $a_n = \left(1 + \dfrac{1}{n}\right)^{2n},\ n \geq 1$

37. Consider the sequence whose first terms are 1, 0, 1, 0, 0, 1, 0, 0, 0, 1, 0, 0, 0, 0, 1, 0, 0, 0, 0, 0, 1, 0 . . . . Suppose that the terms of this sequence continue in this manner. Does the sequence converge?

38. The second edition of a book *Liber abaci* (meaning beyond the abacus), published in 1202 by the Italian mathematician Leonardo de Pisa (better known as Fibonacci, meaning "son of Bonaccio") contains the problem: "How many pairs of rabbits are produced from one pair in a year if every month each pair produces a new pair which from the second month itself becomes productive?" If F_n denotes the number of pairs of rabbits at the end of the nth month, then $F_1 = 1$, $F_2 = 1$, $F_3 = 2$ and, in general, $F_n = F_{n-1} + F_{n-2}$ for $n \geq 2$. Find the number of pairs of rabbits at the end of the first year.

39. It is estimated that 5000 people move into a certain Canadian city each year and that 40% of those who have moved into the city prior to that year leave during that year. Let x_n denote the population due to this migration n years after 1975, the year the migration began. Find a formula for x_n in terms of n and determine $\lim\limits_{n\to\infty} x_n$.

40. The equation in Exercise 39 is called a linear first-order difference equation. The general form of this equation is $x_n = ax_{n-1} + b_n$, where a is a known constant and $\{b_n\}$ a known sequence. Verify that

$$x_n = x_0 a^n + \sum_{m=1}^{n} a^{n-m} b_m$$

satisfies this equation.

41. Show that $\lim\limits_{n\to\infty} r^n = 0$ if $|r| < 1$, and that $\{r^n\}$ diverges if $|r| > 1$.

42. Use Definition 9.2 to prove that if $\lim\limits_{n\to\infty} a_n = L$, then $\lim\limits_{n\to\infty} |a_n| = |L|$.

43. Find a sequence $\{a_n\}$ with the property that $\lim\limits_{n\to\infty} |a_n|$ exists, but $\lim\limits_{n\to\infty} a_n$ does not exist.

44. Use Definition 9.2 to prove that if $\lim\limits_{n\to\infty} a_n = L$ and $\lim\limits_{n\to\infty} b_n = M$, then

(a) $\lim\limits_{n\to\infty} (a_n + b_n) = L + M$;

(b) $\lim\limits_{n\to\infty} ca_n = cL$, for any constant c.

45. Find sequences $\{a_n\}$ and $\{b_n\}$ with the property that $\lim\limits_{n\to\infty} (a_n + b_n)$ exists, but neither $\lim\limits_{n\to\infty} a_n$ nor $\lim\limits_{n\to\infty} b_n$ exists.

46. Show that if $\{a_n\}$ and $\{b_n\}$ are sequences with the property that $\lim_{n \to \infty} (a_n + b_n)$ and $\lim_{n \to \infty} (a_n - b_n)$ both exist, then $\lim_{n \to \infty} a_n$ and $\lim_{n \to \infty} b_n$ both exist.

47. Prove Corollary 9.5 directly using Definition 9.2.

48. Show that if $\{a_n\}$ is a convergent sequence, then for any number $\varepsilon > 0$, there exists a positive integer M such that $|a_n - a_m| < \varepsilon$ whenever $n, m > M$. (The condition described in this problem is known as the *Cauchy condition*. It is proved in advanced calculus books that a sequence is convergent if and only if it satisfies the Cauchy condition.)

49. In Exercise 40 of Section 3.9 it was shown that:
If a function g satisfies the following

    (a) g is differentiable on $[a, b]$,

    (b) a number L exists with $|g'(x)| \leq L < 1$ for all x in $[a, b]$, and

    (c) $a \leq g(x) \leq b$, for all x in $[a, b]$,

then g has a unique fixed point p in $[a, b]$ (i.e. $g(p) = p$).

    Show that if x_0 is chosen arbitrarily in $[a, b]$ and $x_n = g(x_{n-1})$ for each $n \geq 1$, then $\lim_{n \to \infty} x_n = p$.

50. Use Exercise 49 to show that the sequence defined by

$$x_n = \frac{1}{2}\left(x_{n-1} + \frac{2}{x_{n-1}}\right),$$

for $n \geq 1$ converges to $\sqrt{2}$ for any $x_0 > 1$.

51. A *prime number* is a positive integer that can be evenly divided only by one and itself. Euclid demonstrated around 300 B.C. that there is an infinite sequence of prime numbers by assuming that the number of primes $p_1, p_2, \ldots, p_n$ is finite and constructing a new prime $p = p_1 \cdot p_2 \cdots p_n + 1$. Show that p must be prime. Are all prime numbers in the form of p?

52. The sequence $\{F_n\}$ described in Exercise 38 is called a *Fibonacci sequence*. Its terms occur naturally in many botanical species, particularly those with petals or scales arranged in the form of a logarithmic spiral. (Sunflower seed arrangement, scales on pine cones and pineapples, bud formation on alders and birches, are examples of this phenomenon. The study of the arrangement of scales, leaves and such is known as *phyllotaxiology*.) Consider the sequence $\{x_n\}$, where $x_n = \dfrac{F_{n+1}}{F_n}$. Assuming that $\lim_{n \to \infty} x_n = x$ exists, show that $x = \dfrac{1 + \sqrt{5}}{2}$. This number is called the *golden ratio* and, among its many applications, gives the average monthly increase rate in the rabbit population described in Exercise 38. (*Hint:* Divide $F_{n+1} = F_n + F_{n-1}$ by F_n to obtain a relationship between x_n and x_{n-1}. Then show that $x = 1 + \dfrac{1}{x}$ and solve for x.)

53. Inscribe an equilateral polygon with n sides within a unit circle. Show that the perimeter of the polygon is $p_n = 2n \sin \pi/n$. Find $\lim\limits_{n \to \infty} p_n$. (*Hint*: Bisect the angle $2\pi/n$ and find a pair of congruent triangles, one of which clearly has a side with length equal to one half the length of a side of the polygon.)

54. Circumscribe an equilateral polygon with n sides outside a unit circle. Show that the perimeter of the polygon is $p_n = 2n \tan \pi/n$. Find $\lim\limits_{n \to \infty} p_n$.

9.2
INFINITE SERIES

In Examples 2 and 3 of Section 9.1 we considered the sequences $\{1/2^n\}$ and $\left\{\dfrac{2^n - 1}{2^n}\right\}$ and remarked at the time that the terms of $\left\{\dfrac{2^n - 1}{2^n}\right\}$ can be generated by summing the terms of $\{1/2^n\}$. That is,

$$\frac{1}{2} = \frac{1}{2}, \quad \frac{3}{4} = \frac{1}{2} + \frac{1}{4}, \quad \frac{7}{8} = \frac{1}{2} + \frac{1}{4} + \frac{1}{8}, \quad \frac{15}{16} = \frac{1}{2} + \frac{1}{4} + \frac{1}{8} + \frac{1}{16}, \cdots$$

and, in general, for each positive integer n,

$$\frac{2^n - 1}{2^n} = \frac{1}{2} + \frac{1}{2^2} + \cdots + \frac{1}{2^n}.$$

Since $\quad \lim\limits_{n \to \infty} \left(\dfrac{1}{2} + \dfrac{1}{2^2} + \cdots + \dfrac{1}{2^n} \right) = \lim\limits_{n \to \infty} \left(\dfrac{2^n - 1}{2^n} \right) = 1,$

it seems reasonable to define the infinite sum

$$\frac{1}{2} + \frac{1}{2^2} + \frac{1}{2^3} + \cdots$$

to be 1.

In this section, we begin a study of expressions in which the terms of an infinite sequence are summed. An expression of the type

(9.10) $$\sum_{n=1}^{\infty} a_n = a_1 + a_2 + a_3 + \cdots$$

is called an **infinite series** and a_n is called the **nth term of the series**.

(9.11)
DEFINITION

Associated with any sequence $\{a_n\}$ is a sequence $\{S_n\}$ defined by

$$S_1 = a_1, S_2 = S_1 + a_2 = a_1 + a_2, S_3 = S_2 + a_3 = a_1 + a_2 + a_3,$$

and, in general,

$$S_n = S_{n-1} + a_n = a_1 + a_2 + \cdots + a_n = \sum_{i=1}^{n} a_i.$$

The sequence $\{S_n\}$ is called the **sequence of partial sums** of the series $\sum_{n=1}^{\infty} a_n$.

If the sequence $\{S_n\}$ converges to a number L, the infinite series $\sum_{n=1}^{\infty} a_n$ is said to **converge** to L; in this case we write $\sum_{n=1}^{\infty} a_n = L$ and L is called the **sum** of the series. If the sequence $\{S_n\}$ diverges, the infinite series $\sum_{n=1}^{\infty} a_n$ is said to **diverge**.

EXAMPLE 1   Show that the infinite series $\sum_{n=1}^{\infty} \frac{1}{2^n}$ converges and find its limit.

SOLUTION

Let $S_n = \frac{1}{2} + \frac{1}{2^2} + \cdots + \frac{1}{2^n} = \frac{2^n - 1}{2^n}.$

Then

$$\lim_{n \to \infty} S_n = \lim_{n \to \infty} \frac{2^n - 1}{2^n} = 1,$$

so

$$\sum_{n=1}^{\infty} \frac{1}{2^n} = 1. \qquad \square$$

EXAMPLE 2   Find the sequence of partial sums of the series $\sum_{n=1}^{\infty} \frac{1}{n(n + 1)}$ and determine if $\sum_{n=1}^{\infty} \frac{1}{n(n + 1)}$ converges.

SOLUTION

For each n, $S_n = \frac{1}{1 \cdot 2} + \frac{1}{2 \cdot 3} + \frac{1}{3 \cdot 4} + \cdots + \frac{1}{n(n + 1)}.$

The procedure introduced for the partial fractions integration technique (Section 8.4) can be used to show that for any integer n,

$$\frac{1}{n(n + 1)} = \frac{1}{n} - \frac{1}{n + 1}.$$

So
$$S_n = 1 - \frac{1}{2} + \frac{1}{2} - \frac{1}{3} + \cdots + \frac{1}{n-1} - \frac{1}{n} + \frac{1}{n} - \frac{1}{n+1}$$

$$= 1 - \frac{1}{n+1}$$

and $\lim\limits_{n \to \infty} S_n = \lim\limits_{n \to \infty} \left(1 - \dfrac{1}{n+1} \right) = 1$. Consequently, $\sum\limits_{n=1}^{\infty} \dfrac{1}{n(n+1)}$ converges;

in fact, $\sum\limits_{n=1}^{\infty} \dfrac{1}{n(n+1)} = 1$.   □

EXAMPLE 3   Does $\sum\limits_{n=1}^{\infty} (-1)^n$ converge?

SOLUTION

The sequence of partial sums $\{S_n\}$ is given by

$$S_n = \begin{cases} -1 & \text{if } n \text{ is odd} \\ 0 & \text{if } n \text{ is even.} \end{cases}$$

The sequence $\{S_n\}$ does not converge, so $\sum\limits_{n=1}^{\infty} (-1)^n$ diverges.   □

(9.12)
THEOREM

If the infinite series $\sum\limits_{n=1}^{\infty} a_n$ converges, then $\lim\limits_{n \to \infty} a_n = 0$.

PROOF

If $\sum\limits_{n=1}^{\infty} a_n$ converges, a number L exists with $L = \lim\limits_{n \to \infty} S_n$. Since S_n is the sum of the first n terms and S_{n-1} is the sum of the first $n - 1$ terms, $S_n = S_{n-1} + a_n$ and $S_n - S_{n-1} = a_n$.

But, $\lim\limits_{n \to \infty} S_n = \lim\limits_{n \to \infty} S_{n-1} = L,$

so, $\lim\limits_{n \to \infty} a_n = \lim\limits_{n \to \infty} (S_n - S_{n-1}) = \lim\limits_{n \to \infty} S_n - \lim\limits_{n \to \infty} S_{n-1} = 0.$ □

The following corollary is the logical equivalent of Theorem 9.13 and is called the **contrapositive** of the statement in Theorem 9.13. In general, the contrapositive of the statement "p implies q" is "not q implies not p." For example, we know that: "if a function f is differentiable at b, then f is continuous at b." The contrapositive of this statement is: "if a function f is not continuous at b, then f is not differentiable at b."

(9.13)
COROLLARY

If $\{a_n\}$ is a sequence that does not have the limit zero, then $\sum\limits_{n=1}^{\infty} a_n$ diverges.

EXAMPLE 4 Does $\sum\limits_{n=1}^{\infty} \left[1 - \left(\dfrac{1}{2}\right)^n \right]$ converge?

SOLUTION

Since $\lim\limits_{n\to\infty} \left[1 - \left(\dfrac{1}{2}\right)^n \right] = 1$, it follows from Corollary 9.13 that the infinite

series $\sum\limits_{n=1}^{\infty} \left[1 - \left(\dfrac{1}{2}\right)^n \right]$ diverges. $\qquad\qquad$ ☐

The converse of Theorem 9.12 is not true: $\lim\limits_{n\to\infty} a_n = 0$ does not imply that a_n converges.

For example, the **harmonic series**

$$\sum_{n=1}^{\infty} \frac{1}{n}$$

diverges even though $\lim\limits_{n\to\infty} \dfrac{1}{n} = 0$.

To see that this is true, consider the partial sums of the form S_{2^n}:

$$S_2 = 1 + \frac{1}{2} = \frac{3}{2},$$

$$S_4 = S_2 + \frac{1}{3} + \frac{1}{4} > S_2 + \frac{1}{4} + \frac{1}{4} = \frac{3}{2} + \frac{1}{2} = \frac{4}{2},$$

$$S_8 = S_4 + \frac{1}{5} + \frac{1}{6} + \frac{1}{7} + \frac{1}{8} > S_4 + \frac{1}{8} + \frac{1}{8} + \frac{1}{8} + \frac{1}{8} > \frac{4}{2} + \frac{1}{2} = \frac{5}{2},$$

and in general

$$S_{2^n} > \frac{n+2}{2}.$$

These partial sums are unbounded, so $\sum\limits_{n=1}^{\infty} \dfrac{1}{n}$ diverges. This important result will be established using another method in Section 9.3. A third demonstration is considered in Exercise 34.

The following theorems will be used often in our study of series; their proofs follow from the corresponding results for sequences.

(9.14)
THEOREM

If $\displaystyle\sum_{n=1}^{\infty} a_n = L$ and $\displaystyle\sum_{n=1}^{\infty} b_n = M$, then

(i) $\displaystyle\sum_{n=1}^{\infty} a_n \pm b_n = L \pm M$

(ii) $\displaystyle\sum_{n=1}^{\infty} ca_n = cL$, for any constant c.

(9.15)
THEOREM

If an integer M exists with $a_n = b_n$ for all $n \geq M$, then $\displaystyle\sum_{n=1}^{\infty} a_n$ and $\displaystyle\sum_{n=1}^{\infty} b_n$ both converge or both diverge.

EXAMPLE 5

Does $\displaystyle\sum_{n=1}^{\infty} \frac{1}{(n+4)(n+5)}$ converge?

SOLUTION

This infinite series can be alternately expressed, using the change of index $m = n + 4$, as

$$\sum_{n=1}^{\infty} \frac{1}{(n+4)(n+5)} = \sum_{m=5}^{\infty} \frac{1}{m(m+1)}.$$

The series $\displaystyle\sum_{n=1}^{\infty} \frac{1}{n(n+1)}$ was shown in Example 2 to converge to 1 and

$$\sum_{n=1}^{\infty} \frac{1}{n(n+1)} = \sum_{m=5}^{\infty} \frac{1}{m(m+1)} + \frac{1}{(1)(2)} + \frac{1}{(2)(3)} + \frac{1}{(3)(4)} + \frac{1}{(4)(5)}.$$

So the series

$$\sum_{m=5}^{\infty} \frac{1}{m(m+1)} = \sum_{n=1}^{\infty} \frac{1}{(n+4)(n+5)}$$

converges. In fact

$$\sum_{n=1}^{\infty} \frac{1}{(n+4)(n+5)} = \sum_{m=5}^{\infty} \frac{1}{m(m+1)}$$

$$= \sum_{m=1}^{\infty} \frac{1}{m(m+1)} - \frac{1}{(1)(2)} - \frac{1}{(2)(3)} - \frac{1}{(3)(4)} - \frac{1}{(4)(5)}$$

$$= 1 - \frac{1}{2} - \frac{1}{6} - \frac{1}{12} - \frac{1}{20} = \frac{1}{5}. \qquad \square$$

This section concludes by considering a class of series for which the question of convergence can be easily determined. These series are the geometric series, the terms of which are generated by multiplying the previous term by a fixed constant called the *ratio* of the series. Examples of geometric series are

$$\sum_{n=1}^{\infty} \left(\frac{1}{10}\right)^n = \frac{1}{10} + \frac{1}{100} + \frac{1}{1000} + \cdots$$

and

$$\sum_{n=1}^{\infty} 5\left(\frac{1}{3}\right)^n = \frac{5}{3} + \frac{5}{9} + \frac{5}{27} + \cdots.$$

The ratio in the first example is 1/10 while that in the second is 1/3. In general, a **geometric series** has the form

$$\sum_{n=1}^{\infty} ar^{n-1},$$

where $a \neq 0$ is called the first term and r is the ratio.

(9.16)
THEOREM

A geometric series $\sum_{n=1}^{\infty} ar^{n-1}$ is convergent if and only if $|r| < 1$. When $|r| < 1$,

$$\sum_{n=1}^{\infty} ar^{n-1} = a + ar + ar^2 + \cdots = \frac{a}{1-r}.$$

PROOF

For $|r| \geq 1$, $\lim_{n \to \infty} ar^{n-1} \neq 0$ so by Corollary 9.13 the series diverges.

For $|r| < 1$, the nth partial sum S_n of the series can be written as

$$S_n = a + ar + ar^2 + \ldots + ar^{n-2} + ar^{n-1};$$

so

$$rS_n = ar + ar^2 + ar^3 + \ldots + ar^{n-1} + ar^n$$

and

$$rS_n - S_n = ar^n - a.$$

Thus,

$$S_n = \frac{ar^n - a}{r-1} = \frac{a(r^n - 1)}{r-1}$$

and

$$\sum_{n=1}^{\infty} ar^{n-1} = \lim_{n \to \infty} S_n = \lim_{n \to \infty} \frac{a(r^n - 1)}{r-1}.$$

Since $|r| < 1$, $\lim_{n \to \infty} r^n = 0$. So

$$\sum_{n=1}^{\infty} ar^{n-1} = \frac{a(0-1)}{r-1} = \frac{a}{1-r}$$

and the series converges.   □

Theorem 9.16 implies that the series $\sum_{n=1}^{\infty} \left(\frac{1}{10}\right)^n$ and $\sum_{n=1}^{\infty} 5\left(\frac{1}{3}\right)^n$ converge:

$$\sum_{n=1}^{\infty} \left(\frac{1}{10}\right)^n = \frac{\frac{1}{10}}{1 - \frac{1}{10}} = \frac{1}{9} \quad \text{and} \quad \sum_{n=1}^{\infty} 5\left(\frac{1}{3}\right)^n = \frac{5\left(\frac{1}{3}\right)}{1 - \frac{1}{3}} = \frac{5}{2}.$$

EXAMPLE 6   Does the series $\displaystyle\sum_{n=1}^{\infty} 3\left(\frac{4^{n-1}}{5^n}\right)$ converge?

SOLUTION

This series can be rewritten as:

$$\sum_{n=1}^{\infty} 3\left(\frac{4^{n-1}}{5^n}\right) = \sum_{n=1}^{\infty} \frac{3}{5}\left(\frac{4}{5}\right)^{n-1},$$

which is a geometric series with first term $a = 3/5$ and ratio $r = 4/5$. By Theorem 9.16, this series converges; in fact,

$$\sum_{n=1}^{\infty} 3\left(\frac{4^{n-1}}{5^n}\right) = \frac{3/5}{1 - 4/5} = 3.$$ □

EXAMPLE 7   Does the series $\displaystyle\sum_{n=1}^{\infty} \frac{3^n + 2^n}{4^n}$ converge?

SOLUTION

The geometric series

$$\sum_{n=1}^{\infty} \frac{3^n}{4^n} = \sum_{n=1}^{\infty} \left(\frac{3}{4}\right)^n \quad \text{and} \quad \sum_{n=1}^{\infty} \frac{2^n}{4^n} = \sum_{n=1}^{\infty} \left(\frac{2}{4}\right)^n$$

both converge and have sums

$$\frac{3/4}{1 - 3/4} = 3 \quad \text{and} \quad \frac{2/4}{1 - 2/4} = 1$$

respectively. Consequently, Theorem 9.14 implies that

$$\sum_{n=1}^{\infty} \frac{3^n + 2^n}{4^n} = \sum_{n=1}^{\infty} \frac{3^n}{4^n} + \sum_{n=1}^{\infty} \frac{2^n}{4^n} = 3 + 1 = 4.$$ □

EXERCISE SET 9.2

Determine whether the series in Exercises 1 through 8 converge or diverge.

1. $\displaystyle\sum_{n=1}^{\infty} \frac{n}{n+1}$

2. $\displaystyle\sum_{n=2}^{\infty} \frac{2n^2 + 1}{n^2 - 1}$

3. $\displaystyle\sum_{n=1}^{\infty} \left(\frac{2}{3}\right)^n$

4. $\displaystyle\sum_{n=1}^{\infty} \left(\frac{3}{2}\right)^n$

5. $\displaystyle\sum_{n=1}^{\infty} \frac{e^n}{n}$

6. $\displaystyle\sum_{n=1}^{\infty} \sin n\pi$

7. $\displaystyle\sum_{n=1}^{\infty} \frac{1}{(n+5)(n+6)}$

8. $\displaystyle\sum_{n=1}^{\infty} \frac{n^2}{(n+5)(n+6)}$

Find the sum of each infinite series given in Exercises 9 through 22. Use procedures similar to those discussed in the examples.

9. $\displaystyle\sum_{n=1}^{\infty} \left(\frac{1}{3}\right)^n$

10. $\displaystyle\sum_{n=2}^{\infty} \left(\frac{3}{4}\right)^n$

11. $\displaystyle\sum_{n=3}^{\infty} \left(\frac{2}{5}\right)^{n-1}$

12. $\displaystyle\sum_{n=1}^{\infty} \left(-\frac{1}{2}\right)^n$

13. $\displaystyle\sum_{n=0}^{\infty} 3^n \cdot 5^{-n}$

14. $\displaystyle\sum_{n=1}^{\infty} 4\left(\frac{3}{\pi}\right)^n$

15. $\displaystyle\sum_{n=1}^{\infty} \frac{1}{(n+1)(n+2)}$

16. $\displaystyle\sum_{n=1}^{\infty} \frac{1}{(n+2)(n+3)}$

17. $\displaystyle\sum_{n=2}^{\infty} \frac{1}{n^2-1}$

18. $\displaystyle\sum_{n=3}^{\infty} \frac{1}{n^2-4}$

19. $\displaystyle\sum_{n=1}^{\infty} \left(\frac{1}{2^n} - \frac{1}{3^n}\right)$

20. $\displaystyle\sum_{n=1}^{\infty} (-1)^n \, 2^{n+1} \, 3^{-n+2}$

21. $\displaystyle\sum_{n=2}^{\infty} \left[\left(\frac{2}{3}\right)^n + \frac{1}{n^2-1} \right]$

22. $\displaystyle\sum_{n=1}^{\infty} \frac{4^n - 3^n}{12^n}$

23. Show that the series $\displaystyle\sum_{n=1}^{\infty} \ln\left(\frac{n}{n+1}\right)$ diverges.

24. Show that for any positive real number a, the series $\displaystyle\sum_{n=1}^{\infty} a^{1/n}$ diverges.

25. Compare the finite sums $\displaystyle\sum_{n=1}^{4} 2^n$, $\displaystyle\sum_{n=1}^{4} 3^n$, $\displaystyle\sum_{n=1}^{4} 6^n$, and $\displaystyle\sum_{n=1}^{4} \left(\frac{2}{3}\right)^n$. Is there any connection between the product of the sums $\left(\displaystyle\sum_{n=1}^{4} 2^n\right)$ and $\left(\displaystyle\sum_{n=1}^{4} 3^n\right)$ and the sum $\displaystyle\sum_{n=1}^{4} (2^n \cdot 3^n)$? Is there any connection between the quotient of the sums $\left(\displaystyle\sum_{n=1}^{4} 2^n\right)$ and $\left(\displaystyle\sum_{n=1}^{4} 3^n\right)$ and the sum $\displaystyle\sum_{n=1}^{4} \left(\frac{2}{3}\right)^n$?

26. An arithmetic sequence is a sequence of the form $a_n = a + nd$, $n \geq 0$ where a is the first term of the sequence and d is called the *common difference*. Show that the series associated with this sequence diverges except in the case when $a = 0$ and $d = 0$.

27. Find the nth partial sum of the series $1 + 11 + 111 + 1111 + \cdots$.

28. Show that the infinite repeating decimal $.19191919\ldots$ can be expressed in infinite series form as

$$.19191919\ldots = \sum_{n=1}^{\infty} 19 \cdot \left(\frac{1}{100}\right)^n.$$

Use this expression to determine the rational number whose decimal expansion is $.191919\ldots$.

29. Find a pair of integers whose quotient is the rational number that has the infinite repeating decimal expansion .345734573457 . . . .

30. Give an example of infinite series $\sum_{n=1}^{\infty} a_n$ and $\sum_{n=1}^{\infty} b_n$ which both diverge, yet $\sum_{n=1}^{\infty} (a_n + b_n)$ converges.

31. Give an example of an infinite series $\sum_{n=1}^{\infty} a_n$ and a real number c such that $\sum_{n=1}^{\infty} ca_n$ converges and $\sum_{n=1}^{\infty} a_n$ diverges.

32. Prove that if $\sum_{n=1}^{\infty} a_n = L$ and c is a constant, then $\sum_{n=1}^{\infty} ca_n = cL$.

33. Show that $\sum_{n=1}^{\infty} \frac{1}{n^2} < 2$ by grouping the terms of the series in increasing powers of two, that is, consider the first term of the series, then the next two terms, the next four terms, and so on. Show that the mth grouping is less than 2^{1-m}.

34. Complete the steps in the following proof that the harmonic series $\sum_{n=1}^{\infty} 1/n$ diverges. Suppose $\sum_{n=1}^{\infty} \frac{1}{n} = S$. Find $\sum_{n=1}^{\infty} \frac{1}{2n}$ and show that

$$\sum_{n=1}^{\infty} \frac{1}{2n} < \sum_{n=1}^{\infty} \frac{1}{2n - 1}.$$

Use this to construct the contradiction that $S > \frac{1}{2}S + \frac{1}{2}S$.

35. Find a nonzero sequence $\{a_n\}$ with the property that $S_n^2 = S_n$ for each n.

36. A white square is divided into 25 equal squares. One of these squares is blackened. Each of the remaining smaller white squares is then divided into 25 equal squares and one of each is blackened. If this process is continued indefinitely, how many sizes of black squares will be necessary to blacken a total of at least 80% of the original white square?

37. An equilateral triangle has sides of unit length. Another equilateral triangle is constructed within the first by placing its vertices at the midpoints of the sides of the first triangle. A third equilateral triangle is constructed within the second in the same manner, and so on. (a) What is the sum of the perimeters of the triangles? (b) What is the sum of the areas of the triangles?

38. The Rhind papyrus, written in Egypt about 1650 B.C., is a text by the scribe Ahmes that contains 85 problems thought to be transcripted from an earlier Egyptian work. Problem 79 in this work is a forerunner to the old English rhyme: As I was going to St. Ives, I met a man with seven wives; each wife had seven sacks; each sack had seven cats; each cat had seven kits. Kits, cats, sacks, and wives, how many are going to St. Ives? Solve this problem.

39. The Federal government claims to be able to substantially stimulate the economy by giving each taxpayer a $50 rebate. They reason that 90% of this amount will be spent, that 90% of the amount spent will again be spent, and so on. If this is true, how much total expenditure will result from this $50 rebate?

40. Suppose that the government levies a 20% tax on all money spent. How does this tax influence the total expenditure from the tax rebate described in Exercise 39? How much of its rebate will the government eventually recover with this tax?

41. A toy company introduces a new ball called a "Whizball." It has the property that when dropped from a height h onto a hard surface it rebounds approximately 4/5 h. Suppose that this ball is dropped from a point 20 ft above a hard surface. (a) How far would it travel before it hits the surface four times? (b) Assuming that it bounces continually when dropped from this height, what would be the total distance the ball travels?

42. How long does it take the ball described in Exercise 41 to complete bouncing when dropped from the height of 20 feet? (This answer may not agree with your physical intuition.)

43. In actual practice the ball described in Exercise 41 bounces only as long as the force of rebound exceeds the force of friction between the ball and the surface on which it bounces. A ball dropped from 20 feet actually bounces 19 times before coming to rest on a concrete pavement. (a) How far does the ball travel during this time? (b) How long does it take the ball to travel this distance?

44. Suppose that the *Whizball* described in Exercise 41 is dropped from a height of 40 feet rather than from 20 feet and bounces continually. (a) How would this change the distance the ball travels? (b) How would this change the time required to complete the bouncing?

45. An old Greek fable concerns the god Achilles chasing a hare. Both are running in a straight line with the hare preceding Achilles by 100 meters. Achilles runs at a speed of 5 meters per second, twice the speed of the hare. How long does it take Achilles to catch the hare and how far must he run?

46. A famous story about the outstanding mathematician John von Neumann (1903–1957) concerns the following problem: Two bicyclists start 20 miles apart and head toward each other, each going 10 mph. At the same time, a fly traveling 15 mph leaves the front wheel of one bicycle, flies to the front wheel of the other bicycle, turns around and flies back to the wheel of the first bicycle, and so on, continuing in this manner until smashed between the two wheels. What total distance did he fly? There is a quick way to solve this problem. However, von Neumann allegedly solved the problem instantly by summing an infinite series. Solve this problem by both methods. (A very interesting article on John von Neumann written by P. R. Halmos, appears in the April 1973 issue of the *American Mathematical Monthly*, available in most college libraries.)

47. Two students, Pat and Mike, flip a coin to see who buys lunch. The first to flip a head must buy. Mike suggests that Pat flip first. Pat protests saying that one half of the time he will lose on the first flip, but relents when Mike explains to him that he will have this same likelihood of buying if Pat's flip is a tail. Who gets the better of this deal and by how much?

Putnam exercises:

48. $S_1 = \ln a$, and $S_n = \displaystyle\sum_{i=1}^{n-1} \ln (a - S_i)$, $n > 1$. Show that

$$\lim_{n \to \infty} S_n = a - 1.$$

(This exercise was problem 6, part I of the seventeenth William Lowell Putnam examination given on March 2, 1957. The examination and its solution can be found in the November 1957 issue of the *American Mathematical Monthly*, pages 649–654.)

49. If $f(x)$ is a real-valued function defined for $0 < x < 1$, then the formula $f(x) = o(x)$ is an abbreviation for the statement that

$$\frac{f(x)}{x} \to 0 \text{ as } x \to 0.$$

Keeping this in mind, prove the following: if

$$\lim_{x \to 0} f(x) = 0 \qquad \text{and} \qquad f(x) - f\left(\frac{x}{2}\right) = o(x),$$

then $f(x) = o(x)$.

(This exercise was problem 5, part I of the fourteenth William Lowell Putnam examination given on March 6, 1954. The examination and its solution can be found in the October 1954 issue of the *American Mathematical Monthly*, pages 542–549.)

50. If $a_0, a_1, \ldots, a_n$ are real numbers satisfying

$$\frac{a_0}{1} + \frac{a_1}{2} + \cdots + \frac{a_n}{n + 1} = 0,$$

show that the equation $a_0 + a_1 x + a_2 x^2 + \cdots + a_n x^n = 0$ has at least one real root.

(This exercise was problem 1, part I of the eighteenth William Lowell Putnam examination given on February 8, 1958. The examination and its solution can be found in the January 1961 issue of the *American Mathematical Monthly*, pages 18–22.)

9.3
INFINITE SERIES WITH POSITIVE TERMS

The definition of convergence of an infinite series is generally difficult to apply to a specific problem. In Section 9.2 we saw that if $\lim_{n \to \infty} a_n \neq 0$, then the series $\displaystyle\sum_{n=1}^{\infty} a_n$ diverges, and that a geometric series $\displaystyle\sum_{n=1}^{\infty} ar^{n-1}$ diverges if $|r| \geq 1$ and converges if $|r| < 1$. In this and succeeding sections we develop additional results to determine convergence without directly applying the definition.

First we study the class of series with all positive terms. A series of this type has the property that its sequence of partial sums is increasing. Convergence of the series depends entirely on whether the sequence of partial sums is bounded.

(9.17)
THEOREM

Suppose $\{a_n\}$ is a sequence with positive terms and $\{S_n\}$ is the associated sequence of partial sums. The series $\sum_{n=1}^{\infty} a_n$ converges if and only if the sequence $\{S_n\}$ is bounded. If $\{S_n\}$ is bounded and L is its least upper bound, then $\sum_{n=1}^{\infty} a_n = L$.

PROOF

By definition, the series $\sum_{n=1}^{\infty} a_n$ is convergent and $\sum_{n=1}^{\infty} a_n = L$ if and only if $\lim_{n \to \infty} S_n = L$. But $\{S_n\}$ is an increasing sequence, so Theorem 9.9 implies that $\lim_{n \to \infty} S_n = L$ if and only if L is the least upper bound of $\{S_n\}$. $\quad\square$

EXAMPLE 1

Show that the series $\sum_{n=1}^{\infty} \dfrac{1}{(n + 1)!}$ is convergent.

SOLUTION

For a positive integer n we use $n!$, read $n\ factorial$, to represent the product of the first n positive integers.

$$n! = 1 \cdot 2 \cdot 3 \ldots n.$$

We also define $0! = 1$.

Let $a_n = \dfrac{1}{(n + 1)!}$; then, for all n,

$$0 < a_n = \frac{1}{(n + 1)!} = \frac{1}{1 \cdot 2 \cdot 3 \ldots (n + 1)} \leq \frac{1}{1 \cdot 2 \cdot 2 \ldots \cdot 2} = \frac{1}{2^n}.$$

Consequently, for each n,

$$S_n = a_1 + a_2 + \cdots + a_n < \frac{1}{2} + \frac{1}{2^2} + \cdots + \frac{1}{2^n} \leq \sum_{i=1}^{\infty} \frac{1}{2^i}.$$

The series $\sum_{i=1}^{\infty} \dfrac{1}{2^i}$ is a geometric series with $r = 1/2$ and

$$\sum_{i=1}^{\infty} \frac{1}{2^i} = \frac{\dfrac{1}{2}}{1 - \dfrac{1}{2}} = 1.$$

So $S_n \leq 1$ for every positive integer n and $\displaystyle\sum_{n=1}^{\infty} \frac{1}{(n+1)!}$ converges.

In fact, $0 < \displaystyle\sum_{n=1}^{\infty} \frac{1}{(n+1)!} \leq 1$. Theorem 9.17 does not, however, give the precise value. ☐

In Example 1 we implicitly applied the following theorem when comparing the series $\displaystyle\sum_{n=1}^{\infty} \frac{1}{(n+1)!}$ to the series $\displaystyle\sum_{n=1}^{\infty} \frac{1}{2^n}$.

(9.18)
THEOREM

The Comparison Test   Suppose $\displaystyle\sum_{n=1}^{\infty} a_n$ and $\displaystyle\sum_{n=1}^{\infty} b_n$ are series with positive terms and M is a positive integer.

(i)   If $\displaystyle\sum_{n=1}^{\infty} b_n$ converges and $a_n \leq b_n$ for all $n > M$, then $\displaystyle\sum_{n=1}^{\infty} a_n$ converges.

(ii)   If $\displaystyle\sum_{n=1}^{\infty} b_n$ diverges and $a_n \geq b_n$ for all $n > M$, then $\displaystyle\sum_{n=1}^{\infty} a_n$ diverges.

In essence, part (i) of the comparison test states that when the terms of the sequence $\{a_n\}$ do not exceed those of $\{b_n\}$ and the terms of $\{b_n\}$ are small enough to ensure convergence, then the terms of $\{a_n\}$ are also small enough to ensure convergence. Part (ii) states that if the terms of $\{a_n\}$ exceed those of $\{b_n\}$ and the partial sums of $\{b_n\}$ are unbounded, then the partial sums of $\{a_n\}$ will be unbounded as well. The proof of the comparison test follows directly from this logic.

The following is a more useful, but slightly more complicated, version of the comparison test.

(9.19)
THEOREM

The Limit Comparison Test   Suppose $\displaystyle\sum_{n=1}^{\infty} a_n$ and $\displaystyle\sum_{n=1}^{\infty} b_n$ are series with positive terms.

(i)   If $\displaystyle\sum_{n=1}^{\infty} b_n$ converges and $0 \leq \displaystyle\lim_{n \to \infty} \frac{a_n}{b_n} < \infty$, then $\displaystyle\sum_{n=1}^{\infty} a_n$ converges.

(ii)   If $\displaystyle\sum_{n=1}^{\infty} b_n$ diverges and $0 < \displaystyle\lim_{n \to \infty} \frac{a_n}{b_n} \leq \infty$, then $\displaystyle\sum_{n=1}^{\infty} a_n$ diverges.

PROOF

Suppose first that $\displaystyle\lim_{n \to \infty} \frac{a_n}{b_n} = L$ and $0 < L < \infty$. An integer M exists with the property that

$$\left| \frac{a_n}{b_n} - L \right| < \frac{L}{2},$$

whenever $n > M$. Thus, when $n > M$,

$$\frac{L}{2} < \frac{a_n}{b_n} < \frac{3L}{2}$$

and

$$\frac{L}{2} b_n < a_n < \frac{3L}{2} b_n.$$

This inequality together with the comparison test implies that $\sum_{n=1}^{\infty} a_n$ converges if $\sum_{n=1}^{\infty} \frac{3L}{2} b_n$ converges, and diverges if $\sum_{n=1}^{\infty} \frac{L}{2} b_n$ diverges. We know from Theorem 9.14 that $\sum_{n=1}^{\infty} \frac{3L}{2} b_n$ converges when $\sum_{n=1}^{\infty} b_n$ converges and $\sum_{n=1}^{\infty} \frac{L}{2} b_n$ diverges when $\sum_{n=1}^{\infty} b_n$ diverges. Consequently, $\sum_{n=1}^{\infty} a_n$ converges precisely when $\sum_{n=1}^{\infty} b_n$ converges.

If $\sum_{n=1}^{\infty} b_n$ converges and $\lim_{n \to \infty} \frac{a_n}{b_n} = 0$, an integer M exists with $\frac{a_n}{b_n} < 1$ when $n > M$. Consequently, for all $n > M$, $a_n < b_n$. The comparison test implies that $\sum_{n=1}^{\infty} a_n$ converges.

Similarly, if $\sum_{n=1}^{\infty} b_n$ diverges and $\lim_{n \to \infty} \frac{a_n}{b_n} = \infty$, an integer M exists with $\frac{a_n}{b_n} > 1$ and $a_n > b_n$ when $n > M$. By the comparison test, $\sum_{n=1}^{\infty} a_n$ diverges.   □

EXAMPLE 2   Is the series $\sum_{n=1}^{\infty} \frac{3}{2^n + 1}$ convergent or divergent?

SOLUTION

The series $\sum_{n=1}^{\infty} \frac{1}{2^n}$ converges, so the series $\sum_{n=1}^{\infty} \frac{3}{2^n}$ converges as well.

Since

$$\frac{3}{2^n + 1} < \frac{3}{2^n},$$

the comparison test implies that $\sum_{n=1}^{\infty} \frac{3}{2^n + 1}$ converges.   □

EXAMPLE 3   Is the series $\sum_{n=1}^{\infty} \frac{3}{2^n - 1}$ convergent or divergent?

SOLUTION

Noting the similarity between this series and the series considered in Example 2, we would expect that this series is convergent. However, the denominator $2^n - 1$ is less than the denominator 2^n of the series to which we would naturally compare, so we cannot use the comparison test. Instead we employ the limit comparison test. Since $\displaystyle\sum_{n=1}^{\infty} \frac{1}{2^n}$ converges and

$$\lim_{n \to \infty} \frac{\dfrac{3}{2^n - 1}}{1/2^n} = \lim_{n \to \infty} \frac{3 \cdot 2^n}{2^n - 1} = 3,$$

the limit comparison test implies that $\displaystyle\sum_{n=1}^{\infty} \frac{3}{2^n - 1}$ converges.   ☐

Before we can effectively apply the comparison tests, we need a larger collection of convergent and divergent series with which to compare. An important application of the next theorem is to determine convergence of a type of series that can be easily used for comparison.

(9.20)
THEOREM

The Integral Test If f is continuous and decreasing on $[1, \infty)$ and $f(n) = a_n$ for each term in the positive term series $\displaystyle\sum_{n=1}^{\infty} a_n$, then $\displaystyle\sum_{n=1}^{\infty} a_n$ converges if and only if $\displaystyle\int_{1}^{\infty} f(x)dx$ converges.

PROOF

Since f is decreasing on each interval $[i, i + 1]$, the inequalities

$$a_{i+1} = f(i + 1) \le f(x) \le f(i) = a_i$$

hold for each integer i and each x in $[i, i + 1]$.

Consequently, for each integer i,

$$a_{i+1} = \int_{i}^{i+1} a_{i+1}dx \le \int_{i}^{i+1} f(x)dx \le \int_{i}^{i+1} a_i dx = a_i.$$

These inequalities can be expressed geometrically by considering the rectangles R_1 and R_2 with base $[i, i + 1]$ and heights a_{i+1} and a_i shown in Figure 9.6:

area of R_1 ≤ area below the graph f on $[i, i + 1]$ ≤ area of R_2.

For each integer n,

$$\sum_{i=1}^{n} a_{i+1} \le \sum_{i=1}^{n} \int_{i}^{i+1} f(x)dx = \int_{1}^{n+1} f(x)dx \le \sum_{i=1}^{n} a_i.$$

From the inequality,

$$\int_{1}^{n+1} f(x)dx \le \sum_{i=1}^{n} a_i,$$

we see that when $\displaystyle\int_{1}^{\infty} f(x)dx = \lim_{n \to \infty} \int_{1}^{n+1} f(x)dx = \infty$, then $\displaystyle\sum_{i=1}^{\infty} a_i$ is infinite as well. So if $\displaystyle\int_{1}^{\infty} f(x)dx$ diverges, then $\displaystyle\sum_{n=1}^{\infty} a_n$ diverges.

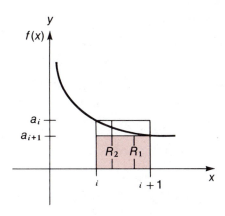

FIGURE 9.6

On the other hand, the series $\sum\limits_{i=1}^{\infty} a_{i+1}$ can be rewritten $\sum\limits_{i=1}^{\infty} a_{i+1} = \sum\limits_{i=2}^{\infty} a_i$. If

$\int_1^{\infty} f(x)dx$ converges, then the inequality

$$\sum_{i=1}^{n} a_{i+1} \leq \int_1^{n} f(x)dx$$

implies that $\sum\limits_{i=1}^{\infty} a_{i+1} = \sum\limits_{i=2}^{\infty} a_i$ converges and hence that $\sum\limits_{i=1}^{\infty} a_i$ converges.   ☐

EXAMPLE 4   Use the integral test to show that the series $\sum\limits_{n=1}^{\infty} \dfrac{1}{n}$ diverges.

SOLUTION

The function defined by $f(x) = \dfrac{1}{x}$ is positive, continuous, and decreasing on the interval $[1, \infty)$.

$$\int_1^{\infty} \frac{1}{x}\, dx = \lim_{M \to \infty} \int_1^{M} \frac{1}{x}\, dx$$

$$= \lim_{M \to \infty} \ln |x| \Big]_1^{M}$$

$$= \lim_{M \to \infty} (\ln M - \ln 1) = \infty.$$

It follows from the integral test that $\sum\limits_{n=1}^{\infty} \dfrac{1}{n}$ diverges.   ☐

EXAMPLE 5   Does the series $\sum\limits_{n=1}^{\infty} \dfrac{n}{e^n}$ converge?

SOLUTION

The function f defined by $f(x) = xe^{-x}$ is positive, continuous, and decreasing on the interval $[1, \infty)$. To evaluate the associated integral, $\int_1^\infty xe^{-x}\, dx$, we use integration by parts. Let $u = x$ and $dv = e^{-x}\, dx$, then $du = dx$ and $v = -e^{-x}$. So

$$\int xe^{-x}\, dx = -xe^{-x} + \int e^{-x}\, dx = -xe^{-x} - e^{-x} + C$$

and

$$\int_1^\infty xe^{-x}\, dx = \lim_{M \to \infty} \int_1^M xe^{-x}\, dx$$

$$= \lim_{M \to \infty} \left[-xe^{-x} - e^{-x} \right]_1^M$$

$$= \lim_{M \to \infty} (-Me^{-M} - e^{-M}) - (-e^{-1} - e^{-1})$$

$$= 2e^{-1} - \lim_{M \to \infty} \frac{M + 1}{e^M}.$$

By L'Hôpital's rule,

$$\lim_{M \to \infty} \frac{M + 1}{e^M} = \lim_{M \to \infty} \frac{1}{e^M} = 0,$$

so

$$\int_1^\infty xe^{-x}\, dx = 2e^{-1} < \infty.$$

By the integral test, $\displaystyle\sum_{n=1}^\infty ne^{-n} = \sum_{n=1}^\infty \frac{n}{e^n}$ converges.   □

An important application of the integral test is to determine the behavior of series of the form

(9.21)
$$\sum_{n=1}^\infty \frac{1}{n^p},$$

where p is a positive real number. These series are called the **p-series** and are the series most frequently used for comparison.

Since the function defined by

$$f(x) = \frac{1}{x^p}$$

is positive, continuous, and decreasing on $[1, \infty)$, the integral test implies that the convergence of a p-series depends on whether f has a finite integral on $[1, \infty)$. We saw in Example 4 that if $p = 1$, then the series $\displaystyle\sum_{n=1}^\infty \frac{1}{n}$ diverges.

When $p \neq 1$,

$$\int_1^\infty \frac{1}{x^p} \, dx = \lim_{M \to \infty} \int_1^M \frac{1}{x^p} \, dx$$

$$= \lim_{M \to \infty} \left(\frac{1}{-p + 1} \right) \frac{1}{x^{p-1}} \Bigg]_1^M$$

$$= \lim_{M \to \infty} \frac{1}{1 - p} \left[\frac{1}{M^{p-1}} - 1 \right].$$

This limit is finite when $p > 1$ (in fact, it is $1/(p - 1)$), but is infinite when $p < 1$. The results of the integral test applied to the p-series are listed in Theorem 9.22.

(9.22)
THEOREM

The p-series

$$\sum_{n=1}^\infty \frac{1}{n^p},$$

converges if $p > 1$ and diverges if $p \leq 1$.

This theorem used in conjunction with the comparison and limit comparison tests is a very powerful tool for determining the behavior of positive term series.

EXAMPLE 6

Does $\displaystyle\sum_{n=1}^\infty \frac{n}{(n + 1)^{5/2}}$ converge?

SOLUTION

For large values of n

$$\frac{n}{(n + 1)^{5/2}} \approx \frac{n}{n^{5/2}} = \frac{1}{n^{3/2}},$$

so a likely p-series with which to compare is $\displaystyle\sum_{n=1}^\infty \frac{1}{n^{3/2}}$. This series converges since $p = 3/2 > 1$.

Using the limit comparison test,

$$\lim_{n \to \infty} \frac{\dfrac{n}{(n + 1)^{5/2}}}{\dfrac{1}{n^{3/2}}} = \lim_{n \to \infty} \frac{n}{(n + 1)^{5/2}} \cdot n^{3/2} = \lim_{n \to \infty} \left(\frac{n}{n + 1} \right)^{5/2} = 1,$$

so the series $\displaystyle\sum_{n=1}^\infty \frac{n}{(n + 1)^{5/2}}$ converges. $\square$

EXAMPLE 7   Does $\displaystyle\sum_{n=1}^{\infty} \frac{5n}{2\sqrt[3]{n^5 + n}}$ converge?

SOLUTION

For large values of n,

$$\frac{5n}{2\sqrt[3]{n^5 + n}} \approx \frac{5n}{2\sqrt[3]{n^5}} = \frac{5}{2n^{2/3}},$$

so we compare this series with the p-series $\displaystyle\sum_{n=1}^{\infty} \frac{1}{n^{2/3}}$, a divergent series.

Applying the limit comparison test:

$$\lim_{n \to \infty} \frac{\dfrac{5n}{2\sqrt[3]{n^5 + n}}}{\dfrac{1}{n^{2/3}}} = \lim_{n \to \infty} \frac{5n}{2\sqrt[3]{n^5 + n}} \cdot \sqrt[3]{n^2} = \lim_{n \to \infty} \frac{5}{2} \frac{\sqrt[3]{n^5}}{\sqrt[3]{n^5 + n}} = \frac{5}{2},$$

so the series $\displaystyle\sum_{n=1}^{\infty} \frac{5n}{2\sqrt[3]{n^5 + n}}$ diverges. $\square$

EXERCISE SET 9.3

Use the comparison and limit comparison tests to determine if the series in Exercises 1 through 18 converge.

1. $\displaystyle\sum_{n=1}^{\infty} \frac{1}{2n + 1}$ **2.** $\displaystyle\sum_{n=1}^{\infty} \frac{2}{3n - 1}$

3. $\displaystyle\sum_{n=1}^{\infty} \frac{1}{n^2 + 2n + 2}$ **4.** $\displaystyle\sum_{n=1}^{\infty} \frac{1}{n^2 - 2n + 2}$

5. $\displaystyle\sum_{n=5}^{\infty} \frac{1}{\sqrt{n} - 2}$ **6.** $\displaystyle\sum_{n=1}^{\infty} \frac{n^3 + 3n^2 + 3}{n^5 + 2n - 5}$

7. $\displaystyle\sum_{n=1}^{\infty} \frac{1}{n(n + 4)}$ **8.** $\displaystyle\sum_{n=1}^{\infty} \frac{1}{n + \sqrt{n}}$

9. $\displaystyle\sum_{n=1}^{\infty} \frac{\sin n\pi + 2}{n^2}$ **10.** $\displaystyle\sum_{n=1}^{\infty} \frac{\sin n\pi}{n^2}$

11. $\displaystyle\sum_{n=1}^{\infty} \frac{1}{\sqrt{n^3 + 4}}$ **12.** $\displaystyle\sum_{n=1}^{\infty} \frac{n}{\sqrt{n^2 + 4}}$

13. $\displaystyle\sum_{n=1}^{\infty} \frac{1}{n2^n}$ **14.** $\displaystyle\sum_{n=1}^{\infty} \frac{1}{\sqrt{n}\, 3^n}$

15. $\displaystyle\sum_{n=1}^{\infty} \frac{n!}{(n+1)!}$ **16.** $\displaystyle\sum_{n=1}^{\infty} \frac{n!}{(n+2)!}$

17. $\displaystyle\sum_{n=1}^{\infty} \frac{n!}{(2n)!}$ **18.** $\displaystyle\sum_{n=1}^{\infty} \frac{(n!)^2}{(2n)!}$

Use the integral test to determine if the series in Exercises 19 through 28 converge.

19. $\displaystyle\sum_{n=1}^{\infty} \frac{1}{n+1}$ **20.** $\displaystyle\sum_{n=1}^{\infty} \frac{1}{(2n-1)^2}$

21. $\displaystyle\sum_{n=1}^{\infty} \frac{\ln n}{n}$ **22.** $\displaystyle\sum_{n=2}^{\infty} \frac{1}{n\sqrt{\ln n}}$

23. $\displaystyle\sum_{n=2}^{\infty} \frac{1}{n \ln n}$ **24.** $\displaystyle\sum_{n=2}^{\infty} \frac{1}{n(\ln n)^2}$

25. $\displaystyle\sum_{n=1}^{\infty} \frac{n}{e^{n^2}}$ **26.** $\displaystyle\sum_{n=1}^{\infty} \frac{1}{1+n^2}$

27. $\displaystyle\sum_{n=1}^{\infty} \frac{n}{(1+n^2)^2}$ **28.** $\displaystyle\sum_{n=3}^{\infty} \frac{1}{n \cdot \ln n \cdot \ln(\ln n)}$

29. Show that for all $x > 0$, $\ln x < x$ and conclude from this that $\displaystyle\sum_{n=2}^{\infty} \frac{1}{\ln n}$ diverges.

30. Determine all values of p for which the series $\displaystyle\sum_{n=2}^{\infty} \frac{1}{n(\ln n)^p}$ converges.

31. Show that $\displaystyle\sum_{n=2}^{\infty} \frac{1}{n \ln n(\ln(\ln n))^p}$ converges if and only if $p > 1$.

32. Does $\displaystyle\sum_{n=1}^{\infty} \frac{1}{n^{1+\frac{1}{n}}}$ converge?

33. A biologist examines a circular plate for a certain type of bacteria by drawing concentric circles of radius n, for positive integers n. The number of bacteria between the $n-1$st circle and the nth circle is inversely proportional to the area of the nth circle. (The constant of proportionality is independent of n.) Show that the number of bacteria on the plate is finite without assuming that the plate has finite radius.

34. Suppose that the biologist in Exercise 33 finds that the number of bacteria between the $n-1$st circle and the nth circle is inversely proportional to the radius of the nth circle, instead of its area. Can it still be deduced that the number of bacteria on the plate is finite, without assuming that the plate has finite radius?

35. (a) Use L'Hôpital's rule to show that for any positive real number r,

$$\lim_{n \to \infty} \frac{\ln n}{n^r} = 0.$$

(b) Show that for any positive real number r the relationship

$$\frac{1}{n^{1+r}} < \frac{1}{n \ln n} < \frac{1}{n}$$

holds for sufficiently large n.

(c) Conclude from (a) and (b) that the convergence or divergence of the series $\sum_{n=2}^{\infty} \dfrac{1}{n \ln n}$ cannot be determined by comparison to any of the p-series.

36. (a) Show that if $\sum_{n=1}^{\infty} a_n$ is a convergent series of positive terms, then

$\sum_{n=1}^{\infty} a_n^2$ is a convergent series.

(b) If $\sum_{n=1}^{\infty} a_n$ is a divergent series of positive terms, must $\sum_{n=1}^{\infty} a_n^2$ also diverge?

37. Use the identity $2a_n b_n = (a_n + b_n)^2 - a_n^2 - b_n^2$ and the result in Exercise 36 to show that if $\sum_{n=1}^{\infty} a_n$ and $\sum_{n=1}^{\infty} b_n$ are both convergent series with positive terms, then $\sum_{n=1}^{\infty} a_n b_n$ converges.

38. Use the limit comparison test and the fact that the terms of a convergent series are bounded to show that if $\sum_{n=1}^{\infty} a_n$ and $\sum_{n=1}^{\infty} b_n$ are convergent series with positive terms, then $\sum_{n=1}^{\infty} a_n b_n$ converges.

39. Suppose that $\sum_{n=1}^{\infty} a_n$ is a convergent series of positive terms and that $\{b_n\}$ is a sequence with $\lim_{n \to \infty} b_n = b > 0$. Show that the series $\sum_{n=1}^{\infty} a_n b_n$ converges.

40. (a) Show that $\sum_{n=1}^{\infty} \tan \dfrac{1}{n}$ diverges.

(b) Show that $\sum_{n=1}^{\infty} \left(\tan \dfrac{1}{n} \right)^2$ converges.

(c) For which values of k does $\left(\sum \tan \dfrac{1}{n} \right)^k$ converge?

41. Show that $\displaystyle\sum_{n=1}^{\infty} \arctan\left[\frac{1}{n^2 + n + 1}\right] = \frac{\pi}{4}$. (*Hint*: Show first that

$\arctan x - \arctan y = \arctan\left(\dfrac{x - y}{1 + xy}\right)$ then cleverly choose x and y.)

42. Evaluate $\displaystyle\sum_{k=0}^{\infty} \frac{k^2 + 3k + 1}{(k + 2)!}$.

43. By examining the proof of the integral test, show that if f is continuous and decreasing and $f(i) = a_i$ for each i, then for $n \geq 2$

$$\int_1^{n+1} f(x)dx \leq \sum_{i=1}^{n} a_i \leq \int_1^{n} f(x) + a_1.$$

44. In Exercise 23 you found that the series $\displaystyle\sum_{i=2}^{\infty} \frac{1}{n \ln n}$ diverges. The number of subatomic particles in the known universe is estimated to be 10^{125}. Use the result in Exercise 43 to find an upper bound for

$$\sum_{n=2}^{10^{125}} \frac{1}{n \ln n}.$$

45. In many applications, it is important to have an approximation for $n!$ when n is large. A crude approximation can be obtained by noting that $\ln n! = \displaystyle\sum_{i=1}^{n} \ln i$ and using an inequality obtained in a similar manner to that in Exercise 43 to deduce that

$$e(n/e)^n < n! < e((n + 1)/e)^{n+1}.$$

Verify this inequality.

46. A better approximation to $n!$ is known as *Stirling's formula*:

$$n! \approx \sqrt{2\pi n}\,(n/e)^n.$$

Show that this approximation is within the upper and lower bounds for $n!$ given in Exercise 45.

47. It seems intuitively reasonable that if one begins stacking cards at the edge of a table at no time can any card in the stack totally extend beyond the edge of the table. Not true. In fact, the cards can be arranged to extend any finite distance beyond the edge of the table. Show that this is true by showing that

(a) The first card can be placed to extend 1/2 its length beyond the table.

(b) The first two cards can be placed to extend a total of 3/4 a card length beyond the table.

(c) The first n cards can be placed to extend a total of $\displaystyle\sum_{i=1}^{n} 1/2i$ card lengths beyond the table.

(*Hint*: Assume that the ith card from the top of the stack extends $1/2i$ times its length beyond the $i + 1$st in the stack. Sum the total amount of card length extending beyond the table and compare to the amount lying above the table.)

Putnam exercises:

48. Suppose that $u_0, u_1, u_2 \cdots$ is a sequence of real numbers such that

$$u_n = \sum_{k=1}^{\infty} u_{n+k}^2 \qquad \text{for } n = 0, 1, 2, \ldots .$$

Prove that if $\Sigma\, u_n$ converges, then $u_k = 0$ for all k.

(This exercise was problem 6, part I of the fourteenth William Lowell Putnam examination given on March 6, 1954. The examination and its solution can be found in the October 1954 issue of the *American Mathematical Monthly*, pages 542–549.)

49. Let $\{a_n\}$ be a sequence of real numbers satisfying the inequalities

$$0 \le a_k \le 100a_n \text{ for } n \le k \le 2n \text{ and } n = 1, 2, \ldots ,$$

and such that the series

$$\sum_{n=0}^{\infty} a_n$$

converges. Prove that

$$\lim_{n \to \infty} na_n = 0.$$

(This exercise was problem 5, part II of the twenty-fourth William Lowell Putnam examination given on December 7, 1963. The examination and its solution can be found in the June–July 1964 issue of the *American Mathematical Monthly*, pages 636–641.)

9.4
ALTERNATING SERIES

In Section 9.3 we discussed series of positive terms. In this section we consider another special type of series: those with terms that alternate in sign, called **alternating series.** For example,

$$\sum_{n=1}^{\infty} (-1)^n \frac{1}{n} \quad \text{and} \quad \sum_{n=1}^{\infty} (-1)^{n+1} \frac{1}{\ln (n + 1)}$$

are alternating series. The convergence test for this class of series is given in the following theorem.

(9.23)
THEOREM

The Alternating Series Test Suppose $\sum_{n=1}^{\infty} (-1)^{n+1} a_n$ is an alternating series with $0 < a_{n+1} < a_n$ for each integer n. The series $\sum_{n=1}^{\infty} (-1)^{n+1} a_n$ converges if and only if $\lim_{n \to \infty} a_n = 0$.

PROOF

It has already been shown in Theorem 9.12 that if $\sum\limits_{n=1}^{\infty} (-1)^n\, a_n$ converges, then $\lim\limits_{n \to \infty} (-1)^n\, a_n = 0$. This implies that $\lim\limits_{n \to \infty} a_n = 0$. We need only show the converse.

To show that if $0 < a_{n+1} < a_n$ and $\lim\limits_{n \to \infty} a_n = 0$, then the series $\sum\limits_{n=1}^{\infty} (-1)^{n+1} a_n$ converges, we consider the sequence of partial sums for this series. Because the signs of the terms alternate, we consider the sequence of partial sums with odd and even indices separately.

For the even-indexed partial sums, a typical term S_{2n} can be written either as

$$S_{2n} = (a_1 - a_2) + (a_3 - a_4) + \cdots + (a_{2n-1} - a_{2n})$$

or as

$$S_{2n} = a_1 - (a_2 - a_3) - (a_4 - a_5) - \cdots - (a_{2n-2} - a_{2n-1}) - a_{2n}.$$

Since all the expressions within parentheses are positive, the first equation implies that the sequence of even partial sums is increasing, while the second equation implies that the sequence is bounded above by a_1. Consequently, by Theorem 9.9, the limit of the sequence of even partial sums exists. We denote this limit by S;

$$S = \lim_{n \to \infty} S_{2n}.$$

If we now consider the odd-indexed partial sums we have

$$S_{2n+1} = S_{2n} + a_{2n+1}$$

and since

$$\lim_{n \to \infty} S_{2n} = S \qquad \text{and} \qquad \lim_{n \to \infty} a_{2n+1} = 0;$$

$$\lim_{n \to \infty} S_{2n+1} = \lim_{n \to \infty} S_{2n} + \lim_{n \to \infty} a_{2n+1} = S + 0 = S.$$

Showing that both $\lim\limits_{n \to \infty} S_{2n} = S$ and $\lim\limits_{n \to \infty} S_{2n+1} = S$ is equivalent to showing that $\lim\limits_{n \to \infty} S_n = S$. Consequently,

$$\sum_{n=1}^{\infty} (-1)^{n+1}\, a_n \text{ converges.} \qquad \square$$

EXAMPLE 1   Is the series $\sum\limits_{n=1}^{\infty} (-1)^n\, \dfrac{1}{\sqrt{n}}$ convergent or divergent?

SOLUTION

The series satisfies the hypothesis of Theorem 9.23 because

$$0 < \frac{1}{\sqrt{n + 1}} < \frac{1}{\sqrt{n}}.$$

Since $\lim\limits_{n \to \infty} a_n = \lim\limits_{n \to \infty} \dfrac{1}{\sqrt{n}} = 0$, the series converges. $\qquad \square$

By examining the alternating behavior of the series $\sum_{n=1}^{\infty} (-1)^{n+1} a_n$ more closely, we can determine a bound for the difference between the nth partial sum and the sum of the series.

(9.24)
COROLLARY

If $\sum_{n=1}^{\infty} (-1)^{n+1} a_n$ is a convergent alternating series with $0 < a_{n+1} < a_n$ for each integer n and S is its limit, then

$$|S_n - S| < a_{n+1}$$

for each integer n.

PROOF

In the proof of Theorem 9.23 we found that $\{S_{2n}\}$ is an increasing sequence and $\lim_{n \to \infty} S_{2n} = S$. Consequently,

$$S_2 < S_4 < \cdots < S_{2n} < S.$$

The sequence of partial sums with odd indices is decreasing because

$$S_{2n+1} = S_{2n-1} - (a_{2n} - a_{2n+1})$$

and $(a_{2n} - a_{2n+1}) > 0$. Since $\lim_{n \to \infty} S_{2n+1} = S$, we have

$$S_2 < S_4 < \cdots < S_{2n} < S < S_{2n+1} < S_{2n-1} < \cdots < S_3 < S_1.$$

If the index is an odd integer of the form $2n - 1$, then

$$|S_{2n-1} - S| = S_{2n-1} - S < S_{2n-1} - S_{2n} = a_{2n}.$$

If the index is an even integer, then

$$|S_{2n} - S| = S - S_{2n} < S_{2n+1} - S_{2n} = a_{2n+1}.$$

In either case, $|S_n - S| < a_{n+1}$. $\square$

EXAMPLE 2

Show that $\sum_{n=1}^{\infty} (-1)^{n+1} \dfrac{1}{n} = 1 - \dfrac{1}{2} + \dfrac{1}{3} - \dfrac{1}{4} + \cdots$ converges, and find n so that

$$|S_n - S| < 10^{-3}.$$

SOLUTION

Since $\lim_{n \to \infty} \dfrac{1}{n} = 0$ and $0 < \dfrac{1}{n+1} < \dfrac{1}{n}$ the series $\sum_{n=1}^{\infty} (-1)^{n+1} \dfrac{1}{n}$ converges. It follows from Corollary 9.24 that for each integer n

$$\left| \sum_{i=1}^{n} (-1)^{i+1} \frac{1}{i} - S \right| < \frac{1}{n+1}.$$

To find n so that $|S_n - S| < 10^{-3}$, it suffices to find n with

$$\frac{1}{n+1} = 10^{-3};$$

that is, $n + 1 = 10^3$, so

$$n = 10^3 - 1 = 999.$$ □

EXAMPLE 3   How many terms of the convergent series $\sum_{n=1}^{\infty} (-1)^{n+1} \dfrac{1}{n!}$ are required to guarantee that $|S_n - S| < 10^{-3}$?

SOLUTION

We need to find n so that

$$\left| \sum_{i=1}^{n} (-1)^{i+1} \frac{1}{i!} - S \right| < \frac{1}{(n+1)!} \le 10^{-3}.$$

In this case, we cannot algebraically solve for n. However, the values in the accompanying table indicate that $n = 6$ is the first integer with the property that $1/(n+1)! \le 10^{-3}$. With six terms, $|S_n - S| < 10^{-3}$. The value of S to three decimal places is .632.

n	1	2	3	4	5	6
$\dfrac{1}{(n+1)!}$	.5000	.1667	.0417	.0083	.0014	.0002
S_n	1.0000	.5000	.6667	.6250	.6333	.6319

□

EXERCISE SET 9.4

Determine whether the series in Exercises 1 through 36 are convergent.

1. $\displaystyle\sum_{n=1}^{\infty} (-1)^{n+1} \frac{1}{n+1}$

2. $\displaystyle\sum_{n=1}^{\infty} \frac{(-1)^n}{3n+2}$

3. $\displaystyle\sum_{n=1}^{\infty} (-1)^{n+1} \frac{1}{\sqrt{n}-2}$

4. $\displaystyle\sum_{n=1}^{\infty} \frac{(-1)^n}{n\sqrt{n}}$

5. $\displaystyle\sum_{n=1}^{\infty} (-1)^{n-1} \frac{n}{n^2+1}$

6. $\displaystyle\sum_{n=1}^{\infty} \frac{(-1)^{n+1}}{n^2}$

7. $\displaystyle\sum_{n=0}^{\infty} (-1)^n \frac{n+2}{4n+5}$

8. $\displaystyle\sum_{n=1}^{\infty} \frac{(-1)^n}{(n+1)\sqrt{n}}$

9. $\displaystyle\sum_{n=1}^{\infty} \frac{(-1)^{n-1} n^2}{n^2+1}$

10. $\displaystyle\sum_{n=1}^{\infty} (-1)^{n+1} \frac{n^{3/2}}{n+4}$

11. $\displaystyle\sum_{n=1}^{\infty} (-1)^n \frac{\ln n}{n}$

12. $\displaystyle\sum_{n=1}^{\infty} \frac{(-1)^n n}{\ln n}$

13. $\displaystyle\sum_{n=2}^{\infty} (-1)^n \frac{1}{\ln n}$

14. $\displaystyle\sum_{n=2}^{\infty} \frac{(-1)^n}{\ln(\ln n)}$

15. $\displaystyle\sum_{n=1}^{\infty} (-1)^{n-1} \frac{n}{e^n}$

16. $\displaystyle\sum_{n=1}^{\infty} (-1)^n \frac{e^n}{n}$

17. $\displaystyle\sum_{n=1}^{\infty} (-1)^n \frac{e^n}{n!}$

18. $\displaystyle\sum_{n=1}^{\infty} (-1)^{n-1} \frac{e^n}{n^e}$

19. $\displaystyle\sum_{n=1}^{\infty} \frac{(-1)^{2n+1}}{n}$

20. $\displaystyle\sum_{n=1}^{\infty} (-1)^{2n} \frac{\sqrt{n+2}}{n^2 + 3n + 1}$

21. $\displaystyle\sum_{n=1}^{\infty} \frac{(-1)^n (1000)^n}{n!}$

22. $\displaystyle\sum_{n=1}^{\infty} \frac{(-1)^{n+1} n!}{(1000)^n}$

23. $\displaystyle\sum_{n=0}^{\infty} (-1)^n \frac{n^2}{\pi^n}$

24. $\displaystyle\sum_{n=0}^{\infty} \frac{(-1)^n e^n}{n^4}$

25. $\displaystyle\sum_{n=1}^{\infty} (-1)^{n-1} \frac{\pi^n}{n e^n}$

26. $\displaystyle\sum_{n=1}^{\infty} (-1)^{n-1} \frac{e^n}{n \pi^n}$

27. $\displaystyle\sum_{n=1}^{\infty} \frac{n(-2)^{2n}}{5^n}$

28. $\displaystyle\sum_{n=1}^{\infty} \frac{n(-4)^{3n}}{5^n}$

29. $\displaystyle\sum_{n=1}^{\infty} (-1)^n \frac{n^3}{n^3 + e^n}$

30. $\displaystyle\sum_{n=1}^{\infty} \frac{(-1)^n 2e^n}{n^3 + e^n}$

31. $\displaystyle\sum_{n=1}^{\infty} (-1)^n \frac{\cos n\pi}{n}$

32. $\displaystyle\sum_{n=1}^{\infty} \frac{(-1)^n \sin \dfrac{n\pi}{2}}{(n^3 + n)^{1/2}}$

33. $\displaystyle\sum_{n=1}^{\infty} \frac{\sin \dfrac{n\pi}{2}}{\sqrt{n}}$

34. $\displaystyle\sum_{n=1}^{\infty} \frac{\cos\left(\dfrac{n\pi}{2}\right)}{n^2}$

35. $\displaystyle\sum_{n=1}^{\infty} \frac{(-1)^n (n!)^2}{(2n)!}$

36. $\displaystyle\sum_{n=1}^{\infty} \frac{(-1)^n (3n)!}{(n!)^3}$

37. Find an approximation to the sum of the series $\displaystyle\sum_{n=1}^{\infty} (-1)^{n+1} \frac{1}{n!}$ that is accurate to within 10^{-5}.

38. How many terms are necessary to find an approximation to the sum of the series $\displaystyle\sum_{n=1}^{\infty} \frac{(-1)^n}{n^2}$ that is accurate to within 10^{-4}?

39. In Exercise 36 of Section 9.3 it was stated that if $\displaystyle\sum_{n=1}^{\infty} a_n$ is a convergent series of positive terms, then $\displaystyle\sum_{n=1}^{\infty} a_n^2$ is also convergent. Show that this need not be true if the terms of $\displaystyle\sum_{n=1}^{\infty} a_n$ are alternating.

9.5
ABSOLUTE CONVERGENCE

In preceding sections we considered special types of series: positive-term series and alternating series. This section is concerned with the convergence of general series and the relationship between a general series $\sum\limits_{n=1}^{\infty} a_n$ and the series of absolute values $\sum\limits_{n=1}^{\infty} |a_n| = |a_1| + |a_2| + |a_3| + \cdots$.

(9.25)
DEFINITION

A series $\sum\limits_{n=1}^{\infty} a_n$ is said to be **absolutely convergent** if the series of the absolute values of the terms,

$$\sum_{n=1}^{\infty} |a_n|,$$

converges. A convergent series $\sum\limits_{n=1}^{\infty} a_n$ that does not converge absolutely is said to be **conditionally convergent**.

Notice that a series with positive terms is absolutely convergent precisely when it is convergent. It is never conditionally convergent.

EXAMPLE 1   Show that the series $\sum\limits_{n=1}^{\infty} (-1)^n \dfrac{1}{n^2}$ converges absolutely.

SOLUTION

$\sum\limits_{n=1}^{\infty} \left| (-1)^n \dfrac{1}{n^2} \right| = \sum\limits_{n=1}^{\infty} \dfrac{1}{n^2}$ is a convergent p-series, so $\sum\limits_{n=1}^{\infty} (-1)^n \dfrac{1}{n^2}$ is absolutely convergent.   □

EXAMPLE 2   Does $\sum\limits_{n=1}^{\infty} (-1)^n \dfrac{1}{n}$ converge absolutely, converge conditionally, or diverge?

SOLUTION

In Example 2 of Section 9.4 we used the alternating series test to show that $\sum\limits_{n=1}^{\infty} (-1)^n \dfrac{1}{n}$ converges. However, the harmonic series $\sum\limits_{n=1}^{\infty} \dfrac{1}{n}$ diverges, so $\sum\limits_{n=1}^{\infty} (-1)^n \dfrac{1}{n}$ is conditionally convergent.   □

The following theorem details an important reason for considering absolute convergence.

(9.26)
THEOREM

If a series is absolutely convergent, then it is convergent.

PROOF

Suppose $\sum_{n=1}^{\infty} |a_n|$ is a convergent series. Then the multiple of this series $\sum_{n=1}^{\infty} 2|a_n|$ is also convergent. Moreover, for each integer n,

$$-|a_n| \le a_n \le |a_n|,$$

so
$$0 \le a_n + |a_n| \le 2|a_n|.$$

By the comparison test, Theorem 9.18, the series

$$\sum_{n=1}^{\infty} (a_n + |a_n|)$$

is convergent. So $\sum_{n=1}^{\infty} a_n$ can be expressed as the difference of two convergent series

$$\sum_{n=1}^{\infty} a_n = \sum_{n=1}^{\infty} [(a_n + |a_n|) + (-|a_n|)] = \sum_{n=1}^{\infty} (a_n + |a_n|) - \sum_{n=1}^{\infty} |a_n|.$$

It follows from Theorem 9.14 that $\sum_{n=1}^{\infty} a_n$ is convergent. ☐

EXAMPLE 3   Is $\sum_{n=1}^{\infty} \dfrac{\cos n}{n^{3/2} + 1}$ a convergent series?

SOLUTION

$$\left| \frac{\cos n}{n^{3/2} + 1} \right| \le \frac{1}{n^{3/2} + 1} \le \frac{1}{n^{3/2}} \quad \text{and} \quad \sum_{n=1}^{\infty} \frac{1}{n^{3/2}} \text{ is a convergent } p\text{-series.}$$

It follows from the comparison test that $\sum_{n=1}^{\infty} \left| \dfrac{\cos n}{n^{3/2} + 1} \right|$ converges. Therefore

$$\sum_{n=1}^{\infty} \frac{\cos n}{n^{3/2} + 1}$$ is absolutely convergent and hence convergent. ☐

A very important test for determining if a series is absolutely convergent uses the ratio of consecutive terms of the series as its guide.

(9.27)
THEOREM

The Ratio Test   Suppose $\displaystyle\sum_{n=1}^{\infty} a_n$ is a series of nonzero terms and

$$\lim_{n \to \infty} \left| \frac{a_{n+1}}{a_n} \right| = L$$

exists. (L may be ∞). Then
(i)   if $0 \le L < 1$, the series converges absolutely,
(ii) if $L > 1$, the series diverges,
(iii) if $L = 1$, nothing can be deduced from this test.

PROOF

(i) For $0 \le L < 1$, let $r = \dfrac{L + 1}{2}$ and $\varepsilon = 1 - r$. Then $0 < r < 1$ and

$\varepsilon > 0$. Since $\displaystyle\lim_{n \to \infty} \left| \frac{a_{n+1}}{a_n} \right| = L$, an integer M exists with

$$\left| \left| \frac{a_{n+1}}{a_n} \right| - L \right| < \varepsilon$$

whenever $n \ge M$. Consequently,

$$\left| \frac{a_{n+1}}{a_n} \right| < L + \varepsilon = L + (1 - r) = L + 1 - \left(\frac{L + 1}{2} \right) = \frac{L + 1}{2} = r,$$

so $\left| a_{n+1} \right| < \left| a_n \right| r$ whenever $n \ge M$.

In particular,

$$\left| a_{M+1} \right| < \left| a_M \right| r,$$
$$\left| a_{M+2} \right| < \left| a_{M+1} \right| r < \left| a_M \right| r^2$$
$$\left| a_{M+3} \right| < \left| a_{M+2} \right| r < \left| a_M \right| r^3$$

and in general,

$$\left| a_{M+n} \right| < \left| a_M \right| r^n.$$

The series $\displaystyle\sum_{n=1}^{\infty} \left| a_M \right| r^n$ is a geometric series with ratio r, $0 < r < 1$, and hence

converges. Therefore, the comparison test implies that $\displaystyle\sum_{n=1}^{\infty} \left| a_{M+n} \right|$ converges. Since

$\displaystyle\sum_{n=1}^{\infty} \left| a_n \right|$ differs from $\displaystyle\sum_{n=1}^{\infty} \left| a_{M+n} \right|$ in only the first M terms, $\displaystyle\sum_{n=1}^{\infty} \left| a_n \right|$ converges.

(ii) This statement is shown in a similar manner by choosing r to be a number between 1 and L. If $\varepsilon = L - r$, an integer M exists with

$$\left| \left| \frac{a_{n+1}}{a_n} \right| - L \right| < \varepsilon,$$

whenever $n \geq M$. So

$$\left| \frac{a_{n+1}}{a_n} \right| > L - \varepsilon = L - (L - r) = r > 1,$$

for $n \geq M$. Thus for all $n \geq M$, $|a_{n+1}| > |a_n| > 0$. Consequently $\lim\limits_{n \to \infty} |a_n| \neq 0$ and $\lim\limits_{n \to \infty} a_n \neq 0$, which implies that the series diverges.

Series that demonstrate the validity of statement (iii) in the theorem are given in Example 6. □

EXAMPLE 4   Does the series $\sum\limits_{n=1}^{\infty} (-1)^n \dfrac{n}{2^n}$ converge?

SOLUTION

Applying the ratio test to the series, we have

$$\lim_{n \to \infty} \left| \frac{a_{n+1}}{a_n} \right| = \lim_{n \to \infty} \left| \frac{(-1)^{n+1} \dfrac{n+1}{2^{n+1}}}{(-1)^n \dfrac{n}{2^n}} \right|$$

$$= \lim_{n \to \infty} \frac{n+1}{2^{n+1}} \cdot \frac{2^n}{n} = \lim_{n \to \infty} \frac{n+1}{n} \cdot \frac{1}{2} = \frac{1}{2},$$

so the series converges absolutely and hence converges. □

EXAMPLE 5   Use the ratio test to discuss the behavior of the series

$$\sum_{n=1}^{\infty} \frac{n!}{n^n}.$$

SOLUTION

The ratio of the terms of the series is

$$\left| \frac{a_{n+1}}{a_n} \right| = \frac{\dfrac{(n+1)!}{(n+1)^{n+1}}}{\dfrac{n!}{n^n}} = \frac{(n+1)!}{(n+1)^{n+1}} \cdot \frac{n^n}{n!}$$

$$= (n+1) \frac{n^n}{(n+1)^{n+1}}$$

$$= \frac{n^n}{(n+1)^n} = \frac{1}{\left(1 + \dfrac{1}{n}\right)^n}.$$

In Section 7.4 we found that $\lim\limits_{x \to \infty} (1 + 1/x)^x$ is an indeterminate form of type 1^∞ and used L'Hôpital's rule to show that the value of this limit is e. Consequently, the series converges, since

$$\lim_{n \to \infty} \left| \frac{a_{n+1}}{a_n} \right| = \lim_{n \to \infty} \frac{1}{\left(1 + \dfrac{1}{n}\right)^n} = \frac{1}{e} < 1.$$

□

EXAMPLE 6   Consider the ratio test applied to the p-series $\sum\limits_{n=1}^{\infty} \dfrac{(-1)^n}{n^p}$, where p is an arbitrary positive real number.

SOLUTION

For any positive p, the ratio test gives

$$\lim_{n \to \infty} \left| \frac{a_{n+1}}{a_n} \right| = \lim_{n \to \infty} \frac{\dfrac{1}{(n+1)^p}}{\dfrac{1}{n^p}} = \lim_{n \to \infty} \left(\frac{n}{n+1} \right)^p = 1.$$

However, for $p = 2$ the series converges absolutely; when $p = 1$ the series converges conditionally; while for $p = -1$ the series diverges. This demonstrates the validity of the statement in (iii) of Theorem 9.27 that no information concerning convergence of a series can be deduced when the limit is one.   □

The following test for convergence is known as the **root test**. It is useful for series $\sum\limits_{n=1}^{\infty} a_n$ when a_n has the form $(b_n)^n$.

(9.28)
THEOREM

The Root Test   Suppose $\sum\limits_{n=1}^{\infty} a_n$ is a series and $\lim\limits_{n \to \infty} |a_n|^{1/n} = L$ exists.

(L may be ∞). Then
(i)   if $0 \leq L < 1$, the series converges absolutely,
(ii) if $L > 1$, the series diverges,
(iii) if $L = 1$, nothing can be deduced from this test.

The proof of Theorem 9.28 is similar to the proof that establishes the ratio test.

EXAMPLE 7   Does the series $\displaystyle\sum_{n=1}^{\infty}\left(-\frac{1}{n}\right)^{n}$ converge?

SOLUTION

We apply the root test:

$$\lim_{n\to\infty}\left|\left(-\frac{1}{n}\right)^{n}\right|^{1/n} = \lim_{n\to\infty}\frac{1}{n} = 0 < 1.$$

The series converges absolutely and hence converges.   □

Elementary properties of algebra ensure that when the terms of a finite sum are rearranged, the value of the sum remains the same. For example,

$$3 = 11 + 3 - 4 - 7 = -4 + 11 + 3 - 7 = -7 + 3 + 11 - 4$$

or any other possible rearrangement of the terms $-7, -4, 3, 11$. This equality under rearrangement is not generally true for infinite series, however, and provides an interesting distinction between conditionally and absolutely convergent series. The results are presented in the following theorem. A proof of these and stronger results can be found in most advanced calculus books.

(9.29)
THEOREM   Suppose $\displaystyle\sum_{n=0}^{\infty} a_n$ is a convergent series.

(i) If the series is absolutely convergent, then any rearrangement of the terms produces a series that is absolutely convergent and sums to the same value.

(ii) If the series is conditionally convergent and L represents either an arbitrary real number, ∞, or $-\infty$, then there is a rearrangement of the terms to produce a series that sums to L.

We close this section with a review of the tests for convergence discussed in Sections 9.2 through 9.5. This review is presented in the flowchart in Figure 9.7, which can also be used as a guide to determine the convergence property of a series. The chart correctly indicates that some questions regarding series convergence have been left unanswered.

Given a series $\sum\limits_{n=1}^{\infty} a_n$:

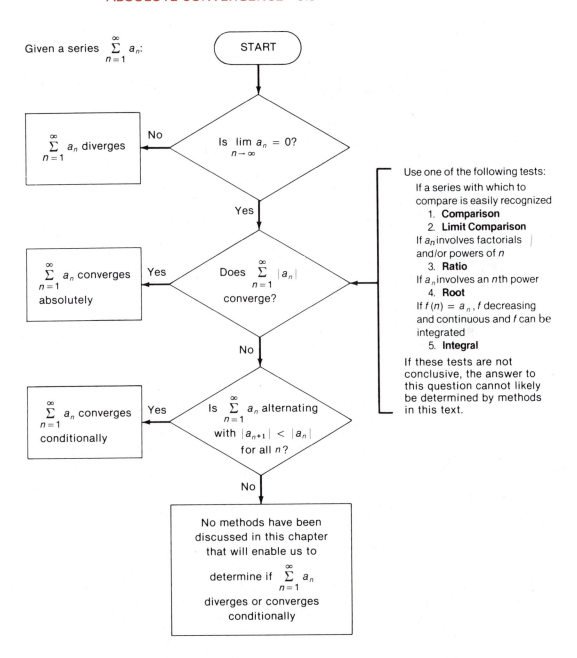

FIGURE 9.7

EXERCISE SET 9.5

Determine whether the series in Exercises 1 through 42 are divergent, conditionally convergent, or absolutely convergent.

1. $\displaystyle\sum_{n=1}^{\infty} (-1)^{n+1} \frac{1}{n+1}$

2. $\displaystyle\sum_{n=1}^{\infty} \frac{(-1)^n}{3n+2}$

3. $\displaystyle\sum_{n=1}^{\infty} (-1)^{n+1} \frac{1}{\sqrt{n}-2}$

4. $\displaystyle\sum_{n=1}^{\infty} \frac{(-1)^n}{n\sqrt{n}}$

5. $\displaystyle\sum_{n=1}^{\infty} (-1)^{n-1} \frac{n}{n^2+1}$

6. $\displaystyle\sum_{n=1}^{\infty} \frac{(-1)^{n+1}}{n^2}$

7. $\displaystyle\sum_{n=0}^{\infty} (-1)^n \frac{n+2}{4n+5}$

8. $\displaystyle\sum_{n=1}^{\infty} \frac{(-1)^n\, n}{(n+1)\sqrt{n}}$

9. $\displaystyle\sum_{n=1}^{\infty} \frac{(-1)^{n-1}\, n^2}{n^2+1}$

10. $\displaystyle\sum_{n=1}^{\infty} (-1)^{n+1} \frac{n^{3/2}}{n+4}$

11. $\displaystyle\sum_{n=1}^{\infty} (-1)^n \frac{\ln n}{n}$

12. $\displaystyle\sum_{n=1}^{\infty} \frac{(-1)^n\, n}{\ln n}$

13. $\displaystyle\sum_{n=2}^{\infty} (-1)^n \frac{1}{\ln n}$

14. $\displaystyle\sum_{n=2}^{\infty} \frac{(-1)^n}{\ln(\ln n)}$

15. $\displaystyle\sum_{n=1}^{\infty} (-1)^{n-1} \frac{n}{e^n}$

16. $\displaystyle\sum_{n=1}^{\infty} (-1)^n \frac{e^n}{n}$

17. $\displaystyle\sum_{n=1}^{\infty} (-1)^n \frac{e^n}{n!}$

18. $\displaystyle\sum_{n=1}^{\infty} (-1)^{n-1} \frac{e^n}{n^e}$

19. $\displaystyle\sum_{n=1}^{\infty} \frac{(-1)^{2n+1}}{n}$

20. $\displaystyle\sum_{n=1}^{\infty} (-1)^{2n} \frac{\sqrt{n+2}}{n^2+3n+1}$

21. $\displaystyle\sum_{n=1}^{\infty} \frac{(-1)^n\, (1000)^n}{n!}$

22. $\displaystyle\sum_{n=1}^{\infty} \frac{(-1)^{n+1}\, n!}{(1000)^n}$

23. $\displaystyle\sum_{n=0}^{\infty} (-1)^n \frac{n^2}{\pi^n}$

24. $\displaystyle\sum_{n=0}^{\infty} \frac{(-1)^n\, e^n}{n^4}$

25. $\displaystyle\sum_{n=1}^{\infty} (-1)^{n-1} \frac{\pi^n}{ne^n}$

26. $\displaystyle\sum_{n=1}^{\infty} (-1)^{n-1} \frac{e^n}{n\pi^n}$

27. $\displaystyle\sum_{n=1}^{\infty} \frac{n(-2)^{2n}}{5^n}$

28. $\displaystyle\sum_{n=1}^{\infty} \frac{n(-4)^{3n}}{5^n}$

29. $\displaystyle\sum_{n=1}^{\infty} (-1)^n \frac{n^3}{n^3+e^n}$

30. $\displaystyle\sum_{n=1}^{\infty} \frac{(-1)^n\, 2e^n}{n^3+e^n}$

31. $\displaystyle\sum_{n=1}^{\infty} (-1)^n \frac{\cos n\pi}{n}$

32. $\displaystyle\sum_{n=1}^{\infty} \frac{(-1)^n \sin (n\pi/2)}{(n^3 + n)^{1/2}}$

33. $\displaystyle\sum_{n=1}^{\infty} \frac{\sin (n\pi/2)}{\sqrt{n}}$

34. $\displaystyle\sum_{n=1}^{\infty} \frac{\cos (n\pi/4)}{n^2}$

35. $\displaystyle\sum_{n=1}^{\infty} \frac{(-1)^n (n!)^2}{(2n)!}$

36. $\displaystyle\sum_{n=1}^{\infty} \frac{(-1)^n (n!)^3}{(3n)!}$

37. $\displaystyle\sum_{n=1}^{\infty} \frac{(-1)^n n^n}{(2n)!}$

38. $\displaystyle\sum_{n=1}^{\infty} (-1)^n \ln\left(\frac{n}{n+1}\right)$

39. $\displaystyle\sum_{n=1}^{\infty} \frac{n!}{1 \cdot 3 \cdot 5 \cdot 7 \cdots (2n-1)}$

40. $\displaystyle\sum_{n=1}^{\infty} \frac{1 \cdot 3 \cdot 5 \cdot 7 \cdots (2n-1)}{2 \cdot 4 \cdot 6 \cdot 8 \cdots (2n)}$

41. $\displaystyle\sum_{n=1}^{\infty} \left(\frac{-n}{n^2 + 1}\right)^n$

42. $\displaystyle\sum_{n=1}^{\infty} \left(\frac{-2n}{n+1}\right)^n$

Use the root test to test the convergence of the series in Exercises 43 through 46.

43. $\displaystyle\sum_{n=2}^{\infty} \frac{1}{(\ln n)^n}$

44. $\displaystyle\sum_{n=1}^{\infty} \frac{1}{(\ln 2)^n}$

45. $\displaystyle\sum_{n=1}^{\infty} \left(\frac{n-1}{3n+1}\right)^n$

46. $\displaystyle\sum_{n=1}^{\infty} \frac{2^n}{n^2}$

47. Show that $\displaystyle\sum_{n=1}^{\infty} (-1)^n \left[\left(1 + \frac{1}{n}\right)^{n+1} - \left(1 + \frac{1}{n}\right)^n\right]$ converges conditionally.

48. Show that if $\displaystyle\sum_{n=1}^{\infty} a_n$ is a series that is absolutely convergent, then $\displaystyle\sum_{n=1}^{\infty} a_n^2$ is also absolutely convergent.

49. Find a series $\displaystyle\sum_{n=1}^{\infty} a_n$ that is convergent but for which $\displaystyle\sum_{n=1}^{\infty} a_n^2$ diverges.

50. Prove that if $\displaystyle\sum_{n=0}^{\infty} a_n$ is an absolutely convergent series with no zero terms, then $\displaystyle\sum_{n=0}^{\infty} \frac{1}{a_n}$ diverges.

51. Suppose that the series $\displaystyle\sum_{n=0}^{\infty} a_n$ is conditionally convergent. What can be said about the series

$$\sum_{n=0}^{\infty} \frac{|a_n| + a_n}{2} \quad \text{and} \quad \sum_{n=0}^{\infty} \frac{|a_n| - a_n}{2} ?$$

9.6
POWER SERIES

A polynomial is a function P with the property that

$$P(x) = a_0 + a_1 x + \cdots + a_n x^n$$

for some integer n and collection of constants $a_0, a_1, \ldots a_n$. We have seen that polynomials are always continuous, differentiable, and integrable. Derivatives and indefinite integrals of polynomials are also polynomials. This section considers an extension of polynomials that results from considering a series with an infinite number of terms of the form $a_n x^n$. This produces an infinite series of constants and powers of x called a **power series**. A power series, then, is an expression of the form

(9.30) $$\sum_{n=0}^{\infty} a_n x^n = a_0 + a_1 x + a_2 x^2 + \cdots + a_n x^n + \cdots,$$

where $a_0, a_1, a_2, \ldots$ represent constants and x represents a variable. Examples of power series are

$$\text{(a)} \sum_{n=0}^{\infty} x^n = 1 + x + x^2 + \cdots$$

and

$$\text{(b)} \sum_{n=0}^{\infty} \frac{(-1)^n x^n}{n+1} = 1 - \frac{1}{2}x + \frac{1}{3}x^2 - \frac{1}{4}x^3 + \cdots.$$

To apply the techniques of calculus to a function that is expressed in power series form,

(9.31) $$f(x) = \sum_{n=0}^{\infty} a_n x^n,$$

we must first consider the domain of f, that is, those values of x for which the series converges. For example, the series in (a) is a geometric series with first term $a = 1$ and ratio $r = x$ and consequently converges precisely when $|x| < 1$. In fact, for $|x| < 1$, $\displaystyle\sum_{n=0}^{\infty} x^n = \frac{1}{1-x}$.

EXAMPLE 1   Use the ratio test to find the values of x for which the power series $\displaystyle\sum_{n=0}^{\infty} \frac{(-1)^n x^n}{n+1}$ converges.

SOLUTION

For a fixed value of x,

$$\lim_{n \to \infty} \left| \frac{\dfrac{(-1)^{n+1} x^{n+1}}{n+2}}{\dfrac{(-1)^n x^n}{n+1}} \right| = \lim_{n \to \infty} \frac{|x^{n+1}|}{n+2} \frac{n+1}{|x^n|} = \lim_{n \to \infty} \frac{n+1}{n+2} |x| = |x|.$$

The ratio test implies that the series converges when $|x| < 1$ and diverges when $|x| > 1$. The question of convergence when $x = 1$ and $x = -1$ remains.

When $x = 1$, the power series is $\displaystyle\sum_{n=0}^{\infty} \frac{(-1)^n}{n+1}$, an alternating series with

$$0 < a_{n+1} = \frac{1}{n+2} < \frac{1}{n+1} = a_n$$

and $\displaystyle\lim_{n\to\infty} \frac{1}{n+1} = 0$. Hence it converges.

When $x = -1$, the power series is

$$\sum_{n=0}^{\infty} \frac{(-1)^n (-1)^n}{n+1} = \sum_{n=0}^{\infty} \frac{1}{n+1} = 1 + \frac{1}{2} + \frac{1}{3} + \cdots,$$

which is the divergent harmonic series. Consequently the power series $\displaystyle\sum_{n=0}^{\infty} \frac{(-1)^n x^n}{n+1}$ converges for $-1 < x \le 1$. The function defined by

$$f(x) = \sum_{n=0}^{\infty} \frac{(-1)^n x^n}{n+1}$$

has $(-1, 1]$ as its domain. $\quad\square$

Unless the domain of a function defined by a power series consists of an interval or collection of intervals we cannot discuss continuity, differentiability, or integrability of the function. Each of the functions defined by the power series $\displaystyle\sum_{n=0}^{\infty} x^n$ and $\displaystyle\sum_{n=0}^{\infty} \frac{(-1)^n x^n}{n+1}$ has an interval for its domain. That this is true in general of functions defined by power series is a result of the following lemma.

(9.32)
LEMMA

Suppose $\displaystyle\sum_{n=0}^{\infty} a_n x^n$ is a power series.

(i) If the series converges at c, then the series converges absolutely at x whenever $|x| < |c|$.

(ii) If the series diverges at b, then the series diverges at x whenever $|x| > |b|$.

PROOF

(i) Since $\displaystyle\sum_{n=0}^{\infty} a_n c^n$ converges, $\displaystyle\lim_{n\to\infty} a_n c^n = 0$.

A convergent sequence is bounded, so a number M exists with $|a_n c^n| \le M$ for all integers $n \ge 0$. Suppose x is arbitrarily chosen with $|x| < |c|$. Then $\left|\dfrac{x}{c}\right| < 1$ and the geometric series with first term M and ratio $\left|\dfrac{x}{c}\right|$,

$$\sum_{n=0}^{\infty} M\left(\frac{x}{c}\right)^n,$$

converges.

However,
$$|a_n x^n| = |a_n c^n| \cdot \left| \frac{x^n}{c^n} \right| \le M \left| \frac{x}{c} \right|^n,$$

so the comparison test implies that

$$\sum_{n=0}^{\infty} |a_n x^n|$$

converges and that the series is absolutely convergent when $|x| < |c|$.

(ii) This part of the lemma can be proved directly, but it is easier to apply part (i).

Suppose the series diverges at b but converges for some x with $|b| < |x|$. This implies, by part (i), that $\sum_{n=0}^{\infty} a_n b^n$ converges absolutely, which contradicts the assumption of divergence at b. Thus, if the series diverges at b and $|b| < |x|$, the series diverges at x.   □

The importance of Lemma 9.32 is that it can be used to partition all power series into three classes. These classes are described by the parts of the following theorem.

(9.33)
THEOREM   If $\sum_{n=0}^{\infty} a_n x^n$ is a power series, then precisely one of the following statements holds:

(i) $\sum_{n=0}^{\infty} a_n x^n$ converges only at $x = 0$.

(ii) $\sum_{n=0}^{\infty} a_n x^n$ converges for all real numbers x.

(iii) A number $R > 0$ exists and

(a) $\sum_{n=0}^{\infty} a_n x^n$ converges absolutely if $|x| < R$.

(b) $\sum_{n=0}^{\infty} a_n x^n$ diverges if $|x| > R$.

The number R given in part (iii) is called the **radius of convergence** of the series. A series of type (ii) is said to have an infinite radius of convergence, while a series of type (i) has radius of convergence zero.

The interval on which the series converges is called the **interval of convergence** for the series. The behavior of the power series at the endpoints of this interval is determined by applying one of the tests for convergence of a series of constants. Since only the constant term a_0 remains when a power series

$$a_0 + a_1 x + a_2 x^2 + \cdots = \sum_{n=0}^{\infty} a_n x^n \text{ is evaluated at zero, all power series converge}$$

at zero.

EXAMPLE 2   Find the radius of convergence for the power series

$$\sum_{n=0}^{\infty} n!x^n = 1 + x + 2x^2 + 6x^3 + \cdots.$$

SOLUTION

For any fixed value of x we can apply the ratio test:

$$\lim_{n\to\infty} \left| \frac{(n+1)!\, x^{n+1}}{n!\, x^n} \right| = \lim_{n\to\infty} (n+1)\,|x| = \infty.$$

Consequently for all $x \neq 0$, this series diverges. The radius of convergence is therefore zero, and the series converges only at $x = 0$. □

EXAMPLE 3   Find the domain of the function defined by the power series

$$f(x) = \sum_{n=0}^{\infty} \frac{3^n}{n^3 + 1} x^n = 1 + \frac{3}{2}x + x^2 + \frac{27}{28}x^3 + \cdots.$$

SOLUTION

The domain of f is the interval of convergence for the power series. To determine this, we apply the ratio test.

$$\lim_{n\to\infty} \left| \frac{\dfrac{3^{n+1}}{(n+1)^3 + 1} x^{n+1}}{\dfrac{3^n}{n^3 + 1} x^n} \right| = \lim_{n\to\infty} \frac{n^3 + 1}{(n+1)^3 + 1}\, 3|x| = 3|x|.$$

Consequently, the series converges when $3|x| < 1$, that is, when $|x| < \frac{1}{3}$, and diverges when $|x| > \frac{1}{3}$. That is, the series converges when x is in the interval $(-1/3, 1/3)$ and diverges when $x > 1/3$ and when $x < -1/3$.

To determine whether the series converges at the endpoints of the interval, we must examine the values individually. For $x = 1/3$,

$$\sum_{n=0}^{\infty} \frac{3^n \left(\frac{1}{3}\right)^n}{n^3 + 1} = \sum_{n=0}^{\infty} \frac{1}{n^3 + 1},$$

which converges by comparison with the p-series $\sum_{n=1}^{\infty} \frac{1}{n^3}$. For $x = -1/3$,

$$\sum_{n=0}^{\infty} \frac{3^n \left(-\frac{1}{3}\right)^n}{n^3 + 1} = \sum_{n=0}^{\infty} \frac{(-1)^n}{n^3 + 1}$$

converges absolutely and hence converges. The interval of convergence of the series, and consequently the domain of f, is $[-1/3, 1/3]$. □

The general form of a power series is

(9.34) $$\sum_{n=0}^{\infty} a_n(x - a)^n,$$

where a is a constant. A power series of this form is said to be *centered* at a. The general form is simply a linear translation of the case when $a = 0$. The last examples of this section illustrate that the method for finding the interval of convergence in case $a \neq 0$ is the same as when $a = 0$.

EXAMPLE 4   Find the interval of convergence of the power series

$$\sum_{n=1}^{\infty} (-1)^{n+1} \frac{(x - 1)^n}{n} = (x - 1) - \frac{1}{2}(x - 1)^2 + \frac{1}{3}(x - 1)^3 - \cdots.$$

SOLUTION

Applying the ratio test to the series:

$$\lim_{n \to \infty} \left| \frac{(-1)^{n+2} \dfrac{(x - 1)^{n+1}}{n + 1}}{(-1)^{n+1} \dfrac{(x - 1)^n}{n}} \right| = \lim_{n \to \infty} \frac{n}{n + 1} |x - 1| = |x - 1|.$$

Thus, the series converges if $|x - 1| < 1$; that is, if x is in $(0, 2)$. The series diverges if $|x - 1| > 1$.

When $x = 0$, the series diverges since

$$\sum_{n=1}^{\infty} (-1)^{n+1} \frac{(-1)^n}{n} = \sum_{n=1}^{\infty} (-1)^{2n+1} \frac{1}{n} = -\sum_{n=1}^{\infty} \frac{1}{n}.$$

When $x = 2$, the series converges since

$$\sum_{n=1}^{\infty} \frac{(-1)^{n+1} (1)^n}{n} = \sum_{n=1}^{\infty} \frac{(-1)^{n+1}}{n}.$$

The interval of convergence is $(0, 2]$. □

EXAMPLE 5   Find the domain of the function defined by

$$f(x) = \sum_{n=0}^{\infty} \frac{(2x - 1)^n}{3^n}.$$

SOLUTION

Since

$$\lim_{n \to \infty} \left| \frac{\dfrac{(2x - 1)^{n+1}}{3^{n+1}}}{\dfrac{(2x - 1)^n}{3^n}} \right| = \lim_{n \to \infty} \frac{|2x - 1|}{3} = \frac{|2x - 1|}{3},$$

the series converges if

$$\frac{|2x - 1|}{3} < 1, \qquad \text{that is, if} \qquad |2x - 1| < 3.$$

The inequality $|2x - 1| < 3$ is equivalent to $-3 < 2x - 1 < 3$, and hence to $-2 < 2x < 4$. The series converges if $-1 < x < 2$ and diverges if $x < -1$ or $x > 2$.
     When $x = -1$,

$$\sum_{n=0}^{\infty} \frac{(-3)^n}{3^n} = \sum_{n=0}^{\infty} (-1)^n$$

diverges.
     When $x = 2$,

$$\sum_{n=0}^{\infty} \frac{3^n}{3^n} = \sum_{n=0}^{\infty} 1$$

diverges. Consequently, the domain of f is $(-1, 2)$.   □

EXERCISE SET 9.6

Find the radius of convergence and interval of convergence of the power series in Exercises 1 through 26.

1. $\displaystyle\sum_{n=0}^{\infty} \frac{x^n}{n + 1}$

2. $\displaystyle\sum_{n=1}^{\infty} \frac{x^n}{\sqrt{n}}$

3. $\displaystyle\sum_{n=1}^{\infty} \frac{(\ln n)\, x^n}{n^2}$

4. $\displaystyle\sum_{n=1}^{\infty} \frac{x^n}{n(n + 1)}$

5. $\displaystyle\sum_{n=0}^{\infty} \frac{x^n}{n!}$

6. $\displaystyle\sum_{n=1}^{\infty} \frac{(1001)^n\, x^n}{n!}$

7. $\displaystyle\sum_{n=1}^{\infty} \frac{x^{n+1}}{n^2}$

8. $\displaystyle\sum_{n=0}^{\infty} \frac{x^{2n}}{n + 1}$

9. $\displaystyle\sum_{n=0}^{\infty} \frac{(-1)^n\, x^n}{n!}$

10. $\displaystyle\sum_{n=0}^{\infty} \frac{(-1)^{n+1}\, x^{2n}}{(2n)!}$

11. $\displaystyle\sum_{n=0}^{\infty} \frac{(-1)^{n+1}\, x^{2n+1}}{(2n + 1)!}$

12. $\displaystyle\sum_{n=2}^{\infty} \frac{x^{2n+1}}{\ln n}$

13. $\displaystyle\sum_{n=1}^{\infty} \frac{2^n\, x^n}{n}$

14. $\displaystyle\sum_{n=0}^{\infty} \frac{n^n\, x^n}{n!}$

15. $\displaystyle\sum_{n=5}^{\infty} \frac{x^{n-4}}{\sqrt{n + 4}}$

16. $\displaystyle\sum_{n=1}^{\infty} \frac{(3x)^n}{\arctan n}$

17. $\displaystyle\sum_{n=0}^{\infty} (x - 2)^n$

18. $\displaystyle\sum_{n=0}^{\infty} \frac{(x - 2)^n}{n + 1}$

19. $\displaystyle\sum_{n=0}^{\infty} (n + 2)^2\, (x - 2)^n$

20. $\displaystyle\sum_{n=0}^{\infty} n!(x - 2)^n$

21. $\displaystyle\sum_{n=1}^{\infty} \frac{(x + 3)^{2n}}{n}$

22. $\displaystyle\sum_{n=1}^{\infty} \frac{\ln n(x + 1)^{3n}}{n(n + 1)}$

23. $\displaystyle\sum_{n=1}^{\infty} \frac{(3x - 1)^n}{n^2}$

24. $\displaystyle\sum_{n=0}^{\infty} \frac{n(2x - 5)^n}{n + 1}$

25. $\displaystyle\sum_{n=1}^{\infty} \frac{(2x + 1)^n}{\sqrt{n}}$

26. $\displaystyle\sum_{n=0}^{\infty} \frac{(3x + 4)^{2n}}{2^n}$

Find the domain of each function described in Exercises 27 through 34.

27. $f(x) = \displaystyle\sum_{n=2}^{\infty} \frac{x^n}{n \ln n}$

28. $f(x) = \displaystyle\sum_{n=1}^{\infty} \frac{(-1)^n x^{2n}}{n\sqrt{n} + 1}$

29. $f(x) = \displaystyle\sum_{n=0}^{\infty} \frac{(x - 1)^n}{n^2 + 2}$

30. $f(x) = \displaystyle\sum_{n=1}^{\infty} \frac{(3x + 1)^n}{\ln(n + 1)}$

31. $f(x) = \displaystyle\sum_{n=0}^{\infty} \frac{(2x - 3)^n}{n^2 + 1}$

32. $f(x) = \displaystyle\sum_{n=1}^{\infty} \frac{\ln n(x + 7)^{2n}}{(n + 1) \arctan n}$

33. $f(x) = \displaystyle\sum_{n=0}^{\infty} x^n (x - 2)^n$

34. $f(x) = \displaystyle\sum_{n=0}^{\infty} (x - 1)^n (x + 1)^n$

Find the values of x for which the series in Exercises 35 through 40 converge.

35. $\displaystyle\sum_{n=0}^{\infty} \frac{1}{x^n}$

36. $\displaystyle\sum_{n=0}^{\infty} \frac{n}{x^n}$

37. $\displaystyle\sum_{n=0}^{\infty} \frac{2^n}{x^n}$

38. $\displaystyle\sum_{n=0}^{\infty} \frac{3^n}{(x - 1)^n}$

39. $\displaystyle\sum_{n=0}^{\infty} (\cos x)^n$

40. $\displaystyle\sum_{n=1}^{\infty} \frac{(\tan x)^n}{n}$

41. Find the interval of convergence for the series $\displaystyle\sum_{n=0}^{\infty} [2 + (-1)^n] \, x^n$.

42. Suppose that the ratio test applied to $\displaystyle\sum_{n=0}^{\infty} a_n x_0{}^n$ gives $\displaystyle\lim_{n\to\infty} \left| \frac{a_{n+1}}{a_n} x_0 \right| < 1$.

Show that $\displaystyle\sum_{n=0}^{\infty} n^k a_n x^n$ also converges absolutely for any integer k and every value of x with $|x| < |x_0|$.

43. Suppose that the power series $\displaystyle\sum_{n=0}^{\infty} a_n x^n$ and $\displaystyle\sum_{n=0}^{\infty} b_n x^n$ have radius of convergence R_1 and R_2 respectively. What can be said about the radius of convergence R of $\displaystyle\sum_{n=0}^{\infty} (a_n + b_n) \, x^n$?

44. Suppose the series $\displaystyle\sum_{n=0}^{\infty} a_n x^n$ has the property that $\dfrac{1}{L} < a_n < L$ for all n, where L is a positive constant. Show that the radius of convergence of this series must be 1. What is the interval of convergence of such a series?

9.7
DIFFERENTIATION AND INTEGRATION OF POWER SERIES

Power series can be thought of as a generalization of polynomials. As we have seen, a polynomial

$$P(x) = a_0 + a_1 x + \cdots + a_m x^m = \sum_{n=0}^{m} a_n x^n$$

can be differentiated and integrated easily:

$$D_x P(x) = a_1 + 2a_2 x + \cdots + m a_m x^{m-1} = \sum_{n=1}^{m} n a_n x^{n-1}$$

and

$$\int_0^x P(t)dt = a_0 x + \frac{a_1 x^2}{2} + \cdots + \frac{a_m x^{m+1}}{m+1} = \sum_{n=0}^{m} \frac{a_n x^{n+1}}{n+1}.$$

An important feature of power series is that they have calculus properties similar to those of polynomials. We can apply differentiation and integration techniques to functions defined by power series, since the domain of these functions is an interval.

(9.35)
THEOREM

If f is a function defined by the power series

$$f(x) = \sum_{n=0}^{\infty} a_n x^n = a_0 + a_1 x + a_2 x^2 + a_3 x^3 + \cdots$$

with radius of convergence R, then
(i) f is differentiable at each x in $(-R, R)$, and

$$D_x f(x) = D_x \sum_{n=0}^{\infty} a_n x^n = \sum_{n=1}^{\infty} n a_n x^{n-1} = a_1 + 2a_2 x + 3a_3 x^2 + \cdots$$

(ii) f is integrable on $(-R, R)$ and

$$\int_0^x f(t)dt = \int_0^x \left[\sum_{n=0}^{\infty} a_n t^n \right] dt = \sum_{n=0}^{\infty} \frac{a_n}{n+1} x^{n+1}$$

$$= a_0 x + \frac{a_1}{2} x^2 + \frac{a_2}{3} x^3 + \cdots .$$

The proof of this theorem, especially part (i), is rather long and somewhat complicated and is deferred to Appendix B. It is not, however, beyond the scope of this course and is instructive for reviewing the definition of the derivative and the fundamental theorem of calculus.

EXAMPLE 1   Use the fact that $\dfrac{1}{1-x} = \sum_{n=0}^{\infty} x^n$ for $|x| < 1$ to find a power series representation for $f(x) = \dfrac{1}{(1-x)^2}$.

SOLUTION

Finding a power series representation for $f(x)$ means finding a series $\sum\limits_{n=0}^{\infty} a_n x^n$ such that $f(x) = \sum\limits_{n=0}^{\infty} a_n x^n$ for each x in the interval of convergence of the power series. Since

$$D_x\left(\frac{1}{1-x}\right) = \frac{1}{(1-x)^2},$$

it follows from part (i) of Theorem 9.33 that

$$\frac{1}{(1-x)^2} = D_x\left(\frac{1}{1-x}\right) = D_x\left[\sum_{n=0}^{\infty} x^n\right] = \sum_{n=1}^{\infty} nx^{n-1}$$

for $|x| < 1$. The series $\sum\limits_{n=1}^{\infty} nx^{n-1}$ diverges at $x = -1$ and at $x = 1$, so the interval of convergence is $(-1, 1)$.   □

EXAMPLE 2   Find a power series representation for $\ln(1 - x)$.

SOLUTION

We again start with the power series

$$\frac{1}{1-x} = \sum_{n=0}^{\infty} x^n, \quad \text{for } |x| < 1.$$

By integration, $-\ln(1 - x) = \int_0^x \frac{1}{1-t}\, dt = \sum\limits_{n=0}^{\infty} \frac{x^{n+1}}{n+1}, \quad \text{for } |x| < 1,$

So $\ln(1 - x) = -\sum\limits_{n=0}^{\infty} \frac{x^{n+1}}{n+1}, \quad \text{for } |x| < 1.$

Note that when $x = -1$, this series is a convergent alternating series, so

$$\ln 2 = -\sum_{n=0}^{\infty} \frac{(-1)^{n+1}}{n+1}.$$

When $x = 1$, the series is the divergent harmonic series, so the interval of convergence is $[-1, 1)$.   □

EXAMPLE 3   Show that $e^x = \sum\limits_{n=0}^{\infty} \frac{x^n}{n!}$.

SOLUTION

Applying the ratio test to $\sum\limits_{n=0}^{\infty} \dfrac{x^n}{n!}$ gives

$$\lim_{n\to\infty} \left| \frac{\dfrac{x^{n+1}}{(n+1)!}}{\dfrac{x^n}{n!}} \right| = \lim_{n\to\infty} \left| \frac{x}{n+1} \right| = 0,$$

independent of x, so the series converges for all values of x. Let

$$f(x) = \sum_{n=0}^{\infty} \frac{x^n}{n!}.$$

We see that $f(0) = 1$. To show that $f(x) = e^x$ it suffices to show that $f'(x) = f(x)$, since the natural exponential function is the only function with these properties. It follows from part (i) of Theorem 9.35 that

$$f'(x) = \sum_{n=1}^{\infty} \frac{nx^{n-1}}{n!}$$

$$= \sum_{n=1}^{\infty} \frac{x^{n-1}}{(n-1)!} = 1 + x + \frac{x^2}{2} + \frac{x^3}{6} + \cdots$$

$$= \sum_{n=0}^{\infty} \frac{x^n}{n!} = f(x),$$

for all values of x. Consequently,

$$e^x = \sum_{n=0}^{\infty} \frac{x^n}{n!},$$

for all real numbers x.   □

EXAMPLE 4   Approximate $\displaystyle\int_0^{.1} e^{-x^2}\, dx$ to within 10^{-8}.

SOLUTION

The function given by $f(x) = e^{-x^2}$ is continuous and hence integrable. We cannot use the fundamental theorem of calculus for the evaluation of its integral since f has no elementary antiderivative. However, by modifying the power series in the previous example, we can express this function as

$$e^{-x^2} = \sum_{n=0}^{\infty} \frac{(-x^2)^n}{n!} = \sum_{n=0}^{\infty} \frac{(-1)^n x^{2n}}{n!}$$

and

$$\int_0^x e^{-t^2}\, dt = \sum_{n=0}^{\infty} \frac{(-1)^n x^{2n+1}}{n!(2n+1)}.$$

Thus
$$\int_0^{.1} e^{-t^2}\, dt = \sum_{n=0}^{\infty} \frac{(-1)^n\,(.1)^{2n+1}}{n!(2n+1)},$$

which is an alternating series. By Corollary 9.24, the mth partial sum approximates the true value with error not exceeding the magnitude of the $(m+1)$st term:

$$\frac{(.1)^{2m+3}}{(m+1)!(2m+3)}.$$

By evaluating this quotient at $m = 1$ and then at $m = 2$, we find that the first integer for which this value does not exceed 10^{-8} is $m = 2$. This approximation is

$$\int_0^{.1} e^{-x^2}\, dx \approx \sum_{n=0}^{2} \frac{(-1)^n\,(.1)^{2n+1}}{n!(2n+1)}$$

$$= .1 - \frac{.001}{3} + \frac{.00001}{10} = .09966767.$$

The error in the approximation is at most the magnitude of the third term of this alternating series

$$\frac{(.1)^{2(2)+3}}{(2+1)!(2\cdot 2 + 3)} = \frac{(.1)^7}{42} \approx 2.4 \times 10^{-9}. \qquad \Box$$

EXERCISE SET 9.7

Find a power series representation for the functions described in Exercises 1 through 20. Use the fact that $\dfrac{1}{1-x} = \displaystyle\sum_{n=0}^{\infty} x^n$, for $|x| < 1$, $e^x = \displaystyle\sum_{n=0}^{\infty} \dfrac{x^n}{n!}$ for all x, and the arithmetic and calculus properties of power series.

1. $f(x) = \dfrac{1}{1+x}$

2. $f(x) = \dfrac{1}{1-2x}$

3. $f(x) = \dfrac{1}{(1+x)^2}$

4. $f(x) = \ln(1+x)$

5. $f(x) = \dfrac{1}{2-x}$

6. $f(x) = \dfrac{1}{3-2x}$

7. $f(x) = \dfrac{1}{1+x^2}$

8. $f(x) = \arctan x$

9. $f(x) = \dfrac{x}{1+x^2}$

10. $f(x) = \ln(1+x^2)$

11. $f(x) = \dfrac{1}{1-x^2}$

12. $f(x) = \dfrac{1+x^2}{1-x^2}$

13. $f(x) = \dfrac{e^x - 1}{x}$

14. $f(x) = \dfrac{x}{1-x}$

15. $f(x) = x \ln(1+x)$

16. $f(x) = \ln((1+x)/(1+x^2))$

17. $f(x) = \dfrac{2x}{1 - 3x^2}$

18. $f(x) = \displaystyle\int_0^x e^{t^2}\, dt$

19. $f(x) = \displaystyle\int_0^x t\, e^{t^3}\, dt$

20. $f(x) = \displaystyle\int_0^x [e^t + \ln(t + 1)]\, dt$

Use a power series representation to approximate the definite integrals in Exercises 21 and 22 to within 10^{-4}.

21. $\displaystyle\int_0^{.2} e^{-x^2}\, dx$

22. $\displaystyle\int_0^{.5} \dfrac{1}{1 + x^3}\, dx$

23. Use integration by parts to find $\int \ln(1 + x)\,dx$ and use this result to find a power series representation for $x \ln(1 + x)$. Compare this result with the result in Exercise 15.

24. Suppose that f is a function with a power series representation of the form
$f(x) = \displaystyle\sum_{n=0}^{\infty} a_n x^n$. If $f(0) = 0$ and $f'(x) = f(x)$ for all x, determine the constants
in the power series representation.

25. Suppose that f is a function with a power series representation of the form
$f(x) = \displaystyle\sum_{n=0}^{\infty} a_n x^n$. If $f(0) = 0, f'(0) = 1$ and $f''(x) = -f(x)$ for all x, determine
the constants in the power series representation.

26. Consider the functions f and g defined by
$$f(x) = \sum_{n=0}^{\infty} \frac{(-1)^{n+1}\, x^{2n+1}}{(2n + 1)!} \quad \text{and} \quad g(x) = \sum_{n=0}^{\infty} \frac{(-1)^n\, x^{2n}}{(2n)!}.$$
Compute $f'(x)$ and $g'(x)$ (and be observant).

27. Consider the series given in Example 2 on p. 552 for $\ln(1 - x)$. Let
$x = \dfrac{1}{2}$, and show that $\ln 2$ also can be expressed as
$$\ln 2 = \sum_{n=0}^{\infty} \frac{1}{2^{n+1}(n + 1)}.$$

28. Since $\tan \pi/4 = 1$, π can be found by evaluating $4 \arctan 1$.
   (a) Find a power series for $\arctan x$ (see Exercise 8) and determine the number of terms of the series that need to be summed to ensure that $|4 \arctan 1 - \pi| < 10^{-3}$.
   (b) The single precision version of the scientific programming language FORTRAN requires the value of π to be within 10^{-7}. How many terms of this series must be summed to obtain this degree of accuracy?

29. Exercise 28 details a rather inefficient means of obtaining an approximation to π. The method can be improved substantially by observing that $\pi/4 = \arctan 1/2 + \arctan 1/3$ and evaluating the series for $\arctan$ at $1/2$ and at $1/3$. Show that this identity holds and determine the number of terms of each series that must be summed to ensure an approximation to π within 10^{-3}.

30. Another formula for computing π can be deduced from the identity $\pi/4 = 4 \arctan 1/5 - \arctan 1/239$. Determine the number of terms of each series that must be summed to ensure an approximation to π within 10^{-3}.

31. The German astronomer and mathematician Friedrich Wilhelm Bessel (1784–1846) used the differential equation $x^2 y'' + xy' + (x^2 - m^2)y = 0$, where m is a positive integer, in his study of planetary motion. The solution to this equation is known as the *Bessel function* of order m, defined by the series

$$J_m(x) = \sum_{n=0}^{\infty} \frac{(-1)^n x^{2n+m}}{n!(m+n)! \, 2^{2n+m}}.$$

Show that J_m satisfies the differential equation. (The Bessel functions are also important in the study of heat flow in cylinders, vibrations in circular membranes, the vibration of chains, and many other mathematical applications.)

Putnam exercises:

32. For $0 < x < 1$, express

$$\sum_{n=0}^{\infty} \frac{x^{2n}}{1 - x^{2n+1}}$$

as a rational function of x.

(This exercise was problem A–4 of the thirty-eighth William Lowell Putnam examination given on December 3, 1977. The examination and its solution can be found in the March 1979 issue of the *American Mathematical Monthly*, pages 170–175.)

33. Evaluate in closed form

$$\sum_{k=1}^{n} \binom{n}{k} k^2.$$

Note:

$$\binom{n}{k} = \frac{n(n-1) \cdots (n-k+1)}{1 \cdot 2 \cdots k}.$$

(This exercise was problem 5, part I of the twenty-third William Lowell Putnam examination given on December 1, 1962. The examination and its solution can be found in the September 1963 issue of the *American Mathematical Monthly*, pages 713–717.)

9.8
TAYLOR POLYNOMIALS AND TAYLOR SERIES

Example 3 in the previous section demonstrated that the natural exponential function can be represented by a power series that converges for all real numbers. In this section we discuss power series representations for other functions. Three questions we will consider are:

(i)   Which functions have a power series representation?

(ii) If such a representation exists, is it unique?

(iii) Is there a direct way to obtain the power series representation?

We first consider question (ii): Suppose we assume that a function f has a power series representation of the form

$$f(x) = \sum_{n=0}^{\infty} a_n x^n = a_0 + a_1 x + a_2 x^2 + \cdots$$

with radius of convergence $R > 0$. We need to determine the constants a_0, a_1, a_2, ...

When f is evaluated at zero, all terms of the series but the first vanish, so

$$a_0 = f(0).$$

Using Theorem 9.35 to differentiate f gives

$$f'(x) = \sum_{n=1}^{\infty} n a_n x^{n-1} = a_1 + 2a_2 x + 3a_3 x^2 + \cdots,$$

so $$a_1 = f'(0).$$

This procedure can be continued by repeated differentiation:

$$f''(x) = \sum_{n=2}^{\infty} n(n-1) a_n x^{n-2},$$

so

$$f''(0) = 2 \cdot 1 \, a_2 \qquad \text{and} \qquad a_2 = \frac{f''(0)}{2};$$

$$f'''(x) = \sum_{n=3}^{\infty} n(n-1)(n-2) a_n x^{n-3},$$

so $$f'''(0) = 3 \cdot 2 \cdot 1 \, a_3 \qquad \text{and} \qquad a_3 = \frac{f'''(0)}{6}.$$

In general,

$$a_n = \frac{f^{(n)}(0)}{n!}$$

for all nonnegative integers n, if we define $0! = 1$ and let $f^{(0)}(x) = f(x)$.

The following theorem is a direct consequence of these observations.

(9.36)
THEOREM

If a function f has a power series representation $\sum_{n=0}^{\infty} a_n x^n$ with radius of convergence $R > 0$, then the power series is unique and is given by

$$f(x) = \sum_{n=0}^{\infty} \frac{f^{(n)}(0)}{n!} x^n,$$

for x in $(-R, R)$.

The power series representation given in Theorem 9.36 is called the **Maclaurin series** for f. The generalization of Theorem 9.36 to power series centered at an arbitrary value a is contained in the following corollary. This series is called the **Taylor series** for f at a. The Maclaurin series for f is consequently the Taylor series centered at $a = 0$. (See the Historical Note.)

(9.37)
COROLLARY

If f has a power series representation at a with radius of convergence $R > 0$, then the power series at a is unique and is given by

$$f(x) = \sum_{n=0}^{\infty} \frac{f^{(n)}(a)}{n!} (x - a)^n$$

for x in $(a - R, a + R)$.

While these results are important, they do not completely answer the questions about functions and their power series representations. They tell us how to find a power series representation when one is known to exist, but not *when* we expect to find such a representation. The following result is more important in this regard, and answers question (iii).

(9.38)
THEOREM

If a function f has $n + 1$ derivatives on an open interval I about a and x is in this interval, then a number ξ between x and a exists with

where
$$f(x) = P_n(x) + R_n(x),$$

and
$$P_n(x) = \sum_{i=0}^{n} \frac{f^{(i)}(a)}{i!} (x - a)^i$$

$$R_n(x) = \frac{f^{(n+1)}(\xi)}{(n + 1)!} (x - a)^{n+1}.$$

PROOF

Assume that $P_n(x)$ is defined as stated in the theorem and that $R_n(x)$ is defined by

$$R_n(x) = f(x) - P_n(x)$$

for an arbitrary but fixed value of x in I. We will show that $R_n(x)$ must have the representation given in the theorem.

HISTORICAL NOTE

Brook Taylor (1685–1731) discussed series of this type in *Methodus incrementorum directa et inversa*, published in 1715. Special cases of this result, perhaps even the result itself, had been known to Newton, Gregory, and others much earlier.

The Scottish mathematician **Colin Maclaurin** (1698–1746) is best remembered for his strong defense of the calculus techniques of Newton, which came under attack in 1734 by a nonmathematician, Bishop George Berkley. The result bearing Maclaurin's name appeared in his work *Treatise on Fluxions* published in 1742, but was known at least as early as Taylor's work in 1715.

Although Maclaurin did not discover the series that bears his name, he did devise, in about 1729, a means of finding solutions to systems of linear equations. The method is known as Cramer's rule, although Cramer did not publish it until 1750.

If $x = a$, $f(x) = P_n(x)$ and $R_n(x) = 0$. If $x \neq a$, the proof proceeds by applying Rolle's theorem to the function defined by

$$g(z) = f(x) - \sum_{i=0}^{n} \frac{f^{(i)}(z)}{i!}(x - z)^i - R_n(x) \frac{(x - z)^{n+1}}{(x - a)^{n+1}}.$$

First note that when $z = a$,

$$g(a) = f(x) - \sum_{i=0}^{n} \frac{f^{(i)}(a)}{i!}(x - a)^i - R_n(x) \frac{(x - a)^{n+1}}{(x - a)^{n+1}}$$

$$= f(x) - P_n(x) - R_n(x) = 0,$$

and when $z = x$,

$$g(x) = f(x) - \sum_{i=0}^{n} \frac{f^{(i)}(x)}{i!}(x - x)^i - R_n(x) \frac{(x - x)^{n+1}}{(x - a)^{n+1}}$$

$$= f(x) - f(x) = 0.$$

The hypothesis that f has $n + 1$ derivatives on I ensures that g is differentiable and hence continuous on the closed interval with endpoints a and x. It follows from Rolle's theorem that a number ξ between x and a exists with $g'(\xi) = 0$.

Keeping in mind that x is fixed, we find that

$$g'(z) = 0 - f'(z) - \sum_{i=1}^{n} \left[\frac{f^{(i+1)}(z)}{i!}(x - z)^i - \frac{f^{(i)}(z)}{(i - 1)!}(x - z)^{i-1} \right]$$

$$+ \frac{R_n(x)(n + 1)(x - z)^n}{(x - a)^{n+1}}.$$

Expanding the series and cancelling terms,

$$g'(z) = -f'(z) - [f''(z)(x - z) - f'(z)]$$

$$- \left[\frac{f'''(z)(x - z)^2}{2} - f''(z)(x - z) \right] - \cdots$$

$$- \left[\frac{f^{(n+1)}(z)(x - z)^n}{n!} - \frac{f^{(n)}(z)(x - z)^{n-1}}{(n - 1)!} \right]$$

$$+ \frac{R_n(x)(n + 1)(x - z)^n}{(x - a)^{n+1}}$$

$$= \frac{-f^{(n+1)}(z)(x - z)^n}{n!} + \frac{R_n(x)(n + 1)(x - z)^n}{(x - a)^{n+1}}$$

$$= -(x - z)^n \left[\frac{f^{(n+1)}(z)}{n!} - R_n(x) \frac{(n + 1)}{(x - a)^{n+1}} \right].$$

Thus,

$$0 = g'(\xi) = -(x - \xi)^n \left[\frac{f^{(n+1)}(\xi)}{n!} - R_n(x) \frac{(x + 1)}{(x - a)^{n+1}} \right],$$

which implies, since $x \neq \xi$, that

$$R_n(x) = \frac{f^{(n+1)}(\xi)}{(n+1)!}(x-a)^{n+1}. \qquad \square$$

The function P_n defined by

$$P_n(x) = \sum_{i=0}^{n} \frac{f^{(i)}(a)}{i!}(x-a)^i$$

is called the **nth degree Taylor polynomial for f at a** and

$$R_n(x) = \frac{f^{(n+1)}(\xi)}{(n+1)!}(x-a)^{n+1}$$

is called the **remainder** for this polynomial. When $a = 0$, P_n is also known as the **nth degree Maclaurin polynomial for f**.

EXAMPLE 1   Find the Maclaurin polynomial of degree two for $f(x) = \cos x$ and use the remainder to find a bound for the error introduced when this polynomial is used to approximate $\cos .01$.

SOLUTION
Since $f(x) = \cos x$, $f'(x) = -\sin x$, $f''(x) = -\cos x$, and $f'''(x) = \sin x$,

$$P_2(x) = f(0) + x f'(0) + \frac{x^2}{2} f''(0)$$

$$= 1 + x \cdot (0) + \frac{x^2}{2} \cdot (-1) = 1 - \frac{x^2}{2}$$

and
$$R_2(x) = \frac{x^3 f'''(\xi)}{6} = \frac{x^3}{6} \sin \xi,$$

for some number ξ between x and zero.

Thus, $\cos .01 \approx P_3(.01) = 1 - \dfrac{(.01)^2}{2} = .99995.$

Since $|\sin \xi| < 1$, the maximum error in this approximation is

$$|R_2(.01)| = \left| \frac{(.01)^3}{6} \sin \xi \right| \le \frac{(.01)^3}{6} \approx 1.67 \times 10^{-7}.$$

Consequently, $\cos .01 \approx .999950$, where all 6 decimal places listed are accurate.

$\square$

EXAMPLE 2   Repeat the calculations in Example 1 using instead the third degree Maclaurin polynomial for $f(x) = \cos x$.

SOLUTION

The third degree Maclaurin polynomial is given by

$$P_3(x) = 1 - \frac{x^2}{2} + \frac{x^3}{6}f'''(0) = 1 - \frac{x^2}{2} + \frac{x^3}{6}(0) = 1 - \frac{x^2}{2},$$

the same expression as $P_2(x)$. Hence,

$$\cos .01 \approx P_3(.01) = .99995.$$

Before dismissing this example as superfluous, however, let us consider the remainder term $R_3(x)$. Since $f^{(4)}(x) = \cos x$ and $|\cos \xi| \le 1$,

$$|R_3(.01)| = \left| \frac{(.01)^4}{24} \cos \xi \right| \le \frac{(.01)^4}{24} \approx 4.17 \times 10^{-10}.$$

This shows that the approximation to $\cos .01$ is much more accurate than deduced in Example 1. In fact,

$$\cos .01 \approx .999950000$$

is accurate to all 9 decimal places listed. $\square$

The following corollary to Theorem 9.38 answers the question of which functions have power series representations.

(9.39)
COROLLARY

If a function f has derivatives of all orders on an open interval about a and if

$$\lim_{n \to \infty} R_n(x) = 0$$

for all x in that interval, then f has the Taylor series representation

$$f(x) = \sum_{n=0}^{\infty} \frac{f^{(n)}(a)}{n!}(x - a)^n$$

for all x in the interval.

EXAMPLE 3

Find the Maclaurin series for the sine and cosine functions.

SOLUTION

Letting $f(x) = \sin x$, we have

$$\begin{array}{ll} f(x) = \sin x & f(0) = 0 \\ f'(x) = \cos x & f'(0) = 1 \\ f''(x) = -\sin x & f''(0) = 0 \\ f'''(x) = -\cos x & f'''(0) = -1 \\ f^{(4)}(x) = \sin x & f^{(4)}(0) = 0 \end{array}$$

In general, $f^{(n)}(0)$ is zero when n is even and is alternately 1 and -1 when n is odd. The Maclaurin series is consequently

$$\sin x = x - \frac{x^3}{3!} + \frac{x^5}{5!} - \frac{x^7}{7!} + \cdots = \sum_{n=0}^{\infty} (-1)^n \frac{x^{2n+1}}{(2n+1)!}$$

a series that converges for all real numbers. (You can use the ratio test to verify this.) To show that this series represents $\sin x$ for all values of x, we show that

$$\lim_{n \to \infty} R_n(x) = 0$$

for all values of x.

For a fixed but arbitrary value of x, there is a number ξ between x and zero with

$$R_n(x) = \frac{x^{n+1} f^{(n+1)}(\xi)}{(n+1)!}.$$

Since $|f^{(n+1)}(\xi)|$ is either $|\cos(\xi)|$ or $|\sin(\xi)|$, $|f^{(n+1)}(\xi)| \leq 1$ and

$$|R_n(x)| \leq \frac{|x|^{n+1}}{(n+1)!}.$$

Since $\displaystyle\sum_{n=0}^{\infty} \frac{x^{n+1}}{(n+1)!}$ converges for all x, $\displaystyle\lim_{n \to \infty} \frac{x^{n+1}}{(n+1)!} = 0$, so

$$\lim_{n \to \infty} |R_n(x)| \leq \lim_{n \to \infty} \frac{|x|^{n+1}}{(n+1)!} = 0.$$

While the Maclaurin series for $\cos x$ can be derived in the same manner as $\sin x$, it is easier to apply Theorem 9.35 once the series representation for $\sin x$ is known.

$$\cos x = D_x \sin x = D_x \left(x - \frac{x^3}{3!} + \frac{x^5}{5!} - \frac{x^7}{7!} + \cdots \right)$$

$$= 1 - \frac{x^2}{2!} + \frac{x^4}{4!} - \frac{x^6}{6!} + \cdots = \sum_{n=0}^{\infty} (-1)^n \frac{x^{2n}}{(2n)!} \qquad \square$$

The power series representations of certain functions will be referred to in this and later mathematics courses. For ease of reference, we list the most frequently used series.

(9.40) $\displaystyle e^x = \sum_{n=0}^{\infty} \frac{x^n}{n!}$,   for all x in $\mathbb{R}$,

(9.41) $\displaystyle \sin x = \sum_{n=0}^{\infty} (-1)^n \frac{x^{2n+1}}{(2n+1)!}$,   for all x in $\mathbb{R}$,

(9.42) $\displaystyle \cos x = \sum_{n=0}^{\infty} (-1)^n \frac{x^{2n}}{(2n)!}$,   for all x in $\mathbb{R}$,

(9.43) $\displaystyle \frac{1}{1-x} = \sum_{n=0}^{\infty} x^n$,   for $|x| < 1$.

EXERCISE SET 9.8

Find the first four terms of the Maclaurin series for the functions described in Exercises 1 through 6 by using Theorem 9.38.

1. $f(x) = e^{2x}$

2. $f(x) = e^{-x}$

3. $f(x) = \ln(x + 1)$

4. $f(x) = \cos 2x$

5. $f(x) = \arctan x$

6. $f(x) = x^3 + 3x + 1$

In Exercises 7 through 16, find the Taylor polynomial of degree four for the function f about the number a. Find the remainder for the Taylor polynomial.

7. $f(x) = e^x, a = 1$

8. $f(x) = e^x, a = -2$

9. $f(x) = \sin x, a = \dfrac{\pi}{2}$

10. $f(x) = \cos x, a = \dfrac{\pi}{2}$

11. $f(x) = \tan x, a = \dfrac{\pi}{4}$

12. $f(x) = \dfrac{1}{x}, a = 1$

13. $f(x) = \sqrt{x}, a = 4$

14. $f(x) = \ln x, a = 1$

15. $f(x) = x^2 + 2 + e^x, a = 1$

16. $f(x) = x^2 + 2x + 3, a = 1$

17. Use the Maclaurin polynomial of degree three to approximate sin .01 and find the maximum error for this approximation.

18. Use the third degree Taylor polynomial about 1 to approximate ln 1.1 and find the maximum error for this approximation.

19. Use a Taylor polynomial for the function $f(x) = \sqrt{x}$ about $a = 4$ to find an approximation to $\sqrt{4.1}$ that is accurate to within 10^{-8}.

20. Use a Taylor polynomial for the function $f(x) = \ln x$ about $a = e$ to find an approximation to ln 3 that is accurate to within 10^{-4}.

21. Use a Taylor polynomial to approximate sin 5° to an accuracy of 10^{-4}.

22. Use a Taylor polynomial to approximate cos 42° to an accuracy of 10^{-4}.

23. Suppose P is a polynomial of degree n. Show that the nth degree Maclaurin polynomial about $a = 0$ is precisely P.

24. Suppose P is a polynomial of degree n. Show that the nth degree Taylor polynomial about a is precisely P regardless of the value chosen for a.

25. The method for completing the square of a quadratic polynomial can be derived by using the result in Exercise 24. Suppose that $f(x) = ax^2 + bx + c$. Expand f in a second degree polynomial about an arbitrary point d and then choose d so that the coefficient of $(x - d)$ is zero. Use this technique to complete the square in the quadratic polynomial $P(x) = 2x^2 + 3x + 1$.

26. Prove that the Maclaurin series of a function has only even powers if and only if the function is even. (*Hint:* First show that the derivative of an even function must be an odd function and that the derivative of an odd function must be even.)

27. Prove that the Maclaurin series of a function has only odd powers if and only if the function is odd.

28. The nth degree Taylor polynomial for a function f at a is also referred to as the polynomial that "best" approximates f near a.

(a) Explain why this description is accurate.

(b) Find the quadratic polynomial that best approximates a function f at $x = 1$, if the function has as its tangent line the line with equation $y = 4x - 1$ when $x = 1$ and has $f''(1) = 6$.

29. A Maclaurin polynomial for e^x is used to give the approximation 2.5 to e. The error in this approximation is estimated to be $E = 1/6$. Find a bound for the error in E.

30. It is important to verify that $\lim\limits_{n \to \infty} R_n(x) = 0$ to ensure that the Taylor series for a function actually represents the function. Consider the function

$$f(x) = \begin{cases} e^{-1/x^2} & \text{if } x \neq 0 \\ 0 & \text{if } x = 0. \end{cases}$$

The Maclaurin series of this function is identically zero, so the series represents f only when $x = 0$. Show that this statement is true by constructing the Maclaurin series in the following manner.

(a) Show that $\lim\limits_{x \to 0} e^{-1/x^2} x^{-n} = 0$ for each n.

(b) Use the definition of the derivative to find $f'(0)$.

(c) Use the definition of the derivative to find $f^{(n)}(0)$ for $n \geq 2$.

(d) Construct the Maclaurin series.

Putnam exercises:

31. Assume that $|f(x)| \leq 1$ and $|f''(x)| \leq 1$ for all x on an interval of length at least 2. Show that $|f'(x)| \leq 2$ on the interval.
(This exercise was problem 4, part I of the twenty-third William Lowell Putnam examination given on December 1, 1962. The examination and its solution can be found in the September 1963 issue of the *American Mathematical Monthly*, pages 713–717.)

32. Let C be a real number, and let f be a function such that

$$\lim_{x \to \infty} f(x) = C, \qquad \lim_{x \to \infty} f'''(x) = 0.$$

Prove that $\lim\limits_{x \to \infty} f'(x) = 0$ and $\lim\limits_{x \to \infty} f''(x) = 0.$

(This exercise was problem 4, part II of the nineteenth William Lowell Putnam examination given on November 22, 1958. The examination and its solution can be found in the January 1961 issue of the *American Mathematical Monthly*, pages 22–27.)

APPLICATIONS OF TAYLOR POLYNOMIALS AND SERIES

In this section we consider a few of the many interesting results connected with Taylor polynomials and Taylor series.

THE BINOMIAL SERIES

In algebra courses we learn that integral powers of a sum can be expanded by using the binomial theorem:

$$(a + b)^k = a^k + ka^{k-1}b + \frac{k(k-1)}{2}a^{k-2}b^2 + \cdots$$

$$+ \frac{k(k-1)\cdots(k-n+1)}{n!}a^{k-n}b^n + \cdots + kab^{k-1} + b^k$$

where a and b are any real numbers and k is a positive integer.

We will consider this expansion when $a = 1$ and $b = x$, and relax the restriction on the power k to allow it to assume any real value. That is, we will consider the expansion of $(1 + x)^k$ for real numbers x and k. To obtain this expansion we use the Maclaurin series representation of the function f where $f(x) = (1 + x)^k$.

$$f(x) = (1 + x)^k, \quad \text{so } f(0) = 1$$

$$f'(x) = k(1 + x)^{k-1}, \quad \text{so } f'(0) = k$$

$$f''(x) = k(k-1)(1 + x)^{k-2}, \quad \text{so } f''(0) = k(k-1)$$

and, in general, for any integer n,

$$f^{(n)}(x) = k(k-1)\ldots(k-n+1)(1+x)^{k-n},$$

so
$$f^{(n)}(0) = k(k-1)\ldots)(k-n+1).$$

Consequently, the Maclaurin series is

$$1 + \sum_{n=1}^{\infty} \frac{k(k-1)\ldots(k-n+1)}{n!}x^n.$$

The series has radius of convergence 1, since

$$\lim_{n\to\infty} \left| \frac{\dfrac{k(k-1)\ldots(k-n)}{(n+1)!}x^{n+1}}{\dfrac{k(k-1)\ldots(k-n+1)}{n!}x^n} \right| = \lim_{n\to\infty} \left| \frac{k-n}{n+1}x \right| = |x|.$$

This series,

(9.44)
$$1 + \sum_{n=1}^{\infty} \frac{k(k-1)\ldots(k-n+1)}{n!}x^n = (1 + x)^k,$$

is called the **binomial series**. To show that it is the power series representation for $f(x) = (1 + x)^k$ when $|x| < 1$ requires showing that $\lim_{n \to \infty} R_n(x) = 0$ for $|x| < 1$, where

$$R_n(x) = \frac{f^{(n+1)}(\xi)}{(n+1)!} x^{n+1} = \frac{k(k-1)\ldots(k-n)}{(n+1)!} (1 + \xi)^{k-n-1} x^{n+1},$$

for some ξ between x and zero. This is a rather difficult demonstration and will not be considered here.

EXAMPLE 1   Find the binomial series for $\sqrt{1 + x}$ and use this series to approximate $\sqrt{1.1}$ to within 10^{-6}.

SOLUTION

$$(1 + x)^{1/2} = 1 + \sum_{n=1}^{\infty} \frac{\frac{1}{2}\left(\frac{1}{2} - 1\right) \cdots \left(\frac{1}{2} - n + 1\right)}{n!} x^n$$

$$= 1 + \frac{1}{2}x + \frac{1}{2}\left(-\frac{1}{2}\right)\frac{x^2}{2} + \frac{1}{2}\left(-\frac{1}{2}\right)\left(-\frac{3}{2}\right)\frac{x^3}{6} + \cdots$$

$$= 1 + \frac{1}{2}x - \frac{1}{8}x^2 + \frac{3}{48}x^3 - \frac{15}{384}x^4 + \cdots.$$

When $x = 1.1$,

$$\sqrt{1.1} = (1 + .1)^{1/2}$$

$$= 1 + \frac{1}{2}(.1) - \frac{1}{8}(.1)^2 + \frac{3}{48}(.1)^3 - \frac{15}{384}(.1)^4 + \cdots.$$

Since this is an alternating series, to find the number of terms to ensure an approximation to within 10^{-6}, we need to find n so that $|a_{n+1}| < 10^{-6}$. By evaluating the terms in the series we see that this first occurs when $n = 4$, and

$$a_{n+1} = a_5 = \frac{\left(\frac{1}{2}\right)\left(-\frac{1}{2}\right)\left(-\frac{3}{2}\right)\left(-\frac{5}{2}\right)\left(-\frac{7}{2}\right)}{120}(.1)^5 \approx 2.73 \times 10^{-7}.$$

Consequently, an approximation to $\sqrt{1.1}$ to within 10^{-6} is

$$\sqrt{1.1} \approx 1 + \frac{1}{2}(.1) - \frac{1}{8}(.1)^2 + \frac{3}{48}(.1)^3 - \frac{15}{384}(.1)^4 \approx 1.048809. \quad \square$$

EXAMPLE 2   Use Theorem 9.35 and the binomial series with $k = -1/2$ to derive a series for arcsin x that is valid when $|x| < 1$.

SOLUTION
   Since

$$\arcsin x = \int_0^x (1 - t^2)^{-1/2} \, dt,$$

we first find the series representation for $(1 - t^2)^{-1/2}$ and then integrate using Theorem 9.35. Since

$$(1 + x)^{-1/2} = 1 - \frac{1}{2}x + \left(-\frac{1}{2}\right)\left(-\frac{3}{2}\right)\frac{x^2}{2!} + \left(-\frac{1}{2}\right)\left(-\frac{3}{2}\right)\left(-\frac{5}{2}\right)\frac{x^3}{3!}$$

$$+ \cdots + \left[\left(-\frac{1}{2}\right)\left(-\frac{3}{2}\right)\cdots\left(\frac{-(2n-1)}{2}\right)\right]\frac{x^n}{n!} + \cdots$$

$$= 1 - \frac{1}{2}x + \frac{1 \cdot 3}{2^2}\frac{x^2}{2!} - \frac{1 \cdot 3 \cdot 5}{2^3}\frac{x^3}{3!} + \cdots$$

$$+ (-1)^n \frac{1 \cdot 3 \cdot 5 \cdots (2n + 1)}{2^n}\frac{x^n}{n!} + \cdots ;$$

replacing x by $-t^2$ gives

$$(1 - t^2)^{-1/2} = 1 + \frac{1}{2}t^2 + \frac{1 \cdot 3}{2^2}\frac{t^4}{2!} + \frac{1 \cdot 3 \cdot 5}{2^3}\frac{t^6}{3!} + \cdots$$

$$= 1 + \sum_{n=1}^{\infty} \frac{1 \cdot 3 \cdot 5 \cdots (2n - 1)}{2^n}\frac{t^{2n}}{n!}.$$

Consequently,

$$\arcsin x = \int_0^x (1 - t^2)^{-1/2} \, dt$$

$$= x + \sum_{n=1}^{\infty} \frac{1 \cdot 3 \cdot 5 \cdots (2n - 1)}{2^n} \frac{x^{2n+1}}{(2n + 1) \, n!}.$$

This series has radius of convergence $R = 1$. (See Exercise 9.) □

NEWTON'S METHOD

Newton's method is a technique used to find approximate solutions to an equation of the form $f(x) = 0$. This method was discussed in Section 4.6 as an application of the derivative. It proceeds by assuming an initial approximation x_0 to the solution of the equation. Subsequent approximations $x_1, x_2, \ldots$ are generated by using the formula

(9.45) $$x_{n+1} = x_n - \frac{f(x_n)}{f'(x_n)}$$

for $n > 0$, provided that $f'(x_n)$ is defined and nonzero. The method often produces exceedingly accurate results. With the aid of Taylor polynomials we can show why this is true.

Suppose $\hat{x}$ denotes the solution to $f(x) = 0$ and f'' exists in an interval containing both $\hat{x}$ and the initial approximation x_0. Expanding f in a first degree

Taylor polynomial at x_0 and evaluating at $x = \hat{x}$ gives

$$0 = f(\hat{x}) = f(x_0) + f'(x_0)(\hat{x} - x_0) + \frac{f''(\xi)}{2}(\hat{x} - x_0)^2,$$

where ξ lies between x_0 and $\hat{x}$. Consequently, if $f'(x_0) \neq 0$,

$$\hat{x} - x_0 + \frac{f(x_0)}{f'(x_0)} = \frac{-f''(\xi)}{2f'(x_0)}(\hat{x} - x_0)^2.$$

Since
$$x_1 = x_0 - \frac{f(x_0)}{f'(x_0)},$$

this implies that

$$\hat{x} - x_1 = -\frac{f''(\xi)}{2f'(x_0)}(\hat{x} - x_0)^2.$$

If a bound M is known for the second derivative in an interval about $\hat{x}$, and x_0 is within this interval, then

$$|\hat{x} - x_1| \leq \frac{M}{|2f'(x_0)|}|\hat{x} - x_0|^2.$$

Since $f'(x_0) \neq 0$, there is a number α with $2|f'(x_0)| \geq \alpha > 0$, so

$$|\hat{x} - x_1| \leq \frac{M}{\alpha}|\hat{x} - x_0|^2.$$

This inequality indicates that Newton's method has the tendency to approximately double the number of significant digits of accuracy with each successive approximation. (If the error in approximating $\hat{x}$ by x_0 is 10^{-n}, then the error in approximating $\hat{x}$ by x_1 is bounded by $M10^{-2n}/\alpha$.)

EXAMPLE 3   Find an approximation to the solution of the equation $x = 3^{-x}$ that is accurate to within 10^{-8}.

SOLUTION

A solution to this equation corresponds to a solution to $f(x) = 0$ where

$$f(x) = x - 3^{-x}.$$

Since $f(0) = -1$ and $f(1) = 2/3$, a solution to the equation lies in the interval $(0, 1)$. As an initial approximation we use $x_0 = .5$. Succeeding approximations are generated by applying the formula

$$x_{n+1} = x_n - \frac{x_n - 3^{-x_n}}{1 + 3^{-x_n} \ln 3}.$$

These approximations are listed in the accompanying table, together with differences between successive approximations.

i	x_i	$\lvert x_i - x_{i-1} \rvert$
0	.500000000	—
1	.547329757	.047329757
2	.547808574	.000478817
3	.547808622	.000000048

The difference in successive approximations leads us to the correct conclusion that x_3 is accurate to the places listed and that the solution to $x = 3^{-x}$ accurate to within 10^{-8} is .54780862. ☐

EULER'S FORMULA

The Maclaurin series representation for the sine, cosine, and natural exponential functions can be used to extend the domain of these functions to the set of complex numbers. By a **complex number** we mean a number of the form

$$z = x + iy,$$

where x and y are real numbers and i is a symbol with the property that $(i)^2 = -1$. The number x is called the **real part** of z and the number y the **imaginary part** of z.

A graphical representation of the complex numbers is called an **Argand diagram**, named for Jean Robert Argand (1768–1822), a Swiss mathematician who, in 1806, published a paper using a graphical representation of complex numbers. In actuality, Caspar Wessel (1745–1818) first used this graphic representation in a paper published by the Danish Academy of Sciences in 1798. The complex plane is also referred to as the Gaussian plane, although Karl Friedrich Gauss did not publish his views on complex numbers until 1828.

The Argand diagram uses a horizontal axis to represent the real part of the number and a vertical line to represent the imaginary part. This is illustrated in Figure 9.8.

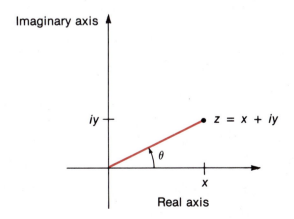

FIGURE 9.8

The magnitude of a complex number $z = x + iy$ is given by $|z| = \sqrt{x^2 + y^2}$. With the notation in Figure 9.8, we see that

$$x = |z| \cos \theta \quad \text{and} \quad y = |z| \sin \theta,$$

so
$$z = x + iy = |z| (\cos \theta + i \sin \theta).$$

For complex numbers z, e^z is defined using the power series representation

$$e^z = \sum_{n=0}^{\infty} \frac{z^n}{n!} = 1 + z + \frac{z^2}{2!} + \frac{z^3}{3!} + \frac{z^4}{4!} + \frac{z^5}{5!} + \cdots.$$

In particular, for real numbers x this implies that

$$e^{ix} = 1 + ix + \frac{(ix)^2}{2!} + \frac{(ix)^3}{3!} + \frac{(ix)^4}{4!} + \frac{(ix)^5}{5!} + \cdots.$$

Since $i^2 = -1$, $i^3 = -i$, $i^4 = 1$, $i^5 = i$, etc.,

$$e^{ix} = 1 + ix - \frac{x^2}{2!} - \frac{ix^3}{3!} + \frac{x^4}{4!} + \frac{ix^5}{5!} + \cdots.$$

However, absolute convergence permits us to rewrite this series as

$$e^{ix} = \left(1 - \frac{x^2}{2!} + \frac{x^4}{4!} - \cdots \right) + i\left(x - \frac{x^3}{3!} + \frac{x^5}{5!} - \cdots \right).$$

Since
$$\cos x = 1 - \frac{x^2}{2!} + \frac{x^4}{4!} - \cdots$$

and
$$\sin x = x - \frac{x^3}{3!} + \frac{x^5}{5!} - \cdots,$$

we have the equation

(9.46) $$e^{ix} = \cos x + i \sin x.$$

This intriguing relation is known as **Euler's formula**. It has a number of interesting applications and is the basis for showing a result known as **de Moivre's formula**. If x is a real number and n is a positive integer, then

$$\begin{aligned}(\cos x + i \sin x)^n &= (e^{ix})^n \\ &= e^{i(nx)} \\ &= \cos(nx) + i \sin (nx),\end{aligned}$$

so for every integer n,

(9.47) $$(\cos x + i \sin x)^n = \cos nx + i \sin nx.$$

You are perhaps familiar with this formula from a precalculus course.

EXAMPLE 4   Use de Moivre's formula to find all complex numbers that are cube roots of 8.

SOLUTION

If we express the real number 8 as

$$8 = 8(\cos 0 + i \sin 0)$$

and assume that, for some value of θ,

$$z = |z|(\cos \theta + i \sin \theta)$$

is a cube root of 8, then

$$8 = 8(\cos 0 + i \sin 0) = |z|^3(\cos \theta + i \sin \theta)^3 = z^3.$$

By de Moivre's formula

$$(\cos \theta + i \sin \theta)^3 = \cos 3\theta + i \sin 3\theta,$$

so we must have

$$8(\cos 0 + i \sin 0) = |z|^3 (\cos 3\theta + i \sin 3\theta).$$

Consequently,

$$|z|^3 = 8, \quad \text{which implies } |z| = 2$$

while

$$\cos 3\theta = \cos 0 \quad \text{and} \quad \sin 3\theta = \sin 0.$$

There is flexibility in the choice of θ since these equations are satisfied whenever $\theta = 2n\pi/3$ for some integer n.

When $n = 0$, $n = 1$, and $n = 2$ we have the cube roots

$$n = 0: z_0 = 2(\cos 0 + i \sin 0) = 2,$$

$$n = 1: z_1 = 2\left(\cos \frac{2\pi}{3} + i \sin \frac{2\pi}{3}\right) = 2\left(-\frac{1}{2} + i\frac{\sqrt{3}}{2}\right) = -1 + i\sqrt{3},$$

and

$$n = 2: z_2 = 2\left(\cos \frac{4\pi}{3} + i \sin \frac{4\pi}{3}\right) = 2\left(-\frac{1}{2} - i\frac{\sqrt{3}}{2}\right) = -1 - i\sqrt{3}.$$

Other integral values of n produce repetitions of these results. The three complex

---

HISTORICAL NOTE   **Abraham de Moivre** (1667–1754) did his most lasting work in the area of probability theory continuing the subject from the work of Pascal and Fermat. These results appeared in a series of papers published during the early eighteenth century and were collected and expanded in a memoir in 1718 entitled *Doctrine of Chance, or a Method of Calculating the Probability of Events in Play*. The formula that bears his name was first presented in a paper in 1722, although the result in another form had been used by Roger Cotes (1682–1716) around 1710.

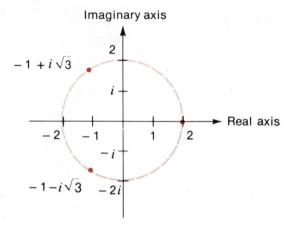

FIGURE 9.9

cube roots of 8 are consequently

$$2, \quad -1 + i\sqrt{3}, \quad \text{and} \quad -1 - i\sqrt{3}.$$

These roots are represented on the Argand diagram in Figure 9.9. They are equally spaced on the circle centered at the origin with radius $2 = 8^{1/3}$.   ☐

EXERCISE SET 9.9

Find a power series representation for each function described in Exercises 1 through 6 and give the radius of convergence of the series.

1. $f(x) = \sqrt[3]{1 + x}$

2. $f(x) = \sqrt{1 + x^2}$

3. $f(x) = \dfrac{1}{\sqrt{1 + x}}$

4. $f(x) = (1 + x)^{-3}$

5. $f(x) = (3x + 2)^{3/2}$

6. $f(x) = \sqrt{9 + 2x}$

Approximate the definite integrals in Exercises 7 and 8 to within 10^{-4}

7. $\displaystyle\int_0^{1/2} \sqrt[3]{1 + x^2}\, dx$

8. $\displaystyle\int_0^{1/2} \dfrac{dx}{\sqrt[4]{1 + x^3}}$

9. Show that the series for arcsin x given in Example 2 has radius of convergence $R = 1$.

10. For the function described by $f(x) = x^3 - 2x - 1$, find an approximation, accurate to within 10^{-5}, to the x-intercept that is between 1 and 2.

11. Find approximations to the solutions of the equation $x^2 - 2x - 2 = 0$ that are accurate to within 10^{-5}.

12. Compute numerically:

(a) e^{13i}

(b) $e^{2 + 13i}$

(c) $e^{-1 + 2i}$

13. Write $e^{2+\pi i}$ in the form $a + bi$ where a and b are real numbers.

14. In each of the following, indicate on an Argand diagram the set of complex numbers $z = x + iy$ that satisfy the given condition.

   (a) (Real part of z) > 0 (b) $|z| \leq 1$

   (c) z is a real number   (d) $|e^z| \geq 1$

15. Find all cube roots of 1 and represent these roots graphically.

16. Find the cube roots of -1 and represent these roots graphically.

17. Find all sixth roots of 64 and represent these roots graphically.

18. Find the sixth roots of i and represent these roots graphically.

19. Solve $x^4 + 4 = 0$. Use the solutions to factor $x^4 + 4$ into two real quadratic factors.

20. Solve $x^4 + 32 = 0$. Use the solutions to factor $x^4 + 32$ into two real quadratic factors.

21. Use Euler's formula to show that

$$\cos x = \frac{e^{ix} + e^{-ix}}{2} \quad \text{and} \quad \sin x = \frac{e^{ix} - e^{-ix}}{2}.$$

22. For which complex numbers $z = x + iy$ does the inequality $|e^{-iz}| < 1$ hold?

23. Show that the square roots of i are $\pm \dfrac{(1 + i)}{\sqrt{2}}$.

24. The accumulated value of a savings account based on regular periodic payments can be determined from the *annuity due* equation

$$A = \frac{P}{r}(1 + r)[(1 + r)^n - 1].$$

   In this equation A is the amount in the account, P is the amount regularly deposited, and r is the rate of interest per period for the n deposit periods.

   An engineer would like to have a savings account valued at \$75,000 upon retirement in 20 years, and can afford to put \$150 per month toward this goal. Use Newton's method to answer the following questions.

   (a) What is the minimal interest rate at which this amount can be deposited, assuming that the interest is compounded quarterly?

   (b) What is the minimal interest rate if the interest is compounded daily? (Assume a 360-day year, the common banking practice.)

25. The amount of money required to pay off a mortgage over a fixed period of time is given by

$$A = \frac{P}{r}[1 - (1 + r)^{-n}],$$

known as an *ordinary annuity* equation. In this equation A is the amount of the mortgage, P is the amount of each payment, and r is the interest rate per period for the n payment periods. Suppose a 30-year home mortgage in the amount of \$50,000 is needed and that the borrower can afford house payments of at most \$450 per month. Use Newton's method to find the maximal interest rate the borrower can afford to pay.

REVIEW EXERCISES

The nth term of a sequence is given in Exercises 1 through 14. Determine if the sequence is convergent. If the sequence is convergent, give its limit.

1. $a_n = \dfrac{3n + 2}{1 - 2n}$

2. $a_n = \dfrac{n\pi + 1}{n}$

3. $a_n = 3$

4. $a_n = \dfrac{\sin n\pi}{n}$

5. $a_n = \sqrt[n]{n + 1}$

6. $a_n = \dfrac{\ln n}{n}$

7. $a_n = \displaystyle\sum_{i=1}^{n} \dfrac{1}{i}$

8. $a_n = \displaystyle\sum_{i=1}^{n} \left(\dfrac{1}{2}\right)^i$

9. $a_n = \dfrac{e^n}{n}$

10. $a_n = \dfrac{n}{e^n}$

11. $a_n = \sqrt{n^2 + 1} - n$

12. $a_n = \dfrac{n^2 - 2n + 5}{3n^2 - 8}$

13. $a_n = \dfrac{1}{\sqrt[n]{2}}$

14. $a_n = \left(\dfrac{1}{n}\right)^{1/n}$

Determine whether the series in Exercises 15 through 34 are divergent, conditionally convergent, or absolutely convergent.

15. $\displaystyle\sum_{n=1}^{\infty} \dfrac{1}{1 + \sqrt{n}}$

16. $\displaystyle\sum_{n=1}^{\infty} \dfrac{1}{n^2} \sin n$

17. $\displaystyle\sum_{n=1}^{\infty} \dfrac{1}{e^n}$

18. $\displaystyle\sum_{n=1}^{\infty} \dfrac{\ln n}{n^3}$

19. $\displaystyle\sum_{n=1}^{\infty} (-1)^{n+1} \dfrac{1 + \sqrt{n}}{2^n}$

20. $\displaystyle\sum_{n=1}^{\infty} \dfrac{\arctan n}{n^2 + 1}$

21. $\displaystyle\sum_{n=1}^{\infty} \dfrac{(-1)^n}{2n + 1}$

22. $\displaystyle\sum_{n=1}^{\infty} \dfrac{1}{\sqrt[n]{2}}$

23. $\displaystyle\sum_{n=1}^{\infty} \dfrac{n!}{n^n}$

24. $\displaystyle\sum_{n=1}^{\infty} (-1)^n \dfrac{n}{5^n}$

25. $\displaystyle\sum_{n=1}^{\infty} \dfrac{(-1)^n (1 + 4n)}{7n^2 - 1}$

26. $\displaystyle\sum_{n=1}^{\infty} (-1)^n \dfrac{2^n}{n!}$

27. $\displaystyle\sum_{n=2}^{\infty} \dfrac{3}{n \ln n}$

28. $\displaystyle\sum_{n=1}^{\infty} (-1)^{n+1} \dfrac{2^n}{e^n}$

29. $\displaystyle\sum_{n=1}^{\infty} \dfrac{(-1)^n \ln n}{n + 1}$

30. $\displaystyle\sum_{n=1}^{\infty} \dfrac{1}{n^n}$

31. $\displaystyle\sum_{n=1}^{\infty} \frac{(-1)^n n}{(2n+1)^2}$

32. $\displaystyle\sum_{n=2}^{\infty} \frac{1}{\ln n}$

33. $\displaystyle\sum_{n=1}^{\infty} (-1)^{n+1} \left(\frac{3}{4n}\right)$

34. $\displaystyle\sum_{n=1}^{\infty} \frac{n^2 - 4n + 4}{9n^2 + 3n - 2}$

Find the radius and interval of convergence of each power series in Exercises 35 through 46.

35. $\displaystyle\sum_{n=1}^{\infty} \frac{(x-1)^n}{n^2}$

36. $\displaystyle\sum_{n=1}^{\infty} n^2 x^n$

37. $\displaystyle\sum_{n=0}^{\infty} (x+2)^n$

38. $\displaystyle\sum_{n=1}^{\infty} \frac{x^n}{(2n+1)^2}$

39. $\displaystyle\sum_{n=0}^{\infty} \frac{n^2}{2^n} (x-2)^n$

40. $\displaystyle\sum_{n=1}^{\infty} \frac{\ln n}{n} x^n$

41. $\displaystyle\sum_{n=0}^{\infty} 3^n (x-3)^n$

42. $\displaystyle\sum_{n=0}^{\infty} \frac{2^n}{n!} (x-3)^n$

43. $\displaystyle\sum_{n=0}^{\infty} \left(\frac{x^2-1}{2}\right)^n$

44. $\displaystyle\sum_{n=1}^{\infty} \left(\frac{x}{n}\right)^n$

45. $\displaystyle\sum_{n=1}^{\infty} \frac{2 \cdot 4 \cdot 6 \cdots (2n)}{3 \cdot 6 \cdot 9 \cdots (3n)} x^n$

46. $\displaystyle\sum_{n=2}^{\infty} \frac{1 \cdot 2 \cdot 5 \cdots (3n-7)}{3 \cdot 6 \cdot 9 \cdots (3n-3)} x^{n-1}$

For the series in Exercises 47 through 49, find an approximation to the sum that is accurate to within 10^{-5}.

47. $\displaystyle\sum_{n=2}^{\infty} \frac{(-1)^{n-1}}{4^n (n-1)}$

48. $\displaystyle\sum_{n=1}^{\infty} \frac{(-1)^{n+1}}{(2n-1)^3}$

49. $\displaystyle\sum_{n=1}^{\infty} \frac{(-1)^{n-1}}{n3^n}$

50. Give an example of a bounded sequence that does not converge.

51. Give an example of a monotonic sequence that does not converge.

52. Show that the sequence $\left\{\dfrac{\ln n}{n+1}\right\}$ is a decreasing sequence for $n > 4$.

Find a power series representation for each function described in Exercises 53 through 58. Use the power series representations (9.40) through (9.44) and the arithmetic and calculus properties of power series.

53. $f(x) = \dfrac{\sin x}{x}$

54. $f(x) = \sqrt{1 + 2x}$

55. $f(x) = \dfrac{1}{\sqrt{1 + 2x}}$

56. $f(x) = \ln \dfrac{1}{1-x}$.

57. $f(x) = \cosh x$

58. $f(x) = \sinh x$

59. Let $x_1 = 1$ and $x_{n+1} = \dfrac{x_n}{2} + \dfrac{3}{2x_n}$. Find the fourth term of this sequence and compare it with $\sqrt{3}$.

60. Find the values of x for which the series $\displaystyle\sum_{n=1}^{\infty} \left(\frac{\cos x}{n}\right)^n$ converges.

61. Find the values of x for which the series $\displaystyle\sum_{n=1}^{\infty} \frac{1}{n}\left(\frac{x-1}{x}\right)^n$ converges.

62. Find the Maclaurin polynomial of degree four for the function described by $f(x) = x^6 - 7x^5 + 4x^2 - 2x - 3$.

63. Find the Maclaurin series for $f(x) = \dfrac{x-1}{x+1}$.

64. Find the Maclaurin series for $\displaystyle\int_0^x \frac{e^t - 1}{t}\, dt$.

65. Find the Taylor series for $f(x) = \ln 2x$ about 1.

66. (a) Find the Taylor polynomial of degree four for the function described by $f(x) = \sec x$ about $\pi/4$.

　(b) Use the polynomial found in (a) to approximate $\sec \pi/4$.

67. Find a series representation for the definite integral $\displaystyle\int_0^1 e^{x^2}\, dx$.

68. Sketch the set of complex numbers z satisfying $|z - 1| \le 1$.

69. Find all fourth roots of 16.

70. Write $e^{\pi i - 2}$ in the form $a + bi$ where a and b are real numbers.

10
POLAR COORDINATES AND PARAMETRIC EQUATIONS

The first portion of this chapter introduces a new method for representing points in the plane and discusses curves that are more easily represented in this new coordinate system. The remainder of the chapter is concerned with the description of general curves in the plane and the connection between this topic and some previously discussed results.

THE POLAR COORDINATE SYSTEM

In a rectangular, or cartesian, coordinate system a point P is located by using directed distances to P from a pair of perpendicular axes, as shown in Figure 10.1.

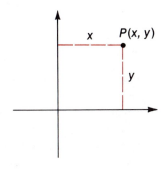

FIGURE 10.1

An alternate method for locating a point P is to specify its direction and distance from a fixed point in the plane, a point we will denote O. We first fix a **ray**, or half line, that originates at O and extends infinitely in one direction. This ray is called the **polar axis** and is generally drawn horizontally to the right of O. The point O is called the **pole.**

Pole ●————————————————→
Polar axis

FIGURE 10.2

Let θ denote the directed angle from the polar axis to the line that joins O and P, and r denote the directed distance along this line to P. Then (r, θ) is called a pair of **polar coordinates** for the point P. The polar coordinate representations of some points are shown in Figure 10.3.

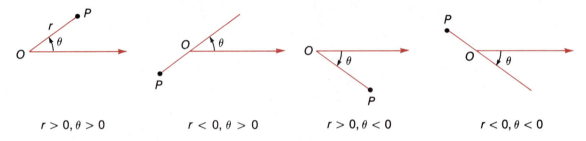

$r > 0, \theta > 0$ $r < 0, \theta > 0$ $r > 0, \theta < 0$ $r < 0, \theta < 0$

FIGURE 10.3

While the polar coordinate system has significant advantages for many applications, it suffers from the complication that a point does not have a unique representation. In fact, the coordinates (r, θ), $(r, 2\pi + \theta)$, $(-r, \pi + \theta)$, $(-r, -3\pi + \theta)$, and an infinite number of other such pairs describe the point shown in Figure 10.4. The pole O is described by $(0, \theta)$ for any angle θ. In most physical applications this multiple representation of points does not cause a problem.

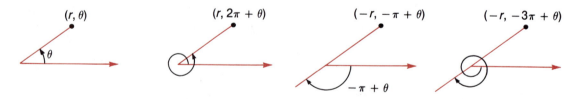

(r, θ) $(r, 2\pi + \theta)$ $(-r, -\pi + \theta)$ $(-r, -3\pi + \theta)$

$-\pi + \theta$

FIGURE 10.4

It is advantageous to have both cartesian and polar coordinates available and be able to change readily from one system to the other. To do this, we use a common scale and orient the systems so that the pole is at the origin and the polar axis and the positive x-axis coincide. The change from one system to the other is immediate for some points. For example, if a point has polar coordinates $(2, \pi/2)$, it lies on the positive y-axis, 2 units from the origin, and has cartesian coordinates $(0, 2)$. (See Figure 10.5.)

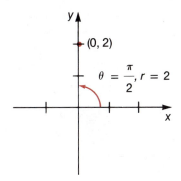

FIGURE 10.5

Some other points for which polar, as well as cartesian coordinates, are relatively simple to supply are illustrated in Figure 10.6.

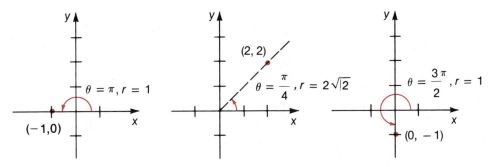

FIGURE 10.6

To derive a general procedure for changing from one system to the other, suppose P is a point with cartesian coordinates (x, y) and polar coordinates (r, θ). Then, as shown in the triangle in Figure 10.7,

$$\cos \theta = \frac{x}{r} \quad \text{and} \quad \sin \theta = \frac{y}{r}.$$

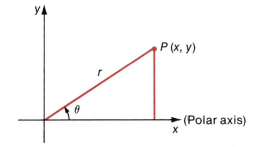

FIGURE 10.7

To change from polar to cartesian coordinates we use

(10.1)
$$x = r \cos \theta$$
$$y = r \sin \theta.$$

Although our figure demonstrates the situation only for P in the first quadrant and $r > 0$, the equations in (10.1) can be verified for any values of r and θ.

To change from cartesian to polar coordinates we modify the equations in (10.1):

$$x^2 + y^2 = r^2 (\cos \theta)^2 + r^2 (\sin \theta)^2 = r^2$$

and

$$\frac{y}{x} = \frac{r \sin \theta}{r \cos \theta} = \tan \theta, \quad \text{provided } x \neq 0.$$

So one set of polar coordinates describing the point with cartesian coordinates (x, y) is given by

(10.2)

$$r = \pm \sqrt{x^2 + y^2}$$

$$\theta = \arctan \left(\frac{y}{x}\right), \quad \text{provided } x \neq 0.$$

The sign for r is chosen to ensure that the point lies in the appropriate quadrant.

In the remainder of this chapter we will assume that both sets of coordinates are available when discussing a problem and change from one system to the other when convenient.

EXAMPLE 1   Find the cartesian coordinates of the points with polar coordinates $\left(2, \dfrac{2\pi}{3}\right)$ and $\left(-3, \dfrac{\pi}{6}\right)$.

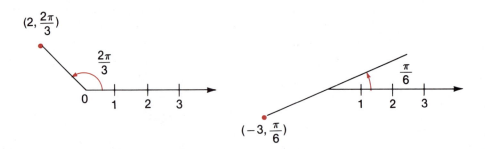

FIGURE 10.8

SOLUTION

These points are shown in Figure 10.8. Using the equations in (10.1), we have

$$x = 2 \cos \frac{2\pi}{3} = 2\left(-\frac{1}{2}\right) = -1$$

and

$$y = 2 \sin \frac{2\pi}{3} = 2\left(\frac{\sqrt{3}}{2}\right) = \sqrt{3};$$

so the cartesian coordinates corresponding to $(2, 2\pi/3)$ are $(-1, \sqrt{3})$. The point with polar coordinates $(-3, \pi/6)$ has cartesian coordinates

and

$$x = -3 \cos \frac{\pi}{6} = -3 \frac{\sqrt{3}}{2}$$

$$y = -3 \sin \frac{\pi}{6} = -3\left(\frac{1}{2}\right) = -\frac{3}{2}. \qquad \square$$

EXAMPLE 2   Find polar coordinates for the point with cartesian coordinates $(\sqrt{3}, 1)$.

SOLUTION

Figure 10.9 indicates one possibility for r and θ:

$$r = \sqrt{(\sqrt{3})^2 + 1^2} = 2$$

and

$$\theta = \arctan\left(\frac{1}{\sqrt{3}}\right) = \frac{\pi}{6},$$

so one pair of polar coordinates of this point is $(2, \pi/6)$.

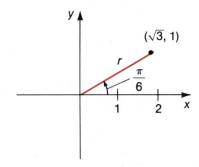

FIGURE 10.9

   Some other pairs of polar coordinates of the point with cartesian coordinates $(\sqrt{3}, 1)$ are $(2, 13\pi/6)$, $(2, -11\pi/6)$, $(-2, 7\pi/6)$ and $(-2, -5\pi/6)$. In fact, any pair of points (r, θ) satisfying

$$r = 2 \qquad \text{and} \qquad \theta = \frac{\pi}{6} + 2n\pi, \ n \text{ an integer}$$

or

$$r = -2 \qquad \text{and} \qquad \theta = \frac{7\pi}{6} + 2n\pi, \ n \text{ an integer}$$

describes the point with cartesian coordinates $(\sqrt{3}, 1)$. $\qquad \square$

EXERCISE SET 10.1

1. Plot the points whose polar coordinates are

(a) $\left(3, \frac{\pi}{6}\right)$,   (b) $\left(1, \frac{4\pi}{3}\right)$,   (c) $\left(0, \frac{\pi}{6}\right)$,   (d) $\left(2, -\frac{\pi}{3}\right)$.

2. Plot the points whose polar coordinates are

   (a) $\left(-2, \dfrac{\pi}{3}\right)$,   (b) $\left(2, \dfrac{4\pi}{3}\right)$,   (c) $\left(-2, -\dfrac{4\pi}{3}\right)$, (d) $\left(-2, -\dfrac{2\pi}{3}\right)$.

3. Plot the points whose polar coordinates are

   (a) $\left(1, \dfrac{375\pi}{2}\right)$,   (b) $(1, 1)$,   (c) $(0, 0)$,   (d) $(\pi, 1)$.

4. Give four other sets of polar coordinates that describe the point with polar coordinates

   (a) $(2, 0)$,   (b) $\left(-1, \dfrac{\pi}{2}\right)$,   (c) $\left(1, \dfrac{3\pi}{4}\right)$,   (d) $(1, -\pi)$.

5. Give four other sets of polar coordinates that describe the points with polar coordinates:

   (a) $\left(2, \dfrac{2\pi}{3}\right)$,   (b) $(1, \pi)$,   (c) $(0, 0)$,   (d) $(1, 1)$.

6. Find the rectangular coordinates of the following points given in polar coordinates:

   (a) $\left(2, \dfrac{\pi}{4}\right)$,   (b) $\left(1, \dfrac{\pi}{2}\right)$,   (c) $\left(-1, \dfrac{\pi}{2}\right)$,   (d) $\left(2, \dfrac{2\pi}{3}\right)$.

7. Find the rectangular coordinates of the following points given in polar coordinates:

   (a) $(-1, \pi)$,   (b) $\left(-3, \dfrac{7\pi}{6}\right)$,   (c) $\left(2, -\dfrac{7\pi}{4}\right)$,   (d) $(-4, \pi)$.

8. Show that the distance between points with polar coordinates (r_1, θ_1) and (r_2, θ_2) is

$$d((r_1, \theta_1), (r_2, \theta_2)) = \sqrt{r_1^2 + r_2^2 - 2r_1r_2 \cos(\theta_2 - \theta_1)}.$$

9. Determine a polar equation for the line passing through the points with polar coordinates (r_1, θ_1) and (r_2, θ_2).

10. Find a cartesian equation corresponding to the polar equation $r = a \sin \theta + b \cos \theta$, where a and b are arbitrary constants. Make a general statement about the graph of an equation of this form.

11. A football field is 100 yards long and 160 feet wide. Place a polar coordinate system with the pole at one corner of the field and the polar axis along the adjacent 100-yard side. What are the polar coordinates of the corner of the field diagonally opposite the pole?

12. A surveyor sets a transit at one corner of a lot with four straight sides. An angle of 0° is set on a line through another corner that is 900 feet away. The two remaining corners have polar coordinates (630, 45°) and (200, 90°). Draw a sketch of the lot and determine its acreage. (1 acre is 43,560 square feet.)

13. A major league baseball field is in the shape of a square with sides 90 feet long. Suppose a polar coordinate system is placed with the pole at home plate and the polar axis parallel to the line from first to third base with positive direction on the first base side. What are the polar coordinates of first, second, and third base?

10.2
GRAPHING IN POLAR COORDINATES

Certain types of curves are more easily described by using polar equations than cartesian equations. The polar equation $r = 2$, for example, describes the circle with radius two whose center is at the origin. (See Figure 10.10.) This equation is considerably simpler than its cartesian counterpart $x^2 + y^2 = 4$.

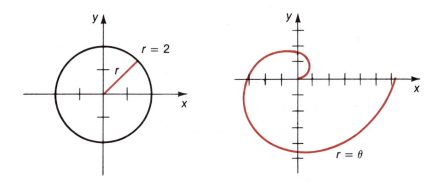

FIGURE 10.10   FIGURE 10.11

The polar equation

$$r = \theta, \quad 0 \le \theta \le 2\pi$$

describes the spiraling curve shown in Figure 10.11. This curve is very difficult to express using cartesian coordinates. For example, a cartesian equation for the portion of the graph lying in the first quadrant can be determined by using the relations between polar and cartesian coordinates discussed in the preceding section. The cartesian equation for this portion of the curve is

$$\sqrt{x^2 + y^2} = \arctan\left(\frac{y}{x}\right), \quad x \ne 0,$$

a considerably more complicated representation.

Although there are exceptions, polar equations are generally appropriate for describing curves that have some systematic rotational behavior centered at the origin. Most other curves are more easily represented by cartesian equations.

Paper specifically designed for sketching graphs in polar coordinates is available in most college bookstores or can be easily constructed with a ruler and compass. Figure 10.16 is sketched on paper of this type. If you have difficulty sketching polar curves it may help to use this special paper.

Many of the graphing techniques used for cartesian equations can be applied to polar equations with only minor modifications. For example, intercepts of the x-, or polar, axis are important. They occur at (r, θ) when θ is any multiple of π. Intercepts of the y-axis or polar line $\theta = \pi/2$, occur at (r, θ) when θ is any odd multiple of $\pi/2$.

Symmetry with respect to the x-axis occurs if the equation is unchanged when (r, θ) is replaced by $(r, -\theta)$, as shown in Figure 10.12.

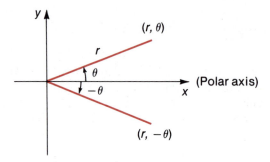

FIGURE 10.12

Symmetry with respect to the y-axis occurs if the equation is unchanged when (r, θ) is replaced by either $(r, \pi - \theta)$ or $(-r, -\theta)$, as shown in Figure 10.13.

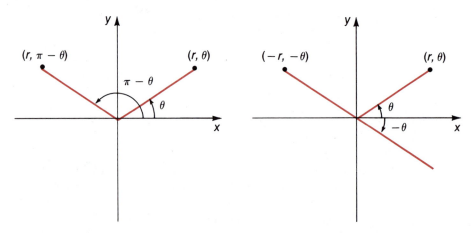

FIGURE 10.13

Symmetry with respect to the origin, or pole, occurs if the equation is unchanged when (r, θ) is replaced by either $(r, \pi + \theta)$ or $(-r, \theta)$. This is illustrated in Figure 10.14.

These symmetry tests are not complete since each point in the plane can be represented in an infinite number of ways; they are, however, the easiest tests to apply and are generally sufficient.

One other useful technique in graphing polar equations is to find the values of θ that correspond to $r = 0$. This determines when the graph intersects the pole and at what angle the graph approaches the pole.

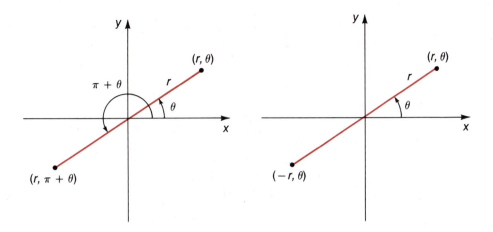

FIGURE 10.14

EXAMPLE 1   Sketch the graph of the polar equation

$$r = 2 \sin \theta$$

and find a corresponding cartesian equation.

SOLUTION

Since $\sin(-\theta) = -\sin \theta$, the equation is unchanged if (r, θ) is replaced by $(-r, -\theta)$. So the graph is symmetric with respect to the y-axis. It intersects the x-, or polar, axis when $\theta = 0$ or $\theta = \pi$. Since $\sin 0 = \sin \pi = 0$, the only x-intercept occurs at the pole. When $\theta = \pi/2$ or $\theta = 3\pi/2$, the graph intersects the y-axis. This gives only one y-intercept, since $(2, \pi/2)$ and $(-2, 3\pi/2)$ describe the same point.

The table on page 586 lists representative values for θ, $-\pi/2 \le \theta \le \pi/2$, and the corresponding values for r. These points give the shape of the graph to the right of the y-axis. Symmetry with respect to the y-axis enables us to sketch the remainder of the graph as shown in Figure 10.15.

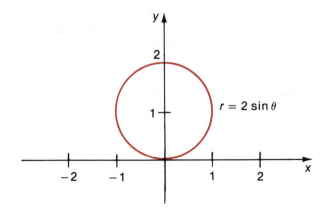

FIGURE 10.15

θ	$-\dfrac{\pi}{2}$	$-\dfrac{\pi}{3}$	$-\dfrac{\pi}{4}$	$-\dfrac{\pi}{6}$	0	$\dfrac{\pi}{6}$	$\dfrac{\pi}{4}$	$\dfrac{\pi}{3}$	$\dfrac{\pi}{2}$
r	-2	$-\sqrt{3}$	$-\sqrt{2}$	-1	0	1	$\sqrt{2}$	$\sqrt{3}$	2

The graph is a circle with center at the point with cartesian coordinates $(0, 1)$ and radius 1.

To find a cartesian equation, first multiply both sides of the equation $r = 2 \sin \theta$ by r to obtain $r^2 = 2r \sin \theta$. Using the relationships, $r^2 = x^2 + y^2$ and $y = r \sin \theta$, we can write the equation in cartesian coordinates:

$$x^2 + y^2 = 2y,$$

which simplifies to

$$x^2 + y^2 - 2y = 0,$$
$$x^2 + y^2 - 2y + 1 = 1,$$

so

$$x^2 + (y - 1)^2 = 1. \qquad \square$$

EXAMPLE 2   Sketch the graph of the polar equation

$$r = 2(1 + \sin \theta).$$

SOLUTION

The equation is changed when (r, θ) is replaced by $(-r, \theta)$; however, when (r, θ) is replaced by $(r, \pi - \theta)$, we have

$$r = 2[1 + \sin(\pi - \theta)]$$
$$= 2[1 + \sin \pi \cos \theta - \sin \theta \cos \pi]$$
$$= 2(1 + \sin \theta).$$

The graph is symmetric with respect to the y-axis but for a different reason than was the equation in Example 1. This symmetry implies that we need only consider θ in the interval $[-\pi/2, \pi/2]$. An x-intercept occurs in this interval at the point with polar coordinates $(2, 0)$ and the y-intercepts have polar coordinates $(0, -\pi/2)$ and $(4, \pi/2)$. A table listing representative values of θ and the corresponding values of r is given below. The graph is shown in Figure 10.16; this graph is called a **cardioid**, a name derived from the Greek word *kardia*, meaning heart.

θ	$-\dfrac{\pi}{2}$	$-\dfrac{\pi}{3}$	$-\dfrac{\pi}{4}$	$-\dfrac{\pi}{6}$	0	$\dfrac{\pi}{6}$	$\dfrac{\pi}{4}$	$\dfrac{\pi}{3}$	$\dfrac{\pi}{2}$
r	0	$2 - \sqrt{3}$	$2 - \sqrt{2}$	1	2	3	$2 + \sqrt{2}$	$2 + \sqrt{3}$	4

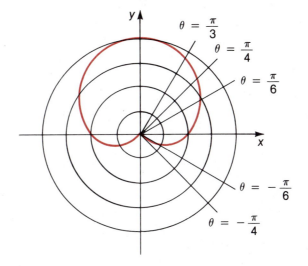

FIGURE 10.16

EXAMPLE 3   Sketch the graph of the polar equation

$$r = \cos 2\theta.$$

SOLUTION

   Since $\cos(-2\theta) = \cos(2\theta)$, this graph is symmetric with respect to the x-, or polar, axis. The graph is also symmetric with respect to the y-axis, since

$$\cos 2(\pi - \theta) = \cos(2\pi - 2\theta) = \cos(-2\theta) = \cos 2\theta.$$

Because of the symmetry, it suffices to determine the behavior of the graph for θ in the interval $[0, \pi/2]$. The x-intercept in this interval occurs at the point with polar coordinates $(1, 0)$ and the y-intercept at $(-1, \pi/2)$. Notice that $r = 0$ when $\theta = \pi/4$. This observation produces the dotted lines shown in Figure 10.17. These lines are rays along which the graph approaches the pole.

   The following table lists some representative values. The values give the portion of the graph of the equation that is shown in Figure 10.17.

θ	0	$\dfrac{\pi}{8}$	$\dfrac{\pi}{6}$	$\dfrac{\pi}{4}$	$\dfrac{\pi}{3}$	$\dfrac{5\pi}{12}$	$\dfrac{\pi}{2}$
r	1	$\dfrac{\sqrt{2}}{2}$	$\dfrac{1}{2}$	0	$-\dfrac{1}{2}$	$-\dfrac{\sqrt{3}}{2}$	-1

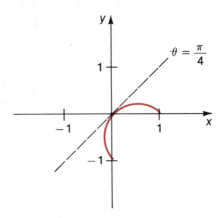

FIGURE 10.17

Symmetry with respect to the x-axis enables us to extend the graph as shown in Figure 10.18 and symmetry with respect to the y-axis gives the completed graph shown in Figure 10.19. The graph is called a **four-leafed rose**.   □

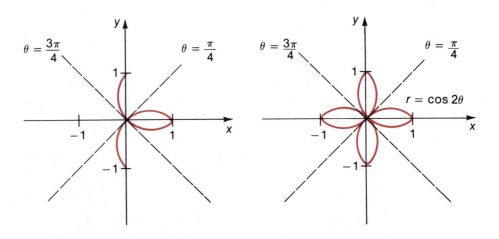

FIGURE 10.18   FIGURE 10.19

EXAMPLE 4   Sketch the graph of the polar equation

$$r^2 = \cos 2\theta.$$

SOLUTION

The analysis of this equation is similar to that of the equation in Example 3, except that values of θ for which $\cos 2\theta < 0$ must be excluded; that is, θ must satisfy $-\dfrac{\pi}{2} < 2\theta < \dfrac{\pi}{2}$ or $\dfrac{3\pi}{2} < 2\theta < \dfrac{5\pi}{2}$. So the graph is determined by values of θ satisfying

$$-\frac{\pi}{4} \le \theta \le \frac{\pi}{4} \qquad \text{or} \qquad \frac{3\pi}{4} \le \theta \le \frac{5\pi}{4}.$$

The graph is a two-leafed rose, called a **lemniscate**, and is shown in Figure 10.20. Again, we have used dotted lines to indicate those rays along which the graph approaches the pole. □

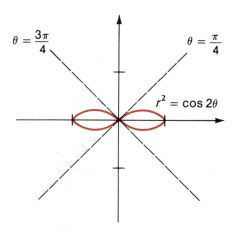

FIGURE 10.20

EXAMPLE 5   Sketch the graph of the polar equation

$$r = 2 \cos \theta + 1.$$

SOLUTION

The graph is symmetric with respect to the x-axis so it suffices to determine the graph for $0 \le \theta \le \pi$. The only y-intercept in this interval occurs at $(1, \pi/2)$. The other intercepts are at the points with polar coordinates $(3, 0)$, $(-1, \pi)$ and $(0, 2\pi/3)$. The rays along which the graph approaches the pole are $\theta = 2\pi/3$ and $\theta = 4\pi/3$. A table of values for θ in the interval $[0, \pi]$ is given on p. 590; these values indicate that the graph appears as shown in Figure 10.21. This figure is known as a **limaçon**.

θ	0	$\dfrac{\pi}{6}$	$\dfrac{\pi}{4}$	$\dfrac{\pi}{3}$	$\dfrac{\pi}{2}$	$\dfrac{2\pi}{3}$	$\dfrac{3\pi}{4}$	$\dfrac{5\pi}{6}$	π
r	3	$1+\sqrt{3}$	$1+\sqrt{2}$	2	1	0	$1-\sqrt{2}$	$1-\sqrt{3}$	-1

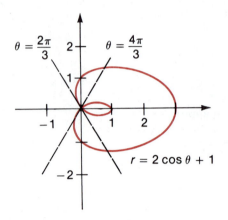

FIGURE 10.21

The curves in Examples 1 through 5 are samples of certain general classes of polar curves, summarized in the table below.

Equation	Curve
$r = a$	Circle with center at $(0, 0)$ and radius $\lvert a \rvert$.
$\begin{aligned} r &= a \pm b\cos\theta \\ r &= a \pm b\sin\theta \end{aligned}$	Limaçon. If $a < b$, the limaçon has a loop. If $a = b$, the limaçon is heart-shaped and called a cardioid. See Figure 10.22.
$\begin{aligned} r^2 &= \pm\, a^2\cos 2\theta \\ r^2 &= \pm\, a^2\sin 2\theta \end{aligned}$	Lemniscate
$\begin{aligned} r &= a\sin n\theta \\ r &= a\cos n\theta \end{aligned}$	n-leaved rose if n is odd, $2n$-leaved rose if n is even
$r = f(\theta)$, where f is a monotonic function	Spiral

TABLE 10.1

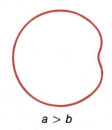

$a > b$

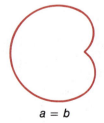

$a = b$

$a < b$

FIGURE 10.22

Listed in Table 10.2 are items that can be helpful for sketching the graph of a polar equation. Not all items will be appropriate for use in every problem, but the table provides a good checklist of techniques that can be applied.

Test	Result
Substitute $(r, -\theta)$ for (r, θ)	Symmetry to x-axis if equation is unchanged
Substitute $(r, \pi - \theta)$ for (r, θ)	Symmetry to y-axis if equation is unchanged
Substitute $(-r, -\theta)$ for (r, θ)	Symmetry to y-axis if equation is unchanged
Substitute $(-r, \theta)$ for (r, θ).	Symmetry to pole if equation is unchanged
Substitute $(r, \pi + \theta)$ for (r, θ)	Symmetry to pole if equation is unchanged
Determine θ when $r = 0$	Lines along which the graph approaches the pole
Determine r when $\theta = n\pi$	Intercepts of the x-axis
Determine r when $\theta = \dfrac{(2n+1)\pi}{2}$	Intercepts of the y-axis
Determine θ when $r'(\theta)$ is zero or does not exist	Candidates at which maximum and minimum values of r can occur

TABLE 10.2

EXERCISE SET 10.2

Sketch the graphs of the polar equations listed in Exercises 1 through 30.

1. $r = 3$

2. $r = \dfrac{5\pi}{4}$

3. $\theta = \dfrac{5\pi}{4}$

4. $\theta = 3$

5. $r = 3 \cos \theta$

6. $r = 6 \sin \theta$

7. $r = -3 \cos \theta$

8. $r = -6 \sin \theta$

9. $r = 1 + \cos \theta$

10. $r = 1 + \sin \theta$

11. $r = 2 + \sin \theta$

12. $r = 1 + 2 \sin \theta$

13. $r = 2 - \sin \theta$

14. $r = 1 - 2 \sin \theta$

15. $r = 3 \sin 3\theta$

16. $r = \cos 3\theta$

17. $r = 4 \sin 4\theta$

18. $r = \cos 4\theta$

19. $r = 2 \sin 2\theta$

20. $r^2 = 4 \sin 2\theta$

21. $r^2 = -9 \cos 2\theta$

22. $r^2 = 16 \cos 4\theta$

23. $r = 3 \sec \theta$

24. $r = -2 \csc \theta$

25. $r = \theta$

26. $r = -2^{\theta}$

27. $r = e^{\theta}$ **28.** $r = \ln \theta$

29. $r = \dfrac{1}{\theta}$ **30.** $r = 2^{-\theta}$

Compare the graphs of the polar equations in each part of Exercises 31 through 37.

31. (a) $r = 1 + \cos \theta$ (b) $r = 1 - \cos \theta$
      (c) $r = \cos \theta - 1$ (d) $r = -\cos \theta - 1$

32. (a) $r = 1 + \sin \theta$ (b) $r = 1 - \sin \theta$
      (c) $r = \sin \theta - 1$ (d) $r = -\sin \theta - 1$

33. (a) $r = \cos \theta$ (b) $r = 1 + \cos \theta$
      (c) $r = 2 + \cos \theta$ (d) $r = 3 + \cos \theta$

34. (a) $r = \sin \theta$ (b) $r = \sin 2\theta$
      (c) $r = \sin 3\theta$ (d) $r = \sin 4\theta$

35. (a) $r = \sin \theta$ (b) $r = 1 + \sin \theta$
      (c) $r = 1 + 2 \sin \theta$ (d) $r = 1 + 3 \sin \theta$

36. (a) $r = \sin \theta$ (b) $r = \cos \theta$
      (c) $r = \csc \theta$ (d) $r = \sec \theta$

37. (a) $r = 2^{\theta}$ (b) $r = 2^{-\theta}$
      (c) $r = -2^{\theta}$ (d) $r = -2^{-\theta}$

38. Suppose f is a differentiable function and $r = f(\theta)$. Find $dy/d\theta$ and $dx/d\theta$ in terms of θ, $f(\theta)$, and $f'(\theta)$. Use these expressions and the chain rule for derivatives to show that the graph of the equation will have

(a) a horizontal tangent at (r, θ) if $r = -f'(\theta) \tan \theta$.

(b) a vertical tangent at (r, θ) if $r = f'(\theta) \cot \theta$.

39. Use the result in Exercise 38 to find the horizontal and vertical tangents of

(a) $r = \cos \theta, 0 \le \theta \le 2\pi$ (b) $r = 2 + 2 \sin \theta, 0 \le \theta \le 2\pi$

(c) $r = 1 - \cos \theta, 0 \le \theta \le 2\pi$ (d) $r = e^{-\theta}, -\pi \le \theta \le \pi$

40. A chambered nautilus (*Nautilus pompilus*) is a type of mollusk found in the Pacific and Indian oceans. The outside of its shell grows in the form of an exponential spiral. A typical equation of such a spiral is $r = 2e^{.2\theta}$. Sketch the graph of this spiral and compare the resulting curve to the curve in the accompanying picture.

41. The lengthwise cross section of an apple can be reasonably approximated using a polar equation. What form would you expect the equation to assume if the polar axis is the stem of the apple and the pole is at the base of the stem?

42. A polar coordinate system on a major league baseball field was described in Exercise 13 of Section 10.1 (p. 583). Use this coordinate system to describe the base paths of the field.

10.3
AREAS OF REGIONS USING POLAR COORDINATES

In Section 10.2 we considered some typical curves that can be expressed using polar coordinates. In this section we derive a result that uses polar coordinate representation to find the area of certain regions of the plane. The area of a region whose boundaries are described using cartesian coordinates was found by summing the areas of approximating rectangles. To find the area of a region bounded by a polar equation, we first divide the region into approximating circular sectors (see Figure 10.23) and then sum the areas of these sectors.

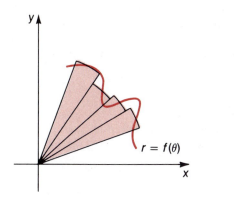

FIGURE 10.23

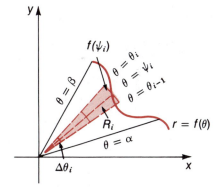

FIGURE 10.24

Suppose a curve is described by $r = f(\theta)$ where f is a continuous, nonnegative function defined on $[\alpha, \beta]$. Let R denote the region bounded by the graph of f and by the rays $\theta = \alpha$ and $\theta = \beta$, where $\alpha < \beta \leq \alpha + 2\pi$. To find an approximation to the area of this region we let

$$\mathcal{P} = \{\theta_0, \theta_1, \ldots, \theta_n\}$$

be a partition of the interval $[\alpha, \beta]$.

We choose ψ_i arbitrarily in the subinterval $[\theta_{i-1}, \theta_i]$ for each $i = 1, 2, \ldots, n$ and let $\Delta\theta_i = \theta_i - \theta_{i-1}$. The partition determines subregions R_i bounded by the rays $\theta = \theta_{i-1}$ and $\theta = \theta_i$ and by the graph of the function. An approximation to the area of R_i is the area of the sector of the circle with radius $f(\psi_i)$ and central angle $\Delta\theta_i$. (See Figure 10.24).

The area of a circle with radius $f(\psi_i)$ is $\pi[f(\psi_i)]^2$. Since the sector contains $\Delta\theta_i/2\pi$ of the circle, the area of the sector is

$$\pi[f(\psi_i)]^2\,\frac{\Delta\theta_i}{2\pi} = \frac{[f(\psi_i)]^2}{2}\,\Delta\theta_i.$$

An approximation to the entire area of the region R is the sum of these approximations

$$\sum_{i=1}^{n}\frac{[f(\psi_i)]^2}{2}\,\Delta\theta_i,$$

a Riemann sum for the function described by $\dfrac{[f(\theta)]^2}{2}$ on the interval $[\alpha, \beta]$.

(10.3)
DEFINITION

If f is a continuous, nonnegative function on the interval $[\alpha, \beta]$ where $\alpha < \beta \le \alpha + 2\pi$, then the area of the region bounded by the rays $\theta = \alpha$ and $\theta = \beta$ and by the graph of $r = f(\theta)$ is

$$A = \frac{1}{2}\int_{\alpha}^{\beta}[f(\theta)]^2 d\theta.$$

EXAMPLE 1   Find the area of the region bounded by the graph of the cardioid

$$r = 2(1 + \sin\theta).$$

SOLUTION
    The graph of this equation was discussed in Example 2 of Section 10.2 and is shown in Figure 10.25, together with a typical sector.

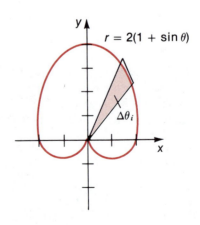

FIGURE 10.25

The area of the region is

$$A = \frac{1}{2} \int_0^{2\pi} [2(1 + \sin \theta)]^2 d\theta$$

$$= 2 \int_0^{2\pi} [1 + 2\sin \theta + (\sin \theta)^2] d\theta$$

$$= 2 \left[\theta - 2\cos\theta \right]_0^{2\pi} + 2 \int_0^{2\pi} \frac{(1 - \cos 2\theta)}{2} d\theta$$

$$= 4\pi + \left[\theta - \frac{\sin 2\theta}{2} \right]_0^{2\pi}$$

$$= 6\pi. \qquad \square$$

EXAMPLE 2   Find the area of the region bounded by one leaf of the four-leafed rose $r = \cos 2\theta$.

SOLUTION

The graph of this equation was discussed in Example 3 of Section 10.2 and is shown in Figure 10.26, together with a typical polar sector.

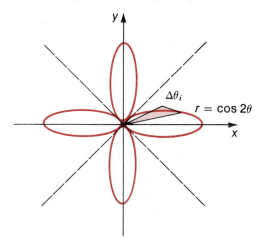

FIGURE 10.26

The leaf that is horizontal and to the right can be described by taking θ in the interval $[-\pi/4, \pi/4]$; the area of this region is

$$A = \frac{1}{2} \int_{-\pi/4}^{\pi/4} [\cos 2\theta]^2 d\theta$$

$$= \frac{1}{2} \int_{-\pi/4}^{\pi/4} \frac{(1 + \cos 4\theta)}{2} d\theta$$

$$= \frac{1}{4} \left[\theta + \frac{\sin 4\theta}{4} \right]_{-\pi/4}^{\pi/4} = \frac{\pi}{8}. \qquad \square$$

EXAMPLE 3   Find the area of the region formed by the intersection of the graphs of the polar equations $r = 1$ and $r = 2 \sin \theta$.

SOLUTION

The first problem is to determine where the curves intersect. We see in Figure 10.27 that there are two points of intersection. If (r, θ) is a point of intersection, then $r = 1$ and $r = 2 \sin \theta$. So

$$1 = 2 \sin \theta \quad \text{or} \quad \sin \theta = \frac{1}{2}$$

and

$$\theta = \frac{\pi}{6} \quad \text{or} \quad \theta = \frac{5\pi}{6}.$$

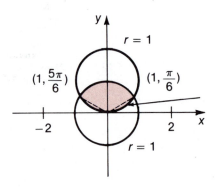

FIGURE 10.27

The region whose area we are to find is bounded by the graph of $r = 2 \sin \theta$ for θ in $[0, \pi/6]$, by the graph of $r = 1$ for θ in $[\pi/6, 5\pi/6]$, and again by the graph of $r = 2 \sin \theta$ in $[5\pi/6, \pi]$.

$$A = \frac{1}{2} \int_0^{\pi/6} [2\sin \theta]^2 d\theta + \frac{1}{2} \int_{\pi/6}^{5\pi/6} (1)^2 d\theta + \frac{1}{2} \int_{5\pi/6}^{\pi} [2\sin \theta]^2 d\theta$$

$$= \frac{1}{2} \int_0^{\pi/6} 4\left(\frac{1 - \cos 2\theta}{2}\right) d\theta + \frac{1}{2} \int_{\pi/6}^{5\pi/6} d\theta + \frac{1}{2} \int_{5\pi/6}^{\pi} 4\left(\frac{1 - \cos 2\theta}{2}\right) d\theta$$

$$= \left[\theta - \frac{\sin 2\theta}{2}\right]_0^{\pi/6} + \frac{1}{2}\left(\frac{5\pi}{6} - \frac{\pi}{6}\right) + \left[\theta - \frac{\sin 2\theta}{2}\right]_{5\pi/6}^{\pi}$$

$$= \frac{\pi}{6} - \frac{\sqrt{3}}{4} + \frac{\pi}{3} + \pi - \frac{5\pi}{6} - \frac{\sqrt{3}}{4}$$

$$= \frac{2\pi}{3} - \frac{\sqrt{3}}{2}.$$

The integration could have been simplified by noting the symmetry with respect to the y-axis. The symmetry implies that the area is also given by

$$A = 2\left[\frac{1}{2}\int_0^{\pi/6} [2\sin\theta]^2 d\theta + \frac{1}{2}\int_{\pi/6}^{\pi/2} (1)^2 d\theta\right]. \qquad \square$$

In Example 3 we found the points of intersection of two curves by solving the equations simultaneously. This method does not necessarily give *all* points of intersection of curves with polar equations. Because a point has an infinite number of sets of polar coordinates, it is possible for the intersection to occur at a point for which no single set of polar coordinates satisfies both equations.

EXAMPLE 4   Find the points of intersection of the cardioids $r = 1 - \cos\theta$ and $r = 1 + \sin\theta$.

SOLUTION

The graphs of the cardioids are given in Figure 10.28 where three points of intersection are indicated. First we solve the equations simultaneously to obtain:

$$1 - \cos\theta = 1 + \sin\theta$$
$$-\cos\theta = \sin\theta.$$

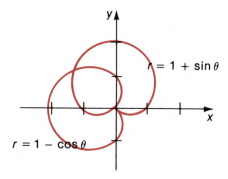

FIGURE 10.28

This equation is satisfied when $\theta = 3\pi/4$ and $\theta = 7\pi/4$, so two points of intersection are $(1 + \sqrt{2}/2, 3\pi/4)$ and $(1 - \sqrt{2}/2, 7\pi/4)$. A third point of intersection is shown in Figure 10.29 to be at the origin. The polar coordinates of the origin that satisfy $r = 1 - \cos\theta$ are $(0, 0)$ while those satisfying $r = 1 + \sin\theta$ are $(0, 3\pi/2)$. $\qquad \square$

The preceding example demonstrates that it is important to sketch the graphs of curves given in polar coordinates when finding points of intersection. It is quite possible for the intersection to occur at a point for which no single set of polar coordinates satisfies both equations.

EXERCISE SET 10.3

Sketch the graphs of the polar equations in Exercises 1 through 16. Find the area of the region bounded by the graph.

1. $r = 2$

2. $r = \dfrac{\pi}{2}$

3. $r = 2 \sin \theta$

4. $r = 4 \cos \theta$

5. $r = -3 \cos \theta$

6. $r = -5 \sin \theta$

7. $r = 1 + \cos \theta$

8. $r = 2 - \sin \theta$

9. $r = 4 - 2 \sin \theta$

10. $r = 3 - 2 \cos \theta$

11. $r = \sin 2\theta$

12. $r = \sin 3\theta$

13. $r^2 = 4 \cos 2\theta$

14. $r^2 = 9 \sin 2\theta$

15. $r = \theta, \ -\dfrac{\pi}{2} \le \theta \le \dfrac{\pi}{2}$

16. $r = \theta^2, \ -\pi \le \theta \le \pi$

Sketch the graphs of the pair of polar equations in Exercises 17 through 24. Find the area of the region inside the graph of the first equation and outside the graph of the second equation.

17. $r = 2, r = 1$

18. $r = 2, r = 1 + \cos \theta$

19. $r = 1 + \cos \theta, r = 1$

20. $r = 1, r = 1 + \cos \theta$

21. $r = 2, r^2 = 4 \cos 2\theta$

22. $r = \sin \theta, r^2 = \sin 2\theta$

23. $r = 1 + \cos \theta, r = 1 + \sin \theta$

24. $r = 1 + \cos \theta, r = 1 - \cos \theta$

Sketch the graphs of the pair of polar equations in Exercises 25 through 30. Find the area of the region that lies inside the graphs of both equations.

25. $r = 1, r = 1 + \cos \theta$

26. $r = 1 + \cos \theta, r = 1 - \cos \theta$

27. $r = 1 + \cos \theta, r = 1 + \sin \theta$

28. $r = 2 + 2 \sin \theta, r = 3$

29. $r = \sin 2\theta, r = \cos 2\theta$

30. $r = 3 + 3 \cos \theta, r = -3 \cos \theta$

Show graphically that the pairs of curves listed in Exercises 31 through 36 have intersections at which there is no single pair (r, θ) of polar coordinates that satisfies both equations. Find polar coordinates for the intersection points.

31. $r = \cos \theta, r = \sin \theta$

32. $r = \cos \theta + 1, r = -\cos \theta - 1$

33. $r = \cos \theta + 1, r = \cos \theta - 1$

34. $r = 1, r = 2 \sin 2\theta$

35. $r = \cos \dfrac{\theta}{2}, r = \sin \dfrac{\theta}{2}$

36. $r = 2 \cos 2\theta, r = 2 \sin 2\theta$

37. In Exercise 40 of Section 10.2, the equation $r = 2e^{.2\theta}$ is used to describe the outside of the shell of the chambered nautilus pictured with that exercise. Suppose that θ_1 is a fixed positive real number and A_n denotes the area enclosed by the spiral from $\theta = (n - 1)\theta_1$ to $\theta = n\theta_1$. Show that A_{n+1}/A_n is a constant depending only on θ_1.

10.4
PARAMETRIC EQUATIONS

In a number of past instances we have expressed a problem involving cartesian coordinates in terms of a third variable. In Section 4.3, for example, we discussed related rate problems in which the variables are not only related to one another, but depend on variations in time. The definition of the basic trigonometric functions makes use of the cartesian equation

$$x^2 + y^2 = 1$$

together with the specifications

$$x = \cos t, \qquad y = \sin t.$$

In this case t can represent any real number.

In the following definition we formally recognize this type of expression and define in general what is meant by a curve in the plane.

(10.4)
DEFINITION

A **curve** C in the plane is a set of points $(f(t), g(t))$ where f and g are continuous functions defined on an interval I.
The equations

$$x = f(t), y = g(t), \text{ for } t \text{ in } I$$

are called **parametric equations** for the curve C in the **parameter** t.

For example, the circle with radius 1 and center $(0, 0)$, shown in Figure 10.29, has parametric equations

$$x = \cos t, \qquad y = \sin t, \qquad 0 \le t \le 2\pi.$$

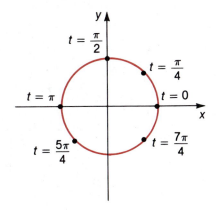

FIGURE 10.29

Another set of parametric equations for the circle is

$$x = \sin t, \qquad y = \cos t, \qquad 0 \le t \le 2\pi,$$

and still another is

$$x = \sin 2t, \qquad y = \cos 2t, \qquad 0 \le t \le \pi.$$

The only conditions that need to be satisfied to describe this curve are that $x^2 + y^2 = 1$ and that each point on the circle is the image of some value of the parameter.

Parametric equations are introduced in this chapter because of the natural relationship between this concept and functions that are defined using polar coordinates. If r is a continuous function on an interval I, then

$$x = r(\theta)\cos \theta$$

and

$$y = r(\theta)\sin \theta$$

are parametric equations with θ as the parameter. We implicitly used this fact in Section 10.2 when sketching the graphs of polar equations in a cartesian coordinate system.

EXAMPLE 1   Describe and sketch the graph of the curve whose parametric equations are

$$x = 2t + 1, \qquad y = 4t^2 - 1, \qquad -1 \le t \le 1.$$

SOLUTION

By eliminating the parameter t we obtain an equation in cartesian coordinates:

$$t = \frac{x - 1}{2}$$

so

$$y = 4\left(\frac{x - 1}{2}\right)^2 - 1 = (x - 1)^2 - 1.$$

This is an equation of a parabola. As t varies from -1 to 1, $P(t)$ moves along the curve from the point $(-1, 3)$ to $(3, 3)$. The graph of the curve is shown in Figure 10.30.

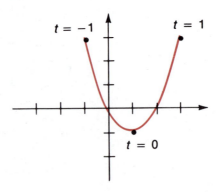

FIGURE 10.30

EXAMPLE 2   Find parametric equations for the straight line through $(1, 2)$ with slope $1/2$.

SOLUTION

A cartesian equation for this line is

$$y - 2 = \frac{1}{2}(x - 1).$$

One set of parametric equations can be found by letting

$$t = x - 1.$$

Then

$$y - 2 = \frac{1}{2}t,$$

so

$$y = \frac{1}{2}t + 2.$$

The line is consequently described by the parametric equations

$$x = t + 1, \qquad y = \frac{1}{2}t + 2, \qquad -\infty < t < \infty.$$

Another set of parametric equations that describe this line can be found by letting

$$x = t, \qquad y = \frac{1}{2}(t - 1) + 2, \qquad -\infty < t < \infty.$$

A third set can be found by letting

$$x = \ln t + 1, \qquad y = \frac{1}{2}\ln t + 2 = \ln\sqrt{t} + 2, \qquad 0 < t < \infty.$$

The only conditions that need be fulfilled are that

$$y - 2 = \frac{1}{2}(x - 1)$$

and that both x and y assume all real values.

Notice that the parametric equations

$$x = t^2, \qquad y = \frac{1}{2}(t^2 + 3), \qquad -\infty < t < \infty,$$

do not describe the entire line. This set of equations describes only the portion of the line when $x \geq 0$ and traces this portion of the line twice for t in $(-\infty, \infty)$, as shown in Figure 10.31.

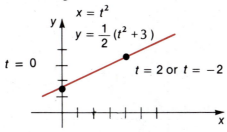

FIGURE 10.31

EXAMPLE 3   A circle of radius R rolls along a horizontal line. Find parametric equations of the curve traced by a fixed point P on the circumference of the circle. This curve is called a **cycloid.** (It is the curve described by the head of a nail embedded in a tire as the tire rolls along the ground.)

SOLUTION

Suppose the horizontal line along which the circle rolls is the x-axis and the point P is originally at the origin. Let O denote the center of the circle and t denote the radian measure of the angle through which the line OP has moved. (If $0 < t < \pi/2$, P has moved to the position shown in Figure 10.32.)

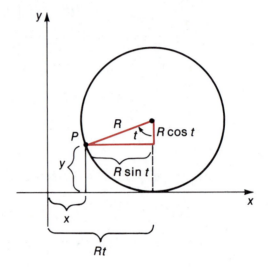

FIGURE 10.32

The new coordinates (x, y) of P are consequently

$$x = Rt - R \sin t,$$
$$y = R - R \cos t.$$

These equations hold for any value of $t \geq 0$. The curve that this set of parametric equations describes is shown in Figure 10.33.

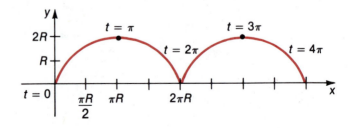

FIGURE 10.33

The cycloid has some interesting geometric properties. The area of the region bounded by the x-axis and one arc of the curve is $3\pi R^2$ (see Exercise 23). So when the circle with radius R has completed half its revolution, the boundary of the circle divides the region into three parts with equal area. (See Figure 10.34.)

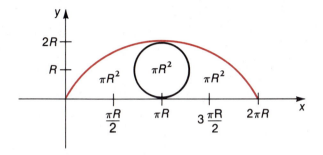

FIGURE 10.34

When the parametric equations describing the cycloid are expressed with $R < 0$, the graph of the cycloid is shown in Figure 10.35.

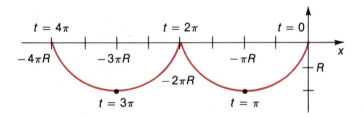

FIGURE 10.35

This curve is the solution to two famous historical problems:

1. The **brachistochrone** or shortest time problem: Find the curve that describes the shape a wire should assume to enable a frictionless bead to slide from a point A to a lower point B in the shortest time. (See Figure 10.36(a).) Johann Bernoulli (1667–1748) showed in 1696 that such a curve is the arc of a cycloid. An interesting feature of this solution is that for some points B, the shortest time is accomplished by sliding to a point lower than B and then to B. (See Figure 10.36(b).)

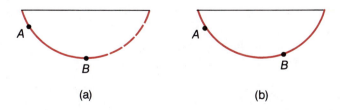

FIGURE 10.36

2. The **tautochrone** or equal time problem: Find the curve that describes the shape a wire should assume if the time it takes a frictionless bead to travel to the bottom of the curve is independent of the point on the curve at which it starts. Three beads starting at the same time from points *A*, *B*, and *C* on the cycloid shown in Figure 10.37 will reach *P* simultaneously.

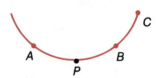

FIGURE 10.37

Jean Bernoulli (1654–1705), brother of Johann, solved this problem in 1690. He published this result in response to a challenge presented by Johann regarding the brachistochrone problem. The paper in which the solution appears contains a number of interesting facts concerning cycloids.

EXERCISE SET 10.4

Sketch the graphs of the curves described by the parametric equations in Exercises 1 through 16. Find a cartesian equation describing each curve.

1. $x = 3t, y = \dfrac{1}{2}t$ 　　　　　　　**2.** $x = 2t + 1, y = 3t - 2$

3. $x = \sqrt{t}, y = t + 1$ 　　　　　　**4.** $x = t + 2, y = -3\sqrt{t}$

5. $x = \sin t, y = (\cos t)^2$ 　　　　**6.** $x = \sec t, y = \tan t$

7. $x = 3 \sin t, y = 4 \cos t$ 　　　　**8.** $x = 3 \sec t, y = 4 \tan t$

9. $x = e^t, y = e^{-t}$ 　　　　　　　　**10.** $x = \ln t, y = \ln \sqrt{t}$

11. $x = \sin t, y = \sec t$ 　　　　　**12.** $x = \sin t, y = \cot t$

13. $x = 4 + 2 \cos t, y = 6 + 2 \sin t$ 　　**14.** $x = \sin t + 1, y = 2 \cos t - 1$

15. $x = 4t^3, y = 2t^2 + 1$ 　　　　**16.** $x = e^{-t} - 2, y = e^{2t} + 3$

In Exercises 17 through 22, the graph of the parametric equations in each part represents a portion of the same curve. Sketch the graph of the parametric equations and label representative values of the parameter.

17. (a) $x = \cos t, y = \sin t; 0 \le t \le 2\pi$

　　(b) $x = \sin t, y = \cos t; 0 \le t \le 2\pi$

　　(c) $x = t, y = \sqrt{1 - t^2}; -1 \le t \le 1$

　　(d) $x = -t, y = \sqrt{1 - t^2}; -1 \le t \le 1$

18. (a) $x = t + 1, y = 4t + 5$

(b) $x = t^2 - \dfrac{1}{4}, y = 4t^2$

(c) $x = \ln t, y = 1 + \ln t^4; 0 < t$

(d) $x = \sin t, y = 4 \sin t + 1; 0 \leq t \leq \pi$

19. (a) $x = t, y = \ln t; 1 < t$

(b) $x = e^t, y = t; 0 < t$

(c) $x = t^2, y = 2 \ln t; 1 < t$

(d) $x = \dfrac{1}{t}, y = -\ln t; 1 < t$

20. (a) $x = t, y = \dfrac{1}{t}; 0 < t$

(b) $x = e^t, y = e^{-t}; 0 < t$

(c) $x = \sin t, y = \csc t; 0 < t < \pi$

(d) $x = \tan t, y = \cot t; 0 < t < \dfrac{\pi}{2}$

21. (a) $x = t, y = t^2$

(b) $x = \sqrt{t}, y = t$

(c) $x = \sin t, y = 1 - (\cos t)^2$

(d) $x = -e^t, y = e^{2t}$

22. (a) $x = t, y = \sqrt{t^2 - 1}; 1 < t$

(b) $x = \sqrt{t^2 + 1}, y = t; 1 < t$

(c) $x = \sec t, y = \tan t$

(d) $x = \sinh t, y = \cosh t$

23. Show that the area of the region bounded by the x-axis and one arc of the cycloid $x = Rt - R \sin t, y = R - R \cos t$ is $3\pi R^2$. (See Figure 10.34.) (*Hint:* Express the integral in terms of t.)

24. Show that when the circle of radius R has completed half its revolution the boundary of the circle divides the region described in Exercise 23 into three parts with equal area. (See Figure 10.34.)

25. Does the curve given parametrically by $x = \dfrac{\sin t}{t}, y = \dfrac{\cos t}{t}$ have a horizontal or vertical asymptote?

26. A 20-foot ladder rests against a vertical wall. A cat starts to climb the ladder at 1 foot per second and the foot of the ladder slides away from the wall at a rate of 1 foot per second. Find:

(a) The equation of the curve described by the cat.

(b) Maximum distance of the cat from the floor.

(c) Maximum distance of the cat from the wall.

(d) Maximum distance of the cat from the corner.

(e) The velocity of the cat at the end of 16 seconds.

10.5
TANGENT LINES TO CURVES

In this section we consider the problem of finding the equation of a line tangent to a plane curve described by parametric equations. Since this is a common problem, the power of parametric representation would be greatly diminished if it was necessary to express the equation in cartesian coordinates before finding the slope of the tangent line. To see that this is unnecessary, suppose C is a curve described by

$$x = f(t), \quad y = g(t), \qquad \text{for } t \text{ in } I,$$

where f and g are differentiable functions. If $f'(t) = dx/dt$ is never zero in I, then f has a continuous inverse on I and $y = g(t) = g(f^{-1}(x))$. So y is a function of x and, by the chain rule,

$$\frac{dy}{dt} = \frac{dy}{dx}\frac{dx}{dt};$$

hence

(10.5)
$$\frac{dy}{dx} = \frac{dy/dt}{dx/dt} = \frac{g'(t)}{f'(t)}.$$

EXAMPLE 1   Find the slope of the line tangent to the curve described by the parametric equations

$$x = 1 + \ln t, \qquad y = 2 + \ln \sqrt{t}, \qquad 0 < t < \infty.$$

SOLUTION
We saw in Example 2 of the previous section that these parametric equations describe a line with slope 1/2. This agrees with the result from equation (10.5) since

$$\frac{dy}{dx} = \frac{dy/dt}{dx/dt} = \frac{\left(\dfrac{1}{\sqrt{t}}\right)\left(\dfrac{1}{2}t^{-1/2}\right)}{\dfrac{1}{t}} = \frac{1}{2},$$

for any positive value of t.   ☐

EXAMPLE 2   Find an equation of the tangent line to the curve described by $x = 3t + t^3$, $y = t^3 - 9t^2$ at the point when $t = 1$.

SOLUTION
The slope of the tangent line at any point is given by the derivative:

$$\frac{dy}{dx} = \frac{3t^2 - 18t}{3 + 3t^2} = \frac{t^2 - 6t}{1 + t^2}.$$

When $t = 1$, the slope is $-5/2$ and the point on the curve is $(4, -8)$. Consequently, an equation of the tangent line is

$$y + 8 = -\frac{5}{2}(x - 4) \qquad \text{or} \qquad 5x + 2y - 4 = 0.$$ ☐

EXAMPLE 3   Sketch the curve described by $x = t^2$, $y = t^3 - 3t$, $-\infty < t < \infty$.

SOLUTION

We will use the first derivative to determine when the curve is increasing and decreasing, and the second derivative to determine the concavity of the curve.

$$\frac{dy}{dx} = \frac{3t^2 - 3}{2t} = \frac{3(t - 1)(t + 1)}{2t}$$

This derivative is zero when $t = 1$ and $t = -1$, and the curve has horizontal tangent lines at these points. When $t = 0$, dy/dx is undefined and the tangent line at this point is vertical. The second derivative,

$$\frac{d^2y}{dx^2} = \frac{d\left(\frac{dy}{dx}\right)}{dx} = \frac{\frac{d(dy/dx)}{dt}}{\frac{dx}{dt}} = \frac{D_t\left(\frac{3t^2 - 3}{2t}\right)}{D_t t^2} = \frac{\frac{3}{2}(1 + t^{-2})}{2t} = \frac{3}{4}\left(\frac{t^2 + 1}{t^3}\right)$$

is positive when $t > 0$ and negative when $t < 0$. The results obtained from the derivatives are summarized below.

When t is in	dy/dx is	d^2y/dx^2 is	The graph is
$(\infty, -1)$	negative	negative	decreasing, concave downward
$(-1, 0)$	positive	negative	increasing, concave downward
$(0, 1)$	negative	positive	decreasing, concave upward
$(1, \infty)$	positive	positive	increasing, concave upward

By setting $y = 0$ and solving for t, we see that the curve intersects the x-axis when $t = 0$ and when $t = \pm\sqrt{3}$. The graph intersects the y-axis only when $t = 0$.

Plotting the points for $t = -2, -1, 0, 1, 2$ and using the information in the table, we obtain the graph shown in Figure 10.38. The arrows on the curve indicate the direction of increasing values of t.

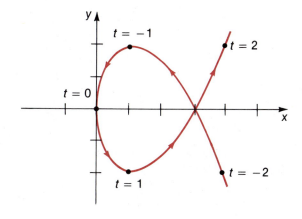

FIGURE 10.38

Suppose parametric equations for a curve are given in terms of polar coordinates,

$$x = r(\theta) \cos \theta \qquad \text{and} \qquad y = r(\theta) \sin \theta.$$

In this case, θ is the parameter and

$$\frac{dy}{dx} = \frac{\dfrac{dy}{d\theta}}{\dfrac{dx}{d\theta}} = \frac{D_\theta[r(\theta) \sin \theta]}{D_\theta[r(\theta) \cos \theta]}$$

$$= \frac{r'(\theta) \sin \theta + r(\theta) \cos \theta}{r'(\theta) \cos \theta - r(\theta) \sin \theta},$$

or, dividing by $\cos \theta$, if $\cos \theta \neq 0$,

(10.6)
$$\frac{dy}{dx} = \frac{\dfrac{dr}{d\theta} \tan \theta + r}{\dfrac{dr}{d\theta} - r \tan \theta}.$$

EXAMPLE 4   Find the slope of the line tangent to the spiral with polar equation

$$r = \ln \theta, \qquad \theta \geq 1,$$

at the end of one rotation about the origin.

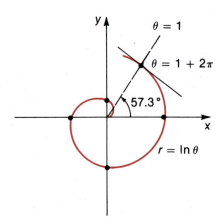

FIGURE 10.39

SOLUTION

The spiral is shown in Figure 10.39. One rotation is completed when $\theta = 1 + 2\pi$ and $r = \ln(1 + 2\pi)$. Since

$$\frac{dr}{d\theta} = \frac{1}{\theta},$$

$$\frac{dy}{dx} = \frac{\dfrac{\tan \theta}{\theta} + r}{\dfrac{1}{\theta} - r \tan \theta} = \frac{\tan \theta + \theta \ln \theta}{1 - \theta(\ln \theta)(\tan \theta)}.$$

When $\theta = 1 + 2\pi$, the slope of the line tangent to the spiral is

$$\frac{dy}{dx} = \frac{\tan(1 + 2\pi) + (1 + 2\pi)\ln(1 + 2\pi)}{1 - (1 + 2\pi)\ln(1 + 2\pi)\tan(1 + 2\pi)} \approx -.74429. \qquad \square$$

EXAMPLE 5   Find a cartesian equation of the line tangent to the cardioid $r = 1 - \cos \theta$ at the point with polar coordinates $(3/2, 2\pi/3)$.

SOLUTION

Using (10.6), the slope of the tangent line at any point (r, θ) is

$$\frac{dy}{dx} = \frac{\sin \theta \tan \theta + 1 - \cos \theta}{\sin \theta - (1 - \cos \theta)\tan \theta}.$$

So the slope at $\left(\dfrac{3}{2}, 2\pi/3\right)$ is

$$\frac{dy}{dx}\left(\text{when } \theta = \frac{2\pi}{3}\right) = \frac{\dfrac{\sqrt{3}}{2}(-\sqrt{3}) + 1 + \dfrac{1}{2}}{\dfrac{\sqrt{3}}{2} - \left(1 + \dfrac{1}{2}\right)(-\sqrt{3})} = 0$$

and the tangent line is horizontal. The cartesian coordinates of the point with polar coordinates $\left(\dfrac{3}{2}, \dfrac{2\pi}{3}\right)$ are

$$x = \frac{3}{2}\cos\frac{2\pi}{3} = -\frac{3}{4}$$

and

$$y = \frac{3}{2}\sin\frac{2\pi}{3} = \frac{3\sqrt{3}}{4}.$$

Consequently, the cartesian equation of the tangent line is

$$y = \frac{3\sqrt{3}}{4}. \qquad \square$$

EXERCISE SET 10.5

In Exercises 1 through 16 find the slope, if it exists, of the line tangent to the curve at the specified value of t. Check this result with the sketch of the curve that was determined in the corresponding exercise in Exercise Set 10.4.

1. $x = 3t, y = \dfrac{1}{2}t, t = 1$ **2.** $x = 2t + 1, y = 3t - 2, t = -2$

3. $x = \sqrt{t}, y = t + 1, t = 4$ **4.** $x = t + 2, y = -3\sqrt{t}, t = 9$

5. $x = \sin t, y = (\cos t)^2, t = \dfrac{\pi}{3}$ **6.** $x = \sec t, y = \tan t, t = \dfrac{\pi}{4}$

7. $x = 3 \sin t, y = 4 \cos t, t = \pi$ **8.** $x = 3 \sec t, y = 4 \tan t, t = \dfrac{\pi}{4}$

9. $x = e^t, y = e^{-t}, t = 0$ **10.** $x = \ln t, y = \ln \sqrt{t}, t = e$

11. $x = \sin t, y = \sec t, t = \dfrac{\pi}{4}$ **12.** $x = \sin t, y = \cot t, t = \dfrac{\pi}{2}$

13. $x = 4 + 2 \cos t, y = 6 + 2 \sin t, t = \dfrac{\pi}{2}$

14. $x = \sin t + 1, y = 2 \cos t - 1, t = \pi$

15. $x = 4t^3, y = 2t^2 + 1, t = -1$

16. $x = e^{-t} - 2, y = e^{2t} + 3, t = 1$

In Exercises 17 through 30 find the slope, if it exists, of the line tangent to the polar curve at the specified value of θ. Find the cartesian equation of the tangent line.

17. $r = 1, \theta = \dfrac{\pi}{4}$ **18.** $r = 2, \theta = \dfrac{\pi}{3}$

19. $r = 2 \sin \theta, \theta = \dfrac{\pi}{4}$ **20.** $r = -3 \cos \theta, \theta = \dfrac{\pi}{4}$

21. $r = 1 + \cos \theta, \theta = \dfrac{\pi}{3}$ **22.** $r = 2 + \sin \theta, \theta = 0$

23. $r = 3 \sin 2\theta, \theta = \dfrac{\pi}{4}$ **24.** $r = 4 \cos 2\theta, \theta = \dfrac{\pi}{4}$

25. $r^2 = 4 \cos 2\theta, \theta = \dfrac{\pi}{6}$ **26.** $r^2 = 16 \sin 4\theta, \theta = \dfrac{\pi}{16}$

27. $r = \theta, \theta = \dfrac{5\pi}{4}$ **28.** $r = \ln \theta, \theta = e$

29. $r = 2^{-\theta}, \theta = -2$ **30.** $r = 3 \sec \theta, \theta = \dfrac{\pi}{4}$

In Exercises 31 through 40 find $D_x^2 y$.

31. $x = 3t, y = \dfrac{1}{2} t$ **32.** $x = 2t + 1, y = 3t - 2$

33. $x = 3 \sin t, y = 4 \cos t$ **34.** $x = \sec t, y = \tan t$

35. $x = e^t, y = e^{-t}$ **36.** $x = 4t^3, y = 2t^2 + 1$

37. $r = \sin \theta$ **38.** $r = 2 + \cos \theta$

39. $r = 2^{-\theta}$ **40.** $r = \ln \theta$

A pair of intersecting polar curves are listed in Exercises 41 through 46. Find equations of the tangent lines to the curves at the points of intersection.

41. $r = 1, r = 1 + \cos \theta$ **42.** $r = 1 + \cos \theta, r = 1 - \cos \theta$

43. $r = 1 + \cos\theta$, $r = 1 + \sin\theta$; $0 \le \theta \le \dfrac{\pi}{2}$

44. $r = 3$, $r = 2 + 2\sin\theta$ **45.** $r = \sin\theta$, $r = \cos\theta$

46. $r = \sin 2\theta$, $r = \cos 2\theta$, $0 \le \theta \le \dfrac{\pi}{2}$

47. Find the values of θ that produce horizontal and vertical tangents to the graph of $r = e^\theta$.

48. The parametric equations

$$x = a\tan\theta, \qquad y = a(\cos\theta)^2$$

describe a curve known as the witch of Agnesi. (See the Historical Note.) Express y in terms of x and sketch this curve.

49. To connect a circular segment of a roadbed for a highway or railroad to a straight segment, engineers often use a cubic spiral of the form $r = a_3\theta^3 + a_2\theta^2 + a_1\theta + a_0$. The constants are chosen so that in the accompanying figure, both the segments and their tangents agree with the connecting piece. Find an equation of the cubic spiral shown in the figure.

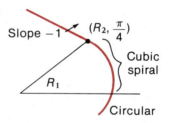

50. Solve the general problem of finding the connecting spiral between a circle of radius R_1 and a line of angle α from the end of the circular segment having a slope m.

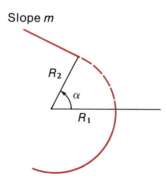

HISTORICAL NOTE   This name is derived from a mistranslation. Maria Gaetana Agnesi (1718–1799) discussed this curve in her calculus book *Instituzoni Analitiche*, published in 1748. This was the first comprehensive calculus text published after the work of L'Hôpital. The curve was called a *versiera*, a word derived from the Latin *vertere*, "to turn," but also an abbreviation for the Italian word *avversiera* meaning "wife of the devil." When Maria's text was translated into English, the word *versiera* was rendered as witch.

10.6
LENGTHS OF CURVES

In Section 6.5, we derived a formula for determining the length of a curve described by $y = f(x)$, $a \le x \le b$, provided the function f has a continuous derivative on $[a, b]$. In this chapter we have seen that many curves, such as cardioids, spirals, and cycloids, can be described by polar and parametric equations when y cannot conveniently be written as a function of x. In this section we derive a formula to find the length of these curves.

Suppose C is a curve described by the parametric equations

$$x = f(t), \qquad y = g(t),$$

where f' and g' are continuous on $[a, b]$ and never simultaneously zero. Such a curve is said to be **smooth**. If $\mathcal{P} = \{t_0, t_1, \ldots, t_n\}$ is a partition of $[a, b]$, then the straight line segments joining the points $(f(t_{i-1}), g(t_{i-1}))$ and $(f(t_i), g(t_i))$ for each $i = 1, 2, \ldots, n$, approximate the curve C, as shown in Figure 10.40. The sum of their lengths approximates the length of C.

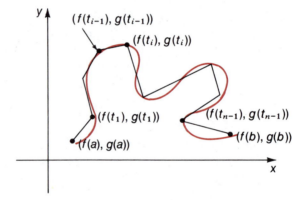

FIGURE 10.40

Since the length of the ith line segment is

$$\sqrt{[f(t_i) - f(t_{i-1})]^2 + [g(t_i) - g(t_{i-1})]^2},$$

the total length of the straight line segments is

$$\Delta L_i = \sum_{i=1}^{n} \sqrt{[f(t_i) - f(t_{i-1})]^2 + [g(t_i) - g(t_{i-1})]^2}$$

$$= \sum_{i=1}^{n} \left\{ \left[\frac{f(t_i) - f(t_{i-1})}{t_i - t_{i-1}} \right]^2 + \left[\frac{g(t_i) - g(t_{i-1})}{t_i - t_{i-1}} \right]^2 \right\}^{1/2} \Delta t_i.$$

Since both f and g satisfy the hypotheses of the mean value theorem, there is a w_i in $[t_{i-1}, t_i]$ such that

$$\frac{f(t_i) - f(t_{i-1})}{t_i - t_{i-1}} = f'(w_i)$$

and there is a z_i in $[t_{i-1}, t_i]$ such that

$$\frac{g(t_i) - g(t_{i-1})}{t_i - t_{i-1}} = g'(z_i).$$

Consequently, the length of the curve is approximated by

$$\sum_{i=1}^{n} \{[f'(w_i)]^2 + [g'(z_i)]^2\}^{1/2} \, \Delta t_i,$$

where w_i and z_i are in $[t_{i-1}, t_i]$ for each i. This is not a Riemann sum because of the possible, and even likely, fact that $w_i \neq z_i$. However, a result known as **Duhamel's principle for integrals** can be used to show that

$$\int_a^b \{[f'(t)]^2 + [g'(t)]^2\}^{1/2} \, dt = \lim_{\Delta t \to 0} \sum_{i=1}^{n} \{[f'(w_i)]^2 + [g'(z_i)]^2\}^{1/2} \, \Delta t_i,$$

provided that f' and g' are continuous on $[a, b]$. (The theorem of G. Bliss referred to in Section 6.5 is a special case of Duhamel's principle.)

(10.7)
DEFINITION

If C is a curve described by

$$x = f(t), \qquad y = g(t), \qquad a \leq t \leq b,$$

then the length of C is

$$L_C = \int_a^b \{[f'(t)]^2 + [g'(t)]^2\}^{1/2} \, dt,$$

provided the integral exists.

When the integral in Definition 10.7 exists, C is called a **rectifiable** curve. It follows from the fundamental theorem of calculus that any smooth curve is rectifiable.

If h is a function with a continuous derivative on $[a, b]$ and $y = h(x)$, then the parametric equations

$$x = t \qquad y = h(t),$$

describe a curve C whose length, by Definition 10.7, is

$$L_C = \int_a^b \{[1]^2 + [h'(t)]^2\}^{1/2} \, dt$$

$$= \int_a^b \sqrt{1 + [h'(x)]^2} \, dx.$$

This shows that Definition 10.7 reduces to the definition of the length of a curve given in Section 6.5 when y is expressed as a function of x.

EXAMPLE 1

Find the length of one arch of the cycloid with parametric equations

$$x = t - \sin t$$
$$y = 1 - \cos t.$$

SOLUTION

The graph of the cycloid is as shown on page 603 as Figure 10.33 with $R = 1$. As t varies from 0 to 2π, the corresponding point $P(t)$ on the cycloid moves from the point $(0, 0)$ to the point $(2\pi, 0)$.

Hence,
$$L = \int_0^{2\pi} \{[D_t(t - \sin t)]^2 + [D_t(1 - \cos t)]^2\}^{1/2} \, dt$$

$$= \int_0^{2\pi} [(1 - \cos t)^2 + (\sin t)^2]^{1/2} \, dt$$

$$= \int_0^{2\pi} [1 - 2\cos t + (\cos t)^2 + (\sin t)^2]^{1/2} \, dt$$

$$= \int_0^{2\pi} [2 - 2\cos t]^{1/2} \, dt = \int_0^{2\pi} 2\sin\left(\frac{t}{2}\right) dt$$

$$= -4\cos\frac{t}{2}\Bigg]_0^{2\pi} = 8. \qquad \Box$$

When C is a curve described by a polar equation
$$r = f(\theta), \qquad \alpha \le \theta \le \beta,$$

the parametric equations
$$x = f(\theta)\cos\theta$$
$$y = f(\theta)\sin\theta, \qquad \alpha \le \theta \le \beta$$

also describe C and

$$L_C = \int_\alpha^\beta \{[D_\theta(f(\theta)\cos\theta)]^2 + [D_\theta(f(\theta)\sin\theta)]^2\}^{1/2} \, d\theta$$

$$= \int_\alpha^\beta \{[f'(\theta)\cos\theta - f(\theta)\sin\theta]^2 + [f'(\theta)\sin\theta + f(\theta)\cos\theta]^2\}^{1/2} \, d\theta$$

$$= \int_\alpha^\beta \{[f'(\theta)]^2 (\cos\theta)^2 - 2f'(\theta)f(\theta)\cos\theta\sin\theta + [f(\theta)]^2 (\sin\theta)^2$$
$$+ [f'(\theta)]^2 (\sin\theta)^2 + 2f'(\theta)f(\theta)\sin\theta\cos\theta + [f(\theta)]^2 (\cos\theta)^2\}^{1/2} \, d\theta$$

$$= \int_\alpha^\beta \{[f'(\theta)]^2 + [f(\theta)]^2\}^{1/2} \, d\theta,$$

or, since $r = f(\theta)$,

(10.8)
$$L_C = \int_\alpha^\beta \left[r^2 + \left(\frac{dr}{d\theta}\right)^2\right]^{1/2} \, d\theta.$$

EXAMPLE 2   Find the length of the cardioid $r = 2(1 + \cos \theta)$.

SOLUTION
    This cardioid, as shown in Figure 10.41, is symmetric with respect to the x-axis, so

$$L = 2 \int_0^\pi \{4(1 + \cos \theta)^2 + (-2 \sin \theta)^2\}^{1/2} \, d\theta$$

$$= 2 \int_0^\pi \{4[1 + 2 \cos \theta + (\cos \theta)^2] + 4(\sin \theta)^2\}^{1/2} \, d\theta$$

$$= 2 \int_0^\pi \{4 + 8 \cos \theta + 4[(\cos \theta)^2 + (\sin \theta)^2]\}^{1/2} \, d\theta$$

$$= 2 \int_0^\pi (8 + 8 \cos \theta)^{1/2} \, d\theta$$

$$= 4\sqrt{2} \int_0^\pi (1 + \cos \theta)^{1/2} \, d\theta = 8 \int_0^\pi \cos \frac{\theta}{2} \, d\theta$$

$$= 16 \sin \frac{\theta}{2} \bigg]_0^\pi = 16. \qquad \square$$

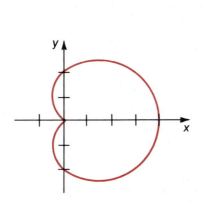

FIGURE 10.41

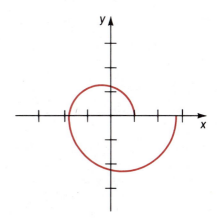

FIGURE 10.42

EXAMPLE 3   Find the length of the spiral

$$r(\theta) = e^{\theta/2\pi}, \qquad 0 \le \theta \le 2\pi.$$

SOLUTION
    The spiral is shown in Figure 10.42.

$$L = \int_0^{2\pi} \left[(e^{\theta/2\pi})^2 + \left(\frac{1}{2\pi} e^{\theta/2\pi} \right)^2 \right]^{1/2} d\theta$$

$$= \int_0^{2\pi} \left[e^{\theta/\pi} + \frac{1}{4\pi^2} e^{\theta/\pi} \right]^{1/2} d\theta$$

$$= \int_0^{2\pi} \left[e^{\theta/\pi} \left(1 + \frac{1}{4\pi^2} \right) \right]^{1/2} d\theta = \left(1 + \frac{1}{4\pi^2} \right)^{1/2} \int_0^{2\pi} e^{\theta/2\pi} d\theta$$

$$= \left[\left(\frac{4\pi^2 + 1}{4\pi^2} \right)^{1/2} (2\pi e^{\theta/2\pi}) \right]_0^{2\pi}$$

$$= (4\pi^2 + 1)^{1/2} (e - 1) \approx 10.93216. \qquad \square$$

Although the integrals in these examples can be easily determined, there are many curves for which the exact value of the length is very difficult to determine. Recall from Section 6.5 that this is also true for curves described by cartesian equations.

EXERCISE SET 10.6

Find the length of each curve described by the parametric equations in Exercises 1 through 10.

1. $x = 3t - 1, y = 2t + 4, 1 \leq t \leq 3$
2. $x = 1 - t, y = 3t + 4, -1 \leq t \leq 5$
3. $x = t^3 + 1, y = 3t^2, 0 \leq t \leq 3$
4. $x = 3t^2 + 2, y = 4t^2 - 1, -1 \leq t \leq 2$
5. $x = t^2 + 1, y = 2t - 3, 0 \leq t \leq 1$
6. $x = \ln \sin 2t, y = 2t + 3, \dfrac{\pi}{6} \leq t \leq \dfrac{\pi}{3}$
7. $x = t + 1, y = \ln \cos t, 0 \leq t \leq \dfrac{\pi}{4}$
8. $x = e^{-t} \cos t, y = e^{-t} \sin t, 0 \leq t \leq 2\pi$
9. $x = e^t \sin t, y = e^t \cos t, 0 \leq t \leq 2\pi$
10. $x = \cos t + t \sin t, y = t \cos t - \sin t, 0 \leq t \leq \pi$

Find the length of each curve described by the polar equations in Exercises 11 through 16.

11. $r = 3, 0 \leq \theta \leq \pi$ 12. $r = 3\theta, 0 \leq \theta \leq \pi$
13. $r = e^{2\theta}, 0 \leq \theta \leq 1$ 14. $r = 3^{-\theta}, -1 \leq \theta \leq 1$
15. $r = 1 - \cos \theta, 0 \leq \theta \leq 2\pi$ 16. $r = 1 + \sin \theta, 0 \leq \theta \leq 2\pi$

17. Sketch the graphs of the parametric equations given below for $0 \leq t \leq \dfrac{\pi}{2}$ and find the length of each curve.
    (a) $x = \cos t, y = \sin t$ (b) $x = (\cos t)^2, y = (\sin t)^2$
    (c) $x = (\cos t)^3, y = (\sin t)^3$

18. Sketch the graphs of the polar equations given below for $0 \le \theta \le \pi$ and find the length of each curve.

(a) $r = \cos \theta$ (b) $r = \left(\cos \dfrac{\theta}{2} \right)^2$ (c) $r = \left(\cos \dfrac{\theta}{3} \right)^3$

19. Find the length of one arch of the cycloid

$$x = R(t - \sin t), \qquad y = R(1 - \cos t).$$

20. In Exercise 40 of Section 10.2 (p. 592), the equation $r = 2e^{.2\theta}$ was used to describe the outside of the shell of the chambered nautilus pictured with that exercise. Find the total length of the outside of the shell as a function of θ.

21. The valve stem of a tire that is 27 inches in diameter is located 7 inches from the center of the tire. How far does the valve stem travel per mile of tire travel?

22. The study of light diffraction at a rectangular aperture involves the Fresnel integrals $C(t) = \int_0^t \cos (\pi w^2/2)dw$ and $S(t) = \int_0^t \sin (\pi w^2/2)dw$, named in honor of the French physicist Augustin Jean Fresnel (1788–1827). These integrals cannot be evaluated directly, but Marie Alfred Cornu (1841–1902) used the parametric representation $x = C(t), y = S(t)$ to produce the curves, known as *Cornu spirals*, shown in the accompanying figure.

(a) Find the length of the Cornu spiral for t in (t_1, t_2).

(b) Find the slope of the line tangent to the spiral at t_1.

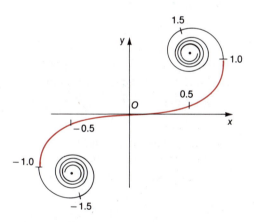

23. How much recording tape of one mil (.001 inch) thickness is required to fill a reel with inside diameter 2 inches and outside diameter 7 inches? (*Hint*: Make the assumption that on the first revolution the thickness varies linearly from zero to its actual thickness .001 inch and then remains .001 inch thick.)

REVIEW EXERCISES

Sketch the graphs of the curves described in Exercises 1 through 9.

1. $r = 3$ **2.** $r = 2 \cos \theta$

3. $r = 3 \sin 2\theta$ **4.** $r^2 = 16 \sin 4\theta$

5. $x = 2 \cos \theta, y = 3 \sin \theta, 0 \le \theta \le 2\pi$ **6.** $x = 2t + 1, y = 3t + 7$

7. $x = e^t, y = 1 + e^{-t}$ **8.** $x = \cos 2t, y = \sin t$

9. $x = 3(\sin t)^2 + 2(\cos t)^2, y = 2(\sin t)^2 + (\cos t)^2, 0 \le t \le \dfrac{\pi}{2}$

Sketch the graph of each curve described in Exercises 10 through 15 and find the slope of the tangent line to the curve at the indicated point.

10. $x = t^2, y = t; t = 2$ **11.** $x = -t^2, y = t + 1; t = -1$

12. $x = \ln t, y = \dfrac{1}{2}\left(t - \dfrac{1}{t}\right); t = 1$ **13.** $x = \ln t, y = \dfrac{1}{2}\left(t + \dfrac{1}{t}\right); t = e$

14. $r = 2 \cos \theta, \left(\sqrt{2}, \dfrac{\pi}{4}\right)$ **15.** $r^2 = \cos \theta, \left(\dfrac{\sqrt{2}}{2}, \dfrac{\pi}{3}\right)$

16. Sketch the graphs of the following polar equations, and find the length of each curve.

   (a) $r = 2 + 2 \cos \theta$ (b) $r = 2 - 2 \cos \theta$

   (c) $r = 2 \cos \theta - 2$ (d) $r = -2 \cos \theta - 2$

17. Find a polar equation for the curve described by the parametric equations:

$$x = \frac{\cos \theta}{\theta}, \qquad y = \frac{\sin \theta}{\theta}.$$

18. Sketch the graphs of the following polar equations.

   (a) $r \sin \theta = 3$ (b) $r \cos \theta = 3$

   (c) $r \sin \theta = -3$ (d) $r \cos \theta = -3$

19. (a) Sketch the graphs of the cardioid $r = 2(1 - \cos \theta)$ and the circle $r = -6 \cos \theta$ and find their points of intersection.

   (b) Find the area of the region inside the cardioid and outside the circle.

   (c) Find the area of the region inside the circle and outside the cardioid.

   (d) Find the area of the region inside both the cardioid and the circle.

20. Find the points on the cardioid $r = 2(1 - \cos \theta)$ at which the tangent line is (a) perpendicular to the polar axis, (b) parallel to the polar axis.

21. (a) Find $\lim\limits_{\theta \to \pi^+/2} \tan \theta$, $\lim\limits_{\theta \to \pi^-/2} \tan \theta$.

   (b) If $r = \tan \theta$, find $dr/d\theta$.

   (c) Sketch the graph of the polar equation $r = \tan \theta$.

22. Give a set of parametric equations that describe the cartesian equation $16x^2 + 9y^2 = 1$.

Sketch the graphs of the equations in Exercises 23 through 26 and find the corresponding cartesian equations.

23. $r = \sec \theta$ **24.** $r = \dfrac{1}{1 + \cos \theta}$

25. $r(4 \cos \theta - 2 \sin \theta) = 6$ **26.** $r = \dfrac{2}{1 - \sin \theta}$

27. The position of a particle at time t is given by $x = \sin t, y = \cos t$. Find the distance the particle moves when $0 \le t \le \pi/2$.

11
CONIC SECTIONS

In this chapter we systematically examine the graphs associated with equations of the form

$$Ax^2 + Bxy + Cy^2 + Dx + Ey + F = 0,$$

where $A, B, C, D, E,$ and F are constants.

We have seen several examples of equations of this type, for example, the parabola with equation $x^2 - y = 0$ and the circle with equation $x^2 + y^2 = 1$. The graphs of these equations are called **conic sections**, or simply **conics**, because they describe figures that result from intersecting a plane with a double-napped right circular cone. A double-napped cone with its component parts labeled is shown in Figure 11.1.

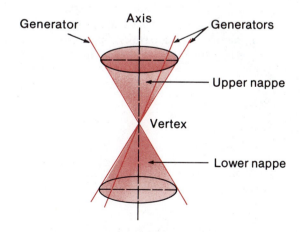

FIGURE 11.1

In general, three distinct figures can be produced by such an intersection: a **parabola**, if the plane intersects only one nappe of the cone and is parallel to a line lying entirely in the cone, called a *generator* of the cone (see Figure 11.2(a)); an **ellipse**, if the plane intersects only one nappe of the cone but is not parallel to a generator (see Figure 11.2(b)); and a **hyperbola,** if the plane intersects both nappes of the cone (see Figure 11.2(c)).

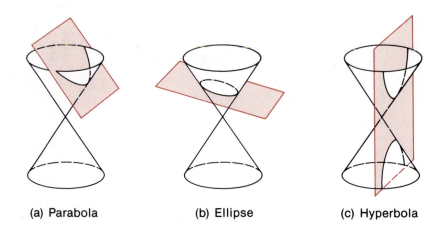

(a) Parabola   (b) Ellipse   (c) Hyperbola

FIGURE 11.2

Figures such as a point, circle, line, or pair of intersecting lines can also be produced, but these can be considered special or *degenerate* cases of the parabola, ellipse, and hyperbola.

We will not use this three-dimensional description of conic sections, but will instead define the conic sections in terms of distances from fixed lines and points.

One reason for discussing conic sections at this time is in preparation for the graphing of solids and surfaces in space. Describing three-dimensional objects with a graph that, by necessity, is two dimensional can be quite difficult. The usual practice is to determine the shape of the cross sections of the objects, and many common three-dimensional objects have cross sections that are conic sections.

11.1 PARABOLAS

(11.1) DEFINITION

A **parabola** is the set of points in a plane that are equidistant from a given point, called the **focus** or **focal point**, and a given line, called the **directrix**.

The line through the focal point of the parabola perpendicular to the directrix is called the **axis** of the parabola. The intersection of the parabola and the axis is called the **vertex**. See Figure 11.3.

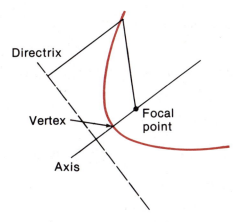

FIGURE 11.3

Since we have discussed parabolas in the past without referring to this definition, it is important to establish that the definition describes the same figures that have been previously discussed. Suppose an xy-coordinate system is superimposed so that the vertex of a parabola is at the origin and the focal point lies on the y-axis at the point $(0, f)$. Then the directrix has equation $y = -f$. Figure 11.4 shows a parabola when $f > 0$ and when $f < 0$.

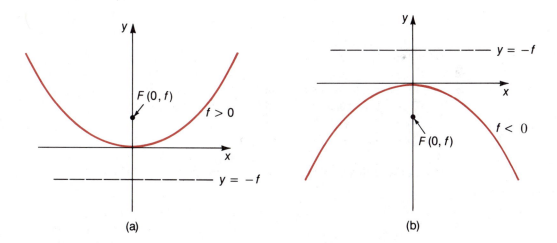

FIGURE 11.4

The distance from an arbitrary point (x, y) on the graph of the parabola to the focal point $(0, f)$ is the same as the distance from (x, y) to the line $y = -f$; that is,

$$d((x, y), (0, f)) = d((x, y), (x, -f)).$$

So
$$\sqrt{(x - 0)^2 + (y - f)^2} = \sqrt{(x - x)^2 + (y + f)^2}$$

and
$$x^2 + (y - f)^2 = (y + f)^2.$$

Simplifying, this becomes

$$x^2 + y^2 - 2yf + f^2 = y^2 + 2yf + f^2,$$

so the parabola with focal point $(0, f)$ and directrix $y = -f$ has equation

(11.2) $$y = \frac{1}{4f}x^2.$$

EXAMPLE 1   Find the focal point and directrix of the parabola with equation $y = x^2$.

SOLUTION

The focal point of this parabola lies at $(0, f)$, where $1/4f = 1$, so $f = 1/4$. The equation of the directrix is consequently $y = -1/4$. The graph of the parabola is given in Figure 11.5.

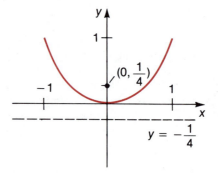

FIGURE 11.5   □

In a similar manner it can be shown that a parabola with focal point at $(f, 0)$ and directrix $x = -f$ is described by the equation

(11.3) $$x = \frac{1}{4f}y^2.$$

Parabolas of this type are shown in Figure 11.6.

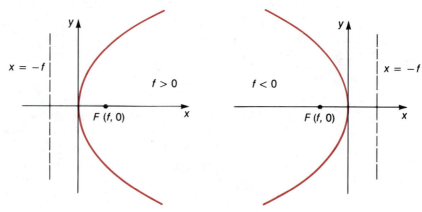

FIGURE 11.6

A parabola with equation either of form (11.2) or of form (11.3) is said to be in **standard position**.

EXAMPLE 2   Find an equation of the parabola with focal point at $(-2, 0)$ and directrix $x = 2$.

SOLUTION

The focal point lies along the x-axis. Since the vertex is the midpoint of the line segment from the focal point perpendicular to the directrix, the vertex is at the origin and the parabola is in standard position. An equation of this parabola is

$$x = \frac{1}{4(-2)} y^2 \qquad \text{or} \qquad y^2 = -8x.$$

The graph of this parabola is shown in Figure 11.7.

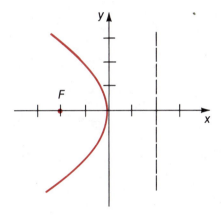

FIGURE 11.7

By applying the graphing techniques illustrated in Chapter 1, translations can be used to graph any parabola whose directrix is parallel to one of the coordinate axes.

EXAMPLE 3   Find the focal point and directrix of the parabola with equation $2y^2 + 8y + x + 11 = 0$.

SOLUTION

To compare this equation with one of the standard forms, we complete the square in y:

$$2(y^2 + 4y + 4) + x + 11 - 8 = 0$$

or $\qquad\qquad 2(y + 2)^2 + x + 3 = 0,$

so $\qquad\qquad x + 3 = -2(y + 2)^2.$

Compare this equation to the equation

$$x = -2y^2,$$

which describes a parabola with focal point at $(-1/8, 0)$ and directrix $x = 1/8$. The graph of $x + 3 = -2(y + 2)^2$ is a translation of the graph of $x = -2y^2$: down 2 units and left 3 units. Consequently, the focal point of the parabola $x + 3 = -2(y + 2)^2$ is at $(-3 - 1/8, -2) = (-25/8, -2)$ and the directrix of this parabola is $x = -3 + 1/8 = -23/8$. Figure 11.8 shows the graph of both parabolas.

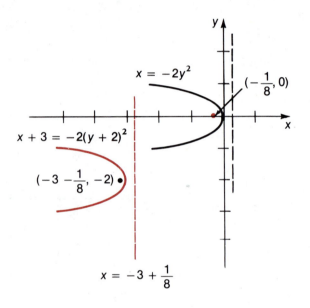

FIGURE 11.8

Find an equation of the parabola with focal point at $(1, 5)$ and directrix $y = -1$.

SOLUTION

The location of the directrix and focal point, as shown in Figure 11.9, indicates that the axis of this parabola is the line $x = 1$. The vertex is the point on the axis midway between the focal point $(1, 5)$ and the point where the directrix and axis of the parabola intersect, $(1, -1)$. Consequently, the vertex is at $(1, 2)$.

If the vertex is translated to the origin, each point is moved down 2 units and to the left 1 unit. Thus, the focal point is translated to $(0, 3)$ and the directrix to $y = -3$. An equation of this "translated" parabola is

$$y = \frac{1}{4(3)} x^2 = \frac{1}{12} x^2.$$

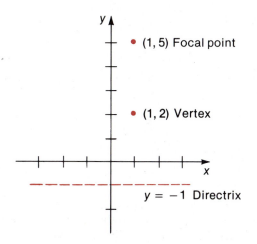

The original parabola has equation

$$(y - 2) = \frac{1}{12}(x - 1)^2 \quad \text{or} \quad y = \frac{1}{12}(x - 1)^2 + 2.$$

Both parabolas are shown in Figure 11.10.

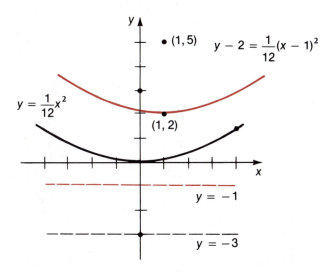

An important application of parabolas follows from a basic law of physics, which states that the angle of reflection of sound, light, or any reflected object hitting a surface, is the same as the angle of incidence of the object. The angles are measured with respect to the tangent line to the surface. (See Figure 11.11.)

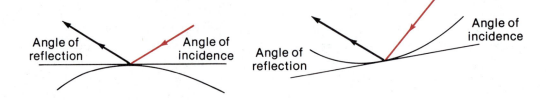

FIGURE 11.11

We will show that when an object is emitted from the focal point of a parabola to the surface of the parabola, the angle of reflection is the same as the angle formed by the tangent line and axis of the parabola. Consequently, an object emitted from the focal point of a parabola to its surface is reflected along a line parallel to the axis of the parabola.

Suppose we consider a parabola with equation $x = \dfrac{1}{4f} y^2$ containing the arbitrary point (x_1, y_1), as shown in Figure 11.12.

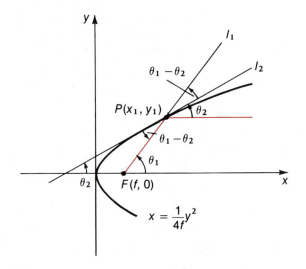

FIGURE 11.12

Let l_1 denote the line joining the focal point $(f, 0)$ and the point (x_1, y_1) and l_2 denote the tangent line to the parabola at (x_1, y_1). If θ_1 is the angle formed by the intersection of l_1 and the x-axis, then

$$\text{slope of } l_1 = \tan \theta_1 = \frac{y_1 - 0}{x_1 - f} = \frac{y_1}{x_1 - f}.$$

If θ_2 is the angle formed by the intersection of l_2 and the x-axis, then

$$\text{slope of } l_2 = \tan \theta_2 = \frac{dy}{dx}(\text{at } (x_1, y_1)).$$

Differentiating $x = \dfrac{y^2}{4f}$, we have

$$1 = \frac{y}{2f}\frac{dy}{dx} \quad \text{so} \quad \frac{dy}{dx} = \frac{2f}{y}$$

and

$$\text{slope of } l_2 = \tan\theta_2 = \frac{2f}{y_1}.$$

To show that the reflection is parallel to the x-axis, we must show that the angle $\theta_1 - \theta_2$ is equal to the angle θ_2. Since both θ_2 and $\theta_1 - \theta_2$ are between 0 and π, we can show that the two angles are equal by showing that $\tan(\theta_1 - \theta_2) = \tan\theta_2$.

$$\tan(\theta_1 - \theta_2) = \frac{\tan\theta_1 - \tan\theta_2}{1 + \tan\theta_1 \tan\theta_2}$$

$$= \frac{\dfrac{y_1}{x_1 - f} - \dfrac{2f}{y_1}}{1 + \dfrac{y_1}{x_1 - f}\dfrac{2f}{y_1}} = \frac{\dfrac{y_1^2 - 2x_1 f + 2f^2}{y_1(x_1 - f)}}{\dfrac{x_1 + f}{x_1 - f}}$$

$$= \frac{y_1^2 - 2x_1 f + 2f^2}{y_1(x_1 + f)} = \frac{4fx_1 - 2x_1 f + 2f^2}{y_1(x_1 + f)}$$

$$= \frac{2f}{y_1} = \tan\theta_2.$$

Consequently, any object emitted from the focal point of a parabola is reflected in a line parallel to the axis of the parabola.

It follows that if a parabola is rotated about its axis to form a hollow shell, an object that strikes the inside of the shell when emitted from the focal point is reflected outward parallel to the axis, as illustrated in Figure 11.13.

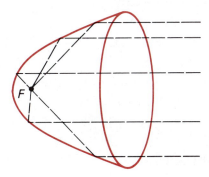

FIGURE 11.13

Reflectors in flashlights and automobile headlights use this property of parabolas. The light source is not always placed precisely at the focal point, but more often is placed between the focal point and the vertex with the location depending on the amount of spread desired for the emitted light.

In the reverse manner, incoming rays of light or sound that are parallel to the axis of a parabola are reflected to the focal point. This property is used in the design of parabolic microphones and solar collectors. For example, the microphone used by television broadcasters to relay the conversation in a football huddle is of parabolic design, as are the receivers used to pick up television signals from orbiting satellites.

A parabola is also the shape a suspended cable assumes when it hangs between two fixed points and supports a uniformly distributed load. Bridges built on this principle are called **suspension bridges**. Most of the world's longest and most famous bridges are of this type: for example, the Golden Gate in San Francisco, Verrazano-Narrows joining Brooklyn and Staten Island in New York, and the Mackinac Straits Bridge in upper Michigan.

The path of an object propelled into the air also assumes the shape of a parabola, a fact that artillery experts have known for centuries. (This topic is considered in more detail in Section 13.4.)

EXERCISE SET 11.1

Sketch the graph of each parabola given in Exercises 1 through 16 and find the vertex, focal point, and equation of the directrix.

1. $y = 2x^2$

2. $16y = 9x^2$

3. $y = -2x^2$

4. $9y = -16x^2$

5. $y^2 = 2x$

6. $16y^2 = 9x$

7. $y^2 = -2x$

8. $9y^2 = -16x$

9. $x^2 + 4x + 4 = 2y$

10. $x^2 + 6x + 9 - y = 0$

11. $y^2 - 8y + 12 = 2x$

12. $y^2 + 6y + 6 - 3x = 0$

13. $2x^2 + 4x - 9y + 20 = 0$

14. $3x^2 - 12x - 4y + 8 = 0$

15. $9y^2 - 36y - 2x + 34 = 0$

16. $4y^2 + 8y - 3x + 10 = 0$

In Exercises 17 through 24, find an equation of a parabola that satisfies the stated conditions.

17. Focus $(-2, 2)$, directrix $y = -2$.

18. Focus $(-2, 2)$, directrix $x = 2$.

19. Focus $(-2, 2)$, directrix $y = -1$.

20. Focus $(-2, 2)$, directrix $x = -4$.

21. Vertex $(-2, 2)$, directrix $x = 4$.

22. Vertex $(-2, 2)$, focus $(-2, 0)$.

23. Vertex $(3, 4)$, focus $(3, 6)$.

24. Vertex $(3, 4)$, directrix $y = 6$.

25. Find an equation of the parabola with axis the y-axis and vertex $(0, 0)$ that passes through the point $(4, 6)$.

26. Find an equation of the parabola with axis parallel to the y-axis and vertex $(1, 2)$ that passes through the point $(5, 8)$. (*Hint*: modify the result obtained in Exercise 25.)

27. Find an equation of the parabola with axis the x-axis and vertex $(0, 0)$ that passes through the point $(4, 6)$.

28. Find an equation of the parabola with axis parallel to the x-axis and vertex $(1, 2)$ that passes through the point $(5, 8)$.

29. Find an equation of the parabola with axis the y-axis and vertex $(0, 0)$ that passes through the point (x_1, y_1), where $x_1 \neq 0$, $y_1 \neq 0$. Use this result to find an equation of the parabola with axis parallel to the y-axis and vertex (h, k) that passes through (x_1, y_1) where $x_1 \neq h$, $y_1 \neq k$.

30. (a) Find an equation of the parabola with axis the x-axis and vertex $(0, 0)$ that passes through the point (x_1, y_1), where $x_1 \neq 0$, $y_1 \neq 0$.

   (b) Use the result of (a) to find an equation of the parabola with axis parallel to the x-axis and vertex (h, k) that passes through (x_1, y_1), where $x_1 \neq h$, $y_1 \neq k$.

31. Show that the vertex of a parabola is the point on the parabola that is closest to the focal point.

32. Find the focal point of the parabola with axis the y-axis, passing through $(-1, 3)$ and having a tangent with slope 2 at this point.

33. Find a general form for the equation of a parabola with axis the y-axis and passing through $(1, 1)$. Of these parabolas, which have the property that the normal line at $(1, 1)$ is tangent to $y = x^2$ at this point?

34. Consider the pair of parabolas $y = ax^2$ and $x = ay^2$. Show that the angle formed by the tangent lines at their intersection is independent of a.

35. Consider the pair of parabolas $y = a - bx^2$ and $y = -a + bx^2$, where a and b are positive constants. What must be true of a and b if the tangent line to one parabola at their point of intersection is normal to the other parabola?

36. The *latus rectum* of a parabola is the line segment that passes through the focus perpendicular to the axis and joins two points on the curve.

   (a) Show that the length of the latus rectum is four times the distance from the vertex to the focal point.

   (b) Show that the area of the region bounded by the graph of the parabola and the latus rectum is one sixth the square of the length of the latus rectum.

37. Find an equation of the parabola with axis parallel to the y-axis and passing through $(1, 0)$, $(0, 1)$ and $(2, 2)$.

38. A driving light has a parabolic cross section with a depth of $2''$ and a cross-section height of $4''$. Where should the light source be placed to produce a parallel beam of light?

39. A ball thrown horizontally from the top edge of a building follows a parabolic curve with vertex at the top edge of the building and axis along the side of the building. The ball passes through a point 100 feet from the building when it is a vertical distance of 16 feet from the top.

(a) How far from the building will the ball land if the building is 64 ft high?

(b) How far will the ball travel before it hits the ground?

(c) Suppose instead that the ball is thrown from the top of the Sears tower in Chicago, the world's tallest building with a height of 1450 feet. Recompute the answers to the questions.

40. The Coliseum at the University of Georgia shown in the accompanying figure encloses six acres of floor area under a roof supported by two diagonal parabolic arches. The arches span 384 feet between fan-shaped columns at the edge of the roof. The buttresses continue beyond these columns for 63.5 feet to their foundation. The clear height of the building at the intersection of the arches is 75 feet and the arches are 3 feet thick at this point. Find the height of the top of the arches at the point where the columns meet the edge of the roof.

41. The world's largest hangar is the Goodyear Airdock, built in 1929 in Akron, Ohio to house and service the rigid airships USS Akron and USS Macon. The main structure of this hangar is a 1175-foot long cylinder whose cross-section is parabolic with height 211 feet and width 325 feet. What is the volume enclosed by this structure?

42. The longest bridge on the North American continent is the Verrazano Narrows bridge in New York City. This is a suspension bridge with a 4260 ft span and twin supporting towers standing about 700 feet above the water. The distance between the towers is 2627 feet and the low point on the supporting cables is 225 feet above the water. What is the length of a supporting cable between the two towers?

43. The George Washington bridge crossing the Hudson river in New York City has a main span of 3500 feet and has cables that make an angle of approximately 20° with its supporting towers. Use this information to approximate the sag in the cables. The sag in the cables is about 5/8 of the height of the towers above the water. Find this height.

44. The world's largest reflecting telescope is called the Hale telescope in honor of the American astronomer George Ellery Hale (1868–1939). It is at the Palomar Mountain Observatory 45 miles northeast of San Diego, California. The main parabolic mirror of this telescope is 200 inches in diameter and has a depth from rim to vertex of 3.75 inches. A small cylindrical platform is located within the tube of the telescope along the axis of the parabola for an observer to view and record the reflection from the telescope. How far from the center of the mirror is the observer's viewing area located?

45. A satellite is placed in a position to make a parabolic flight past the moon with the center of the moon at the focal point of the parabola. When the satellite is 5783 kilometers from the surface of the moon, it makes an angle of 60° with the axis of the parabola. The closest the satellite gets to the surface is 143 kilometers. What is the diameter of the moon? (Assume that the gravitational center of the moon is at its center; this is the focus of the parabola. This assumption is not quite correct since the gravitational center is offset approximately 2 kilometers from the center toward the earth.)

Putnam exercises:

46. Find the length of the shortest chord that is normal to the parabola $y^2 = 2ax$ at one end of the chord. (This is a problem in the first William Lowell Putnam examination given on April 16, 1938.)

47. Consider the two mutually tangent parabolas $y = x^2$ and $y = -x^2$. [These have foci at $(0, 1/4)$ and $(0, -1/4)$, and directrices $y = -1/4$ and $y = 1/4$, respectively.] The upper parabola rolls without slipping around the fixed lower parabola. Find the locus of the focal point of the moving parabola.

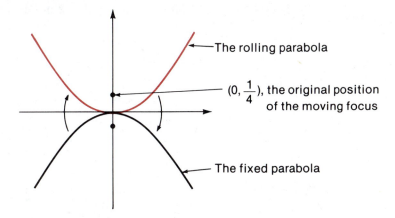

The rolling parabola

$(0, \frac{1}{4})$, the original position of the moving focus

The fixed parabola

(This exercise was problem A–5 of the thirty-fifth William Lowell Putnam examination given on December 7, 1974. The examination and its solution can be found in the November 1975 issue of the *American Mathematical Monthly*, pages 907–912.)

11.2 ELLIPSES

(11.4)
DEFINITION

An **ellipse** is the set of points in the plane the sum of whose distances from two fixed points is a given constant.

The fixed points are called the **focal points** of the ellipse, the line through these points is called the **axis** of the ellipse, and the points where the ellipse intersects the axis are called **vertices** of the ellipse. These are illustrated in Figure 11.14.

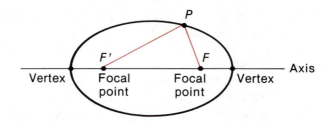

FIGURE 11.14

Standard equations for an ellipse can be derived by placing an xy-coordinate system with the x-axis along the axis of the ellipse and the y-axis along the perpendicular bisector of the line segment joining the focal points. The focal points can then be assigned coordinates $(f, 0)$ and $(-f, 0)$.

For reasons of convenience, the sum of the distances from the focal points to points on the ellipse (shown by dashed lines in Figure 11.15) is denoted by the constant $2a$. Any point (x, y) on the ellipse has the property that

$$d((x, y), (f, 0)) + d((x, y), (-f, 0)) = 2a.$$

So
$$\sqrt{(x - f)^2 + (y - 0)^2} + \sqrt{(x + f)^2 + (y - 0)^2} = 2a$$

and
$$\sqrt{(x - f)^2 + y^2} = 2a - \sqrt{(x + f)^2 + y^2}.$$

Squaring both sides, we have

$$(x - f)^2 + y^2 = 4a^2 - 4a\sqrt{(x + f)^2 + y^2} + (x + f)^2 + y^2,$$

which simplifies to

$$a + \frac{f}{a}x = \sqrt{(x + f)^2 + y^2}.$$

Squaring again gives

$$a^2 + 2fx + \frac{f^2}{a^2}x^2 = (x + f)^2 + y^2,$$

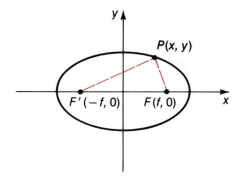

FIGURE 11.15

which can be simplified to

$$x^2 \left[\frac{f^2 - a^2}{a^2} \right] - y^2 = f^2 - a^2.$$

Dividing both sides by $f^2 - a^2$, we have

$$\frac{x^2}{a^2} + \frac{y^2}{a^2 - f^2} = 1.$$

The sum of the lengths of two sides of the triangle shown in Figure 11.15 with vertices $(-f, 0)$, $(f, 0)$ and (x, y) is $2a$. Since $2f$ is the length of the third side, $2a > 2f$ so $a > f$. Consequently, $a^2 - f^2 > 0$. For convenience we replace this constant by a new constant

$$b^2 = a^2 - f^2.$$

The equation of the ellipse then becomes

(11.5) $$\frac{x^2}{a^2} + \frac{y^2}{b^2} = 1.$$

The graph of this equation is symmetric with respect to the x-axis and the y-axis. It has y-intercepts $(0, b)$ and $(0, -b)$ and x-intercepts at the vertices $(a, 0)$ and $(-a, 0)$. It is shown in Figure 11.16. Note that for any ellipse $a > \sqrt{a^2 - f^2} = b$.

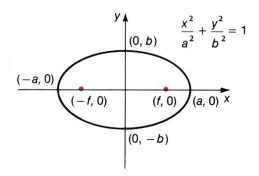

FIGURE 11.16

The line segment from $(-a, 0)$ to $(a, 0)$ is called the **major axis** of the ellipse, the line segment from $(0, -b)$ to $(0, b)$ is called the **minor axis**.

If the focal points are placed along the y-axis at $(0, f)$ and $(0, -f)$, the ellipse has equation

(11.6)
$$\frac{y^2}{a^2} + \frac{x^2}{b^2} = 1.$$

An ellipse with equation in either form (11.5) or (11.6) is said to be in **standard position**.

Sketch the graph of the ellipse with equation $9x^2 + 16y^2 = 144$ and find its focal points.

SOLUTION

Rewriting the equation as

$$\frac{x^2}{16} + \frac{y^2}{9} = 1,$$

we see that $a^2 = 16$ and $b^2 = 9$, so $a = 4$ and $b = 3$. The graph is symmetric with respect to both coordinate axes and has intercepts $(4, 0)$, $(-4, 0)$, $(0, 3)$, and $(0, -3)$. Figure 11.17 shows this graph.

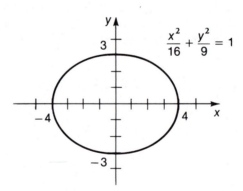

FIGURE 11.17

The focal points of this ellipse are at $(f, 0)$ and $(-f, 0)$, where f satisfies the equation $b^2 = a^2 - f^2$. Hence, $9 = 16 - f^2$ and $f = \sqrt{7}$.   □

EXAMPLE 2   Sketch the graph of the ellipse with equation $16x^2 + 9y^2 = 144$ and find its focal points.

SOLUTION

This equation can be rewritten as

$$\frac{x^2}{9} + \frac{y^2}{16} = 1.$$

This is an ellipse of the form of (11.6), because the denominator of y^2 exceeds that of x^2. As in Example 1, $a^2 = 16$ and $b^2 = 9$, so $a = 4$ and $b = 3$. The intercepts are $(3, 0)$, $(-3, 0)$, $(0, 4)$, and $(0, -4)$ and the focal points lie along the y-axis at $(0, \sqrt{7})$ and $(0, -\sqrt{7})$. The graph is shown in Figure 11.18. Note the similarity between this ellipse and the ellipse in Example 1.

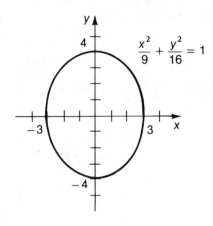

FIGURE 11.18

EXAMPLE 3   Find an equation of the ellipse in standard position that has a vertex at $(5, 0)$ and a focal point at $(3, 0)$.

SOLUTION

Since the ellipse is in standard position, the other focal point must be at $(-3, 0)$ and the other vertex at $(-5, 0)$.

Since $a = 5$, $a^2 = 25$ and $b^2 = a^2 - f^2 = 25 - 9 = 16$,

the ellipse has equation

$$\frac{x^2}{25} + \frac{y^2}{16} = 1 \quad \text{or} \quad 16x^2 + 25y^2 = 400.$$

EXAMPLE 4   Sketch the graph of the equation $9x^2 - 72x + 4y^2 + 16y + 124 = 0$.

SOLUTION

To compare this equation with the equation of an ellipse in standard form, we complete the square in both x and y:

$$9(x^2 - 8x + 16) + 4(y^2 + 4y + 4) + 124 - 9(16) - 4(4) = 0.$$

This simplifies to

$$9(x - 4)^2 + 4(y + 2)^2 = 36 \quad \text{or} \quad \frac{(x - 4)^2}{4} + \frac{(y + 2)^2}{9} = 1.$$

Compare this equation to the equation in standard form

$$\frac{x^2}{4} + \frac{y^2}{9} = 1,$$

which describes an ellipse with intercepts $(2, 0)$, $(-2, 0)$, $(0, 3)$, and $(0, -3)$ and focal points at $(0, \sqrt{5})$ and $(0, -\sqrt{5})$. The ellipse in standard position is shown in Figure 11.19.

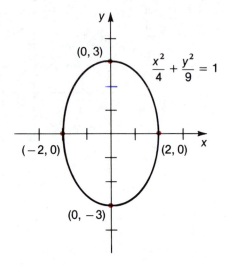

FIGURE 11.19

The graph of the original equation is simply a translation of the graph of the ellipse in standard form, $x^2/4 + y^2/9 = 1$, down 2 units and 4 units to the right. Consequently, the vertices of the ellipse are at $(4, 1)$ and $(4, -5)$ and the focal points are at $(4, -2 + \sqrt{5})$ and $(4, -2 - \sqrt{5})$. The graph is shown in Figure 11.20.

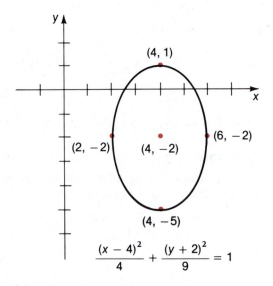

FIGURE 11.20

An ellipse has a reflection property similar to the reflection property of a parabola. In this case, however, any object emitted from one of the focal points is reflected from the curve through the other focal point. (The proof of this fact is considered in Exercise 30.) If the object continues through the second focal point and is again reflected from the curve, it returns to the original focal point.

This reflection property is most evident in "whispering galleries," rooms that have elliptical ceilings. The Mormon Tabernacle in Salt Lake City is built on this design (see Exercise 40), as is Statuary Hall in the United States Capital building in Washington D.C. In fact, markers are placed at the focal points in Statuary Hall and visitors can verify this reflection property. A person standing at one focal point can clearly hear the whisper of a person standing at the other focal point, even when the room is quite crowded.

The reflection property of an ellipse has also been used recently in the design of small billiard tables. The playing surface of the table is in the form of an ellipse with holes located at the focal points. Generally there are obstacles placed in the center of the table so that a bank, or reflection, shot must be performed to go from one end of the table to the hole in the other end. Consider how the intelligent player would design such a shot.

The ellipse also plays a vital role in astronomy, since the orbit of each planet is an ellipse with the sun at one of its focal points. This fact is known as **Kepler's First Law** (the Law of Ellipses). It was first announced by Johann Kepler (1571–1630) in 1609, following a decade of work based on observations performed by the Dane astronomer, Tycho Brahe (1546–1601). There are two other laws of planetary motion that were formulated by Kepler. These laws are considered in Section 13.6.

EXERCISE SET 11.2

Sketch the graph of each ellipse in Exercises 1 through 14 and find the vertices and focal points.

1. $\dfrac{x^2}{4} + \dfrac{y^2}{9} = 1$ **2.** $\dfrac{x^2}{9} + \dfrac{y^2}{4} = 1$

3. $\dfrac{x^2}{25} + \dfrac{y^2}{16} = 1$ **4.** $25x^2 + 16y^2 = 400$

5. $3x^2 + 2y^2 = 6$ **6.** $4x^2 + 3y^2 = 12$

7. $4x^2 + y^2 = 1$ **8.** $x^2 + 4y^2 = 1$

9. $4x^2 + y^2 + 16x + 7 = 0$ **10.** $16x^2 + 9y^2 - 54y - 63 = 0$

11. $x^2 + 4y^2 - 2x - 16y + 13 = 0$

12. $4x^2 + 9y^2 - 16x + 90y + 97 = 0$

13. $3x^2 + 2y^2 - 18x + 4y + 28 = 0$

14. $2x^2 + 5y^2 + 8x - 10y - 27 = 0$

In Exercises 15 through 20, find an equation of the ellipse that satisfies the stated conditions.

15. Foci at $(\pm 2, 0)$, vertices at $(\pm 3, 0)$.

16. Foci at $(\pm 2, 0)$, y-intercepts at $(0, \pm 2)$.

17. Foci at $(0, \pm 1)$, x-intercepts at $(\pm 2, 0)$.

18. Foci at $(3, 0)$ and $(1, 0)$, a vertex at $(0, 0)$.

19. Vertices at $(2, 2)$ and $(6, 2)$, a focal point at $(5, 2)$.

20. Foci at $(3, 3)$ and $(3, -1)$, passing through $(4, 0)$.

21. Find the area bounded by the ellipse $9x^2 + 4y^2 = 36$.

22. Show that the area bounded by the ellipse $x^2/a^2 + y^2/b^2 = 1$ is πab.

23. Find an equation of the line tangent to the ellipse $9x^2 + 4y^2 = 36$ at $\left(1, \dfrac{3\sqrt{3}}{2}\right)$.

24. Show that an equation of the line tangent to the ellipse with equation $x^2/a^2 + y^2/b^2 = 1$ at the point (x_0, y_0) can be written in the form

$$\frac{x\, x_0}{a^2} + \frac{y\, y_0}{b^2} = 1.$$

25. Find an equation of a line tangent to the ellipse $4x^2 + y^2 = 4$ that passes through the point $(3, 0)$.

26. Show that a line tangent to the ellipse with equation $x^2/a^2 + y^2/b^2 = 1$ and passing through $(x_0, 0)$, where $x_0 > a$, intersects the ellipse at $x = a^2/x_0$.

27. Find the area of the largest rectangle with sides parallel to the x- and y-axes that can be inscribed within the ellipse having equation $x^2/a^2 + y^2/b^2 = 1$.

28. Find the area of the largest isosceles triangle with a vertex at $(0, b)$ and base parallel to the x-axis that can be inscribed within the ellipse $x^2/a^2 + y^2/b^2 = 1$.

29. Show that the triangle found in Exercise 28 has the same area as the largest isosceles triangle with a vertex at $(a, 0)$ and base parallel to the y-axis that can be inscribed within $x^2/a^2 + y^2/b^2 = 1$.

30. Show that if an object is emitted from one of the focal points of an ellipse, it is reflected from the curve through the other focal point.

31. A *latus rectum* of an ellipse is a line segment that passes through a focus perpendicular to the major axis and joins the points on the ellipse. Find the length of a latus rectum of the ellipse with equation $x^2/a^2 + y^2/b^2 = 1$.

32. It appears that the closest and farthest points on an ellipse from the center of the ellipse lie at the intersection of the ellipse with the coordinate axes. Use calculus to show that this is true.

33. Consider the equation $Ax^2 + Cy^2 + Dx + Ey + F = 0$, where A and C are positive constants. Find conditions on the constants A, C, D, E, and F that will ensure that this equation describes

(a) an ellipse,

(b) a single point,

(c) no points in the plane.

34. Describe the curve traced by a point 2 feet from the top of a ladder 8 feet long as the bottom of the ladder moves away from a vertical wall.

35. Olympic stadium in Montreal, Canada is constructed in the shape of an ellipse with major and minor axes of 480 and 280 meters respectively. Find an equation of this ellipse. How much area is covered by this stadium?

36. Lou's Knolls Fruit Market sells two varieties of equally delicious watermelons. One type, called the Crimson melon, costs $3.99 and is 14 inches long and 12 inches wide, the other costs $2.99 and is 20 inches long and 9 inches wide. They both have elliptical cross sections when cut lengthwise.

(a) Which melon contains more volume?

(b) Suppose the rind on each melon is 1-inch thick. Which melon is the better buy?

37. The largest rigid airships ever built, the U.S. Navy ships *USS Akron* and *USS Macon*, were 785 feet long and had a gas capacity of 6,500,000 cubic feet. Assuming the cross sections of the ships were elliptical when cut lengthwise, what was their width? How much surface material is contained on these ships?

38. The Goodyear airships *Columbia*, *America*, and *Europa* are each 192 feet long and 50 feet wide. A brochure available from Goodyear Aerospace in Akron, Ohio states that the volume contained in one of the blimps is 202,700 cubic feet. Is this figure reasonable?

39. Halley's comet is named to honor Edmund Halley (1652–1742), the friend and contemporary of Isaac Newton who persuaded him to publish his *Principia Mathematica*, the work in which Newton described the laws of motion. In 1682 Halley determined the orbit of the comet that bears his name, and predicted its return 76 years later. The orbit is elliptical with a major axis of length 36.2 A.U. (A.U. is an abbreviation for Astronomical Unit. 1 A.U. $\approx 9.25 \times 10^7$ mi) and a minor axis of length 9.1 A.U., with the sun at one focus. How close does the comet pass to the sun? (The comet will next be at this point, called the *perihelion* of its orbit, on April 29, 1986. It will be visible from earth during periods before and after this date.)

40. The Mormon Tabernacle in Salt Lake City, Utah was built between 1863 and 1867. The tabernacle is 250 feet long, 150 feet wide and 80 feet high and built in the shape of a whispering gallery with its cross section in the form of half an ellipse. Determine where the focal points of the ellipse lie.

Putnam exercise:

41. Of all ellipses inscribed in a square, show that the circle has the maximum perimeter.
(This exercise was problem A–4 of the thirty-third William Lowell Putnam examination given on December 2, 1972. The examination and its solution can be found in the November 1973 issue of the *American Mathematical Monthly*, pages 1019–1028.

11.3
HYPERBOLAS

(11.7)
DEFINITION

A **hyperbola** is the set of points in the plane, the difference of whose distances from two fixed points has a constant magnitude.

The fixed points are called the **focal points** of the hyperbola; the line through the focal points is called the **axis** of the hyperbola; and the points where the hyperbola intersects the axis are called the **vertices** of the hyperbola. Figure 11.21 illustrates these terms.

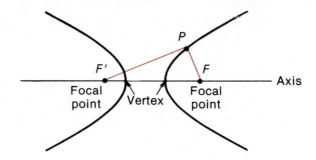

FIGURE 11.21

Standard equations for the hyperbola can be derived in the same manner as for the ellipse. In fact, if we let the x-axis be the axis of the hyperbola with focal points at $(f, 0)$ and $(-f, 0)$, as shown in Figure 11.22, and let the constant magnitude of the difference be denoted by $2a$, the equation derived is the same as that of the ellipse:

$$\frac{x^2}{a^2} + \frac{y^2}{a^2 - f^2} = 1.$$

However, in this case,

$$2a = \left| d((x, y), (f, 0)) - d((x, y), (-f, 0)) \right|.$$

If $(f, 0)$ is the focal point closest to (x, y), then

$$d((x, y), (-f, 0)) > d((x, y), (f, 0))$$

and

$$2a = d((x, y), (-f, 0)) - d((x, y), (f, 0))$$

so

$$2a + d((x, y), (f, 0)) = d((x, y), (-f, 0)).$$

Since $d((x, y), (-f, 0))$ is the length of one side of the triangle shown in Figure 11.22 with vertices $(-f, 0)$, $(f, 0)$, and (x, y) and the sum of the other two sides is $2f + d((x, y), (f, 0))$. This implies that

$$2a + d((x, y), (f, 0)) = d((x, y), (-f, 0) \leq 2f + d((x, y), (f, 0))$$

and that $a < f$. Since $0 < a < f$, $a^2 - f^2 < 0$. Letting

$$b^2 = f^2 - a^2,$$

the equation of a hyperbola in standard form is

(11.8) $$\frac{x^2}{a^2} - \frac{y^2}{b^2} = 1.$$

   This hyperbola is symmetric with respect to both coordinate axes and has
x-intercepts at $(a, 0)$ and $(-a, 0)$. It does not intersect the y-axis. In fact, since

$$y^2 = \frac{b^2}{a^2}(x^2 - a^2),$$

the graph does not intersect the region where $-a < x < a$.

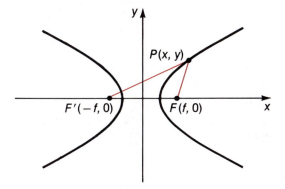

FIGURE 11.22

   Suppose we restrict the graph of the hyperbola in (11.8) to the first quadrant
and consider the difference between the y-coordinate on the line with equation
$y = \dfrac{b}{a}x$ and the y-coordinate on the hyperbola. This difference is

$$\frac{b}{a}x - \frac{b}{a}\sqrt{x^2 - a^2}$$

and

$$\lim_{x \to \infty}\left(\frac{b}{a}x - \frac{b}{a}\sqrt{x^2 - a^2}\right) = \lim_{x \to \infty}\frac{b}{a}(x - \sqrt{x^2 - a^2})\left(\frac{x + \sqrt{x^2 - a^2}}{x + \sqrt{x^2 - a^2}}\right)$$

$$= \lim_{x \to \infty}\frac{b}{a}\left(\frac{x^2 - (x^2 - a^2)}{x + \sqrt{x^2 - a^2}}\right)$$

$$= \frac{b}{a}\lim_{x \to \infty}\frac{a^2}{x + \sqrt{x^2 - a^2}} = 0.$$

Consequently, the hyperbola approaches the line $y = \dfrac{b}{a}x$ as x increases. The

line $y = \dfrac{b}{a}x$ is an **asymptote** of the hyperbola. By the symmetry of the graph of the hyperbola, the graph in the other quadrants must approach either $y = \dfrac{b}{a}x$ or $y = -\dfrac{b}{a}x$. This implies that the rectangle that "surrounds" the ellipse with equation $x^2/a^2 + y^2/b^2 = 1$ can be used to sketch the graph of the hyperbola with equation $x^2/a^2 - y^2/b^2 = 1$, as shown in Figure 11.23.

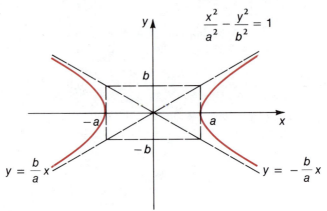

FIGURE 11.23

In a similar manner, the hyperbola with focal points on the y-axis at $(0, f)$ and $(0, -f)$ has equation

(11.9)
$$\frac{y^2}{a^2} - \frac{x^2}{b^2} = 1,$$

has vertices at the y-intercepts $(0, a)$ and $(0, -a)$, and has asymptotes $y = \pm\dfrac{a}{b}x$. Since $x^2 = \dfrac{b^2(y^2 - a^2)}{a^2}$, the graph of this hyperbola does not intersect the region where $-a < y < a$. The graph is shown in Figure 11.24.

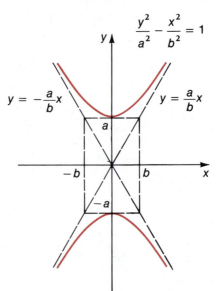

FIGURE 11.24

A hyperbola with equation in either form (11.8) or (11.9) is said to be in **standard position**.

EXAMPLE 1   Sketch the graph of the hyperbola with equation $16x^2 - 9y^2 = 144$ and find its focal points.

SOLUTION

The equation can be rewritten

$$\frac{x^2}{9} - \frac{y^2}{16} = 1,$$

so $a = 3$ and $b = 4$. The graph intersects the x-axis at $(3, 0)$ and $(-3, 0)$ and has asymptotes $y = \frac{4}{3}x$ and $y = -\frac{4}{3}x$. This is sufficient information to sketch the graph shown in Figure 11.25.

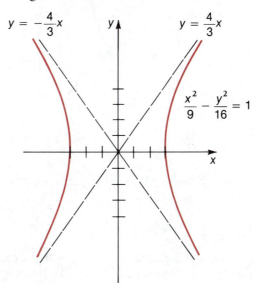

Since $b^2 = f^2 - a^2, f = \sqrt{a^2 + b^2} = \sqrt{9 + 16} = 5$ and the focal points are at $(5, 0)$ and $(-5, 0)$.   □

EXAMPLE 2   Sketch the graph of $9y^2 - 16x^2 = 144$.

SOLUTION

The given equation can be rewritten

$$\frac{y^2}{16} - \frac{x^2}{9} = 1$$

and the asymptotes to the graph are $y = \pm \frac{4}{3}x$, the same as the asymptotes in Example 1. In this example, however, the vertices are on the y-axis at $(0, 4)$ and

$(0, -4)$. The graph is as shown in Figure 11.26 together with the graph of the hyperbola described in Example 1: $x^2/9 - y^2/16 = 1$. These hyperbolas are called **conjugate hyperbolas**.

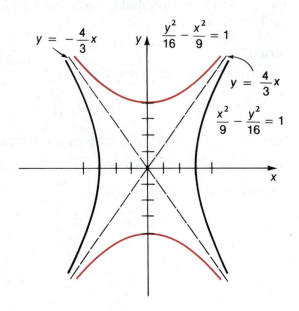

FIGURE 11.26   ☐

EXAMPLE 3   Sketch the graph of the hyperbola with equation

$$y^2 - 2y - 9x^2 + 36x = 39.$$

SOLUTION

To graph the equation, we show that its graph is the translation of the graph of a hyperbola in standard form. The first step is to complete the square in both x and y:

$$(y^2 - 2y + 1) - 9(x^2 - 4x + 4) = 39 + 1 - 36,$$

so

$$(y - 1)^2 - 9(x - 2)^2 = 4,$$

and

$$\frac{(y - 1)^2}{4} - \frac{(x - 2)^2}{\dfrac{4}{9}} = 1.$$

Consequently, $a = 2$, $b = 2/3$ and this hyperbola is a translation of the hyperbola in standard form with equation

$$\frac{y^2}{4} - \frac{x^2}{\dfrac{4}{9}} = 1.$$

The hyperbola in standard form has asymptotes $y = 3x$ and $y = -3x$; its graph is shown in Figure 11.27.

$$\frac{y^2}{4} - \frac{x^2}{\frac{4}{9}} = 1$$

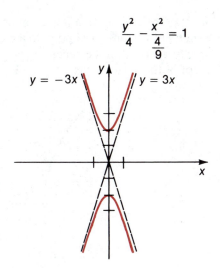

FIGURE 11.27

The graph of

$$\frac{(y-1)^2}{4} - \frac{(x-2)^2}{\frac{4}{9}} = 1$$

is consequently the graph in Figure 11.27 translated up one unit and to the right two units. The asymptotes of

$$\frac{(y-1)^2}{4} - \frac{(x-2)^2}{\frac{4}{9}} = 1$$

are

$$y - 1 = 3(x - 2) \quad \text{and} \quad y - 1 = -3(x - 2),$$

that is

$$y = 3x - 5 \quad \text{and} \quad y = -3x + 7.$$

The graph of $y^2 - 2y - 9x^2 + 36x = 39$ is shown in Figure 11.28.

$$\frac{(y-1)^2}{4} - \frac{(x-2)^2}{\frac{4}{9}} = 1$$

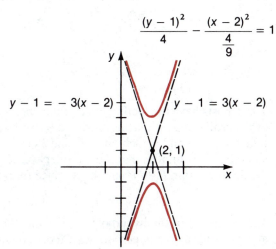

FIGURE 11.28

The hyperbola also has a reflection property involving its focal points. An object emitted from one focal point to the curve is reflected from the curve on a line directly away from the second focal point, as illustrated in Figure 11.29. (The proof of this result is discussed in Exercise 38.)

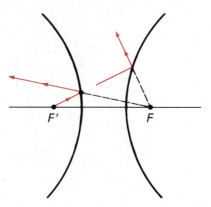

FIGURE 11.29

An interesting application of this reflection property of the hyperbola concerns spotting the position of an object when its relative distance from stationary points is known. For example, suppose an artillery piece is at an unknown location and that observation posts are located at points P_1, P_2, and P_3. (See Figure 11.30.)

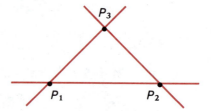

FIGURE 11.30

These posts have the facility to determine precisely the time at which the sound of the artillery piece reaches them, so the difference between the length of time it takes the sound to reach posts P_1 and P_2 is known and is constant in the given situation. This implies that if the artillery piece is located at P, the difference in the distance from P_1 to P and the distance from P_2 to P is constant:

$$d(P_1, P) - d(P_2, P) = c.$$

The definition of the hyperbola implies that the artillery piece must be located along a branch of a hyperbola with focal points P_1 and P_2, as shown in Figure 11.31.

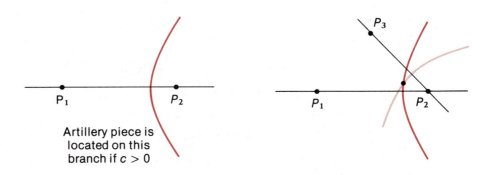

FIGURE 11.31   FIGURE 11.32

In a similar manner, the difference between the length of time it takes the sound to reach the posts P_2 and P_3 fixes the artillery piece along a branch of the hyperbola with focal points P_2 and P_3, as in Figure 11.32. The intersection of these hyperbolas determines the location P of the artillery piece.

The reverse technique is used in air and sea navigation by using three fixed reference points that emit radio signals at precise time intervals. By knowing the speed at which the signal travels, the differences between the times the three signals are received can be used to pinpoint the position of an object. This technique is used in the **LORAN** (**Lo**ng **Ra**nge **N**avigation) system of ship navigation and by the Omega system of air navigation.

EXERCISE SET 11.3

In Exercises 1 through 18, sketch the graph of the hyperbola, find the coordinates of the vertices and focal points, and write equations of the asymptotes.

1. $\dfrac{x^2}{4} - \dfrac{y^2}{9} = 1$

2. $\dfrac{x^2}{9} - \dfrac{y^2}{4} = 1$

3. $\dfrac{y^2}{4} - \dfrac{x^2}{9} = 1$

4. $\dfrac{y^2}{9} - \dfrac{x^2}{4} = 1$

5. $16x^2 - 4y^2 = 64$

6. $16y^2 - 4x^2 = 64$

7. $x^2 - y^2 = 1$

8. $y^2 - 4x^2 = 1$

9. $y^2 - 2x^2 = 8$

10. $x^2 - 4y^2 = 100$

11. $x^2 + 2x - 4y^2 = 3$

12. $9y^2 - 18y - 4x^2 = 27$

13. $3x^2 - y^2 = 6x$

14. $2y^2 + 8y = 9x^2$

15. $9x^2 - 4y^2 - 18x - 8y = 31$

16. $x^2 - 4y^2 - 2x - 16y = 19$

17. $4x^2 - 3y^2 + 8x + 18y = 11$

18. $4y^2 - 4x^2 + 32x + 8y = 56$

In Exercises 19 through 30, find an equation of the hyperbola that satisfies the stated conditions.

19. Foci at $(\pm 5, 0)$, vertices at $(\pm 3, 0)$.

20. Foci at $(0, \pm 13)$, vertices at $(0, \pm 12)$.

21. Foci at $(0, \pm 5)$, vertices at $(0, \pm 4)$.

22. Foci at $(\pm 13, 0)$, vertices at $(\pm 5, 0)$.

23. Foci at $(-1, 4)$ and $(5, 4)$, a vertex at $(0, 4)$.

24. A focus at $(2, 1)$, vertices at $(2, 0)$ and $(2, -3)$.

25. Foci at $(\pm 3\sqrt{2}, 0)$, passing through $(5, 4)$.

26. Vertices at $(2, \pm 3)$, passing through $(8, 8)$.

27. Foci $(\pm 3, 0)$, equations of asymptotes $y = \pm 3x/4$.

28. Vertices at $(\pm 3, 0)$, equations of asymptotes $y = \pm 2x/3$.

29. Vertices at $(6, 1)$ and $(-2, 1)$, equations of asymptotes $y = 3x/4 - 1/2$ and $y = -3x/4 + 5/2$.

30. Foci at $(6, 1)$ and $(-2, 1)$, equations of asymptotes $y = 3x/4 - 1/2$ and $y = -3x/4 + 5/2$.

31. Find an equation of a line tangent to the hyperbola with equation $3x^2 - 4y^2 = 12$ at the point $(2\sqrt{2}, \sqrt{3})$.

32. Show that an equation of the line tangent to the hyperbola with equation $x^2/a^2 - y^2/b^2 = 1$ at the point (x_0, y_0) can be written in the form $x x_0/a^2 - y y_0/b^2 = 1$.

33. Find an equation of a line tangent to the hyperbola with equation $9x^2 - 4y^2 = 36$ that passes through $(1, 0)$.

34. Show that a line tangent to the hyperbola with equation $x^2/a^2 - y^2/b^2 = 1$ and passing through $(x_0, 0)$, when $0 < x_0 < a$, intersects the hyperbola when $x = a^2/x_0$.

35. The *latus rectum* of a hyperbola is a line segment that passes through a focal point perpendicular to the axis and joins two points on the hyperbola. Find the length of a latus rectum of the hyperbola with equation $x^2/a^2 - y^2/b^2 = 1$.

36. Use calculus to show that the point on a hyperbola that is closest to a focal point is not a vertex of the hyperbola.

37. Consider the equation $Ax^2 - Cy^2 + Dx + Ey + F = 0$, where A and C are positive constants. Find conditions on the constants A, C, D, E, and F that will ensure that this equation describes

(a) a hyperbola with axis parallel to the x-axis,

(b) a single point,

(c) a hyperbola with axis parallel to the y-axis.

38. Show that if an object is emitted from one focal point of a hyperbola to the portion of the curve closest to the other focal point, it is reflected from that curve on a line directly away from the second focal point.

39. Three detection stations lie on an east-west line 1150 meters apart. The eastmost station detects a sound from an object on the ground 2 seconds before the westmost station and 1 second before the station in the middle. Can the object emitting the noise be pinpointed? (Assume that the sound travels at 330 meters/second.)

40. A company has two manufacturing plants that produce identical automobiles. Because of differing manufacturing and labor conditions in the plants, it costs $130 more to produce a car in plant A than in plant B. The shipping costs from both plants are the same, $1 per mile, as are the loading and unloading costs, $25 per car. State criteria for determining from which plant a car should be shipped.

41. The prime focus configuration of the 200-inch Hale telescope at the Palomar Mountain Observatory was described in Exercise 44 of Section 11.1. In addition to this configuration there is a Cassegrain configuration that consists of a hyperbolic mirror inserted between the parabolic reflector and the parabola's focal point. This hyperbolic mirror has one focus at the focal point of the parabola and reflects the image back through a hole in the center of the parabolic mirror. From there it goes to the other focal point of the hyperbolic mirror, located 5 feet beyond the surface of the parabolic mirror. What is the depth from rim to vertex of the hyperbolic mirror if it is located 64 inches from the center of the parabolic mirror and has a diameter of 41 inches?

11.4
ROTATION OF AXES

The equation $xy = 1$ describes a hyperbola that has been rotated from standard position by the angle $\theta = \pi/4$. The axis of this hyperbola is the line $y = x$, as shown in Figure 11.33.

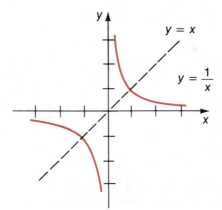

FIGURE 11.33

As the name implies, a rotation shifts a graph through an angle, as opposed to a translation, in which a graph is shifted horizontally and vertically. Rotation is a useful technique to employ when graphing certain second-degree equations that include a product of the variables. As we will see in this section, an appropriate rotation can eliminate the product term from the equation.

The first object is to determine how the rotation of a coordinate system affects an arbitrary point in the plane. Suppose a point P has coordinates (x, y) in one coordinate system and coordinates $(\hat{x}, \hat{y})$ in a second coordinate system, obtained from the first by rotating through an angle θ with the origins of the two systems remaining the same.

Let r denote the distance from the point P to the origin and ψ denote the angle formed by the intersection of the $\hat{x}$ axis and the line from P to the origin as in Figure 11.34.

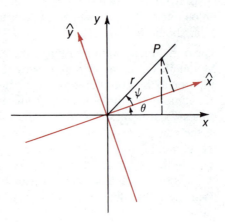

FIGURE 11.34

Since the coordinates of P are (x, y) in the original coordinate system and $(\hat{x}, \hat{y})$ in the rotated system, it can be seen from Figure 11.34 that

$$\cos(\psi + \theta) = \frac{x}{r}, \quad \sin(\psi + \theta) = \frac{y}{r}$$

$$\cos \psi = \frac{\hat{x}}{r}, \quad \sin \psi = \frac{\hat{y}}{r}.$$

Thus, $\quad x = r \cos(\psi + \theta) = r \cos \psi \cos \theta - r \sin \psi \sin \theta$
$$y = r \sin(\psi + \theta) = r \cos \psi \sin \theta + r \sin \psi \cos \theta;$$

so

(11.10) $$x = \hat{x} \cos \theta - \hat{y} \sin \theta$$
$$y = \hat{x} \sin \theta + \hat{y} \cos \theta.$$

EXAMPLE 1   Find the change in the equation $xy = 1$ that occurs when the x- and y- axes are rotated through the angle $\theta = \pi/4$.

SOLUTION
Denoting a point in the rotated system by $(\hat{x}, \hat{y})$, equations (11.10) give:

$$x = \hat{x} \cos \frac{\pi}{4} - \hat{y} \sin \frac{\pi}{4} = \frac{\sqrt{2}}{2} (\hat{x} - \hat{y})$$

and $$y = \hat{x} \sin \frac{\pi}{4} + \hat{y} \cos \frac{\pi}{4} = \frac{\sqrt{2}}{2} (\hat{x} + \hat{y}).$$

Consequently $\quad\quad\quad\quad\quad xy = 1$

becomes $$\frac{\sqrt{2}}{2} (\hat{x} - \hat{y}) \frac{\sqrt{2}}{2} (\hat{x} + \hat{y}) = 1$$

or
$$\frac{\hat{x}^2}{2} - \frac{\hat{y}^2}{2} = 1.$$

This $(\hat{x}, \hat{y})$ coordinate equation is the equation of a hyperbola with asymptotic lines $\hat{y} = \pm\,\hat{x}$, that is, the x- and y-axes. This, of course, agrees with our previous knowledge of the graph of $y = 1/x$. $\square$

EXAMPLE 2   Sketch the graph of the equation $21x^2 + 31y^2 - 10\sqrt{3}\,xy = 144$ by graphing the equation that results from rotating the x- and y-axes through an angle $\theta = \pi/6$.

SOLUTION

Since
$$x = \hat{x}\cos\frac{\pi}{6} - \hat{y}\sin\frac{\pi}{6} = \frac{\sqrt{3}}{2}\,\hat{x} - \frac{1}{2}\,\hat{y}$$

and
$$y = \hat{x}\sin\frac{\pi}{6} + \hat{y}\cos\frac{\pi}{6} = \frac{1}{2}\,\hat{x} + \frac{\sqrt{3}}{2}\,\hat{y},$$

the new equation is

$$144 = 21\left(\frac{\sqrt{3}}{2}\,\hat{x} - \frac{1}{2}\,\hat{y}\right)^2 + 31\left(\frac{1}{2}\,\hat{x} + \frac{\sqrt{3}}{2}\,\hat{y}\right)^2$$

$$- 10\sqrt{3}\left(\frac{\sqrt{3}}{2}\,\hat{x} - \frac{1}{2}\,\hat{y}\right)\left(\frac{1}{2}\,\hat{x} + \frac{\sqrt{3}}{2}\,\hat{y}\right)$$

$$= \frac{21}{4}(3\hat{x}^2 - 2\sqrt{3}\,\hat{x}\hat{y} + \hat{y}^2) + \frac{31}{4}(\hat{x}^2 + 2\sqrt{3}\,\hat{x}\hat{y} + 3\hat{y}^2)$$

$$- \frac{10\sqrt{3}}{4}(\sqrt{3}\,\hat{x}^2 + 2\hat{x}\hat{y} - \sqrt{3}\,\hat{y}^2)$$

$$= \hat{x}^2\left(\frac{63}{4} + \frac{31}{4} - \frac{30}{4}\right) + \hat{y}^2\left(\frac{21}{4} + \frac{93}{4} + \frac{30}{4}\right)$$

$$+ \hat{x}\hat{y}\left(-\frac{42}{4}\sqrt{3} + \frac{62}{4}\sqrt{3} - \frac{20}{4}\sqrt{3}\right).$$

So
$$144 = 16\hat{x}^2 + 36\hat{y}^2$$

or
$$\frac{\hat{x}^2}{9} + \frac{\hat{y}^2}{4} = 1.$$

The graph of the equation
$$21x^2 + 31y^2 - 10\sqrt{3}\,xy = 144$$

is consequently the graph of the ellipse $\dfrac{x^2}{9} + \dfrac{y^2}{4} = 1$ rotated through the angle $\theta = \pi/6$, and is sketched in Figure 11.35. $\square$

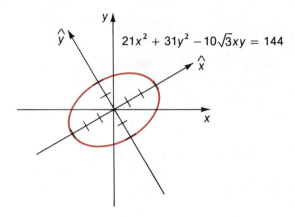

$$21x^2 + 31y^2 - 10\sqrt{3}xy = 144$$

FIGURE 11.35

In the preceding examples, the appropriate angle was given for the elimination of the product term from the second-degree equation. We now show how to find this angle in a general situation.

Suppose that

(11.11) $Ax^2 + Bxy + Cy^2 + Dx + Ey + F = 0$

is a second-degree equation and x and y are defined by the equations

$$x = \hat{x} \cos \theta - \hat{y} \sin \theta$$
$$y = \hat{x} \sin \theta + \hat{y} \cos \theta,$$

where θ is an angle between 0 and $\pi/2$. Introducing the variables $\hat{x}$ and $\hat{y}$ into equation (11.11) gives

$$A(\hat{x} \cos \theta - \hat{y} \sin \theta)^2 + B(\hat{x} \cos \theta - \hat{y} \sin \theta)(\hat{x} \sin \theta + \hat{y} \cos \theta)$$
$$+ C(\hat{x} \sin \theta + \hat{y} \cos \theta)^2 + D(\hat{x} \cos \theta - \hat{y} \sin \theta) + E(\hat{x} \sin \theta + \hat{y} \cos \theta)$$
$$+ F = 0.$$

Multiplying and collecting similar terms, we have

$$\hat{x}^2 [A(\cos \theta)^2 + B \cos \theta \sin \theta + C(\sin \theta)^2] + \hat{x}\hat{y} \{ -2A \cos \theta \sin \theta$$
$$+ B[(\cos \theta)^2 - (\sin \theta)^2] + 2C \sin \theta \cos \theta \}$$
$$+ \hat{y}^2 [A(\sin \theta)^2 - B \sin \theta \cos \theta + C(\cos \theta)^2]$$
$$+ \hat{x}(D \cos \theta + E \sin \theta) + \hat{y}(-D \sin \theta + E \cos \theta) + F = 0.$$

To eliminate the $\hat{x}\hat{y}$-term we choose θ so that the $\hat{x}\hat{y}$-coefficient is zero. This means that

$$0 = -2A \cos \theta \sin \theta + B[(\cos \theta)^2 - (\sin \theta)^2] + 2C \sin \theta \cos \theta$$
$$= (C - A) 2 \sin \theta \cos \theta + B [(\cos \theta)^2 - (\sin \theta)^2]$$
$$= (C - A)\sin 2\theta + B \cos 2\theta,$$

or $$\cot 2\theta = \frac{A - C}{B}.$$

Consequently, the appropriate rotation is through the angle

(11.12)
$$\theta = \frac{1}{2} \text{arccot} \left(\frac{A - C}{B} \right).$$

This rotation changes the equation

$$Ax^2 + Bxy + Cy^2 + Dx + Ey + F = 0$$

into

$$\hat{A}\hat{x}^2 + \hat{C}\hat{y}^2 + \hat{D}\hat{x} + \hat{E}\hat{y} + \hat{F} = 0,$$

where

(11.13)
$$
\begin{aligned}
\hat{A} &= A(\cos\theta)^2 + B\cos\theta\sin\theta + C(\sin\theta)^2 \\
\hat{B} &= 2(C - A)\sin\theta\cos\theta + B[(\cos\theta)^2 - (\sin\theta)^2] \\
\hat{C} &= A(\sin\theta)^2 - B\sin\theta\cos\theta + C(\cos\theta)^2 \\
\hat{D} &= D\cos\theta + E\sin\theta \\
\hat{E} &= -D\sin\theta + E\cos\theta \\
\hat{F} &= F
\end{aligned}
$$

and, because of the choice of θ, $\hat{B} = 0$.

EXAMPLE 3   Use a rotation to sketch the graph of $2x^2 - 72xy + 23y^2 + 25 = 0$.

SOLUTION

To simplify the equation, we rotate through an angle θ, where

$$\cot 2\theta = \frac{A - C}{B} = \frac{2 - 23}{-72} = \frac{7}{24},$$

so

$$\theta = \frac{1}{2}\text{arccot}\left(\frac{7}{24}\right) \approx .6435 \text{ radian.}$$

Although we can determine $\sin\theta$ and $\cos\theta$ directly from the value of θ, it is usually more convenient to use trigonometric identities and the value of $\cot 2\theta$ to find $\sin\theta$ and $\cos\theta$. This approach is similar to that used in integration by trigonometric substitutions (Section 8.3).

Drawing a triangle with $\cot 2\theta = 7/24$ as shown in Figure 11.36, we see that $\cos 2\theta = 7/25$.

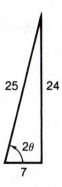

FIGURE 11.36

Since $0 < \theta < \dfrac{\pi}{2}$,

$$\cos \theta = \sqrt{\frac{1 + \cos 2\theta}{2}} = \sqrt{\frac{1 + \dfrac{7}{25}}{2}} = \frac{4}{5}$$

and

$$\sin \theta = \sqrt{\frac{1 - \cos 2\theta}{2}} = \sqrt{\frac{1 - \dfrac{7}{25}}{2}} = \frac{3}{5}.$$

Since

$$\hat{A} = 2\left(\frac{4}{5}\right)^2 - 72\left(\frac{3}{5}\right)\left(\frac{4}{5}\right) + 23\left(\frac{3}{5}\right)^2 - -\frac{625}{25} - -25$$

$$\hat{C} = 2\left(\frac{3}{5}\right)^2 + 72\left(\frac{3}{5}\right)\left(\frac{4}{5}\right) + 23\left(\frac{4}{5}\right)^2 = \frac{1250}{25} = 50$$

$\hat{F} = 25$, and $\hat{B} = \hat{D} = \hat{E} = 0$, the equation after rotation is

$$-\frac{625}{25}\,\hat{x}^2 + \frac{1250}{25}\,\hat{y}^2 + 25 = 0,$$

which simplifies to

$$\hat{x}^2 - 2\hat{y}^2 = 1.$$

This is a hyperbola with asymptotes

$$\hat{y} = \pm\,\frac{\sqrt{2}}{2}\,\hat{x}.$$

The graph of this hyperbola is given in Figure 11.37.   □

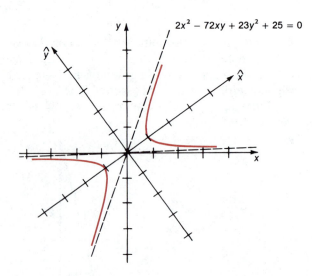

$$2x^2 - 72xy + 23y^2 + 25 = 0$$

FIGURE 11.37

Looking back at the equations in Examples 2 and 3, we see that the equation

$$21x^2 - 10\sqrt{3}yx + 31y^2 - 144 = 0$$

is an ellipse, while the graph of the somewhat similar equation

$$2x^2 - 72xy + 23y^2 + 25 = 0$$

is a hyperbola. There is an interesting formula that permits us to determine the form of the graph by observing the coefficients. If the graph of the equation

$$Ax^2 + Bxy + Cy^2 + Dx + Ey + F = 0$$

is rotated through an angle θ, and the new equation is

$$\hat{A}\hat{x}^2 + \hat{B}\hat{x}\hat{y} + \hat{C}\hat{y}^2 + \hat{D}\hat{x} + \hat{E}\hat{y} + \hat{F} = 0,$$

it can be easily, though rather tediously, shown (see Exercise 26) that

$$B^2 - 4AC = \hat{B}^2 - 4\hat{A}\hat{C}.$$

In particular, when the angle is chosen so that $\hat{B} = 0$, this implies that

$$\hat{A}\hat{x}^2 + \hat{C}\hat{y}^2 + \hat{D}\hat{x} + \hat{E}\hat{y} + \hat{F} = 0$$

and
$$B^2 - 4AC = -4\hat{A}\hat{C}.$$

However, except for degenerate cases, the graph of the equation

$$\hat{A}\hat{x}^2 + \hat{C}\hat{y}^2 + \hat{D}\hat{x} + \hat{E}\hat{y} + \hat{F} = 0$$

is

  (i)   An ellipse if $\hat{A}$ and $\hat{C}$ are both positive or both negative; that is, if $-4\hat{A}\hat{C} < 0$;
  (ii) A hyperbola if one of $\hat{A}$ or $\hat{C}$ is positive and the other is negative; that is, if $-4\hat{A}\hat{C} > 0$;
  (iii) A parabola if $\hat{A} = 0$ or $\hat{C} = 0$; that is, if $-4\hat{A}\hat{C} = 0$.

Consequently, except for degenerate cases, the graph of the equation

$$Ax^2 + Bxy + Cy^2 + Dx + Ey + F = 0$$

is

  (i)   An ellipse if $B^2 - 4AC < 0$;
(11.14) (ii) A hyperbola if $B^2 - 4AC > 0$;
  (iii) A parabola if $B^2 - 4AC = 0$.

In the degenerate cases, the graph is

  (i)   A straight line if $\hat{A} = \hat{C} = 0$ and at least one of $\hat{D}$ or $\hat{E}$ is nonzero.
  (ii) A pair of straight lines if $\hat{D} = \hat{E} = \hat{F} = 0$ and both $\hat{A}$ and $\hat{C}$ are nonzero.
  (iii) A point or pair of points if only one of $\hat{A}$, $\hat{C}$, $\hat{D}$, or $\hat{E}$ is nonzero.

Note that:
The equation $xy = 1$ in Example 1 has

$$B^2 - 4AC = 1^2 - 4(0)(0) = 1 > 0$$

and is a hyperbola.

The equation $21x^2 - 10\sqrt{3}\,xy + 31y^2 - 144 = 0$ in Example 2 has

$$B^2 - 4AC = (-10\sqrt{3})^2 - 4(21)(31) = -2304 < 0$$

and is an ellipse.

The equation $2x^2 - 72xy + 23y^2 + 25 = 0$ in Example 3 has

$$B^2 - 4AC = (-72)^2 - 4(2)(23) = 5000 > 0,$$

and is a hyperbola.

We recommend that you apply this test when a rotation is involved because it provides a check on the final result—a result that generally requires a significant amount of algebraic and trigonometric manipulation, which can be error-producing.

EXERCISE SET 11.4

The graph of each of the equations in Exercises 1 through 20 is a conic section that can be sketched by using a suitable rotation of axes. First use (11.14) to determine if the conic is an ellipse, hyperbola, or parabola. Then perform the rotation, and if necessary a translation, and sketch the graph.

1. $x^2 - xy + y^2 = 2$
2. $x^2 + 4xy + y^2 = 3$
3. $x^2 + 2xy + y^2 + \dfrac{\sqrt{2}}{2}x - \dfrac{\sqrt{2}}{2}y = 0$
4. $14x^2 + 24xy + 7y^2 = 1$
5. $14x^2 - 24xy + 7y^2 = 1$
6. $\sqrt{3}x^2 - 3xy = \sqrt{3}$
7. $4x^2 - 4xy + 7y^2 = 24$
8. $5x^2 - 3xy + y^2 = 5$
9. $17x^2 - 6xy + 9y^2 = 0$
10. $108x^2 - 312xy + 17y^2 + 240x + 320y + 500 = 0$
11. $225x^2 - 1080xy + 1296y^2 - 624x - 260y - 2028 = 0$
12. $21x^2 - 10\sqrt{3}xy + 11y^2 + 6x + 6\sqrt{3}y - 150 = 0$
13. $6x^2 + 4\sqrt{3}xy + 2y^2 - 9x + 9\sqrt{3}y - 63 = 0$
14. $6x^2 - 60xy - 19y^2 + 48\sqrt{13}x - 32\sqrt{13}y + 338 = 0$
15. $484x^2 + 352xy + 64y^2 - 748x - 272y + 289 = 0$
16. $x^2 + 2\sqrt{3}xy + 3y^2 + (4 + 2\sqrt{3})x + (4\sqrt{3} - 2)y + 12 = 0$
17. $16x^2 + 4xy + 19y^2 - 20\sqrt{5}x - 10\sqrt{5}y - 25\sqrt{5} = 0$
18. $97x^2 + 42xy + 153y^2 - 98\sqrt{10}x + 246\sqrt{10}y + 10 = 0$
19. $7x^2 + 48xy - 7y^2 - 10x - 70y - 25 = 0$
20. $7x^2 + 48xy - 7y^2 - 170x + 60y - 125 = 0$

21. Find an equation of the line tangent to the conic section $x^2 - xy + y^2 - 2 = 0$ at the point $(\sqrt{2}, \sqrt{2})$. Sketch the graph of the conic section and the tangent line.

22. Find an equation of the line tangent to the graph of the conic section $x^2 + xy + y^2 - 4\sqrt{2}x - 4\sqrt{2}y = 0$ at the point $(0, 0)$. Sketch the graph of the curve and the tangent line.

23. Show that the graph of $\sqrt{x^2 + (y - 1)^2} + 1 = \sqrt{x^2 + (y + 1)^2}$ is a conic section.

24. Find an equation of the parabola whose axis is $y = x$ and passes through $(1, 0)$, $(0, 1)$ and $(1, 1)$.

25. Find an equation of the ellipse whose vertices lie at $(0, 0)$ and $(6, 8)$ and that passes through the point $(0, 25/4)$.

26. Use the equations in (11.12) and (11.13) to show that if the graph of the equation $Ax^2 + Bxy + Cy^2 + Dx + Ey + F = 0$ is rotated through any angle θ, the coefficients in the new equation $\hat{A}\hat{x}^2 + \hat{B}\hat{x}\hat{y} + \hat{C}\hat{y}^2 + \hat{D}\hat{x} + \hat{E}\hat{y} + \hat{F} = 0$ have the property that $B^2 - 4AC = \hat{B}^2 - 4\hat{A}\hat{C}$.

27. Show that a second-degree equation can represent a circle only if the coefficient of the xy-term is zero.

28. Use the equations in (11.13) to show that A and $C = \hat{A} + \hat{C}$, regardless of the value of θ.

29. Two nonparallel lines l_1 and l_2 are given. Show that the set of all points, the product of whose distances from l_1 and l_2 is constant, must be a hyperbola. Show, in addition, that l_1 and l_2 are asymptotes of the hyperbola.

11.5
POLAR EQUATIONS OF CONIC SECTIONS

Conic sections can be derived using an alternate geometric method. This method involves a constant ratio called the **eccentricity** of the conic. (The eccentricity of a conic has historically been denoted e, a symbol that unfortunately has a dual purpose, since it is also used for describing the base of the natural exponential function.) The following theorem details the role that the eccentricity of a conic plays in determining the behavior of the conic.

(11.15)
THEOREM

Suppose F is a fixed point in the plane, l is a fixed line, and e is a fixed positive constant. The set of points P whose distance from F is the product of e and the distance from P to l,

$$d(P, F) = e\, d(P, l),$$

is a conic section with a focal point at F.
Moreover, except for degenerate cases, the conic section is
(i)   a parabola, if $e = 1$,
(ii) an ellipse, if $e < 1$,
(iii) a hyperbola, if $e > 1$.

The line l is called a **directrix** of the conic. (This coincides with the definition of the directrix of a parabola in the case when $e = 1$.)

PROOF

Suppose we position the point F and the line l so that F is at the origin of a polar coordinate system and l lies perpendicular to the polar axis a distance d to the left of F. (See Figure 11.38.)

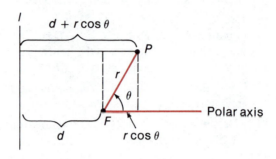

FIGURE 11.38

If P has polar coordinates (r, θ), then $d(P, F) = r$ and $d(P, l) = d + r \cos \theta$. So

$$r = e(d + r \cos \theta),$$

or in cartesian coordinates

$$\sqrt{x^2 + y^2} = e(d + x).$$

Squaring, we have

$$x^2 + y^2 = e^2(d^2 + 2xd + x^2)$$

or $$(1 - e^2)x^2 + y^2 - 2de^2x - e^2d^2 = 0.$$

In (11.14) of Section 11.4 we saw that this is the equation of a conic and is, except for degenerate cases,

    (i)   a parabola if $B^2 - 4AC = -4(1 - e^2) \cdot 1 = 0$; that is, if $e = 1$;
    (ii) an ellipse if $B^2 - 4AC = -4(1 - e^2) \cdot 1 < 0$; that is, if $e < 1$;
    (iii) a hyperbola if $B^2 - 4AC = -4(1 - e^2) \cdot 1 > 0$; that is, if $e > 1$. $\square$

The polar equation, $r = e(d + r \cos \theta)$, describing the conic with directrix a distance d to the left of the focal point F, is usually expressed in the form

(11.16) $$r = \frac{ed}{1 - e \cos \theta}.$$

Modifications of the polar equation of conics occur when the directrix is moved to one of the other standard positions. When the directrix l of a conic is a distance d to the right of the focal point F, the polar equation is

(11.17) $$r = \frac{ed}{1 + e \cos \theta}.$$

When the focal point is at the origin and the directrix is parallel to and a distance d above the polar axis, the equation is

(11.18)
$$r = \frac{ed}{1 + e \sin \theta}.$$

When the directrix is parallel to and a distance d below the polar axis, the equation is

(11.19)
$$r = \frac{ed}{1 - e \sin \theta}.$$

If $e = 1$, the equation of the parabola with focal point at the pole and directrix perpendicular to the polar axis and d units to the left of the pole is

(11.20)
$$r = \frac{d}{1 - \cos \theta},$$

which is undefined when $\theta = 0$. When $\theta = \pi$, $r = d/2$ and when $\theta = \pi/2$ or $\theta = 3\pi/2$, $r = d$. This agrees with our previous knowledge of a parabola. (See Figure 11.39.)

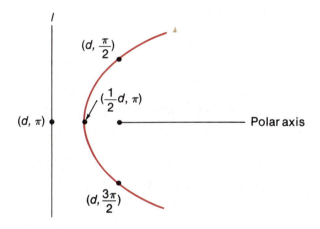

FIGURE 11.39

EXAMPLE 1   Sketch the graph of the parabola whose polar equation is

$$r = \frac{3}{1 - \cos \theta}.$$

SOLUTION
Comparing this equation with equation (11.20), we see that the focal point of this parabola is at the origin and the directrix is the line

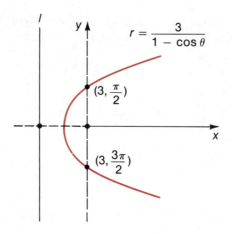

FIGURE 11.40

perpendicular to the polar axis with cartesian equation $x = -3$. When $\theta = \pi/2$ or $\theta = 3\pi/2$, $r = 3$. The graph is shown in Figure 11.40.   ☐

If $e < 1$ (an ellipse), the polar equation $r = \dfrac{ed}{1 - e\cos\theta}$ has a denominator that is always positive and consequently is defined for all values of θ. The vertices of the ellipse occur at the absolute extrema of r. To find the extrema on the interval $[0, 2\pi)$, we find $D_\theta r$:

$$D_\theta r = D_\theta\left(\frac{ed}{1 - e\cos\theta}\right) = \frac{-e^2 d \sin\theta}{(1 - e\cos\theta)^2}.$$

If $\theta = 0$ or if $\theta = \pi$, $D_\theta r = 0$.

For $-\pi/2 < \theta < 0$, $D_\theta r > 0$, and for $0 < \theta < \pi/2$, $D_\theta r < 0$. Consequently, r has an absolute maximum value of $\dfrac{ed}{1 - e}$ when $\theta = 0$.

For $\pi/2 < \theta < \pi$, $D_\theta r < 0$, and for $\pi < \theta < 3\pi/2$, $D_\theta r > 0$. Thus r has an absolute minimum value of $\dfrac{ed}{1 + e}$ when $\theta = \pi$.

The vertices of the ellipse are at

$$\left(\frac{ed}{1 - e}, 0\right) \qquad \text{and} \qquad \left(\frac{ed}{1 + e}, \pi\right)$$

as shown in Figure 11.41.

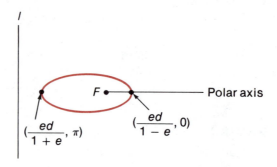

FIGURE 11.41

EXAMPLE 2   Sketch the graph of the conic whose polar equation is

$$r = \frac{16}{5 + 3 \cos \theta}$$

and find the cartesian equation of this conic.

SOLUTION

The standard form for a conic with focal point at the origin and directrix perpendicular to the polar axis d units to the right of the focal point is

$$r = \frac{ed}{1 + e \cos \theta}.$$

Our equation is

$$r = \frac{16}{5 + 3 \cos \theta} = \frac{\dfrac{16}{5}}{1 + \dfrac{3}{5} \cos \theta},$$

so $e = 3/5$ and since $ed = 16/5$, $d = 16/3$. The conic is an ellipse with vertices occurring at $(2, 0)$ and $(8, \pi)$. The graph is shown in Figure 11.42.

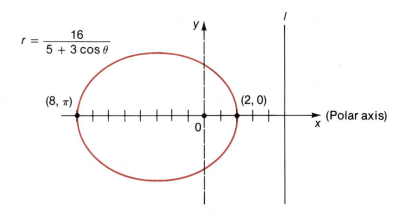

FIGURE 11.42

To find the cartesian equation of the ellipse, we first write the polar equation as

$$5r + 3r \cos \theta = 16,$$

so

$$5\sqrt{x^2 + y^2} + 3x = 16.$$

Rearranging the terms and squaring both sides, we obtain

$$25(x^2 + y^2) = (16 - 3x)^2 = 256 - 96x + 9x^2,$$

or

$$16x^2 + 96x + 25y^2 = 256.$$

Completing the square,

$$16(x^2 + 6x + 9) + 25y^2 = 256 + 144$$

so

$$16(x + 3)^2 + 25y^2 = 400,$$

and

$$\frac{(x + 3)^2}{25} + \frac{y^2}{16} = 1. \qquad \square$$

If $e > 1$ (a hyperbola), the polar equation $r = \dfrac{ed}{1 - e \cos \theta}$ has a denominator that is undefined when $\cos \theta = 1/e$, that is, when $\theta = \pm \arccos 1/e$. The left side of the hyperbola is produced when

$$-\arccos \frac{1}{e} < \theta < \arccos \frac{1}{e}$$

and the right side when

$$\arccos \frac{1}{e} < \theta < 2\pi - \arccos \frac{1}{e}.$$

The vertices occur at $\theta = 0$, $r = \dfrac{ed}{1 - e}$ (a negative number) and at $\theta = \pi$,

$r = \dfrac{ed}{1 + e}$, as shown in Figure 11.43.

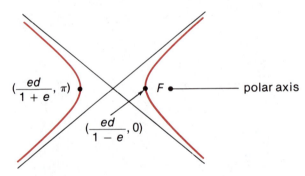

$(\dfrac{ed}{1 + e}, \pi)$ F ———— polar axis

$(\dfrac{ed}{1 - e}, 0)$

FIGURE 11.43

EXAMPLE 3   Sketch the graph of the hyperbola whose polar equation is

$$r = \frac{10}{2 + 5 \sin \theta}.$$

SOLUTION

The directrix for this hyperbola lies parallel to and above the polar axis, so both vertices will lie along the positive y-axis.

$$r = \frac{10}{2 + 5 \sin \theta} = \frac{5}{1 + \dfrac{5}{2} \sin \theta}.$$

The directrix lies above the x-axis a distance

$$d = \frac{5}{e} = \frac{5}{\frac{5}{2}} = 2.$$

The graph is shown in Figure 11.44. The points with polar coordinates $(5, 0)$, $(5, \pi)$, $(10/7, \pi/2)$ and $(-10/3, 3\pi/2)$ are shown on the graph. Additional points can be plotted if more accuracy is desired.   □

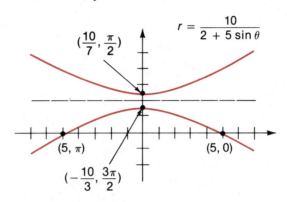

$$r = \frac{10}{2 + 5 \sin \theta}$$

FIGURE 11.44

EXAMPLE 4   Find a polar equation of the hyperbola with cartesian equation $x^2 - y^2 = 1$. (See Figure 11.45.)

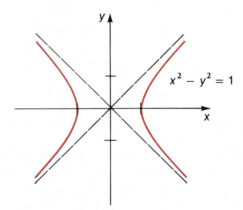

$x^2 - y^2 = 1$

FIGURE 11.45

SOLUTION

The polar equation is

$$(r \cos \theta)^2 - (r \sin \theta)^2 = 1,$$

or

$$r^2 = \frac{1}{(\cos \theta)^2 - (\sin \theta)^2} = \frac{1}{\cos 2\theta}.$$

Note that this is not the "standard" polar equation of a conic, because the hyperbola does not have a focal point at the origin.   □

EXERCISE SET 11.5

In Exercises 1 through 12, sketch the graph of each conic section and determine a corresponding cartesian equation.

1. $r = \dfrac{2}{1 + \cos \theta}$

2. $r = \dfrac{3}{3 - 2 \cos \theta}$

3. $r = \dfrac{3}{3 + 2 \sin \theta}$

4. $r = \dfrac{4}{1 + 2 \cos \theta}$

5. $r = \dfrac{2}{3 - 3 \sin \theta}$

6. $r = \dfrac{5}{2 - 5 \sin \theta}$

7. $r = \dfrac{5}{2 - 3 \cos \theta}$

8. $r = \dfrac{2}{4 + \cos \theta}$

9. $r = \dfrac{3}{1 + 2 \sin \theta}$

10. $r = \dfrac{9}{2 - 6 \cos \theta}$

11. $r = \dfrac{1}{4 + 2 \sin \theta}$

12. $r = \dfrac{14}{7 - 2 \sin \theta}$

In Exercises 13 through 18, find a polar equation of the conic satisfying the given conditions.

13. $e = 2$, the directrix has equation $x = 4$.

14. $e = \dfrac{1}{2}$, the directrix has equation $y = -2$.

15. $e = 1$, the directrix has equation $y = -\dfrac{1}{4}$.

16. $e = \dfrac{1}{3}$, directrix has polar equation $r = \sec \theta$.

17. $e = 3$, directrix has polar equation $r = 2 \csc \theta$.

18. $e = 1$, directrix has polar equation $r = -\dfrac{1}{2} \csc \theta$.

19. Find a polar equation of the ellipse in standard position that has vertices at points with polar coordinates $(1, 0)$ and $(3, \pi)$.

20. Find a polar equation of the parabola with a focus at the pole and a vertex at the point with cartesian coordinates $(-6, 0)$.

Express the cartesian equations in Exercises 21 through 26 as polar equations, and state which of these conics is in standard polar position.

21. $y = 3x^2$

22. $2x^2 + y^2 = 1$

23. $4x^2 + 3y^2 - 2y = 1$

24. $y^2 = x + 1$

25. $x^2 - y^2 - 2y = 4$

26. $y^2 - 3x^2 + 6x = 9$

27. Show that the conic section with polar equation $r = ed/(1 + e \cos \theta)$ can also be expressed as $r = ed/(e \cos \theta - 1)$. [*Hint*: Consider the relationship between (r, θ) and $(-r, \theta + \pi)$.]

28. Show that the conic section with polar equation $r = ed/(1 + e \sin \theta)$ can also be expressed as $r = ed/(e \sin \theta - 1)$.

Use the relationships in Exercises 27 and 28 to sketch the graphs of the polar equations in Exercises 29 through 32.

29. $r = \dfrac{2}{\cos \theta - 1}$

30. $r = \dfrac{3}{\sin \theta - 2}$

31. $r = \dfrac{1}{2 \sin \theta - 3}$

32. $r = \dfrac{4}{2 \cos \theta - 1}$

33. The world's first orbiting satellite, *Sputnik I*, was launched in the Soviet Union on Oct. 4, 1957. Its elliptical orbit reached a maximum height of 560 miles above the earth and a minimum height of 145 miles. Write a polar equation for this orbit, assuming that a focal point is at the pole, which is placed at the center of the earth. Also assume that the earth is spherical with a radius of 4000 miles.

34. The earth moves in an elliptical orbit around the sun with the sun at one focal point and an eccentricity of .017. The major axis of the orbit is approximately 3×10^{11} meters. Write a polar equation of this ellipse if the pole is taken as the center of the sun.

REVIEW EXERCISES

Identify and sketch the graph of each conic section described in Exercises 1 through 30.

1. $y^2 = 4x$

2. $y = -x^2$

3. $x^2 = -2(y - 5)$

4. $y^2 = -16(x - 5)$

5. $(y - 2)^2 = 2(x + 3)$

6. $(x - 1)^2 = 4(y + 3)$

7. $y^2 - 2x + 2y + 7 = 0$

8. $16x^2 + 25y^2 = 400$

9. $16(x - 2)^2 + 25(y - 3)^2 = 400$

10. $16x^2 - 25y^2 = 400$

11. $16(x - 2)^2 - 25(y - 3)^2 = 400$

12. $9(x + 2)^2 - 4(y - 5)^2 = 36$

13. $9x^2 - 16y^2 = 144$

14. $x^2 - 2x - 6y - 7 = 0$

15. $9x^2 + 4y^2 - 90x - 16y + 205 = 0$

16. $4x^2 - 9y^2 - 16x - 90y + 16 = 0$

17. $4x^2 - 9y^2 - 16x - 90y - 210 = 0$

18. $9x^2 + 4y^2 - 90x - 16y - 83 = 0$

19. $4x^2 + 9y^2 - 16x - 90y + 205 = 0$

20. $y^2 - 6y + 9 - 4x = 0$

21. $x^2 + 4xy + y^2 - 16 = 0$

22. $23x^2 + 26\sqrt{3}xy - 3y^2 - 144 = 0$

23. $31x^2 + 10\sqrt{3}xy + 21y^2 - 144 = 0$

24. $x^2 + 3xy + y^2 - 1 = 0$

25. $25x^2 + 14xy + 25y^2 - 288 = 0$

26. $r = \dfrac{1}{1 - \sin\theta}$

27. $r = \dfrac{8}{1 - 4\cos\theta}$

28. $r = \dfrac{2}{3 + \sin\theta}$

29. $r = \dfrac{1}{1 - \cos\theta}$

30. $r = \dfrac{7}{3 + 5\cos\theta}$

Find an equation of each conic described in Exercises 31 through 39.

31. A parabola with focus at $(0, 0)$ and directrix $y = 2$

32. A parabola with focus at $(0, 0)$ and directrix $x = 2$

33. An ellipse with foci at $(0, \pm 1)$ and vertices at $(0, \pm 3)$

34. An ellipse with foci at $(\pm 1, 0)$ and vertices at $(\pm 3, 0)$

35. A hyperbola with foci at $(\pm 3, 0)$ and vertex at $(1, 0)$

36. A parabola with focus at $(0, 1)$ and vertex at $(0, 0)$

37. An ellipse with foci at $(0, \pm 5)$ and passing through the point $(4, 0)$

38. A conic with a focus at the origin, eccentricity 3/4, and directrix $x = 2$

39. A conic with a focus at the origin, eccentricity 3, and directrix $y = -2$

40. Explain why a rotation does not alter the equation $x^2 + y^2 = r^2$.

41. Find an equation of the line tangent to the graph of $16x^2 - 4y^2 = 144$ at $(5, 8)$.

42. Find an equation of the line tangent to the graph of $x^2 + 4y^2 = 40$ at $(2, 3)$.

43. Find an equation (in cartesian coordinates) of a line tangent to $r = 1/(1 - \cos\theta)$ at $(1, \pi/2)$.

12
VECTORS

The first eleven chapters concerned the study of functions whose domain and range are both contained in the set of real numbers. This essentially restricts us to the study of two-dimensional problems. Since we live in three-dimensional space, we need to study more general functions. In this chapter, we lay the groundwork by defining and studying the properties of vectors, which are used for describing direction in multidimensional space.

12.1
THE RECTANGULAR COORDINATE SYSTEM IN SPACE

To analytically describe objects in space, we need to introduce a three-dimensional coordinate system. The natural extension of the rectangular xy-coordinate system in the plane is made by simply adding a third coordinate axis, labeled z, perpendicular to the xy-plane. Theoretically and analytically, this is a satisfactory procedure; graphically, however, problems can arise. The work of M. C. Escher entitled ''Waterfall,'' shown in Figure 12.1 on the next page, illustrates quite clearly the fallacies that can arise from adding a depth perception to a two-dimensional plane. However, since the plane is the only vehicle available to us for graphical representation, we use it as best we can.

To add a dimension of depth to our rectangular coordinate system, we assume that the plane determined by the y- and z-axes, the yz-plane, coincides with the plane in which we draw, and that the positive x-axis is directed toward us. We draw the x-axis as a straight line that has the *appearance* of making an angle of

FIGURE 12.1

135° with both the y- and z-axes. The positive portion of each axis is drawn as a black line, while the negative direction, when needed, is represented by a lighter colored line as illustrated in Figure 12.2.

As an additional aid to the perception of depth, we introduce the standard drafting procedure of using a physical length of approximately 3/4 of one unit along the x-axis for each unit along the y- and z-axes, as shown in Figure 12.3.

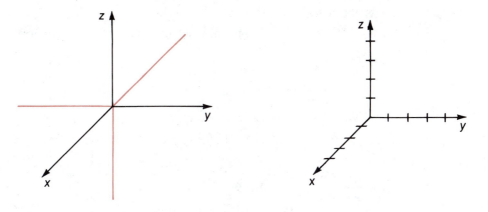

A point in space is represented in this rectangular coordinate system by an **ordered triple** called the coordinates of the point. A point with coordinates (a, b, c) is drawn in the rectangular coordinate system by sketching the parallelepiped shown in Figure 12.4.

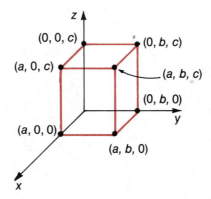

FIGURE 12.4

The point (a, b, c) is the vertex of the parallelepiped that is diagonally opposite the vertex at the origin $(0, 0, 0)$.

The coordinates of a point in the xy-plane have the form $(a, b, 0)$; similarly in the xz-plane, $(a, 0, c)$; and in the yz-plane, $(0, b, c)$. The xy-, yz-, and xz-planes (also known as the **coordinate planes**) divide space into eight sections called octants. The portion of space determined by those points, all of whose coordinates are positive, is called the **first octant.** We do not label the other seven octants.

EXAMPLE 1   Sketch the position of the points $(2, 3, 2)$ and $(4, 4, 3)$ in the rectangular coordinate system.

SOLUTION

These points lie in the first octant and are shown in Figures 12.5(a) and (b).

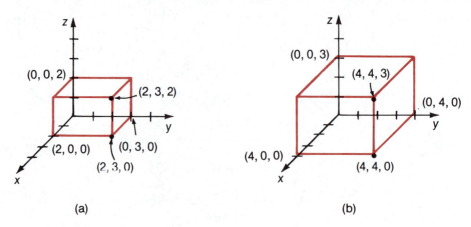

(a)   (b)

FIGURE 12.5

While it is easy to observe the relative position of these two points when their parallelepipeds are drawn, notice that when the parallelepipeds are removed from the sketch, the points are indistinguishable, as shown in Figure 12.6.

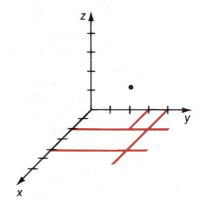

FIGURE 12.6

In fact, the point in this sketch could have coordinates $(0, 2, 1)$, $(-2, 1, 0)$, or even $(-100, -48, -49)$. Although drawing a parallelepiped seems to involve extra work, this example demonstrates that it is the only way to be confident of the graphical representation.   □

The parallelepiped can also be used to find a formula for the distance between two points in space. Suppose $P_1(x_1, y_1, z_1)$ and $P_2(x_2, y_2, z_2)$ are two points in space. Consider the parallelepiped with diagonally opposite vertices at P_1 and P_2 and sides parallel to the coordinate planes, as shown in Figure 12.7.

The line from P_1 to Q is perpendicular to the line from Q to P_2 so the Pythagorean theorem implies that

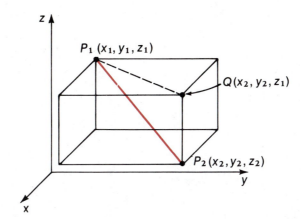

FIGURE 12.7

$$[d(P_1, P_2)]^2 = [d(P_1, Q)]^2 + [d(Q, P_2)]^2,$$

where, as usual, d denotes the distance between the specified points. However,

$$d(Q, P_2) = |z_2 - z_1|$$

and $$d(P_1, Q) = \sqrt{(x_2 - x_1)^2 + (y_2 - y_1)^2},$$

so $$[d(P_1, P_2)]^2 = (x_2 - x_1)^2 + (y_2 - y_1)^2 + (z_2 - z_1)^2.$$

Thus, the distance between $P_1(x_1, y_1, z_1)$ and $P_2(x_2, y_2, z_2)$ is

(12.1) $$d(P_1, P_2) = \sqrt{(x_2 - x_1)^2 + (y_2 - y_1)^2 + (z_2 - z_1)^2}.$$

This is the natural extension of the two-dimensional distance formula.

In a similar manner, the midpoint of the line segment joining P_1 and P_2 has the same form as in the plane. The midpoint of the line segment joining P_1 and P_2 is $\left(\dfrac{x_1 + x_2}{2}, \dfrac{y_1 + y_2}{2}, \dfrac{z_1 + z_2}{2} \right)$.

EXAMPLE 2   Find the distance between the points $P_1(1, -1, 2)$ and $P_2(3, 4, -1)$ and the midpoint of the line segment joining P_1 and P_2.

SOLUTION

This distance is

$$d(P_1, P_2) = \sqrt{(3 - 1)^2 + (4 - (-1))^2 + (-1-2)^2}$$
$$= \sqrt{4 + 25 + 9} = \sqrt{38}.$$

The midpoint of the line segment joining P_1 and P_2 is

$$\left(\frac{1 + 3}{2}, \frac{-1 + 4}{2}, \frac{2 - 1}{2} \right) = \left(2, \frac{3}{2}, \frac{1}{2} \right). \qquad \square$$

EXAMPLE 3   Find an equation of the sphere with center (h, k, l) and radius r that is graphed in Figure 12.8.

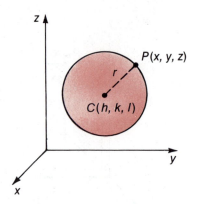

FIGURE 12.8

SOLUTION

   All points on the sphere have the property that their distance from the center (h, k, l) is the constant r. So a point (x, y, z) lies on the sphere if and only if

$$r = d((x, y, z), (h, k, l))$$
$$= \sqrt{(x - h)^2 + (y - k)^2 + (z - l)^2}.$$

Squaring both sides, we have the equation

$$r^2 = (x - h)^2 + (y - k)^2 + (z - l)^2;$$

this is known as the **standard equation of a sphere of radius r with center (h, k, l).**   □

EXAMPLE 4   Describe geometrically the set of points (x, y, z) that satisfy the equation $x^2 + y^2 = 1$.

SOLUTION

   The equation $x^2 + y^2 = 1$ describes a circle in the xy-plane. There is no restriction on z, so z can assume any real number value. The equation describes a right circular cylinder whose center is the z-axis, as shown in Figure 12.9.   □

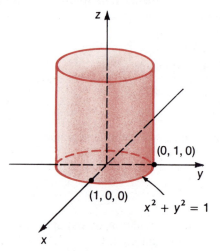

FIGURE 12.9

The graph of an equation in three variables is called a **surface.** Examples of surfaces are the sphere and cylinder discussed in Examples 3 and 4. Surfaces are studied in detail in Chapter 14.

EXERCISE SET 12.1

1. Plot each of the following points and sketch the associated parallelepiped.
   (a) $(1, 3, 4)$ (b) $(1, -3, 4)$ (c) $(2, 4, -3)$
2. Plot each of the following points and sketch the associated parallelepiped.
   (a) $(2, 3, 3)$ (b) $(-2, 3, 3)$ (c) $(-2, -3, 3)$

In Exercises 3 through 8, plot the points A and B and find the distance between them.

3. $A(1, 2, 3)$, $B(-1, 3, 4)$ 4. $A(2, 5, 0)$, $B(3, -3, 1)$
5. $A(-3, 4, 0)$, $B(-3, 4, 2)$ 6. $A(0, 0, 0)$, $B(4, 2, -4)$
7. $A(1, 8, 3)$, $B(0, 1, 0)$ 8. $A(3, 5, 0)$, $B(5, 3, 0)$

9–14. Find the midpoint of the line segment joining points A and B given in Exercises 3 through 8.

In Exercises 15 through 18, the points A and B are the endpoints of a diagonal of a parallelepiped having its faces parallel to the coordinate planes. Sketch the parallelepiped and find the coordinates of the other vertices.

15. $A(0, 0, 0)$, $B(2, 4, 4)$ 16. $A(2, 2, 0)$, $B(3, 5, 4)$
17. $A(3, 0, 0)$, $B(4, 3, 5)$ 18. $A(0, 2, 0)$, $B(2, -2, 4)$

In Exercises 19 through 22, find an equation of the sphere with center C and radius r.

19. $C(0, 0, 0)$, $r = 2$ 20. $C(2, 0, 0)$, $r = 2$
21. $C(2, 3, 4)$, $r = 1$ 22. $C(0, 3, 0)$, $r = 3$

23. Find an equation of the sphere having endpoints of a diameter at $(-2, 1, 1)$ and $(1, 4, 5)$.
24. Find an equation of a sphere having center at $(1, 2, 3)$ and passing through the origin.
25. Find the center and radius of the sphere described by

$$x^2 + y^2 + z^2 + 2x - 4y + 6z = 0.$$

   (*Hint*: Complete the square in each of the variables x, y, and z.)
26. Sketch the parallelepiped consisting of the points (x, y, z) with $2 \le x \le 4$, $2 \le y \le 3$, and $0 \le z \le 5$. Give the coordinates of the eight corners of the parallelepiped.
27. (a) Show that the points $(4, 4, 1)$, $(1, 1, 1)$ and $(0, 8, 5)$ are the vertices of a right triangle.
   (b) Find an equation of the sphere passing through the three points and having a diameter along the hypotenuse of the triangle.

28. Sketch the points (2, 4, 2), (2, 1, 5) and (5, 1, 2) and show that they are the vertices of an equilateral triangle.

29. The floor of a room is 12 feet by 8 feet and the height of the ceiling is 7 feet. Make a representative sketch of the room in a three-dimensional coordinate system and label the points corresponding to the corners of the room.

30. A dome tent has a circular floor with diameter 8 feet. The height of the tent in the center is 6 feet. Make a representative sketch of the tent in a three-dimensional coordinate system.

12.2
VECTORS IN SPACE

Many physical properties can be described by simply stating a magnitude. Some examples are the mass of an object, the area of a triangle, the volume of a sphere, and the distance between a pair of points in space. Other properties, however, can only be described by specifying both a magnitude and a direction. Examples of this type are particularly abundant in physics and include force, velocity, acceleration, and momentum.

There is no difficulty describing direction in a one-dimensional setting. From a fixed point a on a directed line there are only two directions: a positive direction for those points greater than a and a negative direction for those points less than a. When additional dimensions are introduced, the problem of describing direction requires a different method. From a fixed point in the plane or in space the number of directions is infinite. Vectors are used to describe the notion of distance and direction in the plane and in space.

(12.2)
DEFINITION

A **vector** in space is an ordered triple $\mathbf{v} = \langle v_1, v_2, v_3 \rangle$ of real numbers that is used to describe the collection of directed line segments joining any pair of points $P_1(x_1, y_1, z_1)$ and $P_2(x_2, y_2, z_2)$ with the property that

$$v_1 = x_2 - x_1, \qquad v_2 = y_2 - y_1, \qquad \text{and} \qquad v_3 = z_2 - z_1.$$

The points P_1 and P_2 are called **initial** and **terminal points,** respectively, for the vector $\mathbf{v}$; this is denoted by writing $\mathbf{v} = \overrightarrow{P_1 P_2}$. The **length** of the vector $\mathbf{v}$, denoted $\|\mathbf{v}\|$, is defined by

$$\|\mathbf{v}\| = \sqrt{v_1^2 + v_2^2 + v_3^2}.$$

The only vector with length zero, $\mathbf{0} = \langle 0, 0, 0 \rangle$, is assigned no direction.

A boldface type is used to represent vectors so that it is easy to distinguish between scalar (real-valued) and vector quantities. A vector $\mathbf{v}$ is represented geometrically by a directed line segment of a particular length, $\|\mathbf{v}\|$, and a particular direction. The direction is denoted by an arrow. For example, the vector $\mathbf{v} = \langle 1, 2, 3 \rangle$ can be represented by a line segment from any point (x, y, z) to the point $(x + 1, y + 2, z + 3)$. The arrow in Figure 12.10 gives one representation of $\mathbf{v}$.

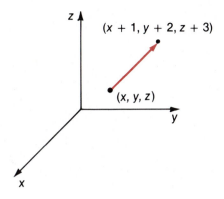

FIGURE 12.10   **FIGURE 12.11**

There is flexibility incorporated into the definition of vectors that makes their geometric representation movable. Since a vector is completely determined by its direction and magnitude, different representations can be given by parallel displacements, as shown in Figure 12.11.

At times it is convenient to position the initial point of a vector $\mathbf{v} = \langle v_1, v_2, v_3 \rangle$ at the origin $(0, 0, 0)$. The terminal point of the vector $\mathbf{v}$ is then located at the point (v_1, v_2, v_3). This representation is called the **position vector** representation of $\mathbf{v}$. There is a one-to-one correspondence between the set of position vectors and the set of points in space.

EXAMPLE 1   Determine the vector describing the line segment joining points $(1, 2, -1)$ and $(2, 3, 4)$. Find the length of this vector.

SOLUTION

The problem is not precisely posed since two such vectors exist.

If $(1, 2, -1)$ is the initial point and $(2, 3, 4)$ is the terminal point, then the vector is $\langle 2 - 1, 3 - 2, 4 - (-1) \rangle = \langle 1, 1, 5 \rangle$. If the initial and terminal points are reversed, the vector is $\langle 1 - 2, 2 - 3, -1 - 4 \rangle$ or $\langle -1, -1, -5 \rangle$. In either case the length is $\sqrt{(1)^2 + (1)^2 + (5)^2} = \sqrt{27}$. The vector with initial point $(1, 2, -1)$ is shown in Figure 12.12. $\square$

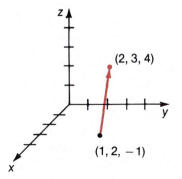

FIGURE 12.12

EXAMPLE 2   What is the terminal point of the vector $\langle 3, 0, 1 \rangle$, if the initial point is
(a) $(-1, 1, 0)$,   (b) $(2, e, \pi)$,   (c) $(0, 0, 0)$?

SOLUTION

(a) The terminal point is (x, y, z) where $\langle x + 1, y - 1, z - 0 \rangle = \langle 3, 0, 1 \rangle$;
so the terminal point is $(2, 1, 1)$. The terminal points in the other cases are (b)
$(5, e, \pi + 1)$ and (c) $(3, 0, 1)$.   □

EXAMPLE 3   Find the position vector representation for the vector with initial point
$(3, 7, -2)$ and terminal point $(-1, 2, 4)$.

SOLUTION

If $\mathbf{v} = \langle v_1, v_2, v_3 \rangle$ denotes this vector,
then
$$v_1 = -1 - 3 = -4, \qquad v_2 = 2 - 7 = -5, \qquad \text{and} \qquad v_3 = 4 - (-2) = 6$$
so
$$\mathbf{v} = \langle -4, -5, 6 \rangle.$$

The position vector representation of $\mathbf{v}$ is the directed line segment with initial
point at $(0, 0, 0)$ and terminal point at $(-4, -5, 6)$, shown in Figure 12.13.□

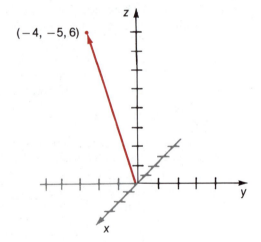

FIGURE 12.13

The preceding example illustrates the difference between the geometric and
analytic aspects of vectors. To establish results concerning vectors it is generally
more convenient to consider the ordered triple definition of a vector. To see
geometrically the application of these results, it is preferable to consider vectors
as movable directed line segments.

Though the definition has been given for vectors in space, it is desirable in
some applications to consider vectors restricted to the xy-plane. These vectors
have the form $\langle v_1, v_2, 0 \rangle$ and are often written simply as $\langle v_1, v_2 \rangle$. Throughout the
study of vectors, keep in mind that the results proved for vectors in space are
equally valid for vectors in the plane. In fact, it is often convenient to use vectors
in the xy-plane to illustrate vector concepts.

EXAMPLE 4   Find the position vector representation for the vector in the *xy*-plane with initial point $(3, 4)$ and terminal point $(-2, 1)$.

SOLUTION

If $\mathbf{v} = \langle v_1, v_2 \rangle$ denotes this vector, then

$$v_1 = -2 - 3 = -5 \quad \text{and} \quad v_2 = 1 - 4 = -3,$$

so $\mathbf{v} = \langle -5, -3 \rangle$. The position vector representation is shown in Figure 12.14.   □

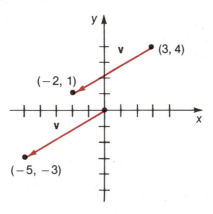

FIGURE 12.14

We continue the study of vectors with the definition of some arithmetic operations.

(12.3)   If $\mathbf{a} = \langle a_1, a_2, a_3 \rangle$ and $\mathbf{b} = \langle b_1, b_2, b_3 \rangle$ are vectors and α is a real number,
DEFINITION   we define

(i) $\mathbf{a} + \mathbf{b} = \langle a_1 + b_1, a_2 + b_2, a_3 + b_3 \rangle$

(ii) $\mathbf{a} - \mathbf{b} = \langle a_1 - b_1, a_2 - b_2, a_3 - b_3 \rangle$

(iii) $\alpha\mathbf{a} = \langle \alpha a_1, \alpha a_2, \alpha a_3 \rangle$.

Definitions (i) and (ii) are called **vector addition** and **vector subtraction,** respectively, while (iii) is known as **scalar multiplication.**

Vector addition can be described geometrically as follows. Suppose **a** and **b** are two vectors drawn so that the terminal point of **a** and initial point of **b** coincide. With the vectors in this position, **a** + **b** is the vector from the initial point of **a** to the terminal point of **b** as illustrated in Figure 12.15.

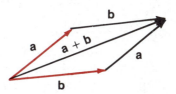

FIGURE 12.15

Since $\mathbf{b} + (\mathbf{a} - \mathbf{b}) = \mathbf{a}$, this description of addition enables us to geometrically represent vector subtraction as shown in Figure 12.16.

Scalar multiplication by a real number α has the effect of compressing or expanding the length of the vector and of reversing its direction in case $\alpha < 0$. (See Figure 12.17.)

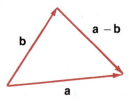

FIGURE 12.16

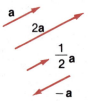

FIGURE 12.17

EXAMPLE 5   Find the sum of the vectors $\mathbf{a} = \langle 3, 1, 0 \rangle$ and $\mathbf{b} = \langle 2, 4, 0 \rangle$ and illustrate this sum geometrically.

SOLUTION

$\mathbf{a} + \mathbf{b} = \langle 3, 1, 0 \rangle + \langle 2, 4, 0 \rangle = \langle 5, 5, 0 \rangle$. Since these vectors all have a representation in the xy-plane, we use only the plane in Figure 12.18.   □

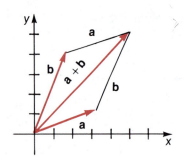

FIGURE 12.18

EXAMPLE 6   Find the result of the indicated operations when applied to the vectors $\mathbf{a} = \langle 3, -1, 2 \rangle$ and $\mathbf{b} = \langle 1, 0, -3 \rangle$.
(a) $\mathbf{a} + \mathbf{b}$ (b) $3\mathbf{a} - 2\mathbf{b}$

SOLUTION

(a) $\mathbf{a} + \mathbf{b} = \langle 3, -1, 2 \rangle + \langle 1, 0, -3 \rangle = \langle 3 + 1, -1 + 0, 2 - 3 \rangle$
    $= \langle 4, -1, -1 \rangle$ (see Figure 12.19(a)).

(b) $3\mathbf{a} - 2\mathbf{b} = 3\langle 3, -1, 2 \rangle - 2\langle 1, 0, -3 \rangle = \langle 9, -3, 6 \rangle - \langle 2, 0, -6 \rangle$
    $= \langle 7, -3, 12 \rangle$ (see Figure 12.19(b)).   □

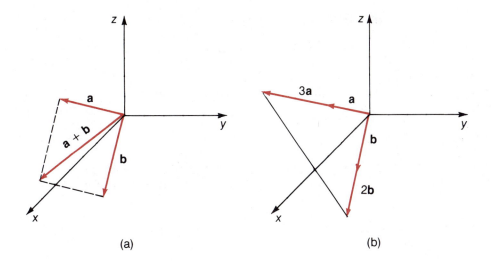

FIGURE 12.19

The next results follow easily from Definition 12.3. Proofs of portions of this theorem are considered in the exercises.

(12.4)
THEOREM

If **a**, **b** and **c** are vectors and α, β are real numbers, then

(i) $\mathbf{a} + \mathbf{b} = \mathbf{b} + \mathbf{a}$ (ii) $\mathbf{a} + (\mathbf{b} + \mathbf{c}) = (\mathbf{a} + \mathbf{b}) + \mathbf{c}$

(iii) $(\alpha\beta)\mathbf{a} = \alpha(\beta\mathbf{a})$ (iv) $(\alpha + \beta)\mathbf{a} = \alpha\mathbf{a} + \beta\mathbf{a}$

(v) $\mathbf{a} + \mathbf{0} = \mathbf{a}$ (vi) $0(\mathbf{a}) = \mathbf{0}$

(vii) $1\,(\mathbf{a}) = \mathbf{a}$ (viii) $\|\alpha\mathbf{a}\| = |\alpha| \cdot \|\mathbf{a}\|$

Scalar multiplication provides a precise manner for defining parallel vectors.

(12.5)
DEFINITION

Nonzero vectors **a** and **b** are called **parallel** if a real number α exists with $\mathbf{a} = \alpha\mathbf{b}$. (We also say that the zero vector is parallel to every vector **a**.)

Vectors with length 1 are called **unit vectors.** Any nonzero vector **v** can be multiplied by the reciprocal of its length, $\dfrac{1}{\|\mathbf{v}\|}$, to produce the unit vector, $\dfrac{\mathbf{v}}{\|\mathbf{v}\|}$, which has the same direction as **v**. The vector $\dfrac{-\mathbf{v}}{\|\mathbf{v}\|}$ is the unit vector in the direction opposite to **v**.

EXAMPLE 7

Find a unit vector that has the same direction as $\mathbf{v} = \langle 1, 1, 1 \rangle$ and a unit vector in the opposite direction.

SOLUTION

Since $\|\mathbf{v}\| = \sqrt{3}$, the vector $\dfrac{\mathbf{v}}{\|\mathbf{v}\|} = \dfrac{1}{\sqrt{3}}\langle 1, 1, 1 \rangle = \left\langle \dfrac{1}{\sqrt{3}}, \dfrac{1}{\sqrt{3}}, \dfrac{1}{\sqrt{3}} \right\rangle$ is the unit

vector with the same direction as $\mathbf{v}$ and $\left\langle \dfrac{-1}{\sqrt{3}}, \dfrac{-1}{\sqrt{3}}, \dfrac{-1}{\sqrt{3}} \right\rangle$ is the unit vector in the

opposite direction.   ☐

Of particular interest are the unit vectors in the positive direction of each of the coordinate axes (see Figure 12.20). These vectors are given special designations:

(12.6) $\mathbf{i} = \langle 1, 0, 0 \rangle, \qquad \mathbf{j} = \langle 0, 1, 0 \rangle, \qquad \mathbf{k} = \langle 0, 0, 1 \rangle.$

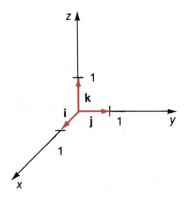

FIGURE 12.20

The results of Theorem 12.4 allow any vector $\mathbf{v} = \langle v_1, v_2, v_3 \rangle$ to be expressed in terms of $\mathbf{i}$, $\mathbf{j}$, and $\mathbf{k}$ as

$$\mathbf{v} = \langle v_1, v_2, v_3 \rangle = v_1\mathbf{i} + v_2\mathbf{j} + v_3\mathbf{k}.$$

The numbers v_1, v_2, and v_3 are called the **components** of $\mathbf{v}$ in the x-, y-, and z- directions, respectively.

EXAMPLE 8   Express the vector $\langle -3, 2, 0 \rangle$ in terms of $\mathbf{i}$, $\mathbf{j}$, and $\mathbf{k}$ and illustrate this representation geometrically.

SOLUTION

$$\langle -3, 2, 0 \rangle = -3\mathbf{i} + 2\mathbf{j} + 0\mathbf{k} = -3\mathbf{i} + 2\mathbf{j}$$

The representation is shown in the xy-plane in Figure 12.21(a). Figure 12.21(b) shows the representation in three-dimensional space.   ☐

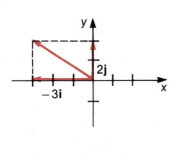

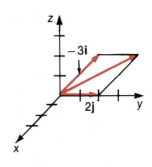

(a)   (b)

FIGURE 12.21

EXERCISE SET 12.2

In Exercises 1 through 6, the initial point A and terminal point B of a vector are given. Find the position vector representation and length of the vector.

1. $A(0, 0, 0)$, $B(1, 5, 3)$

2. $A(0, 2, 0)$, $B(0, 2, 7)$

3. $A(2, 3, 4)$, $B(2, -3, 4)$

4. $A(1, 2, 0)$, $B(3, 5, 0)$

5. $A(-2, 2, 1)$, $B(0, 5, 0)$

6. $A(4, 1, 0)$, $B(1, -2, 0)$

In Exercises 7 through 10, a vector and its initial point A are given. Find the terminal point of the vector.

7. $\mathbf{v} = \langle 1, 3, 4 \rangle$, $A(0, 0, 0)$

8. $\mathbf{v} = \langle 1, 3, 4 \rangle$, $A(2, 1, 0)$

9. $\mathbf{v} = \langle -1, 2, 0 \rangle$, $A(-4, 1, 1)$

10. $\mathbf{v} = \langle 2, -2, 1 \rangle$, $A(0, 3, 3)$

In Exercises 11 through 18, $\mathbf{a} = \langle 1, 1, 0 \rangle$, $\mathbf{b} = \langle -1, 1, 2 \rangle$, and $\mathbf{c} = \langle 2, 3, 4 \rangle$.

11. Find $\mathbf{a} + \mathbf{b}$ and illustrate this sum geometrically.

12. Find $3\mathbf{a}$. Sketch $\mathbf{a}$ and $3\mathbf{a}$ using the same rectangular coordinate system.

13. Find $-2\mathbf{c}$. Sketch $\mathbf{c}$ and $-2\mathbf{c}$ using the same rectangular coordinate system.

14. Find $\mathbf{a} - \mathbf{c}$ and illustrate this difference geometrically.

15. Find $\mathbf{a} + \mathbf{b} + \mathbf{c}$.

16. Find $2\mathbf{a} - 3\mathbf{b} + 4\mathbf{c}$.

17. Find $\|\mathbf{c}\|$ and $\|-2\mathbf{c}\|$.

18. Find $\|\mathbf{a}\|$, $\|\mathbf{b}\|$, and $\|\mathbf{a} + \mathbf{b}\|$.

19. Find a unit vector that has the same direction as

(a) $\langle 2, 3, 4 \rangle$ (b) $\langle -1, 2, 3 \rangle$.

20. Find a unit vector that has direction opposite that of

(a) $\langle 2, 3, 4 \rangle$ (b) $\langle -1, 2, 3 \rangle$.

21. Find a vector parallel to $\langle 3, 4, -1 \rangle$ that has length 3.

22. Find a vector parallel to $\langle 2, 4, 4 \rangle$ that has length 7.

23. Express each of the following vectors in terms of **i**, **j**, and **k** and illustrate the representation geometrically.

(a) $\langle 1, 4, 0 \rangle$ (b) $\langle 1, 4, 5 \rangle$ (c) $\langle 0, 2, 3 \rangle$

24. Determine which of the following pairs of vectors are parallel.

(a) $\langle 1, 0, 1 \rangle, \langle -1, 0, -1 \rangle$ (b) $\langle 1, 0, 0 \rangle, \langle 1, 1, 0 \rangle$
(c) $\langle 2, 3, -1 \rangle, \langle 4, 6, -2 \rangle$ (d) $\langle 3, -1, 2 \rangle, \langle -6, 2, 4 \rangle$

25. Show that if **a** and **b** are vectors and α is a real number then $\alpha(\mathbf{a} + \mathbf{b}) = \alpha\mathbf{a} + \alpha\mathbf{b}$.

26. Show that if **a** is a vector and α is a real number, then $\|\alpha\mathbf{a}\| = |\alpha| \|\mathbf{a}\|$.

27. A canoeist crosses a 300-foot stream in one minute, arriving at a point directly opposite the starting point. The current in the stream is 3 feet per second. In what direction did the canoeist paddle and at what average speed?

28. A steady easterly wind is blowing at 20 ft/sec and pulling on a helium-filled balloon attached to a string. In still air, the balloon would rise vertically at 5 ft/sec. Where is the balloon and how much string has been released when the balloon is 250 feet from the ground?

29. A private pilot flies for one hour and arrives at a point 100 miles due north of his departure point. A steady wind was blowing from the northwest at 20 miles per hour. In what direction did the pilot fly, and at what average speed?

30. Suppose that the pilot of the plane in Exercise 29 flies to a point 100 miles due south instead of due north. Assuming that the other conditions remain the same, in what direction and at what average speed did the plane fly?

12.3
THE DOT PRODUCT OF VECTORS

Addition, subtraction, and scalar multiplication apply quite naturally to vectors because of the convenient geometric representation. In this section we introduce a type of multiplication of two vectors that produces a real number. This operation is known as the **dot product** of two vectors. (It is also referred to as the *inner* or *scalar product*.)

(12.7)
DEFINITION

The **dot product** of the vectors $\mathbf{a} = \langle a_1, a_2, a_3 \rangle$ and $\mathbf{b} = \langle b_1, b_2, b_3 \rangle$ is

$$\mathbf{a} \cdot \mathbf{b} = a_1 b_1 + a_2 b_2 + a_3 b_3.$$

EXAMPLE 1

For vectors $\mathbf{a} = \langle 1, -1, 1 \rangle$ and $\mathbf{b} = \langle -2, 3, 1 \rangle$, find
(a) $\mathbf{a} \cdot \mathbf{b}$ (b) $\mathbf{a} \cdot \mathbf{a}$ (c) $\mathbf{b} \cdot \mathbf{i}$ (d) $\mathbf{b} \cdot \mathbf{j}$

SOLUTION
(a) $\mathbf{a} \cdot \mathbf{b} = \langle 1, -1, 1 \rangle \cdot \langle -2, 3, 1 \rangle = (1)(-2) + (-1)(3) + (1)(1) = -4$
(b) $\mathbf{a} \cdot \mathbf{a} = \langle 1, -1, 1 \rangle \cdot \langle 1, -1, 1 \rangle = 1 + 1 + 1 = 3$
(c) $\mathbf{b} \cdot \mathbf{i} = \langle -2, 3, 1 \rangle \cdot \langle 1, 0, 0 \rangle = -2 + 0 + 0 = -2$
(d) $\mathbf{b} \cdot \mathbf{j} = \langle -2, 3, 1 \rangle \cdot \langle 0, 1, 0 \rangle = 0 + 3 + 0 = 3$ $\square$

Some immediate consequences of the definition of the dot product are listed in the following theorem.

(12.8)
THEOREM

If $\mathbf{a}$, $\mathbf{b}$, and $\mathbf{c}$ **are vectors and** α **is a real number, then**
(i) $\mathbf{a} \cdot \mathbf{b} = \mathbf{b} \cdot \mathbf{a}$ (ii) $(\mathbf{a} + \mathbf{b}) \cdot \mathbf{c} = \mathbf{a} \cdot \mathbf{c} + \mathbf{b} \cdot \mathbf{c}$
(iii) $(\alpha \mathbf{a}) \cdot \mathbf{b} = \alpha(\mathbf{a} \cdot \mathbf{b})$ (iv) $\mathbf{0} \cdot \mathbf{a} = 0$
(v) $\mathbf{a} \cdot \mathbf{a} = \|\mathbf{a}\|^2$
Additionally,
(vi) $\mathbf{i} \cdot \mathbf{i} = \mathbf{j} \cdot \mathbf{j} = \mathbf{k} \cdot \mathbf{k} = 1$ (vii) $\mathbf{i} \cdot \mathbf{j} = \mathbf{i} \cdot \mathbf{k} = \mathbf{j} \cdot \mathbf{k} = 0$

An important consequence of the dot product is that it allows us to compute the angle between two vectors. The **angle between nonzero vectors a and b** is the angle θ in $[0, \pi]$ determined by the position vector representations of $\mathbf{a}$ and $\mathbf{b}$. (See Figure 12.22.) If A denotes the terminal point of $\mathbf{a}$ and B denotes the terminal point of $\mathbf{b}$, then the angle between $\mathbf{a}$ and $\mathbf{b}$ is angle AOB.

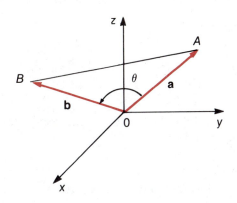

FIGURE 12.22

When **a** and **b** are parallel vectors and $\mathbf{b} = \alpha\mathbf{a}$, we say the angle between **a** and **b** is 0 if $\alpha > 0$ and π if $\alpha < 0$.

(12.9)
THEOREM

If θ is the angle between two vectors **a** and **b**, then

$$\mathbf{a} \cdot \mathbf{b} = \|\mathbf{a}\| \|\mathbf{b}\| \cos \theta.$$

PROOF

Consider the position vector representations of **a** and **b**. By connecting the terminal points of these vectors, a triangle is formed with sides of length $\|\mathbf{a}\|$, $\|\mathbf{b}\|$, and $\|\mathbf{b} - \mathbf{a}\|$ (see Figure 12.23).

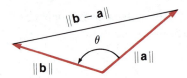

FIGURE 12.23

The law of cosines (Equation A.26) applied to this triangle implies that

$$\|\mathbf{b} - \mathbf{a}\|^2 = \|\mathbf{a}\|^2 + \|\mathbf{b}\|^2 - 2\|\mathbf{a}\| \|\mathbf{b}\| \cos \theta.$$

It follows from Theorem 12.8 that

$$\begin{aligned}\|\mathbf{b} - \mathbf{a}\|^2 &= (\mathbf{b} - \mathbf{a}) \cdot (\mathbf{b} - \mathbf{a}) \\ &= \mathbf{b} \cdot \mathbf{b} - \mathbf{a} \cdot \mathbf{b} - \mathbf{b} \cdot \mathbf{a} + \mathbf{a} \cdot \mathbf{a} \\ &= \|\mathbf{b}\|^2 - 2\mathbf{a} \cdot \mathbf{b} + \|\mathbf{a}\|^2.\end{aligned}$$

So

$$\|\mathbf{b}\|^2 - 2\mathbf{a} \cdot \mathbf{b} + \|\mathbf{a}\|^2 = \|\mathbf{a}\|^2 + \|\mathbf{b}\|^2 - 2\|\mathbf{a}\| \|\mathbf{b}\| \cos \theta$$

and

$$-2\mathbf{a} \cdot \mathbf{b} = -2\|\mathbf{a}\| \|\mathbf{b}\| \cos \theta.$$

Hence,

$$\mathbf{a} \cdot \mathbf{b} = \|\mathbf{a}\| \|\mathbf{b}\| \cos \theta. \qquad \square$$

EXAMPLE 2

Find the dot product of the vectors $\langle 3, 0, -1 \rangle$ and $\langle 2, 1, 2 \rangle$ and determine the angle between them.

SOLUTION

The dot product of these vectors is

$$\langle 3, 0, -1 \rangle \cdot \langle 2, 1, 2 \rangle = 6 + 0 - 2 = 4.$$

Since

$$\|\langle 3, 0, -1 \rangle\| = \sqrt{10} \qquad \text{and} \qquad \|\langle 2, 1, 2 \rangle\| = \sqrt{9} = 3,$$

it follows from Theorem 12.9 that

$$\cos \theta = \frac{\mathbf{a} \cdot \mathbf{b}}{\|\mathbf{a}\| \|\mathbf{b}\|} = \frac{4}{3\sqrt{10}} = \frac{2\sqrt{10}}{15},$$

so

$$\theta = \arccos \left(\frac{2\sqrt{10}}{15} \right) \approx 1.14. \qquad \square$$

EXAMPLE 3   Show that the points $(0, 0, 0)$, $(3, 1, 0)$ and $(-2, 6, 0)$ are vertices of a right triangle.

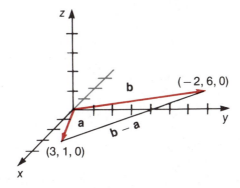

FIGURE 12.24

SOLUTION

The position vector representation of the vectors $\mathbf{a} = \langle 3, 1, 0 \rangle$ and $\mathbf{b} = \langle -2, 6, 0 \rangle$ are two sides of the triangle whose third side is the vector $\mathbf{b} - \mathbf{a}$. See Figure 12.24. To show that this is a right triangle, we find the angle θ between $\mathbf{a}$ and $\mathbf{b}$.

Since

$$\mathbf{a} \cdot \mathbf{b} = \langle 3, 1, 0 \rangle \cdot \langle -2, 6, 0 \rangle = -6 + 6 + 0 = 0,$$

$$\cos \theta = \frac{\mathbf{a} \cdot \mathbf{b}}{\|\mathbf{a}\| \|\mathbf{b}\|} = 0,$$

and hence $\theta = \pi/2$. (An alternative method for solving this problem would be to show that $\|\mathbf{a}\|^2 + \|\mathbf{b}\|^2 = \|\mathbf{a} - \mathbf{b}\|^2$. This is considered in Exercise 34.) $\square$

Motivated by the preceding example, we make the following definition.

(12.10)
DEFINITION
Vectors $\mathbf{a}$ and $\mathbf{b}$ are said to be **orthogonal** (or perpendicular) if θ, the angle between them, satisfies $\cos \theta = 0$. (In addition, we say that $\mathbf{0}$ is orthogonal to every vector.)

A corollary of Theorem 12.9 follows from this definition.

(12.11)
COROLLARY   Vectors **a** and **b** are orthogonal if and only if $\mathbf{a} \cdot \mathbf{b} = 0$.

Three angles of particular interest for a vector $\mathbf{v} = \langle v_1, v_2, v_3 \rangle$ are the angles α between **v** and **i**, β between **v** and **j**, and γ between **v** and **k**. (See Figure 12.25.) These are called the **direction angles** of **v**. The numbers $\cos \alpha$, $\cos \beta$, and $\cos \gamma$ are called the **direction cosines** of **v**.

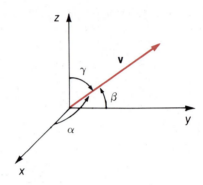

FIGURE 12.25

(12.12)
THEOREM   The direction cosines of a nonzero vector $\mathbf{v} = \langle v_1, v_2, v_3 \rangle$ are

$$\cos \alpha = \frac{v_1}{\|\mathbf{v}\|}, \quad \cos \beta = \frac{v_2}{\|\mathbf{v}\|}, \quad \cos \gamma = \frac{v_3}{\|\mathbf{v}\|}$$

and

$$(\cos \alpha)^2 + (\cos \beta)^2 + (\cos \gamma)^2 = 1.$$

PROOF

$$\cos \alpha = \frac{\mathbf{v} \cdot \mathbf{i}}{\|\mathbf{v}\| \|\mathbf{i}\|} = \frac{v_1}{\|\mathbf{v}\|}, \cos \beta = \frac{\mathbf{v} \cdot \mathbf{j}}{\|\mathbf{v}\| \|\mathbf{j}\|} = \frac{v_2}{\|\mathbf{v}\|}, \text{ and } \cos \gamma = \frac{\mathbf{v} \cdot \mathbf{k}}{\|\mathbf{v}\| \|\mathbf{k}\|} = \frac{v_3}{\|\mathbf{v}\|},$$

so $(\cos \alpha)^2 + (\cos \beta)^2 + (\cos \gamma)^2 = \dfrac{v_1^2}{\|\mathbf{v}\|^2} + \dfrac{v_2^2}{\|\mathbf{v}\|^2} + \dfrac{v_3^2}{\|\mathbf{v}\|^2} = \dfrac{\|\mathbf{v}\|^2}{\|\mathbf{v}\|^2} = 1$ □

EXAMPLE 4   Find the direction cosines and direction angles of the vector $\mathbf{v} = \langle 1, -2, 2 \rangle$.

SOLUTION

$$\|\mathbf{v}\| = \sqrt{1 + 4 + 4} = 3, \text{ so } \cos \alpha = \frac{1}{3}, \cos \beta = -\frac{2}{3}, \text{ and } \cos \gamma = \frac{2}{3}.$$

So $\alpha = \arccos \left(\dfrac{1}{3} \right) \approx 1.23, \beta = \arccos \left(-\dfrac{2}{3} \right) \approx 2.30$, and

$$\gamma = \arccos \left(\frac{2}{3} \right) \approx .84.$$ □

Since any vector $\mathbf{v} = \langle v_1, v_2, v_3 \rangle$ can be written as $\mathbf{v} = v_1\mathbf{i} + v_2\mathbf{j} + v_3\mathbf{k}$, Theorem 12.12 implies that

(12.13) $\mathbf{v} = \|\mathbf{v}\| \, (\cos \alpha \, \mathbf{i} + \cos \beta \, \mathbf{j} + \cos \gamma \, \mathbf{k})$.

If $\mathbf{v} = \langle v_1, v_2 \rangle$ is a vector in the xy-plane, then α is the angle between $\mathbf{v}$ and the positive x-axis and β is the angle between $\mathbf{v}$ and the positive y-axis. (See Figure 12.26.) In this case

(i) $\cos \alpha = \dfrac{v_1}{\|\mathbf{v}\|}, \qquad \cos \beta = \sin \alpha = \dfrac{v_2}{\|\mathbf{v}\|}$

(12.14)

(ii) $v = \|\mathbf{v}\|(\cos \alpha \, \mathbf{i} + \sin \alpha \, \mathbf{j})$.

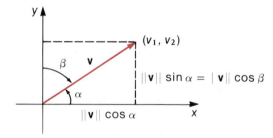

FIGURE 12.26

It follows from Theorem 12.9 that for any vectors $\mathbf{a}$ and $\mathbf{b}$,

(12.15) $|\mathbf{a} \cdot \mathbf{b}| = \|\mathbf{a}\| \, \|\mathbf{b}\| \, |\cos \theta| \leq \|\mathbf{a}\| \, \|\mathbf{b}\|$.

Inequality (12.15), known as the **Cauchy-Buniakowsky-Schwarz inequality,** is used to establish the important result given in Theorem 12.16. This result is known as the **triangle** (or Minkowski) **inequality.**

(12.16)
THEOREM   For any vectors $\mathbf{a}$ and $\mathbf{b}$, $\|\mathbf{a} + \mathbf{b}\| \leq \|\mathbf{a}\| + \|\mathbf{b}\|$.

PROOF
Consider $\|\mathbf{a} + \mathbf{b}\|^2$ and apply Theorem 12.8 (v) and (12.15).

$$\|\mathbf{a} + \mathbf{b}\|^2 = (\mathbf{a} + \mathbf{b}) \cdot (\mathbf{a} + \mathbf{b})$$
$$= \mathbf{a} \cdot \mathbf{a} + \mathbf{b} \cdot \mathbf{a} + \mathbf{a} \cdot \mathbf{b} + \mathbf{b} \cdot \mathbf{b}$$
$$= \|\mathbf{a}\|^2 + 2\mathbf{a} \cdot \mathbf{b} + \|\mathbf{b}\|^2$$
$$\leq \|\mathbf{a}\|^2 + 2|\mathbf{a} \cdot \mathbf{b}| + \|\mathbf{b}\|^2$$
$$\leq \|\mathbf{a}\|^2 + 2\|\mathbf{a}\| \, \|\mathbf{b}\| + \|\mathbf{b}\|^2 = (\|\mathbf{a}\| + \|\mathbf{b}\|)^2.$$

Taking the positive square root of each side produces the desired result.   □

If Theorem 12.16 is interpreted geometrically, it says that the length of one side of a triangle cannot exceed the sum of the lengths of the remaining two sides. (See Figure 12.27.)

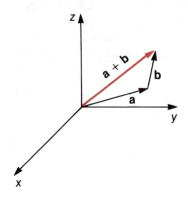

FIGURE 12.27

Certain applications require that a given vector be decomposed into a sum of two vectors of a special type. For any fixed nonzero vector **a**, a vector **b** can be written as the sum of a vector parallel to **a** and another vector orthogonal to **a**. The vector parallel to **a** is called the **orthogonal projection** of **b** onto **a** and denoted **proj$_a$b**.

It can be seen from Figure 12.28 that the length of the orthogonal projection of **b** onto **a** is

$$\|\mathbf{proj_a b}\| = \left| \|\mathbf{b}\| \cos \theta \right| = \left| \|\mathbf{b}\| \frac{\mathbf{a} \cdot \mathbf{b}}{\|\mathbf{a}\| \|\mathbf{b}\|} \right| = \left| \frac{\mathbf{a} \cdot \mathbf{b}}{\|\mathbf{a}\|} \right|.$$

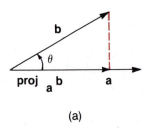

(a)

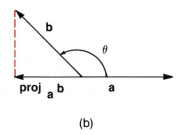

(b)

FIGURE 12.28

The number $\dfrac{\mathbf{a} \cdot \mathbf{b}}{\|\mathbf{a}\|}$ is called the component of **b** in the direction of **a** and denoted comp$_a$**b**.

Since $\dfrac{\mathbf{a}}{\|\mathbf{a}\|}$ is the unit vector in the direction of **a**,

$$\mathbf{proj_a b} = \mathbf{comp_a b} \, \frac{\mathbf{a}}{\|\mathbf{a}\|}.$$

The details of the decomposition of a vector **b** into the sum of two vectors, one parallel to **a**, the other orthogonal to **a**, are given in the following theorem.

(12.17)
THEOREM

If **a** is a nonzero vector, then any vector **b** can be written $\mathbf{b} = \mathbf{b_1} + \mathbf{b_2}$, where

$$\mathbf{b_1} = \mathbf{proj_a b}$$

is parallel to **a** and

$$\mathbf{b_2} = \mathbf{b} - \mathbf{b_1}$$

is orthogonal to **a**.

PROOF

By definition of $\mathbf{proj_a b}$, $\mathbf{b_1}$ is parallel to **a**. With $\mathbf{b_2}$ defined by $\mathbf{b_2} = \mathbf{b} - \mathbf{b_1}$, it is clear that $\mathbf{b} = \mathbf{b_1} + \mathbf{b_2}$. It remains to show that $\mathbf{b_2}$ is orthogonal to **a**, that is, that $\mathbf{a} \cdot \mathbf{b_2} = 0$.

$$\mathbf{a} \cdot \mathbf{b_2} = \mathbf{a} \cdot (\mathbf{b} - \mathbf{b_1})$$
$$= \mathbf{a} \cdot \mathbf{b} - \mathbf{a} \cdot \mathbf{proj_a b}$$
$$= \mathbf{a} \cdot \mathbf{b} - \mathbf{a} \cdot \mathrm{comp_a} \mathbf{b} \left(\frac{\mathbf{a}}{\|\mathbf{a}\|} \right)$$
$$= \mathbf{a} \cdot \mathbf{b} - \mathrm{comp_a} \mathbf{b} \left(\frac{\mathbf{a} \cdot \mathbf{a}}{\|\mathbf{a}\|} \right)$$
$$= \mathbf{a} \cdot \mathbf{b} - \frac{\mathbf{a} \cdot \mathbf{b}}{\|\mathbf{a}\|} \left(\frac{\|\mathbf{a}\|^2}{\|\mathbf{a}\|} \right) = 0. \qquad \square$$

EXAMPLE 5

Decompose $\mathbf{b} = \langle 2, 3, -1 \rangle$ into a sum $\mathbf{b_1} + \mathbf{b_2}$, where $\mathbf{b_1}$ is parallel to $\mathbf{a} = \langle 0, 4, 2 \rangle$ and $\mathbf{b_2}$ is orthogonal to **a**.

SOLUTION

$$\mathrm{comp_a} \mathbf{b} = \frac{\mathbf{a} \cdot \mathbf{b}}{\|\mathbf{a}\|} = \frac{\langle 0, 4, 2 \rangle \cdot \langle 2, 3, -1 \rangle}{\sqrt{0 + 16 + 4}} = \frac{10}{2\sqrt{5}} = \sqrt{5}$$

and

$$\mathbf{proj_a b} = \mathrm{comp_a} \mathbf{b} \left(\frac{\mathbf{a}}{\|\mathbf{a}\|} \right) = \sqrt{5} \frac{\langle 0, 4, 2 \rangle}{2\sqrt{5}} = \langle 0, 2, 1 \rangle.$$

So $\mathbf{b_1} = \langle 0, 2, 1 \rangle$ and $\mathbf{b_2} = \langle 2, 3, -1 \rangle - \langle 0, 2, 1 \rangle = \langle 2, 1, -2 \rangle$. Figure 12.29 shows the decomposition. $\qquad \square$

An application of the orthogonal projection of one vector onto another arises in the definition of the work done by a force in moving an object. Recall from Section 6.7 that if a constant force F is applied in the direction of motion, then the work W done by F in moving an object a distance D is given by

$$W = (\text{force})(\text{distance}) = FD.$$

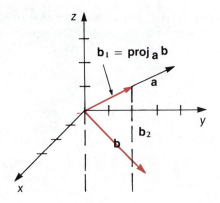

FIGURE 12.29

If a vector force **F** is constant, but is applied at an angle to the direction of motion (as in Figure 12.30), then the work done by **F** is defined by

$$W = \text{(component of force in the direction of motion)(distance)}$$

$$= (\text{comp}_{\vec{PQ}}\, \mathbf{F})\, \|\vec{PQ}\|.$$

Thus,

(12.18) $W = (\|\mathbf{F}\| \cos \theta)\, \|\vec{PQ}\| = \mathbf{F} \cdot \vec{PQ}.$

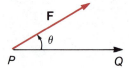

FIGURE 12.30

EXAMPLE 6   A toboggan loaded with camping gear is pulled 500 feet across the snow with a rope that makes an angle of $\pi/4$ with the ground. Find the work done if a force of 20 pounds is exerted in pulling the toboggan.

SOLUTION

The situation is illustrated in Figure 12.31. The component of force in the direction of motion is

$$20\left(\cos \frac{\pi}{4}\right) = 20\, \frac{\sqrt{2}}{2} = 10\sqrt{2} \text{ lbs.}$$

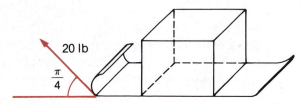

FIGURE 12.31

The work done is

$$W = (10\sqrt{2})(500) = 5000\sqrt{2} \text{ ft lbs.}$$

EXERCISE SET 12.3

Find $\mathbf{a} \cdot \mathbf{b}$ for the pairs of vectors given in Exercises 1 through 8.

1. $\mathbf{a} = \langle 1, 3, 3 \rangle$, $\mathbf{b} = \langle 2, 2, 4 \rangle$ **2.** $\mathbf{a} = \langle -1, -3, 2 \rangle$, $\mathbf{b} = \langle 1, 3, 2 \rangle$

3. $\mathbf{a} = \langle 2, 4, 0 \rangle$, $\mathbf{b} = \langle 3, 7, 0 \rangle$ **4.** $\mathbf{a} = \langle 1, -3, 2 \rangle$, $\mathbf{b} = \langle 2, 2, 2 \rangle$

5. $\mathbf{a} = \langle 0, 4, 6 \rangle$, $\mathbf{b} = \langle 0, -3, 2 \rangle$ **6.** $\mathbf{a} = \langle 1, 5, -3 \rangle$, $\mathbf{b} = \langle 3, 1, 1 \rangle$

7. $\mathbf{a} = \langle 3, 4, \pi \rangle$, $\mathbf{b} = \langle 1, e, 0 \rangle$

8. $\mathbf{a} = \langle -\sqrt{2}, \sqrt{3}, 1 \rangle$, $\mathbf{b} = \langle \sqrt{3}, 0, \sqrt{2} \rangle$

9–16. Find the angle between the pairs of vectors given in Exercises 1 through 8 and determine in each case whether the vectors are orthogonal.

17–24. Find the component of $\mathbf{b}$ in the direction of $\mathbf{a}$ and the orthogonal projection of $\mathbf{b}$ onto $\mathbf{a}$ for the vectors given in Exercises 1 through 8.

Find the direction cosines of each of the vectors in Exercises 25 through 28.

25. $\langle 2, 3, 4 \rangle$ **26.** $\langle -1, 3, -3 \rangle$

27. $\langle 0, 3, 4 \rangle$ **28.** $\langle 2, 3, 0 \rangle$

29. Find a unit vector that has the same direction cosines as the vector $\langle 0, 3, 4 \rangle$.

30. Find a vector lying in the xy-plane that is orthogonal to the vector $\langle 2, 4, 0 \rangle$. Is there more than one such vector?

31. Find a unit vector lying in the xy-plane that is orthogonal to the vector $\langle 2, 4, 0 \rangle$. Is there more than one such vector?

32. Find the component of $\mathbf{b} = \langle b_1, b_2, b_3 \rangle$ in the direction of each of the vectors $\mathbf{i}$, $\mathbf{j}$, and $\mathbf{k}$ and find the orthogonal projection of $\mathbf{b}$ onto each of $\mathbf{i}$, $\mathbf{j}$, and $\mathbf{k}$.

33. Express $\mathbf{b} = \langle -4, 1, -2 \rangle$ as $\mathbf{b} = \mathbf{b}_1 + \mathbf{b}_2$ where $\mathbf{b}_1$ is parallel to $\mathbf{a} = \langle 1, 3, -3 \rangle$ and $\mathbf{b}_2$ is orthogonal to $\mathbf{a}$.

34. Show that $(0, 0, 0)$, $(3, 1, 0)$ and $(-2, 6, 0)$ are vertices of a right triangle by showing that $\|\mathbf{a}\|^2 + \|\mathbf{b}\|^2 = \|\mathbf{a} - \mathbf{b}\|^2$ for appropriate vectors $\mathbf{a}$ and $\mathbf{b}$.

35. Show that $(2, 2, 2)$, $(2, 0, 1)$ and $(4, 1, -1)$ are vertices of a right triangle.

36. Write an equation to describe all points (x, y, z) with the property that the vector $\langle x - 1, y - 2, z - 3 \rangle$ is orthogonal to the vector $\langle 1, -1, 2 \rangle$.

37. Find a vector that is orthogonal to both of the vectors $\mathbf{a} = \langle 1, 2, 4 \rangle$ and $\mathbf{b} = \langle 2, -2, 5 \rangle$. Is there more than one such vector?

38. Prove that if $\mathbf{a}$ and $\mathbf{b}$ are vectors, then $\mathbf{a} \cdot \mathbf{b} = \mathbf{b} \cdot \mathbf{a}$.

39. Prove that if $\mathbf{a}$ is a vector, then $\mathbf{a} \cdot \mathbf{a} = \|\mathbf{a}\|^2$.

40. Prove that if $\mathbf{a}$ and $\mathbf{b}$ are vectors, then $\|\mathbf{a} - \mathbf{b}\| \geq \|\mathbf{a}\| - \|\mathbf{b}\|$. (*Hint*: Write $\mathbf{a}$ as $\mathbf{b} + (\mathbf{a} - \mathbf{b})$ and use the triangle inequality given in Theorem 12.16.)

41. Show that if $\mathbf{v}$ is orthogonal to $\mathbf{a}$ and $\mathbf{b}$, then $\mathbf{v}$ is orthogonal to $c_1\mathbf{a} + c_2\mathbf{b}$ for all real numbers c_1 and c_2.

42. Show that vectors **a** and **b** are orthogonal if and only if

$$\|\mathbf{a} + \mathbf{b}\|^2 = \|\mathbf{a}\|^2 + \|\mathbf{b}\|^2.$$

43. Find the work done by the force $\mathbf{F} = \mathbf{j} + 2\mathbf{k}$ (in pounds) applied to an object that moves two feet along the y-axis.

44. A wagon loaded with groceries is pulled horizontally a half mile by a handle that makes an angle of $\pi/3$ with the horizontal. Find the work done if a force of 20 pounds is exerted on the handle.

45. A block is to be moved 6 feet up a ramp that makes an angle of $\pi/4$ with the horizontal. (a) Find the work done if 20 pounds of force are applied in the direction of the motion. (b) Find the work done if 20 pounds of force are applied at an angle of $\pi/6$ with the horizontal.

46. A 200-pound piece of slate for the top of a pool table is moved along a horizontal surface by a pulling force applied from above. If the force is applied at an angle of $\pi/3$ to the slate, what force must be applied to move it?

12.4
THE CROSS PRODUCT OF VECTORS

In the previous section we examined a product of a pair of vectors that produces a scalar. In this section we consider a different product of vectors. This operation produces a vector and is known as the **cross product** (or *vector product*) of two three-dimensional vectors. Since it is somehwat easier to describe the geometric form of the vector product than the analytic form, we give the geometric description first.

Suppose **a** and **b** are a pair of three-dimensional nonzero vectors that are not scalar multiples of each other. These vectors can be positioned to form a parallelogram as shown in Figure 12.32.

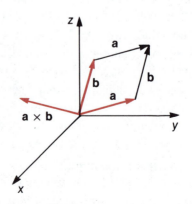

FIGURE 12.32

The cross product of **a** and **b**, written **a** × **b**, is a vector that is orthogonal to both **a** and **b** and has length equal to the area of this parallelogram. There are two vectors in three-dimensional space satisfying this condition: one "above" the plane of the vectors **a** and **b** and one "below" this plane. The vector **a** × **b** is the one whose direction is specified by the following "right-hand rule."

Let θ denote the angle between the vectors **a** and **b** and suppose **a** is rotated through the angle θ to coincide with **b**. If the right hand is positioned so that the wrist is at the common initial point of these vectors and the natural fold of the fingers points in the direction of this rotation, then the thumb will point in the direction of **a** × **b**. (See Figure 12.33.)

Note that this "right-hand" rule indicates that

$$\mathbf{a} \times \mathbf{b} = -(\mathbf{b} \times \mathbf{a}).$$

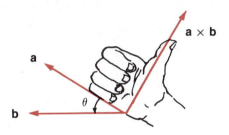

FIGURE 12.33

EXAMPLE 1   Find the cross product **i** × **j**.

SOLUTION
The parallelogram formed by **i** and **j** is a square with area 1.

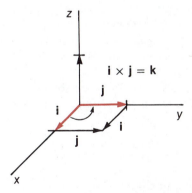

FIGURE 12.34

Figure 12.34 indicates that the direction of **i** × **j** coincides with that of the positive z-axis, so

$$\mathbf{i} \times \mathbf{j} = \mathbf{k}. \qquad \square$$

To calculate the cross product of more general vectors and to establish results about the operation, an analytic definition is needed.

(12.19)
DEFINITION

The **cross** (or *vector*) **product a $\times$ b** of the vectors $\mathbf{a} = \langle a_1, a_2, a_3 \rangle$ and $\mathbf{b} = \langle b_1, b_2, b_3 \rangle$ is

$$\mathbf{a} \times \mathbf{b} = \langle a_2 b_3 - a_3 b_2, a_3 b_1 - a_1 b_3, a_1 b_2 - a_2 b_1 \rangle.$$

EXAMPLE 2

Use Definition 12.19 to

(a) show that $\mathbf{i} \times \mathbf{j} = \mathbf{k}$ and (b) find $\langle 1, 2, 1 \rangle \times \langle 3, 0, 1 \rangle$.

SOLUTION

(a) Since $\mathbf{i} = \langle 1, 0, 0 \rangle$ and $\mathbf{j} = \langle 0, 1, 0 \rangle$,

$$\mathbf{i} \times \mathbf{j} = \langle 0 \cdot 0 - 0 \cdot 1, 0 \cdot 0 - 1 \cdot 0, 1 \cdot 1 - 0 \cdot 0 \rangle = \langle 0, 0, 1 \rangle = \mathbf{k}.$$

(b) $\langle 1, 2, 1 \rangle \times \langle 3, 0, 1 \rangle = \langle 2 \cdot 1 - 1 \cdot 0, 1 \cdot 3 - 1 \cdot 1, 1 \cdot 0 - 2 \cdot 3 \rangle$
$$= \langle 2, 2, -6 \rangle. \qquad \square$$

The analytic definition of $\mathbf{a} \times \mathbf{b}$ must, of course, agree with the geometric definition. To show that $\mathbf{a} \times \mathbf{b}$ is orthogonal to $\mathbf{a}$, we find $\mathbf{a} \cdot (\mathbf{a} \times \mathbf{b})$:

$$\mathbf{a} \cdot (\mathbf{a} \times \mathbf{b}) = \langle a_1, a_2, a_3 \rangle \cdot \langle a_2 b_3 - a_3 b_2, a_3 b_1 - a_1 b_3, a_1 b_2 - a_2 b_1 \rangle$$
$$= a_1 a_2 b_3 - a_1 a_3 b_2 + a_2 a_3 b_1 - a_2 a_1 b_3 + a_3 a_1 b_2 - a_3 a_2 b_1$$
$$= 0.$$

Similarly, $\mathbf{a} \times \mathbf{b}$ is orthogonal to $\mathbf{b}$. At the end of this section, we show that Definition 12.19 also implies that the magnitude of $\mathbf{a} \times \mathbf{b}$ is the area of the parallelogram formed by $\mathbf{a}$ and $\mathbf{b}$.

Although the analytic definition of the vector product appears to be a rather complicated expression, there is a simple mnemonic device for remembering this result. To describe this device, we need determinant notation.

A square array of numbers is an array with the same number of rows and columns. Assigned to each square array is a number called the **determinant** of the array. For an array with two rows and two columns, the determinant is defined by

$$\begin{vmatrix} a & b \\ c & d \end{vmatrix} = ad - bc.$$

The determinant of an array with three rows and three columns is defined by

$$\begin{vmatrix} a & b & c \\ d & e & f \\ g & h & i \end{vmatrix} = a \begin{vmatrix} e & f \\ h & i \end{vmatrix} - b \begin{vmatrix} d & f \\ g & i \end{vmatrix} + c \begin{vmatrix} d & e \\ g & h \end{vmatrix}$$
$$= a(ei - fh) - b(di - fg) + c(dh - eg).$$

Since $\mathbf{a} \times \mathbf{b} = (a_2 b_3 - a_3 b_2)\mathbf{i} - (a_1 b_3 - a_3 b_1)\mathbf{j} + (a_1 b_2 - a_2 b_1)\mathbf{k}$, this product can be written

(12.20) $\mathbf{a} \times \mathbf{b} = \begin{vmatrix} a_2 & a_3 \\ b_2 & b_3 \end{vmatrix}\mathbf{i} - \begin{vmatrix} a_1 & a_3 \\ b_1 & b_3 \end{vmatrix}\mathbf{j} + \begin{vmatrix} a_1 & a_2 \\ b_1 & b_2 \end{vmatrix}\mathbf{k}.$

If we extend the determinant notation to permit the first row of the array to consist of the unit vectors $\mathbf{i}$, $\mathbf{j}$, and $\mathbf{k}$, then the cross product of $\mathbf{a}$ and $\mathbf{b}$ can be written

(12.21) $\mathbf{a} \times \mathbf{b} = \begin{vmatrix} \mathbf{i} & \mathbf{j} & \mathbf{k} \\ a_1 & a_2 & a_3 \\ b_1 & b_2 & b_3 \end{vmatrix}.$

EXAMPLE 3   Find the vector product $\langle 3, 0, 4 \rangle \times \langle 1, 2, 3 \rangle$.

SOLUTION

$$\langle 3, 0, 4 \rangle \times \langle 1, 2, 3 \rangle = \begin{vmatrix} \mathbf{i} & \mathbf{j} & \mathbf{k} \\ 3 & 0 & 4 \\ 1 & 2 & 3 \end{vmatrix}$$

$$= (0 \cdot 3 - 4 \cdot 2)\mathbf{i} - (3 \cdot 3 - 4 \cdot 1)\mathbf{j} + (3 \cdot 2 - 0 \cdot 1)\mathbf{k}$$

$$= -8\mathbf{i} - 5\mathbf{j} + 6\mathbf{k}$$

$$= \langle -8, -5, 6 \rangle. \qquad \square$$

EXAMPLE 4   Find a unit vector perpendicular to both vectors $\mathbf{a} = \langle 3, 2, 4 \rangle$ and $\mathbf{b} = \langle 0, 3, -1 \rangle$.

SOLUTION
By definition, a vector perpendicular to both $\mathbf{a}$ and $\mathbf{b}$ is

$$\mathbf{a} \times \mathbf{b} = \begin{vmatrix} \mathbf{i} & \mathbf{j} & \mathbf{k} \\ 3 & 2 & 4 \\ 0 & 3 & -1 \end{vmatrix}$$

$$= (-2 - 12)\mathbf{i} - (-3 - 0)\mathbf{j} + (9 - 0)\mathbf{k}$$

$$= \langle -14, 3, 9 \rangle.$$

The unit vector in the direction of $\mathbf{a} \times \mathbf{b}$ is

$$\frac{\mathbf{a} \times \mathbf{b}}{\|\mathbf{a} \times \mathbf{b}\|} = \frac{\langle -14, 3, 9 \rangle}{\sqrt{286}}.$$

Another unit vector perpendicular to both $\mathbf{a}$ and $\mathbf{b}$ is

$$\frac{\mathbf{b} \times \mathbf{a}}{\|\mathbf{b} \times \mathbf{a}\|} = \frac{-(\mathbf{a} \times \mathbf{b})}{\|\mathbf{a} \times \mathbf{b}\|} = \frac{\langle 14, -3, -9 \rangle}{\sqrt{286}}. \qquad \square$$

The results below concerning the vector product follow from the analytic definition.

(12.22)
THEOREM

If **a**, **b**, and **c** are vectors and α is a real number, then

(i) $\mathbf{a} \times \mathbf{b} = -(\mathbf{b} \times \mathbf{a})$

(ii) $\mathbf{a} \times \mathbf{0} = \mathbf{0}$

(iii) $\mathbf{a} \times \mathbf{a} = \mathbf{0}$

(iv) $\alpha(\mathbf{a} \times \mathbf{b}) = (\alpha\mathbf{a}) \times \mathbf{b} = \mathbf{a} \times (\alpha\mathbf{b})$

(v) $(\mathbf{a} + \mathbf{b}) \times \mathbf{c} = \mathbf{a} \times \mathbf{c} + \mathbf{b} \times \mathbf{c}$

(vi) $\mathbf{a} \times (\mathbf{b} \times \mathbf{c}) = (\mathbf{a} \cdot \mathbf{c})\mathbf{b} - (\mathbf{a} \cdot \mathbf{b})\mathbf{c}$

(vii) $\mathbf{a} \cdot (\mathbf{b} \times \mathbf{c}) = (\mathbf{a} \times \mathbf{b}) \cdot \mathbf{c}$

(viii) $\|\mathbf{a} \times \mathbf{b}\|^2 = (\|\mathbf{a}\| \|\mathbf{b}\|)^2 - (\mathbf{a} \cdot \mathbf{b})^2$

Part (viii) of Theorem 12.22 can be used to obtain a result about the length of the vector $\mathbf{a} \times \mathbf{b}$ that is similar to the result in Theorem 12.9 concerning the dot product of **a** and **b**.

(12.23)
COROLLARY

If **a** and **b** are vectors and θ is the angle between **a** and **b**, then

$$\|\mathbf{a} \times \mathbf{b}\| = \|\mathbf{a}\| \|\mathbf{b}\| \sin \theta.$$

PROOF

It follows from Theorem 12.22, part (viii) and Theorem 12.9 that

$$
\begin{aligned}
\|\mathbf{a} \times \mathbf{b}\|^2 &= (\|\mathbf{a}\| \|\mathbf{b}\|)^2 - (\mathbf{a} \cdot \mathbf{b})^2 \\
&= (\|\mathbf{a}\| \|\mathbf{b}\|)^2 - (\|\mathbf{a}\| \|\mathbf{b}\| \cos \theta)^2 \\
&= (\|\mathbf{a}\| \|\mathbf{b}\|)^2 [1 - (\cos \theta)^2] \\
&= (\|\mathbf{a}\| \|\mathbf{b}\|)^2 (\sin \theta)^2.
\end{aligned}
$$

Hence, $\|\mathbf{a} \times \mathbf{b}\| = \|\mathbf{a}\| \|\mathbf{b}\| \sin \theta$. $\square$

An important consequence of this corollary characterizes parallel vectors.

(12.24)
COROLLARY

Vectors **a** and **b** are parallel if and only if $\mathbf{a} \times \mathbf{b} = \mathbf{0}$.

PROOF

If **a** and **b** are parallel, then $\mathbf{a} = \alpha\mathbf{b}$, for some real number α and

$$
\begin{aligned}
\mathbf{a} \times \mathbf{b} = (\alpha\mathbf{b}) \times \mathbf{b} &= \alpha(\mathbf{b} \times \mathbf{b}) \\
&= \alpha\mathbf{0} = \mathbf{0}.
\end{aligned}
$$

Conversely, if $\mathbf{a} \times \mathbf{b} = \mathbf{0}$, then $\|\mathbf{a} \times \mathbf{b}\| = 0$. Applying Corollary 12.23, we have

$$0 = \|\mathbf{a} \times \mathbf{b}\| = \|\mathbf{a}\| \|\mathbf{b}\| \sin \theta,$$

where θ is the angle between **a** and **b**. Then either $\mathbf{a} = \mathbf{0}$, $\mathbf{b} = \mathbf{0}$, or $\sin \theta = 0$, which implies that $\theta = 0$ or π. In any case, **a** is parallel to **b**. $\square$

With the result of Corollary 12.23, we can show that the magnitude of $\|\mathbf{a} \times \mathbf{b}\|$ is the area of the parallelogram determined by $\mathbf{a}$ and $\mathbf{b}$. Recall that this is needed to complete the verification that the geometric and analytic definitions of $\mathbf{a} \times \mathbf{b}$ agree.

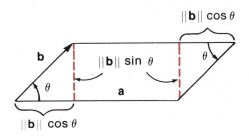

FIGURE 12.35

From Figure 12.35, we see that the area of the parallelogram determined by $\mathbf{a}$ and $\mathbf{b}$ is given by

$$A = 2\left(\frac{1}{2}\, \|\mathbf{b}\|\, \sin\theta\, \|\mathbf{b}\|\, \cos\theta\right) + \|\mathbf{b}\|\, \sin\theta[\|\mathbf{a}\| - \|\mathbf{b}\|\, \cos\theta]$$

$$= \|\mathbf{a}\|\, \|\mathbf{b}\|\, \sin\theta.$$

By Corollary 12.23, $A = \|\mathbf{a} \times \mathbf{b}\|$.

EXERCISE SET 12.4

Find $\mathbf{a} \times \mathbf{b}$ for the vectors given in Exercises 1 through 4.

1. $\mathbf{a} = \langle -1, 3, 5 \rangle, \mathbf{b} = \langle 2, -1, 0 \rangle$

2. $\mathbf{a} = \langle 0, 1, 1 \rangle, \mathbf{b} = \langle 5, 2, -3 \rangle$

3. $\mathbf{a} = -2\mathbf{i} + 3\mathbf{k}, \mathbf{b} = 4\mathbf{i} + \mathbf{j}$

4. $\mathbf{a} = 2\mathbf{i} + \mathbf{j} - 3\mathbf{k}, \mathbf{b} = -\mathbf{i} - \mathbf{j} + \mathbf{k}$

Use the vectors $\mathbf{a} = \langle 1, 2, 0 \rangle, \mathbf{b} = \langle 1, 3, 1 \rangle, \mathbf{c} = \langle 0, 1, 0 \rangle, \mathbf{d} = \langle 3, -2, 2 \rangle$ in Exercises 5 through 16. Find the result or state why the operation is impossible.

5. $\mathbf{a} \times \mathbf{b} + \mathbf{a} \times \mathbf{c}$ **6.** $\mathbf{a} \times (\mathbf{b} + \mathbf{c})$

7. $(\mathbf{a} \times \mathbf{b}) \times \mathbf{c}$ **8.** $(\mathbf{c} \times \mathbf{b}) \times \mathbf{a}$

9. $(\mathbf{a} \times \mathbf{b}) \cdot \mathbf{c}$ **10.** $\mathbf{a} \cdot (\mathbf{b} \times \mathbf{c})$

11. $(\mathbf{a} \cdot \mathbf{b}) \times \mathbf{c}$ **12.** $(\mathbf{a} \times \mathbf{b}) \cdot (\mathbf{c} \times \mathbf{d})$

13. $(\mathbf{a} \times \mathbf{d}) \cdot (\mathbf{c} \times \mathbf{b})$ **14.** $(\mathbf{a} \cdot \mathbf{b}) \times (\mathbf{c} \cdot \mathbf{d})$

15. $\dfrac{\mathbf{a} \times \mathbf{b}}{\mathbf{c} \cdot \mathbf{d}}$ **16.** $\dfrac{\mathbf{a} \cdot \mathbf{b}}{\mathbf{c} \times \mathbf{d}}$

17. Use the cross product to find a vector orthogonal to both $\mathbf{a} = \langle 1, 2, 4 \rangle$ and $\mathbf{b} = \langle 2, -2, 5 \rangle$. Compare this method to the method required in Exercise 37 of Section 12.3. Which method do you prefer?

18. Find a vector orthogonal to both $\mathbf{a} = \langle 3, -4, 0 \rangle$ and $\mathbf{b} = \langle 0, 2, 5 \rangle$.

19. Find the area of the parallelogram determined by the vectors $\langle 4, 3, 0 \rangle$ and $\langle 1, 7, 0 \rangle$.

20. Find the area of the parallelogram determined by the vectors $\langle -1, 1, 2 \rangle$ and $\langle 3, 5, 1 \rangle$.

21. Find the area of the triangle with vertices at $(2, 1, 0)$, $(0, 4, 2)$ and $(2, -3, 2)$.

22. Find the area of the triangle with vertices at $(2, 3, 4)$, $(3, 0, 0)$ and $(1, 7, 2)$.

23. Suppose that vectors $\mathbf{a}$, $\mathbf{b}$, and $\mathbf{c}$ determine the sides of a parallelepiped. Show that the volume of the parallelepiped is $|\mathbf{a} \cdot (\mathbf{b} \times \mathbf{c})|$. (*Hint:* Let the parallelogram determined by $\mathbf{b}$ and $\mathbf{c}$ be the base; its area is $\|\mathbf{b} \times \mathbf{c}\|$. The altitude of the parallelepiped is given by the component of $\mathbf{a}$ in the direction of $\mathbf{b} \times \mathbf{c}$.)

24. A tetrahedron is a solid whose surface is formed by connecting in space four points that do not all lie in the same plane. The surface of a tetrahedron consists of 4 triangles and its volume is (area of a base) · (altitude) 3. Show that the tetrahedron formed by the vectors $\mathbf{a}$, $\mathbf{b}$, $\mathbf{c}$ has volume $|(\mathbf{a} \times \mathbf{b}) \cdot \mathbf{c}|/6$.

In Exercises 25 through 28, use the result in Exercise 23 to find the volume of the parallelepiped determined by the vectors $\mathbf{a}$, $\mathbf{b}$, and $\mathbf{c}$.

25. $\mathbf{a} = \langle 2, 0, 0 \rangle$, $\mathbf{b} = \langle 0, 3, 0 \rangle$, $\mathbf{c} = \langle 0, 0, 4 \rangle$

26. $\mathbf{a} = \langle 0, 0, 2 \rangle$, $\mathbf{b} = \langle 3, 0, 0 \rangle$, $\mathbf{c} = \langle 2, 4, 0 \rangle$

27. $\mathbf{a} = \langle 3, -3, 2 \rangle$, $\mathbf{b} = \langle 0, 2, 2 \rangle$, $\mathbf{c} = \langle 2, 0, 0 \rangle$

28. $\mathbf{a} = \langle -1, 1, 4 \rangle$, $\mathbf{b} = \langle 3, 2, 1 \rangle$, $\mathbf{c} = \langle 1, 0, -1 \rangle$

29–32. Use the result in Exercise 24 to find the volume of the tetrahedron determined by the vectors $\mathbf{a}$, $\mathbf{b}$, and $\mathbf{c}$ given in Exercises 25 through 28.

33. Use Definition 12.19 to show that $\mathbf{a} \times \mathbf{b} = -(\mathbf{b} \times \mathbf{a})$ for any vectors $\mathbf{a}$ and $\mathbf{b}$.

34. Show that for any vector $\mathbf{a}$, $\mathbf{a} \times \mathbf{a} = \mathbf{0}$.

35. Show that for any vectors $\mathbf{a}$, $\mathbf{b}$, and $\mathbf{c}$, $(\mathbf{a} \times \mathbf{b}) \cdot \mathbf{c} = \mathbf{a} \cdot (\mathbf{b} \times \mathbf{c})$.

36. Show that for any pair of vectors $\mathbf{a}$ and $\mathbf{b}$,

$$(\mathbf{a} + \mathbf{b}) \times (\mathbf{a} - \mathbf{b}) = 2(\mathbf{b} \times \mathbf{a}).$$

37. Show that for any pair of vectors $\mathbf{a}$ and $\mathbf{b}$,

$$\|\mathbf{a} \times \mathbf{b}\| = [\|\mathbf{a}\|^2 \|\mathbf{b}\|^2 - (\mathbf{a} \cdot \mathbf{b})^2]^{1/2}.$$

38. Show that for any vectors $\mathbf{a}$, $\mathbf{b}$, and $\mathbf{c}$,

$$\mathbf{a} \times (\mathbf{b} \times \mathbf{c}) = (\mathbf{a} \cdot \mathbf{c})\mathbf{b} - (\mathbf{a} \times \mathbf{b})\mathbf{c}.$$

12.5 PLANES

Planes are the most elementary surfaces in space. To completely describe a plane requires specifying one point in the plane and a direction. The direction is given by a vector orthogonal to all line segments in the plane; such a vector is called a **normal** to the plane. Since only the *direction* of a normal vector is important, any vector parallel to a normal to a plane is also a normal vector to that plane.

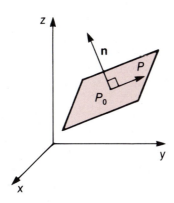

FIGURE 12.36

Suppose $P_0(x_0, y_0, z_0)$ is a point in a plane $\mathcal{P}$ and $\mathbf{n} = \langle a, b, c \rangle$ is a normal vector to $\mathcal{P}$, as illustrated in Figure 12.36. A point $P(x, y, z)$ lies in the plane $\mathcal{P}$ precisely when $\mathbf{n}$ is orthogonal to the vector $\overrightarrow{P_0P} = \langle x - x_0, y - y_0, z - z_0 \rangle$; that is, when

$$\langle a, b, c \rangle \cdot \langle x - x_0, y - y_0, z - z_0 \rangle = 0.$$

Consequently, a plane with normal vector $\mathbf{n} = \langle a, b, c \rangle$ and passing through (x_0, y_0, z_0) has equation

(12.25) $a(x - x_0) + b(y - y_0) + c(z - z_0) = 0,$

or

$$ax + by + cz = d,$$

where $d = ax_0 + by_0 + cz_0.$

EXAMPLE 1   Find an equation of the plane containing the point $(1, 0, 2)$ that has normal vector $\langle 2, 3, 7 \rangle$. Sketch the graph of this equation.

SOLUTION

If (x, y, z) is a point in this plane, then the vector $\langle x - 1, y - 0, z - 2 \rangle$ is orthogonal to $\langle 2, 3, 7 \rangle$. So the plane has equation

$$2(x - 1) + 3(y - 0) + 7(z - 2) = 0,$$

which simplifies to

$$2x + 3y + 7z = 16.$$

To sketch the graph of this equation, we first find the points at which the plane intersects the coordinate axes:

$$\text{when } y = 0 \text{ and } z = 0, \quad 2x = 16, \quad \text{so } x = 8;$$

$$\text{when } x = 0 \text{ and } z = 0, \quad 3y = 16, \quad \text{so } y = \frac{16}{3};$$

$$\text{when } x = 0 \text{ and } y = 0, \quad 7z = 16, \quad \text{so } z = \frac{16}{7}.$$

Thus the points $(8, 0, 0)$, $(0, 16/3, 0)$, and $(0, 0, 16/7)$ lie in the plane. The line segments joining these points will also lie in the plane, so the graph in the first octant is as shown in Figure 12.37. The plane extends linearly in all directions from this triangular segment. □

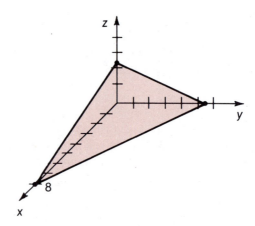

FIGURE 12.37

We have seen that a plane with normal vector $\mathbf{n} = \langle a, b, c \rangle$ and passing through (x_0, y_0, z_0) has equation $ax + by + cz = d$, where $d = ax_0 + by_0 + cz_0$.

On the other hand, if a, b, and c are not all zero, then the graph of an equation of the form

(12.26) $$ax + by + cz = d$$

must be a plane with normal vector $\langle a, b, c \rangle$. To see this, suppose $a \neq 0$. Equation (12.26) can be written as

$$a\left(x - \frac{d}{a}\right) + b(y - 0) + c(z - 0) = 0,$$

which is an equation of the plane that contains the point $(d/a, 0, 0)$ and has normal vector $\langle a, b, c \rangle$.

An equation of the form (12.26) is called a linear equation in three variables. The two-dimensional analogue to (12.26) is the equation of a straight line.

EXAMPLE 2   Sketch the graphs of the equations

(a) $2x + 3y + z = 6$ (b) $x + y = 2$.

SOLUTION
As shown in Example 1, it is convenient to find the points at which the plane intersects the coordinate axes before sketching the graph of the equation of a plane. The plane described in (a) intersects the coordinate axes at $(3, 0, 0)$, $(0, 2, 0)$, and $(0, 0, 6)$. The line segments joining these points lie in the plane, so the graph in the first octant is as shown in Figure 12.38(a).

The plane described in (b) does not intersect the z-axis since both x and y cannot be zero. It intersects the x-axis at $(2, 0, 0)$ and the y-axis at $(0, 2, 0)$. The graph in the first octant is illustrated in Figure 12.38(b).   □

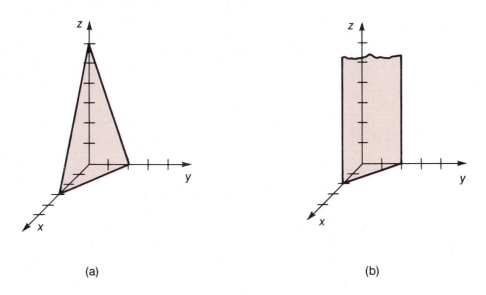

(a)   (b)

FIGURE 12.38

The close relationship between planes and their normal vectors permits us to define the **angle between two planes** as the angle between their normal vectors. **Parallel and orthogonal planes** are planes with parallel and orthogonal normal vectors, respectively. The next theorem follows directly from Corollaries 12.11 and 12.24, which give conditions concerning orthogonal and parallel vectors.

(12.27)
THEOREM
Suppose $\mathcal{P}_1$ and $\mathcal{P}_2$ are planes with normal vectors $\mathbf{n}_1$ and $\mathbf{n}_2$ respectively.
(i) $\mathcal{P}_1$ is orthogonal to $\mathcal{P}_2$ if and only if $\mathbf{n}_1 \cdot \mathbf{n}_2 = 0$.
(ii) $\mathcal{P}_1$ is parallel to $\mathcal{P}_2$ if and only if $\mathbf{n}_1 = \alpha \mathbf{n}_2$, for some constant α, which occurs if and only if $\mathbf{n}_1 \times \mathbf{n}_2 = \mathbf{0}$.

The coordinate planes (the xz-, yz-, and xy- planes) and any planes parallel to these coordinate planes are described by an equation in one variable. For example, the equation of the yz-plane is $x = 0$ and the equation $x = 3$ describes a plane parallel to the yz-plane and passing through the point $(3, 0, 0)$.

A plane orthogonal to the xy-plane has normal vector $\langle a, b, 0 \rangle$ so its equation is of the form

$$ax + by = d.$$

A plane of this type was considered in Example 2(b) and shown in Figure 12.38(b). Similarly, a plane orthogonal to the yz-plane has an equation of the form

$$by + cz = d$$

and a plane orthogonal to the xz-plane has an equation of the form

$$ax + cz = d.$$

Any three points in space that are not collinear (do not lie on the same straight line) determine a unique plane. The equation of this plane can be determined by solving a system of three linear equations. An easier method uses the fact that the line segments joining these points also lie in the plane and that these segments describe vectors. A normal to the plane is the cross product of two such vectors.

EXAMPLE 3   Find an equation of the plane containing the points $(2, 0, 1)$, $(0, 6, -2)$, and $(-2, 3, 1)$.

SOLUTION

The vector with initial point $(2, 0, 1)$ and terminal point $(0, 6, -2)$ is

$$\mathbf{v}_1 = \langle 0 - 2, 6 - 0, -2 - 1 \rangle = \langle -2, 6, -3 \rangle$$

and the vector with the same initial point and terminal point $(-2, 3, 1)$ is

$$\mathbf{v}_2 = \langle -2 - 2, 3 - 0, 1 - 1 \rangle = \langle -4, 3, 0 \rangle.$$

Vectors $\mathbf{v}_1$ and $\mathbf{v}_2$ lie in the plane, so a normal to the plane is

$$\mathbf{v}_1 \times \mathbf{v}_2 = \begin{vmatrix} \mathbf{i} & \mathbf{j} & \mathbf{k} \\ -2 & 6 & -3 \\ -4 & 3 & 0 \end{vmatrix} = 9\mathbf{i} + 12\mathbf{j} + 18\mathbf{k}.$$

The plane has an equation of the form

$$9x + 12y + 18z = d.$$

To determine d, we use the fact that the point $(2, 0, 1)$ lies in the plane. Thus,

$$d = 9 \cdot 2 + 12 \cdot 0 + 18 \cdot 1 = 36$$

and the equation is

$$9x + 12y + 18z = 36$$

or

$$3x + 4y + 6z = 12.$$

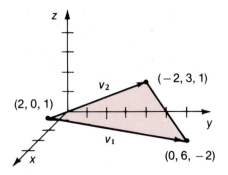

FIGURE 12.39

The graph of this plane is shown in Figure 12.39.   □

The last example in this section shows how the dot product can be used to find the distance from a plane to a point not on the plane.

EXAMPLE 4   Find the distance from point $P(1, 7, -13)$ to the plane $2x + 3y - z = -1$.

SOLUTION

Let $P_0(x_0, y_0, z_0)$ be an arbitrary point in the plane and consider the vector

$$\overrightarrow{P_0P} = \langle 1 - x_0, 7 - y_0, -13 - z_0 \rangle.$$

The absolute value of the component of this vector in the direction of a normal to the plane gives the distance d we are seeking. (See Figure 12.40.)

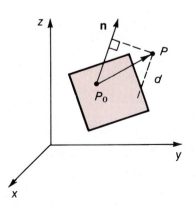

FIGURE 12.40

Since $\mathbf{v} = \langle 2, 3, -1 \rangle$ is normal to the plane, this component is

$$\text{comp}_{\mathbf{v}} \overrightarrow{P_0 P} = \frac{\langle 1 - x_0, 7 - y_0, -13 - z_0 \rangle \cdot \langle 2, 3, -1 \rangle}{\| \langle 2, 3, -1 \rangle \|}$$

$$= \frac{2(1 - x_0) + 3(7 - y_0) - 1(-13 - z_0)}{\sqrt{4 + 9 + 1}}$$

$$= \frac{(2 + 21 + 13) - (2x_0 + 3y_0 - z_0)}{\sqrt{14}}.$$

The point (x_0, y_0, z_0) lies in the plane so

$$2x_0 + 3y_0 - z_0 = -1$$

and the distance from $(1, 7, -13)$ to the plane $2x + 3y - z = -1$ is

$$\frac{36 - (-1)}{\sqrt{14}} = \frac{37}{\sqrt{14}} = \frac{37\sqrt{14}}{14} \approx 9.89. \qquad \square$$

EXERCISE SET 12.5

In Exercises 1 through 6, find an equation of the plane that contains the given point and has normal vector $\mathbf{n}$.

1. $(3, 3, 2)$, $\mathbf{n} = \langle 1, -1, 1 \rangle$ **2.** $(-1, 2, 1)$, $\mathbf{n} = \langle 2, 3, 4 \rangle$

3. $(2, 3, 4)$, $\mathbf{n} = \langle 0, 0, 3 \rangle$ **4.** $(2, -3, 1)$, $\mathbf{n} = \langle 1, 1, 0 \rangle$

5. $(0, 0, 0)$, $\mathbf{n} = \langle 2, 3, 4 \rangle$ **6.** $(0, 0, 2)$, $\mathbf{n} = \langle 0, 1, 1 \rangle$

Sketch the graph of each equation in Exercises 7 through 14.

7. $2x + 3y + 4z = 12$ **8.** $2x + y - z - 4 = 0$

9. $x = 2$ **10.** $z = 3$

11. $2x + 3z = 4$ **12.** $2y - 3z = 4$

13. $z = 0$ **14.** $x + y - 4 = 0$

Find an equation of a plane that satisfies the conditions stated in Exercises 15 through 26.

15. Contains the point $(2, 3, 1)$ and has normal vector $\langle 2, 3, 1 \rangle$.

16. Is parallel to and a distance of 3 units from the xy-plane.

17. Contains the point $(1, 0, -1)$ and is parallel to the plane $x + y - z = 4$.

18. Contains the point $(2, 3, 4)$ and is parallel to the plane $x + y + z = 1$.

19. Contains the point $(1, 2, 3)$ and is parallel to the xy-plane.

20. Contains the point $(2, 3, 4)$ and is parallel to the yz-plane.

21. Contains the points $(1, 2, 3)$ and $(0, 1, 1)$ and is orthogonal to the xy-plane.

22. Contains the points $(-1, 2, -3)$ and $(5, 0, 4)$ and is orthogonal to the *xz*-plane.

23. Contains the points $(1, -1, 4)$, $(0, 2, 3)$, and $(2, 1, 0)$.

24. Contains the points $(3, 2, -1)$, $(2, 3, 5)$, and $(-1, -3, 4)$.

25. Contains the points $(1, 0, -1)$ and $(2, 1, 3)$ and is orthogonal to the plane $2x - y + 3z = 6$.

26. Contains the point $(1, -1, 4)$ and is orthogonal to the planes $x - 2y + z = 2$ and $2x + 2y + z = 1$.

In Exercises 27 through 30, find the distance from the point to the plane.

27. $(0, 0, 0)$, $2x + 3y + 2z = 6$

28. $(0, 0, 3)$, $z = 0$

29. $(1, -2, 3)$, $x + z = 1$

30. $(1, 5, 4)$, $x + y + 2z = 2$

In Exercises 31 through 34, find the distance between the parallel planes.

31. $x - y + 2z = 2$, $x - y + 2z = -2$

32. $2x + y + 3z = 6$, $2x + y + 3z = 1$

33. $2x - 3y + z = 3$, $4x - 6y + 2z = 9$

34. $x - z = 3$, $x - z = 5$

35. Find an equation that describes the graph of all points equidistant from the points $(3, 1, 1)$ and $(7, 5, 6)$.

36. Show that the plane that intersects the coordinate axes at $(a, 0, 0)$, $(0, b, 0)$ and $(0, 0, c)$ has equation

$$\frac{x}{a} + \frac{y}{b} + \frac{z}{c} = 1,$$

provided a, b, and c are all nonzero.

37. Show that the shortest distance from the point (x_0, y_0, z_0) to the plane $Ax + By + Cz + D = 0$ is

$$d = \frac{|Ax_0 + By_0 + Cz_0 + D|}{\sqrt{A^2 + B^2 + C^2}}.$$

▌ 12.6
▌ LINES IN SPACE

Lines are described by specifying a point and a direction. When the line lies in the *xy*-plane, the direction of the line is given by its slope. The slope measures the change in the second or *y*-coordinate of the points on the line relative to the change in the first or *x*-coordinate. Lines in a plane are uniquely determined by the fact that this ratio remains constant for each pair of points on the line.

Since lines in space have three coordinates that can change, we need a different method for indicating direction. A natural tool to use is a vector. Figure 12.41 shows a line l that passes through the point (x_0, y_0, z_0) and is parallel to the vector $\mathbf{v} = \langle v_1, v_2, v_3 \rangle$.

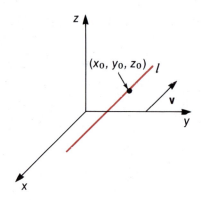

FIGURE 12.41

A point (x, y, z) lies on this line if and only if the vector $\langle x - x_0, y - y_0, z - z_0 \rangle$ is parallel to $\mathbf{v}$, which means there is a real number t such that

$$\langle x - x_0, y - y_0, z - z_0 \rangle = t\langle v_1, v_2, v_3 \rangle.$$

Consequently, all points (x, y, z) on the line passing through (x_0, y_0, z_0) in the direction of $\mathbf{v} = \langle v_1, v_2, v_3 \rangle$ are given by

$$x - x_0 = tv_1, \qquad y - y_0 = tv_2, \qquad z - z_0 = tv_3$$

or

(12.28) $$x = x_0 + tv_1, \qquad y = y_0 + tv_2, \qquad z = z_0 + tv_3$$

for some real number t.

The numbers v_1, v_2, v_3 are called **direction numbers** of l for the natural reason that they indicate its direction. The set of equations in (12.28) is called a set of **parametric equations** for l in the parameter t.

EXAMPLE 1   Find a set of parametric equations for the line passing through the point $(-1, 1, 3)$ and having direction given by $\mathbf{v} = \langle 4, 4, -2 \rangle$.

SOLUTION

Since (x, y, z) lies on this line if and only if

$$\langle x - (-1), y - 1, z - 3 \rangle = t\langle 4, 4, -2 \rangle$$

for some real number t, a set of parametric equations for the line is

$$x = -1 + 4t, \qquad y = 1 + 4t, \qquad z = 3 - 2t.$$

This line is sketched in Figure 12.42.   ☐

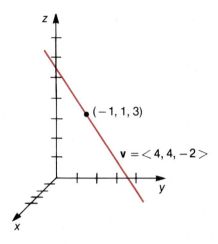

FIGURE 12.42

EXAMPLE 2   Find parametric equations for the line passing through the points $(1, 2, 3)$ and $(0, 1, 3)$.

SOLUTION

A vector **v** that describes the direction of this line is a vector determined by the given points:

$$\mathbf{v} = \langle 1 - 0, 2 - 1, 3 - 3 \rangle = \langle 1, 1, 0 \rangle.$$

Using **v** and the point $(1, 2, 3)$, we obtain the parametric equations:

$$x = 1 + t, \qquad y = 2 + t, \qquad z = 3.$$

Since $z = 3$ regardless of the value of t, this line lies in the plane $z = 3$.   □

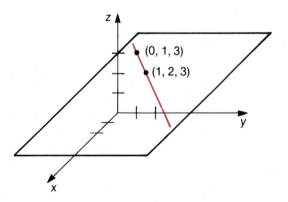

FIGURE 12.43

In Example 2, the line lies in a plane parallel to the xy-plane because the direction number of the line in the z-direction is zero. In general, when one of the direction numbers of a line is zero, the line is parallel to a coordinate plane. When two of the direction numbers of the line are zero, the line is parallel to a coordinate axis.

When a line has all nonzero direction numbers, each of its parametric equations

$$x = x_0 + v_1t, \qquad y = y_0 + v_2t, \qquad z = z_0 + v_3t$$

can be solved for t:

$$\frac{x - x_0}{v_1} = t, \qquad \frac{y - y_0}{v_2} = t, \qquad \frac{z - z_0}{v_3} = t.$$

This gives a set of equations known as **symmetric equations** for the line:

(12.29) $$\frac{x - x_0}{v_1} = \frac{y - y_0}{v_2} = \frac{z - z_0}{v_3}.$$

EXAMPLE 3   Find a set of symmetric equations for the line passing through the point $(1, 0, 2)$ and parallel to the line with parametric equations $x = 2 + t$, $y = 1 + 3t$, and $z = 1 + 4t$.

SOLUTION

The direction of the line with parametric equations $x = 2 + t$, $y = 1 + 3t$, and $z = 1 + 4t$ is given by $\mathbf{v} = \langle 1, 3, 4 \rangle$, which also gives the direction of any line parallel to this line. The symmetric equations for the line parallel to the given line and passing through the point $(1, 0, 2)$ are:

$$\frac{x - 1}{1} = \frac{y - 0}{3} = \frac{z - 2}{4}.$$

Parametric equations for this same line are found by setting each part of the symmetric equations equal to a parameter t and solving for x, y, and z.

$$x = 1 + t, \qquad y = 3t, \qquad z = 2 + 4t. \qquad \square$$

EXAMPLE 4   Find the point of intersection of the xy-plane and the line with symmetric equations

$$\frac{x - 1}{2} = \frac{y + 1}{3} = \frac{z - 2}{-1}.$$

SOLUTION

The z-coordinate of this point of intersection is zero, so

$$\frac{x - 1}{2} = \frac{y + 1}{3} = \frac{0 - 2}{-1} = 2.$$

Solving for x and y, we have

$$\frac{x - 1}{2} = 2, \quad \text{so } x = 5$$

and

$$\frac{y + 1}{3} = 2, \quad \text{so } y = 5.$$

Consequently, the point of intersection of the line and the xy-plane is $(5, 5, 0)$. $\qquad \square$

Suppose a line has symmetric equations

$$\frac{x - x_0}{v_1} = \frac{y - y_0}{v_2} = \frac{z - z_0}{v_3}.$$

By taking the equations in pairs, say

$$\frac{x - x_0}{v_1} = \frac{y - y_0}{v_2} \qquad \text{or} \qquad v_2 x - v_1 y = v_2 x_0 - v_1 y_0$$

and

$$\frac{x - x_0}{v_1} = \frac{z - z_0}{v_3} \qquad \text{or} \qquad v_3 x - v_1 z = v_3 x_0 - v_1 z_0,$$

the line is the intersection of the planes with these equations.

If one of the direction numbers of the line is zero, say $v_3 = 0$, then

$$\frac{x - x_0}{v_1} = \frac{y - y_0}{v_2} \qquad \text{and} \qquad z = z_0,$$

and the line is again expressed as the intersection of two planes.

Intuitively, we expect that whenever two planes intersect, then they either intersect in a straight line or coincide; that is, intersecting planes contain a common straight line. This is shown in the next theorem.

(12.30)
THEOREM

If $\mathcal{P}_1$ and $\mathcal{P}_2$ are distinct planes containing a common point, then they intersect in a line.

PROOF

Refer to Figure 12.44. Suppose (x_0, y_0, z_0) is common to both planes and that $\mathbf{n}_1$ and $\mathbf{n}_2$ are normal vectors to the planes $\mathcal{P}_1$ and $\mathcal{P}_2$ respectively. A directed line segment containing (x_0, y_0, z_0) lies in $\mathcal{P}_1$ if and only if its direction is orthogonal to $\mathbf{n}_1$, and lies in $\mathcal{P}_2$ if and only if its direction is orthogonal to $\mathbf{n}_2$.

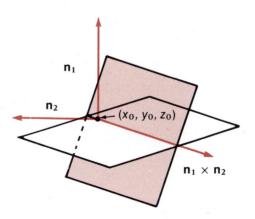

FIGURE 12.44

This implies that a line segment containing (x_0, y_0, z_0) lies in both $\mathcal{P}_1$ and $\mathcal{P}_2$ precisely when its direction is orthogonal to both $\mathbf{n}_1$ and $\mathbf{n}_2$. But $\mathbf{n}_1 \times \mathbf{n}_2$ is orthogonal to both $\mathbf{n}_1$ and $\mathbf{n}_2$, and any vector orthogonal to both $\mathbf{n}_1$ and $\mathbf{n}_2$ is parallel to $\mathbf{n}_1 \times \mathbf{n}_2$. Consequently, a unique line through (x_0, y_0, z_0) exists that is contained in both planes. This line has parametric equations

$$x = x_0 + v_1 t, \qquad y = y_0 + v_2 t, \qquad z = z_0 + v_3 t$$

where

$$\langle v_1, v_2, v_3 \rangle = \mathbf{n}_1 \times \mathbf{n}_2. \qquad \square$$

EXAMPLE 5   Find parametric and symmetric equations for the line described by the intersection of the planes

$$x + 2y + z = 4 \qquad \text{and} \qquad 2x - y + 3z = 3.$$

SOLUTION

We first determine a point of intersection of the planes. In particular, we will find a point of intersection that lies in the xy-plane. If no such point exists we will try to find an intersection point in one of the other coordinate planes. Exercise 34 contains an interesting justification that this procedure is always successful if an intersection point exists.

A point of intersection lying in the xy-plane has the property that $z = 0$, so the equations become:

$$x + 2y = 4 \qquad \text{and} \qquad 2x - y = 3.$$

Solving these equations for x and y, we find that $x = 2$ and $y = 1$. So the point $(2, 1, 0)$ is common to the planes.

The normal vectors to the planes are

$$\mathbf{n}_1 = \langle 1, 2, 1 \rangle \qquad \text{and} \qquad \mathbf{n}_2 = \langle 2, -1, 3 \rangle$$

so the line has direction given by

$$\mathbf{n}_1 \times \mathbf{n}_2 = \begin{vmatrix} \mathbf{i} & \mathbf{j} & \mathbf{k} \\ 1 & 2 & 1 \\ 2 & -1 & 3 \end{vmatrix} = (6 + 1)\mathbf{i} - (3 - 2)\mathbf{j} + (-1 - 4)\mathbf{k}$$

$$= 7\mathbf{i} - \mathbf{j} - 5\mathbf{k}.$$

Parametric equations for the line are

$$x = 2 + 7t, \qquad y = 1 - t, \qquad z = -5t$$

and symmetric equations are

$$\frac{x - 2}{7} = \frac{y - 1}{-1} = \frac{z}{-5}. \qquad \square$$

We now have three ways to describe a line: by parametric equations, by symmetric equations, and as the intersection of two planes.

EXERCISE SET 12.6

In Exercises 1 through 4, find the parametric equations for the line passing through the point and having direction given by **v**.

1. $(2, -1, 2)$, $\mathbf{v} = \langle 1, 1, 1 \rangle$ **2.** $(0, 4, 3)$, $\mathbf{v} = \langle 2, 0, 1 \rangle$

3. $(0, 0, 0)$, $\mathbf{v} = \langle 2, 4, 3 \rangle$ **4.** $(2, 3, 4)$, $\mathbf{v} = \langle 0, -3, 4 \rangle$

In Exercises 5 through 8, find the parametric equations for the line passing through the given points.

5. $(1, 2, 0)$, $(1, -1, 4)$ **6.** $(2, 0, -1)$, $(-1, 2, 1)$

7. $(3, 4, 4)$, $(2, -3, 5)$ **8.** $(0, 0, 2)$, $(3, 0, 0)$

9–16. Find symmetric equations for the lines described in Exercises 1 through 8.

Find the parametric equations of a line that satisfies the conditions stated in Exercises 17 through 22.

17. Passes through the point $(1, 2, 3)$ and is parallel to the line with parametric equations $x = t + 1$, $y = -t$, $z = 2 - t$.

18. Passes through the origin and is parallel to the line with parametric equations $x = 1 - 2t$, $y = 3t + 2$, $z = t - 4$.

19. Passes through the point $(1, -2, 3)$ and is orthogonal to the plane $x + y + z = 1$.

20. Passes through the point $(1, 1, 1)$ and is orthogonal to the plane $x - 2y - 3z = 6$.

21. Passes through $(1, 7, 0)$ and is parallel to the x-axis.

22. Passes through $(2, -2, 5)$ and is parallel to the y-axis.

In Exercises 23 through 26, determine whether each pair of lines is (a) parallel, (b) orthogonal, and (c) find any points of intersection.

23. $\dfrac{x - 1}{2} = y + 2 = \dfrac{z - 3}{2}$, $\dfrac{x}{-2} = \dfrac{y - 1}{4} = \dfrac{z + 2}{2}$

24. $x - 1 = \dfrac{y + 1}{2} = \dfrac{z - 2}{3}$, $x - 1 = \dfrac{y + 2}{-3} = \dfrac{z - 5}{2}$

25. $x - 1 = \dfrac{y - 1}{3} = \dfrac{z - 2}{-1}$, $x - 1 = y - 2 = z - 1$

26. $\dfrac{x}{5} = \dfrac{y - 1}{2} = \dfrac{z + 2}{3}$, $\dfrac{x - 3}{3} = \dfrac{y + 1}{6} = \dfrac{3 - z}{3}$

27. Find the point of intersection of the line with parametric equations $x = t$, $y = 2 - t$, $z = 2t - 3$ and

(a) the yz-plane   (b) the xz-plane   (c) the xy-plane.

28. Find the point of intersection of the line with parametric equations $x = t + 3$, $y = 2t - 1$, $z = t - 4$ and

(a) the yz-plane   (b) the xz-plane   (c) the xy-plane.

29. Show that the lines with parametric equations

$$x = t - 3, y = 2t, z = 1 + t$$

and

$$x = 1 + 2s, y = s - 4, z = 5 + 2s$$

intersect and find the point of intersection.

30. Show that the lines with parametric equations

$$x = 1 + t, \qquad y = 1 + 3t, \qquad z = 2 - t$$

and

$$x = 1 + s, \qquad y = 2 + s, \qquad z = 1 + s$$

intersect and find the point of intersection.

31. Show that the planes $x + 2y + z = 4$ and $x + z = 2$ intersect and find parametric equations of the line of intersection.

32. Find parametric equations of the line that passes through the point $(1, -1, 1)$, is orthogonal to the line $3x = 2y = z$, and parallel to the plane $x + y - z = 0$.

33. Show that any line in space must intersect at least one of the coordinate planes.

34. Show that if two planes intersect then their intersection contains a point in one of the coordinate planes.

35. Suppose l_1 is a line passing through P_1 (x_1, y_1, z_1) with direction $\mathbf{v}_1$ and l_2 is a line through P_2 (x_2, y_2, z_2) with direction $\mathbf{v}_2$. Show that the shortest distance from l_1 to l_2 is $\|\mathbf{proj}_{\mathbf{v}_1 \times \mathbf{v}_2} \overrightarrow{P_1P_2}\|$.

36. Show that the shortest distance from a point P_0 to the line joining P_1 and P_2 is

$$\frac{\|\overrightarrow{P_0P_1} \times \overrightarrow{P_1P_2}\|}{\|\overrightarrow{P_1P_2}\|}.$$

REVIEW EXERCISES

Identify each surface described in Exercises 1 through 12 as a plane, a line, or a sphere. Sketch the graph of the equations on a rectangular coordinate system in space.

1. $z = 3$

2. $x + y = 1$

3. $y = 1 - x, z = 0$

4. $2x + 2y - z = 4$

5. $\dfrac{x - 2}{2} = y - 1 = z - 3$

6. $4x + 9y = 1$

7. $x^2 + y^2 + z^2 = 9$

8. $x = 2$

9. $x + 2z = 4$

10. $x = t + 2, y = t - 3, z = t$

11. $x + y + 2z = 4$

12. $x^2 + y^2 + z^2 - 4z = 0$

In Exercises 13 through 20, find a vector that satisfies the stated condition.

13. Initial point $(1, 0, 0)$, terminal point $(1, 3, 3)$.

14. A unit vector parallel to the vector described in Exercise 13.

15. A unit vector parallel to $\langle 1, -4, -2 \rangle$.

16. Parallel to $\langle 2, -6, 3 \rangle$ with length 14.

17. Normal to the plane $3x - 7y + z = 21$.

18. Orthogonal to $\langle 2, 3, 0 \rangle$ and $\langle 0, 3, 1 \rangle$.

19. Describes the direction of the line normal to the plane $3x - 7y + z = 21$ and passing through the origin.

20. Gives the direction of the line with parametric equations:

$$x = 2 - t, \qquad y = 2t - 3, \qquad z = \frac{t + 2}{5}$$

Use the vectors $\mathbf{a} = \langle 2, 3, 0 \rangle$ and $\mathbf{b} = \langle -2, 1, 0 \rangle$ in Exercises 21 through 28.

21. Sketch the position vector representation of $\mathbf{a}$ and $\mathbf{b}$ in the xy-plane.

22. Sketch $\mathbf{a} + \mathbf{b}$, $\mathbf{a} - \mathbf{b}$, $2\mathbf{a}$, and $-\mathbf{b}$ on the same coordinate system.

23. Find a unit vector that is orthogonal to both $\mathbf{a}$ and $\mathbf{b}$.

24. Find the area of the parallelogram determined by $\mathbf{a}$ and $\mathbf{b}$.

25. Find the angle between $\mathbf{a}$ and $\mathbf{b}$.

26. Find the component of $\mathbf{b}$ in the direction of $\mathbf{a}$.

27. Find the component of $\mathbf{b}$ in the direction of $\mathbf{k}$.

28. Find the component of $\mathbf{a}$ in the direction of $\mathbf{i}$.

Find an equation of each surface described in Exercises 29 through 40.

29. A plane that contains the point $(1, 3, 3)$ and has normal vector $\mathbf{i} + \mathbf{j}$.

30. A plane that contains the point $(3, 5, 0)$ and is parallel to the plane $x - 2y + 3z = 6$.

31. A line that passes through the point $(-2, 1, 4)$ and has direction given by $\mathbf{j}$.

32. A line that passes through the points $(1, -3, 4)$ and $(0, 3, 7)$.

33. A plane that contains the points $(2, 0, 0)$, $(0, 3, 0)$, and $(0, 0, -3)$.

34. A plane that contains the points $(0, 2, 3)$ and $(1, -3, 7)$ and is orthogonal to the plane $2x + y - z = 5$.

35. A sphere with center $(2, 0, 0)$ and radius 2.

36. A sphere that passes through the point $(3, 5, 7)$ and has center at $(1, 3, 4)$.

37. A plane that contains the point $(-3, 2, 4)$ and is parallel to the xy-plane.

38. A plane that passes through the origin and contains the line

$$5x = 3 - y = 5 - z.$$

39. A line that passes through the point $(-2, 1, 2)$ and is parallel to the y-axis.

40. A line that passes through the point $(6, 2, 4)$ and is orthogonal to the x-axis and the y-axis.

41. If $\mathbf{a} = \langle 3, 2, 0 \rangle$ and $\mathbf{b} = \langle 2, k, 0 \rangle$, find k so that

(a) $\mathbf{a}$ and $\mathbf{b}$ are orthogonal;   (b) $\mathbf{a}$ and $\mathbf{b}$ are parallel.

42. For the vectors $\mathbf{a} = 2\mathbf{i} - 3\mathbf{j} + \mathbf{k}$ and $\mathbf{b} = -\mathbf{i} - 2\mathbf{j} + 2\mathbf{k}$, find each of the following:

    (a) $\mathbf{a} \cdot \mathbf{b}$ (b) $\mathbf{a} \times \mathbf{b}$ (c) $\|\mathbf{a}\|$ (d) $\mathbf{a} \cdot \mathbf{i}$

    (e) The direction cosines of $\mathbf{a}$ (f) A unit vector in the direction of $\mathbf{a}$

43. Find the distance from the point $(2, 4, 6)$ to each of the planes described below.

    (a) xy-plane   (b) $2x + y + z = 3$

44. (a) Sketch the triangle with vertices at the points $(4, 9, 1)$, $(-2, 6, 3)$, and $(7, 3, 1)$.

    (b) Show that this triangle is a right triangle.

    (c) Find the area of this triangle.

    (d) Find the interior angles of the triangle.

45. Find the work done if a force $\mathbf{F} = 20(\mathbf{i} + \mathbf{j} + \mathbf{k})$ moves an object from the origin to the point $(5, 5, 0)$.

46. A force of 50 pounds is exerted at an angle of $\pi/3$ to move an object 200 feet. Find the work done.

13

VECTOR-VALUED FUNCTIONS

Vectors and their properties were studied in Chapter 12. In this chapter we develop a theory of calculus for functions whose range is a set of vectors and consider applications of this calculus, particularly applications associated with motion of particles in space.

13.1
DEFINITION OF A VECTOR-VALUED FUNCTION

At the beginning of Chapter 1, we defined a function as describing a rule of correspondence between two sets. The only functions we have studied are those with both domain and range contained in the set of real numbers. In this chapter, such functions will be called real-valued or scalar functions to distinguish them from the vector-valued functions we now define.

(13.1)
DEFINITION

A **vector-valued function** is a function whose domain is a subset of the set of real numbers and whose range is a set of vectors.

Boldface notation, such as **F**, is used to distinguish vector-valued functions from real-valued functions in the same way that boldface notation was used in Chapter 12 to distinguish vectors from scalars. If

$$x = x(t), \qquad y = y(t), \qquad z = z(t)$$

is a set of parametric equations, then

(13.2)
$$\mathbf{F}(t) = x(t)\mathbf{i} + y(t)\mathbf{j} + z(t)\mathbf{k}$$

describes a vector-valued function. Since each vector in space consists of an ordered triple of real numbers, a vector-valued function always has such a parametric representation. The real-valued functions described by $x(t)$, $y(t)$ and $z(t)$ are called the **component functions of F**. Unless specified otherwise, the domain of a vector-valued function is the largest set of real numbers for which all its component functions are defined.

EXAMPLE 1   Determine the domain of the vector-valued function described by

$$\mathbf{F}(t) = t\mathbf{i} + \ln t\mathbf{j} + e^t\mathbf{k}.$$

SOLUTION

The domain of **F** is the set of real numbers t for which $x(t) = t$, $y(t) = \ln t$ and $z(t) = e^t$ are defined. Since $\ln t$ is defined only for $t > 0$ and the others are defined for all values of t, the domain of **F** is $(0, \infty)$.   □

A graphic representation of a vector-valued function **F** can be given by connecting the terminal points of the position vectors $\mathbf{F}(t)$ for t in the domain of **F**, or equivalently by sketching the graph of the associated set of parametric equations. The resulting figure is a curve in space such as that shown in Figure 13.1.

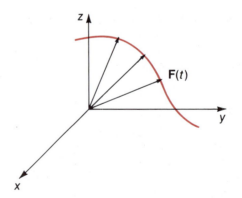

FIGURE 13.1

Since there is a one-to-one correspondence between the set of position vectors and the points in space, a vector-valued function describes a mapping from a portion of the real line into space. This is of particular interest in describing the motion of objects in space: the domain represents time and a vector in the range gives the position of the object at a particular time.

EXAMPLE 2   Sketch the graph of the function described by

$$\mathbf{F}(t) = (4 + 3t)\mathbf{i} + (2 - 2t)\mathbf{j} + (1 - 2t)\mathbf{k}.$$

SOLUTION

The graph of **F** is the same as the graph of the parametric equations:

$$x(t) = 4 + 3t, \qquad y(t) = 2 - 2t, \qquad z(t) = 1 - 2t,$$

which are equations of the line in space that passes through the point $(4, 2, 1)$ and has direction given by $\mathbf{v} = \langle 3, -2, -2 \rangle$. The graph of this line is shown in Figure 13.2. The arrow on the line indicates the direction of increasing values of t. $\quad\square$

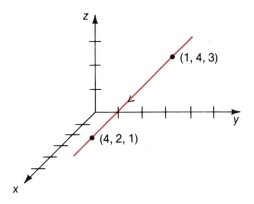

FIGURE 13.2

This gives us four ways to describe lines in space: by a set of parametric equations, by symmetric equations, as the intersection of two planes, and now as the range of a vector-valued function.

EXAMPLE 3   Sketch the graph of $\mathbf{F}(t) = t\mathbf{i} + t^2\mathbf{j}$.

SOLUTION

The graph of $\mathbf{F}$ is the graph of the parametric equations

$$x(t) = t, \qquad y(t) = t^2, \qquad z(t) = 0.$$

Eliminating the parameter t, $y = x^2$, and $z = 0$. The three-dimensional sketch of the graph of $\mathbf{F}$ is shown in Figure 13.3. Since $z = 0$ the graph of $\mathbf{F}$ lies in the xy-plane, as shown in Figure 13.4. $\quad\square$

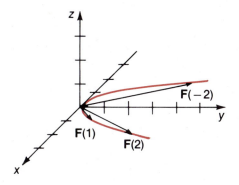

FIGURE 13.3

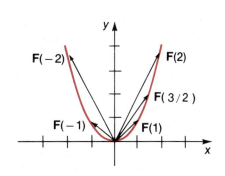

FIGURE 13.4

EXAMPLE 4   Sketch the graph of $\mathbf{F}(t) = \cos t\mathbf{i} + \sin t\mathbf{j} + t\mathbf{k}$ for $t \geq 0$.

SOLUTION

The set of parametric equations

$$x(t) = \cos t, \qquad y(t) = \sin t, \qquad z(t) = 0$$

describes the unit circle in the xy-plane. The graph of $\mathbf{F}$ is similar, but spirals upward in space with increasing values of t, since $z(t) = t$. The curve is called a **circular helix** and is shown in Figure 13.5. It has the appearance of a wire wrapped around the right circular cylinder of radius 1 centered on the z-axis. $\square$

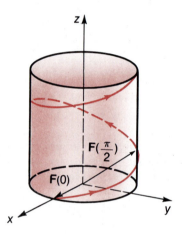

FIGURE 13.5

The range of a vector-valued function is a collection of vectors, a set for which we have established certain arithmetic properties. These properties can in turn be used to establish arithmetic definitions for vector-valued functions.

If $\mathbf{F}$ and $\mathbf{G}$ are vector-valued functions defined by

$$\mathbf{F}(t) = F_1(t)\mathbf{i} + F_2(t)\mathbf{j} + F_3(t)\mathbf{k},$$
$$\mathbf{G}(t) = G_1(t)\mathbf{i} + G_2(t)\mathbf{j} + G_3(t)\mathbf{k},$$

and f is a real-valued function, then we define the functions $\mathbf{F} + \mathbf{G}$, $\mathbf{F} - \mathbf{G}$, $f\mathbf{F}$, $\mathbf{F} \cdot \mathbf{G}$, $\mathbf{F} \times \mathbf{G}$ and $\mathbf{F} \circ f$ by

(i) $(\mathbf{F} + \mathbf{G})(t) = \mathbf{F}(t) + \mathbf{G}(t)$
(ii) $(\mathbf{F} - \mathbf{G})(t) = \mathbf{F}(t) - \mathbf{G}(t)$
(iii) $(f\mathbf{F})(t) = f(t)\,\mathbf{F}(t)$
(iv) $(\mathbf{F} \circ f)(t) = \mathbf{F}(f(t))$
(v) $(\mathbf{F} \times \mathbf{G})(t) = \mathbf{F}(t) \times \mathbf{G}(t)$
(vi) $(\mathbf{F} \cdot \mathbf{G})(t) = \mathbf{F}(t) \cdot \mathbf{G}(t)$

All these functions are vector-valued except $\mathbf{F} \cdot \mathbf{G}$, which is a scalar or real-valued function.

EXAMPLE 5 Describe each of the functions defined in (i) through (vi) if $\mathbf{F}$, $\mathbf{G}$, and f are given by $\mathbf{F}(t) = t^2\mathbf{i} + 2t\mathbf{j} + \sin t\mathbf{k}$, $\mathbf{G}(t) = e^t\mathbf{i} + t\mathbf{j} + \mathbf{k}$ and $f(t) = t^2 - 1$.

SOLUTION

(i) $(\mathbf{F} + \mathbf{G})(t) = (t^2 + e^t)\mathbf{i} + 3t\mathbf{j} + (\sin t + 1)\mathbf{k}$

(ii) $(\mathbf{F} - \mathbf{G})(t) = (t^2 - e^t)\mathbf{i} + t\mathbf{j} + (\sin t - 1)\mathbf{k}$

(iii) $(f\mathbf{F})(t) = (t^2 - 1)[t^2\mathbf{i} + 2t\mathbf{j} + \sin t\mathbf{k}]$
$= (t^4 - t^2)\mathbf{i} + (2t^3 - 2t)\mathbf{j} + (t^2 - 1)\sin t\mathbf{k}$

(iv) $(\mathbf{F} \circ f)(t) = \mathbf{F}(t^2 - 1) = (t^2 - 1)^2\mathbf{i} + 2(t^2 - 1)\mathbf{j} + \sin(t^2 - 1)\mathbf{k}$

(v) $(\mathbf{F} \times \mathbf{G})(t) = \begin{vmatrix} \mathbf{i} & \mathbf{j} & \mathbf{k} \\ t^2 & 2t & \sin t \\ e^t & t & 1 \end{vmatrix}$

$= (2t - t\sin t)\mathbf{i} - (t^2 - e^t\sin t)\mathbf{j} + (t^3 - 2te^t)\mathbf{k}$

(vi) $(\mathbf{F} \cdot \mathbf{G})(t) = (t^2\mathbf{i} + 2t\mathbf{j} + \sin t\mathbf{k}) \cdot (e^t\mathbf{i} + t\mathbf{j} + \mathbf{k})$
$= t^2e^t + 2t^2 + \sin t.$ ☐

EXERCISE SET 13.1

Determine the domain of the vector-valued functions described in Exercises 1 through 6.

1. $\mathbf{F}(t) = t\mathbf{i} + t^2\mathbf{j} + 2\mathbf{k}$

2. $\mathbf{F}(t) = 3\mathbf{i} + \sqrt{1 - t}\,\mathbf{j} + t\mathbf{k}$

3. $\mathbf{F}(t) = \sqrt{t}\mathbf{i} + (t^2 - 2)\mathbf{j} + t\mathbf{k}$

4. $\mathbf{F}(t) = \sqrt{t}\mathbf{i} + (t^2 - 2)\mathbf{j} + \dfrac{1}{t}\mathbf{k}$

5. $\mathbf{F}(t) = \ln t\mathbf{i} + (1 - t^2)\mathbf{j} + \mathbf{k}$

6. $\mathbf{F}(t) = \ln t\mathbf{j} + e^t\mathbf{k}$

Sketch the graphs in the xy-plane of the vector-valued functions described in Exercises 7 through 12 and indicate the direction of increasing t.

7. $\mathbf{F}(t) = t\mathbf{i} + t^2\mathbf{j}$

8. $\mathbf{F}(t) = t\mathbf{i} + \dfrac{1}{t}\mathbf{j}, t > 0$

9. $\mathbf{F}(t) = t\mathbf{i} + \sin t\mathbf{j}$

10. $\mathbf{F}(t) = e^t\mathbf{i} + e^{-2t}\mathbf{j}$

11. $\mathbf{F}(t) = 4t^2\mathbf{j}$

12. $\mathbf{F}(t) = t\mathbf{i} + \ln t\mathbf{j}$

Sketch the graphs in space of the vector-valued functions described in Exercises 13 through 20 and indicate the direction of increasing t.

13. $\mathbf{F}(t) = (t + 1)\mathbf{i} + (2t - 1)\mathbf{j} + (2 - 2t)\mathbf{k}$

14. $\mathbf{F}(t) = (2t + 1)\mathbf{i} + (3 - t)\mathbf{j} + (2 - t)\mathbf{k}$

15. $\mathbf{F}(t) = \cos t\mathbf{i} + \sin t\mathbf{j} + 2\mathbf{k}, 0 \leq t \leq 2\pi$

16. $\mathbf{F}(t) = 2\cos t\mathbf{i} + 3\sin t\mathbf{j}, 0 \leq t \leq 2\pi$

17. $\mathbf{F}(t) = t\mathbf{i} + e^t\mathbf{j} + \mathbf{k}$

18. $\mathbf{F}(t) = t\mathbf{i} + e^t\mathbf{j} + e^t\mathbf{k}$

19. $\mathbf{F}(t) = (200 - 16t^2)\mathbf{k}$

20. $\mathbf{F}(t) = \sin t\mathbf{i} + t\mathbf{j} + t\mathbf{k}$

21. If $\mathbf{F}(t) = t\mathbf{i} + t^{-2}\mathbf{j} + e^t\mathbf{k}$, $\mathbf{G}(t) = \sqrt{t}\mathbf{i} + \sin t\mathbf{k}$ and $f(t) = 2 - t$, describe each of the following functions and determine their domains.

(a) $(\mathbf{F} \cdot \mathbf{G})(t)$ (b) $(\mathbf{F} \circ f)(t)$

(c) $(f\mathbf{F})(t)$ (d) $(\mathbf{F} + \mathbf{G})(t)$

(e) $(\mathbf{F} - \mathbf{G})(t)$ (f) $(\mathbf{F} \times \mathbf{G})(t)$

22. A circle of radius R rolls along a horizontal line. Find a vector-valued function whose graph is the curve traced by a fixed point P on the circumference of the circle. (*Hint:* See Example 3 in Section 10.4.)

23. The motion of a baseball is described by

$$\mathbf{F}(t) = 3.4\mathbf{i} + 5.5\,t\mathbf{j} + (4.9t - 4.9t^2)\mathbf{k}, \quad 0 \le t \le 1,$$

where t is expressed in seconds and the coordinate values are in meters. Sketch a curve that represents the path of this baseball.

24. A 16-foot sailboat is caught in a storm on Lake Muskoka. The motion of the top of the mast is described by

$$\mathbf{F}(t) = (6 \sin t + 1)\mathbf{i} + t\mathbf{j} + (\cos t + 12)\mathbf{k}, \quad t \ge 0.$$

Describe the turbulent motion of this mast top.

13.2
THE CALCULUS OF VECTOR-VALUED FUNCTIONS

The representation of a vector-valued function in terms of its component, or coordinate, functions provides a natural means for defining the various concepts of calculus for these functions.

(13.3)
DEFINITION

Suppose $\mathbf{F}$ is described by $\mathbf{F}(t) = x(t)\mathbf{i} + y(t)\mathbf{j} + z(t)\mathbf{k}$.

(i) If $\lim_{t \to a} x(t)$, $\lim_{t \to a} y(t)$, and $\lim_{t \to a} z(t)$ exist, then the limit of $\mathbf{F}$ at a is said to exist and

$$\lim_{t \to a} \mathbf{F}(t) = \lim_{t \to a} x(t)\mathbf{i} + \lim_{t \to a} y(t)\mathbf{j} + \lim_{t \to a} z(t)\mathbf{k}.$$

(ii) If x, y, and z are continuous at a, then $\mathbf{F}$ is said to be continuous at a.

(iii) If x, y, and z are differentiable at a, then $\mathbf{F}$ is said to be differentiable at a and

$$D_t\mathbf{F}(a) = \mathbf{F}'(a) = x'(a)\mathbf{i} + y'(a)\mathbf{j} + z'(a)\mathbf{k}.$$

(iv) If x, y, and z are integrable on the interval $[a, b]$, then $\mathbf{F}$ is said to be integrable on $[a, b]$ and

$$\int_a^b \mathbf{F}(t)\, dt = \left(\int_a^b x(t)\, dt \right)\mathbf{i} + \left(\int_a^b y(t)\, dt \right)\mathbf{j} + \left(\int_a^b z(t)\, dt \right)\mathbf{k}.$$

We use $\int \mathbf{F}(t)dt$ to represent the indefinite integral of $\mathbf{F}$, defined by

$$\mathbf{F}(t)dt = \left(\int x(t)dt\right)\mathbf{i} + \left(\int y(t)dt\right)\mathbf{j} + \left(\int z(t)dt\right)\mathbf{k}.$$

EXAMPLE 1   Find $D_t\mathbf{F}\left(\dfrac{\pi}{4}\right)$ and $\displaystyle\int_0^{\pi/4} \mathbf{F}(t)\, dt$ for $\mathbf{F}(t) = \cos t\mathbf{i} + \sin t\mathbf{j} + t\mathbf{k}$.

SOLUTION

$$D_t\mathbf{F}(t) = D_t\,(\cos t)\mathbf{i} + D_t\,(\sin t)\mathbf{j} + D_t(t)\mathbf{k}$$
$$= (-\sin t)\mathbf{i} + (\cos t)\mathbf{j} + \mathbf{k},$$

so

$$D_t\mathbf{F}\left(\frac{\pi}{4}\right) = -\frac{\sqrt{2}}{2}\mathbf{i} + \frac{\sqrt{2}}{2}\mathbf{j} + \mathbf{k}.$$

$$\int_0^{\pi/4} \mathbf{F}(t)\, dt = \left(\int_0^{\pi/4} \cos t\, dt\right)\mathbf{i} + \left(\int_0^{\pi/4} \sin t\, dt\right)\mathbf{j} + \left(\int_0^{\pi/4} t\, dt\right)\mathbf{k}$$

$$= \sin t\,\Big]_0^{\pi/4}\, \mathbf{i} - \cos t\,\Big]_0^{\pi/4}\, \mathbf{j} + \frac{t^2}{2}\Big]_0^{\pi/4}\, \mathbf{k}$$

$$= \left(\sin\frac{\pi}{4} - \sin 0\right)\mathbf{i} - \left(\cos\frac{\pi}{4} - \cos 0\right)\mathbf{j} + \left(\frac{\left(\frac{\pi}{4}\right)^2}{2} - 0\right)\mathbf{k}$$

$$= \frac{\sqrt{2}}{2}\mathbf{i} - \left(\frac{\sqrt{2}}{2} - 1\right)\mathbf{j} + \frac{\pi^2}{32}\mathbf{k}. \qquad \square$$

Theorems for real-valued functions concerning these concepts hold for vector-valued functions as well, unless the theorems include operations that are undefined for vector-valued functions, such as the reciprocal and quotient of vectors.

It is instructive to note that the definitions of continuity and differentiability can be expressed in the same context as their real-valued counterparts. The proofs of the following theorems are applications of the definitions.

(13.4)
THEOREM

A function $\mathbf{F}$ is continuous at a if and only if
(i) $\mathbf{F}(a)$ exists,
(ii) $\lim\limits_{t\to a} \mathbf{F}(t)$ exists, and
(iii) $\lim\limits_{t\to a} \mathbf{F}(t) = \mathbf{F}(a)$.

$\square$

(13.5)
THEOREM

A function $\mathbf{F}$ is differentiable at a if and only if $\lim\limits_{t \to a} \dfrac{\mathbf{F}(t) - \mathbf{F}(a)}{t - a}$ exists. In this case,

$$D_t\mathbf{F}(a) = \lim_{t \to a} \frac{\mathbf{F}(t) - \mathbf{F}(a)}{t - a}.$$

EXAMPLE 2   Use Theorem 13.5 to show that $\mathbf{F}$ is differentiable at 1, where

$$\mathbf{F}(t) = t^2\mathbf{i} + 2t\mathbf{j} + \frac{1}{t}\mathbf{k}.$$

SOLUTION

$$D_t\mathbf{F}(1) = \lim_{t \to 1} \frac{\mathbf{F}(t) - \mathbf{F}(1)}{t - 1}$$

$$= \lim_{t \to 1} \frac{t^2 - 1}{t - 1}\mathbf{i} + \lim_{t \to 1} \frac{2t - 2}{t - 1}\mathbf{j} + \lim_{t \to 1} \frac{\frac{1}{t} - 1}{t - 1}\mathbf{k}$$

$$= \lim_{t \to 1} \frac{(t + 1)(t - 1)}{t - 1}\mathbf{i} + \lim_{t \to 1} 2\mathbf{j} + \lim_{t \to 1} \frac{1 - t}{t(t - 1)}\mathbf{k}$$

$$= \lim_{t \to 1} (t + 1)\mathbf{i} + \lim_{t \to 1} 2\mathbf{j} + \lim_{t \to 1} \frac{-1}{t}\mathbf{k}$$

$$= 2\mathbf{i} + 2\mathbf{j} - \mathbf{k}.$$

Note that this agrees with the result obtained by using part (iii) of Definition 13.3:

$$D_t\mathbf{F}(t) = 2t\mathbf{i} + 2\mathbf{j} - \frac{1}{t^2}\mathbf{k},$$

so

$$D_t\mathbf{F}(1) = 2\mathbf{i} + 2\mathbf{j} - \mathbf{k}. \qquad \square$$

Using Theorem 13.5, we can give a geometric interpretation of $D_t\mathbf{F}(a)$. Consider $\mathbf{F}(a)$ for a fixed value a in the domain of $\mathbf{F}$. If t is any other number in the domain of $\mathbf{F}$, the difference

$$\mathbf{F}(t) - \mathbf{F}(a)$$

is a vector describing the secant line through the terminal points of $\mathbf{F}(t)$ and $\mathbf{F}(a)$. The graph of $\mathbf{F}$ and a secant line are shown in Figure 13.6.

The vector

(13.6)
$$\frac{\mathbf{F}(t) - \mathbf{F}(a)}{t - a}$$

is parallel to this secant line. If $t > a$, then $\mathbf{F}(t) - \mathbf{F}(a)$ points in the direction of increasing values of t and so does the vector in (13.6). If $t < a$, then $\mathbf{F}(t) - \mathbf{F}(a)$ points in the direction of decreasing values of t. In this case, the vector in (13.6) has direction opposite that of $\mathbf{F}(t) - \mathbf{F}(a)$ and again points in the direction of increasing values of t. The direction of the vector in (13.6) is always that of increasing values of t. The derivative of $\mathbf{F}$ at a,

$$D_t\mathbf{F}(a) = \lim_{t \to a} \frac{\mathbf{F}(t) - \mathbf{F}(a)}{t - a},$$

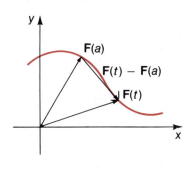

FIGURE 13.6   FIGURE 13.7

is tangent to the graph of $\mathbf{F}$ at a and points in the direction of increasing t. (See Figure 13.7.)

EXAMPLE 3   Let $\mathbf{F}(t) = \cos t\,\mathbf{i} + \sin t\,\mathbf{j} + t\mathbf{k}$. Find $D_t\mathbf{F}\left(\dfrac{\pi}{2}\right)$ and sketch the graph of $\mathbf{F}$ and the tangent vector $D_t\mathbf{F}\left(\dfrac{\pi}{2}\right)$.

SOLUTION
Since $D_t\mathbf{F}(t) = -\sin t\,\mathbf{i} + \cos t\,\mathbf{j} + \mathbf{k},$

$D_t\mathbf{F}\left(\dfrac{\pi}{2}\right) = -\mathbf{i} + \mathbf{k}$. The graph of $\mathbf{F}$ is the circular helix considered in Example 4 of Section 13.1. The vector $D_t\mathbf{F}\left(\dfrac{\pi}{2}\right) = -\mathbf{i} + \mathbf{k}$, sketched with initial point $\left(0, 1, \dfrac{\pi}{2}\right)$ lying on the graph of $\mathbf{F}$ and terminal point $\left(-1, 1, \dfrac{\pi}{2} + 1\right)$, is shown in Figure 13.8.

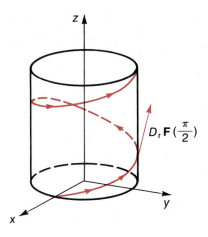

FIGURE 13.8

The study of the calculus of real-valued functions involves the derivatives of various combinations of functions. The next theorem indicates that similar results are valid for vector-valued functions.

(13.7)
THEOREM

If $\mathbf{F}$, $\mathbf{G}$, and f are differentiable, then

(i) $D_t(f\,\mathbf{F})(t) = [D_t f(t)]\mathbf{F}(t) + f(t)[D_t\mathbf{F}(t)]$

(ii) $D_t(\mathbf{F} \times \mathbf{G})(t) = D_t\mathbf{F}(t) \times \mathbf{G}(t) + \mathbf{F}(t) \times D_t\mathbf{G}(t)$

(iii) $D_t(\mathbf{F} \cdot \mathbf{G})(t) = D_t\mathbf{F}(t) \cdot \mathbf{G}(t) + \mathbf{F}(t) \cdot D_t\mathbf{G}(t)$.

If, in addition, g is differentiable at a and $\mathbf{F}$ is differentiable at $g(a)$, then

(iv) $D_t(\mathbf{F} \circ g)(a) = \mathbf{F}'(g(a))\,g'(a)$.

PROOF

(i) To facilitate the proof, we express $\mathbf{F}$ in terms of its components:

$$\mathbf{F}(t) = x(t)\mathbf{i} + y(t)\mathbf{j} + z(t)\mathbf{k}.$$

Then

$$D_t(f\,\mathbf{F})(t) = D_t[f(t)x(t)]\mathbf{i} + D_t[f(t)y(t)]\mathbf{j} + D_t[f(t)z(t)]\mathbf{k}$$

and applying the product rule to each component,

$$\begin{aligned}
D_t(f\,\mathbf{F})(t) &= [f'(t)x(t) + f(t)x'(t)]\mathbf{i} + [f'(t)y(t) + f(t)y'(t)]\mathbf{j} \\
&\quad + [f'(t)z(t) + f(t)z'(t)]\mathbf{k} \\
&= f'(t)x(t)\mathbf{i} + f'(t)y(t)\mathbf{j} + f'(t)z(t)\mathbf{k} \\
&\quad + f(t)x'(t)\mathbf{i} + f(t)y'(t)\mathbf{j} + f(t)z'(t)\mathbf{k} \\
&= [D_t f(t)]\mathbf{F}(t) + f(t)[D_t\mathbf{F}(t)].
\end{aligned}$$

The other parts are proved in a similar manner.   ☐

EXAMPLE 4   Find $D_t(f\,\mathbf{F})(t)$, $D_t(\mathbf{F} \circ f)(t)$, $D_t(\mathbf{F} \times \mathbf{G})(t)$, and $D_t(\mathbf{F} \cdot \mathbf{G})(t)$ if $\mathbf{F}(t) = t\mathbf{i} + \mathbf{j} + e^t\mathbf{k}$, $\mathbf{G}(t) = \mathbf{i} + 2t\mathbf{j} + \sin t\mathbf{k}$, and $f(t) = t^2 + 1$.

SOLUTION

$$\begin{aligned}
D_t(f\,\mathbf{F})(t) &= D_t(t^2 + 1)(t\mathbf{i} + \mathbf{j} + e^t\mathbf{k}) + (t^2 + 1)D_t[t\mathbf{i} + \mathbf{j} + e^t\mathbf{k}] \\
&= 2t(t\mathbf{i} + \mathbf{j} + e^t\mathbf{k}) + (t^2 + 1)(\mathbf{i} + e^t\mathbf{k}) \\
&= (3t^2 + 1)\mathbf{i} + 2t\mathbf{j} + (t^2 + 2t + 1)e^t\mathbf{k}.
\end{aligned}$$

Since $\mathbf{F}'(t) = \mathbf{i} + e^t\mathbf{k}$,

$$\begin{aligned}
D_t(\mathbf{F} \circ f)(t) &= \mathbf{F}'(f(t))\,f'(t) \\
&= (\mathbf{i} + e^{t^2+1}\,\mathbf{k})(2t) \\
&= 2t\mathbf{i} + 2te^{t^2+1}\,\mathbf{k}.
\end{aligned}$$

$$\begin{aligned}
D_t(\mathbf{F} \times \mathbf{G})(t) &= D_t(\mathbf{F}(t)) \times \mathbf{G}(t) + \mathbf{F}(t) \times D_t\mathbf{G}(t) \\
&= (\mathbf{i} + e^t\mathbf{k}) \times (\mathbf{i} + 2t\mathbf{j} + \sin t\mathbf{k}) \\
&\quad + (t\mathbf{i} + \mathbf{j} + e^t\mathbf{k}) \times (2\mathbf{j} + \cos t\mathbf{k}) \\
&= \begin{vmatrix} \mathbf{i} & \mathbf{j} & \mathbf{k} \\ 1 & 0 & e^t \\ 1 & 2t & \sin t \end{vmatrix} + \begin{vmatrix} \mathbf{i} & \mathbf{j} & \mathbf{k} \\ t & 1 & e^t \\ 0 & 2 & \cos t \end{vmatrix} \\
&= [-2te^t\mathbf{i} - (\sin t - e^t)\mathbf{j} + 2t\mathbf{k}] \\
&\quad + [(\cos t - 2e^t)\mathbf{i} - t\cos t\mathbf{j} + 2t\mathbf{k}] \\
&= (\cos t - 2te^t - 2e^t)\mathbf{i} + (e^t - \sin t - t\cos t)\mathbf{j} + 4t\mathbf{k}.
\end{aligned}$$

$$D_t(\mathbf{F} \cdot \mathbf{G})(t) = D_t\mathbf{F}(t) \cdot \mathbf{G}(t) + \mathbf{F}(t) \cdot D_t\mathbf{G}(t)$$
$$= (\mathbf{i} + e^t\mathbf{k}) \cdot (\mathbf{i} + 2t\mathbf{j} + \sin t\mathbf{k}) + (t\mathbf{i} + \mathbf{j} + e^t\mathbf{k}) \cdot (2\mathbf{j} + \cos t\mathbf{k})$$
$$= 1 + e^t\sin t + 2 + e^t \cos t$$
$$= 3 + e^t(\sin t + \cos t). \qquad \square$$

EXERCISE SET 13.2

In Exercises 1 through 6, find $\lim_{t \to a} \mathbf{F}(t)$, if it exists.

1. $\mathbf{F}(t) = \sqrt{t}\,\mathbf{i} + \dfrac{1}{t}\mathbf{j} + t^2\mathbf{k}, \ a = 1$ **2.** $\mathbf{F}(t) = e^t\mathbf{i} + t\mathbf{j} + 3\mathbf{k}, \ a = 0$

3. $\mathbf{F}(t) = \dfrac{t^2 - 4}{t - 2}\mathbf{i} + 3\mathbf{j}, \ a = 2$ **4.** $\mathbf{F}(t) = \mathbf{i} + \dfrac{t^2 - 9}{t - 3}\mathbf{j} + \dfrac{1}{t}\mathbf{k}, \ a = 3$

5. $\mathbf{F}(t) = \dfrac{1}{t}\mathbf{i} + t^3\mathbf{j} + \sin t\mathbf{k}, \ a = 0$ **6.** $\mathbf{F}(t) = \dfrac{1}{t - 2}\mathbf{i} + (t - 2)\mathbf{j}, \ a = 2$

7–12. Find the values of t at which the functions described in Exercises 1 through 6 are continuous.

In Exercises 13 through 16, find $D_t\mathbf{F}(t)$.

13. $\mathbf{F}(t) = \ln t\mathbf{i} + t\mathbf{j} + \mathbf{k}$ **14.** $\mathbf{F}(t) = te^t\mathbf{j} + t\mathbf{k}$

15. $\mathbf{F}(t) = t^2\mathbf{i} + e^{t^2}\mathbf{j} + t\mathbf{k}$ **16.** $\mathbf{F}(t) = (\sec t)^2\mathbf{i} + (\tan t)^2\mathbf{j}$

In Exercises 17 through 22, a vector-valued function $\mathbf{F}$ and a point a are given. Find $D_t\mathbf{F}(a)$, sketch the graph of $\mathbf{F}$ and the tangent vector $D_t\mathbf{F}(a)$.

17. $\mathbf{F}(t) = 2 \cos t\mathbf{i} + 3 \sin t\mathbf{j}, \ a = 0$
18. $\mathbf{F}(t) = e^t\mathbf{i} + e^{-t}\mathbf{j}, \ a = \ln 2$
19. $\mathbf{F}(t) = t\mathbf{i} + e^t\mathbf{j} + 2t\mathbf{k}, \ a = 0$
20. $\mathbf{F}(t) = t\mathbf{i} + e^t\mathbf{j} + \mathbf{k}, \ a = 1$
21. $\mathbf{F}(t) = \sin t\mathbf{i} + \cos t\mathbf{j} + t\mathbf{k}, \ a = 0$
22. $\mathbf{F}(t) = 2\cos t\mathbf{i} + 3\sin t\mathbf{j} + t\mathbf{k}, \ a = \pi/2$

Evaluate the integrals in Exercises 23 through 30.

23. $\displaystyle\int_0^{\pi/2} (\sin t\mathbf{i} + \cos t\mathbf{j} + t\mathbf{k})dt$ **24.** $\displaystyle\int_0^1 (t\mathbf{i} + (t - 1)^2\mathbf{j} + \sqrt{t}\mathbf{k})dt$

25. $\displaystyle\int (te^t\mathbf{i} + t\mathbf{j} + \mathbf{k})dt$ **26.** $\displaystyle\int_1^e (t\mathbf{i} + \ln t\mathbf{j})dt$

27. $\displaystyle\int (t\mathbf{i} + \dfrac{1}{t}\mathbf{j})dt$ **28.** $\displaystyle\int (t \cos t\mathbf{i} + \cos t\mathbf{j} + t\mathbf{k})dt$

29. $\displaystyle\int \|\sin t\mathbf{i} + \cos t\mathbf{j} + \mathbf{k}\|dt$ **30.** $\displaystyle\int_0^1 \|\sin t\mathbf{i} + \cos t\mathbf{j} + t\mathbf{k}\|dt$

In Exercises 31 through 34, determine where the derivative of the function is continuous.

31. $F(t) = t\mathbf{i} + \sin t\mathbf{j}$

32. $F(t) = \mathbf{i} + t\mathbf{j} + t^2\mathbf{k}$

33. $F(t) = \tan t\mathbf{i} + t\mathbf{j} + \mathbf{k}$

34. $F(t) = t\mathbf{j} + \ln t\mathbf{k}$

35. If $F(t) = t\mathbf{i} + \cos t\mathbf{j} + \mathbf{k}$, $G(t) = e^t\mathbf{i} + \mathbf{j} + \sin t\mathbf{k}$, and $f(t) = t^2 - 1$, find

   (a) $D_t(\mathbf{F} \cdot \mathbf{G})(t)$
                   (b) $D_t(\mathbf{F} \circ f)(t)$

   (c) $D_t(\mathbf{F} \times \mathbf{G})(t)$
                  (d) $D_t(\mathbf{F} + \mathbf{G})(t)$

36. If $F(t) = t^2\mathbf{i} + e^t\mathbf{j} + (t + 1)\mathbf{k}$, $G(t) = \sin t\mathbf{i} + t\mathbf{j} + \ln t\mathbf{k}$, and $f(t) = e^{t^2}$, find

   (a) $D_t(\mathbf{F} \cdot \mathbf{G})(t)$
                   (b) $D_t(\mathbf{F} \circ f)(t)$

   (c) $D_t(\mathbf{F} \times \mathbf{G})(t)$
                  (d) $D_t(\mathbf{F} + \mathbf{G})(t)$

37. Show that the line segment joining the points $(1, 3, 1)$ and $(2, 5, 4)$ is described by each of the functions:

$$\mathbf{F}_1(t) = (t + 1)\mathbf{i} + (2t + 3)\mathbf{j} + (3t + 1)\mathbf{k}, \quad 0 \le t \le 1$$
$$\mathbf{F}_2(t) = (t^2 + 1)\mathbf{i} + (2t^2 + 3)\mathbf{j} + (3t^2 + 1)\mathbf{k}, \quad 0 \le t \le 1$$
$$\mathbf{F}_3(t) = (\ln t + 1)\mathbf{i} + (\ln t^2 + 3)\mathbf{j} + (\ln t^3 + 1)\mathbf{k}, \quad 1 \le t \le e$$

Find $\mathbf{F}_1\left(\dfrac{1}{2}\right)$, $\mathbf{F}_2\left(\dfrac{1}{2}\right)$, and $\mathbf{F}_3\left(\dfrac{1 + e}{2}\right)$.

38. Show that the following functions describe the same curve in space.

$$\mathbf{F}_1(t) = t\mathbf{i} + 2t\mathbf{j} + t^2\mathbf{k}, \quad 0 \le t \le 1$$
$$\mathbf{F}_2(t) = t^2\mathbf{i} + 2t^2\mathbf{j} + t^4\mathbf{k}, \quad 0 \le t \le 1$$
$$\mathbf{F}_3(t) = \ln t\mathbf{i} + 2\ln t\mathbf{j} + (\ln t)^2\mathbf{k}, \quad 1 \le t \le e$$

Find $\mathbf{F}_1\left(\dfrac{1}{2}\right)$, $\mathbf{F}_2\left(\dfrac{1}{2}\right)$, and $\mathbf{F}_3\left(\dfrac{1 + e}{2}\right)$.

39. Show that $D_t[c\mathbf{F}(t)] = cD_t\mathbf{F}(t)$, where c is a real number.

40. Show that if $\mathbf{F}$ is continuous at a, then $\|\mathbf{F}\|$ is continuous at a.

41. Find a function $\mathbf{F}$ and a number a with $\|\mathbf{F}\|$ continuous at a, but $\mathbf{F}$ not continuous at a.

42. Prove Theorem 13.7, part (iv): If g is differentiable at a and $\mathbf{F}$ is differentiable at $g(a)$, then

$$D_t(\mathbf{F} \circ g)(a) = \mathbf{F}'(g(a))g'(a).$$

43. Prove Theorem 13.7, part (iii): If $\mathbf{F}$ and $\mathbf{G}$ are differentiable, then

$$D_t(\mathbf{F} \cdot \mathbf{G})(t) = D_t\mathbf{F}(t) \cdot \mathbf{G}(t) + \mathbf{F}(t) \cdot D_t\mathbf{G}(t).$$

44. If $\mathbf{F}$ and $\mathbf{G}$ are integrable on $[a, b]$ and c is a real number, show that

   (a) $\displaystyle\int_a^b [\mathbf{F}(t) + \mathbf{G}(t)]dt = \int_a^b \mathbf{F}(t)dt + \int_a^b \mathbf{G}(t)dt$

   (b) $\displaystyle\int_a^b c\mathbf{F}(t)dt = c\int_a^b \mathbf{F}(t)dt$

13.3
ARC LENGTH AS A PARAMETER; UNIT TANGENT AND UNIT NORMAL VECTORS

A curve in space can be described by many different vector-valued functions. For example, the line segment joining the points $(1, 0, 2)$ and $(4, 5, 7)$ is described by each of the functions:

$$\mathbf{F}_1(t) = (6t + 1)\mathbf{i} + 10t\mathbf{j} + (10t + 2)\mathbf{k}, \quad 0 \le t \le \frac{1}{2}$$

$$\mathbf{F}_2(t) = t^2\mathbf{i} + \frac{5}{3}(t^2 - 1)\mathbf{j} + \frac{5}{3}\left(t^2 + \frac{1}{5}\right)\mathbf{k}, \quad 1 \le t \le 2$$

$$\mathbf{F}_3(t) = (\ln t^3 + 1)\mathbf{i} + \ln t^5\mathbf{j} + (\ln t^5 + 2)\mathbf{k}, \quad 1 \le t \le e.$$

This can be easily verified by observing that the coordinate functions for each satisfy the symmetric equations for the line:

$$\frac{x - 1}{3} = \frac{y}{5} = \frac{z - 2}{5}.$$

Although these functions describe the same line, they describe a motion from $(1, 0, 2)$ to $(4, 5, 7)$ in quite different ways. For example, at the midpoint of the domain of each function, three distinct points on the line are represented (see Figure 13.9):

$\mathbf{F}_1(.25) = 2.5\mathbf{i} + 2.5\mathbf{j} + 4.5\mathbf{k}$, the midpoint of the line segment;

$\mathbf{F}_2(1.5) = 2.25\mathbf{i} + 2.08\mathbf{j} + 4.08\mathbf{k}$, a point closer to $(1, 0, 2)$;

$\mathbf{F}_3\left(\dfrac{e + 1}{2}\right) \approx 2.86\mathbf{i} + 3.10\mathbf{j} + 5.10\mathbf{k}$, a point closer to $(4, 5, 7)$.

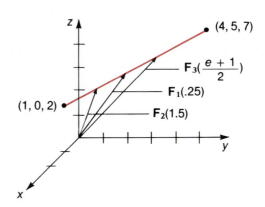

FIGURE 13.9

Although a variety of functions can represent a curve in space, there is a standard method called the **arc length representation**. An arc length representation of a curve of length l can be described geometrically by first fixing one endpoint of the curve and then using the arc length along the curve as the parameter. The point on the curve associated with each value s in $[0, l]$ is the point that is a distance s along the curve from the fixed endpoint.

Suppose **F** is an arbitrary function describing a curve C in space:

$$\mathbf{F}(t) = x(t)\mathbf{i} + y(t)\mathbf{j} + z(t)\mathbf{k}, \qquad t \text{ in an interval } I.$$

The curve C is described by the parametric equations:

(13.8) $x = x(t), \qquad y = y(t), \qquad z = z(t), \qquad t \text{ in } I.$

This curve is called **smooth** if the derivatives x', y', and z' are continuous on I and are not simultaneously zero, except possibly at an endpoint.

If C is a smooth curve defined on $[a, b]$, then the length l of C is given by

$$l = \int_a^b \sqrt{[x'(t)]^2 + [y'(t)]^2 + [z'(t)]^2}\, dt.$$

The derivation of this formula is the same as that of the equivalent formula for parametric equations in the plane. (See Section 10.6.)

For t in the interval $[a, b]$, consider the length $s(t)$ of C from the terminal point of $\mathbf{F}(a)$ to the terminal point of $\mathbf{F}(t)$.

(13.9) $s \equiv s(t) = \int_a^t \sqrt{[x'(\tau)]^2 + [y'(\tau)]^2 + [z'\tau]^2}\, d\tau.$

This describes a one-to-one correspondence between $[a, b]$ and $[0, l]$ and permits the function **F** in the parameter t to be transformed into a new function in the parameter s. The domain of this new function is $[0, l]$. (Figure 13.10 illustrates the lengths l and $s(t)$.)

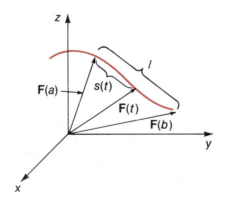

FIGURE 13.10

Let us reconsider the straight line segment joining $(1, 0, 2)$ and $(4, 5, 7)$ to see how the arc length representation evolves. First, we take one of the functions describing the line segment, say $\mathbf{F}_1$:

$$\mathbf{F}_1(t) = (6t + 1)\mathbf{i} + 10t\mathbf{j} + (10t + 2)\mathbf{k}, \quad 0 \le t \le \frac{1}{2}.$$

Then,

$$s(t) = \int_0^t \sqrt{[D_\tau(6\tau + 1)]^2 + [D_\tau 10\tau]^2 + [D_\tau(10\tau + 2)]^2} \, d\tau$$

$$= \int_0^t \sqrt{36 + 100 + 100} \, d\tau = \sqrt{236}\tau \Big]_0^t$$

$$= 2\sqrt{59} \, t.$$

This implies that the length of the line segment from $(1, 0, 2)$ to $(4, 5, 7)$ is $s(1/2) = \sqrt{59}$ and that t and s are related by

$$t = \frac{s}{2\sqrt{59}} = \frac{\sqrt{59}\,s}{118}.$$

Substituting $\sqrt{59}s/118$ for t in the description of $\mathbf{F}_1$, we obtain a vector-valued function $\mathbf{F}$ with arc length as parameter:

$$\mathbf{F}(s) = \left(\frac{3\sqrt{59}}{59}s + 1\right)\mathbf{i} + \frac{5\sqrt{59}}{59}s\mathbf{j} + \left(\frac{5\sqrt{59}}{59}s + 2\right)\mathbf{k}, \quad 0 \le s \le \sqrt{59}.$$

The other functions $\mathbf{F}_2$ or $\mathbf{F}_3$ could have been used to produce this same arc length representation $\mathbf{F}$. In fact, (13.9) can be used to give the arc length representation of a curve whenever any parametric representation is known.

Although the arc length representation is important in theory, arc length often cannot be determined explicitly. Fortunately, an explicit representation is not generally required; we need only know that it exists, and know the method that describes it. The primary reason for introducing arc length parametric representation of a curve at this time is associated with the following definition.

(13.10)
DEFINITION

If a curve C is described by a function $\mathbf{F}$ on $[a, b]$, the **principal unit tangent vector** to C at the terminal point of $\mathbf{F}(t_0)$ is defined by:

$$\mathbf{T}(t_0) = \frac{\mathbf{F}'(t_0)}{\|\mathbf{F}'(t_0)\|},$$

provided $\|\mathbf{F}'(t_0)\| \ne 0$.

EXAMPLE 1

Find the principal unit tangent vector to the curve C in the xy-plane described by $\mathbf{F}(t) = 2\cos t\,\mathbf{i} + 3\sin t\,\mathbf{j}$ when $t = \dfrac{\pi}{4}$.

SOLUTION

According to Definition 13.10,

$$\mathbf{T}(t) = \frac{\mathbf{F}'(t)}{\|\mathbf{F}'(t)\|} = \frac{-2\sin t\,\mathbf{i} + 3\cos t\,\mathbf{j}}{[4(\sin t)^2 + 9(\cos t)^2]^{1/2}}$$

so

$$\mathbf{T}\left(\frac{\pi}{4}\right) = \frac{-2\dfrac{\sqrt{2}}{2}\mathbf{i} + 3\dfrac{\sqrt{2}}{2}\mathbf{j}}{\left[4\left(\dfrac{1}{2}\right) + 9\left(\dfrac{1}{2}\right)\right]^{1/2}} = \frac{-\sqrt{2}\mathbf{i} + \dfrac{3\sqrt{2}}{2}\mathbf{j}}{\dfrac{\sqrt{13}}{\sqrt{2}}}$$

$$= \left(-\sqrt{2}\mathbf{i} + \frac{3}{2}\sqrt{2}\mathbf{j}\right)\frac{\sqrt{2}}{\sqrt{13}} = -\frac{2\sqrt{13}}{13}\mathbf{i} + \frac{3\sqrt{13}}{13}\mathbf{j}.$$

This is illustrated in Figure 13.11. ☐

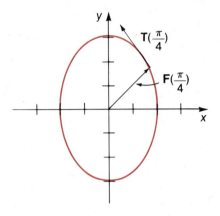

FIGURE 13.11

There are two unit tangent vectors to a curve at a point t_0. A reference to "*the* unit tangent vector" will always mean the principal unit tangent vector $\mathbf{T}(t_0)$, not $-\mathbf{T}(t_0)$. (See Figure 13.12.)

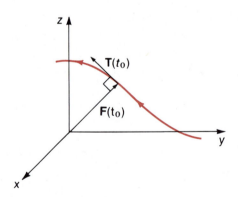

FIGURE 13.12

To see the role played by arc length in Definition 13.10, recall that

$$s = s(t) = \int_a^t \sqrt{[x'(\tau)]^2 + [y'(\tau)]^2 + [z'(\tau)]^2}\, d\tau,$$

where

$$\mathbf{F}(t) = x(t)\mathbf{i} + y(t)\mathbf{j} + z(t)\mathbf{k}, \quad a \leq t \leq b, \quad \text{describes } C.$$

It follows from the fundamental theorem of calculus that

$$\frac{ds}{dt} = \sqrt{[x'(t)]^2 + [y'(t)]^2 + [z'(t)]^2} = \|\mathbf{F}'(t)\|.$$

Consequently, $\mathbf{T}(t)$ can be written

(13.11) $$\mathbf{T}(t) = \frac{\mathbf{F}'(t)}{\|\mathbf{F}'(t)\|} = \frac{\mathbf{F}'(t)}{\dfrac{ds}{dt}} = \frac{d\mathbf{F}(t)}{dt}\frac{dt}{ds} = \frac{d\mathbf{F}(t)}{ds}.$$

Thus, the unit tangent vector is the rate of change of the values of the function **F** relative to the change in arc length along the curve described by **F**.

EXAMPLE 2   Show that the functions

$$\mathbf{F}_1(t) = (6t + 1)\mathbf{i} + 10t\mathbf{j} + (10t + 2)\mathbf{k}, \quad 0 \leq t \leq .5$$

$$\mathbf{F}_2(t) = t^2\mathbf{i} + \frac{5}{3}(t^2 - 1)\mathbf{j} + \frac{5}{3}\left(t^2 + \frac{1}{5}\right)\mathbf{k}, \quad 1 \leq t \leq 2$$

$$\mathbf{F}_3(t) = (\ln t^3 + 1)\mathbf{i} + (\ln t^5)\mathbf{j} + (\ln t^5 + 2)\mathbf{k}, \quad 1 \leq t \leq e$$

give the same unit tangent vector.

SOLUTION

$$\mathbf{F}_1'(t) = 6\mathbf{i} + 10\mathbf{j} + 10\mathbf{k}, \quad \|\mathbf{F}_1'(t)\| = \sqrt{36 + 100 + 100} = 2\sqrt{59}.$$

$$\mathbf{F}_2'(t) = 2t\mathbf{i} + \frac{10}{3}t\mathbf{j} + \frac{10}{3}t\mathbf{k}, \quad \|\mathbf{F}_2'(t)\| = \sqrt{4t^2 + \frac{100}{9}t^2 + \frac{100}{9}t^2} = \frac{2\sqrt{59}}{3}t.$$

$$\mathbf{F}_3'(t) = \frac{3}{t}\mathbf{i} + \frac{5}{t}\mathbf{j} + \frac{5}{t}\mathbf{k}, \quad \|\mathbf{F}_3'(t)\| = \sqrt{\frac{9}{t^2} + \frac{25}{t^2} + \frac{25}{t^2}} = \frac{\sqrt{59}}{t}.$$

So $$\mathbf{T}(t) = \frac{\mathbf{F}_1'(t)}{\|\mathbf{F}_1'(t)\|} = \frac{3}{\sqrt{59}}\mathbf{i} + \frac{5}{\sqrt{59}}\mathbf{j} + \frac{5}{\sqrt{59}}\mathbf{k}$$

$$= \frac{3\sqrt{59}}{59}\mathbf{i} + \frac{5\sqrt{59}}{59}\mathbf{j} + \frac{5\sqrt{59}}{59}\mathbf{k}$$

It is easily seen that $\dfrac{\mathbf{F}_2'(t)}{\|\mathbf{F}_2'(t)\|}$ and $\dfrac{\mathbf{F}_3'(t)}{\|\mathbf{F}_3'(t)\|}$ give the representation. In this case $\mathbf{T}(t)$ is independent of t because the curve is a straight line, in fact, a line in the direction $\mathbf{T}(t)$.   □

The principal unit tangent vector $\mathbf{T}(t)$ to a curve is orthogonal to the vector $\mathbf{T}'(t)$. This is a result of the following theorem.

(13.12)
THEOREM

If $\mathbf{u}(t)$ is a unit vector for each t, then $\mathbf{u}(t) \cdot \mathbf{u}'(t) = 0$.

PROOF

Since $\mathbf{u}(t)$ is of unit length,

$$1 = \|\mathbf{u}(t)\|^2 = \mathbf{u}(t) \cdot \mathbf{u}(t).$$

Using the product rule,

$$0 = D_t(\mathbf{u}(t) \cdot \mathbf{u}(t)) = \mathbf{u}'(t) \cdot \mathbf{u}(t) + \mathbf{u}(t) \cdot \mathbf{u}'(t) = 2\mathbf{u}(t) \cdot \mathbf{u}'(t)$$

and $\qquad\qquad\qquad\qquad \mathbf{u}(t) \cdot \mathbf{u}'(t) = 0.$ $\qquad\qquad$ □

In a plane, the normal line to a curve at a specified point is the unique line perpendicular to the tangent line at that point (see Figure 13.13(a)). For curves in space, this unique line is replaced by a plane passing through the point, as shown in Figure 13.13(b).

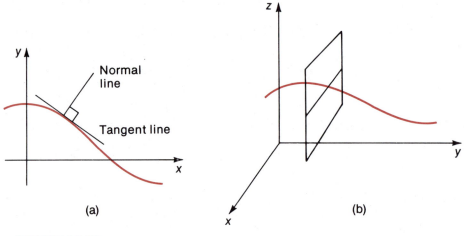

FIGURE 13.13

One particular normal direction is more important than others in applications: the normal in the direction of $\mathbf{T}'(t)$.

(13.13)
DEFINITION

If C is a curve described by the function $\mathbf{F}$ on $[a, b]$ and $\mathbf{T}$ is the function describing the principal unit tangent vector at points on C, then the **principal unit normal vector** at t_0 is the unit vector in the direction of $\mathbf{T}'(t_0)$,

$$\mathbf{N}(t_0) = \frac{\mathbf{T}'(t_0)}{\|\mathbf{T}'(t_0)\|},$$

provided $\|\mathbf{T}'(t_0)\| \neq 0.$

EXAMPLE 3   Find the principal unit tangent and unit normal vectors to the circular helix described by $\mathbf{F}(t) = \cos t\mathbf{i} + \sin t\mathbf{j} + t\mathbf{k}$ at the point $(0, 1, \pi/2)$.

SOLUTION

Since $\mathbf{F}'(t) = -\sin t\mathbf{i} + \cos t\mathbf{j} + \mathbf{k}$, the unit tangent vector is given by:

$$\mathbf{T}(t) = \frac{\mathbf{F}'(t)}{\|\mathbf{F}'(t)\|} = \frac{-\sin t\mathbf{i} + \cos t\mathbf{j} + \mathbf{k}}{\sqrt{(-\sin t)^2 + (\cos t)^2 + 1}} = \frac{\sqrt{2}}{2}[-\sin t\mathbf{i} + \cos t\mathbf{j} + \mathbf{k}].$$

The unit normal vector is

$$\mathbf{N}(t) = \frac{\mathbf{T}'(t)}{\|\mathbf{T}'(t)\|} = \frac{\frac{\sqrt{2}}{2}[-\cos t\mathbf{i} - \sin t\mathbf{j}]}{\frac{\sqrt{2}}{2}\sqrt{(-\cos t)^2 + (-\sin t)^2}} = -\cos t\mathbf{i} - \sin t\mathbf{j}.$$

Since $(0, 1, \pi/2)$ is the terminal point of $\mathbf{F}(\pi/2)$, the principal unit tangent and unit normal vectors at this point are

$$\mathbf{T}\left(\frac{\pi}{2}\right) = \frac{\sqrt{2}}{2}(-\mathbf{i} + \mathbf{k}) \qquad \text{and} \qquad \mathbf{N}\left(\frac{\pi}{2}\right) = -\mathbf{j}.$$

These vectors are shown in Figure 13.14.   □

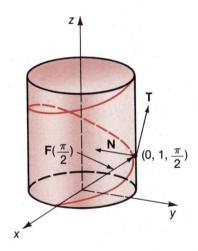

FIGURE 13.14

EXERCISE SET 13.3

Determine the intervals on which the curves described in Exercises 1 through 6 are smooth.

1. $\mathbf{F}(t) = t\mathbf{i} + t^2\mathbf{j}$ **2.** $\mathbf{F}(t) = \sqrt{1 - t^2}\,\mathbf{j} + t\mathbf{k}$

3. $F(t) = i + tj + \sqrt{t}k$

4. $F(t) = tj + |t|\,k$

5. $F(t) = ti + e^t j + tk$

6. $F(t) = ti + \tan tj + 2k$

Find a vector-valued function with arc length as parameter to describe the curves in Exercises 7 through 12.

7. The straight line from $(1, 1, 0)$ to $(2, 5, 4)$.

8. The straight line from $(2, 0, 1)$ to $(1, 2, 3)$.

9. A circle of radius 1 in the xy-plane with center at the origin.

10. A circle of radius 2 in the xz-plane with center at the origin.

11. The circular helix described by

$$F(t) = \sin ti + \cos tj + tk, \quad 0 \le t \le \pi.$$

12. A quarter circle with center $(0, 0, 1)$ that begins at

$$\left(\frac{\sqrt{2}}{2}, \frac{\sqrt{2}}{2}, 1\right) \text{ and ends at } \left(-\frac{\sqrt{2}}{2}, \frac{\sqrt{2}}{2}, 1\right).$$

Find the principal unit tangent vector at t_0 to the curves described in Exercises 13 through 18.

13. $F(t) = (2t - 1)i + (t + 1)j + (1 - 3t)k, \quad t_0 = 2$

14. $F(t) = (t - 2)i + (2 - t)j + (2t - 1)k, \quad t_0 = -1$

15. $F(t) = \cos ti + \sin tj, \quad t_0 = \dfrac{\pi}{2}$

16. $F(t) = \sin ti + \cos tj + tk, \quad t_0 = \pi$

17. $F(t) = t^2 i + \ln tj + tk, \quad t_0 = 1$

18. $F(t) = 2ti + e^t j + t^3 k, \quad t_0 = 0$

19–24. Find the principal unit normal vector at t_0 to the curves described in Exercises 13 through 18.

25. (a) Sketch the curve described by $F(t) = 2i + tj + (t^2 - 2)k$.

(b) Find and sketch the principal unit tangent and unit normal vectors to this curve at the point $(2, 2, 2)$.

(c) Determine the values of t at which the principal unit tangent vector is parallel to the xy-plane. What is the direction of the principal unit normal vector in this case?

26. The vector-valued function $F(t) = 3i + 6tj + (5t - 5t^2)k$ describes the motion of a baseball.

(a) Sketch the curve described by $F(t)$.

(b) Find and sketch the principal unit tangent and unit normal vectors at $t = .5$.

(c) Determine the values of t when the unit tangent vector is parallel to the xy-plane.

27. A 5/8-inch-diameter hex bolt has 200 threads in its 10 inches of length. Grooves have been cut between the threads to a depth of .01 inch. What is the total length of (a) the threads and (b) the grooves?

28. Find the length of the threads and grooves of a bolt or pipe that has diameter D, with n threads in its length l, cut to a depth d.

13.4
VELOCITY AND ACCELERATION OF OBJECTS IN SPACE

In this section we study the motion of an object moving along a curve in space. In so doing we will see applications of the principal unit tangent and unit normal vectors and discover why the directions described by these vectors are called the *principal directions*.

Suppose an object travels through space so that at any time t, its position is described by the vector-valued function

$$\mathbf{r}(t) = x(t)\mathbf{i} + y(t)\mathbf{j} + z(t)\mathbf{k}, \quad a \le t \le b.$$

The motion of the object describes a curve C in space.

As the object moves along C, its speed and direction may vary with time. The speed is the instantaneous rate of change of the arc length traveled with respect to time. The velocity of the object describes both its speed and direction.

The definitions of velocity, speed, and acceleration for an object traveling in space are similar to these definitions for rectilinear motion.

(13.14)
DEFINITION

If the motion of an object through space is described by

$$\mathbf{r}(t) = x(t)\mathbf{i} + y(t)\mathbf{j} + z(t)\mathbf{k}, \qquad a \le t \le b,$$

then the **velocity** of the object is

$$\mathbf{v}(t) = \mathbf{r}'(t) = x'(t)\mathbf{i} + y'(t)\mathbf{j} + z'(t)\mathbf{k}$$

and its **speed** is

$$v(t) = \|\mathbf{v}(t)\| = \sqrt{[x'(t)]^2 + [(y'(t)]^2 + [z'(t)]^2}.$$

The **acceleration** of the object is

$$\mathbf{a}(t) = \mathbf{v}'(t) = \mathbf{r}''(t) = x''(t)\mathbf{i} + y''(t)\mathbf{j} + z''(t)\mathbf{k}.$$

The velocity is the tangent vector to the curve that points in the direction of increasing values of t. This is illustrated in Figure 13.15. Thus the velocity vector does indeed give the direction of motion. Since $\|\mathbf{v}(t)\| = \|D_t\mathbf{r}(t)\| = ds/dt$, the speed gives the rate at which the arc length traveled changes with time.

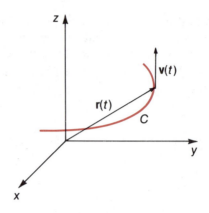

FIGURE 13.15

EXAMPLE 1   An object travels along the curve of a circular helix, such that its position at any time t is at the terminal point of the position vector

$$\mathbf{r}(t) = \cos t\mathbf{i} + \sin t\mathbf{j} + t\mathbf{k}, \qquad 0 \le t \le 2\pi.$$

Find the velocity, acceleration, and speed of this object at any time t in $[0, 2\pi]$.

SOLUTION
     The velocity is

$$\mathbf{v}(t) = \mathbf{r}'(t) = -\sin t\mathbf{i} + \cos t\mathbf{j} + \mathbf{k},$$

the acceleration is

$$\mathbf{a}(t) = \mathbf{v}'(t) = -\cos t\mathbf{i} - \sin t\mathbf{j},$$

and the speed is

$$v(t) = \|\mathbf{v}(t)\| = \sqrt{(-\sin t)^2 + (\cos t)^2 + 1} = \sqrt{2}.$$

Notice that the speed is constant, although both the velocity and acceleration depend on t.   □

     We will now show how vector functions can be used to describe the motion of an object propelled from the earth if the initial velocity $\mathbf{v}_0$ and initial position $\mathbf{r}_0$ are known. Since this is essentially a two-dimensional problem, we assume that the component in the direction of the vector $\mathbf{k}$ is zero and that the object is propelled in the plane shown in Figure 13.16. In this coordinate system, x represents the horizontal distance and y represents the vertical distance.
     Gravity is the only force acting on the object and this force acts in the negative y direction with magnitude 32 ft/sec². This is expressed by

$$\mathbf{a}(t) = -32\mathbf{j} \text{ ft/sec}^2.$$

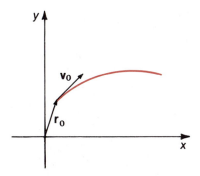

FIGURE 13.16

The velocity of the object is given by

$$\mathbf{v}(t) = \int \mathbf{a}(t)dt = -32t\mathbf{j} + \mathbf{C}_1,$$

for some constant vector $\mathbf{C}_1$. Since

$$\mathbf{v}_0 = \mathbf{v}(0) = -32(0)\mathbf{j} + \mathbf{C}_1 = \mathbf{C}_1,$$
$$\mathbf{v}(t) = -32t\mathbf{j} + \mathbf{v}_0 \text{ ft/sec.}$$

The position of the object at time t is

$$\mathbf{r}(t) = \int \mathbf{v}(t)dt = -16t^2\mathbf{j} + \mathbf{v}_0 t + \mathbf{C}_2,$$

for some constant vector $\mathbf{C}_2$. Since $\mathbf{r}(0) = \mathbf{r}_0$, $\mathbf{C}_2 = \mathbf{r}_0$. Consequently

$$\mathbf{r}(t) = -16t^2\mathbf{j} + \mathbf{v}_0 t + \mathbf{r}_0 \text{ ft.}$$

By using vector-valued functions, we have solved this problem as easily as if the motion had been rectilinear.

Sometimes we need to determine the motion in the horizontal and vertical directions independently. This can be done by decomposing the initial velocity vector $\mathbf{v}_0$ into its horizontal and vertical components.

EXAMPLE 2   A projectile is fired from ground level with an initial speed of 1000 ft/sec and an angle of elevation of $\pi/4$ radians. Describe the subsequent motion of the projectile and find (a) the time when the projectile strikes the ground, (b) the distance from the firing point when the projectile strikes the ground, (c) the maximum height of the projectile, and (d) the speed of the projectile at the instant of impact.

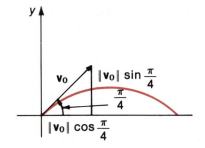

FIGURE 13.17

SOLUTION

Figure 13.17 shows the initial velocity vector $\mathbf{v}_0$ and its horizontal and vertical components. The origin has been placed at the firing point, so $\mathbf{r}_0 = \mathbf{0}$. The subsequent motion of the projectile is along the curve. Since $\|\mathbf{v}_0\| = 1000$ and the angle of elevation is $\pi/4$,

$$\mathbf{v}_0 = 1000 \cos \frac{\pi}{4} \mathbf{i} + 1000 \sin \frac{\pi}{4} \mathbf{j}$$
$$= 500\sqrt{2}\, \mathbf{i} + 500\sqrt{2}\, \mathbf{j}.$$

Since $\mathbf{a}(t) = -32\mathbf{j}$,

$$\mathbf{v}(t) = -32t\mathbf{j} + \mathbf{v}_0 = -32t\mathbf{j} + 500\sqrt{2}\mathbf{i} + 500\sqrt{2}\mathbf{j}$$
$$= 500\sqrt{2}\mathbf{i} + (500\sqrt{2} - 32t)\mathbf{j}.$$

Consequently, the motion of the projectile is described by

$$\mathbf{r}(t) = 500\sqrt{2}\, t\mathbf{i} + (500\sqrt{2}\, t - 16t^2)\mathbf{j}.$$

(a) The time of impact occurs when the y-component of $\mathbf{r}(t)$ is zero, that is, when

$$0 = 500\sqrt{2}\, t - 16t^2 = 4t(125\sqrt{2} - 4t).$$

The time $t = 0$ corresponds to the initial position, so the projectile strikes the ground when $t = 125\sqrt{2}/4 \approx 44.2$ seconds.

(b) Since

$$\mathbf{r}\left(\frac{125\sqrt{2}}{4}\right) = (500\sqrt{2})\left(\frac{125\sqrt{2}}{4}\right)\mathbf{i} = 31250\,\mathbf{i},$$

the projectile strikes the ground 31,250 feet from the firing point.

(c) The maximum height occurs when the y-component of $\mathbf{r}(t)$ is a maximum, that is, when the derivative of this component is zero. This implies that the velocity vector is horizontal. Since

$$\mathbf{v}(t) = \mathbf{r}'(t) = 500\sqrt{2}\, \mathbf{i} + (500\sqrt{2} - 32t)\mathbf{j},$$

the maximum height occurs when

$$0 = 500\sqrt{2} - 32t$$

or

$$t = \frac{500\sqrt{2}}{32} = \frac{125\sqrt{2}}{8}.$$

The maximum height is the y-component of $\mathbf{r}(125\sqrt{2}/8)$,

$$y\left(\frac{125\sqrt{2}}{8}\right) = 500\sqrt{2}\left(\frac{125\sqrt{2}}{8}\right) - 16\left(\frac{125\sqrt{2}}{8}\right)^2$$
$$= (125)^2 - \frac{(125)^2}{2}$$
$$= \frac{1}{2}(125)^2 = 7812.5 \text{ ft.}$$

(d) Since the projectile strikes the ground when $t = 125\sqrt{2}/4$, the speed of the projectile at the instant of impact is given by

$$\left\| \mathbf{v}\left(\frac{125\sqrt{2}}{4} \right) \right\| = \sqrt{(500\sqrt{2})^2 + \left[500\sqrt{2} - (32)\left(\frac{125\sqrt{2}}{4} \right) \right]^2}$$

$$= \sqrt{(500)^2 \cdot 2 + [-500\sqrt{2}]^2}$$

$$= (500)(2) = 1000 \text{ ft/sec},$$

the speed at which it was propelled. ☐

The velocity vectors of an object in space are tangent to the curve traced by the position vectors. Thus an intimate relationship exists between the velocity, speed, and acceleration of the object and the principal unit tangent and unit normal vectors to its curve of motion.

Since the unit tangent vector for $\mathbf{r}(t)$ is

$$\mathbf{T}(t) = \frac{\mathbf{r}'(t)}{\|\mathbf{r}'(t)\|} = \frac{\mathbf{v}(t)}{v(t)},$$

the velocity of an object is

$$\mathbf{v}(t) = v(t)\, \mathbf{T}(t).$$

If $\mathbf{T}'(t) \neq \mathbf{0}$, the acceleration is

$$\mathbf{a}(t) = \mathbf{v}'(t) = v'(t)\mathbf{T}(t) + v(t)\mathbf{T}'(t)$$

$$= v'(t)\mathbf{T}(t) + v(t)\,\|\mathbf{T}'(t)\|\,\frac{\mathbf{T}'(t)}{\|\mathbf{T}'(t)\|},$$

so

(13.15) $$\mathbf{a}(t) = v'(t)\mathbf{T}(t) + v(t)\,\|\mathbf{T}'(t)\|\,\mathbf{N}(t).$$

The reason for choosing the direction of $\mathbf{N}(t)$ to be the *principal* normal direction can now be better appreciated. The acceleration of an object has one component in this direction and one component in the direction of the principal unit tangent vector. See Figure 13.18.

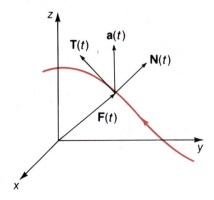

FIGURE 13.18

The tangential component of acceleration is denoted $a_\mathbf{T}$:

(13.16) $$a_\mathbf{T}(t) = v'(t).$$

The normal component of acceleration, also called the **centripetal acceleration,** is denoted a_N:

(13.17) $a_N(t) = v(t) \|\mathbf{T}'(t)\|.$

With this notation, we have:

(13.18) $\mathbf{a}(t) = a_T(t)\mathbf{T}(t) + a_N(t)\mathbf{N}(t).$

The fact that $\mathbf{T}(t)$ and $\mathbf{N}(t)$ are orthogonal unit vectors enables us to express a_N in terms of $\|\mathbf{a}\|$ and $\mathbf{a}_T$.

$$
\begin{aligned}
\|\mathbf{a}(t)\|^2 &= \mathbf{a}(t) \cdot \mathbf{a}(t) \\
&= (a_T\,\mathbf{T}(t) + a_N\,\mathbf{N}(t)) \cdot (a_T\,\mathbf{T}(t) + a_N\,\mathbf{N}(t)) \\
&= (a_T)^2\,\mathbf{T}(t) \cdot \mathbf{T}(t) + 2a_N a_T\,\mathbf{T}(t) \cdot \mathbf{N}(t) + (a_N)^2\,\mathbf{N}(t) \cdot \mathbf{N}(t) \\
&= a_T^2\,\|\mathbf{T}(t)\|^2 + 0 + a_N^2\,\|\mathbf{N}(t)\|^2 \\
&= a_T^2 + a_N^2.
\end{aligned}
$$

Since $a_N = \|\mathbf{r}'(t)\|\,\|\mathbf{T}'(t)\| > 0,$

(13.19) $a_N = \sqrt{\|\mathbf{a}(t)\|^2 - a_T^2}.$

EXAMPLE 3   Find the tangential and normal components of acceleration of a particle whose motion is given by

$$\mathbf{r}(t) = t\mathbf{i} + t\mathbf{j} + t^2\,\mathbf{k}, \qquad 0 \le t \le 1.$$

SOLUTION

The velocity, acceleration, and speed of this object are given by

$$\mathbf{v}(t) = \mathbf{i} + \mathbf{j} + 2t\mathbf{k}, \quad \mathbf{a}(t) = 2\mathbf{k}, \quad \text{and} \quad v(t) = \sqrt{2 + 4t^2}.$$

The tangential component of acceleration is

$$a_T(t) = v'(t) = \frac{4t}{\sqrt{2 + 4t^2}}.$$

To compute the normal component of acceleration, we will use (13.19). Since $\|\mathbf{a}(t)\|^2 = \|2\mathbf{k}\|^2 = 4,$

$$a_N(t) = \sqrt{\|\mathbf{a}(t)\|^2 - a_T^2} = \sqrt{4 - \frac{16t^2}{2 + 4t^2}} = \sqrt{\frac{8}{2 + 4t^2}} = \frac{2}{\sqrt{1 + 2t^2}}. \ \square$$

EXAMPLE 4   An object rotates in a circular orbit of radius R at a constant angular velocity ω radians/second. Show that the acceleration is totally centripetal.

SOLUTION

Assume that the rotation occurs counterclockwise in the xy-plane with the center of the orbit at the origin and that the object is at $(R, 0)$ when $t = 0$. Then the orbit is described

by $\mathbf{r}(t) = R[\cos \omega t\mathbf{i} + \sin \omega t\mathbf{j}],$

so $\mathbf{v}(t) = \omega R[-\sin\omega t\mathbf{i} + \cos\omega t\mathbf{j}]$

and $\mathbf{a}(t) = -\omega^2 R[\cos\omega t\mathbf{i} + \sin\omega t\mathbf{j}] = -\omega^2\,\mathbf{r}(t).$

(See Figure 13.19.) Since $v(t) = \|\mathbf{v}(t)\| = \omega R,$

$$a_{\mathbf{T}}(t) = v'(t) = 0$$

and

$$a_{\mathbf{N}}(t) = \sqrt{\|\mathbf{a}(t)\|^2 - (a_{\mathbf{T}}(t))^2} = \sqrt{\omega^2\|\mathbf{r}(t)\|} = \omega\sqrt{\|\mathbf{r}(t)\|}.$$

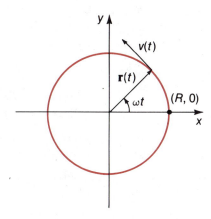

FIGURE 13.19

Consequently, the acceleration of an object rotating in a circular orbit at constant speed is totally centripetal, in the direction toward the center of the circle.   □

EXERCISE SET 13.4

The motion of an object is described in Exercises 1 through 8. Find the velocity, acceleration, and speed of the object at any time t.

1. $\mathbf{r}(t) = \sin t\mathbf{i} + \cos t\mathbf{j} + t\mathbf{k}$

2. $\mathbf{r}(t) = \sin t\mathbf{i} + \cos t\mathbf{j} + e^t\mathbf{k}$

3. $\mathbf{r}(t) = (t - 1)\mathbf{i} + (2t - 1)\mathbf{j} + (3t - 3)\mathbf{k}$

4. $\mathbf{r}(t) = \mathbf{i} + 3t\mathbf{j} + 6t^2\mathbf{k}$

5. $\mathbf{r}(t) = 3\mathbf{i} + 4t\mathbf{j} + (5t - 5t^2)\mathbf{k}, \quad 0 \le t \le 1$

6. $\mathbf{r}(t) = (10\sin t - 1)\mathbf{i} + t\mathbf{j} + (16\cos t - 2)\mathbf{k}, \quad t \ge 0$

7. $\mathbf{r}(t) = \sqrt{3t - 1}\,\mathbf{j} + e^t\mathbf{k}$

8. $\mathbf{r}(t) = t\mathbf{j} + (16t^2 - 10t)\mathbf{k}$

The motion of an object is described in Exercises 9 through 14. Determine the tangential and normal components of acceleration of the object.

9. $\mathbf{r}(t) = \sin t\,\mathbf{i} + \cos t\,\mathbf{j} + t\mathbf{k}$

10. $\mathbf{r}(t) = \cos t\,\mathbf{i} + \sin t\,\mathbf{j} + t\mathbf{k}$

11. $\mathbf{r}(t) = (2t - 1)\mathbf{i} + (2 - t)\mathbf{j} + (t - 3)\mathbf{k}$

12. $\mathbf{r}(t) = 3\mathbf{i} + 2t\mathbf{j} + t^2\mathbf{k}$

13. $\mathbf{r}(t) = t\mathbf{j} + (16t^2 - 10t)\mathbf{k}$

14. $\mathbf{r}(t) = 2\mathbf{i} + 3t\mathbf{j} + e^t\mathbf{k}$

15. The path of a golf ball hit is described by

$$\mathbf{r}(t) = \begin{cases} 110t\mathbf{j} + 50t(2 - t)\mathbf{k}, & 0 \le t \le 2 \\ 220\sqrt{t - 1}\,\mathbf{j}, & 2 \le t \le 2.4, \end{cases}$$

where the time t is given in minutes and the distance in feet.

(a) Sketch the curve that represents the path of the golf ball.

(b) Find the maximum height of the golf ball.

(c) Find the speed of the ball when it strikes the ground.

16. The path of a baseball is described by

$$\mathbf{r}(t) = 2t\left(t^2 - \frac{5}{4}t + \frac{1}{4}\right)\mathbf{i} + 45t\mathbf{j} + (4 + t)\mathbf{k}, \quad 0 \le t \le 1,$$

where the time t is in seconds and the distance in meters. Find

(a) the initial speed of the ball.

(b) the speed of the ball when $t = 1$ second.

(c) the horizontal distance traveled by the ball.

17. A projectile is fired from the ground with an initial speed of 1200 ft/sec and an angle of elevation of $\pi/3$ radians. Find

(a) the subsequent motion of the projectile.

(b) the velocity at time t.

(c) the maximum height of the projectile.

(d) the time when the projectile strikes the ground.

(e) the speed of the projectile at the instant of impact.

18. A quarterback throws a football downfield with an initial speed of 55 ft/sec and at an angle of elevation of $\pi/6$ radians. How far from the quarterback should the receiver be to catch the ball, and how long from the time the ball is thrown does he have to get into this position?

19. In the 1972 Olympics the winning throw in the shotput was for a distance of 21.18 meters. If the shot was launched at an angle of 45° relative to the vertical, what was

(a) the initial speed?   (b) the maximum height?   (c) the time of flight?

20. A baseball hit at an angle of elevation of $\pi/4$ just clears a 21 foot high wall in right field 335 feet from home plate at Three Rivers Stadium in Pittsburgh. If the ball is 4 feet above the ground when hit, find

(a) the initial speed of the ball.

(b) the time it takes to reach the wall.

(c) the velocity as it clears the wall.

(d) the speed as it clears the wall.

21. A center fielder throws a ball toward home plate 330 feet away. The fielder throws at his maximum velocity, 110 feet per second. At what angle should the ball be thrown to hit home plate?

22. A runner tags third base and heads for home at 25 feet per second at precisely the time the fielder in Exercise 21 releases the ball. Is he sure to beat the throw to the plate (90 feet away)?

23. A fifty pound block of ice is placed on a slide 30 feet long that makes an angle of 15° with the horizontal. The end of the slide is 4 feet above the ground and the speed of the block at the end is 20 feet/second. Where will the block land?

24. Find the total horizontal distance a golf ball travels in the air if it leaves a tee at a velocity of 140 ft/sec and an angle of $\pi/6$. Is it possible for the golf ball to hit the branch of a tree that is 90 feet above the fairway?

25. Find the maximum horizontal distance that an object can travel if it is propelled from the ground with initial speed v_0. Show that if an object is propelled from the ground at an angle of θ from the horizontal, then it hits the ground at this same angle.

26. A slob throws an empty can from the window of a truck while the truck is stopped at a traffic light. The can is thrown perpendicular to the side of the truck, the window is 5 feet above the ground and the can lands 25 feet from the truck. Suppose the next empty can is thrown in the same manner, but the truck is traveling at 35 miles per hour. How far from the truck will this can land if the effect of the wind resistance is neglected?

27. Show that the acceleration of an object moving along the helix described by $\mathbf{r}(t) = \cos t\mathbf{i} + \sin t\mathbf{j} + t\mathbf{k}$ is totally centripetal.

28. Suppose an object rotates about the z-axis on a circle of radius a lying in a plane parallel to and b units from the xy-plane. Suppose the angular speed $d\theta/dt$ is a constant ω. The vector $\boldsymbol{\omega} = \omega\mathbf{k}$ is called the angular velocity of the object. Show that the velocity vector $\mathbf{v}(t) = \mathbf{r}'(t)$ is the cross-product of the angular velocity and the position vector $\mathbf{r}(t)$.

29. Show that the speed of an object is constant if and only if the tangential component of acceleration is zero.

30. Show that the speed of an object is constant if and only if the velocity and acceleration vectors are orthogonal.

13.5 CURVATURE

One distinguishing feature of a curve in space is its length, another is the way it changes direction. For example, the direction of a straight line is constant. (See Figure 13.20(a).) For the curve in Figure 13.20(b), the change in direction is very gradual while the change is more rapid for the curve in Figure 13.20(c). The notion of curvature provides a measure of how rapidly a curve is changing direction.

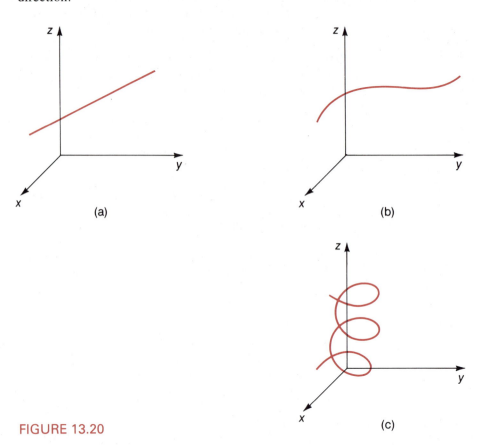

FIGURE 13.20

The unit tangent vector of a curve gives its direction at each point. The rate of change of this vector with respect to arc length, $D_s\,\mathbf{T}(s)$, provides a measure for the way a curve changes direction at the point.

(13.20)
DEFINITION

If C is a curve in space, then the **curvature** of C is defined by
$$K(s) = \|D_s\,\mathbf{T}(s)\|,$$
where the function $\mathbf{T}$ describes the principal unit tangent vector at each point on C.

The vector $D_s\mathbf{T}(s)$ is known as the **curvature vector** of the curve C and denoted $\mathbf{K}(s)$.

EXAMPLE 1   Show that the curvature of the line described by

$$\mathbf{r}(s) = \left(\frac{3\sqrt{59}}{59} s + 1\right)\mathbf{i} + \frac{5\sqrt{59}}{59} s \,\mathbf{j} + \left(\frac{5\sqrt{59}}{59} s + 2\right)\mathbf{k}$$

is zero.

SOLUTION

Since

$$\mathbf{r}'(s) = \frac{3\sqrt{59}}{59}\mathbf{i} + \frac{5\sqrt{59}}{59}\mathbf{j} + \frac{5\sqrt{59}}{59}\mathbf{k},$$

a constant; $\mathbf{T}(s) = \dfrac{\mathbf{r}'(s)}{\|\mathbf{r}'(s)\|}$ is also constant, and $K(s) = \|D_s\, \mathbf{T}(s)\| = 0$.   □

EXAMPLE 2   Show that the curvature of a circle of radius R at any point on the circle is $1/R$.

SOLUTION

A vector-valued function that describes a circle of radius R is

$$\mathbf{r}(t) = R[\cos t\,\mathbf{i} + \sin t\,\mathbf{j}], \quad 0 \le t \le 2\pi.$$

So

$$\mathbf{T}(t) = \frac{\mathbf{r}'(t)}{\|\mathbf{r}'(t)\|} = \frac{R[-\sin t\,\mathbf{i} + \cos t\,\mathbf{j}]}{R} = -\sin t\,\mathbf{i} + \cos t\,\mathbf{j}.$$

Since the arc length s along a circle of radius R from $\mathbf{r}(0)$ to $\mathbf{r}(t)$ is $s = Rt$, the circle has arc length representation

$$\mathbf{r}(s) = R\left[\cos\left(\frac{s}{R}\right)\mathbf{i} + \sin\left(\frac{s}{R}\right)\mathbf{j}\right].$$

The principal unit tangent vector is

$$\mathbf{T}(s) = -\sin\left(\frac{s}{R}\right)\mathbf{i} + \cos\left(\frac{s}{R}\right)\mathbf{j}$$

and $K(s) = \|D_s\, \mathbf{T}(s)\| = \left\|\dfrac{1}{R}\left(-\cos\left(\dfrac{s}{R}\right)\mathbf{i} - \sin\left(\dfrac{s}{R}\right)\mathbf{j}\right)\right\| = \dfrac{1}{R}$.

The curvature of this curve is constant, since the curve changes direction in a uniform manner.   □

The striking aspects of the curvature and curvature vector for a circle of radius R are that the radius of the circle is the reciprocal of the curvature:

$$R = \frac{1}{K(s)}$$

and the curvature vector

$$\mathbf{K}(s) = D_s\,\mathbf{T}(s) = \frac{1}{R}\left(-\cos\left(\frac{s}{R}\right)\mathbf{i} - \sin\left(\frac{s}{R}\right)\mathbf{j}\right) = -\frac{1}{R^2}\,\mathbf{r}(s)$$

indicates the direction from the point on the curve to the center of the circle. This is shown in Figure 13.21.

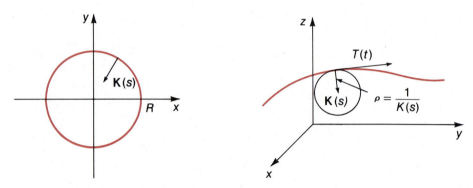

FIGURE 13.21   FIGURE 13.22

To consider the impact of these results on the analysis of the direction of a general curve in space, suppose we fix a point P on a curve C and try to draw a circle through P that "best" approximates C. The "best" circle to approximate the curve C at P will have (see Figure 13.22):

(i) the same unit tangent vector as the curve C at P.
(ii) The same curvature as the curve C at P.

Since the curvature of a circle is the reciprocal of its radius, the radius of the circle that best approximates C at P is the reciprocal of the curvature $K(s)$ at P. We call this value

(13.21) $$\rho = \frac{1}{K(s)},$$

the **radius of curvature** of the curve at P.

Since the arc length of a curve is often very difficult or impossible to determine, the usefulness of curvature would be limited if its calculation required an arc length parametrization of the curve. Fortunately, it does not, as we now illustrate.

Suppose a curve C is described by

$$\mathbf{r}(t) = x(t)\mathbf{i} + y(t)\mathbf{j} + z(t)\mathbf{k}.$$

The curvature vector represented in terms of the parameter t is

$$\mathbf{K}(t) = \frac{d\mathbf{T}(t)}{ds} = \frac{d\mathbf{T}(t)}{dt} \cdot \frac{dt}{ds}.$$

However, if $\mathbf{r}'(t) \neq 0$,

$$\frac{dt}{ds} = \frac{1}{\dfrac{ds}{dt}} = \frac{1}{\|\mathbf{r}'(t)\|} = \frac{1}{v(t)},$$

so the curvature vector is

(13.22) $$\mathbf{K}(t) = \frac{\mathbf{T}'(t)}{\|\mathbf{r}'(t)\|} = \frac{\mathbf{T}'(t)}{v(t)}$$

and the curvature is

(13.23) $$K(t) = \frac{\|\mathbf{T}'(t)\|}{\|\mathbf{r}'(t)\|} = \frac{\|\mathbf{T}'(t)\|}{v(t)}.$$

EXAMPLE 3   Find the radius of curvature of the circular helix described by

$$\mathbf{r}(t) = \cos t\mathbf{i} + \sin t\mathbf{j} + t\mathbf{k}$$

at the point $P(0, 1, \pi/2)$.

SOLUTION
First note that $(0, 1, \pi/2)$ corresponds to $\mathbf{r}(\pi/2)$. Since

$$\mathbf{r}'(t) = -\sin t\mathbf{i} + \cos t\mathbf{j} + \mathbf{k}$$

and

$$\|\mathbf{r}'(t)\| = \sqrt{(-\sin t)^2 + (\cos t)^2 + 1} = \sqrt{2},$$

the unit tangent vector is given by

$$\mathbf{T}(t) = \frac{\mathbf{r}'(t)}{\|\mathbf{r}'(t)\|} = \frac{1}{\sqrt{2}}(-\sin t\mathbf{i} + \cos t\mathbf{j} + \mathbf{k})$$

$$= \frac{\sqrt{2}}{2}(-\sin t\mathbf{i} + \cos t\mathbf{j} + \mathbf{k})$$

and

$$\mathbf{T}'(t) = \frac{\sqrt{2}}{2}(-\cos t\mathbf{i} - \sin t\mathbf{j}).$$

The curvature is

$$K(t) = \frac{\|\mathbf{T}'(t)\|}{\|\mathbf{r}'(t)\|} = \frac{1}{\sqrt{2}}\frac{\sqrt{2}}{2}\sqrt{(-\cos t)^2 + (-\sin t)^2}$$

$$= \frac{1}{2}$$

and the radius of curvature of the helix at $(0, 1, \pi/2)$ is

$$\rho\left(\frac{\pi}{2}\right) = \frac{1}{K\left(\dfrac{\pi}{2}\right)} = \frac{1}{\dfrac{1}{2}} = 2. \qquad \square$$

Equation (13.23) also shows the relationship between the curvature and the normal component of acceleration

(13.24) $$a_N(t) = v(t)\|\mathbf{T}'(t)\| = [v(t)]^2 K(t).$$

Computing $\|\mathbf{T}'(t)\|$ can be quite tedious. However, with a little manipulation, we can deduce a more easily evaluated representation for $K(t)$. Recall from Section 13.4 that the velocity and acceleration can be written in terms of the principal unit tangent and unit normal vectors:

$$\mathbf{v}(t) = \mathbf{r}'(t) = \|\mathbf{r}'(t)\|\mathbf{T}(t) = v(t)\mathbf{T}(t)$$

and

$$\mathbf{a}(t) = v'(t)\mathbf{T}(t) + v(t)\|\mathbf{T}'(t)\|\mathbf{N}(t).$$

Thus,

$$\begin{aligned}
\mathbf{v}(t) \times \mathbf{a}(t) &= [v(t)\mathbf{T}(t)] \times [v'(t)\mathbf{T}(t) + v(t)\|\mathbf{T}'(t)\|\mathbf{N}(t)] \\
&= [v(t)\mathbf{T}(t) \times v'(t)\mathbf{T}(t)] + [v(t)\mathbf{T}(t) \times v(t)\|\mathbf{T}'(t)\|\mathbf{N}(t)] \\
&= v(t)v'(t)[\mathbf{T}(t) \times \mathbf{T}(t)] + [v(t)]^2 \|\mathbf{T}'(t)\|[\mathbf{T}(t) \times \mathbf{N}(t)] \\
&= [v(t)]^2 \|\mathbf{T}'(t)\|[\mathbf{T}(t) \times \mathbf{N}(t)].
\end{aligned}$$

Since $\mathbf{T}(t)$ and $\mathbf{N}(t)$ are orthogonal unit vectors, $\|\mathbf{T}(t) \times \mathbf{N}(t)\| = 1$ and

$$\|\mathbf{v}(t) \times \mathbf{a}(t)\| = [v(t)]^2 \|\mathbf{T}'(t)\|.$$

Equation (13.23) can therefore be rewritten as

(13.25)
$$K(t) = \frac{\|\mathbf{T}'(t)\|}{v(t)} = \frac{\|\mathbf{v}(t) \times \mathbf{a}(t)\|}{[v(t)]^3}.$$

EXAMPLE 4   Find the curvature and components of acceleration, $a_\mathbf{T}$ and $a_\mathbf{N}$, of an object at the point $(1, 2, 1)$ if the position of the object is given by

$$\mathbf{r}(t) = t\mathbf{i} + 2t\mathbf{j} + t^3\mathbf{k}, \qquad 0 \le t \le 2.$$

SOLUTION

The velocity, acceleration, and speed of this object at any time t are given by:

$$\mathbf{v}(t) = \mathbf{i} + 2\mathbf{j} + 3t^2\mathbf{k}, \qquad \mathbf{a}(t) = 6t\mathbf{k}, \qquad \text{and} \cdot v(t) = \sqrt{5 + 9t^4}.$$

Since the curvature vector is not required in this example, it is easier to use formula (13.25) to determine the curvature.

$$\begin{aligned}
K(t) &= \frac{\|\mathbf{v}(t) \times \mathbf{a}(t)\|}{[v(t)]^3} = \frac{\|12t\mathbf{i} - 6t\mathbf{j}\|}{(5 + 9t^4)^{3/2}} \\
&= \frac{\sqrt{180}\, t}{(5 + 9t^4)^{3/2}} = \frac{6\sqrt{5}\, t}{(5 + 9t^4)^{3/2}}.
\end{aligned}$$

The point $(1, 2, 1)$ is determined by $t = 1$, so the curvature at this point is

$$K(1) = \frac{6\sqrt{5}}{(14)^{3/2}} = \frac{3\sqrt{70}}{98}.$$

Substituting into (13.24) gives

$$a_\mathbf{N}(1) = [v(1)]^2 K(1) = (\sqrt{14})^2 \frac{3\sqrt{70}}{98} = \frac{3}{7}\sqrt{70}.$$

To determine $a_T(1)$, we use Equation (13.16) on p. 739.

$$a_T(t) = v'(t) = \frac{18t^3}{\sqrt{5 + 9t^4}}$$

so
$$a_T(1) = \frac{18}{\sqrt{14}} = \frac{9\sqrt{14}}{7}. \qquad \square$$

EXERCISE SET 13.5

In Exercises 1 through 8, find the curvature of the curves at the given point.

1. $\mathbf{r}(t) = \sin t\mathbf{i} + \cos t\mathbf{j} + 2t\mathbf{k}, \quad (1, 0, \pi)$
2. $\mathbf{r}(t) = \sin t\mathbf{i} + \cos t\mathbf{j} + \mathbf{k}, \quad (0, 1, 1)$
3. $\mathbf{r}(t) = (2t - 1)\mathbf{i} + (t - 3)\mathbf{j} + (2 - 3t)\mathbf{k}, \quad (1, -2, -1)$
4. $\mathbf{r}(t) = t\mathbf{i} + t^2\mathbf{j} + 2t\mathbf{k}, \quad (2, 4, 4)$
5. $\mathbf{r}(t) = 2t\mathbf{i} + t^2\mathbf{j} + 5t^3\mathbf{k}, \quad (0, 0, 0)$
6. $\mathbf{r}(t) = t\mathbf{i} + \ln t\mathbf{j} + .5t\mathbf{k}, \quad (2, \ln 2, 1)$
7. $\mathbf{r}(t) = t\mathbf{j} + e^t\mathbf{k}, \quad (0, 0, 1)$
8. $\mathbf{r}(t) = t\mathbf{i} + \sqrt{9 - t^2}\ \mathbf{j}, \quad (3, 0, 0)$

For Exercises 9 through 18, find the curvature and radius of curvature at the indicated point and sketch the curve.

9. $\mathbf{r}(t) = 3\cos t\mathbf{i} + 2\sin t\mathbf{j}, \quad t = 0$

10. $\mathbf{r}(t) = 3\cos t\mathbf{i} + 2\sin t\mathbf{j}, \quad t = \dfrac{\pi}{2}$

11. $\mathbf{r}(t) = t\mathbf{i} + \dfrac{1}{2}t^2\ \mathbf{j}, \quad t = 1$

12. $\mathbf{r}(t) = t\mathbf{i} + \tan t\mathbf{j}, \quad t = \dfrac{\pi}{4}$

13. $\mathbf{r}(t) = 3\cos t\mathbf{i} + 2\sin t\mathbf{j} + t\mathbf{k}, \quad t = \pi/2$
14. $\mathbf{r}(t) = t\mathbf{i} + \sin t\mathbf{j} + t\mathbf{k}, \quad t = \pi/2$
15. $y = \ln x, x = 1$ [*Hint*: $\mathbf{r}(t) = t\mathbf{i} + \ln t\mathbf{j}$]
16. $y = \ln x, \quad x = 3$
17. $y = x^3 - 3x - 2, \quad x = 1$

18. $y = \dfrac{1}{x}, \quad x = 1$

Find any points on the plane curves described in Exercises 19 through 26 at which the curvature is zero.

19. $\mathbf{r}(t) = t\mathbf{i} + 2t^3\mathbf{j}$ 20. $\mathbf{r}(t) = 3t^2\mathbf{i} + 4t^2\mathbf{j}$
21. $\mathbf{r}(t) = 2\cos t\mathbf{i} + 3\sin t\mathbf{j}$ 22. $\mathbf{r}(t) = e^t\mathbf{i} + e^{-t}\mathbf{j}$
23. $y = x^3 - 3x^2 + 2x$ 24. $y = x^3 - 3x^2 + 3x - 1$
25. $y = \sin x + x$ 26. $y = \sin x + \cos x$

27. Show that if a curve lying in the xy-plane is described by $\mathbf{r}(t) = x(t)\mathbf{i} + y(t)\mathbf{j}$, then the curvature is given by

$$K = \frac{|x'(t)y''(t) - x''(t)y'(t)|}{[(x'(t))^2 + (y'(t))^2]^{3/2}}.$$

28. Show that if a curve lies in the xy-plane and is described by $y = f(x)$, then the curvature is given by

$$K = \frac{\left|\dfrac{d^2y}{dx^2}\right|}{\left[1 + \left(\dfrac{dy}{dx}\right)^2\right]^{3/2}}.$$

29–36. Use the results of Exercises 27 and 28 to solve Exercises 19 through 26.

37. Find the points on the ellipse described by $\mathbf{r}(t) = 3\cos t\,\mathbf{i} + 2\sin t\,\mathbf{j}$, $0 \le t \le 2\pi$, at which the curvature is a maximum and the points at which the curvature is a minimum.

38. Show that if the second derivative of a function f exists and is zero at a point, then the curvature of the graph of f at this point is zero.

39. Show that the curvature of any line in space is zero.

40. Show that the curvature of the parabola $y = x^2$ approaches 0 as $x \to \infty$.

13.6
NEWTON'S AND KEPLER'S LAWS OF MOTION

By 1665 the plague that had broken out in London in 1664 had spread to the college at Cambridge, where 23-year-old Isaac Newton (1642-1727) was a scholar at Trinity College. The plague forced the college to close temporarily and Newton returned to his family home at Woolsthorpe in Lincolnshire. During the next two years, Newton made and refined some of the most important mathematical and physical discoveries of his illustrious life.

At the insistence of his friend Edmund Halley (1656-1742) (who predicted the orbit of the comet that bears his name), Newton published in 1689 the 530-page treatise *Philosophiae Naturalis Principia Mathematica*, commonly called simply the *Principia*. At the beginning of this work, he states his three laws of motion in Latin, the scientific language of the day.

L E X I.

Corpus omne perseverare in statu suo quiescendi vel movendi uniformiter in directum, nisi quatenus illud a viribus impressis cogitur statum suum mutare.

L E X II.

Mutationem motus proportionalem esse vi motrici impressæ,
& fieri secundum lineam rectam qua vis illa imprimitur.

L E X III.

Actioni contrariam semper & æqualem esse reactionem : sive
corporum duorum actiones in se mutuo semper esse æquales
& in partes contrarias dirigi.

Translated into English, the laws read:
1. A body continues in its state of rest or uniform motion in a straight line unless an unbalanced force acts on it.
2. The acceleration of a body is directly proportional to the force exerted on the body, is inversely proportional to the mass of the body, and is in the same direction as the force. (This law is usually expressed in the equation form $\mathbf{F} = m\mathbf{a}$, where $\mathbf{F}$ represents the sum of all forces acting on a body, m the mass of the body, and $\mathbf{a}$ the resulting acceleration due to $\mathbf{F}$.)
3. Whenever one body exerts a force upon a second body, the second exerts an equal and opposite force upon the first.

Newton used the first 60 pages of the *Principia* to show that his laws of motion imply some previously assumed facts about the solar system. In particular, he derived Johann Kepler's three laws of planetary motion:
1. Law of Ellipses: The orbit of each planet is an ellipse with the sun at one focus.
2. Law of Areas: The position vector from the sun to a planet sweeps out equal areas in equal intervals of time.
3. Harmonic Law: The square of the period of a planet's orbit about the sun is proportional to the cube of the length of the major axis of the ellipse describing the orbit.

Kepler published the Law of Ellipses and the Law of Areas in 1609 and the Harmonic Law in 1619. He devised these laws after studying an extremely detailed collection of astronomical data about the planet Mars. This data was compiled primarily by the Danish astronomer, Tycho Brahe (1546–1607). Brahe was the premiere astronomer of his day and the last great astronomer to record observations without benefit of a telescope, a device credited to Galileo Galilei (1564–1642) in 1609. Incidentally, the first reflecting telescope, the type used by most modern observatories, was invented by Newton.

In addition to deriving Kepler's laws, Newton used the results in the *Principia* to deduce his law of gravitational attraction:

The force $\mathbf{F}$ between two bodies having masses m_1 and m_2 whose centers of mass are separated by a distance r is an attraction acting along the line joining the centers of mass and having magnitude $\|\mathbf{F}\| = Gm_1m_2/r^2$, where G is a constant called the *universal gravitational constant*.

The derivation of Kepler's laws from Newton's second law of motion and his law of gravitation is an interesting illustration of the power of vector function techniques when applied to motion problems. The remainder of this chapter is devoted to this endeavor.

KEPLER'S FIRST LAW

The orbit of a planet is an ellipse with the sun at a focal point.

We first need to show that the orbit of a planet is confined to a single plane. Let $\mathbf{r}(t)$ denote the vector directed from the sun to the planet at time t. The sun's gravitational force $\mathbf{F}(t)$ is so much larger than all other forces acting on a planet that these other forces produce an almost negligible effect on the planet and are neglected in this argument.

By Newton's law of gravitation, the force $\mathbf{F}(t)$ is directed toward the sun, in the direction of the unit vector $-\mathbf{r}(t)/r(t)$ and has magnitude $GMm/(r(t))^2$, where $r(t) = \|\mathbf{r}(t)\|$ and M and m denote respectively the mass of the sun and planet. So,

$$\mathbf{F}(t) = -\frac{GMm}{(r(t))^2}\frac{\mathbf{r}(t)}{r(t)} = -\frac{GMm}{(r(t))^3}\mathbf{r}(t),$$

and by Newton's second law of motion,

$$\mathbf{F}(t) = m\mathbf{a}(t) = m\frac{d^2\mathbf{r}(t)}{dt^2} = -\frac{GMm}{(r(t))^3}\mathbf{r}(t).$$

This implies that $\mathbf{a}$ is parallel to $\mathbf{r}$ so

$$\mathbf{a} \times \mathbf{r} = \mathbf{0}.$$

Consequently,

$$D_t\left(\mathbf{r}(t) \times \frac{d\mathbf{r}(t)}{dt}\right) = \frac{d\mathbf{r}(t)}{dt} \times \frac{d\mathbf{r}(t)}{dt} + \mathbf{r}(t) \times \frac{d^2\mathbf{r}(t)}{dt} = \mathbf{0} + \mathbf{0} = \mathbf{0}$$

and $\mathbf{r}(t) \times d\mathbf{r}(t)/dt = \mathbf{c}$, where $\mathbf{c}$ is a constant vector. Thus,

$$\mathbf{0} = \mathbf{r}(t) \times \left(\mathbf{r}(t) \times \frac{d\mathbf{r}(t)}{dt}\right) = \mathbf{r}(t) \times \mathbf{c}.$$

Since $\mathbf{r}(t)$ is orthogonal to a constant vector $\mathbf{c}$ for every value of t, $\mathbf{r}(t)$ must lie in a plane that is independent of t.

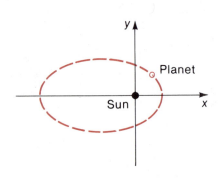

FIGURE 13.23

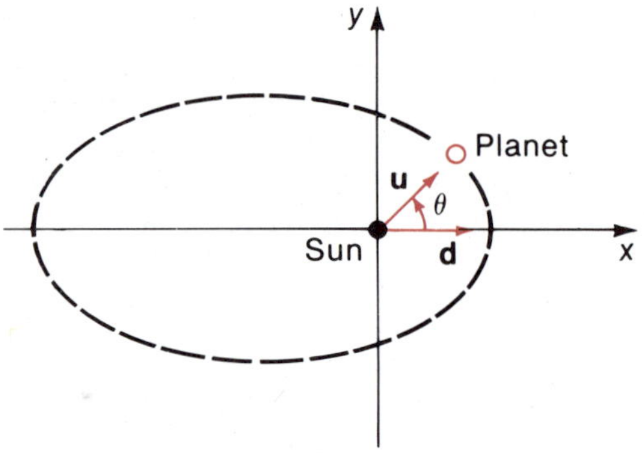

FIGURE 13.24

We will assume that this plane is the xy-plane with the sun at the origin and that $\mathbf{c} = c\mathbf{k}$ for some scalar constant c. (See Figure 13.23.) The position of the axes in the xy-plane will be chosen later.

To derive the equation of the orbit in the xy-plane, let $\mathbf{u}$ denote the unit vector in the direction of $\mathbf{r}$. Then

$$c\mathbf{k} = \mathbf{r} \times \frac{d\mathbf{r}}{dt} = r\mathbf{u} \times \frac{d(r\mathbf{u})}{dt} = r\mathbf{u} \times \left(\frac{dr}{dt}\mathbf{u} + r\frac{d\mathbf{u}}{dt} \right),$$

so

(13.26) $$c\mathbf{k} = r\frac{dr}{dt}(\mathbf{u} \times \mathbf{u}) + r^2\left(\mathbf{u} \times \frac{d\mathbf{u}}{dt} \right) = r^2\left(\mathbf{u} \times \frac{d\mathbf{u}}{dt} \right).$$

From Newton's second law and the law of gravitation,

$$\mathbf{a} = -\frac{GM}{r^3}\mathbf{r} = -\frac{GM}{r^2}\mathbf{u},$$

so by (13.26) $$\mathbf{a} \times c\mathbf{k} = -\frac{GM}{r^2}\mathbf{u} \times \left(r^2\left(\mathbf{u} \times \frac{d\mathbf{u}}{dt} \right) \right)$$

$$= -GM\left(\mathbf{u} \times \left(\mathbf{u} \times \frac{d\mathbf{u}}{dt} \right) \right).$$

By part (vi) of Theorem 12.22, this triple cross-product can be rewritten using dot products and

$$\mathbf{a} \times c\mathbf{k} = -GM\left[\left(\mathbf{u} \cdot \frac{d\mathbf{u}}{dt} \right)\mathbf{u} - (\mathbf{u} \cdot \mathbf{u})\frac{d\mathbf{u}}{dt} \right].$$

Since $\mathbf{u}$ is a unit vector, $\mathbf{u} \cdot \mathbf{u} = 1$. By Theorem 13.12, $\mathbf{u} \cdot \dfrac{d\mathbf{u}}{dt} = 0$, so

$$\mathbf{a} \times c\mathbf{k} = GM\frac{d\mathbf{u}}{dt}$$

and $$D_t\,(\mathbf{v} \times c\mathbf{k}) = \mathbf{a} \times c\mathbf{k} = GM\frac{d\mathbf{u}}{dt}.$$

Thus

$$\mathbf{v} \times c\mathbf{k} = GM\mathbf{u} + \mathbf{d},$$

for some constant vector $\mathbf{d}$. Since $\mathbf{v} \times c\mathbf{k}$ is perpendicular to $\mathbf{k}$ and $\mathbf{u}$ is perpendicular to $\mathbf{k}$, $\mathbf{d}$ must also be perpendicular to $\mathbf{k}$, so $\mathbf{d}$ lies in the xy-plane. We now fix the axes in the xy-plane so that the positive x-axis is the direction of $\mathbf{d}$; that is, $\mathbf{d} = d\mathbf{i}$ for some scalar constant d.

Let θ denote the angle between $\mathbf{u}$ and $\mathbf{d} = d\mathbf{i}$ at any time t. See Figure 13.24. Then the polar coordinates of the planet's position at time t are (r, θ) and

$$\mathbf{r} \cdot \mathbf{d} = \|\mathbf{r}\|\,\|\mathbf{d}\|\cos\theta = rd\cos\theta.$$

So

$$c^2 = c\mathbf{k} \cdot c\mathbf{k} = \left(\mathbf{r} \times \frac{d\mathbf{r}}{dt} \right) \cdot c\mathbf{k} = \mathbf{r} \cdot (\mathbf{v} \times c\mathbf{k})$$

$$= \mathbf{r} \cdot (GM\mathbf{u} + d\mathbf{i}) = GMr + dr\cos\theta.$$

Solving for r implies that

$$r = \frac{c^2}{GM + d\cos\theta} = \frac{c^2/GM}{1 + (d/GM)\cos\theta},$$

which is the equation of a conic section with eccentricity $e = d/GM$ (see Section 11.5). Since the orbit of a planet is a closed curve, the orbit is an ellipse and $0 < d/GM < 1$.

KEPLER'S SECOND LAW

The position vector from the sun to a planet sweeps out area at a constant rate.

The area swept out by the position vectors from

$$\mathbf{r}(t_0) = r(t_0)\mathbf{u}(t_0) = r(t_0)[\cos\theta(t_0)\mathbf{i} + \sin\theta(t_0)\mathbf{j}]$$

to

$$\mathbf{r}(t) = r(t)\mathbf{u}(t) = r(t)[\cos\theta(t)\mathbf{i} + \sin\theta(t)\mathbf{j}]$$

is

$$A = \frac{1}{2}\int_{\theta(t_0)}^{\theta(t)} [r(t)]^2\, d\theta = \frac{1}{2}\int_{t_0}^{t} [r(t)]^2 \frac{d\theta}{dt}\, dt.$$

See Figure 13.25.

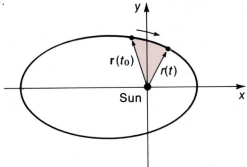

FIGURE 13.25

The rate at which the position vectors sweep out area is consequently

$$\frac{dA}{dt} = \frac{dA}{d\theta}\frac{d\theta}{dt} = \frac{1}{2}[r(t)]^2 \frac{d\theta}{dt}.$$

However $\mathbf{u} = \mathbf{i}\cos\theta + \mathbf{j}\sin\theta$, so

$$\mathbf{u} \times \frac{d\mathbf{u}}{dt} = \mathbf{u} \times \frac{d\mathbf{u}}{d\theta}\frac{d\theta}{dt}$$

$$= \frac{d\theta}{dt}[(\cos\theta\,\mathbf{i} + \sin\theta\,\mathbf{j}) \times (-\sin\theta\,\mathbf{i} + \cos\theta\,\mathbf{j})]$$

$$= \frac{d\theta}{dt}[(\cos\theta)^2 + (\sin\theta)^2](\mathbf{i} \times \mathbf{j}) = \frac{d\theta}{dt}\mathbf{k}.$$

But equation (13.26) implies that

$$c\mathbf{k} = r^2\,\mathbf{u} \times \frac{d\mathbf{u}}{dt},$$

so

$$c\mathbf{k} = r^2 \frac{d\theta}{dt}\mathbf{k}$$

and
$$\frac{dA}{dt} = \frac{1}{2} r^2 \frac{d\theta}{dt} = \frac{c}{2},$$

a constant.

KEPLER'S THIRD LAW

The square of the period of a planet's orbit about the sun is proportional to the cube of the length of the major axis of the ellipse describing the orbit.

In Newton's original language this law is: "the periodic times in ellipses are as the 3/2th power of their greater axes."

Let T denote the period of an elliptical orbit. By Kepler's second law $dA/dt = c/2$, so the area enclosed by an elliptical orbit is $A = cT/2$.

The area of an ellipse is $A = \pi ab$, where $2a$ and $2b$ are respectively the lengths of the major and minor axes. (This is Exercise 22 in Section 11.2.) Since $b = a\sqrt{1 - e^2}$, this implies that

$$T = \frac{2}{c} A = \frac{2}{c} \pi a^2 \sqrt{1 - e^2}.$$

The eccentricity of the ellipse is $e = d/GM$, so the square of the period of orbital rotation is

(13.27) $$T^2 = \frac{4\pi^2}{c^2} a^4 \left(1 - \left(\frac{d}{GM}\right)^2\right) = \frac{4\pi^2 a^4 ((GM)^2 - d^2)}{c^2 (GM)^2}.$$

However, the length of the major axis of the ellipse is the sum of the distances $r(0)$ and $r(\pi)$, so

$$a = \frac{1}{2}[r(0) + r(\pi)] = \frac{1}{2}\left[\frac{c^2}{GM + d} + \frac{c^2}{GM - d}\right] = \frac{c^2 GM}{(GM)^2 - d^2}.$$

Consequently, equation (13.27) can be rewritten

$$T^2 = \frac{4\pi^2 a^4}{GM} \frac{1}{a} = \frac{4\pi^2}{GM} a^3.$$

REVIEW EXERCISES

For the curves described in Exercises 1 through 6, (a) sketch the graph and indicate the direction of increasing t; (b) find and sketch $D_t \mathbf{F}(a)$; (c) find the principal unit tangent and unit normal vectors at a and sketch these vectors; (d) find the curvature at a.

1. $\mathbf{F}(t) = (1 + t)\mathbf{i} + (2 + t)\mathbf{j} + 4\mathbf{k}, a = 0$

2. $\mathbf{F}(t) = (t + 3)\mathbf{i} + (2 - t)\mathbf{j} + 3t\mathbf{k}, a = 1$

3. $\mathbf{F}(t) = \cos t\mathbf{i} + \sin t\mathbf{j} + 3\mathbf{k}, a = \dfrac{\pi}{3}$

4. $\mathbf{F}(t) = \cos t\mathbf{i} + \sin t\mathbf{j} + e^t\mathbf{k}, a = \pi$

5. $\mathbf{F}(t) = t\mathbf{j} + \cos t\mathbf{k}, a = \dfrac{3\pi}{2}$

6. $\mathbf{F}(t) = 2 \sin t\mathbf{i} + 3 \cos t\mathbf{j}, a = \dfrac{\pi}{2}$

Determine the values of t for which each of the functions described in Exercises 7 through 10 is continuous.

7. $\mathbf{F}(t) = \sqrt{1 - t^2}\,\mathbf{i} + t\mathbf{j}$

8. $\mathbf{F}(t) = t\mathbf{j} + \cos t\mathbf{k}$

9. $\mathbf{F}(t) = t\mathbf{i} + \dfrac{1}{t - 1}\mathbf{j} + \mathbf{k}$

10. $\mathbf{F}(t) = e^t\mathbf{i} + t\mathbf{j}$

Evaluate the integral in Exercises 11 through 14.

11. $\displaystyle\int_0^\pi (\cos t\mathbf{i} + \sin t\mathbf{j} + e^t\mathbf{k})\,dt$

12. $\displaystyle\int (t\cos t\mathbf{i} + t\sin t\mathbf{j} + t\mathbf{k})\,dt$

13. $\displaystyle\int_0^1 (te^t\mathbf{i} + \ln(t + 1)\mathbf{k})\,dt$

14. $\displaystyle\int_0^{2\pi} \|3\cos t\mathbf{i} + 3\sin t\mathbf{j}\|\,dt$

15. If $\mathbf{F}(t) = t\mathbf{i} + \ln t\mathbf{j} + t\mathbf{k}$ and $f(t) = t^2$, find

(a) $D_t\mathbf{F}(t)$ (b) $D_t(\mathbf{F}\circ f)(t)$ (c) $D_t(f\mathbf{F})(t)$

16. Why is the curve described by $\mathbf{F}(t) = t\mathbf{i} + |t|\mathbf{j} + \mathbf{k}$ not a smooth curve?

17. Find a vector-valued function with arc length as parameter that describes the straight line from $(3, 0, 0)$ to $(2, 7, 5)$.

18. Sketch the graph of the vector-valued function described by $\mathbf{F}(t) = 2\sin t\mathbf{i} + 5\cos t\mathbf{j}$, $0 \le t \le 2\pi$ and find the radius of curvature at $t = 0$.

19. Find the curvature at $x = 2$ of the curve in the xy-plane described by $y = x^3 - 3x^2 - 2$.

20. Find the point (x, y) on the parabola $y = x^2$ at which the curvature is a maximum. What is the curvature at this point?

21. Find the radius of curvature of the parabola $y = x^2$ at $(0, 0)$.

22. The motion of an object is described by

$$\mathbf{r}(t) = \cos t^2\,\mathbf{i} + \sin t^2\,\mathbf{j}, \quad t \ge 0.$$

Find the velocity, speed, acceleration, and tangential and normal components of acceleration. Sketch the curve representing the path of the object.

23. The motion of an object is described by $\mathbf{r}(t) = t\mathbf{j} + 2|\sin t|\mathbf{k}$.

(a) Sketch the curve that represents the path of the object.

(b) Find the velocity, acceleration, and speed of the object at any time t.

(c) Find the maximum height of the object.

24. Determine the tangential and normal components of acceleration if the motion of an object is described by $\mathbf{r}(t) = \cos t\mathbf{i} + \sin t\mathbf{j} + e^t\mathbf{k}$.

25. A projectile is fired from ground level with an initial speed of 500 ft/sec and an angle of elevation of $\pi/3$ radians. Describe the subsequent motion of the projectile and find (a) the time in the air, (b) the maximum height, (c) the horizontal range, and (d) the speed at the instant of impact.

26. The only significant force acting on an unpowered satellite in a circular orbit about the earth is the gravitational force of the earth. Use the result in Example 4 of Section 13.4 to show that a satellite in a circular orbit x miles above the surface of the earth has a speed of approximately $v = 32\sqrt{165(4000 + x)}$ ft/sec. Find the length of time it takes the satellite to orbit the earth.

14

MULTIVARIATE FUNCTIONS

The functions we studied in Chapter 13 mapped the real numbers into vectors. The other functions studied in the first 13 chapters had domains and ranges that were subsets of $\mathbb{R}$. Many common problems, however, require functions with more than one dimension for their domain. In this chapter we consider the properties of functions whose domain is a subset of either the plane, denoted $\mathbb{R}^2$, or space, denoted $\mathbb{R}^3$, and whose range is a subset of $\mathbb{R}$. In subsequent chapters we will see how the results of calculus can be extended to this type of function.

14.1
FUNCTIONS OF SEVERAL VARIABLES

Suppose n is a positive integer greater than one. By a **function of n variables** we mean a function whose domain is a subset of $\mathbb{R}^n$, the set of all ordered n-tuples of real numbers, and whose range is a subset of the real numbers $\mathbb{R}$. In general, we consider only the case when n is either two or three, but applications frequently occur requiring functions of more than three variables.

Before we begin the study of specific properties of functions of several variables, let us consider some applications of single variable functions and analogous applications of several variable functions.

Functions from $\mathbb{R}$ into $\mathbb{R}$ have been used to describe:

1. The distance from points on a line to a fixed origin on the line.
2. The length of objects.
3. The temperature at each point of a rod.
4. The selling price of an article whose cost depends entirely on the cost of materials.

Analogous functions from $\mathbb{R}^2$ into $\mathbb{R}$ can be used to describe:
1. The distance from points in a plane to a fixed origin in the plane.
2. The area of regions.
3. The temperature at each point on a plate.
4. The selling price of an article whose cost depends on both the cost of materials and the cost of labor.

Similarly, functions from $\mathbb{R}^3$ into $\mathbb{R}$ can be used to describe:
1. The distance from points in space to a fixed origin in space.
2. The volume of a solid.
3. The temperature at each point in a solid, for instance, the temperature within a turkey that is roasting in an oven.
4. The selling price of an article whose cost depends on the cost of materials, the cost of labor, and the cost of overhead.

A function from $\mathbb{R}^n$ to $\mathbb{R}$ could be constructed for any integer n by considering the application listed under (4) and increasing to n the number of variables upon which the selling price depends. It is not uncommon for functions describing economic models to be functions of hundreds of variables. For our purposes, however, three variables are sufficient.

EXAMPLE 1   A flat circular plate is heated so that the temperature at each point on the plate is directly proportional to the distance of the point from the center. Find a function that describes the temperature of the plate.

SOLUTION

We introduce a two-dimensional coordinate system so that the center of the plate is at the origin, as in Figure 14.1.

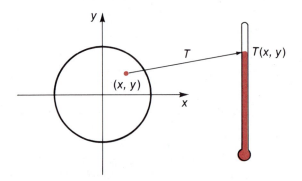

FIGURE 14.1

Since the distance from a point (x, y) to $(0, 0)$ is $\sqrt{x^2 + y^2}$, the temperature is described by

$$T(x, y) = k\sqrt{x^2 + y^2},$$

where k is the constant of proportionality.   □

EXAMPLE 2   For $f(x, y, z) = x^2 + y^2 + z^2$, find
(a) $f(1, 0, -1)$ and $f(2, 1, 3)$, and   (b) the domain and range of f.

SOLUTION

(a) $$f(1, 0, -1) = 1^2 + 0^2 + (-1)^2 = 2$$

and $$f(2, 1, 3) = 2^2 + 1^2 + 3^2 = 14.$$

(b) The domain of f is the set of all (x, y, z) for which f is defined. Consequently, the domain of f is $\mathbb{R}^3$.

Since $x^2 + y^2 + z^2 \geq 0$, the range of f is the set of nonnegative real numbers. $\square$

EXAMPLE 3   Find the domain and range of the function described by

$$f(x, y) = \sqrt{1 - x^2 - y^2}.$$

SOLUTION

The domain of f is the set of ordered pairs (x, y) for which f is defined; that is, when

$$1 - x^2 - y^2 \geq 0 \qquad \text{or} \qquad x^2 + y^2 \leq 1.$$

The range of f is $[0, 1]$. The correspondence can be seen in Figure 14.2. $\square$

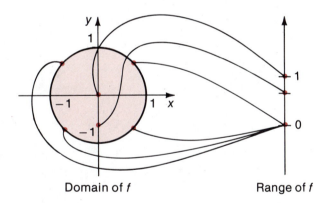

Domain of f Range of f

FIGURE 14.2

The arithmetic properties of functions of several variables are defined in the same manner as for functions of one variable. For example, if f and g are functions of two variables x and y, we define

(i) $(f + g)(x, y) = f(x, y) + g(x, y)$
(ii) $(f - g)(x, y) = f(x, y) - g(x, y)$

(iii) $(f \cdot g)(x, y) = f(x, y)g(x, y)$

and for any constant k

(iv) $(kf)(x, y) = kf(x, y).$

The domain of these functions is the intersection of the domain of f and the domain of g.

Similarly, we define

(v) $\left(\dfrac{f}{g}\right)(x, y) = \dfrac{f(x, y)}{g(x, y)}.$

The domain of f/g is the set of ordered pairs (x, y) that are in both the domain of f and the domain of g and for which $g(x, y) \neq 0$.

The composition of functions of several variables can also be defined; however, a bit of caution is needed because we are mapping from different sets. If g is a function of n variables, then $g: \mathbb{R}^n \rightarrow \mathbb{R}$, so composition is possible with a function $f: \mathbb{R} \rightarrow \mathbb{R}$, and $f \circ g$ is again a function from $\mathbb{R}^n$ into $\mathbb{R}$.

EXAMPLE 4   Suppose $g(x, y) = x + y$ describes a function from $\mathbb{R}^2$ into $\mathbb{R}$ and $f(x) = \sqrt{x}$ describes a function from $\mathbb{R}$ into $\mathbb{R}$. Find $g \cdot g$ and $f \circ g$.

SOLUTION

$$(g \cdot g)(x, y) = g(x, y) \cdot g(x, y) = (x + y)^2$$

and $(f \circ g)(x, y) = f(g(x, y)) = f(x + y) = \sqrt{x + y}.$ $\square$

EXERCISE SET 14.1

1. If $f(x, y) = x^2 + y$, find
   (a) $f(1, 0)$ (b) $f(-1, -1)$ (c) $f(\sqrt{3}, 2)$
2. If $f(x, y, z) = 2xy + 4xz$, find
   (a) $f(2, 1, 3)$ (b) $f(1, 3, 2)$ (c) $f(0, 1, 5)$

Find the domain of the functions described in Exercises 3 through 10.

3. $f(x, y) = \sqrt{1 - x - y}$ 4. $f(x, y) = \tan x + y$

5. $f(x, y) = \dfrac{x}{y}$ 6. $f(x, y) = \sqrt{x^2 + y^2 - 1}$

7. $f(x, y) = \ln\left(\dfrac{y - x}{x}\right)$ 8. $f(x, y) = \dfrac{1}{2 - x - y}$

9. $f(x, y, z) = \sqrt{4 - x^2 - y^2 - z^2}$ 10. $f(x, y, z) = \dfrac{1}{xy(z - 1)}$

For the functions described in Exercises 11 through 14, find

$$\dfrac{f(x + h, y) - f(x, y)}{h} \quad \text{and} \quad \dfrac{f(x, y + k) - f(x, y)}{k}.$$

11. $f(x, y) = 3 - x - y$ 12. $f(x, y) = x^2 + y^2$
13. $f(x, y) = 4 - x^2 - y^2$ 14. $f(x, y) = xy$
15. Find $f \circ g$ if $g(x, y, z) = 2xy + 4xz + 6yz$ and $f(x) = 2x$.
16. Find $f \circ g$ if $g(x, y) = \sqrt{x^2 + y^2}$ and $f(x) = x^2$.

17. Express the volume of a right circular cylinder as a function of its height and radius.

18. Express the height of a right circular cone as a function of its volume and the radius of its base.

19. If A dollars are deposited at a 5% continuous interest rate, then the amount in the account at the end of t years is given by $P(A, t) = Ae^{.05t}$. Find $P(1000, 20)$.

20. Physicians sometimes need to know the body surface area of a patient. For a person who weighs w kilograms and is h centimeters tall, an approximation to the body surface area is given in square meters by $A(w, h) = .00718w^{.425} h^{.725}$. Find $A(55, 170)$.

21. A box is to be built in the form of a rectangular parallelepiped. The bottom of the box can be made from scrap material costing \$1 per square foot but the material for the top and sides costs \$8 per square foot. Find a function that describes the cost of materials for the box.

22. Suppose the builder of the box in Exercise 21 is paid \$7 per hour. Find a function (of four variables) that describes the total cost of constructing the box.

23. Express the cost of painting a rectangular wall as a function of the dimensions if paint costing \$8 per gallon covers 200 square feet, and a painter who paints 150 square feet per hour is paid \$10 per hour.

14.2
FUNCTIONS OF TWO VARIABLES

To sketch the graph of a function of two variables, we introduce a third variable

$$z = f(x, y)$$

and sketch the collection of points (x, y, z) for (x, y) in the domain of f. The resulting figure is called a **surface.** (See Figure 14.3(a).)

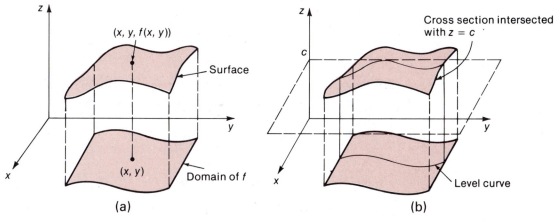

(a)   (b)

FIGURE 14.3

A cross section or **trace** of a surface in a plane is the intersection of the surface with that plane. The set of points (x, y) in the xy-plane that satisfy $c = f(x, y)$ for some constant c is called a **level curve** of the function f. Thus, a level curve is the projection onto the xy-plane of a trace in the plane $z = c$. (See Figure 14.3(b) on the previous page.)

EXAMPLE 1   Describe the level curves of

$$f(x, y) = x^2 + y^2$$

and sketch the graph of the function.

SOLUTION
Setting $z = f(x, y)$, we have the equation

$$z = x^2 + y^2.$$

The level curves of f are determined by setting the variable z equal to a constant. If $z = c > 0$, the equation is

$$c = x^2 + y^2,$$

and the level curve is a circle with center at $(0, 0, c)$ and radius $\sqrt{c}$.

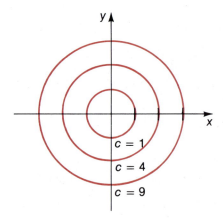

FIGURE 14.4

The level curves for various values of c are shown in the xy-plane in Figure 14.4. The portions of the graph of f corresponding to these level curves are shown in Figure 14.5.

In the xz-plane, $y = 0$, so the graph has equation

$$z = x^2,$$

while in the yz-plane, $x = 0$, and the equation is

$$z = y^2.$$

Consequently, the graph is as shown in Figure 14.6.

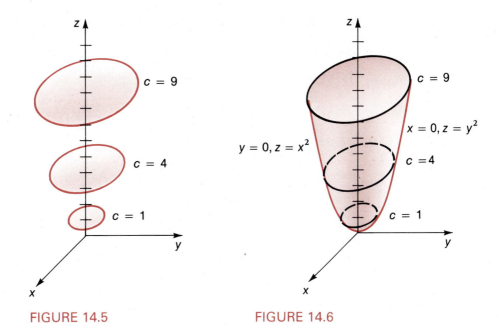

FIGURE 14.5   FIGURE 14.6

The traces parallel to the xy-plane are circular, while traces parallel to both the yz- and xz-planes are parabolas. This surface is called a **circular paraboloid.** Surfaces of this type are studied in more detail in Section 14.4.   □

EXAMPLE 2   Sketch the graph of f if $f(x, y) = 1 - x - y$.

SOLUTION

Setting $z = f(x, y)$, we have

$$z = 1 - x - y \qquad \text{or} \qquad x + y + z = 1.$$

This is an equation of the plane with normal vector $\langle 1, 1, 1 \rangle$ shown in Figure 14.7. The level curves are straight lines with equations $z = c$, $x + y = 1 - c$, for c a constant.   □

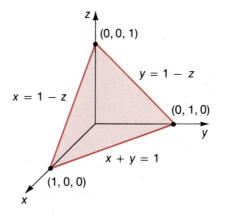

FIGURE 14.7

EXAMPLE 3   Sketch the level curves of the function described by $f(x, y) = xy$.

SOLUTION

The level curves for this function are given by

$$c = xy \qquad \text{or} \qquad y = \frac{c}{x}.$$

The graphs of these level curves for various values of c are shown in Figure 14.8.

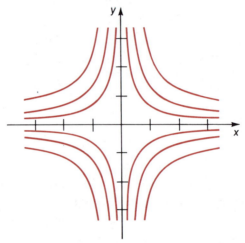

FIGURE 14.8

The surface in the first octant appears as shown in Figure 14.9.   □

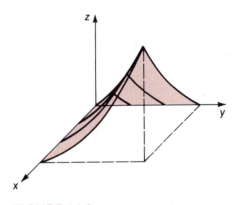

FIGURE 14.9

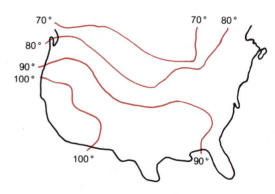

FIGURE 14.10

In Example 3, the level curves in the xy-plane give a better graphical representation for f than does the three-dimensional representation shown in Figure 14.9. Using level curves to present three-dimensional information in a two-dimensional setting is common in many areas. For example, topographical maps use this technique to indicate curves of constant elevation. Weather forecast maps use this same type of representation to show level curves of temperature as well as high and low pressure zones throughout a large area. Figure 14.10 shows a weather map with levels of constant temperature for the United States. Such level curves are called *isothermals*.

One advantage to a level-curve graph is that the graph for one portion of the domain is not obscured by the graph of another portion. The two illustrations of Mt. Rainier shown in Figure 14.11 demonstrate this feature.

(a)

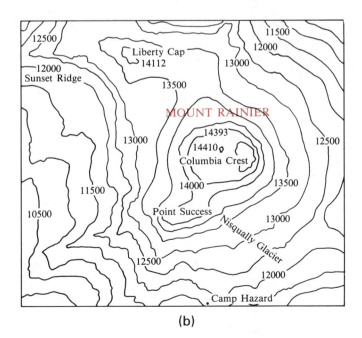

(b)

FIGURE 14.11

Since the photograph in Figure 14.11(a) is taken from northeast of the summit, the elevation at points on the southwest slope are obscured. However, the elevation of all points on the mountain can be quite easily found on the topographical map in Figure 14.11(b).

The disadvantage of the level-curve representation is the lack of depth perception. It is not immediately apparent, for example, that Figure 14.12 illustrates the plane $z = 1 - x - y$.

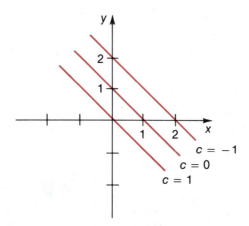

FIGURE 14.12

COMPUTER GRAPHICS

The increasing availability of relatively low-cost computer systems has brought with it an increase in the use of computer graphics systems. A computer graphics system consists of a computer combined with some form of graphic device. The graphic device can be a printer or plotter that produces a paper copy, but is more commonly a cathode ray tube (TV screen) on which a visual image is formed.

The applications of computer graphics range from entertainment, such as computer games and the generation of animated cartoons, to sophisticated simulation techniques that can be used for everything from automobile design to spacecraft navigation.

Most computer graphics systems have the programming capability to allow realistic two-dimensional representation of three-dimensional surfaces. The computer sketches surface representations in the plane just as we do, by drawing representative curves. The advantage that the computer has is the ability to analyze data and plot points very rapidly. We have the advantage of a built-in analyzer that enables us to distinguish the most important features of the surface and to recognize the portions of the surface that are hidden from view.

Even though most calculus students may not have access to computer graphics systems, it is interesting to examine some products of these systems.

Consider the surface shown in Figures 14.13 (a) and (b). This surface is called a monkey saddle and has equation $z = x^3 - 3xy^2$. The name is derived from the fact that if in the unlikely event a monkey were to sit on this surface at the origin with its head in the direction of the positive x-axis, it would have a place to rest its tail (along the negative x-axis) as well as both feet (along the lines $y = x$ and $y = -x$, when $x > 0$). Figure 14.13 (a) shows a computer-generated sketch of the monkey saddle using a grid of curves on the surface that are parallel to the coordinate planes. Figure 14.13 (b) shows computer-generated level curves on the surface.

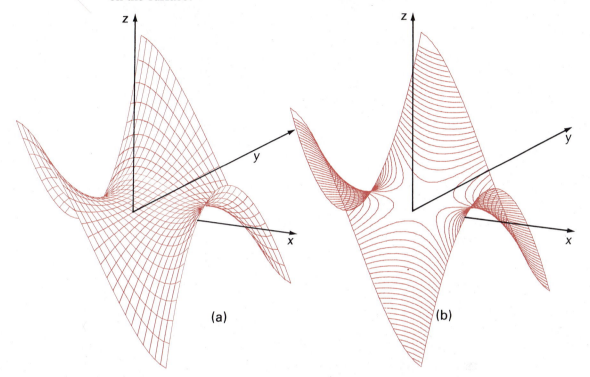

(a)   (b)

FIGURE 14.13

Notice that in each case the drawing consists of a collection of curves in planes parallel to the coordinate planes. The system has been programmed to recognize and delete portions of the curves that are naturally hidden behind other parts of the surface. These graphic representations do not, however, make use of the shading feature that helps identify portions of a surface in artist drawings of a textbook. Because of this, your own graphs will probably have more in common with the computer-generated graphs than those done by an artist.

Some computer graphics systems have the ability to rotate the surface to allow it to be viewed from different perspectives. Figure 14.14 on p. 768 illustrates the use of a rotation in examining the surface whose equation is $z = x^2 + \sin y$. This rotation property is particularly useful for architectural and engineering design applications.

We will present samples of computer drawn graphs as we proceed through this chapter. If you have access to a computer graphics system it would be worthwhile to observe it drawing a graph to see how the features of the graph are generated. Since the graphs are generally formed by drawing one curve at a time, you can see exactly how the graph is obtained.

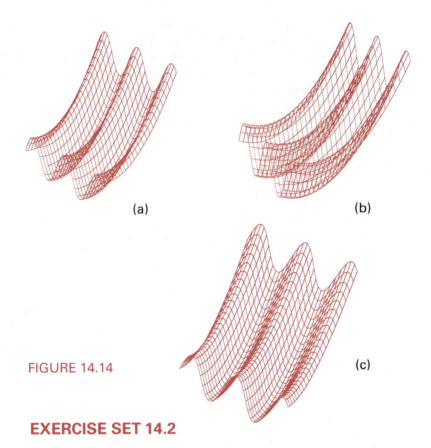

(a)

(b)

FIGURE 14.14

(c)

EXERCISE SET 14.2

In Exercises 1 through 10, sketch the level curves for the given values of c.

1. $f(x, y) = 1 + x - y,\ c = 1, 2, 3$

2. $f(x, y) = 2x - y + 4,\ c = 0, 1, 3, 5$

3. $f(x, y) = 2 - x^2 - y^2,\ c = 0, -2$

4. $f(x, y) = 4x^2 + y^2,\ c = 1, 2, 4$

5. $f(x, y) = x^2 - y,\ c = 0, -1, 1$

6. $f(x, y) = x^2 - y^2,\ c = 1, 4$

7. $f(x, y) = \ln(x + y),\ c = 0, 1$

8. $f(x, y) = \ln x + \ln y,\ c = 0, 1, 2$

9. $f(x, y) = \sin x + y,\ c = 0, 1$

10. $f(x, y) = \dfrac{1}{xy},\ c = -1, -\dfrac{1}{2}, \dfrac{1}{2}, 1$

Sketch the graphs of the functions described in Exercises 11 through 24.

11. $f(x, y) = x + y - 2$ **12.** $f(x, y) = 6 - 2x - 3y$

13. $f(x, y) = x^2 + y^2 + 2$ **14.** $f(x, y) = 4 - x^2 - y^2$

15. $f(x, y) = x^2$ **16.** $f(x, y) = \sin y$

17. $f(x, y) = 3 - x$ **18.** $f(x, y) = 3 - y$

19. $f(x, y) = 4x^2 + 9y^2$ **20.** $f(x, y) = \sqrt{x^2 + y^2}$

21. $f(x, y) = \begin{cases} x^2 + y^2, & \text{if } x^2 + y^2 \le 9 \\ 0 & , \text{if } x^2 + y^2 > 9 \end{cases}$

22. $f(x, y) = \begin{cases} \sqrt{x^2 + y^2}, & \text{if } x^2 + y^2 \le 4 \\ 4 & , \text{if } x^2 + y^2 > 4 \end{cases}$

23. $f(x, y) = \begin{cases} 6 - 2x - 3y, & \text{if } x \ge 0, y \ge 0, 6 - 2x - 3y \ge 0 \\ 0 & , \text{elsewhere} \end{cases}$

24. $f(x, y) = \begin{cases} 2 - x + y, & \text{if } x \ge 0, y \ge 0, 2 - x + y \ge 0 \\ 0 & , \text{elsewhere} \end{cases}$

25. The elevation of a mountain above a point (x, y) in the base plane is described by $E(x, y) = 200 - x^2 - 4y^2$. Sketch the curves of constant elevation for $E = 0, E = 100, E = 150, E = 200$.

26. The temperature at any point (x, y) on a circular plate is given by $T(x, y) = \sqrt{x^2 + y^2}$. Sketch the curves of constant temperature for $T = 0, 3, 5$.

27. The cost of building a box of height h inches with a square base of length l inches is given by $C(l, h) = l^2 + 6lh$ dollars. Sketch the curves of constant cost for $l \ge 0, h \ge 0$ and $C = 4, 9, 16$.

14.3
FUNCTIONS OF THREE VARIABLES: LEVEL SURFACES

In the preceding section we saw that the graph of a function of two variables involves three dimensions: two dimensions for the domain of the function and a third dimension to describe the range. Functions of three variables need three dimensions to describe their domain, so the graph would require four dimensions. For this reason it is impossible to sketch the graph of any function of three variables.

We can, however, represent a function of three variables using level surfaces analogous to the level curves used for functions of two variables. For a function f of three variables and a given constant c, the **level surface** associated with f and c is the surface described by those ordered triples (x, y, z) in the domain of f that satisfy

(14.1) $c = f(x, y, z).$

EXAMPLE 1   Sketch the level surfaces of

$$f(x, y, z) = x^2 + y^2 + z^2$$

corresponding to $c = 0, 1,$ and 3.

SOLUTION

For $c > 0$, the surface

$$c = x^2 + y^2 + z^2$$

is a sphere with center at the origin $(0, 0, 0)$ and radius $\sqrt{c}$. When $c = 0$ we have the equation

$$0 = x^2 + y^2 + z^2,$$

which implies that $x = 0$, $y = 0$, and $z = 0$.

When $c = 1$,

$$1 = x^2 + y^2 + z^2,$$

an equation of a sphere with center $(0, 0, 0)$ and radius 1.

When $c = 3$ the sphere has center $(0, 0, 0)$ and radius $\sqrt{3}$, since in this case

$$3 = x^2 + y^2 + z^2.$$

The portions of these level surfaces lying in the first octant are shown in Figure 14.15. □

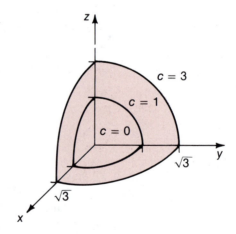

FIGURE 14.15

EXAMPLE 2 Describe the level surfaces of the function defined by

$$f(x, y, z) = x + y + z.$$

SOLUTION

For a constant c, the level surface associated with f and c is a plane with equation

$$c = x + y + z.$$

For various values of c, the level surfaces of f are parallel planes with normal vector $\langle 1, 1, 1 \rangle$. Sketches of the level surfaces for $c = 1$ and $c = 2$ are shown in Figure 14.16. □

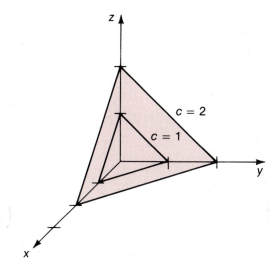

FIGURE 14.16

An important special case of level surface occurs when one of the three variables does *not* occur in the description of the function. For example, the equations of level surfaces of the function described by

$$f(x, y, z) = x^2 + y^2 - 1$$

do not depend on the variable z. For any constant c the level surface for f and c intersects each plane perpendicular to the z-axis in the same curve, the curve with equation

$$x^2 + y^2 = c + 1.$$

For $c = 0$ and $c = 3$, the level curves are shown in Figure 14.17. A surface of this type is called a **cylinder.**

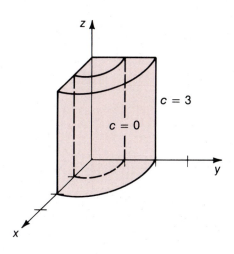

FIGURE 14.17

(14.2)
DEFINITION

Suppose C is a curve lying in a plane and l is a line that does not lie in this plane. The **cylinder** generated by C and l consists of all lines parallel to l that intersect C.

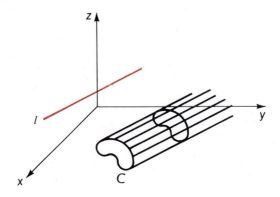

FIGURE 14.18

Figure 14.18 illustrates Definition 14.2. The curve C is called the **directrix** of the cylinder and l is called a **generator.** When l is perpendicular to the plane containing C the surface is called a **right cylinder.** The most common cylinder occurs when the curve C is a circle and the cylinder is a right cylinder, and is called a **right circular cylinder.** (The cylinders shown in Figure 14.17 are right circular cylinders.)

EXAMPLE 3

Sketch the graph of the right cylinder that is perpendicular to the xy-plane and whose equation in that plane is $y = x^2$.

SOLUTION

The graph of the equation in the xy-plane is shown in Figure 14.19. The cylinder is shown in Figure 14.20.   □

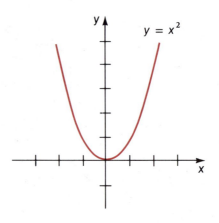

FIGURE 14.19

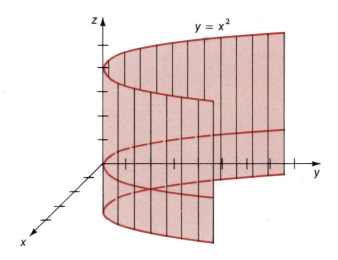

$y = x^2$

FIGURE 14.20

EXAMPLE 4   Sketch the graph of the right cylinder that is perpendicular to the xz-plane and whose equation in that plane is $z = \cos x$, $0 \le x \le 2\pi$.

SOLUTION

This cylinder is shown in Figure 14.21.

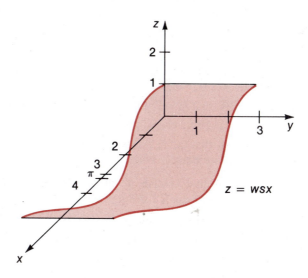

$z = \cos x$

FIGURE 14.21

A primary reason for considering functions of three variables is their relation to surfaces in space. For example, the level surfaces in Examples 1 and 2 are well-known surfaces in space. In succeeding chapters, we will study such topics as tangent planes and normals to surfaces, surface area, and volumes bounded by surfaces. Functions of three variables will be used to handle these topics with precision.

EXERCISE SET 14.3

Describe the level surfaces of the functions given in Exercises 1 through 12.

1. $f(x, y, z) = 2x + 3y + z$ **2.** $f(x, y, z) = x - y - 2z$

3. $f(x, y, z) = z - x - y$ **4.** $f(x, y, z) = x - z$

5. $f(x, y, z) = x^2 + y^2$ **6.** $f(x, y, z) = x^2 - y$

7. $f(x, y, z) = z^2 + y^2$ **8.** $f(x, y, z) = z^2 - y^2$

9. $f(x, y, z) = x^2 + 4y^2 + 9z^2$ **10.** $f(x, y, z) = 4x^2 + 9y^2 + 4z^2$

11. $f(x, y, z) = y - x^2 - z^2 - 1$ **12.** $f(x, y, z) = z - x^2 - y^2$

13. Sketch the graph of the right cylinder perpendicular to the xy-plane whose equation in that plane is $4x^2 + 9y^2 = 36$.

14. Sketch the graph of the right cylinder perpendicular to the yz-plane whose equation in that plane is $z = \sin y$, $0 \le y \le 2\pi$.

15. Sketch the graph of the right cylinder perpendicular to the xz-plane whose equation in that plane is $|z| = |x|$.

16. Sketch the graph of the right cylinder perpendicular to the xy-plane whose equation in that plane is $y = e^x$.

17. Suppose that C is a curve in the xy-plane described by $y = f(x)$, where $x \ge 0$, $y \ge 0$.

    (a) Show that (x, y, z) lies on the surface generated by rotating C about the y-axis precisely when $y = f(\sqrt{x^2 + z^2})$.

    (b) Show that (x, y, z) lies on the surface generated by rotating C about the x-axis precisely when $\sqrt{y^2 + z^2} = f(x)$.

Use the result in Exercise 17 to find an equation of the surface of rotation described in Exercises 18 through 23. Sketch the graph of the surface.

18. $y = 2x + 1$, $x \ge 0$, about the y-axis

19. $y = \sqrt{1 - x^2}$, $x \ge 0$, about the y-axis

20. $y = x^2$, $x \ge 0$, about the x-axis

21. $y = x^2$, $x \ge 0$, about the y-axis

22. $y = 1 + \sqrt{4 - x^2}$, $x \ge 0$, about the x-axis

23. $y = \sqrt{x^2 - 4}$, $x \ge 2$, about the x-axis

24. The temperature at any point (x, y, z) in a room is given in degrees Fahrenheit by $T(x, y, z) = 65 - x^2 - y^2 - z$. Sketch the surfaces of constant temperature for $T = 61, 56, 29$.

14.4
QUADRIC SURFACES

In Chapter 11, we found that the graph of a second-degree equation in two variables

$$Ax^2 + By^2 + Cxy + Dx + Ey + F = 0$$

is a conic section. In general, a conic section assumes the form of either a parabola, an ellipse, or a hyperbola, depending on the values of the constants in its equation.

The graph of a second-degree equation in three variables

$$Ax^2 + By^2 + Cz^2 + Dxy + Exz + Fyz + Gx + Hy + Iz + J = 0$$

is called a **quadric surface.** In this section we consider some common quadric surfaces in standard position, similar to the cartesian coordinate standard position of the conic sections considered in Chapter 11. The names of these surfaces are derived from the names of the conic sections formed by the intersections of the surface with planes parallel to the various coordinate planes. For example, the quadric surface whose equation is

(14.3)
$$\frac{x^2}{a^2} + \frac{y^2}{b^2} + \frac{z^2}{c^2} = 1$$

for positive real numbers a, b, and c, is called an **ellipsoid** because its trace in each of the coordinate planes is an ellipse. Figure 14.22(a) is an artist's rendering of an ellipsoid, while Figure 14.22 (b) shows a computer-generated drawing of the same surface.

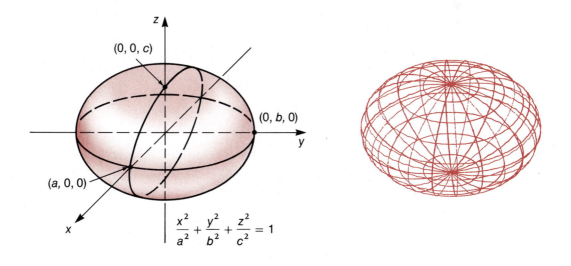

(a)

(b)

FIGURE 14.22

In a similar manner, the graph of the equation

(14.4) $$\frac{x^2}{a^2} + \frac{y^2}{b^2} - \frac{z^2}{c^2} = 1,$$

for positive constants a, b, and c, is called an **elliptic hyperboloid of one sheet.** The trace of this surface in the xy-plane is an ellipse with planar equation

$$\frac{x^2}{a^2} + \frac{y^2}{b^2} = 1,$$

while the traces in the xz- and yz-planes are hyperbolas with planar equations

$$\frac{x^2}{a^2} - \frac{z^2}{c^2} = 1$$

and $$\frac{y^2}{b^2} - \frac{z^2}{c^2} = 1$$

respectively. (See Figures 14.23(a) and (b).)

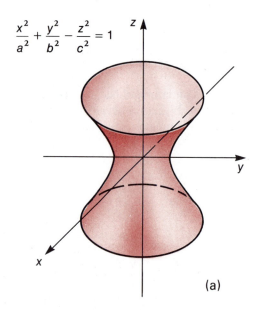

$$\frac{x^2}{a^2} + \frac{y^2}{b^2} - \frac{z^2}{c^2} = 1$$

(a)

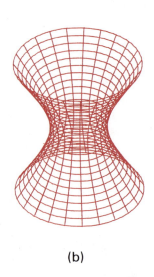

(b)

FIGURE 14.23

The surface with equation of the form (14.4) was called an elliptic hyperboloid of *one sheet* to distinguish it from the type of surface involving ellipses and hyperbolas that we consider next.

The graph of an equation of the form

(14.5) $$\frac{x^2}{a^2} - \frac{y^2}{b^2} - \frac{z^2}{c^2} = 1,$$

for positive constants a, b, and c, is called an **elliptic hyperboloid of two sheets.** The trace of this surface in either the xy- or xz-plane is a hyperbola, but the surface does not intersect the yz-plane. However, the trace in any plane parallel to the yz-plane having equation $x = k$ is an ellipse whose equation in that plane is

$$\frac{y^2}{b^2} + \frac{z^2}{c^2} = \frac{k^2}{a^2} - 1,$$

provided that $|k| > a$. Such a graph is shown in Figures 14.24(a) and (b).

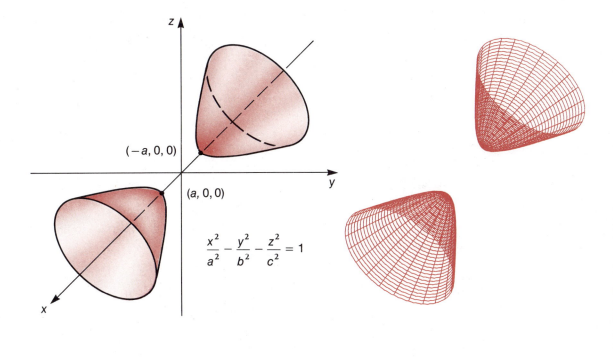

(a)   (b)

FIGURE 14.24

Another commonly occurring quadric surface is the **elliptic paraboloid,** a surface that typically has an equation of the form

(14.6) $$\frac{x^2}{a^2} + \frac{y^2}{b^2} = cz,$$

where a and b are positive constants and $c \neq 0$. For positive values of c, the surface appears as shown in Figures 14.25(a) and (b) and its traces in the xz- and the yz-planes are parabolas. The surface intersects the xy-plane only at the origin. The trace of the surface in a plane of the form $z = k$, for $k > 0$, is an ellipse with equation

$$\frac{x^2}{a^2 ck} + \frac{y^2}{b^2 ck} = 1.$$

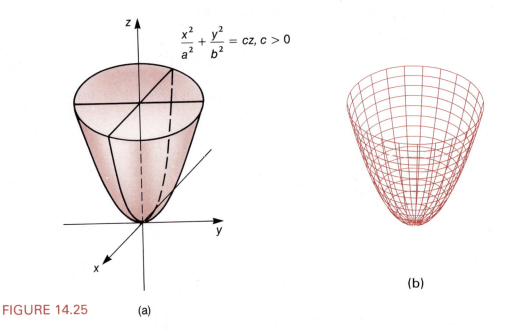

$$\frac{x^2}{a^2} + \frac{y^2}{b^2} = cz, c > 0$$

(b)

FIGURE 14.25   (a)

When $c < 0$, z assumes only nonpositive values and the surface in this case appears as shown in Figure 14.26.

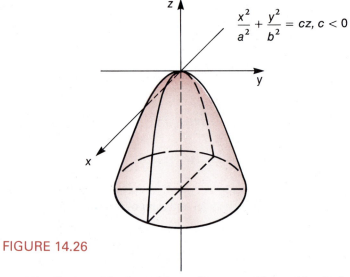

$$\frac{x^2}{a^2} + \frac{y^2}{b^2} = cz, c < 0$$

FIGURE 14.26

The final standard quadric surface we will consider is the **hyperbolic paraboloid.** Surfaces of this type are of particular interest when studying the calculus of multivariate functions. A typical equation for a hyperbolic paraboloid is of the form

(14.7) $$\frac{x^2}{a^2} - \frac{y^2}{b^2} = cz,$$

for positive constants a and b and $c \neq 0$. The trace of this surface in either the

xz- or yz-plane is a parabola, while the trace in the xy-plane is the pair of straight lines with equations

$$y = \pm \frac{b}{a}x.$$

Planes parallel to the xy-plane intersect the surface in a hyperbola. If $c > 0$, the major axis of this hyperbola is the x-axis if $z > 0$ and is the y-axis if $z < 0$. Consequently, the surface looks like a saddle that is infinitely high and deep, as shown in Figures 14.27(a) and (b). In this figure, the xy-plane has been rotated 90° to allow the figure to be more easily illustrated.

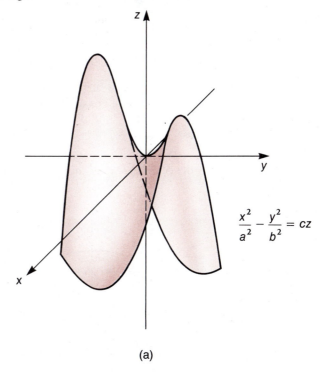

$$\frac{x^2}{a^2} - \frac{y^2}{b^2} = cz$$

(a)

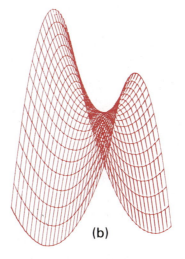

(b)

FIGURE 14.27

EXAMPLE 1   Sketch the graph of the quadric surface having equation

$$\frac{x^2}{4} + \frac{y^2}{9} = z.$$

SOLUTION

This surface is an elliptic paraboloid. When $x = 0$, the surface intersects the yz-plane in the parabola with equation

$$z = \frac{y^2}{9}.$$

When $y = 0$, the surface intersects the xz-plane in the parabola with equation

$$z = \frac{x^2}{4}.$$

The only intersection of the surface with the xy-plane is the point $(0, 0, 0)$. For $z = k$, where k is a positive constant, the intersection is an ellipse with equation

$$\frac{x^2}{4} + \frac{y^2}{9} = k.$$

The trace in the plane $z = 1$ is shown in Figure 14.28.   □

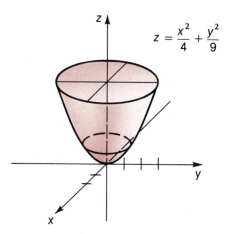

$$z = \frac{x^2}{4} + \frac{y^2}{9}$$

FIGURE 14.28

EXAMPLE 2   Sketch the quadric surface with equation

$$\frac{x^2}{9} + \frac{z^2}{16} - \frac{y^2}{25} = 1.$$

SOLUTION

The trace of this surface in the xz-plane is an ellipse with equation

$$\frac{x^2}{9} + \frac{z^2}{16} = 1.$$

Its intersection with both the *xy*- and *yz*-planes gives hyperbolas with equations

$$\frac{x^2}{9} - \frac{y^2}{25} = 1 \quad \text{and} \quad \frac{z^2}{16} - \frac{y^2}{25} = 1,$$

respectively. The surface is the elliptic hyperboloid of one sheet shown in Figure 14.29. ☐

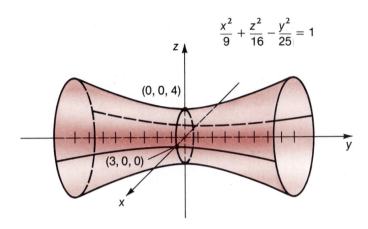

$$\frac{x^2}{9} + \frac{z^2}{16} - \frac{y^2}{25} = 1$$

(0, 0, 4)

(3, 0, 0)

FIGURE 14.29

The graph of a second-degree equation in three variables can also produce some degenerate types of quadric surfaces, such as those shown in Figures 14.30 through 14.35.

Right circular cylinder	Elliptic cone	Parabolic cylinder	Hyperbolic cylinder
$\dfrac{x^2}{a^2} + \dfrac{y^2}{b^2} = 1$	$\dfrac{x^2}{a^2} + \dfrac{y^2}{b^2} - \dfrac{z^2}{c^2} = 0$	$y = ax^2$	$\dfrac{x^2}{a^2} - \dfrac{y^2}{b^2} = 1$

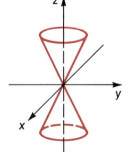

(0, b, 0)

(a, 0, 0)

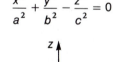

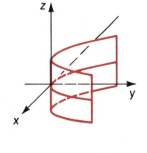

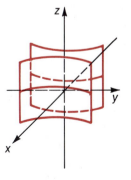

FIGURE 14.30   **FIGURE 14.31**   **FIGURE 14.32**   **FIGURE 14.33**

Perpendicular planes

$$\frac{x^2}{a^2} - \frac{y^2}{b^2} = 1$$

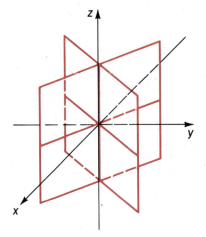

FIGURE 14.34

Parallel planes

$$\frac{x^2}{a^2} = 1$$

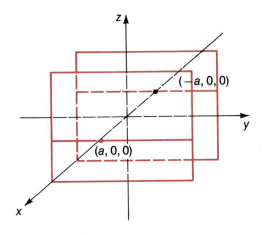

FIGURE 14.35

EXERCISE SET 14.4

Name and sketch the surfaces described in Exercises 1 through 20.

1. $\dfrac{x^2}{9} + \dfrac{y^2}{4} + \dfrac{z^2}{16} = 1$

2. $9x^2 + 4y^2 + z^2 = 36$

3. $\dfrac{x^2}{25} + \dfrac{y^2}{16} - \dfrac{z^2}{9} = 1$

4. $\dfrac{x^2}{25} - \dfrac{y^2}{16} - \dfrac{z^2}{9} = 1$

5. $x^2 - 4y^2 - 9z^2 = 36$

6. $x^2 + 4y^2 - 9z^2 = 36$

7. $z = x^2 + 4y^2$

8. $z = x^2 - 4y^2$

9. $\dfrac{x^2}{9} + \dfrac{y^2}{4} = 1$

10. $x^2 + 16y^2 = 4$

11. $4x^2 + 9y^2 - z^2 = 0$

12. $x^2 + y^2 - z^2 = 0$

13. $x^2 - y^2 = 0$

14. $4x^2 - 9y^2 = 0$

15. $\dfrac{y^2}{16} + \dfrac{z^2}{9} - \dfrac{x^2}{25} = 1$

16. $y = x^2 + 4z^2$

17. $y^2 = x^2 - z^2$

18. $y^2 + z^2 = 1$

19. $y^2 = 9$

20. $x = y^2 + z^2$

Name and sketch the surfaces described in Exercises 21 through 26. (*Hint*: Complete the square.)

21. $x^2 + y^2 + z^2 - 2x - 4z = 0$

22. $x^2 + y^2 - z^2 + 4x - 6y - 3 = 0$

23. $x^2 + y^2 - 6x - 8y - z + 20 = 0$

24. $x^2 - y^2 + z^2 + 4x - 2y - 6 = 0$

25. $x^2 - 2x = 0$

26. $x^2 - z^2 - 4x + 2z + 3 = 0$

27. Sketch the surface $z = x^2 + y^2$ and use the result to sketch the following surfaces:

    (a) $z = -(x^2 + y^2)$ (b) $z = x^2 + y^2 + 2$

    (c) $z = x^2 + y^2 - 4$ (d) $z = x^2 + (y - 1)^2$

28. Name and sketch the surface obtained when the ellipse $9x^2 + 4y^2 = 36$ is rotated about the y-axis.

29. Name and sketch the surface obtained when the parabola $z = y^2$ is rotated about the z-axis.

30. Name and sketch the surface obtained when the line $y = z$ is rotated about the z-axis.

31. Consider the graph of $\dfrac{x^2}{4} + \dfrac{y^2}{8} + \dfrac{z^2}{8} = 1$. Which of the following planes intersect this graph in a circle?

    (a) $x = y$,   (b) $x = 1$,   (c) $y = -z$,   (d) $y = 4$,   (e) $z = 3$.

32. A garland is strung around a Christmas tree seven times in an equally-spaced spiral. The conical tree has a height and diameter of 140 cm. Place a rectangular coordinate system with the origin at the top of the tree and the positive z-axis coinciding with the axis of the cone.

    (a) Find an equation describing the tree.

    (b) Find an equation describing the garland.

    (c) Find the length of the garland.

14.5
CYLINDRICAL AND SPHERICAL COORDINATES IN SPACE

Cylindrical and spherical coordinate systems are both extensions of the polar coordinate system in the plane. The names of these systems indicate the geometric figures that are most often described using these systems. Certain cylinders are easily represented using cylindrical coordinates, while spheres centered at the origin are most easily described using spherical coordinates.

CYLINDRICAL COORDINATE SYSTEM

The cylindrical coordinate system in space is derived from the polar coordinate system in the plane in the same way the rectangular system in space is derived from the rectangular system in the plane. We add a third or z-coordinate direction perpendicular to the xy-plane. Consequently, a point P with rectangular coordinates (x, y, z) will have a cylindrical coordinate representation (r, θ, z) where

(14.8) $x = r \cos \theta, \quad y = r \sin \theta, \quad r^2 = x^2 + y^2, \quad \tan \theta = \dfrac{y}{x},$

and, of course, $z = z$.

Figure 14.36 shows a typical point and its cylindrical coordinates.

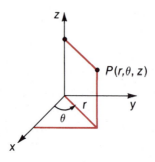

FIGURE 14.36

The cylindrical coordinates representing a point are not unique because of the various ways of describing a point in the plane using polar coordinates. For example, if (r, θ, z) is one representation of P, then for any integer n, $(r, \theta + 2n\pi, z)$ also represents P.

EXAMPLE 1   Find a cylindrical equation corresponding to the three-dimensional rectangular equation $x^2 + y^2 = 1$ and sketch the graph of this equation.

SOLUTION

A cylindrical equation is simply

$$r = \sqrt{x^2 + y^2} = 1$$

and the graph is the right circular cylinder shown in Figure 14.37.   □

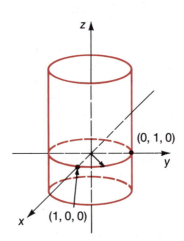

FIGURE 14.37

In general, the cylindrical equation $r = k$ for a constant k produces a right circular cylinder whose axis is the z-axis and whose radius is $|k|$.

EXAMPLE 2   Sketch the graph of the cylindrical equation $\theta = \pi/3$ and determine a corresponding rectangular equation.

SOLUTION

Since there are no restrictions on z and r, the graph of $\theta = \pi/3$ is the plane shown in Figure 14.38.

The rectangular equation for $\theta = \pi/3$ is

$$\frac{y}{x} = \tan\frac{\pi}{3} = \sqrt{3} \qquad \text{or} \qquad y = \sqrt{3}x.$$

This is the equation of the plane containing the z-axis and intersecting the xy-plane at the line given by $y = \sqrt{3}x$. ☐

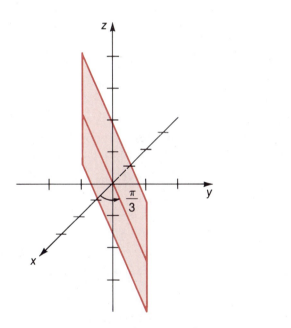

FIGURE 14.38   FIGURE 14.39

In general, the cylindrical equation $\theta = k$, for a constant k, describes a plane containing the z-axis and the line in the xy-plane with equation $y = (\tan k)x$.

EXAMPLE 3   Find a cylindrical equation corresponding to the rectangular equation $z^2 = x^2 + y^2$ and sketch the graph of this equation.

SOLUTION

Since $r^2 = x^2 + y^2$, the cylindrical equation is $z^2 = r^2$ or $z = \pm r$. The graph is the circular cone of two nappes with axis along the z-axis, as shown in Figure 14.39. ☐

SPHERICAL COORDINATE SYSTEM

While the cylindrical system uses one angle and two distances to locate a point in space, the spherical coordinate system uses two angles and one distance.

To describe a point P using spherical coordinates we use the distance ρ between P and the origin O, the angle θ used in the cylindrical coordinate system, and the angle ϕ $(0 \leq \phi \leq \pi)$ formed by the positive z-axis and the line joining P to the origin (see Figure 14.40(a)). The ordered triple (ρ, θ, ϕ) describes the spherical coordinates of P.

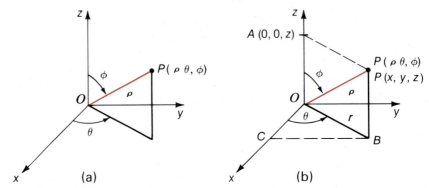

(a)   (b)

FIGURE 14.40

By considering the right triangles PAO, PBO, and BCO in Figure 14.40(b) we can see that the rectangular coordinates (x, y, z) and the cylindrical coordinates (r, θ, z) for P are related to the spherical coordinates (ρ, θ, ϕ) by:

(14.9)

$$z = \rho \cos \phi, \quad r = \rho \cos(\pi/2 - \phi) = \rho \sin \phi$$
$$x = r \cos \theta = \rho \sin \phi \cos \theta$$
$$y = r \sin \theta = \rho \sin \phi \sin \theta$$
$$\rho^2 = r^2 + z^2 = x^2 + y^2 + z^2.$$

EXAMPLE 4   Sketch the graphs of $\rho = k$, $\theta = k$, and $\phi = k$, for k a positive constant.

SOLUTION

The spherical equation $\rho = k$ corresponds to the rectangular equation $k^2 = x^2 + y^2 + z^2$, an equation of the sphere with center at the origin and radius k. The graph in the first octant is shown in Figure 14.41.

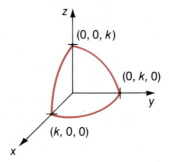

FIGURE 14.41

The spherical equation $\theta = k$ coincides, by definition, with the cylindrical equation $\theta = k$. An equation of this type is discussed in Example 2. The graph is a plane containing the z-axis and intersecting the xy-plane at the line $y = (\tan k)x$.

The spherical equation $\phi = k$, $0 \leq k \leq \pi$, describes a right circular cone of one nappe with axis along the z-axis and vertex at the origin. Figure 14.42(a) shows the cone obtained if $0 < k < \pi/2$, while Figure 14.42(b) shows the sketch for $\pi/2 < k < \pi$. The equation $\phi = 0$ describes the positive z-axis, $\phi = \pi$ describes the negative z-axis, and $\phi = \pi/2$ describes the xy-plane. $\square$

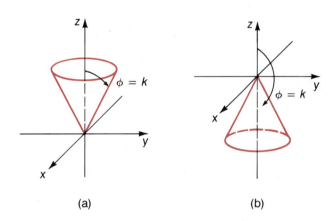

(a)   (b)

FIGURE 14.42

While the spherical coordinate system is introduced primarily for the representation of spheres, the system is useful only for spheres centered at the origin. The last example of this section illustrates how complicated a spherical equation becomes when describing a sphere that is not centered at the origin.

EXAMPLE 5   Find a spherical equation for the sphere that has its center at $(1, 1, 1)$ and passes through the origin.

SOLUTION

Since the distance from $(1, 1, 1)$ to $(0, 0, 0)$ is $\sqrt{3}$, the rectangular equation for this sphere is

$$(x - 1)^2 + (y - 1)^2 + (z - 1)^2 = 3,$$

which simplifies to

$$x^2 + y^2 + z^2 = 2(x + y + z).$$

Thus, $\rho^2 = 2(\rho \sin \phi \cos \theta + \rho \sin \phi \sin \theta + \rho \cos \phi)$

so $\rho = 2(\sin \phi \cos \theta + \sin \phi \sin \theta + \cos \phi).$ $\square$

EXERCISE SET 14.5

Rectangular coordinates of a point are given in Exercises 1 through 6. Determine
(a) cylindrical and (b) spherical coordinates of the point.

1. $(1, 1, 3)$ **2.** $(1, 0, -1)$ **3.** $(\sqrt{2}, \sqrt{2}, 0)$
4. $(1, \sqrt{3}, -2)$ **5.** $(0, 1, 3)$ **6.** $(1, -1, 3)$

Cylindrical coordinates of a point are given in Exercises 7 through 12. Determine
(a) rectangular and (b) spherical coordinates of the point.

7. $\left(3, \dfrac{\pi}{3}, 5\right)$ **8.** $(-2, 0, 2)$ **9.** $\left(4, \dfrac{\pi}{4}, -2\right)$

10. $\left(-1, \dfrac{\pi}{2}, 2\right)$ **11.** $(0, \pi, 1)$ **12.** $(-1, 0, 1)$

Spherical coordinates of a point are given in Exercises 13 through 18. Determine
(a) rectangular and (b) cylindrical coordinates of the point.

13. $\left(1, \dfrac{\pi}{4}, \dfrac{\pi}{3}\right)$ **14.** $\left(2, \dfrac{\pi}{6}, \dfrac{\pi}{4}\right)$ **15.** $\left(2, \dfrac{5\pi}{6}, \dfrac{3\pi}{4}\right)$

16. $\left(3, \dfrac{3\pi}{4}, \dfrac{\pi}{2}\right)$ **17.** $(5, \pi, \pi)$ **18.** $\left(1, \dfrac{7\pi}{4}, \dfrac{\pi}{6}\right)$

Three-dimensional rectangular equations are given in Exercises 19 through 26.
Find corresponding cylindrical equations.

19. $y = x$ **20.** $\sqrt{3}y = x$
21. $x^2 + y^2 = 4$ **22.** $y = -2$
23. $z = x^2 + y^2$ **24.** $z = \sqrt{x^2 + y^2}$
25. $x^2 + y^2 + 4z^2 = 4$ **26.** $x^2 + y^2 + z = 1$

Sketch the graphs of the cylindrical equations given in Exercises 27 through 38.
27. $r = 3$ **28.** $r = -2$

29. $\theta = \dfrac{\pi}{2}$ **30.** $\theta = \dfrac{\pi}{4}$

31. $z = 4r^2$ **32.** $z = 3$
33. $r^2 + z^2 = 9$ **34.** $r^2 + z^2 = 16$
35. $r = \cos \theta$ **36.** $r = \cos \theta + 1$
37. $r = \theta$ **38.** $r^2 = (\cos \theta)^2$

Sketch the graphs of the spherical equations given in Exercises 39 through 46.

39. $\rho = 2$ **40.** $\theta = \dfrac{\pi}{4}$ **41.** $\phi = \dfrac{\pi}{4}$

42. $\rho^2 = 9$ **43.** $\rho = \cos \phi$ **44.** $\phi = \dfrac{\pi}{4}, \theta = \dfrac{\pi}{4}$

45. $\rho = 3, \phi = \dfrac{\pi}{4}$ **46.** $\rho = 3, \theta = \dfrac{\pi}{4}$

Sketch the regions described in Exercises 47 through 54.

47. $1 \le r \le 3, \quad 0 \le z \le 4$

48. $0 \le \theta \le \dfrac{\pi}{2}, \quad 0 \le r \le 1, \quad 0 \le z \le 3$

49. $1 \le r \le 2, \quad \dfrac{\pi}{6} \le \theta \le \dfrac{\pi}{3}, \quad 2 \le z \le 3$

50. $1 \le \rho \le 2$

51. $0 \le \phi \le \dfrac{\pi}{4}, \quad \rho = 3$

52. $0 \le \phi \le \dfrac{\pi}{4}, \quad 0 \le \theta \le \dfrac{\pi}{2}, \quad \rho = 3$

53. $1 \le \rho \le 2, \quad \dfrac{\pi}{4} \le \phi \le \dfrac{\pi}{3}, \quad \dfrac{\pi}{6} \le \theta \le \dfrac{\pi}{3}$

54. $0 \le z \le \sqrt{4 - r^2}, \quad 0 \le r \le 2$

55. Use cylindrical coordinates to describe the region that is inside the sphere $x^2 + y^2 + z^2 = 4$ and outside the cylinder $x^2 + y^2 = 1$.

56. Use cylindrical coordinates to describe the region that is inside both the sphere $x^2 + y^2 + z^2 = 4$ and the cylinder $x^2 + y^2 = 1$.

57. Use spherical coordinates to describe the region that is inside the sphere $x^2 + y^2 + z^2 = 9$ and outside the sphere $x^2 + y^2 + z^2 = 4$.

58. Use spherical coordinates to describe the region inside both the cone $z^2 = x^2 + y^2$ and the sphere $x^2 + y^2 + z^2 = 9$.

59. Find an equation in spherical coordinates for the sphere

$$x^2 + y^2 + (z - 1)^2 = 4.$$

REVIEW EXERCISES

Find the domain of the functions described in Exercises 1 through 6.

1. $f(x, y) = \sqrt{1 - x^2 - y}$

2. $f(x, y) = \dfrac{1}{\sqrt{4 - x^2 - y^2}}$

3. $f(x, y) = \dfrac{\sqrt{1 - x^2 - y^2}}{x^2}$

4. $f(x, y, z) = \sqrt{x^2 + y^2 + z^2 - 1}$

5. $f(x, y, z) = \ln xyz$

6. $f(x, y, z) = \dfrac{1}{x}$

Sketch representative level curves for the functions described in Exercises 7 through 10.

7. $f(x, y) = 2 - x + y$

8. $f(x, y) = 2x + 3y - 6$

9. $f(x, y) = 9x^2 + 4y^2$

10. $f(x, y) = x - y^2$

Sketch the graph in the first octant of representative level surfaces for the functions described in Exercises 11 through 14.

11. $f(x, y, z) = 9x^2 + 4y^2$

12. $f(x, y, z) = 2 - x + y - z$

13. $f(x, y, z) = x^2 + y^2 + z^2$

14. $f(x, y, z) = 4x^2 + 9y^2 + z^2$

Identify and sketch the graphs of the surfaces described in Exercises 15 through 36.

15. $z = 4 - x - 2y$

16. $z = x - 1$

17. $z = x^2 + y^2 - 2$

18. $9x^2 + 4y^2 + z^2 = 36$

19. $y^2 + 4z^2 = 4$

20. $x^2 = 4$

21. $\dfrac{x^2}{9} + \dfrac{y^2}{4} - \dfrac{z^2}{16} = 1$

22. $x^2 + 25y^2 - z^2 = 0$

23. $\dfrac{x^2}{16} - \dfrac{y^2}{25} = 0$

24. $y^2 - z^2 = 0$

25. $x^2 - y^2 = 1$

26. $yz = 1$

27. $x^2 + y^2 + z^2 - 2x - 2y - 2z = 1$

28. $x^2 + 2y^2 + 3z^2 - 8y - 18z = 1$

29. $z^2 - 4z = 0$

30. $z^2 - 4z + 4 = 0$

31. $r = 2 \sin \theta$

32. $\theta = \dfrac{5\pi}{3}$

33. $r^2 = 4 \sin \theta$

34. $\rho = 4 \cos \phi$

35. $\rho \sin \phi = 1$

36. $\rho^2 - 5\rho + 6 = 0$

37. Determine cylindrical and spherical coordinates of the following points given in rectangular coordinates.

(a) $(1, 0, 0)$ (b) $(1, 1, 0)$ (c) $(1, 1, 1)$

38. Determine rectangular and spherical coordinates of the following points given in cylindrical coordinates.

(a) $\left(4, \dfrac{\pi}{4}, 0\right)$ (b) $\left(1, \dfrac{\pi}{2}, 1\right)$ (c) $\left(1, \dfrac{\pi}{6}, 2\right)$

39. Determine rectangular and cylindrical coordinates of the following points given in spherical coordinates.

(a) $\left(1, \dfrac{\pi}{2}, \dfrac{\pi}{2}\right)$ (b) $\left(1, \dfrac{\pi}{4}, \dfrac{\pi}{2}\right)$ (c) $\left(2, \dfrac{\pi}{2}, \dfrac{\pi}{4}\right)$

40. The temperature at any point (x, y) on a circular plate of radius 2 is given by $T(x, y) = 4 - (x^2 + y^2)$. Sketch the curves of constant temperature for $T = 0, 1, 3$.

41. Describe the surface obtained by rotating the curve $y = \sqrt{4 - x^2}$, $z = 0$ about the x-axis and determine an equation of the surface. (See Exercise 17 in Section 14.3.)

42. Describe the surface obtained by rotating the line $y = z$, $x = 0$ about the z-axis and determine an equation of the surface.

43. Describe the surface obtained by rotating the circle $(y - 2)^2 + z^2 = 1$, $x = 0$ about the z-axis.

15

THE DIFFERENTIAL CALCULUS OF MULTIVARIATE FUNCTIONS

The development of the calculus of multivariate functions parallels that for single variable functions. We begin by considering the concepts of the limit and continuity of a multivariate function and proceed to study differentiation and integration. Since the concepts are similar for arbitrary functions of n variables, we will generally simplify the discussion by considering only functions of two variables. Occasionally we will also state results for functions of three variables, primarily to illustrate the similarity of results.

15.1
LIMITS AND CONTINUITY

The concept of the limit of a single variable function, expressed by

$$\lim_{x \to a} f(x) = L,$$

is that $f(x)$ becomes and remains close to L as x becomes close but not equal to a. The definition of the limit of a function makes use of tolerance intervals about a and L to express analytically the concept of "close to." This definition was given in Section 2.2 and is repeated here.

(15.1)
DEFINITION

The limit of the single variable function f at a is L, provided that for every number $\varepsilon > 0$, a number $\delta > 0$ can be found with the property that

$$|f(x) - L| < \varepsilon$$

whenever

$$0 < |x - a| < \delta.$$

The definition of the limit of a function of two variables parallels the single variable definition. Modifications are necessary, however, because the domain of the function is now a subset of $\mathbb{R}^2$.

A point (x, y) is within δ units of a point (a, b) precisely when (x, y) lies in the interior of the circle about (a, b) with radius δ. (See Figure 15.1.) The points in the interior of this circle are those points (x, y) that satisfy the inequality

$$\sqrt{(x - a)^2 + (y - b)^2} < \delta.$$

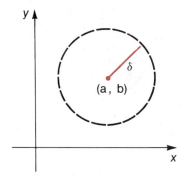

FIGURE 15.1

If, in addition, the point (x, y) differs from (a, b), the inequality becomes

$$0 < \sqrt{(x - a)^2 + (y - b)^2} < \delta.$$

Using this inequality to express the tolerance about the point (a, b) leads to the following definition.

(15.2)
DEFINITION

The limit of a function f of two variables at (a, b) is L, written

$$\lim_{(x,y) \to (a,b)} f(x, y) = L,$$

provided that for every number $\varepsilon > 0$, a number δ can be found with the property that

$$|f(x, y) - L| < \varepsilon$$

whenever

$$0 < \sqrt{(x - a)^2 + (y - b)^2} < \delta.$$

An illustration of the limit of a function of two variables is shown in Figure 15.2.

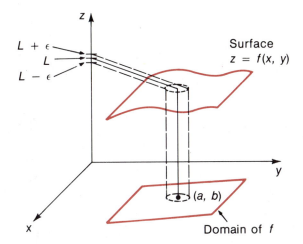

FIGURE 15.2

In a similar manner, the definition of the limit of a function of three variables uses the inequality

$$0 < \sqrt{(x - a)^2 + (y - b)^2 + (z - c)^2} < \delta$$

to express that (x, y, z) is within δ units of, but not equal to, (a, b, c).

(15.3)
DEFINITION

The limit of a function f of three variables at (a, b, c) is L, written

$$\lim_{(x,y,z)\to(a,b,c)} f(x, y, z) = L,$$

provided that for every number $\varepsilon > 0$, a number δ can be found with the property that

$$|f(x, y, z) - L| < \varepsilon$$

whenever

$$0 < \sqrt{(x - a)^2 + (y - b)^2 + (z - c)^2} < \delta.$$

These definitions ensure that when the limit of a multivariate function exists it must be unique.

EXAMPLE 1   Prove that $\lim_{(x,y)\to(1,2)} f(x, y) = 2$, where $f(x, y) = 6 - 2x - y$.

SOLUTION

The graph of f is the plane shown in Figure 15.3.

To prove that $\lim_{(x,y)\to(1,2)} (6 - 2x - y) = 2$, we must show that for an arbitrary $\varepsilon > 0$, a number δ exists with

$$|(6 - 2x - y) - 2| < \varepsilon$$

whenever $$0 < \sqrt{(x - 1)^2 + (y - 2)^2} < \delta.$$

Suppose such a δ exists and (x, y) is restricted so that

$$0 < \sqrt{(x - 1)^2 + (y - 2)^2} < \delta.$$

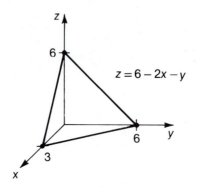

FIGURE 15.3

The fact that both

$$|x - 1| \le \sqrt{(x - 1)^2 + (y - 2)^2}$$

and

$$|y - 2| \le \sqrt{(x - 1)^2 + (y - 2)^2}$$

implies that

$$
\begin{aligned}
|(6 - 2x - y) - 2| &= |-2(x - 1) - (y - 2)| \\
&\le 2|x - 1| + |y - 2| \\
&\le 2\sqrt{(x - 1)^2 + (y - 2)^2} + \sqrt{(x - 1)^2 + (y - 2)^2} \\
&< 3\delta.
\end{aligned}
$$

To ensure that $|(6 - 2x - y) - 2| < \varepsilon$, choose $\delta \le \varepsilon/3$. Since $\varepsilon > 0$ was chosen arbitrarily, this establishes that $\lim_{(x,y)\to(1,2)} (6 - 2x - y) = 2$. ☐

If $\lim_{(x,y)\to(a,b)} f(x, y) = L$, then $f(x, y)$ must approach L as (x, y) approaches (a, b) along every curve in the domain of f that passes through (a, b) To show that a limit of a function of two variables does not exist, it suffices to show that the limiting values differ along two distinct curves.

EXAMPLE 2   Show that $\lim_{(x,y)\to(0,0)} f(x, y)$ does not exist for $f(x, y) = \dfrac{xy}{x^2 + y^2}$.

SOLUTION

Every circle about the origin contains both points on the x-axis and points on the line $y = x$.

For points $(x, 0)$ on the x-axis,

$$f(x, 0) = \frac{x \cdot 0}{x^2 + y^2} = 0,$$

so if this limit exists it must be zero. However, for points (x, x) along the line $y = x$,

$$f(x, x) = \frac{x \cdot x}{x^2 + x^2} = \frac{1}{2}.$$

(See Figure 15.4.) Hence $\lim_{(x,y)\to(0,0)} \dfrac{xy}{x^2 + y^2}$ does not exist. ☐

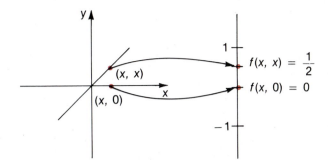

FIGURE 15.4

The condition required for a function of several variables to be continuous at a point is the same as the condition for functions of a single variable: the limit of the function at the point must agree with the value of the function there.

(15.4)
DEFINITION

A function f of two variables is continuous at (a, b) if

(i) $f(a, b)$ exists,
(ii) $\lim\limits_{(x,y)\to(a,b)} f(x, y)$ exists, and
(iii) $\lim\limits_{(x,y)\to(a,b)} f(x, y) = f(a, b)$.

The arithmetic and composition rules for functions of a single variable also hold for functions of several variables when the operations are defined. The results for limits are stated; the analogous continuity results will be used though not stated.

(15.5)
THEOREM

If $\lim\limits_{(x,y)\to(a,b)} f(x, y) = L$ and $\lim\limits_{(x,y)\to(a,b)} g(x, y) = M$, then

(i) $\lim\limits_{(x,y)\to(a,b)} (f + g)(x, y) = L + M,$
(ii) $\lim\limits_{(x,y)\to(a,b)} (f - g)(x, y) = L - M,$
(iii) $\lim\limits_{(x,y)\to(a,b)} (f \cdot g)(x, y) = L \cdot M,$ and
(iv) $\lim\limits_{(x,y)\to(a,b)} \left(\dfrac{f}{g}\right)(x, y) = \dfrac{L}{M},$ provided $M \neq 0.$

(15.6)
THEOREM

If $\lim\limits_{(x,y)\to(a,b)} f(x, y) = L$ and g is a function of one variable that is continuous at L, then

$$\lim\limits_{(x,y)\to(a,b)} g(f(x, y)) = g(L).$$

EXAMPLE 3

Show that $\lim\limits_{(x,y)\to\left(1,\frac{\pi}{2}\right)} \dfrac{x + \cos y}{x^2 + \sin y} = \dfrac{1}{2}.$

SOLUTION

Since

$$\lim_{(x,y)\to\left(1,\frac{\pi}{2}\right)} x = 1 \quad \text{and} \quad \lim_{(x,y)\to\left(1,\frac{\pi}{2}\right)} \cos y = 0,$$

$$\lim_{(x,y)\to\left(1,\frac{\pi}{2}\right)} (x + \cos y) = 1.$$

Similarly,

$$\lim_{(x,y)\to\left(1,\frac{\pi}{2}\right)} (x^2 + \sin y) = \lim_{(x,y)\to\left(1,\frac{\pi}{2}\right)} x^2 + \lim_{(x,y)\to\left(1,\frac{\pi}{2}\right)} \sin y = 1 + 1 = 2.$$

Consequently,

$$\lim_{(x,y)\to\left(1,\frac{\pi}{2}\right)} \frac{x + \cos y}{x^2 + \sin y} = \frac{1}{2}. \qquad \square$$

EXAMPLE 4

Consider the function described by

$$f(x, y) = \begin{cases} \dfrac{xy}{x^2 + y^2} & \text{if } (x, y) \neq (0, 0) \\ 0 & \text{if } (x, y) = (0, 0). \end{cases}$$

Determine the points at which f is continuous.

SOLUTION

We saw in Example 2 that $\displaystyle\lim_{(x,y)\to(0,0)} \dfrac{xy}{x^2 + y^2}$ does not exist, so f is not continuous at the origin. When $(a, b) \neq (0, 0)$, $a^2 + b^2 \neq 0$ and

$$\lim_{(x,y)\to(a,b)} f(x, y) = \lim_{(x,y)\to(a,b)} \frac{xy}{x^2 + y^2} = \frac{ab}{a^2 + b^2} = f(a, b).$$

So f is continuous at every point in the plane except the origin. $\qquad \square$

EXERCISE SET 15.1

In Exercises 1 through 16, determine the limit of f at the given point, if it exists.

1. $f(x, y) = 3 - x - y$, $(1, 1)$ **2.** $f(x, y) = \sqrt{x + y}$, $(2, 2)$

3. $f(x, y) = x^2 + y^2$, $(0, 1)$ **4.** $f(x, y) = \dfrac{1}{x + y}$, $(1, 0)$

5. $f(x, y) = \ln (x + y)$, $(1, 0)$ **6.** $f(x, y) = \dfrac{x^2 + y^2}{xy + 1}$, $(0, 0)$

7. $f(x, y) = \dfrac{x^3 - x^2y + xy^2}{x + y}$, $(1, 1)$ **8.** $f(x, y) = \cos(x + y)$, $\left(\dfrac{\pi}{2}, 0\right)$

9. $f(x, y) = e^{x+y}$, $(1, 0)$ **10.** $f(x, y) = y \ln x$, $(e^2, 2)$

11. $f(x, y) = \dfrac{x^2 - y^2}{x - y}$, $(1, 1)$

12. $f(x, y) = \dfrac{y(x^2 - 4)}{x - 2}$, $(2, 1)$

13. $f(x, y) = \dfrac{\sin(x^2 + y^2)}{x^2 + y^2}$, $(0, 0)$

14. $f(x, y) = \dfrac{1 - \cos(x^2 + y^2)}{x^2 + y^2}$ $(0, 0)$

15. $f(x, y, z) = x^2 + y^2 + z^2$, $(1, -1, 1)$

16. $f(x, y, z) = \dfrac{x(z^2 - 9)}{y(z - 3)}$, $(2, 1, 3)$

Determine the regions of the plane in which the functions described in Exercises 17 through 24 are continuous.

17. $f(x, y) = x^2 + y^2$

18. $f(x, y) = e^{xy}$

19. $f(x, y) = \dfrac{y - x}{x^2 + y^2}$

20. $f(x, y) = \dfrac{1}{1 - x + y}$

21. $f(x, y) = \sqrt{1 - x^2 - y^2}$

22. $f(x, y) = \dfrac{x^2 + y^2}{1 - x^2 - y^2}$

23. $f(x, y) = \sin(x + y)$

24. $f(x, y) = \ln(x + y - 1)$

In Exercises 25 through 32, show that the limits do not exist.

25. $\lim\limits_{(x,y)\to(0,0)} \dfrac{y^2}{x^2 + y^2}$

26. $\lim\limits_{(x,y)\to(0,0)} \dfrac{x}{y}$

27. $\lim\limits_{(x,y)\to(0,0)} \dfrac{y}{x^2 - y}$

28. $\lim\limits_{(x,y)\to(0,0)} \dfrac{x^2 y^2}{x^4 + 3y^4}$

29. $\lim\limits_{(x,y)\to(0,0)} \dfrac{\sin xy}{xy}$

30. $\lim\limits_{(x,y)\to(0,0)} \dfrac{1 - \cos xy}{xy}$

31. $\lim\limits_{(x,y,z)\to(0,0,0)} \dfrac{x^2 + y^2}{x^2 + y^2 + z^2}$

32. $\lim\limits_{(x,y,z)\to(0,0,0)} \dfrac{xy + z}{x + y + z^2}$

33. Sketch the regions in the xy-plane that are described by the following.

(a) $0 < \sqrt{(x - 1)^2 + (y - 2)^2} < .5$

(b) $0 < \sqrt{x^2 + (y + 1)^2} < 1$

(c) $1 < \sqrt{(x + 2)^2 + (y + 3)^2} < 2$

34. Determine the regions in space that are described by the following.

(a) $0 < \sqrt{(x - 1)^2 + (y - 2)^2 + (z - 3)^2} < .5$

(b) $0 < \sqrt{x^2 + y^2 + (z + 2)^2} < .1$

(c) $3 < \sqrt{(x + 1)^2 + (y + 2)^2 + (z + 1)^2} < 4$

35. Use Definition 15.2 to prove that $\lim\limits_{(x,y)\to(0,0)} (3 - x - y) = 3$.

36. Use Definition 15.2 to prove that $\lim\limits_{(x,y)\to(2,1)} (6 - 2x - 3y) = -1$.

37. Use Definition 15.2 to prove that $\lim\limits_{(x,y)\to(0,0)} (x^2 + y^2) = 0$.

38. Use Definition 15.2 to prove that $\lim\limits_{(x,y)\to(0,0)} (4 - x^2 - y^2) = 4$.

39. Use Definition 15.3 to prove that $\lim\limits_{(x,y,z)\to(1,2,3)} (2x - y + z) = 3$.

40. Use Definition 15.3 to prove that $\lim\limits_{(x,y,z)\to(0,0,0)} (x^2 + y^2 + z^2) = 0$.

41. If $f(x, y) = x^2 + y$ and $g(x) = \sin x$, find $\lim\limits_{(x,y)\to(0,\frac{\pi}{2})} g(f(x, y))$.

42. If $f(x, y, z) = x + y + z$ and $g(x) = \cos x$, find $\lim\limits_{(x,y,z)\to(\frac{\pi}{2},\frac{\pi}{2},0)} g(f(x, y, z))$.

43. If $f(x, y, z) = \dfrac{z}{xy}$ and $g(x) = \ln x$, find $\lim\limits_{(x,y,z)\to(1,2,2e)} g(f(x, y, z))$.

44. Consider the function described by

$$f(x, y) = \begin{cases} \dfrac{xy}{x^2 + y^2} & , (x, y) \neq (0, 0) \\ 0 & , (x, y) = (0, 0). \end{cases}$$

(a) Show that the single-variable function described by $f(x, 0)$ is a continuous function.

(b) Show that the single-variable function described by $f(0, y)$ is a continuous function.

45. Show that if $\lim\limits_{(x,y)\to(a,b)} f(x, y) = L$ and $\lim\limits_{(x,y)\to(a,b)} f(x, y) = M$, then $L = M$.

15.2
PARTIAL DERIVATIVES

Suppose (a, b) is in the domain of a function f of two variables x and y. By fixing y at b and allowing x to vary we can define the derivative of f with respect to x at (a, b). By fixing x at a and allowing y to vary we can define the derivative of f with respect to y at (a, b).

The derivatives produced in this manner are called **partial derivatives** since they are defined by allowing only the numbers in a part of the domain of the function to vary. Partial derivatives give the rate of change of the values of a function relative to a variable change along a line parallel to one of the coordinate axes.

A multivariable function has as many partial derivatives at a point in the interior of its domain as it has variables, provided that the required limits exist. For simplicity, we will consider the definition of partial derivatives only for functions of two variables, leaving the analogous discussion involving functions of three or more variables to the examples and exercises.

(15.7)
DEFINITION

Suppose f is a function of two variables x and y. The **first partial derivative of f with respect to x, f_x,** is defined by

$$f_x(x_0, y_0) = \lim_{x \to x_0} \frac{f(x, y_0) - f(x_0, y_0)}{x - x_0},$$

and the **first partial derivative with respect to y, f_y,** by

$$f_y(x_0, y_0) = \lim_{y \to y_0} \frac{f(x_0, y) - f(x_0, y_0)}{y - y_0},$$

provided that the limits exist.

Another common notation for partial derivatives is similar to the Leibniz differential notation for the derivative of a single variable function:

(15.8) $\dfrac{\partial f}{\partial x} \equiv f_x$ and $\dfrac{\partial f}{\partial y} \equiv f_y.$

If $z = f(x, y)$, then $\partial z/\partial x \equiv f_x$ and $\partial z/\partial y \equiv f_y$. A less common, but occasionally used, notation is to assign an integer to each variable, say 1 to x and 2 to y, and let $f_1 \equiv f_x$ and $f_2 \equiv f_y$.

It follows from Definition 15.7 that $f_x(x_0, y_0)$ is the derivative with respect to x of the single variable function $f(x, y_0)$ at $x = x_0$. To find f_x, we simply differentiate $f(x, y)$ with y treated as a constant. Similarly, to find f_y, differentiate $f(x, y)$ with x treated as a constant.

EXAMPLE 1   Find the first partial derivatives of f if

$$f(x, y) = x^2 + xy + y \sin x$$

and evaluate these at $(0, 1)$.

SOLUTION

$$f_x(x, y) = \frac{\partial(x^2 + xy + y \sin x)}{\partial x} = 2x + y + y \cos x,$$

so $f_x(0, 1) = 0 + 1 + 1 \cos 0 = 2$

and $f_y(x, y) = \dfrac{\partial(x^2 + xy + y \sin x)}{\partial y} = 0 + x + \sin x,$

so $f_y(0, 1) = 0.$ ☐

EXAMPLE 2   Find f_x, f_y, and f_z if

$$f(x, y, z) = \sqrt{x} + 2xyz + z^2 e^x \ln y.$$

SOLUTION

Proceeding as with functions of two variables, we consider y and z constant to determine

$$f_x(x, y, z) = \frac{1}{2\sqrt{x}} + 2yz + z^2 e^x \ln y.$$

Fixing x and z, we have

$$f_y(x, y, z) = 0 + 2xz + z^2 e^x \cdot \frac{1}{y},$$

and fixing x and y gives

$$f_z(x, y, z) = 0 + 2xy + 2ze^x \ln y. \qquad \square$$

EXAMPLE 3   Find f_x and f_y if $f(x, y) = \dfrac{y(x^2 - y^2)}{x^2 + y^2}$.

SOLUTION

We apply the quotient rule in each case.

$$f_x(x, y) = \frac{2xy(x^2 + y^2) - y(x^2 - y^2)2x}{(x^2 + y^2)^2} = \frac{4xy^3}{(x^2 + y^2)^2}$$

$$f_y(x, y) = \frac{(x^2 - 3y^2)(x^2 + y^2) - y(x^2 - y^2)2y}{(x^2 + y^2)^2} = \frac{x^4 - 4x^2y^2 - y^4}{(x^2 + y^2)^2}$$

The function f and, consequently, the partial derivatives f_x and f_y are undefined at $(0, 0)$. $\qquad \square$

EXAMPLE 4   Find $f_x(0, 0)$ and $f_y(0, 0)$ if

$$f(x, y) = \begin{cases} \dfrac{xy}{x^2 + y^2}, & \text{if } (x, y) \neq (0, 0) \\ 0 & \text{, if } (x, y) = (0, 0). \end{cases}$$

SOLUTION

In this case, we must apply Definition 15.7 to find the partial derivatives, since the quotient rule is invalid when the denominator is zero.

$$f_x(0, 0) = \lim_{x \to 0} \frac{f(x, 0) - f(0, 0)}{x - 0} = \lim_{x \to 0} \frac{0 - 0}{x} = 0$$

and

$$f_y(0, 0) = \lim_{y \to 0} \frac{f(0, y) - f(0, 0)}{y - 0} = \lim_{y \to 0} \frac{0 - 0}{y} = 0.$$

Notice that both first partial derivatives of f exist at $(0, 0)$, even though Example 4 of Section 15.1 shows that f is not continuous at $(0, 0)$. $\qquad \square$

The first partial derivatives of a function of two variables have an important geometric application. Suppose f is a function whose graph is the surface shown in Figure 15.5, where $z = f(x, y)$.

Fixing y at y_0 and allowing x to vary traces the curve C_1 on the surface. This curve is the intersection of the surface with the plane $y = y_0$ and is a planar curve

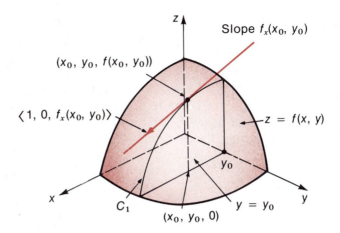

FIGURE 15.5

with equation $z = f(x, y_0)$. The partial derivative $f_x(x_0, y_0)$ is the slope of the tangent line to the curve C_1 at the point $(x_0, y_0, f(x_0, y_0))$. The equation of this tangent line in the plane $y = y_0$ is

$$\frac{z - z_0}{x - x_0} = f_x(x_0, y_0),$$

so symmetric equations of the line are

(15.9) $$y = y_0, \qquad x - x_0 = \frac{z - z_0}{f_x(x_0, y_0)},$$

and the direction vector is $\langle 1, 0, f_x(x_0, y_0)\rangle$.

   Similarly, the curve C_2 shown in Figure 15.6 is the intersection of the surface with the plane $x = x_0$, so $f_y(x_0, y_0)$ is the slope of the tangent line to the curve C_2 at the point $(x_0, y_0, f(x_0, y_0))$. Symmetric equations of this tangent line are

(15.10) $$x = x_0, \qquad y - y_0 = \frac{z - z_0}{f_y(x_0, y_0)}$$

and the direction vector is $\langle 0, 1, f_y(x_0, y_0)\rangle$.

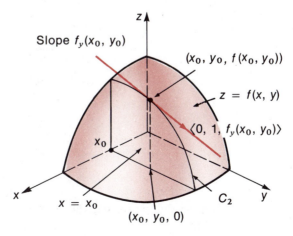

FIGURE 15.6

EXAMPLE 5   Find an equation of the line tangent to the curve of intersection of the surface $z = 4 - x^2 - y^2$ and the plane $x = 1$ at the point $(1, 1, 2)$.

SOLUTION

Let $f(x, y) = 4 - x^2 - y^2$. The curve of intersection of the graph of f and the plane $x = 1$ is shown in Figure 15.7. The slope of the tangent line to this curve at the point $(1, 1, 2)$ is $f_y(1, 1)$. Since $f_y(x, y) = -2y$, $f_y(1, 1) = -2$. Consequently, the equations of the tangent line are:

$$x = 1, \qquad \frac{z - 2}{y - 1} = -2$$

or

$$x = 1, \qquad \frac{z - 2}{-2} = y - 1.$$ □

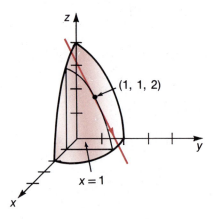

FIGURE 15.7

If f is a function of the variables x and y, the **second partial derivatives** of f are defined by

$$f_{xx} = \frac{\partial^2 f}{\partial x^2} = \frac{\partial}{\partial x}\left(\frac{\partial f}{\partial x}\right)$$

$$f_{yy} = \frac{\partial^2 f}{\partial y^2} = \frac{\partial}{\partial y}\left(\frac{\partial f}{\partial y}\right)$$

$$f_{xy} = \frac{\partial^2 f}{\partial y \partial x} = \frac{\partial}{\partial y}\left(\frac{\partial f}{\partial x}\right)$$

and

$$f_{yx} = \frac{\partial^2 f}{\partial x \partial y} = \frac{\partial}{\partial x}\left(\frac{\partial f}{\partial y}\right).$$

If f and its second partial derivatives are all continuous, it can be shown that $f_{xy} = f_{yx}$, so the order of differentiation is immaterial. A result of this type is also true for functions of more than two variables and for derivatives of order higher than two. In general, the partial derivatives of a function are independent of the order in which the partial differentiation is performed, provided that all partial derivatives of that order are continuous.

EXAMPLE 6   Suppose $f(x, y) = x^3 + 2x^2y + e^{2x} \sin \pi y$. Find the second partial derivatives of f and the third partial derivative f_{xyx}.

SOLUTION

$$\frac{\partial f}{\partial x}(x, y) = 3x^2 + 4xy + 2e^{2x} \sin \pi y$$

and

$$\frac{\partial f}{\partial y}(x, y) = 2x^2 + \pi e^{2x} \cos \pi y,$$

so

$$\frac{\partial^2 f}{\partial x^2}(x, y) = \frac{\partial}{\partial x}\left(\frac{\partial f}{\partial x}\right) = 6x + 4y + 4e^{2x} \sin \pi y,$$

$$\frac{\partial^2 f}{\partial y^2}(x, y) = \frac{\partial}{\partial y}\left(\frac{\partial f}{\partial y}\right) = -\pi^2 e^{2x} \sin \pi y,$$

$$\frac{\partial^2 f}{\partial y \partial x}(x, y) = \frac{\partial}{\partial y}\left(\frac{\partial f}{\partial x}\right) = 4x + 2\pi e^{2x} \cos \pi y,$$

and

$$\frac{\partial^2 f}{\partial x \partial y}(x, y) = \frac{\partial}{\partial x}\left(\frac{\partial f}{\partial y}\right) = 4x + 2\pi e^{2x} \cos \pi y.$$

The third partial derivative f_{xyx} is

$$f_{xyx}(x, y) = \frac{\partial^3 f(x, y)}{\partial x \partial y \partial x} = \frac{\partial}{\partial x}\left(\frac{\partial^2 f}{\partial y \partial x}\right)(x, y) = 4 + 4\pi e^{2x} \cos \pi y. \quad \square$$

EXAMPLE 7   Find f_{xyyz} for

$$f(x, y, z) = \frac{x^2 \ln y}{z}.$$

SOLUTION

$$f_x(x, y, z) = \frac{2x \ln y}{z},$$

$$f_{xy}(x, y, z) = \frac{\partial}{\partial y}\left(\frac{2x \ln y}{z}\right) = \frac{2x}{zy},$$

$$f_{xyy}(x, y, z) = \frac{\partial}{\partial y}\left(\frac{2x}{zy}\right) = -\frac{2x}{zy^2},$$

and

$$f_{xyyz}(x, y, z) = \frac{\partial}{\partial z}\left(-\frac{2x}{zy^2}\right) = \frac{2x}{z^2y^2}. \quad \square$$

EXERCISE SET 15.2

Find the first partial derivatives of the functions described in Exercises 1 through 20.

1. $f(x, y) = x + y + 1$
2. $f(x, y) = 6 - x - 2y$
3. $f(x, y) = x^2 + xy + y^2$
4. $f(x, y) = 5x^3y^2 - 3x^2y + y^4$
5. $f(x, y) = e^{xy}$
6. $f(x, y) = \ln xy$

7. $f(x, y) = \dfrac{xy}{x^2 + y^2}$
8. $f(x, y) = \dfrac{x}{y}$

9. $f(x, y) = x^2 (x + y^2)$
10. $f(x, y) = (x - 2y)(x^2 + y^2)$
11. $f(x, y) = ye^x$
12. $f(x, y) = \sin (x^2 + y^3)$
13. $f(x, y) = \sqrt{x^2 + y^2}$
14. $f(x, y) = (2x - 3y)^3$
15. $f(x, y) = e^x \cos y$
16. $f(x, y) = \tan(x^2 + y)$
17. $f(x, y, z) = 3xz + 4y$
18. $f(x, y, z) = 10xy + 7z$

19. $f(x, y, z) = \cos(2x + 3y + 4z)$
20. $f(x, y, z) = \dfrac{1}{xyz}$

Find the first partial derivatives of f at the given point in Exercises 21 through 28.

21. $f(x, y) = x^2 + y^2$, $(0, 1)$
22. $f(x, y) = \dfrac{1}{x^2 + y^2}$, $(-1, 1)$

23. $f(x, y) = y \cos x$, $(\pi, 1)$
24. $f(x, y) = \ln(x^2 + y^2)$, $(1, 0)$
25. $f(x, y) = \cos x + \sin y$, (π, π)
26. $f(x, y) = \cos x \sin y$, (π, π)
27. $f(x, y, z) = \sqrt{x^2 + y^2 + z^2}$, $(1, 1, 2)$
28. $f(x, y, z) = e^x \sin y + z^2$, $(0, \pi, 3)$

Show that $f_{xy} = f_{yx}$ for the functions described in Exercises 29 through 36.

29. $f(x, y) = x^3 + 2x^2y$
30. $f(x, y) = x^2 + \cos(x + y)$
31. $f(x, y) = e^{xy}$
32. $f(x, y) = ye^x$
33. $f(x, y) = e^y \cos \pi x$
34. $f(x, y) = \ln(x^2 + 1)$
35. $f(x, y, z) = 3xz + 4xy + 2yz$
36. $f(x, y, z) = x^2 + y^2 + z^2$

37. Find f_{xyx} and f_{xyy} if $f(x, y) = x^3 + e^{x^2 + y^2}$.

38. Find f_{xyz} and f_{zyx} if $f(x, y, z) = \cos(x^2 + y^2 + z^2)$.

39. Find $\dfrac{\partial^2 w}{\partial x^2}$ if $w = \sqrt{x^2 + y^2 + z^2}$.

40. Find $\dfrac{\partial^2 w}{\partial x \partial y}$ if $w = \sin xy$.

41. Find $\dfrac{\partial^3 w}{\partial x \partial y \partial z}$ if $w = \sin xyz$.

42. Find $f_x(0, 0)$ and $f_y(0, 0)$ if

$$f(x, y) = \begin{cases} \dfrac{xy^2}{x^2 + y^2}, & \text{for } (x, y) \neq (0, 0) \\ 0, & \text{for } (x, y) = (0, 0). \end{cases}$$

43. Find an equation of the line in the plane $y = 1$ that is tangent to the curve of intersection of this plane and the surface $z = x^2 + y^2$ at $(2, 1, 5)$.

44. Find an equation of the line in the plane $y = 1$ that is tangent to the curve of intersection of this plane and the surface $z = 4 - x^2 - y^2$ at $(1, 1, 2)$.

45. Consider the hyperbolic paraboloid described by

$$f(x, y) = y^2 - x^2.$$

(a) Determine the relative extrema of the curve of intersection of this surface and the plane $x = 0$.

(b) Determine the relative extrema of the curve of intersection of this surface and the plane $y = 0$.

[*Hint*: Apply the calculus of functions of one variable to the function described by $f(0, y)$ in (a) and by $f(x, 0)$ in (b).]

46. An equation of the form

$$\frac{\partial f}{\partial t} = c \frac{\partial^2 f}{\partial x^2}$$

is called a *diffusion equation*. The molecular concentration $C(x, t)$ of a fluid injected into a tube can be described by

$$C(x, t) = t^{-1/2} e^{-x^2/(Kt)}$$

where t denotes time, x denotes the distance from the point at which the fluid was injected and K is a positive constant. Show that C satisfies the diffusion equation with $c = K/4$.

47. An equation of the form

$$\frac{\partial^2 f}{\partial t^2} = \alpha^2 \frac{\partial^2 f}{\partial x^2}$$

is called a *one-dimensional wave equation*. The vibrations of an elastic string are described by such an equation. Show that if $f(x, t) = \sin 2(x + \alpha t)$, then f satisfies a wave equation.

48. Show that if $f(x, t) = \sin x \, [A \cos \alpha t + B \sin \alpha t]$ where A and B are arbitrary constants, then f satisfies the one-dimensional wave equation.

49. If T represents the temperature at a point with rectangular coordinates (x, y, z) at time t and no heat sources are present, then T satisfies the diffusion equation

$$\frac{\partial T}{\partial t} = h^2 \left(\frac{\partial^2 T}{\partial x^2} + \frac{\partial^2 T}{\partial y^2} + \frac{\partial^2 T}{\partial z^2} \right),$$

where h^2 is a physical constant called *thermal diffusivity*. Show that if

$$T = T(x, y, z, t) = e^{-t} [\cos x + \cos y + \cos z],$$

then T satisfies the heat equation with $h^2 = 1$. [In this setting, the diffusion equation is called a *heat equation*.]

50. When a fluid is injected into a tube containing a liquid flowing with constant velocity v, the molecular concentration is described by

$$C(x, t) = t^{-1/2} e^{-(x - vt)^2/(Kt)}.$$

Show that C satisfies the modified diffusion equation

$$\frac{\partial f}{\partial t} = c \frac{\partial^2 f}{\partial x^2} - v \frac{\partial f}{\partial x}$$

with $c = K/4$. (This type of equation describes the diffusion of a serum injected into the bloodstream.)

51. One of the simplest modifications of the one-mole ideal gas law $PV = RT$ is to add a term αP to the right side of the equation, where α is a function of temperature only. Suppose that a gas satisfies this law: $PV = RT + \alpha P$.

    Show that $\dfrac{\partial P}{\partial T} = \dfrac{P}{T} \left[1 + \dfrac{P}{R} \dfrac{\partial \alpha}{\partial T} \right]$.

52. Stokes' law for the velocity of a particle falling in a fluid is given by

    $v = \dfrac{2g(\rho - d)r^2}{9n}$, where g is the acceleration due to gravity, ρ is the density

    of the particle, d is the density of the liquid, r is the radius of the particle,

    and n is the absolute viscosity of the liquid. Find $\dfrac{\partial v}{\partial d}, \dfrac{\partial v}{\partial r}$, and $\dfrac{\partial v}{\partial n}$.

53. A formula, $S = .00718\ W^{.425}\ H^{.725}$ devised by Dubois and Dubois in 1916 relates the surface area S of a human in square meters, to the weight W in kilograms, and height H in centimeters. (This formula was also considered in Exercise 20 of Section 14.1.)

    (a) Determine the rate of change of surface area with respect to the change in weight when the person has a height of 2 meters and weight of 70 kilograms.

    (b) Determine the rate of change of the surface area with respect to the change in weight of a person of your own height and weight.

15.3
DIFFERENTIABILITY OF MULTIVARIATE FUNCTIONS

In Section 15.2 we discussed the partial derivatives of a function of two variables. Partial derivatives provide a measure of the relative change in the values of a function as a fixed point is approached along a line parallel to a coordinate axis. There are many other ways to approach a fixed point in the plane. In order to say that a function is differentiable, we need to consider the relative change in the values of the function regardless of how the point is approached.

Consider the function defined by

$$f(x, y) = \begin{cases} \dfrac{xy}{x^2 + y^2}, & \text{if } (x, y) \neq (0, 0) \\ 0 & , \text{if } (x, y) = (0, 0). \end{cases}$$

This function was considered as Example 4 of Section 15.1, where we saw that f is not continuous at $(0, 0)$. However, it was shown in Example 4 of Section 15.2 that the partial derivatives of f at $(0, 0)$ exist and are:

$$\frac{\partial f}{\partial x}(0, 0) = \lim_{x \to 0} \frac{f(x, 0) - f(0, 0)}{x} = 0$$

and

$$\frac{\partial f}{\partial y}(0, 0) = \lim_{y \to 0} \frac{f(0, y) - f(0, 0)}{y} = 0.$$

We would expect, by analogy with the single variable case, that differentiability of a function at a point would ensure that the function is continuous at the point. The preceding example implies that more must be required for differentiability than simply the existence of the partial derivatives.

To obtain a definition for the differentiability of a multivariate function, we first present an alternate characterization of differentiability for a function of a single variable. While this result is not important in the single variable case, it leads to an appropriate definition for the differentiability of a multivariate function.

A single variable function f is differentiable at a number x_0 if and only if there exists an open interval I about x_0 and a constant m and function ε with

$$\lim_{x \to x_0} \varepsilon(x) = 0$$

and

$$f(x) - f(x_0) = m(x - x_0) + \varepsilon(x)(x - x_0),$$

whenever x is in I. (This result is discussed in Exercise 49 of Section 3.2.)

(15.11)
DEFINITION

A function f of two variables is said to be **differentiable** at (x_0, y_0) if there exists a circle D about (x_0, y_0) and functions ε_1 and ε_2 with

(i) $\quad f(x, y) - f(x_0, y_0) = f_x(x_0, y_0)(x - x_0) + f_y(x_0, y_0)(y - y_0)$
$$+ \varepsilon_1(x, y)(x - x_0) + \varepsilon_2(x, y)(y - y_0)$$

and

(ii) $\quad \lim_{(x,y) \to (x_0, y_0)} \varepsilon_1(x, y) = 0, \qquad \lim_{(x,y) \to (x_0, y_0)} \varepsilon_2(x, y) = 0,$

whenever (x, y) is inside D.

EXAMPLE 1

Show that the function described by

$$f(x, y) = x^2 + y^2$$

is differentiable at every point in $\mathbb{R}^2$.

SOLUTION

To show that f is differentiable at (x_0, y_0), we must find functions ε_1 and ε_2 that satisfy (i) and (ii) in Definition 15.11. Since $f_x(x, y) = 2x$ and $f_y(x, y) = 2y$, (i) becomes:

$$(x^2 + y^2) - (x_0^2 + y_0^2) = 2x_0(x - x_0) + 2y_0(y - y_0)$$
$$+ \varepsilon_1(x, y)(x - x_0) + \varepsilon_2(x, y)(y - y_0).$$

Simplifying this, we obtain:

$$x^2 - 2x_0x + x_0^2 + y^2 - 2y_0y + y_0^2 = \varepsilon_1(x, y)(x - x_0) + \varepsilon_2(x, y)(y - y_0),$$

which factors to:

$$(x - x_0)^2 + (y - y_0)^2 = \varepsilon_1(x, y)(x - x_0) + \varepsilon_2(x, y)(y - y_0).$$

We can choose

$$\varepsilon_1(x, y) = x - x_0 \qquad \text{and} \qquad \varepsilon_2(x, y) = y - y_0$$

to satisfy (i) and (ii). Since it is necessary to find only one such pair of functions, f is differentiable at (x_0, y_0) and hence at every point in $\mathbb{R}^2$.   □

(15.12)
THEOREM

If a function f of two variables is differentiable at (x_0, y_0), then f is continuous at (x_0, y_0).

PROOF

From (i) in Definition 15.11:

$$f(x, y) - f(x_0, y_0) = f_x(x_0, y_0)(x - x_0) + f_y(x_0, y_0)(y - y_0)$$
$$+ \varepsilon_1(x, y)(x - x_0) + \varepsilon_2(x, y)(y - y_0).$$

Since both $\lim\limits_{(x,y) \to (x_0, y_0)} \varepsilon_1(x, y)$ and $\lim\limits_{(x,y) \to (x_0, y_0)} \varepsilon_2(x, y)$ exist, it is clear that

$$\lim\limits_{(x,y) \to (x_0, y_0)} [f(x, y) - f(x_0, y_0)] = 0$$

and

$$\lim\limits_{(x,y) \to (x_0, y_0)} f(x, y) = f(x_0, y_0).$$

This is precisely the condition required for continuity at (x_0, y_0).   □

The following important result gives sufficient conditions on the partial derivatives of a function to ensure differentiability.

(15.13)
THEOREM

If a function f of two variables has partial derivatives at each point of the interior of a circle D about (x_0, y_0) and if f_x and f_y are continuous at (x_0, y_0), then f is differentiable at (x_0, y_0).

PROOF

This proof is a nice application of the mean value theorem for functions of one variable (Theorem 3.34). To use the mean value theorem, we first note that for any point (x, y) in D we can write

(15.14) $f(x, y) - f(x_0, y_0) = f(x, y) - f(x_0, y) + f(x_0, y) - f(x_0, y_0).$

Consider the single-variable function that results from fixing y. Since f_x exists in D, it follows from the mean value theorem that there is a number $\hat{x}$ between x and x_0 such that

(15.15) $f(x, y) - f(x_0, y) = f_x(\hat{x}, y)(x - x_0).$

Similarly, by fixing x at x_0 and allowing y to vary, there is a number $\hat{y}$ between y and y_0 such that

(15.16) $f(x_0, y) - f(x_0, y_0) = f_y(x_0, \hat{y})(y - y_0).$

Substituting (15.15) and (15.16) into equation (15.14) produces

$$f(x, y) - f(x_0, y_0) = f_x(\hat{x}, y)(x - x_0) + f_y(x_0, \hat{y})(y - y_0),$$

which can be rewritten as:

$$\begin{aligned} f(x, y) - f(x_0, y_0) = {} & f_x(x_0, y_0)(x - x_0) + f_y(x_0, y_0)(y - y_0) \\ & + [f_x(\hat{x}, y) - f_x(x_0, y_0)](x - x_0) \\ & + [f_y(x_0, \hat{y}) - f_y(x_0, y_0)](y - y_0). \end{aligned}$$

Let $\varepsilon_1(x, y) = f_x(\hat{x}, y) - f_x(x_0, y_0)$

and $\varepsilon_2(x, y) = f_y(x_0, \hat{y}) - f_y(x_0, y_0).$

Since $\hat{x}$ is between x and x_0, $\hat{y}$ is between y and y_0, and the partial derivatives are continuous at (x_0, y_0),

$$\lim_{(x,y)\to(x_0,y_0)} f_x(\hat{x}, y) = f_x(x_0, y_0)$$

and

$$\lim_{(x,y)\to(x_0,y_0)} f_y(x_0, \hat{y}) = f_y(x_0, y_0).$$

Therefore,

$$\lim_{(x,y)\to(x_0,y_0)} \varepsilon_1(x, y) = 0, \quad \lim_{(x,y)\to(x_0,y_0)} \varepsilon_2(x, y) = 0$$

and f is differentiable at (x_0, y_0). $\square$

In light of the results in Theorems 15.12 and 15.13, let us reconsider the function defined by

$$f(x, y) = \begin{cases} \dfrac{xy}{x^2 + y^2} & , \text{ if } (x, y) \neq (0, 0) \\ 0 & , \text{ if } (x, y) = (0, 0). \end{cases}$$

We have shown in Example 4 of Section 15.1 that f is not continuous at $(0, 0)$. The contrapositive of Theorem 15.12:

> If a function f is not continuous at (x_0, y_0), then f is not differentiable at (x_0, y_0) implies that f is not differentiable at $(0, 0)$.

The notion of the differential of a multivariate function is defined in a manner similar to that in the single variable case.

(15.17)
DEFINITION

Suppose a function f of two variables is differentiable in the interior of a circle D about (x_0, y_0) that contains $(x_0 + \Delta x, y_0 + \Delta y)$. If we let $z = f(x, y)$, then
(i)   the **differential of x** is denoted dx and defined by: $dx = \Delta x$;
(ii) the **differential of y** is denoted dy and defined by: $dy = \Delta y$;
(iii) the **differential of z** is denoted dz and defined by:
$$dz = f_x(x_0, y_0)dx + f_y(x_0, y_0)dy.$$

The differential dz approximates the change in the value of f relative to changes dx and dy in the variables x and y.

EXAMPLE 2

Suppose $z = f(x, y) = x^3 + 2xy$. Find dz as (x, y) changes from $(1, 3)$ to $(1.01, 2.98)$ and use this to approximate $f(1.01, 2.98)$.

SOLUTION

The partial derivatives of f are

$$f_x(x, y) = 3x^2 + 2y \qquad \text{and} \qquad f_y(x, y) = 2x.$$

So the differentials for this problem are

$$dx = 1.01 - 1 = .01, \qquad dy = 2.98 - 3 = -.02$$

and
$$\begin{aligned} dz &= f_x(1, 3)dx + f_y(1, 3)dy \\ &= 9(.01) + 2(-.02) \\ &= .05. \end{aligned}$$

The approximate change is $dz = .05$, and an approximation to $f(1.01, 2.98)$ is $f(1, 3) + dz = 7 + dz = 7.05$. The exact value is $f(1.01, 2.98) = 7.049901$. □

EXAMPLE 3

Let $w = f(x, y, z) = x^2ye^{3z}$. Find the differential dw and use this to determine an approximation for $f(1.01, -1.03, .02)$ if $dx = .01$, $dy = -.03$ and $dz = .02$.

SOLUTION

The definition of the differential dw for a function of three variables is the extension of Definition 15.17 to a third variable:

$$dw = f_x dx + f_y dy + f_z dz.$$

Since $\qquad f_x(x, y, z) = 2xye^{3z}, \qquad f_y(x, y, z) = x^2e^{3z},$

and $\qquad\qquad\qquad f_z(x, y, z) = 3x^2ye^{3z},$

$$dw = f_x(1, -1, 0)dx + f_y(1, -1, 0)dy + f_z(1, -1, 0)dz$$
$$= -2(.01) + 1(-.03) - 3(.02)$$
$$= -.11.$$

Consequently,

$$f(1.01, -1.03, .02) \approx f(1, -1, 0) + dw = -1 - .11 = -1.11. \quad \square$$

EXAMPLE 4   A tin can is constructed in the form of a right circular cylinder with inside height 4 inches and inside radius 1.5 inches. Use differentials to approximate the amount of tin required to construct the can if the tin is .0012 inch thick and the can has both a top and a bottom.

SOLUTION

The amount of tin required is the difference between the volume of a cylinder with dimensions those of the outside of the can, and the volume of a cylinder with dimensions those of the inside of the can. This difference is approximated by dV when $dr = .0012$, the thickness of the side, and $dh = .0024$, the combined thickness of the top and bottom. Since the volume of a cylinder is given by

$$V(r, h) = \pi r^2 h,$$

$$\frac{\partial V}{\partial r} = 2\pi rh \qquad \text{and} \qquad \frac{\partial V}{\partial h} = \pi r^2.$$

So $\qquad\qquad\qquad\qquad dV = 2\pi rh\, dr + \pi r^2 dh$

and the tin required is approximately

$$dV = 2\pi(1.5)(4)(.0012) + \pi(1.5)^2\,(.0024)$$
$$\approx .062 \text{ in}^3. \qquad\qquad\qquad \square$$

EXERCISE SET 15.3

Use Theorem 15.13 to show that the functions in Exercises 1 through 6 are differentiable.

1. $f(x, y) = \sin(x + y)$ $\qquad\qquad$ **2.** $f(x, y) = \cos x - y^2$

3. $f(x, y) = \ln(1 + x^2 + y^2)$ $\qquad$ **4.** $f(x, y) = e^{x+y}$

5. $f(x, y) = ye^x$ $\qquad\qquad\qquad$ **6.** $f(x, y) = e^x \sin y$

Find dz in Exercises 7 through 12.

7. $z = x^2 - 3y^2$ $\qquad\qquad\qquad$ **8.** $z = e^{xy}$

9. $z = \dfrac{x}{y}$ $\qquad\qquad\qquad\qquad$ **10.** $z = \sqrt{x^2 + y^2}$

11. $z = y^2 \cos x$ $\qquad\qquad\qquad$ **12.** $z = \sin(x + 2y)$

Find dw in Exercises 13 through 18.

13. $w = xyz$

14. $w = 2xy + 4xz + yz$

15. $w = \cos xyz$

16. $w = \ln xy$

17. $w = xye^z$

18. $w = \dfrac{3x^2}{y} + e^z y^2$

In Exercises 19 through 24, find functions ε_1 and ε_2 that satisfy Definition 15.11, and show that f is differentiable.

19. $f(x, y) = 3 - x - y$

20. $f(x, y) = x^2$

21. $f(x, y) = x^2 - y^2$

22. $f(x, y) = 4 - x^2 - y^2$

23. $f(x, y) = xy$

24. $f(x, y) = x^2 y$

25. Suppose $z = f(x, y) = e^x \sin y + x^3$. Use differentials to approximate the change in the value of the function as (x, y) changes from $(1, \pi)$ to $(1.01, \pi + .01)$.

26. Suppose $z = f(x, y) = \ln(x^2 + y^2 + 1)$. Use differentials to approximate the change in the value of the function as (x, y) changes from $(1, 1)$ to $(1.02, .98)$.

27. Use differentials to approximate the error in the area measurement of a $12' \times 14'$ rectangular floor if an error of $.1''$ is made in measuring the length and the width.

28. If two capacitors with capacitance C_1 and C_2 are connected in series, then the capacitance of the resulting circuit is $C = \left(\dfrac{1}{C_1} + \dfrac{1}{C_2}\right)^{-1}$. Find the partial derivatives of C with respect to C_1 and C_2, and use these to approximate the change in capacitance that results from changing C_1 from 1 to 1.2 microfarads and C_2 from 3 to 2.8 microfarads.

29. If two resistors with resistance R_1 and R_2 are connected in parallel, then the resistance of the resulting circuit is

$$R = \left(\frac{1}{R_1} + \frac{1}{R_2}\right)^{-1}.$$

Find the partial derivatives of R with respect to R_1 and R_2, and use these to approximate the change in resistance that results from changing R_1 from 3 to 3.2 ohms and R_2 from 8 to 7.5 ohms.

30. A company that spends x thousand dollars on advertisement and y thousand dollars on research and development makes an annual profit of $f(x, y) = 3x + 4y + \dfrac{xy}{100} - \dfrac{x^2}{100} - \dfrac{y^2}{200}$ thousand dollars. Suppose the company currently spends $x_0 = \$500{,}000$ on advertising and $y_0 = \$200{,}000$ on research and development and has an additional \$20,000 to apply to one of these areas. Compute $f_x(x_0, y_0)$ and $f_y(x_0, y_0)$ and use these to determine to which area the additional \$20,000 should be applied.

15.4
THE CHAIN RULE

If g and f are differentiable functions with $y = g(x)$ and $z = f(y) = f(g(x))$, then the chain rule states that

$$\frac{dz}{dx} = \frac{dz}{dy}\frac{dy}{dx}.$$

In this section, we extend the chain rule to multivariate functions.

Suppose f is a function of the two variables x and y, each of which is a function of a third variable t. In effect, f is a function of the single variable t and under certain conditions is differentiable with respect to t. To motivate the chain rule for determining $D_t f(x(t), y(t))$, consider what happens when t changes. A change in t directly affects $x(t)$, which changes the value of f. Also, a change in t directly affects $y(t)$, which again changes the value of f. See Figure 15.8. The derivatives of x and y and the partial derivatives of f are used to quantitatively express this result.

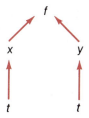

FIGURE 15.8

(15.18) **THEOREM**	Suppose x and y are differentiable functions of t and f is a differentiable function of x and y. Then f is a differentiable function of t and $$\frac{df}{dt}(x(t), y(t)) = f_x(x(t), y(t))\, x'(t) + f_y(x(t), y(t))\, y'(t).$$

PROOF

Let

$$\Delta x = x(t + \Delta t) - x(t) \qquad \text{and} \qquad \Delta y = y(t + \Delta t) - y(t).$$

Since f is differentiable, functions ε_1 and ε_2 exist with

$$\lim_{(\Delta x, \Delta y) \to (0,0)} \varepsilon_1(\Delta x, \Delta y) = \lim_{(\Delta x, \Delta y) \to (0,0)} \varepsilon_2(\Delta x, \Delta y) = 0$$

and

$$f(x(t + \Delta t), y(t + \Delta t)) - f(x(t),\ y(t)) = [f_x(x(t), y(t)) + \varepsilon_1(\Delta x, \Delta y)]\Delta x$$
$$+ [f_y(x(t), y(t)) + \varepsilon_2(\Delta x, \Delta y)]\Delta y.$$

Since x and y are differentiable functions of t, both are continuous so

$$\lim_{\Delta t \to 0} \Delta x = \lim_{\Delta t \to 0} [x(t + \Delta t) - x(t)] = 0$$

and

$$\lim_{\Delta t \to 0} \Delta y = \lim_{\Delta t \to 0} [y(t + \Delta t) - y(t)] = 0.$$

Thus,

$$\frac{df}{dt}(x(t), y(t)) = \lim_{\Delta t \to 0} \frac{f(x(t + \Delta t), y(t + \Delta t)) - f(x(t), y(t))}{\Delta t}$$

$$= \lim_{\Delta t \to 0} [f_x(x(t), y(t)) + \varepsilon_1(\Delta x, \Delta y)] \lim_{\Delta t \to 0} \frac{\Delta x}{\Delta t}$$

$$+ \lim_{\Delta t \to 0} [f_y(x(t), y(t)) + \varepsilon_2(\Delta x, \Delta y)] \lim_{\Delta t \to 0} \frac{\Delta y}{\Delta t}$$

$$= f_x(x(t), y(t)) \, x'(t) + f_y(x(t), y(t)) \, y'(t). \qquad \square$$

The equation in Theorem 15.18 is known as the chain rule, and is often expressed in Leibniz notation as

$$\frac{df}{dt} = \frac{\partial f}{\partial x}\frac{dx}{dt} + \frac{\partial f}{\partial y}\frac{dy}{dt}.$$

EXAMPLE 1   Let $f(x, y) = x^2 \sin y + e^x \cos y$, $x(t) = t^2 + 1$, and $y(t) = t^3$. Find $D_t f(x(t), y(t))$.

SOLUTION

$$D_t f(x(t), y(t)) = f_x(x(t), y(t)) \, x'(t) + f_y(x(t), y(t)) \, y'(t)$$
$$= (2x \sin y + e^x \cos y)(2t)$$
$$+ (x^2 \cos y - e^x \sin y)(3t^2). \qquad \square$$

The chain rule can be generalized to functions of more than two variables and also to the situation when the variables themselves are multivariate functions. The most frequently used results are listed below. In each case it is assumed that the required differentiability conditions are fulfilled. The result in Theorem 15.18 is repeated for ease of reference:

If f is a function of $x(t)$ and $y(t)$, then

(15.19)
$$\frac{df}{dt} = \frac{\partial f}{\partial x}\frac{dx}{dt} + \frac{\partial f}{\partial y}\frac{dy}{dt}.$$

If f is a function of $x(u, v)$ and $y(u, v)$, then

(15.20)
$$\frac{\partial f}{\partial u} = \frac{\partial f}{\partial x}\frac{\partial x}{\partial u} + \frac{\partial f}{\partial y}\frac{\partial y}{\partial u} \quad \text{and} \quad \frac{\partial f}{\partial v} = \frac{\partial f}{\partial x}\frac{\partial x}{\partial v} + \frac{\partial f}{\partial y}\frac{\partial y}{\partial v}.$$

If f is a function of $x(t)$, $y(t)$ and $z(t)$, then

(15.21)
$$\frac{df}{dt} = \frac{\partial f}{\partial x}\frac{dx}{dt} + \frac{\partial f}{\partial y}\frac{dy}{dt} + \frac{\partial f}{\partial z}\frac{dz}{dt}.$$

If f is a function of $x(u, v)$, $y(u, v)$ and $z(u, v)$, then

(15.22) $\dfrac{\partial f}{\partial u} = \dfrac{\partial f}{\partial x}\dfrac{\partial x}{\partial u} + \dfrac{\partial f}{\partial y}\dfrac{\partial y}{\partial u} + \dfrac{\partial f}{\partial z}\dfrac{\partial z}{\partial u}$ and $\dfrac{\partial f}{\partial v} = \dfrac{\partial f}{\partial x}\dfrac{\partial x}{\partial v} + \dfrac{\partial f}{\partial y}\dfrac{\partial y}{\partial v} + \dfrac{\partial f}{\partial z}\dfrac{\partial z}{\partial v}.$

EXAMPLE 2   If $f(x, y) = x^2 y + y^2 \tan x$, $x = u^2 - v^2$ and $y = u^2 + v^2$, find $\dfrac{\partial f}{\partial u}$ and $\dfrac{\partial f}{\partial v}$.

SOLUTION

$$\frac{\partial f}{\partial u} = \frac{\partial f}{\partial x}\frac{\partial x}{\partial u} + \frac{\partial f}{\partial y}\frac{\partial y}{\partial u}$$
$$= [2xy + y^2 (\sec x)^2]\, (2u) + (x^2 + 2y \tan x)\, (2u)$$

and

$$\frac{\partial f}{\partial v} = \frac{\partial f}{\partial x}\frac{\partial x}{\partial v} + \frac{\partial f}{\partial y}\frac{\partial y}{\partial v}$$
$$= [2xy + y^2 (\sec x)^2]\, (-2v) + (x^2 + 2y \tan x)\, (2v). \qquad \square$$

EXAMPLE 3   Suppose f, x, and y are as given in Example 2 and that, in addition, $u = \sin t$ and $v = \cos t$. Find $\dfrac{df}{dt}(\pi/4)$.

SOLUTION

The total dependence of f on the variable t can be seen from the diagram in Figure 15.9.

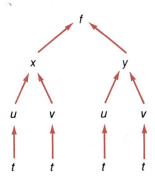

FIGURE 15.9

There are four "paths" from t to f, that is, four ways in which a change in t influences f. Each path contributes to the derivative of f with respect to t. Thus,

$$\frac{df}{dt} = \frac{\partial f}{\partial x}\frac{\partial x}{\partial u}\frac{du}{dt} + \frac{\partial f}{\partial x}\frac{\partial x}{\partial v}\frac{dv}{dt} + \frac{\partial f}{\partial y}\frac{\partial y}{\partial u}\frac{du}{dt} + \frac{\partial f}{\partial y}\frac{\partial y}{\partial v}\frac{dv}{dt}$$
$$= [2xy + y^2 (\sec x)^2](2u)(\cos t) + [2xy + y^2 (\sec x)^2](-2v)(-\sin t)$$
$$+ (x^2 + 2y \tan x)(2u)(\cos t) + (x^2 + 2y \tan x)(2v)(-\sin t).$$

When $t = \dfrac{\pi}{4}$, $u = \sin \dfrac{\pi}{4} = \dfrac{\sqrt{2}}{2}$, $v = \cos \dfrac{\pi}{4} = \dfrac{\sqrt{2}}{2}$, $x = u^2 - v^2 = 0$,

and $y = u^2 + v^2 = 1$.

So $\dfrac{df}{dt}\left(\dfrac{\pi}{4}\right) = [0 + (\sec 0)^2]\ (\sqrt{2})\left(\dfrac{\sqrt{2}}{2}\right) + [0 + (\sec 0)^2](-\sqrt{2})\left(-\dfrac{\sqrt{2}}{2}\right)$

$$+ (0 + 2\tan 0)(\sqrt{2})\left(\dfrac{\sqrt{2}}{2}\right) + (0 + 2\tan 0)(\sqrt{2})\left(-\dfrac{\sqrt{2}}{2}\right)$$

$$= (1)(1) + (1)(1) + 0 + 0 = 2. \qquad \square$$

EXAMPLE 4 Let $f(x, y, z) = xye^z$, $x = \cos t$, $y = \sin t$, and $z = \ln t$. Find $\dfrac{df}{dt}$ (a) by expressing $f(x, y, z)$ in terms of t before differentiating; and (b) by using the chain rule.

SOLUTION

(a) $$f(x(t), y(t), z(t)) = (\cos t)(\sin t)e^{\ln t}$$
$$= t\cos t \sin t.$$

So $$\dfrac{df}{dt} = \cos t \sin t - t(\sin t)^2 + t(\cos t)^2.$$

(b) Using the chain rule produces:

$$\dfrac{df}{dt} = \dfrac{\partial f}{\partial x}\dfrac{dx}{dt} + \dfrac{\partial f}{\partial y}\dfrac{dy}{dt} + \dfrac{\partial f}{\partial z}\dfrac{dz}{dt}$$

$$= ye^z(-\sin t) + xe^z(\cos t) + xye^z \cdot \dfrac{1}{t}.$$

To see that this is the same as the result obtained in (a), we must express x, y, and z in terms of t.

$$\dfrac{df}{dt} = -(\sin t)^2 e^{\ln t} + (\cos t)^2 e^{\ln t} + \sin t \cos t\, e^{\ln t} \cdot \dfrac{1}{t}$$

$$= -t(\sin t)^2 + t(\cos t)^2 + \sin t \cos t. \qquad \square$$

EXAMPLE 5 A conveyor belt is delivering iron ore pellets onto a conical pile. When the height of the pile is 60 ft and the radius of the pile is 40 ft, the height and radius are increasing at 4 ft/sec and 3 ft/sec, respectively. How fast is the conveyor belt delivering ore at this time?

SOLUTION

The conveyor belt is delivering ore at the rate at which the volume of the pile is increasing. If V denotes the volume, h the height, and r the radius, this rate is dV/dt when $h = 60$ and $r = 40$. The problem is concisely expressed as:

Know	Find
1. $V = \dfrac{1}{3}\pi r^2 h$	(a) $\dfrac{dV}{dt}$ when $h = 60$ ft, $r = 40$ ft.
2. $\dfrac{dh}{dt} = 4 \dfrac{\text{ft}}{\text{sec}}$, when $h = 60$ ft, $r = 40$ ft	
3. $\dfrac{dr}{dt} = 3 \dfrac{\text{ft}}{\text{sec}}$, when $h = 60$ ft, $r = 40$ ft	
4. $\dfrac{dV}{dt} = \dfrac{\partial V}{\partial h}\dfrac{dh}{dt} + \dfrac{\partial V}{\partial r}\dfrac{dr}{dt}$	

Since
$$\frac{\partial V}{\partial h} = \frac{1}{3}\pi r^2 \quad \text{and} \quad \frac{\partial V}{\partial r} = \frac{2}{3}\pi rh,$$

$$\frac{dV}{dt} = \frac{1}{3}\pi r^2 \frac{dh}{dt} + \frac{2}{3}\pi rh \frac{dr}{dt}.$$

When $h = 60$ and $r = 40$,

$$\frac{dV}{dt} = \frac{1}{3}\pi (40)^2(4) + \frac{2}{3}\pi (40)(60)(3)$$

$$= \frac{20800}{3}\pi \frac{\text{ft}^3}{\text{sec}}.$$

At this rate a standard railroad ore hopper car could be filled in about one minute. ☐

The chain rule for finding partial derivatives of functions of several variables can be used to expand the technique of implicit differentiation using multivariate functions. The following result forms the basis for this extension.

(15.23)
THEOREM

If $F(x, y) = 0$ implicitly defines y as a differentiable function of x, then
$$\frac{dy}{dx} = \frac{-F_x(x, y)}{F_y(x, y)},$$

provided that $F_y(x, y) \neq 0$.

PROOF

Let $z = F(x, y)$. Since $F(x, y) = 0$, the chain rule implies that
$$0 = \frac{dz}{dx} = \frac{\partial z}{\partial x}\frac{dx}{dx} + \frac{\partial z}{\partial y}\frac{dy}{dx}$$

so
$$0 = \frac{\partial z}{\partial x} + \frac{\partial z}{\partial y}\frac{dy}{dx},$$

and
$$\frac{dy}{dx} = -\frac{\dfrac{\partial z}{\partial x}}{\dfrac{\partial z}{\partial y}} = \frac{-F_x(x, y)}{F_y(x, y)}.$$
☐

EXAMPLE 6   Use Theorem 15.23 to find dy/dx if

$$x^3 + 3xy + y^5 + 4 = 0.$$

SOLUTION

Let $F(x, y) = x^3 + 3xy + y^5 + 4$,

then $F_x(x, y) = 3x^2 + 3y$,   and $F_y(x, y) = 3x + 5y^4$.

So $$\frac{dy}{dx} = \frac{-F_x(x, y)}{F_y(x, y)} = \frac{-(3x^2 + 3y)}{3x + 5y^4}.$$ ☐

EXERCISE SET 15.4

Find $\dfrac{df}{dt}$ in Exercises 1 through 12.

1. $f(x, y) = x^2y^3 + xy^2$, $x(t) = \sin t + t$, $y(t) = t^2 + 1$
2. $f(x, y) = \sqrt{2x - y^2}$, $x(t) = \ln t$, $y(t) = e^t$
3. $f(x, y) = 3x^3 - \sqrt{x + y}$, $x(t) = \ln t$, $y(t) = t^2$
4. $f(x, y) = e^x \sin y + x^2 \cos y$, $x(t) = t^3 + 1$, $y(t) = \sqrt{t}$
5. $f(x, y) = \ln xy + e^{x^2+y^2}$, $x(t) = \sqrt{t + 1}$, $y(t) = e^t$
6. $f(x, y) = e^{x^2} \ln y + e^y \ln x^2$, $x(t) = \sqrt{t}$, $y(t) = e^t$
7. $f(x, y, z) = x^2y + yz^2$, $x(t) = \ln t$, $y(t) = e^t$, $z(t) = t^2$
8. $f(x, y, z) = x^2y^4z^5$, $x(t) = e^t + t^2$, $y(t) = \ln (t^2 + 1)$, $z(t) = t^3 + t$
9. $f(x, y, z) = e^z \cos xy$, $x(t) = \sin t$, $y(t) = t^2 + 1$, $z(t) = e^{t^2}$
10. $f(x, y, z) = \ln (x^2 - 3y^4 + z^3)$, $x(t) = \sqrt{t}$, $y(t) = t^{-2}$, $z(t) = \cos t$
11. $f(x, y, z) = \sin xyz^2$, $x(t) = e^t$, $y(t) = \ln t$, $z(t) = t + \ln t$
12. $f(x, y, z) = \sqrt{x^2 + y^2 + z^2}$, $x(t) = \sin t$, $y(t) = \cos 2t$, $z(t) = \sin 2t$

Find $\dfrac{\partial f}{\partial u}$ and $\dfrac{\partial f}{\partial v}$ in Exercises 13 through 20.

13. $f(x, y) = x^3y^5$, $x(u, v) = u + v$, $y(u, v) = u^2 - v^2$
14. $f(x, y) = 2xy + x^3$, $x(u, v) = v \ln u$, $y(u, v) = e^{u+v}$
15. $f(x, y) = e^x \cos y$, $x(u, v) = ue^v$, $y(u, v) = u + v - 1$
16. $f(x, y) = e^{xy}(\cos xy + \sin xy)$, $x(u, v) = e^{u+v}$, $y(u, v) = uv$
17. $f(x, y) = \tan (x + y)$, $x(u, v) = e^u \cos v$, $y(u, v) = e^u \sin v$
18. $f(x, y) = e^{x^2+y^2}$, $x(u, v) = ve^u$, $y(u, v) = v \ln u$
19. $f(x, y, z) = \ln (x + y + z)$, $x(u, v) = e^{u^2+v^2}$, $y(u, v) = \ln (u^2 + v^2)$, $z(u, v) = u - v$
20. $f(x, y, z) = \dfrac{1}{xyz}$, $x(u, v) = u^2v$, $y(u, v) = u + v$, $z(u, v) = \sqrt{u^2 + v^2}$

In Exercises 21 through 26, use Theorem 15.23 to find $\dfrac{dy}{dx}$.

21. $x^2y^2 - xy + x^3 + y^3 + 6 = 0$

22. $x^5 + 2x^4y + x^2y^3 + y^5 = 0$

23. $x^2y + xy^2 = 6(x^2 + y^2)$

24. $y^2 = \dfrac{x}{xy + 1}$

25. $\sin xy = \cos x$

26. $\sin x + \cos y = 1$

27. Show that if $z = f(x, y)$, $x = r \cos \theta$, and $y = r \sin \theta$, then

(i) $\dfrac{\partial z}{\partial r} = \dfrac{\partial f}{\partial x} \cos \theta + \dfrac{\partial f}{\partial y} \sin \theta$

and

(ii) $\dfrac{\partial z}{\partial \theta} = r\left[\dfrac{-\partial f}{\partial x} \sin \theta + \dfrac{\partial f}{\partial y} \cos \theta\right]$.

28. Show that if $z = f(x - y)$, then

$$\frac{\partial z}{\partial x} + \frac{\partial z}{\partial y} = 0.$$

29. Show that if

$$z = g(y - \alpha x) + f(y + \alpha x),$$

then z satisfies the one-dimensional wave equation

$$\frac{\partial^2 z}{\partial x^2} = \alpha^2 \frac{\partial^2 z}{\partial y^2}.$$

30. The length, width, and depth of a rectangular box are each increasing at the rate of 1 inch per minute. Find the rate at which the volume is increasing when the length is 18 inches, the width 12 inches, and the depth 3 inches.

31. Oil spilling from a tanker into a large lake forms an inverted conical shape. If the oil is spilling at 12π ft^3/min and the radius of the spill is increasing at 1 ft/min, at what rate is the center depth of the spill increasing when the radius is 6 feet and the volume of oil spilled is 24π ft^3?

32. The class of Cobb-Douglass functions for describing the production of a firm's output is based on two variables: x describing property resources (capital goods and natural resources) and y describing human resources (labor and managerial potential). These functions have the form $f(x, y) = a\, x^k\, y^{1-k}$ for a positive constant a and a constant $0 < k < 1$. Show that any Cobb-Douglass function satisfies the equation

$$f(x, y) = xf_x(x, y) + yf_y(x, y).$$

33. Sketch the graph of the x and y values that are possible if the Cobb-Douglass function described by $f(x, y) = 4\, x^{1/3}\, y^{2/3}$ is to assume the value $f(x, y) = 16$. Find the partial derivatives of f when $x = 1$ and $y = 8$ and describe their practical significance.

15.5
DIRECTIONAL DERIVATIVES AND GRADIENTS

Partial derivatives give the rate of change of the values of a function relative to a variable change along a line parallel to one of the coordinate axes. The directional derivative gives the rate of change of the function values along lines not necessarily parallel to a coordinate axis.

(15.24)
DEFINITION

If $\mathbf{u} = u_1\mathbf{i} + u_2\mathbf{j}$ is a unit vector, the **directional derivative** of f at (x, y) in the direction of $\mathbf{u}$ is

$$D_{\mathbf{u}} f(x, y) = \lim_{h \to 0} \frac{f(x + hu_1, y + hu_2) - f(x, y)}{h},$$

provided this limit exists.

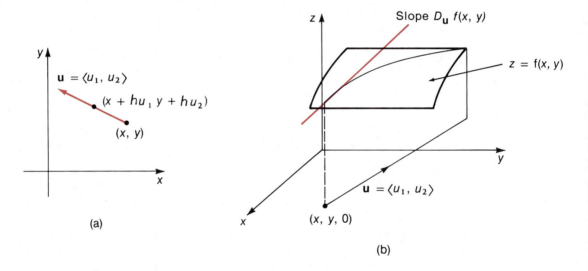

FIGURE 15.10

The directional derivative $D_{\mathbf{u}}(x, y)$ gives the slope of the tangent line to the curve of intersection of the plane and surface illustrated in Figure 15.10.

When $\mathbf{u} = \mathbf{i}$, the directional derivative in the direction of $\mathbf{i}$ is the partial derivative of f with respect to x:

$$D_{\mathbf{i}} f(x, y) = \lim_{h \to 0} \frac{f(x + h, y) - f(x, y)}{h} = f_x(x, y).$$

Similarly, when $\mathbf{u} = \mathbf{j}$,

$$D_{\mathbf{j}} f(x, y) = \lim_{h \to 0} \frac{f(x, y + h) - f(x, y)}{h} = f_y(x, y).$$

Since it is generally cumbersome to use the limit to determine derivatives, the natural question to ask is whether there is an alternative way to find directional derivatives. The following theorem shows that whenever a function is differentiable, the partial derivatives can be used to determine its directional derivatives.

(15.25)
THEOREM

If f is differentiable at (x, y) and $\mathbf{u} = u_1\mathbf{i} + u_2\mathbf{j}$ is a unit vector, then

$$D_{\mathbf{u}}f(x, y) = f_x(x, y)u_1 + f_y(x, y)u_2 = \langle f_x(x, y), f_y(x, y)\rangle \cdot \mathbf{u}.$$

PROOF

To simplify notation, we introduce a single variable function g defined, for fixed values of x, y, u_1, and u_2, by

$$g(h) = f(x + hu_1, y + hu_2).$$

Then

$$D_{\mathbf{u}}f(x, y) = \lim_{h \to 0} \frac{f(x + hu_1, y + hu_2) - f(x, y)}{h}$$

$$= \lim_{h \to 0} \frac{g(h) - g(0)}{h}$$

$$= g'(0).$$

With $u = x + hu_1$ and $v = y + hu_2$, the chain rule implies that

$$g'(h) = \frac{dg}{dh} = \frac{\partial f}{\partial u}\frac{du}{dh} + \frac{\partial f}{\partial v}\frac{dv}{dh} = \frac{\partial f}{\partial u}u_1 + \frac{\partial f}{\partial v}u_2.$$

When $h = 0$, $u = x$ and $v = y$, so

$$D_{\mathbf{u}}f(x, y) = g'(0) = \frac{\partial f}{\partial x}u_1 + \frac{\partial f}{\partial y}u_2 = f_x(x, y)u_1 + f_y(x, y)u_2. \quad \square$$

It follows from Theorem 15.25 that if a function f is differentiable at (x, y), then f has a directional derivative at (x, y) in every direction.

EXAMPLE 1

Let $f(x, y) = 2x^2 - y^2 - 1$ and $\mathbf{v} = 3\mathbf{i} + 4\mathbf{j}$. Find the directional derivative of f at $(2, 1)$ in the direction of $\mathbf{v}$.

SOLUTION

Since $\mathbf{v}$ is not a unit vector, the first task is to find a unit vector in the direction of $\mathbf{v}$:

$$\mathbf{u} = \frac{\mathbf{v}}{\|\mathbf{v}\|} = \frac{3\mathbf{i} + 4\mathbf{j}}{\sqrt{3^2 + 4^2}} = \frac{3}{5}\mathbf{i} + \frac{4}{5}\mathbf{j}.$$

Since

$$f_x(x, y) = 4x \quad \text{and} \quad f_y(x, y) = -2y,$$

$$D_{\mathbf{u}}(x, y) = 4x\left(\frac{3}{5}\right) - 2y\left(\frac{4}{5}\right) = \frac{4}{5}(3x - 2y),$$

and

$$D_{\mathbf{u}}f(2, 1) = \frac{4}{5}(3 \cdot 2 - 2 \cdot 1) = \frac{16}{5}. \quad \square$$

If θ represents the angle formed by a vector $\mathbf{v}$ and the unit vector $\mathbf{i}$, then $\mathbf{v} = \|\mathbf{v}\| \, (\mathbf{i} \cos \theta + \mathbf{j} \sin \theta)$. (See Figure 15.11.) Consequently, the unit vector in the direction of $\mathbf{v}$ is

$$\mathbf{u} = \frac{\mathbf{v}}{\|\mathbf{v}\|} = \mathbf{i} \cos \theta + \mathbf{j} \sin \theta$$

and by Theorem 15.25 the directional derivative of a function f at (x, y) in the direction of $\mathbf{v}$ is

(15.26) $D_{\mathbf{u}} f(x, y) = f_x(x, y) \cos \theta + f_y(x, y) \sin \theta.$

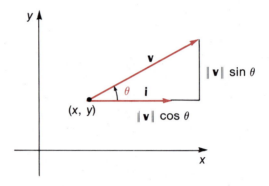

FIGURE 15.11

This representation is more useful theoretically than computationally, since $\cos \theta$ and $\sin \theta$ are not always readily available.

The directional derivative $D_{\mathbf{u}} f(x, y)$ describes the rate of change in the function values at (x, y) in the direction of the unit vector $\mathbf{u}$. Not surprisingly, we are interested in determining which direction produces the maximum and minimum rates of change. To study this problem we need to introduce some additional terminology.

(15.27)
DEFINITION

If f is a function with partial derivatives at (x, y), then the **gradient** of f at (x, y), denoted grad $f(x, y)$ or $\nabla f(x, y)$, is defined by

$$\text{grad } f(x, y) \equiv \nabla f(x, y) = f_x(x, y)\mathbf{i} + f_y(x, y)\mathbf{j}.$$

EXAMPLE 2   Find the gradient of f if $f(x, y) = e^x \sin y$.

SOLUTION

$$\nabla f(x, y) = f_x(x, y)\mathbf{i} + f_y(x, y)\mathbf{j}$$
$$= e^x \sin y \, \mathbf{i} + e^x \cos y \, \mathbf{j}. \qquad \square$$

The gradient is a function that transforms a scalar multivariate function into a vector multivariate function. One of the immediate applications of the gradient is that, for a differentiable function f and a unit vector $\mathbf{u}$,

(15.28) $D_{\mathbf{u}}f(x, y) = f_x(x, y)u_1 + f_y(x, y)u_2 = \nabla f(x, y) \cdot \mathbf{u}.$

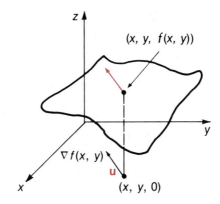

FIGURE 15.12

This important relation is illustrated in Figure 15.12, and is the basis for the following theorem.

(15.29)
THEOREM

If f is differentiable at (x, y), then the maximum value of $D_{\mathbf{u}}f(x, y)$ is $\| \nabla f(x, y)\|$ and occurs when $\mathbf{u}$ is parallel to $\nabla f(x, y)$.

PROOF

For any unit vector $\mathbf{u} = u_1\mathbf{i} + u_2\mathbf{j}$,

$$
\begin{aligned}
D_{\mathbf{u}}f(x, y) &= f_x(x, y)u_1 + f_y(x, y)u_2 \\
&= \nabla f(x, y) \cdot \mathbf{u} \\
&= \|\nabla f(x, y)\| \, \|\mathbf{u}\| \cos \theta \\
&= \|\nabla f(x, y)\| \cos \theta,
\end{aligned}
$$

where θ is the angle between $\mathbf{u}$ and $\nabla f(x, y)$. Hence, the maximum value of $D_{\mathbf{u}}f(x, y)$ is $\|\nabla f(x, y\|$. This occurs when $\cos \theta = 1$, that is, when $\mathbf{u}$ is parallel to $\nabla f(x, y)$.   □

Since $\nabla f(x, y)$ describes the direction and magnitude of maximum rate of change in the values of f at (x, y), it also describes the direction of maximum slope of the surface $z = f(x, y)$ at $(x, y, f(x, y))$. It is this application that gives the gradient its name: the maximum slope of a surface at a point is called the *grade* of the surface at the point.

EXAMPLE 3

If $f(x, y) = x^2 + y^2 - 1$, in what direction is f increasing most rapidly at $(2, 1)$ and what is this maximum rate of increase?

SOLUTION

The graph of $f(x, y)$ is sketched in Figure 15.13. Since $f_x(x, y) = 2x$ and $f_y(x, y) = 2y$, $\nabla f(x, y) = 2x\mathbf{i} + 2y\mathbf{j}$.

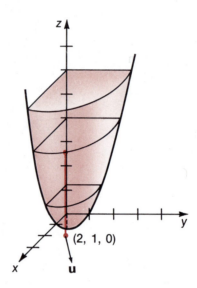

FIGURE 15.13

The direction in which f is increasing most rapidly at $(2, 1)$ is given by the vector

$$\nabla f(2, 1) = 4\mathbf{i} + 2\mathbf{j}.$$

The maximum rate of increase is

$$\|\nabla f(2, 1)\| = \sqrt{20} = 2\sqrt{5}. \qquad \square$$

Directional derivatives and gradients are defined for functions of three variables in a manner similar to the definitions for two variable functions:

If $\mathbf{u} = u_1\mathbf{i} + u_2\mathbf{j} + u_3\mathbf{k}$ is a unit vector, the **directional derivative** of f at (x, y, z) in the direction of $\mathbf{u}$ is

$$D_{\mathbf{u}} f(x, y, z) = \lim_{h \to 0} \frac{f(x + hu_1, y + hu_2, z + hu_3) - f(x, y, z)}{h},$$

provided this limit exists.

If f is differentiable at (x, y, z), then

$$D_{\mathbf{u}} f(x, y, z) = f_x(x, y, z)u_1 + f_y(x, y, z)u_2 + f_z(x, y, z)u_3.$$

The **gradient** of f is defined by

(15.30) $\operatorname{grad} f(x, y, z) \equiv \nabla f(x, y, z) = f_x(x, y, z)\mathbf{i} + f_y(x, y, z)\mathbf{j} + f_z(x, y, z)\mathbf{k},$

so

(15.31) $D_{\mathbf{u}} f(x, y, z) = \nabla f(x, y, z) \cdot \mathbf{u}.$

This leads to a theorem for functions of three variables that is equivalent to the two-variable counterpart, Theorem 15.29.

(15.32)
THEOREM   If f is differentiable at (x, y, z), then the maximum value of $D_{\mathbf{u}} f(x, y, z)$ is $\|\nabla f(x, y, z)\|$ and occurs when $\mathbf{u}$ is parallel to $\nabla f(x, y, z)$.

EXAMPLE 4   The temperature distribution in a solid metal ball of radius 3 inches centered at $(0, 0, 0)$ is given by $T(x, y, z) = 300e^{-(x^2 + y^2 + z^2)/9}$ degrees Celsius. Determine the direction and maximum rate of decrease in temperature at $(1, 1, 0)$.

SOLUTION

$$\nabla T(x, y, z) = 300e^{-(x^2+y^2+z^2)/9} \left(-\frac{1}{9}\right) (2x\mathbf{i} + 2y\mathbf{j} + 2z\mathbf{k})$$

$$= -\frac{200}{3} e^{-(x^2+y^2+z^2)/9} (x\mathbf{i} + y\mathbf{j} + z\mathbf{k}),$$

so the direction of maximum rate of decrease in temperature at $(1, 1, 0)$ is

$$-\nabla T(1, 1, 0) = \frac{200}{3} e^{-2/9} (\mathbf{i} + \mathbf{j}).$$

The rate of decrease in this direction is

$$\|\nabla T(1, 1, 0)\| = \frac{200\sqrt{2}}{3} e^{-2/9} \approx 75.5 \frac{\text{degrees Celsius}}{\text{inch}}.\qquad \square$$

EXERCISE SET 15.5

In Exercises 1 through 10, find the directional derivative of f at P in the direction of $\mathbf{v}$.

1. $f(x, y) = 6 - 2x - 3y, \mathbf{v} = \mathbf{i} + \mathbf{j}, P(0, 0)$

2. $f(x, y) = x^2 - xy + y^2, \mathbf{v} = \mathbf{i} - \mathbf{j}, P(1, 2)$

3. $f(x, y) = x^2 + y^2, \mathbf{v} = \mathbf{i} + \mathbf{j}, P(1, 1)$

4. $f(x, y) = 4 - x^2 - y^2, \mathbf{v} = 2\mathbf{i} + \mathbf{j}, P(-1, -1)$

5. $f(x, y) = \sqrt{36 - 9x^2 - 4y^2}, \mathbf{v} = -\mathbf{i} + \mathbf{j}, P(1, -2)$

6. $f(x, y) = xe^y, \mathbf{v} = 3\mathbf{i} + \mathbf{j}, P(2, 1)$

7. $f(x, y, z) = 2x + y + 3z, \mathbf{v} = -\mathbf{i} + \mathbf{j} + 2\mathbf{k}, P(1, 2, 3)$

8. $f(x, y, z) = ze^{x^2+y^2}, \mathbf{v} = \mathbf{i} + \mathbf{j} + \mathbf{k}, P(1, 1, -3)$

9. $f(x, y, z) = \cos(x + y + z), \mathbf{v} = \mathbf{i} - 2\mathbf{j} + 2\mathbf{k}, P\left(0, \frac{\pi}{4}, \frac{\pi}{2}\right)$

10. $f(x, y, z) = \dfrac{x + y}{x + z}, \mathbf{v} = 2\mathbf{i} + \mathbf{j} - 3\mathbf{k}, P(1, 3, 1)$

In Exercises 11 through 18, find the gradient of the function.

11. $f(x, y) = x^2 - y^2$ **12.** $f(x, y) = \sin x + \cos y$

13. $f(x, y) = ye^x$ **14.** $f(x, y) = ye^x \sin xy$

15. $f(x, y) = \dfrac{xy}{x^2 + y^2}$ **16.** $f(x, y) = \dfrac{x}{y}$

17. $f(x, y, z) = \sin x^2 yz$ **18.** $f(x, y, z) = \ln(x^2 + 2y - z^3)$

In Exercises 19 through 24, find the gradient of the function at the given point.

19. $f(x, y) = x^2 + y^2$, $(1, 2)$ **20.** $f(x, y) = \dfrac{2x - y}{xy + 1}$, $(1, 2)$

21. $f(x, y) = e^x \cos y$, $(1, \pi)$ **22.** $f(x, y) = e^{xy} \cos xy$, $\left(2, \dfrac{\pi}{4}\right)$

23. $f(x, y, z) = \cos xyz$, $\left(1, 2, \dfrac{\pi}{4}\right)$ **24.** $f(x, y, z) = \dfrac{2x}{y} \ln z$, $(2, 2, 2)$

In Exercises 25 through 32, find the direction in which f is increasing most rapidly at P and this maximum rate of increase.

25. $f(x, y) = x^2 + y^2 + 1$, $P(1, 1)$ **26.** $f(x, y) = y - x^2$, $P(0, 0)$

27. $f(x, y) = 9x^2 + 4y^2$, $P(2, 0)$ **28.** $f(x, y) = \sin y$, $P(0, \pi)$

29. $f(x, y) = e^{x^2 + y^2}$, $P\left(\dfrac{\sqrt{2}}{2}, \dfrac{\sqrt{2}}{2}\right)$ **30.** $f(x, y) = e^x \cos xy$, $P(1, \pi)$

31. $f(x, y, z) = e^x (\cos y + \sin z)$, $P\left(1, \pi, \dfrac{\pi}{2}\right)$

32. $f(x, y, z) = \ln(x^2 + y^2 + z)$, $P(1, -1, 1)$

In Exercises 33 through 36, find the directional derivative of f at P in the direction of a vector $\mathbf{v}$ that makes an angle θ with the positive x-axis.

33. $f(x, y) = 9 - 4x^2 - y^2$, $\theta = \dfrac{\pi}{6}$, $P(-1, -2)$

34. $f(x, y) = e^x \sin y$, $\theta = \dfrac{\pi}{4}$, $P\left(1, \dfrac{\pi}{2}\right)$

35. $f(x, y) = x^2 \ln xy$, $\theta = -\dfrac{\pi}{3}$, $P(e, 1)$

36. $f(x, y) = 2 - x - y$, $\theta = \dfrac{\pi}{3}$, $P(0, 1)$

In Exercises 37 through 42, find the directional derivative of the function f at the point P in the direction from P to Q.

37. $f(x, y) = x \cos y$, $P\left(2, \dfrac{\pi}{2}\right)$, $Q\left(3, \dfrac{\pi}{4}\right)$

38. $f(x, y) = ye^{x^2 + y^2}$, $P(-1, 1)$, $Q(1, 3)$

39. $f(x, y) = \dfrac{x^2}{y}$, $P(2, 1)$, $Q(-2, -1)$

40. $f(x, y) = 4x^2 + 9y^2$, $P(-1, -1)$, $Q(-2, -4)$

41. $f(x, y, z) = x^2 + y^2 + z^2$, $P(0, 1, 1)$, $Q(2, -1, 3)$

42. $f(x, y, z) = \sqrt{x}\, e^{yz}$, $P(1, 2, 3)$, $Q(3, 6, -1)$

43. Suppose the height of a mountain is given by $z = 1000 - 4x^2 - y^2$ feet, where (x, y) is a point in the plane at the base of the mountain. If a climber is at $(4, 4, 920)$, in what direction should the climber turn to

(a) ascend the mountain most rapidly?

(b) descend the mountain most rapidly?

44. The temperature at a point (x, y) on a rectangular metal plate in the xy-plane is $T(x, y) = \dfrac{500}{x^2 + y^2}$ degrees Celsius.

(a) Find the rate of change of T at $(3, 4)$ in the direction of the vector $\mathbf{i}$ and in the direction of the vector $\mathbf{i} + \mathbf{j}$.

(b) Find the direction in which T increases most rapidly at $(3, 4)$.

(c) Find the direction in which T decreases most rapidly at $(3, 4)$.

45. The temperature within a solid is given by $T(x, y, z) = x^2 + 3xy + z^2$. An object moves in a straight line from $(1, 2, 3)$ to $(-1, 1, 4)$.

(a) What rate of change of temperature is initially encountered by the object?

(b) What rate of change is finally encountered by the object?

46. Modify the definition of the directional derivative of a function to include vectors that are not unit vectors. State a theorem concerning these more general derivatives that reduces to Theorem 15.25 when the vector is a unit vector.

15.6
TANGENT PLANES AND NORMALS

This section contains an application of the gradient that can be used to describe another important geometric property of a surface. We found in Section 15.5 that if a surface S is described by $z = f(x, y)$, then $\nabla f(x, y)$ gives the direction and maximum rate of change of f at (x, y). As such, $\nabla f(x, y)$ describes the grade, or maximum slope of the surface S at $(x, y, f(x, y))$.

A surface in space can also be described as a level surface of a function of three variables. For example, the surface S described by $z = f(x, y)$ is the zero-level surface of

$$F(x, y, z) = z - f(x, y)$$

or the one-level surface of

$$G(x, y, z) = z + 1 - f(x, y).$$

Functions of three variables can, in fact, represent more general surfaces than functions of two variables. The unit sphere with equation $x^2 + y^2 + z^2 = 1$ cannot be represented in the form $z = f(x, y)$, but can easily be described as the one-level surface of

$$F(x, y, z) = x^2 + y^2 + z^2.$$

It is the level surface representation of a surface that we now consider.

(15.33)
THEOREM
Suppose F has continuous partial derivatives and S is a level surface described by

$$F(x, y, z) = c, \text{ where } c \text{ is a constant.}$$

If $P_0\,(x_0, y_0, z_0)$ lies on S and $\nabla F(x_0, y_0, z_0) \neq \mathbf{0}$, then $\nabla F(x_0, y_0, z_0)$ is normal to every curve on S passing through P_0.

PROOF

Suppose C is an arbitrary smooth curve lying entirely on S and passing through P_0, as shown in Figure 15.14.

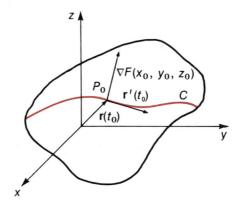

FIGURE 15.14

Let

$$\mathbf{r}(t) = x(t)\mathbf{i} + y(t)\mathbf{j} + z(t)\mathbf{k}, \quad a \le t \le b$$

be a vector-valued function describing C and $\mathbf{r}(t_0) = x_0\mathbf{i} + y_0\mathbf{j} + z_0\mathbf{k}$. The vector $\mathbf{r}'(t_0)$ is tangent to C at P_0, and

$$
\begin{aligned}
\mathbf{r}'(t_0) \cdot \nabla F(x_0, y_0, z_0) &= [x'(t_0)\mathbf{i} + y'(t_0)\mathbf{j} + z'(t_0)\mathbf{k}] \cdot [F_x(x_0, y_0, z_0)\mathbf{i} \\
&\quad + F_y(x_0, y_0, z_0)\mathbf{j} + F_z(x_0, y_0, z_0)\mathbf{k}] \\
&= F_x(x_0, y_0, z_0)x'(t_0) + F_y(x_0, y_0, z_0)y'(t_0) \\
&\quad + F_z(x_0, y_0, z_0)z'(t_0) \\
&= \frac{dF}{dt}(x_0, y_0, z_0).
\end{aligned}
$$

However, F is constant on S, so $\dfrac{dF}{dt}(x_0, y_0, z_0) = 0$. This implies that

$$\mathbf{r}'(t_0) \cdot \nabla F(x_0, y_0, z_0) = 0$$

and that $\nabla F(x_0, y_0, z_0)$ is orthogonal to $\mathbf{r}'(t_0)$. Since C is an arbitrary curve on S passing through $P_0(x_0, y_0, z_0)$, the vector $\nabla F(x_0, y_0, z_0)$ is orthogonal to the tangent vector to any curve in S. Hence $\nabla F(x_0, y_0, z_0)$ is normal to every curve in S passing through P_0. $\square$

Under the conditions of Theorem 15.33, all tangent vectors to the surface S at P_0 lie in the same plane. This plane is called the **tangent plane** to S at P_0. A normal to this plane is said to be **normal** to the surface at P_0. Since a normal to the tangent plane to $F(x, y, z) = c$ at $P_0(x_0, y_0, z_0)$ is $\nabla F(x_0, y_0, z_0) = F_x(x_0, y_0, z_0)\mathbf{i} + F_y(x_0, y_0, z_0)\mathbf{j} + F_z(x_0, y_0, z_0)\mathbf{k}$, an equation of this tangent plane is

(15.34)

$$F_x(x_0, y_0, z_0)(x - x_0) + F_y(x_0, y_0, z_0)(y - y_0) + F_z(x_0, y_0, z_0)(z - z_0) = 0.$$

EXAMPLE 1   Find an equation of the plane tangent to the paraboloid $x^2 + y^2 = z - 3$ at $(1, 1, 5)$.

SOLUTION

If F is defined by $F(x, y, z) = z - x^2 - y^2$, then the paraboloid is described by the level surface $F(x, y, z) = 3$. Since

$$\nabla F(x, y, z) = -2x\mathbf{i} - 2y\mathbf{j} + \mathbf{k},$$

a normal vector to the paraboloid at $(1, 1, 5)$ is

$$\nabla F(1, 1, 5) = -2\mathbf{i} - 2\mathbf{j} + \mathbf{k}.$$

An equation of the tangent plane is

$$-2(x - 1) - 2(y - 1) + (z - 5) = 0$$

or $$2x + 2y - z = -1.$$

This plane is shown in Figure 15.15.   □

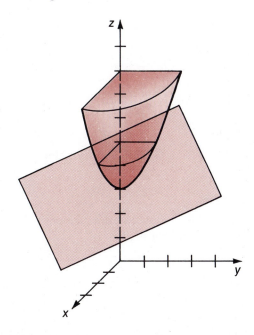

FIGURE 15.15

EXAMPLE 2   Find an equation of the line normal to the surface $z = e^{xy}$ at $(1, 1, e)$.

SOLUTION

If F is defined by $F(x, y, z) = z - e^{xy}$, then the surface is the level surface $F(x, y, z) = 0$. Since $\nabla F(x, y, z) = -ye^{xy}\mathbf{i} - xe^{xy}\mathbf{j} + \mathbf{k}$, the normal line to the surface at $(1, 1, e)$ has direction given by $\langle -e, -e, 1 \rangle$. An equation for the line through $(1, 1, e)$ with direction $\langle -e, -e, 1 \rangle$ is given in parametric form by

$$x = 1 - et, \qquad y = 1 - et, \qquad z = e + t$$

or in symmetric form by

$$\frac{x - 1}{-e} = \frac{y - 1}{-e} = \frac{z - e}{1}.$$

This line is shown in Figure 15.16.

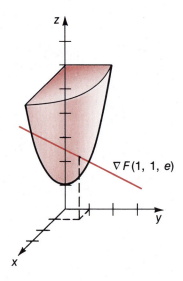

$\nabla F(1, 1, e)$

FIGURE 15.16

EXAMPLE 3   Show that a normal line to a sphere must pass through the center of the sphere.

SOLUTION

It suffices to show that the statement is true for spheres centered at the origin, spheres with equation

$$x^2 + y^2 + z^2 = r^2.$$

Let $F(x, y, z) = x^2 + y^2 + z^2$. Then $F(x, y, z) = r^2$ describes the sphere and

$$\nabla F(x, y, z) = 2x\mathbf{i} + 2y\mathbf{j} + 2z\mathbf{k}$$

is normal to the sphere at (x, y, z). Consequently, the normal line to the sphere at a point (x_0, y_0, z_0) on the sphere has direction given by $\langle x_0, y_0, z_0 \rangle$. Symmetric equations for the normal line are

$$\frac{x - x_0}{x_0} = \frac{y - y_0}{y_0} = \frac{z - z_0}{z_0},$$

and the point $(0, 0, 0)$, the center of the sphere, lies on this line. $\quad\square$

We see from Examples 1 and 2 that if a surface S is described by $z = f(x, y)$, then S is a level surface of

(15.35) $F(x, y, z) = z - f(x, y).$

This implies that a normal to S at any point $(x_0, y_0, z_0) = (x_0, y_0, f(x_0, y_0))$ is given by

$$\nabla F(x_0, y_0, z_0) = F_x(x_0, y_0, z_0)\mathbf{i} + F_y(x_0, y_0, z_0)\mathbf{j} + \mathbf{k}$$

so

(15.36) $\nabla F(x_0, y_0, z_0) = -f_x(x_0, y_0)\mathbf{i} - f_y(x_0, y_0)\mathbf{j} + \mathbf{k}.$

In Section 15.2, we saw that if a surface S is described by $z = f(x, y)$, then the vector $\langle 1, 0, f_x(x_0, y_0) \rangle$ gives the direction of the tangent line to the curve of intersection of the plane $y = y_0$ and the surface S at (x_0, y_0, z_0). In a similar manner, the vector $\langle 0, 1, f_y(x_0, y_0) \rangle$ gives the direction of the tangent line to the curve of intersection of S and the plane $x = x_0$.

Consequently, a normal to the surface S at $(x_0, y_0, f(x_0, y_0))$ is the cross product of these tangent vectors:

$$\begin{aligned}\mathbf{n} &= \langle 1, 0, f_x(x_0, y_0) \rangle \times \langle 0, 1, f_y(x_0, y_0) \rangle \\ &= -f_x(x_0, y_0)\mathbf{i} - f_y(x_0, y_0)\mathbf{j} + \mathbf{k},\end{aligned}$$

which agrees with (15.36).

Theorem 15.33 has an interesting application for intersecting surfaces. Suppose that surfaces S_1 described by $F(x, y, z) = c_1$ and S_2 described by $G(x, y, z) = c_2$ intersect in a curve C. See Figure 15.17.

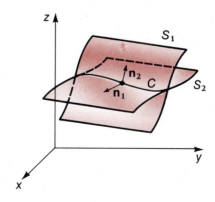

FIGURE 15.17

Since C lies on both S_1 and S_2, the normals $\mathbf{n}_1$ to S_1 and $\mathbf{n}_2$ to S_2 at a point

$P(x, y, z)$ on C are both normal to C at the point. As a consequence, the tangent to C at P has the direction

(15.37) $\mathbf{n}_1 \times \mathbf{n}_2 = \nabla F(x, y, z) \times \nabla G(x, y, z),$

provided that both F and G are differentiable at $P(x, y, z)$ and $\mathbf{n}_1$ is not parallel to $\mathbf{n}_2$.

EXAMPLE 4   The surfaces $x^2 + y^2 + z^2 = 8$ and $z^2 = x^2 + y^2$ intersect in a curve. Find parametric equations for the line tangent to this curve at $(1, \sqrt{3}, 2)$.

SOLUTION

Let $F(x, y, z) = x^2 + y^2 + z^2$ and $G(x, y, z) = z^2 - x^2 - y^2$. Then

$\nabla F(x, y, z) = 2x\mathbf{i} + 2y\mathbf{j} + 2z\mathbf{k}$, so $\nabla F(1, \sqrt{3}, 2) = 2\mathbf{i} + 2\sqrt{3}\mathbf{j} + 4\mathbf{k}$

and

$\nabla G(x, y, z) = -2x\mathbf{i} - 2y\mathbf{j} + 2z\mathbf{k}$, so $\nabla G(1, \sqrt{3}, 2) = -2\mathbf{i} - 2\sqrt{3}\mathbf{j} + 4\mathbf{k}.$

The direction of the tangent line to the curve of intersection is consequently

$$\nabla F(1, \sqrt{3}, 2) \times \nabla G(1, \sqrt{3}, 2) = 16\sqrt{3}\mathbf{i} - 16\mathbf{j} = 16(\sqrt{3}\mathbf{i} - \mathbf{j}).$$

So the tangent line to the curve of intersection of the two surfaces at $(1, \sqrt{3}, 2)$ has direction $\sqrt{3}\,\mathbf{i} - \mathbf{j}$. Parametric equations for this tangent line are

$$x = 1 + \sqrt{3}\,t, \qquad y = \sqrt{3} - t, \qquad z = 2. \qquad \square$$

The final example in this section illustrates how the two applications of the gradient can be used to find information about the same surface.

EXAMPLE 5   Let S be the paraboloid $x^2 + y^2 - z = 1$. Use the gradient to find
(a) the direction in which z increases most rapidly at $(1, 1)$ and this maximum rate of increase, and
(b) the equations of the tangent plane and normal line to S at $(1, 1, 1)$.

SOLUTION

The paraboloid is shown in Figure 15.18.

(a) To solve this part of the problem, we describe S by a function f of two variables

$$f(x, y) = z = x^2 + y^2 - 1.$$

Since

$$\nabla f(x, y) = 2x\mathbf{i} + 2y\mathbf{j}, \qquad \nabla f(1, 1) = 2\mathbf{i} + 2\mathbf{j}.$$

The direction of maximum increase of z is consequently in the direction of

$$\nabla f(1, 1) = 2\mathbf{i} + 2\mathbf{j},$$

and the maximum rate of increase is

$$\|\nabla f(1, 1)\| = \sqrt{8} = 2\sqrt{2}.$$

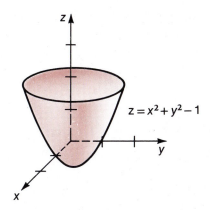

$z = x^2 + y^2 - 1$

FIGURE 15.18

(b) To solve this part of the problem, we describe S as a level surface of a function F of three variables. If

$$F(x, y, z) = x^2 + y^2 - z,$$

then S is the one-level surface of F.

Since

$$\nabla F(x, y, z) = 2x\mathbf{i} + 2y\mathbf{j} - \mathbf{k},$$

the direction of the normal to S at $(1, 1, 1)$ is

$$\nabla F(1, 1, 1) = 2\mathbf{i} + 2\mathbf{j} - \mathbf{k}.$$

An equation of the tangent plane is therefore

$$2(x - 1) + 2(y - 1) - (z - 1) = 0$$

or

$$2x + 2y - z = 3.$$

Parametric equations for the normal line are

$$x = 1 + 2t, \qquad y = 1 + 2t, \qquad z = 1 - t. \qquad \square$$

EXERCISE SET 15.6

In Exercises 1 through 8, find a vector that is normal to the surface at the given point.

1. $x^2 + y^2 + z^2 = 6$, $(1, 1, 2)$
2. $x^2 + y^2 - z = 0$, $(1, -1, 2)$
3. $z = e^y$, $(2, 0, 1)$
4. $z = e^x [\cos y + \sin y]$, $(0, \pi, -1)$
5. $f(x, y) = \ln xy$, $(e, e^2, 3)$
6. $f(x, y) = \sqrt{x^2 + y^2}$, $(3, -4, 5)$
7. $f(x, y) = \sqrt{x} + \sqrt{y}$, $(1, 1, 2)$
8. $f(x, y) = xy^2$, $(2, 1, 2)$

In Exercises 9 through 16, find an equation of the plane tangent to the surface at the given point.

9. $z = x^2 - y^2$, $(2, 1, 3)$

10. $z = \dfrac{x^2}{9} + \dfrac{y^2}{4}$, $(3, -2, 2)$

11. $z^2 = x^2 + y^2$, $(3, -4, -5)$

12. $z = \cos y$, $\left(1, \dfrac{\pi}{2}, 0\right)$

13. $f(x, y) = \ln(x^2 + y^2)$, $(1, 0, 0)$

14. $f(x, y) = ye^x$, $(0, -2, -2)$

15. $f(x, y) = x \tan y$, $\left(2, \dfrac{\pi}{4}, 2\right)$

16. $f(x, y) = \sec(x + y)$, $(0, 0, 1)$

In Exercises 17 through 20, the two surfaces intersect in a curve that contains the given point. Find equations of the line tangent to the curve of intersection at the point.

17. $x^2 + y^2 = 5$, $x^2 + z^2 = 5$; $(1, 2, 2)$

18. $x = z^2$, $x = 4 - y^2$; $(4, 0, 2)$

19. $x = 6 - y^2 - z^2$, $x = 2$; $(2, 1, \sqrt{3})$

20. $z = \sin x$, $x^2 + y^2 = \pi^2$; $\left(\dfrac{\pi}{2}, \dfrac{\sqrt{3}\pi}{2}, 1\right)$

21. Find the points on the paraboloid $z = 4x^2 + y^2$ at which the tangent plane is parallel to the plane $x + 2y + z = 6$.

22. Find the points on the ellipsoid $\dfrac{x^2}{4} + \dfrac{y^2}{9} + \dfrac{z^2}{36} = 1$ at which the tangent plane is parallel to the plane $x + y + z = 0$.

23. If two surfaces have a common tangent plane at a point of intersection, then the surfaces are said to be *tangent* at that point. Show that if the surfaces $F(x, y, z) = c_1$ and $G(x, y, z) = c_2$ are tangent at (x_0, y_0, z_0) then $\nabla F(x_0, y_0, z_0) \times \nabla G(x_0, y_0, z_0) = 0$.

24. Show that if the two surfaces described by $F(x, y, z) = c_1$ and $G(x, y, z) = c_2$ intersect *orthogonally* (have orthogonal tangent planes) at a point (x_0, y_0, z_0) then $\nabla F(x_0, y_0, z_0) \cdot \nabla G(x_0, y_0, z_0) = 0$.

25. Show that the plane tangent to the quadric surface $\dfrac{x^2}{a^2} \pm \dfrac{y^2}{b^2} \pm \dfrac{z^2}{c^2} = 1$ at (x_0, y_0, z_0) has equation $\dfrac{x_0 x}{a^2} \pm \dfrac{y_0 y}{b^2} \pm \dfrac{z_0 z}{c^2} = 1$.

26. Suppose a mountainous region is mapped on the xy-plane and that the height above sea level at each point (x, y) is given by $f(x, y)$. The level curves described by $f(x, y) = c$ are called *contour lines*. Show that the direction of flow of a stream of water at each point is always normal to the contour line at that point.

15.7
EXTREMA OF MULTIVARIATE FUNCTIONS

The extreme value theorem for single variable functions states that a continuous function defined on a closed interval assumes maximum and minimum values on the interval. In studying the derivative in Chapter 3, we found that the maximum and minimum values of a differentiable function occur only at endpoints of the interval or at critical points, numbers in the domain of the function at which the derivative is zero or does not exist.

This section examines some extensions of the single variable results to functions of several variables. In general, we will consider only functions of two variables. Corresponding results for functions of more than two variables are available, but are usually too complicated to apply.

First we need to introduce some terminology so that we can be certain what is meant by extrema, relative extrema, and critical points for functions of two variables.

(15.38)
DEFINITION

A function f of two variables is said to have an **absolute maximum,** or simply a **maximum,** at (x_0, y_0) in the region R if $f(x, y) \leq f(x_0, y_0)$ for all (x, y) in R. An **absolute minimum,** or simply a **minimum,** of f in R occurs at (x_0, y_0) if $f(x, y) \geq f(x_0, y_0)$ for all (x, y) in R.

A **relative maximum** of f occurs at (x_0, y_0) if a circle D about (x_0, y_0) exists with $f(x, y) \leq f(x_0, y_0)$ for all (x, y) in the interior of D. A **relative minimum** of f occurs at (x_0, y_0) if a circle D exists with $f(x, y) \geq f(x_0, y_0)$ for all (x, y) in the interior of D.

Both maxima and minima are referred to as **extrema;** relative maxima and relative minima are referred to as **relative extrema.**

EXAMPLE 1

Show that if $f(x, y) = x^2 + y^2$, then an absolute minimum of f occurs at $(0, 0)$.

SOLUTION

The graph of f is shown in Figure 15.19. Since $f(x, y) = x^2 + y^2 \geq 0$ and $f(0, 0) = 0, f(x, y) \geq f(0, 0)$ for all (x, y). Hence f has an absolute minimum value of 0 at $(0, 0)$. ☐

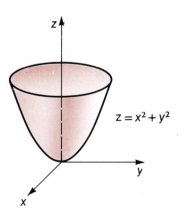

$$z = x^2 + y^2$$

FIGURE 15.19

(15.39)
DEFINITION

A function f of two variables is said to have a **critical point** at (x_0, y_0) in the domain of f if either

(i)   both partial derivatives of f are zero at (x_0, y_0), or

(ii) at least one of the partial derivatives fails to exist at (x_0, y_0).

EXAMPLE 2

Find all critical points of f if $f(x, y) = 4 - 4x^2 - y^2$.

SOLUTION

Since $f_x(x, y) = -8x$ and $f_y(x, y) = -2y$, the partial derivatives of f exist for all values of x and y.

The only critical points occur when both partial derivatives are zero, that is, when

$$f_x(x, y) = -8x = 0 \qquad \text{and} \qquad f_y(x, y) = -2y = 0.$$

The only critical point is $(0, 0)$.

It is easy to see that a maximum of f occurs at $(0, 0)$, since $f(0, 0) = 4 \geq 4 - 4x^2 - y^2$ for all (x, y) in the plane. (See Figure 15.20.)   □

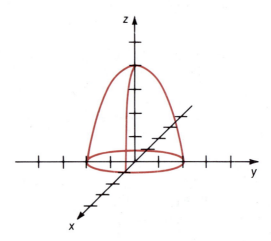

FIGURE 15.20

The following theorems are analogous to results for single variable functions and can be proved using the single variable results. The proof of the first theorem is included; the proof of Theorem 15.41 follows directly from Theorem 15.40.

(15.40)
THEOREM

If a function f has a relative extrema at (x_0, y_0), then (x_0, y_0) is a critical point of f.

PROOF

Suppose f has a relative maximum at (x_0, y_0) and D is a circle about (x_0, y_0) with $f(x, y) \leq f(x_0, y_0)$ for all (x, y) in the interior of D. The single variable functions described by $f(x, y_0)$ and $f(x_0, y)$ also have a relative maximum at (x_0, y_0). It follows from Theorem 3.32, page 146, that if $f_x(x_0, y_0)$ and $f_y(x_0, y_0)$

both exist, then $f_x(x_0, y_0) = f_y(x_0, y_0) = 0$. Consequently, the partial derivatives of f are either both zero at (x_0, y_0) or one of them fails to exist at (x_0, y_0).

The case when f has a relative minimum at (x_0, y_0) is handled similarly. □

(15.41)
THEOREM

If R is a bounded region of the plane and f has an extremum at (x_0, y_0) in R, then either (x_0, y_0) is a critical point of f or (x_0, y_0) is on the boundary of R.

EXAMPLE 3

Find all critical points of $f(x, y) = e^x \sin y$.

SOLUTION

The partial derivatives of f are

$$f_x(x, y) = e^x \sin y \qquad \text{and} \qquad f_y(x, y) = e^x \cos y;$$

both exist for all values of x and y, so critical points occur only when

$$e^x \sin y = 0 \qquad \text{and} \qquad e^x \cos y = 0.$$

The solutions to the first equation are $y = n\pi$ for n an integer. However, if these values are substituted in the second equation,

$$0 = e^x \cos(n\pi) = \pm e^x,$$

which has no solution. Consequently, the partial derivatives are never simultaneously zero, and f has no critical points. It follows from Theorem 15.40 that f has neither a relative maximum nor a relative minimum. □

EXAMPLE 4

Determine whether $f(x, y) = y^2 - x^2$ has absolute extrema.

SOLUTION

The partial derivatives of f are

$$f_x(x, y) = -2x \qquad \text{and} \qquad f_y(x, y) = 2y$$

so $(0, 0)$ is the only critical point of f and the only point at which an extremum can occur. In every circle containing $(0, 0)$, there are points $(x, 0)$ at which $f(x, 0) = -x^2 < 0$ and points $(0, y)$ at which $f(0, y) = y^2 > 0$. Consequently, f has neither a maximum nor a minimum at $(0, 0)$. This can also be seen by looking at the graph of f shown in Figure 15.21, a hyperbolic paraboloid. □

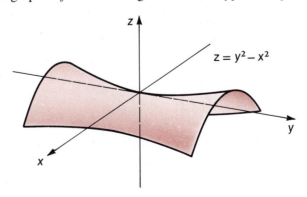

FIGURE 15.21

A critical point at which a relative extremum does not occur is often called a **saddle point** of the graph. In Example 4 this name is quite appropriate, since the graph has the appearance of a saddle and the critical point determines the point of the saddle at which one would sit. The terminology is not always as descriptive, since saddle points often occur when the graph resembles nothing like a saddle.

It is usually more difficult to determine absolute extrema on bounded regions of the plane. However, in applications this is often the problem that needs to be solved.

EXAMPLE 5

Find the absolute maximum and minimum values of $f(x, y) = 4 - 4x^2 - y^2$ in the square R in the xy-plane described by $0 \le x \le 1, 0 \le y \le 1$.

SOLUTION

In Example 2 we found that the absolute maximum for this function occurs at $(0, 0)$ and since $(0, 0)$ is in R we need only find the absolute minimum. Since $(0, 0)$ is the only critical point of f, the minimum must occur on the boundary of R.

The boundary has four portions, described by the equations

$$B_1: y = 0, \quad 0 \le x \le 1; \qquad B_2: x = 0, \quad 0 \le y \le 1;$$
$$B_3: y = 1, \quad 0 \le x \le 1; \qquad B_4: x = 1, \quad 0 \le y \le 1.$$

On B_1, $f(x, 0) = 4 - 4x^2$ is a function of the single variable x. The minimum value of f on B_1 occurs either when $f_x(x, 0) = -8x = 0$, or at the endpoints of B_1, $(0, 0)$, and $(1, 0)$. The minimum on B_1 occurs at $(1, 0)$ and $f(1, 0) = 0$.

On B_2, $f(0, y) = 4 - y^2$, and extreme values occur when $f_y(0, y) = -2y = 0$ or at the endpoints of B_2, $(0, 0)$, and $(0, 1)$. The minimum on B_2 is $f(0, 1) = 3$.

On B_3, $f(x, 1) = 3 - 4x^2$. Extrema can occur when $f_x(x, 1) = -8x = 0$ or at the endpoints of B_3, $(0, 1)$ and $(1, 1)$. The minimum on B_3 is $f(1, 1) = -1$.

On B_4, $f(1, y) = -y^2$ and the minimum is $f(1, 1) = -1$. Therefore, the minimum value on the boundary of R and consequently on R itself is $f(1, 1) = -1$.   □

The graph of f for (x, y) in R is shown in Figure 15.22.

In the previous examples, it was easy to determine whether relative maxima or minima occurred at critical points. In less obvious cases, we can apply the following extension of the second derivative test. The proof of this result is quite complicated and is not presented.

(15.42)
THEOREM

Let f be a function with continuous second partial derivatives and suppose (x_0, y_0) is a critical point of f.
(i)   If $f_{xx}(x_0, y_0) \cdot f_{yy}(x_0, y_0) - [f_{xy}(x_0, y_0)]^2 > 0$ and $f_{xx}(x_0, y_0) < 0$, then f has a relative maximum at (x_0, y_0).
(ii) If $f_{xx}(x_0, y_0) \cdot f_{yy}(x_0, y_0) - [f_{xy}(x_0, y_0)]^2 > 0$ and $f_{xx}(x_0, y_0) > 0$, then f has a relative minimum at (x_0, y_0).
(iii) If $f_{xx}(x_0, y_0) \cdot f_{yy}(x_0, y_0) - [f_{xy}(x_0, y_0)]^2 < 0$, then f has a saddle point at (x_0, y_0).

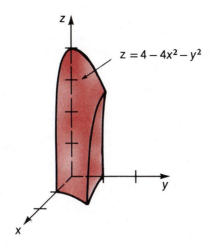

FIGURE 15.22

Notice that no conclusion can be drawn from Theorem 15.42 when $f_{xx}(x_0, y_0) \cdot f_{yy}(x_0, y_0) - [f_{xy}(x_0, y_0)]^2 = 0$. This corresponds to the single variable situation when the second derivative at a critical point is zero. As in that case, it may happen that a relative maximum occurs, a relative minimum occurs, or the point is a saddle point. Examples of such situations are considered in Exercises 37 and 38.

EXAMPLE 6   Determine any relative extrema for $f(x, y) = x^3 - y^3 + 3xy$, using Theorem 15.42.

SOLUTION

Critical points of f occur when both

$$0 = f_x(x, y) = 3x^2 + 3y \quad \text{so} \quad y = -x^2$$

and

$$0 = f_y(x, y) = -3y^2 + 3x \quad \text{so} \quad x = y^2.$$

Thus

$$x = y^2 = (-x^2)^2 = x^4$$

and

$$x = 0, \quad \text{or} \quad x = 1.$$

So critical points of f are at $(0, 0)$ and $(1, -1)$.

The second partial derivatives of f are

$$f_{xx}(x, y) = 6x, \quad f_{yy}(x, y) = -6y, \quad \text{and} \quad f_{xy}(x, y) = 3.$$

At the critical point $(0, 0)$,

$$f_{xx}(0, 0)f_{yy}(0, 0) - [f_{xy}(0, 0)]^2 = (0)(0) - (3)^2 < 0,$$

so a saddle point occurs at $(0, 0)$. At the critical point $(1, -1)$

$$f_{xx}(1, -1)f_{yy}(1, -1) - [f_{xy}(1, -1)]^2 = (6)(6) - (3)^2 > 0.$$

Since $f_{xx}(1, -1) = 6 > 0$, a relative minimum of f occurs at $(1, -1)$. This relative minimum value is $f(1, -1) = -1$.   □

EXAMPLE 7   A box in the form of a rectangular parallelepiped is to be built to enclose 36 ft³. The bottom of the box can be made from scrap material costing $1 per square foot, but the material for the top and sides costs $8 per square foot. Find the most economical dimensions for the box and the corresponding cost of the box.

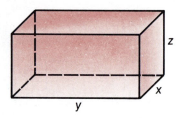

FIGURE 15.23

SOLUTION

By labeling the sides as shown in Figure 15.23, we can express the problem concisely as shown in the table below.

Know	Find
1. Volume $= xyz = 36$ ft³,	(a) The values of x, y, z at which C has a minimum, and
2. Cost $= xy + 8(2xz + 2yz + xy)$, and	
3. $C(x, y) = xy + 8\left[2x\left(\dfrac{36}{xy}\right) + 2y\left(\dfrac{36}{xy}\right) + xy\right]$	(b) the cost corresponding to the values of $x, y,$ and z found in (a).
$\qquad = 9xy + \dfrac{576}{y} + \dfrac{576}{x}.$	

To find the critical points of C we need

$$C_x(x, y) = 9y - \frac{576}{x^2} = \frac{9}{x^2}(yx^2 - 64)$$

and

$$C_y(x, y) = 9x - \frac{576}{y^2} = \frac{9}{y^2}(xy^2 - 64).$$

$C_x(x, y)$ does not exist when $x = 0$ and $C_y(x, y)$ does not exist when $y = 0$. Thus, critical points occur whenever $x = 0$ or $y = 0$, but these do not produce physically reasonable results since no box is formed. The only other critical points occur when

$$yx^2 - 64 = 0 \qquad \text{and} \qquad xy^2 - 64 = 0.$$

Solving these equations simultaneously, we have:

$$y = \frac{64}{x^2} \qquad \text{and} \qquad x\left(\frac{64}{x^2}\right)^2 - 64 = 0$$

so $x^3 = 64$. Thus $x = 4$ and $y = \dfrac{64}{16} = 4$.

The cost corresponding to $x = 4$ and $y = 4$ is

$$C(4, 4) = 9(16) + \frac{576}{4} + \frac{576}{4} = \$432.$$

Using Theorem 15.42 and the fact that the domain of C is not bounded, we can verify that the minimum value of C is \$432:

$$C_{xx}(x, y) = \frac{1152}{x^3}, \quad C_{yy}(x, y) = \frac{1152}{y^3}, \quad C_{xy}(x, y) = 9.$$

At $(4, 4)$,

$$C_{xx}(4, 4) \cdot C_{yy}(4, 4) - [C_{xy}(4, 4)]^2 = (18)(18) - 81 > 0$$

and

$C_{xx}(4, 4) = 18 > 0$ so C has a relative and hence absolute minimum value at $(4, 4)$.

The most economical box costs \$432 and is one with a square base of length 4 feet and height $z = \dfrac{36}{xy} = \dfrac{36}{16} = 2.25$ feet. ☐

EXERCISE SET 15.7

1. Show that if $f(x, y) = 4 - x^2 - y^2$, then f has an absolute maximum at $(0, 0)$.

2. Show that if $f(x, y) = x^2 - y^2$, then f has neither an absolute maximum nor an absolute minimum at $(0, 0)$.

3. Show that if $f(x, y) = \sin y$, $0 \le y \le 2\pi$, then f has an absolute maximum at any point $\left(x, \dfrac{\pi}{2} \right)$ and an absolute minimum at $\left(x, \dfrac{3\pi}{2} \right)$ for any value of x.

4. Show that if $f(x, y) = \cos x$, $0 \le x \le 2\pi$, then f has an absolute maximum at $(0, y)$ and an absolute minimum at (π, y) for any value of y.

In Exercises 5 through 12, find all critical points of f.

5. $f(x, y) = x^2 - xy + y^2$

6. $f(x, y) = x^2 + y^2 - 4x - 2y + 5$

7. $f(x, y) = x^2 + 2xy + 3y^2 - 4$

8. $f(x, y) = \dfrac{3x}{y}$

9. $f(x, y) = x \cos y$

10. $f(x, y) = e^x \sin y$

11. $f(x, y) = xy^2 + x^3 - 4y^2$

12. $f(x, y) = x^2y + x + y$

13–20. Use Theorem 15.42 to determine whether each critical point in Exercises 5 through 12 is a relative maximum, relative minimum, or a saddle point.

In Exercises 21 through 30, find all critical points of f and use Theorem 15.42 to determine any relative extrema for f.

21. $f(x, y) = x^3 + y^3 - 3xy$

22. $f(x, y) = x^4 + y^3 + 32x - 27y$

23. $f(x, y) = x^2y - 5xy + 6y$

24. $f(x, y) = x^2 + y^4 - y^2 - 2xy$

25. $f(x, y) = \cos x + \cos y$

26. $f(x, y) = \sin x + \sin y + y \sin x$

27. $f(x, y) = e^{x^2 + y^2}$

28. $f(x, y) = e^{xy}$

29. $f(x, y) = 16x^2 - 24xy + 9y^2 - 60x - 8y$

30. $f(x, y) = \dfrac{-x}{x^2 + y^2 + 1}$

In Exercises 31 through 36, find the absolute maximum and minimum values of f on the region described by the inequalities.

31. $f(x, y) = x^2 + y^2; x^2 + y^2 \le 1$

32. $f(x, y) = x^2 + y^2; 0 \le x \le 1, 0 \le y \le 2$

33. $f(x, y) = xy; x^2 + 2y^2 \le 4$

34. $f(x, y) = y^2 - x^2; 0 \le x \le 1, 0 \le y \le 1$

35. $f(x, y) = 8xy + y, 0 \le y \le 15 - x, 0 \le x \le 5$

36. $f(x, y) = 2x^2 + 2y^2 + xy, 5 - x \le y \le 10, 0 \le x \le 3$

37. (a) Show that if $f(x, y) = x^4 + y^4$, then $f_{xx}(0, 0) f_{yy}(0, 0) - [f_{xy}(0, 0)]^2 = 0$, so no extrema information can be concluded from Theorem 15.42.

   (b) Show that f has a minimum at $(0, 0)$.

38. (a) Show that if $f(x, y) = x^4 - y^4$, then $f_{xx}(0, 0) f_{yy}(0, 0) - [f_{xy}(0, 0)]^2 = 0$, so no extrema information can be concluded from Theorem 15.42.

   (b) Show that f has a saddle point at $(0, 0)$.

39. Find the absolute minimum of the function f described by $f(x, y, z) = 2x + 3y + z$ if $x, y,$ and z satisfy $z = x^2 + y^2$.

40. Find the point on the plane $2x + 3y + z = 6$ that is nearest the origin.

41. Find the minimum distance between the lines with parametric equations $x = t, y = 3 - 2t, z = 1 + 2t$ and $x = 1 + s, y = -2 - s, z = s$.

42. The trace of $z = f(x, y)$ in the yz-plane is $z = y^2$. The trace in the xz-plane is $z = -x^2$. Can z have a relative extrema at $(0, 0, 0)$?

43. If a cereal box with a top is to have a volume of 80 cubic inches, find the dimensions that will minimize the surface area.

44. A box in the form of a rectangular parallelepiped is to be built at a cost of $500. If the material for the bottom of the box costs $1 per square foot and the material for the top and sides costs $4 per square foot, find the approximate dimensions of the box of greatest volume that can be made.

45. U.S. postal regulations require that in post offices serving more than 600 units the length plus the girth of a package to be mailed cannot exceed 84 inches. Find the dimensions of the rectangular package of greatest volume that can be mailed. [Girth = 2(width) + 2(height)]

46. Method of Least Squares. Given n points (x_1, y_1), (x_2, y_2), . . . , (x_n, y_n) in the xy-plane, it is usually impossible to find a straight line that passes through all the points. The method of least squares is a method for finding a best approximating straight line, $f(x) = ax + b$, by finding values of a and b that minimize the sum of the squares of the differences between the y-values on the approximating line and the given y-values:

$$\sum_{i=1}^{n} [y_i - (ax_i + b)]^2.$$

Show that these values of a and b are

$$a = \frac{n\left(\sum_{i=1}^{n} x_i y_i\right) - \left(\sum_{i=1}^{n} x_i\right)\left(\sum_{i=1}^{n} y_i\right)}{n\left(\sum_{i=1}^{n} x_i^2\right) - \left(\sum_{i=1}^{n} x_i\right)^2}$$

and

$$b = \frac{\left(\sum_{i=1}^{n} x_i^2\right)\left(\sum_{i=1}^{n} y_i\right) - \left(\sum_{i=1}^{n} x_i y_i\right)\left(\sum_{i=1}^{n}\right)}{n\left(\sum_{i=1}^{n} x_i^2\right) - \left(\sum_{i=1}^{n} x_i\right)^2}.$$

In Exercises 47 through 50, use the method of least squares to find the best approximating line for the given collection of points.

47. $(1, 2), (2, 2), (3, 4)$

48. $(0, 0), (1, 3), (2, 1)$

49. $(2, 2), (4, 11), (6, 28), (8, 40)$

50. $(-1, 6), (1, 5), (2, 3), (5, 3), (7, -1)$

51. The following set of data, presented in March 1970 to the Senate Antitrust Subcommittee, shows the comparative crash-survivability characteristics of cars in various classes. Find the least-squares line that approximates this data. (The table shows the percent of accident-involved vehicles in which the most severe injury was fatal or serious.)

Type	Average weight	Percent occurrence
1. Domestic "luxury" regular	4800 lb.	3.1
2. Domestic "intermediate" regular	3700 lb.	4.0
3. Domestic "economy" regular	3400 lb.	5.2
4. Domestic compact	2800 lb.	6.4
5. Foreign compact	1900 lb.	9.6

HISTORICAL NOTE   The method of least squares was first published by **Adrien Marie Legendre** (1752–1833) in 1805 as an addendum to a work on the orbits of comets.

Karl Friedrich Gauss (1777–1855), called by some the greatest mathematician of all time, had, however, been using the method as early as 1794. He used least squares to compute, from a few observations, the orbit of Ceres, an asteroid that was discovered on January 1, 1800. His method is still used to track satellites.

52. To determine a relationship between the number of fish and the number of species of fish in samples taken from a portion of the Great Barrier Reef, P. Sale and R. Dybdahl used the method of least squares and the following collection of data. Let x be the number of fish in the sample, and y be the number of species in the sample.

x	y	x	y	x	y
13	11	29	12	60	14
15	10	30	14	62	21
16	11	31	16	64	21
21	12	36	17	70	24
22	12	40	13	72	17
23	13	42	14	100	23
25	13	55	22	130	34

Determine the least-squares line for this data.

Putnam exercises:

53. Given two points in the plane, P and Q, at fixed distances from a line L, and on the same side of the line as indicated, the problem is to find a third point R so that $PR + RQ + RS$ is a minimum, where RS is perpendicular to L. Consider all cases.

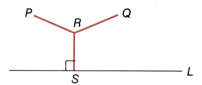

(This exercise was problem 4, part I of the twenty-first William Lowell Putnam examination given on December 3, 1960. The examination and its solution can be found in the September 1961 issue of the *American Mathematical Monthly*, pages 632–637.)

54. Let f be a real-valued function having partial derivatives and which is defined for $x^2 + y^2 \leq 1$ and is such that $|f(x, y)| \leq 1$. Show that there exists a point (x_0, y_0) in the interior of the unit circle such that

$$\left(\frac{\partial f}{\partial x}(x_0, y_0) \right)^2 + \left(\frac{\partial f}{\partial y}(x_0, y_0) \right)^2 \leq 16.$$

(This exercise was problem B–6 of the twenty-eighth William Lowell Putnam examination given on December 2, 1967. The examination and its solution can be found in the September 1968 issue of the *American Mathematical Monthly*, pages 734–739.)

55. Show that if t_1, t_2, t_3, t_4, t_5 are real numbers, then

$$\sum_{j=1}^{5} (1 - t_j) \exp \sum_{k=1}^{j} t_k \leq e^{e^{e^e}}.$$

(This exercise was problem 3, part I of the twenty-first William Lowell Putnam examination given on December 3, 1960. The examination and its solution can be found in the September 1961 issue of the *American Mathematical Monthly,* pages 632–637.)

15.8
LAGRANGE MULTIPLIERS

In Example 7 of Section 15.7 we considered the problem of minimizing the cost of building a box to enclose 36 ft^3 if the bottom costs \$1 per square foot, and the sides and top cost \$8 per square foot. If x and y denote the lengths of the base, z the height of the box, and C is the cost function, the problem can be expressed as:

Minimize $C(x, y, z) = 9xy + 16yz + 16xy$,
subject to the condition that $xyz = 36$.

In this form, the condition $xyz = 36$ is called a **constraint** or **side condition** for the problem. Problems in economics and business often assume this form.

In the previous section we solved the constraint equation for z and then considered C as a function of only two variables x and y. In this section we give an alternate approach to problems of determining extrema of a function subject to one or more constraints. The technique is known as **Lagrange multipliers** because it involves introducing an additional variable, or multiplier, into the problem, and because it was first used in the 18th century by the Italian mathematician Joseph Louis Lagrange (1736–1813).

Our plan is to motivate the method geometrically by considering the solution to the problem:

(15.43)

Maximize $f(x, y, z)$,
subject to $g(x, y, z) = 0$,

where f and g have continuous first partial derivatives.

Consider the surface S described by the constraint equation

$$g(x, y, z) = 0.$$

If problem (15.43) has a solution, then there is at least one constant c such that the level surface T_c described by

$$f(x, y, z) = c$$

intersects S. The maximum of f subject to the constraint occurs when c is the largest constant for which T_c and S have nonempty intersection.

Let us consider a constant c_0 for which T_{c_0} intersects S. The intersection partitions S into two portions:

S_0: when (x, y, z) is in S and $f(x, y, z) > c_0$

and $\hat{S}_0$: when (x, y, z) is in S and $f(x, y, z) \le c_0$.

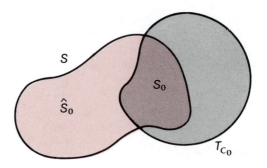

FIGURE 15.24

Figure 15.24 shows a partitioning of S into S_0 and $\hat{S}_0$.

If $c_1 > c_0$ is a constant for which T_{c_1} intersects S, then another partitioning of S is introduced:

$$S_1: \text{when } f(x, y, z) > c_1$$

and $$\hat{S}_1: \text{when } f(x, y, z) \leq c_1.$$

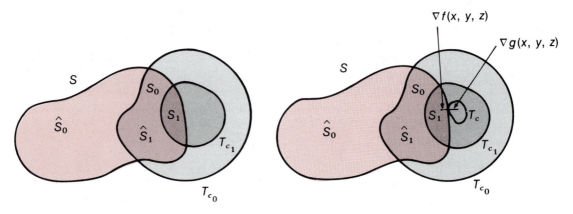

FIGURE 15.25   FIGURE 15.26

Since $c_1 > c_0$, S_1 is contained entirely within S_0. Any larger value of c for which T_c and S have nonempty intersection will induce a similar partition with

$$S_c: \quad \text{when} \quad f(x, y, z) > c$$

contained in S_1. (See Figure 15.25.) The largest constant c for which the intersection of T_c and S is nonempty has the property that the surfaces T_c and S have a common tangent plane. For if they do not have a common tangent plane, then S passes through T_c and intersects level surfaces of f for values both greater and less than c. Consequently, the maximum value of f can occur only when the normal to T_c is parallel to the normal to S. (See Figure 15.26.) The normal to T_c at (x, y, z) is $\nabla f(x, y, z)$ and the normal to S is $\nabla g(x, y, z)$. These vectors are parallel if and only if a constant λ exists with

$$\nabla f(x, y, z) = \lambda \nabla g(x, y, z).$$

The method of Lagrange multipliers proceeds to solve the problem:

(15.44)   Maximize or minimize $f(x, y, z)$,
subject to $g(x, y, z) = 0$

by determining (x, y, z) and λ so that

$$g(x, y, z) = 0 \qquad \text{and} \qquad \nabla f(x, y, z) = \lambda \nabla g(x, y, z).$$

EXAMPLE 1   Apply the method of Lagrange multipliers to the problem:

Minimize $C(x, y, z) = 9xy + 16yz + 16xz$,
subject to $xyz = 36$.

(This minimizes the cost of the box described at the beginning of this section and in Example 7 of Section 15.7.)

SOLUTION
Let $g(x, y, z) = xyz - 36$; then

$$\nabla g(x, y, z) = yz\mathbf{i} + xz\mathbf{j} + xy\mathbf{k}.$$

Also,

$$\nabla C(x, y, z) = (9y + 16z)\mathbf{i} + (9x + 16z)\mathbf{j} + (16y + 16x)\mathbf{k}.$$

The problem is to find (x, y, z) and λ such that

$$(9y + 16z)\mathbf{i} + (9x + 16z)\mathbf{j} + (16y + 16x)\mathbf{k} = \lambda(yz\mathbf{i} + xz\mathbf{j} + xy\mathbf{k}),$$

that is

$$9y + 16z = \lambda yz, \quad 9x + 16z = \lambda xz, \quad 16y + 16x = \lambda xy,$$

and satisfying the constraint equation

$$xyz = 36.$$

Multiplying the first three equations respectively by x, y, and z gives

$$9xy + 16xz = \lambda xyz, \quad 9xy + 16yz = \lambda xyz, \quad \text{and} \quad 16xz + 16yz = \lambda xyz.$$

Thus,

$$9xy + 16xz = 9xy + 16yz, \quad \text{so } x = y,$$

and

$$9xy + 16xz = 16yz + 16xz, \quad \text{so } x = \frac{16}{9}z.$$

The equation $xyz = 36$ implies that

$$(16z/9)^2 z = 36, \quad \text{so } z^3 = 36(81/256) = 729/64$$

and $z = 9/4$. Consequently, the solution is $x = 4$, $y = 4$, $z = 9/4$.   □

EXAMPLE 2   Find the maximum and minimum values of $f(x, y, z) = 2x - 2y + z$ when restricted to the sphere $x^2 + y^2 + z^2 = 9$.

SOLUTION

The problem is to:
Maximize and minimize $f(x, y, z) = 2x - 2y + z$,
subject to $g(x, y, z) = x^2 + y^2 + z^2 - 9 = 0$.

Since $$\nabla f(x, y, z) = 2\mathbf{i} - 2\mathbf{j} + \mathbf{k}$$

and $$\nabla g(x, y, z) = 2x\mathbf{i} + 2y\mathbf{j} + 2z\mathbf{k},$$

the problem reduces to finding (x, y, z) and λ satisfying

$$2 = 2\lambda x, \quad -2 = 2\lambda y, \quad 1 = 2\lambda z$$

and the constraint equation

$$x^2 + y^2 + z^2 = 9.$$

Solving for x, y, and z in the first three equations and substituting into the constraint equation gives

$$x = \frac{1}{\lambda}, \quad y = -\frac{1}{\lambda}, \quad z = \frac{1}{2\lambda}$$

and $$9 = x^2 + y^2 + z^2 = \frac{1}{\lambda^2} + \frac{1}{\lambda^2} + \frac{1}{4\lambda^2} = \frac{9}{4\lambda^2}.$$

Thus, $$\lambda = \frac{1}{2} \quad \text{or} \quad \lambda = -\frac{1}{2}.$$

When $\lambda = \frac{1}{2}$: $\quad x = 2, \quad y = -2, \quad z = 1$

and $$f(2, -2, 1) = 9.$$

When $$\lambda = -\frac{1}{2}: \quad x = -2, \quad y = 2, \quad z = -1$$

and $$f(-2, 2, -1) = -9.$$

The maximum is 9 and occurs at $(2, -2, 1)$; the minimum is -9 and occurs at $(-2, 2, -1)$. $\quad\square$

The method of Lagrange multipliers can easily be modified to handle extrema problems involving more than one constraint. For example, to solve a problem:

(15.45)   Maximize $f(x, y, z)$,
subject to $g(x, y, z) = 0$ and $h(x, y, z) = 0$.

We use two multipliers, λ and μ, chosen so that

$$\nabla f(x, y, z) = \lambda \nabla g(x, y, z) + \mu \nabla h(x, y, z).$$

The algebraic manipulation is often more complicated than in the case of a single constraint, but the technique is the same.

EXAMPLE 3   Consider the intersection of the cone $z^2 = 2x^2 + 2y^2$ and the plane $x + y + z = 1$. (See Figure 15.27.) Find the point in this intersection that is closest to the origin.

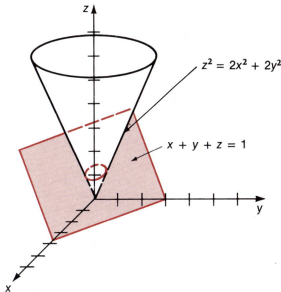

$z^2 = 2x^2 + 2y^2$

$x + y + z = 1$

FIGURE 15.27

SOLUTION

The problem is:

Minimize $D(x, y, z) = \sqrt{x^2 + y^2 + z^2}$,
subject to $2x^2 + 2y^2 - z^2 = 0$ and $x + y + z - 1 = 0$.

The partial derivatives of D are quite complicated. However, minimizing

$$[D(x, y, z)]^2 = x^2 + y^2 + z^2$$

is equivalent to minimizing $D(x, y, z)$, and the partial derivatives are much easier to work with. For this reason, we consider the problem

Minimize $[D(x, y, z)]^2 = x^2 + y^2 + z^2$,
subject to $2x^2 + 2y^2 - z^2 = 0$ and $x + y + z - 1 = 0$.

If the Lagrange multipliers λ and μ are introduced, the problem is to find (x, y, z), λ, and μ satisfying:

$$2x\mathbf{i} + 2y\mathbf{j} + 2z\mathbf{k} = \lambda(4x\mathbf{i} + 4y\mathbf{j} - 2z\mathbf{k}) + \mu(\mathbf{i} + \mathbf{j} + \mathbf{k}),$$

that is,

(1) $2x = 4\lambda x + \mu$,
(2) $2y = 4\lambda y + \mu$,
(3) $2z = -2\lambda z + \mu$,

and the constraint equations. Subtracting equation (2) from (1) yields

$$2(x - y) = 4\lambda(x - y), \text{ which implies } \lambda = 1/2 \text{ or } x = y.$$

Subtracting equation (3) from (1) implies

$$2(x - z) = 2\lambda(2x + z).$$

If $\lambda = 1/2$, $2(x - z) = 2x + z$ and $z = 0$. In this case the constraint equation $z^2 = 2x^2 + 2y^2$ implies that $x = 0$ and $y = 0$. However, $(0, 0, 0)$ does not satisfy the constraint equation $x + y + z = 1$, so $\lambda \neq 1/2$.

The only alternative is that $x = y$. In this case, the constraint equations become

$$4x^2 - z^2 = 0 \qquad \text{and} \qquad 2x + z - 1 = 0.$$

Substituting $z = 1 - 2x$ into $4x^2 - z^2 = 0$, we obtain

$$0 = 4x^2 - (1 - 2x)^2 = 4x^2 - 1 + 4x - 4x^2 = 4x - 1.$$

So $x = 1/4$, $y = x = 1/4$, and $z = 1 - 2x = 1/2$. The point on the intersection of the cone $z^2 = 2x^2 + 2y^2$ and the plane $x + y + z = 1$ that is closest to the origin is $(1/4, 1/4, 1/2)$. ☐

EXERCISE SET 15.8

In Exercises 1 through 12, find the local extrema of f subject to the given constraints.

1. $f(x, y) = x^2 + y^2$, $x + y = 2$

2. $f(x, y) = x^2 - y^2$, $x^2 + y^2 = 2$

3. $f(x, y) = x^2 - 4xy + 4y^2$, $x + y = 1$

4. $f(x, y) = x^2 + 2xy + y^2$, $x + y = 1$

5. $f(x, y, z) = 4x^2 + y^2 + z^2$, $2x + 3y + 4z = 12$

6. $f(x, y, z) = 2x + 3y + z$, $z = x^2 + y^2$

7. $f(x, y, z) = 2xy + 4xz + 6yz$, $xyz = 48$

8. $f(x, y, z) = x^2y^2z^2$, $x + y + z = 1$

9. $f(x, y, z) = x^2 + y^2 + z^2$, $z^2 - xy = 1$

10. $f(x, y, z) = \ln x + \ln y + 3 \ln z$, $x^2 + y^2 + z^2 = 20$

11. $f(x, y, z) = x^2 + y^2 + z^2$, $x^2 + 4y^2 + 4z^2 = 4$, $x - 4y - z = 0$

12. $f(x, y, z) = x^2 + y^2 + (z + 1)^2$, $x^2 - xy + y^2 - z^2 = 1$, $x^2 + y^2 = 1$

13. Find the point on the sphere $x^2 + y^2 + z^2 - 4x - 6y - 8z + 28 = 0$ that is closest to the origin.

14. Find the point on the sphere $x^2 + y^2 + z^2 = 9$ that is closest to the point $(3, 4, -1)$.

15. Use Lagrange multipliers to solve Exercise 43 in Section 15.7.

16. Use Lagrange multipliers to solve Exercise 44 in Section 15.7.

17. Find the rectangular box with maximum volume that can be enclosed by the ellipsoid $\dfrac{x^2}{9} + \dfrac{y^2}{16} + \dfrac{z^2}{25} = 1$ if the sides of the box are parallel to the coordinate planes.

18. A recreation shelter is in the shape of the upper half of an ellipsoid $\dfrac{x^2}{400} + \dfrac{y^2}{2500} + \dfrac{z^2}{100} = 1$. Electrical power is to be provided by an existing transformer. If the power supply is at the point $(0, 75, 20)$, find the point on the shelter closest to the transformer.

19. Suppose a rectangular room is to be built inside the recreation shelter described in Exercise 18. Find the dimensions of the room with maximum volume if the corners of the ceiling are to lie on the ellipsoid.

20. Use Lagrange multipliers to find the dimensions of a 12-ounce cylindrical beer can that can be constructed using the least amount of material. Compare these dimensions to those of a standard can. (Note: one fluid ounce is approximately 1.805 cubic inches.)

21. A standard cylindrical oil drum has a capacity of 55 gallons (approximately 12705 in^3). Use Lagrange multipliers to find the dimensions of the drum that requires the minimal amount of material to construct.

22. An oil drum is usually constructed by crimping the top and bottom to the sides. Suppose that this crimping requires one inch of additional material for the top and the bottom and two inches of additional material for the sides. Find the dimensions of the 55-gallon drum that uses the least material.

23. A company needs a warehouse to contain a million cubic feet. They estimate that the floor and ceiling of the building will cost $3.00 per square foot to construct and the walls will cost $7.00 per square foot. Use Lagrange multipliers to find the cost of the most economical rectangular building.

24. The quantities Q_1 and Q_2 of output of a two-product firm depend on labor l and capital k according to the production functions $Q_1 = 3l_1\sqrt[3]{k_1}$ and $Q_2 = 3l_2\sqrt[3]{k_2}$. Assume that there is a fixed total quantity $L = l_1 + l_2$ of labor and a fixed total quantity $K = k_1 + k_2$ of capital available to the firm and that the respective prices p_1 and p_2 are fixed. Show that the maximum revenue occurs when $\dfrac{l_1}{L} = \dfrac{k_1}{K}$.

25. Consider $f(x, y, z) = x^2 + y^2 + z$ and $g(x, y, z) = x^2 + 4y^2 + 9z^2 - 36$. Show that there are solutions to $\nabla f(x, y, z) = \lambda \nabla g(x, y, z)$ that neither maximize nor minimize $f(x, y, z)$ subject to satisfying $g(x, y, z) = 0$.

REVIEW EXERCISES

1. Find the domain of the function described by $f(x, y) = \sqrt{4 - x^2 - y^2}$. Sketch the graph of f.

2. Find the domain of the function described by $f(x, y) = \sqrt{x^2 + y^2 - 9}$. Sketch the graph of f.

3. Determine any points of discontinuity of the function described by

$$f(x, y) = \begin{cases} x^2 + y^2, & \text{if } x^2 + y^2 \le 1 \\ 0 & , & \text{if } x^2 + y^2 > 1. \end{cases}$$

Sketch the graph of f.

4. Determine any points of discontinuity of the function described by

$$f(x, y) = \begin{cases} 4x^2 + 9y^2, & \text{if } 4x^2 + 9y^2 \leq 36 \\ 0 & , & \text{if } 4x^2 + 9y^2 > 36. \end{cases}$$

Sketch the graph of f.

Find the first partial derivatives of the functions described in Exercises 5 through 12.

5. $f(x, y) = e^{xy}$

6. $f(x, y) = \dfrac{x}{y} + \dfrac{y}{x}$

7. $f(x, y, z) = x^2 + y^2 - 2xy \cos z$

8. $f(x, y) = 3x^2 - 2y^2 + 2xy$

9. $f(x, y) = x^2y + 2xe^{1/y}$

10. $f(x, y, z) = \ln(x^2 + y^2 + z^2)$

11. $f(x, y) = (x^2 + xy)^3$

12. $f(x, y, z) = e^{x+2y-z^2}$

Find all critical points of the functions described in Exercises 13 through 16 and determine whether each critical point is a relative maximum, relative minimum, or a saddle point.

13. $f(x, y) = 4 - x^2 - y^2$

14. $f(x, y) = 4 + x^2 - y^2$

15. $f(x, y) = x^2e^{x+y}$

16. $f(x, y) = (x - 1)\ln y - x^2$

17. If $f(x, y, z) = xe^{yz} + yze^x$, find f_{yxz} and $\dfrac{\partial^3 f}{\partial x^2 \partial y}$.

18. Show that $f_{xy} = f_{yx}$ if $f(x, y) = \dfrac{x}{x + \sin y}$.

In Exercises 19 through 24, find the directional derivative of f at P in the direction of $\mathbf{v}$.

19. $f(x, y) = x^2 + 2xy \cos x$, $\mathbf{v} = \dfrac{\sqrt{3}}{3}\mathbf{i} - \dfrac{\sqrt{6}}{3}\mathbf{j}$, $P(0, 1)$

20. $f(x, y) = x^2 + 2xy + y^2$, $\mathbf{v} = \dfrac{4}{5}\mathbf{i} - \dfrac{3}{5}\mathbf{j}$, $P(1, 1)$

21. $f(x, y) = 7 - x - y$, $\mathbf{v} = \mathbf{i} + \mathbf{j}$, $P(2, 2)$

22. $f(x, y) = x^2 + 2xy - y^2$, $\mathbf{v} = 4\mathbf{i} + 3\mathbf{j}$, $P(2, 4)$.

23. $f(x, y, z) = 5x^2 - 2y^2 + z$, $\mathbf{v} = \dfrac{3}{5}\mathbf{i} - \dfrac{4}{5}\mathbf{k}$, $P(1, 1, 1)$.

24. $f(x, y, z) = e^{-x+y^2+2z}$, $\mathbf{v} = \left(\dfrac{\sqrt{3}}{3}, \dfrac{-\sqrt{3}}{3}, \dfrac{\sqrt{3}}{3} \right)$, $P(1, 1, 0)$

25. If $f(x, y, z) = xy^2 + y^2z^3 + z^3x$, find the directional derivative of f at $(2, -1, 1)$ in the direction toward $(3, 4, 5)$.

26. If $f(x, y) = x^2y + xy^2$, find the directional derivative of f at $(2, 1)$ in the direction toward $(3, 2)$.

27. Find the gradient for the following functions.

(a) $f(x, y) = \ln\sqrt{x^2 + y^2}$ (b) $f(x, y, z) = e^{-x} + e^{-2y} + e^{3z}$

28. Consider the function described by $f(x, y) = 16 - (x - 2)^2 - (y - 3)^2$.

(a) Sketch the graph of f.

(b) Find the directional derivative of f at $(4, 4)$ in the direction toward the point $(2, 3)$.

(c) Find the direction in which the rate of change of f at $(4, 4)$ is a maximum. What is this maximum rate of change?

(d) Find an equation of the tangent plane to the graph of f at the point $(4, 4, 11)$.

(e) Find an equation of the tangent line to the curve of intersection of the graph of f and the plane $y = 4$ at the point $(4, 4, 11)$.

29. Find the slope of the curve at the point $(1, 2, 4)$ formed by the intersection of the surface $z = xy^2$ with a plane perpendicular to the y-axis.

30. If $f(x, y, z) = 3x^2 + xy - 2y^2 - yz + z^2$, find the rate of change of $f(x, y, z)$ at $(1, -2, -1)$ in the direction $\mathbf{v} = 2\mathbf{i} - 2\mathbf{j} - \mathbf{k}$.

In Exercises 31 through 34, find the direction in which the function is increasing most rapidly at P, and find the maximum rate of increase.

31. $f(x, y) = x^2 + xy$, $P(1, -1)$ 32. $f(x, y) = \sqrt{1 - x^2 - y^2}$, $P(0, 1)$

33. $f(x, y) = x^2 + y^2$, $P(-1, 2)$ 34. $f(x, y) = 6 - 2x - 3y$, $P(0, 0)$

Find an equation of the tangent plane to the surfaces described in Exercises 35 through 40 at the given point.

35. $z = x^2 + y^2 - 1$, $(1, 2, 4)$ 36. $z = 2x + 3y + 1$, $(1, -1, 0)$

37. $z = xy \sin xz$, $\left(1, \dfrac{\pi}{2}, \dfrac{\pi}{2}\right)$ 38. $z = x^2 - 2y^2 + 3xy$, $(2, -1, -4)$

39. $z = x^2 + 4y^2$, $(-2, 1, 8)$ 40. $x^2 + y^2 - z^2 = 0$, $(1, 0, 1)$

41. Find $\partial f/\partial r$ and $\partial f/\partial\theta$ if $f(x, y) = xy$, where $x = e^{2r} \cos\theta$, $y = e^r \sin\theta$.

42. If $w = \dfrac{x}{y} + \dfrac{y}{z} + \dfrac{z}{x}$, show that $x\dfrac{\partial w}{\partial x} + y\dfrac{\partial w}{\partial y} + z\dfrac{\partial w}{\partial z} = 0$.

43. Suppose $w = u^2v + uv^2 + 2u - 3v$, $u = \sin(x + y + z)$, and $v = \cos(x + y - z)$. Find $\partial w/\partial x$ when $x = \pi/2$, $y = \pi/4$, $z = \pi/4$.

44. If $z = x^2 + 2xy + y^2$, $x = t \cos t$, and $y = t \sin t$, find dz/dt.

45. In which direction from the point $(1, 1)$ does the directional derivative of $f(x, y) = x^2 + y^2$ vanish?

46. Let

$$f(x, y) = \begin{cases} \dfrac{y^3 - x^3}{y^2 + x^2}, & \text{if } (x, y) \ne (0, 0) \\ 0, & \text{if } (x, y) = (0, 0). \end{cases}$$

Find $f_x(0, 0)$ and $f_y(0, 0)$.

47. Show that $\displaystyle\lim_{(x,y)\to(1,1)} \dfrac{(y - 1)^2}{(x - 1)^2 + (y - 1)^2}$ does not exist.

48. If $x + y + z$ is constant, show that the product xyz is a maximum when $x = y = z$.

49. Find the point on the plane $2x + 2y + z = 4$ that is nearest the origin.

50. Find the maximum value of $P(x, y, z) = 900z - 400x - 100y$ subject to the constraint $z = 5 - \dfrac{1}{x} - \dfrac{1}{y}$, $x > 0$, $y > 0$.

51. Maximize $f(x, y) = x^2 - y^2 - y$ subject to the constraint $x^2 + y^2 - 1 = 0$.

52. The radius of a right circular cone is decreasing at 2 in/min and the height is increasing at 3 in/min. Find the rate of change of the volume when the height is 8 inches and the radius is 4 inches.

53. A closed metal can in the shape of a right circular cylinder is to have an inside height of 10 inches, an inside radius of 2 inches, and a thickness of 0.1 inch. If the cost of the metal to be used is 20 cents per cubic inch, use differentials to find the approximate cost of the metal to be used in manufacturing the can.

16

INTEGRAL CALCULUS OF MULTIVARIATE FUNCTIONS

The definite integral was introduced as a means of determining the area of a region bounded by the x-axis, a pair of lines $x = a$ and $x = b$, and the graph of a continuous nonnegative function f. The procedure was to partition $[a, b]$ into subintervals, evaluate f at a value z_i in each subinterval, multiply this by the width Δx_i of the subinterval, and sum the products. The result is a Riemann sum of the form

$$\sum_{i=1}^{n} f(z_i) \, \Delta x_i,$$

which approximates the area A.

Taking the limit of Riemann sums as the width of each subinterval approaches zero leads to the definite integral:

$$A = \int_{a}^{b} f(x)dx.$$

While this application motivated the study of definite integrals, we later found many other concepts such as volume, moment of inertia, arc length, and surface area, that could be approximated by a Riemann sum. The integration of multivariate functions has a similar development. We begin by discussing a simple geometric application, and once the notion has been presented it can be applied to widely differing problems.

16.1
DOUBLE INTEGRALS

Suppose f is a continuous nonnegative function of two variables defined on a bounded region R in the xy-plane. Our objective is to determine the volume V of the solid bounded by the cylinder with base R and top lying in the surface described by $z = f(x, y)$. See Figure 16.1.

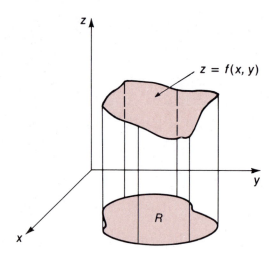

FIGURE 16.1

We begin by drawing a grid of lines through R parallel to the x- and y-axes, as shown in Figure 16.2. Let $R_1, R_2, R_3, \ldots, R_n$ denote those rectangles formed by the grid that lie entirely within R. The collection of these rectangles is denoted $\mathcal{P}$, while $\|\mathcal{P}\|$ denotes the maximum length of the diagonals of the rectangles R_i.

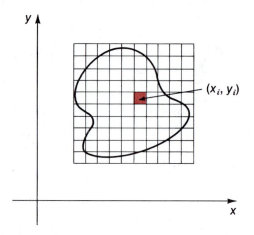

FIGURE 16.2

We arbitrarily choose (x_i, y_i) in R_i for each i and compute $f(x_i, y_i)$. The volume ΔV_i of the parallelepiped with base R_i and height $f(x_i, y_i)$ is

$$\Delta V_i = f(x_i, y_i)\Delta A_i,$$

where ΔA_i denotes the area of the rectangle R_i. A typical parallelepiped is shown in Figure 16.3. The volume of the portion of the solid with base R_i and top lying in the surface $z = f(x, y)$ is approximated by ΔV_i.

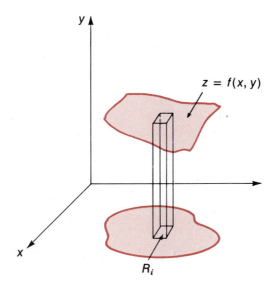

FIGURE 16.3

Summing the volumes of parallelepipeds gives an approximation to the total volume:

(16.1)
$$V \approx \sum_{i=1}^{n} f(x_i, y_i)\Delta A_i.$$

This sum is called a **Riemann sum** for f on the region R.

EXAMPLE 1   A solid has a rectangular base R in the xy-plane bounded by $x = 2$, $y = 1$, $y = 2$ and the y-axis. The top of the solid lies in the surface $z = x^2 + y^2$. Approximate the volume of the solid by using a Riemann sum for $f(x, y) = x^2 + y^2$ on R. Partition the rectangle with the lines $x = 0$, $x = 1$, $x = 2$, $y = 1$, $y = 3/2$, $y = 2$ and choose the point (x_i, y_i) to be the lower left corner of each rectangular element R_i. This partitioning is shown in Figure 16.4 on p. 858.

SOLUTION

$$\sum_{i=1}^{4} f(x_i, y_i)\Delta A_i = f(0, 1)\left(\frac{1}{2}\right) + f(1, 1)\left(\frac{1}{2}\right) + f\left(0, \frac{3}{2}\right)\left(\frac{1}{2}\right) + f\left(1, \frac{3}{2}\right)\left(\frac{1}{2}\right)$$

$$= \frac{1}{2}\left(1 + 2 + \frac{9}{4} + \frac{13}{4}\right) = \frac{17}{4}.$$

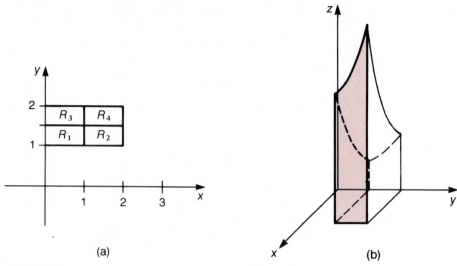

FIGURE 16.4

This Riemann sum gives a lower bound for the volume of the solid, since all the parallelepipeds lie beneath the top of the solid.   □

EXAMPLE 2   Repeat Example 1 using the same partition, but choosing (x_i, y_i) as the upper right corner of R_i.

SOLUTION

In this case,

$$\sum_{i=1}^{4} f(x_i, y_i)A_i = f\left(1, \frac{3}{2}\right)\left(\frac{1}{2}\right) + f\left(2, \frac{3}{2}\right)\left(\frac{1}{2}\right) + f(1, 2)\left(\frac{1}{2}\right) + f(2, 2)\left(\frac{1}{2}\right)$$

$$= \frac{1}{2}\left(\frac{13}{4} + \frac{25}{4} + 5 + 8\right) = \frac{45}{4}.$$

This Riemann sum provides an upper bound for the volume of the solid since the parallelepipeds contain the solid.   □

The volume V of the solid bounded by the cylinder with base R and top lying in the surface $z = f(x, y)$ is the limit of the approximations

$$\sum_{i=1}^{n} f(x_i, y_i)\Delta A_i$$

as the grid lines in both directions become closer together, (that is, as $\|\mathscr{P}\| \to 0$). This produces more parallelepipeds, the sum of whose volumes approaches the volume V. That is,

$$V = \lim_{\|\mathscr{P}\| \to 0} \sum_{i=1}^{n} f(x_i, y_i)\Delta A_i.$$

This procedure for computing volume provides the basic notion of the integral of a function f of two variables over a region R in the plane. The integral is called a double integral and denoted $\iint$ to emphasize that the region of integration is expressed in terms of two variables.

(16.2)
DEFINITION

If a function f of two variables is defined on a region R in the plane, the **double integral of f over R** is defined by

$$\iint_R f(x, y)\,dA = \lim_{\|\mathscr{P}\| \to 0} \sum_{i=1}^{n} f(x_i, y_i)\Delta A_i,$$

provided this limit exists.

Notice that Definition 16.2 does not require f to be nonnegative, although this requirement is needed when applying the double integral to volume. There is also no continuity restriction on f.

When the limit in Definition 16.2 exists, f is said to be **integrable** over R. It can be shown that if f is continuous on R, then f is integrable over R. This proof is included in most advanced calculus textbooks.

The limit in Definition 16.2 is taken over every collection of grid lines and every choice (x_i, y_i) in R_i, $i = 1, 2, \ldots, n$. To be precise,

$$\lim_{\|\mathscr{P}\| \to 0} \sum_{i=1}^{n} f(x_i, y_i)\Delta A_i = L$$

means that for every number $\varepsilon > 0$, there exists a number $\delta > 0$ such that if grid lines produce a set of rectangles $\mathscr{P} = \{R_1, \ldots, R_n\}$, with $\|\mathscr{P}\| < \delta$, then the Riemann sum $\sum_{i=1}^{n} f(x_i, y_i)\,\Delta A_i$ has the property that

$$\left| \sum_{i=1}^{n} f(x_i, y_i)\Delta A_i - L \right| < \varepsilon,$$

for any collection (x_i, y_i) in R_i, $i = 1, \ldots, n$.

EXAMPLE 3

Evaluate $\iint_R \sqrt{16 - x^2 - y^2}\, dA$ where R is the disk $x^2 + y^2 \leq 16$ in the xy-plane.

SOLUTION

It would be tedious to evaluate this integral using the definition. However, since $\sqrt{16 - x^2 - y^2} \geq 0$, this double integral is the volume of the solid with base R and top lying in the surface $z = \sqrt{16 - x^2 - y^2}$. This surface is the

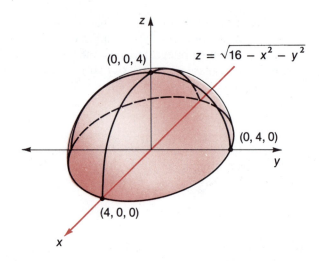

FIGURE 16.5

upper part of the sphere $x^2 + y^2 + z^2 = 16$ shown in Figure 16.5. Consequently, the volume is

$$\frac{1}{2}\left(\frac{4}{3}\pi r^3\right) = \frac{4}{6}\pi(4)^3 = \frac{128}{3}\pi,$$

and

$$\iint_R \sqrt{16 - x^2 - y^2}\, dA = \frac{128}{3}\pi.$$

□

The double integral satisfies the usual arithmetic properties associated with the definite integral. For example,

$$\iint_R (f \pm g)(x, y)dA = \iint_R f(x, y)dA \pm \iint_R g(x, y)dA$$

$$\iint_R cf(x, y)dA = c\iint_R f(x, y)dA,$$

provided that the integrals exist and c is a constant. In addition, if R is a region composed of disjoint subregions R_1 and R_2 on which the integral of f exists, then

$$\iint_R f(x, y)dA = \iint_{R_1} f(x, y)dA + \iint_{R_2} f(x, y)dA.$$

EXERCISE SET 16.1

1. Find the value of the Riemann sum for the function and partition described in Example 1 if (x_i, y_i) in R_i is chosen to be

   (a) the lower right corner of R_i;   (b) the upper left corner of R_i;

   (c) the midpoint of R_i.

2. Suppose $f(x, y) = x^2y$, R is the rectangle in the xy-plane bounded by $x = 0$, $y = 0$, $x = 3$ and $y = 2$, and $\mathcal{P}$ is the partition defined by the lines bounding R together with the lines $x = 1$, $x = 2$ and $y = 1$. Find the value of the Riemann sum if (x_i, y_i) is chosen to be

   (a) the lower left corner of R_i;   (b) the upper right corner of R_i;

   (c) the midpoint of R_i.

3. Suppose $f(x, y) = y^2 - x^2$, R is the triangle in the xy-plane with vertices $(1, 1)$, $(5, 4)$ and $(7, 0)$, and $\mathcal{P}$ is the partition defined by the lines at integral values of x and y. Find the value of the Riemann sum if (x_i, y_i) is chosen to be

   (a) the lower left corner of R_i;   (b) the upper right corner of R_i;

   (c) the midpoint of R_i.

4. Find the value of the Riemann sum for the function described in Exercise 3 if the partition is enlarged to include those rectangles bounded by the lines $x = 3/2$, $x = 11/2$ and $y = 1/2$.

5. Suppose $f(x, y) = 3 - x - y$, R is the triangle in the xy-plane bounded by $x = 0$, $y = 0$, $y = 3 - x$, and $\mathcal{P}$ is the partition defined by the lines $x = 0$, $x = 1$, $x = 2$, $y = 0$, $y = 1$, and $y = 2$. Find the value of the Riemann sum if (x_i, y_i) is chosen to be the upper left corner of R_i.

6. Use the definition of the double integral to show that if f is integrable over R and c is a constant, then cf is integrable over R and

$$\iint_R cf(x, y)dA = c \iint_R f(x, y)dA.$$

7. Use the definition of the double integral to show that if f and g are both integrable over R and $f(x, y) \geq g(x, y)$ on R, then

$$\iint_R f(x, y)dA \geq \iint_R g(x, y)dA.$$

8. Use the definition of the double integral to show that if f is the constant function $f(x, y) = k$ and R is the rectangular region bounded by $x = a$, $x = b$, $y = c$, and $y = d$, with $a < b$ and $c < d$ then

$$\iint_R f(x, y)dA = k(b - a)(d - c).$$

9. Use the method in Example 3 to evaluate $\iint_R \sqrt{9 - x^2 - y^2}\, dA$ where R is the disk $x^2 + y^2 \leq 9$ in the xy-plane.

10. Evaluate $\iint_R 4\, dA$ where R is the disk $x^2 + y^2 \leq 9$ in the xy-plane.

16.2
ITERATED INTEGRALS

The double integral of a function of two variables is defined as the limit of Riemann sums.

You will recall that very little could be accomplished with the definite integral of a single variable function until the fundamental theorem of calculus provided a means to determine integrals without using the limit process. The same problem occurs with double integrals; in fact, it has been compounded. Until we determine a method for evaluating double integrals without resorting to the limit process, double integration is theoretically interesting but not often applicable. Fortunately, a double integral can often be converted into what can be considered a pair of single integrals.

Suppose the region R in the xy-plane is described by

$$a \le x \le b, \qquad g_1(x) \le y \le g_2(x),$$

where g_1 and g_2 are continuous functions. See Figure 16.6.

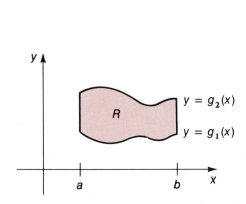

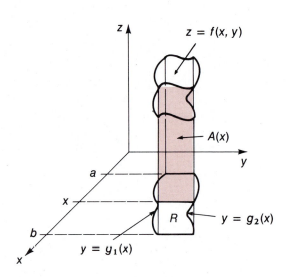

FIGURE 16.6

FIGURE 16.7

Consider the double integral of a continuous nonnegative function f over R, that is,

$$\iint_R f(x, y)dA.$$

The geometric interpretation of this double integral is the volume of the solid bounded by the cylinder with base R and top lying in the surface $z = f(x, y)$. For each x in $[a, b]$, let $A(x)$ denote the area of the region formed by intersecting the solid with the plane through $(x, 0, 0)$ perpendicular to the x-axis, as shown in Figure 16.7.

If we view this region projected into the yz-plane, we see from Figure 16.8 that the cross-sectional area $A(x)$ of our solid is given by

$$A(x) = \int_{g_1(x)}^{g_2(x)} f(x, y)dy.$$

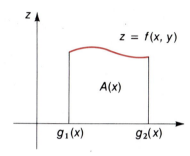

FIGURE 16.8

The volume of a solid with known cross section was shown in Section 6.4 to be $\int_a^b A(x)dx$. Consequently, the volume of this solid is

$$\int_a^b A(x)dx = \int_a^b \left[\int_{g_1(x)}^{g_2(x)} f(x, y)dy\right]dx.$$

Since this volume is also $\iint_R f(x, y)dA$,

(16.3)
$$\iint_R f(x, y)dA = \int_a^b \left[\int_{g_1(x)}^{g_2(x)} f(x, y)dy\right]dx.$$

The expression on the right side is called an **iterated integral**. It is generally expressed with the brackets deleted.

EXAMPLE 1   Find the volume of the solid with rectangular base in the xy-plane bounded by the planes $x = 2$, $y = 1$, $y = 2$, and the y-axis. The top of the solid lies in the surface $z = x^2 + y^2$.

SOLUTION
The volume to be calculated is shown in Figure 16.9. The base region R is described by

$$0 \leq x \leq 2, \qquad 1 \leq y \leq 2$$

so $g_1(x) = 1$, $g_2(x) = 2$, and $A(x) = \int_1^2 (x^2 + y^2)dy$.

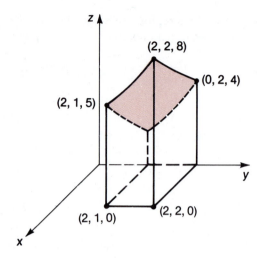

FIGURE 16.9

Thus,

$$V = \iint_R (x^2 + y^2)dA = \int_0^2 A(x)dx$$

$$= \int_0^2 \left[\int_1^2 (x^2 + y^2)dy \right] dx.$$

In the integral $\int_1^2 (x^2 + y^2)dy$, x is considered as a constant. Consequently,

$$\int_1^2 (x^2 + y^2)dy = \left[x^2 y + \frac{y^3}{3} \right]_{y=1}^{y=2} = \left(2x^2 + \frac{8}{3} \right) - \left(x^2 + \frac{1}{3} \right) = x^2 + \frac{7}{3},$$

and

$$V = \int_0^2 \left(x^2 + \frac{7}{3} \right) dx = \frac{x^3}{3} + \frac{7}{3}x \bigg]_0^2 = \frac{8}{3} + \frac{14}{3} = \frac{22}{3}. \qquad \square$$

EXAMPLE 2   Find the volume of the solid bounded above by the graph of $f(x, y) = xy$, below by the xy-plane, and by the cylinders $y = x^2$ and $y = x$.

SOLUTION

The base region of the solid in the xy-plane is shown in Figure 16.10. The graphs of $y = x^2$ and $y = x$ intersect at $(0, 0)$ and $(1, 1)$, so R is described by

$$0 \leq x \leq 1, \qquad x^2 \leq y \leq x.$$

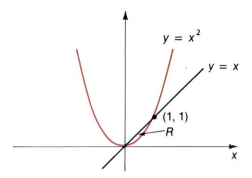

FIGURE 16.10

Since the graph of f lies above R, the volume is

$$V = \int_0^1 \int_{x^2}^x xy \, dy \, dx$$

$$= \int_0^1 \frac{xy^2}{2} \Bigg]_{y=x^2}^{y=x} dx$$

$$= \frac{1}{2} \int_0^1 (x^3 - x^5) dx$$

$$= \frac{1}{2} \left[\frac{x^4}{4} - \frac{x^6}{6} \right]_0^1 = \frac{1}{2} \left(\frac{1}{4} - \frac{1}{6} \right) = \frac{1}{24}. \qquad \square$$

EXAMPLE 3   Find the volume of the solid in the first octant that is bounded below by the plane $x + z = 1$ and above by the plane $4x + y + z = 4$.

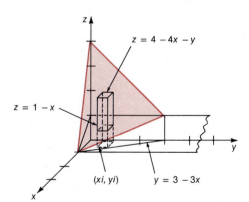

FIGURE 16.11

SOLUTION

   The solid is sketched in Figure 16.11. An element of volume for this solid has the form

$$\Delta V_i = [(4 - 4x_i - y_i) - (1 - x_i)] \Delta A_i.$$

To find the total volume using a double integral, we must know the region R in the xy-plane over which to integrate. This is obtained by finding the intersection of the two surfaces and projecting this intersection into the xy-plane. Solving the equations $z = 4 - 4x - y$ and $z = 1 - x$ simultaneously, we have

$$1 - x = 4 - 4x - y, \qquad \text{so } y = 3 - 3x.$$

The region R in the first octant is described by $0 \le y \le 3 - 3x$, $0 \le x \le 1$. The volume of the solid is

$$V = \iint\limits_R [(4 - 4x - y) - (1 - x)]dA = \int_0^1 \int_0^{3-3x} (3 - 3x - y)\, dy\, dx$$

$$= \int_0^1 \left[3y - 3xy - \frac{y^2}{2} \right]_{y=0}^{y=3-3x} dx$$

$$= \int_0^1 \left[3(3 - 3x) - 3x(3 - 3x) - \frac{(3 - 3x)^2}{2} \right] dx$$

$$= \frac{9}{2} \int_0^1 (1 - x)^2 dx = \left[-\frac{9}{2}\frac{(1 - x)^3}{3} \right]_0^1 = \frac{9}{6} = \frac{3}{2}. \qquad \square$$

The connection between double integrals and iterated integrals is not limited to the case when the integrand is nonnegative. It can be established for arbitrary functions by using Riemann sums, but the process is quite laborious. The theorem relating these concepts is stated below.

(16.4)
THEOREM

Suppose f is a continuous function on a region R in the xy-plane.
(i) If R is described by $a \le x \le b$, $g_1(x) \le y \le g_2(x)$,

where g_1 and g_2 are continuous functions, then

$$\iint\limits_R f(x, y)dA = \int_a^b \int_{g_1(x)}^{g_2(x)} f(x, y)dy\, dx.$$

(ii) If R is described by $c \le y \le d$, $h_1(y) \le x \le h_2(y)$,

where h_1 and h_2 are continuous functions, then

$$\iint\limits_R f(x, y)dA = \int_c^d \int_{h_1(y)}^{h_2(y)} f(x, y)dx\, dy.$$

EXAMPLE 4

Evaluate $\displaystyle\iint\limits_R (x^2 - 2y)dA$ where R is the region bounded by the x-axis, the line $x = 1$ and the graph of $y = 2x^2$.

SOLUTION

The region can be described by fixing x in $[0, 1]$ and observing that (x, y) is in R precisely when $0 \le y \le 2x^2$. See Figure 16.12.

Thus,
$$\iint\limits_{R} (x^2 - 2y)dA = \int_0^1 \int_0^{2x^2} (x^2 - 2y)dy\, dx$$

$$= \int_0^1 \left[x^2 y - y^2 \right]_{y=0}^{y=2x^2} dx$$

$$= \int_0^1 (2x^4 - 4x^4)dx$$

$$= -\frac{2}{5}x^5 \bigg]_0^1 = -\frac{2}{5}.$$

Alternatively, R can be described by fixing y in $[0, 2]$ and observing that (x, y) is in R precisely when $\sqrt{\dfrac{y}{2}} \le x \le 1$. See Figure 16.13.

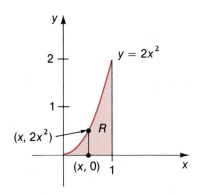

FIGURE 16.12

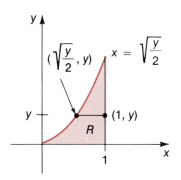

FIGURE 16.13

Using this representation,

$$\iint\limits_{R} (x^2 - 2y)dA = \int_0^2 \int_{\sqrt{\frac{y}{2}}}^1 (x^2 - 2y)dx\, dy$$

$$= \int_0^2 \left[\frac{x^3}{3} - 2xy \right]_{x=\sqrt{\frac{y}{2}}}^{x=1} dy$$

$$= \int_0^2 \left[\left(\frac{1}{3} - 2y \right) - \left(\frac{y^{3/2}}{6\sqrt{2}} - \frac{2y^{3/2}}{\sqrt{2}} \right) \right] dy$$

$$= \int_0^2 \left(\frac{1}{3} - 2y + \frac{11}{6\sqrt{2}} y^{3/2} \right) dy$$

$$= \left[\frac{1}{3}y - y^2 + \frac{11}{6\sqrt{2}} y^{5/2} \cdot \frac{2}{5} \right]_0^2$$

$$= \frac{2}{3} - 4 + \frac{11}{15\sqrt{2}} (4\sqrt{2}) = -\frac{2}{5}. \qquad \square$$

Example 4 demonstrates that for certain regions, $\iint_R f(x,\ y)dx$ can be expressed as either

$$\int_a^b \int_{g_1(x)}^{g_2(x)} f(x,\ y)dy\ dx$$

or as

$$\int_c^d \int_{h_1(y)}^{h_2(y)} f(x,\ y)dx\ dy.$$

Changing from one iterated integral form to the other is called **reversing the order of integration**. The following example shows that the technique of reversing the order of integration is sometimes essential to the evaluation of an iterated integral.

EXAMPLE 5   Evaluate the iterated integral

$$\int_0^1 \int_x^1 e^{y^2}\ dy\ dx.$$

SOLUTION

Since there is no elementary antiderivative for the function described by $f(y) = e^{y^2}$, the fundamental theorem of calculus cannot be used to determine $\int e^{y^2}\ dy$. Suppose, however, we reexpress this iterated integral as a double integral. The region of integration R is described by

$$0 \le x \le 1, \qquad x \le y \le 1$$

and shown in Figure 16.14. This region can also be described by

$$0 \le y \le 1, \qquad 0 \le x \le y.$$

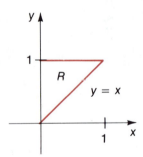

FIGURE 16.14

Thus,

$$\int_0^1 \int_x^1 e^{y^2}\ dy\ dx = \iint_R e^{y^2}\ dA = \int_0^1 \int_0^y e^{y^2}\ dx\ dy$$

$$= \int_0^1 xe^{y^2}\Big]_{x=0}^{x=y}\ dy = \int_0^1 ye^{y^2}\ dy$$

$$= \frac{1}{2}\,e^{y^2}\Big]_0^1 = \frac{1}{2}(e - 1).\qquad\square$$

The area A of a region R in the plane has the same numerical value as the volume of the solid with base R and constant height one. See Figure 16.15.

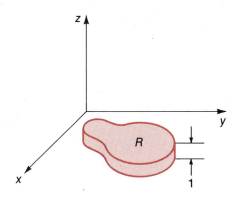

FIGURE 16.15

Thus,
$$A = \iint_R 1 \, dA.$$

In particular, if R is a region bounded by the lines $x = a$, $x = b$, and the graphs of the continuous functions f and g, where $g(x) \leq f(x)$ for x in $[a, b]$, then

$$A = \iint_R 1 \, dA = \int_a^b \int_{g(x)}^{f(x)} dy \, dx = \int_a^b [f(x) - g(x)]dx.$$

This is the formula given in Section 6.1 for the area of a region bounded between two curves.

EXAMPLE 6   Find the area of the region bounded by the curves $y = \sin x$ and $y = \cos x$ for $0 \leq x \leq \pi/4$.

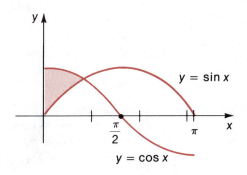

FIGURE 16.16

SOLUTION

From Figure 16.16, we see that $\cos x \geq \sin x$ in the interval $[0, \pi/4]$. So

$$A = \iint\limits_{R} dA = \int_{0}^{\pi/4} \int_{\sin x}^{\cos x} dy \, dx$$

$$= \int_{0}^{\pi/4} y \Big]_{y=\sin x}^{y=\cos x} dx$$

$$= \int_{0}^{\pi/4} (\cos x - \sin x) dx$$

$$= (\sin x + \cos x) \Big]_{0}^{\pi/4} = \frac{\sqrt{2}}{2} + \frac{\sqrt{2}}{2} - 1 = \sqrt{2} - 1. \quad \square$$

EXERCISE SET 16.2

Evaluate the integrals in Exercises 1 through 14.

1. $\displaystyle\int_{1}^{2} \int_{3}^{4} xy^2 \, dy \, dx$ **2.** $\displaystyle\int_{1}^{2} \int_{3}^{4} xy^2 \, dx \, dy$

3. $\displaystyle\int_{-\pi}^{3\pi/2} \int_{0}^{2\pi} (y \sin x + x \cos y) dy \, dx$ **4.** $\displaystyle\int_{0}^{1} \int_{\pi}^{2\pi} (x + \sin y) dy \, dx$

5. $\displaystyle\int_{-1}^{3} \int_{-2}^{1} xe^{xy} \, dy \, dx$ **6.** $\displaystyle\int_{-1}^{1} \int_{0}^{1} xy \, e^{x^2+y^2} \, dx \, dy$

7. $\displaystyle\int_{0}^{2} \int_{x}^{2x} (x^2 + y^3) dy \, dx$ **8.** $\displaystyle\int_{0}^{2} \int_{y}^{2y} (x^2 + y^3) dx \, dy$

9. $\displaystyle\int_{0}^{\pi} \int_{0}^{x} \cos x \, dy \, dx$ **10.** $\displaystyle\int_{0}^{\pi} \int_{0}^{y} \cos x \, dx \, dy$

11. $\displaystyle\int_{0}^{1} \int_{0}^{x} (x^2 + \sqrt{y}) \, dy \, dx$ **12.** $\displaystyle\int_{0}^{\pi} \int_{x/2}^{x} (\sin x + \cos y) dy \, dx$

13. $\displaystyle\int_{1}^{e} \int_{1}^{x} \ln xy \, dy \, dx$ **14.** $\displaystyle\int_{0}^{\pi/2} \int_{0}^{\sin y} \frac{1}{\sqrt{1 - x^2}} dx \, dy$

15–28. Sketch the graphs of the regions over which the integration in Exercises 1 through 14 is performed, and express the integral as a double integral that reverses the order of integration. Evaluate the double integral obtained and compare the result to the previously obtained value.

Exercises 29 through 32 describe the boundaries of a solid whose base lies in the xy-plane. Use double integrals to compute (a) the area of the base of the solid, and (b) the volume of the solid.

29. $x = 0, x = 1, y = 0, y = 2, z = 0, z = x^2y$

30. $x = -1, x = 2, y = -1, y = 3, z = 0, z = 4 + 2x + 2y$

31. $x = 2, y = 1, y = x, z = 0, z = x + y^2$

32. $x = -1, x = 1, y = -x, y = x, z = 0, z = x^2 + y^2$

In Exercises 33 through 40, use double integration to find the volume of the solid.

33. The solid in the first octant bounded by the plane $z = 4 - x - y$.

34. The solid bounded by the cylinder $x^2 + y^2 = 1$ and the planes $z = 0$ and $z = 1 - x$.

35. The solid bounded above by the paraboloid $z = 4 - x^2 - y^2$ and below by the xy-plane.

36. The solid in the first octant bounded by the cylinder $x^2 + z^2 = 4$ and the plane $z = 3 - y$.

37. The wedge cut from the cylinder $x^2 + y^2 = 9$ by the planes $z = 0$ and $z = y$, when $y \geq 0$.

38. The wedge cut from the cylinder $x^2 + y^2 = 9$ by the planes $z = 0$ and $z = 2y$, when $y \geq 0$.

39. The solid bounded by the cylinder $9x^2 + 4y^2 = 36$ and the planes $z = 0$ and $x + y + 2z = 6$.

40. The solid in the first octant bounded above by planes $z = y$ and $x + y + z = 3$.

41. Suppose f is continuous and g is defined by

$$g(t) = \int_0^t \int_0^x f(x, y)\, dy\, dx.$$

Determine $g'(t)$ and $g''(t)$.

Putnam exercises:

42. Show that the integral equation

$$f(x, y) = 1 + \int_0^x \int_0^y f(u, v)\, du\, dv$$

has at most one solution continuous for $0 \leq x \leq 1, 0 \leq y \leq 1$. (This exercise was problem 5, part I of the eighteenth William Lowell Putnam examination given on February 8, 1958. The examination and its solution can be found in the January 1961 issue of the *American Mathematical Monthly*, pages 18–22.)

43. If f is a positive, monotone decreasing function defined on $0 \leq x \leq 1$, prove that

$$\frac{\int_0^1 x(f(x))^2 dx}{\int_0^1 xf(x)dx} \leq \frac{\int_0^1 (f(x))^2 dx}{\int_0^1 f(x)dx}.$$

(This exercise was problem 3, part II of the seventeenth William Lowell Putnam examination given on December 6, 1956. The examination and its solution can be found in the November 1957 issue of the *American Mathematical Monthly*, pages 649–654.)

16.3
DOUBLE INTEGRALS IN POLAR COORDINATES

In Chapter 10 we saw that certain regions in the plane can be more naturally represented by polar, rather than cartesian, equations. Regions bounded by circles or spirals centered at the origin fall into this category, as do those bounded by lemniscates, cardioids, and so on. In this section, we will describe the double integral of a function over a region represented in polar coordinates, following as closely as possible the development of double integrals using cartesian coordinates.

To begin, suppose R is a region in the plane and f is a continuous nonnegative function on R. The volume of the solid bounded by the cylinder with base R and above by the graph of f is

$$V = \iint\limits_{R} f(x, y)dA,$$

or, if f is expressed using polar coordinates,

$$V = \iint\limits_{R} f(r, \theta)dA.$$

When R is described using cartesian equations, this double integral is equivalent to an iterated integral with dA replaced by $dy\,dx$ or $dx\,dy$, since an element of area ΔA_i is given by $\Delta x_i \Delta y_i$.

To derive an equivalent iterated integral using polar equations, we first place a polar grid over the region R. This grid consists of lines through the origin and arcs of circles centered at the origin. These curves are the graphs of equations in which one of the polar coordinates is set to a constant value. See Figure 16.17.

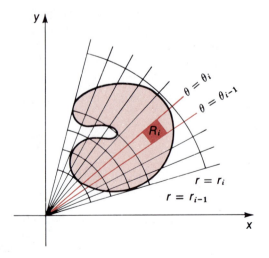

FIGURE 16.17

As in the case of cartesian regions, we label the "polar rectangles" formed by the grid and lying totally inside R by $R_1, R_2, \ldots, R_n$. The set of these polar rectangles is denoted by $\mathcal{P}$ and $\|\mathcal{P}\|$ denotes the length of the longest diagonal of the polar rectangles in $\mathcal{P}$. Let $\theta = \theta_{i-1}$ and $\theta = \theta_i$ be the lines through the origin that form the boundaries of R_i. The boundaries formed by the circular arcs are denoted by $r = r_{i-1}$ and $r = r_i$.

In Section 10.3 we found that the area of the sector of the circle bounded by $\theta = \theta_{i-1}$, $\theta = \theta_i$, and $r = r_i$ is

$$\frac{1}{2} r_i^2 (\theta_i - \theta_{i-1}) = \frac{1}{2} r_i^2 \, \Delta\theta_i.$$

Similarly, the area bounded by $\theta = \theta_{i-1}$, $\theta = \theta_i$, and $r = r_{i-1}$ is $\frac{1}{2} r_{i-1}^2 \, \Delta\theta_i$.

Since R_i has boundaries $\theta = \theta_{i-1}$, $\theta = \theta_i$, $r = r_{i-1}$, and $r = r_i$, the area of R_i is

$$\Delta A_i = \frac{1}{2} r_i^2 \, \Delta\theta_i - \frac{1}{2} r_{i-1}^2 \, \Delta\theta_i$$

$$= \frac{1}{2}(r_i + r_{i-1})(r_i - r_{i-1})\Delta\theta_i = \frac{r_i + r_{i-1}}{2} \, \Delta r_i \, \Delta\theta_i.$$

To approximate the volume ΔV_i bounded by the cylinder with base R_i and above by the graph of f, we evaluate f at a point in R_i and multiply by ΔA_i.

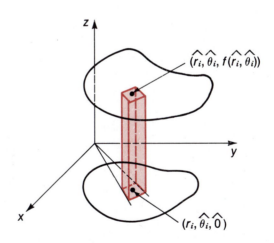

FIGURE 16.18

In this case, we evaluate f at $(\hat{r}_i, \hat{\theta}_i)$ (see Figure 16.18), where

$$\hat{r}_i = \frac{r_i + r_{i-1}}{2} \qquad \text{and} \qquad \hat{\theta}_i = \frac{\theta_i + \theta_{i-1}}{2}.$$

Then
$$\Delta V_i = f(\hat{r}_i, \hat{\theta}_i)\Delta A_i$$

$$= f(\hat{r}_i, \hat{\theta}_i) \frac{1}{2}(r_i + r_{i-1})\Delta r_i \, \Delta\theta_i$$

$$= f(\hat{r}_i, \hat{\theta}_i)\hat{r}_i \, \Delta r_i \, \Delta\theta_i.$$

The total volume is approximated by the Riemann sum

$$V \approx \sum_{i=1}^{n} \Delta V_i = \sum_{i=1}^{n} f(\hat{r}_i, \hat{\theta}_i) \hat{r}_i \, \Delta r_i \, \Delta \theta_i$$

and

$$\iint\limits_{R} f(r, \theta) dA = V = \lim_{\|\mathcal{P}\| \to 0} \sum_{i=1}^{n} f(\hat{r}_i, \hat{\theta}_i) \hat{r}_i \, \Delta r_i \, \Delta \theta_i.$$

This motivation leads to the following theorem. As in the case of cartesian coordinates, it is not necessary that the function f be nonnegative.

(16.5)
THEOREM

Suppose f is a continuous function on a region R in the xy-plane.
(i) If continuous functions g_1 and g_2 exist with R described by

$$\alpha \le \theta \le \beta, \qquad g_1(\theta) \le r \le g_2(\theta),$$

then

$$\iint\limits_{R} f(r, \theta) dA = \int_{\alpha}^{\beta} \int_{g_1(\theta)}^{g_2(\theta)} f(r, \theta) r \, dr \, d\theta.$$

(ii) If continuous functions h_1 and h_2 exist with R described by

$$a \le r \le b, \qquad h_1(r) \le \theta \le h_2(r),$$

then

$$\iint\limits_{R} f(r, \theta) dA = \int_{a}^{b} \int_{h_1(r)}^{h_2(r)} f(r, \theta) r \, d\theta \, dr.$$

EXAMPLE 1

Find $\displaystyle\iint\limits_{R} (3x + y) dA$ where R is the region in the first quadrant that lies inside the circle $x^2 + y^2 = 4$ and outside the circle $x^2 + y^2 = 1$.

SOLUTION

The region R is sketched in Figure 16.19. It is easily expressed in polar coordinates:

$$0 \le \theta \le \frac{\pi}{2}, \qquad 1 \le r \le 2.$$

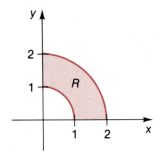

FIGURE 16.19

Since the relationship between rectangular and polar coordinates is: $x = r \cos \theta$ and $y = r \sin \theta$,

$$\iint\limits_{R} (3x + y)dA = \int_0^{\pi/2} \int_1^2 (3r \cos \theta + r \sin \theta)r \, dr \, d\theta$$

$$= \int_0^{\pi/2} \left[r^3 \cos \theta + \frac{r^3}{3} \sin \theta \right]_{r=1}^{r=2} d\theta$$

$$= \int_0^{\pi/2} \left[\left(8 \cos \theta + \frac{8}{3} \sin \theta \right) - \left(\cos \theta + \frac{1}{3} \sin \theta \right) \right] d\theta$$

$$= \int_0^{\pi/2} \left(7 \cos \theta + \frac{7}{3} \sin \theta \right) d\theta$$

$$= \left[7 \sin \theta - \frac{7}{3} \cos \theta \right]_0^{\pi/2} = 7 + \frac{7}{3} = \frac{28}{3}. \qquad \square$$

EXAMPLE 2   A silo has the shape of a right circular cylinder with a hemispherical top. The cylinder has a height of 40 feet and the radius of the cylinder and hemisphere is 5 feet. How much grain will the silo hold?

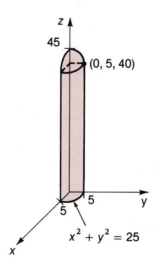

FIGURE 16.20

SOLUTION

Suppose a cartesian coordinate system is placed with the xy-plane at the base of the silo and the z-axis through the center of the silo, as in Figure 16.20. The equation of the hemispherical top of the silo is

$$x^2 + y^2 + (z - 40)^2 = 5^2, \qquad 40 \le z \le 45$$

or
$$z = 40 + \sqrt{25 - x^2 - y^2}.$$

Consequently, the volume of the silo is

$$V = \iint\limits_{R} (40 + \sqrt{25 - x^2 - y^2})\,dA,$$

where R is the region in the xy-plane described by

$$x^2 + y^2 \leq 5.$$

This volume is much easier to compute using polar coordinates.

$$V = \int_0^{2\pi} \int_0^5 [40 + \sqrt{25 - r^2}]r\,dr\,d\theta$$

$$= \int_0^{2\pi} \left[20r^2 - \frac{1}{3}(25 - r^2)^{3/2} \right]_0^5 d\theta$$

$$= \int_0^{2\pi} \left(500 + \frac{125}{3} \right) d\theta$$

$$= \frac{1625}{3}\,\theta \bigg]_0^{2\pi} = \frac{3250}{3}\,\pi\ \text{ft}^3.$$

This is equivalent to approximately 2735 bushels of grain. □

Iterated integrals involving polar coordinates can also be used to find volumes bounded between two surfaces. However, we must be able to project the intersection of the surfaces into a region of the plane that can be described in one of the forms of Theorem 16.5.

EXAMPLE 3   A hen's egg is six centimeters in length and has a surface roughly described by the equations $z = \dfrac{x^2 + y^2}{2}$ and $z = 6 - x^2 - y^2$. Find the volume of the egg.

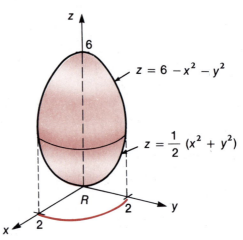

FIGURE 16.21

SOLUTION

   The graphs of the equations $z = (x^2 + y^2)/2$ and $z = 6 - x^2 - y^2$ intersect when $z = 2$, $x^2 + y^2 = 4$. The surface in the first octant is shown in Figure 16.21. Consequently, the projection of the intersection of the two surfaces into the xy-plane is the circle, $x^2 + y^2 = 4$ or $r = 2$. Thus,

$$V = \iint\limits_{R} \left[(6 - x^2 - y^2) - \frac{1}{2}(x^2 + y^2) \right] dA$$

$$= \int_0^{2\pi} \int_0^2 \left[(6 - r^2) - \frac{1}{2}r^2 \right] r \, dr \, d\theta$$

$$= \int_0^{2\pi} \int_0^2 \left(6r - \frac{3}{2}r^3 \right) dr \, d\theta$$

$$= \int_0^{2\pi} \left[3r^2 - \frac{3}{8}r^4 \right]_{r=0}^{r=2} d\theta$$

$$= \int_0^{2\pi} (12 - 6)d\theta = 60 \Big]_0^{2\pi} = 12\pi \text{ cm}^3. \qquad \square$$

   At times it is convenient to change an iterated integral expressed in cartesian coordinates to one in polar coordinates before evaluating the integral. The following example gives an illustration of the technique.

EXAMPLE 4   Evaluate $\displaystyle\int_0^2 \int_0^{\sqrt{4-x^2}} e^{-x^2} e^{-y^2} \, dy \, dx$.

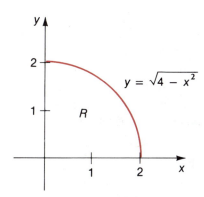

FIGURE 16.22

SOLUTION

   The region described by this iterated integral is shown in Figure 16.22. It can be expressed in polar coordinates by

$$0 \leq \theta \leq \frac{\pi}{2}, \qquad 0 \leq r \leq 2.$$

Thus,
$$\int_0^2 \int_0^{\sqrt{4-x^2}} e^{-x^2} e^{-y^2} \, dy \, dx = \int_0^2 \int_0^{\sqrt{4-x^2}} e^{-(x^2+y^2)} \, dy \, dx$$

$$= \int_0^{\pi/2} \int_0^2 e^{-r^2} r \, dr \, d\theta$$

$$= \int_0^{\pi/2} \left. -\frac{1}{2} e^{-r^2} \right]_{r=0}^{r=2} d\theta$$

$$= \left. \frac{1}{2}(1 - e^{-4})\theta \right]_0^{\pi/2} = \frac{\pi}{4}(1 - e^{-4}). \qquad \square$$

EXERCISE SET 16.3

Use double integrals in polar coordinates to find the volume of the solids described in Exercises 1 through 12.

1. Bounded by $z = 0$, $z = r$, and $r = 3$
2. Bounded by $z = 0$, $z = r$, and $r = \sin\theta$, $0 \le \theta \le \pi$
3. Bounded by the cylinders $x^2 + y^2 = 4$, $x^2 + y^2 = 16$ and the planes $z = 0$, $z = 4$
4. Inside the cylinder $x^2 + y^2 = 9$ and outside the cone $z^2 = x^2 + y^2$
5. Inside both the sphere $x^2 + y^2 + z^2 = 4$ and the cylinder $x^2 + y^2 = 1$
6. Inside the sphere $x^2 + y^2 + z^2 = 16$ and outside the cylinder $x^2 + y^2 = 9$
7. Inside the sphere $x^2 + y^2 + z^2 = 4$ and inside the cone $z^2 = x^2 + y^2$
8. Inside the sphere $x^2 + y^2 + z^2 = 4$ and outside the cone $z^2 = x^2 + y^2$
9. Inside the cylinder $r = \sin\theta$, above the paraboloid $z = x^2 + y^2 - 4$, and below the xy-plane.
10. Outside the cylinder $r = \sin\theta$, above the paraboloid $z = x^2 + y^2 - 4$ and below the xy-plane
11. Inside the cylinder formed by the intersection of $r = \sin\theta$ and $r = \cos\theta$, bounded above by the paraboloid $z = x^2 + y^2$ and below by $z = 0$
12. Inside the cylinder bounded by $r = \cos\theta$ when $0 \le \theta \le \pi/2$, bounded above by $y = x \tan z$ and below by $z = 0$

Use double integrals to find the area of the regions described in Exercises 13 through 22.

13. Inside $r = 2\sin\theta$
14. Inside $r = 1 + \cos\theta$
15. Inside $r = 3 - 2\cos\theta$
16. Inside $r^2 = 4\cos 2\theta$
17. Inside $r = 1 + \cos\theta$ and outside $r = 1$
18. Inside $r = 1 + \cos\theta$ and outside $r = 1 + \sin\theta$
19. Inside $r = \sin\theta$ and outside $r^2 = (\sin 2\theta)/2$
20. Inside $r = 2 + 2\sin\theta$ and inside $r = 3$
21. Inside $r = \sin 2\theta$ and inside $r = \cos 2\theta$
22. Between $r = \theta$ and $r = e^\theta$ for $0 \le \theta \le 2\pi$

Evaluate the integrals in Exercises 23 through 26 by changing to a double integral in polar coordinates.

23. $\displaystyle\int_{-1}^{1}\int_{0}^{\sqrt{1-x^2}} \frac{y}{\sqrt{x^2+y^2}}\, dy\, dx$ **24.** $\displaystyle\int_{-3}^{3}\int_{-\sqrt{9-x^2}}^{\sqrt{9-x^2}} e^{x^2} e^{y^2}\, dy\, dx$

25. $\displaystyle\int_{0}^{1}\int_{\sqrt{3}\,x/3}^{x} \frac{1}{\sqrt{x^2+y^2}}\, dy\, dx$ **26.** $\displaystyle\int_{0}^{2}\int_{-\sqrt{2x-x^2}}^{\sqrt{2x-x^2}} \sqrt{x^2+y^2}\, dy\, dx$

Evaluate the integrals in Exercises 27 through 30 by changing to a double integral in cartesian coordinates.

27. $\displaystyle\int_{0}^{\pi/4}\int_{0}^{\sec\theta} r^3\,(\sin\theta)^2\, dr\, d\theta$ **28.** $\displaystyle\int_{0}^{\pi/4}\int_{\sec\theta}^{2\sec\theta} r^2\,\sqrt{\sin 2\theta}\, dr\, d\theta$

29. $\displaystyle\int_{\pi/4}^{3\pi/4}\int_{0}^{3\csc\theta} r\, e^{r^2\,(\sin\theta)^2}\, dr\, d\theta$ **30.** $\displaystyle\int_{5\pi/4}^{3\pi/2}\int_{0}^{-2\csc\theta} r\,\ln(r^2\,\sin 2\theta)\, dr\, d\theta$

31. Organisms are released from a point taken as the pole of a polar coordinate system. The density of the organisms as a function of the time after release and is

$$f(r, \theta, t) = \frac{1}{\pi t} e^{-r^2/t}.$$

Show that $\displaystyle\int_{0}^{\infty}\int_{0}^{2\pi} f(r, \theta, t)r\, d\theta\, dr = 1$. What physical interpretation does this result have?

32. The energy generated by a 100-megaton nuclear bomb is approximately 4.18×10^{24} ergs. Suppose that such a bomb is detonated at a height H and radiates energy equally in all directions. How much of this energy reaches a circle of radius R on the level ground if the center of the circle is directly below the point of detonation?

33. A liquid flowing through a cylindrical tube has varying velocity depending on its distance r from the center of the tube:

$$v(r) = k(R^2 - r^2),$$

where R is the radius of the tube and k is a constant that depends on the length of the tube and the velocity of the fluid at the ends. Suppose $R = 3.5$ cm and $k = 1$. Use double integrals to find the rate of flow of the liquid through the tube.

16.4
CENTER OF MASS AND MOMENTS OF INERTIA

In Section 6.8 the definite integral of a single variable function was used to determine the moments and center of mass of a plane lamina with a constant density ρ. By using double integrals, we can extend these concepts to plane lamina that do not have constant density.

Suppose a region R in the xy-plane describes a lamina S with density at (x, y) given by $\rho(x, y)$, where ρ is a continuous function. Place a grid over R and let $R_1, R_2, \ldots, R_n$ denote those rectangles formed by the grid that lie entirely within R. See Figure 16.23. Let ΔA_i denote the area of R_i. If (x_i, y_i) is chosen arbitrarily in R_i, then an approximation to the mass of the lamina described by R_i is $\rho(x_i, y_i)\Delta A_i$. An approximation to the total mass of S is

$$M \approx \sum_{i=1}^{n} \rho(x_i, y_i)\Delta A_i.$$

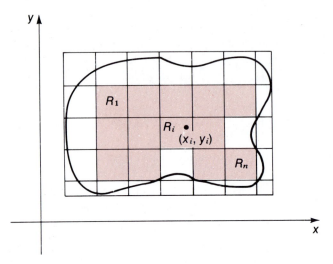

FIGURE 16.23

The approximation improves as the grid lines in both directions get closer together and leads to the following definition of mass and moments.

(16.6)
DEFINITION

Suppose R is a region in the xy-plane describing a plane lamina S with density at each point (x, y) given by $\rho(x, y)$, where ρ is a continuous function. The **mass of S** is

$$M = \iint_R \rho(x, y)dA.$$

The **moment of S about the x-axis** is

$$M_x = \iint_R y\,\rho(x, y)dA$$

and the **moment of S about the y-axis** is

$$M_y = \iint_R x\,\rho(x, y)dA.$$

The **center of mass of S** has coordinates $(\bar{x}, \bar{y})$ where

$$\bar{x} = \frac{M_y}{M} \quad \text{and} \quad \bar{y} = \frac{M_x}{M}.$$

This definition seems natural if you recall the physical concept expressed by a moment. The moment of a mass concentrated at a point about a line is the product of the mass and the directed distance from the line. For the moment about the x-axis,

$$M_x = \iint\limits_R y\,\rho(x, y)\,dA,$$

y gives the directed distance from the x-axis and $\rho(x, y)\,dA$ represents an element of mass.

Suppose ρ is a constant function and R is a region bounded by $x = a$, $x = b$, and the graphs of f_1 and f_2 where $f_1(x) \le f_2(x)$. Then the results in Definition 16.6 reduce to those given in Section 6.8. For example, in this case:

$$M_y = \iint\limits_R x\,\rho\,dA = \int_a^b \int_{f_1(x)}^{f_2(x)} x\,\rho\,dy\,dx$$

$$= \rho \int_a^b x[f_2(x) - f_1(x)]\,dx.$$

EXAMPLE 1   Compute the mass, moments, and center of mass of a plane lamina in the shape of the region R described by

$$0 \le x \le 4, \qquad 0 \le y \le \sqrt{x},$$

if the density is given by $\rho(x, y) = xy$.

SOLUTION

$$M = \iint\limits_R \rho(x, y)\,dA = \int_0^4 \int_0^{\sqrt{x}} xy\,dy\,dx$$

$$= \int_0^4 \frac{xy^2}{2}\Bigg]_{y=0}^{y=\sqrt{x}} dx$$

$$= \int_0^4 \frac{x^2}{2}\,dx = \frac{x^3}{6}\Bigg]_0^4 = \frac{64}{6} = \frac{32}{3}.$$

$$M_x = \iint\limits_R y(xy)\,dA = \int_0^4 \int_0^{\sqrt{x}} xy^2\,dy\,dx$$

$$= \int_0^4 \frac{xy^3}{3}\Bigg]_0^{\sqrt{x}} dx$$

$$= \frac{1}{3}\int_0^4 x^{5/2}\,dx = \frac{2}{21}x^{7/2}\Bigg]_0^4 = \frac{256}{21},$$

and

$$M_y = \iint\limits_R x(xy)\,dA = \int_0^4 \int_0^{\sqrt{x}} x^2 y\,dy\,dx = \int_0^4 \frac{x^2 y^2}{2}\Bigg]_0^{\sqrt{x}} dx$$

$$= \frac{1}{2}\int_0^4 x^3\,dx = \frac{x^4}{8}\Bigg]_0^4 = 32.$$

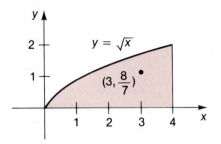

FIGURE 16.24

The center of mass is $(\bar{x}, \bar{y})$, where

$$\bar{x} = \frac{M_y}{M} = \frac{32}{\dfrac{32}{3}} = 3 \qquad \text{and} \qquad \bar{y} = \frac{M_x}{M} = \frac{\dfrac{256}{21}}{\dfrac{32}{3}} = \frac{8}{7}.$$

This is shown in Figure 16.24.   □

The moment M_x of a plane lamina about the x-axis describes the tendency of the lamina to rotate about this axis. Similarly, M_y describes the tendency of the lamina to rotate about the y-axis. If the mass M of the lamina is concentrated at the center of mass $(\bar{x}, \bar{y})$, then the resulting system has the same moments and rotation tendencies. While these moments give valuable information about the gravitational center of the lamina, they give no indication about the way the region is distributed about this center. For example, if the regions in Figure 16.25 both have $\rho = 1$, then they have the same moments M_x and M_y and the same center of mass $(0, 0)$.

To describe the spread or distribution of a region about an axis we must introduce a concept known as the second moment about the axis and called the moment of inertia about the axis.

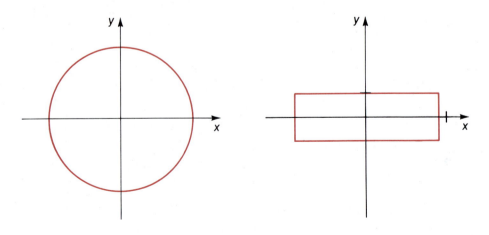

FIGURE 16.25

(16.7)
DEFINITION

Suppose R is a region in the plane describing a plane lamina S with density at each point (x, y) given by $\rho(x, y)$, where ρ is a continuous function on R. Then the **moment of inertia of S about the x-axis** is

$$I_x = \iint_R y^2 \rho(x, y)dA;$$

the **moment of inertia of S about the y-axis** is

$$I_y = \iint_R x^2 \rho(x, y)dA.$$

The second moment can also be used to describe the distribution about a point. The second moment about the origin is called the **polar moment of inertia of S** and is given by

$$I_0 = I_x + I_y = \iint_R (x^2 + y^2) \, \rho(x, y)dA.$$

Analogous to the center of mass is a point $(\hat{x}, \hat{y})$ with the property that if the total mass is concentrated at $(\hat{x}, \hat{y})$, then the resulting system has the same moments of inertia as the original system. This means that $\hat{x}^2 M = I_y$ and $\hat{y}^2 M = I_x$, so the coordinates of this point are given by

$$\hat{x} = \sqrt{\frac{I_y}{M}}, \text{ called the \textbf{radius of gyration} about the } x\text{-axis,}$$

(16.8)

$$\hat{y} = \sqrt{\frac{I_x}{M}}, \text{ called the \textbf{radius of gyration} about the } y\text{-axis.}$$

Notice that by virtue of their definitions, the moments of inertia as well as the radii of gyration must be nonnegative.

EXAMPLE 2

Find the moments of inertia and radii of gyration of the plane lamina in the shape of the region R described by

$$0 \le x \le 4, \qquad 0 \le y \le \sqrt{x}$$

if the density is given by $\rho(x, y) = xy$.

SOLUTION

The region is the same as that described in Example 1 and shown in Figure 16.24. Hence,

$$I_x = \iint_R y^2 \, (xy)dA = \int_0^4 \int_0^{\sqrt{x}} xy^3 \, dy \, dx$$

$$= \int_0^4 \frac{xy^4}{4} \Bigg]_0^{\sqrt{x}} dx$$

$$= \frac{1}{4} \int_0^4 x^3 \, dx = \frac{x^4}{16} \Bigg]_0^4 = 16$$

and
$$I_y = \iint\limits_R x^2\,(xy)dA = \int_0^4 \int_0^{\sqrt{x}} x^3 y\,dy\,dx$$

$$= \int_0^4 \frac{x^3 y^2}{2}\Big]_0^{\sqrt{x}}\,dx$$

$$= \frac{1}{2}\int_0^4 x^4\,dx = \frac{x^5}{10}\Big]_0^4 = \frac{512}{5}.$$

Also,
$$I_0 = I_x + I_y = 16 + \frac{512}{5} = \frac{592}{5}.$$

Since M was found in Example 1 to be $32/3$,

$$\hat{x} = \sqrt{\frac{I_y}{M}} = \sqrt{\frac{\dfrac{512}{5}}{\dfrac{32}{3}}} = \sqrt{\frac{48}{5}} = \frac{4\sqrt{15}}{5} \approx 3.1$$

and
$$\hat{y} = \sqrt{\frac{I_x}{M}} = \sqrt{\frac{16}{\dfrac{32}{3}}} = \sqrt{\frac{48}{32}} = \sqrt{\frac{3}{2}} \approx 1.2. \qquad \square$$

To demonstrate that the moments of inertia and consequently the radii of gyration do give a measure of the spread of a lamina, consider the case of two circular laminae shown in Figure 16.26.

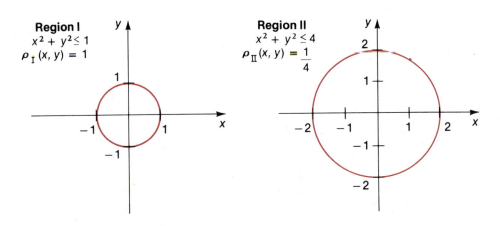

FIGURE 16.26

Both of these circular laminae have mass π and center of mass $(0, 0)$. Consider, however, their moments of inertia.

For Region I:

$$I_x = \iint\limits_{R_{\mathrm{I}}} y^2\, \rho(x, y)dA = \iint\limits_{R_{\mathrm{I}}} y^2 \cdot 1\, dA$$

$$= \int_0^{2\pi}\int_0^1 (r\sin\theta)^2\, r\, dr\, d\theta$$

$$= \int_0^{2\pi}\int_0^1 (\sin\theta)^2\, r^3\, dr\, d\theta = \int_0^{2\pi}(\sin\theta)^2\, \frac{r^4}{4}\Bigg]_0^1 d\theta$$

$$= \frac{1}{4}\int_0^{2\pi}(\sin\theta)^2\, d\theta = \frac{1}{4}\int_0^{2\pi}\frac{1-\cos 2\theta}{2}\, d\theta$$

$$= \frac{1}{8}\left[\theta - \frac{1}{2}\sin 2\theta\right]_0^{2\pi} = \frac{\pi}{4},$$

and by symmetry, $I_y = \pi/4$.
For Region II:

$$I_x = \iint\limits_{R_{\mathrm{II}}} y^2\, \rho(x, y)dA = \iint\limits_{R_{\mathrm{II}}} y^2 \cdot \frac{1}{4}\, dA$$

$$= \frac{1}{4}\int_0^{2\pi}\int_0^2 (r\sin\theta)^2\, r\, dr\, d\theta = \frac{1}{4}\int_0^{2\pi}\int_0^2 r^3\, (\sin\theta)^2\, dr\, d\theta$$

$$= \frac{1}{4}\int_0^{2\pi}(\sin\theta)^2\, \frac{r^4}{4}\Bigg]_0^2 d\theta = \frac{1}{4}\int_0^{2\pi} 4(\sin\theta)^2\, d\theta$$

$$= \int_0^{2\pi}\frac{1-\cos 2\theta}{2}\, d\theta = \frac{1}{2}\left[\theta - \frac{1}{2}\sin 2\theta\right]_0^{2\pi} = \pi.$$

Again, by symmetry, $I_y = \pi$.
   The difference between the moments of inertia in regions I and II correctly describes the fact that the lamina in II is more widely distributed from its center of mass than is the lamina in I.

EXERCISE SET 16.4

Find the mass, moments about the axes, and center of mass of the lamina bounded by the curves described in Exercises 1 through 8 and having density ρ.

1. $y = x^2$, $y = \sqrt{x}$, $\rho(x, y) = 2$
2. $y = x^2$, $y = 2 - x^2$, $\rho(x, y) = 4$
3. $y = 1 - x$, $y = 0$, $x = 0$, $\rho(x, y) = x$
4. $y = 1 - x$, $y = 0$, $x = 0$, $\rho(x, y) = y$
5. $y = 1 - x^2$, $y = 0$, $\rho(x, y) = x^2$
6. $y = 1 - x^2$, $y = 0$, $\rho(x, y) = y$

7. $y = x$, $y = 1$, $x = 0$, $\rho(x, y) = x + y$

8. $y = \dfrac{1}{x}$, $y = 0$, $x = 1$, $x = 4$, $\rho(x, y) = x + y$

9–16. Find the moments of inertia about the axes and the origin and the radii of gyration for the laminas described in Exercises 1 through 8.

17. Show that the formulas for the mass, moments about the axes and center of mass $(\bar{r}, \bar{\theta})$ for a plane lamina with density $\rho(r, \theta)$ and shape that of a region R can be described in polar coordinates by

$$M = \iint\limits_R r\,\rho(r, \theta)dr\,d\theta; \qquad M_x = \iint\limits_R r^2 \sin\theta\,\rho(r, \theta)dr\,d\theta,$$

$$M_y = \iint\limits_R r^2 \cos\theta\,\rho(r, \theta)dr\,d\theta \qquad \text{and} \qquad \bar{r}^2 = \frac{M_x^2 + M_y^2}{M^2}, \ \tan\bar{\theta} = \frac{M_x}{M_y}.$$

18. Show that the formulas for the moments of inertia of a plane lamina can be described using polar coordinates by

$$I_x = \iint\limits_R r^3 (\sin\theta)^2 \,\rho(r, \theta)dr\,d\theta,$$

$$I_y = \iint\limits_R r^3 (\cos\theta)^2 \,\rho(r, \theta)dr\,d\theta$$

and

$$I_0 = \iint\limits_R r^3 \,\rho(r, \theta)dr\,d\theta.$$

Use the formulas in Exercises 17 and 18 to find the mass, moments about the axes, and moments of inertia for the plane laminas whose boundaries and density functions are described in Exercises 19 through 22.

19. $r = 4$, $\rho(r, \theta) = r$ **20.** $r = 4$, $\rho(r, \theta) = \theta$

21. $r = \sin\theta$, $\rho(r, \theta) = r$ **22.** $r = \cos\theta$, $\rho(r, \theta) = r$

Use the formulas in Exercise 17 to find the center of mass of the plane laminas whose boundaries and density functions are described in Exercises 23 and 24.

23. $r = 1 - \sin\theta$, $\rho(r, \theta) = r$ **24.** $r = 1 + \cos\theta$, $\rho(r, \theta) = r$

25. Find the center of mass of a lamina that has the shape of the region bounded by $x^2 + y^2 = 4$ if the density at any point (x, y) is the square of the distance from the center of the region.

26. Find the center of mass of the lamina that has the shape of the region bounded by $x = 0$, $y = x$ and $y = 3 - x$ if the density at any point (x, y) is twice the distance from the y-axis.

27. A lamina with constant density has the shape of an isosceles right triangle with equal sides of length 3. Find
    (a) the center of mass.
    (b) the moment of inertia about the lines $x = \bar{x}$ and $y = \bar{y}$.
    (c) the moment of inertia about the hypotenuse.

16.5
TRIPLE INTEGRALS

The development of the triple integral for functions of three variables over regions in space parallels the development of the double integral of functions of two variables over regions in the plane. Suppose a function f is defined on a bounded region D in space. Since D is bounded, D is contained within some parallelepiped P. We begin by forming a grid through D consisting of planes through P parallel to the xy-, yz-, and xz-coordinate planes. This forms a set of rectangular parallelepipeds that cover D. Let $D_1, D_2, \ldots, D_n$ denote those parallelepipeds that are contained entirely in D. The collection of all these parallelepipeds is denoted by $\mathscr{P}$ and $\|\mathscr{P}\|$ denotes the maximum length of the diagonals of the D_i. We choose (x_i, y_i, z_i) arbitrarily within D_i. See Figure 16.27.

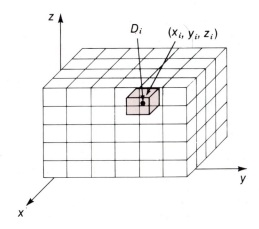

FIGURE 16.27

A Riemann sum can be formed with respect to f, the grid, and the choice of points within each D_i:

$$\sum_{i=1}^{n} f(x_i, y_i, z_i)\Delta V_i,$$

where ΔV_i denotes the volume of the parallelepiped D_i. Taking the limit of the Riemann sums over all possible grid constructions and choices of points within the grid elements gives the definition of the triple integral of f over D.

(16.9)
DEFINITION

If a function f of three variables is defined on a region D in space, the **triple integral** of f over D is defined by

$$\iiint_{D} f(x, y, z)\,dV = \lim_{\|\mathscr{P}\| \to 0} \sum_{i=1}^{n} f(x_i, y_i, z_i)\Delta V_i,$$

provided that this limit exists.

This definition satisfies the usual properties associated with the integral. For example,

$$(16.10) \qquad \iiint_D (f \pm g)(x, y, z)dV = \iiint_D f(x, y, z)dV \pm \iiint_D g(x, y, z)dV$$

$$(16.11) \qquad \iiint_D cf(x, y, z)dV = c \iiint_D f(x, y, z)dV,$$

provided that the integrals exist and c is a constant. In addition, if D is composed of subregions D_1 and D_2 on which the integral f exists, then

$$(16.12) \qquad \iiint_D f(x, y, z)dV = \iiint_{D_1} f(x, y, z)dV + \iiint_{D_2} f(x, y, z)dV.$$

As with double integrals, the triple integral can often be expressed by a sequence of single or iterated integrals. When the region D is itself a parallelepiped with boundaries parallel to the coordinate planes, the iterated integrals are the successive integrals of the function over each of the coordinate intervals describing D.

EXAMPLE 1   Evaluate the triple integral of $f(x, y, z) = xy \sin z$ over the region D in space described by $0 \leq x \leq 2, 1 \leq y \leq 2, 0 \leq z \leq \pi$.

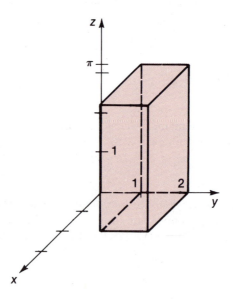

FIGURE 16.28

SOLUTION

The region is shown in Figure 16.28 and

$$\iiint\limits_{D} f(x, y, z)\,dV = \int_0^2 \int_1^2 \int_0^\pi xy \sin z \, dz \, dy \, dx$$

$$= \int_0^2 \int_1^2 -xy \cos z \, \Big]_{z=0}^{z=\pi} dy \, dx$$

$$= \int_0^2 \int_1^2 2xy \, dy \, dx$$

$$= \int_0^2 xy^2 \Big]_{y=1}^{y=2} dx = \int_0^2 3x \, dx = \frac{3}{2} x^2 \Big]_0^2 = 6. \qquad \square$$

The integral in Example 1 can be expressed as an iterated integral in five other ways by interchanging the order of the variables. For example,

$$\iiint\limits_{D} xy \sin z \, dV = \int_0^2 \int_0^\pi \int_1^2 xy \sin z \, dy \, dz \, dx = \int_0^\pi \int_1^2 \int_0^2 xy \sin z \, dx \, dy \, dz.$$

Reversing the order of integration in this example causes no difficulty, since the variables describing D are independent of one another.

The triple integral of a function f over a region D can in general be expressed as an iterated integral involving integration, first with respect to z, then with respect to y, and finally, with respect to x (written $dz\,dy\,dx$), provided that D can be described by

$$a \le x \le b, \quad g_1(x) \le y \le g_2(x), \quad h_1(x, y) \le z \le h_2(x, y),$$

where g_1, g_2, h_1, and h_2 are continuous functions. A region of this type is shown in Figure 16.29.

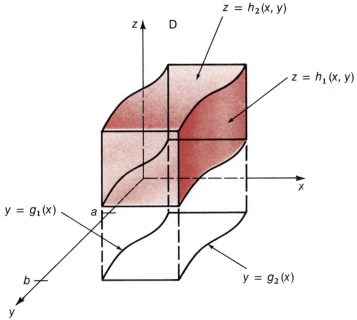

FIGURE 16.29

In this situation,

(16.13) $$\iiint_D f(x, y, z)dV = \int_{x=a}^{x=b} \int_{y=g_1(x)}^{y=g_2(x)} \int_{z=h_1(x,y)}^{z=h_2(x,y)} f(x, y, z)dz\, dy\, dx.$$

The region D can be thought of as being formed of "slices" from $x = a$ to $x = b$. For each fixed value of x, the slice has a pair of straight lateral boundaries described by $y = g_1(x)$ and $y = g_2(x)$. Each line in the slice parallel to the lateral boundaries is bounded by $z = h_1(x, y)$ and $z = h_2(x, y)$. See Figure 16.30.

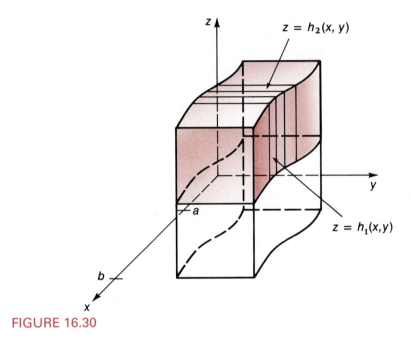

FIGURE 16.30

Determining the functions that describe the boundaries of the region can be difficult, since these boundaries are usually described in terms of intersecting surfaces.

EXAMPLE 2   Find $\iiint_D z\, dV$ where D is the region in the first octant bounded by the planes

$x = y$, $z = 0$, $y = 0$ and the cylinder $x^2 + z^2 = 1$.

SOLUTION

The region D is bounded below by the xy-plane. In the first octant, the right circular cylinder $x^2 + z^2 = 1$ intersects the xy-plane in the line $x = 1$, so the base of D can be described by $0 \le x \le 1$, $0 \le y \le x$. To describe the other boundaries of D, first draw a plane parallel to the yz-plane through D at an arbitrary point between $x = 0$ and $x = 1$. The solid and an intersecting plane parallel to the yz-plane are shown in Figure 16.31. By drawing a line from the top to the bottom of this slice of D we see that D is described by

$$0 \le x \le 1, \quad 0 \le y \le x, \quad 0 \le z \le \sqrt{1 - x^2}.$$

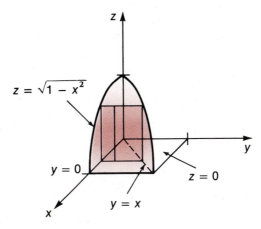

FIGURE 16.31

So $\qquad \displaystyle\iiint\limits_{D} z\,dV = \int_{x=0}^{x=1} \int_{y=0}^{y=x} \int_{z=0}^{z=\sqrt{1-x^2}} z\,dz\,dy\,dx$

$$= \int_0^1 \int_0^x \frac{z^2}{2}\,\bigg]_0^{\sqrt{1-x^2}} dy\,dx = \int_0^1 \int_0^x \frac{1-x^2}{2}\,dy\,dx$$

$$= \int_0^1 \frac{1-x^2}{2}\,y\,\bigg]_0^x dx$$

$$= \int_0^1 \frac{x-x^3}{2}\,dx = \frac{x^2}{4} - \frac{x^4}{8}\,\bigg]_0^1 = \frac{1}{8}. \qquad \square$$

EXAMPLE 3   Evaluate $\displaystyle\iiint\limits_{D} x^2 y\,dV$ where D is the region in the first octant bounded by

the plane $z = 4$ and the cone $z^2 = x^2 + y^2$.

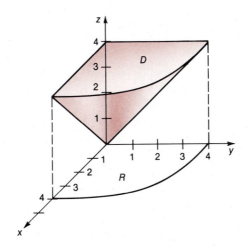

FIGURE 16.32

SOLUTION

The region D is shown in Figure 16.32. The plane $z = 4$ and the cone

$z^2 = x^2 + y^2$ intersect in the circle $x^2 + y^2 = 16$ lying in the plane $z = 4$. The projection of this intersection to the xy-plane is the circle $x^2 + y^2 = 16$. Consequently, D can be described by

$$0 \le x \le 4, \quad 0 \le y \le \sqrt{16 - x^2}, \quad \sqrt{x^2 + y^2} \le z \le 4.$$

A typical defining slice and line are shown in Figure 16.33.

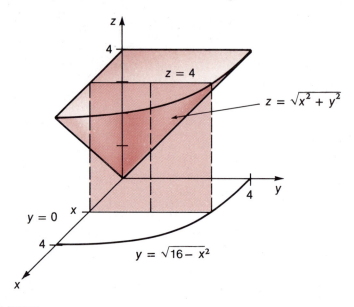

FIGURE 16.33

Hence,

$$\iiint_D x^2 y \, dV = \int_0^4 \int_0^{\sqrt{16-x^2}} \int_{\sqrt{x^2+y^2}}^4 x^2 y \, dz \, dy \, dx = \int_0^4 \int_0^{\sqrt{16-x^2}} x^2 y z \bigg]_{z=\sqrt{x^2+y^2}}^{z=4} dy \, dx$$

$$- \int_0^4 \int_0^{\sqrt{16-x^2}} (4x^2 y - x^2 y \sqrt{x^2 + y^2}) \, dy \, dx$$

$$= \int_0^4 \left[2x^2 y^2 - \frac{x^2}{3} (x^2 + y^2)^{3/2} \right]_{y=0}^{y=\sqrt{16-x^2}} dx$$

$$= \int_0^4 \left[2x^2 (16 - x^2) - \frac{x^2}{3} (x^2 + 16 - x^2)^{3/2} + \frac{x^5}{3} \right] dx.$$

$$= \int_0^4 \left[\frac{32}{3} x^2 - 2x^4 + \frac{x^5}{3} \right] dx = \left[\frac{32}{9} x^3 - \frac{2}{5} x^5 + \frac{x^6}{18} \right]_0^4 = \frac{2048}{45}. \quad \square$$

Triple integrals can be used to give an alternate expression for the volume of a solid. When the solid S has the shape of a region D described by

$$a \le x \le b, \quad g_1(x) \le y \le g_2(x), \quad h_1(x, y) \le z \le h_2(x, y),$$

the double integration formula for volume implies that

$$\text{Volume of } S = \int_{x=a}^{x=b} \int_{y=g_1(x)}^{y=g_2(x)} [h_2(x, y) - h_1(x, y)]dy\, dx$$

$$= \int_{x=a}^{x=b} \int_{y=g_1(x)}^{y=g_2(x)} \int_{z=h_1(x,y)}^{z=h_2(x,y)} dz\, dy\, dx = \iiint_D dV.$$

EXAMPLE 4   Determine the volume of the solid with base in the xy-plane and bounded by the planes $x = 0$, $y = 0$, and $x + 2y + 2z = 4$.

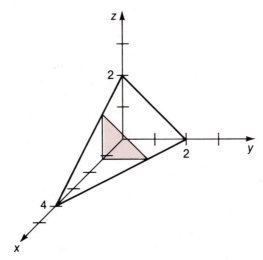

FIGURE 16.34

SOLUTION

The solid is shown in Figure 16.34. The plane $x + 2y + 2z = 4$ intersects the xy-plane in the line $x + 2y = 4$, so the solid has the shape of the region described by:

$$0 \le x \le 4, \quad 0 \le y \le \frac{4 - x}{2}, \quad 0 \le z \le \frac{4 - x - 2y}{2}.$$

$$V = \int_0^4 \int_0^{(4-x)/2} \int_0^{(4-x-2y)/2} dz\, dy\, dx$$

$$= \int_0^4 \int_0^{(4-x)/2} \frac{1}{2}(4 - x - 2y)dy\, dx$$

$$= \frac{1}{2} \int_0^4 \left[4y - xy - y^2 \right]_0^{(4-x)/2} dx$$

$$= \frac{1}{2} \int_0^4 \left[2(4 - x) - \frac{x}{2}(4 - x) - \frac{1}{4}(4 - x)^2 \right] dx$$

$$= \frac{1}{2} \int_0^4 \left[8 - 4x + \frac{1}{2}x^2 - \frac{1}{4}(4 - x)^2 \right] dx$$

$$= \frac{1}{2} \left[8x - 2x^2 + \frac{1}{6}x^3 + \frac{1}{12}(4 - x)^3 \right]_0^4 = \frac{8}{3}. \qquad \square$$

The regions considered in the previous examples have had representations that permitted iterated integration to be performed in the order $dz\ dy\ dx$. The boundaries in the x-direction have been constant, those in the y-direction depended only on x, and those in the z-direction depended on both x and y. Many regions, however, can be more easily described by a different ordering of variables. When the region is described by

$$a \le x \le b, \quad g_1(x) \le z \le g_2(x), \quad h_1(x, z) \le y \le h_2(x, z),$$

the integral is

(16.14) $$\iiint_D f(x, y, z)dV = \int_{x=a}^{x=b} \int_{z=g_1(x)}^{z=g_2(x)} \int_{y=h_1(x,z)}^{y=h_2(x,z)} f(x, y, z)dy\ dz\ dx.$$

A region of this type is shown in Figure 16.35.

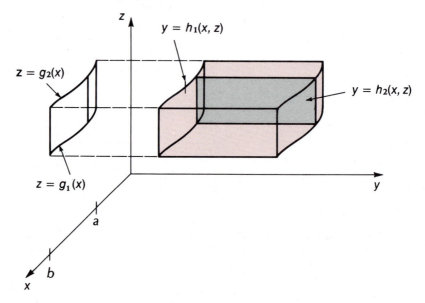

FIGURE 16.35

Similarly, when a region is represented by

$$a \le z \le b, \quad g_1(z) \le y \le g_2(z), \quad h_1(y, z) \le x \le h_2(y, z),$$

(see Figure 16.36), the integral has the form

(16.15) $$\iiint_D f(x, y, z)dV = \int_{z=a}^{z=b} \int_{y=g_1(z)}^{y=g_2(z)} \int_{x=h_1(y,z)}^{x=h_2(y,z)} f(x, y, z)dx\ dy\ dz.$$

There are three other variable orderings that can be used: $dx\ dz\ dy$, $dy\ dx\ dz$, and $dz\ dx\ dy$.

Often a region can be expressed as an iterated integral in more than one way. When this occurs, the order of integration depends upon the ease of representation as well as on the order in which the function can be most readily integrated. To make an intelligent choice requires experience. The rather extensive list of exercises at the end of this section is designed to provide this experience.

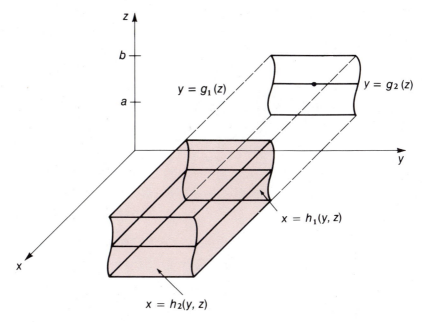

FIGURE 16.36

EXAMPLE 5   Express $\iiint\limits_D z\,dV$ in the six iterated integral forms, where D is the region in the first octant bounded by the planes $x = y$, $z = 0$, $y = 0$ and the cylinder $x^2 + z^2 = 1$.

SOLUTION

This triple integral was expressed in Example 2 as

(1) $\displaystyle\int_0^1 \int_0^x \int_0^{\sqrt{1-x^2}} z\,dz\,dy\,dx,$

seen in Figure 16.31. It could alternately be expressed as

(2) $\displaystyle\int_0^1 \int_0^{\sqrt{1-x^2}} \int_0^x z\,dy\,dz\,dx,$

(see Figure 16.37),

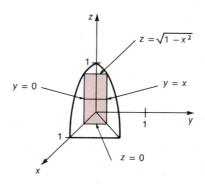

FIGURE 16.37

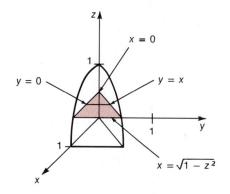

FIGURE 16.38

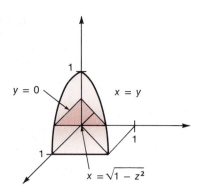

FIGURE 16.39

(3) $\displaystyle\int_0^1 \int_0^{\sqrt{1-z^2}} \int_0^x z\,dy\,dx\,dz,$

(see Figure 16.38),

(4) $\displaystyle\int_0^1 \int_0^{\sqrt{1-z^2}} \int_y^{\sqrt{1-z^2}} z\,dx\,dy\,dz,$

(see Figure 16.39),

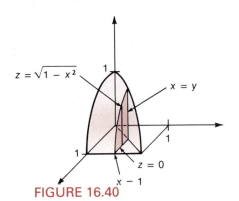

FIGURE 16.40

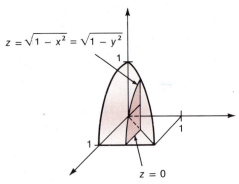

FIGURE 16.41

(5) $\displaystyle\int_0^1 \int_y^1 \int_0^{\sqrt{1-x^2}} z\,dz\,dx\,dy,$

(see Figure 16.40),

or

(6) $\displaystyle\int_0^1 \int_0^{\sqrt{1-y^2}} \int_y^{\sqrt{1-z^2}} z\,dx\,dz\,dy,$

(see Figure 16.41). $\qquad\qquad\qquad\qquad\qquad\qquad\square$

EXAMPLE 6   Determine the volume of the solid lying in the first octant that is bounded above by the sphere $x^2 + y^2 + z^2 = 2$ and below by the paraboloid $z = x^2 + y^2$.

SOLUTION

To find the curve of intersection of the sphere and paraboloid, we solve the equations $x^2 + y^2 + z^2 = 2$ and $z = x^2 + y^2$ simultaneously:

$$2 = x^2 + y^2 + z^2 = z + z^2.$$

So $z^2 + z - 2 = 0$ and $z = -2$ or $z = 1$.

Since the region lies above the xy-plane, $z = 1$ and the curve of intersection is $x^2 + y^2 = 1$, $z = 1$. Projecting the curve of intersection into the xy-plane determines the limits of integration on x and y

$$0 \le x \le 1, \qquad 0 \le y \le \sqrt{1 - x^2}.$$

The solid and an intersecting plane parallel to the yz-plane are shown in Figure 16.42.

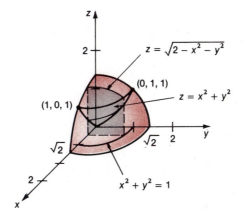

FIGURE 16.42

$$V = \int_0^1 \int_0^{\sqrt{1-x^2}} \int_{x^2+y^2}^{\sqrt{2-x^2-y^2}} dz \, dy \, dx$$

$$= \int_0^1 \int_0^{\sqrt{1-x^2}} z \Big]_{x^2+y^2}^{\sqrt{2-x^2-y^2}} dy \, dx$$

$$= \int_0^1 \int_0^{\sqrt{1-x^2}} [\sqrt{2 - x^2 - y^2} - (x^2 + y^2)] dy \, dx.$$

This double integral is much easier to evaluate using polar coordinates.

$$V = \int_0^{\pi/2} \int_0^1 (\sqrt{2 - r^2} - r^2) r \, dr \, d\theta$$

$$= \int_0^{\pi/2} \left[-\frac{(2 - r^2)^{3/2}}{3} - \frac{r^4}{4} \right]_0^1 d\theta$$

$$= \int_0^{\pi/2} \left(-\frac{1}{3} - \frac{1}{4} + \frac{2\sqrt{2}}{3} \right) d\theta$$

$$= \int_0^{\pi/2} \frac{4(\sqrt{2})^3 - 7}{12} d\theta = \frac{8\sqrt{2} - 7}{24} \pi \approx .56. \qquad \square$$

EXERCISE SET 16.5

Evaluate the triple integrals in Exercises 1 through 8.

1. $\displaystyle\int_0^1 \int_{x^2}^x \int_{x-y}^{x+y} y\,dz\,dy\,dx$

2. $\displaystyle\int_0^1 \int_{x^2}^x \int_{x-z}^{x+z} y\,dy\,dz\,dx$

3. $\displaystyle\int_0^1 \int_{x^2}^x \int_{x-y}^{x+y} z\,dz\,dy\,dx$

4. $\displaystyle\int_0^1 \int_{x^2}^x \int_{x-z}^{x+z} z\,dy\,dz\,dx$

5. $\displaystyle\int_0^\pi \int_0^z \int_0^{xz} \frac{1}{x}\sin\left(\frac{y}{x}\right)dy\,dx\,dz$

6. $\displaystyle\int_{-1}^0 \int_1^2 \int_{1/y}^1 xy\sin\pi yz\,dz\,dy\,dx$

7. $\displaystyle\int_0^1 \int_{-1}^1 \int_{-xy}^{xy} e^{x^2+y^2}\,dz\,dx\,dy$

8. $\displaystyle\int_0^1 \int_0^1 \int_0^{xy} e^{x^2+y^2}\,dz\,dx\,dy$

Rewrite the integrals in Exercises 9 through 12 in five alternative ways. Evaluate the triple integrals in two ways.

9. $\displaystyle\int_0^1 \int_0^{2-2x} \int_0^{3-3x-3y/2} (2x+y)dz\,dy\,dx$

10. $\displaystyle\int_0^1 \int_y^1 \int_0^x x^2z\,dz\,dx\,dy$

11. $\displaystyle\int_0^2 \int_0^{3\sqrt{4-z^2}/2} \int_0^{\sqrt{36-4y^2-9z^2}} z\,dx\,dy\,dz$

12. $\displaystyle\int_0^1 \int_{-\sqrt{1-z}}^{\sqrt{1-z}} \int_0^{\sqrt{1-y^2-z}} y\,dx\,dy\,dz$

Use triple integrals to find the volume of the solids whose boundaries are described in Exercises 13 through 22.

13. $x = 0,\ y = 0,\ x+y+z = 2,\ x+y-z = 2$

14. $x+y+z = 2,\ x = 1,\ y = 0,\ z = 0$

15. $z = x^2 + y^2,\ z = 1,\ z = 4$

16. $z = y^2,\ z = 1,\ x = 0,\ x = 1$

17. $y = \sin x,\ y = \cos x,\ z = \cos x,\ x = 0,\ z = 0,\ x = \dfrac{\pi}{4}$

18. $y = \sin x,\ y = \cos x,\ z = \sin x,\ x = 0,\ z = 0,\ x = \dfrac{\pi}{4}$

19. $4x^2 + 9y^2 + z^2 = 36$

20. $4x^2 + 9y^2 - z^2 = 36,\ z = 0,\ z = 4$

21. $x^2 + y^2 + z^2 = 6,$ inside $z = x^2 + y^2$

22. $x^2 + y^2 = 1,\ x^2 + z^2 = 1$

In Exercises 23 through 26, sketch the solid whose volume is given by the triple integral.

23. $\displaystyle\int_{-2}^2 \int_0^3 \int_0^y dz\,dy\,dx$

24. $\displaystyle\int_{-2}^2 \int_{-\sqrt{4-x^2}}^{\sqrt{4-x^2}} \int_{x^2+y^2}^{8-x^2-y^2} dz\,dy\,dx$

25. $\displaystyle\int_0^3 \int_{z-3}^0 \int_{-x-3}^0 dy\,dx\,dz$ **26.** $\displaystyle\int_{-1}^1 \int_0^3 \int_0^{\sqrt{1-z^2}} dx\,dy\,dz$

Putnam exercise:

27. Evaluate

$$\int_0^1 \int_0^1 \int_0^1 \left[\cos\left(\frac{\pi}{6}(x+y+z)\right) \right]^2 dx\,dy\,dz.$$

(This exercise was derived from problem B–1 of the twenty-sixth William Lowell Putnam examination given on November 20, 1965. The examination and its solution can be found in the September 1966 issue of the *American Mathematical Monthly*, pages 727–732.)

16.6
TRIPLE INTEGRALS IN CYLINDRICAL AND SPHERICAL COORDINATES

Example 6 of Section 16.5 shows the need for evaluating triple integrals using the cylindrical coordinate system. In this example we first integrated with respect to the variable z and then changed the remaining double integral to polar coordinates. To use cylindrical coordinates initially on a problem of this type requires that the region D be expressed in terms of r, θ, and z as

$$\alpha \le \theta \le \beta, \quad g_1(\theta) \le r \le g_2(\theta), \quad h_1(r,\theta) \le z \le h_2(r,\theta).$$

Then

(16.16) $\displaystyle\iiint\limits_{D} f(r,\theta,z)\,dV = \int_{\theta=\alpha}^{\theta=\beta} \int_{r=g_1(\theta)}^{r=g_2(\theta)} \int_{z=h_1(r,\theta)}^{z=h_2(r,\theta)} f(r,\theta,z)\,dz\,r\,dr\,d\theta$

$$= \int_\alpha^\beta \int_{g_1(\theta)}^{g_2(\theta)} \int_{h_1(r,\theta)}^{h_2(r,\theta)} f(r,\theta,z)\,r\,dz\,dr\,d\theta.$$

EXAMPLE 1   Use the cylindrical representation to find the volume of the region in the first octant that is inside the cylinder $r = 1$, bounded below by the xy-plane and above by the plane $3x + 2y + 6z = 6$.

SOLUTION

Referring to Figure 16.43, we see that the solid is bounded above by the plane $3x + 2y + 6z = 6$. In cylindrical coordinates this plane is described by

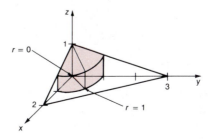

FIGURE 16.43

$$z = 1 - \frac{x}{2} - \frac{y}{3} = 1 - \frac{r}{2}\cos\theta - \frac{r}{3}\sin\theta, \text{ so}$$

$$V = \int_{\theta=0}^{\theta=\pi/2} \int_{r=0}^{r=1} \int_{z=0}^{z=1-(r\cos\theta)/2-(r\sin\theta)/3} r \, dz \, dr \, d\theta$$

$$= \int_0^{\pi/2} \int_0^1 r\left(1 - \frac{r}{2}\cos\theta - \frac{r}{3}\sin\theta\right) dr \, d\theta$$

$$= \int_0^{\pi/2} \left[\frac{r^2}{2} - \frac{r^3}{6}\cos\theta - \frac{r^3}{9}\sin\theta\right]_0^1 d\theta$$

$$= \int_0^{\pi/2} \left(\frac{1}{2} - \frac{1}{6}\cos\theta - \frac{1}{9}\sin\theta\right) d\theta = \left[\frac{\theta}{2} - \frac{1}{6}\sin\theta + \frac{1}{9}\cos\theta\right]_0^{\pi/2}$$

$$= \left(\frac{\pi}{4} - \frac{1}{6}\right) - \frac{1}{9} = \frac{\pi}{4} - \frac{5}{18}. \qquad \square$$

EXAMPLE 2   What is the maximum amount of ice cream that can be packed in a cone with diameter 2″ and height 3″? (Assume that the top of the filled ice cream cone is hemispherical with diameter 2″.)

SOLUTION

Place a three-dimensional rectangular coordinate system so that the bottom of the cone is at the origin and the center of the cone is along the z-axis. The cone can then be described by the equation $9(x^2 + y^2) = z^2$ and the top of the ice cream by $x^2 + y^2 + (z - 3)^2 = 1$, $z \geq 3$. (See Figure 16.44.) Hence the

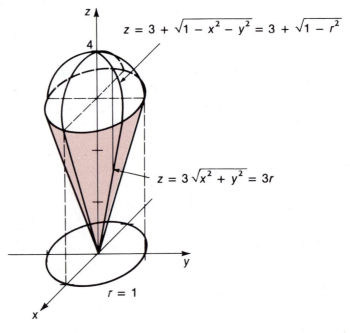

FIGURE 16.44

maximum volume of ice cream that can be packed in the cone is

$$V = \int_{\theta=0}^{\theta=2\pi} \int_{r=0}^{r=1} \int_{z=3r}^{z=3+\sqrt{1-r^2}} r \, dz \, dr \, d\theta$$

$$= \int_0^{2\pi} \int_0^1 (3 + \sqrt{1 - r^2} - 3r)r \, dr \, d\theta$$

$$= \int_0^{2\pi} \int_0^1 (3r + r\sqrt{1 - r^2} - 3r^2) dr \, d\theta$$

$$= \int_0^{2\pi} \left[\frac{3}{2} r^2 - \frac{1}{3}(1 - r^2)^{3/2} - r^3 \right]_0^1 d\theta$$

$$= \int_0^{2\pi} \left[\left(\frac{3}{2} - 0 - 1 \right) - \left(0 - \frac{1}{3} - 0 \right) \right] d\theta$$

$$= \int_0^{2\pi} \frac{5}{6} \, d\theta = \frac{5}{6} \, \theta \Big]_0^{2\pi} = \frac{5}{3} \, \pi \text{ in}^3.$$

(Approximately 44 cones of this size can be made from 1 gallon of ice cream. If the ice cream is not packed into the cone but is left spherical, 11 more cones can be made.) □

The cylindrical coordinate system in space is an extension of polar coordinates in the plane in the same way that the rectangular coordinate system in space is an extension of the rectangular system in the plane. In each case, a third dimension described by the z-coordinate direction is introduced perpendicular to the original coordinate plane. A typical volume element in the cylindrical coordinate system is shown in Figure 16.45. The volume of the element is

$$\Delta V = \Delta A \Delta z = r\Delta r\Delta \theta \Delta z,$$

since the base was shown in Section 16.3 to have area $r\Delta r\Delta \theta$ and the height is Δz.

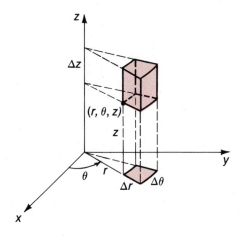

FIGURE 16.45

The spherical coordinate system in space is also an extension of the polar coordinate system in the plane, but the typical volume element in the spherical coordinate system has a much different form.

To approximate the volume of a typical element in spherical coordinates, consider the distances along the edges of the volume element: D_ρ from (ρ, θ, ϕ) to $(\rho + \Delta\rho, \theta, \phi)$, D_θ from (ρ, θ, ϕ) to $(\rho, \theta + \Delta\theta, \phi)$, and D_ϕ from (ρ, θ, ϕ) to $(\rho, \theta, \phi + \Delta\phi)$. The volume of the element shown in Figure 16.46 is approximated by

$$\Delta V = D_\rho\, D_\theta\, D_\phi.$$

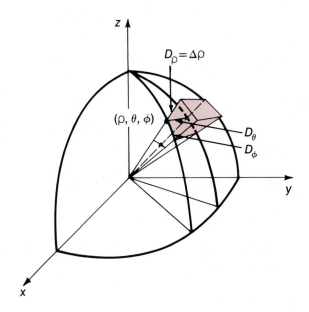

FIGURE 16.46

The distance D_ρ is

$$D_\rho = \Delta\rho.$$

The distance D_θ is the portion of the arc of the circle of radius $\rho \sin \phi$ determined by the angle $\Delta\theta$. See Figure 16.47(a).

$$D_\theta = \rho \sin \phi\, \Delta\theta.$$

The distance D_ϕ is the portion of the arc of the circle of radius ρ determined by the angle $\Delta\phi$. See Figure 16.47(b).

$$D_\phi = \rho\Delta\phi.$$

Consequently,

(16.17) $\Delta V \approx D_\rho\, D_\theta\, D_\phi = (\Delta\rho)(\rho \sin \phi\, \Delta\theta)(\rho\Delta\phi)$
$$= \rho^2 \sin \phi\, \Delta\rho\, \Delta\phi\, \Delta\theta.$$

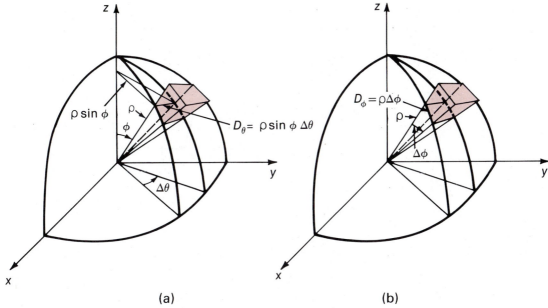

(a)   (b)

FIGURE 16.47

This result implies that if a region D in space can be described by

$$\alpha \le \theta \le \beta, \quad g_1(\theta) \le \phi \le g_2(\theta), \quad h_1(\theta, \phi) \le \rho \le h_2(\theta, \phi),$$

then

(16.18) $\displaystyle\iiint_D f(\rho, \theta, \phi)dV = \int_{\theta=\alpha}^{\theta=\beta} \int_{\phi=g_1(\theta)}^{\phi=g_2(\theta)} \int_{\rho=h_1(\theta,\phi)}^{\rho=h_2(\theta,\phi)} f(\rho, \theta, \phi)\rho^2\sin\phi\, d\rho\, d\phi\, d\theta.$

EXAMPLE 3   Use an iterated integral in spherical coordinates to determine the volume between the concentric spheres $x^2 + y^2 + z^2 = 2$ and $x^2 + y^2 + z^2 = 1$.

SOLUTION

Refer to Figure 16.48 on page 904, where the portion in the first octant is shown. From geometry we know that this volume should be

$$V = \frac{4\pi}{3}(\sqrt{2})^3 - \frac{4\pi}{3}(1)^3 = \frac{4\pi}{3}(2\sqrt{2} - 1).$$

To obtain this result by iterated integration in spherical coordinates, we evaluate

$$V = 2\int_{\theta=0}^{\theta=2\pi} \int_{\phi=0}^{\phi=\pi/2} \int_{\rho=1}^{\rho=\sqrt{2}} \rho^2 \sin\phi\, d\rho\, d\phi\, d\theta$$

$$= 2\int_0^{2\pi} \int_0^{\pi/2} \frac{\rho^3}{3}\Bigg]_1^{\sqrt{2}} \sin\phi\, d\phi\, d\theta$$

$$= \frac{2}{3}\int_0^{2\pi} \int_0^{\pi/2} ((\sqrt{2})^3 - 1)\sin\phi\, d\phi\, d\theta = \frac{2}{3}\int_0^{2\pi} (2\sqrt{2} - 1)(-\cos\phi)\Bigg]_0^{\pi/2} d\theta$$

$$= \frac{2}{3}\int_0^{2\pi} (2\sqrt{2} - 1)(-0 + 1)d\theta = \frac{2}{3}(2\sqrt{2} - 1)\cdot 2\pi = \frac{4\pi}{3}[2\sqrt{2} - 1].$$

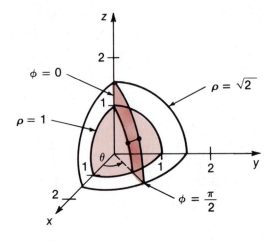

FIGURE 16.48

This volume can also be calculated by using cylindrical coordinates, but the computation requires evaluating two iterated integrals:

$$V = 2 \int_0^{2\pi} \int_0^1 \int_{\sqrt{1-r^2}}^{\sqrt{2-r^2}} r \, dz \, dr \, d\theta + 2 \int_0^{2\pi} \int_1^2 \int_0^{\sqrt{2-r^2}} r \, dz \, dr \, d\theta. \qquad \Box$$

EXAMPLE 4   Evaluate the integral of $f(x, y, z) = x^2 + y^2 + z$ over the region S bounded below by the cone $z = \sqrt{x^2 + y^2}$ and above by the sphere $x^2 + y^2 + z^2 = 9$.

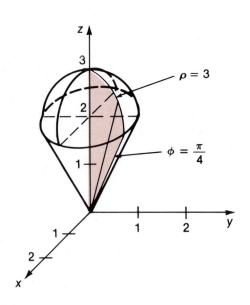

FIGURE 16.49

SOLUTION

The region is shown in Figure 16.49. The integral can be more easily evaluated by translating the problem into spherical coordinates, since S can be described in spherical coordinates by $0 \leq \rho \leq 3$ and $0 \leq \theta < 2\pi$. To find the limits on ϕ, observe that the cone $z = x^2 + y^2$ intersects the yz-plane in the line $z^2 = y^2$, so $0 \leq \phi \leq \pi/4$. Since

$$x^2 + y^2 + z = (\rho \sin \phi \cos \theta)^2 + (\rho \sin \phi \sin \theta)^2 + \rho \cos \phi$$
$$= \rho^2 (\sin \phi)^2 + \rho \cos \phi,$$

$$\iiint_S f(x, y, z)dv = \int_0^{2\pi} \int_0^{\pi/4} \int_0^3 [\rho^2 (\sin \phi)^2 + \rho \cos \phi]\rho^2 \sin \phi \, d\rho \, d\phi \, d\theta$$

$$= \int_0^{2\pi} \int_0^{\pi/4} \left[\frac{\rho^5}{5} (\sin \phi)^3 + \frac{\rho^4}{4} \cos \phi \sin \phi \right]_{\rho=0}^{\rho=3} d\phi \, d\theta$$

$$= \int_0^{2\pi} \int_0^{\pi/4} \left[\frac{243}{5} (1 - (\cos \phi)^2) \sin \phi + \frac{81}{4} \cos \phi \sin \phi \right] d\phi \, d\theta$$

$$= \int_0^{2\pi} \left[\frac{243}{5} \left(-\cos \phi + \frac{(\cos \phi)^3}{3} \right) + \frac{81}{8} (\sin \phi)^2 \right]_{\phi=0}^{\phi=\pi/4} d\theta$$

$$= \int_0^{2\pi} \frac{81}{80} (37 - 20\sqrt{2})d\theta = \frac{81\pi}{40} (37 - 20\sqrt{2}) \approx 55.5. \quad \square$$

EXERCISE SET 16.6

Evaluate the triple integrals in Exercises 1 through 6.

1. $\displaystyle\int_{-\pi}^{\pi} \int_0^1 \int_0^{\sqrt{1-r^2}} rz \sin \theta \, dz \, dr \, d\theta$

2. $\displaystyle\int_{-\pi}^{\pi} \int_0^1 \int_0^{\sqrt{1-r^2}} rz \sin \theta \, dz \, dr \, d\theta$

3. $\displaystyle\int_0^{\pi} \int_0^{\pi/2} \int_0^{\sin \phi} \rho^2 \sin \phi \cos \phi \, d\rho \, d\phi \, d\theta$

4. $\displaystyle\int_0^{\pi} \int_0^{\pi/2} \int_0^{\sin \phi} \rho^2 \sin \phi \cos \phi \, d\rho \, d\theta \, d\phi$

5. $\displaystyle\int_0^{2\pi} \int_0^{\sin \theta} \int_0^r r^2 \cos \theta \, dz \, dr \, d\theta$

6. $\displaystyle\int_0^1 \int_0^{\arccos \theta} \int_0^{\cos \phi} \rho \sin \phi \, e^\theta \, d\rho \, d\phi \, d\theta$

Use either cylindrical coordinates or spherical coordinates to reexpress and evaluate the integrals in Exercises 7 through 12.

7. $\displaystyle\int_0^2 \int_0^{\sqrt{4-x^2}} \int_0^{\sqrt{4-x^2-y^2}} \frac{z}{\sqrt{x^2 + y^2}} \, dz \, dy \, dx$

8. $\displaystyle\int_0^2 \int_{-\sqrt{4-y^2}}^{\sqrt{4-y^2}} \int_0^{\sqrt{4-x^2-y^2}} z\sqrt{x^2 + y^2} \, dz \, dx \, dy$

9. $\displaystyle\int_0^2 \int_{-\sqrt{4-y^2}}^{\sqrt{4-y^2}} \int_0^{\sqrt{4-x^2-y^2}} z\sqrt{x^2 + y^2 + z^2} \, dz \, dx \, dy$

10. $\displaystyle\int_0^2 \int_0^{\sqrt{4-x^2}} \int_0^{\sqrt{4-x^2-y^2}} \frac{z}{\sqrt{x^2 + y^2 + z^2}} \, dz \, dy \, dx$

11. $\displaystyle\int_0^1 \int_0^{\sqrt{1-x^2}} \int_{\sqrt{1-x^2-y^2}}^{1-x^2-y^2} \arctan\left(\frac{y}{x}\right) \, dz \, dy \, dx$

12. $\displaystyle\int_{-1}^2 \int_0^2 \int_0^{\sqrt{4-x^2}} e^{x^2+y^2} \, dy \, dx \, dz$

Use either cylindrical or spherical coordinates to find the volume of the solid described in Exercises 13 through 20.

13. Inside both $z^2 = x^2 + y^2$ and $x^2 + y^2 + z^2 = 4$

14. Inside $x^2 + y^2 + z^2 = 4$ and outside $z^2 = x^2 + y^2$

15. Bounded by $z = x^2 + y^2$, $z = 1$, and $z = 4$

16. Bounded by $z^2 = x^2 + y^2$ and $x^2 + y^2 = 4$

17. With boundaries $z = \sqrt{x^2 + y^2}$, $z = \sqrt{3x^2 + 3y^2}$, and $z = 9$

18. With boundaries $z = \sqrt{x^2 + y^2}$, $z = \sqrt{3x^2 + 3y^2}$, and $\sqrt{x^2 + y^2 + z^2} = 3$

19. With boundaries $z = \sqrt{x^2 + y^2}$, $\sqrt{x^2 + y^2 + z^2} = 2$, and $\sqrt{x^2 + y^2 + z^2} = 3$

20. With boundaries $z = \sqrt{x^2 + y^2}$, $x^2 + y^2 = 4$, $x^2 + y^2 = 9$, and $z = 0$

In Exercises 21 through 24, sketch the solid whose volume is given by the triple integral.

21. $\displaystyle\int_0^{2\pi} \int_0^2 \int_0^{r^2} r \, dz \, dr \, d\theta$ **22.** $\displaystyle\int_0^\pi \int_1^3 \int_0^4 r \, dz \, dr \, d\theta$

23. $\displaystyle\int_0^{\pi/2} \int_0^{\pi/4} \int_0^2 \rho^2 \sin\phi \, d\rho \, d\phi \, d\theta$ **24.** $\displaystyle\int_0^{\pi/2} \int_0^{\pi/4} \int_0^2 \rho^2 \sin\theta \, d\rho \, d\theta \, d\phi$

In Exercises 25 through 27, set up the integral(s) in (a) rectangular, (b) cylindrical, and (c) spherical coordinates to evaluate a function f over the solid described.

25. Inside $x^2 + y^2 + z^2 = a^2$

26. Inside $x^2 + y^2 = a^2$, bounded below by $z = 0$ and above by $z = a$

27. Bounded by $x = a$, $y = a$, $z = a$, and the coordinate planes

28. A conical dunce cap of diameter 5 inches and height 15 inches sits upon the (spherical with radius 3.5 inches) head of the class dullard. Find the volume of air in the cap.

29. A hole of radius 1 inch is drilled through the center of a sphere of radius 6 inches. Find the volume of the solid removed.

30. A hole drilled through the center of a sphere of radius R removes half the volume of the sphere. What is the radius of the hole?

16.7
APPLICATIONS OF TRIPLE INTEGRALS

The concepts of mass, moments, and inertia for solids in space are direct extensions of these concepts for plane lamina. The mass of a solid described by the region D in space with density function ρ is

(16.19)
$$M = \iiint_D \rho(x, y, z)dV.$$

Since the distances of a point (x, y, z) from the xy-, yz-, and xz-planes are z, x, and y respectively, the moments with respect to these coordinate planes are

(16.20)
$$M_{xy} = \iiint_D z\,\rho(x, y, z)dV,$$

(16.21)
$$M_{yz} = \iiint_D x\,\rho(x, y, z)dV,$$

and

(16.22)
$$M_{xz} = \iiint_D y\,\rho(x, y, z)dV.$$

The center of mass $(\bar{x}, \bar{y}, \bar{z})$ is

(16.23)
$$\bar{x} = \frac{M_{yz}}{M}, \qquad \bar{y} = \frac{M_{xz}}{M}, \qquad \bar{z} = \frac{M_{xy}}{M}.$$

EXAMPLE 1   Find the center of mass of the ice cream cone described in Example 2 of Section 16.6. The cone is topped with a hemisphere of ice cream. The height of the cone is $3''$ and the diameter of the top of the cone and of the hemisphere of ice cream is $2''$. Assume that the density of the ice cream is a constant ρ.

SOLUTION
The cone is described by $z^2 = 9(x^2 + y^2)$ and the top of the ice cream by $x^2 + y^2 + (z - 3)^2 = 1, z \geq 3$. We found that the volume of the cone is $5\pi/3$ (inches)^3. Since the density ρ is constant, the mass of the ice cream cone is

$$M = \frac{5\pi}{3}\rho.$$

The cone is symmetric with respect to both the xz- and yz- planes, so the center of mass lies along the z-axis; that is, $\bar{x} = 0$ and $\bar{y} = 0$. To find $\bar{z}$ requires finding

$$M_{xy} = \iiint_D z\rho\, dV.$$

By referring again to Example 2 in Section 16.6, we see that

$$M_{xy} = \rho \int_{\theta=0}^{\theta=2\pi} \int_{r=0}^{r=1} \int_{z=3r}^{z=3+\sqrt{1-r^2}} z \, r \, dz \, dr \, d\theta$$

$$= \rho \int_0^{2\pi} \int_0^1 \frac{z^2 r}{2} \bigg]_{3r}^{3+\sqrt{1-r^2}} dr \, d\theta$$

$$= \rho \int_0^{2\pi} \int_0^1 \frac{r}{2} [(3 + \sqrt{1 - r^2})^2 - (3r)^2] dr \, d\theta$$

$$= \rho \int_0^{2\pi} \int_0^1 (5r + 3r\sqrt{1 - r^2} - 5r^3) dr \, d\theta$$

$$= \rho \int_0^{2\pi} \left[\frac{5r^2}{2} - (1 - r^2)^{3/2} - \frac{5r^4}{4} \right]_0^1 d\theta$$

$$= \rho \int_0^{2\pi} \frac{9}{4} \, d\theta = \frac{9\pi}{2} \rho.$$

Thus,

$$\bar{z} = \frac{M_{xy}}{M} = \frac{\frac{9\pi}{2} \rho}{\frac{5\pi}{3} \rho} = 2.7 \text{ inches.}$$

The center of mass lies at $(0, 0, 2.7)$. $\square$

To describe the moment of inertia of a solid about an axis, we first need to consider the distance of an arbitrary point from the axis. Suppose we want to determine the moment of inertia I_z of a solid region D with density ρ. The distance from an arbitrary point (x, y, z) to the z-axis is $\sqrt{x^2 + y^2}$. (See Figure 16.50.)

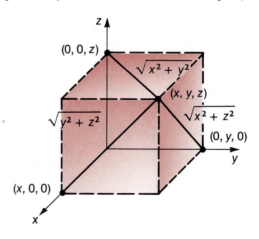

FIGURE 16.50

Consequently,

$$I_z = \iiint_D (\sqrt{x^2 + y^2})^2 \, \rho(x, y, z) \, dV = \iiint_D (x^2 + y^2)\rho(x, y, z) \, dV.$$

Similarly,

$$I_x = \iiint\limits_{D} (y^2 + z^2)\rho(x, y, z)\, dV$$

and

$$I_y = \iiint\limits_{D} (x^2 + z^2)\rho(x, y, z)\, dV.$$

The radii of gyration $\hat{x}$, $\hat{y}$, $\hat{z}$ with respect to the x-, y-, and z-axes can be expressed in a manner similar to the case of double integrals:

$$\hat{x} = \sqrt{\frac{I_x}{M}}, \qquad \hat{y} = \sqrt{\frac{I_y}{M}}, \qquad \hat{z} = \sqrt{\frac{I_z}{M}}.$$

EXAMPLE 2   Find the radius of gyration about the z-axis for the solid described by $x^2 + y^2 + z^2 = 1$ with density at (x, y, z) given by $x^2 + y^2 + z^2$.

SOLUTION

The mass

$$M = \iiint\limits_{D} (x^2 + y^2 + z^2)dV$$

is most easily evaluated by using spherical coordinates.

$$M = \int_{\theta=0}^{\theta=2\pi} \int_{\phi=0}^{\phi=\pi} \int_{\rho=0}^{\rho=1} \rho^2\, (\rho^2 \sin\phi)d\rho\, d\phi\, d\theta$$

$$= \int_0^{2\pi} \int_0^{\pi} \frac{\rho^5}{5} \sin\phi \bigg]_0^1 d\phi\, d\theta = \frac{1}{5}\int_0^{2\pi} -\cos\phi \bigg]_0^{\pi} d\theta$$

$$= \frac{2}{5}\int_0^{2\pi} d\theta = \frac{4\pi}{5}.$$

The moment of inertia about the z-axis is

$$I_z = \iiint\limits_{D} (x^2 + y^2 + z^2)(x^2 + y^2)dV.$$

Because of the product of $x^2 + y^2$ terms, we use cylindrical coordinates:

$$I_z = 2\int_0^{2\pi} \int_0^1 \int_0^{\sqrt{1-r^2}} (r^2 + z^2)r^2\, r\, dz\, dr\, d\theta$$

$$= 2\int_0^{2\pi} \int_0^1 r^5 z + \frac{r^3 z^3}{3} \bigg]_0^{\sqrt{1-r^2}} dr\, d\theta$$

$$= 2\int_0^{2\pi} \int_0^1 \left[r^5 \sqrt{1 - r^2} + r^3 \frac{(1 - r^2)^{3/2}}{3} \right] dr\, d\theta$$

$$= \frac{2}{3}\int_0^{2\pi} \int_0^1 (2r^5 + r^3)\sqrt{1 - r^2}\, dr\, d\theta.$$

Using the trigonometric substitution $r = \sin \psi$, $dr = \cos \psi \, d\psi$ yields:

$$I_z = \frac{2}{3} \int_0^{2\pi} \int_{r=0}^{r=1} [2(\sin \psi)^5 + (\sin \psi)^3](\cos \psi)^2 \, d\psi \, d\theta$$

$$= \frac{2}{3} \int_0^{2\pi} \int_{r=0}^{r=1} [2(1 - (\cos \psi)^2)^2 + 1 - (\cos \psi)^2](\cos \psi)^2 \sin \psi \, d\psi \, d\theta$$

$$= \frac{2}{3} \int_0^{2\pi} \int_{r=0}^{r=1} [3(\cos \psi)^2 - 5(\cos \psi)^4 + 2(\cos \psi)^6]\sin \psi \, d\psi \, d\theta$$

$$= \frac{2}{3} \int_0^{2\pi} \left[-(\cos \psi)^3 + (\cos \psi)^5 - \frac{2}{7}(\cos \psi)^7 \right]_{r=0}^{r=1} d\theta$$

$$= \frac{2}{3} \int_0^{2\pi} \left[-(\sqrt{1 - r^2})^3 + (\sqrt{1 - r^2})^5 - \frac{2}{7}(\sqrt{1 - r^2})^7 \right]_{r=0}^{r=1} d\theta$$

$$= \frac{2}{3} \int_0^{2\pi} \frac{2}{7} \, d\theta = \frac{4}{21}(2\pi) = \frac{8}{21} \pi.$$

Consequently,

$$\hat{z} = \sqrt{\frac{I_z}{M}} = \sqrt{\frac{\dfrac{8}{21}\pi}{\dfrac{4}{5}\pi}} = \sqrt{\frac{10}{21}} \approx .7. \qquad \square$$

In Section 5.4 we defined the average value of a function of one variable with respect to x by:

Average value of f over $[a, b]$ = $\dfrac{1}{b - a} \displaystyle\int_a^b f(x) \, dx$.

If f is a continuous function of two variables x and y, the average value of f over a region R in the xy-plane is:

Average value of f over R = $\dfrac{1}{\text{area of } R} \displaystyle\iint_R f(x, y) \, dA = \dfrac{\displaystyle\iint_R f(x, y) dA}{\displaystyle\iint_R dA}$.

Similarly, if f is a continuous function of three variables, then the average value of f over a region D in space is

Average value of f over D = $\dfrac{1}{\text{volume of } D} \displaystyle\iiint_D f(x, y, z) dV$

$$= \dfrac{\displaystyle\iiint_D f(x, y, z) \, dV}{\displaystyle\iiint_D dV}.$$

EXAMPLE 3   A dome tent has the shape of a hemisphere with radius 4 feet. The temperature on the floor is 55°F and at the top of the dome is 65°F. Find the average temperature in the tent under the assumption that the temperature at any point is proportional to its distance from the floor.

SOLUTION

We first place a rectangular coordinate system so that the *xy*-plane coincides with the tent's floor with the center at the origin and the positive *z*-axis directed toward the top of the dome, as in Figure 16.51.

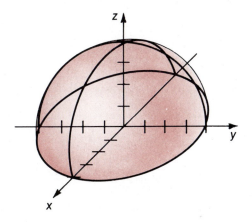

The outside of the tent can then be described by

$$x^2 + y^2 + z^2 = 16, \qquad z \geq 0$$

or in cylindrical coordinates by

$$z = \sqrt{16 - r^2}.$$

Since the temperature varies linearly with height from 55° at the floor to 65° at the top of the dome, the temperature at any point in the tent is described by

$$T(x, y, z) = \left(55 + \frac{5}{2} z \right)^{\circ}.$$

Since the volume of the hemispherical tent is $\dfrac{1}{2}\left(\dfrac{4}{3}\pi 4^3 \right) = \dfrac{128\pi}{3},$

$$\text{Average temperature} = \frac{3}{128\pi} \iiint_D \left(55 + \frac{5}{2}z\right) dV$$

$$= \frac{3}{128\pi} \int_0^{2\pi} \int_0^4 \int_0^{\sqrt{16-r^2}} \left(55 + \frac{5}{2}z\right) r \, dz \, dr \, d\theta$$

$$= \frac{15}{128\pi} \int_0^{2\pi} \int_0^4 \left[\left(11z + \frac{z^2}{4}\right)r\right]_0^{\sqrt{16-r^2}} dr \, d\theta$$

$$= \frac{15}{128\pi} \int_0^{2\pi} \int_0^4 \left(11r\sqrt{16-r^2} + 4r - \frac{r^3}{4}\right) dr \, d\theta$$

$$= \frac{15}{128\pi} \int_0^{2\pi} \left[\frac{-11}{3}(16-r^2)^{3/2} + 2r^2 - \frac{r^4}{16}\right]_0^4 d\theta$$

$$= \frac{15}{128\pi} \int_0^{2\pi} \left(32 - 16 + \frac{11}{3}(64)\right) d\theta$$

$$= \frac{15}{128\pi} \left(\frac{752}{3}\right)(2\pi) = 58.75°\text{F.} \qquad \square$$

EXERCISE SET 16.7

Find the mass of the solids described in Exercises 1 through 6.

1. Bounded by the coordinate planes and the planes $x = 1$, $y = 2$, and $z = 3$. Density described by $\rho(x, y, z) = 3z + 1$.

2. Bounded by the coordinate planes and the plane $x + y + z = 1$. Density of the solid is constant.

3. Bounded by the cylinder $x^2 + y^2 = 4$ and the planes $z = 1$ and $z = 5$. Density described by $\rho(x, y, z) = kz$.

4. Bounded by the cone $z = \sqrt{x^2 + y^2}$ and the plane $z = 2$. Density function described by $\rho(x, y, z) = k\sqrt{x^2 + y^2}$.

5. Bounded by the hemisphere of radius 3 centered at $(0, 0, 0)$ and lying above the xy-plane. Density proportional to the square of the distance from the base.

6. Bounded by the xy-plane, the cone $z = \sqrt{x^2 + y^2}$ and the hemisphere $z = \sqrt{1 - x^2 - y^2}$. Density proportional to $\sqrt{x^2 + y^2 + z^2}$.

7–12. Find the moments about the coordinate planes and center of mass of the solids and density functions described in Exercises 1 through 6.

13–18. Find the moments of inertia about the coordinate axes and radii of gyration of the solids and density functions described in Exercises 1 through 6.

19. Find the average height of that portion of the plane $3x + 2y + z = 6$ that lies in the first octant.

20. A solid of constant density has the shape of a cube with sides of length 4. Find the moment about a diagonal plane of the cube.

21. Set up the integrals required to find the moments of inertia about the x- and y-axes for the solid and density function described in Example 2 of Section 16.7. Determine the values of these moments. (*Hint:* Think carefully about the situation before performing the integrations.)

22. Recall the connection between the average value of a single variable function and the mean value theorem for integrals discussed in Section 5.4. State a corresponding theorem for triple integrals.

23. A conical teepee has a base of radius 4 feet and height 6 feet. Use the same temperature assumptions as described in Example 3 to find the average temperature inside the teepee. Which tent is cooler?

24. A 16-oz can of beer is sitting on a block of ice. Suppose an xyz-coordinate system is introduced with the center of the bottom of the can at the origin and the positive z-axis passing through the center of the can. The temperature of the liquid can be described by $T(x, y, z) = 35 + z\,[3 + 2\sqrt{x^2 + y^2}]$ degrees Fahrenheit, for $0 \le z \le 6$ and $0 \le x^2 + y^2 \le (1.25)^2$. Find the average temperature of the beer.

25. Paint in a one-gallon cylindrical can of height 6 inches and diameter 7 inches has a density that varies linearly from 55.3 lb/ft^3 at the top of the can to 59.4 lb/ft^3 at the bottom. How much does the paint weigh and where is the center of mass of the paint?

26. Modern lumber designed for outdoor use is pressure-treated with a liquid that penetrates the wood to prevent rot and insect damage. Suppose that an 8-foot 4-by-4 (actual dimensions 3.5 inches by 3.5 inches) originally weighing 40 lb is pressure-treated to absorb 5 lb of preservative. The liquid penetrates the wood so that the preservative is linearly distributed from zero at the center of the wood to a maximum on a pair of parallel outside edges. Show that the center of mass of the wood is unchanged, but that the moment of inertia about a line placed perpendicular to the base of the wood is increased.

27. A *pousse-cafe* is a (generally alcoholic) drink made by gently pouring colored liquids with varying density into a glass so that the liquids do not mix. Suppose that 1/4 oz of each of the following liqueurs are poured into a conical glass of height 4 inches and diameter 2 inches in order of decreasing density and remain in layers in the glass. What is the center of mass of the liquid? Suppose the contents of the glass are stirred to make a homogeneous liquid. Does the center of mass of the liquid stay the same?

Liqueur	Specific Gravity (Density)
Benedictine	1.150
Drambuie	1.180
Liquore Galliano	1.193
Grand Marnier	1.070
Irish Mist	1.063

REVIEW EXERCISES

1. Suppose $f(x, y) = 4 - x - 2y$, R is the triangle in the xy-plane bounded by $x = 0$, $y = 0$, and $x = 4 - 2y$, and $\mathcal{P}$ is the partition defined by the lines $x = 0$, $x = 1$, $x = 2$, $x = 3$, $x = 4$, $y = 0$, $y = \dfrac{1}{2}$, $y = 1$, $y = \dfrac{3}{2}$, and $y = 2$. Find the value of the Riemann sum if the R_i are the rectangles that lie entirely within R and (x_i, y_i) is chosen to be the upper left corner of R_i.

2. Repeat Exercise 1 if (x_i, y_i) is chosen to be the lower right corner of R_i.

3. Find the volume of the solid bounded by the coordinate planes and above by the plane $x + 2y + z = 4$. Find the area of the face of this solid lying in the xy-plane.

4. Find the volume of the solid bounded by the coordinate planes and by the plane $3z - 4x - 2y = 12$. Find the area of the face of this solid lying in the xy-plane.

Sketch the graphs of the regions in the xy-plane over which the integration in Exercises 5 through 10 is taken. Reverse the order of integration and evaluate the double integral obtained.

5. $\displaystyle\int_0^4 \int_x^{2\sqrt{x}} dy\, dx$

6. $\displaystyle\int_0^1 \int_0^{\sqrt{y}} (2 - x - y)dx\, dy$

7. $\displaystyle\int_0^4 \int_{-\sqrt{y}}^{\sqrt{y}} (4 - y)dx\, dy$

8. $\displaystyle\int_0^2 \int_{-2}^2 y^3 e^{4x}\, dx\, dy$

9. $\displaystyle\int_0^1 \int_y^1 ye^{-x^3}\, dx\, dy$

10. $\displaystyle\int_0^4 \int_y^{2\sqrt[3]{2y}} xy\, dx\, dy$

In Exercises 11 through 14, the boundaries of a solid with base in the xy-plane are described. Use double integrals to find the area of the base and the volume of the solid.

11. $x = y^2$, $x = 8 - y^2$, $z = 4x$

12. $x = 0$, $y = 1$, $y = x^3$, $z = 2 - x - y$

13. $r = 1 - \sin\theta$, $x = 0$, $z = 4$

14. $r = \cos\theta$, $x = 0$, $z = xy$

In Exercises 15 through 18, sketch a solid whose volume is represented by the integral. Find the volume.

15. $\displaystyle\int_0^1 \int_0^{\sqrt{1-y^2}} 4y\, dx\, dy$

16. $\displaystyle\int_0^2 \int_0^{\sqrt{4-x^2}} \int_0^{2-y} dz\, dy\, dx$

17. $\displaystyle\int_0^{\pi/2} \int_0^{2\sqrt{2}} \int_{r^2/8}^{\sqrt{9-r^2}} r\, dz\, dr\, d\theta$

18. $\displaystyle\int_0^{\pi/2} \int_0^{\pi/2} \int_0^2 \rho^2 \sin\phi\, d\rho\, d\theta\, d\phi$

In Exercises 19 through 28, find the volume of each solid.

19. The solid bounded below by the paraboloid $z = x^2 + y^2$ and above by the plane $z = 4$.

20. The solid bounded below by the paraboloid $z = x^2 + y^2 - 1$ and above by the plane $z = 4$.

21. The solid bounded by the cylinder $r = 2 - 2 \cos \theta$ and the planes $z = 0$ and $z = 4$.

22. The solid bounded by the cylinder $r = \sin \theta$, the xy-plane, and $z = r^2$.

23. The solid bounded below by the paraboloid $z = x^2 + y^2$ and above by the sphere $x^2 + y^2 + z^2 = 2$.

24. The solid bounded below by the cone $z^2 = x^2 + y^2$ and above by the sphere $x^2 + y^2 + (z - 2)^2 = 4$.

25. The solid in the first octant bounded by the cylinder $y^2 + z^2 = 9$ and the plane $z = 3 - x$.

26. The portion of the cylinder $9x^2 + 4y^2 = 36$ bounded below by the xy-plane and above by the plane $x + y + 2z = 6$.

27. The solid in the first octant bounded above by the planes $z = x/3$ and $x + 2y + 3z = 6$.

28. The solid bounded below by the xy-plane, above by the plane $z = 4 - y$, and by the cylinder $y = x^2$.

In Exercises 29 through 34, find the mass, moments about the axes, and center of mass of the lamina that has the shape of the region bounded by the curves and has the indicated density.

29. $y = e^x$, $y = 0$, $x = 0$, $x = 2$, $\rho(x, y) = 1$

30. $y = \sin x$, $y = \cos x$, $y = 0$, $x = 0$, $x = \dfrac{\pi}{2}$, $\rho(x, y) = 3$

31. $y = e^x$, $y = e^{-x}$, $y = 0$, $x = -1$, $x = 1$, $\rho(x, y) = |xy|$

32. $y = 1/x$, $y = 0$, $x = 1$, $x = 2$, $\rho(x, y) = x^2 + y^2$

33. $x = 0$, $y = 0$, $y = \sqrt{9 - x^2}$, $\rho(x, y) = x^2 + y^2$

34. $y = 1$, $y = \sqrt{2 - x^2}$, $\rho(x, y) = x^2 + y^2$

35. Find the mass of a sphere of radius a if the density at every point is twice the distance from the center.

36. Find the mass of a plate in the shape of a cardioid $r = 1 - \sin \theta$ if the density is proportional to the distance from the origin.

37. Find the radius of gyration with respect to the z-axis of the cube bounded by the coordinate planes and the planes $x = a$, $y = a$, and $z = a$, if the density at any point (x, y, z) is given by $x + y + z$.

38. Show that
$$\int_0^\pi \int_0^\pi \sin x \sin y \, dx \, dy = \left(\int_0^\pi \sin x \, dx \right)^2,$$
but
$$\int_0^\pi \int_0^\pi (\sin x)^2 \, dx \, dy \neq \left(\int_0^\pi \sin x \, dx \right)^2.$$

17

LINE AND SURFACE INTEGRALS

The definite integral has been defined for a function whose domain is an interval of the real line, a region in the plane, or a solid in space. In this chapter we extend the definition of the definite integral to include functions whose domain consists of a curve or surface in space. These definitions lead to results that have important applications in engineering and the physical sciences.

17.1
VECTOR FIELDS

The gradient of a scalar function f defined in Section 15.5 is the vector function described by

$$\nabla f(x, y, z) = \frac{\partial f(x, y, z)}{\partial x} \mathbf{i} + \frac{\partial f(x, y, z)}{\partial y} \mathbf{j} + \frac{\partial f(x, y, z)}{\partial z} \mathbf{k}.$$

By definition, ∇f is a function that associates a vector with each point (x, y, z) in a region in space. We have seen applications of the gradient in determining the direction and magnitude of the maximum and minimum rates of change along a surface and in finding the normal to a surface at a point.

In this section, we begin a study of general functions that associate each point in a region in space with a vector, and begin laying the groundwork for the remainder of the chapter.

If to each point (x, y, z) a vector $\mathbf{F}(x, y, z)$ is assigned, the collection of all vectors $\mathbf{F}(x, y, z)$ is called a **vector field.** When $\mathbf{F}(x, y, z)$ describes velocity, the vector field is called a **velocity field.** (A velocity field is illustrated in Figure

17.1.) Similarly, if $\mathbf{F}(x, y, z)$ describes force, the field is called a **force field** and if $\mathbf{F}$ is the gradient of some function, the vector field is called a **gradient field.**

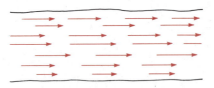

Velocity field of a body of water

FIGURE 17.1

For convenience of notation, we use the term vector field to describe either the collection of vectors in the field or the function used to describe this collection.

EXAMPLE 1   Describe the vector field if

$$\mathbf{F}(x, y, z) = -y\mathbf{i} + x\mathbf{j} \quad \text{for } (x, y, z)$$

satisfying $x^2 + y^2 \le 4, 0 \le z \le 1$.

SOLUTION
The vector field at various points is illustrated in Figure 17.2.

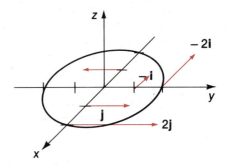

FIGURE 17.2

This field can be thought of as the velocity field of a rotating disk of radius 2 with the z-axis as the axis of rotation. □

Associated with every differentiable vector field are two functions: the *divergence*, a scalar function and the *curl*, a vector function.

(17.1)
DEFINITION
Suppose $\mathbf{F}(x, y, z) = M(x, y, z)\mathbf{i} + N(x, y, z)\mathbf{j} + P(x, y, z)\mathbf{k}$ is a vector function and M, N, and P are differentiable. The **divergence** of $\mathbf{F}$ is the scalar function

$$\text{div } \mathbf{F} = \frac{\partial M}{\partial x} + \frac{\partial N}{\partial y} + \frac{\partial P}{\partial z}$$

and the **curl** of **F** is the vector function

$$\text{curl } \mathbf{F} = \left(\frac{\partial P}{\partial y} - \frac{\partial N}{\partial z} \right) \mathbf{i} + \left(\frac{\partial M}{\partial z} - \frac{\partial P}{\partial x} \right) \mathbf{j} + \left(\frac{\partial N}{\partial x} - \frac{\partial M}{\partial y} \right) \mathbf{k}.$$

The names *divergence* and *curl* are derived from the application of these concepts to fluid flow. Later in this chapter we will see that the divergence of the velocity field of a moving fluid describes the rate of change of the fluid mass per unit volume. As such, the divergence describes how the fluid diverges at the point. We will also see that the curl of the velocity field of the fluid describes how the fluid rotates, or curls, about a point. European writers use the word *rotation* instead of curl to describe this concept and write rot **F** in place of curl **F**. This terminology is perhaps more descriptive.

A notation similar to that used for the gradient of a function f, ∇f, can be used to represent the divergence and curl. Let

(17.2)
$$\nabla = \frac{\partial}{\partial x} \mathbf{i} + \frac{\partial}{\partial y} \mathbf{j} + \frac{\partial}{\partial z} \mathbf{k}$$

be a function that maps a scalar function f into a vector function ∇f

$$\nabla : f \rightarrow \frac{\partial f}{\partial x} \mathbf{i} + \frac{\partial f}{\partial y} \mathbf{j} + \frac{\partial f}{\partial z} \mathbf{k} = \nabla f.$$

With this definition,

(17.3)
$$\text{div } \mathbf{F} = \nabla \cdot \mathbf{F}$$

and

(17.4)
$$\text{curl } \mathbf{F} = \nabla \times \mathbf{F}.$$

EXAMPLE 2   Find the divergence and curl of

$$\mathbf{F}(x, y, z) = x^2 y \,\mathbf{i} + xyz \,\mathbf{j} + (y^2 + z^2) \mathbf{k}.$$

SOLUTION

$$\text{div } \mathbf{F}(x, y, z) = \nabla \cdot \mathbf{F}(x, y, z)$$

$$= \left(\frac{\partial}{\partial x} \mathbf{i} + \frac{\partial}{\partial y} \mathbf{j} + \frac{\partial}{\partial z} \mathbf{k} \right) \cdot (x^2 y \,\mathbf{i} + xyz \,\mathbf{j} + (y^2 + z^2) \mathbf{k})$$

$$= \frac{\partial}{\partial x}(x^2 y) + \frac{\partial}{\partial y}(xyz) + \frac{\partial}{\partial z}(y^2 + z^2)$$

$$= 2xy + xz + 2z.$$

$$\text{curl } \mathbf{F}(x, y, z) = \nabla \times \mathbf{F}(x, y, z)$$

$$= \begin{vmatrix} \mathbf{i} & \mathbf{j} & \mathbf{k} \\ \dfrac{\partial}{\partial x} & \dfrac{\partial}{\partial y} & \dfrac{\partial}{\partial z} \\ x^2 y & xyz & y^2 + z^2 \end{vmatrix}$$

$$= \left[\frac{\partial}{\partial y}(y^2 + z^2) - \frac{\partial}{\partial z}(xyz) \right]\mathbf{i} - \left[\frac{\partial}{\partial x}(y^2 + z^2) - \frac{\partial}{\partial z}(x^2 y) \right]\mathbf{j}$$

$$+ \left[\frac{\partial}{\partial x}(xyz) - \frac{\partial}{\partial y}(x^2 y) \right]\mathbf{k}$$

$$= (2y - xy)\mathbf{i} + (yz - x^2)\mathbf{k}. \qquad \square$$

The operations of divergence and curl satisfy the usual arithmetic properties: if $\mathbf{F}$ and $\mathbf{G}$ are vector functions and c is a constant, then

(17.5) $\text{div}(\mathbf{F} + \mathbf{G}) = \text{div } \mathbf{F} + \text{div } \mathbf{G}, \qquad \text{div}(c\mathbf{F}) = c(\text{div } \mathbf{F})$

(17.6) $\text{curl}(\mathbf{F} + \mathbf{G}) = \text{curl } \mathbf{F} + \text{curl } \mathbf{G}, \qquad \text{curl}(c\mathbf{F}) = c(\text{curl } \mathbf{F}).$

In addition, the divergence and curl satisfy interesting and useful relationships when applied to various combinations of scalar and vector functions. Some of these are listed below, others are given in the exercises.

If $\mathbf{F}$ and $\mathbf{G}$ are vector functions and f is a scalar function and the various partial derivatives are continuous, then

(17.7) $\text{div}(f\mathbf{F}) = f(\text{div } \mathbf{F}) + (\text{grad } f) \cdot \mathbf{F}$

(17.8) $\text{curl}(\text{grad } f) = \mathbf{0}$

(17.9) $\text{div}(\text{curl } \mathbf{F}) = 0$

(17.10) $\text{curl}(f\mathbf{F}) = f(\text{curl } \mathbf{F}) + (\text{grad } f) \times \mathbf{F}$

(17.11) $\text{div}(\mathbf{F} \times \mathbf{G}) = (\text{curl } \mathbf{F}) \cdot \mathbf{G} - \mathbf{F} \cdot (\text{curl } \mathbf{G})$

(17.12) $\text{div}(\text{grad } f) = \nabla^2 f = \dfrac{\partial^2 f}{\partial x^2} + \dfrac{\partial^2 f}{\partial y^2} + \dfrac{\partial^2 f}{\partial z^2}.$

These identities can be verified by directly applying the definitions. For example, to verify (17.7), let $\mathbf{F} = M\mathbf{i} + N\mathbf{j} + P\mathbf{k}$. Then $f\mathbf{F} = fM\mathbf{i} + fN\mathbf{j} + fP\mathbf{k}$, so

$$\text{div}(f\mathbf{F}) = \nabla \cdot (f\mathbf{F}) = \frac{\partial(fM)}{\partial x} + \frac{\partial(fN)}{\partial y} + \frac{\partial(fP)}{\partial z}$$

$$= M\frac{\partial f}{\partial x} + f\frac{\partial M}{\partial x} + N\frac{\partial f}{\partial y} + f\frac{\partial N}{\partial y} + P\frac{\partial f}{\partial z} + f\frac{\partial P}{\partial z}$$

$$= f\left[\frac{\partial M}{\partial x} + \frac{\partial N}{\partial y} + \frac{\partial P}{\partial z} \right] + \left[M\frac{\partial f}{\partial x} + N\frac{\partial f}{\partial y} + P\frac{\partial f}{\partial z} \right]$$

$$= f(\nabla \cdot \mathbf{F}) + (\nabla f) \cdot \mathbf{F}$$

$$= f(\text{div } \mathbf{F}) + \text{grad } f \cdot \mathbf{F}.$$

Identity (17.12) is a particularly useful result that has application in many areas of applied mathematics because of its relation to **Laplace's equation:**

(17.13)
$$\nabla^2 f = \frac{\partial^2 f}{\partial x^2} + \frac{\partial^2 f}{\partial y^2} + \frac{\partial^2 f}{\partial z^2} = 0.$$

Functions satisfying this equation are called **harmonic functions** and occur naturally in the study of such areas as steady-state heat flow, electrostatics, and gravitational attraction. The heat flow application is considered in Section 17.8.

EXAMPLE 3   Show that $f(x, y, z) = \dfrac{1}{\sqrt{x^2 + y^2 + z^2}}$ satisfies Laplace's equation.

SOLUTION

$$\nabla f(x, y, z) = \left(\frac{\partial}{\partial x} \mathbf{i} + \frac{\partial}{\partial y} \mathbf{j} + \frac{\partial}{\partial z} \mathbf{k} \right)(x^2 + y^2 + z^2)^{-1/2}$$

$$= -(x^2 + y^2 + z^2)^{-3/2} (x\mathbf{i} + y\mathbf{j} + z\mathbf{k}).$$

So $\nabla^2 f(x, y, z) = -\dfrac{\partial}{\partial x}[(x^2 + y^2 + z^2)^{-3/2} x] - \dfrac{\partial}{\partial y}[(x^2 + y^2 + z^2)^{-3/2} y]$

$$- \frac{\partial}{\partial z}[(x^2 + y^2 + z^2)^{-3/2} z]$$

$$= -(x^2 + y^2 + z^2)^{-3/2} \left[\left(1 - \frac{3x^2}{x^2 + y^2 + z^2} \right) \right.$$

$$\left. + \left(1 - \frac{3y^2}{x^2 + y^2 + z^2} \right) + \left(1 - \frac{3z^2}{x^2 + y^2 + z^2} \right) \right]$$

$$= 0. \qquad \qquad \square$$

EXERCISE SET 17.1

Find the divergence and curl of the vector-valued functions defined in Exercises 1 through 6.

1. $\mathbf{F}(x, y, z) = x^2yz\mathbf{i} + xy^2z\mathbf{j} + xyz^2\mathbf{k}$

2. $\mathbf{F}(x, y, z) = \sqrt{yz}\,\mathbf{i} + \sqrt{xz}\,\mathbf{j} + \sqrt{xy}\,\mathbf{k}$

3. $\mathbf{F}(x, y, z) = xy \sin z\,\mathbf{i} + x^2 \cos y\,\mathbf{j} + z\sqrt{xy}\,\mathbf{k}$

4. $\mathbf{F}(x, y, z) = e^x \sin y\,\mathbf{i} + e^y \cos z\,\mathbf{j} + e^{xyz}\,\mathbf{k}$

5. $\mathbf{F}(x, y, z) = e^z \ln \dfrac{x}{y}\mathbf{i} + e^x \ln \dfrac{y}{z}\mathbf{j} + e^y \ln \dfrac{z}{x}\mathbf{k}$

6. $\mathbf{F}(x, y, z) = \tan xz\,\mathbf{i} + \sec yx\,\mathbf{j} + \arcsin yz\,\mathbf{k}$

Determine if the functions in Exercises 7 through 12 are harmonic.

7. $f(x, y) = x^2y - y^2x$

8. $f(x, y) = \dfrac{1}{\sqrt{x^2 + y^2}}$

9. $f(x, y) = \sqrt{x^2 + y^2}$

10. $f(x, y, z) = \ln|x^2 + y^2|$

11. $f(x, y, z) = 2x^2 - y^2 - z^2$

12. $f(x, y, z) = x^2 + y^2 + z^2$

Find the values of the operations, if possible, in Exercises 13 through 20 where the functions are defined by $f(x, y, z) = x^2 + yz$, $\mathbf{F}(x, y, z) = 2xy\mathbf{i} + (x^2 + z^2)\mathbf{j} + yz\mathbf{k}$.

13. $\nabla \cdot (f\mathbf{F})$

14. $\nabla f \cdot \mathbf{F}$

15. $\nabla \times (\nabla \mathbf{F})$

16. $\nabla \times (\nabla f \cdot \mathbf{F})$

17. $\nabla \times (\nabla \cdot \mathbf{F})$

18. $\nabla \cdot (\nabla f \cdot \mathbf{F})$

19. $\nabla f + \nabla \times \mathbf{F}$

20. $\nabla \cdot \nabla f + \nabla \cdot (\nabla \times \mathbf{F})$

Verify the identities in Exercises 21 through 28, assuming the required differentiability on the vector functions $\mathbf{F}$ and $\mathbf{G}$ and the scalar functions f and g.

21. $\nabla \cdot (\mathbf{F} + \mathbf{G}) = \nabla \mathbf{F} + \nabla \mathbf{G}$

22. $\nabla \times (\mathbf{F} + \mathbf{G}) = \nabla \times \mathbf{F} + \nabla \times \mathbf{G}$

23. $\nabla \cdot (f\mathbf{F}) = f(\nabla \cdot \mathbf{F}) + \nabla f \cdot \mathbf{F}$

24. $\nabla \cdot (\mathbf{F} \times \mathbf{G}) = (\nabla \times \mathbf{F}) \cdot \mathbf{G} - \mathbf{F} \cdot (\nabla \times \mathbf{G})$

25. $\nabla \times (\nabla f) = \mathbf{0}$

26. $\nabla \cdot (\nabla \times \mathbf{F}) = 0$

27. $\nabla \cdot (f \nabla g) = f \nabla^2 g + \nabla f \cdot \nabla g$

28. $\nabla \times (f\mathbf{F}) = f(\nabla \times \mathbf{F}) + (\nabla f \times \mathbf{F})$

29. Let $\mathbf{F} = M\mathbf{i} + N\mathbf{j} + P\mathbf{k}$. Show that

$$\nabla \times (\nabla \times \mathbf{F}) = \nabla(\nabla \cdot \mathbf{F}) - (\nabla^2 M\mathbf{i} + \nabla^2 N\mathbf{j} + \nabla^2 P\mathbf{k}).$$

30. Suppose $\mathbf{F}$ is a gradient field and has continuous second partial derivatives. Show that curl $\mathbf{F} = 0$.

31. Let (r, θ, z) be the cylindrical representation of Q, $\mathbf{a}$ the unit vector in the positive direction of r at Q, and $\mathbf{b}$ the unit vector in the direction of the tangent to the circle of radius r parallel to the xy-plane at Q. Show that if f is a scalar function and $\mathbf{F} = M\mathbf{a} + N\mathbf{b} + P\mathbf{k}$ is a vector function described in cylindrical coordinates, then

(a) $\nabla f = \dfrac{\partial f}{\partial r}\mathbf{a} + \dfrac{1}{r}\dfrac{\partial f}{\partial \theta}\mathbf{b} + \dfrac{\partial f}{\partial z}\mathbf{k}$

(b) $\nabla \cdot \mathbf{F} = \dfrac{1}{r}\dfrac{\partial}{\partial r}(rM) + \dfrac{1}{r}\dfrac{\partial N}{\partial \theta} + \dfrac{\partial P}{\partial z}$

(c) $\nabla^2 f = \dfrac{1}{r}\dfrac{\partial}{\partial r}\left(r\dfrac{\partial f}{\partial r}\right) + \dfrac{1}{r^2}\dfrac{\partial^2 f}{\partial \theta^2} + \dfrac{\partial^2 f}{\partial z^2}.$

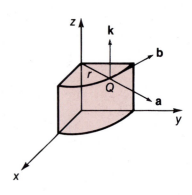

17.2
LINE INTEGRALS

In this section we define a **line integral,** which is a natural generalization of the definite integral. For the definite integral, $\int_a^b f(x)dx$, we integrate a function defined at each value between a and b along the x-axis. For a line integral we integrate along a curve in space and the integrand is a function defined at each point on the curve.

Suppose f is a continuous function defined on a subset D of $\mathbb{R}^3$ and C is a smooth curve in D described parametrically by

(17.14) $x = x(t), \qquad y = y(t), \qquad z = z(t), \qquad a \le t \le b.$

(A smooth curve in $\mathbb{R}^3$ is one for which x', y', and z' are continuous and not simultaneously zero.) The curve C is given an orientation by assigning a positive direction, generally that of increasing t. If the initial point $(x(a), y(a), z(a))$ and terminal point $(x(b), y(b), z(b))$ coincide, C is called a **closed** curve. If, in addition, there are no other points of intersection, C is called a **simple closed** curve. Figure 17.3 shows some types of closed curves.

Smooth closed curve   Smooth simple closed curve   Simple closed curve
  that is not simple   that is not smooth

FIGURE 17.3

Two types of line integrals will be considered: those of the type

(I) $\displaystyle\int_C f(x,\, y,\, z)dx, \qquad \int_C f(x,\, y,\, z)dy, \qquad \int_C f(x,\, y,\, z)dz,$

where the integration is with respect to a coordinate direction, and those of the type

(II) $\displaystyle\int_C f(x,\, y,\, z)ds,$

where the integration is with respect to the arc length of the curve.

Both types of line integrals are defined in terms of Riemann sums and, under the conditions on f and C stated at the beginning of the section, both can be reduced to Riemann integrals in the parameter t.

First we consider the line integrals of type (I). Since the derivation of each of these integrals is the same, only integrals of the form $\int_C f(x,\, y,\, z)dx$ will be discussed.

$$\int_C f(x, y, z)dx$$

Let $\mathcal{P} = \{t_0, t_1, \ldots, t_n\}$ be a partition of $[a, b]$

and $$P_i = P(t_i) = (x_i, y_i, z_i)$$

be the points along C determined by $\mathcal{P}$. The line integral $\int_C f(x, y, z)dx$ is defined as

(17.15) $$\int_C f(x, y, z)dx = \lim_{\|\mathcal{P}\| \to 0} \sum_{i=1}^{n} f(P(\tau_i))\Delta x_i,$$

where, for each i, $\Delta x_i = x_i - x_{i-1}$ and τ_i is an arbitrary point in $[t_{i-1}, t_i]$. Since x' exists on $[a, b]$, it follows from the mean value theorem that the Riemann sum on the right side of (17.15) can be rewritten

$$\sum_{i=1}^{n} f(P(\tau_i))\Delta x_i = \sum_{i=1}^{n} f(P(\tau_i)) \frac{x(t_i) - x(t_{i-1})}{t_i - t_{i-1}} \Delta t_i$$

$$= \sum_{i=1}^{n} f(P(\tau_i)) \, x'(w_i)\Delta t_i,$$

for some w_i in (t_{i-1}, t_i). The theorem of Bliss, Theorem 6.19, can be employed to deduce that

(17.16)
$$\int_C f(x, y, z)dx = \lim_{\|\mathcal{P}\| \to 0} \sum_{i=1}^{n} f(P(\tau_i))x'(w_i)\Delta t_i$$

$$= \int_a^b f(x(t), y(t), z(t)) \, x'(t)dt.$$

Notice that if the curve C happens to be the interval $[a, b]$ on the x-axis and f is a function of x only, then C is described by $x = t$, $a \le t \le b$, and this line integral reduces to the definite integral $\int_a^b f(x)dx$.

The line integrals $\int_C f(x, y, z)dy$ and $\int_C f(x, y, z)dz$ can be similarly defined and shown to satisfy

(17.17) $$\int_C f(x, y, z)dy = \int_a^b f(x(t), y(t), z(t)) \, y'(t)dt$$

and

(17.18) $$\int_C f(x, y, z)dz = \int_a^b f(x(t), y(t), z(t)) \, z'(t)dt.$$

EXAMPLE 1   Find $\int_C xydx$ where C is described by $x = 1 + 2t$, $y = t$, $z = 0$, for $0 \le t \le 2$.

SOLUTION

Since $dx = 2dt$,

$$\int_C xydx = \int_0^2 (1 + 2t)t \, 2dt = 2\int_0^2 (t + 2t^2)dt$$

$$= 2\left(\frac{t^2}{2} + \frac{2t^3}{3}\right)\Bigg]_0^2 = 2\left(2 + \frac{16}{3}\right) = \frac{44}{3}. \qquad \square$$

EXAMPLE 2   Find $\int_C (x + y + z)dy$, where C is the curve of intersection of the upper half of the sphere $x^2 + y^2 + z^2 = 1$ and the plane $x = 0$ with initial point $(0, 1, 0)$.

SOLUTION

A sketch of C is shown in Figure 17.4.

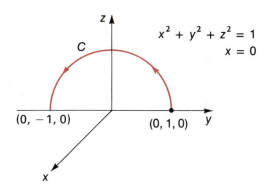

FIGURE 17.4

Since C can be represented parametrically by

$$x = 0, \qquad y = \cos t, \qquad z = \sin t, \qquad 0 \le t \le \pi,$$

$$\int_C (x + y + z)dy = \int_0^\pi (0 + \cos t + \sin t)(-\sin t)dt$$

$$= \int_0^\pi [\cos t(-\sin t) - (\sin t)^2]dt$$

$$= \frac{(\cos t)^2}{2}\Bigg]_0^\pi - \int_0^\pi \frac{1 - \cos 2t}{2} dt$$

$$= \left(\frac{1}{2} - \frac{1}{2}\right) - \left(\frac{1}{2}t - \frac{1}{4}\sin 2t\right)\Bigg]_0^\pi = -\frac{\pi}{2}. \qquad \square$$

Because of the relationship expressed in equations (17.16), (17.17), and (17.18), many of the properties of definite integrals also hold for these line integrals. For example,

(17.19) $$\int_C (f \pm g)(x, y, z)dx = \int_C f(x, y, z)dx \pm \int_C g(x, y, z)dx$$

(17.20) $$\int_C k f(x, y, z)dx = k \int_C f(x, y, z)dx, \text{ for any constant } k.$$

If $C_1 + C_2$ denotes the union of the curves C_1 and C_2 when the terminal point of C_1 and the initial point of C_2 coincide (see Figure 17.5), then

(17.21) $$\int_{C_1+C_2} f(x, y, z)dx = \int_{C_1} f(x, y, z)dx + \int_{C_2} f(x, y, z)dx.$$

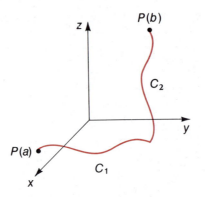

FIGURE 17.5

Finally, if $-C$ denotes the curve that passes through the same points as C but is traversed in the opposite direction, then

(17.22)
$$\int_{-C} f(x, y, z)dx = -\int_{C} f(x, y, z)dx.$$

Many applications of line integrals involve integrals of the form

$$\int_{C} M(x, y, z)dx + N(x, y, z)dy + P(x, y, z)dz.$$

This is simply an alternative notation for the sum of the line integrals

$$\int_{C} M(x, y, z)dx + \int_{C} N(x, y, z)dy + \int_{C} P(x, y, z)dz.$$

The term $M(x, y, z)dx + N(x, y, z)dy + P(x, y, z)dz$ is called a **differential form.** (A particularly important type of differential form is discussed in Section 17.4.)

EXAMPLE 3   Evaluate $\int_{C} x^2 y \, dx + (y - z)dy + xz \, dz$ where
(a) C is the curve described by

$$\mathbf{r}(t) = \mathbf{i} + t^2 \mathbf{j} + t\mathbf{k}, \qquad 0 \le t \le 2,$$

(b) C is the straight line joining $(1, 0, 0)$ and $(1, 4, 2)$.

SOLUTION

(a) With $\mathbf{r}(t) = \mathbf{i} + t^2 \mathbf{j} + t\mathbf{k}$, $\dfrac{d\mathbf{r}(t)}{dt} = 2t\mathbf{j} + \mathbf{k}$, so $\dfrac{dx}{dt} = 0, \dfrac{dy}{dt} = 2t,$

and $\dfrac{dz}{dt} = 1$. Thus,

$$\int_{C} x^2 y \, dx + (y - z)dy + xz \, dz = \int_{0}^{2} (1^2 \cdot t^2 \cdot 0dt + (t^2 - t)2t \, dt + 1 \cdot t \, dt)$$

$$= \int_{0}^{2} (2t^3 - 2t^2 + t)dt$$

$$= \frac{1}{2}t^4 - \frac{2}{3}t^3 + \frac{1}{2}t^2 \Bigg]_{0}^{2}$$

$$= 8 - \frac{16}{3} + 2 = \frac{14}{3}.$$

(b) A position vector representation for the straight line from $(1, 0, 0)$ to $(1, 4, 2)$ is

$$\mathbf{r}(t) = \mathbf{i} + 4t\mathbf{j} + 2t\mathbf{k}, \qquad 0 \le t \le 1.$$

Since $\dfrac{d\mathbf{r}(t)}{dt} = 4\mathbf{j} + 2\mathbf{k}, \qquad \dfrac{dx}{dt} = 0, \qquad \dfrac{dy}{dt} = 4, \qquad$ and $\qquad \dfrac{dz}{dt} = 2.$

Consequently,

$$\int_C x^2 y \, dx + (y - z)dy + xz \, dz = \int_0^1 1^2 \cdot 4t \cdot 0 \, dt + (4t - 2t)4 \, dt + 2t \cdot 2 \, dt$$

$$= \int_0^1 12t \, dt = 6t^2 \Big]_0^1 = 6.$$

Notice that even though the two curves in (a) and (b) begin and end at the same points (see Figure 17.6), the values of the integrals differ significantly. □

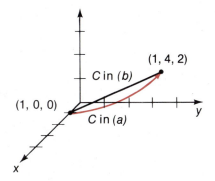

FIGURE 17.6

$\displaystyle\int_C f(x, y, z)ds$

To define $\int_C f(x, y, z)ds$, the line integral of f with respect to arc length, we again begin with a partition $\mathscr{P} = \{t_0, t_1, \ldots, t_n\}$ of $[a, b]$. Since x', y', and z' are continuous on $[a, b]$, the arc length from $P(t_{i-1})$ to $P(t_i)$ is given by

$$\Delta s_i = s(t_i) - s(t_{i-1}) = \int_{t_{i-1}}^{t_i} \sqrt{[x'(t)]^2 + [y'(t)]^2 + [z'(t)]^2}\, dt.$$

The line integral of f over C with respect to arc length is defined by

(17.23)
$$\int_C f(x, y, z)ds = \lim_{\|\mathscr{P}\| \to 0} \sum_{i=1}^n f(P(\tau_i))\Delta s_i,$$

where τ_i is an arbitrary point in $[t_{i-1}, t_i]$.

Since s is differentiable on $[a, b]$, it follows from the mean value theorem that

$$\Delta s_i = s(t_i) - s(t_{i-1}) = s'(w_i)\Delta t_i,$$

for some w_i in $[t_{i-1}, t_i]$ so

$$\Delta s_i = \sqrt{[x'(w_i)]^2 + [y'(w_i)]^2 + [z'(w_i)]^2}\, \Delta t_i.$$

By using the theorem of Bliss, Theorem 6.19, it can be shown that

(17.24) $$\int_C f(x, y, z)ds = \int_a^b f(x(t), y(t), z(t))\sqrt{[x'(t)]^2 + [y'(t)]^2 + [z'(t)]^2} \, dt.$$

If $\mathbf{r}(t) = x(t)\mathbf{i} + y(t)\mathbf{j} + z(t)\mathbf{k}$ describes C, the integral can also be expressed by

(17.25) $$\int_C f(x, y, z)ds = \int_a^b f(x, y, z) \, \|\mathbf{r}'(t)\| \, dt.$$

Note that $\int_C ds$ gives the arc length of C from $P(a)$ to $P(b)$. (See Equation 13.9 on page 728.)

EXAMPLE 4   Determine $\int_C (x^2 + y^2 + z^2)ds$ where C is the portion of the helix shown in Figure 17.7 beginning at $(1, 0, 0)$ and ending at $(1, 0, 2\pi)$.

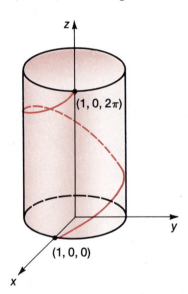

FIGURE 17.7

SOLUTION

A parametric representation for C is

$$x = \cos t, \qquad y = \sin t, \qquad z = t, \qquad 0 \le t \le 2\pi.$$

It follows from (17.24) that

$$\int_C (x^2 + y^2 + z^2)ds = \int_0^{2\pi} [(\cos t)^2 + (\sin t)^2 + t^2] \cdot \sqrt{(-\sin t)^2 + (\cos t)^2 + 1} \, dt$$

$$= \int_0^{2\pi} (1 + t^2)\sqrt{2} \, dt$$

$$= \sqrt{2}\left(t + \frac{t^3}{3}\right)\Bigg]_0^{2\pi}$$

$$= \sqrt{2}\left(2\pi + \frac{8\pi^3}{3}\right) = 2\sqrt{2}\pi\left(1 + \frac{4\pi^2}{3}\right). \qquad \square$$

EXAMPLE 5   Determine $\int_C (x^2 + y^2 + z^2)ds$ where C is the straight line segment from $(1, 0, 0)$ to $(1, 0, 2\pi)$.

SOLUTION

A parametric representation for C is

$$x = 1, \qquad y = 0, \qquad z = t, \qquad 0 \leq t \leq 2\pi.$$

Thus,

$$\int_C (x^2 + y^2 + z^2)ds = \int_0^{2\pi} (1^2 + 0^2 + t^2)\sqrt{0^2 + 0^2 + 1^2}\, dt$$

$$= \int_0^{2\pi} (1 + t^2)dt$$

$$= t + \frac{t^3}{3} \Bigg]_0^{2\pi}$$

$$= 2\pi + \frac{8\pi^3}{3} = 2\pi\left(1 + \frac{4\pi^2}{3}\right). \qquad \square$$

Even though the two curves in Examples 4 and 5 begin and end at the same points, the values of the integrals are not equal.

EXAMPLE 6   Determine $\int_{-C} (x^2 + y^2 + z^2)ds$, where C is the curve described in Example 5.

SOLUTION

Recall that $-C$ denotes the curve that passes through the same points as C but is traversed in the opposite direction. Thus a parametric representation for $-C$ is

$$x = 1, \qquad y = 0, \qquad z = 2\pi - t, \qquad 0 \leq t \leq 2\pi$$

and

$$\int_C f(x, y, z)ds = \int_0^{2\pi} [1^2 + 0^2 + (2\pi - t)^2]\sqrt{0^2 + 0^2 + (-1)^2}\, dt$$

$$= \int_0^{2\pi} [1 + (2\pi - t)^2]dt$$

$$= t - \frac{(2\pi - t)^3}{3} \Bigg]_0^{2\pi}$$

$$= 2\pi + \frac{8\pi^3}{3} = 2\pi\left(1 + \frac{4\pi^2}{3}\right). \qquad \square$$

Notice that the results in Example 5 and 6 are the same, that is,

(17.26) $$\int_C f(x, y, z)ds = \int_{-C} f(x, y, z)ds.$$

This fact, true in general for line integrals with respect to arc length, occurs because $\Delta s_i > 0$ regardless of the orientation of C. On the other hand, Δx_i, Δy_i, and Δz_i change sign when the orientation of C reverses, which causes the line integral with respect to dx, dy, and dz to change sign. This was seen in (17.22).

The result concerning the integral over the union of curves also fails to hold for line integrals involving arc length. The other integral properties do hold, namely

(17.27) $$\int_C (f \pm g)(x, y, z)ds = \int_C f(x, y, z)ds \pm \int_C g(x, y, z)ds$$

and

(17.28) $$\int_C kf(x, y, z)ds = k \int_C f(x, y, z)ds, \text{ for any constant } k.$$

EXERCISE SET 17.2

Determine which of the curves described in Exercises 1 through 4 are closed curves?

1. $\mathbf{r}(t) = \cos t\mathbf{i} + \sin t\mathbf{j} + \mathbf{k}, 0 \le t \le 2\pi$
2. $\mathbf{r}(t) = t\mathbf{i} + t^2\mathbf{j} + (2t + 1)\mathbf{k}, 0 \le t \le 1$
3. $\mathbf{r}(t) = \sin \pi t\mathbf{i} + \cos 2\pi t\mathbf{j} + (t^2 - t + 2)\mathbf{k}, 0 \le t \le 1$
4. C is the set of points $(\cos t, \sin t, t), 0 \le t \le 2\pi$

Evaluate the line integrals in Exercises 5 through 20.

5. $\int_C xyz \, dx$, where C is described by
$$\mathbf{r}(t) = t\mathbf{i} + (t + 2)\mathbf{j} + (2t - 1)\mathbf{k}, 0 \le t \le 1.$$

6. $\int_C (x + y + z)dz$, where C is described by
$$\mathbf{r}(t) = t^2\mathbf{i} + (t^{3/2} + 1)\mathbf{j} + t\mathbf{k}, 0 \le t \le 4.$$

7. $\int_C (xy + x^2 + y^2)dx$, where C is the semicircle in the xy-plane described by $y = \sqrt{4 - x^2}$, beginning at $(2, 0)$ and ending at $(-2, 0)$.

8. $\int_C (x^2 + y^2)dz$, where C is the portion of the helix shown in Figure 17.7 beginning at $(1, 0, 0)$ and ending at $(1, 0, 2\pi)$.

9. $\int_C y \sec x \, dx$, where C is the curve $y = \sec x$ in the xy-plane from $x = 0$ to $x = \pi/4$.

10. $\int_C y \sec x \, dy$, where C is the curve $y = \sec x$ in the xy-plane from $x = 0$ to $x = \pi/4$.

11. $\int_C (x^2 + y^2)dx + 2xy \, dy$, where C is the circle $x^2 + y^2 = 4$ in the xy-plane, oriented clockwise, beginning and ending at $(2, 0)$.

12. $\int_C \dfrac{1}{1 + x^2}\,dx + \dfrac{1}{1 + y^2}\,dy$, where C is the straight line in the xy-plane from $(1, 1)$ to $(2, 2)$.

13. $\int_C (x^3 + y^3)\,ds$, for C as described in Exercise 11.

14. $\int_C (y - z^2)\,ds$, where C is described by $\mathbf{r}(t) = \mathbf{i} + t^2\mathbf{j} + t\mathbf{k}$, $1 \le t \le 2$.

15. $\int_C (x^2 + y^2 + z^2)\,ds$, where C is the straight line joining $(0, 0, 0)$ and $(1, 2, 3)$.

16. $\int_C (x^2 + y^2 + z^2)\,ds$, where C is the arc of the circular helix described by $\mathbf{r}(t) = \cos t\,\mathbf{i} + \sin t\,\mathbf{j} + 2t\mathbf{k}$ from $(1, 0, 0)$ to $(1, 0, 4\pi)$.

17. $\int_C \dfrac{y + z}{x}\,ds$, where C is the arc of the circle $x^2 + y^2 = 1$ in the xy-plane counterclockwise from $(1, 0)$ to $\left(-\dfrac{\sqrt{2}}{2}, \dfrac{\sqrt{2}}{2}\right)$.

18. $\int_C (x + y^{3/2})\,ds$, where C is the parabola $y = x^2$ in the xy-plane from $(0, 0)$ to $(2, 4)$.

19. $\int_C x \cos y\,dx + y\,dy + z \sin y\,dz$, where C is the helix described by $\mathbf{r}(t) = \sin t\,\mathbf{i} + t\mathbf{j} + \cos t\mathbf{k}$, $0 \le t \le \pi$.

20. $\int_C x\,e^{xy}\,dx + y\,e^{yz}\,dy + z\,dz$, where C is described by $\mathbf{r}(t) = t\mathbf{i} + \mathbf{j} + \ln t\mathbf{k}$, $1 \le t \le 2$.

21. Evaluate $\int_C xy^2\,dx + x^2y\,dy$ along the following four curves in the xy-plane from $(1, 0)$ to $(-1, 0)$:
(a) C is the upper unit semicircle.
(b) C is the lower unit semicircle.
(c) C is a segment of the x-axis from $x = 1$ to $x = -1$.
(d) C is a part of the curve $y = x^3 - x$.

22. Evaluate $\int_C x^2y\,dx + (x - y)\,dy$ along the following four curves in the xy-plane from $(0, 0)$ to $(1, 1)$:
(a) C consists of a segment of the x-axis from $x = 0$ to $x = 1$ and a segment of the line $x = 1$ from $y = 0$ to $y = 1$.
(b) C is the straight line joining $(0, 0)$ to $(1, 1)$.
(c) C consists of a segment of the y-axis from $y = 0$ to $y = 1$ and a segment of the line $y = 1$ from $x = 0$ to $x = 1$.
(d) C is a part of the parabola $y = x^2$.

23. Evaluate $\int_C yx\,dx + x\,dy$ over each of the following simple closed curves:

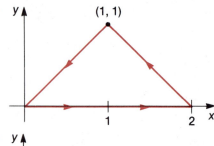

24. Evaluate $\int_C x\,dx + yx\,dy$ over the simple closed curves shown in Exercise 23.

25. Suppose C is a smooth curve described by $\mathbf{r}(t) = x(t)\mathbf{i} + y(t)\mathbf{j} + z(t)\mathbf{k}$, $a \le t \le b$ and $\mathbf{F}$ is a continuous vector-valued function defined in a region containing C. Show that if $\mathbf{T}(x, y, z)$ is the unit tangent vector to C at (x, y, z), then

$$\int_C \mathbf{F}(x, y, z) \cdot \mathbf{T}(x, y, z)ds = \int_a^b \mathbf{F}(x, y, z) \cdot \mathbf{r}'(t)dt.$$

26. Use Exercise 25 to find $\int_C \mathbf{F}(x, y, z) \cdot \mathbf{T}(x, y, z)ds$, where $\mathbf{F}(x, y, z) = x^2\mathbf{i} + y^2\mathbf{j} + z^2\mathbf{k}$, and $\mathbf{T}(x, y, z)$ is the unit tangent vector to C at (x, y, z) if

(a) C is the straight line segment from $(0, 0, 0)$ to $(1, 2, 3)$.

(b) C is described by $\mathbf{r}(t) = t\mathbf{i} + 2t^2\mathbf{j} + 3t^3\mathbf{k}, 0 \le t \le 1$.

27. Consider the integral $\int_C M\,dx + N\,dy + P\,dz$, where $M(x, y, z) = y^2 \sin z$, $N(x, y, z) = 2xy \sin z$, and $P(x, y, z) = xy^2 \cos z$.

(a) Show that $\dfrac{\partial M}{\partial y} = \dfrac{\partial N}{\partial x}, \dfrac{\partial M}{\partial z} = \dfrac{\partial P}{\partial x}$, and $\dfrac{\partial N}{\partial z} = \dfrac{\partial P}{\partial y}$.

(b) Evaluate the integral if C is the straight line joining $(0, 0, 0)$ and $(1, 4, 2)$.

(c) Evaluate the integral if C is the curve described by

$$\mathbf{r}(t) = \frac{t}{2}\mathbf{i} + t^2\mathbf{j} + t\mathbf{k}, 0 \le t \le 2.$$

(d) Evaluate the integral if C is the circle $x^2 + y^2 = 4$ in the plane $z = 2$ (counterclockwise).

28. (a) Find $\int_C (x^2y\mathbf{i} + xy^2\mathbf{j}) \cdot \mathbf{r}'(t)\, dt$, where C is the boundary of the rectangle shown below.

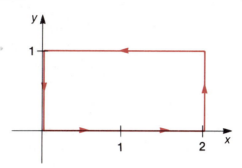

(b) Find $\displaystyle\iint\limits_R \left[\frac{\partial}{\partial x}(xy^2) - \frac{\partial}{\partial y}(x^2y) \right] dA$, where R is the rectangle enclosed by C.

29. The total force on a particle with charge Q moving with velocity $\mathbf{v}$ in the presence of an electric field $\mathbf{E}$ and a magnetic field $\mathbf{B}$ is given by the *Lorentz force* $\mathbf{F} = Q(\mathbf{E} + \mathbf{v} \times \mathbf{B})$. Show that Newton's second law of motion can be used to generate the *power equation* for the particle:

$$m\frac{d\mathbf{v}}{dt} \cdot \mathbf{v} = Q\mathbf{E} \cdot \frac{d\mathbf{r}}{dt}$$

and consequently the *energy equation*

$$Q\int \mathbf{E} \cdot d\mathbf{r} = \frac{1}{2}m\mathbf{v}^2 + \mathbf{C},$$

where $\mathbf{C}$ is a constant (m represents the mass of the particle).

17.3
PHYSICAL APPLICATIONS OF LINE INTEGRALS

WORK

The concept of work was first introduced in Section 6.7, where we found that the work done by a variable force F moving an object from P to Q along an x-coordinate line is

(17.29) $$W = \int_P^Q F(x)\,dx.$$

In Section 12.3 we found that the work done by a constant vector force $\mathbf{F}$ moving an object along the line from P to Q is

(17.30) $$W = \mathbf{F} \cdot \overrightarrow{PQ}.$$

With the introduction of line integrals, we can consolidate our knowledge of work and extend the concept to the most useful situation: work done by a variable vector force moving an object along a curve in space.

Suppose C is a smooth curve described by

$$\mathbf{r}(t) = x(t)\mathbf{i} + y(t)\mathbf{j} + z(t)\mathbf{k}, \qquad a \le t \le b,$$

and $\mathbf{F}$ is a continuous vector function defined in a region containing C. Let $\mathscr{P} = \{t_0, t_1, \ldots, t_n\}$ be a partition of $[a, b]$, $P_i = P(t_i) = (x(t_i), y(t_i), z(t_i))$, and Δs_i be the arc length of that portion of C between P_{i-1} and P_i for each $i = 0, 1, \ldots, n$. The first step is to approximate the work ΔW_i done by $\mathbf{F}$ as an object moves from P_{i-1} to P_i along C. To do this, choose τ_i in $[t_{i-1}, t_i]$, let $Q_i = P(\tau_i)$, and compute the work done by the constant force $\mathbf{F}(Q_i)$ moving an object along the straight line of length Δs_i tangent to C at Q_i.

From Equation (13.11) in Section 13.3, we know that the unit tangent vector to C at Q_i is

$$\mathbf{T}(Q_i) = \frac{\mathbf{r}'(\tau_i)}{\|\mathbf{r}'(\tau_i)\|}.$$

So the vector of length Δs_i tangent to the curve at Q_i is $\mathbf{T}(Q_i)\Delta s_i$, as shown in Figure 17.8. Consequently, by (17.30),

$$\Delta W_i \approx \mathbf{F}(Q_i) \cdot \mathbf{T}(Q_i)\Delta s_i.$$

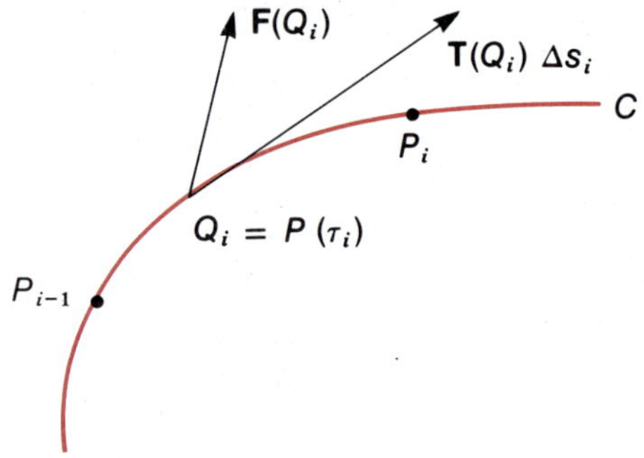

FIGURE 17.8

An approximation to the total work done by the force in moving the object along C is

$$W \approx \sum_{i=1}^{n} \mathbf{F}(Q_i) \cdot \mathbf{T}(Q_i)\Delta s_i.$$

Taking the limit as $\|\mathscr{P}\| \to 0$ produces the definition for the work done by a variable force moving an object along a curve in space.

(17.31)
DEFINITION

Suppose C is a smooth curve in space and an object is moved along C by a force $\mathbf{F}$. If $\mathbf{F}$ is continuous on a region containing C, **the work done by $\mathbf{F}$ along C** is

$$W = \int_C \mathbf{F}(x, y, z) \cdot \mathbf{T}(x, y, z)\,ds,$$

where $\mathbf{T}(x, y, z)$ is the unit tangent vector to C at (x, y, z).

Since
$$\mathbf{F}(x, y, z) \cdot \mathbf{T}(x, y, z) = \mathbf{F}(x, y, z) \cdot \frac{\mathbf{r}'(t)}{\|\mathbf{r}'(t)\|}$$

and
$$ds = \sqrt{[x'(t)]^2 + [y'(t)]^2 + [z'(t)]^2} \, dt = \|\mathbf{r}'(t)\|dt,$$

(17.31) $$W = \int_C \mathbf{F}(x, y, z) \cdot \mathbf{T}(x, y, z)ds = \int_a^b \mathbf{F}(x, y, z) \cdot \mathbf{r}'(t)dt.$$

Since $\mathbf{r}'(t) = \dfrac{d\mathbf{r}}{dt}$, this integral is often written in the compact form

(17.32) $$W = \int_C \mathbf{F} \cdot d\mathbf{r},$$

a form very similar in appearance to the integral in (17.29).

A word of caution must be made about using the line integral notation in (17.32). This integral depends on the orientation of the curve C since a specific parameterization is used to express the Riemann integral $\int_a^b \mathbf{F} \cdot \mathbf{r}' \, dt$. In this sense it differs from the line integral with respect to arc length, which we found to be independent of the orientation of the curve.

EXAMPLE 1   Find the work done by the force $\mathbf{F}(x, y, z) = x^2y\mathbf{i} + y^2x\mathbf{j} + z^2\mathbf{k}$ in moving an object along the straight line from $(0, 0, 0)$ to $(1, 2, 3)$.

SOLUTION

This line can be described by

$$\mathbf{r}(t) = t\mathbf{i} + 2t\mathbf{j} + 3t\mathbf{k}, \qquad 0 \leq t \leq 1.$$

Thus,
$$\begin{aligned}
W &= \int_0^1 \mathbf{F}(x, y, z) \cdot \frac{d\mathbf{r}}{dt} \, dt \\
&= \int_0^1 (2t^3\mathbf{i} + 4t^3\mathbf{j} + 9t^2\mathbf{k}) \cdot (\mathbf{i} + 2\mathbf{j} + 3\mathbf{k})dt \\
&= \int_0^1 (2t^3 + 8t^3 + 27t^2)dt = \frac{5}{2}t^4 + 9t^3 \Big]_0^1 = \frac{23}{2}.
\end{aligned}$$

EXAMPLE 2   Find the work done by the force $\mathbf{F}(x, y, z) = x^2y\mathbf{i} + y^2x\mathbf{j} + z^2\mathbf{k}$ in moving an object from $(0, 0, 0)$ to $(1, 2, 3)$ along the curve described by

$$\mathbf{r}(t) = t\mathbf{i} + 2t^2\mathbf{j} + 3t^3\mathbf{k}, \qquad 0 \leq t \leq 1.$$

SOLUTION

The work done along this curve is

$$\begin{aligned}
W &= \int_0^1 (2t^4\mathbf{i} + 4t^5\mathbf{j} + 9t^6\mathbf{k}) \cdot (\mathbf{i} + 4t\mathbf{j} + 9t^2\mathbf{k})dt \\
&= \int_0^1 (2t^4 + 16t^6 + 81t^8)dt \\
&= \frac{2}{5}t^5 + \frac{16}{7}t^7 + 9t^9 \Big]_0^1 = \frac{409}{35}.
\end{aligned}$$

Notice that the work done by the same force along the curves in Examples 1 and 2 is not the same, even though the curves have the same endpoints.

Moments

Suppose a thin wire in space has the shape of a smooth curve C described by

$$\mathbf{r}(t) = x(t)\mathbf{i} + y(t)\mathbf{j} + z(t)\mathbf{k}, \qquad a \le t \le b$$

and that ρ is a continuous nonnegative function describing the density of the wire at points along C. Line integrals can be used to find the mass and moments of the wire about the coordinate planes.

Let $\mathscr{P} = \{t_0, t_1, \ldots, t_n\}$ be a partition of $[a, b]$, $P_i = P(t_i) = (x(t_i), y(t_i), z(t_i))$, Δs_i be the arc length of that portion of C between P_{i-1} and P_i, and τ_i be an arbitrary point in $[t_{i-1}, t_i]$. An approximation to the mass of the wire between P_{i-1} and P_i is given by

$$\rho(P(\tau_i))\Delta s_i,$$

and the moments about the coordinate planes are approximated by:

$$\text{moment about the } yz\text{-plane} \approx x(\tau_i)\, \rho(P(\tau_i))\Delta s_i,$$
$$\text{moment about the } xy\text{-plane} \approx z(\tau_i)\, \rho(P(\tau_i))\Delta s_i,$$
$$\text{moment about the } xz\text{-plane} \approx y(\tau_i)\, \rho(P(\tau_i))\Delta s_i.$$

Summing these approximations and taking the limit as $\|\mathscr{P}\| \to 0$ leads to the following definition.

(17.33)
DEFINITION

Suppose a thin wire in space can be described by a smooth curve C, and ρ is a continuous nonnegative function defined on a region containing C that describes the density of the wire. The **mass** M and **moments** M_{yz}, M_{xy}, and M_{xz} about the coordinate planes are given by:

$$M = \int_C \rho(x, y, z)\,ds, \qquad M_{yz} = \int_C x\rho(x, y, z)\,ds$$

$$M_{xy} = \int_C z\rho(x, y, z)\,ds, \qquad M_{xz} = \int_C y\rho(x, y, z)\,ds.$$

The **center of mass** of the wire is $(\bar{x}, \bar{y}, \bar{z}) = \left(\dfrac{M_{yz}}{M}, \dfrac{M_{xz}}{M}, \dfrac{M_{xy}}{M} \right)$.

EXAMPLE 3

A wire is bent in the shape of a semicircle with its ends at $(1, 0)$ and $(-1, 0)$ in the xy-plane. The mass of the wire varies linearly with the x-coordinate of the point on the wire from 1 at $(-1, 0)$ to 2 at $(1, 0)$. Find the total mass and center of mass of the wire.

SOLUTION

Suppose the wire has been placed in the xy-plane as shown in Figure 17.9. Then the curve C made by the wire can be represented by

$$\mathbf{r}(t) = \cos t\mathbf{i} + \sin t\mathbf{j}, \qquad 0 \le t \le \pi.$$

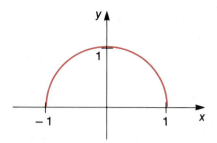

FIGURE 17.9

The mass is given by a linear equation $\rho(x, y) = ax + b$, where

$$\rho(-1, 0) = -a + b = 1 \quad \text{and} \quad \rho(1, 0) = a + b = 2.$$

Solving these equations simultaneously for a and b implies that $\rho(x, y) = x/2 + 3/2$. Consequently,

$$M = \int_C \rho(x, y)ds$$

$$= \int_0^\pi \left(\frac{1}{2}x + \frac{3}{2}\right)\sqrt{[x'(t)]^2 + [y'(t)]^2}\, dt$$

$$= \int_0^\pi \left[\frac{\cos t}{2} + \frac{3}{2}\right]\sqrt{(-\sin t)^2 + (\cos t)^2}\, dt$$

so
$$M = \frac{1}{2}\int_0^\pi (\cos t + 3)dt = \frac{1}{2}(\sin t + 3t)\Big]_0^\pi = \frac{3}{2}\pi.$$

Also,
$$M_{yz} = \int_C x\rho(x, y)ds = \frac{1}{2}\int_0^\pi \cos t(\cos t + 3)dt$$

$$= \frac{1}{2}\int_0^\pi [(\cos t)^2 + 3\cos t]dt$$

$$= \frac{1}{2}\int_0^\pi \left[\left(\frac{1 + \cos 2t}{2}\right) + 3\cos t\right]dt$$

$$= \frac{1}{2}\left(\frac{1}{2}t + \frac{1}{4}\sin 2t + 3\sin t\right)\Big]_0^\pi = \frac{\pi}{4},$$

and
$$M_{xz} = \int_C y\rho(x, y)ds = \frac{1}{2}\int_0^\pi \sin t(\cos t + 3)dt$$

$$= \frac{1}{2}\left[\frac{(\sin t)^2}{2} - 3\cos t\right]_0^\pi = 3.$$

Since the wire lies in the xy-plane, $M_{xy} = 0$.

The center of mass of the wire is at $(\bar{x}, \bar{y}, \bar{z})$, where

$$\bar{x} = \frac{M_{yz}}{M} = \frac{\dfrac{\pi}{4}}{\dfrac{3}{2}\pi} = \frac{1}{6}, \qquad \bar{y} = \frac{M_{xz}}{M} = \frac{3}{\dfrac{3}{2}\pi} = \frac{2}{\pi}, \qquad \bar{z} = \frac{M_{xy}}{M} = 0. \qquad \square$$

EXERCISE SET 17.3

In Exercises 1 through 8, find the work done by the force **F** in moving an object along the given curve.

1. $\mathbf{F}(x, y, z) = x\mathbf{i} + y\mathbf{j} + z\mathbf{k}$, along the parabola $y = x^2$ in the xy-plane from $(1, 1)$ to $(2, 4)$.

2. $\mathbf{F}(x, y, z) = (2x - y)\mathbf{i} + (y - z)\mathbf{j} + 2y\mathbf{k}$ along the helix $\mathbf{r}(t) = \cos t\, \mathbf{i} + \sin t\, \mathbf{j} + 3t\mathbf{k}$ from $(1, 0, 0)$ to $(1, 0, 6\pi)$.

3. $\mathbf{F}(x, y, z) = x^2 z\mathbf{i} - yx^2\mathbf{j} + 3xz\mathbf{k}$ along the straight line from $(0, 0, 0)$ to $(2, 3, 4)$.

4. $\mathbf{F}(x, y, z) = x\mathbf{i} - z\mathbf{j} + 2y\mathbf{k}$ along the y-axis from 0 to 2.

5. $\mathbf{F}(x, y, z) = y^2\mathbf{i} + x^2\mathbf{j}$ along the parabola $y = x^2 + 1$ in the plane $z = 2$, from $(0, 1, 2)$ to $(-1, 2, 2)$.

6. $\mathbf{F}(x, y, z) = \dfrac{1}{1 + x^2}\mathbf{i} + \dfrac{1}{1 + y^2}\mathbf{j}$ along the straight line segment described by $\mathbf{r}(t) = t\mathbf{i} + (t + 1)\mathbf{j} + 2\mathbf{k}$ from $(0, 1, 2)$ to $(2, 3, 2)$.

7. $\mathbf{F}(x, y, z) = (y^2 + xz)\mathbf{i} + xy\mathbf{j} + (x^2 - y)\mathbf{k}$ along the elliptical helix $\mathbf{r}(t) = 2\cos t\mathbf{i} + 3\sin t\mathbf{j} + t\mathbf{k}$ from $(2, 0, 0)$ to $(-2, 0, 3\pi)$.

8. $\mathbf{F}(x, y, z) = y\sin x\, \mathbf{i} + ye^x\, \mathbf{j}$ along the curve $\mathbf{r}(t) = t\mathbf{i} + \cos t\mathbf{j} + \mathbf{k}$, for $0 \le t \le \pi/2$.

9. Find the work done by the force $\mathbf{F}(x, y, z) = yz\mathbf{i} + xz\mathbf{j} + xy\mathbf{k}$ in moving an object from $(0, 0, 0)$ to $(1, 2, 3)$ along the curve $\mathbf{r}(t) = t\mathbf{i} + 2t\mathbf{j} + 3t\mathbf{k}$, $0 \le t \le 1$.

10. Find the work done by the force described in Exercise 9 along the curve $\mathbf{r}(t) = t\mathbf{i} + 2t^2\mathbf{j} + 3t^3\mathbf{k}$, $0 \le t \le 1$.

11. Find the work done by the force $\mathbf{F}(x, y, z) = (x^2 - y)\mathbf{i} + (y^2 - z)\mathbf{j} + (z^2 - x)\mathbf{k}$ in moving an object along each of the following curves.

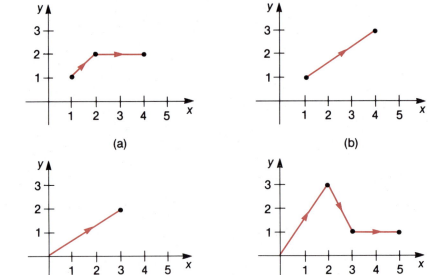

12. Find the work done by the force $\mathbf{F}(x, y, z) = e^{x+y}\,\mathbf{i} - x^2y\mathbf{j} + \mathbf{k}$ in moving an object along the curve that follows.

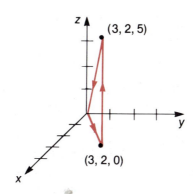

In Exercises 13 through 18, find the total mass and center of mass of a wire described by the curve C and having density ρ.

13. C is the straight line from $(1, 0, 0)$ to $(1, 4, 7)$; $\rho(x, y, z) = xyz$.

14. C is the semicircle $x^2 + y^2 = 1$, $y \geq 0$, in the xy-plane; $\rho(x, y, z) = x + y$.

15. C is the upper half of the circle $x^2 + y^2 = 4$ in the xy-plane; $\rho(x, y, z) = 8$.

16. C is the triangle with vertices $(0, 0, 0)$, $(1, 1, 0)$, $(1, 1, 1)$; $\rho(x, y, z) = \cos x + \cos y + \cos z$.

17. C is the intersection of the ellipsoid $\dfrac{x^2}{9} + \dfrac{y^2}{16} + \dfrac{z^2}{16} = 1$ and the plane $x = 2$; $\rho(x, y, z) = x$.

18. C is the rectangle in the xy-plane with vertices $(0, 0)$, $(2, 0)$, $(2, 3)$, $(0, 3)$;
$$\rho(x, y, z) = \begin{cases} x \text{ from } (0, 0) \text{ to } (2, 0) \text{ and from } (0, 3) \text{ to } (2, 3) \\ 3 \text{ from } (2, 0) \text{ to } (2, 3) \text{ and from } (0, 0) \text{ to } (0, 3). \end{cases}$$

17.4
LINE INTEGRALS INDEPENDENT OF PATH

Examples 3, 4, and 5 of Section 17.2 demonstrate that the line integral of a function generally depends on the curve over which the integration is taken, and not only on the endpoints of the curve. However, certain functions have the property that their line integrals do depend only on the endpoints of the curve.

When the value of the line integral of a function depends only on the endpoints of the curve for all curves in a region D, the line integral is said to be **independent of path** in D. In this section, we discuss the distinguishing properties of functions whose line integrals are independent of path.

(17.34)
DEFINITION

A function $\mathbf{F}$ described by
$$\mathbf{F}(x, y, z) = M(x, y, z)\mathbf{i} + N(x, y, z)\mathbf{j} + P(x, y, z)\mathbf{k}$$

is called **conservative** in a region D in $\mathbb{R}^3$ if a differentiable function f exists with $\nabla f = \mathbf{F}$. In this case, the differential form

$$M(x, y, z)dx + N(x, y, z)dy + P(x, y, z)dz$$

is said to be **exact** in D.

(17.35)
THEOREM

Suppose $\mathbf{F}$ is continuous on a region D in $\mathbb{R}^3$. The line integral $\int_C \mathbf{F} \cdot d\mathbf{r}$ is independent of path in D if and only if $\mathbf{F}$ is conservative in D.

PROOF

If $\mathbf{F}$ is conservative, there is a function f with $\nabla f = \mathbf{F}$. So, using the chain rule,

$$\int_C \mathbf{F} \cdot d\mathbf{r} = \int_a^b \left(\frac{\partial f}{\partial x}\mathbf{i} + \frac{\partial f}{\partial y}\mathbf{j} + \frac{\partial f}{\partial z}\mathbf{k} \right) \cdot \left(\frac{dx}{dt}\mathbf{i} + \frac{dy}{dt}\mathbf{j} + \frac{dz}{dt}\mathbf{k} \right) dt$$

$$= \int_a^b \left(\frac{\partial f}{\partial x}\frac{dx}{dt} + \frac{\partial f}{\partial y}\frac{dy}{dt} + \frac{\partial f}{\partial z}\frac{dz}{dt} \right) dt$$

$$= \int_a^b \frac{df}{dt}dt$$

$$= f(x(b), y(b), z(b)) - f(x(a), y(a), z(a)),$$

and $\int_C \mathbf{F} \cdot d\mathbf{r}$ depends only on the endpoints of C.

Suppose, on the other hand, that $\int_C \mathbf{F} \cdot d\mathbf{r}$ is independent of path in D. For a fixed point (x_0, y_0, z_0) in D, define

$$f(x, y, z) = \int_{(x_0,y_0,z_0)}^{(x,y,z)} \mathbf{F} \cdot d\mathbf{r}$$

$$= \int_{(x_0,y_0,z_0)}^{(x,y,z)} [M(x, y, z)dx + N(x, y, z)dy + P(x, y, z)dz].$$

Let C_1 be the straight line from (x_0, y_0, z_0) to (x_0, y_0, z), C_2 be the straight line from (x_0, y_0, z) to (x_0, y, z) and C_3 be the straight line from (x_0, y, z) to (x, y, z) and $C = C_1 + C_2 + C_3$. See Figure 17.10.

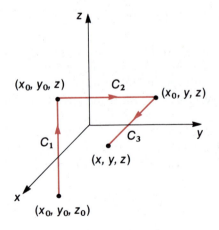

FIGURE 17.10

Then

$$f(x, y, z) = \int_C [M\,dx + N\,dy + P\,dz]$$

$$= \int_{C_1} [M\,dx + N\,dy + P\,dz] + \int_{C_2} [M\,dx + N\,dy + P\,dz]$$

$$+ \int_{C_3} [M\,dx + N\,dy + P\,dz].$$

The first two integrals are independent of x, so

$$\frac{\partial}{\partial x} \left[\int_{C_1} [M\,dx + N\,dy + P\,dz] + \int_{C_2} [M\,dx + N\,dy + P\,dz] \right] = 0.$$

In the third integral, $dy = 0$ and $dz = 0$, so

$$\frac{\partial}{\partial x} \left[\int_{C_3} [M\,dx + N\,dy + P\,dz] \right] = \frac{\partial}{\partial x} \int_{x_0}^{x} M\,dx = M(x, y, z).$$

Thus,
$$\frac{\partial f(x, y, z)}{\partial x} = M(x, y, z).$$

Similarly,
$$\frac{\partial f(x, y, z)}{\partial y} = N(x, y, z) \quad \text{and} \quad \frac{\partial f(x, y, z)}{\partial z} = P(x, y, z).$$

Thus $\nabla f = \mathbf{F}$ and $\mathbf{F}$ is conservative. □

The following corollaries are immediate consequences of Theorem 17.35 and its proof.

(17.36)
COROLLARY
A continuous function $\mathbf{F}$ is conservative in D if and only if $\int_C \mathbf{F} \cdot d\mathbf{r} = 0$ for every simple closed curve C in D.

(17.37)
COROLLARY
Suppose $\mathbf{F}$ is a continuous function and f is a function with $\nabla f = \mathbf{F}$ on a region D in $\mathbb{R}^3$. If C is a curve in D with initial point P and terminal point Q, then

$$\int_C \mathbf{F} \cdot d\mathbf{r} = f(Q) - f(P).$$

Note the close analogy between Corollary 17.37 and the fundamental theorem of calculus. Here f plays the role of the antiderivative of the integrand $\mathbf{F}$, and the endpoints P and Q of the curve C correspond to the endpoints of the interval of integration.

EXAMPLE 1
Find $\int_C \mathbf{F} \cdot d\mathbf{r}$, where $\mathbf{r}(t) = t\mathbf{i} + (t^2 - 2)\mathbf{j} + \dfrac{\pi t^2}{2}\mathbf{k}$, where $0 \le t \le 1$, and
$\mathbf{F}(x, y, z) = y^2 \sin z\,\mathbf{i} + 2xy \sin z\,\mathbf{j} + xy^2 \cos z\,\mathbf{k}$.

SOLUTION
Notice that if $f(x, y, z) = xy^2 \sin z$, then

$$\nabla f(x, y, z) = y^2 \sin z\,\mathbf{i} + 2xy \sin z\,\mathbf{j} + xy^2 \cos z\,\mathbf{k} = \mathbf{F}(x, y, z).$$

Consequently, **F** is conservative and the integral is independent of path. Since

$$\mathbf{r}(1) = \mathbf{i} - \mathbf{j} + \frac{\pi}{2}\mathbf{k} \text{ and } \mathbf{r}(0) = -2\mathbf{j},$$

it follows from Corollary 17.37 that

$$\int_C \mathbf{F} \cdot d\mathbf{r} = f\left(1, -1, \frac{\pi}{2}\right) - f(0, -2, 0)$$

$$= 1 \cdot (-1)^2 \cdot \sin\frac{\pi}{2} - 0 \cdot (-2)^2 \sin 0 = 1. \qquad \square$$

The evaluation of the line integral in the preceding example was simplified by recognizing a function f with $\nabla f = \mathbf{F}$. The natural questions to ask are:

(i) Are there easily verifiable conditions that will ensure that **F** is conservative?

(ii) If **F** is conservative, how do we find a function f with $\nabla f = \mathbf{F}$?

Before presenting answers to these questions, we need a definition.

(17.38)
DEFINITION

A region D (either in the plane or in space) is **simply connected** if every simple closed curve in D can be shrunk to a point without leaving D.

In the plane, the interior of a circle or rectangle is a simply connected region (Figures 17.11(a) and (b)); the region between concentric circles is not simply connected (Figure 17.11(c)).

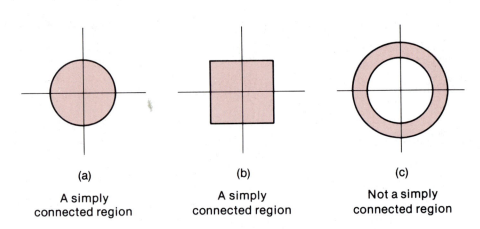

(a)

A simply
connected region

(b)

A simply
connected region

(c)

Not a simply
connected region

FIGURE 17.11

In space, the interior of a sphere with finitely many points removed is simply connected (see Figure 17.12(a). The region bounded by a torus, the doughnut-shaped region shown in Figure 17.12(b), is not simply connected since the curve C shown in the figure cannot be shrunk to a point without leaving the torus.

In the next theorem, we need the concept of an *open* region. A region in the plane is open provided that each of its points is the center of a circle that lies entirely in the region. A region in space is open provided that each of its points is the center of a sphere that lies entirely in the region.

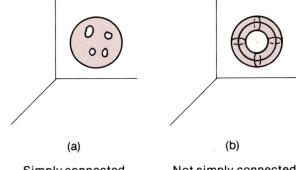

(a)   (b)

Simply connected   Not simply connected

FIGURE 17.12

(17.39)
THEOREM

Suppose $\mathbf{F}(x, y, z) = M(x, y, z)\mathbf{i} + N(x, y, z)\mathbf{j} + P(x, y, z)\mathbf{k}$, where M, N, and P have continuous first partial derivatives in an open simply connected region D. Then $\mathbf{F}$ is conservative in D if and only if

$$\frac{\partial M}{\partial y} = \frac{\partial N}{\partial x}, \qquad \frac{\partial M}{\partial z} = \frac{\partial P}{\partial x}, \qquad \text{and} \qquad \frac{\partial N}{\partial z} = \frac{\partial P}{\partial y}.$$

PROOF

We will show that if $\mathbf{F}$ is conservative in D, then the conditions are satisfied. The proof that the conditions imply $\mathbf{F}$ is conservative is deferred to Section 17.7, where it is an easy consequence of Stokes' theorem, a major result of that section.

If $\mathbf{F}$ is conservative, then there is a function f with

$$\mathbf{F} = \nabla f = \frac{\partial f}{\partial x}\mathbf{i} + \frac{\partial f}{\partial y}\mathbf{j} + \frac{\partial f}{\partial z}\mathbf{k}.$$

Thus, $$M = \frac{\partial f}{\partial x}, \qquad N = \frac{\partial f}{\partial y}, \qquad \text{and} \qquad P = \frac{\partial f}{\partial z}.$$

Since M and N have continuous first partial derivatives,

$$\frac{\partial M}{\partial y} = \frac{\partial^2 f}{\partial y \partial x} \qquad \text{and} \qquad \frac{\partial N}{\partial x} = \frac{\partial^2 f}{\partial x \partial y}$$

are continuous. Consequently

$$\frac{\partial^2 f}{\partial y \partial x} = \frac{\partial^2 f}{\partial x \partial y},$$

which implies

$$\frac{\partial M}{\partial y} = \frac{\partial N}{\partial x}.$$

In a similar manner,

$$\frac{\partial M}{\partial z} = \frac{\partial P}{\partial x} \qquad \text{and} \qquad \frac{\partial N}{\partial z} = \frac{\partial P}{\partial y}. \qquad \square$$

Theorems 17.35 and 17.39 can be combined to give the following result.

(17.40)
COROLLARY

Suppose $\mathbf{F}(x, y, z) = M(x, y, z)\mathbf{i} + N(x, y, z)\mathbf{j} + P(x, y, z)\mathbf{k}$, where M, N, and P have continuous first partial derivatives in an open simply connected region D. The line integral $\int_C \mathbf{F} \cdot d\mathbf{r}$ is independent of path in D if and only if

$$\frac{\partial M}{\partial y} = \frac{\partial N}{\partial x}, \qquad \frac{\partial M}{\partial z} = \frac{\partial P}{\partial x}, \quad \text{and} \quad \frac{\partial N}{\partial z} = \frac{\partial P}{\partial y}.$$

EXAMPLE 2

Find $\int_C \mathbf{F} \cdot d\mathbf{r}$ if C is described by $\mathbf{r}(t) = \sqrt{t}\,\mathbf{i} + t^2\mathbf{j} + \sin \frac{\pi}{2}\, t\mathbf{k}$, $0 \le t \le 1$, and $\mathbf{F}(x, y, z) = 2x\mathbf{i} + \tan z\mathbf{j} + y(\sec z)^2\mathbf{k}$.

SOLUTION

Direct integration requires evaluating the formidable integral

$$\int_0^1 \left[1 + 2t \tan\left(\sin \frac{\pi}{2}\, t\right) + \frac{\pi}{2}\, t^2 \left(\sec\left(\sin \frac{\pi}{2}\, t\right)\right)^2 \left(\cos \frac{\pi}{2}\, t\right) \right] dt.$$

However, $\mathbf{F}$ is conservative since

$$\frac{\partial}{\partial y}(2x) = 0 = \frac{\partial}{\partial x}(\tan z), \qquad \frac{\partial}{\partial z}(2x) = 0 = \frac{\partial}{\partial x}(y(\sec z)^2)$$

and

$$\frac{\partial}{\partial z}(\tan z) = (\sec z)^2 = \frac{\partial}{\partial y}(y(\sec z)^2).$$

Consequently, $\int_C \mathbf{F} \cdot d\mathbf{r}$ is independent of path. Instead of integrating along the given curve C from $(0, 0, 0)$ to $(1, 1, 1)$, we will integrate along the curve $C_1 + C_2 + C_3$, shown in Figure 17.13, where

$$C_1: \mathbf{r}(t) = t\mathbf{i}, \qquad 0 \le t \le 1;$$
$$C_2: \mathbf{r}(t) = \mathbf{i} + t\mathbf{j}, \qquad 0 \le t \le 1;$$

and

$$C_3: \mathbf{r}(t) = \mathbf{i} + \mathbf{j} + t\mathbf{k}, \qquad 0 \le t \le 1.$$

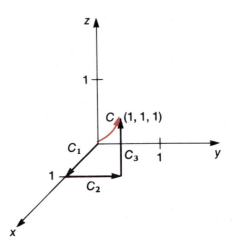

FIGURE 17.13

Although more integrals are involved, the problem is simplified because only one of the coordinate variables changes on each curve.

$$\int_C \mathbf{F} \cdot d\mathbf{r} = \int_{C_1} \mathbf{F} \cdot d\mathbf{r} + \int_{C_2} \mathbf{F} \cdot d\mathbf{r} + \int_{C_3} \mathbf{F} \cdot d\mathbf{r}$$

$$= \int_0^1 [2t\mathbf{i} + (\tan 0)\mathbf{j} + 0(\sec 0)^2\mathbf{k}] \cdot \mathbf{i}\, dt$$

$$+ \int_0^1 [2\mathbf{i} + (\tan 0)\mathbf{j} + t(\sec 0)^2\mathbf{k}] \cdot \mathbf{j}\, dt$$

$$+ \int_0^1 [2\mathbf{i} + (\tan t)\mathbf{j} + (\sec t)^2\mathbf{k}] \cdot \mathbf{k}\, dt$$

$$= \int_0^1 2t\, dt + \int_0^1 0\, dt + \int_0^1 (\sec t)^2 dt$$

$$= t^2 + \tan t \Big]_0^1 = 1 + \tan 1. \qquad \square$$

We now have conditions that determine when a function $\mathbf{F}$ is conservative. The remaining question to be answered is how to determine a function f whose gradient is $\mathbf{F}$. The answer is illustrated in the next example.

EXAMPLE 3   Show that the function described by

$$\mathbf{F}(x, y, z) = 2xy\mathbf{i} + (x^2 + \sin z)\mathbf{j} + (y \cos z + 2)\mathbf{k}$$

is the gradient of some function f, and find f.

SOLUTION
Since

$$\frac{\partial}{\partial y}(2xy) = 2x = \frac{\partial}{\partial x}(x^2 + \sin z), \qquad \frac{\partial}{\partial z}(2xy) = 0 = \frac{\partial}{\partial x}(y \cos z + 2)$$

and $\dfrac{\partial}{\partial z}(x^2 + \sin z) = \cos z = \dfrac{\partial}{\partial y}(y \cos z + 2),$

$\mathbf{F}$ is conservative. Hence there is a function f with

$$\mathbf{F} = \nabla f = \frac{\partial f}{\partial x}\mathbf{i} + \frac{\partial f}{\partial y}\mathbf{j} + \frac{\partial f}{\partial z}\mathbf{k}.$$

To find f, we first equate the components in $\mathbf{F}(x, y, z)$ and $\nabla f(x, y, z)$ to obtain:

(a) $\dfrac{\partial f}{\partial x}(x, y, z) = 2xy$ \qquad\qquad (b) $\dfrac{\partial f}{\partial y}(x, y, z) = x^2 + \sin z$

(c) $\dfrac{\partial f}{\partial z}(x, y, z) = y \cos z + 2$

Integrating (a) with respect to x, we have

(d) $f(x, y, z) = \displaystyle\int 2xy\, dx = x^2 y + g(y, z),$

where $g(y, z)$ is the constant of integration with respect to x, that is,

$$\frac{\partial g}{\partial x}(y, z) = 0.$$

Next, we take the partial derivative of f in (d) with respect to y and set this equal to the expression in (b).

$$\frac{\partial f}{\partial y}(x, y, z) = \frac{\partial}{\partial y}(x^2 y + g(y, z)) = x^2 + \frac{\partial}{\partial y}g(y, z).$$

So

$$x^2 + \frac{\partial}{\partial y}g(y, z) = x^2 + \sin z$$

and

(e) $\dfrac{\partial}{\partial y}g(y, z) = \sin z.$

Integrating both sides of (e) with respect to y gives

$$g(y, z) = \int \sin z\, dy = y \sin z + k(z).$$

Combining this with (d) produces

(f) $f(x, y, z) = x^2 y + y \sin z + k(z).$

Finally, to evaluate $k(z)$ we take the partial derivative of f as given in (f) with respect to z and set this equal to the expression in (c).

$$\frac{\partial f}{\partial z} = \frac{\partial}{\partial z}(x^2 y + y \sin z + k(z)) = y \cos z + k'(z)$$

so

$$y \cos z + k'(z) = y \cos z + 2$$

and $k'(z) = 2$. It follows that

$$k(z) = 2z + C$$

for any constant C and that

$$f(x, y, z) = x^2 y + y \sin z + 2z + C$$

has the property that $\nabla f = \mathbf{F}$. $\square$

EXERCISE SET 17.4

1. Evaluate $\displaystyle\int_C yz\, dx + xz\, dy + xy\, dz$ where

    (a) C is the straight line from $(0, 0, 0)$ to $(1, 1, 1)$.

    (b) C is described by $\mathbf{r}(t) = t\mathbf{i} + t^2\mathbf{j} + t^3\mathbf{k},\ 0 \le t \le 1$.

2. Evaluate $\displaystyle\int_C y \sin z\, dx + x \sin z\, dy + xy \cos z\, dz$, where

    (a) C is the straight line from $(1, 0, 0)$ to $(-1, 0, \pi)$.

    (b) C is the helix described by $\mathbf{r}(t) = \cos t\mathbf{i} + \sin t\mathbf{j} + t\mathbf{k},\ 0 \le t \le \pi$.

In Exercises 3 through 8, determine if the differential form is exact.

3. $y \sin z \, dx + x \sin z \, dy + xy \cos z \, dz$

4. $y^3 \cos x \, dx - 3y^2 \sin x \, dy$

5. $(y + z)dx + (x + z)dy + (x + y)dz$

6. $yz \, dx + xz \, dy + xy \, dz$

7. $x^2 \, dx + y^2 \, dy + z^2 \, dz$

8. $y^2 \, dx + z^2 \, dy + x^2 \, dz$

In Exercises 9 through 16, determine if $\mathbf{F}$ is conservative.

9. $\mathbf{F}(x, y) = e^{-y} \cos x\mathbf{i} - e^{-y} \sin x\mathbf{j}$

10. $\mathbf{F}(x, y) = (2x - y)\mathbf{i} + (2y - x)\mathbf{j}$

11. $\mathbf{F}(x, y, z) = \cos yz\mathbf{i} - xy \sin yz\mathbf{j} - xy \sin yz\mathbf{k}$

12. $\mathbf{F}(x, y) = \dfrac{-x}{x^2 + y^2}\mathbf{i} + \dfrac{y}{x^2 + y^2}\mathbf{j}$

13. $\mathbf{F}(x, y, z) = 2xy^3z\mathbf{i} + 3x^2y^2z\mathbf{j} + x^2y^3\mathbf{k}$

14. $\mathbf{F}(x, y, z) = x^2y^3\mathbf{i} + x^2y^3z\mathbf{j} + \mathbf{k}$

15. $\mathbf{F}(x, y, z) = (z \cos x - y \sin x)\mathbf{i} + (\cos x + z)\mathbf{j} + (\sin x + y - 2z)\mathbf{k}$

16. $\mathbf{F}(x, y, z) = e^{-z}\mathbf{i} + 2y\mathbf{j} + xe^{-z}\mathbf{k}$

17–24. For each of the conservative functions $\mathbf{F}$ in Exercises 9 through 16, find a function f with $\operatorname{grad} f = \mathbf{F}$.

In Exercises 25 through 30, show that the line integral is independent of path and evaluate the integral.

25. $\displaystyle\int_C (2x + y)dx + (2y + x)dy + 2z \, dz$, where C is the circle $x^2 + y^2 = 4$ in the xy-plane.

26. $\displaystyle\int_C e^x \sin y \, dx + e^x \cos y \, dy$, where C is the curve in the xy-plane described by $\mathbf{r}(t) = t\mathbf{i} + 2t\mathbf{j}$, $0 \le t \le 1$.

27. $\displaystyle\int_C y \ln z \, dx + x \ln z \, dy + \dfrac{xy}{z} \, dz$, where C is the helix described by $\mathbf{r}(t) = \cos t\mathbf{i} + \sin t\mathbf{j} + t\mathbf{k}$ from $(1, 0, 2\pi)$ to $(1, 0, 4\pi)$.

28. $\displaystyle\int_C (y^2 + 2xz)dx + (2xy - z)dy + (x^2 - y)dz$, where C is described by $\mathbf{r}(t) = \left(\sin \dfrac{\pi t}{2}\right)^3 \mathbf{i} + (t \cos \pi t)\mathbf{j} + t^2(\cos \pi t)^2\mathbf{k}$, $0 \le t \le 1$.

29. $\displaystyle\int_C \dfrac{1}{x} \ln xy \, dx + \left(\dfrac{1}{y} \ln xy + ze^{yz}\right)dy + ye^{yz} \, dz$, where C is described by $\mathbf{r}(t) = t\mathbf{i} + t^2\mathbf{j} + t^3\mathbf{k}$ from $(1, 1, 1)$ to $(2, 4, 8)$.

30. $\displaystyle\int_C 2xe^{2y}dx + 2x^2e^{2y}dy + (\sec z)^2 \, dz$, where C is any piecewise smooth curve joining $(1, 1, 3)$ and $(1, -2, 3)$.

31. Near the earth's surface, the force exerted by gravity is $-g\mathbf{k}$. Show that if an object of mass m moves from a height h_1 to a height h_2 along a smooth curve, the work done by gravity on this object is $mg(h_1 - h_2)$.

17.5
GREEN'S THEOREM

In Section 17.2 we defined a simple closed curve as one whose endpoints coincide and are the only points of intersection. Green's theorem concerns the relation between a line integral over a simple closed curve in the plane and a double integral over the region bounded by the curve.

Before presenting the theorem, we need to discuss in more detail the orientation of a simple closed curve. A simple closed curve in the plane, described by $\mathbf{r}(t) = x(t)\mathbf{i} + y(t)\mathbf{j}$, $a \leq t \leq b$, is said to have **positive orientation** if the region to the left of the curve is bounded as the curve is traced by $\mathbf{r}(t)$. A curve oriented in the opposite direction is naturally said to have **negative orientation.** See Figure 17.14.

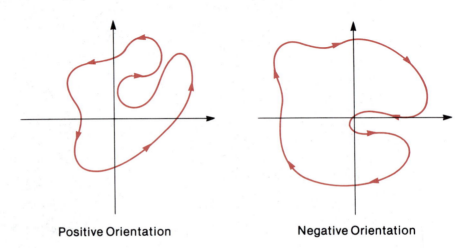

Positive Orientation　　　　　　　Negative Orientation

FIGURE 17.14

(17.41)
THEOREM

Green's Theorem　Suppose C is a positively oriented simple closed curve in the plane and R is the region bounded by C. If

$$\mathbf{F}(x, y) = M(x, y)\mathbf{i} + N(x, y)\mathbf{j}$$

is a vector-valued function with the partial derivatives of M and N continuous on R, then

$$\int_C \mathbf{F} \cdot d\mathbf{r} = \iint_R \left(\frac{\partial N}{\partial x} - \frac{\partial M}{\partial y} \right) dA.$$

PROOF
We will prove the theorem only for the special case when the region R can

---

HISTORICAL NOTE　**Green's theorem** was first published in 1828 in an article on electromagnetism by George Green (1793–1841). The article was privately published and not widely distributed. This theorem was not generally known until it was rediscovered by Lord Kelvin, born William Thompson (1824–1907), in 1845. The result was independently discovered by Michel Ostrogradski (1801–1861) and in Russia is known as *Ostrogradski's theorem.*

be represented both as

(I) $\qquad\qquad\qquad a \le x \le b, \qquad f_1(x) \le y \le f_2(x),$

for appropriate functions f_1 and f_2 (see Figure 17.15), and as

(II) $\qquad\qquad\qquad c \le y \le d, \qquad g_1(y) \le x \le g_2(y),$

for appropriate functions g_1 and g_2 (see Figure 17.16).

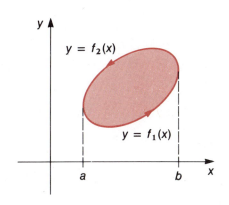

FIGURE 17.15

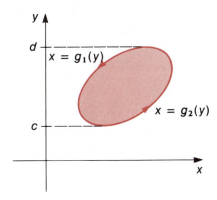

FIGURE 17.16

After giving an example, we will indicate how the result can be extended to arbitrary regions.

The proof of the theorem for this special region relies on the fact that the simple closed curve C can be expressed both as $C_1 - C_2$, where

$$C_1: \mathbf{r}(t) = t\mathbf{i} + f_1(t)\mathbf{j}, \qquad a \le t \le b;$$
$$C_2: \mathbf{r}(t) = t\mathbf{i} + f_2(t)\mathbf{j}, \qquad a \le t \le b,$$

and as $C_3 - C_4$, where

$$C_3: \mathbf{r}(t) = g_2(t)\mathbf{i} + t\mathbf{j}, \qquad c \le t \le d;$$
$$C_4: \mathbf{r}(t) = g_1(t)\mathbf{i} + t\mathbf{j}, \qquad c \le t \le d.$$

See Figures 17.17 and 17.18.

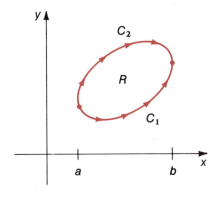

FIGURE 17.17

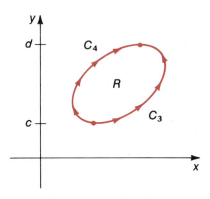

FIGURE 17.18

Thus,

$$\iint_R \left(\frac{\partial N}{\partial x} - \frac{\partial M}{\partial y}\right) dA = \iint_R \frac{\partial N}{\partial x} \, dA - \iint_R \frac{\partial M}{\partial y} \, dA$$

$$= \int_c^d \int_{g_1(y)}^{g_2(y)} \frac{\partial N}{\partial x} \, dx \, dy - \int_a^b \int_{f_1(x)}^{f_2(x)} \frac{\partial M}{\partial y} \, dy \, dx$$

$$= \int_c^d [N(g_2(y), y) - N(g_1(y), y)] dy$$

$$- \int_a^b [M(x, f_2(x)) - M(x, f_1(x))] dx$$

$$= \int_{C_3} N(x, y) dy - \int_{C_4} N(x, y) dy - \int_{C_2} M(x, y) dx$$

$$= \int_{C_3 - C_4} N(x, y) dy + \int_{C_1 - C_2} M(x, y) dx$$

$$= \int_C N(x, y) dy + \int_C M(x, y) dx$$

$$= \int_C (M(x, y)\mathbf{i} + N(x, y)\mathbf{j}) \cdot (dx\mathbf{i} + dy\mathbf{j})$$

$$= \int_C \mathbf{F} \cdot d\mathbf{r}. \qquad \square$$

EXAMPLE 1   Find $\int_C (x^2 y\mathbf{i} + xy^2\mathbf{j}) \cdot d\mathbf{r}$ where C is the positively oriented boundary of the rectangle shown in Figure 17.19.

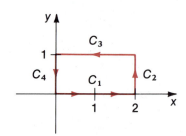

FIGURE 17.19

SOLUTION

To compute this integral directly we express C as $C = C_1 + C_2 + C_3 + C_4$ where

$$C_1: \mathbf{r}(t) = t\mathbf{i}, \qquad 0 \le t \le 2$$
$$C_2: \mathbf{r}(t) = 2\mathbf{i} + t\mathbf{j}, \qquad 0 \le t \le 1$$
$$C_3: \mathbf{r}(t) = (2 - t)\mathbf{i} + \mathbf{j}, \qquad 0 \le t \le 2$$
$$C_4: \mathbf{r}(t) = (1 - t)\mathbf{j}, \qquad 0 \le t \le 1.$$

On C_1, y is zero and on C_4, x is zero. Hence on both C_1 and C_4, $x^2 y\mathbf{i} + xy^2\mathbf{j}$ is zero, as are the integrals over these curves. Consequently,

$$
\begin{aligned}
\int_C (x^2 y\mathbf{i} + xy^2\mathbf{j}) \cdot d\mathbf{r} &= \int_{C_2} (x^2 y\mathbf{i} + xy^2\mathbf{j}) \cdot d\mathbf{r} + \int_{C_3} (x^2 y\mathbf{i} + xy^2\mathbf{j}) \cdot d\mathbf{r} \\
&= \int_0^1 (4t\mathbf{i} + 2t^2\mathbf{j}) \cdot \mathbf{j}\,dt + \int_0^2 [(2-t)^2\mathbf{i} + (2-t)\mathbf{j}] \cdot (-\mathbf{i})\,dt \\
&= \int_0^1 2t^2\,dt + \int_0^2 -(2-t)^2\,dt \\
&= \left. \frac{2t^3}{3} \right]_0^1 + \left. \frac{(2-t)^3}{3} \right]_0^2 = \frac{2}{3} - \frac{8}{3} = -2.
\end{aligned}
$$

Computing the integral using Green's theorem,

$$
\begin{aligned}
\int_C (x^2 y\mathbf{i} + xy^2\mathbf{j})d\mathbf{r} &= \int_0^2 \int_0^1 \left[\frac{\partial(xy^2)}{\partial x} - \frac{\partial(x^2 y)}{\partial y} \right] dy\,dx \\
&= \int_0^2 \int_0^1 (y^2 - x^2)\,dy\,dx \\
&= \int_0^2 \left[\frac{y^3}{3} - x^2 y \right]_0^1 dx = \int_0^2 \left(\frac{1}{3} - x^2 \right) dx \\
&= \left. \frac{x}{3} - \frac{x^3}{3} \right]_0^2 = \frac{2}{3} - \frac{8}{3} = -2. \qquad \square
\end{aligned}
$$

Green's theorem is proved for more general regions in the plane by partitioning the region into subregions that can be expressed in the forms (I) and (II). For example, the region R in Figure 17.20(a) is partitioned into the regions R_1 and R_2 shown in Figure 17.20(b).

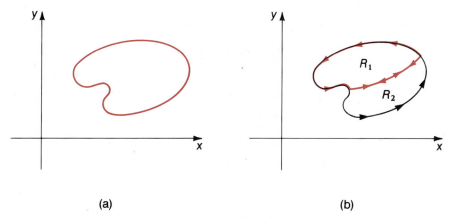

(a)   (b)

FIGURE 17.20

Each of the regions R_1 and R_2 can be represented in forms (I) and (II) with positively oriented boundaries C_1 and C_2. (See Figure 17.21.)

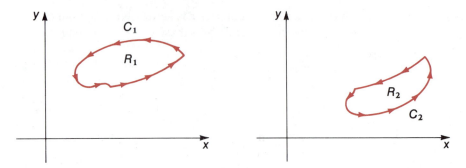

FIGURE 17.21

Since the partitioning line is traversed twice but in opposite directions, the line integrals on these portions of C_1 and C_2 cancel. Consequently,

$$\int_C \mathbf{F} \cdot d\mathbf{r} = \int_{C_1} \mathbf{F} \cdot d\mathbf{r} + \int_{C_2} \mathbf{F} \cdot d\mathbf{r}$$

$$= \iint_{R_1} \left(\frac{\partial N}{\partial x} - \frac{\partial M}{\partial y} \right) dA + \iint_{R_2} \left(\frac{\partial N}{\partial x} - \frac{\partial M}{\partial y} \right) dA$$

$$= \iint_{R} \left(\frac{\partial N}{\partial x} - \frac{\partial M}{\partial y} \right) dA.$$

This technique can even be extended to regions in the plane that have "holes," such as the region R shown in Figure 17.22.

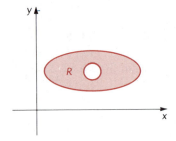

FIGURE 17.22

This region can be partitioned as shown in Figure 17.23(a) with the boundaries of the partitions R_1 and R_2 denoted by $C_1{}'$ and $C_2{}'$, as in Figures 17.23(b) and (c).

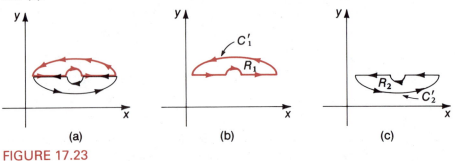

(a)   (b)   (c)

FIGURE 17.23

Then,

$$\iint_R \left(\frac{\partial N}{\partial x} - \frac{\partial M}{\partial y}\right) dA = \iint_{R_1} \left(\frac{\partial N}{\partial x} - \frac{\partial M}{\partial y}\right) dA + \iint_{R_2} \left(\frac{\partial N}{\partial x} - \frac{\partial M}{\partial y}\right) dA$$

$$= \int_{C_1'} M dx + N dy + \int_{C_2'} M dx + N dy,$$

and since the line integrals over the common boundary lines are in opposite directions, these cancel to yield

$$\iint_R \left(\frac{\partial N}{\partial x} - \frac{\partial M}{\partial y}\right) dA = \int_C M dx + N dy = \int_C \mathbf{F} \cdot d\mathbf{r}.$$

EXAMPLE 2   Use Green's theorem to determine

$$\int_C (e^x - 3xy) dx + (\sin y + 6x^2) dy$$

where C is described by $\mathbf{r}(t) = \cos t\mathbf{i} + 2 \sin t\mathbf{j}, 0 \le t \le 2\pi$.

SOLUTION

The curve C is an ellipse with equation $4x^2 + y^2 = 4$, shown in Figure 17.24.

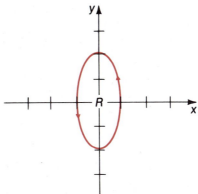

FIGURE 17.24

Thus,

$$\int_C (e^x - 3xy) dx + (\sin y + 6x^2) dy = \iint_R \left[\frac{\partial}{\partial x}(\sin y + 6x^2) - \frac{\partial}{\partial y}(e^x - 3xy)\right] dA$$

$$= \int_{-1}^{1} \int_{-2\sqrt{1-x^2}}^{2\sqrt{1-x^2}} (12x + 3x) dy\, dx$$

$$= \int_{-1}^{1} 15xy \Big]_{-2\sqrt{1-x^2}}^{2\sqrt{1-x^2}} dx = \int_{-1}^{1} 60x\sqrt{1-x^2}\, dx$$

$$= -30(1-x^2)^{3/2} \cdot \frac{2}{3}\Big]_{-1}^{1} = 0. \qquad \square$$

To gain further appreciation for Green's theorem, you might try evaluating the integral in Example 2 directly.

EXAMPLE 3   Evaluate $\displaystyle\int_C [(2x \sin y + x^2 y^2)\mathbf{i} + (x^3 y + x^2 \cos y)\mathbf{j}] \cdot d\mathbf{r}$ along

$$C: \mathbf{r}(t) = t\mathbf{i} + t^2\mathbf{j}, \qquad 0 \le t \le 2.$$

SOLUTION

Instead of evaluating this integral directly, we show how Green's theorem can be used to simplify the problem.

Consider the simple closed curve $\hat{C} = C_1 + C_2 - C$, where

$$C_1: \mathbf{r}(t) = t\mathbf{i}, \qquad 0 \le t \le 2 \quad \text{and} \quad C_2: \mathbf{r}(t) = 2\mathbf{i} + t\mathbf{j}, \qquad 0 \le t \le 4.$$

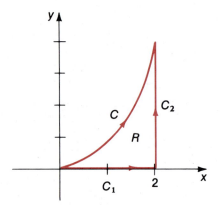

FIGURE 17.25

Then,

$$\int_C \mathbf{F} \cdot d\mathbf{r} = \int_{C_1} \mathbf{F} \cdot d\mathbf{r} + \int_{C_2} \mathbf{F} \cdot d\mathbf{r} - \int_{\hat{C}} \mathbf{F} \cdot d\mathbf{r}.$$

But $\hat{C}$ is a positively oriented curve enclosing the region R shown in Figure 17.25, so it follows from Green's theorem that

$$\int_{\hat{C}} [(2x \sin y + x^2 y^2)\mathbf{i} + (x^3 y + x^2 \cos y)\mathbf{j}] \cdot d\mathbf{r} = \iint_R [(3x^2 y + 2x \cos y)$$

$$- (2x \cos y + 2x^2 y)]\,dA$$

$$= \int_0^2 \int_0^{x^2} x^2 y \, dy \, dx$$

$$= \int_0^2 \left. \frac{x^2 y^2}{2} \right]_0^{x^2} dx$$

$$= \int_0^2 \frac{x^6}{2} \, dx = \left. \frac{x^7}{14} \right]_0^2 = \frac{64}{7}.$$

On C_1, $y = 0$ so $dy = 0$ and

$$\int_{C_1} (2x \sin y + x^2 y^2)dx + (x^3 y + x^2 \cos y)dy = 0.$$

On C_2, $x = 2$ and $y = t$ so $dx = 0$, $dy = dt$, and

$$\int_{C_2} (2x \sin y + x^2 y^2)dx + (x^3 y + x^2 \cos y)dy = \int_0^4 (8t + 4 \cos t)dt$$

$$= 4t^2 + 4 \sin t \Big]_0^4$$

$$= 64 + 4 \sin 4.$$

Consequently,

$$\int_C (2x \sin y + x^2 y^2)dx + (x^3 y - x^2 \cos y)dy = 0 + 64 + 4 \sin 4 - \frac{64}{7}$$

$$= \frac{384}{7} + 4 \sin 4.$$

While this may seem a complicated procedure, it is considerably simpler than the direct method of evaluating this integral. □

   Green's theorem can be used in a somewhat reverse manner to determine the areas of plane regions whose boundaries are described by parametric equations. If R is a region whose boundary is a positively oriented curve C, the area of R can be written

$$\textbf{(17.42)} \quad A = \iint_R dy\, dx = \iint_R \left[\frac{\partial x}{\partial x} - \frac{\partial 0}{\partial y} \right] dA = \int_C [0dx + x\, dy] = \int_C x\, dy$$

and also

$$\textbf{(17.43)} \qquad A = \iint_R dy\, dx = \iint_R \left[\frac{\partial 0}{\partial y} - \frac{\partial(-y)}{\partial y} \right] dA = \int_C -y\, dx.$$

EXAMPLE 4   Determine the area inside the ellipse described by

$$\frac{x^2}{4} + \frac{y^2}{9} = 1.$$

SOLUTION

The ellipse has a parametric representation $x(t) = 2 \cos t$, $y(t) = 3 \sin t$, $0 \leq t \leq 2\pi$. Hence

$$A = \int_C x \, dy = \int_0^{2\pi} (2 \cos t)(3 \cos t \, dt)$$

$$= 6 \int_0^{2\pi} (\cos t)^2 \, dt = 6 \int_0^{2\pi} \frac{1 + \cos 2t}{2} \, dt$$

$$= 3 \left[t + \frac{1}{2} \sin 2t \right]_0^{2\pi} = 3[2\pi] = 6\pi. \qquad \square$$

Green's theorem can also be used to verify a change of variables formula for double integrals. Suppose that $x = f(u, v)$ and $y = g(u, v)$ describe a transformation of variables and R is a simply connected region. Then

(17.44) $$\iint_R G(x, y) dy \, dx = \iint_S G(f(u, v), g(u, v)) \, J(u, v) du \, dv,$$

where $J(u, v)$ is the **Jacobian** of x and y with respect to u and v defined by

$$J(u, v) = \frac{\partial x}{\partial u} \frac{\partial y}{\partial v} - \frac{\partial y}{\partial u} \frac{\partial x}{\partial v},$$

and S is a region in the uv-plane that is mapped in a one-to-one manner onto R by the transformation.

We will show that the formula holds in the special case when $F(x, y) = 1$. The general case can be derived from this result by using the definition of the double integral.

To show that

$$\iint_R dy \, dx = \iint_S J(u, v) du \, dv,$$

let $M(x, y) = 0$ and $N(x, y) = x$. If C is the boundary of the region R, then Green's theorem implies that

$$\iint_R dy \, dx = \iint_R \left(\frac{\partial N}{\partial x} - \frac{\partial M}{\partial y} \right) dy \, dx = \int_C M \, dx + N \, dy = \int_C x \, dy.$$

However, $x = f(u, v)$ and $y = g(u, v)$,

so $$dy = \frac{\partial g}{\partial u}(u, v) du + \frac{\partial g}{\partial v}(u, v) dv$$

and
$$\iint_R dy\, dx = \int_C f(u, v)\left(\frac{\partial g}{\partial u}(u, v)du + \frac{\partial g}{\partial v}(u, v)dv\right)$$

$$= \int_C f\frac{\partial g}{\partial u}\, du + f\frac{\partial g}{\partial v}\, dv$$

$$= \iint_S \left[\frac{\partial}{\partial u}\left(f\frac{\partial g}{\partial v}\right) - \frac{\partial}{\partial v}\left(f\frac{\partial g}{\partial u}\right)\right] du\, dv$$

$$= \iint_S \left[\frac{\partial f}{\partial u}\frac{\partial g}{\partial v} + f\frac{\partial^2 g}{\partial u\,\partial v} - \frac{\partial f}{\partial v}\frac{\partial g}{\partial u} - f\frac{\partial^2 g}{\partial v\,\partial u}\right] du\, dv$$

$$= \iint_S \left[\frac{\partial f}{\partial u}\frac{\partial g}{\partial v} - \frac{\partial f}{\partial v}\frac{\partial g}{\partial u}\right] du\, dv$$

$$= \iint_S J(u, v)du\, dv.$$

EXAMPLE 5   Use the result in equation (17.44) to verify the double integral change of variable formula from rectangular to polar coordinates.

SOLUTION
Since $x = r\cos\theta$, $y = r\sin\theta$,

$$J(r, \theta) = \frac{\partial x}{\partial r}\frac{\partial y}{\partial \theta} - \frac{\partial y}{\partial r}\frac{\partial x}{\partial \theta} = \cos\theta(r\cos\theta) - \sin\theta(-r\sin\theta)$$

$$= r[(\cos\theta)^2 + (\sin\theta)]^2 = r.$$

Thus if R is a simply connected region,

$$\iint_R G(x, y)dy\, dx = \iint_S G(r, \theta)r\, dr\, d\theta$$

where S is the same region R, but described in polar coordinates.   □

EXERCISE SET 17.5

1. Verify Green's theorem for $\mathbf{F}(x, y) = x\mathbf{i} + xy\mathbf{j}$, where C is positively oriented and is the boundary of the square with vertices at $(0, 0)$, $(1, 0)$, $(1, 1)$ and $(0, 1)$.

2. Verify Green's theorem for $\mathbf{F}(x, y) = x^2\mathbf{i} + y^2\mathbf{j}$ on the unit circle $x^2 + y^2 = 1$.

In Exercises 3 through 6, use Green's theorem to find $\int_C \mathbf{F} \cdot d\mathbf{r}$, where C is positively oriented.

3. $\mathbf{F}(x, y) = 3y(\sin x)^3\mathbf{i} + (\cos x)^3\mathbf{j}$, C is the closed curve determined by $y = x^2$ and $y = x$.

4. $\mathbf{F}(x, y) = 6x^2y^3\mathbf{i} - 6xy\mathbf{j}$, C is the triangle determined by $y = 0$, $x = -1$, and $x = y$.

5. $\mathbf{F}(x, y) = \cos x \sin y\,\mathbf{i} + \sin x \cos y\,\mathbf{j}$, C is the ellipse $9x^2 + 4y^2 = 36$.

6. $\mathbf{F}(x, y) = xy^2\mathbf{i} + 2x^2y\mathbf{j}$, C is the cardioid described by $r = 1 + \cos \theta$.

In Exercises 7 through 14, use Green's theorem to evaluate the line integrals assuming C is positively oriented.

7. $\int_C (x^2 + xy)dx + (y^2 + xy)dy$, C is the boundary of the square with vertices $(0, 0)$, $(2, 0)$, $(2, 2)$ and $(0, 2)$.

8. $\int_C x^2y\,dx + (x^3 - y^3)dy$, C is the triangle with vertices $(0, 0)$, $(2, 0)$ and $(1, 1)$.

9. $\int_C (y^2 + x)dx + (x^2 + y)dy$, C is the closed curve determined by $y = x^2$ and $x = y^2$.

10. $\int_C xe^y\,dx + xy\,dy$, C is the closed curve determined by $y = x^2$ and $y = x$.

11. $\int_C (x - y^3)dx + x^3\,dy$, C is the unit circle with center at the origin.

12. $\int_C (x + y)dx + xy\,dy$, C is the cardioid $r = 1 + \cos \theta$.

13. $\int_C \frac{y}{x^2 + y^2}\,dx - \frac{x}{x^2 + y^2}\,dy$, C is the circle with equation $(x - 2)^2 + (y - 2)^2 = 1$.

14. $\int_C \frac{1}{2y}e^{x^2+y^2}\,dx + \frac{1}{x}e^{x^2+y^2}\,dy$, C is the boundary of the region in the first quadrant enclosed between the two circles with radii 1 and 3 and centered at the origin.

In Exercises 15 through 18, use Green's theorem to find the area of the given region.

15. Inside the circle $x^2 + y^2 = 16$.

16. Inside the ellipse $4x^2 + y^2 = 4$.

17. Inside the curve that is described by $x = (\cos t)^3$, $y = (\sin t)^3$, $0 \leq t \leq 2\pi$.

18. Inside the curve described by $x = \cos t$, $y = (\sin t)^3$, $0 \le t \le 2\pi$.

19. Suppose R is a region whose boundary is a positively oriented curve C. Conclude from Equations (17.42) and (17.43) that the area of R is also given by

$$\frac{1}{2} \int_C x \, dy - y \, dx.$$

20. Use the conclusion of Exercise 19 to show that if a region D is described in polar coordinates by $x = r \cos \theta$, $y = r \sin \theta$, $0 \le r \le f(\theta)$, $\alpha \le \theta \le \beta$, then the area of D is

$$A = \frac{1}{2} \int_\alpha^\beta [f(\theta)]^2 \, d\theta.$$

21. Explain why Green's theorem cannot be used to evaluate

$$\int_C \frac{y}{x^2 + y^2} \, dx - \frac{x}{x^2 + y^2} \, dy,$$

where C is the circle $x^2 + y^2 = 1$.

22. Use Green's theorem to prove Corollary 17.36: A continuous function $\mathbf{F}$ is conservative on D if and only if $\int_C \mathbf{F} \cdot d\mathbf{r} = 0$ for every simple closed curve C in D.

23. Assuming the hypothesis of Green's theorem, show that the formula in the theorem can be written as

$$\int_C \mathbf{F} \cdot \mathbf{n} \, ds = \iint_R \text{div } \mathbf{F} \, dA,$$

where $\mathbf{n}$ is the outward unit normal to C.

17.6
SURFACE INTEGRALS AND SURFACE AREA

A surface integral is an extension of the double integral over a region in a coordinate plane. The domain of integration for a surface integral is a surface in space. The particular surfaces over which we define surface integrals are those that can be conveniently described in terms of regions in one of the coordinate planes. In this way, a surface integral can be converted into a double integral over a region in a coordinate plane.

Suppose S is a surface described by $z = f(x, y)$, where (x, y) is in R, a bounded, simply connected region in the xy-plane, and f has continuous partial derivatives on R. Such a surface is called **smooth,** and, as shown in Section 15.6, has a tangent plane at each point of the surface. (See Figure 17.26). Suppose, in addition, that g is a continuous function defined on S. The object of this section is to produce a workable definition for the surface integral of g over the smooth surface S, denoted by

$$\iint_S g(x, y, z)d\sigma.$$

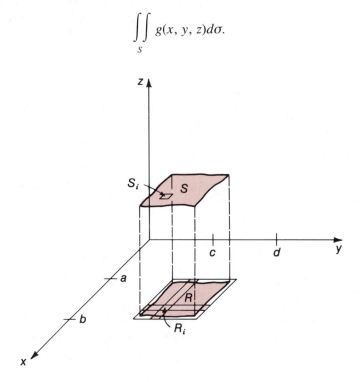

FIGURE 17.26

Since R is bounded, R is contained in a rectangle determined by an x-coordinate interval $[a, b]$ and a y-coordinate interval $[c, d]$. A partition of each of these intervals produces a grid in the xy-plane. Let $R_1, R_2, \ldots, R_n$ denote the elements of this grid that lie entirely inside R. To define the integral of g over S, we must first determine the area ΔS_i of S_i, the portion of the surface above the rectangle R_i.

We choose an arbitrary point (x_i, y_i) in the rectangle R_i and let Δx_i and Δy_i denote the dimensions of R_i. The plane tangent to S at $(x_i, y_i, f(x_i, y_i))$ intersects the right cylinder with base R_i in a parallelogram with sides described by

$$\mathbf{u} = \Delta x_i \mathbf{i} + f_x(x_i, y_i)\Delta x_i \mathbf{k}$$

and

$$\mathbf{v} = \Delta y_i \mathbf{j} + f_y(x_i, y_i)\Delta y_i \mathbf{k}.$$

See Figure 17.27. Since a parallelogram with sides $\mathbf{u}$ and $\mathbf{v}$ has area $\|\mathbf{u} \times \mathbf{v}\|$,

$$\begin{aligned}
\Delta S_i &= \|(\Delta x_i \mathbf{i} + f_x(x_i, y_i)\Delta x_i \mathbf{k}) \times (\Delta y_i \mathbf{j} + f_y(x_i, y_i)\Delta y_i \mathbf{k})\| \\
&= \|-f_x(x_i, y_i)\Delta x_i \, \Delta y_i \mathbf{i} - f_y(x_i, y_i)\Delta x_i \, \Delta y_i \mathbf{j} + \Delta x_i \, \Delta y_i \mathbf{k}\| \\
&= \{[-f_x(x_i, y_i)\Delta x_i \, \Delta y_i]^2 + [-f_y(x_i, y_i)\Delta x_i \, \Delta y_i]^2 + [\Delta x_i \, \Delta y_i]^2\}^{1/2}. \\
&= \{[f_x(x_i, y_i)]^2 + [f_y(x_i, y_i)]^2 + 1\}^{1/2} \, \Delta x_i \, \Delta y_i.
\end{aligned}$$

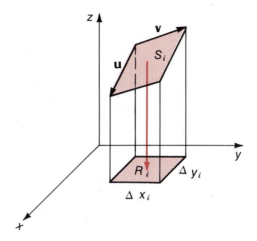

FIGURE 17.27

The definition for the surface integral $\iint_S g(x, y, z)d\sigma$ is produced by summing the products $g(x_i, y_i, f(x_i, y_i)) \, \Delta S_i$ and allowing ΔS_i to approach zero uniformly:

(17.45)

$$\iint_S g(x, y, z)d\sigma = \lim_{\Delta S_i \to 0} \sum_{i=1}^{n} g(x_i, y_i, f(x_i, y_i))\{[f_x(x_i, y_i)]^2 + [f_y(x_i, y_i)]^2 + 1\}^{1/2} \, \Delta x_i \, \Delta y_i,$$

for each i.

While this is not a Riemann sum in x and y, a theorem named to honor the 19th century mathematical physicist Jean Marie Constant Duhamel (1797–1892) can be used to reduce this sum, and consequently the surface integral, to the ordinary double integral.

If S is the smooth surface described by $z = f(x, y)$ and g is continuous on S, then the **surface integral** of g over S is

(17.46)

$$\iint_S g(x, y, z)d\sigma = \iint_R g(x, y, f(x, y))\{[f_x(x, y)]^2 + [f_y(x, y)]^2 + 1\}^{1/2} \, dA.$$

EXAMPLE 1   Determine $\iint_S z \, d\sigma$, where S is described by $z = \sqrt{4 - x^2 - y^2}$.

SOLUTION

The surface S is the hemisphere having radius 2 and lying above the xy-plane as shown in Figure 17.28. The corresponding region R in the xy-plane is the circular region $x^2 + y^2 \leq 4$.

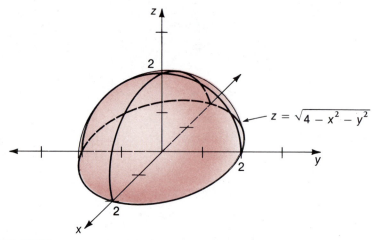

FIGURE 17.28

Consequently,

$$\iint_S z \, d\sigma = \iint_R \sqrt{4 - x^2 - y^2} \left\{ \left(\frac{\partial \sqrt{4 - x^2 - y^2}}{\partial x} \right)^2 + \left(\frac{\partial \sqrt{4 - x^2 - y^2}}{\partial y} \right)^2 + 1 \right\}^{1/2} dA.$$

$$= \int_{-2}^{2} \int_{-\sqrt{4-x^2}}^{\sqrt{4-x^2}} \sqrt{4 - x^2 - y^2} \left[\frac{x^2}{4 - x^2 - y^2} + \frac{y^2}{4 - x^2 - y^2} + 1 \right]^{1/2} dy \, dx$$

$$= \int_{-2}^{2} \int_{-\sqrt{4-x^2}}^{\sqrt{4-x^2}} 2 \, dy \, dx.$$

Using cylindrical coordinates simplifies the integration to

$$\iint_S z \, d\sigma = 2 \int_0^{2\pi} \int_0^2 r \, dr \, d\theta = \int_0^{2\pi} r^2 \bigg]_0^2 d\theta = \int_0^{2\pi} 4 \, d\theta = 4\theta \bigg]_0^{2\pi} = 8\pi. \quad \square$$

An important special case of surface integrals occurs when $g(x, y, z) \equiv 1$. In this case, we see that the sum in (17.45) becomes $\sum_{i=1}^{n} \Delta S_i$, which is an approximation to the area of S. This can be seen in Figure 17.29.

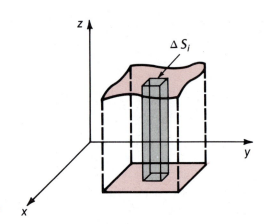

ΔS_i

FIGURE 17.29

For this reason, the special case $g \equiv 1$ is known as the **surface area** of S:

(17.47) $$\iint_S d\sigma = \iint_R \{[f_x(x, y)]^2 + [f_y(x, y)]^2 + 1\}^{1/2} \, dA.$$

EXAMPLE 2   Find the area of that portion of the surface described by $f(x, y) = x^2 + y + 1$ that lies above the region in the xy-plane bounded by the triangle with vertices $(0, 0, 0)$, $(1, 0, 0)$, and $(1, 1, 0)$.

SOLUTION

The surface is shown in Figure 17.30(a) and the region R in the xy-plane in Figure 17.30(b).

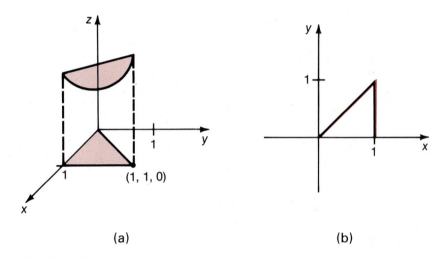

(a)   (b)

FIGURE 17.30

$$A = \iint_R \left\{ \left[\frac{\partial}{\partial x}(x^2 + y + 1) \right]^2 + \left[\frac{\partial}{\partial y}(x^2 + y + 1) \right]^2 + 1 \right\}^{1/2} dA$$

$$= \int_0^1 \int_0^x \sqrt{(2x)^2 + 1^2 + 1} \, dy \, dx = \int_0^1 \int_0^x \sqrt{4x^2 + 2} \, dy \, dx$$

$$= \int_0^1 x\sqrt{4x^2 + 2} \, dx$$

$$= \frac{1}{8}(4x^2 + 2)^{3/2} \cdot \frac{2}{3} \Big]_0^1 = \frac{1}{12}[(6)^{3/2} - (2)^{3/2}] = \frac{\sqrt{2}}{6}[3\sqrt{3} - 1]. \quad \square$$

EXAMPLE 3   Find the area of that portion of the sphere $x^2 + y^2 + z^2 = 4$ that lies above the region bounded by $x^2 + y^2 = 1$.

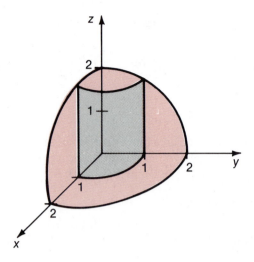

FIGURE 17.31

SOLUTION

The portion of the surface in the first octant is shown in Figure 17.31. This constitutes one fourth of the area, so

$$A = 4 \int_0^1 \int_0^{\sqrt{1-x^2}} \left\{ [f_x(x, y)]^2 + [f_y(x, y)]^2 + 1 \right\}^{1/2} dy \, dx.$$

Since $f(x, y) = z = \sqrt{4 - x^2 - y^2}$,

$$A = 4 \int_0^1 \int_0^{\sqrt{1-x^2}} \left\{ \left(\frac{-x}{\sqrt{4 - x^2 - y^2}} \right)^2 + \left(\frac{-y}{\sqrt{4 - x^2 - y^2}} \right)^2 + 1 \right\}^{1/2} dy \, dx$$

$$= 4 \int_0^1 \int_0^{\sqrt{1-x^2}} \left\{ \frac{x^2 + y^2}{4 - (x^2 + y^2)} + 1 \right\} dy \, dx.$$

This integral calls for a change of variables to cylindrical coordinates:

$$A = 4 \int_0^{\pi/2} \int_0^1 \left\{ \frac{r^2}{4 - r^2} + 1 \right\}^{1/2} r \, dr \, d\theta$$

$$= 4 \int_0^{\pi/2} \int_0^1 \left\{ \frac{4}{4 - r^2} \right\}^{1/2} r \, dr \, d\theta$$

$$= 8 \int_0^{\pi/2} \int_0^1 (4 - r^2)^{-1/2} r \, dr \, d\theta$$

$$= 8 \int_0^{\pi/2} -(4 - r^2)^{1/2} \Big]_0^1 \, d\theta$$

$$= 8(2 - \sqrt{3})\theta \Big]_0^{\pi/2} = 4(2 - \sqrt{3})\pi. \qquad \square$$

Surface integrals are analogously defined for continuous functions over smooth surfaces described using the yz- and xz-planes.

EXAMPLE 4   Find $\displaystyle\iint\limits_{S} xy \sin z \, d\sigma$, where S is the surface of the right circular cylinder $x^2 + y^2 = 4$ in the first octant with $0 \le z \le \pi$.

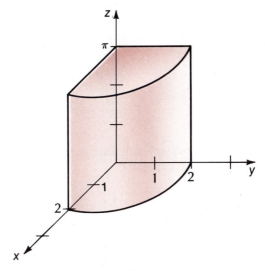

FIGURE 17.32

SOLUTION

This surface, sketched in Figure 17.32, cannot be described by projecting it onto the xy-plane. It can, however, be projected onto the xz-plane. For $0 \le x \le 2$ and $0 \le z \le \pi$, a point (x, y, z) is on the surface S precisely when $y = \sqrt{4 - x^2}$.

Hence,

$$\iint\limits_{S} xy \sin z \, d\sigma = \iint\limits_{R} x\sqrt{4 - x^2} \sin z \left[\left(\frac{\partial(\sqrt{4 - x^2})}{\partial x} \right)^2 + \left(\frac{\partial(\sqrt{4 - x^2})}{\partial z} \right)^2 + 1 \right]^{1/2} dA$$

$$= \int_0^{\pi} \int_0^2 x\sqrt{4 - x^2} \sin z \left[\frac{x^2}{4 - x^2} + 0 + 1 \right]^{1/2} dx \, dz$$

$$= \int_0^{\pi} \int_0^2 2x \sin z \, dx \, dz = \int_0^{\pi} (x^2 \sin z) \Big]_0^2 \, dz$$

$$= \int_0^{\pi} 4 \sin z \, dz = -4 \cos z \Big]_0^{\pi} = 4 - (-4) = 8. \qquad \square$$

Surface integrals occur often in applications in the form

(17.48)
$$\iint\limits_{S} \mathbf{F} \cdot \mathbf{n} \, d\sigma,$$

where $\mathbf{F}(x, y, z) = M(x, y, z)\mathbf{i} + N(x, y, z)\mathbf{j} + P(x, y, z)\mathbf{k}$ and $\mathbf{n} = \mathbf{n}(x, y, z)$ is a unit normal to S at (x, y, z). In many instances a surface integral of this form can be reduced to a double integral. For example, suppose S is a smooth surface described by $z = f(x, y)$ for (x, y) in R, a bounded simply connected region in the xy-plane. Suppose, in addition, that f has continuous partial derivatives on R. Then S can also be described by

$$g(x, y, z) = z - f(x, y) = 0.$$

Consequently, the vector

$$\nabla g(x, y, z) = -f_x(x, y)\mathbf{i} - f_y(x, y)\mathbf{j} + \mathbf{k}$$

is normal to S at (x, y, z) and unit normals to S at (x, y, z) are

(17.49) $\mathbf{n}_1 = \dfrac{-f_x\mathbf{i} - f_y\mathbf{j} + \mathbf{k}}{\sqrt{f_x^2 + f_y^2 + 1}}$ and $\mathbf{n}_2 = \dfrac{f_x\mathbf{i} + f_y\mathbf{j} - \mathbf{k}}{\sqrt{f_x^2 + f_y^2 + 1}}.$

If $\mathbf{n}$ is directed upward, that is, has positive $\mathbf{k}$ component, then $\mathbf{n} = \mathbf{n}_1$ and (17.48) can be expressed:

$$\iint_S \mathbf{F} \cdot \mathbf{n}\, d\sigma = \iint_S (M\mathbf{i} + N\mathbf{j} + P\mathbf{k}) \cdot \left(\dfrac{-f_x\mathbf{i} - f_y\mathbf{j} + \mathbf{k}}{\sqrt{f_x^2 + f_y^2 + 1}}\right) d\sigma$$

$$= \iint_S (-Mf_x - Nf_y + P)\dfrac{1}{\sqrt{f_x^2 + f_y^2 + 1}}\, d\sigma$$

or, using (17.46),

(17.50) $$\iint_S \mathbf{F} \cdot \mathbf{n}\, d\sigma = \iint_R (-Mf_x - Nf_y + P)\, dA.$$

If $\mathbf{n}$ is directed downward, then $\mathbf{n} = \mathbf{n}_2$ and the result is

(17.51) $$\iint_S \mathbf{F} \cdot \mathbf{n}\, d\sigma = \iint_R (Mf_x + Nf_y - P)\, dA.$$

EXAMPLE 5 Determine $\displaystyle\iint_S \mathbf{F} \cdot \mathbf{n}\, d\sigma$ if S is the hemisphere of radius 1 described by

$x^2 + y^2 + z^2 = 1, z \geq 0$, and $\mathbf{F}(x, y, z) = x\mathbf{i} + y\mathbf{j} + z\mathbf{k}$ with $\mathbf{n}$ directed upward.

SOLUTION

The surface S is shown in Figure 17.33. It can be described by

$$z = f(x, y) = \sqrt{1 - x^2 - y^2}.$$

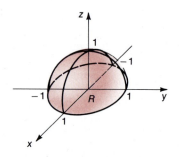

FIGURE 17.33

So by (17.15),

$$\iint_S \mathbf{F} \cdot \mathbf{n} \, d\sigma = \iint_R (-xf_x(x, y) - yf_y(x, y) + z)dA$$

$$= \iint_R \left[-x\left(\frac{-x}{\sqrt{1 - x^2 - y^2}}\right) - y\left(\frac{-y}{\sqrt{1 - x^2 - y^2}}\right) + \sqrt{1 - x^2 - y^2} \right]dA$$

$$= \iint_R \frac{x^2 + y^2 + (1 - x^2 - y^2)}{\sqrt{1 - x^2 - y^2}} \, dA$$

$$= \iint_R \frac{1}{\sqrt{1 - x^2 - y^2}} \, dA.$$

Since R is the region with $x^2 + y^2 \le 1$,

$$\iint_S \mathbf{F} \cdot \mathbf{n} \, d\sigma = \int_{-1}^{1} \int_{-\sqrt{1-x^2}}^{\sqrt{1-x^2}} \frac{1}{\sqrt{1 - x^2 - y^2}} \, dy \, dx.$$

Changing to cylindrical coordinates,

$$\iint_S \mathbf{F} \cdot \mathbf{n} \, d\sigma = \int_0^{2\pi} \int_0^1 \frac{1}{\sqrt{1 - r^2}} r \, dr \, d\theta$$

$$= \int_0^{2\pi} \left. -(1 - r^2)^{1/2} \right|_0^1 \, d\theta$$

$$= \int_0^{2\pi} d\theta = \theta \Big|_0^{2\pi} = 2\pi. \qquad \square$$

EXERCISE SET 17.6

In Exercises 1 through 8, evaluate the surface integral $\iint_S g(x, y, z)d\sigma$.

1. $g(x, y, z) = \dfrac{x}{\sqrt{x^2 + y^2}}$; S is the portion of the paraboloid $z = 4 - x^2 - y^2$ that lies in the first octant.

2. $g(x, y, z) = x^2y^2z^2$; S is described by $z = 3$, $0 \le x \le 2$, $-2 \le y \le 2$.

3. $g(x, y, z) = 8$; S is the portion of the cone $z = \sqrt{x^2 + y^2}$ that lies inside the cylinder $x^2 + y^2 = 1$.

4. $g(x, y, z) = xyz$; S is the portion of the plane $x + y - 2z = 6$ that lies below the triangle in the xy-plane with vertices $(2, 0, 0)$, $(0, 1, 0)$ and $(0, 0, 0)$.

5. $g(x, y, z) = x$; S is the portion above the xy-plane of the cylinder $x^2 + z^2 = 4$, when $0 \le y \le 1$.

6. $g(x, y, z) = xy$; S is described by $y^2 + z^2 = 4$, $0 \le x \le 4$, $z \ge 0$.

7. $g(x, y, z) = x^2 + y^2 + z^2$; S is the hemisphere $z = \sqrt{9 - x^2 - y^2}$.

8. $g(x, y, z) = z(x^2 + y^2)$; S is the cube determined by the planes $x = -1$, $x = 1$, $y = -1$, $y = 1$, $z = -1$, and $z = 1$.

In Exercises 9 through 18, find the surface area of the given surface.

9. The portion of the plane $x + y + z = 3$ that lies in the first octant.

10. The part of the surface $z = x^2$ that lies above the rectangular region in the xy-plane bounded by $x = 0$, $y = 0$, $x = 2$, $y = 3$.

11. The part of the paraboloid $z = x^2 + y^2$ that lies inside the cylinder $x^2 + y^2 = 4$.

12. The portion of the parabolic cylinder $z = y^2$ that lies above the rectangle in the xy-plane with vertices $(0, 1, 0)$, $(2, 1, 0)$, $(2, -1, 0)$, and $(0, -1, 0)$.

13. The portion of the cylinder $y^2 + z^2 = 4$ that lies above the xy-plane between $x = 0$ and $x = 3$.

14. The portion of the plane $x + y + z = 1$ that lies inside the cylinder $x^2 + z^2 = 1$.

15. The portion of the sphere $x^2 + y^2 + z^2 = 6$ that is inside the paraboloid $z = x^2 + y^2$.

16. The parabolic sheet described by $y - x^2 = 1$ for $1 \leq y \leq 2$ and $-1 \leq z \leq 1$.

17. The surface formed by intersecting the cylinders $x^2 + y^2 = 4$ and $x^2 + z^2 = 4$.

18. The portion of the hyperbolic paraboloid $z = y^2 - x^2$ that lies in the first octant inside the cylinder $x^2 + y^2 = 1$.

In Exercises 19 through 22, determine $\iint\limits_{S} \mathbf{F} \cdot \mathbf{n} \, d\sigma$.

19. $\mathbf{F}(x, y, z) = x\mathbf{i} + y\mathbf{j} + z\mathbf{k}$; S is the portion of the surface $z = xy + 1$ that lies above the square $0 \leq x \leq 1$, $0 \leq y \leq 1$ in the xy-plane; $\mathbf{n}$ is directed upward.

20. $\mathbf{F}(x, y, z) = yz\mathbf{i} + xz\mathbf{j} + xy\mathbf{k}$; S is the portion of the paraboloid $z = x^2 + y^2$ that lies inside the cylinder $x^2 + y^2 = 1$; $\mathbf{n}$ is directed downward.

21. $\mathbf{F}(x, y, z) = x^2\mathbf{i} + y^2\mathbf{j} + z\mathbf{k}$; S is the portion of the plane $z = x + y + 1$ that lies above the triangle $0 \leq x \leq 1$, $0 \leq y \leq x$; $\mathbf{n}$ is directed downward.

22. $\mathbf{F}(x, y, z) = x\mathbf{i} - y\mathbf{j} + z^2\mathbf{k}$; S is the sphere $x^2 + y^2 + z^2 = 4$; $\mathbf{n}$ is directed outward from the sphere.

23. When a gas or fluid flows with velocity $\mathbf{v}$ through a surface S, the volume flow and mass flow rates through S are given by $\iint\limits_{S} \mathbf{v} \cdot \mathbf{n} \, d\sigma$ and $\iint\limits_{S} \rho\mathbf{v} \cdot \mathbf{n} \, d\sigma$, respectively, where ρ is the constant density of the gas or fluid and $\mathbf{n}$ is the unit vector in the direction of the flow. Find these rates for a conical shrimp net of radius 1 foot and length 3 feet that is being brought to the surface of the water from a depth of 50 feet at 3 feet per second.

17.7
THE DIVERGENCE THEOREM AND STOKES' THEOREM

Two extensions of Green's theorem are widely used in applications. The divergence theorem of Gauss relates a triple integral over a region in space to a surface integral over the surface of the region. The other extension is Stokes' theorem, which establishes a connection between a surface integral over a surface bounded by a simple closed curve and a line integral over the curve.

The divergence theorem and Stokes' theorem both require the introduction of an orientation on a surface. The idea we wish to express is that a surface is orientable if it has two distinguishable sides, sides that can be painted different colors, say red and white. An orientation on an orientable surface is a means of distinguishing the red side, or alternatively, the white side. (See Figure 17.34(a)). Another way to think of an orientable surface is as a pincushion with the pins as normal vectors. The heads of the pins indicate one orientation of the surface while the points describe the opposite orientation. (See Figure 17.34(b).

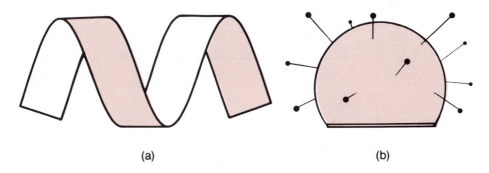

(a)   (b)

FIGURE 17.34

For mathematical purposes we need a nongeometric description. We say that a surface is **orientable** if there is a unit normal vector function

$$\mathbf{n} = \mathbf{n}(x, y, z)$$

that is defined and continuous at every point of S.

Common surfaces such as the surface of a sphere or cube are orientable. For example, the sphere

$$x^2 + y^2 + z^2 = 1$$

is orientable with

$$\mathbf{n} = \mathbf{n}(x, y, z) = x\mathbf{i} + y\mathbf{j} + z\mathbf{k}.$$

These normal vectors point outward from the surface of the sphere, as shown in Figure 17.35. The normal $\mathbf{n}$ is called the outward normal to the surface of the sphere with respect to the solid region enclosed. The opposite orientation is given by $-\mathbf{n}$, which points inward from the surface of the sphere.

One surface that is not orientable is the Möbius strip, a surface named after the German geometer Augustus Ferdinand Möbius (1790–1868). This is a one-

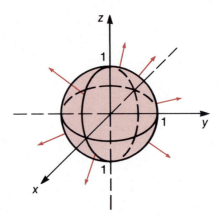

FIGURE 17.35

sided surface that can be constructed by taking a strip of paper, giving one end a twist, and taping the ends of the strip together, as shown in Figure 17.36.

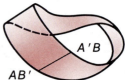

FIGURE 17.36

M. C. Escher's print, Möbius Strip II, shown in Figure 17.37, illustrates the one-sided nature of the surface. Try following the path of an individual ant and you will find that all surfaces of the strip are traversed.

(17.52)
THEOREM

Divergence Theorem   Suppose **F** has component functions with continuous partial derivatives on a solid region D, bounded by a piecewise smooth surface S. If **n** denotes the outward unit normal, that is, the normal to S in the outward direction from D, then

$$\iint_S \mathbf{F} \cdot \mathbf{n}\, d\sigma = \iiint_D \operatorname{div} \mathbf{F}\, dV.$$

PROOF

If we let

$$\mathbf{F}(x, y, z) = M(x, y, z)\mathbf{i} + N(x, y, z)\mathbf{j} + P(x, y, z)\mathbf{k}$$

and write **n** in terms of its direction cosines,

$$\mathbf{n} = \cos \alpha\, \mathbf{i} + \cos \beta\, \mathbf{j} + \cos \gamma\, \mathbf{k},$$

the equation in Theorem 17.52 can be expressed as

(17.53) $$\iint_S (M \cos \alpha + N \cos \beta + P \cos \gamma)\,d\sigma = \iiint_D (M_x + N_y + P_z)\,dV.$$

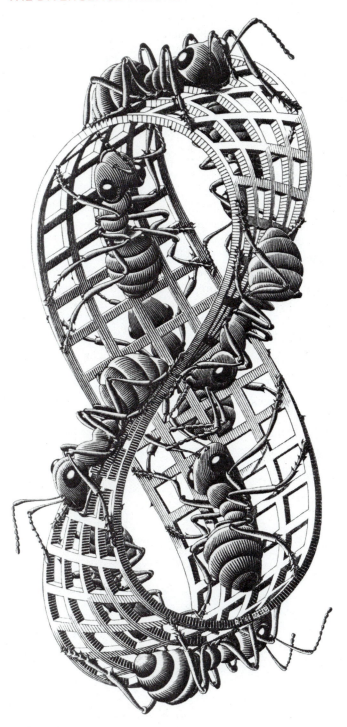

FIGURE 17.37

The proof consists of showing that

$$\iint\limits_S M \cos \alpha \, d\sigma = \iiint\limits_D M_x \, dV, \qquad \iint\limits_S N \cos \beta \, d\sigma = \iiint\limits_D N_y \, dV$$

and

(17.54)
$$\iint\limits_S P \cos \gamma \, d\sigma = \iiint\limits_D P_z \, dV.$$

We will prove (17.54) for regions described by

(17.55)
$$f_1(x, y) \leq z \leq f_2(x, y),$$

for (x, y) in a simply connected region R in the xy-plane with boundary C, and for functions f_1 and f_2 with continuous partial derivatives on R. The proof of the other equalities is given by assuming that D has a similar projection representation in the yz- and xz-planes respectively. Once the result is established for this case it can be extended to a more general solid by decomposing the solid into subregions of this form.

The surface of the solid region described in (17.55) can be decomposed into three smooth surfaces.

the bottom—S_1: $z = f_1(x, y)$, for (x, y) in R

the top—S_2: $z = f_2(x, y)$, for (x, y) in R

the sides—S_3: $f_1(x, y) \leq z \leq f_2(x, y)$ for (x, y) on C

See Figure 17.38.

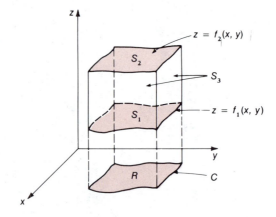

FIGURE 17.38

On S_3 the normal vectors are always parallel to the xy-plane, so $\gamma = \pi/2$, $\cos \gamma = 0$, and

$$\iint\limits_{S_3} P \cos \gamma \, d\sigma = 0.$$

On S_2 the normal vectors are directed upward, so it follows from (17.49) in Section 17.6 that

$$\cos \gamma = \frac{1}{\sqrt{f_x^2 + f_y^2 + 1}}$$

and
$$\iint_{S_2} P \cos \gamma \, d\sigma = \iint_R P \cos \gamma \sqrt{f_x^2 + f_y^2 + 1} \, dA$$

$$= \iint_R P(x, y, f_2(x, y)) dA.$$

On S_1 the normal vectors are directed downward, so it follows from (17.49) that

$$\cos \gamma = \frac{-1}{\sqrt{f_x^2 + f_y^2 + 1}}$$

and
$$\iint_{S_1} P \cos \gamma \, dA = \iint_R -P(x, y, f_1(x, y)) dA.$$

Consequently,

$$\iint_S P \cos \gamma \, d\sigma = \iint_{S_1} P \cos \gamma \, d\sigma + \iint_{S_2} P \cos \gamma \, d\sigma + \iint_{S_3} P \cos \gamma \, d\sigma$$

$$= \iint_R [P(x, y, f_2(x, y) - P(x, y, f_1(x, y))] dA.$$

However,

$$\iiint_D P_z \, dV = \iint_R \left[\int_{z=f_1(x,y)}^{z=f_2(x,y)} P_z \, dz \right] dA$$

$$= \iint_R [P(x, y, f_2(x, y)) - P(x, y, f_1(x, y))] dA,$$

so
$$\iint_S P \cos \gamma \, d\sigma = \iiint_D P_z \, dV. \qquad \square$$

EXAMPLE 1   Use the divergence theorem to evaluate $\iint_S \mathbf{F} \cdot \mathbf{n} \, d\sigma$, where S is the surface of the sphere $x^2 + y^2 + z^2 = 1$, $\mathbf{F}(x, y, z) = x\mathbf{i} + y\mathbf{j} + z\mathbf{k}$, and $\mathbf{n}$ is the outward normal to S.

SOLUTION

By the divergence theorem,

$$\iint_S \mathbf{F} \cdot \mathbf{n} \, d\sigma = \iiint_D \left[\frac{\partial x}{\partial x} + \frac{\partial y}{\partial y} + \frac{\partial z}{\partial z} \right] dV = \iiint_D 3 \, dV$$

$$= \int_{-1}^{1} \int_{-\sqrt{1-x^2}}^{\sqrt{1-x^2}} \int_{\sqrt{1-x^2-y^2}}^{\sqrt{1-x^2-y^2}} 3 \, dz \, dy \, dx.$$

Since the bounding surface is a sphere centered at the origin we change to spherical coordinates for evaluation,

$$\iint_S \mathbf{F} \cdot \mathbf{n}\, d\sigma = \int_0^\pi \int_0^{2\pi} \int_0^1 3\rho^2 \sin\phi\, d\rho\, d\theta\, d\phi = \int_0^\pi \int_0^{2\pi} \rho^3 \sin\phi \Big]_0^1 d\theta\, d\phi$$

$$= \int_0^\pi \theta \sin\phi \Big]_0^{2\pi} d\phi = \Big[-2\pi \cos\phi \Big]_0^\pi = 4\pi.$$

Notice that in this example S is a surface of the type considered in the proof of the theorem with S_3 reducing to a curve, the unit circle in the xy-plane. □

The orientation of an orientable surface bounded by a simple closed curve induces a natural orientation on the curve. The boundary orientation will be *positive* with respect to the surface orientation if the head of a person walking along the boundary of the surface points in the direction of the orientation. This establishes a right-hand rule between the surface orientation and the orientation of its boundary: if the fingers of the right hand are moved in the direction of the positive orientation on the boundary, the extended thumb points in the direction of the orientation of the surface. See Figure 17.39.

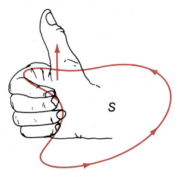

FIGURE 17.39

(17.56)
THEOREM

Stokes' Theorem   Let S be a smooth oriented surface whose boundary C is a smooth simple closed curve with positive orientation with respect to S. Suppose $\mathbf{F}$ has component functions with continuous partial derivatives. Then

$$\int_C \mathbf{F} \cdot d\mathbf{r} = \iint_S \operatorname{curl} \mathbf{F} \cdot \mathbf{n}\, d\sigma,$$

where $\mathbf{n}$ denotes the unit normal to S in the direction of the orientation on S and $\mathbf{r}(t) = x(t)\mathbf{i} + y(t)\mathbf{j} + z(t)\mathbf{k}$, $a \le t \le b$, describes C.

PROOF

Suppose $\mathbf{F}$ is given by

$$\mathbf{F}(x, y, z) = M(x, y, z)\mathbf{i} + N(x, y, z)\mathbf{j} + P(x, y, z)\mathbf{k},$$

then $$\operatorname{curl} \mathbf{F} = (P_y - N_z)\mathbf{i} + (M_z - P_x)\mathbf{j} + (N_x - M_y)\mathbf{k}.$$

If **n** is expressed in terms of its direction cosines,

$$\mathbf{n} = \cos \alpha\, \mathbf{i} + \cos \beta\, \mathbf{j} + \cos \gamma\, \mathbf{k},$$

then the equation in Theorem 17.56 can be rewritten as

(17.57)

$$\int_C M\, dx + N\, dy + P\, dz = \iint_S [(P_y - N_z)\cos \alpha + (M_z - P_x)\cos \beta + (N_x - M_y)\cos \gamma]dA.$$

We will show that

(17.58) $$\int_C M\, dx = \iint_S (M_z \cos \beta - M_y \cos \gamma)dA,$$

when S can be described by $z = f(x, y)$ for (x, y) in a simply connected region R in the xy-plane bounded by the curve $\hat{C}$. ($\hat{C}$ is the projection of C into the xy-plane with orientation inherited from C.)

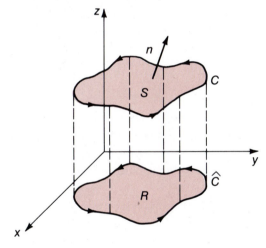

The other equalities

$$\int_C N\, dy = \iint_S (N_x \cos \gamma - N_z \cos \alpha)dA$$

and $$\int_C P\, dz = \iint_S (P_y \cos \alpha - P_x \cos \beta)dA,$$

are proved by assuming that S has a similar projection representation in the yz- and xz-planes. The result can be extended to more general surfaces by decomposing the surface into surfaces of this type.

Suppose the unit normal **n** is directed upward, as shown in Figure 17.40. It follows from (17.49) in Section 17.6 that

$$\cos \beta = \frac{-f_y}{\sqrt{f_x^2 + f_y^2 + 1}} \quad \text{and} \quad \cos \gamma = \frac{1}{\sqrt{f_y^2 + f_y^2 + 1}}.$$

So

$$\iint_S (M_z \cos \beta - M_y \cos \gamma)d\sigma = \iint_S -(M_z f_y + M_y)\frac{1}{\sqrt{f_x^2 + f_y^2 + 1}}\, d\sigma$$

$$= \iint_R -\left[\frac{\partial M}{\partial z}(x, y, z)\frac{\partial z}{\partial y} + \frac{\partial M}{\partial y}(x, y, z)\right]dA.$$

In this last expression M is treated as a function of three variables x, y, and z. If M is considered as a function of two variables x and y with $z = f(x, y)$, then the chain rule implies that

$$\frac{\partial M}{\partial y}(x, y, f(x, y)) = \frac{\partial M}{\partial z}(x, y, z)\frac{\partial z}{\partial y} + \frac{\partial M}{\partial y}(x, y, z).$$

So

(17.59) $$\iint_S (M_z \cos \beta - M_y \cos \alpha)d\sigma = \iint_R -\frac{\partial M}{\partial y}(x, y, f(x, y))dA.$$

Since the orientation of C corresponds to the positive orientation of $\hat{C}$, Green's theorem in the plane can be applied with $N = 0$ to yield

$$\iint_R -\frac{\partial M}{\partial y}(x, y, f(x, y))dA = \int_{\hat{C}} M(x, y, f(x, y))dx = \int_C M(x, y, z)dx.$$

Consequently,

$$\iint_S (M_z \cos \alpha - M_y \cos \gamma)d\sigma = \int_C M(x, y, z)dx.$$

When **n** is directed downward,

$$\cos \beta = \frac{f_y}{\sqrt{f_x^2 + f_y^2 + 1}} \quad \text{and} \quad \cos \gamma = \frac{-1}{\sqrt{f_x^2 + f_y^2 + 1}}.$$

With the same reasoning as that used when **n** is directed upward, it can be shown that:

$$\iint_S (M_z \cos \beta - M_y \cos \gamma)d\sigma = \iint_S (M_z f_y + M_y)\frac{1}{\sqrt{f_x^2 + f_y^2 + 1}}\, d\sigma$$

$$= \iint_R \left[M_z(x, y, z)\frac{\partial z}{\partial y} + M_y(x, y, z)\right]dA$$

$$= \iint_R M_y(x, y, f(x, y))dA$$

$$= -\int_{\hat{C}} M(x, y, f(x, y))dA.$$

However, in this case the positive orientation of $\hat{C}$ corresponds to the negative orientation on C, so

(17.60) $$\int_{\hat{C}} M(x, y, f(x, y))dA = \int_{-C} M(x, y, z)dx = -\int_C M(x, y, z)dx.$$

Once again,

$$\iint_S (M_z \cos \beta - M_y \cos \gamma)d\sigma = \int_C M(x, y, z)dx. \qquad \square$$

If S is a surface in the xy-plane that is oriented by $\mathbf{n} = \mathbf{k}$ and $\mathbf{F}(x, y) = M(x, y)\mathbf{i} + N(x, y)\mathbf{j}$, then Stokes' theorem reduces to Green's theorem:

$$\int_C M\, dx + N\, dy = \iint_S (N_x - M_y)dA.$$

EXAMPLE 2   Verify the result of Stokes' theorem if $\mathbf{F}(x, y, z) = x^2\mathbf{i} + (x + z)\mathbf{j} + yz\mathbf{k}$ and S is that portion of the paraboloid $z = 4 - x^2 - y^2$ that lies above the xy-plane.

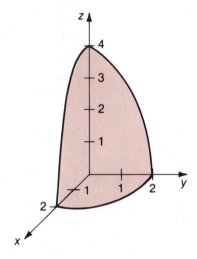

FIGURE 17.41

SOLUTION

The portion of the surface S that lies in the first octant is shown in Figure 17.41 and described by

$$f(x, y) = 4 - x^2 - y^2, \qquad x^2 + y^2 \le 4$$

and

$$\text{curl } \mathbf{F} = \begin{vmatrix} \mathbf{i} & \mathbf{j} & \mathbf{k} \\ \dfrac{\partial}{\partial x} & \dfrac{\partial}{\partial y} & \dfrac{\partial}{\partial z} \\ x^2 & x + z & yz \end{vmatrix}$$

$$= (z - 1)\mathbf{i} + \mathbf{k}$$

$$= (3 - x^2 - y^2)\mathbf{i} + \mathbf{k}.$$

It follows from (17.49) in Section 17.6 that

$$\iint_S \text{curl } \mathbf{F} \cdot \mathbf{n}\, d\sigma = \iint_S [(3 - x^2 - y^2)\mathbf{i} + \mathbf{k}] \cdot \left(\dfrac{-f_x\mathbf{i} - f_y\mathbf{j} + \mathbf{k}}{\sqrt{f_x^2 + f_y^2 + 1}}\right) d\sigma.$$

Using (17.46) to convert to an integral over R,

$$\iint\limits_{S} \text{curl } \mathbf{F} \cdot \mathbf{n} \, d\sigma = \iint\limits_{R} [(3 - x^2 - y^2)\mathbf{i} + \mathbf{k}] \cdot [-f_x \mathbf{i} - f_y \mathbf{j} + \mathbf{k}] dA$$

$$= \iint\limits_{R} [-(3 - x^2 - y^2)f_x + 1] dA$$

$$= \iint\limits_{R} [-(3 - x^2 - y^2)(-2x) + 1] dA$$

$$= \int_0^{2\pi} \int_0^2 [-(3 - r^2)(-2r \cos \theta) + 1] r \, dr \, d\theta$$

$$= \int_0^{2\pi} \int_0^2 [6r^2 \cos \theta - 2r^4 \cos \theta + r] dr \, d\theta$$

$$= \int_0^{2\pi} \left[\left(2r^3 - \frac{2}{5} r^5\right) \cos \theta + \frac{r^2}{2} \right]_0^2 d\theta$$

$$= \int_0^{2\pi} \left[\left(16 - \frac{64}{5}\right) \cos \theta + 2 \right] d\theta$$

$$= \frac{16}{5} \sin \theta + 2\theta \Big]_0^{2\pi} = 4\pi.$$

The boundary of S is a circle C in the xy-plane represented by

$$\mathbf{r}(t) = 2 \cos t \, \mathbf{i} + 2 \sin t \, \mathbf{j}.$$

Since, on C,

$$\mathbf{F}(t) = 4(\cos t)^2 \mathbf{i} + 2 \cos t \, \mathbf{j},$$

$$\int_C \mathbf{F} \cdot d\mathbf{r} = \int_0^{2\pi} [4(\cos t)^2 \mathbf{i} + 2 \cos t \mathbf{j}] \cdot [-2 \sin t \mathbf{i} + 2 \cos t \mathbf{j}] dt$$

$$= \int_0^{2\pi} [-8(\cos t)^2 \sin t + 4(\cos t)^2] dt$$

$$= \frac{8(\cos t)^3}{3} \Big]_0^{2\pi} + 4 \int_0^{2\pi} \frac{1 + \cos 2t}{2} dt$$

$$= \frac{8}{3} - \frac{8}{3} + 2t + \sin 2t \Big]_0^{2\pi} = 4\pi.$$

This shows that

$$\int_C \mathbf{F} \cdot d\mathbf{r} = \iint\limits_{S} \text{curl } \mathbf{F} \cdot \mathbf{n} \, d\sigma. \qquad \square$$

Stokes' theorem can be used to complete the proof to Theorem 17.39.

(17.39)
THEOREM   Suppose $\mathbf{F}(x, y, z) = M(x, y, z)\mathbf{i} + N(x, y, z)\mathbf{j} + P(x, y, z)\mathbf{k}$, where M, N, and P have continuous first partial derivatives in an open simply connected

region D. Then F is conservative on D if and only if

$$\frac{\partial M}{\partial y} = \frac{\partial N}{\partial x}, \qquad \frac{\partial M}{\partial z} = \frac{\partial P}{\partial x}, \qquad \text{and} \qquad \frac{\partial N}{\partial z} = \frac{\partial P}{\partial y}.$$

PROOF

In Section 17.4 we proved that when $\mathbf{F}$ is conservative, the equations in Theorem 17.39 are true. To show the converse, suppose the equations hold. We will show that $\mathbf{F}$ is conservative by applying Corollary 17.36 and showing that $\int_C \mathbf{F} \cdot d\mathbf{r} = 0$ for every simple closed curve C in D.

Suppose C is such a curve and S is any smooth surface whose boundary is C. It follows from Stokes' theorem that

$$\int_C \mathbf{F} \cdot d\mathbf{r} = \iint_S \text{curl } \mathbf{F} \cdot \mathbf{n} \, d\sigma$$

$$= \iint_S [(P_y - N_z)\mathbf{i} + (M_z - P_x)\mathbf{j} + (N_x - M_y)\mathbf{k}] \cdot \mathbf{n} \, d\sigma$$

$$= \iint_S 0 \cdot \mathbf{n} \, d\sigma = 0. \qquad \square$$

EXERCISE SET 17.7

In Exercises 1 through 8, use the divergence theorem to evaluate $\iint_S \mathbf{F} \cdot \mathbf{n} \, d\sigma$, where $\mathbf{n}$ is the outward unit normal to S.

1. $\mathbf{F}(x, y, z) = x\mathbf{i} + y\mathbf{j} + z\mathbf{k}$; S is the cylinder $x^2 + y^2 = 4, 0 \le z \le 4$.

2. $\mathbf{F}(x, y, z) = xz\mathbf{i} + yz\mathbf{j} + xy\mathbf{k}$; S is the cylinder $y^2 + z^2 = 1, 0 \le x \le 2$.

3. $\mathbf{F}(x, y, z) = \mathbf{i} + \mathbf{j} + z\mathbf{k}$; S is the boundary of the region inside the paraboloid $z = x^2 + y^2$ and below the plane $z = 4$.

4. $\mathbf{F}(x, y, z) = x^3\mathbf{i} + y^3\mathbf{j} + z^3\mathbf{k}$; S is the hemisphere $x^2 + y^2 + z^2 = 1$, $z \ge 0$.

5. $\mathbf{F}(x, y, z) = e^x\mathbf{i} - ye^x\mathbf{j} + z^2\mathbf{k}$; S is the boundary of the region in the intersection of the cylinders $x^2 + y^2 = 4$, $y^2 + z^2 = 4$.

6. $\mathbf{F}(x, y, z) = x \sin y\mathbf{i} + e^y\mathbf{j} + z \cos y \, \mathbf{k}$; S is the boundary of the cube described by $0 \le x \le 2, 0 \le y \le 2, 0 \le z \le 2$.

7. $\mathbf{F}(x, y, z) = (x + e^y)\mathbf{i} + (xy + \sin z)\mathbf{j} - (yz + \sqrt{xy})\mathbf{k}$; S is the boundary of the tetrahedron bounded by the planes $x = 0, y = 0, z = 0$, and $2x + 3y + z = 6$.

8. $\mathbf{F}(x, y, z) = yz\mathbf{i} + xy\mathbf{j} + (1 - z - xz)\mathbf{k}$; S is the boundary of the region inside the paraboloid $z = 1 - x^2 - y^2$ and above the xy-plane.

In Exercises 9 through 16, use Stokes' theorem to evaluate $\int_C \mathbf{F} \cdot d\mathbf{r}$, where C has a positive orientation with respect to the surface of which it is the boundary.

9. $\mathbf{F}(x, y, z) = 2y\mathbf{i} + z\mathbf{j} + 3y\mathbf{k}$; C is the intersection of the sphere $x^2 + y^2 + z^2 = 1$ and the plane $z = 0$.

10. $\mathbf{F}(x, y, z) = -y\mathbf{i} + x\mathbf{j} + z\mathbf{k}$; C is the circle $x^2 + y^2 = 1$ in the plane $z = 1$.

11. $\mathbf{F}(x, y, z) = x^2\mathbf{i} + y^2\mathbf{j} + z^2\mathbf{k}$; C is the intersection of the sphere $x^2 + y^2 + z^2 = 1$ and the parabolic cylinder $z = y^2$.

12. $\mathbf{F}(x, y, z) = y^2\mathbf{i} + x^2\mathbf{j} + x\mathbf{k}$; C is the boundary of the portion of the plane $2x + y + 3z = 6$ in the first octant.

13. $\mathbf{F}(x, y, z) = x\mathbf{i} + y^2\mathbf{j} + z\mathbf{k}$; S is the portion of the paraboloid $z = 4 - x^2 - y^2$ above the xy-plane; $\mathbf{n}$ is directed upward.

14. $\mathbf{F}(x, y, z) = zy\mathbf{i} + xz\mathbf{j} + yx\mathbf{k}$; S is the portion of the plane $x + y + z = 1$ in the first octant; $\mathbf{n}$ is directed upward.

15. $\mathbf{F}(x, y, z) = e^z\mathbf{i} + \sin x\mathbf{j} + \sin y\mathbf{k}$; S is the surface whose boundary is the triangle with vertices $(3, 0, 0)$, $(0, 2, 0)$, $(0, 0, 5)$; $\mathbf{n}$ is directed downward.

16. $\mathbf{F}(x, y, z) = z\mathbf{i} + x^2\mathbf{j} - e^x \ln x\mathbf{k}$; S is the part of the paraboloid $z = 4 - x^2 - y^2$ that lies inside the cylinder $x^2 + y^2 = 1$; $\mathbf{n}$ is directed downward.

17. Let D be a solid with unit density and S be the surface of D with outward normal $\mathbf{n}$. Show that

$$\iint_S x^2\mathbf{i} \cdot \mathbf{n} \, d\sigma = 2 M_{yz},$$

where M_{yz} is the moment with respect to the yz-plane. Express M_{xy} and M_{xz} as surface integrals.

18. Let D be a solid in space and S the boundary of D. Show that the volume of D is given by the surface integral

$$\iint_S \frac{1}{3}(x\mathbf{i} + y\mathbf{j} + z\mathbf{k}) \cdot \mathbf{n} \, d\sigma.$$

19. If f and g are twice continuously differentiable on a solid D with boundary S, show that

(i) $\displaystyle\iint_S f\nabla g \cdot \mathbf{n} \, d\sigma = \iiint_D [f\nabla^2 g + \nabla f \cdot \nabla g] dV$

(ii) $\displaystyle\iint_S (f\nabla g - g\nabla f) \cdot \mathbf{n} \, d\sigma = \iiint_D (f\nabla^2 g - g\nabla^2 f) dV.$

20. Let S be a piecewise smooth surface that bounds a region D in space. Show that if a function f of three variables has continuous partial derivatives on D, then

$$\iiint_D \left(\frac{\partial^2 f}{\partial x^2} + \frac{\partial^2 f}{\partial y^2} + \frac{\partial^2 f}{\partial z^2}\right) dV = \iint_S D_\mathbf{n} f \, d\sigma,$$

where $\mathbf{n}$ denotes the outward unit normal to S and $D_\mathbf{n} f$ is the directional derivative of f. Conclude that if f is harmonic throughout D, then

$$\iint_S D_\mathbf{n} f \, d\sigma = 0. \quad (\textit{Hint}: \text{Consider } \mathbf{F} = \nabla f.)$$

21. Let S be a piecewise smooth surface that bounds a region D in space. Show that if a function f of three variables is harmonic throughout D, then

$$\iiint_D \left[\left(\frac{\partial f}{\partial x} \right)^2 + \left(\frac{\partial f}{\partial y} \right)^2 + \left(\frac{\partial f}{\partial z} \right)^2 \right] dV = \iint_S f D_{\mathbf{n}} f \, d\sigma,$$

where $\mathbf{n}$ is the outward unit normal of S and $D_{\mathbf{n}} f$ is the directional derivative of f. Conclude that if f is zero on S, then f must be zero in D.

22. Use Exercise 21 to show that if two harmonic functions agree on the boundary of a solid, then they must agree at each point within the solid.

23. What must be true about f if $\nabla^2 (f^2) = 2 f \nabla^2 f$?

17.8
APPLICATIONS OF THE DIVERGENCE AND STOKES' THEOREMS

FLUID FLOW

Suppose a fluid with velocity described by the vector field $\mathbf{v}(x, y, z)$ flows through a surface S in space. (See Figure 17.42.) The rate at which the fluid flows through S, called the **flux** through S, is found by integrating over S the function described by the component of $\mathbf{v}$ normal to the surface S:

$$\text{Flux through } S = \iint_S \mathbf{v} \cdot \mathbf{n} \, d\sigma.$$

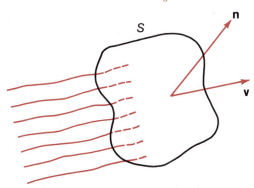

FIGURE 17.42

If S is a closed surface bounding a region D in space, and $\mathbf{n} \equiv \mathbf{n}(x, y, z)$ denotes the outward unit normal to D, then

(17.61)
$$\iint_S \mathbf{v} \cdot \mathbf{n} \, d\sigma$$

gives the net rate of fluid flowing from D. When this integral is positive, fluid is being produced in D and the region D is called a **source** for the fluid. When this integral is negative, fluid is being absorbed in D and D is called a **sink** for the fluid. The divergence theorem can be used to determine precisely which points within D are responsible for this fluid production or absorption. The theorem implies that the net rate of fluid flowing from D is

$$\iint\limits_{S} \mathbf{v} \cdot \mathbf{n} \, d\sigma = \iiint\limits_{D} \operatorname{div} \mathbf{v} \, dV.$$

If V represents the volume of D, then the rate of fluid flowing from D per unit of volume is

$$\frac{1}{V} \iint\limits_{S} \mathbf{v} \cdot \mathbf{n} \, d\sigma = \frac{1}{V} \iiint\limits_{D} \operatorname{div} \mathbf{v} \, dV.$$

An extension of the mean value theorem for integrals can be used to show that for some $(\hat{x}, \hat{y}, \hat{z})$ in D

$$\frac{1}{V} \iiint\limits_{D} \operatorname{div} \mathbf{v} \, dV = \operatorname{div} \mathbf{v}(\hat{x}, \hat{y}, \hat{z}),$$

so

(17.62) $$\frac{1}{V} \iint\limits_{S} \mathbf{v} \cdot \mathbf{n} \, d\sigma = \operatorname{div} \mathbf{v}(\hat{x}, \hat{y}, \hat{z}).$$

Suppose we apply this reasoning when the surface is a sphere S_r of radius r about a fixed point P. The volume bounded by S_r is $4\pi r^3/3$, so (17.62) implies that the rate of flow from inside S_r per unit volume is

(17.63) $$\frac{3}{4\pi r^3} \iint\limits_{S_r} \mathbf{v} \cdot \mathbf{n} \, d\sigma = \operatorname{div} \mathbf{v}(\hat{x}, \hat{y}, \hat{z})$$

for some $(\hat{x}, \hat{y}, \hat{z})$ within S_r. As the radius r approaches zero, the sphere S_r shrinks to the point P. So (17.63) implies that

(17.64) $$\operatorname{div} \mathbf{v}(P) = \lim_{r \to 0} \frac{3}{4\pi r^3} \iint\limits_{S_r} \mathbf{v} \cdot \mathbf{n} \, d\sigma$$

represents the rate of flow per unit volume at the point P. A source occurs at P when $\operatorname{div} \mathbf{v}(P) > 0$; a sink occurs at P when $\operatorname{div} \mathbf{v}(P) < 0$. No fluid is produced or absorbed at P when $\operatorname{div} \mathbf{v}(P) = 0$.

If the fluid has constant density, called an **incompressible** fluid, then $\operatorname{div} \mathbf{v}(P) \equiv 0$ on a region D precisely when there are no sinks or sources within D. Consequently, the only vector fields $\mathbf{v}$ that can be used to represent the flow of a fluid through a region without sinks or sources must satisfy the equation

(17.65) $\operatorname{div} \mathbf{v} \equiv 0.$

This is called the *equation of continuity for fluid flow.* A vector field $\mathbf{v}$ satisfying (17.65) is called **solenoidal.** These concepts and terms are also used in the study of the flow of electrical current.

Stokes' theorem also has an important application to fluid flow. Suppose C is a smooth closed curve lying in a region through which a fluid flows with velocity described by $\mathbf{v}(x, y, z)$. A measure of the tendency of the fluid to propel a particle located at P on C in the positively oriented direction of C is given by $\mathbf{v}(P) \cdot \mathbf{T}(P)$, where $\mathbf{T}(P)$ is the unit tangent vector to C at P. See Figure 17.43.

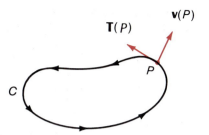

FIGURE 17.43

The **circulation** of the fluid around C is defined by:

$$\text{circulation around } C = \int_C \mathbf{v} \cdot \mathbf{T} \, ds,$$

and measures the tendency of the fluid to propel a particle around C. If C has the parametric representation described by $\mathbf{r}(t)$, the circulation around C can be expressed

$$\int_C \mathbf{v} \cdot \mathbf{T} \, ds = \int_C \mathbf{v} \cdot d\mathbf{r}.$$

Stokes' theorem implies that the circulation around C is

(17.66) $$\int_C \mathbf{v} \cdot \mathbf{T} \, ds = \iint_S \text{curl } \mathbf{v} \cdot \mathbf{n} \, d\sigma,$$

where S is any smooth surface bounded by C and $\mathbf{n}$ is the positively oriented (by C) unit normal to S. If A denotes the area of S, then an extension of the mean value theorem for integrals can be used to show that for some $(\hat{x}, \hat{y}, \hat{z})$ on S

$$\frac{1}{A} \iint_S \text{curl } \mathbf{v} \cdot \mathbf{n} \, d\sigma = \text{curl } \mathbf{v}(\hat{x}, \hat{y}, \hat{z}) \cdot \mathbf{n}(\hat{x}, \hat{y}, \hat{z}),$$

so

$$\frac{1}{A} \int_C \mathbf{v} \cdot \mathbf{T} \, ds = \text{curl } \mathbf{v}(\hat{x}, \hat{y}, \hat{z}) \cdot \mathbf{n}(\hat{x}, \hat{y}, \hat{z}).$$

Suppose P is a fixed point in the region of fluid flow and that for each $r > 0$, C_r represents a circle of radius r centered at P. Then C_r lies in a plane that contains P. Let $\mathbf{n}$ be the normal vector to this plane. Since $A = \pi r^2$,

$$\frac{1}{\pi r^2} \int_{C_r} \mathbf{v} \cdot \mathbf{T} \, ds = \text{curl } \mathbf{v}(\hat{x}, \hat{y}, \hat{z}) \cdot \mathbf{n}(\hat{x}, \hat{y}, \hat{z})$$

for some $(\hat{x}, \hat{y}, \hat{z})$ inside C_r. As r approaches zero, C_r shrinks to the point P, so

(17.67) $$\text{curl } \mathbf{v}(P) \cdot \mathbf{n} = \lim_{r \to 0} \frac{1}{\pi r^2} \int_{C_r} \mathbf{v} \cdot \mathbf{T} \, ds.$$

Equation (17.67) implies that curl $\mathbf{v}(P) \cdot \mathbf{n}$ can be regarded as the circulation per unit area of the flow of the fluid at P in the direction of the vector $\mathbf{n}$. Consequently, curl $\mathbf{v}(P) \cdot \mathbf{n}$ gives a measure for the tendency of a particle placed at P to rotate in the plane perpendicular to $\mathbf{n}$.

The terms divergence and curl have their origin in these applications to fluid flow. The divergence of $\mathbf{v}$ at P measures the amount of change, or divergence, of a fluid at P. The curl of $\mathbf{v}$ at P measures the rotational tendency, or curl, of the fluid at P. The relations expressed in equations (17.64) and (17.67) are often used in advanced textbooks to define the divergence and curl of a vector field. Defined in this way, the divergence and curl are coordinate independent, an advantage when more than one coordinate system is used. Fixing a coordinate system reduces these definitions to those presented in Section 17.1.

HEAT FLOW

One of the basic principles in thermodynamics, the study of heat flow, is that heat flows from an object in the direction of decreasing temperature at a rate proportional to the difference in temperature between the object and its surrounding medium. This is known as **Newton's law of cooling.** Applications of this principle were considered in Section 7.5.

Suppose $T(x, y, z, t)$ represents the temperature of an object at (x, y, z) at time t. The change in temperature is described by grad $T(x, y, z, t)$, so it follows from Newton's law of cooling that the vector field describing the heat flow at (x, y, z) at time t is given by

$$\mathbf{v}(x, y, z, t) = -k \text{ grad } T(x, y, z, t).$$

The positive constant k is called the **thermal conductivity** of the object.

If D is a region in space with a smooth boundary S, then the total amount of heat leaving D through S per unit time is given by

(17.68) $$\iint_S \mathbf{v} \cdot \mathbf{n} \, d\sigma = \iint_S -k \text{ grad } T(x, y, z, t) \cdot \mathbf{n} \, d\sigma.$$

The vector $\mathbf{n}$ is the outward normal to S, as shown in Figure 17.44.

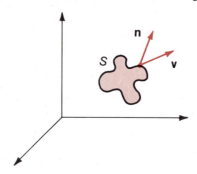

FIGURE 17.44

Another expression for the amount of heat leaving D per unit time can be derived by considering the total amount of heat within D. If the material in D has a constant density ρ and a specific heat σ, then the total heat H in D is

$$H = \iiint_D \rho\sigma \, T(x, y, z, t) \, dV.$$

Consequently, the total amount of heat leaving D per unit time is the negative of the change in H with respect to time

(17.69) $-\dfrac{\partial H}{\partial t} = -\dfrac{\partial}{\partial t}\left[\iiint\limits_{D} \rho\sigma\, T(x, y, z, t)\right]dV = -\iiint\limits_{D} \rho\sigma\, T_t(x, y, z, t)\, dV.$

Equating the expressions (17.68) and (17.69) we have

$$-\iiint\limits_{D} \rho\sigma\, T_t(x, y, z, t)\, dV = \iint\limits_{S} -k\, \mathrm{grad}\, T(x, y, z, t) \cdot \mathbf{n}\, d\sigma.$$

Applying the divergence theorem to the right side of this equation implies that

$$\iiint\limits_{D} \rho\sigma\, T_t(x, y, z, t)\, dV = \iiint\limits_{D} k\, \mathrm{div}\, (\mathrm{grad}\, T(x, y, z, t))\, dV$$

$$= \iiint\limits_{D} k[T_{xx}(x, y, z, t) + T_{yy}(x, y, z, t) + T_{zz}(x, y, z, t)]\, dV$$

so

$$\iiint\limits_{D} [\rho\sigma\, T_t - k(T_{xx} + T_{yy} + T_{zz})]\, dV = 0.$$

Since this equation holds for all regions D, the integrand must be zero. Thus, any function T that describes the temperature within a solid of constant density ρ and specific heat σ must satisfy the equation

(17.70) $\rho\sigma\, T_t = k(T_{xx} + T_{yy} + T_{zz}).$

This equation is known as the **heat equation.**

Heat conduction on a plate is governed by a similar equation

$$\rho\sigma\, T_t = k(T_{xx} + T_{yy}),$$

which is produced from (17.70) by assuming no dependence on the third dimension z. In a similar manner, one-dimensional heat flow in a rod is described by an equation of the form

$$\rho\sigma\, T_t = kT_{xx}.$$

In solutions to physical problems involving the heat equations, the temperature function T must satisfy the appropriate heat equation as well as some specifications regarding the temperature on the boundary of the region. Problems of this type are known as *boundary-value* problems.

MAXWELL'S EQUATIONS

James Clark Maxwell (1831–1879) assembled and refined the laws that detail the connection between electric current and magnetism, a science known as *electromagnetism*. The four equations defining these laws are collectively known as Maxwell's equations, although with each law is associated the name of the person first describing it.

In 1831 Michael Faraday (1791–1867) discovered that an electric current can be produced by varying the magnetic flux through a conducting loop. His experiments showed that the electromotive force of the field is described by

(17.71) $\displaystyle\int_C \mathbf{E} \cdot d\mathbf{r} = -\iint\limits_{S} \dfrac{\partial \mathbf{B}}{\partial t} \cdot \mathbf{n}\, d\sigma,$

where C is the curve described by the loop, S is any surface with boundary C, and $\mathbf{E}$ represents the electric field produced by the magnetic induction $\mathbf{B}$. By applying Stokes' theorem to the left side of equation (17.71), Faraday's law can be transformed into a differential expression. Stokes' theorem implies that

$$\int_C \mathbf{E} \cdot d\mathbf{r} = \iint_S \operatorname{curl} \mathbf{E} \cdot \mathbf{n} \, d\sigma,$$

so

$$\iint_S \operatorname{curl} \mathbf{E} \cdot \mathbf{n} \, d\sigma = \iint_S -\frac{\partial \mathbf{B}}{\partial t} \cdot \mathbf{n} \, d\sigma.$$

Since S is an arbitrary surface bounded by C, the integrands must agree and consequently

(17.72)
$$\operatorname{curl} \mathbf{E} = -\frac{\partial \mathbf{B}}{\partial t}.$$

Maxwell's correction of a law attributed to Andre Marie Ampére (1775–1836) describes empirically how a magnetic field is associated with a time-varying electric current. This connection is expressed by

(17.73)
$$\int_C \mathbf{B} \cdot d\mathbf{r} = \iint_S \mu_0 \left(\mathbf{J} + \varepsilon_0 \frac{\partial \mathbf{E}}{\partial t} \right) \cdot \mathbf{n} \, d\sigma,$$

where $\mathbf{J}$ is the current per unit cross-sectional area, $\varepsilon_0 \, \partial \mathbf{E}/\partial t$ is the displacement current correction needed to ensure the conservation of charge, μ_0 is the magnetic force scale factor, and S is any surface bounded by C, the closed curve describing the path of the electric current. The constant ε_0 is called the *permittivity constant:* it is dependent only on the physical units involved and scales the electrical units so that the forces produced are represented in correct mechanical units.

Applying Stokes' theorem to the left side of equation (17.73) gives

$$\int_C \mathbf{B} \cdot d\mathbf{r} = \iint_S \operatorname{curl} \mathbf{B} \cdot \mathbf{n} \, d\sigma,$$

so

$$\iint_S \mu_0 \left(\mathbf{J} + \varepsilon_0 \frac{\partial \mathbf{E}}{\partial t} \right) \cdot \mathbf{n} \, d\sigma = \iint_S \operatorname{curl} \mathbf{B} \cdot \mathbf{n} \, d\sigma.$$

Since S is an arbitrary surface bounded by C, the integrands agree and the differential expression of Ampéres law is

(17.74)
$$\operatorname{curl} \mathbf{B} = \mu_0 \left(\mathbf{J} + \varepsilon_0 \frac{\partial \mathbf{E}}{\partial t} \right).$$

The other two Maxwell's equations are named for the outstanding mathematician Carl Friedrich Gauss (1777–1855). One law states, in essence, that no magnetic field is produced by a single magnetic pole. This can be expressed mathematically by declaring that

(17.75)
$$\iint_S \mathbf{B} \cdot \mathbf{n} \, d\sigma = 0$$

for all magnetic fields $\mathbf{B}$ and closed surfaces S (there is no magnetic flux through a closed surface). Applying the divergence theorem to this equation, we have

$$0 = \iint_S \mathbf{B} \cdot \mathbf{n}\, d\sigma = \iiint_D \text{div } \mathbf{B}\, dV,$$

where D is the solid bounded by S. Since S and, hence D, are arbitrary, this implies that the differential expression of Gauss' law for magnetism is

(17.76)   div $\mathbf{B} = 0$.

Gauss' law for electricity states that the flux through a closed surface is the net charge enclosed by the surface. If q denotes the total charge enclosed by the surface S, the relationship is expressed by

(17.77) $\iint_S \mathbf{E} \cdot \mathbf{n}\, d\sigma = \dfrac{q}{\varepsilon_0}$,

where ε_0 is the permittivity constant. Applying the divergence theorem to the left side of equation (17.77) implies that

$$\iint_S \mathbf{E} \cdot \mathbf{n}\, d\sigma = \iiint_D \text{div } \mathbf{E}\, dV,$$

where D is the solid bounded by S. The right side of equation (17.77) can also be expressed as a triple integral over D by defining a function ρ to represent the charge per unit volume within D. Then

$$\frac{q}{\varepsilon_0} = \iiint_D \frac{\rho}{\varepsilon_0}\, dV$$

and $$\iiint_D \text{div } \mathbf{E}\, dV = \iiint_D \frac{\rho}{\varepsilon_0}\, dV.$$

Since D is arbitrary, the differential expression of Gauss' law for electricity is

(17.78)   div $\mathbf{E} = \dfrac{\rho}{\varepsilon_0}$.

Collectively, equations (17.71), (17.73), (17.75), and (17.77) are known as the integral form of Maxwell's equations and (17.72), (17.74), (17.76), and (17.78) are known as the differential form. These equations form the backbone of the classical theory of electromagnetism.

EXERCISE SET 17.8

1. Let S be the surface of the sphere $x^2 + y^2 + z^2 = 1$. Assume that the velocity of a fluid flowing through S is given by

$$\mathbf{v}(x, y, z) = x\mathbf{i} + y\mathbf{j} + z\mathbf{k}.$$

(a) Find div $\mathbf{v}(x, y, z)$.

(b) Find $\dfrac{3}{4\pi} \iint_S \mathbf{v} \cdot \mathbf{n}\, d\sigma$, where $\mathbf{n}$ is the outward unit normal to S, and

compare the result with the answer obtained in (a).

2. Let S be the boundary of the region D described by $x^2 + y^2 \leq z \leq 4$. Assume that the velocity of a fluid flowing through S is given by

$$\mathbf{v}(x, y, z) = x^3\mathbf{i} + y^3\mathbf{j} + z^2\mathbf{k}.$$

(a) Find div $\mathbf{v}(x, y, z)$.   (b) Find the volume V of D.

(c) Find the flux, $\displaystyle\iint_S \mathbf{v} \cdot \mathbf{n}\, d\sigma$, through S.

(d) Verify that for some point $(\hat{x}, \hat{y}, \hat{z})$ in D,

$$\frac{1}{V}\iint_S \mathbf{v} \cdot \mathbf{n}\, d\sigma = \text{div } \mathbf{v}(\hat{x}, \hat{y}, \hat{z}).$$

3. Suppose the velocity of a fluid is described by

$$\mathbf{v}(x, y, z) = x^2\mathbf{i} + y^2\mathbf{j} + (z^2 - z)\mathbf{k}.$$

Determine at which points sources occur and at which points sinks occur.

4. Show that if the velocity field $\mathbf{v}(x, y, z) = -x^3\mathbf{i} - y^3\mathbf{j} + 3z\mathbf{k}$ describes the velocity of a fluid, then a source occurs at every point inside the cylinder $x^2 + y^2 = 1$.

5. A fluid of constant density rotates about the z-axis with velocity $\mathbf{v}(x, y, z) = y\mathbf{i} - x\mathbf{j}$.

(a) Find curl $\mathbf{v}$.

(b) Find $\displaystyle\int_{C_r} \mathbf{v} \cdot \mathbf{T}\, ds$, where C_r is a circle of radius r in a plane parallel to the xy-plane.

(c) Show that $\mathbf{v}$ satisfies equation (17.66).

6. If $\mathbf{v}$ is a vector function with curl $\mathbf{v} \equiv \mathbf{0}$, then $\mathbf{v}$ is called *irrotational*. Show that any irrotational function is conservative.

7. A fluid of constant density flows through a region about the origin with velocity given by

$$\mathbf{v}(x, y, z) = x\mathbf{i} + y\mathbf{j} + z\mathbf{k}.$$

Show that $\mathbf{v}$ is irrotational.

8. A fluid of constant density flows through a region D with velocity given by

$$\mathbf{v}(x, y, z) = yz\mathbf{i} + xz\mathbf{j} + xy\mathbf{k}.$$

Show that $\mathbf{v}$ is solenoidal.

9. Suppose $\mathbf{F}(x, y, z) = M(x, y, z)\mathbf{i} + N(x, y, z)\mathbf{j} + P(x, y, z)\mathbf{k}$. Show that if the second partial derivatives of M, N, and P are continuous in a region D, then curl $\mathbf{F}$ is solenoidal.

10. Suppose f is a function of three variables with continuous second partial derivatives. Show that the gradient of f is irrotational.

11. Suppose that a cup of coffee is stirred with a velocity that can be approximated by

$$\mathbf{v}(x, y, z) = -\omega y\mathbf{i} + \omega x\mathbf{j},$$

where ω is the constant angular speed of the rotation. Show that $\mathbf{v}$ is not irrotational.

12. Suppose $T(x, y, z, t)$ represents the temperature at (x, y, z) in D at time t, where D is a region in space with a smooth boundary S. Show that if T is harmonic in D, then the temperature is time independent.

REVIEW EXERCISES

Find the divergence and curl of the vector-valued functions described in Exercises 1 through 4.

1. $\mathbf{F}(x, y, z) = y \sin z\mathbf{i} + x \sin z\mathbf{j} + xy \cos z\mathbf{k}$

2. $\mathbf{F}(x, y, z) = yz\mathbf{i} + xz\mathbf{j} + xy\mathbf{k}$

3. $\mathbf{F}(x, y, z) = x^3 y\mathbf{i} + (x + z)\mathbf{j} + xyz\mathbf{k}$

4. $\mathbf{F}(x, y, z) = \dfrac{x}{y^2 + z^2}\mathbf{i} + \dfrac{y}{x^2 + z^2}\mathbf{j} + \dfrac{z}{x^2 + y^2}\mathbf{k}$

Show that the functions described in Exercises 5 through 8 are harmonic.

5. $f(x, y) = x^3 - 3xy^2 + y$ **6.** $f(x, y) = e^x \cos y$

7. $f(x, y) = xy - x^2 + y^2$ **8.** $f(x, y) = \arctan \dfrac{y}{x}$

9. Which of the curves described below are closed curves? Which are simple closed curves?

(a) $\mathbf{r}(t) = 2 \cos t\mathbf{i} + 3 \sin t\mathbf{j} + \mathbf{k}, 0 \le t \le 2\pi$

(b) $\mathbf{r}(t) = \cos t^2\mathbf{i} + \sin t^2\mathbf{j}, 0 \le t \le 2\pi$

10. Which of the following shaded two-dimensional regions are simply connected?

(a) 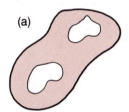   (b) 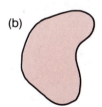   (c)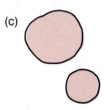

Evaluate the integrals in Exercises 11 through 22.

11. $\displaystyle\int_C \dfrac{-y}{x^2 + y^2}\,dx + \dfrac{x}{x^2 + y^2}\,dy$, where C is the unit circle $x^2 + y^2 = 1$ oriented in the counterclockwise direction.

12. $\displaystyle\int_C x^2 y\,dy$, where C is the straight line segment joining $(1, 1, 0)$ and $(1, 4, 2)$.

13. $\displaystyle\int_C xy\,dx + (x - z)\,dy + yz\,dz$, where C is described by

$$\mathbf{r}(t) = t\mathbf{i} + t^2\mathbf{j} + t^3\mathbf{k}, \qquad 0 \le t \le 1.$$

14. $\int_C x\,dx - z\,dy + 2y\,dz$, where C is the curve $z = y^4$, $x = 1$ from $(1, 0, 0)$ to $(1, 1, 1)$.

15. $\int_C (3x^2 + y)dx + 4y^2 dy$, where C is the curve $y = x^3$, $z = 2$ from $(1, 1, 2)$ to $(2, 8, 2)$.

16. $\int_C \dfrac{1}{1 - x^2}\,dx + \dfrac{1}{xy}\,dy$, where C is the straight line segment in the xy-plane from $(2, 1)$ to $(4, 3)$.

17. $\int_C e^x\,dx + xy\,dy + \ln xy\,dz$, where C is described by

$$\mathbf{r}(t) = t\mathbf{i} + e^t\mathbf{j} + t^2\mathbf{k}, \qquad 1 \le t \le 3.$$

18. $\int_C e^x\,dx + e^y\,dy + e^z\,dz$, where C is described by

$$\mathbf{r}(t) = \left(1 + \frac{t}{2}\right)\mathbf{i} + \frac{3t^2}{4}\mathbf{j} + 5\sin\frac{\pi}{4}t\mathbf{k}, \qquad 0 \le t \le 2.$$

19. $\int_C e^x \sin y\,dx + e^x \cos y\,dy$, C is the closed curve determined by $y = x$, $y = -x$ and $y = 2$.

20. $\int_C y^3\,dx - x^3\,dy$, C is the boundary of the region enclosed between the two circles with radii 1 and 2 and center at the origin.

21. $\int_C \mathbf{F} \cdot d\mathbf{r}$, where $\mathbf{F}(x, y) = \dfrac{y^2}{1 + x^2}\mathbf{i} + 2y \arctan x\,\mathbf{j}$ and C is the ellipse $x^2 + 4y^2 = 1$.

22. $\int_C \mathbf{F} \cdot d\mathbf{r}$, where $\mathbf{F}(x, y) = xy\mathbf{i} + (x^2 + y^2)\mathbf{j}$ and C is the closed curve determined by $y = 6 - x^2$ and $y = 3 - 2x$.

Show that the line integrals in Exercises 23 through 25 are independent of path, and evaluate them.

23. $\int_C yz\,dx + xz\,dy + xy\,dz$, where C is the helix

$$\mathbf{r}(t) = \cos t\mathbf{i} + \sin t\mathbf{j} + t\mathbf{k}, \qquad 0 \le t \le 2\pi.$$

24. $\int_C 2xy\,dx + x^2\,dy$, where C is the segment of the curve $y = x^2 - 1$ from $(-1, 0)$ to $(2, 3)$.

25. $\int_C (\sec x)^2\,dx + z\,dy + (y + 2z)dz$, where C is the elliptical helix

$$\mathbf{r}(t) = 2\cos t\mathbf{i} + 3\sin t\mathbf{j} + t\mathbf{k}, \; 0 \le t \le \frac{3\pi}{2}.$$

In Exercises 26 and 27, find the work done by the force $\mathbf{F}$ in moving an object along the given curve.

26. $\mathbf{F}(x, y, z) = (y + z)\mathbf{i} + (x + z)\mathbf{j} + (x + y)\mathbf{k}$ along the curve described by $\mathbf{r}(t) = t\mathbf{i} + t^2\mathbf{j} + t^3\mathbf{k}$ from $(0, 0, 0)$ to $(1, 1, 1)$.

27. $\mathbf{F}(x, y, z) = xy\mathbf{i} + x^2\mathbf{j}$ counterclockwise along the unit circle in the xy-plane beginning and ending at $(1, 0, 0)$.

Evaluate the integrals in Exercises 28 through 32.

28. $\displaystyle\iint_S (x^2 + y^2)d\sigma$; S is the portion of the plane $x + 2y + 3z = 6$ in the first octant.

29. $\displaystyle\iint_S (x + y)d\sigma$; S is the portion of the plane $x + y + z = 1$ that lies inside the cylinder $x^2 + y^2 = 9$.

30. $\displaystyle\iint_S (x^2 + y^2)z\,d\sigma$; S is the hemisphere $z = \sqrt{1 - (x^2 + y^2)}$.

31. $\displaystyle\iint_S y^2\,d\sigma$; S is the portion of the cylinder $x^2 + y^2 = 1$ in the first octant bounded above by $z = 1$ and below by $z = 0$.

32. $\displaystyle\iint_S (x - y)d\sigma$; S is the portion of the plane $x + y + z = 1$ in the first octant.

In Exercises 33 through 37, determine $\displaystyle\iint_S \mathbf{F} \cdot \mathbf{n}\,d\sigma$.

33. $\mathbf{F}(x, y, z) = 3x^2\mathbf{i} + xy\mathbf{j} + z\mathbf{k}$; S is the portion of the plane $x + y + z = 1$ in the first octant; $\mathbf{n}$ is directed upward.

34. $\mathbf{F}(x, y, z) = (y - z)\mathbf{i} + (z - x)\mathbf{j} + (x - y)\mathbf{k}$; S is the portion of the plane $x + y + 2z = 4$ that lies in the first octant; $\mathbf{n}$ is directed upward.

35. $\mathbf{F}(x, y, z) = \mathbf{i} + 2\mathbf{j} + 3\mathbf{k}$; S is the hemisphere, $z = \sqrt{1 - x^2 - y^2}$; $\mathbf{n}$ is directed upward.

36. $\mathbf{F}(x, y, z) = \sin x\mathbf{i} + (2 - \cos x)y\mathbf{j} + z\mathbf{k}$; S is the parallelepiped described by $0 \le x \le \pi$, $0 \le y \le 2$, $0 \le z \le 1$; $\mathbf{n}$ is directed outward from S.

37. $\mathbf{F}(x, y, z) = x\mathbf{i} + y\mathbf{j} + z\mathbf{k}$; S is the boundary of the region in the first octant bounded by the plane $x + y + z = 3$; $\mathbf{n}$ is directed outward from S.

In Exercises 38 through 41, evaluate $\displaystyle\int_C \mathbf{F} \cdot d\mathbf{r}$, where C has positive orientation.

38. $\mathbf{F}(x, y, z) = (y + z)\mathbf{i} - (x + z)\mathbf{j} + (x + y)\mathbf{k}$; C is the triangle with vertices $(3, 0, 0)$, $(0, 3, 0)$ and $(0, 0, 3)$.

39. $\mathbf{F}(x, y, z) = (y^2 - x^2)\mathbf{i} + 2xy\mathbf{j}$; C is the curve described by $\mathbf{r}(t) = \sec t\mathbf{i} + \tan t\mathbf{j} + \mathbf{k}$, $0 \le t \le \dfrac{\pi}{4}$.

40. $\mathbf{F}(x, y, z) = (y^2 - e^x)\mathbf{i} + (x^2 - \sin y)\mathbf{j} + (z^2 - \sin z)\mathbf{k}$; C is the intersection of the paraboloid $z = x^2 + y^2$ and the plane $2x + 2y + z = 4$.

41. $\mathbf{F}(x, y, z) = e^y\mathbf{i} + e^z\mathbf{j} + e^x\mathbf{k}$; C is the boundary of the rectangle described by $0 \leq x \leq 3$ and $0 \leq z \leq 5$, with positive direction from $(0, 0, 0)$ to $(0, 0, 5)$.

42. Find the total mass and center of mass of the helix $\mathbf{r}(t) = \cos t\mathbf{i} + \sin t\mathbf{j} + 4t\mathbf{k}$, $0 \leq t \leq \pi/2$, if the density is given by $\rho(x, y, z) = x^2 + y^2 + z^2$.

43. Find the total mass and center of mass of the line segment $\mathbf{r}(t) = \mathbf{i} + 2t\mathbf{j} + \pi t/2\ \mathbf{k}$, $0 \leq t \leq 1$, if the density is given by $\rho(x, y, z) = xy \sin z$.

44. Find the surface area of the portion of the cone $x^2 + y^2 - z^2 = 0$ that lies between the planes $z = -2$ and $z = 3$.

45. Find the surface area of the portion of the sphere $x^2 + y^2 + z^2 = 9$ that is inside the cylinder $x^2 + y^2 = 1$.

46. Use surface integrals to show that the surface area of a sphere with radius r is $4\pi r^2$.

47. Use surface integrals to show that the surface area of a cone with height h and radius r is $\pi r\sqrt{r^2 + h^2}$.

48. Use Green's theorem to find the area bounded above by one arch of the cycloid described by $x = t - \sin t$, $y = 1 - \cos t$, and below by the x-axis.

49. Use Green's theorem to find the area bounded above by the curve that describes the path of a projectile:

$$x(t) = (v_0 \cos \theta)t, \qquad y(t) = (v_0 \sin \theta)t - 16t^2,$$

and below by the x-axis if (a) $\theta = \pi/6$, and (b) $\theta = \pi/3$.

50. Verify Stokes' theorem when $\mathbf{F}(x, y, z) = 2z\mathbf{i} + 3x\mathbf{j} + 4y\mathbf{k}$ and S is that portion of the paraboloid $z = 9 - x^2 - y^2$ that lies above the xy-plane.

51. Evaluate $\displaystyle\iint\limits_{S} \text{curl } \mathbf{F} \cdot \mathbf{n} \, d\sigma$, where $\mathbf{F}(x, y, z) = (y^2 + z^2)\mathbf{i} + (x^2 + z^2)\mathbf{j}$

$+ (x^2 + y^2)\mathbf{k}$; S is the surface of the ellipsoid $4x^2 + 9y^2 + z^2 = 1$ above the xy-plane; and $\mathbf{n}$ is directed upward.

18

ORDINARY DIFFERENTIAL EQUATIONS

A natural sequel to differential and integral calculus is the study of differential equations. A **differential equation** is an equation that involves a function and one or more of its derivatives. This final chapter is devoted to an introduction to the methods of solution and applications of some common differential equations. The exposition presented here is brief; for a more complete discussion we recommend that you consult a text on differential equations or advanced engineering mathematics.

18.1 INTRODUCTION

There are two types of differential equations: **ordinary differential equations,** which involve derivatives of a single-variable function and **partial differential equations,** which involve the partial derivatives of a function of several variables. When we speak of a differential equation in this chapter we mean an ordinary differential equation. The topic of partial differential equations is too involved for the brief summary of differential equations presented here.

Some examples of differential equations are

(i) $\quad y' = x^2$

(ii) $\quad y' = \dfrac{2y}{x}$

(iii) $y'' = 2y' - y$

(iv) $y''' = 2$

(v) $\quad y'' = 2y' + 3y - 10 \sin x + 2e^x$

(vi) $y' = 5 - \dfrac{4y}{x}.$

These are all ordinary differential equations, since the solution $y \equiv y(x)$ is a single-variable function. In fact, equations (i) and (ii) are separable equations, a type considered in Section 7.6.

The complete or **general solution** to each of the differential equations given in (i) through (vi) is listed below. In each case, y represents a function of x and $C, C_1, C_2,$ and C_3 denote arbitrary constants.

(i) $y(x) = \dfrac{x^3}{3} + C$ (ii) $y(x) = Cx^2$

(iii) $y(x) = C_1 e^x + C_2 x e^x$ (iv) $y(x) = \dfrac{1}{3} x^3 + C_1 x^2 + C_2 x + C_3$

(v) $y(x) = C_1 e^{3x} + C_2 e^{-x} - \dfrac{1}{2} e^x + 2 \sin x - \cos x$

(vi) $y(x) = x + \dfrac{C}{x^4}$

EXAMPLE 1   Show that $y(x) = C_1 e^x + C_2 x e^x$ is a solution to $y'' = 2y' - y$ for any constants C_1 and C_2.

SOLUTION

Since $y(x) = C_1 e^x + C_2 x e^x$, $y'(x) = C_1 e^x + C_2 e^x + C_2 x e^x$ and $y''(x) = C_1 e^x + 2 C_2 e^x + C_2 x e^x$. Consequently

$$2y' - y = 2(C_1 e^x + C_2 e^x + C_2 x e^x) - (C_1 e^x + C_2 x e^x)$$
$$= C_1 e^x + 2C_2 e^x + C_2 x e^x = y''. \qquad \square$$

While there is no universal method for solving a differential equation, there are standard methods for solving some types that commonly occur in applications. Before discussing these methods, we must classify the equations with respect to order. A differential equation is said to be of **order** n if the highest derivative that appears in the equation is of order n. The equations in (i), (ii), and (vi) are of order one and are called first-order equations; equations (iii) and (v) are second-order equations: equation (iv) is a third-order equation. Notice that in each case the number of arbitrary constants in the general solution agrees with the order of the equation, that is, the general solution of an nth-order differential equation involves n arbitrary constants.

When specific values for the constants are substituted into the general solution the result is called a **particular solution.** Often conditions are specified that a particular solution to a differential equation must satisfy. For example, the specification

$$y(1) = 2$$

added to the differential equation $y' = 2y/x$ implies that the constant C in the general solution $y(x) = Cx^2$ must be chosen so that

$$2 = y(1) = C \cdot 1, \qquad \text{and} \qquad y(x) = 2x^2.$$

A **boundary-value** problem is a problem involving an nth order differential equation together with n conditions that the solution must satisfy. The conditions are the boundary values. A first-order boundary value problem is also known as an **initial-value** problem and the condition is an **initial condition**. The initial-value terminology is also used for higher-order problems when the boundary values depend on only one value of the independent variable.

Geometrically, the general solution to a differential equation is represented by a family of curves. Graphs of such a family are illustrated in Figure 18.1, where the general solution to $y' = 2y/x$ is represented.

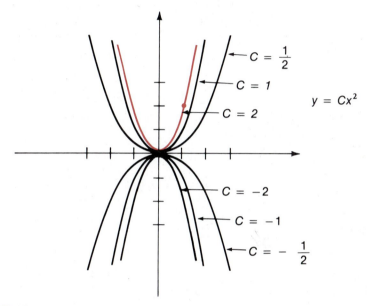

FIGURE 18.1

Giving an initial condition leads to a particular solution and is equivalent to specifying a particular curve in the family. For example, the equation $y' = 2y/x$ and the initial condition $y(1) = 2$ implied that $C = 2$, and produced the particular solution $y = 2x^2$. The graph of this particular solution is shown in color in Figure 18.1.

In a reverse manner, a family of curves described by an arbitrary constant determines a particular differential equation. For example, a differential equation associated with the family of curves

$$y = Ce^{-x},$$

can be found by first solving for C,

$$e^x y = C,$$

and then differentiating:

$$e^x y' + e^x y = 0.$$

The family $y = Ce^{-x}$ determines the first-order differential equation

$$y' = -y.$$

Many applications require the construction of a family of curves whose members are mutually orthogonal to a given family, that is, a member of this new family intersects each curve of the original family at a right angle. In this case the families are said to be **orthogonal trajectories** of one another. For example, the latitudinal lines on a map are orthogonal trajectories of the longitudinal lines on the map.

To determine the orthogonal trajectories to the family $y = Ce^{-x}$, consider its differential equation $y' = -y$. The slope of the tangent line to a curve $y = Ce^{-x}$ at a point (x, y) must have the value $y'(x) = -y(x)$. Two nonvertical lines that intersect orthogonally have slopes whose product is -1. So the orthogonal trajectories must be a family of curves that have a slope at (x, y) given by $-1/(-y(x)) = 1/y(x)$. The differential equation of the orthogonal trajectories to $y = Ce^{-x}$ is therefore

$$y' = \frac{1}{y}.$$

This equation is separable and can be solved using the techniques discussed in Section 7.6. Since

$$yy' = 1,$$

we can integrate both sides to obtain

$$\frac{y^2}{2} = x + C,$$

so

$$y^2 = 2x + C.$$

This is a family of parabolas. Representative members of this family are shown in Figure 18.2 together with graphs of members of the original family $y = Ce^{-x}$.

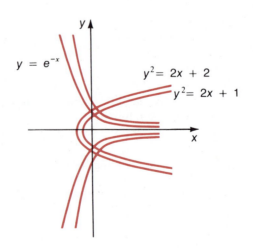

$y = e^{-x}$

$y^2 = 2x + 2$

$y^2 = 2x + 1$

FIGURE 18.2

EXERCISE SET 18.1

In Exercises 1 through 16, verify that the function $y \equiv y(x)$ is the general solution to the accompanying differential equation.

1. $2xy' + y = 0; y = Cx^{-1/2}$

2. $xy' - 3y = 0; y = Cx^3$

3. $2xy' + y = 1; y = Cx^{-1/2} + 1$

4. $xy' - 3y = 1; y = Cx^3 - 1/3$

5. $y' + 2xy = 0; y = Ce^{-x^2}$

6. $xy' - y = xy; y = Cxe^x$

7. $x^2 y' - y^2 - 2xy = 0; y = \dfrac{Cx^2}{1 - Cx}$

8. $(x^2 + 1)y' + y^2 + 1 = 0; y = \dfrac{C - x}{1 + Cx}$

9. $y'' + y = 0; y = C_1 \sin x + C_2 \cos x$

10. $y'' - y = 0; y = C_1 e^x + C_2 e^{-x}$

11. $y'' - y = 0; y = C_1 \sinh x + C_2 \cosh x$

12. $y'' + y = x^2; y = C_1 \sin x + C_2 \cos x + x^2 - 2$

13. $y'' - y = x^2; y = C_1 e^x + C_2 e^{-x} - x^2 - 2$

14. $y''' - 3y'' + 2y' = 6e^{-x}; y = C_1 + C_2 e^x + C_3 e^{2x} - e^{-x}$

15. $yy' + x = 0; x^2 + y^2 = C$

16. $yy' - 4x = 1; y^2 - 4x^2 = 2x + C$

Determine a first-order differential equation whose solution is described in Exercises 17 through 24.

17. $y = Cx$ **18.** $y = 4x + C$

19. $y = \dfrac{C}{x}$ **20.** $y = Ce^x$

21. $y = Ce^{-2x}$ **22.** $x^2 - 2y^2 = C$

23. $y = \sqrt{x + C}$ **24.** $y = Ce^{x^2}$

25–32. Find orthogonal trajectories to the curves described in Exercises 17 through 24 and sketch representative graphs of the curves and the orthogonal trajectories.

18.2
HOMOGENEOUS DIFFERENTIAL EQUATIONS

The general form of a first-order differential equation is

(18.1) $$y'(x) = f(x, y(x)).$$

The most elementary first-order equations occur when f is a continuous function involving only the single variable x. To find a general solution in this case, we apply the fundamental theorem of calculus, that is,

$$y'(x) = f(x) \quad \text{implies} \quad y(x) = \int f(x)dx.$$

We have considered differential equations of this type in previous work. For example, in Section 4.4, an equation describing the motion of a falling object was derived from the knowledge of the object's acceleration due to gravity.

Another type of differential equation, the *separable equation,* was discussed in Section 7.6. Separable equations are differential equations that can be expressed in the form

$$y' = \frac{p(x)}{q(y)}.$$

These equations can be solved by separating the variables and integrating.

A first-order equation of the form

(18.2) $$y' = f(x, y), \quad \text{where } f(x, y) = g\left(\frac{y}{x}\right)$$

is called a **homogeneous** differential equation. An example of such an equation is

$$y' = 1 - \frac{y}{x}.$$

A homogeneous equation can be reduced to a separable equation by using the variable substitution $v = y/x$. With this substitution,

$$y(x) = xv(x)$$

so $$y'(x) = v(x) + xv'(x).$$

Consequently, equation (18.2) can be reexpressed as

$$v + xv' = g(v),$$

or in the separable form

$$v' = \frac{g(v) - v}{x}.$$

This separable equation can then be solved for v and hence for $y(x) = xv(x)$.

EXAMPLE 1   Find the general solution to the differential equation

$$y' = \frac{x + y}{x - y}.$$

SOLUTION

Since $\dfrac{x + y}{x - y} = \dfrac{1 + y/x}{1 - y/x}$, the equation is homogeneous. The substitution $v = \dfrac{y}{x}$ implies $y = xv$ and $y' = v + xv'$. This changes the original differential equation into the separable differential equation

$$v + xv' = \frac{1 + v}{1 - v}.$$

Thus $$xv' = \frac{1 + v}{1 - v} - v = \frac{1 + v^2}{1 - v}$$

$$\frac{1 - v}{1 + v^2} v' = \frac{1}{x},$$

and $$0 = \frac{1}{x} - \frac{1 - v}{1 + v^2} \frac{dv}{dx}.$$

Integrating this equation produces

$$C = \int \frac{1}{x} dx - \int \frac{1 - v}{1 + v^2} dv$$

$$= \ln |x| - \int \frac{1}{1 + v^2} dv + \int \frac{v}{1 + v^2} dv$$

$$= \ln |x| - \arctan v + \frac{1}{2} \ln |1 + v^2|$$

$$= \ln |x \sqrt{1 + v^2}| - \arctan v.$$

Since $v = y/x$, the general solution has the form

$$C = \ln \left| x \sqrt{1 + \left(\frac{y^2}{x^2}\right)} \right| - \arctan \frac{y}{x}$$

$$= \ln \sqrt{x^2 + y^2} - \arctan \frac{y}{x}. \qquad \square$$

EXAMPLE 2   Find the solution to the initial-value problem

$$y' = \frac{y^2 - x^2}{2xy}, \qquad y(1) = 1.$$

SOLUTION

The equation can be rewritten

$$y' = \frac{y^2 - x^2}{2xy}$$

$$= \frac{1}{2} \left[\left(\frac{y}{x}\right) - \left(\frac{x}{y}\right) \right],$$

so the substitution $v = y/x$ reduces the problem to a separable equation. Since

$$y = xv, \qquad y' = xv' + v$$

and the differential equation has the form

$$xv' + v = \frac{1}{2} \left(v - \frac{1}{v} \right) = \frac{v^2 - 1}{2v}.$$

Thus, $$xv' = \frac{v^2 - 1}{2v} - v = -\left(\frac{1 + v^2}{2v} \right),$$

which can be rewritten

$$-\frac{2v}{1+v^2}\,v' = \frac{1}{x}$$

or

$$0 = \frac{1}{x} + \frac{2v}{1+v^2}\frac{dv}{dx}.$$

Integrating this equation produces

$$C = \int \frac{1}{x}\,dx + \int \frac{2v}{1+v^2}\,dv$$
$$= \ln|x| + \ln|1+v^2| = \ln|x(1+v^2)|$$

so

$$e^C = x(1+v^2) = x\left(1 + \frac{y^2}{x^2}\right).$$

Since $y = 1$ when $x = 1$, $e^C = 2$ and the solution is

$$2 = x + \frac{y^2}{x} \qquad \text{or} \qquad y^2 = 2x - x^2. \qquad \square$$

EXERCISE SET 18.2

Find the general solution to the differential equations in Exercises 1 through 6.

1. $y' = \dfrac{y}{x+y}$

2. $y' = \dfrac{2x+y}{x+2y}$

3. $y' = \dfrac{x-y}{x+y}$

4. $y' = \dfrac{2xy+y^2}{x^2}$

5. $y' = \dfrac{x^2+xy+y^2}{x^2}$

6. $y' = \dfrac{x^2+3xy+y^2}{x^2}$

Solve the initial value problems in Exercises 7 through 10.

7. $y' = \dfrac{x+y}{x-y}$; $y(1) = \sqrt{3}$

8. $y' = \dfrac{2x}{x+0}$; $y(1) = 2$

9. $y' = \dfrac{2x+y}{x+2y}$; $y(1) = 1$

10. $y' = \dfrac{4x^2+2y^2}{xy}$; $y(1) = 2$

11. Find the general solution to the differential equation

$$y' = \frac{2x+y+2}{x+2y-1}$$

by first making a substitution $x = u - a$ and $y = w - b$ where u and w are variables and a and b are appropriate constants. Then let $v = w/u$ and obtain a separable equation.

12. Use the technique described in Exercise 11 to solve the differential equation

$$y' = \frac{y+x+5}{x-y-4}.$$

13. Recall from Theorem 15.23 of Section 15.4 that if $F(x, y) = c$, then $\dfrac{dy}{dx} = \dfrac{-F_x(x, y)}{F_y(x, y)}$. Use this fact to show that if the orthogonal trajectories to $F(x, y) = C$ are described by $G(x, y) = C$, then

$$F_x(x, y)\, G_x(x, y) + F_y(x, y)\, G_y(x, y) = 0.$$

14. Suppose that a family of curves is described by $F(x, y) = C$. Show that no member of the family can be mutually orthogonal to itself unless all intersections between members of the family are vertical and horizontal.

18.3
EXACT DIFFERENTIAL EQUATIONS

In discussing line integrals in Section 17.4 we introduced the idea of an exact differential form. A differential form in two variables x and y is an expression

$$f(x, y)dx + g(x, y)dy.$$

The differential form is said to be exact if a function u exists with

$$du = f(x, y)dx + g(x, y)dy.$$

Since
$$du = \frac{\partial u}{\partial x}\, dx + \frac{\partial u}{\partial y}\, dy,$$

the differential form is exact if and only if

$$\frac{\partial u}{\partial x}(x, y) = f(x, y) \quad \text{and} \quad \frac{\partial u}{\partial y}(x, y) = g(x, y).$$

An **exact differential equation** is a first-order differential equation produced when an exact differential form is set equal to zero; that is, a differential equation of the form

(18.3) $$f(x, y)dx + g(x, y)dy = 0,$$

or equivalently,

(18.4) $$\frac{dy}{dx} = -\frac{f(x, y)}{g(x, y)},$$

where a function u exists with

$$\frac{\partial u}{\partial x} = f \quad \text{and} \quad \frac{\partial u}{\partial y} = g.$$

If the function u is known, the solution to this differential equation is immediate, since

$$du = f(x, y)dx + g(x, y)dy = 0$$

implies that

$$u(x, y) = C.$$

EXAMPLE 1   Solve the differential equation

$$\frac{dy}{dx} = - \frac{2xy + y^2}{x^2 + 2xy}.$$

SOLUTION

The substitution $v = y/x$ will reduce this equation to a separable equation, since

$$- \frac{2xy + y^2}{x^2 + 2xy} = \frac{-2(y/x) - (y/x)^2}{1 + 2(y/x)}.$$

This method of solution is quite complicated, however. (See Exercise 17 on p. 1006.)

Instead, suppose we notice that for $u(x, y) = x^2y + y^2x$,

$$\frac{\partial u}{\partial x}(x, y) = 2xy + y^2 \quad \text{and} \quad \frac{\partial u}{\partial y}(x, y) = x^2 + 2yx.$$

Then the differential equation is exact and its general solution can be written

$$x^2y + y^2x = C. \qquad \square$$

The previous example brings out two important questions.
(i) How can we determine whether a differential equation is exact without knowing the function u?
(ii) When the equation is exact, how do we find u?

The answer to the first question is found in the following result, which is a special case of Theorem (17.39).

(18.5)
THEOREM

If $\dfrac{\partial f}{\partial y}$ and $\dfrac{\partial g}{\partial x}$ are continuous, then the differential form

$$f(x, y)dx + g(x, y)dy$$

is exact if and only if

$$\frac{\partial f}{\partial y} = \frac{\partial g}{\partial x}.$$

The theorem gives an easy test to apply. For example, the differential equation in Example 1,

$$\frac{dy}{dx} = \frac{-(2xy + y^2)}{x^2 + 2xy}$$

can be written

$$(2xy + y^2)dx + (x^2 + 2xy)dy = 0$$

and is exact because

$$\frac{\partial}{\partial y}(2xy + y^2) = 2x + 2y = \frac{\partial}{\partial x}(x^2 + 2yx).$$

Theorem (18.5) also provides the answer to question (ii), although this answer is not as immediate.

Suppose the differential equation

$$f(x, y)dx + g(x, y)dy = 0$$

is exact and (a, b) is an arbitrary point in the domain of u, where $\partial u/\partial x = f$ and $\partial u/\partial y = g$. Since $\partial u/\partial x = f$, integrating with respect to x produces

$$u(x, y) = \int_a^x f(x, y)dx + C(y).$$

The integration "constant" $C(y)$ does not depend on x, but may involve y.

In addition, $\partial u/\partial y = g$, so

$$g(x, y) = \frac{\partial u}{\partial y}(x, y) = \frac{\partial}{\partial y}\left[\int_a^x f(x, y)dx\right] + C'(y)$$

$$= \int_a^x \frac{\partial f}{\partial y}(x, y)dx + C'(y).$$

However, the differential equation is exact, so

$$\frac{\partial f}{\partial y}(x, y) = \frac{\partial g}{\partial x}(x, y)$$

and

$$g(x, y) = \int_a^x \frac{\partial g}{\partial x}(x, y)dx + C'(y)$$

$$= g(x, y) - g(a, y) + C'(y).$$

This implies that

$$C'(y) = g(a, y)$$

so

$$C(y) = \int_b^y g(a, y)dy$$

and

$$u(x, y) = \int_a^x f(x, y)dx + \int_b^y g(a, y)dy.$$

Consequently, the general solution to the exact differential equation

$$\frac{dy}{dx} = \frac{-f(x, y)}{g(x, y)}$$

can be expressed as

(18.6)
$$C = \int_a^x f(x, y)dx + \int_b^y g(a, y)dy.$$

This form is especially convenient when an initial condition of the form $y(a) = b$ is specified. In this case the particular solution satisfying the initial condition is

$$\int_a^x f(x, y)dx + \int_b^y g(a, y)dy = 0.$$

EXAMPLE 2   Find the general solution to the differential equation

$$(2xy + e^y)dx + (x^2 + xe^y)dy = 0.$$

SOLUTION

This equation is exact because

$$\frac{\partial}{\partial y}(2xy + e^y) = 2x + e^y = \frac{\partial}{\partial x}(x^2 + xe^y).$$

The choice of (a, b) is arbitrary; to simplify the integration we choose $(a, b) = (0, 0)$. The solution is

$$C = \int_0^x (2xy + e^y)dx + \int_0^y (0^2 + 0 \cdot e^y)dy = x^2y + xe^y. \qquad \square$$

EXAMPLE 3   Find the solution to the initial value problem

$$\frac{dy}{dx} = \frac{2xy^2 + y \sin x}{\cos x - 2x^2y}, \qquad y(0) = 1.$$

SOLUTION

The equation can be rewritten as

$$-(2xy^2 + y \sin x)dx + (\cos x - 2x^2y)dy = 0,$$

and is exact because

$$\frac{\partial}{\partial y}[-(2xy^2 + y \sin x)] = -4xy - \sin x = \frac{\partial}{\partial x}(\cos x - 2x^2y).$$

The particular solution is therefore

$$0 = \int_0^x -(2xy^2 + y \sin x)dx + \int_1^y (\cos (0) - 2 \cdot 0^2 \cdot y)dy$$

$$= (-x^2y^2 + y \cos x)\Big]_{x=0}^{x=x} + (y)\Big]_{y=1}^{y=y}$$

$$= (-x^2y^2 + y \cos x - y) + (y - 1)$$

or $$y \cos x - x^2y^2 = 1. \qquad \square$$

EXAMPLE 4   Solve the differential equation

$$\frac{dy}{dx} = \frac{y}{xy - x}.$$

SOLUTION

This differential equation is not exact because

$$\frac{\partial}{\partial y}(-y) = -1 \neq y - 1 = \frac{\partial}{\partial x}(xy - x).$$

However, the differential equation can be converted into an exact equation by multiplying the numerator and denominator by e^{-y}. The differential equation

$$\frac{dy}{dx} = \frac{ye^{-y}}{xye^{-y} - xe^{-y}}$$

is exact, since

$$\frac{\partial}{\partial y}(-ye^{-y}) = -e^{-y} + ye^{-y} = \frac{\partial}{\partial x}(xye^{-y} - xe^{-y}).$$

The general solution to this equation, and hence to the original problem, can be expressed as

$$C = \int_0^x -ye^{-y}\,dx + \int_0^y (0 \cdot ye^{-y} - 0 \cdot e^{-y})dy$$

$$= -xye^{-y}. \qquad \qquad \Box$$

A term (such as e^{-y} in Example 4) that converts a nonexact equation into an exact equation is one example of what is known as an **integrating factor.** In general, an integrating factor converts a differential equation into a form that can be solved by a simple integration. Every differential equation with a solution of the form $u(x, y) = C$ has an infinite number of integrating factors that make the problem exact. However, in general it is as difficult to find an integrating factor that makes the problem exact as it is to solve the original equation. Exercises 20 and 22 consider two special cases when an integrating factor can be quite readily obtained.

When a problem has been converted to an exact equation by using an integrating factor, the solutions should be verified in the original differential equation because extraneous solutions can occur.

EXERCISE SET 18.3

Determine whether the equations in Exercises 1 through 12 are exact and solve the exact equations.

1. $y' = \dfrac{3y + 2x}{y - 3x}$

2. $y' = \dfrac{y - 4x}{y - x}$

3. $y' = \dfrac{3y - 2x}{y + 3x}$

4. $y' = \dfrac{ax - by}{bx - ay}$

5. $y' = \dfrac{x^3 - xy^4}{2x^2y^3 + 5}$

6. $y' = \dfrac{e^x \cos y - y}{e^x \sin y + x}$

7. $y' = \dfrac{x - y^2}{2xy}$ **8.** $y\,dx + dy = 0$

9. $ye^x dx + e^x\,dy = 0$ **10.** $(6x + y^2)dx + (2xy - 3y^2)dy = 0$

11. $(\cot x - \cos x \cos y)dx + \sin x \sin y\,dy = 0$

12. $\sinh x \cos y\,dx - \sin y \cosh x\,dy = 0$

Solve the initial-value problems in Exercises 13 through 16.

13. $(2x - 3y)dx + (4y - 3x)dy = 0$, $y(0) = 1$

14. $(x^2 + 2xy)dx + (x^2 - y^2)dy = 0$, $y(1) = 1$

15. $\sin x \sin y\,dx - \cos x \cos y\,dy = 0$, $y\left(\dfrac{\pi}{4}\right) = \dfrac{\pi}{2}$

16. $ye^{xy}dx + (2y + xe^{xy})dy = 0$, $y(0) = 1$

17. Solve the differential equation in Example 1 by using the fact that it is homogeneous.

18. Show that every separable equation is an exact equation.

19. Show that the equation $6x^2y\,dx + (x^3 + y)dy = 0$ is not exact, but that $y^{-1/2}[6x^2y\,dx + (x^3 + y)dy] = 0$ is exact. Solve this equation.

20. If the equation $M(x, y)dx + N(x, y)dy = 0$ has the property that $\dfrac{N_x - M_y}{M}$ is a function of y only, then the equation

$$I(y)[M(x, y)\,dx + N(x, y)dy] = 0$$

is exact, where

$$I(y) = \exp\left[\int \frac{N_x - M_y}{M}\,dy\right].$$

Show that this statement holds for the equation considered in Exercise 19.

21. Show that the equation $(\cos y + x)dx + x \sin y\,dy = 0$ is not exact, but that $x^{-2}[(\cos y + x)dx + x \sin y\,dy] = 0$ is exact. Solve this equation.

22. If the equation $M(x, y)dx + N(x, y)dy = 0$ has the property that $\dfrac{M_y - N_x}{N}$ is a function of x only, then the equation

$$I(x)[M(x, y)dx + N(x, y)dy] = 0$$

is exact, where

$$I(x) = \exp\left[\int \frac{M_y - N_x}{N}\,dx\right].$$

Show that this statement holds for the equation considered in Exercise 21.

18.4
LINEAR FIRST-ORDER DIFFERENTIAL EQUATIONS

A **linear** first-order differential equation has the form

(18.7)
$$\frac{dy}{dx} + P(x)y = Q(x).$$

The technique for solving this type of equation involves multiplying the differential equation by an integrating factor $I(x)$ that changes equation (18.7) into an equation whose left side can be expressed as the derivative of the product $I(x)y$. To determine this integrating factor, consider the differential equation

$$I(x)\frac{dy}{dx} + I(x)P(x)y = I(x)Q(x).$$

To choose $I(x)$ so that

$$D_x[I(x)y] = I(x)\frac{dy}{dx} + I(x)P(x)y$$

requires that

$$I(x)\frac{dy}{dx} + I'(x)y = I(x)\frac{dy}{dx} + I(x)P(x)y.$$

This implies that

(18.8)
$$I'(x) = I(x)P(x).$$

Equation (18.8) is a separable equation that can be solved by using an elementary integration method discussed in Section 7.6. The solution is $I(x) = Ce^{\int P(x)dx}$.

Since we need only one integrating factor, we can let $C = 1$ and choose

(18.9)
$$I(x) = e^{\int P(x)dx}.$$

EXAMPLE 1   Solve the linear differential equation

$$y' + \frac{1}{x}y = x^2, \qquad \text{for } x > 0.$$

SOLUTION

An integrating factor for this equation is

$$I(x) = e^{\int 1/x \, dx} = e^{\ln x} = x.$$

Multiplying both sides of the differential equation by x produces

$$xy' + y = x^3.$$

Since
$$D_x(xy) = xy' + y,$$

it follows that

$$D_x(xy) = x^3$$

and
$$xy = \int x^3 dx = \frac{x^4}{4} + C.$$

The general solution is therefore

$$y(x) = \frac{x^3}{4} + \frac{C}{x}. \qquad \square$$

EXAMPLE 2   Solve the initial-value problem

$$(\cos x + y)dx + \cot x \, dy = 0, \qquad y(0) = 2 \qquad \text{for } -\frac{\pi}{2} < x < \frac{\pi}{2}.$$

SOLUTION

The equation can be expressed

$$\cot x \frac{dy}{dx} + y = -\cos x.$$

Before determining the integrating factor we must divide by $\cot x$, so that the coefficient of dy/dx will be 1, as in (18.7):

$$\frac{dy}{dx} + (\tan x)y = -\frac{\cos x}{\cot x} = -\sin x.$$

The integrating factor for this equation is
$$I(x) = e^{\int \tan x \, dx} = e^{\ln |\sec x|} = \sec x.$$

The equation becomes

$$\sec x \frac{dy}{dx} + (\sec x \tan x)y = -\sec x \sin x = -\tan x.$$

Since

$$D_x((\sec x)y) = \sec x \frac{dy}{dx} + (\sec x \tan x)y,$$
$$D_x((\sec x)y) = -\tan x$$

and

$$(\sec x)y = -\int \tan x \, dx + C = -\ln |\sec x| + C = \ln |\cos x| + C$$

so

$$y(x) = \cos x(\ln |\cos x| + C).$$

Since $y(0) = 2$,

$$2 = \cos 0 \, (\ln |\cos 0| + C) = C$$

and the solution to the initial-value problem is

$$y(x) = \cos x(\ln |\cos x| + 2).$$

The restriction on x is necessary in this problem so that the integrating factor and solution are defined. $\qquad \square$

Bernoulli differential equations have the form

(18.10)
$$\frac{dy}{dx} + P(x)y = Q(x)y^n.$$

A Bernoulli equation is linear when $n = 0$ or $n = 1$. For other values of n, a substitution of the form $w = y^{1-n}$ converts the Bernoulli equation in y into a linear differential equation in w. First divide equation (18.10) by y^n:

$$y^{-n}\frac{dy}{dx} + y^{1-n}P(x) = Q(x).$$

Since $w = y^{1-n}$, $\frac{dw}{dx} = (1 - n)y^{-n}\frac{dy}{dx}$, and the equation becomes

$$\frac{1}{1-n}\frac{dw}{dx} + P(x)w = Q(x),$$

or

(18.11)
$$\frac{dw}{dx} + (1 - n)P(x)w = (1 - n)Q(x).$$

Equation (18.11) is a linear equation that can be solved for $w(x)$ by using the integrating factor

$$I(x) = e^{\int(1-n)P(x)dx}.$$

The solution $y(x)$ to equation (18.10) can then be found from

$$y(x) = [w(x)]^{1/(1-n)}.$$

EXAMPLE 3   Solve the Bernoulli differential equation

$$\frac{dy}{dx} + y = xy^3.$$

SOLUTION

Let $w = y^{1-3} = y^{-2}$, then $dw/dx = -2y^{-3}\,dy/dx$. The substitution of w and dw/dx is easier if the equation is first divided by y^3:

$$y^{-3}\frac{dy}{dx} + y^{-2} = x.$$

Then

$$-\frac{1}{2}\frac{dw}{dx} + w = x \quad \text{or} \quad \frac{dw}{dx} - 2w = -2x.$$

The integrating factor

$$I(x) = e^{\int -2dx} = e^{-2x}$$

produces

$$e^{-2x}\frac{dw}{dx} - 2e^{-2x}w = -2xe^{-2x}.$$

Since
$$D_x(e^{-2x} w) = e^{-2x} \frac{dw}{dx} - 2e^{-2x} w,$$

$$D_x(e^{-2x} w) = - 2xe^{-2x}$$

and
$$e^{-2x} w = -2 \int xe^{-2x} dx + C.$$

Integrating by parts gives

$$\int xe^{-2x} dx = -\frac{1}{2} xe^{-2x} + \frac{1}{2} \int e^{-2x} dx = -\frac{1}{2} xe^{-2x} - \frac{1}{4} e^{-2x} + C$$

so

$$e^{-2x} w = xe^{-2x} + \frac{1}{2} e^{-2x} + C,$$

and
$$w(x) = e^{2x} \left[xe^{-2x} + \frac{1}{2} e^{-2x} + C \right]$$

$$= x + \frac{1}{2} + Ce^{2x} = \frac{2x + 1 + Ce^{2x}}{2}.$$

Since $w = y^{-2}$,

$$\frac{1}{y^2} = \frac{2x + 1 + Ce^{2x}}{2}$$

and
$$y^2 = \frac{2}{2x + 1 + Ce^{2x}}. \qquad \square$$

EXERCISE SET 18.4

Find a general solution to the differential equations in Exercises 1 through 20.

1. $y' - 2y = e^x$

2. $y' + 5y = e^{-x}$

3. $y' + 3y = xe^{-x}$

4. $y' + 3y = x + 1$

5. $y' + y = \sin x$

6. $y' + \frac{2}{x} y = 6x$

7. $y' + \frac{3}{x} y = e^x$

8. $y' + 3xy = 6x$

9. $y' + \frac{1}{x^2} y = \frac{1}{x^2}$

10. $y' + \frac{1 + x}{x} y = \frac{2}{x}$

11. $y' + y \tan x = \cos x$

12. $y' + y \tan x = \sec x$

13. $y' + y \tan x = x \cos x$

14. $y' + y \tan x = x \sec x$

15. $(x^2 + x)y' + (2x + 1)y = x^2 - 1$

16. $\frac{(x^2 + 1)}{2x} y' + y = e^{x^2}$

17. $y' + y = \frac{x}{y}$

18. $y' + xy = \frac{x}{y}$

19. $y' + \frac{y}{x} = \frac{y^2}{x}$

20. $y' = y(a - by)$

21. Show that if y_1 and y_2 are solutions to the linear equation $y' + P(x)y = Q(x)$, then for any constant C, $C(y_1 + y_2)$ is a solution to the linear equation $y' + P(x)y = 0$.

22. An equation of the form $y' = P(x)y^2 + Q(x)y + R(x)$ is called a *Ricatti equation*. Suppose that y_1 is a particular solution to this equation. Show that a general solution to the Ricatti equation has the form $y = y_1 + 1/v$, where v is a solution to the linear equation

$$v' + [Q(x) + 2P(x)y_1]v = -P(x).$$

23. A particular solution to the Ricatti equation $y' = x^2 + y^2 - 2xy + 1$ is $y_1(x) = x$. Use the method described in Exercise 22 to find a solution satisfying $y(2) = 1$.

18.5
SECOND-ORDER LINEAR EQUATIONS: HOMOGENEOUS TYPE

An nth-order linear differential equation has the form

(18.12)
$$y^{(n)} + P_1(x)y^{(n-1)} + P_2(x)y^{(n-2)} + \cdots + P_{n-1}(x)y' + P_n(x)y = Q(x).$$

Two separate cases are distinguished. When $Q(x) \equiv 0$ the equation is called **homogeneous**; when $Q(x) \not\equiv 0$ the equation is called **nonhomogeneous.** Homogeneous equations are considered in this section, nonhomogeneous equations in Section 18.6.

Homogeneous linear differential equations have the property that for any two solutions y_1 and y_2, and any pair of constants C_1 and C_2, the function described by

$$y(x) = C_1 y_1(x) + C_2 y_2(x)$$

is also a solution of the homogeneous equation. (This result is discussed for the case $n = 2$ in Exercise 35.)

There is no standard elementary technique for solving *general* homogeneous nth-order linear differential equations. The method we discuss is for the special case when the functions $P_1, P_2, \ldots, P_n$ are all constants. A linear equation of this type is called a **linear equation with constant coefficients.** To simplify matters even further, we consider only the case when $n = 2$. The method of solution when $n > 2$ is similar, but often more complicated.

In summary, this section is concerned with solving differential equations of the form

(18.13)
$$y'' + py' + qy = 0,$$

where p and q are constants.

The methods of Section 18.4 can be used to show that

$$y(x) = e^{-qx}$$

is a solution to the first-order linear equation

$$y' + qy = 0.$$

This provides motivation for the conjecture that the second-order linear equation (18.13) has solutions of the form

$$y(x) = e^{mx}.$$

for properly chosen constants m. To test this conjecture, we compute the derivatives of y:

$$y'(x) = me^{mx} \quad \text{and} \quad y''(x) = m^2 e^{mx}$$

and determine any constants m that ensure that (18.13) is satisfied. Thus,

$$\begin{aligned}
0 &= y'' + py' + qy \\
&= m^2 e^{mx} + pme^{mx} + qe^{mx} \\
&= (m^2 + pm + q)e^{mx}.
\end{aligned}$$

Since $e^{mx} > 0$, the equation is satisfied precisely when

(18.14) $0 = m^2 + pm + q.$

Equation (18.14) is called the **characteristic equation** of equation (18.13) and has solutions

$$m_1 = \frac{-p}{2} + \frac{\sqrt{p^2 - 4q}}{2} \quad \text{and} \quad m_2 = \frac{-p}{2} - \frac{\sqrt{p^2 - 4q}}{2}.$$

The solutions to (18.13) assume three different forms depending on the value of $p^2 - 4q$.

CASE I: $p^2 - 4q > 0.$

When $p^2 - 4q > 0$, m_1 and m_2 are distinct real numbers. Since

(18.15) $y_1(x) = e^{m_1 x} \quad \text{and} \quad y_2(x) = e^{m_2 x}$

are both solutions of the homogeneous equation (18.13), this implies that for any constants C_1 and C_2,

(18.16) $y(x) = C_1 e^{m_1 x} + C_2 e^{m_2 x},$

is also a solution to (18.13). In fact, when $p^2 - 4q > 0$, all solutions to (18.13) must be of this form.

EXAMPLE 1   Solve the differential equation

$$y'' - 5y' + 6y = 0.$$

SOLUTION

If $y(x) = e^{mx}$, then

$$\begin{aligned}
y'' - 5y' + 6y &= m^2 e^{mx} - 5me^{mx} + 6e^{mx} \\
&= (m^2 - 5m + 6)e^{mx} \\
&= (m - 3)(m - 2)e^{mx}.
\end{aligned}$$

37. $x = 1, x = -1, y = 1,$ $y = -1$

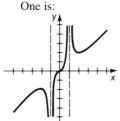

39. $y = 0$

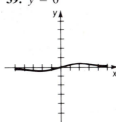

41. There are many examples. One is:

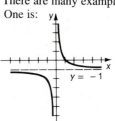

$y = -1$

43. There are many examples. One is:

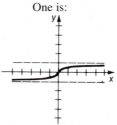

45. There are many examples. One is:

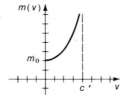

47. (a) $0 < x < .01$
   (b) $0 < x < .0001$

49. There are many examples. One is $f(x) = \dfrac{1}{x}, g(x) = x.$

51. There are many examples. One is $f(x) = \dfrac{1}{x^2}, g(x) = x.$

53. There are many examples. One is $g(x) = x, h(x) = -x.$

55. There are many examples. One is $g(x) = x^2, h(x) = -x.$

57. $\lim\limits_{x \to a^+} f(x) = -\infty$ provided that for any number M, a number $\delta > 0$ can be found with the property that $f(x) < M$ whenever $0 < x - a < \delta.$

59. Because $\lim\limits_{x \to 0} \dfrac{1}{x}$ does not exist, so Theorem 2.24 does not apply.

61. As $V \to 0, T \to \infty.$ As $V \to \infty, T \to \infty.$ In fact, however, in neither case does the equation remain valid.

63. $\lim\limits_{v \to c^-} m(v) = \infty.$

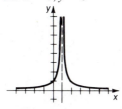

REVIEW EXERCISES , PAGE 94

1. $5/2$ **3.** 0 **5.** 0 **7.** 1 **9.** $\dfrac{1}{4}$ **11.** $\dfrac{-3\sqrt{10}}{10}$ **13.** 0 **15.** ∞ **17.** $-\infty$

19. $-\dfrac{1}{2x^{3/2}}$ **21.** $-2, 2, f(-2), f(2)$ are undefined. **23.** none

25. $3, f(3)$ is undefined. 4, neither $f(4)$ nor $\lim\limits_{x \to 4} f(x)$ exists. **27.** $\left(-\infty, \dfrac{3}{4}\right)$, $\left(\dfrac{3}{4}, \infty\right)$

29. $(-\infty, -1]$, $[1, \infty)$

31. $x = 1, y = 0$

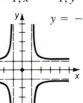

33. $x = 0, x = 2, y = 1$

35. $y = 3$

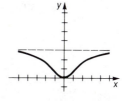

37. $x = 0, y = 0$

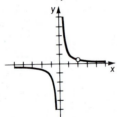

39. Let $f(x) = 2x^3 - x - 2$ and consider $f(-1)$ and $f(2)$.

41. Given $\epsilon > 0$, choose $\delta = \dfrac{\epsilon}{3}$.

43. (a) ∞ (b) $-\infty$ (c) ∞ (d) 0 **45.** $-\dfrac{3}{2}$

47. No. Consider, for example,

$$f(x) = g(x) = \begin{cases} 1, & if\ x < 0 \\ -1, & if\ x \geq 0 \end{cases}$$
$$a = 0.$$

49.

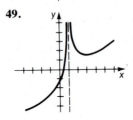

51. (a) $x > 10$ (b) $x > \sqrt{1000}$

CHAPTER 3

EXERCISE SET 3.1, PAGE 101

1.

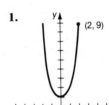

(b) 8
(c) $y = 8x - 7$
(d) $y = -\dfrac{1}{8}x + \dfrac{37}{4}$

3.

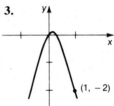

(b) -5
(c) $y = -5x + 3$
(d) $y = \dfrac{1}{5}x - \dfrac{11}{5}$

5.

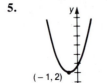

(b) 0
(c) $y = 2$
(d) $x = -1$

7.

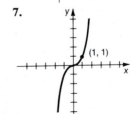

(b) 3
(c) $y = 3x - 2$
(d) $y = -\dfrac{1}{3}x + \dfrac{4}{3}$

9.

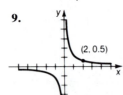

(b) $-\dfrac{1}{4}$
(c) $y = -\dfrac{1}{4}x + 1$
(d) $y = 4x - \dfrac{15}{2}$

11.

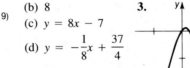

(b) $2a + 2$
(c) $y = (2a + 2)x - a^2 + 2$
(d) $y = \dfrac{-x}{(2a + 2)} + \dfrac{a}{2a + 2} + a^2 + 2a + 2$

13. $\left(-\dfrac{1}{2}, \dfrac{5}{4}\right)$ **15.** $(-1, -1)$ **17.** $(-1, -1)$ **19.** $a = 0$ or $a = -2$

EXERCISE SET 3.2, PAGE 109

1. 3 **3.** 7 **5.** 4 **7.** 0 **9.** 48 **11.** $-\dfrac{1}{16}$ **13.** $-\dfrac{1}{16}$ **15.** $\dfrac{15}{4}$ **17.** 7 **19.** 0

21. $-\dfrac{1}{16}$

23. $f'(x) = 4$

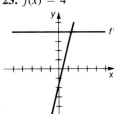

25. $f'(x) = 2x + 2$

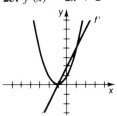

27. $f'(x) = 3x^2$

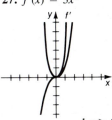

29. $f'(x) = 3(x + 1)^2$

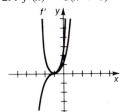

31. $f'(x) = -\dfrac{1}{x^2}$

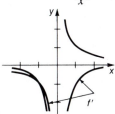

33. $f'(x) = \begin{cases} 1, x > 0 \\ \text{DNE; if } x = 0 \\ -1, x < 0 \end{cases}$

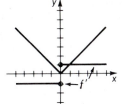

35. 3   **37.** $2\pi r$ **39.** $f'_+ (3) = f'_- (3) = 7$ **41.** $f'_+ (0) = 1, f'_- (0) = -1$

45. $P'(t) = kP(t)$ **47.** y-intercept is $f(a) - af'(a)$; x-intercept is $a - \dfrac{f(a)}{f'(a)}$

EXERCISE SET 3.3, PAGE 120

1. $2x - 2$ **3.** $x + 4$ **5.** $527x^{30} + 406\, x^{28} + 4x^3$ **7.** $2\pi t + 3 + 4\sqrt{2}$ **9.** $4x^3$

11. $2t(2t^2 - 3)(3t^4 - t^{-2}) + (t^4 - 3t^2 + 5)(12t^3 + 2t^{-3})$ **13.** $\dfrac{2}{3}x - \dfrac{6}{x^3}$ **15.** $(z + 1)^{-2}$

17. $\dfrac{-15x^2 + 3 - 40x}{(5x^2 + 1)^2}$ **19.** $\dfrac{4t^2 + 4t + 22}{(t^2 + 3t - 4)^2}$ **21.** $\dfrac{2s^5 + 8s^3 + 10s}{(s^2 + 2)^2}$ **23.** $\dfrac{x^4 + 4x^2 - 1}{(x^2 + 1)^2}$

25. $3x^2 + 12x + 11$ **27.** $4x^3 - 9x^2 - 18x - 5$ **29.** $\left(-3/2, -\dfrac{25}{4}\right)$

31. $y = \dfrac{8x}{25} - \dfrac{1}{25}$ **33.** (a) 150 gal/min   (b) 100 gal/min

35. $P(0) = 1000, P'(t) = 9000/(t + 1)^2$ **37.** $.014\,\dfrac{\text{cal}}{(\text{deg})^2\,\text{mole}}$

39. (a) $\dfrac{3}{2}x^{1/2}$ (b) $\dfrac{5}{2}x^{3/2}$ (c) $-\dfrac{1}{2}x^{-3/2}$ **41.** $\dfrac{-n}{2}x^{-n/2-1}$ **43.** $8(2x + 1)^3$

EXERCISE SET 3.4, PAGE 127

1. $\cos x + (\sec x)^2$ **3.** $3x^2 + \sin x$ **5.** $3t^2 \cos t - t^3 \sin t$ **7.** $(\cos x)^2 - (\sin x)^2$

9. $\tan t \cot t = 1$, so $r'(t) = 0$ **11.** $2 \sin x \cos x$ **13.** $2 \csc x - 2x \csc x \cot x$

15. $\tan x + x(\sec x)^2 + 2x \cot x - x^2(\csc x)^2$ **17.** $\dfrac{2\theta + 2\theta \cos \theta + \theta^2 \sin \theta}{(1 + \cos \theta)^2}$

19. $h(\theta) = \csc \theta + \cot \theta$ so $h'(\theta) = -\cot \theta \csc \theta - (\csc \theta)^2$

21. $\dfrac{\tan t + (\sec t)^2 \sin t\,(\sin t + \cos t)}{(\sin t + \cos t)^2}$ **23.** $\cos x - \sin x$

25. $(\tan x)^2 + 2x \tan x\,(\sec x)^2$ **27.** $\dfrac{1}{2}$ **29.** 0   **31.** 1

35. $y = \dfrac{\sqrt{3}}{2} - \dfrac{\pi}{6} + \dfrac{1}{2}x$ **37.** $\left(\dfrac{\pi}{4}, \sqrt{2}\right)$ **41.** $y = 10 \csc \theta, \dfrac{dy}{d\theta} = -10\sqrt{2}$

EXERCISE SET 3.5, PAGE 133

1. $4(x + 1)^3$ **3.** $7(2x - 3)(x^2 - 3x + 4)^6$ **5.** $-12(6x + 4)^{-3}$

7. $-4(6x - 1)(6x^2 - 2x + 1)^{-3}$ **9.** $\dfrac{24x}{5}(3x^2 + 1)^3$ **11.** $-42x(3x^2 + 13)^{-8}$

13. $6(x - 1)^2/(x + 1)^4$ **15.** $(x - 1)(5 - x)/(x + 1)^4$ **17.** $\pi \cos \pi x$

19. $-3 \sin x (\cos x - 1)^2$ **21.** $-3 \sin x (\cos x)^2$ **23.** $\dfrac{-2}{(x - 1)^2}\left[\sec\left(\dfrac{x + 1}{x - 1}\right)\right]^2$

25. $2x \cos(x^2 + 1) - 2x^3 \sin(x^2 + 1)$ **27.** $(1 - x^2)/(x^2 + 1)^2$

29. $6 (\tan 2x \sec 2x)^2 \sin(1 - x^2) - 2x(\tan 2x)^3 \cos(1 - x^2)$

31. $6(\sin x)^2 \cos x[(\sin x)^3 + 1]$ **33.** (a) $12(4x - 5)^2$ (b) $12(4x - 5)^2 + 2$

(c) $2[(4x - 5)^3 + 2x + 1][12(4x - 5)^2 + 2]$

(d) $4\{[(4x - 5)^3 + 2x + 1]^2 - 7x\}^3 \{2[(4x - 5)^3 + 2x + 1][12(4x - 5)^2 + 2] - 7\}$

35. (a) $2x - 1$ (b) $(2x - 1) \cos(x^2 - x)$

(c) $2[x^3 + \sin(x^2 - x)] [3x^2 + (2x - 1) \cos(x^2 - x)]$

37. $256y - 9x + 14 = 0$ **39.** $y = 4x + 1$ **41.** $8\pi t(t^2 + 3)^2$

43. $10\pi r \dfrac{cm^2}{sec}, 300\pi \dfrac{cm^2}{sec}$ **45.** $3\pi h^2 \dfrac{in^3}{sec}$ **47.** $\dfrac{500}{(2 + t)^2}$

49. $\begin{cases} 2 & , \text{if } x > 0 \\ \text{undefined, if } x = 0 \\ -2 & , \text{if } x < 0 \end{cases}$ **51.** $\begin{cases} 2x & , \text{if } x > 1 \text{ or } x < -1 \\ \text{undefined, if } x = \pm 1 \\ -2x & , \text{if } -1 < x < 1 \end{cases}$

55. $y = ax^3 + bx^2$ where $a = \dfrac{m \cos \alpha - 2 \sin \alpha}{R_2^2 (\cos \alpha)^3}$ and $b = \dfrac{3 \sin \alpha - m \cos \alpha}{R_2(\cos \alpha)^2}$

EXERCISE SET 3.6, PAGE 140

1. $-x^2/y^2$ **3.** $-y/x$ **5.** 1 **7.** $\dfrac{4x - 2xy^2}{2x^2y + 3y^{-4}}$ **9.** $\dfrac{\cos x}{\sin y}$ **11.** $\dfrac{\cos x}{2y}$ **13.** $\dfrac{1}{3}x^{-2/3}$

15. $\dfrac{12}{5}x^{-1/5}$ **17.** $\dfrac{2}{3}(x + 1)^{-1/3} + 1$ **19.** $\dfrac{1}{2}(x + 1)(x^2 + 2x + 2)^{-3/4}$

21. $\dfrac{2}{3}(6x + 4)(3x^2 + 4x + 1)^{-1/3}$ **23.** $\dfrac{1}{2}x^{-1/2} + \dfrac{3}{4}x^{-1/4} - \dfrac{6}{5}x^{-11/5}$

25. $\dfrac{2}{3}x(x^2 + 2)^{-2/3}(x^2 - 2)^{1/4} + \dfrac{1}{2}x(x^2 - 2)^{-3/4}(x^2 + 2)^{1/3}$

27. $12x - \dfrac{3}{x^2} + \dfrac{8}{9}x(2x^2)^{-4/3}$

29. $\dfrac{1}{2}\left[x^2 + \sqrt{x^2 + \sqrt{x + 1}}\right]^{-1/2}\left[2x + \dfrac{1}{2}(x^2 + \sqrt{x + 1})^{-1/2}\left(2x + \dfrac{1}{2}(x + 1)^{-1/2}\right)\right]$

31. $-\dfrac{1}{2}\sqrt[3]{\dfrac{y}{x^2}}$ **33.** $(2y(xy + 1)^2 + x^2)^{-1}$ **35.** $\dfrac{y \cos (x/y) - y^2}{x \cos (x/y)}$ **37.** $\dfrac{y[\sec(xy)]^2}{1 - x[\sec(xy)]^2}$

39. $y = x + 2$ **41.** -2 **43.** $y = 2 - x$ **45.** $-x_0/y_0$ **47.** $y = 4 - \sqrt{3}x$

49. $y = \dfrac{\sqrt{3}}{2} + \sqrt{3}x$ **51.** $r > 1$ **53.** $-\dfrac{7}{26}$

EXERCISE SET 3.7, PAGE 143

1. $6x + 4, 6$ **3.** $4x^3 - 4x^{-5}, 12x^2 + 20x^{-6}$ **5.** $\dfrac{3}{2}x^{-1/2} + 4x, \dfrac{-3}{4}x^{-3/2} + 4$

7. $2 \cos 2x, -4 \sin 2x$ **9.** $-(2x + 5)^{-3/2}, 3(2x + 5)^{-5/2}$ **11.** $(x + 1)^{-2}, -2(x + 1)^{-3}$

13. $x^{-1/2}(\sqrt{x} + 1)^{-2}, -x^{-1}(\sqrt{x} + 1)^{-3} - \dfrac{1}{2}x^{-3/2}(\sqrt{x} + 1)^{-2}$

15. $2x \cos(x^2 + 1), 2 \cos(x^2 + 1) - 4x^2 \sin(x^2 + 1)$

17. $\dfrac{1}{2}x^{-1/2} + \dfrac{3}{2}x^{1/2}, \dfrac{-1}{4}x^{-3/2} + \dfrac{3}{4}x^{-1/2}$ **19.** $2(\sec 2x)^2, 8(\sec 2x)^2 \tan 2x$

21. $\dfrac{1}{2}(x + x^3)^{-1/2}(1 + 3x^2), 3x(x + x^3)^{-1/2} - (1 + 3x^2)^2(x + x^3)^{-3/2}/4$

23. $\dfrac{3}{8}x^{-5/2} - \dfrac{3}{8}x^{-3/2}, \dfrac{-15}{16}x^{-7/2} + \dfrac{9}{16}x^{-5/2}$

25. $32(\sec 2x)^2(\tan 2x)^2 + 16(\sec 2x)^4 = 48(\sec 2x)^4 - 32(\sec 2x)^2,$
    $64 \tan 2x (\sec 2x)^2 (6(\sec 2x)^2 - 1)$

27. $(-1)^n n! (x + 1)^{-(n+1)}$ **29.** $(-1)^n 2^n n! (2x + 1)^{-(n+1)}$

31. $(-1)^{n+1} \dfrac{3}{2^n}((2n - 5)(2n - 7) \cdots 1)x^{-(2n-3)/2}$

33. $f^{(2n)}(x) = (-1)^n \sin x, f^{(2n+1)}(x) = (-1)^n \cos x$ **35.** $-\dfrac{(y^2 - x^2)}{y^2}$ **37.** 0

39. $\dfrac{2(x + y + 1)}{(1 - x)^2}$ **41.** $\dfrac{2y^{3/2}(5y^{1/2} - 8xy)}{(1 - 2xy^{1/2})^3}$ **43.** $h'' = f'(g) \cdot g'' + f''(g) \cdot (g')^2$

45. $P(0) = a_0, P'(0) = a_1,$ and $P''(0) = 2a_2$ **47.** $a_k = \dfrac{P^{(k)}(0)}{k!}$

EXERCISE SET 3.8, PAGE 151

1. 2 **3.** $1, 3$ **5.** $2, -4$ **7.** $\dfrac{1}{3}, -1$ **9.** none   **11.** none   **13.** $2, -2$ **15.** none

17. -2 **19.** $\dfrac{\pi}{4} + n\pi$ **21.** $n\pi, \pm \dfrac{\pi}{3} + 2n\pi$ for any integer n

23. $t = n\pi$ for any integer n or $\cos t = -\dfrac{1}{4}$.

25.   **27.**   **29.** $1, -2$; abs. min at $x = -2$

3; abs. min. at $x = 3$ 3; abs. max at $x = 3$

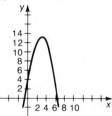
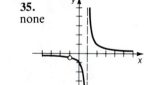

31. none   **33.** $1, -1$; abs. min. at $x = 1$ and at $x = -1$ **35.** none

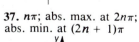
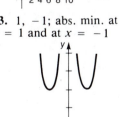
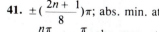

37. $n\pi$; abs. max. at $2n\pi$; abs. min. at $(2n + 1)\pi$ **39.** $n\pi$ **41.** $\pm (\dfrac{2n + 1}{8})\pi$; abs. min. at $x = \dfrac{n\pi}{2} - \dfrac{\pi}{8}$; abs. max. at $x = \dfrac{n\pi}{2} + \dfrac{\pi}{8}$

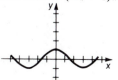
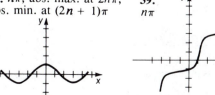
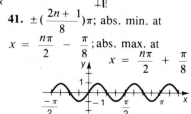

43. 0, 2, −2; abs. max. at $x = 0$, abs. min. at $x = -2$ and $x = 2$

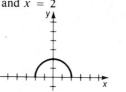

45. 0, 1, $\frac{2}{3}$; abs. min. at $x = 0$ and $x = 1$

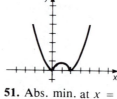

47. abs. min. value at $x = 2$. Abs. max. at $x = 0$ and $x = 4$

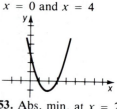

49. Abs. min. at $x = -2$, abs. max. at -4

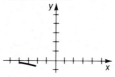

51. Abs. min. at $x = 5$. Abs. max. at $x = 4$.

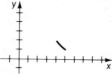

53. Abs. min. at $x = 2$. Abs. max. at $x = 0$.

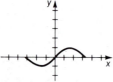

55. Absolute maximum is 1 at $x = \pi/2$. Absolute minimum is -1 at $x = -\pi/2$.

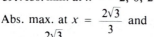

57. Absolute maximum is 1 at $x = 0$. Absolute minimum is $\cos 1 \approx .54$ at $x = \pm 1$.

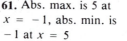

59. Abs. min. at $x = -2, 0, 2$. Abs. max. at $x = \dfrac{2\sqrt{3}}{3}$ and $x = -\dfrac{2\sqrt{3}}{3}$.

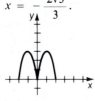

61. Abs. max. is 5 at $x = -1$, abs. min. is -1 at $x = 5$

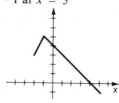

63. 2, when the number is one

67. Some examples are
   (a) $f(x) = 1/x$
   (b) $f(x) = -1/x$
   (c) $f(x) = x$
   (d) $f(x) = \dfrac{1}{x} \sin \dfrac{1}{x}$

69. Absolute maximum is $C(15) = 105$
   Absolute minimum is $C(5) = 5$

71. 17.93°C   **73.** $\dfrac{S_1}{2}$

75. 1.53×10^{-3} moles per liter

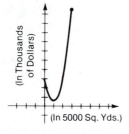

(In Thousands of Dollars)

(In 5000 Sq. Yds.)

EXERCISE SET 3.9, PAGE 159

1. $\dfrac{3}{2}$ **3.** $\dfrac{2 \pm \sqrt{7}}{3}$ **5.** f is not continuous on $[-1, 1]$. Hypotheses not satisfied.

7. $\dfrac{\pi}{2}$ **9.** $\dfrac{3\pi}{4}$ **11.** $\dfrac{1}{2}$ **13.** $\sqrt{2}$ **15.** $\dfrac{32}{9}$ **17.** 0   **19.** $1 - \sqrt{2}$

21. $\cos c = \dfrac{2}{\pi}$, $c \approx .88$ **23.** $f(x) = 3$ **25.** $f(x) = x^2 + 5$ **27.** $f(x) = \sin x + 2$

29. $f(x) = x^2 + 3x + 2$

31. (a) $f'(0)$ is not defined   (b) f is not continuous at $\dfrac{\pi}{2}$ (c) f is not continuous at 1
(d) f is not continuous at $\dfrac{\pi}{2}$

35. One example is $f(x) = 1/x$

EXERCISE SET 3.10, PAGE 165

1. Decreasing on $(-\infty, -2) \cup (1, 3)$; increasing on $(-2, 1) \cup (3, 8)$; relative maximum at 1; relative minimum at -2, absolute minimum of -3 at 3. $f'(-2) = f'(1) = f'(3) = 0$

3. Constant on $(-\infty, 0]$; inc. on $(0, 2)$, decreasing on $(2, \infty)$; relative maximum at 2; absolute maximum at each number in $(-\infty, 0]$. $f'(2) = 0$, $f'(0)$ DNE, $f'(x) = 0$ for all x in $(-\infty, 0)$

5. Decreasing on $(-3, -2) \cup (0, 1) \cup (2, 3)$; increasing on $(-\infty, -3) \cup (-1, 0)$ $\cup (1, 2) \cup (3, \infty)$; relative maximum at -3, at 0 and at 2; relative minimum at 1, at 3, and at each number in $[-2, -1]$. No absolute maximum or minimum. $f'(-3) = f'(0) = f'(2) = 0$. $f'(x)$ does not exist when $x = \pm 1$, $x = -2$, or $x = 3$.

7. Increasing on $(-\infty, \infty)$

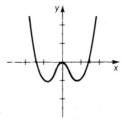

9. Increasing on $(2, \infty)$; decreasing on $(-\infty, 2)$; relative minimum at $(2, 0)$

11. Increasing on $\left(\dfrac{9}{4}, \infty\right)$; decreasing on $\left(-\infty, \dfrac{9}{4}\right)$; relative minimum at $\left(\dfrac{9}{4}, -\dfrac{57}{8}\right)$

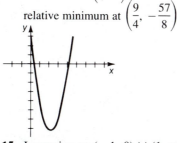

13. Increasing on $(-\infty, -1)$ $\cup (1, \infty)$; decreasing on $(-1, 1)$; relative maximum at $(-1, 2)$; relative minimum at $(1, -2)$

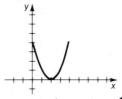

15. Increasing on $(-1, 0) \cup (1, \infty)$; decreasing on $(-\infty, -1) \cup (0, 1)$; relative minimum at $(-1, -1)$ and $(1, -1)$; relative maximum at $(0, 0)$

17. Increasing on $\left(-\infty, \dfrac{1}{2}\right)$ $\cup \left(\dfrac{1}{2}, \infty\right)$; decreasing on $\left(-\dfrac{1}{2}, 0\right) \cup \left(0, \dfrac{1}{2}\right)$; relative maximum at $x = -\dfrac{1}{2}$; relative minimum at $x = \dfrac{1}{2}$.

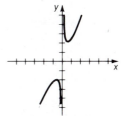

19. Increasing on $\left(-\infty, \dfrac{3}{4}\right)$; decreasing on $\left(\dfrac{3}{4}, \infty\right)$; relative maximum at $\left(\dfrac{3}{4}, \dfrac{27}{256}\right)$

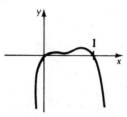

21. Increasing on $(\pi/2 + (2n - 1)\pi,$ $\pi/2 + 2n\pi)$; decreasing on $(\pi/2 + 2n\pi, \pi/2 + (2n + 1)\pi)$; relative maximum at $x = \dfrac{\pi}{2} + 2n\pi$; relative minimum at $x = \dfrac{\pi}{2} + (2n + 1)\pi$

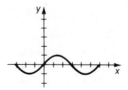

23. Always increasing

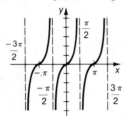

25. Increasing on $(-1, 0) \cup (1, \infty)$; decreasing on $(-\infty, -1) \cup (0, 1)$; relative minimum at -1 and at 1

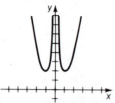

27. Decreasing on $(-\infty, 1) \cup (1, \infty)$

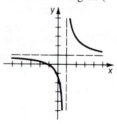

29. Increasing on $(-\infty, -1) \cup (-1, 0)$ decreasing on $(0, 1) \cup (1, \infty)$; relative maximum at 0

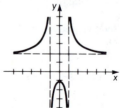

31. Increasing on $\left(\dfrac{-3\sqrt{2}}{2}, \dfrac{3\sqrt{2}}{2}\right)$; decreasing on $\left(-3, \dfrac{-3\sqrt{2}}{2}\right) \cup \left(\dfrac{3\sqrt{2}}{2}, 3\right)$; relative minimum at $-\dfrac{3\sqrt{2}}{2}$; relative maximum at $\dfrac{3\sqrt{2}}{2}$

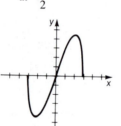

33. Decreasing on $(-\infty, -1) \cup \left(\dfrac{1}{2}, 2\right)$; increasing on $\left(-1, \dfrac{1}{2}\right) \cup (2, \infty)$; relative minimum at $-1, 2$; relative maximum at $\dfrac{1}{2}$

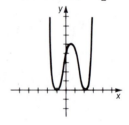

The general solution is therefore

$$y(x) = C_1 e^{3x} + C_2 e^{2x}.$$ □

EXAMPLE 2   Solve the initial-value problem

$$y'' - 3y' - 4y = 0, \qquad y(0) = 0, \qquad y'(0) = 1.$$

SOLUTION

If $y(x) = e^{mx}$,

$$y'' - 3y' - 4y = (m^2 - 3m - 4)e^{mx}$$
$$= (m - 4)(m + 1)e^{mx},$$

so the general solution is

$$y(x) = C_1 e^{4x} + C_2 e^{-x}.$$

Since $y(0) = 0$,

$$0 = C_1 e^0 + C_2 e^0 = C_1 + C_2$$

Since $y'(0) = 1$ and $y'(x) = 4C_1 e^{4x} - C_2 e^{-x}$,

$$1 = 4C_1 e^0 - C_2 e^0 = 4C_1 - C_2.$$

Adding the equations gives

$$1 = 5C_1, \text{ so } C_1 = \frac{1}{5}.$$

Since $C_1 + C_2 = 0$,

$$C_2 = -C_1 = -\frac{1}{5}.$$

The particular solution to the initial-value problem is

$$y(x) = \frac{1}{5} e^{4x} - \frac{1}{5} e^{-x}.$$ □

CASE II: $p^2 - 4q = 0$.

When $p^2 - 4q = 0$, the characteristic equation has only one root (a double root) and only one solution to (18.13) is of the form $y(x) = e^{mx}$. This solution is

$$y_1(x) = e^{m_1 x} = e^{-px/2}.$$

It can be easily verified (see Exercise 36) that in this case an additional solution to (18.13) is

$$y_2(x) = xe^{m_1 x} = xe^{-px/2}.$$

The general solution to equation (18.13) when $p^2 - 4q = 0$ is

(18.17) $$y(x) = C_1 e^{m_1 x} + C_2 x e^{m_1 x}.$$

EXAMPLE 3   Find the solution to the initial-value problem

$$y'' - 4y' + 4y = 0, \qquad y(0) = 1, \qquad y'(0) = 0.$$

SOLUTION

The characteristic equation $m^2 - 4m + 4 = 0$ has the double root 2. Hence, the general solution is

$$y(x) = C_1 e^{2x} + C_2 x e^{2x}.$$

Satisfying the initial condition $y(0) = 1$ requires that

$$1 = C_1 e^0 + C_2 \cdot 0 \cdot e^0 = C_1.$$

Since $y'(x) = 2C_1 e^{2x} + C_2 e^{2x} + 2C_2 x e^{2x}$, satisfying $y'(0) = 0$ requires that

$$0 = 2C_1 e^0 + C_2 e^0 + 2C_2 \cdot 0 e^0 = 2C_1 + C_2.$$

Thus, $C_2 = -2C_1 = -2$ and the particular solution is

$$y(x) = e^{2x} - 2x e^{2x}. \qquad \square$$

CASE III: $p^2 - 4q < 0.$

When $p^2 - 4q < 0$, the roots m_1 and m_2 are distinct but complex numbers. To simplify notation, let $r = -p/2$ and $s = \sqrt{4q - p^2}/2$. Then m_1 and m_2 can be expressed as

$$m_1 = -\frac{p}{2} + \frac{\sqrt{p^2 - 4q}}{2} = -\frac{p}{2} + \frac{\sqrt{4q - p^2}}{2} i = r + is$$

and

$$m_2 = -\frac{p}{2} - \frac{\sqrt{p^2 - 4q}}{2} = r - is.$$

As mentioned in Section 9.9, the symbol i is used to distinguish the nonreal, or imaginary, part of a complex number and has the property that $i^2 = -1$. With this notation, the general solution can be expressed as

$$\begin{aligned}
y(x) &= C_1 e^{m_1 x} + C_2 e^{m_2 x} \\
&= C_1 e^{(r + is)x} + C_2 e^{(r - is)x} \\
&= C_1 e^{rx} e^{sxi} + C_2 e^{rx} e^{-sxi} \\
&= e^{rx} (C_1 e^{sxi} + C_2 e^{-sxi}).
\end{aligned}$$

To simplify this result we use a result that was presented as Equation (9.46) in Section 9.9. This result is called Euler's equation and has the form

$$e^{ix} = \cos x + i \sin x.$$

Then

$$y(x) = e^{rx} [C_1(\cos sx + i \sin sx) + C_2(\cos(-sx) + i \sin(-sx))]$$

or, since $\cos(-sx) = \cos sx$, and $\sin(-sx) = -\sin sx$,

$$y(x) = e^{rx} [(C_1 + C_2)\cos sx + i(C_1 - C_2)\sin sx].$$

The general solution can therefore be expressed as

(18.18) $y(x) = e^{rx} (K_1 \cos sx + K_2 \sin sx),$

where K_1 and K_2 are arbitrary constants. While K_1 and K_2 appear to be complex numbers, they will always be real numbers if the coefficients p and q are real.

EXAMPLE 4   Find the general solution to the differential equation

$$y'' + 25y = 0.$$

SOLUTION

The characteristic equation $m^2 + 25 = 0$ has roots

$$m_1 = 5i \quad \text{and} \quad m_2 = -5i$$

so $r = 0 \quad \text{and} \quad s = 5.$

The general solution is consequently

$$y(x) = e^{0x} (K_1 \cos 5x + K_2 \sin 5x)$$
$$= K_1 \cos 5x + K_2 \sin 5x. \qquad \square$$

EXAMPLE 5   Solve the initial-value problem

$$y'' - 2y' + 10y = 0, \qquad y\left(\frac{\pi}{2}\right) = 0, \qquad y'\left(\frac{\pi}{2}\right) = 1.$$

SOLUTION

The characteristic equation $m^2 - 2m + 10 = 0$ has roots

$$m_1 = \frac{2 + \sqrt{-36}}{2} = 1 + 3i \quad \text{and} \quad m_2 = \frac{2 - \sqrt{-36}}{2} = 1 - 3i,$$

so $r = 1$ and $s = 3$.
The general solution is

$$y(x) = e^x (K_1 \cos 3x + K_2 \sin 3x).$$

Since $y(\pi/2) = 0$,

$$0 = e^{\pi/2} \left(K_1 \cos \frac{3\pi}{2} + K_2 \sin \frac{3\pi}{2} \right) = e^{\pi/2} K_2(-1),$$

which implies $K_2 = 0$

and $y(x) = e^x (K_1 \cos 3x).$

Since $y'\left(\frac{\pi}{2}\right) = 1 \quad \text{and} \quad y'(x) = e^x (K_1 \cos 3x - 3K_1 \sin 3x),$

$$1 = e^{\pi/2} \left(K_1 \cos \frac{3\pi}{2} - 3K_1 \sin \frac{3\pi}{2} \right)$$
$$= e^{\pi/2} (-3K_1)(-1) = 3e^{\pi/2} K_1.$$

Thus,
$$K_1 = \frac{1}{3} e^{-\pi/2}$$

and the particular solution to the initial value problem is

$$y(x) = \frac{1}{3} e^{-\pi/2} e^x \cos 3x = \frac{1}{3} e^{(x-\pi/2)} \cos 3x. \qquad \square$$

EXERCISE SET 18.5

Solve the differential equations in Exercises 1 through 24.

1. $y'' - 9y = 0$

2. $y'' - 6y' + 4y = 0$

3. $y'' + 6y' + 9y = 0$

4. $y'' - 2y' + y = 0$

5. $y'' - 4y' - 5y = 0$

6. $y'' + 5y' + 4y = 0$

7. $y'' - 2y' + 5y = 0$

8. $y'' + 4y' + 13y = 0$

9. $y'' - 2y' = 0$

10. $y'' + 2y' = 0$

11. $y'' + 4y' + 2y = 0$

12. $y'' - 6y' + 9y = 0$

13. $y'' - 7y' - 8y = 0$

14. $y'' + 2y' + 2y = 0$

15. $y'' + 6y' + 5y = 0$

16. $y'' + 8y' + 4y = 0$

17. $y'' + 9y = 0$

18. $y'' + 3y' + 4y = 0$

19. $2y'' + 3y' + y = 0$

20. $2y'' - 2y' + y = 0$

21. $4y'' + 9y = 0$

22. $4y'' - 9y = 0$

23. $5y'' + 6y' + 9y = 0$

24. $3y'' + 7y' - 6y = 0$

Find the particular solutions to the initial-value problems in Exercises 25 through 30.

25. $y'' - 4y' + 4y = 0$, $y(0) = 0$, $y'(0) = 1$.

26. $y'' - 4y' + 3y = 0$, $y(0) = 0$, $y'(0) = 1$.

27. $y'' + 6y' + 13y = 0$, $y(0) = 1$, $y'(0) = 0$.

28. $y'' - 3y' - 4y = 0$, $y(0) = 1$, $y'(0) = 4$.

29. $8y'' - 14y' + 3y = 0$, $y(0) = 1$, $y'(0) = 3$.

30. $2y'' - 6y' + 9y = 0$, $y(\pi) = 0$, $y'(\pi) = e^{3\pi}$.

Find the particular solutions to the boundary-value problems in Exercises 31 through 34.

31. $y'' + y = 0$, $y(0) = 1$, $y\left(\dfrac{\pi}{2}\right) = -1$.

32. $y'' - 2y' + 5y = 0$, $y(0) = 1$, $y\left(\dfrac{\pi}{3}\right) = 0$.

33. $y'' + y' = 0$, $y(0) = 2$, $y(1) = 1$.

34. $3y'' - 5y' + 2y = 0$, $y(0) = 1$, $y(\ln 8) = 2$.

35. Suppose that $y_1(x)$ and $y_2(x)$ are solutions to

$$y'' + P_1(x)y' + P_2(x)y = 0.$$

Show that for any constants C_1 and C_2,

$$y(x) = C_1 y_1(x) + C_2 y_2(x)$$

is also a solution to this equation.

36. Consider the differential equation $y'' + py' + qy = 0$ where $p^2 - 4q = 0$. Show that $y(x) = xe^{-px/2}$ is a solution to this equation.

37. Show that the general solution to $y'' - y = 0$ can be expressed either as

$$y = C_1 e^x + C_2 e^{-x}$$

or as $\qquad\qquad y = K_1 \sinh x + K_2 \cosh x.$

Find the constants C_1, C_2 and K_1, K_2 that give the particular solution satisfying $y(0) = 1$ and $y'(0) = 0$. Show that the two forms of the solutions are equivalent.

38. An equation of the form $y'' + ay' = 0$ can be solved by solving the two first-order equations $w' + aw = 0$ and $y' = w$, since $w' = C_1 e^{-ax}$ and $y = \int w \, dx$. Use this method to solve the differential equations in Exercises 9 and 10.

A simple model for a vibrating system consisting of a mass m, a spring with spring constant k, a shock absorber with damping constant c, and no externally imposed force, is described by the initial-value problem

$$m\,y''(t) + c\,y'(t) + k\,y(t) = 0, \qquad y(0) = y_0, \qquad y'(0) = y_0'.$$

In this equation, $y(t)$ is the distance of the mass below its equilibrium position at time t and describes the motion of the mass, called *harmonic motion*.

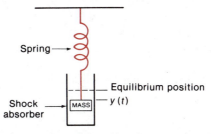

Find an expression for the motion of the vibrating systems described in Exercises 39 through 42.

39. $m = 1$ slug, $k = 16$ lb/ft, $c = 0$, $y_0 = 0$, $y_0' = 0$

40. $m = 4$ kg, $k = 4$ newtons/meter, $c = 0$, $y_0 = 0$, $y_0' = 1$

41. $m = 1$ slug, $k = 16$ lb/ ft , $c = 4$ lb sec/ft, $y_0 = 0$, $y_0' = 1$

42. $m = 4$ kg, $k = 4$ newtons/meter, $c = 3$ dyne sec/meter, $y_0 = 1$, $y_0' = 0$
(1 dyne $= 10^{-5}$ newtons)

A simple model for an electrical circuit containing, in series, an inductance L, a resistance R, a capacitance C, and no impressed voltage is described by the initial value problem

$$L I'(t) + R I(t) + \frac{1}{C} Q(t) = 0, \qquad Q(0) = Q_0, \qquad I(0) = I_0.$$

In this equation, $Q(t)$ is the charge on the capacitor at time t and $I(t) = Q'(t)$ is the current at time t. (The close relationship between the electric circuit differential equation and the differential equation for vibrating systems and other physical phenomena is the basis for analog computers.)

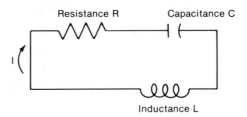

Find an expression for the charge and current in the electrical systems described in Exercises 43 through 46.

43. $L = .25$ henry, $R = 5$ ohms, $C = 10^{-5}$ farads, $I_0 = 0$, $Q_0 = 0$

44. $L = .5$ henry, $R = 4$ ohms, $C = 2 \times 10^{-6}$ farads, $I_0 = 0$, $Q_0 = 0$

45. $L = .25$ henry, $R = 5$ ohms, $C = 10^{-5}$ farads, $I_0 = 0$, $Q_0 = 10^{-6}$ coulombs

46. $L = .5$ henry, $R = 4$ ohms, $C = 2 \times 10^{-6}$ farads, $I_0 = 0$, $Q_0 = 10^{-6}$ coulombs

18.6
SECOND-ORDER LINEAR EQUATIONS: NONHOMOGENEOUS TYPE

In Section 18.5 we discussed the various forms that can be assumed by solutions of the second-order linear homogeneous equation

(18.19)
$$y''(x) + py'(x) + qy(x) = 0,$$

when p and q are constants. In this section we discuss methods for solving the nonhomogeneous counterpart to this equation:

(18.20) $y''(x) + py'(x) + qy(x) = Q(x),$ $Q(x) \neq 0.$

The following result indicates the reason for first considering the homogeneous problem.

(18.21)
THEOREM

If $y_c(x) = C_1 y_1(x) + C_2 y_2(x)$ is the general solution to the homogeneous problem (18.19) and y_p is any solution to the nonhomogeneous problem (18.20), then the general solution to (18.20) has the form

$$y(x) = C_1 y_1(x) + C_2 y_2(x) + y_p(x).$$

The proof of this theorem follows by observing that y is a solution to (18.20) if and only if $y - y_p$ is a solution to (18.19).

The methods of Section 18.5 can be used to find the general solution to the homogeneous equation. Consequently, Theorem 18.21 implies that the problem of finding a general solution to the nonhomogeneous equation is reduced to finding a *single* particular solution.

We present two methods for finding a particular solution to the nonhomogeneous equation. The first technique is called the **variation of parameters** method. In theory, this method can be used to solve any nonhomogeneous problem, regardless of the form of $Q(x)$. In practice, the method often involves integrals that can be very difficult or impossible to evaluate. The second technique is called the **method of undetermined coefficients.** This method is readily applied, but can be used only when $Q(x)$ is the sum and/or product of sine, cosine, polynomial, or exponential functions. This feature limits its use, although many of the nonhomogeneous problems most important for applications have $Q(x)$ in this form.

VARIATION OF PARAMETERS

The variation of parameters method assumes there is a particular solution to the nonhomogeneous problem (18.20) of the form

(18.22)
$$y_p(x) = u_1(x)y_1(x) + u_2(x)y_2(x)$$

where

(18.23)
$$u_1'(x)y_1(x) + u_2'(x)y_2(x) \equiv 0$$

and $C_1y_1(x) + C_2y_2(x)$ is the general solution to the corresponding homogeneous equation (18.19). Differentiating y_p and using condition (18.23) implies that

(18.24)
$$y_p'(x) = u_1'(x)\,y_1(x) + u_1(x)\,y_1'(x) + u_2'(x)\,y_2(x) + u_2(x)\,y_2'(x)$$
$$= u_1(x)\,y_1'(x) + u_2(x)\,y_2'(x),$$

so

(18.25)
$$y_p''(x) = u_1'(x)\,y_1'(x) + u_1(x)\,y_1''(x) + u_2'(x)\,y_2'(x) + u_2(x)\,y_2''(x).$$

Substituting (18.24) and (18.25) into Equation (18.20) and using the fact that y_1 and y_2 are solutions to the homogeneous problem produces

$$Q(x) = y_p''(x) + p\,y_p'(x) + q\,y_p(x)$$
$$= (u_1'\,y_1' + u_1\,y_1'' + u_2'\,y_2' + u_2\,y_2'') + p(u_1u_1' + u_2\,y_2')$$
$$+ q(u_1y_1 + u_2y_2)$$
$$= u_1(y_1'' + py_1'' + qy_1) + u_2\,(y_2'' + py_2' + qy_2) + u_1'y_1' + u_2'y_2'$$
$$= u_1\,0 + u_2\,0 + u_1'y_1' + u_2'y_2'.$$

Thus

(18.26)
$$u_1'(x)\,y_1'(x) + u_2'(x)\,y_2'(x) = Q(x).$$

To solve equations (18.23) and (18.26) for $u_1'(x)$, multiply equation (18.23) by $y_2'(x)/y_2(x)$ and subtract from equation (18.26) to give

$$u_1'(x)\,[y_1'(x) - y_1(x)y_2'(x)/y_2(x)] = Q(x).$$

So

(18.27)
$$u_1'(x) = \frac{Q(x)\,y_2(x)}{y_1'(x)y_2(x) - y_1(x)y_2'(x)}.$$

In a similar manner,

(18.28)
$$u_2'(x) = \frac{Q(x)y_1(x)}{y_1(x)y_2'(x) - y_1'(x)y_2(x)}.$$

A particular solution y_p to the nonhomogeneous equation can be found by using (18.27) and (18.28) to find u_1 and u_2 and then substituting these in (18.22). The general solution to the nonhomogeneous equation (18.20) is then

(18.29)
$$y(x) = C_1y_1(x) + C_2y_2(x) + y_p(x).$$

EXAMPLE 1   Find the solution to the initial-value problem

$$y'' - 2y' + y = \frac{e^x}{x^2}, \qquad y(1) = 0, \qquad y'(1) = 0, \qquad x > 0.$$

SOLUTION

The first step is to determine the solution to the homogeneous equation

$$y'' - 2y' + y = 0,$$

whose characteristic equation is

$$0 = (m^2 - 2m + 1) = (m - 1)^2.$$

Consequently, the general solution to the homogeneous equation is

$$y(x) = C_1e^x + C_2xe^x.$$

A particular solution of the nonhomogeneous equation is assumed to have the form

$$y_p(x) = u_1(x)e^x + u_2(x)xe^x.$$

Equation (18.23) implies that

$$u_1'(x)e^x + u_2'(x)xe^x = 0.$$

Differentiating y_p and substituting into the nonhomogeneous equation leads to the result from equation (18.26):

$$u_1'(x)e^x + u_2'(x)(e^x + xe^x) = \frac{e^x}{x^2}.$$

The equations can then be solved for u_1' and u_2' or we can use Equations (18.27) and (18.28) to obtain

$$u_1'(x) = \frac{\left(\dfrac{e^x}{x^2}\right)xe^x}{e^xxe^x - e^x(e^x + xe^x)} = -\frac{1}{x}$$

and

$$u_2'(x) = \frac{\left(\dfrac{e^x}{x^2}\right)e^x}{e^x(e^x + xe^x) - e^xxe^x} = \frac{1}{x^2}.$$

So
$$u_1(x) = -\int \frac{1}{x}\, dx = -\ln x + K_1,$$

and
$$u_2(x) = \int \frac{1}{x^2}\, dx = -\frac{1}{x} + K_2.$$

Since we are interested in only one particular solution, we can take the constants K_1 and K_2 to be zero.

Thus,
$$y_p(x) = u_1(x)y_1(x) + u_2(x)y_2(x)$$
$$= (-\ln x)e^x - \frac{1}{x}xe^x$$
$$= -e^x - e^x \ln x.$$

The general solution to the differential equation
$$y'' - 2y' + y = \frac{e^x}{x^2}$$

is therefore
$$y(x) = C_1e^x + C_2xe^x - e^x - e^x \ln x = (C_1 - 1)e^x + C_2xe^x - e^x \ln x.$$

Since $C_1 - 1$ is an arbitrary constant, we replace $C_1 - 1$ simply by C_1 and write the solution as
$$y(x) = C_1e^x + C_2xe^x - e^x \ln x.$$

The initial condition $y(1) = 0$ implies that
$$0 = C_1 + C_2.$$

Since
$$y'(x) = C_1e^x + C_2e^x + C_2xe^x - e^x\ln x - e^x \frac{1}{x},$$

$y'(1) = 0$ implies that $\qquad 0 = C_1e + 2C_2e - e.$

Thus, $\qquad\qquad\qquad C_2 = 1 \quad \text{and} \quad C_1 = -1$

and the solution to the initial-value problem is
$$y(x) = xe^x - e^x - e^x \ln x. \qquad\qquad \square$$

UNDETERMINED COEFFICIENTS

The method of undetermined coefficients is more difficult to describe than to use. The basis for the technique lies in the fact that certain common classes of functions have derivatives and integrals belonging to that same class. For example, differentiating or integrating a polynomial always produces a polynomial, differentiating or integrating an exponential function always produces the same type of exponential function. In a similar manner, the derivative or integral of a sine or cosine function can produce only another sine or cosine function.

Combining these observations implies that if $Q(x)$ in the equation
$$y'' + py' + qy = Q(x)$$

has a term of the form
$$(a_nx^n + a_{n-1}x^{n-1} + \cdots + a_1x + a_0)e^{rx} \begin{Bmatrix} \sin kx \\ \text{or} \\ \cos kx \end{Bmatrix},$$

it can only be produced from the differential equation if the particular solution has a term of the form

$$(A_n x^n + A_{n-1} x^{n-1} + \cdots + A_1 x + A_0) e^{rx} \sin kx +$$
$$(B_n x^n + B_{n-1} x^{n-1} + \cdots + B_1 x + B_0) e^{rx} \cos kx,$$

for some collection of constants $A_0, A_1, \ldots, A_n, B_0, B_1, \ldots, B_n$. The form must be modified by multiplying by powers of x if the term of $Q(x)$ happens to be a solution to the homogeneous equation. Examples 2 and 3 illustrate the method when this is not the case.

EXAMPLE 2   Find a particular solution to

$$y'' - y' - 2y = e^x - 2x^2.$$

SOLUTION

To produce the term e^x, $y_p(x)$ must consist in part of a term of the form Ae^x for some constant. Similarly, to produce $-2x^2$ requires that a term of the form $B_2 x^2 + B_1 x + B_0$ be a part of $y_p(x)$. So we assume that $y_p(x)$ has the form

$$y_p(x) = Ae^x + B_2 x^2 + B_1 x + B_0$$

for some collection of constants A, B_0, B_1, and B_2.

Substituting y_p and its derivatives into the differential equation implies that

$$e^x - 2x^2 = y_p''(x) - y_p'(x) - 2y_p(x)$$
$$- (Ae^x + 2B_2) - (Ae^x + 2B_2 x + B_1) - 2(Ae^x + B_2 x^2 + B_1 x + B_0)$$
$$= -2Ae^x - 2B_2 x^2 + (-2B_2 - 2B_1)x + (2B_2 - B_1 - 2B_0).$$

Comparing terms on the two sides of the equation, we see that

$$e^x: 1 = -2A; \quad \text{so } A = -\frac{1}{2}$$

$$x^2: -2 = -2B_2; \quad \text{so } B_2 = 1$$

$$x^1: 0 = -2B_2 - 2B_1; \quad \text{so } B_1 = -B_2 = -1$$

$$x^0: 0 = 2B_2 - B_1 - 2B_0; \quad \text{so } B_0 = \frac{1}{2}(2B_2 - B_1) = \frac{3}{2}$$

and

$$y_p(x) = x^2 - x + \frac{3}{2} - \frac{1}{2}e^x. \qquad \square$$

EXAMPLE 3   Solve the initial-value problem

$$y'' - 4y' + 4y = e^x \sin x, \qquad y(0) = 0, \qquad y'(0) = 0.$$

SOLUTION

First we find the general solution to the homogeneous problem. The characteristic equation is $0 = m^2 - 4m + 4 = (m - 2)^2$, so $m = 2$. The general solution to the homogeneous problem is

$$y(x) = C_1 e^{2x} + C_2 x e^{2x}.$$

The form of the particular solution is dictated by the term $e^x \sin x$ to be

$$y_p(x) = A e^x \sin x + B e^x \cos x.$$

Differentiating produces

$$y_p'(x) = (A - B)e^x \sin x + (A + B)e^x \cos x$$

and

$$y_p''(x) = -2B e^x \sin x + 2A e^x \cos x.$$

So

$$e^x \sin x = y_p''(x) - 4y_p'(x) + 4y_p(x)$$
$$= -2B e^x \sin x + 2A e^x \cos x - 4(A - B)e^x \sin x$$
$$\quad\quad - 4(A + B)e^x \cos x + 4A e^x \sin x + 4B e^x \cos x$$
$$= 2B e^x \sin x - 2A e^x \cos x.$$

Consequently, $B = 1/2$, $A = 0$, and a particular solution is

$$y_p(x) = \frac{1}{2} e^x \cos x.$$

The general solution is

$$y(x) = C_1 e^{2x} + C_2 x e^{2x} + \frac{1}{2} e^x \cos x.$$

The initial condition $y(0) = 0$ implies that

$$0 = C_1 + 1/2 \quad \text{or} \quad C_1 = -1/2.$$

Since $y'(x) = 2C_1 e^{2x} + C_2 e^{2x} + 2C_2 x e^{2x} + \frac{1}{2} e^x \cos x - \frac{1}{2} e^x \sin x$, the condition $y'(0) = 0$ implies that

$$0 = 2C_1 + C_2 + 1/2 \text{ and } C_2 = -2C_1 - 1/2 = 1/2.$$

Consequently, the solution to the initial-value problem is

$$y(x) = \frac{1}{2}(x e^{2x} - e^{2x} + e^x \cos x). \qquad \square$$

A modification of the form of the particular solution is necessary when a term of $Q(x)$ is a solution to the homogeneous equation. Consider, for example, the nonhomogeneous equation

$$y'' - 4y' + 4y = x + e^{2x}.$$

Ordinarily, the particular solution would have the form

$$y_p(x) = A_1 x + A_0 + B e^{2x}.$$

However, we found in Example 3 that the solution to the homogeneous equation

$$y'' - 4y' + 4y = 0$$

is

$$y(x) = C_1 e^{2x} + C_2 x e^{2x}.$$

This implies that the normal form for the particular solution, Be^{2x}, is a solution to the homogeneous equation. Consequently, this part of the particular solution must be modified. This modification involves multiplying Be^{2x} by the lowest integral power of x that produces a term that does not satisfy the homogeneous equation. Since xe^x is also a solution to $y'' - 4y' + 4y = 0$, the lowest power is two and the particular solution has the form

$$y_p(x) = A_1 x + A_0 + Bx^2 e^{2x}.$$

From this point, the problem is solved as in the preceding examples,

$$y_p'(x) = A_1 + 2Bxe^{2x} + 2Bx^2 e^{2x}$$
$$y_p''(x) = 2Be^{2x} + 8Bxe^{2x} + 4Bx^2 e^{2x}$$

so

$$x + e^{2x} = 2Be^{2x} + 8Bxe^{2x} + 4Bx^2 e^{2x} - 4(A_1 + 2Bxe^{2x} + 2Bx^2 e^{2x})$$
$$+ 4(A_1 x + A_0 + Bx^2 e^{2x})$$
$$= 2Be^{2x} + 4A_1 x + (-4A_1 + 4A_0).$$

Comparing the coefficients on each side:

$$e^{2x}: 1 = 2B, \quad \text{so } B = \frac{1}{2}$$

$$x^1: 1 = 4A_1, \quad \text{so } A_1 = \frac{1}{4}$$

$$x^0: 0 = -4A_1 + 4A_0, \quad \text{so } A_0 = A_1 = \frac{1}{4}.$$

This implies that a particular solution is

$$y_p(x) = \frac{1}{4}(x + 1) + \frac{1}{2}x^2 e^{2x}.$$

The general solution to this problem has the form

$$y(x) = C_1 e^{2x} + C_2 x e^{2x} + \frac{1}{4}(x + 1) + \frac{1}{2}x^2 e^{2x}.$$

EXERCISE SET 18.6

Solve the differential equations in Exercises 1 through 10 by using the method of variation of parameters.

1. $y'' + y = e^x$

2. $y'' + 2y' + y = 2e^x$

3. $y'' + 3y' - 4y = \sin x$

4. $y'' - 2y' + 5y = e^x$

5. $y'' + y = \tan x, \ 0 < x < \pi/2$

6. $y'' + y = \sec x, \ 0 < x < \pi/2$

7. $y'' + y = \csc x, \ 0 < x < \pi/2$

8. $y'' + y = \dfrac{1}{1 + \cos x}, \ -\pi < x < \pi$

9. $y'' + 4y' + 4y = \dfrac{e^{-2x}}{x^2}, \; 0 < x < \dfrac{\pi}{2}$

10. $y'' + 3y' + 2y = e^{-x}$

Solve the differential equations in Exercises 11 through 16 by using the method of undetermined coefficients.

11. $y'' + y = e^x$ **12.** $y'' + 2y' + y = 2e^x$

13. $y'' + 3y' - 4y = \sin x$ **14.** $y'' - 2y' + 5y = \cos 2x + e^x$

15. $y'' + 4y = x \sin 2x + \dfrac{1}{4} \cos 2x$ **16.** $y'' + 4y' + 4y = e^{-2x} + \cos x$

Solve the initial-value problems in Exercises 17 through 20.

17. $y'' + y' - 2y = 2x^2 - 3, \; y(0) = 0, \; y'(0) = 0$

18. $y'' + y = 4xe^x, \; y(0) = -2, \; y'(0) = 0$

19. $y'' - y = e^x + e^{-x}, \; y(0) = 1, \; y'(0) = 2$

20. $y'' - 2y' + y = 4xe^x, \; y(0) = 0, \; y'(0) = 1$

The harmonic motion of the vibrating system described before Exercises 39 through 42 in Section 18.5 (p. 1017) is called a *free oscillation* because no external force is imposed on the system. When an external force $F(t)$ is present the motion is called a *forced oscillation* and is described by the initial-value problem

$$m\,y''(t) + c\,y'(t) + k\,y(t) = F(t), \qquad y(0) = y_0, \qquad y'(0) = y_0'.$$

Find an expression for the motion of the vibrating systems described in Exercises 21 through 26.

21. $m = 1$ slug, $k = 16$ lb/ft, $c = 0$, $F(t) = \sin t$ lb, $y_0 = 0$, $y_0' = 0$

22. $m = 4$ kg, $k = 4$ newtons/meter, $c = 0$, $F(t) = \cos 2t$ newtons, $y_0 = 0$, $y_0' = 0$

23. $m = 1$ slug, $k = 16$ lb/ft, $c = 0$, $F(t) = \sin 4t$ lb, $y_0 = 0$, $y_0' = 0$

24. $m = 4$ kg, $k = 4$ newtons/meter, $c = 0$, $F(t) = \cos t$ newtons, $y_0 = 0$, $y_0' = 0$

25. $m = 1$ slug, $k = 16$ lb/ft, $c = 4$ lb sec/ft, $F(t) = \sin 4t$ lb, $y_0 = 0$, $y_0' = 1$

26. $m = 4$ kg, $k = 4$ newtons/meter, $c = 3$ dyne sec/meter, $F(t) = \cos t$ newtons, $y_0 = 1$, $y_0' = 0$

The initial-value problem describing an electrical circuit that contains, in series, an inductance L, a resistance R, a capacitance C, and an impressed voltage $E(t)$ is

$$L\,I'(t) + R\,I(t) + \dfrac{1}{C}\,Q(t) = E(t), \qquad Q(0) = Q_0, \qquad I(0) = I_0.$$

Find an expression for the charge and current as a function of time in the electrical systems described in Exercises 27 and 28.

27. $L = .25$ henry, $R = 5$ ohms, $C = 10^{-5}$ farads, $E = 110 \sin 60t$ volts, $I_0 = 0$, $Q_0 = 0$

28. $L = .5$ henry, $R = 4$ ohms, $C = 2 \times 10^{-6}$ farads, $E = 110 \sin 60t$ volts, $I_0 = 0$, $Q_0 = 0$

REVIEW EXERCISES

Find the general solution to the differential equations in Exercises 1 through 20.

1. $y' = 2y$

2. $y' = 3x^2 - 2x + 1$

3. $y' = \dfrac{x + y}{x - y}$

4. $y' = \dfrac{x^2 + 2xy + y^2}{x^2}$

5. $e^x \sin y \, dx + e^x \cos y \, dy = 0$

6. $(x^2 + y^2)dx + 2xy \, dy = 0$

7. $y' = 2xy$

8. $(y^2 - 1)dx + (3x^2 - 2xy)dy = 0$

9. $y' = x^3 - 2xy$

10. $y' = \dfrac{xy}{1 + x^2}$

11. $xy' + 2y = x^2$

12. $xy \, dx + (x^2 + y^2 + 1)dy = 0$

13. $xy' - 3y + x^4y^2 = 0$

14. $y'' - 4y = 0$

15. $y'' - 2y' - 3y = 0$

16. $y'' - y' - 2y = 0$

17. $y'' + y = \sin x$

18. $y'' - 4y = \sin 2x$

19. $y'' - 3y' + 2y = e^{3x}$

20. $y'' - 4y' + 4y = 0$

Find the particular solution to the initial-value problems in Exercises 21 through 28.

21. $y' = x^3 e^{-2y}, \; y(1) = 0$

22. $y' + y \tan x = 0, \; y(0) = 2$

23. $y'' + 5y' + 6y = 0, \; y(0) = 1, \; y'(0) = 2$

24. $y'' - y = 0, \; y(0) = 1, \; y'(0) = 1$

25. $y'' + y = 2 \cos x, \; y(0) = 5, \; y'(0) = 2$

26. $y'' + y = e^x, \; y(0) = 0, \; y'(0) = 1$

27. $y'' - 4y' + 4y = xe^{2x} + e^{2x}, \; y(0) = 1, \; y'(0) = 0$

28. $y'' - y = x, \; y(0) = 0, \; y'(0) = 0$

APPENDIX A: REVIEW MATERIAL

A.1
THE REAL LINE

The set N of **natural numbers** consists of the numbers $1, 2, 3, \ldots$. Two natural numbers can be added or multiplied and the result is another natural number. The set Z of **integers** consists of the natural numbers, the negative of each natural number, and the number zero. Addition, multiplication, or subtraction of two integers results in an integer.

The set Q of **rational numbers** is introduced to ensure that division is well-defined. A rational number is one of the form p/q, where p and q are integers and $q \neq 0$. The rational numbers satisfy all the necessary arithmetic properties, but fail to have a property mathematicians call **completeness,** because there are gaps or numbers missing from the set. (The precise definition of completeness is discussed in Chapter 9.)

Greek mathematicians knew as early as 400 B.C. that $\sqrt{2}$ is not a rational number, even though it is the length of a diagonal of a unit square. Some other nonrational (called **irrational**) numbers are $\sqrt{5}$, π, and $-\sqrt{3} + 1$. The set $\mathbb{R}$ of **real** numbers consists of the rational numbers together with the irrational numbers, and can be described by considering the set of all possible infinite decimal expansions. The rational numbers are those with expansions that eventually repeat in sequence, such as $1/2 = .50000 \ldots$, $1/3 = .3333 \ldots$, $16/11 = 1.454545 \ldots$, or $123/13 = 9.461538461538 \ldots$. The real numbers that do not have this property are irrational.

The decimal expansion property provides a means for associating each real number with a distinct point on a given line. We call one point on the line the origin and associate with the origin the real number 0. We then associate a point to the right of the origin with the real number 1. The positive integers are marked with equal spacing consecutively to the right of 0 and the negative integers are marked consecutively with the same spacing to the left of 0. Nonintegral real numbers are marked on the line according to their decimal expansions. Figure A.1 shows the number line and points associated with some real numbers.

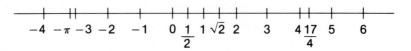

FIGURE A.1

For convenience, we do not generally distinguish between the points on the line and the real numbers they represent. Both are called the set of real numbers and denoted $\mathbb{R}$. The line is called a **coordinate line.**

The relative position of two points on a coordinate line can be used to define an inequality relationship on the set of real numbers. We say a is less than b, written $a < b$, or b is greater than a, written $b > a$, provided that a lies to the left of b on the coordinate line. The notation $a \leq b$ or $b \geq a$ is used to express that a is less than or equal to b. It can be verified that, for any real numbers a, b, and c,

(i)   Precisely one of $a < b$, $b < a$, or $a = b$, holds.
(ii) If $a > b$, then $a + c > b + c$.
(iii) If $a > b$ and $c > 0$, then $ac > bc$.
(iv) If $a > b$ and $c < 0$, then $ac < bc$.

These rules are used frequently to solve problems involving inequality relations.

EXAMPLE 1   Find all values of x satisfying $2x - 1 < 4x + 3$.

SOLUTION

Since $2x - 1 < 4x + 3$,

$$2x - 1 - 3 < 4x,$$

so $-4 < 4x - 2x$ and $-4 < 2x$.

Multiplying both sides of the inequality by 1/2 gives

$$-2 < x. \qquad \square$$

EXAMPLE 2   Find all values of x satisfying $-1 < 2x + 3 < 5$.

SOLUTION

This inequality relation is a compact way of expressing that

$$-1 < 2x + 3 \quad \text{and} \quad 2x + 3 < 5,$$

so $\qquad\qquad\qquad -4 < 2x \quad \text{and} \quad 2x < 2.$

Thus $\qquad\qquad\qquad -2 < x \quad \text{and} \quad x < 1,$

which can be expressed $-2 < x < 1$. $\qquad \square$

EXAMPLE 3   Find all values of x satisfying $x^2 - 4x + 5 > 2$.

SOLUTION

$x^2 - 4x + 5 > 2$ implies $x^2 - 4x + 3 > 0$. Writing $x^2 - 4x + 3$ as $(x - 3)(x - 1)$ produces

$$(x - 3)(x - 1) > 0.$$

This product is positive precisely when both factors are positive or both are negative. Figure A.2 indicates that both are negative if $x < 1$ and that both are positive if $x > 3$.

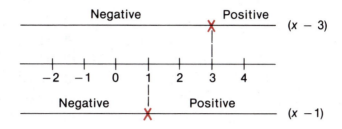

FIGURE A.2

Consequently, $x^2 - 4x + 5 > 2$ precisely when $x < 1$ or $x > 3$.   □

  The distance between two points on the real number line is described by using absolute values. The absolute value of a real number a is

(A.1)
$$|a| = \begin{cases} a, \text{ if } a \geq 0 \\ -a, \text{ if } a < 0. \end{cases}$$

The distance from point a to point b is

(A.2)
$$d(a, b) = |a - b|.$$

Notice that this definition implies that $d(a, b) = d(b, a)$ and that $d(a, 0) = |a|$. Because of this distance relationship, the absolute value frequently occurs in inequality relationships. Some basic properties of absolute values that are useful for working with absolute value inequalities are:

(A.3)
$$|ab| = |a|\,|b|.$$

(A.4)
$$\left|\frac{a}{b}\right| = \frac{|a|}{|b|}, \quad \text{if } b \neq 0.$$

(A.5)
$$|a + b| \leq |a| + |b|.$$

(A.6) $\qquad |x| < a$ if and only if $-a < x < a.$

(A.7) $\quad |x| > a, \quad a \geq 0, \quad$ if and only if $x < -a \quad$ or $\quad x > a.$

EXAMPLE 4   Find all values of x satisfying $|2x - 1| < 3$.

SOLUTION
$|2x - 1| < 3$ if and only if $-3 < 2x - 1 < 3$, that is

$$-3 < 2x - 1 \quad \text{and} \quad 2x - 1 < 3.$$

This implies that $-2 < 2x$ and $2x < 4$, so $-1 < x < 2$.   □

EXERCISE SET A.1

Find all values of x that satisfy the inequalities in Exercises 1 through 16.

1. $x + 4 < 7$

2. $2x + 3 \geq 4$

3. $-3x + 4 < 5$

4. $-(3x + 4) \geq 5$

5. $-2 < 2x + 9 < 5$

6. $2 - x < x < 4 - x$

7. $-2 < 2x + 9 < 5 + x$

8. $x - 2 < 2x + 9 < 5 + x$

9. $x^2 + 2x + 1 \geq 1$

10. $x^3 - 6x^2 + 8x < 0$

11. $|x - 4| \leq 1$

12. $|2x - 3| < 5$

13. $|3x - 1| < .01$

14. $|3x - 2| < .01$

15. $|x^2 - 4| \leq 0$

16. $|x^2 - 2x + 5| \geq 0$

Prove that the rules listed in Exercises 17 through 20 are true for all real numbers.

17. $|ab| = |a|\,|b|$

18. $\left|\dfrac{a}{b}\right| = \dfrac{|a|}{|b|}$ if $b \neq 0$

19. $|x| < a$ if and only if $-a < x < a$

20. $|x| > a$ if and only if $x < -a$ or $x > a$

21. Show that for any pair of real numbers

$$|a + b| \leq |a| + |b|.$$

What must be true about a and b if the equality holds?

22. Show that for any pair of real numbers

$$|a| - |b| \leq |a - b|.$$

What must be true about a and b if equality holds?

A.2
THE COORDINATE PLANE

An ordered pair (a, b) of real numbers can be associated with a point in a plane in a manner similar to the association of a real number with a point on a line. An arbitrary point in the plane is designated the origin and associated with $(0, 0)$. Horizontal and vertical lines are drawn intersecting at the origin. The horizontal line is called the first-coordinate- or x-axis and the vertical line is called the second-coordinate- or y-axis. See Figure A.3.

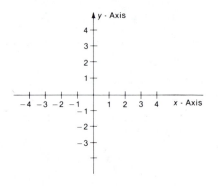

FIGURE A.3

A scale is placed on both axes. The x-axis has positive numbers to the right of the origin and negative numbers to the left. The y-axis has positive numbers above the origin and negative numbers below. The ordered pair (a, b) is associated with the point of intersection of the vertical line drawn through the point a on the x-axis and the horizontal line drawn through b on the y-axis. See Figure A.4.

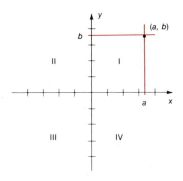

FIGURE A.4

The x- and y-axes divide the plane into four regions, or **quadrants.** These quadrants are labeled as in Figure A.4. The set of all ordered pairs of real numbers is denoted $\mathbb{R} \times \mathbb{R}$ or $\mathbb{R}^2$. The plane determined by the x- and y-axes is called the xy-plane.

EXAMPLE 1   Sketch the points in the coordinate plane associated with the ordered pairs $(1, 2)$, $(-1, 3)$, $(-2, -\pi)$ and $(\sqrt{2}, -\sqrt{3})$.

SOLUTION

These points are shown in Figure A.5.

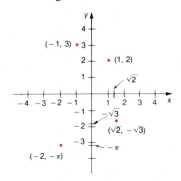

FIGURE A.5

HISTORICAL
NOTE

The coordinate plane is often referred to as the cartesian plane, and the rectangular coordinate system called the cartesian coordinate system. This terminology honors the French mathematician and philosopher **Rene Descartes** (1596–1650) who, in an appendix to his work *La Géométrie*, introduced the mathematical world to the subject of analytic geometry. Descartes was a powerful intellectual who made contributions to nearly every area of knowledge studied in his day.

The distance between two ordered pairs of real numbers (x_1, y_1) and (x_2, y_2) is the distance between the points the pairs represent. (In general, we will not distinguish between an ordered pair and the point it represents in a coordinate plane.) This distance is found by considering the point (x_2, y_1) and applying the Pythagorean theorem. Refer to Figure A.6.

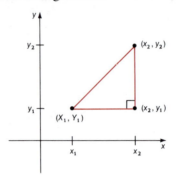

FIGURE A.6

Since on the x-axis $d(x_1, x_2) = |x_1 - x_2|$,

$$d((x_1, y_1), (x_2, y_1)) = |x_1 - x_2|.$$

Similarly, $d((x_2, y_1), (x_2, y_2)) = |y_1 - y_2|$

so

$$d((x_1, y_1), (x_2, y_2)) = \sqrt{[d((x_1, y_1), (x_2, y_1))]^2 + [d((x_2, y_1), (x_2, y_2))]^2}$$
$$= \sqrt{|x_1 - x_2|^2 + |y_1 - y_2|^2}$$

or

(A.8) $d((x_1, y_1), (x_2, y_2)) = \sqrt{(x_1 - x_2)^2 + (y_1 - y_2)^2}.$

. EXAMPLE 2   Find the distance between the points $(1, 2)$ and $(-2, 6)$.

SOLUTION

$$d((1, 2), (-2, 6)) = \sqrt{(1 - (-2))^2 + (2 - 6)^2}$$
$$= \sqrt{9 + 16} = \sqrt{25} = 5. \qquad \square$$

EXAMPLE 3   Show that the midpoint of the line segment joining (x_1, y_1) and (x_2, y_2) is $\left(\dfrac{x_1 + x_2}{2}, \dfrac{y_1 + y_2}{2}\right).$

SOLUTION

$$d\left((x_1, y_1), \left(\frac{x_1 + x_2}{2}, \frac{y_1 + y_2}{2}\right)\right) = \sqrt{\left(x_1 - \frac{x_1 + x_2}{2}\right)^2 + \left(y_1 - \frac{y_1 + y_2}{2}\right)^2}$$
$$= \sqrt{\left(\frac{x_1 - x_2}{2}\right)^2 + \left(\frac{y_1 - y_2}{2}\right)^2}$$

$$= \frac{1}{2} \sqrt{(x_1 - x_2)^2 + (y_1 - y_2)^2}$$

$$= \frac{1}{2} d((x_1, y_1), (x_2, y_2)).$$

Similarly,

$$d\left(\left(\frac{x_1 + x_2}{2}, \frac{y_1 + y_2}{2}\right), (x_2, y_2)\right) = \frac{1}{2} d((x_1, y_1), (x_2, y_2)).$$

Since

$$d((x_1, y_1), (x_2, y_2)) = d\left((x_1, y_1), \left(\frac{x_1 + x_2}{2}, \frac{y_1 + y_2}{2}\right)\right)$$
$$+ d\left(\left(\frac{x_1 + x_2}{2}, \frac{y_1 + y_2}{2}\right), (x_2, y_2)\right),$$

$\left(\dfrac{x_1 + x_2}{2}, \dfrac{y_1 + y_2}{2}\right)$ must be on the line segment and is consequently the midpoint.

$\square$

EXERCISE SET A.2

In Exercises 1 through 4, sketch all the listed points on the same coordinate plane.

1. $(1, 0), (0, 1), (-1, 0), (0, -1)$

2. $(2, 3), (3, 2), (-2, 3), (-3, -2)$

3. $(2, 3), (-2, -3), (2, -3), (-2, 3)$

4. $(5, -10), (10, 20), (-20, 10), (30, 40)$

In Exercises 5 through 8, find the distance between the pairs of points.

5. $(2, 4), (-1, 3)$ **6.** $(-1, 5), (7, 9)$

7. $(\pi, 0), (-1, 2)$ **8.** $(\sqrt{3}, \sqrt{2}), (\sqrt{2}, \sqrt{3})$

In Exercises 9 through 24, indicate on an xy-plane those points (x, y) for which the statement holds.

9. $x = 3$ **10.** $y = -2$

11. $x \geq 0$ and $y \geq 0$ **12.** $x \geq 0$ and $y \leq 0$

13. $-1 \leq x \leq 1$ **14.** $2 < y < 3$

15. $-1 \leq x \leq 1$ and $2 < y < 3$ **16.** $-1 \leq x \leq 1$ or $2 < y < 3$

17. $4 \leq |x|$ **18.** $|y| \leq 2$

19. $|x - 1| < 3$ **20.** $|y + 1| < 2$

21. $|x - 1| < 3$ and $|y + 1| < 2$ **22.** $|x| + |y| > 0$

23. $|x - 1| + |y + 1| = 0$ **24.** $|x| > |y|$

25. Find the distance between the points $(-1, 4)$, $(-3, -4)$ and $(2, -1)$ and show that they are vertices of a right triangle.

26. Show that the points $(2, 1)$, $(-1, 2)$, and $(2, 6)$ are vertices of an isosceles triangle.

27. Find a fourth point, which added to the points in Exercise 25, will form the vertices of a rectangle. Is the point unique?

28. Find a fourth point, which added to the points in Exercise 26, will form the vertices of a parallelogram. Is the point unique?

A.3
TRIGONOMETRIC FUNCTIONS

The trigonometric functions have many applications. They are used to describe the behavior of such diverse topics as sound waves, vibrations, the motion of an automobile on a bumpy road, and the oscillations of a pendulum. In fact, trigonometric functions are commonly required to describe the motion of any object that behaves in a circular, oscillating, or periodic manner.

For calculus purposes it is most useful to define these functions relative to the circle centered at $(0, 0)$ with radius one, called the **unit circle** and denoted U. The trigonometric functions can also be introduced by using the sides and angles of a right triangle. This approach, however, is directed more toward the computational aspects of trigonometry than the functional aspects that are required for the study of calculus. In the next few pages a brief review is given of the definitions and properties of trigonometric functions.

The unit circle U in the xy-plane has equation $x^2 + y^2 = 1$. Our first step is to describe a function P that maps the real line $\mathbb{R}$ onto U in the following manner:

(i) If t is a positive real number, $P(t)$ is the point on the unit circle for which the length of the arc of the circle from $(1, 0)$ to $P(t)$ is t units, measured in the *counterclockwise* direction from $(1, 0)$. (See Figure A.7.)

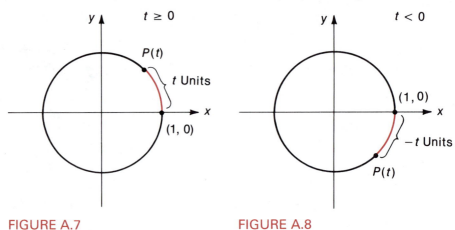

FIGURE A.7   FIGURE A.8

(ii) If t is a negative real number, $P(t)$ is the point on the unit circle for which the length of the arc of the circle from $(1, 0)$ to $P(t)$ is $-t$ units, measured in the *clockwise* direction from $(1, 0)$. (See Figure A.8.)

The mapping of t into $P(t)$ is described geometrically by positioning the real line vertically, with $t = 0$ coinciding with the point $(1, 0)$ on the circle. (See Figure A.9.) For any real number t, the point $P(t)$ is obtained by "wrapping" the line around the circle and marking the point $P(t)$ on the circle that corresponds to the position of t.

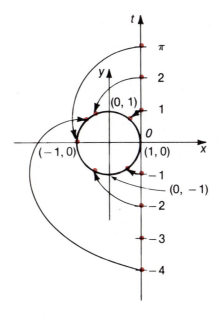

FIGURE A.9

The smallest positive real number that is mapped onto $(-1, 0)$ is called π, an irrational number whose value is approximately 3.1416. Because of the symmetry of the circle, the real number $\pi/2$ is associated with the point $(0, 1)$, $3\pi/2$ with the point $(0, -1)$, and 2π with $(1, 0)$.

Corresponding to each real number t is a pair $(x(t), y(t))$ of xy-coordinates describing the point $P(t)$ on the unit circle U. The basic trigonometric functions are defined in terms of these coordinates.

(A.9)
DEFINITION

If the coordinates of $P(t)$ are $(x(t), y(t))$, then the **sine** of t, written $\sin t$, is defined by

$$\sin t = y(t)$$

and the **cosine** of t, written $\cos t$, by

$$\cos t = x(t).$$

The **tangent** of t, written $\tan t$, is defined by the quotient

$$\tan t = \frac{\sin t}{\cos t}, \qquad \text{provided } \cos t \neq 0.$$

Figure A.10 illustrates these definitions.

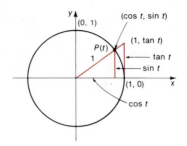

FIGURE A.10

Certain properties of these trigonometric functions follow immediately from the definition.

(A.10) $-1 \leq \sin t \leq 1; \qquad -1 \leq \cos t \leq 1$

(A.11) $(\sin t)^2 + (\cos t)^2 = 1$

(A.12) $\sin(-t) = -\sin t; \qquad \cos(-t) = \cos t; \qquad \tan(-t) = -\tan t$

(A.13) $\sin(t + 2n\pi) = \sin t; \qquad \cos(t + 2n\pi) = \cos t;$

$\tan(t + n\pi) = \tan t, \qquad$ for any integer n.

The result in (A.13) implies that sine and cosine are *periodic* functions with period 2π while the tangent function is periodic with period π. In general, a function f is **periodic** with **period** a if a is the smallest positive real number with $f(x + a) = f(x)$ for all x in the domain of f.

There are a number of values of t for which $\sin t$, $\cos t$, and $\tan t$ can be determined from geometric properties of the unit circle. Some of the more common values are included for reference in Table A.1.

t	$\sin t$	$\cos t$	$\tan t$
0	0	1	0
$\dfrac{\pi}{6}$	$\dfrac{1}{2}$	$\dfrac{\sqrt{3}}{2}$	$\dfrac{\sqrt{3}}{3}$
$\dfrac{\pi}{4}$	$\dfrac{\sqrt{2}}{2}$	$\dfrac{\sqrt{2}}{2}$	1
$\dfrac{\pi}{3}$	$\dfrac{\sqrt{3}}{2}$	$\dfrac{1}{2}$	$\sqrt{3}$
$\dfrac{\pi}{2}$	1	0	not defined
$\dfrac{2\pi}{3}$	$\dfrac{\sqrt{3}}{2}$	$-\dfrac{1}{2}$	$-\sqrt{3}$
$\dfrac{3\pi}{4}$	$\dfrac{\sqrt{2}}{2}$	$-\dfrac{\sqrt{2}}{2}$	-1
π	0	-1	0
$\dfrac{3\pi}{2}$	-1	0	not defined
2π	0	1	0

TABLE A.1

The results in Table A.1 and the properties listed in (A.10) through (A.13) can be used to sketch the graphs of the sine, cosine, and tangent functions. These graphs are shown in Figures A.11, A.12, and A.13.

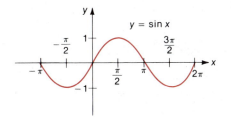

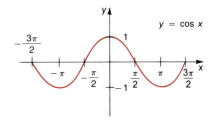

FIGURE A.11   FIGURE A.12

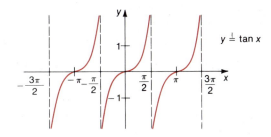

FIGURE A.13

A number of identities are useful in studying trigonometric functions. Two of the most basic are:

(A.14) $\cos(t_1 \pm t_2) = \cos t_1 \cos t_2 \mp \sin t_1 \sin t_2$

(A.15) $\sin(t_1 \pm t_2) = \sin t_1 \cos t_2 \pm \cos t_1 \sin t_2.$

Additional identities having frequent application can be derived from properties (A.10) through (A.15).

(A.16) $\sin\left(\dfrac{\pi}{2} - t\right) = \cos t; \qquad \cos\left(\dfrac{\pi}{2} - t\right) = \sin t$

(A.17) $\sin 2t = 2 \sin t \cos t; \qquad \cos 2t = (\cos t)^2 - (\sin t)^2$

(A.18) $\left(\sin t\right)^2 = \dfrac{1 - \cos 2t}{2}; \qquad \left(\cos t\right)^2 = \dfrac{1 + \cos 2t}{2}$

(A.19) $\tan(t_1 \pm t_2) = \dfrac{\tan t_1 \pm \tan t_2}{1 \pm \tan t_1 \tan t_2}$

(A.20) $\tan t = \dfrac{\sin 2t}{1 + \cos 2t}$

EXAMPLE 1   Verify the identity

$$\sin\left(\frac{\pi}{2} - t\right) = \cos t.$$

SOLUTION

It follows from (A.15) that

$$\sin\left(\frac{\pi}{2} - t\right) = \sin\frac{\pi}{2}\cos t - \cos\frac{\pi}{2}\sin t.$$

From Table A.1, $\sin\dfrac{\pi}{2} = 1$ and $\cos\dfrac{\pi}{2} = 0$ so

$$\sin\left(\frac{\pi}{2} - t\right) = 1 \cdot \cos t - 0 \cdot \sin t = \cos t. \qquad\qquad \square$$

EXAMPLE 2   Find all values of x in $[0, 2\pi)$ that satisfy the equation $\sin x + \cos x = 1$.

SOLUTION

If $\sin x + \cos x = 1$

then $\sin x = 1 - \cos x$

and $(\sin x)^2 = 1 - 2\cos x + (\cos x)^2.$

Since $(\sin x)^2 = 1 - (\cos x)^2,$

$$1 - (\cos x)^2 = 1 - 2\cos x + (\cos x)^2.$$

So $0 = 2(\cos x)^2 - 2\cos x = 2\cos x(1 - \cos x).$

Solutions to this equation occur precisely when $\cos x = 0$, which implies

$$x = \pi/2 \text{ or } x = 3\pi/2$$

or when $\cos x = 1$, which implies $x = 0$. Checking these values in our original equation, we see that the solutions are $x = \pi/2$ and $x = 0$. The extraneous value $x = 3\pi/2$ was introduced when the equation was squared. $\square$

Three other trigonometric functions exist, all definable in terms of the sine and cosine functions.

(A.21)
DEFINITION

For a real number t, the **cosecant** of t, written csc t, is defined by

$$\csc t = \frac{1}{\sin t}, \qquad \text{provided } \sin t \neq 0;$$

the **secant** of t, written sec t, by

$$\sec t = \frac{1}{\cos t}, \qquad \text{provided } \cos t \neq 0;$$

the **cotangent** of t, written cot t, by

$$\cot t = \frac{\cos t}{\sin t}, \qquad \text{provided } \sin t \neq 0.$$

Figures A.14, A.15, and A.16 show the graphs of these functions.

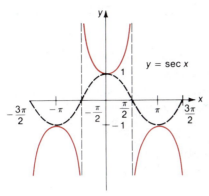

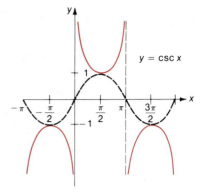

FIGURE A.14   FIGURE A.15

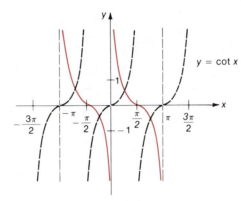

FIGURE A.16

EXAMPLE 3   If $\sin t = \dfrac{2}{3}$ and $0 < t < \dfrac{\pi}{2}$, find the values of each of the other trigonometric functions at t.

SOLUTION

Since $(\sin t)^2 + (\cos t)^2 = 1$, we see that

$$(\cos t)^2 = 1 - (\sin t)^2 = 1 - \frac{4}{9} = \frac{5}{9}.$$

Hence, $\cos t = \dfrac{\sqrt{5}}{3}$. The other trigonometric functions depend on the values of $\cos t$ and $\sin t$ and are given as:

$$\tan t = \frac{\sin t}{\cos t} = \frac{2}{\sqrt{5}} = \frac{2\sqrt{5}}{5}$$

$$\sec t = \frac{1}{\sin t} = \frac{3}{2}$$

$$\csc t = \frac{1}{\cos t} = \frac{3}{\sqrt{5}} = \frac{3\sqrt{5}}{5}$$

and $$\cot t = \frac{\cos t}{\sin t} = \frac{\sqrt{5}}{2}.$$ □

Using the definitions and property (A.11), it can be shown that

(A.22) $(\tan t)^2 + 1 = (\sec t)^2;$ $(\cot t)^2 + 1 = (\csc t)^2.$

There are many other identities involving the trigonometric functions. Some of these are listed in Exercises 34 through 38.

Geometric and computational problems often require that the trigonometric functions be defined in terms of angle measures. An **angle** θ is generated by rotating a **ray,** or half line, l_1 about its endpoint O, in the plane, to a new position determined by a ray l_2 with endpoint O. The ray l_1 is known as the **initial side** of the angle, l_2 the **terminal side,** and O the **vertex.**

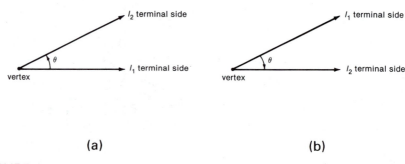

(a)   (b)

FIGURE A.17

If the generating rotation is counterclockwise as in Figure A.17(a), the angle θ is considered a positive angle. If the rotation is clockwise as in Figure A.17(b), θ is considered negative.

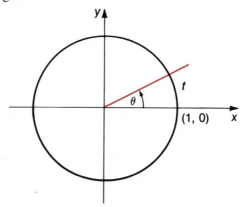

FIGURE A.18

An angle is said to be in **standard position** when its initial side lies along the positive x-axis and its vertex is at the origin. The angle θ shown in Figure A.18 is in standard position. It is said to have **radian measure** t, since the terminal side of θ intersects the unit circle U at $P(t)$.

The trigonometric functions of an angle are defined using the radian measure. If θ is an angle with radian measure t, then $\sin \theta = \sin t$, $\cos \theta = \cos t$, and so on.

EXAMPLE 4   Find the sine, cosine, and tangent of the angle θ in the triangle shown in Figure A.19.

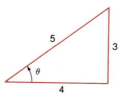

FIGURE A.19

SOLUTION

We superimpose an xy-coordinate system onto the triangle with the vertex of θ at the origin and positive x-axis along the side of the triangle with length 4. Let U be the unit circle in this coordinate system and $P = P(t)$, with coordinates $(x(t), y(t))$, be the point where the terminal side of θ intersects the unit circle (see Figure A.20).

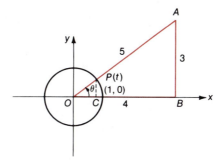

FIGURE A.20

Using similar triangles (triangles AOB and POC) implies that

$$\frac{y(t)}{1} = \frac{3}{5} \quad \text{and} \quad \frac{x(t)}{1} = \frac{4}{5}.$$

Thus, $$\sin \theta = y(t) = \frac{3}{5}, \qquad \cos \theta = x(t) = \frac{4}{5}$$

and $$\tan \theta = \frac{\sin \theta}{\cos \theta} = \frac{3}{4}. \qquad \square$$

An angle that subtends an arc of length $\pi/2$ is called a **right angle** and any triangle with a right angle is called a **right triangle.** For example, the triangle given in Example 4 is a right triangle. The rays determining a right angle are perpendicular.

Notice that in Example 4 the values of the trigonometric functions for the angle θ are quotients of the sides of the right triangle. In general, if θ is the angle in the right triangle shown in Figure A.21, then

(A.23) $$\sin \theta = \frac{b}{c}, \qquad \cos \theta = \frac{a}{c}, \qquad \tan \theta = \frac{b}{a},$$

(A.24) $$\csc \theta = \frac{c}{b}, \qquad \sec \theta = \frac{c}{a}, \qquad \cot \theta = \frac{a}{b},$$

provided, of course, that these quotients are defined.

FIGURE A.21   FIGURE A.22

Although the natural method of measuring angles for calculus purposes is the radian measure, angles can also be measured in **degrees.** An angle formed by rotating a ray counterclockwise from an initial position back to the initial position so the terminal and initial sides coincide has a measure of 360 degrees, written 360°. (See Figure A.22.)

Since the circumference (arc length) of the unit circle is 2π radians,

$$360 \text{ degrees} = 2\pi \text{ radian}$$

so

$$1 \text{ degree} = \frac{\pi}{180} \text{ radians} \approx .0175 \text{ radian}$$

and

$$1 \text{ radian} = \frac{180}{\pi} \text{ degrees} \approx 57.3°.$$

EXAMPLE 5   Express 135° and 405° in radians and $\frac{\pi}{4}$ radians in degrees.

SOLUTION

$$135 \text{ degrees} = 135\left(\frac{\pi}{180} \text{ radians}\right) = \frac{3}{4}\pi \text{ radian}$$

$$405° = 405\left(\frac{\pi}{180}\right) \text{radians} = \frac{9}{4}\pi \text{ radians}.$$

$$\frac{\pi}{4} \text{ radians} = \frac{\pi}{4}\left(\frac{180}{\pi}\right) \text{ degrees} = 45 \text{ degrees}. \qquad \square$$

The final results that will be needed from trigonometry are the law of sines and the law of cosines. These results are useful for determining the unknown sides and angles of triangles that are not right triangles. Using the notation in the triangle given in Figure A.23, the **law of sines** is

(A.25) $$\frac{\sin \alpha}{a} = \frac{\sin \beta}{b} = \frac{\sin \gamma}{c},$$

provided a, b, and c are all nonzero, and the **law of cosines** is

(A.26)

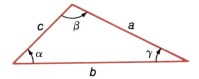

FIGURE A.23

EXERCISE SET A.3

1. Locate the points $P(t)$ on a unit circle that correspond to the real numbers t given below and give the coordinates $(x(t), y(t))$ of the points.

(a) $t = \dfrac{\pi}{6}$

(b) $t = \dfrac{\pi}{4}$

(c) $t = \dfrac{13\pi}{4}$

(d) $t = \dfrac{7\pi}{3}$

(e) $t = \dfrac{-5\pi}{6}$

(f) $t = \dfrac{-5\pi}{4}$

2. Find $\sin t$ and $\tan t$ for the real numbers t given in Exercise 1.

3. Find all values of x in the interval $[0, 2\pi]$ that satisfy the following equations.

(a) $\cos x = 1$

(b) $\sin x = -1$

(c) $\sin x = \dfrac{1}{2}$

(d) $\cos x = \dfrac{-1}{2}$

(e) $\sin x = \dfrac{-\sqrt{3}}{2}$

(f) $\tan x = 1$

(g) $\cot x = -1$

(h) $\sec x = \dfrac{2\sqrt{3}}{3}$

(i) $\csc x = 1$

(j) $\cot x = -2$

4. Use Table A.1 on p. A–10 and identities (A.14) and (A.15) to determine the following values.

(a) $\sin \dfrac{5\pi}{12}$ $\left(Hint:\ \dfrac{5\pi}{12} = \dfrac{\pi}{6} + \dfrac{\pi}{4}\right)$

(b) $\cos \dfrac{7\pi}{12}$

(c) $\tan \dfrac{\pi}{12}$

(d) $\cot\left(\dfrac{-\pi}{12}\right)$

5. If $\sin t = 4/5$ and $0 < t < \pi/2$, find $\cos t$ and $\tan t$.

6. If $\sin t = .1$ and $\pi/2 < t < \pi$, find $\cos t$ and $\tan t$.

7. If $\sin t = 3/5$, find all possible values for each of the other trigonometric functions.

8. If $\cos t = 2/3$, find all possible values for each of the other trigonometric functions.

9. If $\tan t = 2$ and $0 < t < \pi/2$, find the value for each of the other trigonometric functions.

10. If $\sec t = -2$ and $\pi/2 < t < 3\pi/2$, find the value for each of the other trigonometric functions.

11. Find the degree measure of the angles with the following radian measurements.

(a) $\dfrac{3\pi}{4}$

(b) $\dfrac{-5\pi}{6}$

(c) π

(d) $\dfrac{7\pi}{3}$

(e) -4π

(f) $\dfrac{-7\pi}{8}$

12. Find the radian measure of the angles with the following degree measurements.

   (a) 75°   (b) −135°   (c) 240°

   (d) 40°   (e) 1500°   (f) −315°

13. Find the following values.

   (a) $\cot \dfrac{13}{6}\pi$ (b) $\sec \dfrac{322}{3}\pi$ (c) $\csc \dfrac{-107}{3}\pi$

   (d) $\sin(1200°)$ (e) $\cos(-585°)$ (f) $\tan \dfrac{135}{4}\pi$

14. Use the graphs given in this section and the techniques discussed in Chapter 1 to sketch the graph of each function described below.

   (a) $f(x) = 2 \sin x$ (b) $f(x) = \sin x + 1$

   (c) $f(x) = \cos\left(x - \dfrac{\pi}{2}\right)$ (d) $f(x) = 3 \cos x + 2 \sin x$

   (e) $f(x) = \sin\left(x + \dfrac{\pi}{2}\right)$ (f) $f(x) = \cos\left(x - \dfrac{\pi}{2}\right) - \sin\left(x + \dfrac{\pi}{2}\right)$

15. Sketch the graphs of the functions described in the following.

   (a) $f(x) = \sin 2x$ (b) $f(x) = \sin .5x$

   (c) $f(x) = \cos 3x$ (d) $f(x) = \cos \dfrac{1}{4}x$

Find all values of x in the interval $[0, 2\pi)$ that satisfy the equations in Exercises 16 through 21.

16. $\sin 2x = \sin x$ 17. $\sin 2x = \cos x$

18. $|\tan x| = 1$ 19. $\sin x \cos x - \sin x - \cos x + 1 = 0$

20. $\tan x + \cot x = \dfrac{1}{2}\sin 2x$ 21. $\sin\left(x - \dfrac{\pi}{2}\right) = \cos(\pi - x)$

22. The definitions of even and odd functions are given in Exercise 9 of Section 1.1. Use these definitions to determine which of the six trigonometric functions are even and which are odd.

Use the identities listed in this section to verify the identities given in Exercises 23 through 33.

23. $[1 - (\cos \theta)^2](\sec \theta)^2 = (\tan \theta)^2$

24. $\dfrac{\cos x}{1 - \tan x} + \dfrac{\sin x}{1 - \cot x} = \sin x + \cos x$

25. $\cot x + \tan x = \sec x \csc x$ 26. $\dfrac{\sin t}{1 - \cos t} = \cot t + \csc t$

27. $\cot x - \tan x = 2 \cot 2x$ 28. $\dfrac{\sin(b - a)}{\sin(b + a)} = \dfrac{\cot a - \cot b}{\cot a + \cot b}$

29. $(\sin x + \cos x)^2 = 1 + \sin 2x$

30. $\cos x = \sin x \sin 2x + \cos x \cos 2x$

31. $(\tan x)^2 - (\sin x)^2 = (\tan x)^2 (\sin x)^2$

32. $\sec x - \cos x = \sin x \tan x$

33. $\cos x (\cot x + \tan x) = \csc x$

Use identities (A.14) and (A.15) to establish the identities listed in Exercises 34 through 38.

34. $\sin t_1 \sin t_2 = \dfrac{1}{2}[\cos(t_1 - t_2) - \cos(t_1 + t_2)]$

35. $\cos t_1 \cos t_2 = \dfrac{1}{2}[\cos(t_1 - t_2) + \cos(t_1 + t_2)]$

36. $\sin t_1 \cos t_2 = \dfrac{1}{2}[\sin(t_1 - t_2) + \sin(t_1 + t_2)]$

37. $\tan(t_1 + t_2) = \dfrac{\tan t_1 + \tan t_2}{1 - \tan t_1 \tan t_2}$

38. $\tan(t_1 - t_2) = \dfrac{\tan t_1 - \tan t_2}{1 + \tan t_1 \tan t_2}$

39. Four ships A, B, C, and D are at sea in the following relative positions. B is due north of D, D is due west of C, and B is on a line between A and C. The distance between B and D is two miles. The angle BDA measures $40°$ and the angle BCD measures $25°$. What is the distance between A and D? See the accompanying figure.

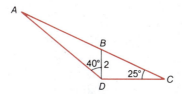

40. A certain biological rhythm can be approximately described by the equation

$$y = 3 + 2 \cos \frac{\pi}{12}(t - 5),$$

where t represents time given in hours. Sketch the graph of this equation.

41. An amateur climber needs to estimate the height of a cliff. She stands at a point on the ground 35 feet from the base of the cliff and estimates that the angle from the ground to a line extending from her feet to the top of the cliff is approximately $60°$. Assuming that the cliff is perpendicular to the ground, find the approximate height of the cliff.

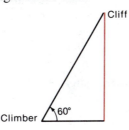

42. A gutter is to be made from a long strip of tin by bending up the sides, so that the base and the sides have the same length b. Express the area of a cross section of the gutter as a function of the angle θ the sides make with the base.

43. Two balls are against the rail on opposite ends of a billiard table. One of the balls lies 2 feet from the right side of the table and the other lies 1 foot from this same side. If the table is 8 feet long, where should the first ball hit the right side of the table in order to hit the second ball?

44. The National Forest Service maintains observation towers to check for the outbreak of forest fires. Suppose there are two towers with the same elevation, one at point A and another 10 miles due west at point B. The ranger at A spots a fire whose line of sight makes an angle of $63°$ with the line between the towers. He immediately contacts the ranger at B, who locates the fire along a line of sight that makes a $50°$ angle with the line between the towers. How far is the fire from tower B?

45. Suppose l_1 and l_2 are two nonvertical, nonperpendicular lines. Show that if l_1 and l_2 have slopes m_1 and m_2 respectively and if θ is the angle from l_1 to l_2, then

$$\tan \theta = \frac{m_2 - m_1}{1 + m_1 m_2}.$$

46. Use the result in Exercise 45 to show that the slope of a nonvertical line l is $\tan \alpha$, where α is the angle of inclination of l.

47. Use the result in Exercise 45 to show that two nonvertical lines l_1 and l_2 with slopes m_1 and m_2 respectively are perpendicular if and only if $m_1 m_2 = -1$.

48. Simplify $\tan t/2 + \cot t/2$.

49. Show that for all t, $|\tan t + \cot t| \geq 2$.

50. To test whether the cut of a circle is a diameter of the circle, a carpenter takes a right angle square and checks to see if it touches the curve at the three points, as shown in the accompanying figure. Show that this test is valid. Find a variation of this test that can be used to locate the center of a given circle.

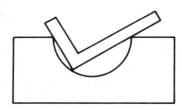

51. Another carpenters' test involving the right angle square is to determine if a closed curve is actually circular. If a carpenters square is placed outside the curve, the curve is circular if the sides of the square intersect the curve at two points that are the same distance from the vertex. Why is this test valid?

52. Suppose that $\sin t_1 + \sin t_2 = \alpha$ and $\cos t_1 + \cos t_2 = \beta$. Find $\sin (t_1 + t_2)$ $\cos (t_1 + t_2)$.

A.4
MATHEMATICAL INDUCTION

It is often necessary to show that a particular mathematical expression is true for all natural numbers. For example, formulas presented in Chapter 5 state that for every natural number n:

$$1 + 2 + \cdots + n = \frac{n(n + 1)}{2},$$

$$1^2 + 2^2 + \cdots + n^2 = \frac{n(n + 1)(2n + 1)}{6},$$

and
$$1^3 + 2^3 + \cdots + n^3 = \frac{n^2 (n + 1)^2}{4}.$$

As another example of this type of expression,

$$\log(a_1 a_2 \ldots a_n) = \log a_1 + \log a_2 + \cdots \log a_n$$

is true for any collection of n positive numbers. These expressions all follow a recursive pattern and can be established by using the principle of mathematical induction. This method of proof was first formally described by Augustus DeMorgan (1807–1871) in 1838. The underlying principle, however, was used by Pascal, Fermat, and others before 1650, and is evident before that time in the work of the Italian mathematician Francesco Maurolico (1494–1575). The method is based on the following axiom.

(A.27)
Axiom of Mathematical Induction

If S is a set of natural numbers that satisfies the conditions
(i) 1 belongs to S,
(ii) whenever an integer k belongs to S, the integer $k + 1$ also belongs to S,
then $S = N$.

To see how this axiom is applied, let us consider some examples.

EXAMPLE 1   Use the axiom of mathematical induction to show that

$$1 + 2 + \cdots + n = \frac{n(n + 1)}{2}$$

for all natural numbers n.

SOLUTION
Let S be the set of natural numbers for which this statement is true.

(i) 1 is in S because $1 = \dfrac{1(1 + 1)}{2}$.

(ii) Suppose the integer k is in S; that is, $1 + 2 + \cdots + k = \dfrac{k(k + 1)}{2}$.

Then

$$1 + 2 + \cdots + (k + 1) = (1 + 2 + \cdots + k) + (k + 1)$$

$$= \frac{k(k + 1)}{2} + (k + 1) = \frac{k(k + 1) + 2(k + 1)}{2}$$

$$= \frac{(k + 1)(k + 2)}{2}.$$

However, this is the formula $1 + 2 + \cdots + n = \dfrac{n(n + 1)}{2}$ when $n = k + 1$, so $k + 1$ is in S.

By the axiom of mathematical induction, $S = N$ and

$$1 + 2 + \cdots + n = \frac{n(n + 1)}{2}$$

for every natural number n.   □

EXAMPLE 2   Show that $2^n > n$ for each natural number n.

SOLUTION

Let S be the set of natural numbers n for which $2^n > n$.
(i)   1 is in S, since $2^1 = 2 > 1$.
(ii) Suppose k is in S; that is, $2^k > k$. Then

$$2^{k+1} = 2^k \, 2 > k \cdot 2 = k + k \geq k + 1.$$

Consequently, $k + 1$ is also in S.
By the axiom of mathematical induction, $S = N$ and $2^n > n$ for each natural number n.   □

EXAMPLE 3   Prove that the rule $\log (a \cdot b) = \log a + \log b$ for positive numbers a and b implies that

$$\log(a_1 a_2 \ldots a_n) = \log a_1 + \log a_2 + \cdots + \log a_n$$

for every collection of n positive numbers.

SOLUTION

Let S be the set of natural numbers n for which

$$\log(a_1 a_2 \ldots a_n) = \log a_1 + \log a_2 + \cdots + \log a_n$$

for every collection of n positive numbers.
(i)   1 is in S, since $\log a_1 = \log a_1$.
(ii) Suppose k is in S; that is, for any collection of k positive numbers

$$\log(a_1 a_2 \ldots a_k) = \log a_1 + \log a_2 + \cdots + \log a_k.$$

Then for any collection $a_1, a_2, \ldots, a_k, a_{k+1}$ of $k + 1$ positive numbers

$$\log(a_1 a_2 \ldots a_k a_{k+1}) = \log((a_1 a_2 \ldots a_k)a_{k+1})$$
$$= \log(a_1 a_2 \ldots a_k) + \log a_{k+1}$$
$$= \log a_1 + \log a_2 + \cdots + \log a_k + \log a_{k+1},$$

so $k + 1$ is in S.
By the axiom of mathematical induction, $S = N$ and

$$\log(a_1 a_2 \ldots a_n) = \log a_1 + \log a_2 + \cdots + \log a_n$$

for any collection of positive numbers $a_1, a_2, \ldots, a_n$, regardless of the value of n. □

EXERCISE SET A.4

Use the axiom of mathematical induction to show that the expressions in Exercises 1 through 17 are true for every natural number n.

1. $1 + 3 + 5 + \cdots + (2n - 1) = n^2$

2. $2 + 4 + 6 + \cdots + 2n = n(n + 1)$

3. $3 + 6 + 9 + \cdots + 3n = \dfrac{2n(n + 1)}{3}$

4. $\dfrac{1}{2} + \dfrac{1}{4} + \dfrac{1}{8} + \cdots + \dfrac{1}{2^n} = 1 - \dfrac{1}{2^n}$

5. $1 - \dfrac{1}{2} - \dfrac{1}{4} - \cdots - \dfrac{1}{2^n} = \dfrac{1}{2^n}$

6. $1 + 4 + 9 + \cdots + n^2 = \dfrac{n(n + 1)(2n + 1)}{6}$

7. $1 + 9 + 27 + \cdots + n^3 = \dfrac{n^2(n + 1)^2}{4}$

8. $2 + 6 + 12 + \cdots + n(n + 1) = \dfrac{n(n + 1)(n + 2)}{3}$

9. $\dfrac{1}{2} + \dfrac{1}{6} + \dfrac{1}{12} + \cdots + \dfrac{1}{n(n + 1)} = \dfrac{n}{n + 1}$

10. $\dfrac{1}{2} + \dfrac{1}{2} + \dfrac{3}{8} + \cdots + \dfrac{n}{2^n} = 2 - \dfrac{n + 2}{2^n}$

11. $3 + 8 + 15 + \cdots + n(n + 2) = \dfrac{n(n + 1)(2n + 7)}{6}$

12. $\dfrac{1}{3} + \dfrac{1}{15} + \dfrac{1}{35} + \cdots + \dfrac{1}{4n^2 - 1} = \dfrac{n}{2n + 1}$

13. $1 - \dfrac{1}{2} - \dfrac{1}{6} - \cdots - \left(\dfrac{1}{n} - \dfrac{1}{n + 1}\right) = \dfrac{1}{n + 1}$

14. $n < \left(\dfrac{3}{2}\right)^n$

15. $1 + na \le (1 + a)^n$, for any $a \ge 0$.

16. $a + ar + ar^2 + \cdots + ar^n = \dfrac{a(1 - r^n)}{1 - r}$, for any a and any $r \ne 1$

17. $a + (a + d) + (a + 2d) + \cdots + (a + nd) = \dfrac{(n + 1)(2a + nd)}{2}$, for any a and d

18. Prove that $x - y$ is a factor of $x^n - y^n$ for any natural number n.

19. Prove that $x + y$ is a factor of $x^{2n-1} - y^{2n-1}$ for any natural number n.

20. Prove that for any natural number n and any $a > 1$, $1 < a^n < a^{n+1}$.

21. Prove that for any natural number n and any $0 < a < 1$, $0 < a^{n+1} < a^n < 1$.

22. Use the axiom of mathematical induction to prove that the following laws of exponents hold for every pair of natural numbers m and n.

(a) $a^n b^n = (ab)^n$

(b) $\dfrac{a^n}{b^n} = \left(\dfrac{a}{b}\right)^n$, provided $b \ne 0$

(c) $a^m a^n = a^{mn}$

(d) $\dfrac{a^n}{a^m} = a^{n-m}$, provided $a \ne 0$

(e) $(a^m)^n = a^{mn}$

[*Hint*: In parts (c), (d) and (e), assume that m is a fixed, but arbitrary natural number.]

23. Let $a_0 = 1$ and $a_{n+1} = \dfrac{1}{2}\left(a_n + \dfrac{2}{a_n}\right)$ for each $n \ge 0$. Show that $1 \le a_n \le 2$ for all $n \ge 0$.

24. Let $a_0 = 1$, $0 < M$, and $a_n = \dfrac{1}{2}\left(a_n + \dfrac{M}{a_n}\right)$ for each $n \ge 0$. Show that $1 \le a_n \le M$ if $M \ge 1$ and $M \le a_n \le 1$ if $M \le 1$.

25. Prove that any set with n elements has 2^n distinct subsets.

26. The Tower of Hanoi problem is shown in the accompanying figure. Four discs are stacked on post A. The problem is to find the number of moves required to move all discs from post A to post C. A move is defined to be the movement of one disc from one pin to another. You can remove only one disc at a time and a larger disc can never rest on top of a smaller disc.

Use mathematical induction to show that the number of moves required to move n discs is $2^n - 1$.

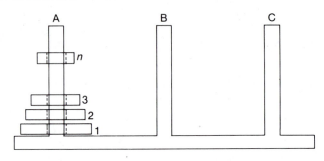

APPENDIX B: ADDITIONAL CALCULUS THEOREMS AND PROOFS

This Appendix contains the proofs of some of the basic theorems in calculus. The motivation and discussion of these results was presented in text and is not repeated here. The theorems have been numbered with their original text references and are presented in the order in which they were originally discussed.

(2.8) THEOREM

If f and g are functions with $\lim\limits_{x \to a} f(x) = L$ and $\lim\limits_{x \to a} g(x) = M$, then

(a) $\lim\limits_{x \to a} [f(x) + g(x)] = L + M$,

(b) $\lim\limits_{x \to a} [f(x) - g(x)] = L - M$,

(c) $\lim\limits_{x \to a} [f(x)\, g(x)] = L \cdot M$,

(d) $\lim\limits_{x \to a} cf(x) = cL$, for any constant c,

(e) $\lim\limits_{x \to a} \dfrac{1}{f(x)} = \dfrac{1}{L}$, provided $L \neq 0$,

(f) $\lim\limits_{x \to a} \dfrac{g(x)}{f(x)} = \dfrac{M}{L}$, provided $L \neq 0$,

(g) $\lim\limits_{x \to a} \sqrt[n]{f(x)} = \sqrt[n]{L}$ if n is an odd positive integer or if n is an even positive integer and $L > 0$.

PROOF

(a) Let $\varepsilon > 0$ be given. Since $\lim\limits_{x \to a} f(x) = L$ and $\lim\limits_{x \to a} g(x) = M$, there exist

A–25

numbers δ_1 and δ_2 with the property that

$$|f(x) - L| < \frac{\varepsilon}{2}, \qquad \text{whenever } 0 < |x - a| < \delta_1$$

and

$$|g(x) - M| < \frac{\varepsilon}{2}, \qquad \text{whenever } 0 < |x - a| < \delta_2.$$

Choose $\delta = $ minimum (δ_1, δ_2) and suppose that $0 < |x - a| < \delta$. Then both $|f(x) - L| < \frac{\varepsilon}{2}$ and $|g(x) - M| < \frac{\varepsilon}{2}$,

so

$$|[f(x) + g(x)] - [L + M]| = |(f(x) - L) + (g(x) - M)|$$

$$\leq |f(x) - L| + |g(x) - M| < \frac{\varepsilon}{2} + \frac{\varepsilon}{2} = \varepsilon.$$

Thus, $\lim_{x \to a} [f(x) + g(x)] = L + M$.

(d) If $c = 0$, then $c f(x) = 0$, $cL = 0$, and the result holds by Theorem 2.6.
Suppose $c \neq 0$ and $\varepsilon > 0$ is given. Let δ be such that

$$|f(x) - L| < \frac{\varepsilon}{|c|}, \qquad \text{whenever } 0 < |x - a| < \delta.$$

Then

$$|c f(x) - cL| = |c| \, |f(x) - L| < |c| \frac{\varepsilon}{|c|} = \varepsilon$$

and $\lim_{x \to a} c f(x) = cL$.

(b) Since $\lim_{x \to a} g(x) = M$, (d) implies that $\lim_{x \to a} -g(x) = -M$.
By (a) $\lim_{x \to a} [f(x) - g(x)] = \lim_{x \to a} [f(x) + (-g(x))] = L - M$.

Before showing the remainder of the results we will prove the following lemma.

LEMMA   If $\lim_{x \to a} f(x) = L$, then numbers $m > 0$ and δ exist with $|f(x)| \leq m$ for $0 < |x - a| < \delta$.

PROOF

Let $\varepsilon = 1$ and δ be such that

$$|f(x) - L| < \varepsilon = 1 \qquad \text{whenever } 0 < |x - a| < \delta.$$

Then $-1 < f(x) - L < 1$, so $L - 1 < f(x) < L + 1$ and

$$|f(x)| \leq \text{maximum } \{|L + 1|, |L - 1|\}.$$

Choosing $m = \max \{|L + 1|, |L - 1|\}$ gives the result.   □

(c) Let $\varepsilon > 0$ be given and δ_1 and $m > 0$ be such that

$$|f(x)| \leq m \text{ whenever } 0 < |x - a| < \delta_1.$$

Let δ_2 and δ_3 be such that

$$|f(x) - L| < \frac{\varepsilon}{2(|M| + 1)} \qquad \text{whenever } 0 < |x - a| < \delta_2$$

and $\qquad |g(x) - M| < \dfrac{\varepsilon}{2m} \qquad \text{whenever } 0 < |x - a| < \delta_3.$

Choose $\delta = \text{minimum } \{\delta_1, \delta_2, \delta_3\}$ and let $0 < |x - a| < \delta$.
Then

$$|f(x)g(x) - L \cdot M| = |f(x)g(x) - f(x)M + f(x)M - L \cdot M|$$

$$\leq |f(x)|\,|g(x) - M| + |f(x) - L|\,|M|$$

$$< m\left(\frac{\varepsilon}{2m}\right) + \frac{\varepsilon}{2(|M| + 1)}|M| < \frac{\varepsilon}{2} + \frac{\varepsilon}{2} = \varepsilon.$$

Thus $\lim\limits_{x \to a} f(x)g(x) = L \cdot M$.

(e) Since $L \neq 0$ there is a number δ_1 such that

$$|f(x) - L| < \frac{|L|}{2} \qquad \text{whenever } 0 < |x - a| < \delta_1.$$

Thus, for $0 < |x - a| < \delta_1$,

$$\left| |f(x)| - |L| \right| \leq |f(x) - L| \leq \frac{|L|}{2}$$

and $\qquad \dfrac{-|L|}{2} < |f(x)| - |L| < \dfrac{|L|}{2}.$

This implies that

$$\frac{|L|}{2} \leq |f(x)| \leq \frac{3|L|}{2}.$$

Let $\varepsilon > 0$ and choose $0 < \delta$ to be a number less than δ_1 with the property that

$$|f(x) - L| < \frac{L^2\varepsilon}{2} \qquad \text{whenever } 0 < |x - a| < \delta.$$

Then

$$\left|\frac{1}{f(x)} - \frac{1}{L}\right| = \left|\frac{L - f(x)}{f(x)\,L}\right| = \frac{|f(x) - L|}{|f(x)|\,|L|}.$$

But since $0 < |x - a| < \delta \leq \delta_1$, $|f(x)| \geq \dfrac{|L|}{2}$, and $\dfrac{1}{|f(x)|} \leq \dfrac{2}{|L|}.$

So $\qquad \left|\dfrac{1}{f(x)} - \dfrac{1}{L}\right| = \dfrac{|f(x) - L|}{|L|\,|f(x)|} \leq \dfrac{2|f(x) - L|}{|L|^2} < \dfrac{2}{|L|^2}\dfrac{L^2\varepsilon}{2} = \varepsilon.$

Thus $\lim\limits_{x \to a} \dfrac{1}{f(x)} = \dfrac{1}{L}$.

(f) By applying (c) and (e),

$$\lim_{x \to a} \frac{g(x)}{f(x)} = \lim_{x \to a} g(x) \frac{1}{f(x)} = \lim_{x \to a} g(x) \lim_{x \to a} \frac{1}{f(x)} = M \frac{1}{L} = \frac{M}{L}.$$

(g) We prove this result only for the case $n = 2$ and $L > 0$. The remainder of the proof is difficult and best postponed to a course in advanced calculus. A proof can be constructed along the lines of the proof with $n = 2$, but it requires the use of the axiom of mathematical induction discussed in Section A.4.

Let $\varepsilon > 0$ be given and choose δ so that

$$|f(x) - L| < \sqrt{L}\, \varepsilon \qquad \text{whenever } 0 < |x - a| < \delta.$$

Since $f(x) - L = (\sqrt{f(x)} - \sqrt{L})(\sqrt{f(x)} + \sqrt{L})$,

$$|\sqrt{f(x)} - \sqrt{L}| = \left| \frac{f(x) - L}{\sqrt{f(x)} + \sqrt{L}} \right| = \frac{|f(x) - L|}{\sqrt{f(x)} + \sqrt{L}} \le \frac{|f(x) - L|}{\sqrt{L}} < \frac{\sqrt{L}\,\varepsilon}{\sqrt{L}} = \varepsilon$$

whenever $0 < |x - a| < \delta$. Thus $\lim\limits_{x \to a} \sqrt{f(x)} = \sqrt{L}$.   □

(3.25)
THEOREM

Chain Rule   If g is differentiable at a and f is differentiable at $g(a)$, then $f \circ g$ is differentiable at a and $(f \circ g)'(a) = f'(g(a))g'(a)$.

Before showing this result we prove the following lemma (the property expressed in this lemma provides the motivation for the definition of differentiability for functions of several variables).

LEMMA

If $f'(b)$ exists, then a function ε exists with

$$f(t) = f(b) + [f'(b) + \varepsilon(t)](t - b),$$

where $\lim\limits_{t \to b} \varepsilon(t) = 0$.

PROOF OF LEMMA

Define ε by

$$\varepsilon(t) = \begin{cases} \dfrac{f(t) - f(b)}{t - b} - f'(b), & \text{if } t \text{ is in the domain of } f \\ & \text{and } t \ne b. \\ 0, & \text{if } t = b. \end{cases}$$

It is easily verified that ε has the required properties.   □

PROOF OF THEOREM

Let $t = g(x)$ and $b = g(a)$. Since f is differentiable at $g(a)$, the lemma implies that a function ε exists with

$$f(g(x)) = f(g(a)) + [f'(g(a)) + \varepsilon(g(x))][g(x) - g(a)]$$

and

$$\lim_{g(x) \to g(a)} \varepsilon(g(x)) = 0.$$

But g is differentiable and hence continuous at a, so $\lim\limits_{x \to a} g(x) = g(a)$ and

$$\lim_{x \to a} \varepsilon(g(x)) = \lim_{g(x) \to g(a)} \varepsilon(g(x)) = 0.$$

Consequently,

$$
\begin{aligned}
(f \circ g)'(a) &= \lim_{x \to a} \frac{f(g(x)) - f(g(a))}{x - a} \\
&= \lim_{x \to a} \frac{[f'(g(a)) + \varepsilon(g(x))][g(x) - g(a)]}{x - a} \\
&= \lim_{x \to a} [f'(g(a)) + \varepsilon(g(x))] \lim_{x \to a} \frac{g(x) - g(a)}{x - a} \\
&= f'(g(a))\, g'(a). \qquad \qquad \square
\end{aligned}
$$

(5.14)
THEOREM

If f is integrable on $[a, b]$ and k is a constant, then kf is also integrable on $[a, b]$ and

$$\int_a^b kf(x)dx = k \int_a^b f(x)dx.$$

PROOF

If $k = 0$ the proof follows from Theorem 5.13. Suppose $k \neq 0$, and $\varepsilon > 0$ is given. Since f is integrable on $[a, b]$, a number $\delta > 0$ exists with the property that for any partition $\mathcal{P} = \{x_0, x_1, \ldots, x_n\}$, with $\|\mathcal{P}\| < \delta$,

$$\left| \int_a^b f(x)\, dx - \sum_{i=1}^{n} f(z_i) \Delta x_i \right| < \frac{\varepsilon}{|k|}$$

whenever z_i is in $[x_{i-1}, x_i]$.
Consequently,

$$\left| k \int_a^b f(x)\, dx - \sum_{i=1}^{n} k f(z_i) \Delta x_i \right| = \left| k \int_a^b f(x)\, dx - k \sum_{i=1}^{n} f(z_i) \Delta x_i \right| < |k| \frac{\varepsilon}{|k|} = \varepsilon.$$

This implies that the function kf is integrable and that

$$\int_a^b kf(x)\, dx = k \int_a^b f(x)\, dx. \qquad \qquad \square$$

(5.15)
THEOREM

If f and g are integrable on $[a, b]$, then

(i) $f + g$ is integrable on $[a, b]$ and

$$\int_a^b (f + g)(x)dx = \int_a^b f(x)dx + \int_a^b g(x)dx,$$

(ii) $f - g$ is integrable on $[a, b]$ and

$$\int_a^b (f - g)(x)dx = \int_a^b f(x)dx - \int_a^b g(x)dx.$$

PROOF

(i) Let $\varepsilon > 0$ be given. Since f and g are both integrable on $[a, b]$ there exist numbers δ_f and δ_g with the property that if $\mathcal{P}$ is any partition of $[a, b]$ with $\|\mathcal{P}\| < \delta_f$ and $\|\mathcal{P}\| < \delta_g$, then

(B.1)
$$\left| \sum_{i=1}^{n} f(z_i)\Delta x_i - \int_a^b f(x)\,dx \right| < \frac{\varepsilon}{2}$$

and

(B.2)
$$\left| \sum_{i=1}^{n} g(z_i)\Delta x_i - \int_a^b g(x)\,dx \right| < \frac{\varepsilon}{2}$$

whenever z_i is in $[x_{i-1}, x_i]$. Let $\delta = \text{minimum } (\delta_f, \delta_g)$. If $\|\mathcal{P}\| < \delta$, then the inequalities in equations (B.1) and (B.2) both hold. Consequently

$$\left| \sum_{i=1}^{n} [f(z_i) + g(z_i)]\Delta x_i - \left[\int_a^b f(x)\,dx + \int_a^b f(x)\,dx \right] \right|$$

$$= \left| \left[\sum_{i=1}^{n} f(z_i)\Delta x_i - \int_a^b f(x)\,dx \right] + \left[\sum_{i=1}^{n} g(z_i)\Delta x_i - \int_a^b g(x)\,dx \right] \right|$$

$$\leq \left| \sum_{i=1}^{n} f(z_i)\Delta x_i - \int_a^b f(x)\,dx \right| + \left| \sum_{i=1}^{n} g(z_i)\Delta x_i - \int_a^b g(x)\,dx \right|$$

$$< \frac{\varepsilon}{2} + \frac{\varepsilon}{2} = \varepsilon$$

whenever z_i is in $[x_{i-1}, x_i]$. This implies that

$$\lim_{\|\mathcal{P}\| \to 0} \sum_{i=1}^{n} (f + g)(z_i)\Delta x_i = \int_a^b (f + g)(x)\,dx$$

exists and that

$$\int_a^b (f + g)(x)\,dx = \int_a^b f(x)\,dx + \int_a^b g(x)\,dx.$$

(ii) Theorem 5.14 implies that

$$\int_a^b - g(x)\,dx = - \int_a^b g(x)\,dx.$$

Combining this result with (i) gives

$$\int_a^b (f - g)(x)\,dx = \int_a^b (f + (-g))(x)\,dx = \int_a^b f(x)\,dx + \int_a^b - g(x)\,dx$$

$$= \int_a^b f(x)\,dx - \int_a^b g(x)\,dx. \qquad \square$$

(5.16)
THEOREM

If f is integrable on the intervals $[a, c]$ and $[c, b]$, then f is integrable on $[a, b]$ and

$$\int_a^b f(x)\,dx = \int_a^c f(x)\,dx + \int_c^b f(x)\,dx.$$

PROOF

We need to show that for a given $\varepsilon > 0$, a number $\delta > 0$ exists with the property that for any partition $\mathcal{P} = \{x_0, x_1, \ldots, x_n\}$ of $[a, b]$, with $\|\mathcal{P}\| < \delta$,

$$\left| \int_a^c f(x)\, dx + \int_c^b f(x)\, dx - \sum_{i=1}^n f(z_i)\Delta x_i \right| < \varepsilon$$

whenever z_i is in $[x_{i-1}, x_i]$.

First choose δ_1 and δ_2 so that if $\mathcal{P}_1$ and $\mathcal{P}_2$ are partitions of $[a, c]$ and $[c, b]$, respectively, with $\|\mathcal{P}_1\| < \delta_1$ and $\|\mathcal{P}_2\| < \delta_2$ and R_1 and R_2 are any Riemann sums associated with $\mathcal{P}_1$ and $\mathcal{P}_2$, respectively, then

$$\left| \int_a^c f(x)\, dx - R_1 \right| < \frac{\varepsilon}{3} \quad \text{and} \quad \left| \int_c^b f(x)\, dx - R_2 \right| < \frac{\varepsilon}{3}.$$

Suppose, in addition, that M_1 and M_1 are bounds for f on $[a, c]$ and $[c, b]$, respectively. These bounds are ensured by the definition of the integrability of f on these intervals.

Let $M = \max \{M_1, M_2\}$, $\delta_3 = \dfrac{\varepsilon}{6M}$, and $\delta = \min \{\delta_1, \delta_2, \delta_3\}$.

Suppose that $\mathcal{P} = \{x_0, x_1, \ldots, x_n\}$ is a partition of $[a, b]$ with $\|\mathcal{P}\| < \delta$. Since $a < c < b$, a unique index k exists, $1 \leq k \leq n$, with $x_{k-1} < c \leq x_k$, and for any collection of z_i in $[x_{i-1}, x_i]$,

$$\sum_{i=1}^n f(z_i)\Delta x_i = \sum_{i=1}^{k-1} f(z_i)\Delta x_i + f(z_k)\Delta x_k + \sum_{i=k+1}^n f(z_i)\Delta x_i.$$

But $f(z_k)\Delta x_k = f(c)\Delta x_k + [f(z_k) - f(c)]\Delta x_k$

$$= f(c)(x_k - c + c - x_{k-1}) + [f(z_k) - f(c)]\Delta x_k$$

$$= f(c)(x_k - c) + f(c)(c - x_{k-1}) + [f(z_k) - f(c)]\Delta x_k,$$

so

$$\sum_{i=1}^n f(z_i)\Delta x_i = \left[\sum_{i=1}^{k-1} f(z_i)\Delta x_i + f(c)(c - x_{k-1}) \right]$$

$$+ \left[f(c)(x_k - c) + \sum_{i=k+1}^n f(z_i)\Delta x_i \right] + [f(z_k) - f(c)]\Delta x_k.$$

However,

$$\sum_{i=1}^{k-1} f(z_i)\Delta x_i + f(c)(c - x_{k-1})$$

and

$$f(c)(x_k - c) + \sum_{i=k+1}^n f(z_i)\Delta x_i$$

are Riemann sums for $[a, c]$ and $[c, b]$, respectively, on partitions with norms bounded by $\delta \leq \min(\delta_1, \delta_2)$. Consequently,

$$\left| \int_a^c f(x)\,dx - \sum_{i=1}^{k-1} f(z_i)\Delta x_i - f(c)(c - x_{k-1}) \right| < \frac{\varepsilon}{3},$$

and

$$\left| \int_c^b f(x)\,dx - f(c)(x_k - c) - \sum_{i=k+1}^{n} f(z_i)\Delta x_i \right| < \frac{\varepsilon}{3},$$

so

$$\left| \int_a^c f(x)\,dx + \int_c^b f(x)\,dx - \sum_{i=1}^{n} f(z_i)\Delta x_i \right| < \frac{\varepsilon}{3} + \frac{\varepsilon}{3} + |f(z_k) - f(c)|\Delta x_k$$

$$\leq \frac{2\varepsilon}{3} + [|f(z_k)| + |f(c)|]\Delta x_k$$

$$\leq \frac{2\varepsilon}{3} + (M + M)\frac{\varepsilon}{6M} = \frac{2\varepsilon}{3} + \frac{\varepsilon}{3} = \varepsilon. \qquad \square$$

(5.18)
THEOREM

If f is integrable on $[a, b]$ and $f(x) \geq 0$ for each x in $[a, b]$, then

$$\int_a^b f(x)\,dx \geq 0.$$

PROOF

Suppose that $\int_a^b f(x)\,dx < 0$. Then $\varepsilon = -\int_a^b f(x)\,dx$ is a positive real number. Since f is integrable, there exists a number δ with the property that whenever $\mathcal{P}$ is a partition of $[a, b]$ and $\|\mathcal{P}\| < \delta$,

$$\left| \sum_{i=1}^{n} f(z_i)\Delta x_i - \int_a^b f(x)\,dx \right| < \varepsilon = -\int_a^b f(x)\,dx.$$

But

$$\sum_{i=1}^{n} f(z_i)\Delta x_i - \int_a^b f(x)\,dx \leq \left| \sum_{i=1}^{n} f(z_i)\Delta x_i - \int_a^b f(x)\,dx \right|.$$

So

$$\sum_{i=1}^{n} f(z_i)\Delta x_i - \int_a^b f(x)\,dx < -\int_a^b f(x)\,dx$$

and

$$\sum_{i=1}^{n} f(z_i)\Delta x_i < 0,$$

whenever $\|\mathcal{P}\| < \delta$ and z_i is in $[x_{i-1}, x_i]$. This is clearly false, since $f(z_i) \geq 0$ and $\Delta x_i > 0$ for each i. Consequently, $\int_a^b f(x)\,dx \geq 0$. $\qquad \square$

(5.19)
THEOREM

If f and g are both integrable on $[a, b]$ and $f(x) \geq g(x)$ for each x in $[a, b]$, then

$$\int_a^b f(x)\, dx \geq \int_a^b g(x)\, dx.$$

PROOF

Let $h(x) = f(x) - g(x)$. Then Theorem 5.18 implies that $\int_a^b h(x)\, dx \geq 0$. Applying Theorem 5.15,

$$\int_a^b f(x)\, dx - \int_a^b g(x)\, dx = \int_a^b [f(x) - g(x)]\, dx = \int_a^b h(x)\, dx \geq 0$$

and

$$\int_a^b f(x)\, dx \geq \int_a^b g(x)\, dx. \qquad \square$$

(7.5)
THEOREM

If f is increasing and continuous on $[a, b]$, then f^{-1} is increasing and continuous on $[f(a), f(b)]$.

PROOF

Theorem 7.3 implies that f^{-1} is increasing on $[f(a), f(b)]$. It remains to show that f^{-1} is continuous on $[f(a), f(b)]$. Suppose y_0 is arbitrarily chosen in $(f(a), f(b))$ and that $y_0 = f(x_0)$. Let $\varepsilon > 0$ be given. We need to show that a number δ exists with $|f^{-1}(y) - f^{-1}(y_0)| < \varepsilon$, whenever $|y - y_0| < \delta$.

Let $\hat{\varepsilon} = $ minimum $\{b - x_0, x_0 - a, \varepsilon\}$. This will ensure that $x_1 = x_0 - \hat{\varepsilon}$ and $x_2 = x_0 + \hat{\varepsilon}$ are both contained in $[a, b]$. Since f is increasing and

$$a < x_1 < x_0 < x_2 < b,$$

$$f(a) < f(x_1) < f(x_0) = y_0 < f(x_2) < f(b).$$

Choose $\delta = $ minimum $\{y_0 - f(x_1), f(x_2) - y_0\}$, and assume that $|y - y_0| < \delta$. (See Figure B.1) Since $f(x_1) < y < f(x_2)$, the intermediate value theorem implies that a number x, $x_1 < x < x_2$ exists with $f(x) = y$. Consequently, $x = f^{-1}(y)$ and

$$|f^{-1}(y) - f^{-1}(y_0)| = |x - x_0|.$$

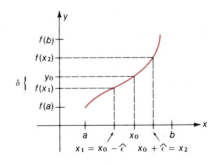

FIGURE B.1

Since $x_0 - \hat{\varepsilon} = x_1 < x < x_2 = x_0 + \hat{\varepsilon}$,

$$-\hat{\varepsilon} < x - x_0 < \hat{\varepsilon}.$$

So, $|x - x_0| < \hat{\varepsilon} < \varepsilon$ and

$$|f^{-1}(y) - f^{-1}(y_0)| < \varepsilon$$

whenever $|y - y_0| < \delta$. This implies that f^{-1} is continuous at each y_0 in $(f(a), f(b))$.

The proof that f^{-1} is continuous from the right at $f(a)$ and continuous from the left at $f(b)$ is similar. $\qquad\square$

(9.35)
THEOREM

If f is a function defined by the power series

$$f(x) = \sum_{n=0}^{\infty} a_n x^n$$

with radius of convergence R, then
(i) f is differentiable at each x in $(-R, R)$, and

$$D_x f(x) = D_x \sum_{n=0}^{\infty} a_n x^n = \sum_{n=1}^{\infty} n a_n x^{n-1}$$

(ii) f is integrable on $(-R, R)$ and

$$\int_0^x f(t)dt = \int_0^x \left[\sum_{n=0}^{\infty} a_n t^n \right] dt = \sum_{n=0}^{\infty} \frac{a_n}{n+1} x^{n+1}.$$

Before proving this theorem we establish the following lemma.

LEMMA

If $\sum_{n=0}^{\infty} a_n x^n$ has radius of convergence R, then the series $\sum_{n=1}^{\infty} n a_n x^{n-1}$ and $\sum_{n=2}^{\infty} n(n-1) a_n x^{n-2}$ have radius of convergence R.

PROOF OF LEMMA

Let x be fixed in $(-R, R)$ and z be such that $|x| < |z| < R$. By Lemma 9.32, $\sum_{n=0}^{\infty} a_n z^n$ is absolutely convergent, so $\lim_{n \to \infty} |a_n z^n| = 0$. Let N be such that if $n > N$, then $|a_n z^n| < 1$. Let

$$M = \max \{|a_1 z|, |a_2 z^2|, \ldots, |a_N z^N|, 1\}.$$

This choice of M ensures that $|a_n z^n| \leq M$, and consequently that

$$\left| n a_n x^{n-1} \right| = \left| n a_n x^{n-1} \frac{z^n}{z^n} \right| = n \frac{|a_n z^n|}{|z|} \left| \frac{x}{z} \right|^{n-1} \leq \frac{nM}{|z|} \left| \frac{x}{z} \right|^{n-1}$$

for each n. Applying the ratio test to the series $\sum_{n=1}^{\infty} \frac{nM}{|z|} \left| \frac{x}{z} \right|^{n-1}$ implies that this series converges, since

$$\lim_{n \to \infty} \frac{(n + 1)M \left|\dfrac{x}{z}\right|^n}{\dfrac{nM}{|z|} \left|\dfrac{x}{z}\right|^{n-1}} = \lim_{n \to \infty} \frac{n + 1}{n} \left|\frac{x}{z}\right| = \left|\frac{x}{z}\right| < 1.$$

By the comparison test, $\displaystyle\sum_{n=1}^{\infty} |n\, a_n x^{n-1}|$ also converges. Since x was arbitrarily

chosen in $(-R, R)$, the radius of convergence $\hat{R}$ of $\displaystyle\sum_{n=1}^{\infty} n\, a_n x^{n-1}$ satisfies $\hat{R} \geq R$.

Now suppose that a number z exists with $|z| > R$ and that $\displaystyle\sum_{n=1}^{\infty} |n\, a_n z^{n-1}|$

converges. When $n > z$, $|n\, a_n z^{n-1}| > |a_n z^n|$ and by the comparison test, the

convergence of $\displaystyle\sum_{n=1}^{\infty} |n\, a_n z^{n-1}|$ implies the convergence of $\displaystyle\sum_{n=1}^{\infty} |a_n z^n|$. This contra-

dicts $|z| > R$. Thus $\hat{R} = R$.

Applying this result to the series $\displaystyle\sum_{n=1}^{\infty} n\, a_n x^{n-1}$ shows that the radius of con-

vergence of $\displaystyle\sum_{n=2}^{\infty} n(n - 1)a_n x^{n-2}$ is also R. $\square$

PROOF OF THEOREM

(i) Let $g(x) = \displaystyle\sum_{n=1}^{\infty} n\, a_n x^{n-1}$ for x in $(-R, R)$. We want to show that

$\displaystyle\lim_{h \to 0} \frac{f(x + h) - f(x)}{h} = g(x)$. Fix x in $(-R, R)$ and restrict h so that

$|x + h| < R$. Consider

$$\left|\frac{f(x + h) - f(x)}{h} - g(x)\right| = \left|\frac{\displaystyle\sum_{n=0}^{\infty} a_n(x + h)^n - \sum_{n=0}^{\infty} a_n x^n}{h} - \sum_{n=1}^{\infty} n\, a_n x^{n-1}\right|.$$

Since all the series converge absolutely and $(x + h)^0 - x^0 = 0$,

$$\left|\frac{f(x + h) - f(x)}{h} - g(x)\right| = \left|\sum_{n=1}^{\infty} a_n\left[\frac{(x + h)^n - x^n}{h} - n\, x^{n-1}\right]\right|.$$

By the mean value theorem, there exists for each n a number ξ_n, ξ_n between x

and $x + h$, with $\dfrac{(x + h)^n - x^n}{h} = n\, \xi_n^{n-1}$. Thus,

$$\left|\frac{f(x + h) - f(x)}{h} - g(x)\right| = \left|\sum_{n=1}^{\infty} a_n[n\, \xi_n^{n-1} - n\, x^{n-1}]\right|$$

$$\leq \sum_{n=1}^{\infty} |n\, a_n|\, |\xi_n^{n-1} - x^{n-1}|.$$

Since $\xi_1^0 - x^0 = 1 - 1 = 0$,

$$\left| \frac{f(x + h) - f(x)}{h} - g(x) \right| \le \sum_{n=2}^{\infty} |n \, a_n| \, |\xi_n^{n-1} - x^{n-1}|.$$

Again, applying the mean value theorem to $\xi_n^{n-1} - x^{n-1}$ implies that there exists for each n a number ζ_n, between x and ξ_n, with $\dfrac{\xi_n^{n-1} - x^{n-1}}{\xi_n - x} = (n - 1)\zeta_n^{n-2}$. Since $|\xi_n - x| < |h|$, this implies that

$$\left| \frac{f(x + h) - f(x)}{h} - g(x) \right| \le \sum_{n=2}^{\infty} |n \, a_n| \, |\xi_n^{n-1} - x^{n-1}|$$

$$= \sum_{n=2}^{\infty} |n \, a_n| \, |(n - 1) \, \zeta_n^{n-2} \, (\xi_n - x)|$$

$$\le |h| \sum_{n=2}^{\infty} |n(n - 1)a_n \, \zeta_n^{n-2}|.$$

But ζ_n is between x and ξ_n, so ζ_n is between x and $x + h$. By the lemma $\sum_{n=2}^{\infty} |n(n - 1)a_n \, \zeta_n^{n-2}|$ converges. Hence

$$\lim_{h \to 0} |h| \sum_{n=2}^{\infty} |n(n - 1)a_n \, \zeta_n^{n-2}| = 0,$$

so

$$\lim_{h \to \infty} \left| \frac{f(x + h) - f(x)}{h} - g(x) \right| = 0.$$

This implies that $f'(x) = g(x)$, that is,

$$D_x \sum_{n=0}^{\infty} a_n x^n = \sum_{n=1}^{\infty} n \, a_n x^{n-1}.$$

(ii) Let $g(x) = \displaystyle\sum_{n=0}^{\infty} \frac{a_n}{n + 1} x^{n+1} = \sum_{n=1}^{\infty} \frac{a_{n-1}}{n} x^n$. By part (i),

$$g'(x) = \sum_{n=1}^{\infty} a_{n-1} x^{n-1} = \sum_{n=0}^{\infty} a_n x^n$$

has radius of convergence R, so $g(x)$ is defined when x is in $(-R, R)$ and $g'(x) = f(x)$. Part (i) also implies that $g''(x) = f'(x)$ exists on $(-R, R)$, so f is continuous on $(-R, R)$. Since g is an antiderivative of the continuous function f, the fundamental theorem of calculus implies that for any x in $(-R, R)$

$$\int_0^x f(t)dt = g(x) - g(0) = g(x)$$

and

$$\int_0^x \left[\sum_{n=0}^{\infty} a_n t^n \right] dt = \sum_{n=0}^{\infty} \frac{a_n}{n + 1} x^{n+1}.$$

$\square$

APPENDIX C: TABLES

TABLE I
Trigonometric Functions
(in degrees)

Degrees	Sine	Tangent	Cotangent	Cosine	
0	0	0	—	1.0000	90
1	0.0175	0.0175	57.290	0.9998	89
2	0.0349	0.0349	28.636	0.9994	88
3	0.0523	0.0524	19.081	0.9986	87
4	0.0698	0.0699	14.301	0.9976	86
5	0.0872	0.0875	11.430	0.9962	85
6	0.1045	0.1051	9.5144	0.9945	84
7	0.1219	0.1228	8.1443	0.9925	83
8	0.1392	0.1405	7.1154	0.9903	82
9	0.1564	0.1584	6.3138	0.9877	81
10	0.1736	0.1763	5.6713	0.9848	80
11	0.1908	0.1944	5.1446	0.9816	79
12	0.2079	0.2126	4.7046	0.9781	78
13	0.2250	0.2309	4.3315	0.9744	77
14	0.2419	0.2493	4.0108	0.9703	76
15	0.2588	0.2679	3.7321	0.9659	75
16	0.2756	0.2867	3.4874	0.9613	74
17	0.2924	0.3057	3.2709	0.9563	73
18	0.3090	0.3249	3.0777	0.9511	72
19	0.3256	0.3443	2.9042	0.9455	71
20	0.3420	0.3640	2.7475	0.9397	70
21	0.3584	0.3839	2.6051	0.9336	69
22	0.3746	0.4040	2.4751	0.9272	68
23	0.3907	0.4245	2.3559	0.9205	67
24	0.4067	0.4452	2.2460	0.9135	66
25	0.4226	0.4663	2.1445	0.9063	65

TABLE 2
Trigonometric Functions
(in radians)

Radians	Sine	Cosine	Tangent
0.00	0.0000	1.0000	0.0000
0.05	0.0500	0.9988	0.0500
0.10	0.0998	0.9950	0.1003
0.15	0.1494	0.9888	0.1511
0.20	0.1987	0.9801	0.2027
0.25	0.2474	0.9689	0.2553
0.30	0.2955	0.9553	0.3093
0.35	0.3429	0.9394	0.3650
0.40	0.3894	0.9211	0.4228
0.45	0.4350	0.9004	0.4831
0.50	0.4794	0.8776	0.5463
0.55	0.5227	0.8525	0.6131
0.60	0.5646	0.8253	0.6841
0.65	0.6052	0.7961	0.7602
0.70	0.6442	0.7648	0.8423
0.75	0.6816	0.7317	0.9316
0.80	0.7174	0.6967	1.0296
0.85	0.7513	0.6600	1.1383
0.90	0.7833	0.6216	1.2602
0.95	0.8134	0.5817	1.3984
1.00	0.8415	0.5403	1.5574
1.05	0.8674	0.4976	1.7433
1.10	0.8912	0.4536	1.9648
1.15	0.9128	0.4085	2.2345
1.20	0.9320	0.3624	2.5722
1.25	0.9490	0.3153	3.0096

TABLE I (cont.)

Degrees	Sine	Tangent	Cotangent	Cosine	
26	0.4384	0.4877	2.0503	0.8988	64
27	0.4540	0.5095	1.9626	0.8910	63
28	0.4695	0.5317	1.8807	0.8829	62
29	0.4848	0.5543	1.8040	0.8746	61
30	0.5000	0.5774	1.7321	0.8660	60
31	0.5150	0.6009	1.6643	0.8572	59
32	0.5299	0.6249	1.6003	0.8480	58
33	0.5446	0.6494	1.5399	0.8387	57
34	0.5592	0.6745	1.4826	0.8290	56
35	0.5736	0.7002	1.4281	0.8192	55
36	0.5878	0.7265	1.3764	0.8090	54
37	0.6018	0.7536	1.3270	0.7986	53
38	0.6157	0.7813	1.2799	0.7880	52
39	0.6293	0.8098	1.2349	0.7771	51
40	0.6428	0.8391	1.1918	0.7660	50
41	0.6561	0.8693	1.1504	0.7547	49
42	0.6691	0.9004	1.1106	0.7431	48
43	0.6820	0.9325	1.0724	0.7314	47
44	0.6947	0.9657	1.0355	0.7193	46
45	0.7071	1.0000	1.0000	0.7071	45
	Cosine	Cotangent	Tangent	Sine	Degrees

TABLE 2 (cont.)

Radians	Sine	Cosine	Tangent
1.30	0.9636	0.2675	3.6021
1.35	0.9757	0.2190	4.4552
1.40	0.9854	0.1700	5.7979
1.45	0.9927	0.1205	8.2381
1.50	0.9975	0.0707	14.1014
1.55	0.9998	0.0208	48.0785
$\pi/2$	1.0000	0.0000	—

TABLE 3
Exponential Functions

x	e^x	e^{-x}	x	e^x	e^{-x}
0.00	1.0000	1.0000	1.5	4.4817	0.2231
0.01	1.0101	0.9901	1.6	4.9530	0.2019
0.02	1.0202	0.9802	1.7	5.4739	0.1827
0.03	1.0305	0.9704	1.8	6.0496	0.1653
0.04	1.0408	0.9608	1.9	6.6859	0.1496
0.05	1.0513	0.9512	2.0	7.3891	0.1353
0.06	1.0618	0.9418	2.1	8.1662	0.1225
0.07	1.0725	0.9324	2.2	9.0250	0.1108
0.08	1.0833	0.9331	2.3	9.9742	0.1003
0.09	1.0942	0.9139	2.4	11.023	0.0907
0.10	1.1052	0.9048	2.5	12.182	0.0821
0.11	1.1163	0.8958	2.6	13.464	0.0743
0.12	1.1275	0.8869	2.7	14.880	0.0672
0.13	1.1388	0.8781	2.8	16.445	0.0608
0.14	1.1503	0.8694	2.9	18.174	0.0550
0.15	1.1618	0.8607	3.0	20.086	0.0498
0.16	1.1735	0.8521	3.1	22.198	0.0450
0.17	1.1853	0.8437	3.2	24.533	0.0408
0.18	1.1972	0.8353	3.3	27.113	0.0369
0.19	1.2092	0.8270	3.4	29.964	0.0334
0.20	1.2214	0.8187	3.5	33.115	0.0302
0.21	1.2337	0.8106	3.6	36.598	0.0273
0.22	1.2461	0.8025	3.7	40.447	0.0247
0.23	1.2586	0.7945	3.8	44.701	0.0224
0.24	1.2712	0.7866	3.9	49.402	0.0202
0.25	1.2840	0.7788	4.0	54.598	0.0183
0.30	1.3499	0.7408	4.1	60.340	0.0166
0.35	1.4191	0.7047	4.2	66.686	0.0150
0.40	1.4918	0.6703	4.3	73.700	0.0136
0.45	1.5683	0.6376	4.4	81.451	0.0123
0.50	1.6487	0.6065	4.5	90.017	0.0111
0.55	1.7333	0.5769	4.6	99.484	0.0101
0.60	1.8221	0.5488	4.7	109.95	0.0091
0.65	1.9155	0.5220	4.8	121.51	0.0082
0.70	2.0138	0.4966	4.9	134.29	0.0074
0.75	2.1170	0.4724	5.0	148.41	0.0067
0.80	2.2255	0.4493	5.5	244.69	0.0041
0.85	2.3396	0.4274	6.0	403.43	0.0025
0.90	2.4596	0.4066	6.5	665.14	0.0015
0.95	2.5857	0.3867	7.0	1096.6	0.0009
1.00	2.7183	0.3679	7.5	1808.0	0.0006
1.10	3.0042	0.3329	8.0	2981.0	0.0003
1.20	3.3201	0.3012	8.5	4914.8	0.0002
1.30	3.6693	0.2725	9.0	8103.1	0.0001
1.40	4.0552	0.2466	10.0	22026	0.00005

TABLE 4
Natural Logarithms

x	$\ln x$	x	$\ln x$	x	$\ln x$
		4.5	1.5041	9.0	2.1972
0.1	-2.3026	4.6	1.5261	9.1	2.2083
0.2	-1.6094	4.7	1.5476	9.2	2.2192
0.3	-1.2040	4.8	1.5686	9.3	2.2300
0.4	-0.9163	4.9	1.5892	9.4	2.2407
0.5	-0.6931	5.0	1.6094	9.5	2.2513
0.6	-0.5108	5.1	1.6292	9.6	2.2618
0.7	-0.3567	5.2	1.6487	9.7	2.2721
0.8	-0.2231	5.3	1.6677	9.8	2.2824
0.9	-0.1054	5.4	1.6864	9.9	2.2925
1.0	0.0000	5.5	1.7047	10	2.3026
1.1	0.0953	5.6	1.7228	11	2.3979
1.2	0.1823	5.7	1.7405	12	2.4849
1.3	0.2624	5.8	1.7579	13	2.5649
1.4	0.3365	5.9	1.7750	14	2.6391
1.5	0.4055	6.0	1.7918	15	2.7081
1.6	0.4700	6.1	1.8083	16	2.7726
1.7	0.5306	6.2	1.8245	17	2.8332
1.8	0.5878	6.3	1.8405	18	2.8904
1.9	0.6419	6.4	1.8563	19	2.9444
2.0	0.6931	6.5	1.8718	20	2.9957
2.1	0.7419	6.6	1.8871	25	3.2189
2.2	0.7885	6.7	1.9021	30	3.4012
2.3	0.8329	6.8	1.9169	35	3.5553
2.4	0.8755	6.9	1.9315	40	3.6889
2.5	0.9163	7.0	1.9459	45	3.8067
2.6	0.9555	7.1	1.9601	50	3.9120
2.7	0.9933	7.2	1.9741	55	4.0073
2.8	1.0296	7.3	1.9879	60	4.0943
2.9	1.0647	7.4	2.0015	65	4.1744
3.0	1.0986	7.5	2.0149	70	4.2485
3.1	1.1314	7.6	2.0281	75	4.3175
3.2	1.1632	7.7	2.0412	80	4.3820
3.3	1.1939	7.8	2.0541	85	4.4427
3.4	1.2238	7.9	2.0669	90	4.4998
3.5	1.2528	8.0	2.0794	100	4.6052
3.6	1.2809	8.1	2.0919	110	4.7005
3.7	1.3083	8.2	2.1041	120	4.7875
3.8	1.3350	8.3	2.1163	130	4.8675
3.9	1.3610	8.4	2.1282	140	4.9416
4.0	1.3863	8.5	2.1401	150	5.0106
4.1	1.4110	8.6	2.1518	160	5.0752
4.2	1.4351	8.7	2.1633	170	5.1358
4.3	1.4586	8.8	2.1748	180	5.1930
4.4	1.4816	8.9	2.1861	190	5.2470

ANSWERS TO ODD-NUMBERED EXERCISES

CHAPTER 1

EXERCISE SET 1.1, PAGE 7

1. (a) 21 (b) 17 (c) $33 + 16\sqrt{3}$ (d) 38

3. $\mathbb{R}$ **5.** $x \neq 0, 1, -1$ **7.** Yes **9.** No **11.** No

13. $x^2 + 2, -x^2 - 2, (1 + 2x^2)/x^2, 1/(x^2 + 2), x + 2, \sqrt{x^2 + 2}$

15. $-1/x, -1/x, x, x, 1/\sqrt{x}, 1/\sqrt{x}$, **17.** $x = \pm 1$ **19.** $x = \pm \sqrt{6}/2$ **21.** $x = -1$

23. $2(x + h) - 4, 2$ **25.** $\frac{3}{2}(x + h) - \frac{1}{4}, 3/2$ **27.** $(x + h)^2, 2x + h$

29. $2(x + h)^2 - (x + h), 4x + 2h - 1$

31. (a) even (b) odd (c) neither (d) even (e) neither (f) odd

33. $R(x) = \begin{cases} 10x & \text{, if } 0 \leq x \leq 10 \\ (12.5 - .25x)\,x & \text{, if } 11 \leq x \leq 24 \\ 6.25x & \text{, if } 25 \leq x \end{cases}$

35. $d(t) = \sqrt{(10t)^2 + (15 + 15t)^2}$ **37.** $A(w) = w\sqrt{100 - w^2}$, domain of A is $(0, 10)$

39. $S(r) = \dfrac{1800}{r} + 2\pi r^2$ **41.** $A(d) = -\dfrac{\pi}{4}d^2 + \dfrac{1}{2}d$

EXERCISE SET 1.2, PAGE 20

1. (a) $m = 2$ (b) $m = -1$ (c) $m = 1$ (d) $m = 3/4$

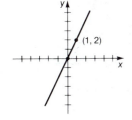
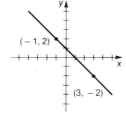
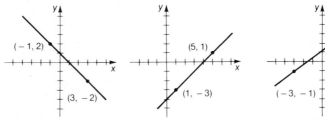
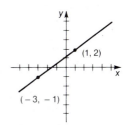

3. Pairs of perpendicular lines are: (a) and (b), (d) and (i), (e) and (h).

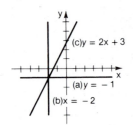
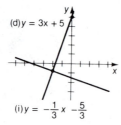
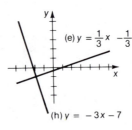
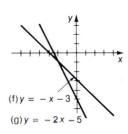

5. (a) $y = 2x$ (b) $y = -\dfrac{1}{2}x$ **7.** (a) $y = 3x - 1$ (b) $y = -\dfrac{1}{3}x + \dfrac{7}{3}$

9. (a) $y = -2x$ (b) $y = \dfrac{1}{2}x + \dfrac{5}{2}$ **11.** (a) $y = -x$ (b) $y = x$

13. Range is $[1, \infty)$ **15.** Range is $[0, \infty)$ **17.** Range is $[0, \infty)$

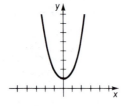
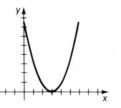
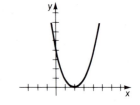

19. Range is $[-1, \infty)$ **21.** Range is $[-4, \infty)$ **23.** Range is $(-\infty, 0]$

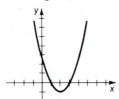
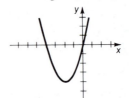
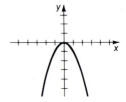

25. Range is $(-\infty, -1]$ **27.** Range is $[0, \infty)$ **29.** Range is $[0, \infty)$

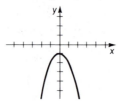
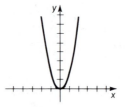
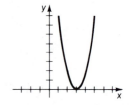

31. Range is $[-2, \infty)$ **33.** Range is $[0, \infty)$ **35.** Range is $[0, \infty)$

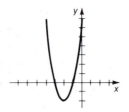
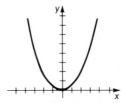
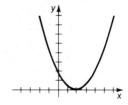

37.

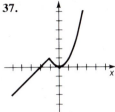

39. Domain of s is $[0, 2\sqrt{10}]$; range of s is $[0, 640]$; $2\sqrt{10}$ seconds

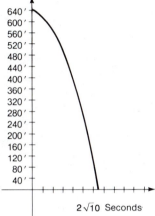

2√10 Seconds

41. Average cost is $-.00125b + 3.25$, where b is the number of bushels

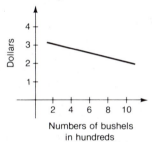

Dollars

Numbers of bushels in hundreds

43. (a)   (b) $TW(n) = 500n - .5n^2$ (c)   No fish remain

In grams

Number of fish in hundreds

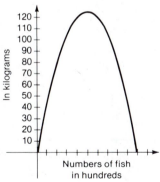

In kilograms

Numbers of fish in hundreds

45. $f(x) \equiv 0$.

EXERCISE SET 1.3, PAGE 27

1. $\sqrt{5}$

3.

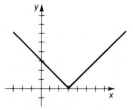

5.

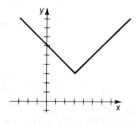

7.

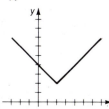

9.

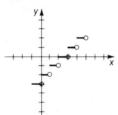

11.

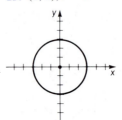

13.

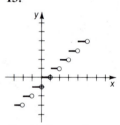

15. $(0, 0)$, $r = 3$

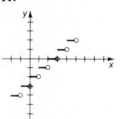

17. $(0, 1)$, $r = 1$

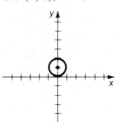

19. $(2, 1)$, $r = 3$

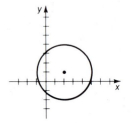

21. This circle is the same as the one in Exercise 19.

23. This circle is the same as the one in Exercise 15, $x^2 + y^2 = 9$.

25. $(x + 1)^2 + (y - 2)^2 = \dfrac{1}{4}$ or $x^2 + 2x + y^2 - 4y + \dfrac{19}{4} = 0$

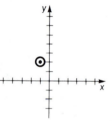

27.

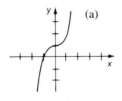 (a)

(b)

(c)

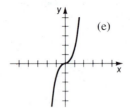 (d) (e) (f)

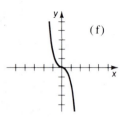

29. $(4, 3)$, $(-4, 3)$, $(4, -3)$, $(-4, -3)$ **31.** $\sqrt{x^4 - x^2 + 1}$

33. $c(x) = \begin{cases} \$5 & \text{, if } 0 < x < 2 \\ \$5 + \$1 \, [\![x - 1]\!], & \text{if } x \geq 2 \end{cases}$ **35.** $d(x) = \begin{cases} |x - 75| & \text{, if } 0 < x \leq 117.5 \\ |x - 160|, & \text{if } 117.5 < x \leq 241 \end{cases}$

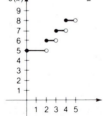

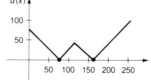

EXERCISE SET 1.4, PAGE 33

1. (a) $(f + g)(x) = 2x + 1$ (b) $(f - g)(x) = -1$ (c) $(f \cdot g)(x) = x^2 + x$

(d) $\left(\dfrac{f}{g}\right)(x) = \dfrac{x}{x + 1}$ (e) $(f \circ g)(x) = x + 1$ (f) $(g \circ f)(x) = x + 1$

Domain for (a), (b), (c), (e), and (f) is $\mathbb{R}$; domain of $\left(\dfrac{f}{g}\right)$ is $\{x \mid x \neq -1\}$.

3. (a) $(f + g)(x) = \dfrac{1}{x} + \sqrt{x - 1}$ (b) $(f - g)(x) = \dfrac{1}{x} - \sqrt{x - 1}$

(c) $(f \cdot g)(x) = \dfrac{\sqrt{x - 1}}{x}$ (d) $\left(\dfrac{f}{g}\right)(x) = \dfrac{1}{x\sqrt{x - 1}}$

(e) $(f \circ g)(x) = \dfrac{1}{\sqrt{x - 1}}$ (f) $(g \circ f)(x) = \sqrt{\dfrac{1 - x}{x}}$

Domain for parts (a), (b), and (c) is $[1, \infty)$; domain for (d) and (e) is $(1, \infty)$; Domain for (f) is $(0, 1]$

5. (a) $(f + g)(x) = \dfrac{x}{x - 1} + \sqrt{x}$ (b) $(f - g)(x) = \dfrac{x}{x - 1} - \sqrt{x}$

(c) $(f \cdot g)(x) = \dfrac{x\sqrt{x}}{x - 1}$ (d) $\left(\dfrac{f}{g}\right)(x) = \dfrac{x}{(x - 1)\sqrt{x}}$

(e) $(f \circ g)(x) = \dfrac{\sqrt{x}}{\sqrt{x} - 1}$ (f) $(g \circ f)(x) = \sqrt{\dfrac{x}{x - 1}}$

Domain for parts (a), (b), (c), and (e) is $[0, 1) \cup (1, \infty)$; domain for (d) is $(0, 1) \cup (1, \infty)$; domain for (f) is $(-\infty, 0] \cup (1, \infty)$

7. (a) $(f + g)(x) = x^2 + x - 1$ (b) $(f - g)(x) = x^2 - x + 3$

(c) $(f \cdot g)(x) = x^3 - 2x^2 + x - 2$ (d) $\left(\dfrac{f}{g}\right)(x) = \dfrac{x^2 + 1}{x - 2}$

(e) $(f \circ g)(x) = x^2 - 4x + 5$ (f) $(g \circ f)(x) = x^2 - 1$

Domain for (a), (b), (c), (e), and (f) is $\mathbb{R}$; domain for (d) is $\{x \mid x \neq 2\}$

9.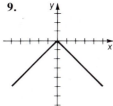

11.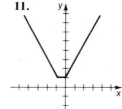

13.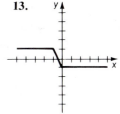

15.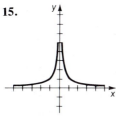

17.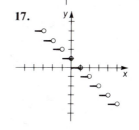

19.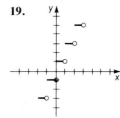

21.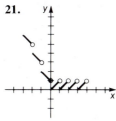

23. $F(x) = (g \circ h)(x)$ where $h(x) = x^2 + 2x$ and $g(x) = \sqrt{x}$.

25. $F(x) = (f \circ h \circ g)(x)$ where $g(x) = x^2 + 1$, $h(x) = \sqrt{x} + 2$, and $f(x) = x^3$.

31. (a) Polynomial of degree at most n.   (b) Polynomial of degree at most n.

(c) Polynomial of degree $2n$.   (d) Polynomial of degree at most n.

(e) Polynomial of degree $n + 1$.

33. (a) 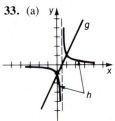   (b) 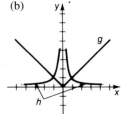   (c)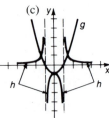

(d)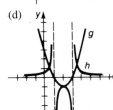

35. $V(t) = \dfrac{4}{3}\pi(3 + .01t)^3 \text{ cm}^3$

37. $\bar{s}(t) = 640 - 16(t - 2)^2$, domain $[2, 2 + 2\sqrt{10}]$, range $[0, 640]$

39. $V(t) = \dfrac{4}{3}\pi(3\sqrt{t} + 5)^3 \text{ cm}^3$,

$s(t) = 4\pi(3\sqrt{t} + 5)^2 \text{ cm}^2$

EXERCISE SET 1.5, PAGE 42

1. $(0, 0)$ **3.** no intercepts **5.** no intercepts **7.** Symmetric with respect to both axes and origin.   **9.** Symmetric with respect to both axes and origin.

11. Symmetric with respect to both axes and origin.   **13.** Symmetric with respect to the x-axis.   **15.** Symmetric with respect to both axes and origin.

17. Symmetric with respect to the origin.   **19.** Symmetric with respect to the origin.

21. Symmetric with respect to the origin.

23. 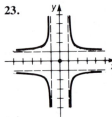   **25.** 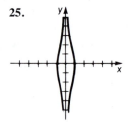   **27.**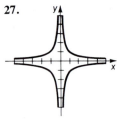

29. 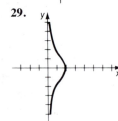   **31.** 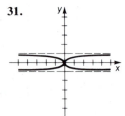   **33.**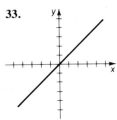

35. 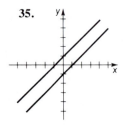   **37.**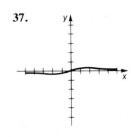

39. (a) Cosine and secant functions are symmetric with respect to the *y*-axis.

(b) Sine, cosecant, tangent, and cotangent are symmetric with respect to the origin.

41. (a) 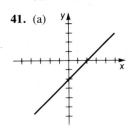 (b) 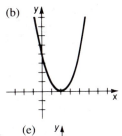 (c)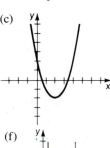

(d) 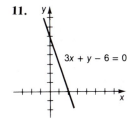 (e) 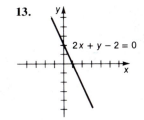 (f)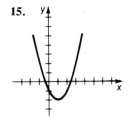

43. $g_1(x) = x^2$, $g_2(x) = x - 2$, $g_3(x) = |x|$, $g_4(x) = \dfrac{1}{x}$, $g_5(x) = 2x$

45. $g_1(x) = x^3$, $g_2(x) = x - 1$, $g_3(x) = |x|$, $g_4(x) = \dfrac{1}{x}$, $g_5(x) = -2x$.

REVIEW EXERCISES, PAGE 43

1. (a) $2\sqrt{2} - 3$ (b) $2\sqrt{2} - 1$ (c) $2\sqrt{2} - 4$ (d) $2\sqrt{x} - 3$ (e) $\sqrt{2x - 3}$ (f) 2

3. (a) $12 - 7\sqrt{2}$ (b) $6 - 5\sqrt{2}$ (c) $16 - 7\sqrt{2}$

(d) $x - 7\sqrt{x} + 10$ (e) $\sqrt{x^2 - 7x + 10}$ (f) $2x - 7 + h$

5. $x + y - 2 = 0$

7.

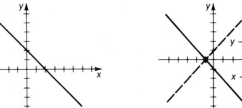

$y - x - 1 = 0$

$x + y + 1 = 0$

9.

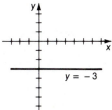

$y = -3$

11.

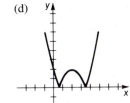

$3x + y - 6 = 0$

13.

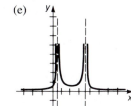

$2x + y - 2 = 0$

15.

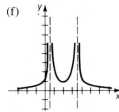

17.

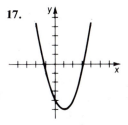

19.

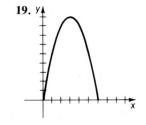

21.

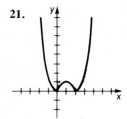

23.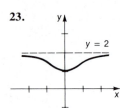

25.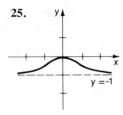

27.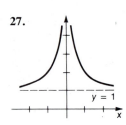

29. $(f \circ g)(x) = \sqrt{\dfrac{1 - 4x^2}{x^2}}$, Domain of $f \circ g$ is $(-.5, 0) \cup (0, .5)$

$(g \circ f)(x) = \dfrac{1}{\sqrt{x^2 - 4}}$, Domain of $g \circ f$ is $(-\infty, -2) \cup (2, \infty)$

31. $(f \circ g)(x) = \dfrac{1}{x^2 + 2x}$, Domain of $f \circ g$ is $(-\infty, -2) \cup (-2, 0) \cup (0, \infty)$

$(g \circ f)(x) = \dfrac{1}{x^2 - 1} + 1$, Domain of $g \circ f$ is $(-\infty, -1) \cup (-1, 1) \cup (1, \infty)$

33. $f(x) = (h \circ g)(x)$, where $g(x) = x^2 - 7x + 1$

35.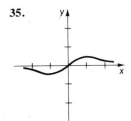

37. Symmetric to both axes. Intercepts at $(\pm 3, 0)$, and $(0, \pm 3)$. A circle with center $(0, 0)$ and radius 3.

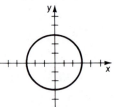

39.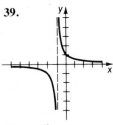

41.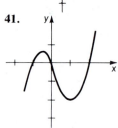

43. Symmetric with respect to the origin. Only intercept at $(0, 0)$.

$-.5 \leq x \leq .5$

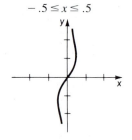

45.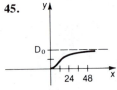

CHAPTER 2

DNE = does not exist.

EXERCISE SET 2.1, PAGE 53

1. 1 **3.** 3 **5.** 4 **7.** 6 **9.** -8 **11.** 6 **13.** 3 **15.** 5 **17.** 1 **19.** DNE
21. DNE **23.** DNE **25.** 1 **27.** 1 **29.** 0 **31.** 1 **33.** 0 **35.** -1
37. (a) $\dfrac{x^2 - 4}{x - 2}$ (b) 4 **39.** (a) $\dfrac{x^3 - 1}{x - 1}$ (b) 3 **41.** 1000

EXERCISE SET 2.2, PAGE 60

1. .5 **3.** .005 **5.** .05 **7.** .05 **9.** .01 **11.** .1/7 **25.** yes **29.** no
31. (a) The "line" $y = x$ with holes at each irrational number and the "line" $y = 1 - x$
with holes at each rational number.   (b) $a = \frac{1}{2}$
33. Hint: If $\lim_{x \to a} f(x) = L$, consider $\epsilon = L/2$.

EXERCISE SET 2.3, PAGE 68

1. (b), (c)   **3.** 1, $f(1)$ DNE   **5.** 1, $\lim_{x \to 1} f(x) \neq f(1)$

7. 2, $f(2)$ DNE   **9.** 0, $f(0)$ and $\lim_{x \to 0} f(x)$ DNE   **11.** 2 and 3, $f(2) = 0 \neq \lim_{x \to 2} f(x) = 4$,

$f(3) = 1 \neq \lim_{x \to 3} f(x) = 5$

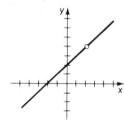 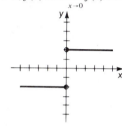 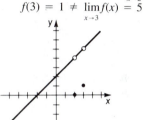

13. $x = (2n + 1)\frac{\pi}{2}$ for any integer n, $\tan\left((2n + 1)\frac{\pi}{2}\right)$ DNE
15. $(f \circ g)(x) = 1/(x - 2), x \neq 2, (g \circ f)(x) = 1/x - 2, x \neq 0$
17. $(f \circ g)(x) = (\sqrt[4]{x} - 1)/(\sqrt[4]{x} + 1), x \geq 0$ $(g \circ f)(x) = \sqrt[4]{\dfrac{x - 1}{x + 1}}, x \geq 1$ or $x < -1$
19. 223, yes **21.** 38, yes **23.** 1, yes **25.** $-5/7$, yes **27.** 1/2, yes **29.** 3, yes
31. -1 **33.** 0 **35.** DNE **37.** 13/41 **39.** 4 **41.** $-1/4$ **43.** DNE **45.** 0
47. (a) 0 (b) 4 (c) 2 **49.** $a = 7, b = -6$ **51.** $g(a) \neq 0, h(a) \neq 0$
57. One example is

$$f(x) = \begin{cases} 1, \text{ if } x \geq 1 \\ -1, \text{ if } x < 1 \end{cases} \quad \text{and} \quad g(x) = \begin{cases} -1, \text{ if } x \geq 1 \\ 1, \text{ if } x < 1 \end{cases}$$

59. The functions given for Exercise 57 will work.

EXERCISE SET 2.4, PAGE 76

1. (a) 2 (b) 1 (c) DNE **3.** DNE **5.** 0 **7.** 0 **9.** $\sqrt{5}$ **11.** 1 **13.** DNE
15. 0 **17.** -1 **19.** 0 **21.** 0 **23.** $(-\infty, -1)$, $[-1, 1)$, $[1, \infty)$
25. $(-\infty, -2)$, $(-2, 2)$, $(2, \infty)$ **27.** $(-\infty, \infty)$ **29.** $(-\infty, 0)$, $(0, \infty)$
31. $[3, \infty)$ **33.** $(-\infty, \infty)$ **35.** $[n, n + 1)$, n an integer

37. (a)

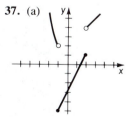

(b) 4, 1, continuous from the left at 2, not continuous at 2 (c) $-5, 2$, continuous from the
right at -1, not continuous at -1 (d) $(-\infty, -1)$, $[-1, 2]$, $(2, \infty)$ **39.** 19/7 **41.** 2
43. $f(1) = -3, f(2) = 1$ and $-3 < 0 < 1 \Rightarrow$ there is at least one number c in $(1, 2)$ with
$f(c) = 0$.

45. $f(0) = 1, f\left(\dfrac{\pi}{2}\right) = -\dfrac{\pi}{2}$ and $\dfrac{-\pi}{2} < 0 < 1 \Rightarrow$ there exists at least one number c in $\left(0, \dfrac{\pi}{2}\right)$ with $f(c) = 0$ **47.** $f(x) = \begin{cases} 1, & \text{if } x \geq 1 \\ -1, & \text{if } x < 1 \end{cases}$

49. One example is $f(x) = \dfrac{2x - 1}{x^2 - x}$; another is $f(x) = \cot \pi x$

EXERCISE SET 2.5, PAGE 84

1. 2/3   **3.** 0   **5.** 0   **7.** 1   **9.** 3/4   **11.** 1   **13.** 0   **15.** 0   **17.** 1   **19.** -1
21. 1/5   **23.** 2   **25.** DNE   **27.** 1/9   **29.** 1/3   **31.** 0   **33.** 1/2   **35.** 1   **37.** 1

39. $y = 0$

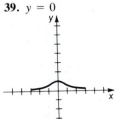

41. $y = 1, y = -1$

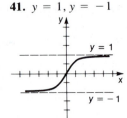

43. (a) 10   (b) 100   **45.** For $\epsilon > 0$, let $M = \dfrac{1}{\sqrt{\epsilon}}$.

49. $\lim\limits_{x \to \infty} f(x) = \dfrac{a_n}{b_n} = \lim\limits_{x \to -\infty} f(x)$

53.

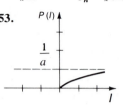

51. $10 , the limiting price

EXERCISE SET 2.6, PAGE 91

1. ∞ **3.** DNE   **5.** ∞ **7.** ∞ **9.** 2   **11.** DNE   **13.** ∞ **15.** $-\infty$ **17.** ∞ **19.** ∞

21. $x = (2n + 1)\dfrac{\pi}{2}$, n an integer   **23.** $x = (2n + 1)\dfrac{\pi}{2}$, n an integer

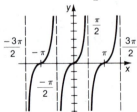

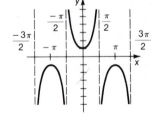

25. $x = -1, y = 1$ **27.** $y = 0$ **29.** $x = 2, x = -2, y = 1, y = -1$

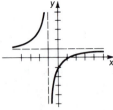

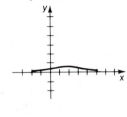

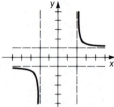

31. $x = 1, x = -1$ **33.** $x = 1, y = 1$ **35.** $x = 3, x = -3, y = -4$

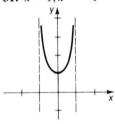

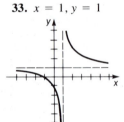

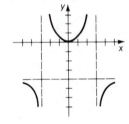

35. Increasing on $(-\infty, 1) \cup \left(2 - \dfrac{\sqrt{3}}{3}, 2 + \dfrac{\sqrt{3}}{3}\right) \cup (3, \infty)$; decreasing on $\left(1, 2 - \dfrac{\sqrt{3}}{3}\right) \cup \left(2 + \dfrac{\sqrt{3}}{3}, 3\right)$; relative maximum at 1 and at $2 + \dfrac{\sqrt{3}}{3}$; relative minimum at $2 - \dfrac{\sqrt{3}}{3}$ and at 3

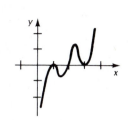

37. Increasing on $(-\sqrt{2}, 0) \cup (\sqrt{2}, \infty)$; decreasing on $(-\infty, -\sqrt{2}) \cup (0, \sqrt{2})$; relative minimum at $-\sqrt{2}$ and at $\sqrt{2}$; relative maximum at 0

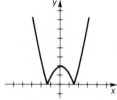

39. Decreasing on $(0, 1)$; increasing on $(1, \infty)$; relative minimum at 1

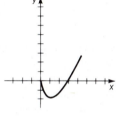

41. Increasing on $\left(0, \dfrac{49}{100}\right)$; decreasing on $\left(\dfrac{49}{100}, \infty\right)$; relative maximum at $\dfrac{49}{100}$

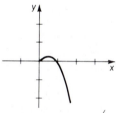

43. Decreasing on $(-\infty, -\sqrt{2})$; increasing on $(\sqrt{2}, \infty)$

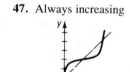

45. Decreasing on $\left(\dfrac{\pi}{3} + 2n\pi, \dfrac{5\pi}{3} + 2n\pi\right)$; increasing on $\left(-\dfrac{\pi}{3} + 2n\pi, \dfrac{\pi}{3} + 2n\pi\right)$; relative maximum at $\dfrac{\pi}{3} + 2n\pi$; relative minimum at $-\dfrac{\pi}{3} + 2n\pi$

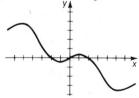

47. Always increasing

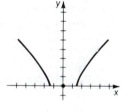

49. Increasing when $500 \le x < 3750$. Decreasing when $3750 < x \le 4000$. Maximum cost is \$28125 when $x = 3750$. Minimum cost is \$7000 when $x = 500$.

51. Impossible **53.** $a \ne 0$, arbitrary, $b = 0$, $c = -3a$.

EXERCISE SET 3.11, PAGE 175

1. Relative minimum when $x = 2$.   **3.** No relative extrema.

5. Relative maximum when $x = -4$; relative minimum when $x = 0$.

7. Relative maximum when $x = \dfrac{\pi}{2} + 2n\pi$; relative minimum when $x = \dfrac{3\pi}{2} + 2n\pi$.

9. Relative maximum when $x = -1$; relative minimum when $x = 1$.

11. Relative minimum when $x = 1$.

13. Concave upward on $(-\infty, -1) \cup (2, \infty)$, Concave downward on $(-1, 2)$, points of inflection $(-1, -10)$, $(2, -40)$.

15. Concave upward on $(-\infty, 1)$; concave downward on $(1, \infty)$

17. Concave upward on $\left(n\pi, (2n + 1)\,\dfrac{\pi}{2} \right)$; concave downward on $\left((2n - 1)\,\dfrac{\pi}{2}, n\pi \right)$; points of inflection $(n\pi, 0)$ where n is an integer

19. Concave downward on $(-\infty, 1)$, concave upward on $(1, \infty)$

21. Concave downward on $(-\infty, -\sqrt{6}) \cup (2, \sqrt{6})$; concave upward on $(-\sqrt{6}, -2) \cup (\sqrt{6}, \infty)$, points of inflection at $(-\sqrt{6}, -2\sqrt{3})$ and $(\sqrt{6}, 2\sqrt{3})$.

23. Concave upward on $(-\infty, -\sqrt{3})$ and $(\sqrt{3}, \infty)$; concave downward on $(-\sqrt{3}, \sqrt{3})$; points of inflection at $(\sqrt{3}, 36)$ and $(-\sqrt{3}, 36)$.

25. Concave upward on $\left(-\infty, \dfrac{1}{2}(1 - \sqrt{3}) \right) \cup \left(\dfrac{1}{2}(1 + \sqrt{3}), \infty \right)$; concave downward on $\left(\dfrac{1}{2}(1 - \sqrt{3}), \dfrac{1}{2}(1 + \sqrt{3}) \right)$; points of inflection $\left(\dfrac{1}{2}(1 - \sqrt{3}), 9/4 \right)$, $\left(\dfrac{1}{2}(1 + \sqrt{3}), 9/4 \right)$

27. Concave downward on $(2n\pi, (2n + 1)\pi)$; concave upward on $((2n - 1)\pi, 2n\pi)$, points of inflection $(2n\pi, 0)$, $((2n + 1)\pi, 0)$, where n is an integer.

29. Critical points at $x = 1$, $x = 3$; increasing on $(-\infty, 1)$ and $(3, \infty)$; relative maximum at $x = 1$; relative minimum at $x = 3$; concave upward on $(2, \infty)$; point of inflection at $(2, -2)$.

31. Critical points at $x = \pm 1$, $x = 0$; increasing on $(-1, 0)$ and $(1, \infty)$; relative minimum at $x = \pm 1$; relative maximum at $x = 0$; concave upward on $\left(-\infty, -\dfrac{\sqrt{3}}{3} \right)$ and $\left(\dfrac{\sqrt{3}}{3}, \infty \right)$; points of inflection $\left(-\dfrac{\sqrt{3}}{3}, \dfrac{-5}{9} \right)$ and $\left(\dfrac{\sqrt{3}}{3}, -\dfrac{5}{9} \right)$.

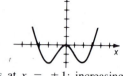

33. Critical points at $x = 2/3$, $x = 2$; increasing on $(-\infty, 2/3)$ and $(2, \infty)$; relative maximum at $x = 2/3$; relative minimum at $x = 2$; concave upward on $(4/3, \infty)$; point of inflection at $(4/3, 8/27)$.

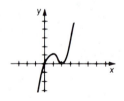

35. Critical points at $x = \pm 1$; increasing on $(-\infty, -1)$ and $(1, \infty)$; relative minima at $x = \pm 1$; concave upward on $(-\infty, 0)$ and $(0, \infty)$; no points of inflection.

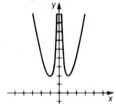

37. No critical points; increasing on $(1, \infty)$; concave upward on

$(4/3, \infty)$; point of inflection at $\left(4/3, \dfrac{4\sqrt{3}}{9} \right)$.

39. Critical points at $x = -1$, $x = 2$, $x = 1/2$; increasing on $(-1, 1/2)$ and $(2, \infty)$; relative maximum at $x = 1/2$; relative minimum at $x = -1$ and $x = 2$; concave upward on

$\left(-\infty, \dfrac{1 - \sqrt{3}}{2} \right)$ and $\left(\dfrac{1 + \sqrt{3}}{2}, \infty \right)$; points

of inflection at $\left(\dfrac{1 - \sqrt{3}}{2}, \dfrac{9}{4} \right)$ and

$\left(\dfrac{1 + \sqrt{3}}{2}, \dfrac{9}{4} \right)$.

41. Critical points at $x = 0$, $x = -2/5$; increasing on $(-\infty, -2/5)$ and $(0, \infty)$; relative maximum at $x = -2/5$ relative minimum at $x = 0$; maximum at $x = -2/5$ relative minimum at $x = 0$; concave upward on $(1/5, \infty)$; concave downward on $(-\infty, 1/5)$; point of inflection at $(1/5, 6/(5^{5/3}))$.

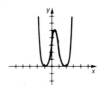

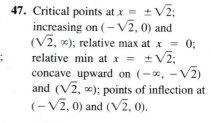

43. Critical points at $x = \pm\sqrt{3}$; increasing on $(-\sqrt{3}, 0)$ and $(0, \sqrt{3})$; relative minimum at $x = -\sqrt{3}$; relative maximum at $x = \sqrt{3}$; concave upward on $(-\sqrt{6}, 0)$ and $(\sqrt{6}, \infty)$; points of inflection at $(-\sqrt{6}, -5\sqrt{6}/36)$ and $(\sqrt{6}, 5\sqrt{6}/36)$.

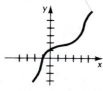

45. Critical points at $x = \dfrac{\pi}{2} + 2n\pi$; increasing on $(-\infty, \infty)$; no relative extrema; concave upward on $\left(\dfrac{\pi}{2} + 2n\pi, \dfrac{3\pi}{2} + 2n\pi \right)$; points of inflection at $\left(\dfrac{\pi}{2} + 2n\pi, \dfrac{\pi}{2} + 2n\pi \right)$ and $\left(\dfrac{3\pi}{2} + 2n\pi, \dfrac{3\pi}{2} + 2n\pi \right)$.

47. Critical points at $x = \pm\sqrt{2}$; increasing on $(-\sqrt{2}, 0)$ and $(\sqrt{2}, \infty)$; relative max at $x = 0$; relative min at $x = \pm\sqrt{2}$; concave upward on $(-\infty, -\sqrt{2})$ and $(\sqrt{2}, \infty)$; points of inflection at $(-\sqrt{2}, 0)$ and $(\sqrt{2}, 0)$.

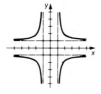

49. Horizontal asymptotes occur at $y = \pm 1$; vertical asymptotes occur at $x = \pm 1$; the graph is concave downward in the third and fourth quadrants and concave upward in the first and second quadrants.

51. Horizontal asymptotes occur at $y = \pm 1$; the graph is concave upward in the third and fourth quadrants and concave downward in the first and second quadrants.

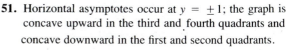

53. A horizontal asymptote occurs at $y = 0$.

55. A vertical asymptote occurs at $x = 0$; the graph is increasing on $(-\infty, -1)$ and $(1, \infty)$; a relative maximum occurs at $x = -1$; a relative minimum occurs at $x = 1$; the graph is concave upward on $(0, \infty)$ and concave downward on $(-\infty, 0)$. This graph is the reflection about the line $y = x$ of the graph in Exercise 53.

61. Any polynomial of the form $f(x) = ax^3 + cx$, where $a < 0$ and $c = -3a$.   **63.** $f'(0)$ does not exist.

REVIEW EXERCISES, PAGE 177

1. $12x^3 - 6x^2 + 1$ **3.** $16 + \dfrac{32}{t^2}$ **5.** $2x(x^2 - 7)^2(5x^4 - 14\,x^2 + 3)$

7. $5[(x^2 + 1)^3 - 7x]^4[6x(x^2 + 1)^2 - 7]$ **9.** $3u^2 + \dfrac{2}{u^3} - \dfrac{1}{u^2}$ **11.** $\dfrac{12x^4 + 12x^2 - 3x}{\sqrt{3x^2 + 4}}$

13. $\dfrac{5(x^3 - 8)^4(x^4 + 12x^2 + 16x)}{(x^2 + 4)^6}$ **15.** $\dfrac{4(x^3 + 1)^3(2x^3 - 1)}{x^5}$

17. $2(\sin x + \cos x)(\cos x - \sin x) = 2\cos 2x$ **19.** $2(\sec 2t)^2 + 3\sec 3t \tan 3t$

21. $36x^2 - 12x, 72x - 12$ **23.** $-64t^{-3}, 192t^{-4}$ **27.** $y = 2x$ **29.** $y = 0$

31. $\dfrac{y\sqrt{x^2 + y^2} - x}{y - x\sqrt{x^2 + y^2}}$ **33.** $-y/x$ **35.** $\dfrac{dy}{dx} = -1, \dfrac{d^2y}{dx^2} = 0$

37. When x is in the interval

	$f'(x)$ is	$f''(x)$ is	The graph of f is
$(-\infty, -1)$	$+$	$-$	increasing; concave downward
$\left(-1, \dfrac{1}{2}\right)$	$-$	$-$	decreasing; concave downward
$\left(\dfrac{1}{2}, 2\right)$	$-$	$+$	decreasing; concave upward
$(2, \infty)$	$+$	$+$	increasing; concave upward

39. When x is in the interval

	$f'(x)$ is	$f''(x)$ is	The graph of f is
$(0, 1)$	$-$	$+$	decreasing; concave upward
$(1, \infty)$	$+$	$+$	increasing; concave upward

41. When x is in the interval

	$f'(x)$ is	$f''(x)$ is	The graph of f is
$(0, 1)$	$+$	$-$	increasing; concave downward
$(1, \infty)$	$-$	$-$	decreasing; concave downward

43.

When x is in the interval	$f'(x)$ is	$f''(x)$ is	The graph of f is
$(-\infty, 0)$	$+$	$+$	increasing; concave upward
$(0, 1)$	$+$	$-$	increasing; concave downward
$(1, 2)$	$-$	$-$	decreasing; concave downward
$(2, \infty)$	$-$	$+$	decreasing; concave upward

45.

When x is in the interval	$f'(x)$ is	$f''(x)$ is	The graph of f is
$(1, 3)$	$-$	$+$	decreasing; concave upward
$(3, 7)$	$+$	$+$	increasing; concave upward
$(7, \infty)$	$+$	$-$	increasing; concave downward

47. Maximum of 2 at $x = 1$; minimum of 1 at $x = 0$ and $x = 2$.

51.   **53.** $f(x) = x^3 + 1$

CHAPTER 4

EXERCISE SET 4.1, PAGE 187

1. 256, 64   **3.** $\dfrac{3}{4}, \dfrac{1}{4}$ **5.** $\dfrac{2}{3}, -\dfrac{2}{9}$ **7.** $\dfrac{-1}{\pi + 1}, \dfrac{2}{(\pi + 1)^2}$

9.
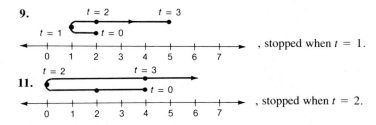
, stopped when $t = 1$.

11.

, stopped when $t = 2$.

13.
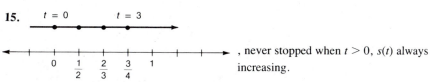
, stopped when $t = 1$ and when $t = 2$.

15.

, never stopped when $t > 0$, $s(t)$ always increasing.

17. It reaches the ground when $t = 5$ sec; $v(5) = -40$ ft/sec, $a(5) = -8$ ft/sec^2. It is 50 feet above the ground when $t = 5\sqrt{2}/2$ sec; $v(5\sqrt{2}/2) = -20\sqrt{2}$ ft/sec, $a(5\sqrt{2}/2) = -8$ ft/sec^2

19. $s(t) = 1600 - 16t^2$ **21.** $v_0 = 14$ ft/sec **23.** $8\sqrt{5}$ ft/sec **25.** 96 ft, 12 seconds

27. 1936 ft **29.** $4.2 \dfrac{\text{yds}}{\text{sec}^2}$, 2.76 seconds **31.** 2.5 sec **33.** 1492 feet

EXERCISE SET 4.2, PAGE 198

1. 1 **3.** $l = w = \sqrt{432}$ ft **5.** $l = 24\sqrt{.6}$, $w = 30\sqrt{.6}$ **7.** $l = 30$, $w = 10$, $h = 7.5$

9. 5488 in^3, 9259.26 in^3 **11.** $\cos\theta = \dfrac{\sqrt{3}}{3}$ or $\theta \approx .955$ radians **13.** $156.25

15. $h = 2r = 2\sqrt[3]{\dfrac{450}{\pi}}$ **17.** $h = 4/3$, $l = w = 16/3$ **21.** $l = w = \sqrt{2}r$

23. Equilateral triangle with sides of length $\sqrt{3}r$. **25.** 7.15 in by 9.53 in

27. $\dfrac{4\sqrt{3}}{9 + 4\sqrt{3}} \approx .43$ ft **29.** $\dfrac{4}{9}\sqrt{3}\pi$ **31.** $\dfrac{2\sqrt{3}}{3}r$ in by $\dfrac{2}{3}\sqrt{6}r$ in

33. $80\sqrt{5}$ ft from end of driveway **35.** 111.5 ft from light with most intensity

37. At the southern factory **39.** $b \approx 5.37$ ft, $h \approx 4.65$ ft

EXERCISE SET 4.3, PAGE 209

1. $360\,\pi\dfrac{\text{mm}^3}{\text{sec}}$ **3.** $16\dfrac{\text{in}^2}{\text{min}}$ **5.** $96\dfrac{\text{in}^3}{\text{min}}$ **7.** $2\sqrt{3}\dfrac{\text{cm}^2}{\text{sec}}$ **9.** $144\,\pi$ in^2, $72\pi\dfrac{\text{in}^2}{\text{sec}}$

11. $\dfrac{1}{18\sqrt[3]{\pi}} \approx .038\dfrac{\text{ft}}{\text{min}}$ **13.** $2\pi r$ **15.** $380\dfrac{\text{mi}}{\text{hr}}$ **17.** $1.2\dfrac{\text{meters}}{\text{sec}}$ **19.** $.5\dfrac{\text{cm}}{\text{sec}}$

21. At the point (x, y) on the circle, the y-coordinate increases at a rate of $\dfrac{x}{y}$ cm/sec.

23. 4000 min, $.002\dfrac{\text{ft}}{\text{min}}$ **25.** $\dfrac{-20\sqrt{91}}{91}\dfrac{\text{ft}}{\text{sec}}$ **27.** $-\sqrt{39/5}\dfrac{\text{ft}}{\text{sec}}$, $-\sqrt{3}\dfrac{\text{ft}}{\text{sec}}$

29. $12.5\dfrac{\text{yd}}{\text{sec}}$ **31.** $.195\,\pi\dfrac{\text{cm}^3}{\text{sec}}$ **33.** $\dfrac{-\sqrt{3}}{20}\dfrac{\text{ft}}{\text{hr}}$, $-\dfrac{\sqrt{6}}{40}\dfrac{\text{ft}}{\text{hr}}$

35. $-9.29\dfrac{\text{lb}}{\text{sec}}$ **37.** (a) $\dfrac{50}{3\pi}\dfrac{\text{cm}}{\text{sec}}$ (b) $\dfrac{25\sqrt{\pi}}{6\pi}\dfrac{\text{cm}}{\text{sec}}$

EXERCISE SET 4.4, PAGE 217

1. $dy = 3\Delta x$, $\Delta y = 3\Delta x$ **3.** $dy = (2x - 2)\Delta x$, $\Delta y = 2x\Delta x - 2\Delta x + (\Delta x)^2$

5. $dy = (3x^2 - 1)\Delta x$, $\Delta y = (\Delta x)^3 + 3x(\Delta x)^2 + (3x^2 - 1)\Delta x$

7. $dy = \dfrac{\Delta x}{2\sqrt{x + 1}}$, $\Delta y = \dfrac{\Delta x}{\sqrt{x + \Delta x + 1} + \sqrt{x + 1}}$

9. $dy = \left(1 - \dfrac{1}{x^2}\right)\Delta x$, $\Delta y = \dfrac{1}{x + \Delta x} - \dfrac{1}{x} + \Delta x$

11. $dy = (\sec x)^2\,\Delta x$, $\Delta y = \tan(x + \Delta x) - \tan x$

13. (a) 0, .01 (b) 1.2, 1.24 **15.** (a) 1, 1.061 (b) $-.2$, $-.328$

17. $dy = 0$, $\Delta y = \dfrac{\sqrt{2}}{2} - 1$ **19.** $dy = .02$, $\Delta y \approx .0204$ **21.** .005, .004996 **23.** 3.005

25. 13121/162   **27.** $\dfrac{2198}{169}$ **29.** 599/1800   **31.** $\pm\,.004\pi$

33. (a) $.016\pi$ ft^3 (b) $.016\pi$ ft^2 **35.** $.086$ in^3 **37.** $\dfrac{300\Delta r}{r}$ **39.** 3.36πcm^3

41. $1.8°$ F   **43.** $(11)\cdot$ (height of stick), $\dfrac{\text{height of stick}}{12}$ feet

EXERCISE SET 4.5, PAGE 226

1. $\dfrac{1}{2}$ **3.** 67/11   **5.** $\dfrac{19}{13}$ **7.** 0   **9.** $\dfrac{2}{3}$ **11.** $-\dfrac{1}{16}$ **13.** $\dfrac{1}{2}$ **15.** 0   **17.** $\dfrac{1}{3}$ **19.** $\dfrac{3}{2}$

21. $-\dfrac{3}{7}$ **23.** 0   **25.** 1   **27.** $-\dfrac{1}{6}$ **29.** $\dfrac{4}{9}$ **31.** 0   **33.** 0   **35.** 0   **37.** ∞ **39.** -1

41. $\dfrac{3}{2}$ **43.** $g'(0) = 0$, so hypotheses of theorem are not satisfied.   **45.** -6

47. This limit was needed to find $D_x \sin x$.   **49.** One pair is: $f(x) = x$, $g(x) = x + \cos x$

EXERCISE SET 4.6, PAGE 232

1. 2.6906   **3.** -1.0905 **5.** 8.8467   **7.** .7391   **9.** .5798
11. 2.2361   **13.** 5.62341   **15.** (.590, .348)   **17.** 809.45 cm^2 **19.** 1.88

EXERCISE SET 4.7, PAGE 239

1. (a) $c(x) = \dfrac{1500}{x} + 2 - .0003x$ (b) $C'(x) = 2 - .0006x$

   (c) $c'(x) = -\dfrac{1500}{x^2} - .0003$ (d) 3.2   (e) 1.40

3. (a) $c(x) = \dfrac{15{,}500}{x} + 77 - .00001x^2$ (b) $C'(x) = 77 - .00003x^2$

   (c) $c'(x) = -\dfrac{15{,}500}{x^2} - .00002x$ (d) 82.5   (e) $\simeq 47$

5. $R'(x) = -.0002$ **7.** $R'(x) = -500\left(\dfrac{x^2 + 7.6x - 300}{(x^2 + 300)^2}\right)$

9. $C'(x) = .76 + .0002x$, $R'(x) = 2.85 - .00016x$
   $P'(x) = 2.09 - .00036x$

11. (a) 5805 or 5806   (b) \$4366.81
13. There is no change. Maximum profit is \$4366.81 with 5805 or 5806 produced.
15. Producing 6055 or 6056 wastebaskets gives a profit of \$4200.56; less profit is produced
   and the work has increased.
17. $R(x) = 1000x$, $P(x) = 700x - 500$, $P'(x) = 700$

19. (a) $\dfrac{600 - x^2}{2x^2}$ (b) 14.14   (c) Increase, decrease   **21.** 4000

REVIEW EXERCISES, PAGE 241

1. (a) $f'(x) = x^2 - 2x - 3$ (b) $v(x) = x^2 - 2x - 3$
   (c) $P'(x) = x^2 - 2x - 3$ (d) $dy = (x^2 - 2x - 3)dx$
3. $(11 + 2\sqrt{31})/4 \approx 5.53$ seconds   **5.** $r = h = (900/\pi)^{1/3}$ **7.** 4

9. $r = h = 4$ in., $V = 64\pi$ in^3 **11.** $\dfrac{\sqrt[3]{4}}{2\pi} \approx .25$ ft/min   **13.** $\left(\dfrac{1}{2}, \dfrac{1}{4}\right)$ **15.** $.6\pi$ ft^3

17. One approximate root is 4.49341   **19.** 0   **21.** $\dfrac{1}{24}$ **23.** 0   **25.** -1 **27.** 0

29. 12 feet from the tree tied at 20 feet above the ground.

CHAPTER 5

EXERCISE SET 5.1, PAGE 254

1. 18   **3.** 105   **5.** 195   **7.** 3   **9.** 203   **11.** $9i + 29k$ **13.** 638

15. $n(n + 1)(n + 2)/3$ **17.** 39,490   **19.** $\sum_{i=0}^{7} (2 + 3i)$ **21.** $\sum_{i=2}^{8} i^2$ **23.** $\sum_{i=1}^{11} (-1)^{i+1} \dfrac{1}{i}$

25. 6   **27.** $\dfrac{1}{3}$ **29.** 74/3   **31.** 8/3   **33.** 16/3   **35.** 6

37. (a) 6   (b) 4   (c) 10   (d) 12   **41.** (a) 60   (b) $(n)(n + 1)/2$

EXERCISE SET 5.2, PAGE 263

1. 15   **3.** $\dfrac{7}{32}$ **5.** $\dfrac{3}{8}$ **7.** $\dfrac{29}{32}$ **9.** $\dfrac{(7\sqrt{2} + 5\sqrt{3})}{24}\pi$ **11.** 2   **13.** $\dfrac{1}{3}$ **15.** $\dfrac{1}{4}$ **17.** 0

19. $\dfrac{1}{4}$ **21.** 2   **23.** 4   **25.** $f(x) = \pi x^2$ **27.** $f(x) = (\cos x)^2$ **29.** $g(x) = \sqrt{1 + [f'(x)]^2}$

31. (a) 20   (b) 20   (c) 20

EXERCISE SET 5.3, PAGE 269

1. 4   **3.** -6 **5.** $4c$ **7.** 8   **9.** -8 **11.** 53/12   **13.** 7/3

15. (a) 4   (b) -2 (c) $2 - 3\pi$ (d) -2

17. 2   **19.** $2 + \pi$ **21.** 1

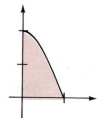   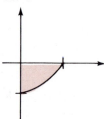

23. (a) 25/2   (b) -25 (c) 75/2   (d) 53/2   **29.** $f(x) = 2$, $g(x) = x$, $[0, 2]$

EXERCISE SET 5.4, PAGE 278

1. 8   **3.** -8 **5.** 21   **7.** 2   **9.** 18   **11.** 1   **13.** 1   **15.** 2   **17.** 9

19. 6   **21.** $x\sqrt{x^2 + 9}$ **23.** 0   **25.** $\int_{1}^{x} \sqrt{t^2 + 9}\, dt + x\sqrt{x^2 + 9}$

27. 12   **29.** 4   **31.** 36

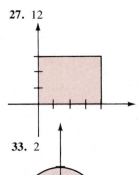   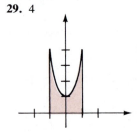   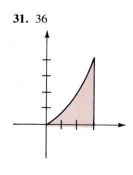

33. 2

35. 1   **37.** 4   **39.** 1   **41.** 5

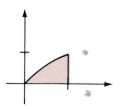

45. 1/3   **47.** (a) $2f(a)$ (b) 0   **53.** 5.62 cal/deg-mole

EXERCISE SET 5.5, PAGE 286

1. $x^4 + C$ **3.** $x^4/4 + C$ **5.** $x^4/2 + x^3 + C$ **7.** $x^3 + 2x^2 - 2x + C$

9. $\dfrac{3x^{5/3}}{5} + \dfrac{9x^{4/3}}{4} + \dfrac{4x^3}{3} + C$ **11.** $\sin x + x + C$ **13.** $-\cot u - u + C$

15. $\dfrac{3t^5}{5} + \dfrac{3t^2}{2} + C$ **17.** $9x^3 + 9x^2 + 3x + C$ **19.** $\dfrac{5t^{6/5}}{6} - \dfrac{5t^{7/5}}{7} + C$

21. $\dfrac{y^2}{2} - 3y + C$ **23.** $-\dfrac{1}{x} + \dfrac{1}{x^2} - \dfrac{1}{x^4} + C$ **25.** $\dfrac{x^7}{7} - \dfrac{2x^9}{9} + \dfrac{x^{11}}{11} + C$ **27.** $\dfrac{284}{3}$

29. 400   **31.** 2   **33.** $2\sqrt{2} - 1$ **35.** 1   **37.** $f(x) = 3x + 2$ **39.** $f(x) = \dfrac{x^2}{2} - \dfrac{1}{2}$

41. $f(x) = \dfrac{x^3}{3} - \dfrac{x^2}{2}$ **43.** $f(x) = -\cos x - \dfrac{x^2}{2}$ **45.** $y = \dfrac{x^2}{2} + 2x$

47. $y = x^3 - x^2 + x + 2$

49. $P(x) = 2.8x - .002x^3 - 1000$

51. 12 cm/sec   **53.** 121 ft; -8 ft/sec $= \dfrac{-60}{11}$ mph

55. 4.75 sec

EXERCISE SET 5.6, PAGE 292

1. $(2x - 3)^3/3 + C$ **3.** $\dfrac{(t^3 + 4)^2}{2} + C$ **5.** $\dfrac{2(t^3 + 4)^{3/2}}{3} + C$ **7.** $\dfrac{(4x - 5)^{3/2}}{6} + C$

9. $-\dfrac{\cos \pi x}{\pi} + C$ **11.** $\dfrac{(\sin x)^{11}}{11} + C$ **13.** $\dfrac{2}{3}(\sqrt{x} - 1)^3 + C$ **15.** $\dfrac{(x^3 + x^2)^5}{5} + C$

17. $\dfrac{6}{25}$ **19.** $-\dfrac{(\cos 3x)^4}{12} + C$ **21.** $(\tan x^2)/2 + C$ **23.** 1/2

25. $-\dfrac{\left(1 + \dfrac{1}{t}\right)^4}{4} + C$ **27.** $(13\sqrt{13} - 1)/3$ **29.** $4\sqrt{3t - 7}/3 + C$

31. $2(z - 1)^{5/2}/5 + 4(z - 1)^{3/2}\big/3 + C$ **33.** $4/3 - 2\sqrt{2}/3$

35. $4[(\sqrt{2} + 1)^{3/2} - 2\sqrt{2}]/3$

37. $2(x - 2)^{7/2}/7 + 8(x - 2)^{5/2}/5 + 10(x - 2)^{3/2}/3 + C$ **39.** $x + 1/(x + 1) + C$

41. $-1/(x + 3) + C$ **43.** $3471/2048$ **45.** $1/(3(2 + 3 \cos x)) + C$

47. $(\sec x)^4/4 - (\sec x)^2/2 + C$. The difference is a constant.

49. $(\sin x)^2/2$ **51.** $(x^2 + 4x)^{3/2}/3$

EXERCISE SET 5.7, PAGE 301

1. $\dfrac{2x}{x^2 + 3}$ **3.** $\dfrac{1}{4x}$ **5.** $1/(x \ln x)$ **7.** $(\sqrt{x^2 - 1} + x)/(x\sqrt{x^2 - 1} + x^2 - 1)$

9. $2/(1 - x^2)$ **11.** $\left[\ln\left(\dfrac{1 + x}{1 - x}\right)\right]^{-1/2} \Big/ (1 - x^2)$ **13.** $\cot x$

15. $\ln(x^2 + 1) + \dfrac{2x^2}{x^2 + 1}$ **17.** $(2x^2 \ln x - x^2 - 1)/(x(\ln x)^2)$ **19.** $\dfrac{2}{2x + 1}$

21. $3x[\ln\sqrt{x^2 + 2}]^2/(x^2 + 2)$ **23.** $-\sin(\ln x)/x$ **25.** $\dfrac{-y}{x}$

27. $\dfrac{1}{(x + y - 1)}$ **29.** $-y(2x^2 + 1)/(x(2y^2 - 1))$ **31.** $\ln|x - 3| + C$

33. $\ln 2 - \ln 4 = -\ln 2$ **35.** $\dfrac{\ln|x^2 + 2x - 3|}{2} + C$ **37.** $[(\ln 3)^2 - (\ln 2)^2]/2$

39. $(\ln x^2)^4/8 + C = 2(\ln x)^4 + C$ **41.** 2 **43.** $\ln|1 + \sin x| + C$

45. $y' = 3(11x^2 + 14x + 40)(x^2 + 8)^2(3x + 7)^4$

47. $y' = (7x + 5)\sqrt{x}\sqrt{(x + 1)\sqrt{x}}/(8x^2 + 8x)$

49. $y' = (x^2 + 1)^{-1/2}(x^3 - 1)^4(2x^2 + 1)^6(60x^6 + 74x^4 - 30x^3 + 15x^2 - 29x)$

51.

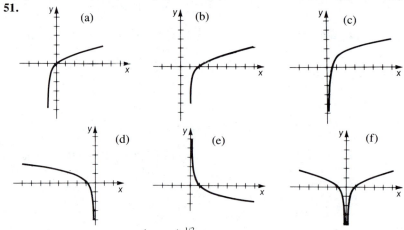

53. $2 \ln 2 \approx 1.39$ **55.** $\ln\left(\dfrac{1 + x}{1 - x}\right)^{1/2} + C$ **57.** 1.118 **59.** $(5.319 \times 10^3)P/T^2$

EXERCISE SET 5.8, PAGE 309

1. Diverges   **3.** $\dfrac{1}{6}$ **5.** $1/\ln 3$ **7.** $1/16$ **9.** Diverges   **11.** $3/2$

13. Diverges   **15.** Diverges   **17.** Diverges   **19.** Diverges   **21.** Diverges

23. Diverges   **25.** $[(\sqrt[3]{2})^2 - 1]/2$ **27.** 0 **29.** Diverges   **31.** Diverges

33. Diverges   **35.** It is not finite   **37.** $1/2$

REVIEW EXERCISES, PAGE 312

1. 30 **3.** 40 **5.** $(3n^2 - 7n)/2$ **9.** (a) $\displaystyle\int_0^2 3x^2\, dx$, 8 (b) $\displaystyle\int_0^2 (2x + 1)dx$, 6

11. $\dfrac{x^4}{4} + C$ **13.** 312 **15.** $\dfrac{(4x + 7)^4}{4} + C$ **17.** $\dfrac{2}{5}x^{5/2} + \ln|x| + C$ **19.** $6x^{11/3}/11 + C$

21. $-\dfrac{1}{3x^3} + \dfrac{4}{3}x^{3/4} + C$ **23.** $-\dfrac{\pi^2}{2}$ **25.** $-\sqrt{4 - x^2} + C$

27. $\frac{1}{5}(w^2 + 1)^{5/2} - \frac{1}{3}(w^2 + 1)^{3/2} + C$ **29.** $\frac{2}{1 - \sqrt{x}} + C$ **31.** $\ln|x - 2| + C$

33. $\frac{1}{2}\ln|x^2 - 6x + 5| + C$ **35.** $\frac{104}{9}$ **37.** $\frac{11}{450}$ **39.** $x + \ln|x^2 - 2x + 2| + C$

41. $\frac{1}{2}\sin 2x + C$ **43.** $\frac{1}{4}\ln|\sec 4x| + C$ **45.** $\frac{1}{3}(\sin x)^3 + C$ **47.** $\sin(\ln x) + C$

49. 3   **51.** 1   **53.** $\frac{1}{\ln 2}$ **55.** Diverges   **57.** 0   **59.** 9   **61.** Diverges   **63.** $\frac{3}{x}(\ln x)^2$

65. $\frac{3x^2 - 7}{x^3 - 7x + 1}$ **67.** $\frac{x^{-2/3}\left[\ln(x^{1/3} + 1)\right]^{-2/3}}{9(x^{1/3} + 1)}$ **69.** $\int_1^x \ln t \, dt + x \ln x$ **71.** $x^2 - x - 12$

73. 9   **75.** 4

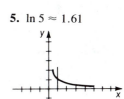

CHAPTER 6
EXERCISE SET 6.1, PAGE 324

1. $\frac{20}{3}$

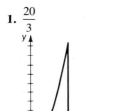

3. $\frac{4}{3}$

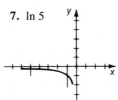

5. $\ln 5 \approx 1.61$

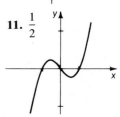

7. $\ln 5$

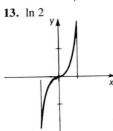

9. $\frac{1}{3}$

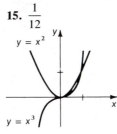

11. $\frac{1}{2}$

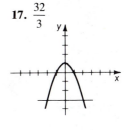

13. $\ln 2$

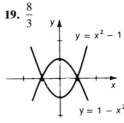

15. $\frac{1}{12}$

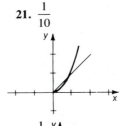

17. $\frac{32}{3}$

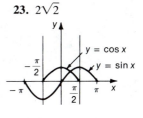

19. $\frac{8}{3}$

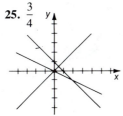

21. $\frac{1}{10}$

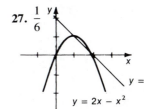

23. $2\sqrt{2}$

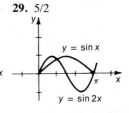

25. $\frac{3}{4}$

27. $\frac{1}{6}$

29. $5/2$

31. 1/2

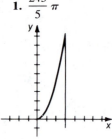

$y = x^3 - x$

$y = 2x^3 - 2x$

33. 2

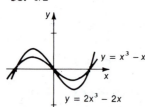

$y = |x|$

$y = x + 2$ $y = 2 - x$

35. $\dfrac{9}{2}$

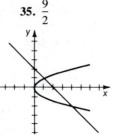

37. $\dfrac{9}{2}$

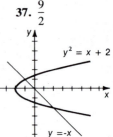

$y^2 = x + 2$

$y = -x$

39. $\dfrac{1}{2}$ **41.** $3\sqrt[3]{6}$ **43.** $8a^2$ **45.** 17

EXERCISE SET 6.2, PAGE 334

1. $\dfrac{243}{5}\pi$

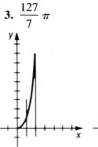

3. $\dfrac{127}{7}\pi$

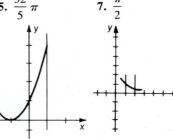

5. $\dfrac{32}{5}\pi$

7. $\dfrac{\pi}{2}$

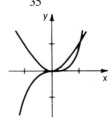

9. $\dfrac{2}{35}\pi$

11. $\dfrac{120}{7}\pi$

13. $\dfrac{64}{5}\pi$

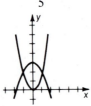

15. $\dfrac{1}{30}\pi$

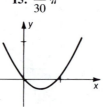

17. 2π

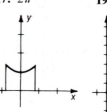

19. $\pi^2/2$

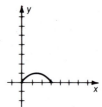

21. $\pi^2/2$

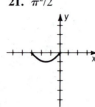

23. $\dfrac{5\pi}{6}$

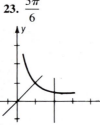

25. 12π

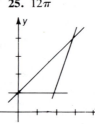

27. $\dfrac{3\pi}{5}$ **29.** $\dfrac{\pi}{6}$ **31.** $\dfrac{\pi}{2}$

33. $\dfrac{8\pi}{15}$ **35.** $\dfrac{64}{27}\pi$

41. (a) $\dfrac{5\pi}{6}$ (b) $\dfrac{8\pi}{15}$ (c) $\dfrac{11\pi}{6}$ (d) $\dfrac{28\pi}{15}$ **43.** f is continuous.

EXERCISE SET 6.3, PAGE 340

1. $\dfrac{\pi}{2}$

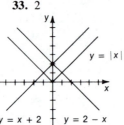

3. $\dfrac{62\pi}{5}$

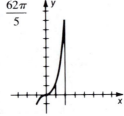

5. $\dfrac{5\pi}{14}$

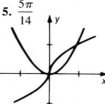

7. π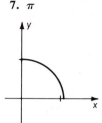

9. $\dfrac{8\pi}{15}$

11. 5π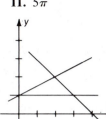

13. $\dfrac{74\pi}{9}$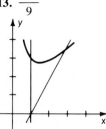

15. $\dfrac{128}{5}\pi$ **17.** 2π **19.** $2\pi/15$ **21.** $2\pi/35$ **23.** $\pi/3$ **25.** $\dfrac{9}{5}\pi$

27. $\dfrac{27\pi}{16}$ **29.** $\pi r^2 h$ **31.** $\dfrac{1}{3}\pi r^2 h$ **33.** (a) $\dfrac{5\pi}{6}$ (b) $\dfrac{8\pi}{15}$ (c) $\dfrac{11\pi}{6}$ (d) $\dfrac{28\pi}{15}$

EXERCISE SET 6.4, PAGE 346

1. $(16/3)r^3$ **3.** $\dfrac{16}{15}$ **5.** $\displaystyle\int_{-r}^{r}\dfrac{r\sqrt{r^2-x^2}}{2}\,dx$ **7.** $\displaystyle\int_{-r}^{r}2h\sqrt{r^2-x^2}\,dx$

9. $\displaystyle\int_{3}^{9}(x-3)\sqrt{81-x^2}\,dx$ **11.** $\dfrac{2}{3}$ **13.** $\dfrac{64}{15}\pi$ **15.** $\dfrac{4\pi}{3}(64-7\sqrt{7})$

17. $r=[1-(.5)^{2/3}]^{1/2}\,R$

EXERCISE SET 6.5, PAGE 354

1. $2\sqrt{10}\approx 6.32$ **3.** $\dfrac{1}{27}(139\sqrt{139}-31\sqrt{31})\approx 54.3$ **5.** $\dfrac{(85)^{3/2}-8}{243}\approx 3.19$ **7.** 6

9. $\dfrac{17}{6}$ **11.** $\dfrac{22}{3}$ **13.** $\dfrac{8(10)^{3/2}-13^{3/2}}{27}\approx 7.63$ **15.** $\dfrac{9}{8}$ for $y\geq 0$. Total length $\dfrac{9}{4}$.

17. $\dfrac{13^{3/2}-8}{27}$ **19.** $\dfrac{(730^{3/2}-10^{3/2})\pi}{27}\approx 2291.26$ **21.** $8\pi\sqrt{2}$ **23.** $\dfrac{49\pi}{3}$

25. $\dfrac{47}{16}\pi$ **27.** $\pi r\sqrt{r^2+h^2}$ **29.** $x\approx .57, y\approx .43$ **31.** $\dfrac{297\sqrt{34}}{25}\pi$ ft^2

33. Hint: Both are equal to $\pi/2$. One is an area, the other is an arc length.

EXERCISE SET 6.6, PAGE 360

1. Same for all three, 499.2 $\dfrac{\text{lb}}{\text{ft}^2}$; 2995.2 lb **3.** 2433.6 lb **5.** 7800 lb

7. 4992 lb **9.** 2928 lb **11.** $\dfrac{80}{9}\sqrt{3}(62.4)$ lb

13. 8424 and 59,904 lb on the ends, 55,640 on the sides
15. 514,800 lb **17.** (a) 42.67 lb (b) 512 lb **19.** 2.6π lb

EXERCISE SET 6.7, PAGE 367

1. 150 ft lb **3.** $(3256)(458)$ mi lb $\approx 7.87\times 10^9$ ft lb
5. 4.8 newton-meters **7.** 40 cm, .4 newton-meters, 2 newton-meters **9.** 62.5 in lb
11. 2808.48 ft lb **13.** 4058.48 ft lb **15.** 3237 ft lb
17. (a) 2,995,200 ft lb (b) 5,241,600 ft lb

19. (a) 3960π ft lb (b) $\displaystyle\int_{-2}^{2}2(55)\,6\sqrt{4-y^2}\,(2-y)\,dy$ **21.** 3,213,600 ft lb

23. Fill as often as possible to minimize the work. The shape of the tank is important.

EXERCISE SET 6.8, PAGE 378

1. $22, \dfrac{11}{3}$ **3.** $17, \dfrac{17}{11}$ **5.** $M_x=11, M_y=5, \left(\dfrac{5}{6},\dfrac{11}{6}\right)$

7. $M_x = -4, M_y = 31, \left(\dfrac{31}{10}, -\dfrac{2}{5}\right)$ **9.** $(0, 1)$ **11.** $(-8, -22)$ **13.** 11.2 lb

15. $\left(0, \dfrac{8}{5}\right)$ **17.** $\left(0, \dfrac{3}{5}\right)$ **19.** $\left(\dfrac{8}{15}, \dfrac{8}{21}\right)$ **21.** $\left(\dfrac{9}{20}, \dfrac{9}{20}\right)$ **23.** $\left(\dfrac{4 \ln 4}{3}, 7/32\right)$

25. $\left(2, \dfrac{2}{3}\right)$ **27.** (a) $\dfrac{4}{3}$ (b) $\dfrac{3}{2}$ (c) $\dfrac{13}{16}$ (d) $\dfrac{1}{2}$ **29.** 7.41 feet from the large end

33. $2\pi^2 r^2 R$ **37.** If the rod is described by $y = \sqrt{1 - x^2}$, the centroid is at $(0, 2/\pi)$

41. Hint: Any line through the centroid of a homogeneous lamina divides the lamina into equal parts.   **43.** 2/9

REVIEW EXERCISES, PAGE 381

1. $\dfrac{5}{12}$ **3.** $\dfrac{125}{6}$ **5.** $\dfrac{64}{3}$ **7.** $\dfrac{27}{2}$ **9.** 9π **11.** 32π **13.** $\dfrac{1024}{15}\pi$

15. $\dfrac{80}{3}\pi$ **17.** $12\displaystyle\int_0^5 \sqrt{25 - x^2}\, dx$ **19.** $\ln(3 + 2\sqrt{2})$ **21.** $\dfrac{2}{3}\pi(2\sqrt{2} - 1)$

23. $(\bar{x}, \bar{y}) = \left(\dfrac{6}{5}, 3\right)$ **25.** 500 in/lb   **27.** (a) $(62.5)(156\pi)$ (b) $(62.5)(441\pi)$

CHAPTER 7
EXERCISE SET 7.1, PAGE 390

1. $f^{-1}(x) = \dfrac{x - 4}{3}$; $\mathbb{R}$ **3.** $f^{-1}(x) = \sqrt[3]{x}$; $\mathbb{R}$ **5.** No inverse function

7. $f_1(x) = x^2 - 4, x \geq 0$ **9.** $f_1(x) = x + 1, x \geq -1$ **11.** $f_1(x) = \sin x,$

or $f_2(x) = x^2 - 4, x \leq 0$ or $f_2(x) = -x - 1, x < -1$ x in $\left[-\dfrac{\pi}{2}, \dfrac{\pi}{2}\right]$

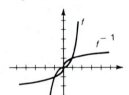

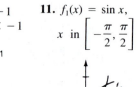

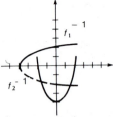

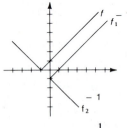

13. $f^{-1}(x) = \dfrac{1}{3}(x + 1), (-\infty, \infty); (f^{-1})'(x) = \dfrac{1}{3}$

15. $f^{-1}(x) = x^{1/3}, [1, 2]; (f^{-1})'(x) = \dfrac{1}{3}x^{-2/3}$

17. $f^{-1}(x) = \dfrac{1}{x}, (0, \infty); (f^{-1})'(x) = -\dfrac{1}{x^2}$

19. $f^{-1}(x) = \dfrac{1}{x - 1}, \left[\dfrac{3}{2}, 2\right]; (f^{-1})'(x) = -\dfrac{1}{(x - 1)^2}$

21. $\dfrac{1}{4}$ **23.** 4   **25.** $-\dfrac{2}{\sqrt{3}}$

33. If $f = f^{-1}$ and (a, b) is on the graph of f then (b, a) is also on the graph of f. The graph is symmetric with respect to the line $y = x$.

35. $a = -1$, b arbitrary, or $a = 1$, $b = 0$

37. $f^{-1}(x) = \begin{cases} \dfrac{1}{\sqrt{x}}, & \text{if } x < 0 \\ 0, & \text{if } x = 0 \\ -\dfrac{1}{\sqrt{x}}, & \text{if } x > 0 \end{cases}$

39. $T(x) = \dfrac{5}{9}(x - 32)$; $T^{-1}(x) = \dfrac{9}{5}x + 32$. T^{-1} converts Celsius readings into Fahrenheit. T' and $(T^{-1})'$ represent, respectively, the rates of change of the Celsius scale with respect to the Fahrenheit, and the change of the Fahrenheit scale with respect to the Celsius.

45. $f(x) = \sin x$ or $f(x) = x^3 - x$ **47.** $f(x) = \dfrac{1}{x}$

EXERCISE SET 7.2, PAGE 400

1. $\dfrac{\pi}{4}$ **3.** $\dfrac{\pi}{3}$ **5.** $\dfrac{\pi}{4}$ **7.** π **9.** $\dfrac{12}{13}$ **11.** 0   **13.** $\dfrac{3 + 4\sqrt{3}}{10}$

15. (a) 1   (b) 0   (c) undefined   (d) 0   **17.** (a) $-\dfrac{\sqrt{2}}{2}$ (b) $\dfrac{\sqrt{2}}{2}$ (c) -1 (d) $-\sqrt{2}$

19. $\dfrac{1}{\sqrt{-x^2 - 2x}}$ **21.** $\dfrac{2x - 1}{x^4 - 2x^3 + 7x^2 - 6x + 10}$ **23.** $\dfrac{\sqrt{1 - x^2}\,\arccos x - x}{\sqrt{1 - x^2}\,\sqrt{1 - (x \arccos x)^2}}$

25. $\cos x \arcsin x + \dfrac{\sin x}{\sqrt{1 - x^2}}$ **27.** $\dfrac{1}{6}\arctan\dfrac{x}{6} + C$ **29.** $\dfrac{\pi}{6}$

31. $\dfrac{1}{6}\arctan\dfrac{2x}{3} + C$ **33.** $\operatorname{arcsec}|4x| + C$ **35.** $\dfrac{1}{2}\operatorname{arcsec} x^2 + C$ **37.** $\dfrac{\pi}{6}$

39. $\dfrac{1}{2}\ln 2 - \dfrac{\pi}{4}$ **41.** $5\arcsin\left(\dfrac{x}{5}\right) + \sqrt{25 - x^2} + C$ **43.** $\dfrac{1}{2}\arctan\left(\dfrac{x + 2}{2}\right) + C$

49.   **51.**   **53.**   **55.**

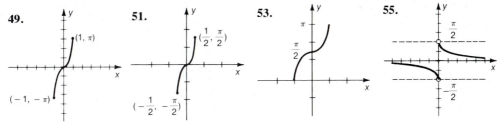

57. 1   **59.** 1   **61.** (a) $-\dfrac{1}{1 + x^2}$ (c) $\dfrac{\pi}{2}$ **65.** 3 feet above the floor   **67.** $\dfrac{\pi}{3}$

69. Hint: Find areas that the integrals represent.

EXERCISE SET 7.3, PAGE 410

1. $3e^{3x}$ **3.** $\dfrac{e^{\sqrt{x}}}{2\sqrt{x}}$ **5.** $e^{-x}(1 - x)$ **7.** $e^x(\ln x + x^{-1})$ **9.** $2e^x\cos x$ **11.** $x\,2^{x^2 + 1}\ln 2$

13. $\cos x\, e^{\sin x}$ **15.** $(3 + \pi)^x \ln(3 + \pi)$ **17.** $2x$ **19.** $(\pi + \sqrt{2})\, x^{\pi + \sqrt{2} - 1}$

21. $x^{\sin x}\left[\cos x \ln x + \dfrac{\sin x}{x}\right]$ **23.** $-\left(\dfrac{e^y + ye^x}{e^x + xe^y}\right)$ **25.** $\dfrac{1 - xye^{xy}}{x^2 e^{xy}}$

27. $\dfrac{e^{3x-2}}{3} + C$ **29.** $\dfrac{e^2 - 1}{2e}$ **31.** $\ln(e^x + 1) + C$ **33.** $\ln|e^x - e^{-x}| + C$

35. $\dfrac{e^{7x}}{7} + C$ **37.** $\arcsin e^x + C$

39.

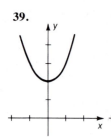

41.

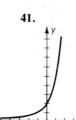

43.

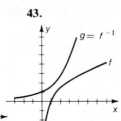

45.

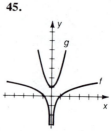

47.

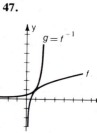

49. 1   **51.** 0   **53.** Does not exist   **55.** $y = e^3(x - 3)$ **57.** $e - 1$

59. $\arctan e^a - \dfrac{\pi}{4}$ **61.** $\dfrac{\pi}{e}(e - 1)$ **63.** $2(2x)^{2x}(1 + \ln 2x)$ **65.** No

71. (a) 230   (b) 29   (c) in 4.99 days   **73.** .0173 cm²/week.

EXERCISE SET 7.4, PAGE 419

1. $\dfrac{2x}{(x^2 + 1)\ln 10}$ **3.** $\dfrac{2x}{(x^2 + 1)(x^2 + 2)\ln 2}$ **5.** $\dfrac{1}{x \ln x \ln 10}$ **7.** $\dfrac{2x + 3}{2|x^2 + 3x - 1|\ln \pi}$

9. $-\dfrac{(x^5 + 5x^4 + 15x + 15)}{3x(x + 2)(x^4 - 5)\ln 4}$ **11.** $\dfrac{xy \ln 10 \log_{10} y - y^2}{xy \ln 10 \log_{10} x - x^2}$ **13.** 1

15. 1   **17.** e^2 **19.** 1   **21.** 1   **23.** 0   **25.** 1

27.

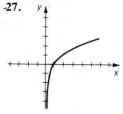

29.

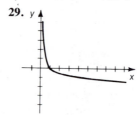

31.

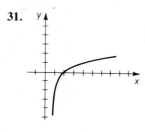

33. e **35.** $\sqrt{e}$ **39.** (a) $f'(x) = -\dfrac{(\log_x a)^2}{x \ln a}$

(b)

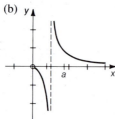

(c)

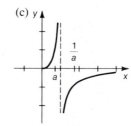

41.

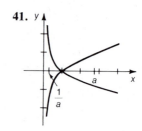

EXERCISE SET 7.5, PAGE 426

1. (a) $(50)2^{-t/590}$ (b) 41 mg   **3.** 69.4 mg   **5.** (a) $(1000)2^{t/4}$ (b) 3364

7. The second bank   **9.** (a) 196,181 thousand   (b) 281,407 thousand

11. 11.6 more minutes   **13.** No. Higher   **15.** 9.8 mins   **17.** 13.92°C

19. 8.64 hours   **21.** Approximately 24,000 years

EXERCISE SET 7.6, PAGE 434

1. $y = Ce^{7x}$ **3.** $y = \tan(x + C)$ **5.** $y - \sin y = \cos x - \dfrac{x^3}{x} + C$

7. $y = \dfrac{1}{\sin x + C}$ **9.** $e^{-x} + e^{-y} = C$ **11.** $\ln|y| = e^{-x} + C$

13. $y = 2e^{x^2/2} - 1$ **15.** $\sin x + e^{-y} + 1 = 0$ **17.** About 2.5 quarts

19. (a) 171,865 thousand   (b) 272,691 thousand

21. 3:34 AM   **23.** $R = k \ln\left(\dfrac{S}{S_0}\right)$ **25.** $S = Ce^{-(A/B)t}$ **27.** ae^{-b}, a

29. (a) 184,531 thousand   (b) 271,006 thousand

EXERCISE SET 7.7, PAGE 446

1. $2x \cosh(x^2)$ **3.** $\dfrac{x+1}{\sqrt{x^2+2x}} \cosh \sqrt{x^2+2x}$ **5.** $\dfrac{\sinh(\ln x)}{x}$

7. $-e^{x^3+3x}(3x^2+3)\text{sech}\,(e^{x^3+3x})\tanh(e^{x^3+3x})$ **9.** $\dfrac{3x^2}{\sqrt{1+x^6}}$ **11.** $\dfrac{1}{2\sqrt{x^2-x}}$

13. $2x\left(\cosh x^2 \text{arcsinh}\, x^2 + \dfrac{\sinh x^2}{\sqrt{1+x^4}}\right)$ **15.** $\dfrac{(\sec x)^2}{1-(\tan x)^2}$

17. $\dfrac{\sqrt{x^2-1}\,\text{arccosh}\, x - \sqrt{1+x^2}\,\text{arcsinh}\, x}{\sqrt{x^4-1}\,(\text{arccosh}\, x)^2}$ **19.** $2 \sinh \sqrt{x} + C$

21. $\dfrac{\tanh 2x}{2}+C$ **23.** $\dfrac{(\sinh x)^2}{2}+C$ **25.** $-\text{csch}\, x + C$

27. $\text{arcsinh}\left(\dfrac{4}{3}\right)$ **29.** $\dfrac{1}{2}\text{arcsinh}\left(\dfrac{x^2}{3}\right)+C$ **31.** $\dfrac{1}{3}\text{arccosh}\,\dfrac{3x}{2}+C$

33. $\text{arctanh}\, e^x + C$, if $x < 0$ or $\text{arccoth}\, e^x + C$ if $x > 0$ **35.** $\dfrac{1}{2}\text{arctanh}\left(\dfrac{x-2}{2}\right)+C$

51. (a) $\dfrac{\pi[\sinh a \cosh a - a]}{2}$ (b) $2\pi[a\cosh a - \sinh a]$

57. (a) $L = \dfrac{2T_0}{w}\sinh\dfrac{50w}{T_0}$ (b) 100.66 feet

REVIEW EXERCISES, PAGE 448

1. $\dfrac{1}{\sqrt{9-x^2}}$ **3.** $3x^2 e^{x^3}$ **5.** $3(\ln 2)x^2 2^{x^3}$ **7.** $3x^2$ **9.** $e^{\tan x} + x(\sec x)^2 e^{\tan x}$

11. $\dfrac{1-2x-x^2}{(x+1)(x^2+1)\ln 10}$ **13.** $3x^2$ **15.** $\dfrac{1}{1+e^x}$ **17.** ex^{e-1}

19. $e^{3x}\text{sech}\, x\,(3-\tanh x) - \text{sech}\, x \tanh x$ **21.** $e^x[\sinh(2x-1)+2\cosh(2x-1)]$

23. $x^{\cos x}\left(\dfrac{\cos x}{x} - \sin x \ln x\right)$ **25.** $\dfrac{xy-x^2}{2x^2+y^2+xy}$ **27.** $\dfrac{1}{3}\arctan\left(\dfrac{x}{3}\right)+C$

29. $\dfrac{1}{3}(e^8-e)$ **31.** $\arcsin\left(\dfrac{x}{3}\right)+C$ **33.** $\arcsin e^x + C$ **35.** $\arctan(x-1)+C$

37. $\dfrac{3}{2\ln 2}+C$ **39.** $\cosh(\ln x)+C$ **41.** (a) $-\dfrac{\pi}{6}$ (b) $-\dfrac{\pi}{4}$ (c) $\dfrac{\pi}{6}$ **43.** ∞

45. 1 **47.** 1 **49.** ∞

51. **53.** **55.**

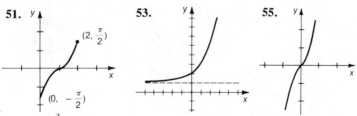

57. $y = \dfrac{x^3}{3}+x^2+C$ **59.** $y = -\ln|C - e^x|$ **61.** $y = \tan(\arctan x + C)$

63. $y = 5e^{x^2}$ **65.** $x^2 = y - \ln|y+1| + 1$

67. (a) $A(t) = A(0)e^{(\ln 3/3)t}$ (b) $A(0) \approx 693$ **69.** $7.88 \times 10^{-7}\,g$ **71.** 11.55 years

CHAPTER 8
EXERCISE SET 8.1, PAGE 457

1. $-e^{-x}(x+1)+C$ **3.** $-\dfrac{\pi}{2}$ **5.** $\dfrac{1}{4}(e^2+1)$ **7.** $x \arcsin x + \sqrt{1-x^2}+C$

9. $x \arccos x - \sqrt{1 - x^2} + C$　**11.** $e^x(x^2 - 2x + 2) + C$　**13.** $\frac{1}{4}(8 - 8\pi - \pi^2)$

15. $2x(\ln x - 1) + C$　**17.** $\frac{x^4}{8}(4 \ln x - 1) + C$　**19.** $\frac{298}{15}$

21. $x \tan x - \ln |\sec x| + C$　**23.** $-\frac{1}{3}(x^2 + 2)\sqrt{1 - x^2} + C$

25. $e^x(\sin x - \cos x)/2 + C$　**27.** $\frac{1}{13}(2e^{3x}\sin 2x + 3e^{3x}\cos 2x) + C$

29. $\frac{1}{2}(\sec x \tan x + \ln |\sec x + \tan x|) + C$　**31.** $\frac{1}{5}$

33. $2(\sqrt{x} \sin \sqrt{x} + \cos \sqrt{x}) + C$　**35.** $2e^{\sqrt{x}}(\sqrt{x} - 1) + C$　**37.** $\pi/4 - 1/2$

39. $3^x(x/\ln 3 - 1/(\ln 3)^2) + C$　**41.** $\frac{2^x}{(\ln 2)^3}[(\ln 2)^2x^2 - 2(\ln 2)x + 2] + C$

43. $\frac{x}{2}[\sin (\ln x) - \cos (\ln x)] + C$　**45.** $x \ln (x + \sqrt{x^2 + 1}) - \sqrt{x^2 + 1} + C$

47. 2π　**49.** $\pi(e - 2)$　**51.** $\frac{3}{2} - \frac{e}{2}$　**53.** $\frac{1}{4}$

57. (a) $\frac{a}{s^2 + a^2}$　(b) $\frac{s}{s^2 + a^2}$　(c) $\frac{1}{s - a}$

EXERCISE SET 8.2, PAGE 463

1. $\frac{1}{3}(\sin x)^2 + C$　**3.** 0　**5.** $-\frac{1}{15}(\cos 3x)^5 + C$　**7.** $\frac{\pi}{96}$　**9.** $3\pi/8$

11. $\frac{2}{3}(\sin x)^{3/2} - \frac{2}{7}(\sin x)^{7/2} + C$　**13.** $\frac{10}{3}$　**15.** $\frac{1}{4}(\tan x)^4 + \frac{1}{6}(\tan x)^6 + C$

17. $1 - \frac{\pi}{4}$　**19.** $-\frac{1}{4}(\csc x)^4 + C$　**21.** $-\frac{1}{5}(\csc x)^5 + \frac{2}{3}(\csc x)^3 - \csc x + C$

23. $\frac{1}{2} \ln |\csc 2x| - \frac{1}{4}(\cot 2x)^2 + C$　**25.** $\tan x - \frac{1}{2}x - 2 \cos x + \frac{1}{4} \sin 2x + C$

27. $\left(\sin x - \frac{(\sin x)^3}{3}\right) \ln(\sin x) - \sin x + \frac{(\sin x)^3}{9} + C$

29. $e^{\sin x}[2 \sin x - (\sin x)^2 - 1] + C$　**31.** $\frac{(\sin x)^2}{2} + C$　**33.** $-\cos x + C$

35. $\ln |\csc x - \cot x| + C$　**37.** $\frac{1}{2}(\tan x)^2 + \ln |\cos x| + C$

39. $\frac{1}{4}(\sin x)^4 + C$　**41.** $\left(\frac{1}{2} + \frac{3\sqrt{3}}{4\pi}\right)$ feet　**43.** $\frac{\pi}{4}(4 - \pi)$　**45.** $\frac{1}{2}\cos x - \frac{1}{6}\cos 3x + C$

47. $\frac{1}{2}\cos x - \frac{1}{10}\cos 5x + C$　**49.** $\frac{1}{4}\sin 2x + \frac{1}{16}\sin 8x + C$

51. (a) $-\frac{1}{2}\left[\frac{\cos(a - b)x}{a - b} + \frac{\cos(a + b)x}{a + b}\right] + C$

(b) $\frac{1}{2}\left[\frac{\sin(a - b)x}{a - b} + \frac{\sin(a + b)x}{a + b}\right] + C$

(c) $\frac{1}{2}\left[\frac{\sin(a - b)x}{a - b} - \frac{\sin(a + b)x}{a + b}\right] + C$

EXERCISE SET 8.3, PAGE 471

1. $\frac{x}{2}\sqrt{1 - x^2} + \frac{1}{2} \arcsin x + C$　**3.** 4π　**5.** $\arctan \frac{x}{3} + C$　**7.** $\ln|\sqrt{16 + x^2} + x| + C$

9. $1 - \dfrac{\pi}{4}$ **11.** $\sqrt{3}$. No trigonometric substitution needed in this problem.

13. $\dfrac{-\pi}{3} - \sqrt{3}$ **15.** $\dfrac{16}{3}\arcsin 3/4 + \sqrt{7}$

17. $\dfrac{x+1}{2}\sqrt{4-(x+1)^2} + 2\arcsin\dfrac{x+1}{2} + C$ **19.** $\dfrac{1}{3}\arctan\dfrac{x+1}{3} + C$

21. $\dfrac{x-1}{2}\sqrt{2x-x^2} + \dfrac{1}{2}\arcsin(x-1) + C$ **23.** $2\arctan\sqrt{x} + C$

25. $\dfrac{e^x}{2}\sqrt{1-e^{2x}} + \dfrac{1}{2}\arcsin e^x + C$ **27.** $2\arctan e^{x/2} + C$ **29.** $2\arctan 2$

31. 12π **33.** $\sqrt{e^2+1} - \sqrt{2} + \ln\left|\dfrac{1-\sqrt{e^2+1}}{e(1-\sqrt{2})}\right|$ **35.** 2π

39. $\dfrac{25}{4}\pi$ cm   **43.** 1 **45.** $\ln|1 + \tan x/2| + C$

47. $\dfrac{1}{2}\left[\ln\left|\tan\dfrac{x}{2}\right| - \dfrac{1}{2}\left(\tan\dfrac{x}{2}\right)^2\right] + C$

EXERCISE SET 8.4, PAGE 479

1. $\dfrac{1}{2}\ln\left|\dfrac{x-1}{x+1}\right| + C$ **3.** $2\ln 2 - \ln 3$ **5.** $\dfrac{1}{2}\ln|x^2-1| + C$ **7.** $\dfrac{5}{2}\ln 3 - 2$

9. $\dfrac{x^2}{2} - x + \dfrac{1}{x+1} + 2\ln|x+1| + C$

11. $-\dfrac{2}{3}\ln|x+1| + \dfrac{5}{12}\ln|x-2| + \dfrac{5}{4}\ln|x+2| + C$ **13.** $\ln\left|\dfrac{x^3}{x-2}\right| - \dfrac{1}{x-2} + C$

15. $\dfrac{1}{4}\ln\left|\dfrac{(x-2)(x+1)^2}{(x+2)(x-1)^2}\right| + C$ **17.** $\arctan(x+1) + C$

19. $\dfrac{1}{4}\ln\dfrac{(x+1)^2}{(x^2+1)} + \dfrac{1}{2}\arctan x + C$ **21.** $\ln\dfrac{42}{13}$ **23.** $\dfrac{1}{2}\left(\arctan x - \dfrac{x}{x^2+1}\right) + C$

25. $\dfrac{x}{x^2+x+1} + \dfrac{2\sqrt{3}}{3}\arctan\dfrac{\sqrt{3}}{3}(2x+1) + C$

27. $\ln\left(\dfrac{2x^2+5}{x^2+3}\right) - \dfrac{3}{5}\sqrt{10}\arctan\dfrac{\sqrt{10}\,x}{5} + \dfrac{3\sqrt{3}}{3}\arctan\dfrac{\sqrt{3}\,x}{3} + C$

29. $\ln\left|\dfrac{\sin x - 1}{\sin x}\right| + C$ **31.** $\ln 2$ **33.** $\dfrac{\pi}{25}[\dfrac{4}{5}\ln 4 + \dfrac{3}{2}]$ **35.** $\dfrac{2\pi}{5}\ln\dfrac{27}{4}$

37. 93.3% **39.** $2\sin 2t + \dfrac{3}{2}\cos 2t + e^{-2t} - 2e^t$ **41.** $\dfrac{6\sqrt{5}}{25}\arctan\left(\dfrac{\sqrt{5}}{5}\right) - \dfrac{2}{15}$

43. $12\left[\dfrac{1}{8}x^{2/3} - \dfrac{1}{7}x^{7/12} + \dfrac{1}{6}x^{1/2} - \dfrac{1}{5}x^{5/12} + \dfrac{1}{4}x^{1/3} - \dfrac{1}{3}x^{1/4} + \dfrac{1}{2}x^{1/6}\quad x^{1/12} + \ln|x^{1/12}+1|\right] + C$

45. $6e^{-x/6} - 4e^{-x/4} - 12e^{-x/12} + 12\ln|1 + e^{-x/12}| + C$

EXERCISE SET 8.5, PAGE 491

1. Trapezoidal rule 1.28201, Simpson's rule 1.29532, actual value 1.29584

3. (a) 1.49068   (b) 1.46371   **5.** (a) .70711   (b) .94281

7. (a) .65916   (b) .65933   **9.** 1.91010

11. 1.91014   **13.** .68016   **15.** $n = 26$ **17.** $n = 6$ **21.** 20.2

23. (a) .683   (b) .341   (c) .273   (d) .067

REVIEW EXERCISES, PAGE 493

1. $-xe^{-x} - e^{-x} + C$ **3.** $-\dfrac{(\cos 2x)^3}{6} + \dfrac{(\cos 2x)^5}{10} + C$ **5.** $\dfrac{x^3 \ln x}{3} - \dfrac{x^3}{9} + C$

7. $\dfrac{1}{2} x \sin(\ln x) - \dfrac{1}{2} x \cos (\ln x) + C$ **9.** $-\ln|x| + 2 \ln|x + 1| + C$

11. $\dfrac{x2^x}{\ln 2} - \dfrac{2^x}{(\ln 2)^2} + C$ **13.** $\dfrac{1}{7}(16 - x^2)^{7/2} - \dfrac{16}{5}(16 - x^2)^{5/2} + C$

15. $\dfrac{1}{2} \ln(x^2 + 6x + 10) - 3 \arctan(x + 3) + C$ **17.** $\dfrac{(\sin x)^{e+1}}{e + 1} - \dfrac{(\sin x)^{e+3}}{e + 3} + C$

19. $x(1 - x^2)^{-1/2} - \arcsin x + C$ **21.** $\dfrac{1}{18}(1 - 2x^3)^{3/2} - \dfrac{1}{6}(1 - 2x^3)^{1/2} + C$

23. $\arcsin \sqrt{x} - \sqrt{x - x^2} + C$ **25.** $\ln(1 + (\sin x)^2) + C$

27. $\dfrac{1}{8}(\tan x)^8 + \dfrac{1}{6}(\tan x)^6 + C$ **29.** $\dfrac{-2}{3} \ln|x| - \dfrac{1}{12} \ln|x + 3| + \dfrac{3}{4} \ln|x - 1| + C$

31. $\sqrt{x^2 - 1} - \text{arcsec } x + C$ **33.** $\sin x \ln(\sin x) - \sin x + C$

35. $\dfrac{4}{5}x(x + 1)^{5/2} - \dfrac{4}{3}(x + 1)^{3/2} + C$ **37.** $\ln(1 + e^x) + C$

39. $\dfrac{-\cos 5x}{10} - \dfrac{\cos x}{2} + C$ **41.** $2 \ln|\sqrt{x} - 1| + C$ **43.** $e^{\tan x} + C$

45. $\arcsin\left(\dfrac{x + 3}{4}\right) + C$ **47.** $\tan x + \sec x + C$

49. $\dfrac{-4}{3} \ln\left(\dfrac{2 - \sqrt{3x}}{2}\right) - \dfrac{2}{3}\sqrt{3x} + C$ **51.** $\pi + \dfrac{\pi^2}{2}$ **53.** $4\pi e^3$

55. $\left(\dfrac{1}{4} (e^2 + 1), \dfrac{1}{2} e - 1\right)$ **57.** (a) .44909   (b) .46788

59. (a) 2.79199   (b) 2.82881

61. 3.82028

CHAPTER 9
EXERCISE SET 9.1, PAGE 505

1. (a) $2, 1, \dfrac{2}{3}, \dfrac{1}{2}$ (b) Yes, 0   (c) Yes   **3.** (a) $-1, \dfrac{1}{2}, -\dfrac{1}{3}$ (b) Yes, 0   (c) Yes

5. (a) $0, \dfrac{1}{5}, \dfrac{1}{3}, \dfrac{3}{7}$ (b) Yes, 1   (c) Yes   **7.** (a) $\dfrac{3}{4}, 3, \dfrac{11}{2}$ (b) Yes, 1   (c) Yes

9. (a) $1, \dfrac{5}{2}, \dfrac{11}{3}$ (b) No   (c) No   **11.** (a) $\pi, \pi/2, \pi/3$ (b) Yes, 0   (c) Yes

13. (a) $e, \dfrac{e^2}{2}, \dfrac{e^3}{3}$ (b) No   (c) No

15. (a) $\dfrac{3}{\sqrt{2}}, -\dfrac{4}{\sqrt[4]{8}}, \dfrac{7}{\sqrt[4]{540}}, -\dfrac{12}{\sqrt[4]{19768}}$ (b) Yes, 0   (c) Yes

17. (a) $0, 1, \sqrt{3}/2$ (b) Yes, 0   (c) Yes   **19.** (a) $-1, \dfrac{e}{3}, \dfrac{e^2}{9}, \dfrac{e^3}{17}$ (b) No   (c) No

21. (a) $\dfrac{1}{2}, \dfrac{1}{6}, \dfrac{1}{12}$ (b) Yes, 0   (c) Yes

23. (a) $1, \sqrt{2} - 1, \sqrt{5} - 2, \sqrt{10} - 3$ (b) Yes, 0   (c) Yes

25. (a) $1, \dfrac{1}{\sqrt{2} - 1}, \dfrac{1}{\sqrt{5} - 2}, \dfrac{1}{\sqrt{10} - 3}$ (b) No (c) No

27. (a) $0, \dfrac{1}{2}, \sqrt{3}/6$ (b) Yes, 0 (c) Yes

29. This is the same as Exercise 3, since $\cos n\pi = (-1)^n$

31. (a) $0, -\dfrac{1}{2}, \dfrac{2}{5}, -\dfrac{3}{10}$ (b) Yes, 0 (c) Yes

33. (a) $\dfrac{1}{2}, \dfrac{e}{1 + e}, \dfrac{e^2}{1 + e^2}, \dfrac{e^3}{1 + e^3}$ (b) Yes, 1 (c) Yes

35. (a) $2, 9/4, 64/27$ (b) Yes, e (c) Yes **37.** No

39. $x_n = 5000(1 + .6n),\ \lim\limits_{n\to\infty} x_n = \infty$ **43.** $a_n = (-1)^n$

45. $a_n = (-1)^n,\ b_n = (-1)^{n+1}$ **51.** No, for example 5, 11, and 13 are not of this form.
53. 2π

EXERCISE SET 9.2, PAGE 514

1. Diverges **3.** Converges **5.** Diverges **7.** Converges **9.** $\dfrac{1}{2}$

11. $\dfrac{4}{15}$ **13.** 5/2 **15.** 1/2 **17.** 3/4 **19.** 1/2 **21.** 25/12

25. $\sum\limits_{n=1}^{4} 2^n = 30, \sum\limits_{n=1}^{4} 3^n = 120, \sum\limits_{n=1}^{4} 6^n = 1554,$

$\sum\limits_{n=1}^{4} \left(\dfrac{2}{3}\right)^n = \dfrac{130}{81}, \sum\limits_{n=1}^{4} (2^n \cdot 3^n) \neq \left(\sum\limits_{n=1}^{4} 2^n\right)\left(\sum\limits_{n=1}^{4} 3^n\right), \sum\limits_{n=1}^{4} \dfrac{2^n}{3^n} \neq \left(\sum\limits_{n=1}^{4} 2^n\right) \div \left(\sum\limits_{n=1}^{4} 3^n\right)$

27. $S_n = \sum\limits_{i=0}^{n} (n - i)10^i$ **29.** $\dfrac{3457}{9999}$

31. This is true for any divergent series if and only if $c = 0$.

35. $a_n = (-1)^{n+1}$ **37.** (a) 6 (b) $\dfrac{\sqrt{3}}{3}$ **39.** \$450

41. (a) 98.08 feet (b) 180 feet **43.** (a) 177.69 feet (b) 17.79 seconds
45. 40 seconds, 200 meters **47.** Mike should win 2 out of every 3 times.

EXERCISE SET 9.3, PAGE 526

1. No **3.** Yes **5.** No **7.** Yes **9.** Yes **11.** Yes **13.** Yes
15. No **17.** Yes **19.** No **21.** No **23.** No **25.** Yes **27.** Yes

EXERCISE SET 9.4, PAGE 533

1. Yes **3.** Yes **5.** Yes **7.** No **9.** No **11.** Yes **13.** Yes
15. Yes **17.** Yes **19.** No. This series is not alternating.
21. Yes **23.** Yes **25.** No **27.** Yes. This series is not alternating.
29. Yes **31.** No. This series is not alternating. **33.** Yes

35. Yes **37.** .632118 **39.** Consider $\sum\limits_{n=1}^{\infty} (-1)^n + \dfrac{1}{\sqrt{n}}$

EXERCISE SET 9.5, PAGE 542

1. Conditionally **3.** Conditionally **5.** Conditionally **7.** Divergent
9. Divergent **11.** Conditionally **13.** Conditionally **15.** Absolutely

17. Absolutely   **19.** Divergent   **21.** Absolutely   **23.** Absolutely
25. Divergent   **27.** Absolutely   **29.** Absolutely   **31.** Divergent
33. Conditionally   **35.** Absolutely   **37.** Absolutely   **39.** Absolutely
41. Absolutely   **43.** Absolutely   **45.** Absolutely   **49.** $\displaystyle\sum_{n=1}^{\infty} \frac{(-1)^n}{\sqrt{n}}$ **51.** Both diverge

EXERCISE SET 9.6, PAGE 549

1. $R = 1, [-1, 1)$ **3.** $R = 1, [-1, 1]$ **5.** $R = \infty, (-\infty, \infty)$
7. $R = 1, [-1, 1]$ **9.** $R = \infty, (-\infty, \infty)$ **11.** $R = \infty, (-\infty, \infty)$

13. $R = \dfrac{1}{2}, \left[-\dfrac{1}{2}, \dfrac{1}{2}\right)$ **15.** $R = 1, [-1, 1)$ **17.** $R = 1, (1, 3)$

19. $R = 1, (1, 3)$ **21.** $R = 1, (-4, -2)$ **23.** $R = \dfrac{1}{3}, \left[0, \dfrac{2}{3}\right]$

25. $R = \dfrac{1}{2}, [-1, 0)$ **27.** $[-1, 1)$ **29.** $[0, 2]$ **31.** $[1, 2]$

33. $(1 - \sqrt{2}, 1) \cup (1, 1 + \sqrt{2})$ **35.** $(-\infty, -1) \cup (1, \infty)$ **37.** $(-\infty, -2) \cup (2, \infty)$

39. $x \neq \dfrac{(2n + 1)\pi}{2}$ for any integer n **41.** $(-1, 1)$ **43.** $R = $ minimum $\{R_1, R_2\}$

EXERCISE SET 9.7, PAGE 554

1. $\displaystyle\sum_{n=0}^{\infty} (-1)^n x^n$ **3.** $\displaystyle\sum_{n=0}^{\infty} (-1)^n (n + 1)x^n$ **5.** $\displaystyle\sum_{n=0}^{\infty} \frac{x^n}{2^{n+1}}$ **7.** $\displaystyle\sum_{n=0}^{\infty} (-1)^n x^{2n}$

9. $\displaystyle\sum_{n=0}^{\infty} (-1)^n x^{2n+1}$ **11.** $\displaystyle\sum_{n=0}^{\infty} x^{2n}$ **13.** $\displaystyle\sum_{n=1}^{\infty} \frac{x^{n-1}}{n!}$ **15.** $\displaystyle\sum_{n=0}^{\infty} \frac{(-1)^n x^{n+2}}{n + 1}$ **17.** $\displaystyle\sum_{n=0}^{\infty} 2 \cdot 3^n x^{2n+1}$

19. $\displaystyle\sum_{n=0}^{\infty} \frac{x^{3n+2}}{(3n + 2)n!}$ **21.** $.1973$ **23.** $\displaystyle\int \ln(1 + x)\, dx = x \ln(1 + x) - x + \ln(1 + x) + C$

25. $f(x) = \displaystyle\sum_{n=0}^{\infty} \frac{(-1)^n x^{2n+1}}{(2n + 1)!}$ **27.** Let $x = -1$ **29.** 3 terms

EXERCISE SET 9.8, PAGE 563

1. $1 + 2x + 2x^2 + \dfrac{4}{3}x^3$ **3.** $x - \dfrac{1}{2}x^2 + \dfrac{1}{3}x^3$ **5.** $x - \dfrac{1}{3}x^3$

7. $P_4(x) = e\left[1 + (x - 1) + \dfrac{1}{2}(x - 1)^2 + \dfrac{1}{6}(x - 1)^3 + \dfrac{1}{24}(x - 1)^4\right];$

$R_4(x) = \dfrac{1}{120} e^{\xi_x}(x - 1)^5$

9. $P_4(x) = 1 - \dfrac{1}{2}\left(x - \dfrac{\pi}{2}\right)^2 + \dfrac{1}{24}\left(x - \dfrac{\pi}{2}\right)^4; \ R_4(x) = \dfrac{1}{120} \cos \xi_x \left(x - \dfrac{\pi}{2}\right)^5$

11. $P_4(x) = 1 + 2\left(x - \dfrac{\pi}{4}\right) + 2\left(x - \dfrac{\pi}{4}\right)^2 + \dfrac{8}{3}\left(x - \dfrac{\pi}{4}\right)^3 + \dfrac{10}{3}\left(x - \dfrac{\pi}{4}\right)^4;$

$R_4(x) = \dfrac{1}{15}\left[2 + 15(\tan \xi_x)^2 + 15(\tan \xi_x)^4\right](\sec \xi_x)^2 \left(x - \dfrac{\pi}{4}\right)^5$

13. $P_4(x) = 2 + \dfrac{1}{4}(x - 4) - \dfrac{1}{64}(x - 4)^2 + \dfrac{1}{512}(x - 4)^3 - \dfrac{5}{16384}(x - 4)^4;$

$R_4(x) = \dfrac{7}{256}(\xi_x)^{-9/2} (x - 4)^5$

15. $P_4(x) = 3 + e + (2 + e)(x - 1) + \frac{1}{2}(2 + e)(x - 1)^2 + \frac{1}{6}(x - 1)^3 + \frac{1}{24}(x - 1)^4$;

$R_4(x) = \frac{1}{120}e^{\xi x}(x - 1)^5$

17. $.0099998333, 4.2 \times 10^{-10}$ **19.** 2.02484567 **21.** $.0872$

25. $P(x) = 2\left(x + \frac{3}{4}\right)^2 - \frac{1}{8}$ **29.** $\frac{1}{24}$

EXERCISE SET 9.9, PAGE 572

1. $1 + \sum_{n=1}^{\infty} \frac{1}{n!}\left[\left(\frac{1}{3}\right)\left(-\frac{2}{3}\right)\left(-\frac{5}{3}\right) \cdots \left(\frac{1}{3} - n + 1\right)\right]x^n, R = 1$

3. $1 + \sum_{n=1}^{\infty} \frac{1}{n!}\left[\left(-\frac{1}{2}\right)\left(-\frac{3}{2}\right)\left(-\frac{5}{2}\right) \cdots \left(-\frac{1}{2} - n + 1\right)\right]x^n, R = 1$

5. $2\sqrt{2}\left\{1 + \sum_{n=1}^{\infty} \frac{3^n}{2^n n!}\left(\frac{3}{2}\right)\left(\frac{1}{2}\right)\left(-\frac{1}{2}\right) \cdots \left(\frac{3}{2} - n + 1\right)\right]x^n\right\}, R = \frac{2}{3}$

7. $.5132$ **11.** $2.73205, -.73205$ **13.** $-e^2$

15.   **17.**

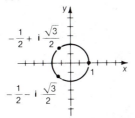   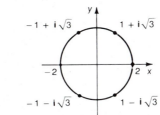

19. $\pm 1 \pm i, x^4 + 4 = (x^2 - 2x + 2)(x^2 + 2x + 2)$ **25.** 10.3%

REVIEW EXERCISES, PAGE 574

1. $-\frac{3}{2}$ **3.** 3 **5.** 1 **7.** Divergent **9.** Divergent **11.** 0 **13.** 1

15. Divergent **17.** Absolutely **19.** Absolutely **21.** Conditionally

23. Absolutely **25.** Conditionally **27.** Divergent **29.** Conditionally **31.** Conditionally

33. Conditionally **35.** $[0, 2]$ **37.** $(-3, -1)$ **39.** $(0, 4)$ **41.** $\left(\frac{8}{3}, \frac{10}{3}\right)$

43. $(-\sqrt{3}, \sqrt{3})$ **45.** $\left(-\frac{3}{2}, \frac{3}{2}\right)$ **47.** $.05579$ **49.** $.28768$ **51.** $a_n = n$

53. $\sum_{n=1}^{\infty} (-1)^n \frac{x^{2n}}{(2n + 1)!}$ **55.** $1 + \sum_{n=1}^{\infty} \frac{\left(-\frac{1}{2}\right)\left(-\frac{3}{2}\right) \cdots \left(-\frac{1}{2} - n + 1\right)}{n!}(2x)^n$

57. $\sum_{n=0}^{\infty} \frac{x^{2n}}{(2n)!}$ **59.** $\frac{97}{56}$ **61.** $\left[\frac{1}{2}, \infty\right)$ **63.** $1 - 2\sum_{n=0}^{\infty} (-1)^n x^n$

65. $\ln 2 + \sum_{n=0}^{\infty} (-1)^{n-1}\frac{(x - 1)^n}{n}$ **67.** $\sum_{n=0}^{\infty} \frac{1}{(2n + 1)n!}$ **69.** $\pm 2, \pm 2i$

CHAPTER 10
EXERCISE SET 10.1, PAGE 581

1.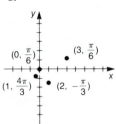

3.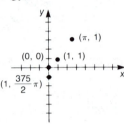

5. (a) $\left(2, \dfrac{8\pi}{3}\right), \left(2, -\dfrac{4\pi}{3}\right)\left(-2, -\dfrac{\pi}{3}\right), \left(-2, \dfrac{5\pi}{3}\right)$

  (b) $(1, 3\pi), (1, -\pi), (-1, 0), (-1, 2\pi)$

  (c) $(0, \pi/2), (0, \pi), \left(0, \dfrac{3\pi}{2}\right), (0, -\pi/4)$

  (d) $(1, 1 + 2\pi), (1, 1 - 2\pi), (-1, 1 - \pi), (-1, \pi + 1)$

7. (a) $(1, 0)$,   (b) $(3\sqrt{3}/2, 3/2)$ (c) $(\sqrt{2}, \sqrt{2})$ (d) $(4, 0)$

9. $r = \dfrac{r_1 r_2 \sin(\theta_2 - \theta_1)}{r_2 \sin(\theta_2 - \theta) - r_1 \sin(\theta_1 - \theta)}$ **11.** $\left(340, \arctan\left(\dfrac{8}{15}\right)\right)$

13. $\left(90, \dfrac{\pi}{4}\right), \left(90\sqrt{2}, \dfrac{\pi}{2}\right), \left(90, \dfrac{3\pi}{4}\right)$

EXERCISE SET 10.2, PAGE 591

1.

3.

5.

7.

9.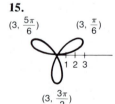

11.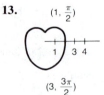

13.

15.

17.

19. $\left(-2, \dfrac{7\pi}{4}\right)$ $\left(2, \dfrac{\pi}{4}\right)$ **21.**   **23.**

$\left(2, \dfrac{5\pi}{4}\right)$ $\left(-2, \dfrac{3\pi}{4}\right)$

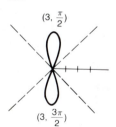

$\left(3, \dfrac{\pi}{2}\right)$

$\left(3, \dfrac{3\pi}{2}\right)$

25.   **27.**   **29.**

31.   **33.**   **35.**

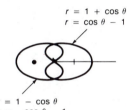

$r = 1 + \cos\theta$
$r = \cos\theta - 1$

$r = 1 - \cos\theta$
$r = -\cos\theta - 1$

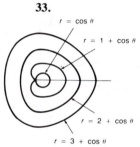

$r = \cos\theta$
$r = 1 + \cos\theta$
$r = 2 + \cos\theta$
$r = 3 + \cos\theta$

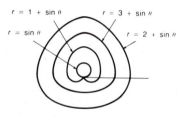

$r = 1 + \sin\theta$ $r = 3 + \sin\theta$
$r = \sin\theta$ $r = 2 + \sin\theta$

37. $r = 2^{\theta}$ $r = -2^{-\theta}$

$r = 2^{-\theta}$ $r = -2^{\theta}$

39. (a) Horizontal tangent when $\theta = \dfrac{\pi}{4}, \dfrac{3\pi}{4}, \dfrac{5\pi}{4}, \dfrac{7\pi}{4}$

Vertical tangent when $\theta = 0, \dfrac{\pi}{2}, \pi, \dfrac{3\pi}{2}, 2\pi$

(b) Horizontal tangent when $\theta = \dfrac{7\pi}{6}, \dfrac{11\pi}{6}$

Vertical tangent when $\theta = \dfrac{\pi}{6}, \dfrac{5\pi}{6}, \dfrac{3\pi}{2}$

(c) Horizontal tangent when $\theta = 0, \dfrac{2\pi}{3}, \dfrac{4\pi}{3}, 2\pi$

Vertical tangent when $\theta = \dfrac{\pi}{3}, \pi, \dfrac{5\pi}{3}$

(d) Horizontal tangent when $\theta = -\dfrac{3\pi}{4}, \dfrac{\pi}{4}$

Vertical tangent when $\theta = -\dfrac{\pi}{4}, \dfrac{3\pi}{4}$

41. $r = a(1 - \cos\theta)$ for some $a > 0$.

EXERCISE SET 10.3, PAGE 598

1.   Area $= 4\pi$

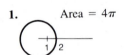

3.   Area $= \pi$

5.   Area $= \dfrac{9}{4}\pi$

7.   Area $= \dfrac{3}{2}\pi$

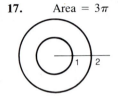

9.   Area $= 18\pi$

11.   Area $= \dfrac{\pi}{2}$

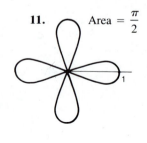

13.   Area $= 4$

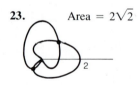

15.   Area $= \dfrac{\pi^3}{24}$

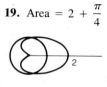

17.   Area $= 3\pi$

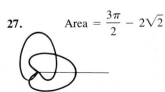

19. Area $= 2 + \dfrac{\pi}{4}$

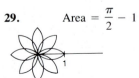

21. Area $= 4\pi - 4$

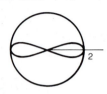

23.   Area $= 2\sqrt{2}$

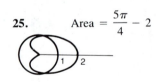

25.   Area $= \dfrac{5\pi}{4} - 2$

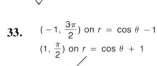

27.   Area $= \dfrac{3\pi}{2} - 2\sqrt{2}$

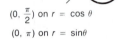

29.   Area $= \dfrac{\pi}{2} - 1$

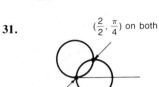

31. $(\frac{2}{2}, \frac{\pi}{4})$ on both

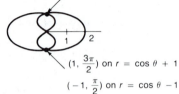

$(0, \frac{\pi}{2})$ on $r = \cos\theta$

$(0, \pi)$ on $r = \sin\theta$

33. $(-1, \frac{3\pi}{2})$ on $r = \cos\theta - 1$

$(1, \frac{\pi}{2})$ on $r = \cos\theta + 1$

$(1, \frac{3\pi}{2})$ on $r = \cos\theta + 1$

$(-1, \frac{\pi}{2})$ on $r = \cos\theta - 1$

35. The graphs of $r = \cos\dfrac{\theta}{2}$ and $r = \sin\dfrac{\theta}{2}$ are the same.

(r, θ) on graph of $r = \cos\dfrac{\theta}{2}$

$(-r, \ 3\pi + \theta)$ on graph of $r = \sin\dfrac{\theta}{2}$

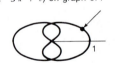

EXERCISE SET 10.4, PAGE 604

1. $y = \dfrac{1}{6}x$ **3.** $x = \sqrt{y-1}$ **5.** $y = 1 - x^2, y \ge 0$

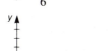

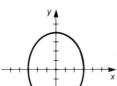   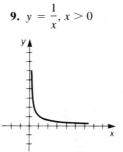   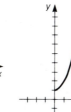

7. $\dfrac{x^2}{9} + \dfrac{y^2}{16} = 1$ **9.** $y = \dfrac{1}{x}, x > 0$

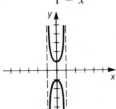   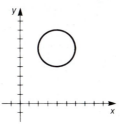

11. $y^2 = \dfrac{1}{1-x^2}$ **13.** $(x-4)^2 + (y-6)^2 = 4$ **15.** $y = \dfrac{\sqrt[3]{4}}{2}x^{2/3} + 1$

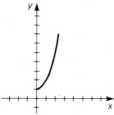   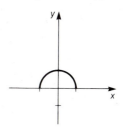   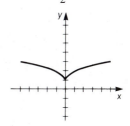

17. (a)   (b)   (c)   (d)

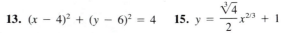

19. (a)   (b)   (c)   (d)

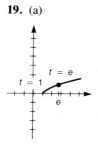   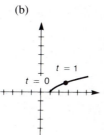   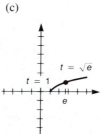   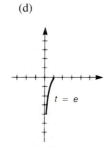

21. (a) 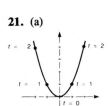 (b) 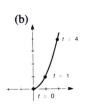 (c) 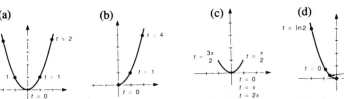 (d)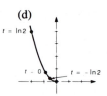

25. A vertical asymptote at $x = 1$, since $\lim\limits_{t \to 0} \dfrac{\sin t}{t} = 1$. No horizontal asymptotes.

EXERCISE SET 10.5, PAGE 609

1. $\dfrac{1}{6}$ **3.** 4   **5.** $-\sqrt{3}$ **7.** 0   **9.** -1 **11.** 1   **13.** 0   **15.** $-\dfrac{1}{3}$

17. $-1; y = -x + \sqrt{2}$ **19.** undefined; $x = 1$ **21.** $1; y = x + 1$

23. $-1; y = -x + 3\sqrt{2}$ **25.** $0; y = \dfrac{\sqrt{2}}{2}$

27. $\dfrac{4 + 5\pi}{4 - 5\pi}; y = \dfrac{4 + 5\pi}{4 - 5\pi}\left(x + \dfrac{5\sqrt{2}\pi}{8}\right) - \dfrac{5\sqrt{2}\pi}{8}$

29. $\dfrac{\tan 2 \ln 2 + 1}{\tan 2 - \ln 2}; y = \left(\dfrac{\tan 2 \ln 2 + 1}{\tan 2 - \ln 2}\right)(x - 4 \cos 2) - 4 \sin 2$ **31.** 0

33. $-\dfrac{4}{9}(\sec t)^3$ **35.** $2e^{-3t}$ **37.** $2(\sec 2\theta)^3$ **39.** $2^\theta \left[\dfrac{1 + (\ln 2)^2}{(\cos \theta \ln 2 + \sin \theta)^3}\right]$

41. At $\left(1, \dfrac{\pi}{2}\right): y = x + 1$ for $r = 1 + \cos \theta$ and $y = 1$ for $r = 1$.

At $\left(1, -\dfrac{\pi}{2}\right): y = -x - 1$ for $r = 1 + \cos \theta$ and $y = -1$ for $r = 1$.

43. At $\left(\dfrac{1}{2}(2 + \sqrt{2}), \dfrac{\pi}{4}\right): y = (1 - \sqrt{2})x + 1 + \dfrac{1}{2}\sqrt{2}$ for $r = 1 + \cos \theta$ and

$y = -(1 + \sqrt{2})x + 2 + \dfrac{3}{2}\sqrt{2}$ for $r = 1 + \sin \theta$.

45. At $\left(\dfrac{\sqrt{2}}{2}, \dfrac{\pi}{4}\right): x = \dfrac{1}{2}$ for $r = \sin \theta$ and $y = \dfrac{1}{2}$ for $r = \cos \theta$.

At the origin: $y = 0$ for $r = \sin \theta$ and $x = 0$ for $r = \cos \theta$.

47. Horizontal tangents when $\theta = \dfrac{3\pi}{4} + n\pi$, Vertical tangents when $\theta = \dfrac{\pi}{4} + n\pi$, for any integer n.

49. $r(\theta) = \dfrac{128}{\pi^3}(R_1 - R_2)\theta^3 + \dfrac{48}{\pi^2}(R_2 - R_1)\theta^2 + R_1$

EXERCISE SET 10.6, PAGE 616

1. $2\sqrt{13}$ **3.** $13\sqrt{13} - 8$ **5.** $\sqrt{2} + \ln(1 + \sqrt{2})$ **7.** $\ln(1 + \sqrt{2})$

9. $\sqrt{2}(e^{2\pi} - 1)$ **11.** 3π **13.** $\dfrac{\sqrt{5}}{2}(e^2 - 1)$ **15.** 8

17. (a) Length $\dfrac{\pi}{2}$ (b) Length $\sqrt{2}$ (c) Length $\dfrac{3}{2}$

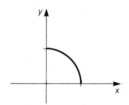 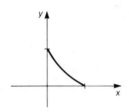 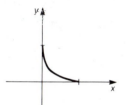

19. $8R$

21. $\dfrac{14}{27}$ mile. (No calculus is needed to solve this problem, but arc length can be used.)

23. Approximately 2945 feet

REVIEW EXERCISES, PAGE 617

1. 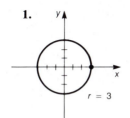 $r = 3$

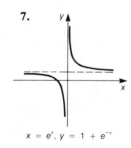

5. 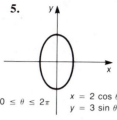 $0 \leq \theta \leq 2\pi$ $\begin{aligned} x &= 2\cos\theta \\ y &= 3\sin\theta \end{aligned}$

7. 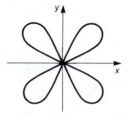 $x = e^t, \, y = 1 + e^{-t}$

9. 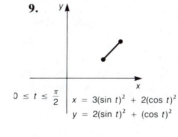 $0 \leq t \leq \dfrac{\pi}{2}$ $\begin{aligned} x &= 3(\sin t)^2 + 2(\cos t)^2 \\ y &= 2(\sin t)^2 + (\cos t)^2 \end{aligned}$

11. $\dfrac{1}{2}$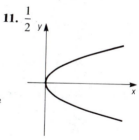

13. $\dfrac{1}{2}\left(e - \dfrac{1}{e}\right)$

15. $\dfrac{\sqrt{3}}{9}$

17. $r = \dfrac{1}{\theta}$

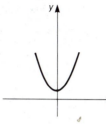

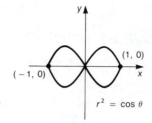 $r^2 = \cos\theta$ $(1, 0)$ $(-1, 0)$

19. (a) $\left(3, \dfrac{2\pi}{3}\right)$ $\left(3, \dfrac{9\pi}{3}\right)$ (b) π (c) 4π (d) 5π

21. (a) $-\infty, \infty$ (b) $(\sec\theta)^2$ (c)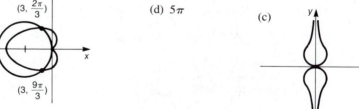

23. $x = 1$

25. $2x - y = 3$

27. $\dfrac{\pi}{2}$

CHAPTER 11
EXERCISE SET 11.1, PAGE 628

1. Vertex $(0, 0)$,
    Focal point $(1, 1/8)$
    Directrix $y = -1/8$

3. Vertex $(0, 0)$,
    focal point $(0, -1/8)$,
    directrix $y = 1/8$

5. Vertex $(0, 0)$,
    focal point $(1/2, 0)$,
    directrix $x = -1/2$

7. Vertex $(0, 0)$,
    focal point $(-1/2, 0)$,
    directrix $x = 1/2$

9. Vertex $(-2, 0)$,
    focal point $(-2, 1/2)$,
    directrix $y = -1/2$

11. Vertex $(-2, 4)$,
    focal point $(-3/2, 4)$,
    directrix $x = -5/2$

13. Vertex $(-1, 2)$,
    focal point $(-1, 25/8)$,
    directrix $y = 7/8$

15. Vertex $(-1, 2)$,
    focal point $(-17/18, 2)$,
    directrix $x = -19/8$

17. $8y = (x + 2)^2$ **19.** $6y - 3 = (x + 2)^2$ **21.** $x + 2 = -(y - 2)^2/24$

23. $y - 4 = \dfrac{1}{8}(x - 3)^2$ **25.** $y = \dfrac{3}{8}x^2$ **27.** $x = \dfrac{1}{9}y^2$

29. (a) $y = \dfrac{y_1}{x_1^2}x^2$ (b) $y - k = \dfrac{y_1 - k}{(x_1 - h)^2}(x - h)^2$

33. $y = ax^2 + 1 - a$, only $y = -\dfrac{1}{4}x^2 + \dfrac{5}{4}$ **35.** $a = \dfrac{1}{4b}$ **37.** $y + \dfrac{1}{24} = \dfrac{3}{2}\left(x - \dfrac{5}{6}\right)^2$

39. (a) 200 feet (b) 212.9 feet (c) 952 feet, 1813 feet

41. Approximately 53.7×10^6 ft^3

43. 318.5 feet, 509.5 feet **45.** Approximately 3473 kilometers

EXERCISE SET 11.2, PAGE 637

1. Vertices $(0, \pm 3)$
    Focal points $(0, \pm\sqrt{5})$

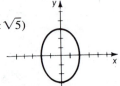

3. Vertices $(\pm 5, 0)$
Focal points $(\pm 3, 0)$

5. Vertices $(0, \pm\sqrt{3})$
Focal points $(0, \pm 1)$

7. Vertices $(0, \pm 1)$
Focal points $\left(0, \pm\dfrac{\sqrt{3}}{2}\right)$

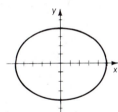

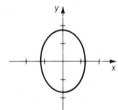

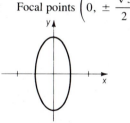

9. Vertices $(-2, \pm 3)$
Focal points $\left(-2, \pm\dfrac{3\sqrt{3}}{2}\right)$

11. Vertices $(-1, 2), (3, 2)$
Focal points $(1 \pm \sqrt{3}, 2)$

13. Vertices $\left(3, -1 \pm\dfrac{\sqrt{2}}{2}\right)$
Focal points $\left(3, -1 \pm\dfrac{\sqrt{6}}{6}\right)$

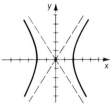

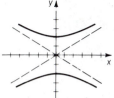

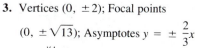

15. $5x^2 + 9y^2 = 45$

17. $5x^2 + 4y^2 = 20$

19. $3(x - 4)^2 + 4(y - 2)^2 = 12$ **21.** 6π **23.** $y = -\dfrac{\sqrt{3}}{2}x + 2\sqrt{3}$

25. $y = \pm\dfrac{\sqrt{2}}{2}(x - 3)$ **27.** $2ab$ **31.** $\dfrac{2b^2}{a}$

33. If $D^2 + E^2 - 4ACF$ (a) is positive (b) is zero (c) is negative

35. $x^2/(240)^2 + y^2/(40)^2 = 1, 33{,}600\pi$ ft^2

37. 125.75 ft, 1.47×10^7 ft^2 **39.** Approximately 53.6×10^6 miles

EXERCISE SET 11.3, PAGE 647

1. Vertices $(\pm 2, 0)$; Focal points
$(\pm\sqrt{13}, 0)$; Asymptotes $y = \pm\dfrac{3}{2}x$

3. Vertices $(0, \pm 2)$; Focal points
$(0, \pm\sqrt{13})$; Asymptotes $y = \pm\dfrac{2}{3}x$

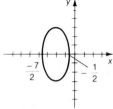

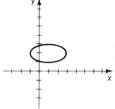

5. Vertices $(\pm 2, 0)$; Focal points
$(\pm 2\sqrt{5}, 0)$; Asymptotes $y = \pm 2x$

7. Vertices $(\pm 1, 0)$; Focal points
$(\pm\sqrt{2}, 0)$; Asymptotes $y = \pm x$

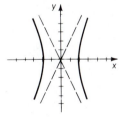

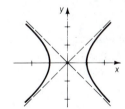

9. Vertices $(0, \pm 2\sqrt{2})$; Focal points $(0, \pm 2\sqrt{3})$; Asymptotes $y = \pm\sqrt{2}x$

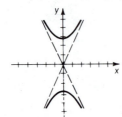

11. Vertices $(-3, 0)$, $(1, 0)$; Focal points $(-1 \pm \sqrt{5}, 0)$; Asymptotes $y = \pm \dfrac{1}{2}(x + 1)$

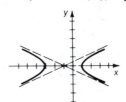

13. Vertices $(0, 0)$, $(2, 0)$; Focal points $(3, 0)$, $(-1, 0)$; Asymptotes $y = \pm\sqrt{3}(x - 1)$

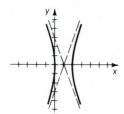

15. Vertices $(-1, -1)$, $(3, -1)$; Focal points $(1 \pm \sqrt{13}, -1)$; Asymptotes $y + 1 = \pm \dfrac{3}{2}(x - 1)$

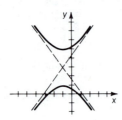

17. Vertices $(-1, 1)$, $(-1, 5)$; Focal points $(-1, 3 \pm\sqrt{7})$; Asymptotes $y - 3 = \pm\dfrac{2\sqrt{2}}{3}(x - 1)$

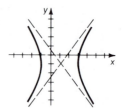

19. $16x^2 - 9y^2 = 144$ **21.** $9y^2 - 16x^2 = 144$

23. $5(x - 2)^2 - 4(y - 4)^2 = 20$

25. $x^2 - y^2 = 9$ **27.** $225x^2 - 400y^2 = 1296$

29. $9(x - 2)^2 - 16(y - 1)^2 = 144$

31. $y = \dfrac{\sqrt{6}}{2}x - \sqrt{3}$ **33.** $y = \pm\sqrt{3}(x - 1)$ **35.** $2b^2/a$

37. If $D^2C - E^2A - 4ACF$, (a) is positive, (b) is zero, (c) is negative

39. No, two points satisfy these conditions **41.** .67 inches

EXERCISE SET 11.4, PAGE 656

1. Ellipse

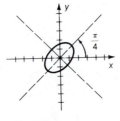

3. Parabola

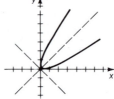

5. Hyperbola

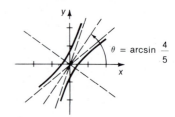

7. Ellipse

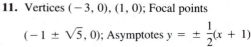

9. (11.14) indicates an ellipse. The graph is the single point $(0, 0)$.

11. Parabola

13. Parabola

15. (11.14) indicates a parabola. The graph is a line with equation
$$x = 17\sqrt{137}/274$$

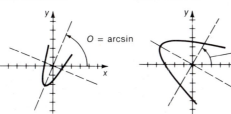

$O = \arcsin$

17. Ellipse

19. A hyperbola.

21.

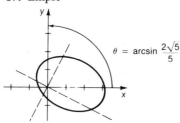

$\theta = \arcsin \dfrac{2\sqrt{5}}{5}$

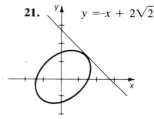

$y = -x + 2\sqrt{2}$

25. $337x^2 - 168xy + 288y^2 - 1350x - 1800y = 0$

EXERCISE SET 11.5, PAGE 664

1. $x - 2 = -\dfrac{y^2}{2}$

3. $\dfrac{9}{5}x^2 + \left(y + \dfrac{6}{5}\right)^2 = \dfrac{81}{25}$

5. $(3y + 1) = \dfrac{9}{4}x^2$

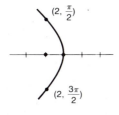

$(2, \frac{\pi}{2})$

$(2, \frac{3\pi}{2})$

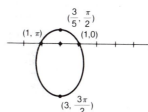

$(\frac{3}{5}, \frac{\pi}{2})$

$(1, \pi)$ $(1, 0)$

$(3, \frac{3\pi}{2})$

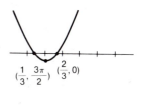

$(\frac{1}{3}, \frac{3\pi}{2})$ $(\frac{2}{3}, 0)$

7. $5(x + 3)^2 - 4y^2 = 20$

9. $3(y - 2)^2 - x^2 = 3$

11. $16x^2 + 12y^2 + 4y = 1$

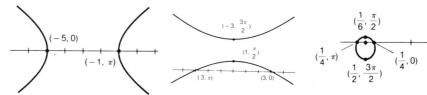

$(-5, 0)$

$(-1, \pi)$

$(-3, \frac{3\pi}{2})$

$(1, \frac{\pi}{2})$

$(3, \pi)$ $(3, 0)$

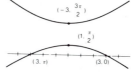

$(\frac{1}{6}, \frac{\pi}{2})$

$(\frac{1}{4}, \pi)$ $(\frac{1}{4}, 0)$

$(\frac{1}{2}, \frac{3\pi}{2})$

13. $r = \dfrac{8}{1 + 2\cos\theta}$

15. $r = \dfrac{1}{4 - 4\sin\theta}$

17. $r = \dfrac{6}{1 + 3\sin\theta}$

19. $r = \dfrac{3}{2 + \cos\theta}$

21. $r = \dfrac{1}{3}\tan\theta\sec\theta$. Not in standard polar position.

23. $r^2(\cos\theta)^2 - 2r\sin\theta + 2 = 0$. Not in standard polar position.

25. $r^2[1 - 2(\sin\theta)^2] - 2r\sin\theta = 4$. Not in standard polar position.

29.

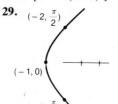

$(-2, \frac{\pi}{2})$

$(-1, 0)$

$(-2, \frac{\pi}{2})$

31.

$(-\frac{1}{5}, \frac{3\pi}{2})$

$(-\frac{1}{3}, 0)$ $(-\frac{1}{3}, \pi)$

$(-1, \frac{\pi}{2})$

33. $r = \dfrac{7560480}{1741 - 83\cos\theta}$

REVIEW EXERCISES, PAGE 665

1.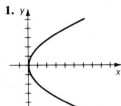

3.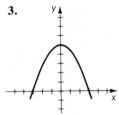

5.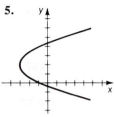

7.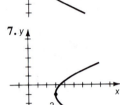

9.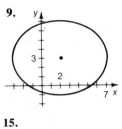

11.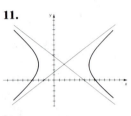

13.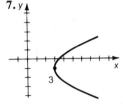

15.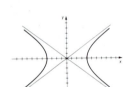

17.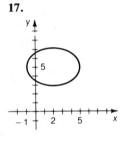

19.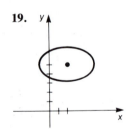

21. $3\hat{x}^2 - \hat{y}^2 = 16$

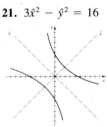

23. $9\hat{x}^2 + 4\hat{y}^2 = 36$

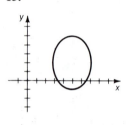

25. $16\hat{x}^2 + 9\hat{y}^2 = 144$

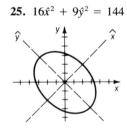

27.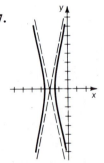

29.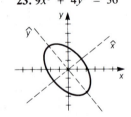

31. $y = 1 - \dfrac{1}{4}x^2$ **33.** $\dfrac{x^2}{8} + \dfrac{y^2}{9} = 1$ **35.** $x^2 - \dfrac{y^2}{8} = 1$

37. $\dfrac{x^2}{16} + \dfrac{y^2}{41} = 1$ **39.** $r = \dfrac{6}{1 - 3\sin\theta}$ **41.** $5x - 2y - 9 = 0$ **43.** $y = x + 1$

CHAPTER 12
EXERCISE SET 12.1, PAGE 673

1. (a)   (b)   (c)

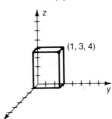

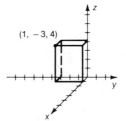

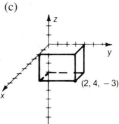

3.   **5.**   **7.**

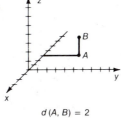

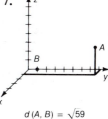

$d(A, B) = \sqrt{6}$ $d(A, B) = 2$ $d(A, B) = \sqrt{59}$

9. $\left(0, \dfrac{5}{2}, \dfrac{7}{2}\right)$ **11.** $(-3, 4, 1)$ **13.** $\left(\dfrac{1}{2}, \dfrac{9}{2}, \dfrac{3}{2}\right)$

15.   **17.**   **19.** $x^2 + y^2 + z^2 = 4$

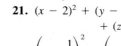

21. $(x - 2)^2 + (y - 3)^2$
$\qquad\qquad + (z - 4)^2 = 1$

23. $\left(x + \dfrac{1}{2}\right)^2 + \left(y - \dfrac{5}{2}\right)^2$
$\qquad\qquad + (z - 3)^2 = \dfrac{34}{4}$

25. Center $(-1, 2, -3)$, radius $\sqrt{14}$ **27.** (b) $\left(x - \dfrac{1}{2}\right)^2 + \left(y - \dfrac{9}{2}\right)^2 + (z - 3)^2 = \dfrac{33}{2}$

29.

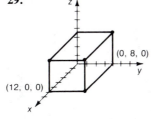

EXERCISE SET 12.2, PAGE 681

1. $\langle 1, 5, 3\rangle$; $\sqrt{35}$ **3.** $\langle 0, -6, 0\rangle$; 6   **5.** $\langle 2, 3, -1\rangle$; $\sqrt{14}$ **7.** $(1, 3, 4)$ **9.** $(-5, 3, 1)$

11.   **13.** $\langle -4, -6, -8\rangle$

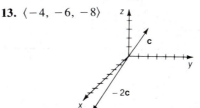

15. $\langle 2, 5, 6 \rangle$

17. $\|\mathbf{c}\| = \sqrt{29}, \|-2\mathbf{c}\| = 2\sqrt{29}$

19. (a) $\left\langle \dfrac{2\sqrt{29}}{29}, \dfrac{3\sqrt{29}}{29}, \dfrac{4\sqrt{29}}{29} \right\rangle$ (b) $\left\langle -\dfrac{\sqrt{14}}{14}, \dfrac{2\sqrt{14}}{14}, \dfrac{3\sqrt{14}}{14} \right\rangle$

21. $\dfrac{3\langle 3, 4, -1 \rangle}{\|\langle 3, 4, -1 \rangle\|} = \left\langle \dfrac{9\sqrt{26}}{26}, \dfrac{6\sqrt{26}}{13}, -\dfrac{3\sqrt{26}}{26} \right\rangle$

23. (a) $\mathbf{i} + 4\mathbf{j}$ (b) $\mathbf{i} + 4\mathbf{j} + 5\mathbf{k}$ (c) $2\mathbf{j} + 3\mathbf{k}$

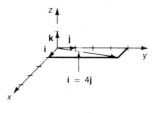 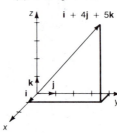 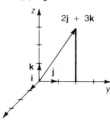

27. If $\mathbf{j}$ points upstream and $\mathbf{i}$ points across the stream, the direction is $300\mathbf{i} - 180\mathbf{j}$. The average speed is 5.8 ft/sec.

29. If $\mathbf{j}$ points north and $\mathbf{i}$ points east, the pilot flies in the direction of $-10\sqrt{2}\,\mathbf{i} + (100 + 10\sqrt{2})\mathbf{j}$, and averages 115 miles per hour.

EXERCISE SET 12.3, PAGE 691

1. 20 **3.** 34 **5.** 0 **7.** $3 + 4e$ **9.** $\arccos \dfrac{5\sqrt{114}}{57} \approx .358$

11. $\arccos \dfrac{17\sqrt{290}}{290} \approx .059$ **13.** $\dfrac{\pi}{2}$, $\mathbf{a}$ and $\mathbf{b}$ are orthogonal

15. $\arccos \dfrac{3 + 4e}{\sqrt{(25 + \pi^2)(1 + e^2)}} \approx .625$ **17.** $\dfrac{20}{19}\sqrt{19}; \dfrac{20}{19}\langle 1, 3, 3 \rangle$

19. $\dfrac{17}{5}\sqrt{5}; \left\langle \dfrac{17}{5}, \dfrac{34}{5}, 0 \right\rangle$ **21.** $0; \mathbf{0}$ **23.** $\dfrac{3 + 4e}{\sqrt{25 + \pi^2}}; \dfrac{3 + 4e}{25 + \pi^2}\langle 3, 4, \pi \rangle$

25. $\dfrac{2\sqrt{29}}{29}, \dfrac{3\sqrt{29}}{29}, \dfrac{4\sqrt{29}}{29}$ **27.** $0, \dfrac{3}{5}, \dfrac{4}{5}$ **29.** $\langle 0, 3/5, 4/5 \rangle$ or $\langle 0, -3/5, -4/5 \rangle$

31. There are two, $\left\langle -\dfrac{2\sqrt{5}}{5}, \dfrac{\sqrt{5}}{5}, 0 \right\rangle$ and $\left\langle \dfrac{2\sqrt{5}}{5}, -\dfrac{\sqrt{5}}{5}, 0 \right\rangle$

33. $\mathbf{b}_1 = \dfrac{1}{19}\langle 5, 15, -15 \rangle$, $\mathbf{b}_2 = \dfrac{1}{19}\langle -81, 4, -23 \rangle$

37. $\langle 6c, 3c, -2c \rangle$ for any constant c.

43. 2 ft/lb **45.** (a) 120 ft/lb (b) 115.91 ft/lb

EXERCISE SET 12.4, PAGE 697

1. $\langle 5, 10, -5 \rangle$ **3.** $\langle -3, 12, -2 \rangle$ **5.** $\langle 2, -1, 2 \rangle$ **7.** $\langle -1, 0, 2 \rangle$ **9.** -1

11. Not possible **13.** 12 **15.** $\dfrac{1}{2}\langle -2, 1, -1 \rangle$ **17.** $\langle 6c, c, -2c \rangle$ for any constant c.

19. 25 **21.** $\sqrt{69}$ **25.** 24 **27.** 20 **29.** 4 **31.** 10/3

EXERCISE SET 12.5, PAGE 704

1. $x - y + z = 2$ **3.** $z = 4$ **5.** $2x + 3y + 4z = 0$

7.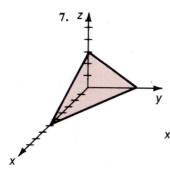

9.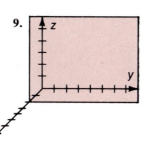

11.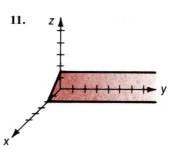

13. The xy-plane.

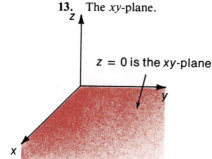

$z = 0$ is the xy-plane

15. $2x + 3y + z = 14$
17. $x + y - z = 2$ **19.** $z = 3$
21. $y - x = 1$
23. $2x + y + z = 5$
25. $7x + 5y - 3z = 10$
27. $\dfrac{6\sqrt{17}}{17}$ **29.** $\dfrac{3\sqrt{14}}{14}$ **31.** $\dfrac{2}{3}\sqrt{6}$ **33.** $\dfrac{3}{7}\sqrt{14}$
35. $8x + 8y + 10z = 99$

EXERCISE SET 12.6, PAGE 711

1. $x = 2 + t, y = -1 + t,$ **3.** $x = 2t, y = 4t,$ **5.** $x = 1, y = 2 - 3t, z = 4t$
$z = 2 + t$ $z = 3t$

7. $x = 3 + t, y = 4 + 7t,$ **9.** $x - 2 = y + 1 = z - 2$ **11.** $\dfrac{x}{2} = \dfrac{y}{4} = \dfrac{z}{3}$
$z = 4 - t$

13. $x = 1, \dfrac{y - 2}{-3} = \dfrac{z}{4}$ **15.** $x - 3 = \dfrac{y - 4}{7} = \dfrac{z - 4}{-1}$ **17.** $x = t + 1, y = 2 - t,$
$z = 3 - t$

19. $x = 1 + t, y = -2 + t, z = 3 + t$ **21.** $x = 1 + t, y = 7, z = 0$

23. Not orthogonal or parallel. Do not intersect.

25. Not orthogonal or parallel. Intersect at $\left(\dfrac{3}{2}, \dfrac{5}{2}, \dfrac{3}{2}\right)$

27. (a) $(0, 2, -3)$ (b) $(2, 0, 1)$ (c) $\left(\dfrac{3}{2}, \dfrac{1}{2}, 0\right)$

29. $(-7, -8, -3)$ **31.** $x = 2t, y = 1, z = 1 - 2t$

REVIEW EXERCISES, PAGE 712

1. Plane 3 units above and parallel to the xy-plane.   **3.** Line in the xy-plane.   **5.** Line passing through the point $(2, 1, 3)$ and having direction given by $\langle 2, 1, 1 \rangle$.

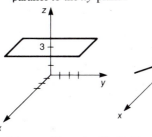

7. Sphere with center $(0, 0, 0)$ and radius 3.

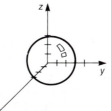

9. Plane orthogonal to the zx-plane.   **11.** Plane

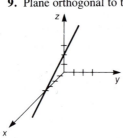

13. $\langle 0, 3, 3 \rangle$ **15.** $\dfrac{\sqrt{21}}{21}\langle 1, -4, -2 \rangle$ or $\dfrac{\sqrt{21}}{21}\langle -1, 4, 2 \rangle$

17. $\langle 3, -7, 1 \rangle$ **19.** $\langle 3, -7, 1 \rangle$ (it doesn't matter that is passes through the origin)

21.

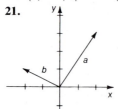

23. $\pm\mathbf{k}$ **25.** $\arccos(-\sqrt{65}/65) \approx 97°$ **27.** 0

29. $x + y = 4$ **31.** $x = -2, y = 1 + t, z = 4$

33. $3x + 2y - 2z = 6$ **35.** $(x-2)^2 + y^2 + z^2 = 4$

37. $z = 4$ **39.** $x = -2, y = 1 + t, z = 2$ **41.** (a) -3

(b) $\dfrac{4}{3}$ **43.** (a) 6   (b) $\dfrac{11\sqrt{6}}{6}$ **45.** 200

CHAPTER 13
EXERCISE SET 13.1, PAGE 719

1. $(-\infty, \infty)$ **3.** $[0, \infty)$ **5.** $(0, \infty)$ **7.**   **9.**

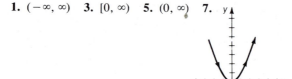

11.   **13.**   **15.**

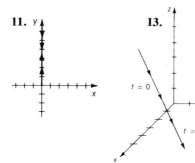

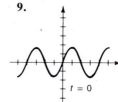

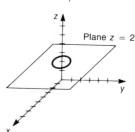

17.   **19.**

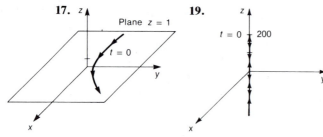

21. (a) $(0, \infty)$; $t^{3/2} + e^t\sin t$ (b) $t \neq 2$; $(2 - t)\mathbf{i} + (2 - t)^{-2}\mathbf{j} + e^{2-t}\mathbf{k}$

(c) $t \neq 0$; $t(2 - t)\mathbf{i} + t^{-2}(2 - t)\mathbf{j} + (2 - t)e^t\mathbf{k}$

(d) $(0, \infty)$; $(t + \sqrt{t})\mathbf{i} + t^{-2}\mathbf{j} + (e^t + \sin t)\mathbf{k}$

(e) $(0, \infty)$; $(t - \sqrt{t})\mathbf{i} + t^{-2}\mathbf{j} + (e^t - \sin t)\mathbf{k}$

(f) $(0, \infty)$; $t^{-2}\sin t\,\mathbf{i} + (e^t\sqrt{t} - t\sin t)\mathbf{j} - t^{-3/2}\mathbf{k}$

23.

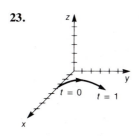

EXERCISE SET 13.2, PAGE 725

1. $\mathbf{i} + \mathbf{j} + \mathbf{k}$ **3.** $4\mathbf{i} + 3\mathbf{j}$ **5.** Does not exist.   **7.** $0 < t < \infty$ **9.** $t \neq 2$

11. $t \neq 0$ **13.** $\frac{1}{t}\mathbf{i} + \mathbf{j}$ **15.** $2t\mathbf{i} + 2te^{t^2}\mathbf{j} + \mathbf{k}$ **17.** $3\mathbf{j}$ **19.** $\mathbf{i} + \mathbf{j} + 2\mathbf{k}$

21. $\mathbf{i} + \mathbf{k}$

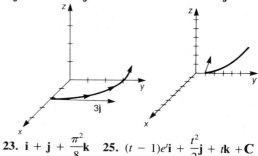

23. $\mathbf{i} + \mathbf{j} + \frac{\pi^2}{8}\mathbf{k}$ **25.** $(t - 1)e^t\mathbf{i} + \frac{t^2}{2}\mathbf{j} + t\mathbf{k} + \mathbf{C}$

27. $\frac{t^2}{2}\mathbf{i} + \ln t\mathbf{j} + \mathbf{C}$ **29.** $\sqrt{2t} + C$ **31.** $(-\infty, \infty)$ **33.** $t \neq k\pi + \pi/2$, k an integer

35. (a) $e^t(t + 1) - \sin t + \cos t$ (b) $2t\mathbf{i} - 2t\sin(t^2 - 1)\mathbf{j}$
   (c) $\cos(2t)\mathbf{i} + [e^t - \sin t - t\cos t]\mathbf{j} + (1 + e^t(\sin t - \cos t))\mathbf{k}$
   (d) $(1 + e^t)\mathbf{i} - \sin t\mathbf{j} + \cos t\mathbf{k}$

37. $\mathbf{F}_1\left(\frac{1}{2}\right) = \frac{3}{2}\mathbf{i} + 4\mathbf{j} + \frac{5}{2}\mathbf{k}$, $\mathbf{F}_2\left(\frac{1}{2}\right) = \frac{5}{4}\mathbf{i} + \frac{7}{2}\mathbf{j} + \frac{7}{4}\mathbf{k}$,

   $\mathbf{F}_3\left(\frac{1 + e}{2}\right) \approx .62\mathbf{i} + 4.24\mathbf{j} + 2.86\mathbf{k}$

41. One example is $\mathbf{F}(t) = \begin{cases} t\mathbf{i}, & \text{if } t \geq 0 \\ -t\mathbf{i}, & \text{if } t < 0. \end{cases}$ $\mathbf{F}$ is discontinuous at $a = 0$, but $\|F(t)\| = 1$ is
   always continuous.

EXERCISE SET 13.3, PAGE 733

1. $(-\infty, \infty)$ **3.** $(0, \infty)$ **5.** $(-\infty, \infty)$

7. $\mathbf{F}(s) = \left(1 + \frac{\sqrt{33}}{33}s\right)\mathbf{i} + \left(1 + \frac{4\sqrt{33}}{33}s\right)\mathbf{j} + \frac{4\sqrt{33}}{33}s\mathbf{k}$ **9.** $\mathbf{F}(s) = \cos s\mathbf{i} + \sin s\mathbf{j}$

11. $\mathbf{F}(s) = \sin\frac{\sqrt{2}}{2}s\mathbf{i} + \cos\frac{\sqrt{2}}{2}s\mathbf{j} + \frac{\sqrt{2}}{2}s\mathbf{k}$, $0 \leq s \leq \sqrt{2}\pi$

13. $\frac{\sqrt{14}}{14}(2\mathbf{i} + \mathbf{j} - 3\mathbf{k})$ **15.** $-\mathbf{i}$ **17.** $\frac{\sqrt{6}}{6}(2\mathbf{i} + \mathbf{j} + \mathbf{k})$

19. $\mathbf{T}'(t_0) = 0$, no unit normal vector exists.   **21.** $-\mathbf{j}$ **23.** $\frac{\sqrt{14}}{7}\left[\mathbf{i} - \frac{3}{2}\mathbf{j} - \frac{1}{2}\mathbf{k}\right]$

25. (b) $\mathbf{T}(2) = \frac{\sqrt{17}}{17}(\mathbf{j} + 4\mathbf{k})$; $\mathbf{N}(2) = \frac{\sqrt{17}}{17}(-4\mathbf{j} + \mathbf{k})$ (c) $t = 0, \mathbf{k}$

27. (a) $\sqrt{15625\pi^2 + 100} \approx 393$ inches.
   (b) $\sqrt{15129\pi^2 + 100} \approx 380$ inches.

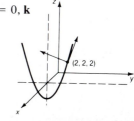

EXERCISE SET 13.4, PAGE 741

1. $\mathbf{v}(t) = \cos t\mathbf{i} - \sin t\mathbf{j} + \mathbf{k}$, $\mathbf{a}(t) = -\sin t\mathbf{i} - \cos t\mathbf{j}$, $v(t) = \sqrt{2}$

3. $\mathbf{v}(t) = \mathbf{i} - 2\mathbf{j} + 3\mathbf{k}$, $\mathbf{a}(t) = \mathbf{0}$, $v(t) = \sqrt{14}$

5. $\mathbf{v}(t) = 4\mathbf{j} + (5 - 10t)\mathbf{k}$, $\mathbf{a}(t) = -10\mathbf{k}$, $v(t) = \sqrt{100t^2 - 100t + 41}$

7. $\mathbf{v}(t) = \dfrac{3}{2}(3t - 1)^{-1/2}\mathbf{j} + e^t\mathbf{k}$, $\mathbf{a}(t) = -\dfrac{9}{4}(3t - 1)^{-3/2}\mathbf{j} + e^t\mathbf{k}$,

$v(t) = \sqrt{\dfrac{9}{4(3t - 1)} + e^{2t}}$

9. $a_\mathbf{T}(t) = 0$, $a_\mathbf{N}(t) = 1$　　**11.** $a_\mathbf{T}(t) = 0$, $a_\mathbf{N}(t) = 0$

13. $a_\mathbf{T}(t) = \dfrac{64(16t - 5)}{\sqrt{1 + (32t - 10)^2}}$, $a_\mathbf{N}(t) = \dfrac{32}{\sqrt{1 + (32t - 10)^2}}$

15. (a)　　(b) 50　　(c) $\approx 149 \dfrac{\text{ft}}{\text{sec}}$

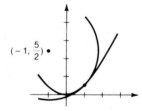

17. (a) $\mathbf{r}(t) = 600t\mathbf{i} + (600\sqrt{3}t - 16t^2)\mathbf{j}$　(b) $\mathbf{v}(t) = 600\mathbf{i} + (600\sqrt{3} - 32t)\mathbf{j}$

(c) 16.875 feet　(d) $\dfrac{75}{2}\sqrt{3}$ seconds　(e) $1200 \dfrac{\text{ft}}{\text{sec}}$

19. (a) 14.41 meters/second　(b) 5.30 meters　(c) 2.08 seconds　**21.** 30.39°

23. The center of the block will land **approximately 7 feet** in front of the slide.　**25.** $v_0^2/32$

EXERCISE SET 13.5, PAGE 749

1. $\dfrac{1}{5}$　**3.** 0　**5.** $\dfrac{1}{2}$　**7.** $\dfrac{1}{4}\sqrt{10}$　**9.** $K(0) = \dfrac{3}{4}$, $\rho(0) = \dfrac{4}{3}$　**11.** $K(1) = \dfrac{\sqrt{2}}{4}$, $\rho(1) = 2\sqrt{2}$

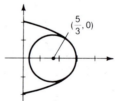

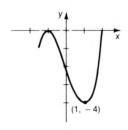

13. $K\left(\dfrac{\pi}{2}\right) = \dfrac{1}{5}$, $\rho\left(\dfrac{\pi}{2}\right) = 5$　**15.** $K(1) = \dfrac{\sqrt{2}}{4}$,　　**17.** $K(1) = 6$, $\rho(1) = \dfrac{1}{6}$

$\rho(1) = 2\sqrt{2}$

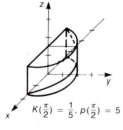

$K\left(\dfrac{\pi}{2}\right) = \dfrac{1}{5}$, $\rho\left(\dfrac{\pi}{2}\right) = 5$

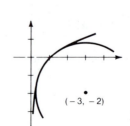

$(-3, -2)$

19. (0, 0)　**21.** None exist.　**23.** (1, 0)　**25.** $(n\pi, n\pi)$ for any integer n.

37. Minimum at $(0, \pm 2)$, maximum at $(\pm 3, 0)$.

REVIEW EXERCISES, PAGE 755

1. (a)

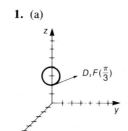

3. (a)

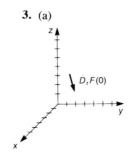

(b) $D_t\mathbf{F}\left(\dfrac{\pi}{3}\right) = -\dfrac{\sqrt{3}}{2}\mathbf{i} + \dfrac{1}{2}\mathbf{j}$

(c) $\mathbf{T}\left(\dfrac{\pi}{3}\right) = -\dfrac{\sqrt{3}}{2}\mathbf{i} + \dfrac{1}{2}\mathbf{j}, \mathbf{N}\left(\dfrac{\pi}{3}\right) = -\dfrac{1}{2}\mathbf{i} - \dfrac{\sqrt{3}}{2}\mathbf{j}$

(d) 1

(b) $D_t\mathbf{F}(0) = \mathbf{i} + \mathbf{j}$

(c) $\mathbf{T}(0) = \dfrac{\sqrt{2}}{2}\mathbf{i} + \dfrac{\sqrt{2}}{2}\mathbf{j}, \mathbf{N}(0) = \mathbf{0}$

(d) 0

5. (a)

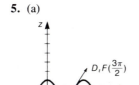

(b) $D_t\mathbf{F}\left(\dfrac{3\pi}{2}\right) = \mathbf{j} + \mathbf{k}$

(c) $\mathbf{T}\left(\dfrac{3\pi}{2}\right) = \dfrac{\sqrt{2}}{2}\mathbf{j} + \dfrac{\sqrt{2}}{2}\mathbf{k}, \mathbf{N}\left(\dfrac{3\pi}{2}\right) = \mathbf{0}$

$= 0$ does not exist

(d) 0

7. $[-1, 1]$ **9.** $t \neq 1$ **11.** $2\mathbf{j} + (e^{\pi} - 1)\mathbf{k}$ **13.** $\mathbf{i} + (2 \ln 2 - 1)\mathbf{k}$

15. (a) $\mathbf{i} + \dfrac{1}{t}\mathbf{j} + \mathbf{k}$ (b) $2t\mathbf{i} + \dfrac{2}{t}\mathbf{j} + 2t\mathbf{k}$ (c) $3t^2\mathbf{i} + (2 \ln t + 1)t\mathbf{j} + 3t^2\mathbf{k}$

17. $\mathbf{F}(s) = \left(3 - \dfrac{\sqrt{3}}{15}s\right)\mathbf{i} + \dfrac{7\sqrt{3}}{15}s\mathbf{j} + \dfrac{\sqrt{3}}{3}s\mathbf{k}, 0 \leq s$ $5\sqrt{3}$ **19.** 6   **21.** $\dfrac{1}{2}$

23. (a)

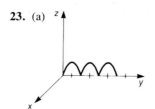

(b) $\mathbf{v}(t) = \mathbf{j} + \dfrac{2 \sin t \cos t}{|\sin t|}\mathbf{k},$

$\mathbf{a}(t) = \dfrac{-2(\sin t)^2}{|\sin t|}\mathbf{k},$

$v(t) = \sqrt{1 + 4(\cos t)^2}$

(c) 2

25. $\mathbf{r}(t) = 250t\mathbf{i} + (250\sqrt{3}t - 16t^2)\mathbf{j}$

(a) $\dfrac{250\sqrt{3}}{16} \approx 27.06$ seconds   (b) 2929.6875 feet

(c) ≈ 6765.82 feet   (d) 500 ft/sec

CHAPTER 14
EXERCISE SET 14.1, PAGE 760

1. (a) 1   (b) 0   (c) 5   **3.** $x + y \leq 1$ **5.** $y \neq 0$ **7.** $x > 0$ and $y > x$ or $x < 0$ and $y < x$.

9. $x^2 + y^2 + z^2 \leq 4$ **11.** $-1, -1$ **13.** $-2x - h, -2y - k$

15. $2(2xy + 4xz + 6yz)$ **17.** $V(h, r) = \pi r^2 h$ **19.** $1000e$

21. $C(x, y, z) = (9xy + 16xy + 16yz)$ dollars   **23.** $C(x, y) = \dfrac{8}{75}xy$ dollars

EXERCISE SET 14.2, PAGE 768

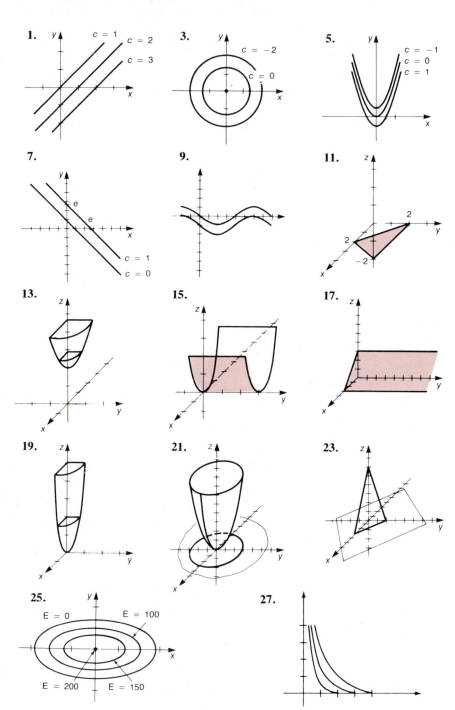

1.

3.

5.

7.

9.

11.

13.

15.

17.

19.

21.

23.

25.

27.

EXERCISE SET 14.3, PAGE 774

1. Parallel planes with normal vector $\langle 2, 3, 1 \rangle$

3. Parallel planes with normal vector $\langle -1, -1, 1 \rangle$

5. Right circular cylinders about the z-axis

7. Right circular cylinders about the x-axis

9. Each level surface is centered at the origin and will intersect each of the coordinate planes in an ellipse.

11. Each level surface is centered at the origin. It will intersect the xz-plane in a circle and the other coordinate planes in ellipses.

13.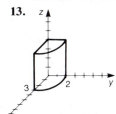

15.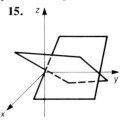

19. $x^2 + y^2 + x^2 = 1$

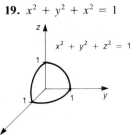

21. $y = x^2 + z^2$

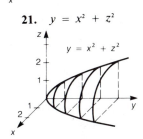

23. $x^2 - y^2 - z^2 = 4, x \geq 2$

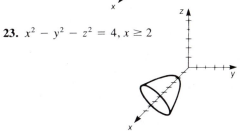

EXERCISE SET 14.4, PAGE 782

1. Ellipsoid

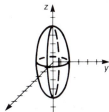

3. Hyperboloid of one sheet

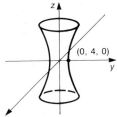
(0, 4, 0)

5. Hyperboloid of two sheets

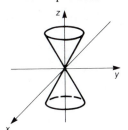
(6, 0, 0)

7. Paraboloid

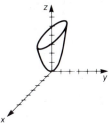

9. Elliptic cylinder

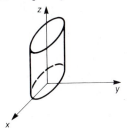

11. Elliptic cone

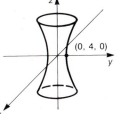

13. Intersecting planes

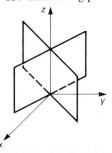

15. Hyperboloid
of one sheet

17. Circular
cone

19. Parallel planes

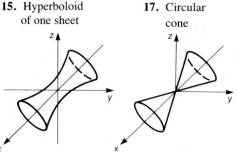 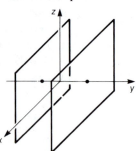

21. A sphere with radius $\sqrt{5}$ centered at $(1, 0, 2)$

23. A paraboloid with vertex at $(3, 4, -5)$ and axis parallel to the *z-axis*

25. Parallel planes $x = 0$ and $x = 2$.

27.

29.

A paraboloid with equation $z = x^2 + y^2$

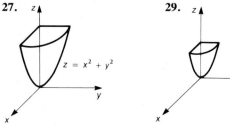 

(a) The original surface is rotated so that it opens downward rather than upward

(b) The original surface is raised two units on the z-axis

(c) The original surface is lowered four units on the z-axis

(d) The original surface is moved one unit to the right on the y-axis

31. (b) and (c)

EXERCISE SET 14.5, PAGE 788

1. (a) $\left(\sqrt{2}, \frac{\pi}{4}, 3 \right)$ (b) $\left(\sqrt{11}, \frac{\pi}{4}, \arccos \frac{3\sqrt{11}}{11} \right)$ **3.** (a) $\left(2, \frac{\pi}{4}, 0 \right)$ (b) $\left(2, \frac{\pi}{4}, \frac{\pi}{2} \right)$

5. (a) $\left(1, \frac{\pi}{2}, 3 \right)$ (b) $\left(\sqrt{10}, \frac{\pi}{2}, \arccos \frac{3\sqrt{10}}{10} \right)$

7. (a) $\left(\frac{3}{2}, \frac{3\sqrt{3}}{2}, 5 \right)$ (b) $\left(\sqrt{34}, \frac{\pi}{3}, \arccos \frac{5\sqrt{34}}{34} \right)$

9. (a) $(2\sqrt{2}, 2\sqrt{2}, -2)$ (b) $\left(\sqrt{20}, \frac{\pi}{4}, \arccos \frac{-2\sqrt{5}}{5} \right)$

11. (a) $(0, 0, 1)$ (b) $(1, \theta, 0)$ for any angle θ

13. (a) $\left(\frac{\sqrt{2}}{4}, \frac{\sqrt{2}}{4}, \frac{\sqrt{3}}{2} \right)$ (b) $\left(\frac{1}{2}, \frac{\pi}{4}, \frac{\sqrt{3}}{2} \right)$

15. (a) $\left(\frac{-\sqrt{2}}{2}, \frac{\sqrt{2}}{2}, -\sqrt{3} \right)$ (b) $\left(1, \frac{5\pi}{6}, -\sqrt{3} \right)$

17. (a) $(0, 0, -5)$ (b) $(0, \theta, -5)$ for any angle θ **19.** $\theta = \frac{\pi}{4}$ **21.** $r = 2$ **23.** $z = r^2$

25. $r^2 = 4(1 - z^2)$ **27.**   **29.** The yz-plane   **31.**

 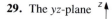 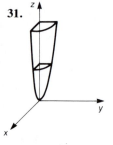

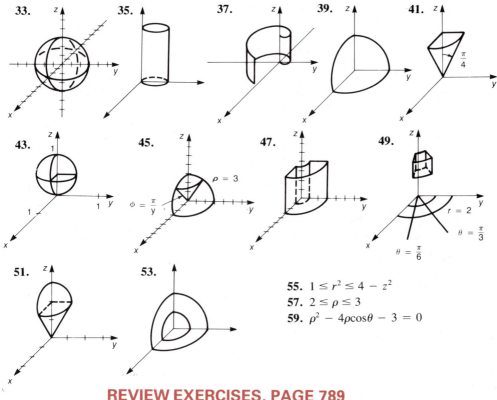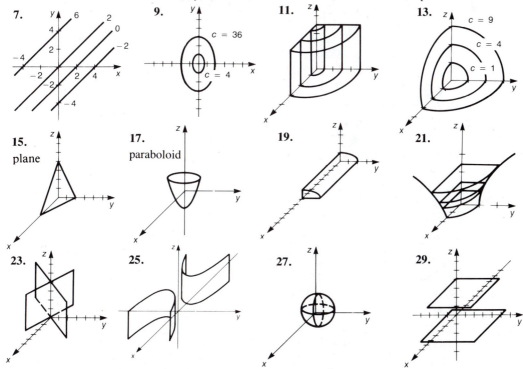

33.

35.

37.

39.

41. $\dfrac{\pi}{4}$

43.

45. $\rho = 3$ $\phi = \dfrac{\pi}{y}$

47.

49. $r = 2$ $\theta = \dfrac{\pi}{3}$ $\theta = \dfrac{\pi}{6}$

51.

53.

55. $1 \le r^2 \le 4 - z^2$

57. $2 \le \rho \le 3$

59. $\rho^2 - 4\rho\cos\theta - 3 = 0$

REVIEW EXERCISES, PAGE 789

1. Values of (x, y) with $y \le 1 - x^2$

3. Values of (x, y) within the unit circle centered at the origin, but excluding points on the y-axis.

5. Values of (x, y, z) in the first octant and not on a coordinate plane.

7.

9. $c = 36$ $c = 4$

11.

13. $c = 9$ $c = 4$ $c = 1$

15. plane

17. paraboloid

19.

21.

23.

25.

27.

29.

31. 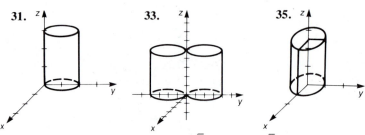 **33.** **35.**

37. (a) $(1, 0, 0)$, $(1, 0, \pi/2)$ (b) $(\sqrt{2}, \pi/4, 0)$, $(\sqrt{2}, \pi/4, \pi/2)$
(c) $(\sqrt{2}, \pi/4, 1)$, $(\sqrt{3}, \pi/4, \pi/4)$

39. (a) $(0, 1, 0)$, $(1, \pi/2, 0)$ (b) $\left(\dfrac{\sqrt{2}}{2}, \dfrac{\sqrt{2}}{2}, 0 \right)$, $(1, \pi/4, 0)$
(c) $(0, \sqrt{2}, \sqrt{2})$, $(\sqrt{2}, \pi/2, \sqrt{2})$

41. Sphere with radius 2 and centered at $(0, 0, 0)$; $x^2 + y^2 + z^2 = 4$.

43. Torus (doughnut) with inside radius 1, outside radius 3, and height 1.

CHAPTER 15
EXERCISE SET 15.1, PAGE 796

1. 1 **3.** 1 **5.** 0 **7.** 0 **9.** e **11.** 2 **13.** 1 **15.** 3 **17.** The entire plane
19. $(x, y) \neq 1$ **21.** $x^2 + y^2 \leq 1$ **23.** The entire plane
33. (a) (b) (c) **41.** 1 **43.** 1

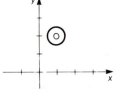 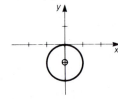 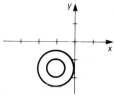

EXERCISE SET 15.2, PAGE 804

1. $f_x = 1, f_y = 1$ **3.** $f_x = 2x + y, f_y = x + 2y$ **5.** $f_x = ye^x, f_y = xe^y$
7. $f_x = \dfrac{y^3 - x^2 y}{(x^2 + y^2)^2}, f_y = \dfrac{x^3 - y^2 x}{(x^2 + y^2)^2}$ **9.** $f_x = 3x^2 + 2xy^2, f_y = 2x^2 y$
11. $f_x = ye^x, f_y = e^x$ **13.** $f_x = x(x^2 + y^2)^{-1/2}, f_y = y(x^2 + y^2)^{-1/2}$
15. $f_x = e^x \cos y, f_y = -e^x \sin y$ **17.** $f_x = 3z, f_y = 4, f_z = 3x$
19. $f_x = -2 \sin (2x + 3y + 4z), f_y = -3 \sin (2x + 3y + 4z),$
    $f_z = -4 \sin (2x + 3y + 4z)$
21. $f_x = 0, f_y = 2$ **23.** $f_x = 0, f_y = -1$ **25.** $f_x = 0, f_y = -1$
27. $f_x = \dfrac{\sqrt{6}}{6}, f_y = \dfrac{\sqrt{6}}{6}, f_z = \dfrac{\sqrt{6}}{3}$
37. $f_{xyx} = 4ye^{x^2+y^2} (1 + 2x^2), f_{xyy} = 4xe^{x^2+y^2} (1 + 2y^2)$
39. $(y^2 + z^2)(x^2 + y^2 + z^2)^{-3/2}$ **41.** $(1 - x^2 y^2 z^2) \sin (xyz) + 3xyz \cos (xyz)$
43. $z = 4x - 3$ **45.** (a) A minimum at $(0, 0)$ (b) A maximum at $(0, 0)$
53. $\dfrac{\partial S}{\partial w} = .00718(.425)(70)^{-.575}(200)^{.725} \approx .012 \dfrac{m^2}{lb}$

EXERCISE SET 15.3, PAGE 811

7. $2x \, dx - 6y \, dy$ **9.** $(y \, dx - x \, dy)/y^2$ **11.** $-y^2 \sin x \, dx + 2y \cos x \, dy$
13. $yz \, dx + xz \, dy + xy \, dz$ **15.** $-(yz \, dx + xz \, dy + xy \, dz) \sin (xyz)$
17. $ye^z dx + xe^z dy + xye^z dz$ **19.** $\varepsilon_1 = 0, \varepsilon_2 = 0$ **21.** $\varepsilon_1 = x - x_0, \varepsilon_2 = y_0 - y$
23. $\varepsilon_1 = y - y_0, \varepsilon_2 = 0$ **25.** $.01(3 - e)$ **27.** 31.2 in^2
29. $\dfrac{\partial R}{\partial R_1} = \dfrac{1}{R_1^2} \left(\dfrac{1}{R_1} + \dfrac{1}{R_2} \right)^{-2} ; \dfrac{\partial R}{\partial R_2} = \dfrac{1}{R_2^2} \left(\dfrac{1}{R_1} + \dfrac{1}{R_2} \right)^{-2} ; \dfrac{83}{1210}$

EXERCISE SET 15.4, PAGE 818

1. $(2xy^3 + y^2)(\cos t + 1) + (3x^2y^2 + 2xy)(2t)$ **3.** $[(1/t) - ye^t](2x - y^2)^{-1/2}$

5. $\left(\dfrac{1}{x} + 2xe^{x^2+y^2}\right)\dfrac{1}{2\sqrt{t+1}} + \left(\dfrac{1}{y} + ye^{x^2+y^2}\right)e^t$ **7.** $\dfrac{2xy}{t} + (x^2 + z^2)e^t + 4yzt$

9. $e^x(2te^{t^2}\cos xy - y\sin t\sin xy - 2xt\sin xy)$

11. $\left(yz^2e^t + \dfrac{xz^2}{t} + 2xyz\left(1 + \dfrac{1}{t}\right)\right)\cos(xyz^2)$

13. $f_u = 3x^2y^5 + 10x^3y^4u, f_v = 3x^2y^5 - 10x^3y^4v$

15. $f_u = e^{x+v}\cos y - e^x\sin y, f_v = ue^{x+v}\cos y - e^x\sin y$

17. $f_u = (e^u\cos v + e^u\sin v)\,[\sec(x+y)]^2, f_v = (e^u\cos v - e^u\sin v)\,[\sec(x+y)]^2$

19. $f_u = \left(2ue^{u^2+v^2} + \dfrac{2u}{u^2+v^2} + 1\right)(x+y+z)^{-1},$

$\quad f_v = \left(2ve^{u^2+v^2} + \dfrac{2v}{u^2+v^2} - 1\right)(x+y+z)^{-1}$

21. $-(2xy^2 - y + 3x^2)/(2x^2y - x + 3y^2)$ **23.** $\dfrac{12x - 2xy - y^2}{x^2 + 2xy - 12y}$

25. $\dfrac{\sin x}{x\cos xy} - \dfrac{y}{x}$ **31.** 1/3 ft/min.

33.

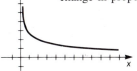

$f_x = \dfrac{16}{3}$ is the rate at which the production changes relative to the change in property resources at $(1, 8)$

$f_y = \dfrac{4}{3}$ is the rate at which the production changes relative to the change in human resources at $(1, 8)$

EXERCISE SET 15.5, PAGE 825

1. $-5\sqrt{2}/2$ **3.** $2\sqrt{2}$ **5.** $17\sqrt{22}/22$ **7.** $\dfrac{5\sqrt{6}}{6}$ **9.** $-\sqrt{2}/6$ **11.** $2x\mathbf{i} - 2y\mathbf{j}$

13. $ye^x\mathbf{i} + e^x\mathbf{j}$ **15.** $[(y^3 - x^2y)\mathbf{i} + (x^3 - xy^2)\mathbf{j}]/(x^2 + y^2)^2$

17. $\cos(x^2yz)[2xyz\mathbf{i} + x^2z\mathbf{j} + x^2y\mathbf{k}]$ **19.** $2\mathbf{i} + 4\mathbf{j}$ **21.** $-e\mathbf{i}$

23. $-\dfrac{\pi}{2}\mathbf{i} - \dfrac{\pi}{4}\mathbf{j} - 2\mathbf{k}$ **25.** $\mathbf{i} + \mathbf{j}, \sqrt{8}$ **27.** $\mathbf{i}, 36$ **29.** $\mathbf{i} + \mathbf{j}, 2e$

31. $2e\mathbf{i}, 2e$ **33.** $2 + 4\sqrt{3}$ **35.** $\dfrac{3}{2}e - \dfrac{\sqrt{3}}{2}e^2$ **37.** $\pi\sqrt{4 + \pi^2}/(4 + \pi^2)$

39. $-\dfrac{4\sqrt{5}}{5}$ **41.** 0   **43.** (a) $4\mathbf{i} + \mathbf{j}$ (b) $-4\mathbf{i} - \mathbf{j}$ **45.** (a) -19 (b) -9

EXERCISE SET 15.6, PAGE 833

1. $\mathbf{i} + \mathbf{j} + 2\mathbf{k}$ **3.** $2\mathbf{i} + 2\mathbf{j} - \mathbf{k}$ **5.** $\dfrac{1}{e}\mathbf{i} + \dfrac{1}{e^2}\mathbf{j} - \mathbf{k}$ **7.** $\dfrac{1}{2}\mathbf{i} + \dfrac{1}{2}\mathbf{j} - \mathbf{k}$

9. $4x - 2y - z = 3$ **11.** $3x - 4y + 5z = 0$ **13.** $2x - z = 2$ **15.** $x + 4y - z = \pi$

17. $x = 1 + 2t, y = 2 - t, z = 2 - t$ **19.** $x = 2, y = 1 - 2\sqrt{3}t, z = \sqrt{3} - 2t$

21. Only at $\left(-\dfrac{1}{8}, -1, \dfrac{17}{16}\right)$

EXERCISE SET 15.7, PAGE 841

5. $(0, 0)$ **7.** $(0, 0)$ **9.** $(0, n\pi + \pi/2)$ for any integer n **11.** $(0, 0)$

13. Saddle point   **15.** Relative minimum   **17.** All saddle points   **19.** Saddle point

21. $(0, 0)$ saddle point; $(1, 1)$ relative minimum **23.** $(0, 0)$ and $(5, 0)$ saddle points

25. $(\pm n\pi, \pm m\pi)$, where n and m are nonnegative integers. Relative maximum when both m and n are even. Relative minimum when both are odd.

27. $(0, 0)$ relative minimum

29. None   **31.** Absolute minimum at $(0, 0)$; absolute maximum whenever $x^2 + y^2 = 1$

33. Absolute mimima at $(\sqrt{2}, 1)$ and $(-\sqrt{2}, -1)$. Absolute maxima at $(\sqrt{2}, -1)$ and $(-\sqrt{2}, 1)$.

35. Absolute maximum at $(5, 10)$. Absolute minimum whenever $y = 0$.

39. $f(-1, -3/2, 13/4) = -13/4$ **41.** $3\sqrt{2}$ **43.** All sides $2\sqrt[3]{10}$ inches

45. Height = width = 14 inches, length = 28 inches   **47.** $y = x + \dfrac{2}{3}$

49. $y = 6.55x - 12.5$ **51.** $y = -.0023x + 13.1$

EXERCISE SET 15.8, PAGE 850

1. $(1, 1)$ **3.** $\left(\dfrac{2}{3}, \dfrac{1}{3}\right)$ **5.** $\left(\dfrac{3}{13}, \dfrac{18}{13}, \dfrac{24}{13}\right)$ **7.** $(6, 4, 2)$

9. $\left(\dfrac{2}{9}\sqrt{306}, \dfrac{8}{153}\sqrt{306}, \dfrac{2}{153}\sqrt{306}\right)$ **11.** $(0, 0, 1)$ and $(0, 0, -1)$

13. $\left(2 - \dfrac{2\sqrt{29}}{29}, 3 - \dfrac{3\sqrt{29}}{29}, 4 - \dfrac{4\sqrt{29}}{29}\right)$ **15.** All sides $2\sqrt[3]{10}$ inches

17. The sides have length $\sqrt{3}, \dfrac{4}{3}\sqrt{3}$, and $\dfrac{5}{3}\sqrt{3}$.

19. The base will have dimensions $\dfrac{20\sqrt{3}}{3}$ feet by $\dfrac{50\sqrt{3}}{3}$ feet. The height is $\dfrac{10\sqrt{3}}{3}$ feet.

21. $h = 2r \approx 25.3$ inches

23. Length and width of the floor should be the same, approximately 132.6 feet. The height of the room should be $\dfrac{3}{7}$ of the length.

REVIEW EXERCISES, PAGE 851

1. $x^2 + y^2 \le 4$

3. Discontinuous if $x^2 + y^2 = 1$.

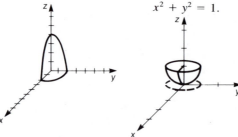

5. $f_x(x, y) = ye^{xy}, f_y(x, y) = xe^{xy}$

7. $f_x(x, y, z) = 2x - 2y\cos z$,
$f_y(x, y, z) = 2y - 2x\cos z$,
$f_z(x, y, z) = 2xy\sin z$

9. $f_x(x, y) = 2xy + 2e^{1/y}$,
$f_y(x, y) = x^2 - \dfrac{2x}{y^2}e^{1/y}$

11. $f_x(x, y) = 3(x^2 + xy)^2(2x + y)$,
$f_y(x, y) = 3x(x^2 + xy)^2$

13. $(0, 0)$, relative maximum

15. $(0, y)$, for any y gives a relative minimum

17. $f_{yxz}(x, y, z) = e^{yz} + zye^{yz} + e^x, \dfrac{\partial^3 f}{\partial x^2 \partial y}(x, y, z) = ze^x$ **19.** $\dfrac{2\sqrt{3}}{3}$ **21.** $-\sqrt{2}$

23. $\dfrac{26}{5}$ **25.** -8 **27.** (a) $\dfrac{x\mathbf{i} + y\mathbf{j}}{x^2 + y^2}$ (b) $-e^{-x}\mathbf{i} - 2e^{-2y}\mathbf{j} + 3e^{3z}\mathbf{k}$

29. Slope 4, equation $\dfrac{z - 4}{4} = x - 1, y = 2$ **31.** $\mathbf{i} + \mathbf{j}; \sqrt{2}$ **33.** $-\mathbf{i} + 2\mathbf{j}; 2\sqrt{5}$

35. $2x + 4y - z = 6$ **37.** $\dfrac{\pi}{2}x + y - z = \dfrac{\pi}{2}$ **39.** $4x - 8y + x = -8$

41. $2ye^{2r}\cos\Theta + xe^r\sin\Theta; -ye^{2r}\sin\Theta + xe^r\cos\Theta$ **43.** 1   **45.** $\mathbf{i} - \mathbf{j}$

49. $(16/9, 16/9, -28/9)$ **51.** Maximum of 9/8 occurs at $(\pm\sqrt{15}/4, -1/4)$ **53.** 96π

CHAPTER 16
EXERCISE SET 16.1, PAGE 860

1. (a) $\dfrac{33}{4}$ (b) $\dfrac{29}{4}$ (c) $\dfrac{57}{8}$ **3.** (a) -59 (b) -81 (c) -70 **5.** 4 **9.** 18π

EXERCISE SET 16.2, PAGE 870

1. $\dfrac{37}{2}$ **3.** $-2\pi^2$ **5.** $e^3 + \dfrac{e^{-6}}{2} - e^{-1} - \dfrac{e^2}{2}$ **7.** 28 **9.** -2 **11.** $\dfrac{31}{60}$ **13.** $e-1$

15. $\displaystyle\int_3^4 \int_1^2 xy^2\, dx\, dy$

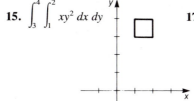

17. $\displaystyle\int_0^{2\pi} \int_{-\pi}^{3\pi/2} (y\sin x + x\cos y)dx\, dy$

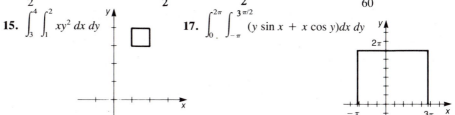

19. $\displaystyle\int_{-2}^1 \int_{-1}^3 xe^{xy}\, dx\, dy$

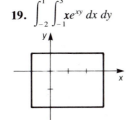

21. $\displaystyle\int_0^2 \int_{y/2}^y (x^2 + y^3)dx\, dy + \int_2^4 \int_{y/2}^2 (x^2 + y^3)dx\, dy$

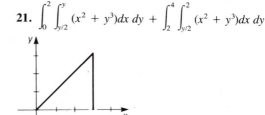

23. $\displaystyle\int_0^\pi \int_y^\pi \cos x\, dx\, dy$

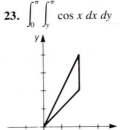

25. $\displaystyle\int_0^1 \int_y^1 (x^2 + \sqrt{y})\, dx\, dy$

27. $\displaystyle\int_1^e \int_y^e \ln xy\, dx\, dy$

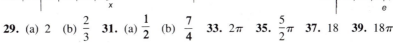
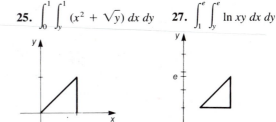

29. (a) 2 (b) $\dfrac{2}{3}$ **31.** (a) $\dfrac{1}{2}$ (b) $\dfrac{7}{4}$ **33.** 2π **35.** $\dfrac{5}{2}\pi$ **37.** 18 **39.** 18π

41. $g'(t) = \displaystyle\int_0^t f(t, y)dy$ $\quad g''(t) = \displaystyle\int_0^t \dfrac{\partial f(t, y)}{\partial t}\, dy + f(t, t)$

EXERCISE SET 16.3, PAGE 878

1. 18π **3.** 48π **5.** $\dfrac{4\pi}{3}(8 - 3\sqrt{3})$ **7.** $\dfrac{4\pi}{3}(8 - 4\sqrt{2})$ **9.** $\dfrac{29}{32}\pi$ **11.** $\dfrac{3\pi}{64} - \dfrac{1}{8}$

13. π **15.** 11π **17.** $2 + \dfrac{\pi}{4}$ **19.** $\dfrac{3\pi}{16}$ **21.** $\dfrac{\pi}{2} - 1$ **23.** 0

25. $\ln\left(\dfrac{\sqrt{3}(\sqrt{2} + 1)}{3}\right)$ **27.** $\dfrac{1}{12}$ **29.** $e^9 - 1$ **33.** $171.5\pi/3$

EXERCISE SET 16.4, PAGE 885

1. $M = \dfrac{2}{3}, M_x = \dfrac{3}{10}, M_y = \dfrac{3}{10}; \left(\dfrac{9}{20}, \dfrac{9}{20}\right)$ **3.** $M = \dfrac{1}{6}, M_x = \dfrac{1}{24}, M_y = \dfrac{1}{12}; \left(\dfrac{1}{2}, \dfrac{1}{4}\right)$

5. $M = \dfrac{4}{15}, M_x = \dfrac{8}{105}, M_y = 0 ; \left(0, \dfrac{2}{7}\right)$ **7.** $M = \dfrac{1}{2}, M_x = \dfrac{3}{8}, M_y = \dfrac{5}{24}; \left(\dfrac{5}{12}, \dfrac{3}{4}\right)$

9. $I_x = \dfrac{6}{35}, I_y = \dfrac{6}{35}, I_0 = \dfrac{12}{35}, \hat{x} = \dfrac{3\sqrt{35}}{35}, \hat{y} = \dfrac{3\sqrt{35}}{35}$

11. $I_x = \dfrac{1}{60}, I_y = \dfrac{1}{20}, I_0 = \dfrac{1}{15}, \hat{x} = \dfrac{\sqrt{30}}{10}, \hat{y} = \dfrac{\sqrt{10}}{10}$

13. $I_x = \dfrac{32}{945}, I_y = \dfrac{4}{35}, I_0 = \dfrac{28}{189}, \quad \hat{x} = \sqrt{21}/7, \hat{y} = 2\sqrt{14}/21$

15. $I_x = \dfrac{3}{10}, I_y = \dfrac{7}{60}, I_0 = \dfrac{5}{12}, \hat{x} = \dfrac{\sqrt{210}}{30}, \hat{y} = \dfrac{\sqrt{15}}{5}$

19. $M = \dfrac{128\pi}{3}, M_x = 0, M_y = 0, I_x = \dfrac{1024}{5}\pi, I_y = \dfrac{1024}{5}\pi, I_0 = \dfrac{2048}{5}\pi$

21. $M = 4/9, M_x = 4/15, M_y = 0, I_x = \dfrac{32}{175}, I_y = \dfrac{16}{525}, I_0 = \dfrac{16}{75}$

23. $\bar{r} = \dfrac{21}{20}, \theta = -\dfrac{\pi}{2}$ **25.** $(0, 0)$ **27.** (a) $(1, 1)$, (b) $I_{x=1} = \dfrac{9}{4}, I_{y=1} = \dfrac{19}{4}$ (c) 0

EXERCISE SET 16.5, PAGE 898

1. $\dfrac{3}{42}$ **3.** $\dfrac{1}{12}$ **5.** $2 + \dfrac{\pi^2}{2}$ **7.** 0

9. $\displaystyle\int_0^1 \int_0^{-3x+3} \int_0^{2-2x-2/3z} (2x + y)\,dy\,dz\,dx = \int_0^2 \int_0^{-(1/2)y+1} \int_0^{3-3x-(3/2)y} (2x + y)\,dz\,dx\,dy$

$\displaystyle\text{\textbf{\textit{1}}} = \int_0^2 \int_0^{-(3/2)+3} \int_0^{1-(3/2)z-(1/2)y} (2x + y)\,dx\,dz\,dy = \int_0^3 \int_0^{-(1/3)z+1} \int_0^{2-2x-(2/3)z} (2x + y)\,dy\,dx\,dz$

$\displaystyle = \int_0^3 \int_0^{-(2/3)z+2} \int_0^{1-(1/3)z-1y/2} (2x + y)\,dx\,dy\,dz = 1$

11. $\displaystyle\int_0^2 \int_0^{3\sqrt{4-z^2}} \int_0^{(1/2)\sqrt{36-x^2-9z^2}} z\,dy\,dx\,dz = \int_0^3 \int_0^{(2/3)\sqrt{9-y^2}} \int_0^{\sqrt{36-4y^2-9z^2}} z\,dx\,dz\,dy$

$\displaystyle = \int_0^3 \int_0^{2\sqrt{9-y^2}} \int_0^{(1/3)\sqrt{36-x^2-4y^2}} z\,dz\,dx\,dy = \int_0^6 \int_0^{3\sqrt{1-x^2}} \int_0^{(1/3)\sqrt{36-x^2-4y^2}} z\,dz\,dy\,dx$

$\displaystyle = \int_0^6 \int_0^{(1/3)\sqrt{36-x^2}} \int_0^{(1/2)\sqrt{36-x^2-9z^2}} z\,dy\,dz\,dx = \dfrac{9\pi}{2}$ **13.** $\dfrac{8}{3}$ **15.** $\dfrac{15\pi}{2}$ **17.** $\dfrac{\pi}{8}$

19. 48π **21.** $\dfrac{2\pi}{3}(6\sqrt{6} - 1)$ **23.**

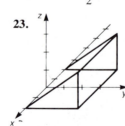

25.

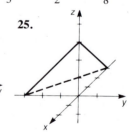

EXERCISE SET 16.6, PAGE 905

1. 0 **3.** $\dfrac{\pi}{15}$ **5.** 0 **7.** $\dfrac{4\pi}{3}$ **9.** $\dfrac{16}{5}\pi$ **11.** $-\dfrac{\pi^2}{96}$ **13.** $\dfrac{16}{3}(2 - \sqrt{2})\pi$

15. $\dfrac{15}{2}\pi$ **17.** 162π **19.** $\dfrac{19}{4}\pi(2 - \sqrt{2})$

21.

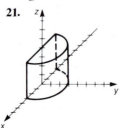

23.

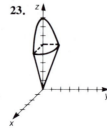

25. $\displaystyle\int_0^{2\pi}\int_0^{\pi/2}\int_0^a \rho^2\sin\phi\,d\rho\,d\phi\,d\theta = \int_0^{2\pi}\int_0^a\int_{\sqrt{a^2-r^2}}^{\sqrt{a^2-r^2}} r\,dz\,dr\,d\theta$

$\displaystyle = \int_{-a}^a\int_{-\sqrt{a^2-x^2}}^{\sqrt{a^2-x^2}}\int_{-\sqrt{a^2-x^2-y^2}}^{\sqrt{a^2-x^2-y^2}} dz\,dy\,dx$

27. $\displaystyle\int_0^a\int_0^a\int_0^a dz\,dy\,dx = \int_0^{\pi/4}\int_0^{a\sec\theta}\int_0^a r\,dz\,dr\,d\theta + \int_{\pi/4}^{\pi/2}\int_0^{a\csc\theta}\int_0^a r\,dz\,dr\,d\theta$

$\displaystyle = \int_0^{\pi/4}\int_0^{\arctan(\sec\theta)}\int_0^{a\sqrt{1+(\sec\theta)^2}} \rho^2\sin\phi\,d\rho\,d\phi\,d\theta$

$\displaystyle + \int_0^{\pi/4}\int_{\arctan(\sec\theta)}^{\pi/2}\int_0^{a\sec\theta\csc\phi} \rho^2\sin\phi\,d\rho\,d\phi\,d\theta$

$\displaystyle + \int_{\pi/4}^{\pi/2}\int_0^{\arctan(\csc\theta)}\int_0^{a\sqrt{1+(\csc\theta)^2}} \rho^2\sin\phi\,d\rho\,d\phi\,d\theta$

$\displaystyle + \int_{\pi/4}^{\pi/2}\int_{\arctan(\csc\theta)}^{\pi/2}\int_0^{a\csc\theta\csc\phi} \rho^2\sin\phi\,d\rho\,d\phi\,d\theta$

29. $4\pi\left(2\sqrt6 - \dfrac{5\sqrt5}{3}\right)$

EXERCISE SET 16.7, PAGE 912

1. 33 **3.** $48k\pi$ **5.** $\dfrac{162}{5}k\pi$

7. $M_{xy} = 63,\ M_{yz} = \dfrac{33}{2},\ M_{xz} = 33\ (\bar x,\bar y,\bar z) = \left(\dfrac12, 1, \dfrac{21}{11}\right)$

9. $M_{xy} = \dfrac{496}{3}k\pi,\ M_{yz} = 0,\ M_{xz} = 0\ (\bar x,\bar y,\bar z) = \left(0,0,\dfrac{31}{9}\right)$

11. $M_{xy} = \dfrac{243}{4}k\pi,\ M_{yz} = 0,\ M_{xz} = 0\ (\bar x,\bar y,\bar z) = \left(0,0,\dfrac{15}{8}\right)$

13. $I_x = \dfrac{367}{2},\ I_y = \dfrac{1011}{6},\ I_z = \dfrac{110}{2},\ \hat x = \sqrt{\dfrac{367}{66}},\ \hat y = \sqrt{\dfrac{1011}{198}},\ \hat z = \sqrt{\dfrac{110}{66}}$

15. $I_x = 624k\pi,\ I_y = 624k\pi,\ I_z = 96k\pi,\ \hat x = \sqrt{13},\ \hat y = \sqrt{13},\ \hat z = \sqrt2$

17. $I_x = \dfrac{4374}{35}k\pi,\ I_y = \dfrac{4374}{35}k\pi,\ I_z = \dfrac{2916}{35}k\pi,\ \hat x = \dfrac37\sqrt{21},\ \hat y = \dfrac37\sqrt{21},\ \hat z = \dfrac37\sqrt{14}$ **19.** 2

21. In spherical coordinates

$\displaystyle I_x = \int_0^{2\pi}\int_0^{\pi}\int_0^1 \rho^6\sin\phi[1 - (\sin\phi)^2(\cos\theta)^2]d\rho\,d\phi\,d\theta$

$\displaystyle I_y = \int_0^{2\pi}\int_0^{\pi}\int_0^1 \rho^6\sin\phi\,[(\sin\phi)^2\,(\cos\theta)^2 + (\cos\phi)^2]\,d\rho\,d\phi\,d\theta$

Because of symmetry, $I_x = I_y = I_z = \dfrac{8}{21}\pi$

23. The conical tent is cooler, averaging $57.5°$

25. $M = \dfrac{4215.225}{1728}\pi \approx 7.66$ lb, center of mass is in the center of the can approximately 2.25 inches from the bottom.

27. The center of mass is in the center of the glass approximately 1.40 inches from the bottom. The center of mass rises after the liquid is **stirred,** provided that the liquid interaction does not change the volume.

REVIEW EXERCISES, PAGE 914

1. 5 **3.** 4

5. $\dfrac{8}{3}$ **7.** $\dfrac{256}{15}$ **9.** $\dfrac{e-1}{6e}$

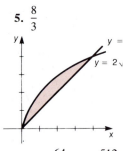   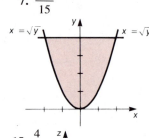   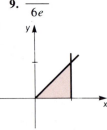

11. $A = \dfrac{64}{3}, V = \dfrac{512}{3}$ **15.** $\dfrac{4}{3}$ **17.** $\dfrac{10\pi}{3}$

13. $A = \dfrac{3}{2}\pi, V = 6\pi$

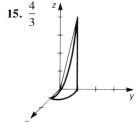

19. 8π **21.** 24π **23.** $\left(\dfrac{8\sqrt{2}-7}{6}\right)\pi$ **25.** $27\left(\dfrac{\pi}{4}-\dfrac{1}{3}\right)$ **27.** 3

29. $M = e^2 - 1, M_x = \dfrac{1}{4}(e^4 - 1), M_y = e^2 + 1, (\bar{x}, \bar{y}) = \left(\dfrac{e^2 + 1}{e^2 - 1}, \dfrac{e^2 + 1}{4}\right)$

31. $M = \dfrac{1}{4}(e^2 + 3e^{-2}), M_x = \dfrac{4}{27}(e^3 + 2e^{-3}), M_y = 0, (\bar{x}, \bar{y}) = \left(0, \dfrac{16(e^3 + 2e^{-3})}{27(e^2 + 3e^{-2})}\right)$

33. $M = \dfrac{81\pi}{8}, M_x = \dfrac{243}{5}, M_y = \dfrac{243}{5}, (\bar{x}, \bar{y}) = \left(\dfrac{24}{5\pi}, \dfrac{24}{5\pi}\right)$

35. πa^4 **37.** $(\hat{x}, \hat{y}, \hat{z}) = \left(\dfrac{7}{9}a^2, \dfrac{7}{9}a^2, \dfrac{7}{9}a^2\right)$

CHAPTER 17
EXERCISE SET 17.1, PAGE 921

1. div $\mathbf{F} = 6xyz$, curl $\mathbf{F} = x(z^2 - y^2)\mathbf{i} + y(x^2 - z^2)\mathbf{j} + z(y^2 - x^2)\mathbf{k}$

3. div $\mathbf{F} = y \sin z - x^2 \sin y + \sqrt{xy}$,

curl $\mathbf{F} = \dfrac{z}{2}\sqrt{\dfrac{x}{y}}\,\mathbf{i} + \left(xy \cos z - \dfrac{z}{2}\sqrt{\dfrac{y}{x}}\right)\mathbf{j} + (2x \cos y - x \sin z)\mathbf{k}$

5. div $\mathbf{F} = \dfrac{e^z}{x}\mathbf{i} + \dfrac{e^x}{y}\mathbf{j} + \dfrac{e^y}{z}\mathbf{k}$, curl $\mathbf{F} = \left[e^y\ \ln\dfrac{z}{x} + \dfrac{e^x}{z}\right]\mathbf{i} + \left[e^z \ln\dfrac{x}{y} + \dfrac{e^y}{x}\right]\mathbf{j} + \left[e^x \ln\dfrac{y}{z} + \right.$

7. No   **9.** No   **11.** Yes   **13.** $7x^2y + 4y^2z + x^2z + z$ **15.** Not possible
17. Not possible   **19.** $(2x - z)\mathbf{i} + z\mathbf{j} + y\mathbf{k}$

EXERCISE SET 17.2, PAGE 930

1. Closed   **3.** Closed   **5.** $1/2$ **7.** -16 **9.** 1 **11.** 0 **13.** 0 **15.** $\dfrac{14}{3}\sqrt{14}$

17. $\ln\sqrt{2}$ **19.** $\dfrac{\pi^2}{2} + \dfrac{2}{3}$ **21.** (a) 0 (b) 0 (c) 0 (d) 0

23. (a) $2/3$ (b) $1/2$ (c) 12 (d) 0 **27.** (b) $16 \sin 2$ (c) $16 \sin 2$ (d) 0

EXERCISE SET 17.3, PAGE 938

1. 9 **3.** 27 **5.** $-\dfrac{41}{30}$ **7.** $9(\pi - 2)$ **9.** 6

11. (a) $107/6$ (b) $275/12$ (c) $26/3$ (d) 35

13. $M = \dfrac{28}{3}\sqrt{65}$, $(\bar{x}, \bar{y}, \bar{z}) = (1, 3, 21/4)$ **15.** $M = 16\pi$, $(\bar{x}, \bar{y}, \bar{z}) = (0, 4/\pi, 0)$

17. $M = \dfrac{16\pi}{3}\sqrt{5}$, $(\bar{x}, \bar{y}, \bar{z}) = (2, 0, 0)$

EXERCISE SET 17.4, PAGE 946

1. (a) 1 (b) 1   **3.** Yes   **5.** Yes   **7.** Yes   **9.** Yes   **11.** No   **13.** Yes
15. Yes   **17.** $e^{-y}\sin x$ **21.** x^2y^3z **23.** $y\cos x + z\sin x + yz - z^2$
25. 0   **27.** 0   **29.** $e^{32} - e + \dfrac{9}{2}(\ln 2)^2$

EXERCISE SET 17.5, PAGE 957

3. $3(\sin 1 + 2\cos 1 - 2)$ **5.** 0   **7.** 0   **9.** 0   **11.** $\dfrac{3\pi}{2}$ **13.** 0   **15.** 15π

17. $\dfrac{3\pi}{8}$ **21.** M_y and N_x are discontinuous at $(0, 0)$

EXERCISE SET 17.6, PAGE 967

1. $\dfrac{(17)^{3/2}}{12}$ **3.** $8\sqrt{2}\pi$ **5.** 0   **7.** 162π **9.** $\dfrac{9}{2}\sqrt{3}$ **11.** $\dfrac{\pi}{6}[(17)^{3/2} - 1]$ **13.** 6π

15. $4\pi(3 - \sqrt{6})$ **17.** 64   **19.** $\dfrac{3}{4}$ **21.** $-\dfrac{2}{3}$

23. $-3\pi\dfrac{\text{ft}^3}{\text{sec}}$; $-187.2\pi\dfrac{\text{lb}}{\text{sec}}$, assuming $\rho = 62.4\dfrac{\text{lb}}{\text{ft}^2}$

EXERCISE SET 17.7, PAGE 979

1. 48π **3.** 8π **5.** 0   **7.** $\dfrac{-21}{2}$ **9.** -2π **11.** 0   **13.** 0

15. $\dfrac{3}{5}e^5 - \dfrac{2}{3}\cos 3 - \dfrac{5}{2}\cos 2 - \dfrac{13}{30}$ **23.** f is constant.

EXERCISE SET 17.8, PAGE 987

1. (a) div $\mathbf{V} = 3$ (b) The results are the same

3. Sources occur when $z > \dfrac{1}{2} - x - y$, sinks occur when $z < \dfrac{1}{2} - x - y$

5. (a) $-2\mathbf{k}$ (b) $-2\pi r^2$

REVIEW EXERCISES, PAGE 989

1. div $\mathbf{F} = 0$, curl $\mathbf{F} = \mathbf{0}$
3. div $\mathbf{F} = 3x^2 + xy$, curl $\mathbf{F} = (xz - 1)\mathbf{i} - yz\mathbf{j} + (1 - x^3)\mathbf{k}$
9. (a) is a simple closed curve   (b) is not a closed curve and is not simple
11. 2π **13.** $\dfrac{107}{120}$ **15.** $\dfrac{8305}{12}$ **17.** $\dfrac{5}{4}e^6 + e^3 - \dfrac{1}{2}e^2 - e + 9\ln 3 + \dfrac{40}{3}$

19. 0   **21.** 0   **23.** 0   **25.** $\dfrac{9\pi}{2}\left(\dfrac{\pi}{2} - 1\right) + \tan 2$ **27.** 0   **29.** 0   **31.** $\dfrac{\pi}{4}$

33. $\dfrac{11}{24}$ **35.** 3π **37.** $\dfrac{27}{2}$ **39.** $\dfrac{1 + \sqrt{2}}{3}$ **41.** $5(1 - e^3)$

43. $M = \dfrac{4\sqrt{16 + \pi^2}}{\pi^2}$, $(\bar{x}, \bar{y}, \bar{z}) = \left(1, \dfrac{4(\pi - 2)}{\pi}, \pi - 2\right)$ **45.** $12\pi(3 - \sqrt{8})$

49. (a) $\dfrac{\sqrt{3}v_0^4}{24576}$ (b) $\dfrac{\sqrt{3}v_0^4}{8192}$ **51.** 0

CHAPTER 18
EXERCISE SET 18.I, PAGE 997

17. $y' = \dfrac{y}{x}$ **19.** $y' = \dfrac{-y}{x}$ **21.** $y' = -2y$ **23.** $y' = \dfrac{1}{2y}$

25. $x^2 + y^2 = C$ **27.** $x^2 - y^2 = C$ **29.** $y = \sqrt{x + C}$ **31.** $y = Ce^{-2x},\, C \geq 0$

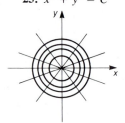
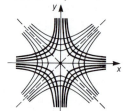
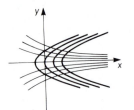
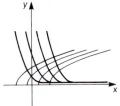

EXERCISE SET 18.2, PAGE 1000

1. $Cy = e^{x/y}$ **3.** $C = y^2 + 2xy - x^2$ **5.** $y = x \tan(\ln |x| + C)$

7. $\ln\dfrac{\sqrt{x^2 + y^2}}{2} = \arctan\dfrac{y}{x} - \dfrac{\pi}{3}$ **9.** $y = x$ **11.** $\dfrac{(x - y + 3)^3}{3x - 3y + 1} = C$

EXERCISE SET 18.3, PAGE 1005

1. Exact; $2x^2 + 6xy - y^2 = C$ **3.** Not exact   **5.** Exact; $x^4 - x^2 y^4 - 20y = C$
7. Exact; $x^2 - 2xy^2 = C$ **9.** Exact; $y = Ce^{-x}$
11. Exact; $\ln|\sin x| - \sin x \cos y = C$ **13.** $x^2 - 3xy + 2y^2 = 2$

15. $\cos x \sin y = \dfrac{\sqrt{2}}{2}$ **19.** $\sqrt{y}\,(3x^3 + y) = C$ **21.** $\ln x - \dfrac{\cos y}{x} = C$

EXERCISE SET 18.4, PAGE 1010

1. $y = Ce^{2x} - e^x$ **3.** $y = \dfrac{1}{4}e^{-x}(2x - 1) + Ce^{-3x}$

5. $y = \dfrac{1}{2}(\sin x - \cos x) + Ce^{-x}$ **7.** $y = e^x x^{-3}(x^3 - 3x^2 + 6x - 6) + Cx^{-3}$

9. $y = 1 + Ce^{1/x}$ **11.** $y = (x + C)\cos x$

13. $y = \left(\dfrac{x^2}{2} + C\right)\cos x$ **15.** $y = \dfrac{x^3 - 3x + C}{3(x^2 + x)}$

17. $y^2 = x - \dfrac{1}{2} + Ce^{-2x}$ **19.** $y = (1 + Cx)^{-1}$ **23.** $y = x + \dfrac{1}{1 - x}$

EXERCISE SET 18.5, PAGE 1016

1. $y = C_1 e^{3x} + C_2 e^{-3x}$ **3.** $y = C_1 e^{-3x} + C_2 x e^{-3x}$
5. $y = C_1 e^{5x} + C_2 e^{-x}$ **7.** $y = e^x(K_1 \cos 2x + K_2 \sin 2x)$
9. $y = C_1 + C_2 e^{2x}$ **11.** $y = C_1 e^{(-2 + \sqrt{2})x} + C_2 e^{(-2 - \sqrt{2})x}$
13. $y = C_1 e^{8x} + C_2 e^{-x}$ **15.** $y = C_1 e^{-x} + C_2 e^{-5x}$
17. $y = K_1 \cos 3x + K_2 \sin 3x$ **19.** $y = C_1 e^{-x} + C_2 e^{-1/2x}$

21. $y = K_1 \cos \dfrac{3}{2}x + K_2 \sin \dfrac{3}{2}x$ **23.** $y = e^{-(3/5)x}\left(K_1 \cos \dfrac{6}{5}x + K_2 \sin \dfrac{6}{5}x\right)$

25. $y = xe^{2x}$ **27.** $y = e^{-3x}\left(\cos 2x + \dfrac{3}{2}\sin 2x\right)$

29. $y = \dfrac{1}{5}(11e^{(3/2)x} - 6e^{(1/4)x})$ **31.** $y = \cos x - \sin x$

33. $y = \dfrac{1}{e-1}[e - 2 + e^{1-x}]$ **37.** $C_1 = C_2 = \dfrac{1}{2}$; $K_1 = 0$, $K_2 = 1$

39. $y = 0$ **41.** $y = (\sqrt{3}\,e^{-2t}\sin 2\sqrt{3}\,t)/6$ **43.** $Q(t) = 0$, $I(t) = 0$

45. $Q(t) = e^{-10t}(10^{-6}\cos 632.38t + 1.58 \times 10^{-8}\sin 632.38t)$
   $I(t) = -6.32 \times 10^{-4}\,e^{-10t}\sin 632.38t$

EXERCISE SET 18.6, PAGE 1024

1. $y = K_1\cos x + K_2\sin x + \dfrac{1}{2}e^x$ **3.** $y = C_1 e^x + C_2 e^{-4x} - \left(\dfrac{5}{34}\sin x + \dfrac{3}{34}\cos x\right)$

5. $y = [K_1 - \ln(\tan x + \sec x)]\cos x + K_2\sin x$

7. $y = (K_1 - x)\cos x + (K_2 + \ln|\sin x|)\sin x$

9. $y = (C_1 - \ln x)e^{-2x} + (C_2 x - 1)e^{-2x}$ **11.** $y = K_1\cos x + K_2\sin x + \dfrac{1}{2}e^x$

13. $y = C_1 e^x + C_2 e^{-4x} - \left(\dfrac{5}{34}\sin x + \dfrac{3}{34}\cos x\right)$

15. $y = K_1\cos 2x + K_2\sin 2x + \dfrac{x}{8}(\sin 2x - x\cos 2x)$

17. $y = \dfrac{1}{3}e^x - \dfrac{1}{3}e^{-2x} - x^2 - x$ **19.** $y = \dfrac{1}{2}e^x(3 + x) - \dfrac{1}{2}e^{-x}(1 + x)$

21. $y = (-\sin 4t)/60 + (\sin t)/15$ **23.** $y = (\sin 4t)/32 - (t\cos 4t)/8$

25. $y = [e^{-2t}(3\cos 2\sqrt{3}\,t + \sqrt{3}\sin 2\sqrt{3}\,t) - 3\cos 4t]/48$

27. $Q(t) = e^{-10t}(.18\cos 62.45t + .56\sin 62.45t) + (11/20)\sin 60t - (11/60)\cos 60t$
   $I(t) = e^{-10t}(33.17\cos 62.45t - 16.84\sin 62.45t) + 33\cos 60t + 11\sin 60t$

REVIEW EXERCISES, PAGE 1026

1. $y = Ce^{2x}$ **3.** $\arctan\dfrac{y}{x} = \ln\sqrt{y^2 + x^2} + C$ **5.** $e^x\sin y = C$ **7.** $y = Ce^{x^2}$

9. $y = \dfrac{x^2 - 1}{2} + Ce^{-x^2}$ **11.** $y = \dfrac{x^2}{4} + \dfrac{C}{x^2}$ **13.** $y = \left(\dfrac{1}{7}x^4 + Cx^{-3}\right)^{-1}$

15. $y = C_1 e^{3x} + C_2 e^{-x}$ **17.** $y = K_1\cos x + K_2\sin x - \dfrac{1}{2}x\cos x$

19. $y = C_1 e^{2x} + C_2 e^x + \dfrac{1}{2}e^{3x}$ **21.** $y = \dfrac{1}{2}\ln((x^4 + 1)/2)$

23. $y = 5e^{-2x} - 4e^{-3x}$ **25.** $y = (x + 2)\sin x + 5\cos x$

27. $y = \dfrac{1}{6}x^3 e^{2x} + \dfrac{1}{2}x^2 e^{2x} + e^{2x} - 2xe^{2x}$

EXERCISE SET A.1, PAGE A-4

1. $x < 3$ **3.** $x > -\dfrac{1}{3}$ **5.** $-\dfrac{11}{2} < x < -2$ **7.** $-\dfrac{11}{2} < x < -3$

9. $x \le -3$ or $x \ge 1$ **11.** $3 \le x \le 5$ **13.** $.33 < x < \dfrac{1.01}{3}$

15. $x = 2$ or $x = -2$ **21.** a and b must have the same sign.

EXERCISE SET A.2, PAGE A-7

1.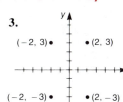

3.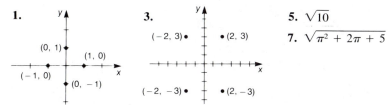

5. $\sqrt{10}$

7. $\sqrt{\pi^2 + 2\pi + 5}$

9.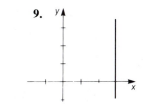

11. 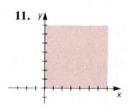   **13.** 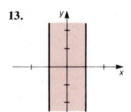   **15.**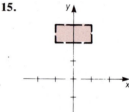

17. 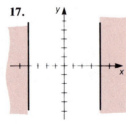   **19.** 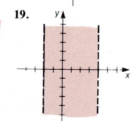   **21.**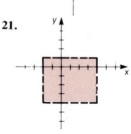

23.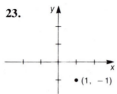

25. $d((-1, 4), (-3, -4)) = 2\sqrt{17}, d((-3, -4), (2, -1)) = \sqrt{34}, d((-1, 4), (2, -1)) = \sqrt{34}$

27. $(-6, 1)$, yes

EXERCISE SET A.3, PAGE A-17

1. (a) $\left(\dfrac{\sqrt{3}}{2}, \dfrac{1}{2}\right)$ (b) $\left(\dfrac{\sqrt{2}}{2}, \dfrac{\sqrt{2}}{2}\right)$ (c) $\left(-\dfrac{\sqrt{2}}{2}, -\dfrac{\sqrt{2}}{2}\right)$ (d) $\left(\dfrac{1}{2}, \dfrac{\sqrt{3}}{2}\right)$

   (e) $\left(-\dfrac{\sqrt{3}}{2}, -\dfrac{1}{2}\right)$ (f) $\left(-\dfrac{\sqrt{2}}{2}, \dfrac{\sqrt{2}}{2}\right)$

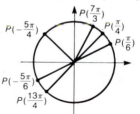

3. (a) $0, 2\pi$ (b) $\dfrac{3\pi}{2}$ (c) $\dfrac{\pi}{6}, \dfrac{5\pi}{6}$ (d) $\dfrac{2\pi}{3}, \dfrac{4\pi}{3}$ (e) $\dfrac{4\pi}{3}, \dfrac{5\pi}{3}$ (f) $\dfrac{\pi}{4}, \dfrac{5\pi}{4}$

   (g) $\dfrac{3\pi}{4}, \dfrac{7\pi}{4}$ (h) $\dfrac{\pi}{6}, \dfrac{11\pi}{6}$ (i) $\dfrac{\pi}{2}$ (j) $\dfrac{5\pi}{6}, \dfrac{11\pi}{6}$

5. $\dfrac{3}{5}, \dfrac{4}{3}$ **7.** $\cos t = \pm\dfrac{4}{5}, \tan t = \pm\dfrac{3}{4}, \cot t = \pm\dfrac{4}{3}, \csc t = \dfrac{5}{3}, \sec t = \pm\dfrac{5}{4}$

9. $\sin t = \dfrac{2\sqrt{5}}{5}, \cos t = \dfrac{\sqrt{5}}{5}, \sec t = \sqrt{5}, \csc t = \dfrac{\sqrt{5}}{2}, \cot t = \dfrac{1}{2}$

11. (a) $135°$ (b) $-150°$ (c) $180°$ (d) $420°$ (e) $-720°$ (f) $-157.5°$

13. (a) $\sqrt{3}$ (b) -2 (c) $\dfrac{2\sqrt{3}}{3}$ (d) $\dfrac{\sqrt{3}}{2}$ (e) $-\dfrac{\sqrt{2}}{2}$ (f) -1

15. (a)

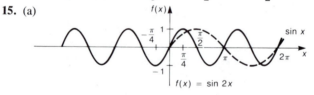

(b)

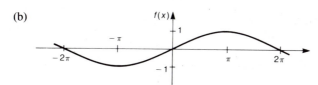

(c)

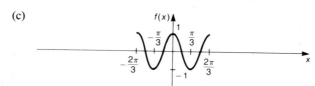

(d)

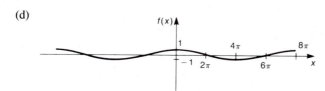

17. $\dfrac{\pi}{2}, \dfrac{3\pi}{2}, \dfrac{\pi}{6}, \dfrac{5\pi}{6}$ **19.** $0, \dfrac{\pi}{2}$ **21.** $\dfrac{\pi}{2}, \dfrac{3\pi}{2}$ **39.** $\dfrac{2}{\tan 25°} \approx 4.29$

41. $35\sqrt{3}$ feet ≈ 61 ft **43.** $5\dfrac{1}{3}$ ft from the end where the first ball lies

INDEX